AF327482

Amorphous and Heterogeneous Silicon Thin Films: Fundamentals to Devices—1999

MATERIALS RESEARCH SOCIETY
SYMPOSIUM PROCEEDINGS VOLUME 557

Amorphous and Heterogeneous Silicon Thin Films: Fundamentals to Devices—1999

Symposium held April 5–9, 1999, San Francisco, California, U.S.A.

EDITORS:

Howard M. Branz
National Renewable Energy Laboratory
Golden, Colorado, U.S.A.

Robert W. Collins
The Pennsylvania State University
University Park, Pennsylvania, U.S.A.

Hiroaki Okamoto
Osaka University
Osaka, Japan

Subhendu Guha
United Solar Systems Corp.
Troy, Michigan, U.S.A.

Ruud Schropp
Utrecht University
Utrecht, The Netherlands

Materials Research Society
Warrendale, Pennsylvania

Single article reprints from this publication are available through
University Microfilms Inc., 300 North Zeeb Road, Ann Arbor, Michigan 48106

CODEN: MRSPDH

Published by:

Materials Research Society
506 Keystone Drive
Warrendale, PA 15086
Telephone (724) 779-3003
Fax (724) 779-8313
Web site: http://www.mrs.org/

Library of Congress Cataloging-in-Publication Data

Amorphous and heterogeneous silicon thin films: fundamentals to devices—1999 :
    symposium held April 5–9, 1999, San Francisco, California, U.S.A. / editors,
    Howard M. Branz, Robert W. Collins, Hiroaki Okamoto, Subhendu Guha,
    Ruud Schropp
        p.cm.—(Materials Research Society symposium proceedings,
        ISSN 0272-9172 ; v. 557)
        Includes bibliographical references and indexes.
        ISSN 1072-3587
        ISBN 1-55899-464-5
        I. Branz, Howard M. II. Collins, Robert W. III. Okamoto, Hiroaki
    IV. Guha, Subhendu V. Schropp, Ruud VI. Materials Research Society symposium
    proceedings ; v. 557
1999   CIP

Manufactured in the United States of America

CONTENTS

PART I: <u>GROWTH AND PROPERTIES</u>

*Invited Paper

PART II: <u>HIGH-RATE DEPOSITION</u>

*Invited Paper

PART III: <u>RECRYSTALLIZATION, AMORPHIZATION AND POROUS SILICON</u>

*Invited Paper

PART V: <u>METASTABILITY</u>

*Invited Paper

PART VII: <u>HETEROGENEOUS MATERIALS AND DEVICES</u>

*Invited Paper

PART IX: <u>SOLAR CELLS</u>

*Invited Paper

*Invited Paper

PREFACE

"Amorphous and Heterogeneous Silicon Thin Films: Fundamentals to Devices—1999," Symposium A at the 1999 MRS Spring Meeting in San Francisco, California, was the seventeenth MRS Spring Meeting symposium to focus on hydrogenated amorphous silicon (a-Si:H), and the third of these symposia with an equal emphasis on microcrystalline thin films. Applications requiring large-area semiconductor coverage rely increasingly on amorphous and heterogeneous silicon materials because they can be deposited at low cost on a variety of substrates. The symposium covered a wide range of topics, from fundamental research on materials to their many device applications. Two joint sessions on thin-film transistors and displays were held with Symposium B, "Flat Panel Displays and Sensors—Principles, Materials and Processes." The papers from these joint sessions are divided between this volume and Volume 558 of the MRS Symposium Proceedings series.

The symposium featured a special session on medium-range order; it has become clear that ordering correlates with the electronic quality of a-Si:H films. Murray Gibson of Argonne National Laboratory reported multi-atom correlations extending to about 1.6 nm in a-Si:H, observed by variable-coherence transmission electron microscopy. Amorphous silicon may resemble paracrystalline structures more than the continuous random networks that have long been used for modeling; the boundary between amorphous and microcrystalline silicon is not so abrupt as once believed. Jeffrey Yang of United Solar Systems Corp. reported a world-record 10.5%-efficient, 920-cm^2, a-Si:H photovoltaic panel, which utilizes amorphous films deposited by plasma CVD at growth parameters just short of the threshold for a transition to microcrystallinity. Don Williamson of the Colorado School of Mines described a narrowing of the first x-ray diffraction peak in these amorphous films that also suggests ordering at 10–30 nm length scales.

Important new experimental observations on metastable effects in a-Si:H were reported. Shuichi Nonomura of Gifu University (Japan) described a metastable volume expansion of about 4 ppm caused by illumination and measured with a sensitive optical-lever bending technique. Stephen Heck of the National Renewable Energy Laboratory showed pulsed-illumination photoconductivity decays that suggest metastable degradation is delayed by about 1 ms after illumination begins. Discussions about H configurations in a-Si:H were spirited. Rafaella Borzi of Washington University showed deuteron magnetic resonance spectra suggesting up to 40% of the H in a-Si:H is found as H_2 molecules in interstitial sites, while Tining Su of the University of Utah reported NMR measurement indicating only about 10%.

Exciting new devices and processing strategies were reported. Kiran Pangal of Princeton University won a Graduate Student Award for fabricating amorphous and microcrystalline thin-film transistors in a single Si film. He exploited the reduction in crystallization time caused by local H plasma treatment. Ruud Schropp of the

University of Utrecht (The Netherlands), in collaboration with a consortium of Dutch universities and Akzo Nobel Chemicals and Coatings, is developing a temporary-superstrate technique for a-Si:H photovoltaic manufacture. In a particularly entertaining presentation, Joao Conde of the Instituto Superior Tecnico (Portugal) showed dramatic micrographs of 20-micron-long bridges, cantilevers and even working MEMS transistors fabricated from a-Si:H.

On behalf of all participants, the organizers thank the financial sponsors of this symposium:

> Akzo Nobel Chemicals and Coatings
> Energy Conversion Devices, Inc.
> Fuji Electric Company
> MVSystems, Inc.
> National Renewable Energy Laboratory
> Sanyo Electric Company
> Solarex Corporation
> SONY Corporation
> United Solar Systems Corp.
> Voltaix, Inc.

The organizing committee is grateful to Peter Fedders, Ping Mei, Fujio Okumura, Milan Vanecek, and Sigurd Wagner for their important help with program selection. Craig Taylor provided vital assistance to the symposium at key moments. Finally, we are deeply grateful to Mary Woolf of the University of Utah for her highly professional organizational assistance, which made the symposium and these proceedings a great success.

> Howard M. Branz
> Robert W. Collins
> Hiroaki Okamoto
> Subhendu Guha
> Ruud Schropp

> October 1999

MATERIALS RESEARCH SOCIETY SYMPOSIUM PROCEEDINGS

Volume 535— III-V and IV-IV Materials and Processing Challenges for Highly Integrated Microelectonics and Optoelectronics, S.A. Ringel, E.A. Fitzgerald, I. Adesida, D. Houghton, 1999, ISBN: 1-55899-441-6

Volume 536— Microcrystalline and Nanocrystalline Semiconductors—1998, L.T. Canham, M.J. Sailor, K. Tanaka, C-C. Tsai, 1999, ISBN: 1-55899-442-4

Volume 537— GaN and Related Alloys, S.J. Pearton, C. Kuo, A.F. Wright, T. Uenoyama, 1999, ISBN: 1-55899-443-2

Volume 538— Multiscale Modelling of Materials, V.V. Bulatov, T. Diaz de la Rubia, R. Phillips, E. Kaxiras, N. Ghoniem, 1999, ISBN: 1-55899-444-0

Volume 539— Fracture and Ductile vs. Brittle Behavior—Theory, Modelling and Experiment, G.E. Beltz, R.L. Blumberg Selinger, K-S. Kim, M.P. Marder, 1999, ISBN: 1-55899-445-9

Volume 540— Microstructural Processes in Irradiated Materials, S.J. Zinkle, G.E. Lucas, R.C. Ewing, J.S. Williams, 1999, ISBN: 1-55899-446-7

Volume 541— Ferroelectric Thin Films VII, R.E. Jones, R.W. Schwartz, S.R. Summerfelt, I.K. Yoo, 1999, ISBN: 1-55899-447-5

Volume 542— Solid Freeform and Additive Fabrication, D. Dimos, S.C. Danforth, M.J. Cima, 1999, ISBN: 1-55899-448-3

Volume 543— Dynamics in Small Confining Systems IV, J.M. Drake, G.S. Grest, J. Klafter, R. Kopelman, 1999, ISBN: 1-55899-449-1

Volume 544— Plasma Deposition and Treatment of Polymers, W.W. Lee, R. d'Agostino, M.R. Wertheimer, B.D. Ratner, 1999, ISBN: 1-55899-450-5

Volume 545— Thermoelectric Materials 1998—The Next Generation Materials for Small-Scale Refrigeration and Power Generation Applications, T.M. Tritt, M.G. Kanatzidis, G.D. Mahan, H.B. Lyon, Jr., 1999, ISBN: 1-55899-451-3

Volume 546— Materials Science of Microelectromechanical Systems (MEMS) Devices, A.H. Heuer, S.J. Jacobs, 1999, ISBN: 1-55899-452-1

Volume 547— Solid-State Chemistry of Inorganic Materials II, S.M. Kauzlarich, E.M. McCarron III, A.W. Sleight, H-C. zur Loye, 1999, ISBN: 1-55899-453-X

Volume 548— Solid-State Ionics V, G-A. Nazri, C. Julien, A. Rougier, 1999, ISBN: 1-55899-454-8

Volume 549— Advanced Catalytic Materials—1998, P.W. Lednor, D.A. Nagaki, L.T. Thompson, 1999, ISBN: 1-55899-455-6

Volume 550— Biomedical Materials—Drug Delivery, Implants and Tissue Engineering, T. Neenan, M. Marcolongo, R.F. Valentini, 1999, ISBN: 1-55899-456-4

Volume 551— Materials in Space—Science, Technology and Exploration, A.F. Hepp, J.M. Prahl, T.G. Keith, S.G. Bailey, J.R. Fowler, 1999, ISBN: 1-55899-457-2

Volume 552— High-Temperature Ordered Intermetallic Alloys VIII, E.F. George, M.J. Mills, M. Yamaguchi, 1999, ISBN: 1-55899-458-0

Volume 553— Quasicrystals, J-M. Dubois, P.A. Thiel, A-P. Tsai, K. Urban, 1999, ISBN: 1-55899-459-9

Volume 554— Bulk Metallic Glasses, W.L. Johnson, C.T. Liu, A. Inoue, 1999, ISBN: 1-55899-460-2

Volume 555— Properties and Processing of Vapor-Deposited Coatings, R.N. Johnson, W.Y. Lee, M.A. Pickering, B.W. Sheldon, 1999, ISBN: 1-55899-461-0

Volume 556— Scientific Basis for Nuclear Waste Management XXII, D.J. Wronkiewicz, J.H. Lee, 1999, ISBN: 1-55899-462-9

Volume 557— Amorphous and Heterogeneous Silicon Thin Films: Fundamentals to Devices—1999, H.M. Branz, R.W. Collins, H. Okamoto, S. Guha, R. Schropp, 1999, ISBN: 1-55899-464-5

MATERIALS RESEARCH SOCIETY SYMPOSIUM PROCEEDINGS

Part I

Growth and Properties

THE FORMATION AND BEHAVIOR OF PARTICLES IN SILANE DISCHARGES

Alan Gallagher[*]
JILA, University of Colorado and National Institute for Standards and Technology
Boulder, CO 80309-0440

ABSTRACT

Particle growth in silane RF discharges, and the incorporation of particles into hydrogenated-amorphous-silicon (a-Si:H) devices is described. These particles have a structure similar to a-Si:H, but their incorporation into the device is believed to yield harmful voids and interfaces. Measurements of particle density and growth in a silane RF plasma, for particle diameters of 8-50 nm, are described. This particle growth rate is very rapid, and decreases in density during the growth indicate a major flux of these size particles to the substrate. Particle densities are a very strong function of pressure, film growth rate and electrode gap, increasing orders of magnitude for small increases in each parameter. A full plasma- chemistry model for particle growth from SiH_m radicals and ions has been developed, and is outlined. It yields particle densities and growth rates, as a function of plasma parameters, which are in qualitative agreement with the data. It also indicates that, in addition to the diameter >2 nm particles that have been observed in films, a very large flux of Si_xH_m molecular radicals with $x \gg 1$ also incorporate into the film. It appears that these large radicals yield more than 1% of the film for typical device-deposition conditions, so this may have a serious effect on device properties.

INTRODUCTION

In silane (SiH_4) and SiH_4-H_2 discharges, a-Si:H film deposition is initiated by electron collisions that dissociate SiH_4. This yields primarily neutral-radical fragments, but accompanied by a small fraction (typically 3-5%) of positive ions (cations) and a very small fraction (typically 0.5-1%) of negative ions (anions). The neutral radicals Si, SiH, SiH_2, and H rapidly react with SiH_4 to form stable molecules or SiH_3, while SiH_3 reacts only with the a-Si:H film. Thus, film growth is primarily due to the reactions of SiH_3 radicals with the surface. Cations that drift out of the plasma induce only a few percent of the growth, but as they typically strike the surface with energies in the 1-30 eV range, this can be important. The anions are produced in much smaller quantity, and they can not even reach the growing film due to sheath fields. Thus, for many years these were neglected as a contributor to film electronic properties. In recent years, however, it has become clear that these anions can profoundly influence the film, for they cause Si particles to grow in the plasma. Since anions ($Si_xH_n^-$) are trapped in the plasma, there is plenty of time for SiH_3 - anion reactions to induce their growth, and this can lead to enormous values of x, the number of Si atoms. Once x exceeds perhaps 30, these anions are believed to be relatively spherical with a structure similar to that of the a-Si:H film; clustered Si at near-crystalline density with occasional H inside and an H terminated surface. Thus, these larger clusters are generally described as "particles," although they can also be described as Si_xH_n with charge z.

From standard "particle-in-plasma" theories, particles with x>200 (radius R_p >1 nm) are expected to be negatively charged, and consequently trapped in the plasma by the sheath fields.[1] (The number of negative charges increases linearly with R_p.) With gas flow, these charged particles are dragged away to the pump once they grow to 0.1-1 μm size ($x=10^8$-10^{11}), and the film does not suffer. (Fig. 1 indicates the forces on negatively charged particles suspended in a rf plasma.) However, we discovered several years ago that particles of 2-15

Mat. Res. Soc. Symp. Proc. Vol. 557 © **1999 Materials Research Society**

nm radius escape the plasma and incorporate into the films in copious quantities.[2] These particles constituted 10^{-3}-10^{-4} of the film volume, and their surface bonds to the surrounding film exceed the electronic defect density of a-Si:H by many orders of magnitude. Here I will report that those observations were just the tip of the iceberg; when smaller particles not visible in that experiment are included, particles probably constitute 1-10% of the film volume, and their bonds to the surrounding film are a significant fraction of all Si-Si bonds within the film. Thus, these particles have become a major concern, as it appears likely that voids and strained bonds occur at particle-film interfaces, particularly below the particles where subsequent growth cannot easily penetrate. Here I will describe what is now understood about this growth and incorporation of particles into the growing a-Si:H film, and mitigation methods that are suggested by this understanding.

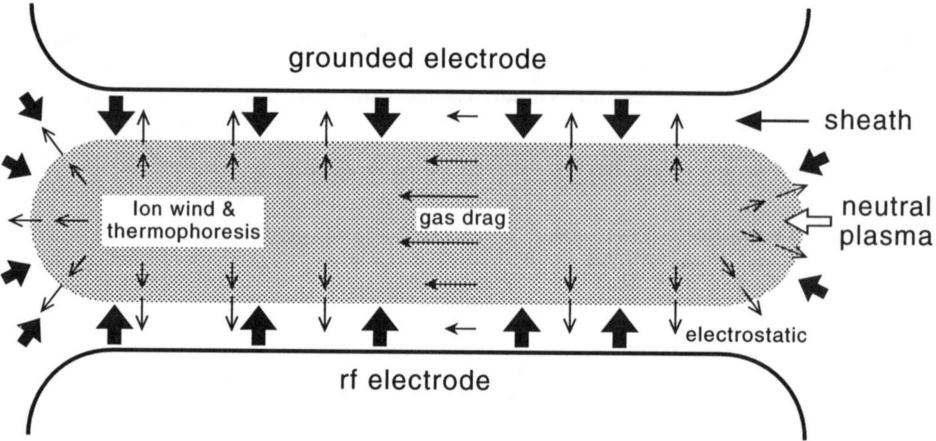

Fig. 1. Diagram of forces on negatively charged particles in a RF, capacitively coupled discharge. Electrostatic forces due to the sheath fields are indicated by large arrows, as these are the strongest forces for small particles. Ion drag results from cation-particle scattering while cations drift toward the electrodes. The gas-drag force increases as R_p^2 whereas the other forces have slower R_p dependence. Thus, large particles are carried away to the pump.

MEASUREMENTS OF PARTICLES IN SILANE PLASMAS

Since the original detection of particles in silane plasmas by Roth et al.,[3] the most common method of measuring particle suspended in plasmas is still light scattering. However, important insights have also been gained from measurements of negative ions escaping the plasma afterglow,[4] and from scanning tunneling microscope[2] and transmission electron microscope (TEM) measurements of particles collected below the electrodes.[5] Unfortunately, most of these measurements were not carried out in pure SiH_4 or SiH_4-H_2, or for conditions that yield device-quality a-Si:H films. Thus, to understand particle growth under device production conditions, I have utilized primarily the insights, not the particle data, from these experiments. In my laboratory we have concentrated on the conditions and gases used to produce devices, so I will primarily discuss our data. Initially, we studied the accumulation of larger particles (R_p >50 nm) at the downstream end of the RF discharge, and their escape to the pumps.[6] These particles can influence the discharge operating conditions, and their exact location indicates an important consideration that will be discussed below in the "particle-mitigation" section: the plasma potential varies within the mid-plane between the electrodes. However, these larger particles are not suspended below the substrate and they do not incorporate into devices. Thus, in the last few years we have studied[7] the very small particles that form directly below the substrate and are indicative of those that incorporate into films.

It is quite difficult to study the very small particles (R_p <20 nm) that incorporate into films, because their light scattering cross section (Q_p), proportional to R_p^6, is very small. Furthermore, it is not straightforward to determine the size of these very small particles, yet this must be done since the scattering signal only establishes $n_p R_p^6$ where n_p is the particle density. We have developed a method based on the particle diffusion, utilizing the fact that neutral particles diffuse out of the discharge afterglow in a time proportional to R_p^{-2}. An example of the scattering signal versus time is shown in Fig. 2, for a discharge that was turned on for 1.8 s. The exponential decay well after discharge termination results from fundamental-spatial-mode diffusion of 20 nm particles. With R_p thus determined, the size of the scattering signal yields n_p. This method works for 4nm<R_p<25nm, where the upper limit is due to the effect of gravity on the motion of larger particles, and the lower limit results from insufficient scattering signal.[7]

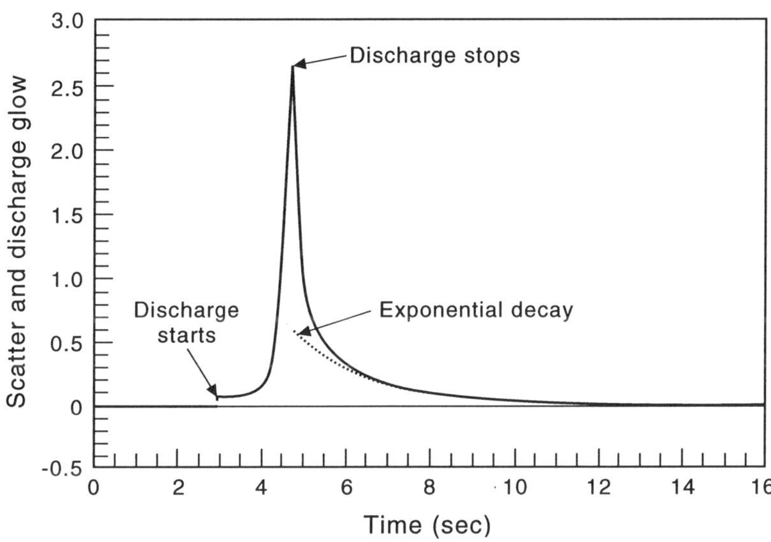

Fig. 2. Time dependence of particle scattering of a 3 W Ar$^+$ laser beam, for a discharge in 0.3 Torr SiH$_4$ that was turned on and off as indicated.[7] The discharge RF amplitude was 82V and the film deposition rate was 0.18 nm/s.

By varying the discharge operating time, t_{on}, we measure the particle size versus discharge time, finding that R_p grows approximately linearly with time within the 4-25 nm range of the data. This can be understood from growth due to radical (primarily SiH$_3$) reactions on the particle surface, which is the same phenomenon that causes film growth on the substrate. In essence, for a spherical particle with silicon density ρ, $x = 4\pi R_p^3/3\rho$, $dx/dt = 4\pi R_p^2 \rho^{-1} dR_p/dt$, while the radical attachment rate dx/dt varies as the cross section πR_p^2. The growth is only approximately linear because the data indicates a more rapid initial growth to $R_p \approx 2$ nm, followed by essentially constant dR_p/dt. This is consistent with a small-particle cross section larger than $4\pi R_p^2$, as is expected for SiH_3 collisions with small, charged particles due to polarization forces. For a more quantitative comparison, the SiH_3 density can be established from the film growth rate, and used to calculate an expected particle growth rate due to SiH_3 collisions. The result agrees with the measurements at low pressures (P) and film growth rate (G), but the measured growth rate is as much as a factor of 2 higher than calculated at high P and G. This increased growth is tentatively attributed to heavier radicals and cations attaching to the particles, perhaps aided by an increased radical reactivity with the particle surface. This overall nearly linear growth result is thus quite reasonable, and in contrast to most previous data, demonstrating the importance of studying the low power conditions and gases appropriate for device production.

A second result is that the measured particle density decreases with discharge operating time, as the particles grow from $R_p = 4$ to 25 nm. A calculation of particle agglomeration rates demonstrates that these are orders of magnitude too small to explain this data, so particles must be escaping the plasma to the electrodes. Indeed, the escape rate implied by the data yields particle densities of $R_p = 4$-8 nm particles in the a-Si:H film that are consistent with those measured in Ref. 2. This rate of particle escape requires that a major fraction of particles are neutral, so that they can escape the plasma. To understand how and why this happens, as well as particle formation and growth, we have developed a full plasma-chemistry model that is outlined in the next section.

A MODEL FOR PARTICLE GROWTH AND ESCAPE TO THE ELECTRODES

In this steady-state model, SiH_3, SiH_n^+ and SiH_3^- are formed by electron collisions, typically in the ratio 100/4/0.8. These radicals grow and alter their charge due to collisions between themselves and with electrons and SiH_4. Most of this growth results from consecutive, one-at-a-time additions of Si atoms. During this growth, the negatively charged complexes, $Si_xH_m^-$, are trapped in the plasma, but neutral and positively charged Si_xH_m radicals can diffuse to the electrodes. Thus, there is a continual attrition of complexes as they grow in size. When x exceeds ~30 and the complexes are largely cross-linked Si with a largely H-terminated surface, we think of it as a particle. The model only keeps track of x and charge z, and makes no distinction between a molecular complex and a particle, so for ease "particle" will be used to describe all $Si_xH_m^z$ radicals or particles. With increasing x or size, particle-diffusion slows and the cross section for adding a Si atom increases. The particles can also accommodate increasing numbers of negative charge as x increases. Thus, diffusive loss is most severe for small x values, and the density of visible $(x>10^4)$ particles depends strongly on the fraction of small-x particles that are lost to diffusion before they can grow to this "large" size. The competition between growth and diffusive loss depends strongly on discharge conditions. At higher powers and pressures, the radical density is high and particle growth is rapid, while diffusion slows with increasing pressure and discharge gap (L). Thus, at high P and G many particles remain in the plasma and visible particle densities are high. On this basis, one might wish to reduce particle incorporation into films by simply lowering P and L. However, at low P and L ion bombardment of the film increases in energy, and this can also cause loss of film quality.

I will now describe the model in more detail, using the notation n(x,z) for the density of particles with charge z and containing x Si atoms. The first requirement is to establish the densities n(1,z) of SiH_m radicals and ions with charge z=0,1, and -1. Since SiH_3 diffusing to the surface produces most of the film growth rate (G), this establishes the SiH_3 production rate

(F_0), and the density $n(1,0)$ then follows from the SiH_3 diffusion coefficient and the electrode gap. Next, based on electron collision coefficients I assume that SiH_3, SiH_m^+ and SiH_3^- are produced in the ratio $F_0/F_1/F_{-1} = 125/5/1$. This yields $n(1,1) = F_1/R(1,1)$ and $n(1,-1) = F_{-1}/R(1,-1)$, where $R(x,z)$ is the loss rate for $Si_xH_m^z$, due to transforming into other x,z species and diffusion to the electrodes. Next, I calculate the particle densities $n(2,z)$ from these $n(1,z)$, based on the rate of Si atom addition from collisions with SiH_m^+, SiH_3 and SiH_4. Here I use available or estimated rate coefficients for collisions between these $x=1$ molecules and electrons, radicals, ions, and silane. The diffusive loss rate for these $x=1$ molecules is included in this $x=1$ to 2 step; it competes with the growth rate to $x=2$, causing some loss of $x=1$ particles. Next, this process is repeated for growth from $x=2$ to $x=3$, again with some loss of $x=2$ particles to the electrodes. This step-by-step ($x \rightarrow x+1$) particle growth is repeated to $x=10^4$ or 10^5, a visible-particle size where comparison to experiment can be made. At appropriate x_i values where Si particles carrying $i>1$ negative charges become stable, these additional charge states are included. Iteratively calculating only $x \rightarrow x+1$ steps is an approximation, as some collisions add more than one Si atom. However, it greatly simplifies the calculation and is justified because particle growth is dominated by collisions with SiH_4 for $x<100$, and by collisions with SiH_3 for $x>100$. An example result from the calculation is shown in Fig.3 for discharge conditions that are typical of those in Ref. 7.

In Fig. 3, first consider the neutral particles that are labeled $n(z=0)$ and flux(0), equivalent to $n(x,z=0)$ and $F(x,0)$ in the more general notation for particle density and flux to

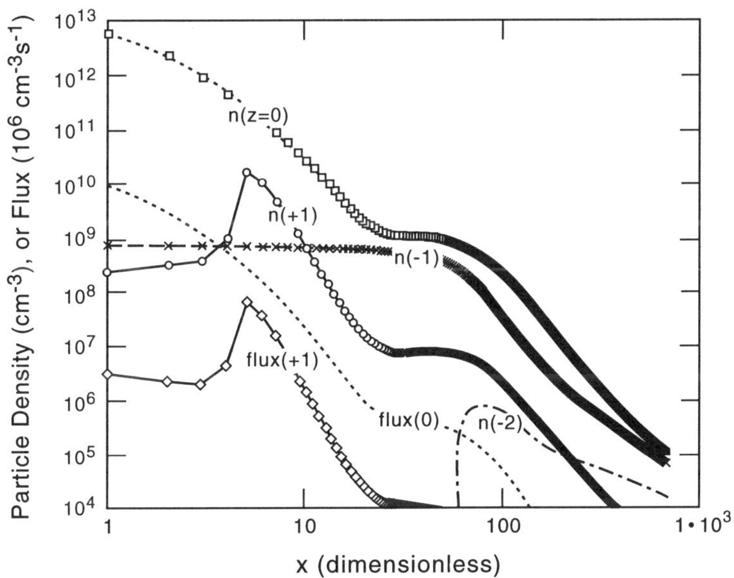

Fig. 3. Densities of particles containing x Si atoms and z charges, and flux of z=0 and +1 particles to the electrodes, in units of 10^6/s and cm^3 of plasma volume. The RF plasma conditions are 0.3 Torr pure SiH_4, T=300K, 0.3 nm/s film growth rate, and 1.3 cm electrode gap, as in Ref. 7. To simplify the figure only the x=1 to 730 range is shown here; for x>730, particles with z = -3 also occur. The charged particle densities are $n_e = 3 \times 10^8$ cm^{-3}, $n_- \approx n_+ = 3.6 \times 10^{10}$ cm^{-3}.

the surfaces with any charge z. These neutral particles are molecular radicals, with $x=1$ representing SiH_3. The $x=2$ radicals result from collisions between a pair of SiH_3, and the $x=3$ primarily result from collisions between $x=1$ and $x=2$ radicals. As x increases from 1 to 2, $n(x,z=0)$ and $F(x,0)$ drop a factor of ~2.5 and ~3.5 respectively. As x further increases, a smaller decrease per $x \rightarrow x+1$ step occurs, due to an increase in the rates of growth/diffusive loss. (Collision cross sections increase with x, and this increases growth rates and decreases diffusion rates.) The $x<20$ data in Fig. 3 can be approximated as $F(x,0) \propto x^{-3}$. If one assumes that x Si atoms are incorporated into the film by each Si_xH_m radical that escapes the plasma, then 50 % of the film results from $x>1$ radicals, 33% from $x>2$ radicals, etc. This result involves assumed radical-radical reaction cross sections for $x>2$, but it is unlikely that these are far off as these are highly exothermic reactions that are stabilized by H release. Thus, even before considering the larger particles, this heavy-radical incorporation could have significant effects on film properties. (This has been pointed out and modeled previously.[8]) This rate of $F(x,0)$ falloff versus x is very sensitive to plasma parameters; it slows if G, L or P is increased and visa versa. For $x>20$, $n(x,0)$ begins to follow the negative ion density, $n(x,-1)$, as these populations become strongly coupled by electron and cation collisions. Since the negative ions grow rapidly for $x<50$, both populations falloff very slowly in the $x=20-50$ range. For $x>50$ this negative ion growth rate decreases, and diffusive loss of neutral particles again causes $n(x,0) \propto x^{-n}$ with $n=3-4$.

Turning now to the anions, labeled $n(-1)$ in Fig.3, their density is almost independent of x for $x<50$ due to very rapid growth and only minor loss by cation collisions. Reactions with silane cause this rapid growth, in a process that has been observed[4] to occur for $x<\sim30$. The measurements do not extend past $x \cong 30$, and are also somewhat ambiguous for $x>6$, but it is clear that at some x value (x_c) the presence of a negative charge can no longer cause a reaction that does not occur on a neutral a-Si:H surface. We guess that this transition occurs when the polarization energy at the surface equals the thermal energy, yielding $x_c \cong 50$. We make x_c an adjustable parameter in the model, while $x_c = 50$ was used for the calculation of Fig. 3. As already noted, for $x>100$ or actually $x \gg x_c$, this anion growth rate slows and diffusive loss of neutral particles causes the coupled densities of $z=0$ and -1 particles to decrease. Due to the much larger ratio of (growth rate)/(diffusion rate) in this $x>100$ region, the attrition per $x \rightarrow x+1$ step is much smaller than in the $x=1-20$ region discussed in the previous paragraph. However, there are many more steps for $x>100$ in Fig. 3, a fact that tends to be hidden by the $log(x)$ scale used in the plot.

Considering next the cations, labeled $n(+1)$ and flux(+1) in Fig. 3, the density of $x=1-3$ is severely lowered by very rapid growth reactions with SiH_4. However, these reactions slow at $x=3$ and essentially terminate by $x=6$, and only slower reactions with radicals cause further cation growth. The result is a peak in $n(x,1)$ at $x=5$, followed by a drop off to $x=20$ as cation diffusive loss competes with growth. For $x>30$, the cations also become closely connected to the $z=-1$ and 0 particles by charge-changing collisions, and all three densities fall together. Note that, for all x values, the cation flux to surfaces is always much smaller than the neutral flux. The cation-silane reaction rates for $x<8$, which lead to this $x=5$ peak in the $F(x,1)$ flux, are measured values[9] that inevitably yield this peak at $x=5\&6$. However, this is in severe contradiction with mass-spectrometer measurements through the substrate electrode of silane RF discharges.[4,10] Those experiments typically observed $F(x+1,1)/F(x,1)$ ratios of ~1/3 for the entire $x=1-6$ range; the opposite of an increase with increasing x. The most likely explanation for this discrepancy is that the larger-x ions exist in the neutral plasma region, but breakup while traversing the sheath due to energetic collisions with SiH_4. This is indirectly supported by comparing mass spectrometer observations from our laboratory, taken at the side of the neutral plasma region and through the cathode of a silane discharge.[11] Although these measurements were in a dc discharge, we measured for x=1-6, $F(x,1) \cong$ constant when sampling at the plasma, but $F(x+1,1)/F(x,1) \cong 1/4$ through the cathode.

There is one additional trace in Fig. 3, corresponding to particles carrying two negative charges. We calculate that this becomes possible at x=60, so n(x,-2) appears for x>60. As can be seen, it plays a minor role for these conditions and range of x values. This, as well as the fact that n(x,0)/n(x,1)>1 for x>100 in Fig. 3, is due to $n_+ \gg n_e$, where n_+ is the total cation density and n_e the electron density. This is a neutral plasma, but a charge $n_- \gg n_e$ is carried by the negatively-charged particles, mostly by the anions with x<50 in Fig. 3. The ratio n(x,0)/n(x,-1) is set by a balance between electron attachment to n(x,0) and cation attachment to n(x,-1), so it is a strong function of n_+/n_e. If $n_+/n_e \cong 1$, as in plasmas without negative ions, this balance would yield n(x,0)/n(x,-1) $\ll$ 1 because electron velocities are much higher than cation velocities. This has a major effect on particle loss to the electrodes, because for each x only the neutral and positive particles can escape. Thus, the large fraction of particles that are neutral in Fig. 3 yields a very large loss flux to the electrodes. This unusually large neutral fraction results from $n_+/n_e \cong 100$ for this plasma.

Figure 4 again shows particle densities and (total) flux to the electrodes for the plasma conditions of Fig. 3. The x range is extended to 10^4 ($R_p = 3.6$ nm), higher charge states are included, and the total density is also plotted as a density of particles per nm of radius. (All other densities are per unit x.) Charge states -5 and -6 also occur in this x range, but are unimportant for these conditions and are not plotted. Note that the particle density per nm of radius is approaching a constant at large x. This is the steady-state result of a constant dR_p/dt growth rate, which occurs as described below Fig. 2 when diffusive loss is minor compared to growth. Very large particles thereby accumulate in this homogeneous plasma model, whereas

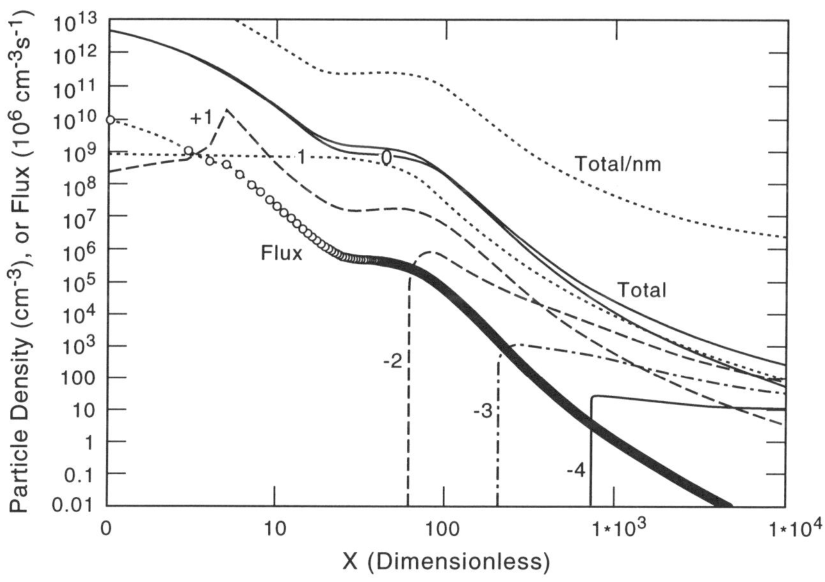

Fig. 4. Calculated particle densities, n(x,2), as a function of x, the number of Si atoms, and 2, the charge. Also shown is density per nm of particle radius and particle flux to surfaces.

in the flowing-gas plasmas used to produce devices these would be normally be carried to the downstream end. However, particles may trap and grow to large size above irregularities in the electrode surfaces, since these can produce a local maxima of the plasma potential that traps negative particles. To estimate the particle sizes that occur below substrates without such irregularities, the typical measured R_p growth rates from the experiment described above should be multiplied by gas dwell times in the reactor. This might typically yield 10nm/s times 0.5s = 5 nm for a device reactor.

The particle density at $x>10^4$ is a very sensitive function of P,G, and L. This is shown in Fig. 5, where n_4, the total particle density per nm of radius at $x=10^4$, is plotted. Here I have used one set of assumed collisional rate coefficients and, as almost all of these are must be estimated from chemical and physical principles, this is not a definitive result. However, the trends seen in Fig. 5 are followed for many different sets of rate coefficients, and I believe it is realistic. The n_4 density tends to limit out at 10^{10}-10^{11}cm^{-3}nm^{-1}, which corresponds to about n_+, the total cation charge. Since the total particle negative charge cannot exceed the cation density, most particles become neutral once the particle density exceeds n_+. Particles then escape the plasma with a high probability, so this sets a limit on total particle density. The data is plotted versus $P \bullet L^{1.8}$ because it is found that this reduces results at different P and L values to a relatively common curve. For each G value, as $P \bullet L^{1.8}$ is reduced n_4 decreases by many orders of magnitude, and at the left side of each curve n_4 varies more than 5 orders of magnitude for a factor of 2 in this parameter or in G. This extreme sensitivity to the plasma parameters results from the multiple-step particle growth, with diffusive loss competing with growth at each step. It is consistent with the results of the experiment described above, as is the range of P,L and G that yield the measured particle densities. However, the exact values of P,L and G for which the model yields these densities is not meaningful, as many collisional rates must be assumed. It is interesting that the sensitivity to L is much stronger than to P.

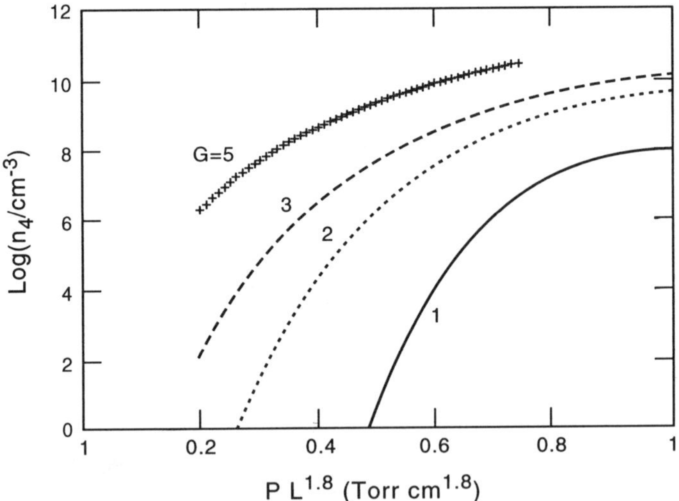

Fig. 5. $Log_{10}(n_4/cm^{-3})$ as a function of RF discharge parameters. G=film growth rate in Å/s, L = electrode gap in cm, P= silane pressure in Torr at 300 K, and n_4 is the density of 3.6 nm radius ($x=10^4$) particles per nm of radius.

Since the discharge sustaining voltage and ion bombardment energies scale more closely with PL, this suggests that for a given ion bombardment one can operate at higher G if L is decreased.

CONCLUSIONS

In order to mitigate particle incorporation into a-Si:H devices, we first set out to understand why this incorporation occurs. In the process of achieving some of this understanding we have discovered that the incorporation of detectable particles into the film is just the tip of the iceberg. Our model has shown that, in addition to the 2-15 nm radius particles which can incorporate at 10^{-3} to 10^{-4} of the film density, large Si_xH_m complexes with x>10 incorporate into films as more than 1% of the volume. It seems likely that some of the interfaces between these particles or large complexes and the remaining film are not well structured or passivated.

I think that we now understand reasonably well why and how the particles grow in the plasma, and their growth appears to be an inevitable consequence of the strong electron-attachment energies of Si-containing molecules and particles. The SiH_3^- production and growth into $Si_{100}H_m^-$ (R_p=0.8 nm) occurs within ms of discharge initiation, and cannot realistically be avoided. Particle growth from x=100 to 10^4 (R_p = 3.6 nm) or larger can be mitigated by using low pressure (P), discharge gap (L) and growth rate (G), but even this is achieved by allowing the growing particles to escape to the electrodes before they reach this "large" size. Our model does suggest that improvements in G can be achieved by decreasing L keeping P•L constant, but this will at best gain as $G \propto L^{-1}$, and it only refers to the incorporation of the larger, visible particles into the film. As there is not much one can do to prevent the growth of particles, it is important to ask why they incorporate into the film, and if this might be avoided. Alternatively, if one wishes to incorporate crystalline silicon particles into the film, can we control this?

Only neutral or positively charged particles diffuse to the electrodes, so it is important to understand why a large fraction of the particles are neutral or positive. (If this were a small fraction, almost all particles would be trapped in the plasma until they are swept to the downstream end of the reactor and finally swept away to the pumps.) The reason a large fraction of the particles are neutral is that the cation and anion densities (n_+ and n_-) are much larger than the electron density (n_e). And the reason for this is that anions are trapped, so n_- and n_+ build up to a value where mutual neutralization (cation-anion) collisions balance the rate of producing anions. When the rate coefficients for electron attachment and ionization, and for mutual neutralization are put into this requirement, $n_+/n_e \cong 100$ results. The model always yields a ratio near this, for the entire range of discharge parameters studied. Thus, there doesn't seem to be much one can do about this in SiH_4 or SiH_4-H_2 gases; a large fraction of particles will be neutral and will escape to the surrounding surfaces before being dragged to the downstream end of the reactor.

One probably cannot prevent particle escape, but there may be something one can do about where the particles escape to. Although they will escape to nearby surfaces, it need not be equally to all surfaces or regions of the electrodes. One very interesting possibility is to drive particles to the RF electrode with the thermophoretic force. This force is proportional to R_p^2, as is gas drag, so it causes all size particles to drift toward colder regions with a velocity proportional to the temperature gradient. (The very small ones partially back diffuse against this drift.) Since the substrate is normally heated to ~240C, this temperature gradient would naturally occur if the RF electrode were cooled. A second method of controlling where particles escape is to design advantageous variations in plasma potential. The plasma is positively biased by a "plasma potential" relative to the substrate, typically by $V_{RF}/2 \sim 30V$ where a V_{RF} is amplitude of the RF voltage. If the discharge runs harder in some regions between the electrodes, this plasma potential may be several volts larger in these regions. This yields a potential well a few eV deep in the plane between the electrodes, and for the particles that have kT of kinetic energy this is a substantial potential gradient, or lateral force. Thus, particles are driven towards regions of higher plasma potential, where they can be trapped for long times.[14] This trapping can occur at steps or protrusions in either electrode surface, so the design of substrate holders may be very important. It may also be possible to guide particles out to the plasma edges with certain electrode structures.[12] For most parallel-plate RF discharges with flat electrodes, the plasma potential is somewhat higher near the outer edges of the RF electrode and particle densities are higher there. This effect is normally not strong enough to drive all particles out of the central region of a discharge, but with some redesign of the RF electrode it might be. As another example, tapering the gap toward the downstream end might draw particles more quickly away from the substrate.

ACKNOWLEDGEMENTS

I wish to thank M. A. Childs and A. V. Phelps for many valuable ideas and suggestions. This work has been supported in part by the National Renewable Energy Laboratory under contract DAD-8-18653-01.

REFERENCES

1. J. E. Daugherty, R. K. Perteous, M. D. Kilgore, and D. B. Graves, J. Appl. Phys. **72**, 3934 (1992).
2. D. M. Tanenbaum, A. L. Laracuente, and A. Gallagher, Appl. Phys. Lett. **68**, 1705 (1996).
3. R. M. Roth, K. G. Spears, G. D. Stein, and G. Wong, Appl. Phys. Lett. **46**, 253 (1985).
4. A. A. Howling, L. Sansonnens, J.-L. Dorier, and Ch. Hollenstein, J. Appl. Phys. **75**, 1340 (1994) .
5. L. Boufendi, A. Plain, J. Ph. Blondeau, A. Bouchoule, C. Laure, and T. Toogood, Appl. Phys. Lett. **60**, 169 (1992).
6. B. M. Jelenkovic and A. Gallagher, J. Appl. Phys. **82**, 1546 (1997).
7. A. Gallagher, S. Barzen, M. Childs, and A. Laraquente, "Atomic Scale Characterization of Hydrogenated Amorphous Silicon Films and Devices" NREL/SR-520-24760, Golden, Colorado (June 1998); M. A. Childs and A. Gallagher, in preparation.
8. A. Gallagher, J. Appl. Phys. **63**, 2406 (1988).
9. J. Perrin, O. Leroy, and M. C. Bordage, Plasma Phys. **36**, 1 (1996).
10. I. Haller, J. Vac. Sci. Technol. **A1**, 1376 (1983).
11. A. Gallagher and J. Scott, "Diagnostics of a Glow Discharge Used To Produce Hydrogenated Amorphous Silicon Films," Annual Report to Solar Energy Research Institute (now NREL) for Contract XJ-0-9053-1, Golden, Colorado (April 1981).

* Member, Quantum Physics Division, NIST. E-mail address: alang@jila.colorado.edu

THE ROLE OF H IN THE GROWTH MECHANISM OF PECVD a-Si:H

M.C.M. van de Sanden, W.M.M. Kessels, A.H.M. Smets, B.A. Korevaar, R.J. Severens[*] and
D.C. Schram
Eindhoven University of Technology, Department of Applied Physics, P.O.Box 513, 5600 MB,
Eindhoven, Netherlands, email: m.c.m.v.d.sanden@phys.tue.nl
[*]AKZO-NOBEL Central Research, P.O.Box 9300, 6800 SB, Arnhem, Netherlands

ABSTRACT

This paper describes an extension of the silyl radical based kinetic growth model by atomic
hydrogen induced surface hydrogen abstraction processes. It is shown that by including this
direct abstraction process several problems of the SiH_3 based model are resolved. The defect
density can be predicted with the proper temperature dependence and order of magnitude. The
implications for high rate deposition of a-Si:H are discussed

INTRODUCTION

The growth mechanism of a-Si:H by means of plasma decomposition of SiH_4 has been the
subject of many research studies. It is widely believed that insight into the deposition mechanism
of a-Si:H opens up the possibility to minimize its defect density and instabilities and possibly
shows ways to increase the growth rate above the growth rate of typically a few A/s obtained
using PECVD in silane. Many experiments have shed light on the main species present during
PECVD and their possible contribution to a-Si:H film formation. In most growth models the role
of the silyl radical is dominant, i.e., SiH_3 is assumed to be responsible for both growth (sticking)
and the creation of growth sites by surface hydrogen abstraction [1,2,3]. The reasoning is based
on the widespread consensus that the silyl density is the highest radical density in the gas phase.
Moreover, the flux of atomic hydrogen on the growth surface is considered to be small due to the
short reactive lifetime of H in the abundant SiH_4 gas. However, in contrast, during deposition of
microcrystalline silicon the H-flux is considered to be important. Especially the transition
between a-Si:H and microcrystalline silicon is assumed to be induced by atomic hydrogen
making it unlikely that atomic hydrogen does not play a role during a-Si:H deposition. Recent
measurements have indicated an atomic hydrogen density of similar order as the silyl density
during PECVD of a-Si:H [4]. Furthermore, isotope studies during a-Si:H deposition from an
intense source of H and SiH_3 radicals revealed directly the influence of atomic hydrogen on the
growth of a-Si:H [5]. Therefore, it is important to address the role of H during a-Si:H deposition
systematically and determine its influence on the quality of PECVD a-Si:H.

In this contribution we will first briefly summarize the kinetic growth model based on the
silyl radical [1,2,3]. We will propose an extension of the model by atomic hydrogen processes
and show that this extension resolves problems of the growth model as discussed by Robertson et
al. [6]. Finally, we discuss the implications for fast deposition of high quality a-Si:H.

SURFACE GROWTH MODEL

A rather successful growth model in explaining the specifics of the growth kinetics of a-
Si:H was formulated by Matsuda et al. [1], Gallagher et al. [2] and Perrin et al. [3] in a number of
papers (hereafter referred to as the MGP-model). The basis of this model is the assumed long
diffusion length of the weakly adsorbed SiH_3 radical on the almost fully hydrogenated growth
surface before sticking. To summarize the MGP-model we will use a notation similar to Perrin et

Mat. Res. Soc. Symp. Proc. Vol. 557 © 1999 Materials Research Society

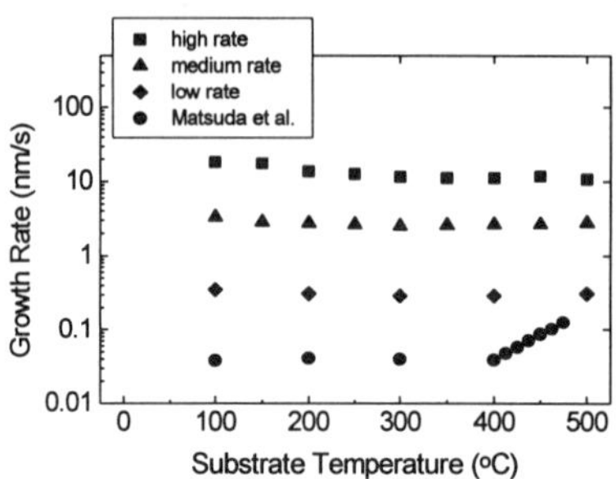

Figure 1: Growth rate as a function of substrate temperature using an intense source of SiH₃ and H for three different plasma conditions [5,12]. The data of Matsuda et al. is shown as reference [1]

al. [3]. All different silicon-hydrogen bonds present on the surface ($\equiv$SiH, $=$SiH$_2$, $-$SiH$_3$) are represented by the total surface hydrogen coverage θ_H. The surface coverage of dangling bonds and weakly adsorbed SiH$_3$ radicals (denoted by $\equiv$Si$-$ and $\equiv$SiHSiH$_3$) are denoted by θ_{DB} and θ_3. Furthermore, the sum of the coverages equals unity or $\theta_H+\theta_{DB}+\theta_3=1$. The different reactions are summarized by Perrin et al. [3] and we will not repeat them here.

To solve for the three surface coverages two additional equations are required. The first equation describes the weakly adsorbed silyl species dwelling on the surface and abstracting, recombining, desorbing and passivating dangling bonds:

$$d\theta_3/dt = \Phi_3 P_{ads3}\theta_H - \theta_3(\nu_{abs}\theta_H + \nu_{des} + 2\nu_{rec3-3}\theta_3 + \nu_{hop3}\theta_{DB}) = 0 \tag{1}$$

Here Φ_3 is the SiH$_3$ flux per surface site, P_{ads3} the probability that SiH$_3$ becomes weakly adsorbed, ν the attempt frequencies for abstraction of chemisorbed hydrogen (ν_{abs}), for desorption into the gas phase (ν_{des}), for recombination of two surface diffusing SiH$_3$ radicals (ν_{rec3-3}) and for hopping to an adjacent hydrogen site (ν_{hop3}). Note that usually desorption is neglected in comparison with the other surface processes (yielding $P_{ads3} = \beta_3$, the surface loss probability) [1,2,3]. In Eq. (1) it is assumed that once a surface diffusing SiH$_3$ encounters a surface dangling bond it sticks with unity probability. The attempt frequencies ν_n usually are assumed to have the generalized form $\nu_n = \nu_{n0}\exp(-E_n/k_bT)$, with E_n the energy barrier for the specific surface process. As discussed by Robertson et al. [6] the parameters ν_{n0} and E_n in practice are strongly constrained by physical requirements, but unfortunately the values are largely unknown.

The second equation describes the surface coverage of dangling bonds. Apart from thermal desorption at high substrate temperature, the MGP-model only includes indirect processes which interact with the dangling bonds on the growth surface. Hence

$$d\theta_{DB}/dt = \theta_3\nu_{abs}\theta_H - \theta_3\nu_{hop3}\theta_{DB} + 2\nu_{H2}\theta_H^2 = 0 \tag{2}$$

The last term on the right hand side describes the formation of H$_2$ from two adjacent chemisorbed hydrogen atoms which desorb into the gas phase. Usually thermal desorption is unimportant for low substrate temperatures (< 400 °C) because of its high activation energy [7].

Since $\theta_{DB}\ll1$ this implies $\nu_{hop3}\gg\nu_{abs}$ (typ. $\nu_{hop3}\approx10^9$-10^{13} Hz depending on temperature [1,3,6]). The MGP-model therefore explains the observation of a reasonable sticking probability of SiH$_3$ (on the order van 0.1-0.2) and a near to unity hydrogenation of the growth surface [1,3]. Furthermore, the MGP-model predicts the growth rate as function of substrate temperature. As can be seen in fig. 1 for low growth rates the growth rate increases as function of the substrate temperature above 400 °C. The latter is caused by the onset of hydrogen desorption from the growth surface [1,3]. This onset of hydrogen desorption was corroborated by hydrogen coverage measurements by Toyoshima et al. [8]. Figure 1 also shows results from an intense source of SiH$_3$/H radicals. Whereas at low growth rates the growth rate increases above 400 °C, the growth

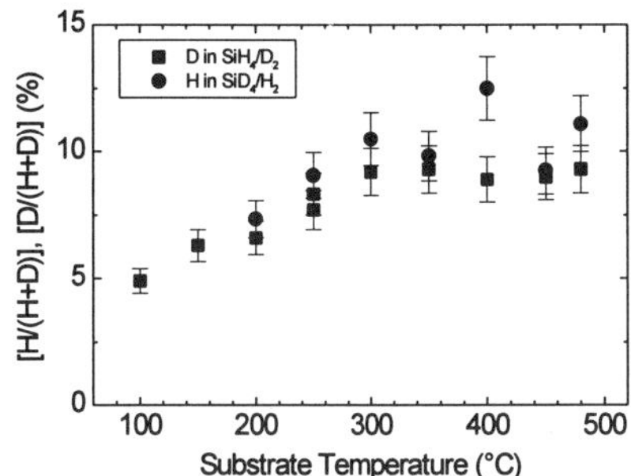

Figure 2: The percentage of H or D originating from gas phase atomic hydrogen/deuterium as measured in bulk a-Si:H:D as a function of substrate temperature deposited using an intense source of H and SiH$_3$ radicals [5,11,12].

rate does not show an increase above a certain temperature for larger growth rates. This can be explained on basis of Eq. (2): the thermal desorption of H$_2$ becomes relatively unimportant in comparison with the abstraction by means of surface diffusing SiH$_3$ radicals.

Another general observation from a-Si:H growth studies is the weak temperature dependence of the growth rate (fig. 1) below the onset of thermal desorption. Within the MGP-model this is explained on the basis of the temperature dependence of β_3. However, the weak temperature dependence of the surface loss probability only leads to a weak temperature dependence of the growth rate if bi-radical recombination on the growth surface is also weakly temperature dependent or can be neglected. The only sticking probability measurement available in the literature and deduced from surface loss probability measurements suggests a constant silyl sticking probability [1]. Unfortunately this measurement is in error since it neglects the increase in mass density with increasing substrate temperature due to hydrogen elimination.

Another problem concerns one of the basic assumptions of the MGP-model and has become clear in recent experiments by Von Keudell et al. [9]. They demonstrated that the dominant interaction of SiH$_3$ with an a-Si:H surface is not abstraction but a self-limiting insertion reaction in strained bonds followed by subsequent elimination of H$_2$. Ramalingam et al. additionally has shown by means of molecular dynamics calculations that the abstraction reaction by means of SiH$_3$ is direct instead of indirect [10]. Moreover, the assumption that only SiH$_3$ abstracts surface hydrogen seems unlikely as other reactive species which can abstract are also present in a plasma like atomic hydrogen [4]. A particular example is shown in fig. 2 and 3. Under similar conditions as for fig. 1 isotope studies were performed [5,11,12]. The presence of both atomic hydrogen-deuterium in the bulk of a-Si:H revealed (cf. fig. 2) that gas phase atomic hydrogen-deuterium takes part in abstraction-passivation reactions on the growth surface. From these experiments the atomic hydrogen abstraction flux is obtained assuming that H is the dominant abstraction reaction (cf. fig. 3). As can be seen the abstraction process is weakly temperature dependent in agreement with hydrogen abstraction-passivation experiment on c-Si and a-Si:H surfaces [13]. Therefore, it seems straightforward to infer that abstraction of chemisorbed hydrogen by atomic hydrogen from the gas phase contributes significantly to the creation of dangling bonds on the a-Si:H growth surface [5]. This assumption naturally explains the temperature dependence of the growth rate since in this case abstraction limits the growth rate. Moreover, including direct abstraction processes by atomic hydrogen has important implications for the problems of the MGP-model as addressed by Robertson et al. [6].

Apart from aditional reactions in Eqs. (1) and (2) to include atomic hydrogen interactions an additional equation is needed which describes the weakly adsorbed atomic hydrogen state (denoted by $\equiv$SiH$-$H) [14]. Thus,

$$d\theta_h/dt = \Phi_H\beta_H\theta_H - \theta_h(2\nu_{recH\text{-}H}\theta_h + \nu_{recH\text{-}3}\theta_3 + \nu_{hopH}\theta_{DB}) = 0 \tag{3}$$

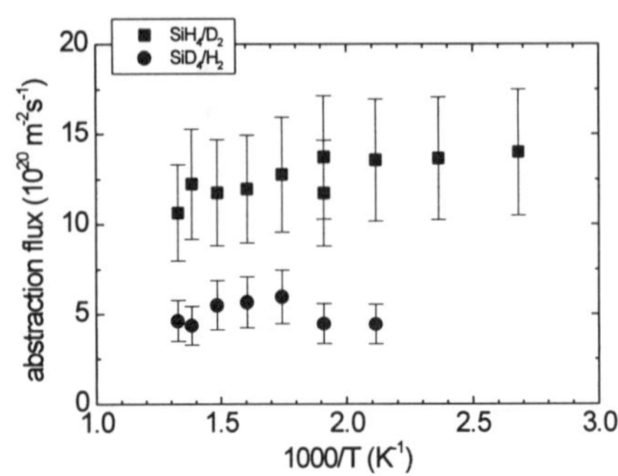

Fig. 3 The atomic hydrogen abstraction flux as obtained from isotope experiments using an intense source of H and SiH₃ [5].

Here θ_h is the fractional coverage of weakly adsorbed hydrogen, Φ_H is the H flux per surface site, β_H the probability that H becomes weakly adsorbed (again desorption of atomic H from the weakly adsorbed state is neglected), ν_{recH-H} the attempt frequency for recombination of two surface diffusing H, ν_{recH-3} the attempt frequency for recombination of surface diffusing H and SiH₃ radicals and ν_{hopH} the attempt frequency for hopping to an adjacent hydrogen site. Direct interaction of gas phase species with weakly adsorbed species is neglected, since, except for θ_H, all surface coverages are small. Equation (1) now becomes

$$d\theta_3/dt = \Phi_3\beta_3\theta_H - \theta_3(\nu_{recH-3}\theta_h + 2\nu_{rec3-3}\theta_3 + \nu_{hop3}\theta_{DB}) = 0 \qquad (4)$$

where we neglected indirect abstraction processes (although this is not essential). In Eq. (2) we explicitly introduce the direct abstraction process by means of atomic hydrogen

$$d\theta_{DB}/dt = \Phi_H P_{abs}\theta_H - \theta_3\nu_{hop3}\theta_{DB} - \theta_h\nu_{hopH}\theta_{DB} + 2\nu_{H2}\theta_H^2 = 0 \qquad (5)$$

Note that Eq. (5) combines the concept of direct abstraction by atomic hydrogen and passivation of dangling bonds by weakly adsorbed SiH₃ and H radicals, similarly to the observations made by Widdra et al. during hydrogen abstraction-passivation experiments [14].

In the limit that H₂ thermal desorption can be neglected (low temperatures or high H and SiH₃ fluxes) and using the fact that $\theta_H \approx 1$, neglecting mutual recombination of two weakly adsorbed SiH₃ or H radicals, it is straightforward to find a solution of Eqs. (3)-(5). The result for θ_{DB} is given by

$$\theta_{DB}=(\nu_{recH-3,0}/(2\nu_{hopH,0}\nu_{hop3,0}))^{1/2}(((\Phi_H P_{abs})^2-(\Phi_3\beta_3-\Phi_H\beta_H)^2)/(\Phi_3\beta_3+\Phi_H\beta_H-\Phi_H P_{abs}))^{1/2} \qquad (6)$$

Equation (6) clearly shows a part depending on the fluxes (and which is weakly temperature dependent through P_{abs}, β_H and β_3) and a temperature dependent part which sets the order of magnitude of the dangling bond coverage. We want to stress here that including direct abstraction processes does not necessary lead to an increase of the surface coverage of dangling bonds. The reason is that the latter is mainly ruled by the interplay between the magnitude of the hopping frequencies, the abstracting flux and the desorption frequency of surface bonded hydrogen. The nature of the weakly adsorbed state, with its high surface diffusivity, automatically garantees a low surface dangling bond density. This reasoning is in agreement with the results of Widdra et al. [14].

Equation (6) also resolves the problem connected to the defect incorporation addressed by Robertson et al. [6]. To make this clear we must return to the MGP-model and its extensions to include a-Si:H bulk properties. The MGP-model, although it explains e.g. the growth rate vs. temperature, it does not predict the a-Si:H defect density nor the hydrogen incorporation. Therefore additional assumptions are necessary. In a previous paper we proposed a statistical incorporation of hydrogen if cross-linking directly following the chemisorption of SiH_3 does not occur [5,15]. Similarly, and in accordance with models by Ganguly et al. [16] and Robertson et al. [6], a statistical incorporation of dangling bonds can occur. In that case the dangling bond density in the bulk scales with the surface dangling bond density

$$N_D = \rho_{a\text{-Si:H}}\theta_{DB} \tag{7}$$

Here N_D and $\rho_{a\text{-Si:H}}$ are the bulk defect density and density of a-Si:H (5×10^{22} cm^{-3}) respectively. The fraction of defects in the bulk is thus equal to the dangling bond coverage (at optimal conditions $N_D \approx 5\times10^{15}$ cm^{-3} or $\theta_{DB} \approx 10^{-7}$). If we now consider the extended MGP-model the dangling bond density for low substrate temperatures is given simply by the ratio of v_{abs}/v_{hop3} (cf. Eq. (2)). It has been pointed out by Robertson et al. [6] that if the statistical incorporation of dangling bonds is valid, the MGP-model is flawed since it leads to unphysically low values for v_{abs} of 10^2-10^6 Hz at reasonable values for v_{hop3}. On the other hand, using ordinary values for the attempt frequencies in the temperature dependent factor, Eq. (6) leads straightforwardly to surface dangling bond densities with the right order of magnitude. Moreover, the temperature dependent factor scales exponentially with temperature (assuming that P_{abs}, β_H and β_3 are weakly temperature dependent) with an activation energy $E_{act} = (E_{hop3} + E_{hopH} - E_{recH\text{-}3})/2$. Since the bulk dangling bond density at constant growth rate decreases with increasing substrate temperature $E_{act} > 0$. In the MGP-model this is obtained if $E_{abs} < E_{hop3}$ a result which seems to be unphysical [1,3]. In the extension of the MGP-model presented here only $E_{recH\text{-}3} < E_{hop3} + E_{hopH}$ which is reasonable since the recombination involves two weakly adsorbed radicals and the barrier for recombination will be likely to be small, or even absent. Since $E_{act}\approx0.3$ eV [1], using $E_{hop3}\approx0.2$ eV [6] and negligible $E_{recH\text{-}3}$ we find $E_{hopH}\approx0.4$ eV.

To get a better insight into Eq. (6) we introduce the following relations between the fluxes and surface loss and sticking probabilities,

$$\Phi_H P_{abs} = \Phi_H s_H + \Phi_3 s_3 \tag{8a}$$

$$\Phi_H \gamma_H = \Phi_3 \gamma_3 \tag{8b}$$

with γ_H, γ_3 the fraction of the flux Φ_H, Φ_3 which is converted into SiH_4 and s_H, s_3 the sticking probability of H and SiH_3. By definition $\beta_3 = \gamma_3 + s_3$ and $\beta_H = \gamma_H + s_H$ (note that we did not include P_{abs} in β_H) and substitution of Eqs. (8) in Eq. (6) leads to

$$\theta_{DB}=(v_{rec,H\text{-}3}/(2v_{hop,H}v_{hop,3}))^{1/2}(s_3 s_H \Phi_H/\gamma_3)^{1/2} \tag{9}$$

Equation (9) has important implications. The flux $s_H\Phi_H$ represents the amount of atomic hydrogen, originating from gas phase atomic hydrogen, which is incorporated and can be determined from isotope experiments (cf. fig. 2). This flux could be actually zero if $s_H = 0$ which would lead to zero defect density in the bulk. In this case SiH_3 scavenges all the weakly adsorbed atomic hydrogen on the growth surface. Furthermore, Eq. (9) states that the defect density does not scale with the growth rate but with the atomic hydrogen incorporated. This implies that high growth rate deposition is possible and will not affect the defect density negatively as long as s_H can be kept small. Using this concept we can explain the defect densities as obtained under high growth conditions (7-10 nm/s) from an intense source of SiH_3 and H (cf. figs. 1-3). The defect density at 400 °C is comparable with PECVD material at 250 °C and deposited at 1-3 Å/s [11,12]: $s_H\Phi_H$ is only about 5 to 10% of the growth flux $s_3\Phi_3$ leading effectively to a lower defect density than would be expected based on the much higher growth rate. The dependence of

θ_{DB} on the atomic hydrogen flux, which implicitly shows the dependence on the growth rate, explains the dependence of the defect density on both temperature and growth rate [6].

CONCLUSIONS

Including direct hydrogen abstraction processes in the MGP-model leads to several interesting features of this extended model. The observed weak temperature dependence of the a-Si:H growth rate is explained in a natural way by the weak temperature dependence of the abstraction process, characteristic for direct gas phase-surface processes. Furthermore, in the extended model the order of magnitude of the defect density is related to the attempt frequencies for the different reactions of the weakly adsorbed species. It is shown that if the defect incorporation mechanism proposed is valid, the defect density is of the proper order of magnitude and scales with gas phase atomic hydrogen incorporated in a-Si:H.

ACKNOWLEDGEMENTS

We gratefully acknowledge financial support from NOVEM, NWO, FOM and AKZO-NOBEL.

REFERENCES

[1] A. Matsuda, K. Nomoto, Y. Takeuchi, A. Suzuki, A. Yuuki, J. Perrin, Surf. Sci. **22**, 50 (1990), A. Matsuda, T. Goto, Mat. Res. Soc. Symp. Proc. **164**, 3 (1990), J.L. Guizot, K. Nomoto, A. Matsuda, Surf. Sci. **244**, 22 (1991)

[2] J.R. Doyle, D.A. Doughty, A. Gallagher, J. Appl. Phys. **68**, 4375 (1990), J.R. Doyle, D.A. Doughty, A. Gallagher, J. Appl. Phys. **71**, 4727 (1992), J.R. Doyle, D.A. Doughty, A. Gallagher, J. Appl. Phys. **71**, 4771 (1992)

[3] J. Perrin, Y. Takeda, N. Hitano, Y. Takeuchi, A. Matsuda, Surf. Sci. **210**, 114 (1989), J. Perrin, M. Shiritani, P. Kae-Nune, H. Videlot, J. Jolly, J. Guillon, J. Vac. Sci. Technol. A **16**, 278 (1998)

[4] K. Miyazaki, Y. Mishiro, T. Kajiwara, K. Uchino, K. Muraoka, T. Okada, M. Maeda, J. Vac. Sci. Technol. A **17**, 155 (1999)

[5] M.C.M. van de Sanden, R.J. Severens, W.M.M. Kessels, F. van de Pas, L.J. van IJzendoorn, D.C. Schram, Mat. Res. Soc. Symp. Proc. **467**, 621 (1997), M.C.M. van de Sanden, R.J. Severens, W.M.M. Kessels, D.C. Schram, Plasma Phys. Control. Fusion **41**, A365 (1999)

[6] J. Robertson, M.J. Powell, Mat. Res. Soc. Symp. Proc. **507**, 897 (1998)

[7] see e.g. K. Sinniah, M.G. Sherman, L.B. Lewis, W.H. Weinberg, J.T. Yates and K.C. Janda, Phys. Rev. Lett. **62**, 567 (1989)

[8] Y. Toyoshima, K. Arai, A. Matsuda, K. Tanaka, J. Appl. Phys. **56**, 1540 (1990)

[9] A. von Keudell, J.R. Abelson, Phys. Rev. B **59**, 5791 (1999)

[10] S. Ramalingam, D. Maroudas, E. Aydill, Mat. Res. Soc. Symp. Proc. **507**, 673 (1998)

[11] W.M.M. Kessels, M.C.M. van de Sanden, R.J. Severens, L.J. van IJzendoorn, D.C. Schram, Mat. Res. Soc. Symp. Proc. **507**, 529 (1998)

[12] W.M.M. Kessels, A.H.M. Smets, B.A. Korevaar, G.J. Adriaenssens, M.C.M. van de Sanden, D.C. Schram, these Proceedings

[13] see e.g. D.D. Koleske, S.M. Gates, B. Jackson, J. Chem. Phys. **101**, 3301 (1994)

[14] W.Widdra, S.I. Yi, R. Maboudian, G.A.D. Briggs, W.H. Weinberg, Phys. Rev. Lett. **74**, 2074 (1995)

[15] W.M.M. Kessels, R.J. Severens, M.C.M. van de Sanden, D.C. Schram, J. Non-Cryst. Solids **227-230**, 133 (1998)

[16] G. Ganguly, A. Matsuda, Phys Rev. B **47**, 3661 (1993)

DEVELOPMENT OF ULTRA-CLEAN PLASMA DEPOSITION PROCESS

Toshihiro KAMEI and Akihisa MATSUDA
Thin Film Silicon Solar Cells Superlab, Electrotechnical Laboratory, 1-1-4 Umezono Tsukuba 3058568, Japan

ABSTRACT

We have developed a new type of ultra-high vacuum plasma-enhanced chemical vapor deposition (UHV/PECVD) system. According to high sensitivity secondary ion mass spectrometry, device quality hydrogenated amorphous silicon (a-Si:H) films deposited at 250°C at a deposition rate of 1 Å/s contains 10^{15} cm^{-3} of O, 10^{15} cm^{-3} of C, and 10^{14} cm^{-3} of N impurities, while low defect hydrogenated microcrystalline silicon (μc-Si:H) films deposited at 200°C at a very low rate of 0.1 Å/s include 10^{16} cm^{-3} of O, 10^{15} cm^{-3} of C and 10^{16} cm^{-3} of N. These are the lowest concentrations of atmospheric contaminants for these kinds of materials observed so far. The essential features of the present UHV/PECVD system are an extremely low outgassing rate of 8×10^{-9} Torr·ℓ/s, extremely low partial pressure of contaminant gas species $<10^{-12}$ Torr, and purification of feed gas SiH$_4$ at "point of use". These efforts are quite important not only for clarifying the microscopic mechanism of photo-induced degradation in a-Si:H, but also for enlarging the crystalline grain size in μc-Si:H. μc-Si:H with a grain size of ≈1000 Å as determined by Scherrer's formula can be obtained at the higher rate of 1.5 Å/s by utilizing a VHF (Very High Frequency) plasma. The specific origins of impurities in the films are also discussed.

INTRODUCTION

Impurities are easily incorporated into the films in the plasma deposition process which is typically used to grow high quality hydrogenated amorphous silicon (a-Si:H) films. Even the films made by an ultra high vacuum (UHV) system contain 10^{18} cm^{-3} of O, 10^{17} - 10^{18} cm^{-3} of C, and 10^{16} - 10^{17} cm^{-3} of N [1, 2, 3]. In particular, the O content had not been reduced below 1×10^{18} cm^{-3} in spite of several attempts to reduce the overall impurity contents. The convergence of minimum O contents to 10^{18} cm^{-3} as observed by these research groups suggests an underlying mechanism which determines the impurity levels. Reactive species in the plasma (e.g., H) pick up the impurity atoms from the walls into the gaseous phase. The impurity atoms (O) are converted into the gaseous species (H$_2$O) which is decomposed in the plasma and then reincorporated into the film. This is the impurity-incorporation mechanism due to the plasma-wall interaction process. It is often assumed that the plasma-wall interaction process mediated by reactive species (H) is the limiting process for reduction of impurity contents, especially for O contents below 10^{18} cm^{-3}.

Recently, however, Kroll et al. managed to reduce the O content to 5×10^{17} cm^{-3} with a high growth rate process ~10 Å/s using a VHF (Very High Frequency) plasma and a silane gas purifier [4]. We have also demonstrated that impurity levels in a-Si:H deposited at 1 Å/s can be reduced to 10^{15} cm^{-3} both for O and C, and 10^{14} cm^{-3} for N, and found that the saturation value of light induced defects (LID) 10^{17} cm^{-3} exceeds any atmospheric contaminant concentrations [5]. These experimental findings exclude any models which assume one-to-one correlation between LID and impurity atoms [6, 7, 8]. Moreover, as we will show below, such a ultra-clean process environment is favorable for obtaining large crystalline grain μc-Si:H films [9].

In this paper, the essential features of our UHV plasma enhanced chemical vapor deposition (PECVD) system are described and the origins of impurities incorporated in the films are identified. In addition, we demonstrate that μc-Si:H with a grain size of ≈1000 Å as

Mat. Res. Soc. Symp. Proc. Vol. 557 © **1999 Materials Research Society**

determined by Scherrer's formula can be made at a reasonably high rate of 1.5 Å/s by utilizing a VHF plasma.

ULTRA-CLEAN PLASMA DEPOSITION PROCESS

We consider four origins of impurities in the plasma deposited materials: 1) desorption of contaminant gas adsorbing on the reactor wall, e.g., H_2O, O_2, CO, CO_2, N_2; 2) oil backdiffusion from vacuum pumps; 3) contamination of the feed gas SiH_4, e.g., H_2O, siloxane, CO, CO_2, N_2, hydrocarbons, chlorosilane, and desorption of contaminant gas adsorbing on the gas line; 4) impurity-incorporation due to the plasma-wall interaction process. We adopt the following methods to suppress impurity contents from these four origins.

(1) We performed a thorough baking to reduce outgassing of the reactor wall. Tanaka et al. reported that the O concentration in the film is well correlated with the reactor buildup rate (=leakage + outgassing rate of the reactor) [3], as shown in Figure 1. Since the reactor buildup rate simply represents an outgassing rate without leakage, it follows from Figure 1 that O impurity levels above 10^{18} cm^{-3} mainly originate from contaminant gas desorbing from the walls. Therefore, a baking process of the reactor is essential. In addition, we emphasize that baking should be homogeneous to reduce the adsorbing gas efficiently; otherwise the transfer of adsorbing gas occurs from higher to lower temperature regions in the chamber. In particular, the homogeneity is very important for the surface exposed to the plasma (inner wall of shroud), since the repeated process of partial peeling-off and film-deposition on the walls increases the surface area significantly. For this purpose, our reactor chamber is equipped with a shroud through which silicone oil heated to 250°C is circulated (Figure 2). This method ensures homogeneity of baking. The load-lock system for loading/unloading substrates is also essential to avoid the air-exposure of the reactor which allows the contaminant gas adsorbing on the wall.

(2) Oil-backdiffusion from vacuum pumps is a source of C impurities; oil molecules not only from the rotary vane pump but also from the conventional turbo molecular pump (TMP) diffuse

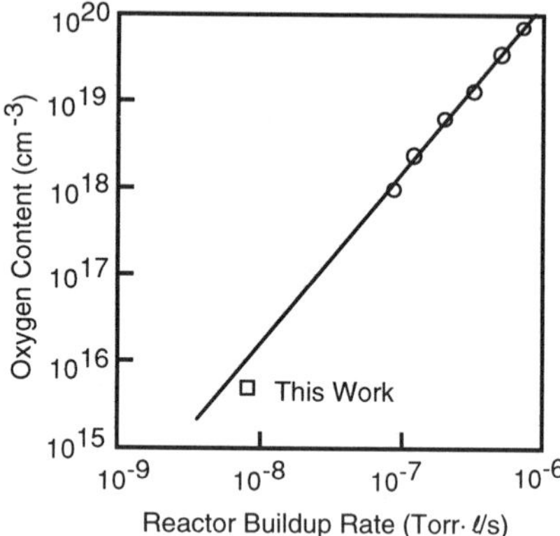

Figure 1. Correlation between buildup rate of reactor chamber and oxygen content in a-Si: H. Data points denoted by open circles are taken from ref. [3].

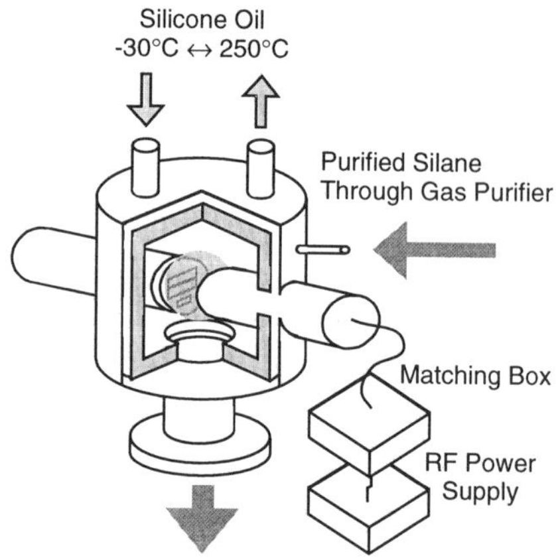

Figure 2. Baking and cooling system flowing the silicone oil through the shroud equipped with the reactor chamber.

back to the reactor and then incorporate into films through plasma decomposition. To avoid this, a rotary vane pump is used only to evacuate the outlet of magnetic bearing TMP without lube oil, and a dry pump is used specifically for rough pumping of the chamber from atmospheric pressure; this inexpensive dry pump cannot be used to evacuate reactive gas. Since oil from the rotary vane pump cannot diffuse back because of an extremely high compressive ratio of the TMP for heavy molecules such as oil molecules, this pumping system is substantially oil free. In particular, evacuation of the reactor chamber is done by a tandem magnetic bearing TMP (STP-X1000A/SEIKO SEIKI).

(3) Contaminant gas present in the feed gas in the absence of air leaks originates from the gas cylinder itself and from adsorption on the gas line; the feed gas is contaminated by outgassing of the gas line before it is introduced from the gas cylinder into the reactor. The contaminant gas concentrations in the SiH_4 gas (99.999%) cylinder are <1 ppm for H_2O, CO, CO_2, and N_2, <0.1 ppm for hydrocarbons, and <2 ppm for chloride. The oxide and chlorinated contamination (H_2O, CO, CO_2, siloxane, chlorosilane, etc.) are reduced below the ppb range by the gas purifier (Waferpure Micro/Millipore) just before it enters the reactor (purification at "point of use"). This purifier, however, cannot eliminate contamination of nitrogen and hydrocarbons. Therefore, efforts are still needed to reduce the outgassing of the gas line. Most components of the gas line, i.e., the stainless steel tube, mass flow controller, and valve etc., are accommodated in a heater box which makes possible homogeneous baking at 150°C by hot air.

(4) An attempt to reduce impurity incorporation due to the plasma-wall interaction process was done by refrigerating the walls; the circulating silicone oil was cooled down to -30°C during the growth, as shown in Figure 2. To some extent, this prevents impurity-incorporation into the gaseous phase which is chemically or thermally enhanced by plasma, i.e., ions, radicals, and light. A temperature as low as -30°C traps H_2O, possibly the main source of O impurity, at the shroud without condensation of SiH_4 molecules.

Based on these concepts, the new type of UHV/PECVD system is designed. A 13.56 -100 MHz power supply is used to produce a glow discharge in a capacitively coupled reactor. Once the reactor chamber is exposed to air, baking is done for a week: 250°C for the shroud, 350-400°C for the substrate unit, 300°C for the cathode, 250°C for the reactor chamber wall, and 150°C for the whole gas line. In the daily procedure, the substrates are baked at 360°C for several hours in the loading chamber before being transferred reactor chamber. The details of deposition procedures were reported elsewhere [5, 9, 10].

A residual gas analysis is carried out by a XHV compatible quadrupole mass spectrometer with a separate ion source (XHV Q-mass/ULVAC Japan). The XHV Q-mass can measure partial pressures of 10^{-13} Torr without being affected by its own outgassing. This aspect makes it different from the conventional Q-mass. The O, C, and N contents in the film are analyzed by secondary ion mass spectrometry (SIMS). For SIMS measurements, films on float zone crystalline Si substrates are used.

RESULTS AND DISCUSSIONS

Prior to a-Si:H deposition, the base pressure and the buildup rate of the reactor chamber are 1.7×10^{-10} Torr and 8×10^{-9} Torr·l/s, respectively. It is noted that the buildup rate is one order of magnitude lower compared to the previous UHV system [2, 3]. Moreover, a major residual gas is hydrogen which is not a contaminant gas for the deposition of a-Si:H, and partial pressures of contaminant gas species, i.e., methane, water, and carbon monoxide are extremely low $<10^{-12}$ Torr, as shown in Figure 3(a). This residual gas analysis suggests that the desorption of contaminant gas is much lower than the observed total buildup rate of 8×10^{-9} Torr·l/s. It is also noted that no broad envelope of hydrocarbon cracking from oil is observed. This confirms that oil-back diffusion does not take place from the vacuum pumps and, thus, the present evacuating system is substantially oil-free, as designed. After deposition, the total residual pressure is higher, but this is due to an increasing partial pressure of hydrogen (Figure 3(b)). Partial pressures of contaminant gas, i.e., methane and water, are still kept very low, but carbon monoxide might be increased to a partial pressure of ~10^{-11} Torr, judging from the cracking pattern of SiH_4. The impurity levels in the resulting a-Si:H films are 5×10^{15} cm^{-3} for O, 7×10^{15} cm^{-3} for C, and 5×10^{14} cm^{-3} for N (Table 1) [5]. It is surprising that the O content is 2-3 orders of magnitude lower than that in the material made by earlier UHV system [1-4].

The O content of 5×10^{15} cm^{-3} achieved in this work is comparable to the value estimated from extrapolation of the linear correlation between the reactor buildup rate and O content in the films, as shown in Figure 1. This linear relationship indicates that impurity levels involved in feed gas SiH_4 are low enough (by use of the gas purifier), and that O impurities in the present pure films are coming from the outgassing of the reactor wall.

We also investigated the

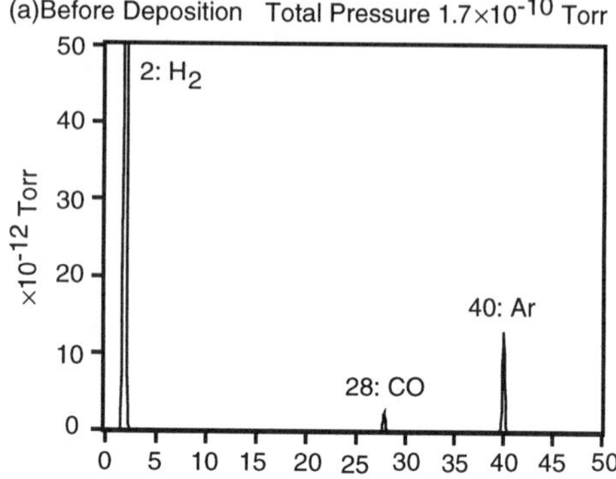

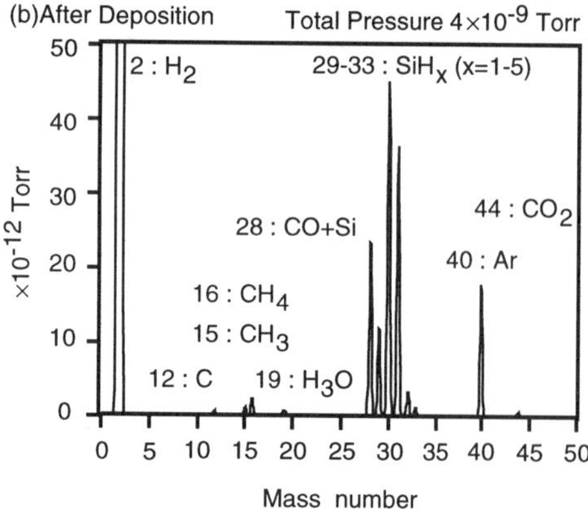

Figure 3.   Residual gas analysis by a XHV quadrupole mass spectrometer (the peak with mass number 1 is omitted) (a) before (a) after deposition.

Table 1 Atmospheric impurity concentrations in a-Si:H as well as μc-Si:H deposited with various conditions. The excitation frequency of the plasma is 13.56 MHz.

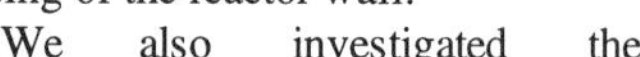

	T_{sub}	GP	Shroud Temp.	O [cm^{-3}]	C [cm^{-3}]	N [cm^{-3}]	Notes
a-Si:H	250°C	Used	-30°C	5×10^{15}	7×10^{15}	5×10^{14}	
a-Si:H	250°C	Used	-30°C	3×10^{15}	6×10^{15}	$< 2\times10^{15}$	
a-Si:H	250°C	Used	+20°C	8×10^{15}	6×10^{15}	$< 2\times10^{15}$	
a-Si:H	250°C	Used	-30°C	3×10^{15}	$< 9\times10^{15}$	$< 2\times10^{15}$	
a-Si:H	250°C	Not used	-30°C	3×10^{16}	$< 9\times10^{15}$	$< 2\times10^{15}$	
a-Si:H	250°C	Not used	+20°C	5×10^{16}	$< 9\times10^{15}$	$< 2\times10^{15}$	
μc-Si:H	200°C	Used	+20°C	5×10^{16}	6×10^{15}	3×10^{16}	Low Defect Density
μc-Si:H	350°C	Used	+20°C	5×10^{17}	1×10^{16}	5×10^{16}	Large Crystalline Grain

T_{sub}: Substrate temperature   Deposition Rate: 1Å/s for a-Si:H; 0.1Å/s for μc-Si:H

contribution of contaminant gas arising from the feed gas and the effect of shroud cooling during the deposition. The results are summarized in Table 1. The contaminant gas present in the unpurified feed gas only contributes to O contents of 3×10^{16} cm^{-3} and 4×10^{16} cm^{-3} at -30°C and 20°C shroud temperatures, respectively. The absence of changes in the C and N contents are due to incapability of the gas purifier to eliminate hydrocarbons and nitrogen. Earlier attempts at reducing impurity contents were less successful not due to impurity incorporation process via the plasma-wall interaction process but rather due to insufficient baking of previous systems. H_2O is the main origin of oxygen impurities in earlier studies [11]. On the contrary, in the present pure films, CO adsorbing on the walls might be the main origin of O impurities, since it is now the major oxygen related gas species (Figure 3(a), (b)). The C impurity in the films may also come from CO adsorbing on the wall or from hydrocarbons (not removed by the gas purifier) in the feed gas.

The effect of shroud cooling is dependent on the relative contaminant gas concentration c_i in the reactor. The shroud cooling from 20°C to -30°C can reduce the O content only by 5×10^{15} cm^{-3} with the gas purifier (for low c_i), while 2×10^{16} cm^{-3} otherwise (for high c_i). No appreciable dependence of C and N concentrations on the shroud temperature can be detected. When SiH_4 gas is purified, c_i is low enough so that cooling of the shroud has only a minor effect on the reduction of the O content. For high c_i, the cooling shroud to trap the contaminant gas, especially water, becomes effective.

Impurity concentrations in μc-Si:H are also reduced drastically down to 5×10^{16} cm^{-3} for O, 6×10^{15} cm^{-3} for C and 3×10^{16} cm^{-3} for N (Table 1). The film was deposited at 200°C where the lowest spin density is usually obtained. These values are the lowest among μc-Si:H films reported so far, but are higher compared to those of their amorphous counterparts because of the low deposition rate of 0.1 Å/s in the μc-Si:H films. When the deposition temperature is increased to 350°C, the crystalline grain size is enlarged to ≈1000 Å, as estimated from the X-ray diffraction pattern using Scherrer's formula [9]. The O, C, and N contents in this film are 5×10^{17} cm^{-3}, 1×10^{16} cm^{-3}, and 5×10^{16} cm^{-3}, respectively (Table 1). An increase in the O content by elevating the temperature from 200°C to 350°C might be ascribed to either an increase in the incorporation probability of O-contaminant species, or to a smaller difference between predeposition baking temperature and deposition temperature. These films were made from rf (13.56 MHz) plasma.

To pursue larger crystalline grains or to clarify which factors are limiting the crystalline grain size, we should reduce ion bombardment as well as impurity concentration while optimizing the deposition temperature [9]. It is expected that the VHF plasma has an advantage for higher growth rate processing with reduced ion damage so that above two factors may be satisfied at the same time [12]. In fact, as shown in Figure 4, μc-Si:H made from a VHF plasma has comparable

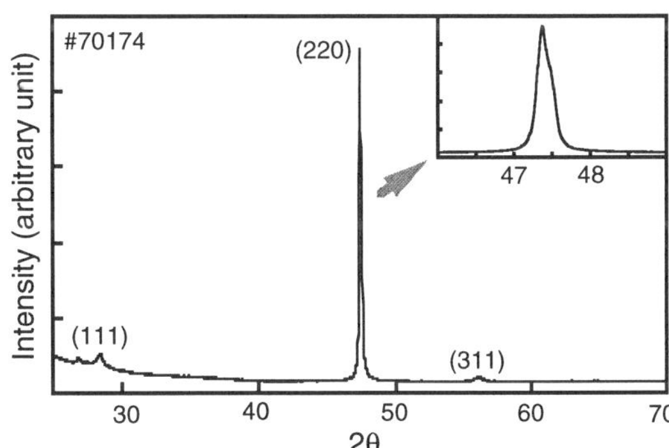

Figure 4. X-ray diffraction pattern of μc-Si:H grown on quartz substrates made from VHF plasma. The inset shows the magnification of (220) diffraction peak. The deposition conditions are following: 60MHz, deposition temperature of 350°C, 300 mTorr, SiH_4 flow rate of 5 sccm, H_2 flow rate of 120 sccm, which give rise to deposition rate of 1.5 Å/s. The film thickness is 1.17 μm.

crystalline grain size of ≈ 1000 Å as determined by Scherrer's formula and enhanced X-ray diffraction intensity at the relatively high rate of 1.5 Å/s.

CONCLUSIONS

We addressed the essential features of our ultra-clean plasma deposition process. The most important is the extremely low outgassing of the reactor walls. The contaminants in unpurified feed gas only contribute $\sim 10^{16}$ cm^{-3} of O impurities in the present films. The shroud cooling during the deposition is not effective as long as the relative contaminant gas concentration c_i in the reactor is kept low. The utilization of the VHF plasma in the ultra-clean process is favorable for obtaining large crystalline grain size (≈ 1000 Å) μc-Si:H at relatively high rates of 1.5 Å/s.

ACKNOWLEDGMENT

The authors acknowledge valuable discussions with Drs. K. Arai, S. Hara, and P. Stradins.

REFERENCES

1. C. C. Tsai, J. C. Knights, and M. J. Thompson, J. Non-Cryst. Solids **6 6**, 45 (1984).
2. S. Tsuda, T. Takahama, M. Isomura, H. Tarui, Y. Nakashima, Y. Hishikawa, N. Nakamura, T. Matsuoka, H. Nishiwaki, S. Nakano, M. Ohnishi, and Y. Kuwano, Jap. J. Appl. Phys. **26**, 33 (1987).
3. H. Tanaka, M. Koyama, K. Miyachi, Y. Ashida, Y. Ohashi, N. Fukuda, and A. Nitta, in Proceedings of the 4th International Photovoltaic Science and Engineering Conference (The Institution of Radio and Radio Electronics Engineers Australia, New South Wales, Australia, 1989).
4. U. Kroll, J. Meier, H. Keppner, A. Shah, S. D. Littlewood, I. E. Kelly, and P. Giannoules, J. Vac. Sci. Technol. **A13**, 2742 (1995).
5. T. Kamei, N. Hata, A. Matsuda, T. Uchiyama, S. Amano, K. Tsukamoto, Y. Yoshioka, and T. Hirao, Appl. Phys. Lett. **6 8**, 2380 (1996).
6. D. Redfield and R. H. Bube, Phys. Rev. Lett. **6 5**, 464 (1990).
7. D. Redfield and R. H. Bube, J. Non-Cryst. Solids **137 & 138**, 215 (1991).
8. H. Fritzsche, J. Non-Cryst. Solids **190**, 180 (1995).
9. T. Kamei, M. Kondo, and A. Matsuda, Jpn. J. Appl. Phys. **3 7**, L265 (1998).
10. T. Kamei and A. Matsuda, J. Vac. Sci. Tech. **A 1 7**, 113 (1999).
11. J. C. Knights, in The *Physics of Hydrogenated Amorphous Silicon I*, edited by J. D. Joannopoulos and G. Lucovsky (Springer, Berlin, 1984), Vol. 55, p. 5.
12. S. Oda, J. Noda, and M. Matsumura, Jpn. J. Appl. Phys. **29**, 1889 (1990).

REMOTE SILANE PLASMA CHEMISTRY EFFECTS AND THEIR CORRELATION WITH a-Si:H FILM PROPERTIES

W.M.M. KESSELS, A.H.M. SMETS, B.A. KOREVAAR, G.J. ADRIAENSSENS[*], M.C.M. VAN DE SANDEN, AND D.C. SCHRAM
Dept. of Appl. Physics, Eindhoven University of Technology, P.O. Box 513, 5600 MB Eindhoven, The Netherlands, w.m.m.kessels@phys.tue.nl
[*]Semiconductor Physics Laboratory, Katholieke Universiteit Leuven, Celestijnenlaan 200 D, B3001 Heverlee-Leuven, Belgium

ABSTRACT

A remote silane plasma, capable of depositing solar grade a-Si:H at a rate of 10 nm/s and with an up to ten times higher hole drift mobility than standard a-Si:H, has been investigated by means of several plasma diagnostics. The creation of the different reactive species in the plasma and their contribution to film growth has been analyzed and is correlated with the film properties obtained under various conditions. Furthermore, the first results on a n-i-p solar cell with the intrinsic a-Si:H film deposited by this remote plasma are presented.

INTRODUCTION

An increase of the a-Si:H growth rate can stimulate its application in thin film solar cells as a more competitive position in the energy market can be obtained. Remote plasmas are one of the candidates to meet this goal as they allow a better and independent optimization of the plasma conditions. The so-called 'expanding thermal plasma' (ETP) is such a plasma which has proven to be able to deposit solar grade a-Si:H at a rate of 10 nm/s. Additionally, the larger freedom in plasma parameters can reveal new insights in a-Si:H film properties as will be shown in this paper. Furthermore, remote plasmas facilitate the study of the plasma and surface chemistry during deposition which can lead to a better understanding of a-Si:H deposition when a correlation with the film properties is made. An onset to the latter will be given in this contribution.

EXPERIMENTAL SETUP

The expanding thermal plasma deposition technique (cf. Fig. 1) is based on generation of an Ar-H_2 plasma in a thermal plasma source (cascaded arc), which is subsequently expanded in a low pressure chamber where it dissociates SiH_4. The dc plasma source is operated at a current and voltage of respectively 45 A and about 180 V. The Ar flow used is 55 sccs (standard cubic centimeter per second) while the H_2 flow is varied between 0 - 15 sccs and the source pressure is about 400 mbar. Pure SiH_4 is admixed in the low pressure (0.2 mbar) deposition chamber just behind the source exit by means of an injection ring and at a fixed flow of 10 sccs. A substrate holder with accurate substrate temperature control (100 - 500 °C) is positioned at 38 cm from the plasma source. The plasma chemistry has been studied by replacing the substrate holder by a mass spectrometer (cf. Fig. 1), capable of analyzing the flux of neutrals, radicals and ions arriving at the substrate. Furthermore, a Langmuir probe and optical system have been used to obtain absolute and spatially resolved ion fluxes [1] and to monitor the plasma emission respectively. A residual gas analyzer is used to determine the consumption of SiH_4 [2]. Films have been deposited on 2.5×2.5 cm^2 p-type Si(111) substrates (10 - 20 Ωcm) and on Corning

Mat. Res. Soc. Symp. Proc. Vol. 557 © 1999 Materials Research Society

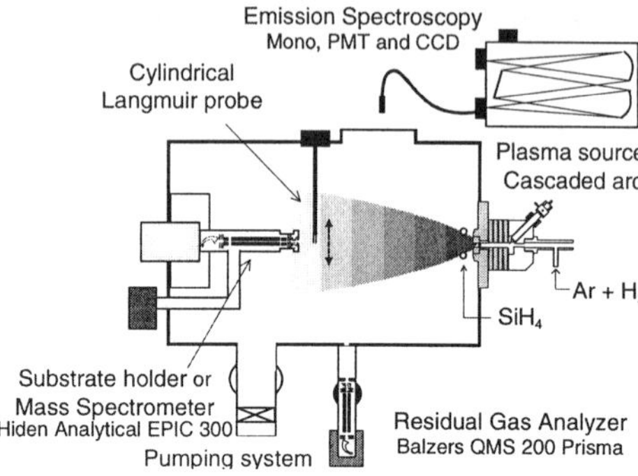

Fig. 1: Expanding thermal plasma deposition setup with plasma diagnostics

7059 glass. Refractive index, growth rate and hydrogen content have been obtained by in-situ ellipsometry (632.8 nm) and ex-situ FTIR transmission spectroscopy. The photo-conductivity has been obtained under AM1.5 (100 mW/cm^2) illumination using coplanar Al contacts and the dark conductivity has been determined after 30 min. anneal of the films at 160 °C. The electron and hole drift mobilities have been obtained by standard time-of-flight (TOF) experiments [3] using 2-4 µm thick a-Si:H films sandwiched between a bottom Cr substrate and a top semi-transparent Cr contact.

PLASMA CHEMISTRY

A detailed study of the plasma emanating from the plasma source has revealed that the downstream electron temperature is relatively low (0.1 - 0.3 eV) due to the fact that no power is coupled into the expanding plasma [1,2]. SiH$_4$ dissociation is therefore dominated by interactions with reactive heavy particles from the source. When no H$_2$ is admixed these are mainly Ar$^+$ ions, while H$^+$ ions prevail for larger H$_2$ flows. At the same time, admixing of H$_2$ leads to a drastic decrease of the total ion flux emanating from the source (from ~3 sccs for no H$_2$ to 0.08 sccs for 10 sccs H$_2$ [1,2]) and the plasma source acts mainly as an atomic hydrogen source.

Consequently at no or low H$_2$ flows ion-molecule reactions dominate the SiH$_4$ dissociation:

$$Ar^+ (H^+) + SiH_4 \;\rightarrow\; SiH_n^+ + pH_2 + qH + Ar \quad (n{\leq}3) \tag{1}$$

which are followed immediately by dissociative recombination reactions with electrons as long as the electron density is high ($>10^{17}$ m^{-3}):

$$SiH_n^+ + e \;\rightarrow\; SiH_m + pH + qH_2 \quad (n{\leq}2) \tag{2}$$

This leads mainly to silane radicals like Si, SiH and SiH$_2$ with a very high reactivity both at the surface and in the plasma (e.g., with SiH$_4$). When the electron density is reduced to a sufficient extent sequential ion-molecule reactions can compete with recombination:

$$Si_nH_m^+ + SiH_4 \;\rightarrow\; Si_{n+1}H_p^+ + qH_2 \tag{3}$$

This leads to large, hydrogen poor cationic clusters containing up to ten silicon atoms [4].

For larger H$_2$ flows SiH$_4$ dissociation is dominated by the hydrogen abstraction reaction by atomic hydrogen:

$$H + SiH_4 \;\rightarrow\; SiH_3 + H_2 \tag{4}$$

leading to a large SiH$_3$ flux. The downstream electron density is already low due to the reduced total ion flux from the source, therefore no strong dissociative recombination takes place and cationic clusters are created by sequential ion-molecule reactions (eq. (3)).

Experimental evidence for this reaction scheme is given in Fig. 2. Figure 2a shows the film growth rate as a function of H$_2$ in the source. It strongly reduces when a significant amount of H$_2$ is admixed due to the lower reactivity of H with SiH$_4$ than Ar$^+$, yet the deposition rate is still around 10 nm/s for higher H$_2$ flows. The growth rate depends weakly on the substrate temperature and this is mainly due to differences in film density (cf. next section). Furthermore, it is interesting to mention that the SiH$_4$ consumption as determined by the residual gas analyzer is well correlated with the growth flux (growth rate multiplied by film density) indicating that SiH$_4$ is mainly converted into a-Si:H and not into polysilanes [1].

Information on the radicals produced is obtained by appearance potential mass spectrometry and optical emission spectroscopy. The SiH$_3$ flux normalized to the SiH$_4$ consumption or growth flux (cf. Fig. 2b) increases with increasing H$_2$ flow as expected. Figure 2c indicates that the generation of reactive radicals per dissociated SiH$_4$ molecule decreases as a function of H$_2$ flow as the SiH* ($A^2\Delta$-$X^2\Pi$, around 414 nm) and Si* ($4s^1P^0$-$3p^{21}S$, 390.6 nm) emission normalized to the SiH$_4$ consumption decreases. It is expected that their emission is an appropriate indication of recombination of silane ions: reaction (2) partly leads to these (excited) radicals and electron induced excitation (dissociation) of ground state radicals (SiH$_4$) is excluded as energetic electrons are absent. It is therefore not surprising that the total SiH* and Si* emission (i.e., before

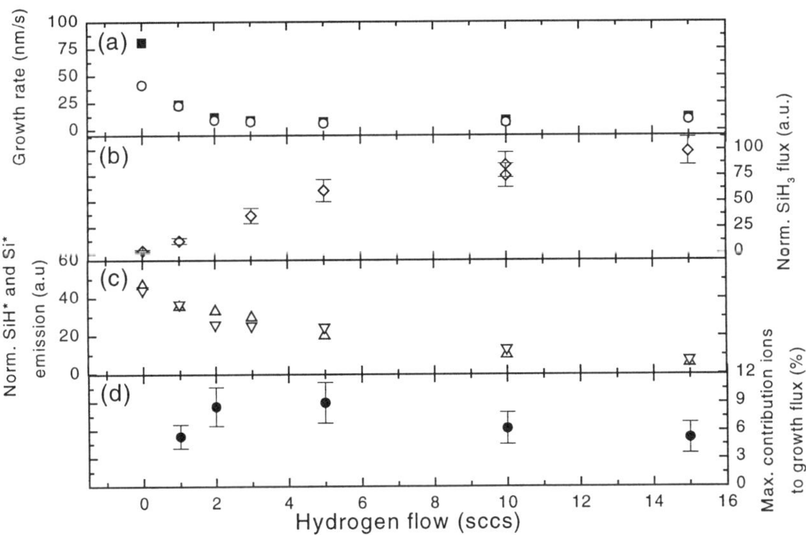

Fig. 2: Growth rate at substrate temperatures of 250 (■) and 400 °C (o) (a), SiH$_3$ flux normalized to growth flux in arbitrary units (b), SiH* and Si* emission normalized to growth flux in arbitrary units (c) and maximum contribution of cluster ions to film growth (d) as a function of H$_2$ flow in the arc plasma source.

normalization) is well correlated with the total ion flux emanation from the plasma source.

The contribution of the hydrogen poor cationic silicon clusters to film growth (cf. Fig. 2d), as determined from a combination of mass spectrometry and Langmuir probe measurements, is in the range of 5 to 9 % when unity sticking probability of the cluster ions is assumed [1]. This rather H_2 flow independent contribution and the increasing contribution of SiH_3 with increasing H_2 flow, suggest again that at low H_2 flows the deposition is governed mainly by reactive silane radicals (SiH_n, $n \leq 2$) and presumably reactive polysilane radicals (e.g., Si_2H_n, $n \leq 4$). The latter can be created in reactions of reactive silane radicals with SiH_4.

FILM PROPERTIES

Films have been deposited under the conditions of Fig. 2 at a substrate temperature of 250 and 400 °C. The growth rates have already been given in Fig. 2a. In ref. 5 it has been shown that the substrate temperature is a key parameter in obtaining dense a-Si:H films as the fraction of hydrogen built in the material is strongly dependent on this temperature. A high substrate temperature (>350 °C) is needed for films deposited by the ETP technique to obtain dense, purely amorphous, films (i.e., refractive index of 4.3 at 632.8 nm) [5]. The hydrogen content of films deposited at 250 °C is 18 % and the hydrogen is mainly bonded as SiH_2 and/or SiH on voids leading to a very high micro-structure parameter of 0.8. For films deposited at 400 °C the hydrogen content is 7 %. This is somewhat lower than for solar grade films deposited with conventional techniques (at lower substrate temperatures), but the micro-structure parameter is still high (0.2). The lower hydrogen content of the latter films leads to a relatively low Tauc band gap of 1.66 eV.

When these parameters are considered for various H_2 flows in the arc, it is striking that the hydrogen content and micro-structure are not significantly depending on this flow. This indicates that these structural parameters are mainly determined by the substrate temperature and that the type of species contributing to growth do not have a significant influence. Only for very low H_2 flows (<2 sccs) there is a significant decrease of the refractive index and the hydrogen content, indicating that these films contain a considerable amount of voids.

The opto-electronic properties of the films show more dependence on the H_2 flow used as shown in Fig. 3. Especially the photoconductivity increases with increasing H_2 admixture and it is higher for the films deposited at 400 °C than at 250 °C. The dark conductivity shows only a decrease for smaller H_2 flows at 250 °C, but as already mentioned the properties of these low density films are inferior. Compared to rf PECVD a-Si:H the photo-response of the films is rather low ($\sim 10^3$) and it is even lower for the films deposited at 400 °C than at 250 °C. This can be attributed to two facts: firstly by the high hole drift mobility, as will discussed below, and secondly by the relatively low hydrogen content of the films deposited at 400 °C corresponding to a relatively small band gap, a relatively small activation energy of the dark conductivity and a relatively high dark conductivity itself. The activation energy of the dark conductivity is 0.76 eV for the films deposited at 400 °C and 0.90 eV for films deposited at 250 °C, which explains the higher dark conductivity for films deposited at 400 °C. The activation energy does not show a clear trend for the different H_2 flows used. Of importance is the fact that the cubic band gap of the material deposited at 400 °C is 1.51 eV, about twice the activation energy, showing that the material does not suffer from impurities [6].

The best film quality has been obtained under the condition with 10 sccs H_2 at 400 °C (and 55 sccs Ar, 10 sccs SiH_4, 45 A arc current and 20 Pa downstream pressure) in which the deposition

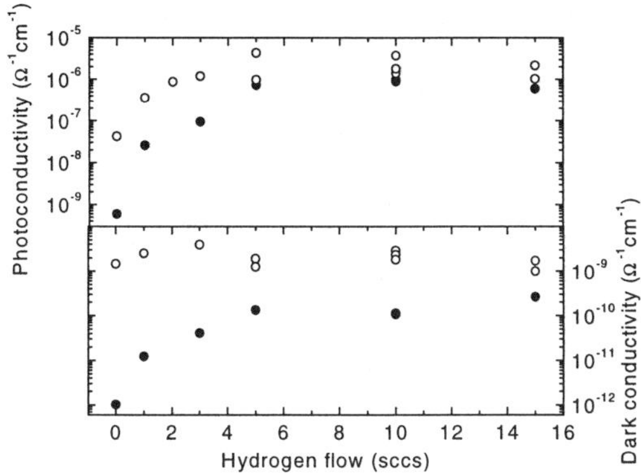

Fig. 3: AM1.5 photo and dark conductivity for films deposited at 250 (●) and 400 °C (o).

is dominated by SiH_3 and the contribution of very reactive radicals is much less significant. The corresponding growth rate and SiH_4 consumption are respectively 10 nm/s and 12%. This condition has been used in the experiments discussed next.

That the conductivity measurements and the corresponding photo-response can be mis-leading in the judgment of a material's electronic properties is shown in Fig. 4, where the hole and electron drift mobilities are shown for an electric field of 10^4 Vcm^{-1}. The hole mobility of the material deposited by the expanding thermal plasma is about one order of magnitude larger than for a-Si:H deposited by conventional techniques [7]. However, the electron drift mobility is about a factor 3 to 6 smaller, in agreement with the lower photoconductivity (cf. Fig. 3) and lower photo-response. The lower photoconductivity is furthermore directly linked to the relatively high hole drift mobility [8]. The hole drift mobility, which correlates with the hole diffusion length being usually the limiting parameter in the application of a-Si:H in thin film solar cells, suggests that a-Si:H deposition by the ETP technique at a rate of 10 nm/s is promising. It is worthy to mention that the ETP a-Si:H shows some similarities with HWCVD a-Si:H, which is also deposited at higher substrate temperatures (>350 °C) and which shows a relatively high ambipolar diffusion length [9].

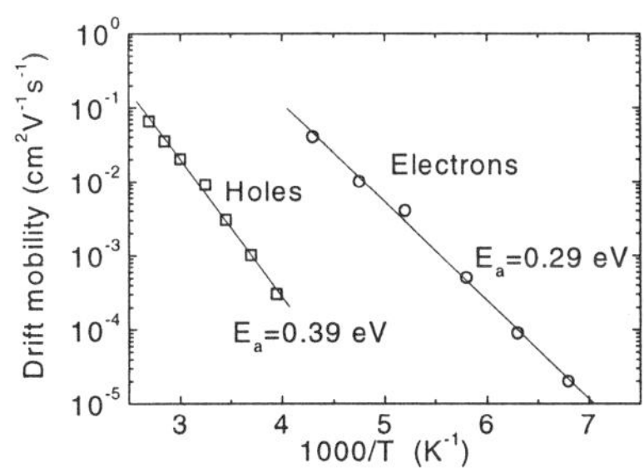

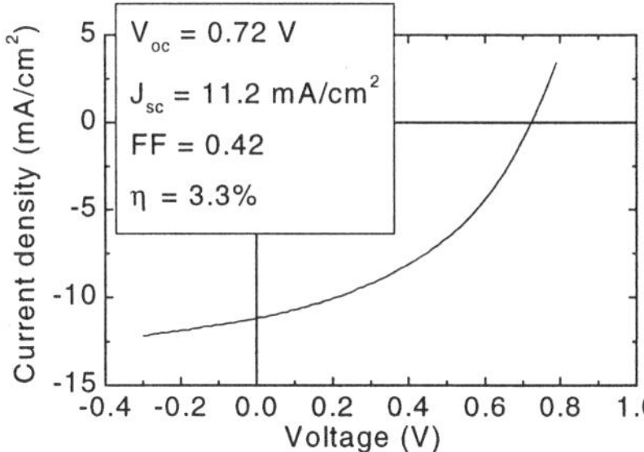

Fig. 4: Hole and electron drift mobility as determined by time-of-flight measurements.

Fig. 5: IV-curve of n-i-p solar cell with intrinsic a-Si:H film deposited by the ETP.

As the best quality a-Si:H is obtained at substrate temperatures in the range of 400 - 450 °C, a n-i-p solar cell is produced at 400 °C using the thermal stable n-layer concept as also applied in HWCVD deposited solar cells [10]. An intrinsic a-Si:H film of 400 nm has been deposited on a 75 nm thick rf PECVD deposited n-layer on stainless steel at a substrate temperature of 430 °C. Afterwards the cell is finished by an rf PECVD deposited 30 nm thick p-layer and a 40 nm thick ITO top contact. The IV-curve of the solar cell obtained is given in Fig. 5, together with its properties. The efficiency of this first solar cell with an intrinsic layer deposited by the ETP technique (at almost 10 nm/s) is still relatively low (3.3%). Yet, the thickness of the intrinsic a-Si:H film has not been optimized. Probably a thinner intrinsic film can be used as the light absorption of the material is larger ($\alpha = 3.8\times10^4$ cm^{-1} at 2.0 eV) due to the lower bandgap. Furthermore, the ITO top contact was not optimized for optical transmission.

CONCLUSION

It has been shown that the (electronic) quality of films deposited by this remote plasma increases with increasing contribution of SiH_3 to deposition, whereas inferior film properties are obtained in conditions with a significant production of very reactive radicals. Moreover, it has been shown that films with good solar grade properties, as concluded from the high hole drift mobility but not from the photo-response, can be obtained by this ETP technique. Possible reasons for the higher hole drift mobility are (a combination of) the high growth rate, the high substrate temperature, the difference in hydrogen bonding (high micro-structure parameter) and the hydrogen poor cationic clusters contributing to growth.

ACKNOWLEDGMENTS

The authors greatly acknowledge Ruud Schropp and Karine van de Werf from Utrecht University for the production and testing of the solar cell. Erik Jan Geluk and Rene van Swaaij of the Delft University of Technology for the bandgap measurements. This work has been financially supported by NOVEM, NWO-prioriteit and FOM-Rolling Grant.

REFERENCES

[1] W.M.M. Kessels, C.M. Leewis, A. Leroux, M.C.M. van de Sanden, D.C. Schram, to appear in J. Vac. Sci. Technol. A 17, Jul/Aug (1999).
[2] M.C.M. van de Sanden, R.J. Severens, W.M.M. Kessels, R.F.G. Meulenbroeks, and D.C. Schram, J. Appl. Phys. **84**, 2426 (1998).
[3] T. Tiedje, J.M. Cebulka, D.L. Morel, and B. Abeles, Phys. Rev. Lett. **46**, 1425 (1981).
[4] W.M.M. Kessels, M.C.M. van de Sanden, D.C. Schram, Appl. Phys. Lett. **72**, 2397 (1998).
[5] W.M.M. Kessels, M.C.M. van de Sanden, R.J. Severens, L.J. Van IJzendoorn, and D.C. Schram, Mater. Res Soc. Symp. Proc. **507**, 529 (1998).
[6] Ruud E.I. Schropp and Miro Zeman, Amorphous and Microcrystalline Silicon Solar Cells: Modeling, Materials and Device Technology, Kluwer Academic Publishers, Boston (1998).
[7] H.-Z. Song, G.J. Adriaenssens, A.H.M. Smets, B.A. Korevaar, W.M.M. Kessels, M.C.M. van de Sanden, and D.C. Schram (to be published).
[8] H. Dersch, L. Schweitzer, and J. Stuke, Phys. Rev. B **28**, 4678 (1983).
[9] A.H. Mahan, J. Carapella, B.P. Nelson, R.S. Crandall, I. Balberg, J. Appl. Phys. **69**, 6728 (1991).
[10] K.F. Feenstra, J.K. Rath, C.H.M. van der Werf, Z. Hartman, R.E.I. Schropp, in Proceedings of the 2nd World Conference on Photovoltaic Energy Conversion, Vienna, Austria, p. 956 (1998).

INTRINSIC, n- AND p-DOPED a-Si:H THIN FILMS GROWN BY DC MAGNETRON SPUTTERING WITH DOPED TARGETS

Å.A. JOHANSSON[*], K. JÄRRENDAHL[*], J. BIRCH[*], B. HJÖRVARSSON[**], H. ARWIN[*]
[*] Dept. of Physics and Measurement Technology, Linköping University, SE-581 83 Linköping, Sweden
[**] Dept. of Physics, Royal Inst of Technology, SE-100 44 Stockholm, Sweden

ABSTRACT

Intrinsic, n- and p-type a-Si:H films were deposited by dc magnetron sputtering and analyzed with several techniques. The films were synthesized in a reactive Ar-H_2 atmosphere giving H contents in the range of 3-20 at %. The films were sputtered from pure silicon targets and doped silicon targets with 1 at % B or P. Doping by co-sputtering from composite Si/B_4C targets was also explored. The doping concentrations were $3 \times 10^{20} - 2 \times 10^{21}$ cm^{-3} for the p-type films and 2.6-2.9 $\times 10^{19}$ cm^{-3} for the n-type films. The conductivity was in the range 10^{-2}-10^{-4} Ω^{-1}cm^{-1} for p-doped films and 10^{-5} Ω^{-1}cm^{-1} for the best n-doped films. Band gap estimations were obtained from dielectric function data and showed an increase with hydrogen content. A comparison to device quality PECVD-samples was also made.

INTRODUCTION

Hydrogenated amorphous silicon (a-Si:H) is used in a number of different applications, such as solar cells, thin film transistors for liquid crystal displays, image sensors for document scanning and medical X-ray imaging [1, 2]. Plasma enhanced chemical vapor deposition (PECVD) is the most frequently used technique for fabricating thin films of a-Si:H. However physical vapor deposition (PVD) has also successfully been used [3-6]. In many cases PVD is a superior technique due to the possibility to deposit films on substrates which must not be heated. PVD also allows modification of the growth conditions far from thermodynamic equilibrium.

This work investigates the possibilities to fabricate device quality intrinsic, n- and p-type a-Si:H by magnetron sputtering with the intention of making photo diodes for digital cameras. The n- and p-type doping is obtained by sputtering from P- and B-doped targets, respectively. Doping by co-sputtering from composite Si/B_4C targets was also explored. A comparison to device quality PECVD samples from Xerox PARC was made.

EXPERIMENTAL DETAILS

All films were grown in a planar magnetron sputtering system. The system consists of two 3 inch magnetrons placed in the top lid of a cylindrical chamber. An electrically isolated plate is placed between the magnetrons to avoid cross-contamination. There are shutters to both the magnetrons and the rotating substrate table. The samples were loaded through a load-lock. For the co-sputtered films a 2 inch magnetron was used.

Mat. Res. Soc. Symp. Proc. Vol. 557 © 1999 Materials Research Society

Before the films were grown, the system was evacuated to 5×10^{-7} Torr, using a turbomolecular pump and a titanium sublimation pump (TSP). Then the substrates were heated to 250°C, by a boron-nitride covered graphite heater placed in the substrate table. Thereafter, H_2 was introduced into the chamber to the desired partial pressure, and followed by 1.5 mTorr Ar, as measured by a capacitance manometer. The target current was kept constant at 0.2 A, giving a target voltage that varied between 450-500 V depending on target and H_2 pressure. The substrates were kept at a floating potential. During the deposition the TSP was operated in a continuous mode, to obtain more uniform hydrogen pumping speed. The deposition rate of the a-Si:H films was 33-100 Å/min depending on the hydrogen partial pressure.

A polycrystalline target of purity 99.999% was used for the intrinsic films. The partial pressure of the 99.9996% pure hydrogen gas was varied between 0.4 - 0.7 mTorr.

B- and P-doped targets of hot-pressed silicon with 1 at % B or P, respectively, were used for the doped films. The co-sputtered film was grown using a pure silicon target with a sector replaced by a piece of B_4C. The hydrogen partial pressure for both the doped series was varied in the range of 0.5 - 6 mTorr. The hydrogen concentration depth profiles were determined by nuclear reaction analysis (NRA) utilizing the $^1H(^{15}N, \alpha\gamma)^{12}C$ nuclear resonance reaction using the resonance at 6.385 MeV. The B- and P-concentration depth profiles were determined by secondary-ion mass-spectroscopy (SIMS) using calibrated standards as a reference.

Film thickness values were measured both by a Jeol JSM-T220A scanning electron microscope, a Dektak 3030 profilometer and Rutherford back scattering spectroscopy. The surface structure was analyzed by a Nanoscope IIIa atomic force microscope system, from Digital Instruments, operated in tapping mode using silicon cantilevers. The scanned area was typically 1 μm^2 from which the surface root-mean-square (rms) roughness was determined. The film microstructure was analyzed by a Philips CM 20 UT electron microscope operated at 200 kV, equipped with a LaB_6 filament, with cleaving as sample preparation method. The optical properties were determined by variable angle spectroscopic ellipsometry (SE) (VASE, J. A. Woollam Co.) in the photon energy range 1.24-5 eV.

The films analyzed with SIMS and SE were grown on float zone silicon wafers. For the electrical measurements oxidized silicon wafers were used. The P-doped films were also deposited on substrates with an aluminum stripe to ensure that the films have good adhesion to a metal contact. The PECVD material was grown on the same types of substrates (sent to the lab at Xerox PARC), as the sputtered films.

RESULTS AND DISCUSSION

The TEM picture in Fig. 1 shows one of the intrinsic films and displays that the film is dense and homogeneous, with no resolved microstructure. Contrary to the intrinsic film, the P-doped film in Fig. 2 shows a columnar structure and is micro crystalline. The TEM analysis of the P-doped film also showed that voids are present at the substrate/film interface.

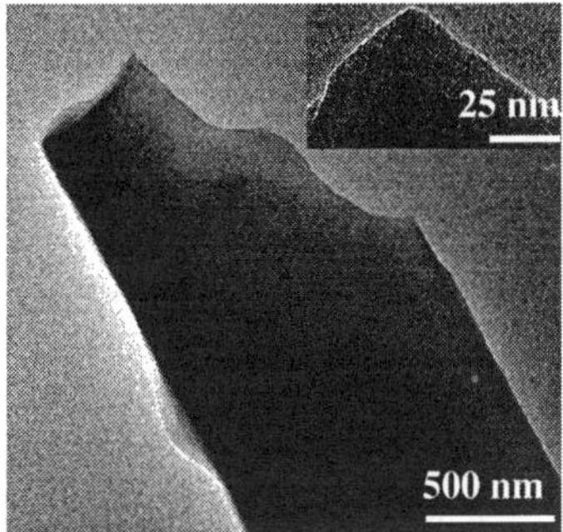

Figure 1. TEM image of intrinsic a-Si:H. The film thickness is 1.25 μm. The substrate is not visible in the picture.

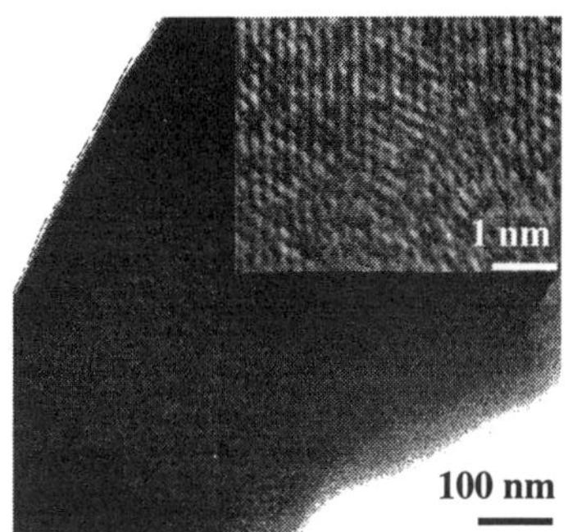

Figure 2. TEM image of P-doped a-Si:H. The film thickness is 0.53 μm. The film is grown on an oxidized silicon wafer and the silicon dioxide is visible to the right in the picture.

The surface structure of an intrinsic a-Si:H film, displayed in the AFM image in Fig. 3 shows a smooth film with an rms roughness of 2.6 nm for a film thickness of 1100 nm. The film is considered to be grown with small thickness variations, since the substrate is oxidized silicon with an rms roughness of 0.21 nm. The sputtered film does not differ significantly from the 600 nm thick PECVD-film in Fig. 4 which has an rms roughness of 1.4 nm. The thickness variations increase with thickness for the sputtered films while the roughness of the films is considered to be of comparable magnitude. The sputtered B-doped films are also smooth with rms roughness values of 0.8-1.1 nm, for 220-1100 nm thick films (Fig. 5). Both the intrinsic and B- doped films have a long range smoothness with peak-to-peak values of maximum 10 nm for B-doped samples and 25 nm for intrinsic a-Si:H.

The sputtered P-doped films, however, have in general larger rms roughness values in the range 3-6 nm for similar film thickness (Fig. 6). For some of the P-doped samples there are larger bumps with heights up to 60 nm separated by 100-200 nm. These thickness variations are considered to be connected to the columnar growth shown in the TEM picture in Fig. 2. The AFM results also indicate a connection of this type of growth to the hydrogen pressure, since the film grown at the lowest hydrogen partial pressure is more smooth.

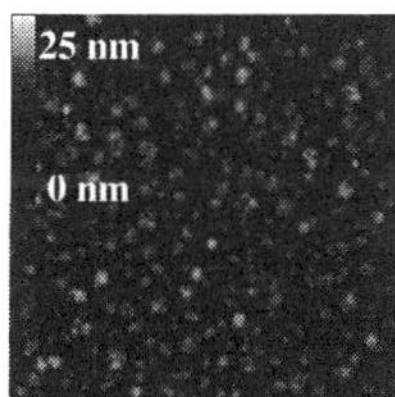

Figure 3. Sputtered intrinsic a-Si:H. Film thickness 1100 nm, rms roughness 2.6 nm.

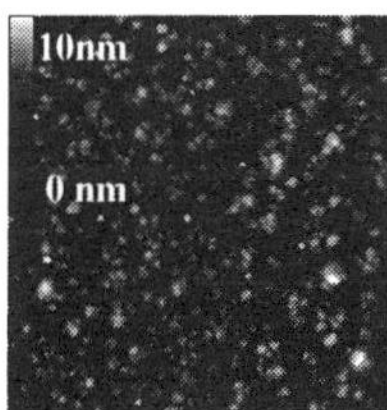

Figure 4. PECVD grown intrinsic a-Si:H. Film thickness 600nm, rms roughness 1.4 nm.

Figure 5. B-doped sputtered a-Si:H. Film thickness 1100 nm and rms roughness 1.1 nm.

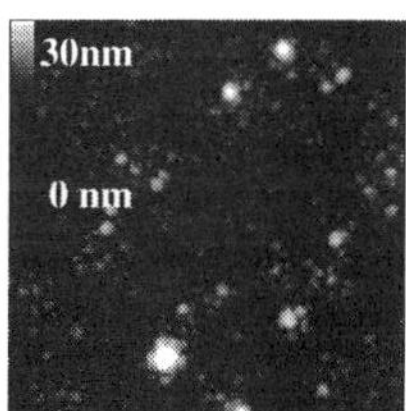

Figure 6. P-doped sputtered a-Si:H. Film thickness 500 nm, rms roughness 3.4 nm.

The NRA studies show that the intrinsic films have a hydrogen content of 10-30%. The B-doped and P-doped films have a hydrogen content of 2-6% and 11-18%, respectively, for similar hydrogen partial pressures.

The highest boron concentration was 2×10^{21} cm^{-3} and found in a co-sputtered film with a hydrogen content of 2.3%. Sputtering from a B-doped target yielded B-concentrations of $2.8 - 3.0 \times 10^{20}$ cm^{-3}. The doping content in the P-doped films was $2.6 - 2.9 \times 10^{19}$ cm^{-3}. The B- or P-contents in the measured films were not affected by the hydrogen partial pressures.

Results from the four-point probe measurements are displayed in Table I. The highest conductivity was 4.1×10^{-2} ohm^{-1}cm^{-1} for one of the B-doped films sputtered from a doped target. The co-sputtered B-doped film with the higher B-concentration has a conductivity of 1.2×10^{-2} ohm^{-1}cm^{-1}, which can be explained by lower doping efficiency for highly doped films [4]. The measurements were not made in total darkness. A very weak background light was present. However, all films show higher conductivity when illuminated by a 20 W microscopy lamp focused on the samples. The intrinsic and P-doped samples show very low conductivity (only one P-doped sample had conductivity within the measurement range of the four-point probe). One reason for the low doping effect in the P-doped samples may be that most P-atoms become threefold-coordinated in the a-Si:H network [7]. This energetically favored configuration is not electrically active. Another possible explanation for the low conductivity in the P-doped samples is the columnar microstructure which can be seen in the TEM image in Fig. 2.

Table I: Results from measurements with four-point-probe.

	Conductivity (ohm^{-1}cm^{-1})
B-doped sputtered (doped target)	4.1×10^{-2}
B-doped co-sputtered	1.2×10^{-2}
B-doped PECVD material	3.9×10^{-3}
P-doped sputtered (doped target)	2.0×10^{-5}
P-doped PECVD material	4.0×10^{-2}

From the spectroscopic ellipsometry data the dielectric function ε of the a-Si:H films was obtained. From ε the optical band gaps were calculated using Tauc plots [8]. Using the AFM rms roughness and peak-to-peak values, a correction for over-layers and film thickness variations was made before making the band gap estimations. A more detailed study of the optical properties for a P-doped film resulted in the real and imaginary parts of the dielectric function shown in Fig. 7. The parameters were extracted from the measured pseudo-dielectric function using a three layer model (over-layer/a-Si:H/SiO$_2$ oxide layer) where the layer thickness values and the dielectric function of a-Si:H were free parameters. The over-layer represents surface oxide and roughness and was modeled with a Bruggeman effective medium approximation mixing 50% a-Si:H and 50% voids. Solutions of the dielectric function of the a-Si:H films were determined for different values of the layer thickness. The physically most correct solution presented in Fig. 7 corresponds to the thickness values for which the dielectric function shows no features corresponding to the interference seen in the pseudo-dielectric function [9]. Some of the optical spectra for P-doped films have features corresponding to the E_1 and E_2 transitions in crystalline silicon [10] which is due to the fact that these films are micro crystalline. The optical properties of the film shown in Fig. 7 did not show effects of these transitions. The general finding is that the band gap increased with hydrogen content, as shown by others.

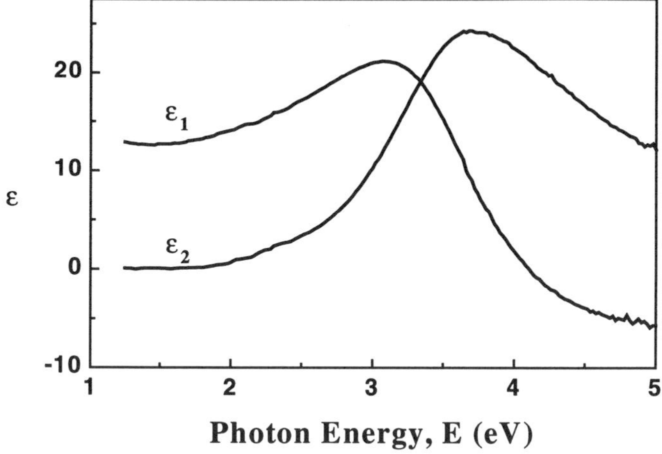

Figure 7. The complex dielectric function,
$\varepsilon = \varepsilon_1 + i\varepsilon_2$ of a P-doped a-Si:H film.

CONCLUSIONS

Successful p-type doping of a-Si:H films has been achieved by incorporation of boron in the films from a doped target as well as by co-sputtering of silicon and B_4C. Further work is needed for phosphorous doped n-type films to remove the columnar growth and increase the conductivity. The intrinsic films are highly resistive as expected.

ACKNOWLEDGEMENTS

Robert A. Street at Xerox PARC, USA, is acknowledged for supplying the PECVD material and for valuable discussions. We also would like to thank Joe Greene and Hyungjun Kim at the University of Illinois for the SIMS measurements and analysis, and Per Persson at Linköping University for help with the TEM handling. We also appreciate the valuable discussions with Christer Svensson, Göran Hansson and Jan-Eric Sundgren. Financial support was obtained from the Swedish Research Council for Engineering Sciences (TFR) and Integrated Vision Products AB. The ion beam analysis was performed at the Tandem Beam Accelerator in Uppsala, Sweden, granted by the Swedish Natural Science Research Council (NFR).

REFERENCES

1. R. A. Street, S. Nelson, L. Antonuk and V. P. Mendez, Mat. Res. Soc. Symp. Proc., **192**, pp. 441-452, (1990).

2. R. A. Street, X. D. Wu, R. Weisfield, S. Ready, R. Apte, M. Ngyuen, W. B. Jackson and P. Nylen, Journal of Non-Crystalline Solids, **198-200**, pp. 1151-1154, (1996).

3. T. D. Moustakas, Solar Energy Materials, **13**, pp. 373-384, (1986).

4. D. E. Carlson, Critical Reviews in Solid State and Materials Science, **16**, pp. 417-435, (1990).

5. Y. H. Liang, S.-Y. Yang, A. Nuruddin and J. R. Abelson, Mat. Res. Soc. Symp. Proc., **336**, pp. 589-94, (1994).

6. M. J. Thompson, "Sputtered Material," in *The physics of Hydrogenated Amorphous Silicon I*, vol. 55, *Topics in Applied Physics*, *edited by:* J. D. Joannopoulus and G. Lucovsky, Springer-Verlag, Berlin Heidelberg, 1984, pp. 119-175.

7. P. A. Fedders, Phys. Rev. B, **58**, pp. 7020-7023, (1998).

8. J. Tauc, R. Grigorovici and A. Vancu, Phys. Stat. Sol., **15**, pp. 627-637, (1966).

9. D. E. Aspnes, A. A. Studna and E. Kinsbron, Phys. Rev. B, **58**, pp. 768-779, (1984).

10. M. Cardona, *Modulation Spectroscopy*, Academic, New York, 1969, pp. 55-65.

SIMULATION OF QUANTUM EFFICIENCY SPECTROSCOPY FOR AMORPHOUS SILICON P-I-N JUNCTIONS

RODNEY ESTWICK AND VIKRAM L. DALAL
Dept. of Electrical and Computer Engineering, Iowa State University, Ames, Iowa 50011

ABSTRACT

Quantum efficiency(QE) spectroscopy of amorphous silicon and alloy solar cells has been used for many years now to determine the mobility-lifetime products for minority carriers. Similarly, matching of I(V) curves, assuming a linear model for collection as a function of applied voltage, has been used to quantify the effects of degradation on cell performance by estimating changes in the collection length [or range] of holes. In this paper, we do a numerical simulation of these techniques, using the AMPS 1-D model developed by Fonash and his coworkers. The simulation shows that neither the lifetime nor the electric field in the devices is constant as a function of position. Nor is the electric field a linear function of applied voltage, particularly when the voltage exceeds about half the built-in voltage. The uniformity of the lifetime depends on the applied bias and on the defect densities in the material. This variation in electric field and lifetime and nonlinearity with applied voltage makes questionable some of the conclusions drawn from fitting device I(V) curves, particularly under forward bias. However, when one uses only a limited range of forward bias, or, preferably, make measurements in cells with thicker i layers under reverse bias, one can make reasonable estimates of the hole mobility-lifetime($\mu\tau$) product or the collection length. The simulations also show that indeed, it is the hole $\mu\tau$ product which is the limiting parameter.

Introduction

It is well known that in materials such as a-Si:H and its alloys, the high mid-gap defect density leads to rather low diffusion lengths for minority carriers. Since devices such as solar cells depend on efficient collection of minority carriers for their performance, most solar cells in these materials use a built-in electric field to create a drift-aided diffusion of carriers to achieve good performance. Extended field regions are achieved in the solar cell by using a nominally undoped(i) material for the base layer of the cell. For such cases, the critical parameter which determines collection efficiency is the range, $\mu\tau E$, where μ is the mobility of the carier, τ is the lifetime and E is the electric field[1]. A high ratio of range/i layer thickness assures a good fill factor for the cell.[2]. Since electric field is a function of applied voltage, the range, and therefore, quantum efficiency(QE), should be a function of the applied voltage. By using such QE-voltage spectroscopy, Crandall[3], and Dalal et al[4,5], were able to estimate the mobility-lifetime products in a-Si solar cells. Crandall[3] assigned the estimated value to the sum of electron and hole mobilities, whereas Dalal and Alvarez[4] claimed that the estimated value represented just the minority carrier $\mu\tau$ product.

These previous workers all assumed a uniform electric field, and more critically, a uniform lifetime of carriers in the material. However, since the earliest simulations of the cell [6,7,8], it has been known that the field in the i layer is not uniform. The first numerical simulations by Hack and Shur[8] showed that the electric field profile was not uniform; rather, it was sharply peaked near both p-i and i-n interfaces. Similarly, intuition suggests that the lifetime

Mat. Res. Soc. Symp. Proc. Vol. 557 © 1999 Materials Research Society

cannot be uniform either. See Fig. 1. Near the p-i interface, where the quasi-Fermi levels are near the valence band, there are many more trapped holes in the midgap and valence band tail states compared to trapped electrons. Therefore, from the Simmons and Taylor analysis of recombination kinetics in a multi-trap system[9], the electron lifetime in this region is likely to be low. In contrast, near the i-n interface, the quasi-Fermi levels are near the conduction band edge, and therefore, hole lifetime is likely to be much lower. Over the middle of the i layer of the device, because of the valence band tail is wider than the conduction band tail, and because the material is always accidentally doped with small amounts of oxygen donors, the hole lifetime is generally much lower than the electron lifetime. [10].

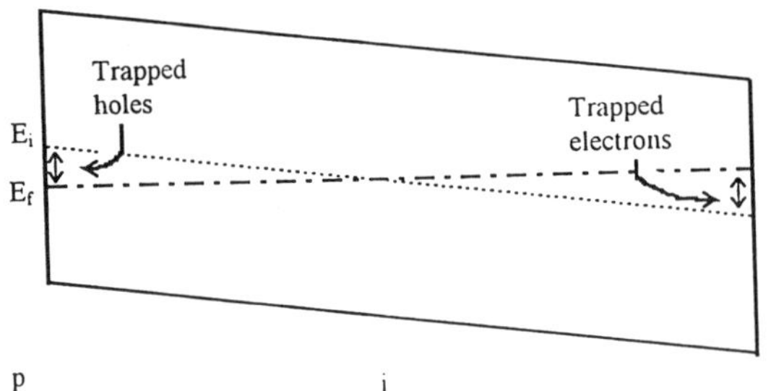

Fig. 1 Expected Trapped electrons and hole distribution in the device

From the above considerations, the question that naturally arises is : when one does the QE vs. voltage spectroscopy, what is one measuring? And what is the accuracy of the measurement? What experimental conditions must be used to obtain reliable measurements of fundamental properties such as hole $\mu\tau$ product? This paper attempts to answer these questions by doing a numerical analysis of QE spectroscopy technique for a-Si cells with different defect densities, using the AMPS 1D simulation package developed by Professor Fonash and his group at Pennsylvania State University.[11].

Cell Design simulations.

The cells that we simulated were of the p-i-n type with light incident from the p side. The p layer was chosen to be an amorphous (Si,C):H layer, and the back n layer was a-Si:H. A graded-gap buffer a-(Si,C):H layer of 10 nm thickness was introduced between the p and the i layers.[12]. The thickness and the defect density in the i layer were varied, while keeping the standard AMPS 1-D Gaussian profiles. The magnitude of the defect density was varied by adjusting the peak of the Gaussian defect density profile.

Results

1.Electric field profiles

The results of the field profile simulations are shown in Fig. 2-5. Fig. 2 shows the electric field profile in the device for different applied voltages for the typical device thickness of 0.4 micrometers and an integrated defect density of $1\times10^{16}/cm^3$. The figure clearly shows that as the applied voltage is increased beyond about 0.5 V in the forward direction, the electric field profile changes dramatically, becoming rather non-uniform. This indicates that the QE measurements must be restricted to a forward voltage not exceeding 0.5 V for a built-in voltage of 0.95 volts for this defect density. In Fig. 3, we show what happens to the electric field as the defect density increases to $5x10^{16}$ /cm^3. Then, the field becomes nonuniform at even a lower voltage, 0.3V. volts.In Fig. 4, we plot the "uniform" field in the center of the cell as a function of applied voltage in the cell for two defect densities, $1x10^{16}$/cm^3 and 5E16/cm^3, and we see that beyond about 0.5 volts applied voltage for the lower defect density, and 0.3V for the higher value, the linearity assumption is violated. That means that using the simple formula for QE, which assumes that the field changes linearly with voltage, is no longer valid beyond about 0.5 V forward bias. And yet, this is precisely the range of voltages used by Crandall[3] and by Hegedus[2] to model the cell and deduce the collection lengths. So, clearly, some question arises about the meaning of collection lengths deduced using such fits of I(V) curves in this voltage range.

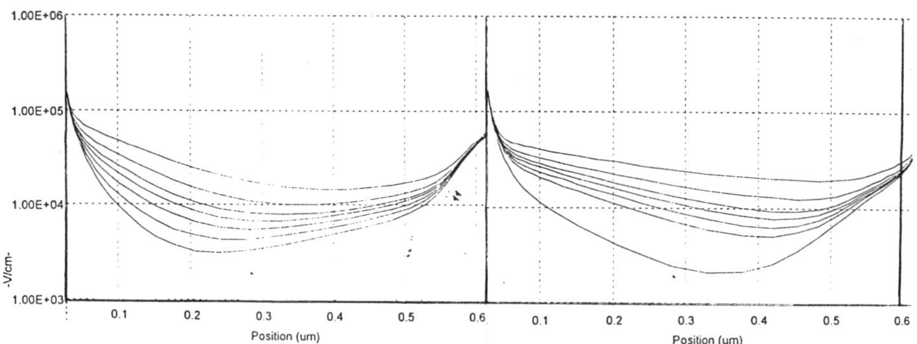

Fig. 2 Field profile as a function of voltage for defect density of 1E16/cm^3.

Fig. 3 Field profile as a function of voltage For defect density of 5E16/cm^3.

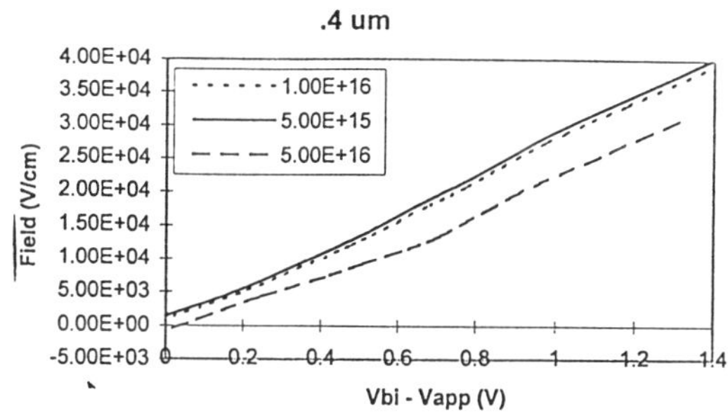

Fig. 4 Non-Linearity of electric field as a function of applied voltage

39

2. Lifetime profiles

We next show, in Fig.5 and 6, the simulation of lifetime in the cell as a function of position for different applied voltages, again for two defect densities, 1E16/cm³ and 5E16/cm³. Once again, as the device goes into forward bias, the lifetimes become rather nonuniform, particularly in the back regions of the cell. In contrast, the lifetimes are much more uniform at reverse voltages, even for the larger i layer thickness, 0.8 micrometer.

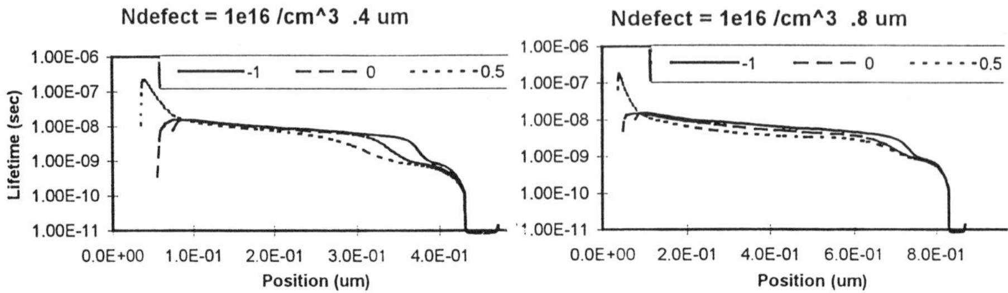

<u>Fig. 5</u> Lifetime vs. position for different applied voltages for two different I layer thicknesses

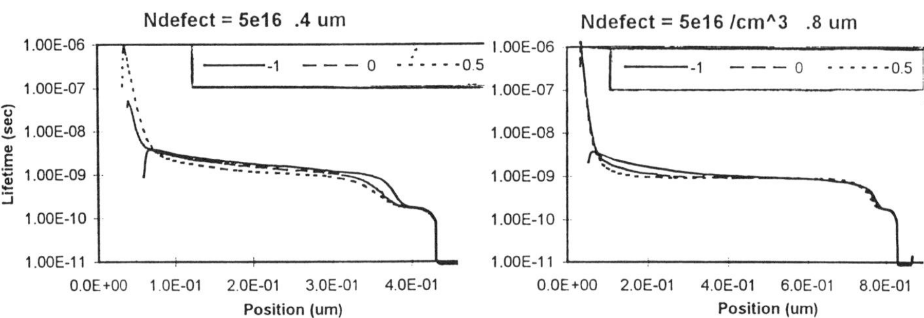

<u>Fig. 6</u> Lifetime vs. position for different applied voltages for two different I layer thicknesses

From Fig. 5 and 6, we can deduce that the hole lifetime increases sharply in the i layer near the p-i interface, as expected from our initial considerations regarding excess of hole trapping in this region. That means that electron lifetime as a function of position must sharply decrease inthe same region. Indeed, numerical simulations bear this out, with the electron lifetime decreasing to below 1E-8 sec. in this region. Typically, this low electron lifetime region is very narrow, only about 0.05 micrometer for the typical defect density of 1E16/cm³ An important point to note from this discussion is that near the p-i interface, the electron lifetime is much lower than in the bulk, and indeed, is lower than the hole lifetime. Except for this region, however, the hole lifetime is much lower than the electron lifetime. What this means is that in Crandall's analysis of device transport[3], the mobility-lifetime product of electrons that ought to be used is the one that corresponds to this low value, and not the bulk value. Therefore, the transport over most of the cell, except right near the p-i interface, is indeed limited by holes as suggested by Dalal and Alvarez[4].

Discussion

Now we are in a position to answer the question about how to use quantum efficiency vs. voltage spectroscopy correctly to estimate the hole $\mu\tau$ product in the bulk of the i layer. The answer is to use reverse and moderate values of forward voltages, preferably less than 0.5 x built-in voltage and use thicker I layers, the thickness depending upon the defect density. Under such conditions, the electric field and lifetimes are more uniform, the lifetime does not depend significantly on the applied voltage, and the field changes linearly with applied bias, a necessary condition for obtaining $\mu\tau$ product from QE vs. voltage measurements. In contrast, when one tries to model the device I(V) curve using a collection length model, one necessarily intrudes into the range of forward voltages where lifetime changes with electric field, and the electric field is not a linear function of applied voltage. Therefore, one must question the accuracy of the collection length derived from such measurements.

Conclusions

In conclusion, we have used AMPS modeling of devices to show that care must be exercised when using the quantum efficiency spectroscopy to deduce hole $\mu\tau$ products. Similarly, its is debatable whether a measurement of device I(V) curve can yield a reliable estimate of collection length or range of holes in the material. The transport over most of the device is limited by hole transport and not by the sum of electron and hole $\mu\tau$ products. To estimate this quantity with reasonable accuracy, it is preferable to use reverse and moderate forward bias voltages (about half the built-in voltage) while doing the QE spectroscopy measurements, and not go into the region of high forward bias. It may be necessary in such cases to use cells with thicker i layers so that QE can vary in response to reverse bias.

Acknowledgment

We acknowledge with thanks the support of NREL for part of this work. We also thank Professor Fonash and his students for their help in learning how to use AMPS and for many useful discussions.

References

1. See, for example, R. Smith, "Semiconductors", (Dover, 1987)

2. S. Hegedus, Progress in Photovoltaics, 5, #3(1997)

3. R. Crandall, J. Appl. Phys., 54,7176(1983)

4. V. L. Dalal and F. Alvarez, J. de Phsyics(Paris), 22, C-4,491(1981)

5. V. L. Dalal et al, Proc. of 18th. IEEE Photovoltaic Spec. Conf., p.837(1985)

6. V. L. Dalal, Solar Cells,2, 261(1980)

7. P. Sichanugrist et al., J. Appl. Phys., <u>54</u>, 6705(1983)

8. M. Hack and M. Shur, J. Appl. Phys., <u>58</u>,997(1985)

9. J. G. Simmons and G. W. Taylor, Phys. Rev. <u>B-4</u>, 501(1971)

10. R. A. Street, "Hydrogenated a-Si", (Cambridge, 1991),p.312

11. P. J> McElheny et al. J.Appl. Phys.,<u>64</u>,1254(1988)

12. A. Vaseashta et al, Proc. of 18th. IEEE Photovolt. Conf.,p.523(1985)

THE PROPERTIES OF a-SiC:H AND a-SiGe:H FILMS DEPOSITED BY 55 kHz PECVD

B.G.BUDAGUAN[*], A.A.SHERCHENKOV[*], A.E.BERDNIKOV[**], J.W.METSELAAR[***], A.A.AIVAZOV[****]

[*]Institute of Electronic Technology, Moscow 103498 RUSSIA, budaguan@ms.miee.ru
[**]Institute of Microelectronics of Russian Academy of Science, Yaroslavl 150007 RUSSIA
[***]University of Technology, Mekelweg 4, 2628 CDDelft, NETHERLANDS
[****]UniSil Corp. 401 National Av, Mountain View, CA, 94043

ABSTRACT

The deposition processes and the properties of a-SiC:H and a-SiGe:H films in 55 kHz glow discharge were investigated. The analysis of deposition rate and RBS measurements showed that the chemical reactions between SiH_n spices and CH_4 control the incorporation of C in a-SiC:H films. High deposition rates of a-SiC:H and a-SiGe:H films fabricated by 55 kHz PECVD is caused by the increase of radical fluxes to the growth surface. The specific features of a-SiC:H and a-SiGe:H microstructure were revealed by IR and AFM analysis. In a-SiC:H films the islands of low size were distinguished on the surfaces of large islands. The large variation of the total hydrogen content in a-SiGe:H did not affect the optical bandgap, while the hydrogen related microstructure controlled the electronic properties such as dark conductivity, $\eta\mu\tau$ product, defect density and Urbach slope.

The results of optoelectronic properties and SW effect measurements of 55 kHz a-SiC:H and a-SiGe:H films demonstrated the increased stability in comparison with a-Si:H.

INTRODUCTION

The a-SiC:H and a-SiGe:H films are currently widely used as a high and a low bandgap optoelectronic materials for a tandem and a triplet solar cell fabrication. The use of these materials increases the utilization of solar spectrum and stabilized efficiency of solar cells, which is the key problem of these devices. However, the properties of these materials are highly sensitive to the fabrication technology. The main drawbacks of a-SiC:H alloys are a low deposition rate ($\sim$3A/c) and a complex microstructure highly sensitive to deposition condition.
For a-SiGe:H alloys the deterioration of electronic properties with the addition of Ge to a-Si:H is observed. It is not now understood yet whether this deterioration of electronic properties is due to the inherent property of this material, or it is the consequence of imperfect technology.

In this work we investigated the deposition and the properties of a-SiC:H and a-SiGe:H alloys fabricated by 55 kHz PECVD at the different amount of the methane and germane content in a gas mixture.

EXPERIMENT

The a-SiC:H and a-SiGe:H films were fabricated by 55 kHz PECVD [1,2]. For deposition of a-SiC:H and a-SiGe:H with various composition the gas mixtures with different contents of methane $R_C=100\cdot[CH_4]/[SiH_4+CH_4]\%$, and germane $R_{Ge}=100\cdot[GeH_4]/[SiH_4+GeH_4]\%$, were used (see Table 1). The substrate temperature, the LF power and the total gas pressure were kept at constant values of 320^0C, 200W and 90 Pa, and 225^0C, 150W and 70 Pa for the deposition of a-SiC:H and a-SiGe:H films, respectively. The films were deposited on c-Si and Corning 7059 glass substrates.

Mat. Res. Soc. Symp. Proc. Vol. 557 © 1999 Materials Research Society

Table 1.

The deposition rate of a-SiC:H and a-SiGe:H films fabricated by 55 kHz PECVD at different gas mixture compositions.

Sample, N	R_C, %	Deposition rate, A/sec	Sample, N	R_{Ge}, %	Deposition rate, A/sec
YC5 17	0	14.4	YG6 12	0	11.1
YC5 23	20	11.1	YG6 23	9.1	8.9
YC5 31	40	10.0	YG6 33	16.7	12.5
YC5 41	60	8.9	YG6 44	27.5	12.5
YC5 59	80	5.3	YG6 57	37.5	9,4
YC5 69	100	0.19	YG6 68	44.5	11,4

The elemental compositions of the a-SiC:H films were determined by Rutherford Backscattering Spectroscopy (BRS). The incorporation of Ge into the a-SiGe:H films was analyzed by the secondary ion mass spectroscopy (SIMS). The surface morphology of the films was investigated by atomic force microscopy (AFM) on P4-SPM-MDT scanning probe microscope. The IR transmission spectra of the films on c-Si substrate were measured on a double beam spectrometer (SPECORD M-80) for the analysis of the chemical composition, concentration and distribution of hydrogen and its bonding forms.

The spectra of optical absorption obtained from the measurements of transmission coefficient as well as the data of constant photocurrent method were used to determine the Tauc optical gap, E_g, and the Urbach energy, E_0. The dark and photoconductivity measurements were carried out in the temperature range of 20-250 ^{0}C. For photoconductivity measurements He-Ne laser (λ=633 nm) with the illumination intensity of $9.6 \cdot 10^{16}$ cm^{-2} c^{-1} was used. The measurements of Staebler-Wronskii effect were carried out at 80 ^{0}C under He-Ne laser illumination.

RESULTS

a-SiC:H films

As it is seen from Table 1 the deposition rate of a-SiC:H films becomes lower than that of a-Si:H films but is higher than typically published values (~3A/c). The considerable decrease of deposition rate is observed for the pure a-C:H films. The dependence of carbon content in the films determined from RBS measurements on the concentration of methane is shown in Fig.1. As it is seen, there is no efficient incorporation of carbon during the growth. Moreover the carbon content in the film does not exceed 50% at high contents of CH_4 up to 80%. This fact with the particularly zero deposition rate for a-C:H films indicate that the used technological

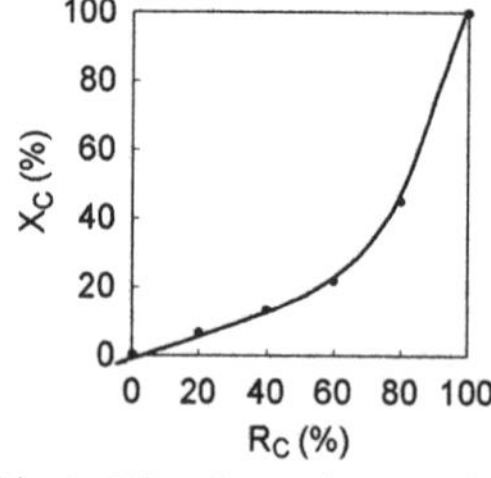

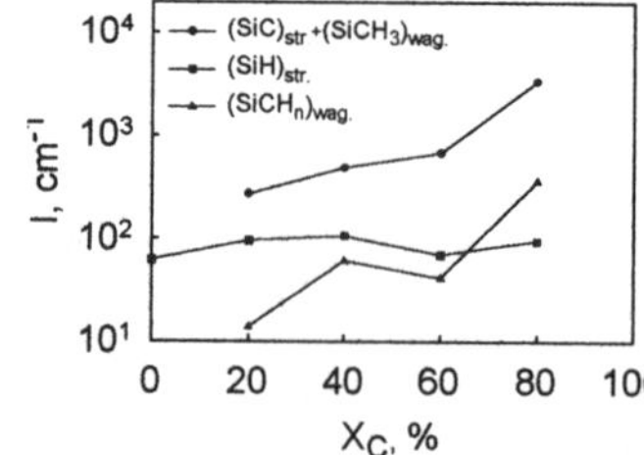

Fig.1. The dependence of C content in films on R_C. The solid line is a guide for eye.

Fig.2. The dependencies of IR peak intensities on X_C. Solid lines are guides for eye.

Fig.3. AFM image of 55 kHz a-SiC:H film at R_C =20%.

conditions allowed to establish "low power" regime of a-SiC:H film deposition [3,4]. On the other hand the growth rate of our films is higher than can be expected from this technological conditions. According to [3] the chemical reactions between $Si-H_n$ spices and CH_4 are responsible for the growth of a-SiC:H films in "low power" regime. It was shown [1,2] that in 55 kHz PECVD the shift of SiH_4 decomposition region to the electrodes and correspondent increase of $Si-H_n$ radical flux to the growth surface is responsible for the increase of a-Si:H deposition rate. The region of chemical reaction of CH_4 with the active radicals is determined by the region of $Si-H_n$ generation. So, as in the case of a-Si:H the increase of deposition rate of 55 kHz a-SiC:H films is caused by the close position of a-SiC:H precursors formation and the growth surface.

The incorporation of carbon into a-SiC:H films has been investigated by IR analysis. The features on IR spectra of a-SiC:H films were identified according to [5]. The shift of Si-H stretching mode to higher wave numbers (from 2007 to 2074 cm^{-1}) and negligible absorption at 890 cm^{-1} ($Si-H_2$ bending) indicates the increase of carbon content in Si-H nearest sides. Figure 2 shows the trends of the integrated intensities of the IR peaks at 780-800 cm^{-1}, attributed to the mixing of $Si-CH_3$ wagging and Si-C stretching vibrations, at 960-990 cm^{-1}, attributed to the Si-CH_n wagging vibrations [5], and at 2007-2074 cm^{-1} (Si-H stretching) as a functions of methane content in a gas mixture. Simultaneous increase of the intensities of absorption band at 780-800 cm^{-1} and Si-CH_n wagging mode indicates on preferable incorporation of carbon in the form of Si-CH_3. The incorporation of C in the form of methyl groups (-CH_3) confirms the "low power" regime at our deposition conditions [3].

The morphology of a-SiC:H films was investigated by AFM analysis. It is seen from Fig. 3 that there is a specific feature on AFM image of films deposited at 20% CH_4 in the gas mixture. Namely the islands of low size are distinguished on the surface of large islands. The distribution of islands was approximated by log-normal dependence and then the average islands size, D_{av}, was calculated [6]. The increase of carbon content in films decreases the average island sizes in a-SiC:H films from 140 nm for R_C =0 to 47 nm for R_C =100%. According to IR data this decrease is accompanied by the increase of methyl group concentration (see Fig. 2). The lower size of carbon atom in comparison with silicon is suggested to be responsible for the decrease of D_{av} at higher methane content in a gas mixture.

The dependence of the optical bandgap on the methane content in the mixture is shown in Fig. 4. The data of Solomon et al. [3] for E_{04} are also presented. It is seen that there are a linear relation between optical bandgap and carbon content and a good correspondence with the literature data. The comparative analysis of optical band gap (see Fig. 4) and IR peak intensities (see Fig. 2) as a function of methane content in a gas mixture has shown that E_g is particularly independent on Si-H stretching peak intensity.

The dependence of dark conductivity at room temperature, σ_{d300}, and activation energy, E_a, on E_g is shown in Fig. 5. The data on σ_{d300} of G.Ambrasone et all. [4] are also presented. It is seen that the activation energy increases with increase of optical bandgap while the room temperature conductivity decreases in a good correspondence with the literature data. At E_g=2.26 eV the room conductivity increases. For these samples the weak temperature dependence of conductivity near the room temperature was observed indicating on the appearance of additional higher conductive channel for electronic transport. It means that at X_C =42.5% a graphite-like microstructure (sp^2-hybrid sites) is formed in 55 kHz a-

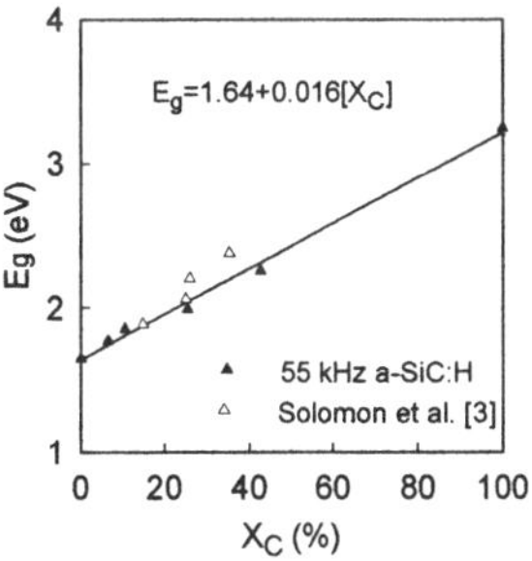

Fig.4. The dependence of E_g on X_C in a-SiC:H films.

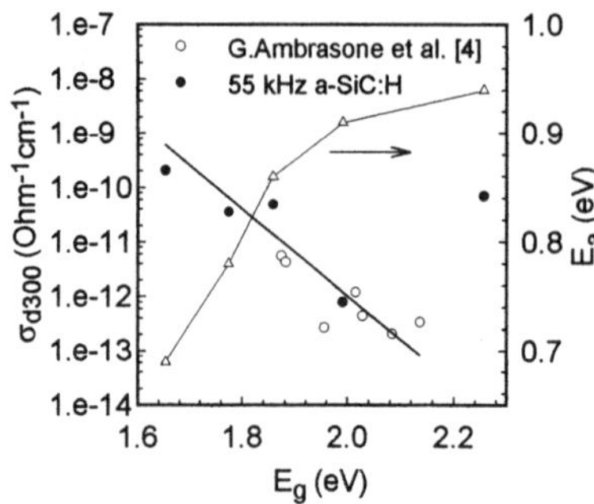

Fig.5. The dependence of σ_{d300} and E_a on Tauc bandgap, E_g, for a-SiC:H films.

SiC:H films.

The lower value of Tauc band gap at $X_C \geq$ 42.5% in comparison with data of work [3], where the onset of diamond-like Si-Si bonds at this carbon concentration was observed also indicates the formation of sp^2-sites of graphite-like microstructure. The formation of sp^2 graphite-like C-C bonds is usually results from clustering of C atoms at high carbon content. According to IR data the concentration of $C-H_n$ configurations increases (see Fig.2). Taking into account the island type microstructure of a-SiC:H films revealed by AFM it is suggested that the clustering of C atoms occurs at the island surfaces in the form of $C-H_n$ bonds.

The increase of $C-H_n$ configurations on islands surfaces at high carbon content facilitate the relaxation of rigid a-SiC:H network through formation of flexible sp^2-sites. So, according to Figs. 2 and 4 the incorporation of Si-C bonds in islands volume determine the optical band gap while CH_n and SiH are located at the islands surface.

The result of Staebler-Wronskii effect measurements for 55 kHz a-SiC:H films deposited at 20%CH_4 content in a gas mixture is presented in Table 2. It is seen that the relative decrease of conductivity under laser illumination during 10^5 sec is by approximately 3 times lower than for undoped a-Si:H. It means that stability of a-SiC:H films increases in comparison with a-Si:H which is important for device application.

a-SiGe:H films.

As it is seen from Table 1 the growth rate of a-SiGe:H is nearly the same as for a-Si:H and doesn't noticeably change with the increase of GeH_4 content. Due to the lower binding energy of Ge-H in comparison with Si-H there is an effective incorporation of Ge into the a-SiGe:H. So, the high deposition rate of a-SiGe:H in 55 kHz PECVD is caused by the close position of the radical generation region to the electrode as in the case of a-Si:H [1,2]. The AFM image of a-SiGe:H films, deposited at 9% GeH_4 in the gas mixture is presented in. Fig.6. The morphology of this film is similar to a-Si:H films with a slightly increase of average islands size from 81 nm for R_{Ge} =0 to 108 nm for R_{Ge} =44.5%. The difference in D_{av} for pure a-SiH between two samples R_{Ge}= 0 and R_C= 0 is related to the difference in other technological parameters such as T_s, W, and P. The increase of D_{av} with the increase of R_{Ge} is suggested to be connected with the higher size of Ge related structural configurations.

Table 2.

The results of Staebler-Wronskii effect measurements under He-Ne laser illumination in a-SiC:H and a-SiGe:H films fabricated by 55 kHz PECVD.

Sample, N	R_C, %	$[\sigma(0)-\sigma(t)]/\sigma(0)$, 80^0C, $t=10^5$sec	Sample, N	R_{Ge}, %	$[\sigma(0)-\sigma(t)]/\sigma(0)$, 80^0C, $t=10^5$sec
YC5 17	0	0.57	YG6 12	0	0.65
YC5 23	20	0.19	YG6 23	9.1	0.12

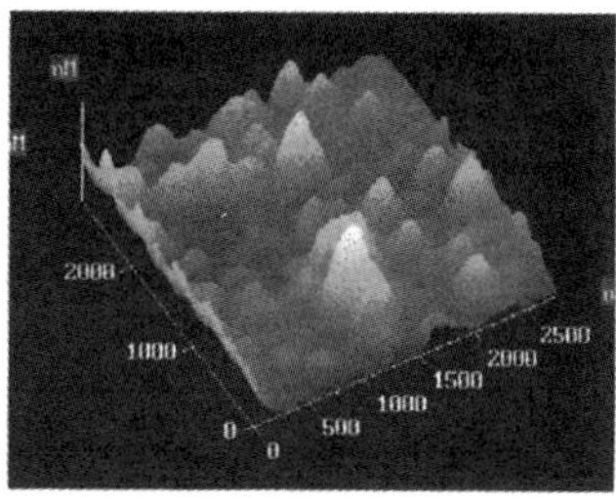

Fig.6. The AFM image of 55 kHz a-SiGe:H film.

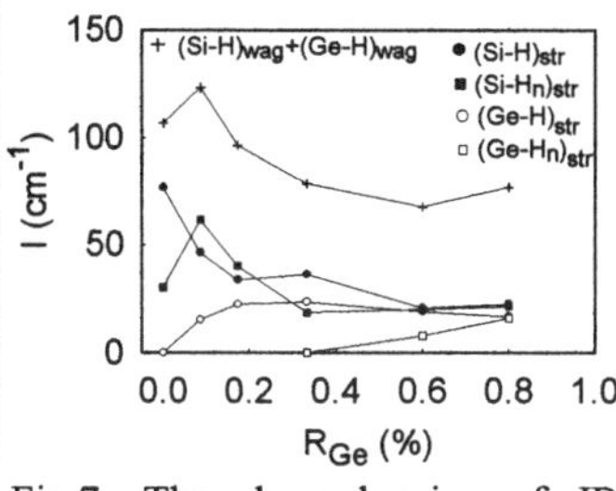

Fig.7. The dependencies of IR peak intensities on R_{Ge}.

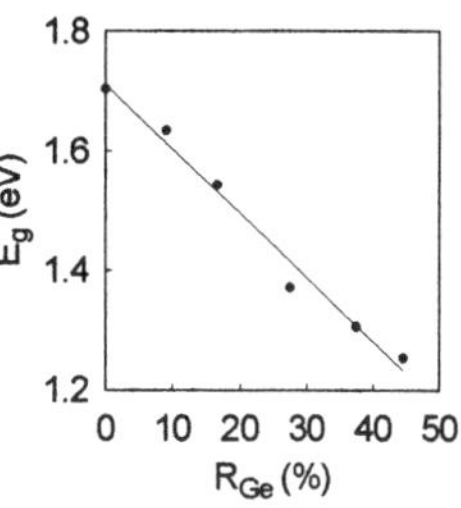

Fig. 8. The dependence of Tauc bandgap on R_{Ge}.

The Si-H_n stretching band at 2000-2063 cm^{-1} on IR spectra was decomposed into two gaussians to determine the contribution of monohydrides Si-H (2000-2004 cm^{-1}) and Si-H_n clusters (2042-2068 cm^{-1}) in the films microstructure. The intensities of these peaks as well as intensities of Si-H and Ge-H wagging vibrations at 605-634 cm^{-1} and Ge-H stretching at 1866-1892 cm^{-1} corresponding to the total hydrogen content and Ge-H bonds, respectively [7], as a function of R_{Ge} are shown in Fig. 7. It is seen that the total hydrogen content first increases with the adding of 9% GeH$_4$ in a gas mixture and then gradually decreases. This variation of total hydrogen content with the R_{Ge} is mainly determined by the amount of Si-H_n configuration. At the same time the amount of monohydride Si-H bonds decreases in the whole range of GeH$_4$ variation. So, the results presented in Fig. 7 show that the adding of GeH$_4$ in a gas mixture leads to substituting of Si-H bonds by Ge-H bonds while the Si-H_n related microstructure dominates at the R_{Ge} up to 27.5%. At higher GeH$_4$ content Ge-H_n configurations appear indicating the formation of Ge related microstructure. According to AFM data the increase of Ge-H bonds and Ge-H_n related microstructure is responsible for the increase of average island size in a-SiGe:H morphology.

The dependence of E_g on R_{Ge} is presented in Fig. 8. It is seen that there is a linear relation between optical band gap and germanium content that is usually reported in literature [8]. Moreover the E_g is particularly independent on the total hydrogen content. Taking into account the island type microstructure of our a-SiGe:H films it is suggested that Si-H_n and Ge-H_n configurations clustered on the island surfaces and doesn't influence the optical band gap while the Ge-Si bonds formed in the island bulk lead to the decrease of E_g.

The dependencies of dark and photoconductivity as well as activation energy on Tauc bandgap are shown in Fig.9. The first decrease of E_g due to increase of GeH$_4$ content is accompanied by the decrease of σ_{d300} and corresponding increase of E_a . The further decrease of

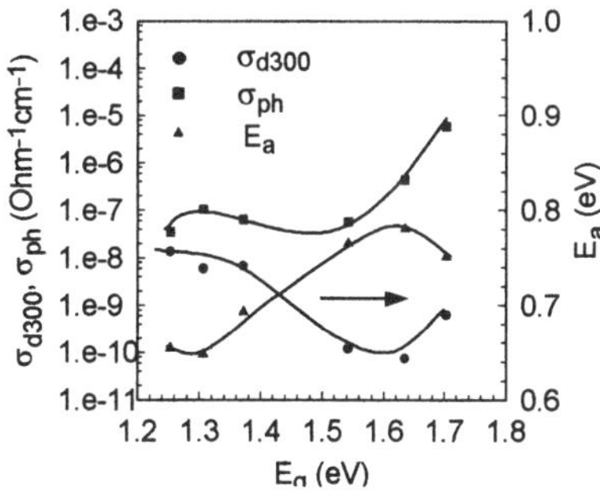

Fig.9. The dependencies of σ_{d300}, σ_{ph} and E_a on Tauc bandgap, E_g.

E_g leads to reciprocal behavior of these parameters. The change in the electronic parameters is connected with the change of the total hydrogen content, particularly with Si-H_n bonds. It means that Si-H_n related microstructure controls the density of states (DOS) distribution in the midgap and the Fermi level position. The dependence of Urbach energy and $\eta\mu\tau$ product on E_g is shown in Fig. 10. The decrease of E_g with germanium content is accompanied with the decrease of $\eta\mu\tau$ product and the increase of Urbach energy. The same dependencies were published in the literature [8]. According to IR data the increase of E_0 and the decrease of $\eta\mu\tau$ is caused by the substitution of Si-

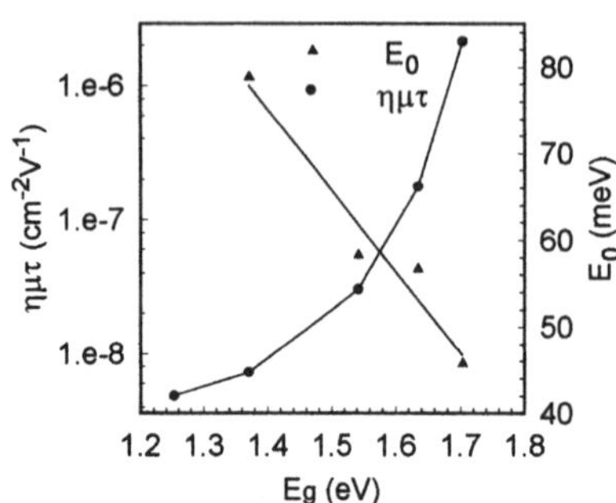

Fig.10. The dependencies of N_d and E_0 on E_g a-SiGe:H films.

H bonds by Ge-H bonds (see Fig.7). This in turn leads to the formation of Ge related defects [7]. The difference in the dependencies of E_a, σ_{d300} (see Fig.9) and E_0, $\eta\mu\tau$ (see Fig.10) on E_g indicates that the Si-H_n related microstructure controls the Fermi level position while Ge-related defects determine the recombination centers and photosensitivity in a-SiGe:H films deposited by 55 kHz PECVD. So, the tunable control of microstructure in a-SiGe:H films deposited by 55 kHz PECVD by choosing of appropriate technological parameters allows to monitor the DOS distribution and photosensitivity of the materials. The detailed analysis of DOS origin in a-SiGe:H films on the basis of photoconductivity modeling is under the progress.

The results of SW effect measurements for 55 kHz a-SiGe:H films (see Table 2) show that degradation of conductivity under He-Ne laser illumination decreases in comparison with a-Si:H which is one of the important advantage of this material.

CONCLUSIONS

The properties of a-SiC:H and a-SiGe:H deposited by 55 kHz PECVD in the wide range of CH_4 and GeH_4 content in the gas mixture were investigated. It was shown that the deposited films possess a comparable with literature data optoelectronic properties while their microstructure is characterized by island type morphology. The stability of a-SiC:H and a-SiGe:H films at R_C=20 and R_{Ge}=9.1%, respectively increases in comparison with a-Si:H which is important for device fabrication.

ACKNOLEDGMENTS

The authors are very grateful to A. Gologanov, S. Nesterov, and S. Lemeshko for the help in the AFM measurements. This work was supported by Netherlands Organization for Scientific Research (project NWO-047.005.09.96).

REFERENCES

1. B.G. Budaguan, A.A. Aivazov, A. Yu Sazonov, A. A. Popov, and A. E. Berdnikov, Mat. Res. Soc. Symp. Proc. **467**, p. 585 (1997).

2. B.G.Budaguan, et all. A.A.Sherchenkov, D.A.Stryahilev, A.Yu.Sazonov, A.G.Radoselsky, V.D.Chernomordik, A.A.Popov, and J.W.Metselaar, J. Electrochem. Soc. **145**, 2508 (1998).

3 . I.Solomon and L.R Tessler, Mat. Res. Soc. Symp. Proc. **336,** p. 505 (1994).

4 . G.Ambrosone S.Catalanotti, U.Coscia, S.Mormone, A.Cutolo, G.Breglio, Proc. 2nd WCPSEC, Vienna , **1**, p. 770 (1998).

5 . I. Pereyra, M.N.P. Carreno, M.H. Tabacniks, R.J. Prado, and M.C.A.Fantini, J. Appl. Phys. **84**, p. 2371 (1998).

6 B.G.Budaguan, A.A.Aivazov, Mat. Res. Soc. Symp. Proc. **513**, p. 387 (1998).

7. Y-P.Chou, S-C.Lee, J. Appl. Phys. **83**, p. 4111 (1998).

8. J. Folsch, F.Finger, T.Kulessa, F.Siebke, W.Beyer, H.Wagner, Mat.Res.Soc.Symp. Proc. **377**, p.517 (1995).

STABLE AMORPHOUS SILICON AND IMPROVED MICROCRYSTALLINE SILICON BY PHOTON-ASSISTED ELECTRON CYCLOTRON RESONANCE CHEMICAL VAPOR DEPOSITION

Young J. Song and Wayne A. Anderson, State University of New York at Buffalo, Dept. of Electrical Engineering, Amherst, NY 14260

ABSTRACT

Amorphous (a-Si) and microcrystalline silicon (μc-Si) are widely used in photovoltaics and thin film transistors. These products suffer from instability and less than desired electrical properties. We have utilized photon-assisted electron cyclotron resonance chemical vapor deposition (PA-ECRCVD) resulting in great improvement in both areas. For example, PA-ECRCVD compared with conventional ECR-CVD gives carrier lifetime of 1.35 μs compared to 0.17 μs, and photovoltaic solar cell efficiency of 10 % compared to 5.9 %. Moreover, the PA-ECRCVD cell only degraded to 9.8 % compared to the ECR-CVD cell degradation to 5.5 %, under long term exposure to tungsten lamp illumination. In addition, PA-ECRCVD gives much enhanced crystallinity in μc-Si as revealed by atomic force microscopy and Raman spectroscopy.

INTRODUCTION

Innumerable efforts to improve the quality of hydrogenated amorphous (a-Si:H) and microcrystalline silicon (μc-Si) films have introduced various growth techniques other than standard rf-driven plasma enhanced chemical vapor deposition (PECVD). Since the hydrogen related profile in the films (e.g. hydrogen content and hydrogen-silcon chemical bonding) plays a key role in electro-optical and structural properties as well as stability of the films [1-2], most of the growth techniques have been developed to provide an optimal hydrogen configuration. Among the techniques, hot wire chemical vapor deposition (HWCVD) [3], which utilizes a thermal decomposition of silane (SiH_4), and high density plasma deposition techniques, such as very high frequency plasma CVD (VHF-CVD) [4] and electron cyclotron resonance CVD (ECR-CVD) [5] have been known to provide better control of hydrogen evolution in the film. The silicon film deposited by these methods normally reveals superior properties due to better microstructure and lower hydrogen content. In this paper, we have introduced a photon-assist (PA) process into the ECR-CVD for the purpose of further control of hydrogen because the PA process during the deposition may modify both gas phase in the plasma and chemical reaction mechanism in the growth zone. We have studied the effects of the PA process both on a-Si:H and μc-Si films, in terms of electro-optical, structural and photovoltaic properties as well as stability.

EXPERIMENT

All silicon films used in this work were deposited by microwave (2.45 GHz) ECR-CVD both with and without the PA process. Its schematic diagram is described elsewhere [6]. The reactant gas was a 2% SiH_4/He mixture, while a minimum amount of Ar was used for plasma generation. The deposition of the a-Si:H film was done without H_2 dilution at a substrate temperature of 250 °C, a chamber pressure of 10 mTorr, an input power of 400 W, and a deposition rate of 70 - 80 Å/min. Meanwhile, the μc-Si film was grown with H_2 dilution (R_H = 0.33, where R_H = H_2/(2% SiH_4/He + H_2 + Ar)) in the conditions of a substrate temperature of 400 °C, a chamber pressure of 1mTorr, and an input power of 350 W. The deposition rate of the μc-

49

Mat. Res. Soc. Symp. Proc. Vol. 557 © 1999 Materials Research Society

Si film was in the range of 30 - 40 Å/min. Both flat Corning glass (for conductivity vs. temperature, carrier lifetime, light absorption coefficient vs. wavelength, Raman and AFM measurements) and HF-treated p-type 1-10 Ω-cm (100) silicon wafer (for FTIR and solar cells) were used as substrates. For the PA process, a commercially available tunsten-halogen lamp was used for illuminating the substrate with 1 - 10 W/cm^2 during the deposition. The solar cell structure used in this study was an undoped a-Si:H/p-type crystalline silicon heterojunction (cell area of 0.28 cm^2). In the FTIR study, 5000 Å of a-Si:H films were deposited both with and without PA. For other measurements, a fixed thickness of silicon film (1000 Å) was deposited. Both photoconductivity and photovoltaic response tests were done under 100 mW/cm^2 AM1.0 spectrum from a tungsten halogen lamp calibrated with a solar cell previously tested at NREL. The dark and photoconductivity measurements were carried out using a planar metal contact geometry with a 120 μm gap. Regarding the metallization for top contacts, a thermally evaporated Mg/Al double layer was used (100 Å/1000 Å for solar cells and 100 Å/500 Å for conductivity measurement). The carrier lifetime measurement of a-Si;H films was done at NREL by the rf conducting method at 532 nm wavelength for excitation.

RESULTS AND DISCUSSION

PA Effects on a-Si:H films

PA effects on the chemical bonding configuration between hydrogen and silicon atoms in the a-Si:H film has been examined by Fourier transform infrared spectroscopy (FTIR). Fig. 1 illustrates the infrared absorption spectra (stretching mode) of the a-Si:H films deposited both with and without PA. According to the early research groups, it was known that the dihydride bonding (=Si-H$_2$) degrades the general semiconductor property [1,7] (e.g. lower photoconductivity and mobility) of a-Si:H. They observed that the percentage of dihydride bonding in the film increases as the deposition temperature decreases below 200 °C, due to less decomposition of hydrogen molecules by thermal energy. As seen in Fig. 1, comparing to the a-Si:H film deposited without PA, the a-Si:H film grown with PA shows a weaker peak at around 2090 cm^{-1}, representing a lesser amount of dihydride bonding in the film. The photons emitted from the tungsten-halogen lamp (i.e. photon energy) in the PA process during the deposition seem to enhance the breakage of dihydride bonding in the growth zone, even at the

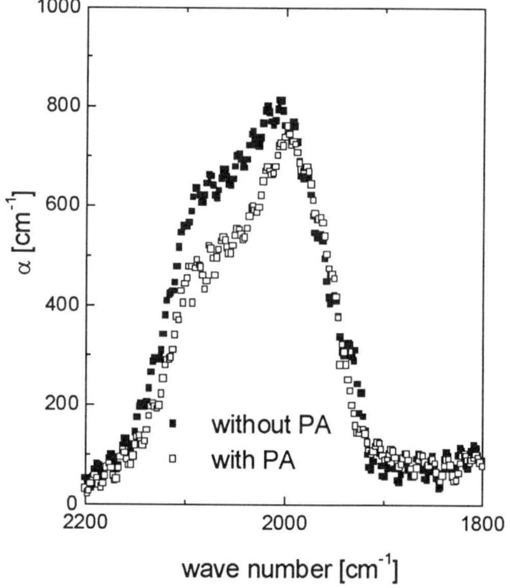

Fig. 1 FTIR spectra (stretching mode) of the a-Si:H films deposited with and without PA.

same substrate temperature (250 °C) because the dihydride bonds are broken more easily than monohydride ($\equiv$Si-H) bonds. This process may accelerate hydrogen removal in the film, resulting in a relatively lower hydrogen content in the a-Si:H film deposited with PA (from the smaller integrated area in the figure). This lower hydrogen content in the PA processed film is related to less degradation after light-soaking than the film grown without PA.

Table I. As-deposited and light-soaked properties of a-Si:H deposited with and without PA.

	As-deposited				After light soaking	
	Carrier lifetime [μsec]	σ_d/σ_p [S/cm]	E_a [eV]	E_{opt} [eV]	σ_d/σ_p [S/cm]	E_a [eV]
with PA	1.35	$8 \times 10^{-9}/1 \times 10^{-4}$	0.94	1.73	$2 \times 10^{-9}/7 \times 10^{-5}$	0.97
without PA	<0.17	$3 \times 10^{-9}/1 \times 10^{-5}$	0.72	1.84	$2 \times 10^{-9}/7 \times 10^{-6}$	0.83

As-deposited and light-soaked properties of a-Si:H processed with and without PA are compared in Table I. As seen in the table, the PA processed a-Si:H film generally shows better semiconductor characteristics. As expected from the previous FTIR result, the film deposited with PA reveals a higher photoconductivity ($\sigma_p \sim 10^{-4}$ S/cm) than that of the film grown without PA ($\sim 10^{-5}$ S/cm), by a factor of about 10. On the other hand, both films have quite comparable dark conductivity (σ_d) values (order of 10^{-9} S/cm). The measured larger carrier lifetime (1.35 μsec) and smaller optical gap (E_{opt}=1.73 eV)) of the PA processed film explain its superior photoconductivity nature. The activation energy (E_a) of the sample processed without PA undergoes a relatively large variation, from 0.72 to 0.83 eV, under the 2 week illumination. In contrast, the E_a of the sample deposited with PA shows a relatively large initial value (0.94 eV) and then it reveals a slight change to 0.97 eV after the illumination. This result reflects the better stability of the film because there is a general agreement that the increase in E_a after light-soaking is mainly due to the introduction of light-induced metastable defects in to the gap of a-Si:H. This is believed to result in the displacement of the Fermi level from the conduction band edge.

As illustrated in Fig. 2, the a-Si:H layer thickness values providing the highest efficiency (η) are quite different between the a-Si:H/c-Si heterojunction solar cells fabricated with and without PA. For the cell fabricated with PA, the maximum efficiency was 10.2% at a thickness of 850 Å. On the other hand, the highest efficiency of 5.9% was observed at a thickness of 200-250 Å for the cell fabricated without PA. The greatly improved carrier lifetime of the PA processed film translates to a greater permitted a-Si:H layer thickness. The photovoltaic data (Fig. 3 shows the corresponding J-V photovoltaic curves of the cells) indicates a much better efficiency for the PA processed cell due to higher short circuit current density J_{sc} (from better photon absorption), higher fill factor FF (from higher photoconductivity), and larger open circuit voltage V_{oc} (since the PA process is likely to reduce defects at the interface between a-Si:H and c-Si). Furthermore, the PA processed cell exhibited a more stable efficiency (less than an average 2% degradation even after 3 week exposure to light) than the cell fabricated without PA (about 6% degradation observed) in spite of the thicker a-Si:H layer of the PA processed cell.

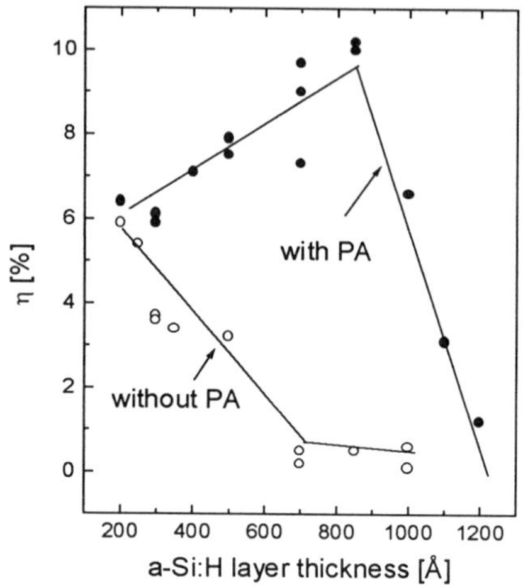
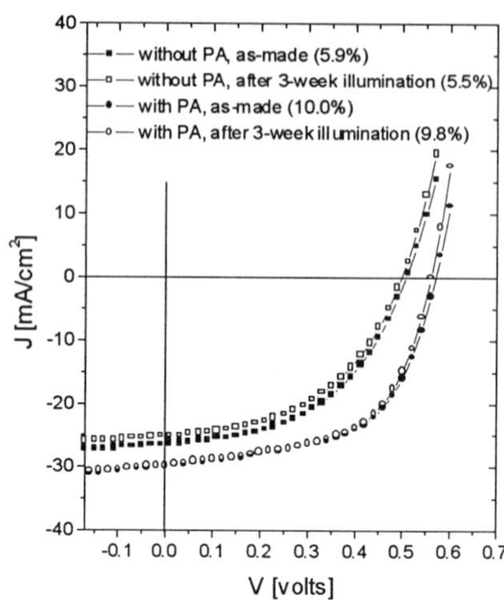

Fig. 2   a-Si:H layer thickness dependence on the efficiency of the cells fabricated with and without PA

Fig. 3 Photovoltaic responses of the cells fabricated with and without PA. The responses after 3 week illumination are also compared.

PA Effects on μc-Si films

The enhanced hydrogen removal capability of the PA process could be applied to μc-Si growth for a higher degree of crystallinity, because the breakage of silicon-hydrogen bonding (by photon absorption), followed by the hydrogen removal, may result in the favored production of silicon-silicon bonding. Fig. 4 demonstrates the effect of PA process on the Raman spectra of μc-Si films, grown with and without PA. From the figure, it is evident that the film deposited with PA has a better crystallinity, even at the same deposition conditions. As expected, the largest grains up to 800 Å are found in the AFM image of the PA processed μc-Si film (Fig. 5 (b)). In contrast, grain size of only 200-300 Å are observed for the μc-Si film deposited without PA (Fig. 5 (a)).

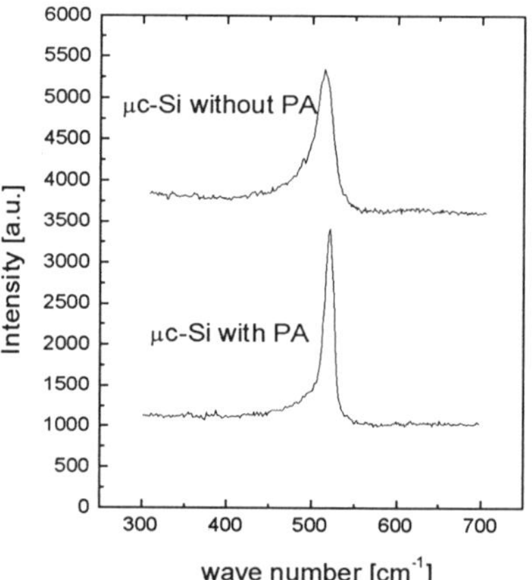

Fig. 4 Raman spectra of the μc-Si deposited with and without PA

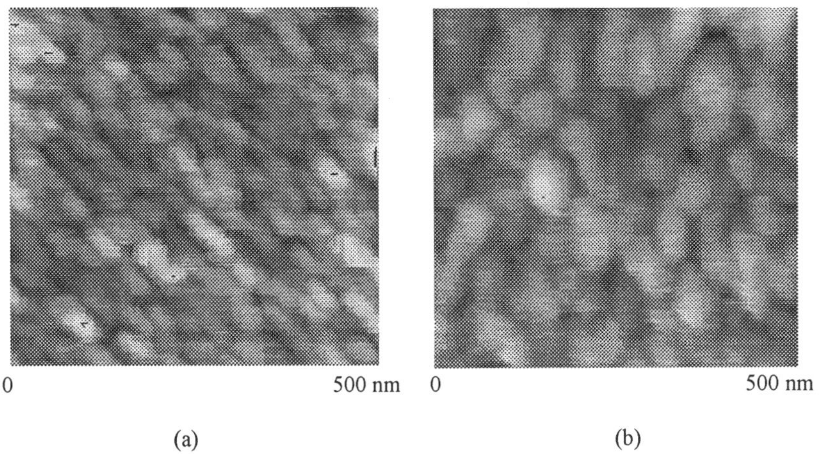

0	500 nm	0	500 nm
	(a)		(b)

Fig. 5   AFM images of the μc-Si films on glass substrates
deposited with and without PA. (a) without PA (b) with PA.

Table II. Structural and electrical properties of the μc-Si film grown with and without PA.

	Raman peak		grain size [Å]	σ_d [S/cm]	E_a [eV]
	FWHM [cm^{-1}]	peak position [cm^{-1}]			
with PA	17	520	500 – 800	4.3×10^{-3}	0.13
without PA	27	513	200 – 300	1.9×10^{-5}	0.35

The electrical conductivity and E_a of both films turned out to be quite different (Table II). For the film grown with PA, a much higher dark conductivity (σ_d) of 4.3×10^{-3} S/cm with smaller E_a of 0.13 eV was measured. The μc-Si film deposited without PA has a σ_d of 1.9×10^{-5} S/cm with E_a of 0.35 eV. The smaller E_a of the film deposited with PA could be explained by the larger amount of defect states in the energy gap, resulting from the dangling bonds created by photon absorption. However, this film still has a potential of higher carrier mobility after conventional hydrogenation.

CONCLUSION

In conclusion, we have investigated the possibility of using the photon assist (PA) process for advantages in the growth of silicon thin films by ECR-CVD. Compared to the a-Si:H film without the PA process, the PA processed film exhibits superior electro-optical and photovoltaic properties as well as stability due to its improved hydrogen-related chemical structure in the film. For µc-Si films, a higher degree of crystallinity has been achieved by the PA process, from the enhanced hydrogen removal during the deposition. The introduction of the PA process into the thin silicon film growth has a potential for increased applications for low-cost, low-temperature solar cells and thin film transistors.

REFERENCES

1. F. R. Jeffrey, H. R. Shanks, and G. C. Danielson, J. Appl. Phys. **50**, 7034 (1979).

2. W. Paul, Solid State Comm. **34**, 283 (1980).

3. X. Liu, E. Iwaniczko, R. O. Pohl, and R. S. Crandall, Mat. Res. Soc. Symp. Proc. **507**, 595 (1998).

4. F. Finger, P. Hapke, M. Luyberg, R. Carius, H. Wager, and M. Scheib, Appl. Phys. Lett. **65**, 2588 (1994).

5. M. Kitagawa, S. Ishihara, K. Setsune, Y. Manabe, and T. Hirao, Jpn. J. Appl. Phys. **26**, L231 (1987).

6. B. Jagannathan, R. L. Wallace, and W. A. Anderson, J. Vac. Sci. and Tech. A **16**, 2751 (1998).

7. P. J. Zanzucchi, C. R. Wronski, and D. E. Carlson, J. Appl. Phys. **48**, 5227 (1977).

AN OPTICAL GAP CALIBRATION APPLIED TO THE CASE OF HYDROGENATED AMORPHOUS SILICON

D. E. SWEENOR*, S. K. O'LEARY**, and B. E. FOUTZ***
* Department of Physics, Applied Physics, and Astronomy, Rensselaer Polytechnic Institute, Troy, New York 12180-3590
** Faculty of Engineering, University of Regina, Regina, Saskatchewan, Canada S4S 0A2
*** School of Electrical Engineering, Cornell University, Ithaca, New York 14853

ABSTRACT

There are many different empirical means whereby the optical gap of an amorphous semiconductor may be defined. We analyze some hydrogenated amorphous silicon data with respect to a number of these empirical measures for the optical gap. By plotting these various gap measures as a function of the breadth of the optical absorption tail, we provide a means of relating these disparate measures of the optical gap. The applicability of this calibration to another set of hydrogenated amorphous silicon data is investigated.

INTRODUCTION

In order to characterize the optical absorption spectrum associated with an amorphous semiconductor it has proven instructive to devise empirical measures for the optical gap. Unfortunately, there is no pronounced feature of this spectrum which can be directly related to such a gap. Instead of terminating abruptly at the fundamental gap, as in the case of a defect-free crystalline semiconductor, the absorption spectrum associated with an amorphous semiconductor exhibits a tail which encroaches into the gap region [1, 2]. While various empirical measures for the optical gap of an amorphous semiconductor have emerged, thus far it has proven difficult to meaningfully compare data which has been interpreted using disparate measures of this gap.

In this paper, we hope to shed light on how these different measures of the optical gap are related to each other. In particular, we analyze some hydrogenated amorphous silicon data of Cody *et al.* [1] with respect to a number of empirical measures of the optical gap, the data of Cody *et al.* [1] being from a study on how the disorder influences the optical absorption spectrum of hydrogenated amorphous silicon. By plotting these various gap measures as a function of the breadth of the optical absorption tail, we provide a means of relating these disparate measures of the optical gap. The applicability of this calibration to another set of hydrogenated amorphous silicon data is also investigated. Further details of this study will be reported in an upcoming publication of Sweenor *et al.* [3].

ON DEFINING THE OPTICAL GAP

For many years, the Tauc model [4] has served as the standard empirical model whereby the optical gap of an amorphous semiconductor may be defined. Assuming square-root distributions of conduction band and valence band states, assuming that the momentum matrix element is independent of $\hbar\omega$, and assuming that the disorder characteristic of amorphous semiconductors relaxes the momentum conservation rules, Tauc *et al.* [4] suggested that an extrapolation of the essentially linear functional dependence of $\sqrt{\alpha(\hbar\omega)\,\hbar\omega}$, observed in

Mat. Res. Soc. Symp. Proc. Vol. 557 © 1999 Materials Research Society

amorphous semiconductors at sufficiently large $\hbar\omega$, allows an empirical optical gap to be defined. Cody *et al.* [5] hypothesized that it is the dipole matrix element instead of the momentum matrix element which is independent of $\hbar\omega$. Thus, they asserted that one should instead extrapolate $\sqrt{\frac{\alpha(\hbar\omega)}{\hbar\omega}}$ to define the optical gap. Sokolov *et al.* [6], while assuming that the momentum matrix element is independent of $\hbar\omega$, argued that the disorder characteristic of amorphous semiconductors will modify the basic form of the optical absorption spectrum. They claimed that an extrapolation of $\sqrt[3]{\alpha(\hbar\omega)\,\hbar\omega}$ is even more appropriate.

An alternative empirical measure for the optical gap of an amorphous semiconductor, which is less sensitive to the interpretational difficulties which plague linear extrapolation methods [7], is the isoabsorption gap. This gap is defined as the point at which the optical absorption spectrum achieves a certain threshold value, i.e., $\alpha(\hbar\omega) = \alpha_\alpha$, where the threshold value, α_α, is usually selected to be 10^3 cm^{-1} or 10^4 cm^{-1} [8, 9]. This isoabsorption analysis can be used in cases where only a limited range of optical absorption data is available, linear extrapolations being even more uncertain in such cases. Freeman and Paul [8] find that the isoabsorption gap, defined where $\alpha_\alpha = 10^3$ cm^{-1}, is in approximate agreement with the Tauc gap.

ANALYSIS

We first determine the breadth of the optical absorption tail. Following the usual procedure, we fit the optical absorption data of Cody *et al.* [1] to an exponential functional dependence, of the form

$$\alpha(\hbar\omega) = \alpha_o\,\exp\left(\frac{\hbar\omega}{E_o}\right), \tag{1}$$

where α_o is a pre-exponential constant and where E_o denotes the breadth of the optical absorption tail. Contrary to the approach taken by Cody *et al.* [1], we do not look for an Urbach focus. Rather, we follow the approach suggested by Persans *et al.* [9], and fit, in a least-squares sense, the optical absorption data, over the range 5×10^2 cm$^{-1} \leq \alpha(\hbar\omega) \leq 5 \times 10^3$ cm^{-1}, to Eq. (1). Some of the fits to Eq. (1) are shown in Figure 1.

We now determine the optical gap using various empirical measures. Following the approach of Cody *et al.* [1], for the linear extrapolations we focus only on the data where $\alpha(\hbar\omega) > 10^4$ cm^{-1}. For the determination of the Tauc gap, we follow the analysis of Tauc *et al.* [4] and fit the relevant experimental data to

$$\sqrt{\alpha(\hbar\omega)\,\hbar\omega} = C_{\text{Tauc}}\left(\hbar\omega - E_{g\text{Tauc}}\right), \tag{2}$$

where C_{Tauc} denotes the slope of the Tauc extrapolation and $E_{g\text{Tauc}}$ represents the corresponding Tauc gap. The linear extrapolations, corresponding to the data depicted in Figure 1, are shown in Figure 2, these extrapolations being performed in the least-squares sense.

The Cody and Sokolov gaps are determined in a similar manner. In particular, the Cody gap is obtained by fitting the experimental data to

$$\sqrt{\frac{\alpha(\hbar\omega)}{\hbar\omega}} = C_{\text{Cody}}\left(\hbar\omega - E_{g\text{Cody}}\right), \tag{3}$$

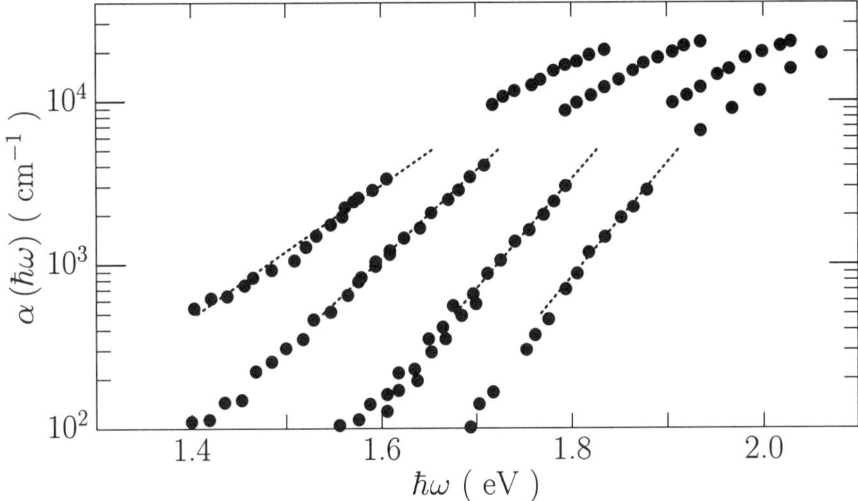

Figure 1: The breadth of the optical absorption tail. We depict some of the hydrogenated amorphous silicon data from Cody *et al.* [1] with the solid points. The fits to Eq. (1), for the data between 5×10^2 cm$^{-1} \leq \alpha(\hbar\omega) \leq 5 \times 10^3$ cm^{-1}, are depicted with the dotted lines. These fits are determined in a least-squares sense. A tabulation of the fitting parameters corresponding to Eq. (1) is found in Table 1 of Sweenor *et al.* [3]. Figure taken after Figure 1 of Sweenor *et al.* [3].

while the Sokolov gap is obtained by fitting the experimental data to

$$\sqrt[3]{\alpha(\hbar\omega)\ \hbar\omega} = C_{\text{Sokolov}}\left(\hbar\omega - E_{g_{\text{Sokolov}}}\right), \tag{4}$$

where C_{Cody} and C_{Sokolov} denote the slopes of the Cody and Sokolov extrapolations, respectively, $E_{g_{\text{Cody}}}$ and $E_{g_{\text{Sokolov}}}$ representing the corresponding Cody and Sokolov gaps. As with the Tauc extrapolations, all linear extrapolations are determined in a least-squares sense, the only data considered being that for which $\alpha(\hbar\omega) > 10^4$ cm^{-1}. The Cody and Sokolov extrapolations, corresponding to the data depicted in Figure 1, are shown in Figures 3 and 4 of Sweenor *et al.* [3], respectively. The fitting parameters corresponding to these linear extrapolations are tabulated in Table 2 of Sweenor *et al.* [3]. The isoabsorption results are tabulated in Table 3 of Sweenor *et al.* [3].

In Figure 3, we plot these empirical measures of the optical gap as a function of the optical absorption tail breadth parameter, E_o. We note that the isoabsorption gap, defined where $\alpha_\alpha = 10^4$ cm^{-1}, $E_{g_{\alpha_\alpha = 10^4}}$, forms an upper bound to all other gap measures, this gap usually exceeding the isoabsorption gap, defined where $\alpha_\alpha = 10^3$ cm^{-1}, $E_{g_{\alpha_\alpha = 10^3}}$, by about 0.2 eV. It is interesting to note that the Tauc gap is in almost exact accordance with $E_{g_{\alpha_\alpha = 10^3}}$, as suggested by earlier by Freeman and Paul [8]. We see, for any given determination of the optical gap, that the Cody gap is always lower than the Tauc gap, and that the Sokolov gap is always lower than the Cody gap. The total range in the optical gap, for any given determination of the optical gap, is found to be between 0.25 and 0.4 eV, this range increasing with E_o.

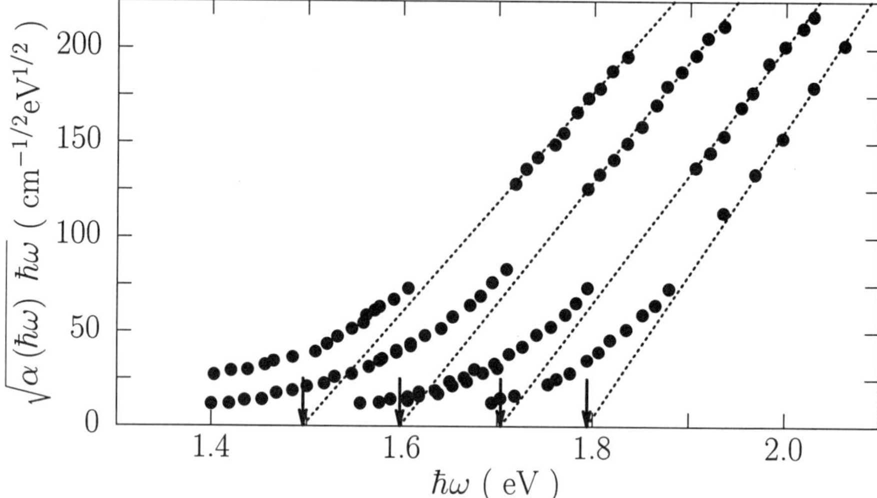

Figure 2: The Tauc gap. We show the Tauc extrapolations corresponding to the hydrogenated amorphous silicon data shown in Figure 1. The data is depicted with the solid points, the least-squares fits to Eq. (2) being indicated with the dotted lines. The extrapolated Tauc gaps are marked with the arrows. A tabulation of the fitting parameters corresponding to Eq. (2) is found in Table 2 of Sweenor *et al.* [3]. Figure taken after Figure 2 of Sweenor *et al.* [3].

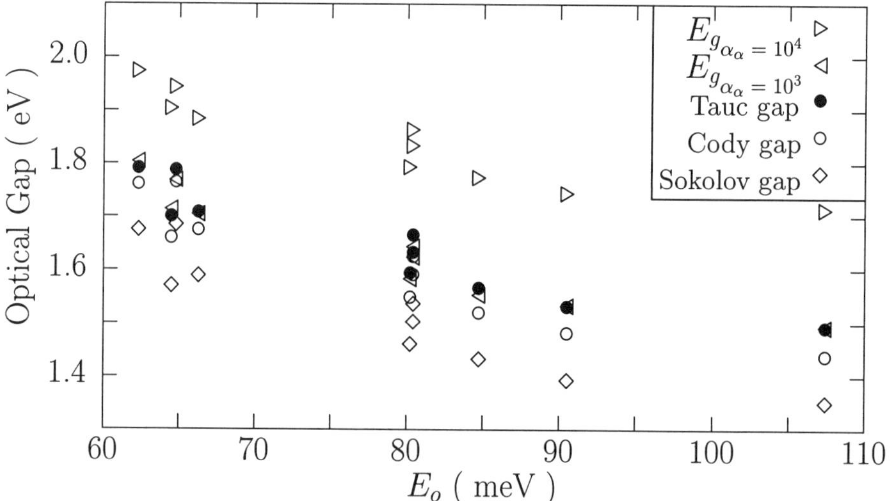

Figure 3: The various measures of the optical gap as a function of the breadth of the optical absorption tail. Figure taken after Figure 5 of Sweenor *et al.* [3].

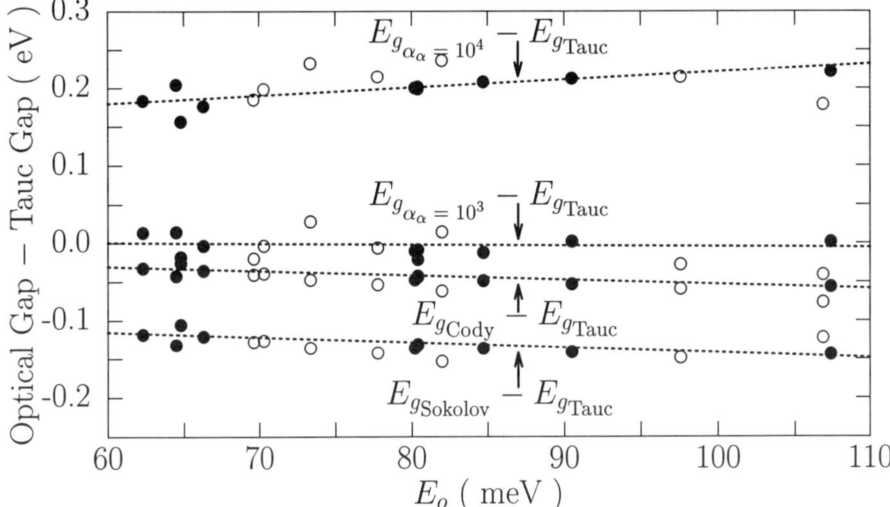

Figure 4: The energy difference between the various measures of the optical gap and the Tauc gap, as a function of the breadth of the optical absorption tail. We plot the experimental data of Cody et al. [1] with the solid points while the experimental data of Viturro and Weiser [11] is depicted with the open points. The least-squares fits corresponding to this data, Eqs. (5), (6), (7), and (8), are depicted with the dotted lines. Figure taken after Figure 6 of Sweenor et al. [3].

To determine how the various measures of the optical gap are related to each other, in Figure 4 we plot, as a function of the breadth of the optical absorption tail, E_o, the energy difference between these measures of the optical gap and the Tauc gap [10]. The Tauc gap is used as a reference in this analysis as the Tauc gap is the most common means whereby the optical gap of an amorphous semiconductor is determined. Performing a series of least-squares linear fits on the data of Cody et al. [1], we find, within the range of experimental error, that

$$E_{g\mathrm{Cody}} = E_{g\mathrm{Tauc}} - 0.57E_o + 4 \text{ meV}, \tag{5}$$

$$E_{g\mathrm{Sokolov}} = E_{g\mathrm{Tauc}} - 0.66E_o - 76 \text{ meV}, \tag{6}$$

$$E_{g_{\alpha_\alpha = 10^4}} = E_{g\mathrm{Tauc}} + 1.04E_o + 118 \text{ meV}, \tag{7}$$

$$E_{g_{\alpha_\alpha = 10^3}} = E_{g\mathrm{Tauc}} - 0.12E_o + 8 \text{ meV}. \tag{8}$$

The error in all cases is found to be of the order of 30 meV. These equations allow one to relate the various disparate measures of the optical gap of hydrogenated amorphous silicon, these measures for the optical gap being determined in the manner previously outlined in this paper.

In order to demonstrate that this calibration is not simply unique to the data of Cody et al. [1], we have also performed an analysis on the hydrogenated amorphous silicon data of Viturro and Weiser [11], the data of Viturro and Weiser [11] being from a study on how the bonded hydrogen content influences the optical absorption spectrum of hydrogenated amorphous silicon. In Figure 4, we also plot the data corresponding to the results of Viturro and Weiser [11]. We find that this data follows essentially the same trends, within the range of experimental error. This suggests that the calibration curve depicted in Figure 4 has a broader range of applicability than just the experimental data of Cody *et al.* [1].

CONCLUSIONS

In conclusion, we have analyzed some hydrogenated amorphous silicon data with respect to a number of empirical measures for the optical gap. By plotting these gap measures as a function of the breadth of the optical absorption tail, we have provided a means whereby one can relate the disparate measures of the optical gap of hydrogenated amorphous silicon.

ACKNOWLEDGEMENTS

The authors wish to thank L. Deutscher, F. Gaspari, J. M. Perz, T. Tiedje, and S. Zukotynski, for many helpful discussions. Financial assistance from the Natural Sciences and Engineering Research Council of Canada is gratefully acknowledged.

REFERENCES

[1] G. D. Cody, T. Tiedje, B. Abeles, B. Brooks, and Y. Goldstein, Phys. Rev. Lett. **47**, 1480 (1981).

[2] G. D. Cody, in *Hydrogenated Amorphous Silicon*, Vol. 21B of *Semiconductors and Semimetals*, edited by J. I. Pankove (Academic, New York, 1984), p. 11.

[3] D. E. Sweenor, S. K. O'Leary, and B. E. Foutz, Solid State Commun. (in press).

[4] J. Tauc, R. Grigorovici, and A. Vancu, Phys. Stat. Sol. **15**, 627 (1966).

[5] G. D. Cody, B. G. Brooks, and B. Abeles, Solar Energy Mater. **8**, 231 (1982).

[6] A. P. Sokolov, A. P. Shebanin, O. A. Golikova, and M. M. Mezdrogina, J. Phys. Cond. Matt. **3**, 9887 (1991).

[7] R. M. Dawson, Y. M. Li, M. Gunes, D. Heller, S. Nag, R. W. Collins, C. R. Wronski, M. Bennett, and Y. M. Li, Mater. Res. Soc. Symp. Proc. **258**, 595 (1992).

[8] E. C. Freeman and W. Paul, Phys. Rev. B **20**, 716 (1979).

[9] P. D. Persans, A. F. Ruppert, S. S. Chan, and G. D. Cody, Solid State Commun. **51**, 203 (1984).

[10] Cody [2] points out that changes in the optical gap of an amorphous semiconductor may be more easily determined than the absolute magnitude of the gap itself. It is for this reason that the calibration curve depicted in Figure 4 is cast relative to the Tauc gap.

[11] R. E. Viturro and K. Weiser, Philos. Mag. B **53**, 93 (1986).

DEPOSITION OF a-Si:H DEVICES IN A RTR SYSTEM FOR PHOTOVOLTAIC AND MACROELECTRONIC APPLICATIONS

M. Scholz, D. Peros and M. Böhm
Institut für Halbleiterelektronik (IHE), Universität GH-Siegen, Hölderlinstr. 3, D-57068 Siegen, Germany

ABSTRACT

This work presents first results of potential manufacturing processes for integrated series connected hydrogenated amorphous silicon (a-Si:H) thin film solar modules and/or pin-diode/TFT based macroelectronic circuits on flexible tapes. A RTR (Reel-To-Reel) deposition system on laboratory scale has been built. The system consists of seven metal sealed UHV stainless steel chambers to obtain ultra high vacuum as a basis for high quality a-Si:H layers. In order to support continuous movement of the tape in the RTR process the chambers cannot be isolated from each other. The necessary pressure difference between the sputtering chambers and the PECVD (Plasma Enhanced Chemical Vapor Deposition) chambers is provided by pressure stages. They are optimized for high molecular flow resistance without any influence on the moving substrate tape. The back metal contacts and the semitransparent TCO (Transparent Conductive Oxide) contacts are deposited by rf magnetron sputtering, the a-Si:H film system is deposited by PECVD. Parallel to the film deposition a Nd:YAG laser patterning system is coupled into one chamber. This allows for instance a total manufacturing of integrated series connected solar modules in one system without breaking the vacuum. Our present investigations focus on the deposition of doped and intrinsic high quality a-Si:H based layers in neighboring chambers. The quality of semiconducting films deposited in adjacent chambers is studied with regard to potential contamination effects.

INTRODUCTION

During the last decade many photovoltaic R&D efforts have been focusing on the optimization of amorphous silicon solar cells and modules. In comparison to mono- and poly-crystalline devices the amorphous material shows better potential for the production of large area low cost PV modules. Despite this fact thin film solar cells have not yet gained considerable market share until now. They contribute only about one fifth to the total shipment of solar cells [1]. One reason is the fact that in general there has been no significant increase in the demand for photovoltaic products. On the other hand the public attention is dominated and perhaps too much focused on lower efficiencies and stability problems of commercially available a-Si:H devices. In spite of the remarkable progress in the cell efficiency and stability the necessary transfer of results from research labs into production is insufficient and production rates for a-Si:H devices are relatively low. Furthermore, there are only a few companies worldwide working on related subjects [e.g. 2,3,4]. Therefore innovative features and new products of a-Si:H thin film technology such as integrated series connection, flexible devices, macroelectronic, sensor and memory applications have not received the appropriate attention.

In order to transfer the R&D results into production, this work aims at potential a-Si:H device manufacturing processes rather than at new world records on cell efficiencies. The design and construction of a seven chamber RTR deposition system on laboratory scale including in-line laser processing of the single layers into integrated a-Si:H modules on flexible metal tapes

Mat. Res. Soc. Symp. Proc. Vol. 557 © 1999 Materials Research Society

is discussed. Results concerned with the vacuum characterization, i.e. system base pressure, capacity of the pump system, pressure gradients between the process chambers etc. are presented as well as first results on cross contamination control in adjacent chambers.

EXPERIMENT

<u>Architecture of the RTR-System</u>

Fig. 1: Overview of the seven chamber RTR lab scale deposition machine with gas supply.

Fig. 2: Sputter chambers connected with slit.

Fig. 3: View of the seven chamber RTR lab scale deposition machine.

The seven chamber laboratory scale RTR machine is shown in figs. 1 to 3. Numerous non-standard vacuum components had to be designed and fabricated in order to meet the special requirements of the RTR deposition technology. The outer vacuum chambers (loading/unloading chambers) contain the substrate reels. These reels can be rotated by mechanical high vacuum feedthroughs. This enables their separation from the driving motors, driving pulley and electronics. The five inner chambers are designed for processing. A sputter region consisting of two chambers is separated from the PECVD zone by the laser chamber (fig. 4). The depositions of the amorphous and contact layers as well as the laser patterning are performed downwards. The main features of the concept under investigation are: a) the laser patterning as part of the flow line process, b) the integration of PECVD and sputtering processes in one vacuum line and c) the simultaneous processing. Since individual chambers cannot be isolated by gate valves the development of pressure stages which enable differential pumping is required. Additionally, for an optimized in-line fabrication process, each process step should have the same duration. These requirements have not been considered for the lab scale prototype. Moreover, at this stage the processing is performed in à stop-and-go mode. Nevertheless, the development is conducted such that an extension towards future continuous device manufacturing is possible. This could be realized by upscaling of the deposition zones. The overall length of the vacuum line is about 5m, the width of the metal tape is 9cm. All process chambers are mounted on adjustable axles which are mechanically operated. This construction allows a very precise alignment of the chambers with respect to each other so that problems associated with the transport of a thin (0.1mm) flexible metal tape can be minimized. The size and geometry of all chambers is identical. Only the built in parts are process specific. To reduce cross contamination and to allow different pressures (typical: PECVD 1mbar, sputtering 0.005mbar) in adjacent chambers, narrow and specially designed slits with high molecular flow resistance are used. These pressure stages allow a quasi-separation of the processes without using excessive pump capacity. In addition, the laser chamber serves as a buffer between PECVD and sputter zones. Differential pumping and purge gas injection is therefore possible [5].

A compact diode pumped Nd:YAG laser operating simultaneously at 1064nm and 532nm is coupled into the vacuum system through a standard 100mm viewport. A CCD video system is provided for positioning of scribing patterns for different layers. The deflection of the laser beam is accomplished by a galvo head. To obtain material selective patterning parameters such as power, Q-switch frequency, focus, wavelength and scan speed are adjusted.

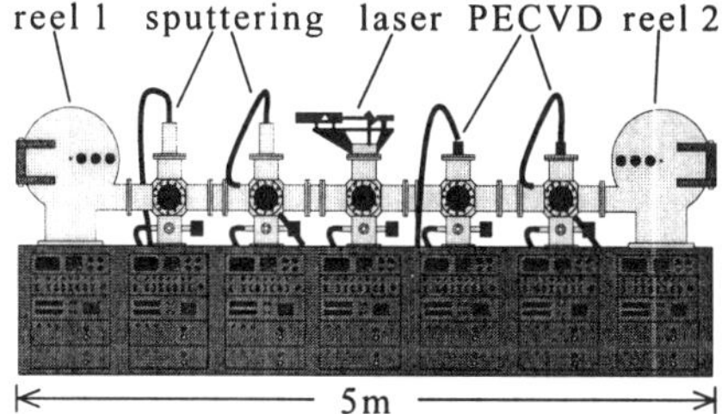

Fig. 4: Schematic drawing of the RTR deposition system.

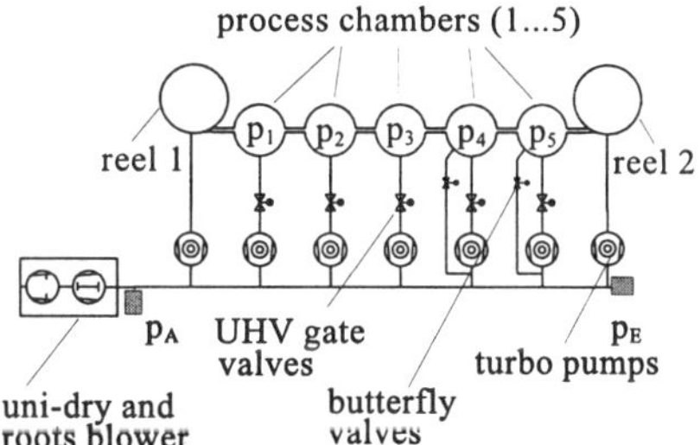

Fig. 5: Schematic layout of the pumping arrangement.

<u>Features of the Vacuum System</u>

Fig. 5 shows the schematic diagram of the vacuum system under investigation. The base pressures (p_1..p_5) of the process chambers are well below 10^{-8}mbar at 22°C.

Detailed calculations have been performed to define the pump capacity of the whole system and to predict pressure differences between neighboring chambers as a function of the slit geometry (fig. 6). These calculations have been verified by measurements. In order to investigate the pressure gradients between two process chambers, a test configuration was designed and built. A test slit of 30cm length, 0.5mm height and 9cm width with five flanges for pressure measurements was used. Fig. 7 shows the experimentally determined pressure gradients along the 30cm test slit. An effective pressure separation of the PECVD and sputter processes extending over more than three orders of magnitude has been verified.

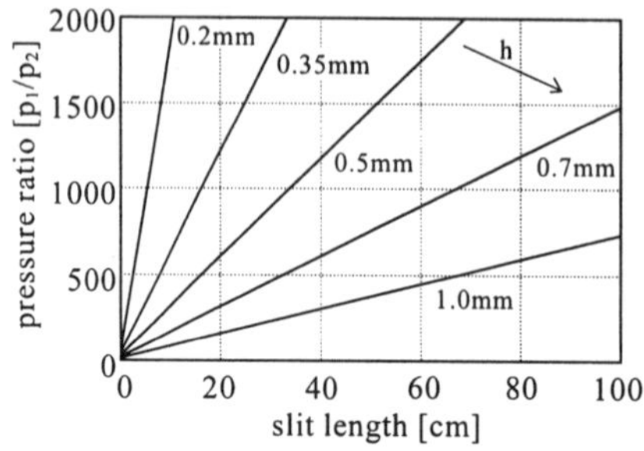

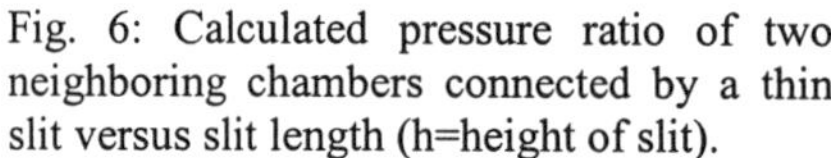

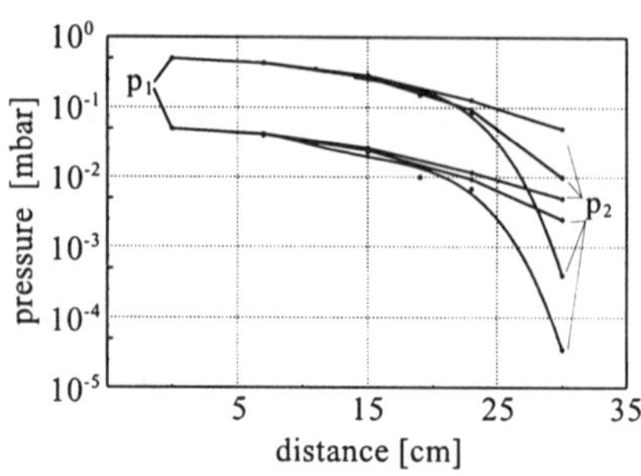

Fig. 6: Calculated pressure ratio of two neighboring chambers connected by a thin slit versus slit length (h=height of slit).

Fig. 7: Experimentally determined pressure gradients between neighboring chambers.

Cross Contamination Control

Two chambers are used for PECVD processing. The chambers are connected with the slit previously described. One chamber is designed for the deposition of intrinsic layers from pure silane with optional hydrogen dilution. The other chamber is designed for the deposition of doped layers using silane with a phospine/hydrogen gas mix (3%) for n-layers or a diborane/helium gas mix (3%) for p-layers. Dilution with methane or hydrogen is also possible. Due to the nearly equal average speed of the gas molecules of phosphine (427m/s) and diborane (473m/s at 20°C) phosphine has been chosen for the following investigations [6]. The contamination level of the doping gas in the intrinsic layers is estimated from the variations of the specific resistance of a-Si:H as a function of the doping gas mixture flow rates [7]. During the deposition of an intrinsic layer in one chamber, the phosphine mix flows under process conditions in the other chamber. No plasma is ignited in the phosphine chamber. Experiments with the above described direct chamber gas feed show that a pressure gradient of 0.3mbar between neighboring chambers (p(SiH$_4$)=0.7mbar, p(PH$_3$ mix)=0.4mbar) is insufficient to produce high quality intrinsic a-Si:H layers. In that case the specific resistance of a-Si:H is only in the range of 10^2-$10^4\Omega$cm, whereas without doping gas flow a specific resistance of more than $2 \cdot 10^9\Omega$cm is measured.

In another experiment a silane gas feed tube is mounted in the middle of the slit. The pressures in the neighboring chambers are kept equal. Fig. 8 shows the pressure at the feed point as function of the silane flow. The pressure in the middle of the slit is higher than inside the chambers. Fig. 9 shows the specific resistance of the i-layer versus the silane flow. The measurements reveal a perfect quasi separation of the two connected chambers. Even a pressure difference of 0.3mbar supporting the dopants' diffusion has no influence on the constantly high specific resistance (fig. 10). To check the influence of doping gas concentration the phosphine flow is varied (fig. 11). It is shown that the specific resistance does not change even if the phosphine flow exceeds the silane flow in either chamber. Investigations with diborane/helium gas mix (3%) show nearly similar effects on the specific resistance of the i-layers. At very small doping gas concentrations a slight increase of the specific resistance of the i-layer is observed. This is due to the compensation of the weak n-conductivity of undoped a-Si:H films.

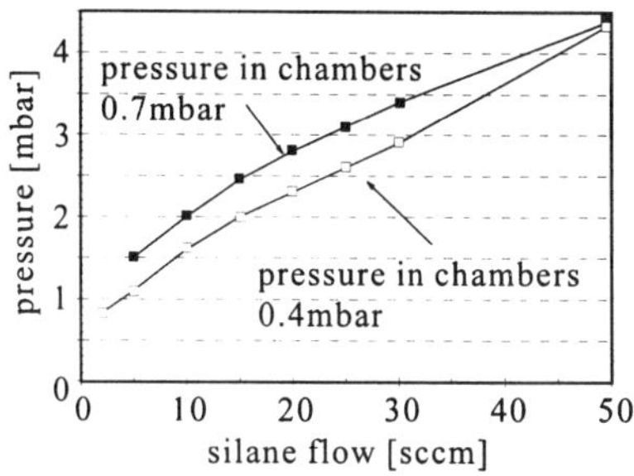
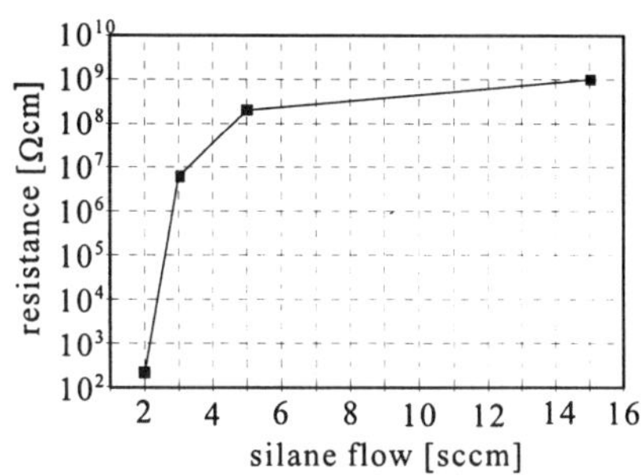

Fig. 8: Experimentally determined pressure in the middle of the slit versus silane flow.

Fig. 9: Experimentally determined specific resistance of the i-layer versus silane flow in the slit.

Above results are explained in terms of the increased gas speed inside the slit. With regard to the Maxwell/Boltzmann distribution the diffusion of dopants becomes negligible if the gas speed in the slit is at least twice as high as the diffusion speed of the dopants. Since, however, some molecules always move much faster than the average diffusion speed, it is impossible to completely eliminate cross contamination in an open system architecture. We find that the gas speed inside our slit is high enough to achieve high quality intrinsic a-Si:H layers even with doping gas flowing in adjacent chambers.

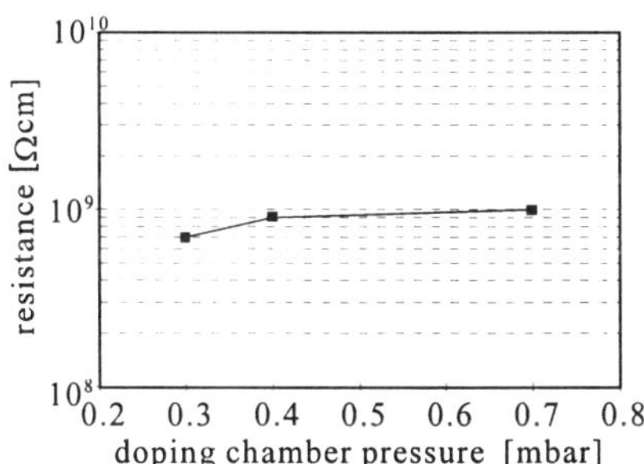
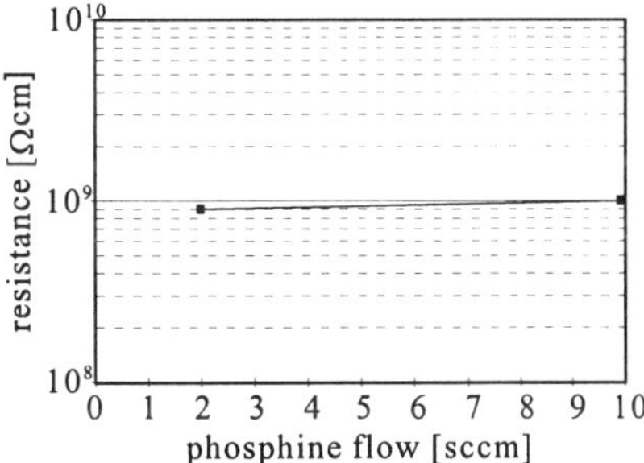

Fig. 10: Experimentally determined specific resistance of the i-layer versus doping chamber pressure (silane flow 15sccm, pressure of i-chamber 0.4mbar).

Fig. 11: Experimentally determined specific resistance of the i-layer versus phosphine flow in the doping chamber (silane flow 2·7,5sccm).

CONCLUSIONS

A RTR deposition system for a-Si:H solar module or macroelectronic device fabrication on laboratory scale has been developed. Pressure differences between adjacent chambers of three orders of magnitude are observed. Parallel sputtering and PECVD is therefore possible. The cross contamination is under control. The developed system allows a perfect quasi separation of the connected PECVD chambers. This has been reached without any influence on the process conditions or the ability of the continuous move of the substrate. A single chamber solar cell deposition with an efficiency of 4.2% (STC) has already been obtained. Further investigations focus on the parallel processing of solar cells and modules on continuous metal tapes.

ACKNOWLEDGMENT

This work was partially supported by the Ministerium für Wissenschaft und Forschung des Landes Nordrhein-Westfalen, project no. 25400392 and industrial partners.

REFERENCES

1. W. Schock, A. Shah, *Proceedings of the 14th EC Photovoltaic Solar Energy Conference*, p. 2000, (1997).

2. S. Guha, X. Xu, J. Yang, A. Banerjee, *Mat. Res. Soc. Symp. Proc.* **377**, p. 621, (1995).

3. S. Fujikake, K. Tabuchi, T. Yoshida, Y. Ichikawa, H. Sakai, *Mat. Res. Soc. Symp. Proc.* **377** p. 609, (1995).

4. Y. Kishi, H. Inoue, H. Tanaka, S. Kouzuma, K. Murata, S. Sakai, M. Nishikuni, K. Wakisaka, H. Shibuya, H. Nishiwaki, A. Takeoka, Y. Kuwano, *Proc. of the 22nd IEEE Photov. Spec. Conf.*, New York, p. 1213, (1991).

5. M. Scholz, D. Peros, M. Wagner, M. Böhm, *Proceedings of the 2nd World Conference on Photovoltaic Solar Energy Conversion*, p. 898, (1998).

6. A. Smith, J. D´ans, *Allgemeine und anorganische Chemie*, Verlag G. Braun, (1931).

7. W. Heywang, *Amorphe und polykristalline Halbleiter*, Springer Verlag, (1984).

CAT-CVD PROCESS AND ITS APPLICATION TO PREPARATION OF Si-BASED THIN FILMS

Hideki Matsumura, Atsushi Masuda and Akira Izumi
JAIST (Japan Advanced Institute of Science and Technology), Tatsunokuchi, Ishikawa 923-1292, Japan, TEL (+81)-761-51-1560, FAX (+81)-761-51-1149, E-mail; h-matsu@jaist.ac.jp.

ABSTRACT

This is to review the present understanding on Cat-CVD (catalytic chemical vapor deposition) or hot wire CVD. Firstly, the deposition mechanism in Cat-CVD process is briefly mentioned along with key issues such as the effect of heat radiation and a method to avoid contamination from the catalyzer. Secondly, the properties of Cat-CVD Si-based thin films such as amorphous silicon (a-Si), polycrystalline silicon (poly-Si) and silicon nitride (SiN_x) films are demonstrated, and finally, the feasibility of such films for industrial application is discussed.

1. INTRODUCTION

Plasma enhanced chemical vapor deposition (PECVD) has been recognized in the semiconductor industry as one of the most useful technologies to obtain thin films at low substrate temperatures. It is widely used in fabricating ultra-large scale integrated circuits (ULSIs) or amorphous silicon (a-Si) devices such as solar cells and thin film transistors (TFTs). However, recently, some other requirements begin to be imposed in obtaining these films. The plasma induced damage by PECVD has to be removed from the surface of substrates. The hydrogen (H) content, which is usually around 10 at. % in PECVD films, should be reduced to around 1 at. %.

A new method which supplies device-quality thin films with low H content at low substrate temperatures without using plasma has been developed in this decade. It is catalytic chemical vapor deposition (Cat-CVD) method, in which deposition gases are decomposed by the catalytic cracking reactions with a heated catalyzer placed near substrates, and thus, the films are deposited at low temperatures without using plasma. Since a tungsten wire is often used as a catalyzer, this method is often called "hot wire CVD (HWCVD)". The method was originally developed to obtain device quality hydro-fluorinated a-Si by our group [1], just after 6 years later from Wiesmann's first report [2] on silane (SiH_4) cracking by a heated tungsten. Then, it is applied to obtain hydrogenated a-Si [3], amorphous silicon-germanium [4], silicon-nitride (SiN_x) [5] and polycrystalline silicon (poly-Si) films [6].

The method is new compared with 30 years history of PECVD technology although the number of researchers is increasing since some useful results such as Mahan's attractive one on low H content a-Si [7] were reported. The feasibility of the method is now extensively studied by many research groups, and recently, promising results for device application begin to be reported [8-10].

This paper is to review the present understanding on Cat-CVD method. The deposition mechanism in Cat-CVD process is briefly mentioned along with key issues of the method such as the effect of heat radiation and a method to avoid contamination from the catalyzer. The properties of Cat-CVD Si-based thin films such as a-Si, poly-Si and SiN_x are demonstrated, and finally, the feasibility of industrial application of such films is discussed.

Mat. Res. Soc. Symp. Proc. Vol. 557 © 1999 Materials Research Society

2. FUNDAMENTALS OF CAT-CVD SYSTEM

(2-1) Deposition Apparatus

An example of Cat-CVD apparatus is schematically illustrated in Fig. 1. In this case, deposition gases are introduced from a shower head placed at top of the chamber and flow down to substrates through a heated catalyzer. A tungsten (W) wire is used as a catalyzer, and it is spread in parallel to the substrates.

The catalyzer is heated by supplying electric power directly on it. The temperature of catalyzer (T_{cat}) is measured by an electronic infrared thermometer through a quartz window at a chamber wall and it is also checked by the temperature dependence of electric resistivity of W. It is easily known that the temperature of thermometer is in good agreement with that obtained by the resistivity for the emissivity of about 0.4. The temperature of substrates, actually the substrate-holder temperature (T_{sh}), is measured by a thermocouple attached there.

Typical deposition conditions are summarized in Table I. In the Table, FR(X), PW_{cat}, S_{cat}, L and P_g refer to the flow rate of gas X, the electric power supply to heat up the catalyzer, the surface area of the catalyzer, the distance between the catalyzer and substrates and the gas pressure during deposition, respectively.

(2-2) Contamination from Catalyzer

In Cat-CVD system, other refractory metals apart from W can be used as a catalyzer, however, W appears the best choice among other metals, since the melting temperature is as high as 3400 ℃ and the vapor pressure at temperatures lower than 1900 ℃ is as low as 10^{-10} Torr. This low vapor pressure suggests that the contamination in the films due to spontaneous evaporation of W during deposition should be less than 10^{15} cm^{-3} when T_{cat} is kept at about 1700 ℃ and the films are formed with deposition rates larger than 2 Å/s. This value of W contamination is acceptable for the present and even future ULSI industry.

The W contamination is usually observed when it is oxidized. When W is heated at the temperature from about 1000 ℃ to 1400 ℃ in poor vacuum or in residual water vapor, the surface of W is easily oxidized, and since the melting temperature of this oxidized W is lower than 1500 ℃, the W is likely to contaminate the deposited films. One of key requirements in Cat-CVD is to avoid this oxidation of W by introducing reduced gas such as H_2 gas during heating of W. The other requirement is to avoid the formation of silicide due to the reaction

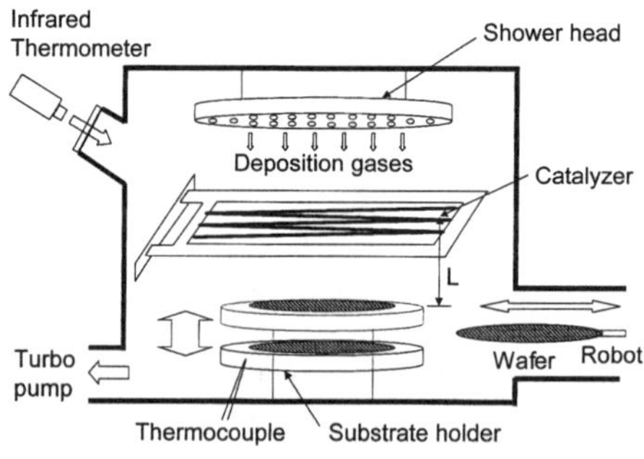

Fig. 1.   Schematic diagram of Cat-CVD apparatus.

Table I.   Typical deposition conditions for Cat-CVD a-Si, poly-Si and SiN_x films.

	a-Si	poly-Si	SiNx
Tcat	1500 -1900 $^{\circ}$C	1500-1900 $^{\circ}$C	1500-1900 $^{\circ}$C
FR(SiH4)	10 - 20 sccm	0.5 – 10 sccm	0.5 – 5 sccm
FR(H2)	0 - 40 sccm	0 – 200 sccm	————
FR(NH3)	————	————	50 – 200 sccm
FR(H2)/FR(SiH4)	0 – 2	0 – 100	————
FR(NH3)/FR(SiH4)	————	————	10 – 100
PWcat	100 – 600 W	800 – 1500 W	300 – 800 W
Scat	5 – 30 cm^2	10 – 60 cm^2	5 – 30 cm^2
L	4 – 8 cm	4 – 6 cm	4 – 6 cm
Pg	0.001 – 0.1 Torr	0.001 – 3 Torr	0.001 – 0.1 Torr
Tsh	150 – 300 $^{\circ}$C	300 – 450 $^{\circ}$C	200 – 300 $^{\circ}$C

of W with SiH_4 gas.   Since the melting temperature of W-silicide is higher than 2400 $^{\circ}$C, the silicide formation does not cause the contamination, however, it changes the resistivity of W wire and degrades the stability and reproducibility of deposition.   When whole parts of the W wire are kept over 1500 $^{\circ}$C, the silicide formation can be avoided [11].

The W contamination has been studied in various groups, and the results are summarized in Fig. 2 [12,13].   The W content in Si films is measured by secondary ion mass spectrometry (SIMS) and Rutherford backscattering spectroscopy (RBS).   All data show that the W content in the films is lower than detectable limit of measurements.

Detailed analysis of impurity contamination in the films demonstrates that the films are often contaminated from iron (Fe), other heavy metals and alkali metals such as potassium (K). For instance, Fe of the order of 10^{17} cm^{-3} is sometimes detected in Si films.   All these impurities are usually come from the W wire itself, since commercially available W wires contain Fe, K and other metals.   However, it is not difficult to reduce the contamination and to obtain "pure films" with the metal contamination lower than 10^{15} cm^{-3}, if a purified W is used. This lowered value of contamination appears acceptable in ULSI fabrication

(2-3) Influence of Heat from Catalyzer

In Cat-CVD method, one may be worried about the influence of thermal radiation from the heated catalyzer on the real substrate temperatures.   The real substrate temperatures (T_{sr}) are measured by special thermocouples embedded inside Si wafers or glass substrates.   The discrepancy between T_{sr} and T_{sh} is usually less than several-tens $^{\circ}$C.   The effect of thermal radiation is not so large when T_{cat} ranges from 1600 $^{\circ}$C to 1800 $^{\circ}$C.   For instance, when S_{cat} is 50 cm^2 and PW_{cat} is about 800 W for T_{cat} of 1800 $^{\circ}$C, only 30 % of the total power supplied onto catalyzer is dissipated by thermal radiation and the rest part of the power is used for heat transport by heated gases.

One of the most unique points in Cat-CVD is the existence of high temperature species.

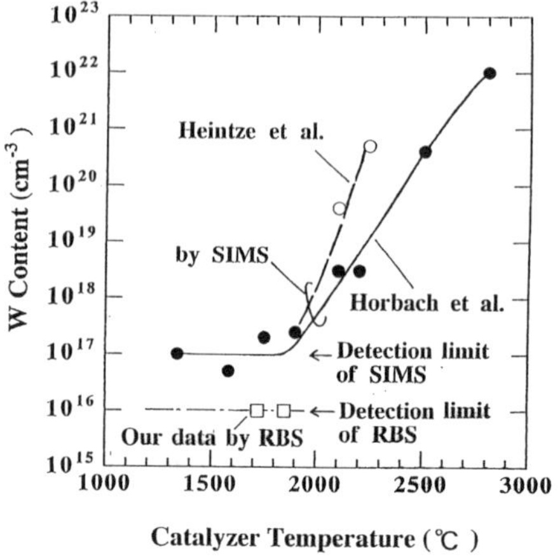

Fig. 2.   W content in Cat-CVD Si films as a function of the catalyzer temperature.

When decomposed species are thermally released from the surface of a hot catalyzer, they receive heat energies in forms of vibration or rotation, etc.   When these heated species reach to the surface of substrates, they start to reconstruct the atomic array of the surface and vibrate atoms there.   It is easily imagined that the heated species induce a local heating, that is, some local regions of growing surface are heated at temperatures between T_{sr} and T_{cat} by heated species when the mean free path of species is long enough, but the substrate itself is kept at T_{sr} because of its large heat capacity.   However, the temperature of such local heating is not known at the present moment although we can guess it indirectly from some experimental results.

In spite of this speculation for local heating, the value of T_{sr} is still an useful measure for actual application.   For instance, when SiN_x films are deposited on gallium arsenide (GaAs) devices, if the surface temperature is higher than 350 ℃ the device performance is degraded due to intermixing of electrode metal with GaAs surface.   However, when T_{sr} indicates the value lower than 350 ℃, the device performance is not surely degraded.

3. PHYSICAL IMAGE OF FILM FORMATION IN CAT-CVD
(3-1) Collision with Heated Catalyzer

Gaseous molecules introduced in a Cat-CVD chamber are travelling around inside the chamber and after a certain residence time they are evacuated by vacuum pumps.   During this travel, these molecules collide with the catalyzer and are cracked there.   The number of molecules colliding with the catalyzer is simply proportional to the gas pressure and also inversely proportional to a root of gas temperature.   It becomes about 10^{17} to 10^{18} cm^{-2}s^{-1} for P_g = 0.001 Torr and the gas temperatures from 0 ℃ to 2000 ℃.   When the gas molecules reach to the surface of the catalyzer, some of them are decomposed and chemically adsorbed but immediately released thermally into space.   Therefore, the number of cracked species is dependent on S_{cat} and also on chemical property of ad-species with catalyzing materials.

Figure 3 shows the deposition rate of Si films as a function of $1/T_{cat}$ for $S_{cat} = 1$ cm^2 and 3 cm^2 of W and molybdenum (Mo) catalyzers. The results are plotted from reported data of Horbach *et al.* for $P_g = 0.01$ Torr [14]. It is clearly shown from the figure that the deposition rate is proportional to S_{cat} but it is likely to saturate for T_{cat} over about 1700 ℃ to 1800 ℃. It is also demonstrated that the activation energy, that is, the slope of plots, appears to depend on the catalyzing materials. Since the number of sites on the catalyzer surface is limited, the number of decomposed species is automatically limited but dependent on S_{cat}. When T_{cat} is elevated, the period for the stay of adsorbed species on the catalyzer surface becomes very short, and the number of active site appearing in a unit time is likely to increase. Thus, for T_{cat} over a certain temperature the decomposition is saturated since the number of molecules colliding with the catalyzer becomes smaller than that of remaining active sites on the catalyzer surface. The results shown in Fig. 3 support that the generation mechanism of decomposed species is catalytic.

The period for stay of ad-species on the catalyzer surface is not clear, but it is believed to be very short and the order of nanoseconds to picoseconds [15]. It is also believed that a SiH$_4$ molecule is adsorbed on the W surface by forming –SiH$_3$ and –H for low T_{cat} [16], but becomes –SiH$_2$ and –2H for T_{cat} over 1000 ℃ [17]. Additionally, there is a claim that a SiH$_4$ is decomposed to –Si and –4H for T_{cat} over 1500 ℃ [18]. However, the decomposed form is dependent on the period from the start of adsorption to thermal release, and since such a period is ambiguous as mentioned above, the real form of decomposed species can not be mentioned clearly at the present moment.

The residence time of the molecules is 0.1 s to 100 s for the common deposition conditions, and during this time they collide with other molecules, chamber walls and the catalyzer. As average, one molecule has a chance to collide with the catalyzer by 1 to 1000 times and thus most of molecules are decomposed at least once during their stay in the chamber. One of major differences in Cat-CVD from PECVD is that higher density

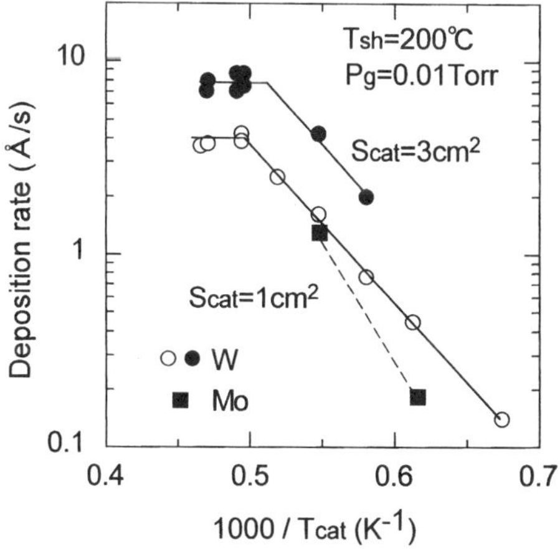

Fig. 3.   Deposition rate of Cat-CVD Si films as a function of
the reciprocal of the catalyzer temperature [14].

decomposed species are generated than PECVD case. The 3-dimensional collision of molecules with electrons is a key phenomenon in PECVD, but the collision of molecules with 2-dimensional solid is a key in Cat-CVD. Thus, the generation of high density species in Cat-CVD is reasonable results.

(3-2) Film Formation

Film formation process appears substantially similar to that in PECVD although the density of deposition precursors is much higher than that in the case of PECVD. Additionally, although the story is limited for the case of L comparable to the mean free path of precursors, the formation of higher silane which causes the degradation of film quality appears to be suppressed due to the heat decomposition of such species at the heated catalyzer or maybe even at the locally heated substrate.

The H content in the Cat-CVD film is usually quite low. The reason for it is not clear. However, the following reasons can be speculated. That is, 1) high density H atoms in Cat-CVD process may play an important role to subtract H from the surface of growing films, 2) the generation of higher silane which causes to bring about large amount of H on the growing surface is likely to be suppressed as mentioned above, and 3) the local heating on the sample surface causes the desorption of H. However, the lack of knowledge about film deposition mechanism in Cat-CVD is obvious.

4. PROPERTIES OF CAT-CVD Si-BASED FILMS
(4-1) Formation of a-Si

As mentioned above, the H content in Cat-CVD films can be kept very low. For instance, the H content in Cat-CVD a-Si can be lowered at less than 1 at. % for the deposition at T_{sr} lower than 400 °C as shown in Fig. 4. The H content in Cat-CVD a-Si can be adjusted by PW_{cat}, L and the ratio of $FR(H_2)$ to $FR(SiH_4)$ when a SiH_4 and H_2 gas mixture is used as

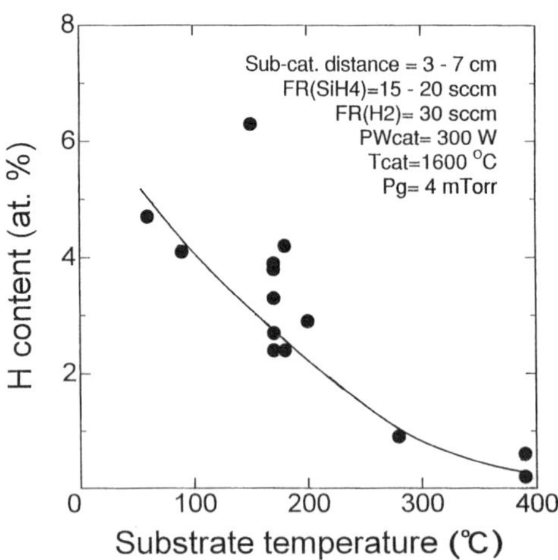

Fig. 4.   H content in Cat-CVD a-Si films as a function of the substrate temperature.

deposition gases. This low H content is one of attractive points for Cat-CVD films. For instance, it can be used directly as a starting material for excimer laser annealing (ELA) without particular degassing process to fabricate TFTs.

In Cat-CVD, the film can be deposited with high deposition rate. Figure 5 demonstrates the photosensitivity of a-Si films, measured under the light of AM-1, 100 mW/cm^2, as a function of the deposition rate. Major deposition conditions are described inside the figure. It is known from the figure that the deposition with the rate as high as 120 Å/s is possible although the film quality is a little degraded. However, the photosensitivity around 10^3, which is realized at the deposition rates over 100 Å /s, may be still acceptable for the application as photoreceptors for copy machines. Actually, although it is hard to adjust the electrical contacts on many aluminum cylinder drums as the base of photoreceptors at one time in the case of PECVD, there is no such limitation in Cat-CVD in addition to high rate deposition. The freedom from electrical contacts is one of major advantages in Cat-CVD.

(4-2) Formation of Poly-Si

Poly-Si films are also easily prepared by Cat-CVD method. When SiH_4 is diluted by H_2, the Si films are easily crystallized even if T_{sr} is lower than 400 °C. When the gas pressure during deposition increases, the films are also simply crystallized without H_2 dilution [6,19]. This increase of P_g is the easiest way to obtain poly-Si. Figure 6 shows the deposition rate of Si films and the crystalline fraction evaluated from Raman spectra, as a function of P_g for 2 types of Cat-CVD apparatuses, A and B. The chamber of the apparatus A has a size of 20 cm diameter and 20 cm height and evacuated by a diffusion pump, and that of B has 35 cm diameter and 35 cm height and evacuated by a turbo-molecular pump. It is found that the crystallinity and the deposition rate are both apparently improved as P_g increases. The Raman spectrum for such a poly-Si film obtained at high P_g exhibits similar sharpness to that of ELA poly-Si [20].

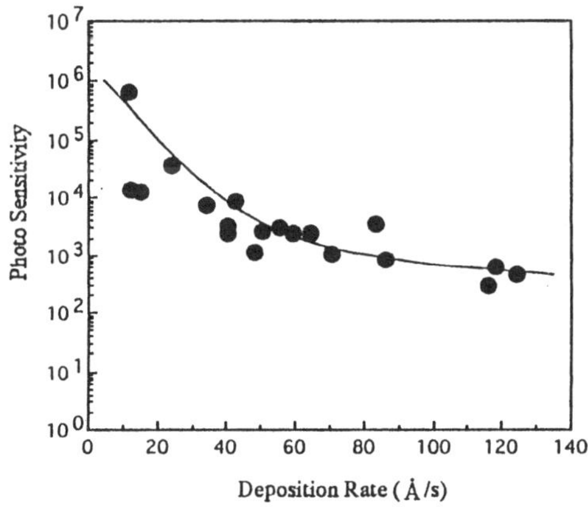

Fig. 5. Photosensitivity for Cat-CVD a-Si films as a function of the deposition rate.

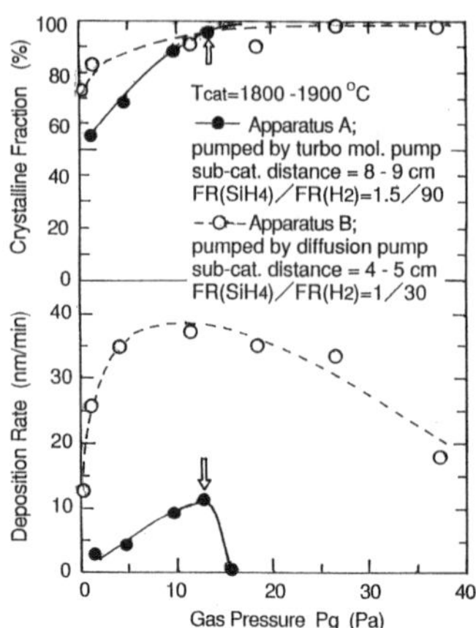

Fig. 6.   Deposition rate and crystalline fraction of Cat-CVD poly-Si films as a function of the gas pressure.

One of care points of poly-Si formation by Cat-CVD method is to suppress oxygen (O) inclusion.   The O atoms are included during deposition, but quite often they are included after air exposure.   Such poly-Si films are likely to be peeled off and not suitable for device fabrication.   The H_2 dilution sometimes causes to make such O contaminated films when P_g is high.

(4-3) Formation of SiN$_x$

SiN$_x$ films are easily obtained by Cat-CVD method when SiH_4 and ammonia (NH_3) gases are used as deposition gases.   Figure 7 shows the refractive index (n) of SiN$_x$ films deposited with various FR(NH_3) from 10 sccm to 60 sccm for fixed FR(SiH_4) of 1.1 sccm.   For this variation of flow rate, P_g is varied from 3 mTorr to 8 mTorr.   By increasing FR(NH_3), n decreases and reach to 2.0 which is often observed in stoichiometric Si_3N_4 films.

Figure 8 demonstrates the ratio of component elements in the films such as Si or nitrogen (N) atoms, measured by the X-ray photoemission spectroscopy (XPS), when the samples are etched.   At the initial step of etching, O atoms are observed due to the surface absorption of water.   However, it is clear that the ratio of N/Si for the film of n = 2.0 is 1.33, that is, the film of n = 2.0 is stoichiometric, however, that the film of n = 2.5 is apparently Si rich one.   In the case of PECVD SiN$_x$ films, device quality films are usually N rich and the H content is over 10 at. %, however, the H content in such a stoichiometric Cat-CVD Si_3N_4 film is lower than a few at. % or sometimes lower than 1 at. %.

Typical properties of SiN$_x$ films obtained by thermal CVD at temperatures higher than 800 ℃, PECVD at 300 ℃, photo-CVD at 300 ℃ and by Cat-CVD at less than 300 ℃ are summarized in Table II.   Cat-CVD SiN$_x$ is a low stressed but high density film since the etching rate by BHF solution is very low and equivalent to that of high temperature thermal

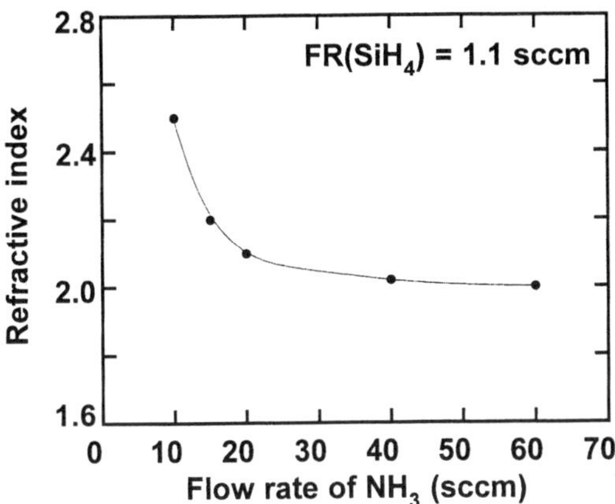

Fig. 7.　Refractive index of Cat-CVD SiN$_x$ films as a function of the flow rate of NH$_3$.

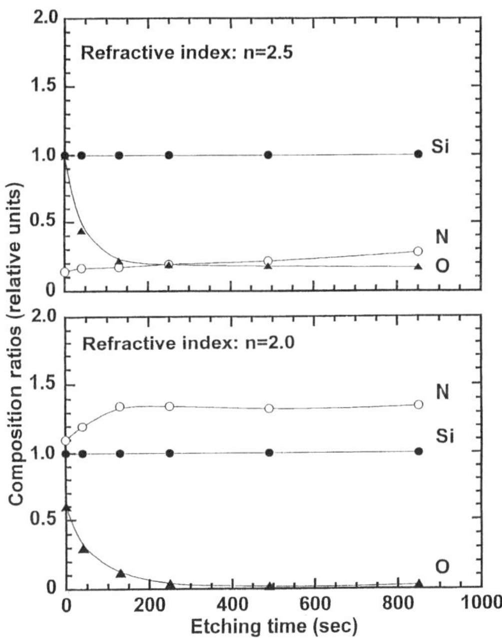

Fig. 8.　Depth profile of composition ratio of Cat-CVD SiN$_x$ films
with the refractive indices of 2.0 and 2.5.

Table II. Typical properties of SiN_x films prepared by various deposition methods.

	Thermal CVD	PECVD	Photo-CVD	Cat-CVD
Deposition gases	SiH_4, NH_3	SiH_4, NH_3	SiH_4, NH_3	SiH_4, NH_3
Refractive index	$2.0 - 2.1$	$1.9 - 2.0$	$1.9 - 2.0$	$1.9 - 2.0$
Deposition rate (nm/min)	30	$30 - 100$	$1 - 10$	$5 - 100$
Breakdown field (MV/cm)	10	$5 - 10$	$5 - 10$	$5 - 10$
H content (at. %)	$2 - 3$	$10 - 20$	$10 - 20$	< 3
Stress (dyn/cm^2)	$10\text{-}20 \times 10^9$	4×10^9	1×10^9	$0.7\text{-}1 \times 10^9$
Etching rate by BHF (nm/min)	2	$20 - 50$	100	$2 - 5$

CVD film. The Cat-CVD SiN_x films are now expected as a novel material for electronic devices because of their unique properties.

5. FEASIBILITY OF INDUSTRIAL APPLICATION OF CAT-CVD FILMS

Electronic devices such as solar cells and TFTs have been already fabricated by many research groups [8-10]. In Fig. 9, recently reported results on the efficiency of solar cells and the mobility of TFTs, which are both made of Cat-CVD a-Si films, are plotted as a function of years. The attempt to fabricate these devices has been just started within this few years. Some results appear preliminary at the present moment, however, they are revealing feasibility of Cat-CVD films.

In a solar cell, only middle i-layer is made of Cat-CVD a-Si and top p-layer and bottom n-layer are both produced in different apparatuses. Both sides of i-layer are exposed in air. In spite of this hard condition, the efficiency around 10 % has been achieved already.

The mobility of a-Si TFT has been already reached to 1 cm^2/Vs, which is equivalent to that of the conventional PECVD a-Si TFT. In Cat-CVD, theoretically, there is no limit in size of deposition area, although in radio-frequency (RF) PECVD the large expansion of deposition area induces a new technical difficulty due to standing wave of RF power. Additionally, the H content in Cat-CVD a-Si films can be lowered less than 1 at. %, and such a-Si films can be used as starting material for ELA without any degassing process as already mentioned above. Therefore, the application to TFT fabrication by using Cat-CVD is also expected.

On the other hand, there is a wide variety of expectation for Cat-CVD SiN_x films. Table III shows the device performance of GaAs high-electron mobility transistor (HEMT) for the samples using a Cat-CVD SiN_x film and a PECVD SiN_x film as surface passivation films [21]. It is clear that a noise figure (NF) at 12 GHz operation in Cat-CVD sample is much lower than that of PECVD sample. This improvement is attributed to damage-free deposition of SiN_x by Cat-CVD. The Cat-CVD SiN_x films are also attempted to be used in ULSI process and some successful results have been obtained as a preliminary step of research [22].

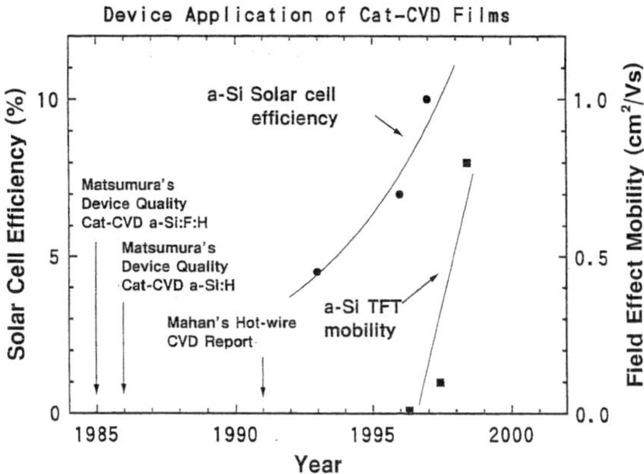

Fig. 9.   Efficiency of solar cells and mobility of TFTs, both made of Cat-CVD
a-Si films, as a function of year.

Table III.   Device performance of GaAs HEMT using Cat-CVD and PECVD
SiN$_x$ passivation films.

	with Cat-CVD films	with PECVD films
Change of active layer resistance after SiNx depo.	1.5 %	12.4 %
Mutual conductance, g_m	75.2 mS	70.7 mS
NF at 12 GHz operation	0.44	0.53

6. CONCLUSIONS

Here, recent situation of Cat-CVD research and application are summarized as follows:
1) By Cat-CVD, device quality Si-based thin films with H contents as low as a few at. % can
be obtained at substrate temperatures around 300 ℃.
2) The efficiency of Cat-CVD a-Si solar cell reaches to 10 %, and the mobility of Cat-CVD
TFT also reaches to 1 cm^2/Vs.   Device application of Cat-CVD films appears promising.
3) The Cat-CVD SiN$_x$ films used in a GaAs device appear superior to PECVD films.   The
Cat-CVD SiN$_x$ films appear also promising.

ACKNOWLEDGEMENTS

From 1998 fiscal year, this Cat-CVD work is supported from NEDO by "Cat-CVD

Fabrication Processes for Semiconductor Devices" through Ishikawa Sunrise Industries Creation Organization. The authors are very grateful to Mr. A. Heya for his help of drawing figures.

REFERENCES

1. H. Matsumura, H. Ihara and H. Tachibana, Proc. 18th IEEE Photovoltaic Specialists Conf., Las Vegas, 1985 (IEEE, New York, 1985), p. 1277.
2. H. Wiesmann, A. K. Ghosh, T. McMahon and M. Strongin, J. Appl. Phys. **50**, 3752 (1979).
3. H. Matsumura, Jpn. J. Appl. Phys. **25**, L949 (1986).
4. H. Matsumura, Appl. Phys. Lett. **51**, 804 (1987).
5. H. Matsumura, J. Appl. Phys. **66**, 3612 (1989).
6. H. Matsumura, Jpn. J. Appl. Phys. **30**, L1522 (1991).
7. A. H. Mahan, J. Carapella, B. P. Nelson, R. S. Crandall and I. Balberg, J. Appl. Phys. **69**, 6728 (1991).
8. S. Bauer, W. Herbst, B. Schroder and H. Oechsner, Conf. Record 26th IEEE Photovoltaic Specialists Conf. -1997, Anaheim, 1997 (IEEE, New York, 1997) p. 719.
9. H. Meiling, A. M. Brockhoff, J. K. Rath and R. E. I. Schropp, Mater. Res. Soc. Symp. Proc. **507**, (1998) (in press).
10. H. Matsumura, K. Mimura and H. Makino, Dig. of Tech. Papers of 1995 Int. Workshop on Active-Matrix Liquid-Crystal Displays, Osaka, 1995 (Business Center for Academic Societies Japan, Tokyo, 1995) p. 89.
11. P. Brogueira, J. P. Conde, S. Arekat and V. Chu, J. Appl. Phys. **79**, 8748 (1996).
12. M. Heintze, R. Zedlitz, H. N. Wanka and M. B. Schubert, J. Appl. Phys. **79**, 2699 (1996).
13. C. Horbach, W. Beyer and H. Wagner, J. Non-Cryst. Solids **137&138**, 661 (1991).
14. C. Horbach, W. Beyer and H. Wagner, J. Non-Cryst. Solids **114**, 187 (1989).
15. G. Ehrlich, Adv. Catal. **14**, 255 (1963).
16. A. G. Sault and D. W. Goodman, Surf. Sci. **235**, 28 (1990).
17. B. A. Scott, Semiconductors and Semimetals (Academic Press, Inc., 1984) Vol. 21A, Chap. 7, p. 123.
18. J. Doyle, R. Robertson, G. H. Lin, M. Z. He and A. Gallagher, J. Appl. Phys. **64**, 3215 (1988).
19. M. Ichikawa, J. Takeshita, A. Yamada and M. Konagai, Jpn. J. Appl. Phys. **38**, L24 (1999).
20. H. Matsumura, Dig. of Tech. Papers of Society for Information Display Int. Symp., Vol. 29, Anaheim, 1998 (Society for Information Display, Santa Ana, 1998) p. 1189.
21. R. Hattori, G. Nakamura, S. Nomura, T. Ichise, A. Masuda and H. Matsumura, Tech. Dig. of 19th IEEE Gallium Arsenide Integrated Circuit Symp., Anaheim, 1997, p. 78.
22. H. Sato, A. Izumi and H. Matsumura, Abst. Mater. Res. Soc. 1999 Spring Meeting, Symposium R, Ultrathin SiO_2 and High-K Materials for ULSI Gate Dielectrics, San Francisco, 1999, R. 6.13.

A NUMERICAL MODEL FOR HOT-WIRE CHEMICAL VAPOR DEPOSITION OF AMORPHOUS SILICON

D. G. GOODWIN
Division of Engineering and Applied Science, California Institute of Technology, Pasadena, CA 91125, dgoodwin@caltech.edu

ABSTRACT

A Direct Simulation Monte Carlo numerical model for hot-wire CVD is described that includes detailed gas-phase chemistry and accurately models gas-phase transport, from the low-pressure ballistic regime to the high-pressure diffusive regime. Model predictions for an idealized reactor geometry are shown to agree qualitatively with experimental trends.

INTRODUCTION

Numerical modeling can be a very useful complement to experiment to help understand and optimize complex processes such as hot-wire chemical vapor deposition (HWCVD) of amorphous silicon. For a realistic simulation of HWCVD, gas-phase chemistry must be taken into account, since chemistry is needed to convert highly-reactive atomic silicon leaving the wire into a less-reactive precursor, which is essential for high-quality films. A realistic simulation of chemistry requires, in turn, accurate treatment of energy transport from the wire, since rates of activated reactions depend exponentially on the local gas temperature. This is not easy for typical hot-wire CVD conditions, since the pressure is too low to assume diffusive transport near the wire, where the gas mean free path is larger than the wire diameter. On the other hand, transport to the substrate is not fully ballistic either, since radicals desorbing from the wire must experience at least a few collisions before reaching the substrate if chemistry is to occur. In such transitional regimes, where the mean free path is neither small nor large compared to dimensions of interest, the most effective strategy is usually direct numerical particle-based simulation.

Two studies have previously modeled aspects of hot-wire CVD of amorphous silicon [1, 2]. Molenbroek *et al.* calculated temperature and concentration profiles for 1% silane in helium at 40 Pa by solving continuum diffusion equations, and also carried out single-species random-walk simulations to compute film thickness spatial profiles. Tsuji, Akiyama, and Komiyama [2] have simulated time-dependent gas-phase kinetics in a uniform gas, with added terms representing diffusion-limited silane decomposition on the wire. In both cases, the models were designed to be relatively simple ones useful for representing specific experiments.

The goal of the present work is to formulate a fairly general model, which applies for all pressures of interest and will eventually be used to simulate hot-wire deposition of amorphous, microcrystalline, or single-crystal silicon. Only the amorphous case is considered here. The model is a particle-based, self-consistent simulation of the HWCVD gaseous environment, including chemistry on the wire and film surface. At present, surface chemistry is limited to specified reactive sticking probabilities, but will eventually be coupled to a kinetic Monte Carlo model of film growth. In this paper, the model will be briefly described and applied to simulate typical experimental conditions in pure silane and silane highly-diluted in helium, in particular the experiments of Molenbroek *et al.* [1, 3].

Mat. Res. Soc. Symp. Proc. Vol. 557 © 1999 Materials Research Society

MODEL DESCRIPTION

For simplicity, only the simplest possible hot-wire reactor is considered here: a long, straight wire enclosed by a concentric cylinder that serves as the substrate. End effects are neglected, and therefore spatial variations in mean gas properties only occur in the radial direction.

The gas is simulated with the Direct Simulation Monte Carlo (DSMC) method [4]. The DSMC method is a direct, physical simulation technique, in which simulation particles representing groups of physical molecules are allowed to move, collide, react, and interact with surfaces according to rules designed to produce statistically correct collision and reaction rates. Mean flow properties are determined by partitioning the domain into cells and sampling the number, identity, momentum, and energy of particles in each cell, averaging over many samples once steady state has been achieved. The technique has been tested and validated in many different low-density gas flows with good results [4].

CHEMISTRY

Si and H Production on the Wire

The primary chemistry on the wire is conversion of incident silicon hydride species and molecular hydrogen into atomic Si and H [5]. The wire is assumed to atomize incident Si_nH_m species with probability 0.7 [1]. For H_2, the dissociation is modeled as an activated process with an activation energy of 50 kcal/mol, resulting in a probability of reaction of 0.14 at 2000 °C.

Film Growth Chemistry

Film growth is modeled approximately, using measured or estimated reactive sticking probabilities [1, 6]. Since the total simulation time required for the gas to come to steady state is well below one monolayer coverage time, no attempt is made to dynamically compute the state of the surface. Instead, surface chemistry is modeled with global reactions. Based on the measurements of [6], atomic H has unity sticking probability on the surface, and SiH_3 can either abstract a surface hydrogen with probability 0.16, or stick to the surface with probability 0.10. Radicals that can insert into silicon-hydrogen bonds (Si, SiH, SiH_2, Si_2, Si_3, H_3SiSiH) are expected to be highly reactive on the surface, and are assigned a sticking probability of 0.7 [1]. At each timestep, the molecules that stick are removed from the simulation, and reaction products, if any, are emitted from the surface. Reactions involving a surface hydrogen are balanced by simply adjusting the H_2 flux back to the gas to insure overall H balance. No attempt is made to simulate hydrogen incorporation into the film.

Gas-Phase Reactions

The gas-phase mechanism consists of 18 reversible reactions among the species H, H_2, Si, Si_2, SiH, SiH_2, SiH_3, SiH_4, Si_2H_2, H_2SiSiH_2, H_3SiSiH, Si_2H_6, and Si_3H_8. The reaction rates are taken from literature sources. The silane pyrolysis mechanism of Ho, Coltrin, and Breiland [7] is used, with additional radical chemistry taken primarily from Woike, Catoire, and Roth [8]. In practice, bimolecular radical chemistry is the most significant, since the pressure is too low for most unimolecular pyrolysis reactions to proceed, or for the products of termolecular association reactions to be stabilized.

The most important gas-phase reaction is the one that converts Si into a less reactive species, which experimental film growth results suggest can only be a fast reaction between Si and SiH_4 [1]. There is still uncertainty about the nature of this reaction. Molenbroek *et al.* [1] discuss several possibilities, and argue that the most likely is

$$Si + SiH_4 \rightleftharpoons H_3SiSiH, \tag{1}$$

followed rapidly by

$$H_3SiSiH \rightleftharpoons H_2SiSiH_2. \tag{2}$$

However, this appears unlikely in light of the rate constant for reaction 1 estimated in Ref. [7] based on an RRKM analysis. At a pressure of 10 Pa, this reaction is far into the falloff regime, with a rate coefficient less than 10^7 cm^3/mol-s, seven orders of magnitude below the gas-kinetic limit. The collision rate at 10 Pa is inadequate to stabilize this highly exothermic reaction. Instead, it seems likely that the excited H_3SiSiH^* or $H_2SiSiH_2^*$ will dissociate into either $Si + SiH_4$, resulting in no reaction, or $Si_2H_2 + H_2$. The latter is energetically favored, since

$$Si + SiH_4 \rightleftharpoons Si_2H_2 + H_2 \tag{3}$$

is exothermic by 110 kJ/mol. If the barrier for H_2 elimination from H_2SiSiH_2 is close to that computed by Dickinson, O'Neil, and Ring [9] (85 kJ/mol) or assumed by Becerra and Walsh [10] (136 kJ/mol), then the exothermicity of reaction (1) (169 kJ/mol) is ample for this reaction to occur. In fact, a recent experimental report states that it proceeds at gas-kinetic rates (4×10^{14} cm^3/mol-s) with no barrier [8], although this value was determined indirectly.

Disilyne is a closed-shell molecule with several isomers. The lowest-energy one has hydrogens in bond-centered positions [$Si(H_2)Si$] [11], and should be far less reactive than Si. It is conceivable that it could play a role in film growth similar to SiH_3. But Si_2H_2 could also lead to highly-reactive intermediates, through the isomerization $Si(H_2)Si \rightleftharpoons H_2SiSi$, since H_2SiSi can insert into silane or into Si-H bonds on the surface [10]. In the present paper, only limited Si_2H_2 chemistry is considered, ignoring possible isomerization reactions.

In the calculations discussed below, reaction (3) is assumed to occur at the gas-kinetic rate of Ref. [8]. On the surface, incident Si_2H_2 is assumed to require a dangling bond to react, and therefore to have low reactivity. For the purposes of the simulation, a reaction probability of 0.03 is chosen, although this must be considered extremely uncertain. Si_2H_2 molecules which do not react to contribute to the film are assumed to recombine to disilane and desorb. Neither assumption is critical to the main results below, since with this assumed reaction rate SiH_3 remains the dominant low-reactivity precursor.

RESULTS

The results presented here aim to simulate approximately the conditions used in the experimental studies of Molenbroek *et al.* [1, 3]. All results shown here are for a 0.5 mm diameter wire at 2000 °C, and a wire-substrate separation of 1.5 cm.

Computed temperature profiles are shown in Fig. 1 for both pure silane and 1% silane in helium. In both cases, the gas at the wire is subtantially colder than the wire temperature of 2273 K. Such temperature discontinuities at boundaries are commonly observed in rarefied gas flows, in which the mean free path is not small compared to relevant length scales

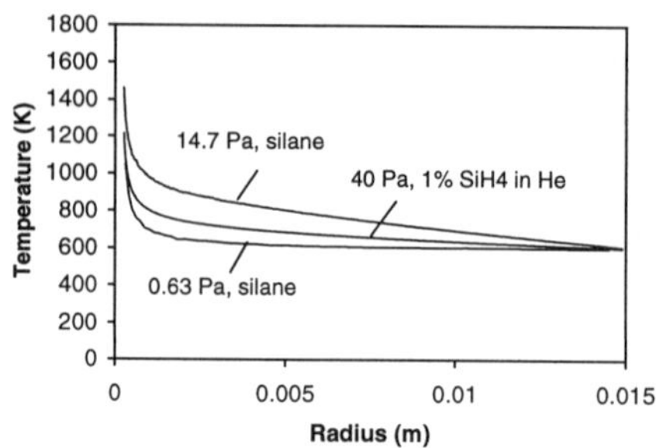

Figure 1: Gas temperature profiles for a wire temperature of 2273 K.

(here the wire diameter). (But note that the temperature is defined as an average over all molecules, independent of direction. Those desorbing from the wire *do* come off with energies characteristic of the wire temperature, but in determining the gas temperature are averaged with low-energy ones heading toward the wire.)

Profiles of the mole fractions of several important species are shown in Fig. 2 for pure silane conditions. Si is seen to be strongly depleted at the surface as the pressure increases, while H at the surface is almost independent of pressure. SiH_3 and most other Si_nH_m radicals have greatest concentration near the wire where they are created (for example by $SiH_4 + H \rightleftharpoons SiH_3 + H_2$), and decrease toward the substrate where they are destroyed.

The predicted growth rate for pure silane conditions is shown in Fig. 3 as a function of pressure. The contributions of low-reactivity species (SiH_3, Si_2H_2, and H_2SiSiH_2) and high-reactivity species (primarily Si, SiH, and SiH_2) are shown separately. At the lowest pressures, Si atoms can reach the substrate without reacting with silane, and are the major contributors to film growth. As the pressure increases, the Si flux to the substrate is exponentially damped, and the contributions of SiH_3 and Si_2H_2 rise, as SiH_3 is created by $H + SiH_4 \rightleftharpoons H_2 + SiH_3$ and Si_2H_2 by $Si + SiH_4 \rightleftharpoons Si_2H_2 + H_2$. At all pressures, the SiH_3 contribution to growth is greater than that of Si_2H_2. The basic trends would not be affected if growth from Si_2H_2 were neglected entirely. As the pressure increases further, the SiH_3 concentration builds to the point that SiH_2 production via $SiH_3 + SiH_3 \rightleftharpoons SiH_2 + SiH_4$ becomes significant, and SiH_2 begins to contribute to growth, causing the second rise in the high-reactivity growth rate.

Since highly reactive species are believed to degrade film quality, the ratio of the growth rate due to high-reactivity species (G_{high}) to that for low reactivity species (G_{low}) should scale inversely with film quality. An analogous experimental measure R' is defined in Ref. [3] based on infrared absorbance in two bands. In Fig. 4, the ratio G_{high}/G_{low} vs. pressure is shown along with R' values from Ref. [3] for films grown under similar conditions. Both show a minimum at a pressure of a few Pa, where the substrate is being effectively shielded from Si but radical-radical reactions have not yet become significant.

Finally, in Fig. 5 mole fraction profiles are shown for a case with silane heavily diluted in helium. To try to achieve sufficient statistics for the trace species, over 300,000 particles were used, and the simulation required approximately 20 hours. However, SiH_3 is still poorly resolved. The residence time was held at the value for the pure silane cases, which results in a significantly lower silane depletion (4%) and therefore a low growth rate (0.43 A/s).

At this pressure, the continuum approximation holds everywhere but near the wire. A

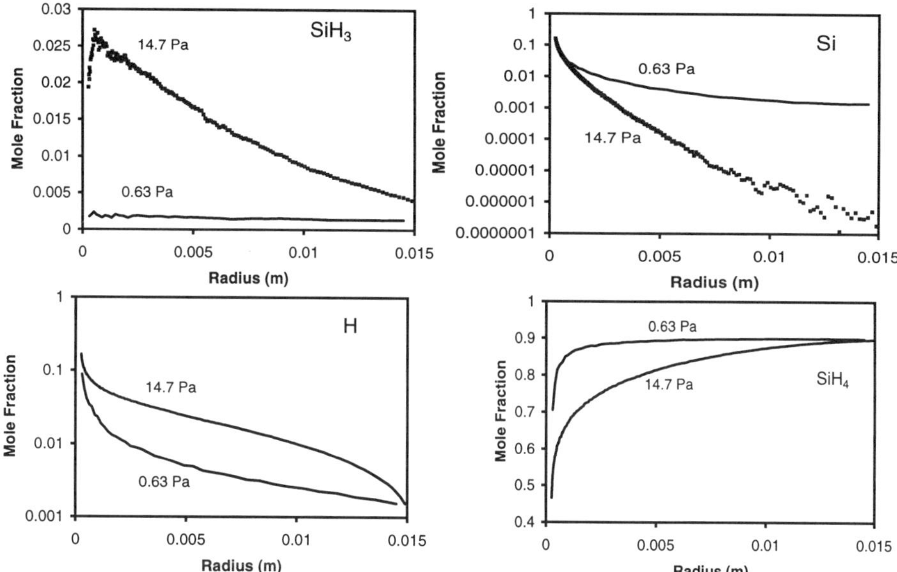

Figure 2: Mole fraction profiles of SiH_4, Si, H, and SiH_3 for pure silane conditions at two pressures.

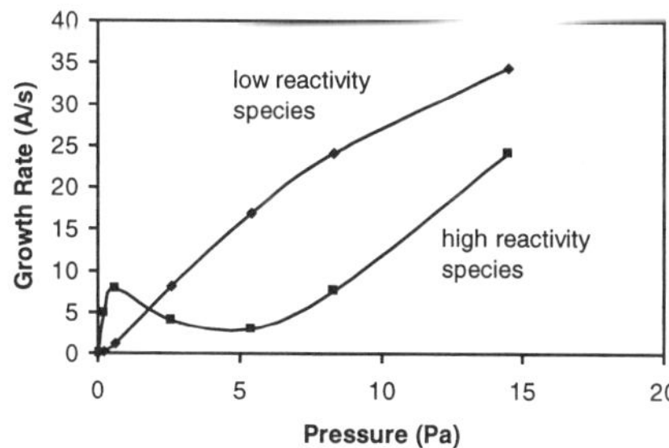

Figure 3: Growth rate due to high-reactivity and low-reactivity species.

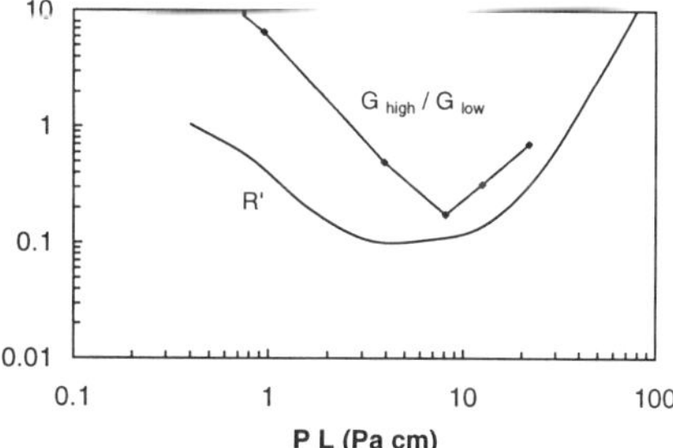

Figure 4: Comparison of the growth rate ratio G_{high}/G_{low} computed here to R' measured in Ref. [3]

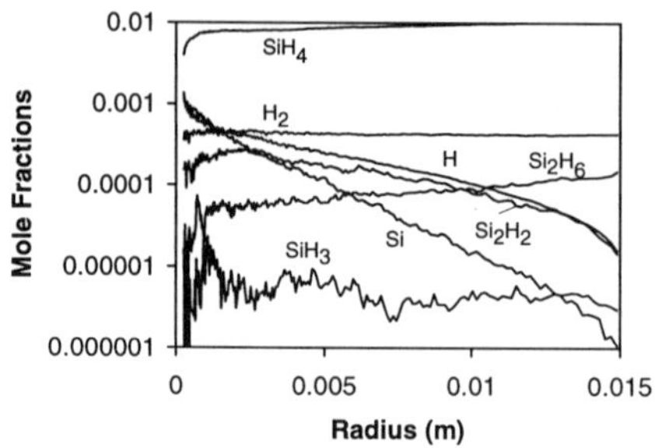

Figure 5: Selected species profiles in 1 % silane / helium at 40 Pa.

much more effective computational approach is to solve continuum equations where they are valid, and limit DSMC simulations to the vicinity of the wire. The capability to do this is present in the model, and is a subject of current work.

ACKNOWLEDGMENTS

This work is supported in part by grant CTS-9529309 from the National Science Foundation. I would also like to acknowledge Prof. Harry Atwater, co-investigator on this program, for enlightening discussions.

REFERENCES

1. E. C. Molenbroek, A. H. Mahan, E. J. Johnson, and A. C. Gallagher, J. Appl. Phys. **79**, 7278 (1996).

2. N. Tsuji, T. Akiyama, and H. Komiyama, J. Non-Cryst. Solids **198–200**, 1054 (1996).

3. E. C. Molenbroek, Mahan A. H., and Alan Gallagher, J. Appl. Phys. **82**, 1909 (1997).

4. G. A. Bird, *Molecular Gas Dynamics and the Direct Simulation of Gas Flows* (Oxford University Press, New York, 1994).

5. J. Doyle, R. Robertson, G. H. Lin, M. Z. He, and A. Gallagher, J. Appl. Phys. **64**, 3215 (1988).

6. J. Perrin, M. Shiratani, P Kae-Nune, H. Videlot, J. Jolly, and J. Guillon, J. Vac. Sci. Technol. A **16**, 278 (1998).

7. P. Ho, M.E. Coltrin, and W.G. Breiland, J. Phys. Chem. **98**, 10138 (1994).

8. D. Woiki, L. Catoire, and P. Roth, AIChE Journal **43**, 2670 (1997).

9. A.P. Dickinson, H.E. O'Neil, and M.A. Ring, Organometallics **10**, 3513 (1991).

10. R. Becerra and R. Walsh, J. Phys. Chem. **96**, 10856 (1992).

11. B. T. Colegrove and H. F. Schaefer III, J. Phys. Chem. **94**, 5593 (1990).

DEPOSITION OF HIGH QUALITY AMORPHOUS SILICON BY A NEW "HOT WIRE" CVD TECHNIQUE

Scott Morrison, Ken Coates, Jianping Xi, Arun Madan

MVSystems Inc., 327 Lamb Lane, Golden, Colorado 80401, or 17301 W. Colfax Ave., Suite 305, Golden, CO 80401, USA.

ABSTRACT

For the "Hot Wire" chemical vapor deposition technique (HWCVD) method to be applicable for photovoltaic applications, certain critical technical issues need to be addressed and resolved such as: lifetime of the filaments, reproducibility, large area demonstration of the material and stable devices. We have developed a new approach (patent applied for) which addresses some of these problems, specifically longevity of the filaments and reproducibility of the materials produced. The new filament material used has so far shown no appreciable degradation even after deposition of >200 μm of amorphous silicon (a-Si). We report that this can produce "state-of-the-art" a-Si with a dark conductivity of $<10^{-10}$ (Ohm*cm)$^{-1}$ and photoconductivity of $>10^{-5}$ (Ohm*cm)$^{-1}$: this material can also be doped p- or n-type. We also provide data using XRD as well as the Raman spectra. These materials have been incorporated into simple Schottky barrier structures. The development of microcrystalline silicon materials is also discussed.

INTRODUCTION

"Hot Wire" chemical vapor deposition (HWCVD) technique has attracted a considerable amount of attention in the last few years because of, (1) the possibility of producing a low concentration of H in the SiH alloys with a view towards reducing or eliminating the instabilities in the material, (2) the possibility of obtaining a high deposition rate and (3), the fabrication of low temperature poly-crystalline silicon. The technique was first demonstrated by Matsumura [1], in which SiH_4 and H_2 gasses were decomposed by catalytic or pyrolytic reactions with a heated catalyzer placed near the substrate.

For the technique to be successful requires further developmental work, specifically in the following areas. (a) Filament: at present, the filament (usually W or Ta) diameter used is on the order of about 1 mm and its usefulness seems to be limited to about 20 μm of a-Si deposition and often less. (b) Embrittlement of filament: this could be an issue as there is a large amount of hydrogen present in the process chamber during deposition which, via reactions, could affect its longevity. This is an obvious concern in production where regular schedule would need to be determined for filament replacement. Another possibility is that silicides can be formed thus weakening the wire. (c) Reproducibility: its lack could result from the inevitable sag of the filaments. Further, the surface conditions and the temperature of the "Hot Wire" elements can be altered during deposition which in turn will alter the temperature of substrates unless rigorous controls are implemented. These effects, amongst others compromise reproducibility.

We have developed a new HWCVD technique (2) which addresses some of the above concerns, and have so far deposited over 200 μm of silicon films without a noticeable change in the physical properties of the "Hot Wire" source. High quality a-Si films (intrinsic, n- and p- type) and preliminary intrinsic μc-Si films have been produced. Using SIMS technique, no significant contamination from the "Hot Wire" source has been found in the films. In this paper, we will describe the preparation, properties of the films and fabrication of preliminary devices.

Mat. Res. Soc. Symp. Proc. Vol. 557 © 1999 Materials Research Society

EXPERIMENTAL DETAILS

Amorphous silicon (a-Si) films were produced in a commercially available PECVD/HWCVD system specifically designed for the thin film semiconductor market and manufactured by MVSystems,Inc. This chamber is capable of depositing in either the PECVD or the HWCVD mode of operation without breaking vacuum on 10 cm X 10 cm substrates situated on either the anode side of the RF electrode assembly (for the PECVD) or above the moveable HWCVD assembly.

In the HWCVD deposition mode, a-Si films were deposited with a silane gas flow rate of 20-100 sccm, deposition pressure in the range of 5-60 mTorr, and a substrate temperature of 250C. The n-type films were deposited by mixing 0.05% of PH_3 with SiH_4 and the p^+ films were deposited by mixing 5% of B_2H_6 with SiH_4. Amorphous to micro-crystalline Si transition was achieved with the use of hydrogen dilution (H_2 to SiH_4 dilution of 10:1). The "Hot Wire" elements were arranged in a grid fashion and the total power dissipated was between 450 and 900 watts. The filament temperature was measured using an optical pyrometer.

Simple solar cell structures of Schottky barrier (SB) configuration of the type: Glass / SnO2 / n+ / i / Pd were employed. The n-type material was fabricated by the PECVD technique, using a SiH_4 and PH_3 gas mixture and exhibiting a conductivity of 2 X 10^{-3} $(Ohm-cm)^{-1}$. The solar cell devices (with i layer fabricated using either the PECVD or the HWCVD technique) were deposited onto Asahi tin oxide coated glass with a transmission >80%. For both devices, n-layer thickness was kept at 200 A, while the intrinsic layer thickness used was 3000 A. The final contacts were opaque palladium, deposited by thermal evaporation.

The opto-electronic, structure, and impurity properties were characterized by photo- (σ_{ph}) and dark conductivity (σ_d), XRD, FTIR, Raman scattering, and SIMS techniques. The average crystal size of the micro-crystalline films was determined using the X-ray diffraction technique. The full width at half maximum (FWHM) of the (111) silicon peak was used to compute the average crystal size via Scherrer's formula.

RESULTS AND DISCUSSION

By using our new HWCVD technique, we have successfully and reproducibly fabricated high quality intrinsic a-Si films, at a deposition rate of about 5 Å/sec. Figure 1 shows photo- and dark conductivity and the gamma factor as a function of deposition pressure. We note that, over the pressures studied, that there is a strong dependence of σ_d, σ_{ph} and γ. The best film was deposited at a pressure of 45 mTorr, power of 450 watts and exhibited, σ_{ph} / σ_d >10^5. It is probable that this material would exhibit a low density of states (DOS) and is supported by the γ data, where the 45 mTorr produced film exhibited a high value (~0.92). (γ is derived from, $\ln(\sigma_{ph})$ α γ $\ln(F)$, where F is the intensity of illumination. It should be noted that the recombination kinetics is dictated by the DOS and as the DOS increases, γ decreases and vice versa [4,5]). Measurement of σ_d as a function of temperature revealed an activation energy of 0.87 eV which locates the Fermi level at the middle of the band gap; infrared absorption measurement (FTIR) showed Si-H bonding and without any discernible peak corresponding to unfavorable bonding configurations, such as $Si-H_2$. The Hydrogen concentration was found to be approximately 4.5%. Doped a-Si samples have been fabricated, with σ_d ~$3x10^{-3}$ $(\Omega \cdot cm)^{-1}$ for the n+ layer with a conductivity activation energy of 0.32 eV, while for the p+ layers, the respective parameters were ~$1.0x10^{-5}$ $(\Omega \cdot cm)^{-1}$, and 0.46 eV.

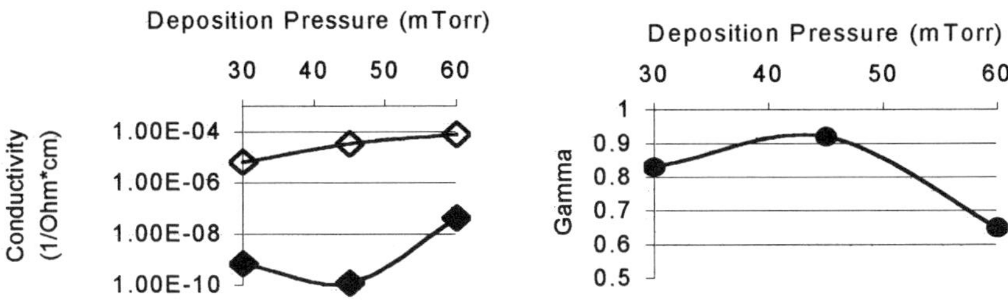

Fig. 1: Photo- and dark conductivity and the gamma factor as a function of deposition pressure for a-Si films deposited using the new HWCVD technique. (♦ = dark conductivity, ◊ = photoconductivity, • = gamma factor)

Longevity of filament

A notable feature of the present approach is that, so far, we have made deposits with a total thickness exceeding 200 μm, without any discernible change in the "Hot Wire" source. We expect that the total thickness deposited may even exceed 500μm before replacement. This is a major departure from the conventional technique, whose lifetime is short (<20μm of deposit).

Reproducibility

Another concern is the irreproducibility of the material produced, which is inherent in the conventional technique due to the use of thin wires and the inevitable sag that would occur during the deposition. In this context, it should be mentioned that the substrate/filament distance is crucial in determining the properties of the films (3). The new technique and the inherent design assembly (a grid pattern) eliminates the sag issue. Further the substrate can be oscillated above this grid, thus opening the way for large area material development. Fig. 2 shows the gamma factor of a-Si films produced, as a function of the total lifetime of the filament, and we note if anything, the material shows an improvement.

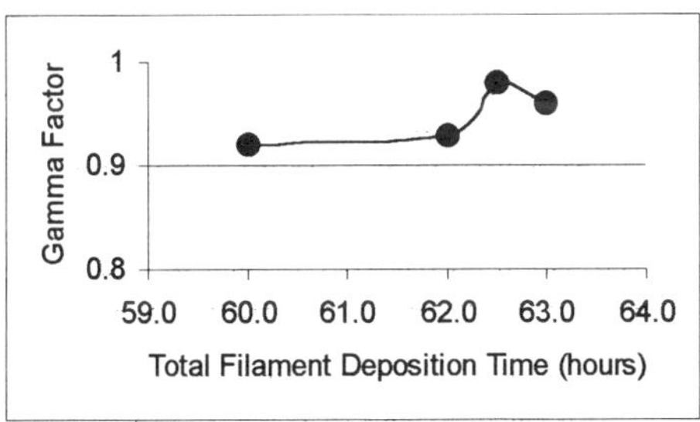

Fig. 2: Gamma factor of a-Si films as a function of total filament deposition time

<u>Preliminary Device Fabrication</u>

We now turn to the perfomance of the Schottky barrier (SB) type device with the i-layer fabricated using either the HWCVD or the PECVD deposition technique. The properties of the corresponding i-layers used are shown in Table I. Comparing the two, it seems that the HWCVD sample may very well possess a higher DOS as the photo-conductivity is lower and gamma value is reduced.

Table I: Properties of HWCVD and PECVD a-Si

<u>SN</u>	<u>Type</u>	<u>Method</u>	<u>Dark cond.</u> $(\Omega.cm)^{-1}$	<u>Photocond.</u> $(\Omega.cm)^{-1}$	<u>Gamma</u>
797	I-layer -amorphous	PECVD	$5x10^{-10}$	$8x10^{-5}$	0.95
807	I-layer-amorphous	HWCVD	$1.2x10^{-10}$	$1.3x10^{-5}$	0.91

Table II shows the device parameters while Figures 3 and 4 show the light and dark J-V curves measured for the two types of SB's. We note the following: (a) both types of SB's show a rectification ratio, $> 10^3$ for voltage of 0.5 volts; (b) the diode quality factor for the PECVD device was found to be 1.34 with a barrier height of approximately 1.00 eV. (In Table II, the diode quality factor, n, is calculated from, $J = Jo[e^{eV/nkT}-1]$. The built-in potential, ϕ_B, is estimated from $J_0 = e\mu g(E_C)F_{MAX}\exp(-e\phi_B/kT)$ where μ is the carrier mobility, $g(E_C)$ is the density of states at the conduction band, and F_{MAX} is the electric field at the semiconductor-metal interface. It should be noted, the expressions are derived using diffusion theory, which is applicable for low mobility materials such as amorphous silicon. In order to evaluate the barrier height, we have estimated the following parameters: $g(E_C)\sim10^{20}/cm^3$, $\mu \sim 5$ $cm^2s^{-1}V^{-1}$, and $F_{MAX} = 4*10^4$ V/cm.). Under white light illumination, the current density is $\sim13mA/cm^2$, The light enters through the glass/SnO2 and hence its intensity, over the visible part of the spectra, reaching the junction is reduced by ~15-20%. Therefore, the current density for this type of device may very well be on the order of 15 mAcm^-2. These results are not too dissimilar from the results, which can be found in the literature (1).

The diode quality factor for the HWCVD device is 1.70 and is larger than found in the corresponding PECVD device discussed above; the device exhibits a reduced barrier height of approximately 0.94 eV. The light characteristics indicate that the current density to be ~11mA cm^-2. (The current density for this type of device may very well be on the order of 13 mAcm^-2, in view of the comments made above). These results coupled with lower FF's suggest that the DOS may be higher in the HWCVD material and supported by the material results discussed above. In addition to this, it is entirely conceivable that the interface DOS may be larger resulting in losses at the n/i and the i/Pd junction.

<u>Microcrystalline Silicon</u>

We have made a brief attempt to fabricate intrinsic micro-crystalline silicon films by using a hydrogen dilution ratio of about 10. The resulting deposition rate is 2.3 Å/sec. Fig. 5 shows the X-Ray Diffraction (XRD) of a µc-Si material deposited on Corning glass and on crystalline silicon. As shown, the material exhibits a (220) orientation, especially on the glass substrate. From this peak, the average grain size of the material was calculated to be 390 Angstroms. Fig. 6 shows the Raman spectrum, with a large sharp peak at 520 cm^{-1} indicative of crystalline structure ; a detailed. quantitative analysis is in progress. The films exhibited, photo- and dark conductivity of $3.2x10^{-6}$ $(\Omega\cdot cm)^{-1}$ and $2.7x10^{-6}$ $(\Omega\cdot cm)^{-1}$, respectively.

Table II: Comparison of a Schottky Barrier Device with i-layer fabricated using the PECVD and HWCVD technique.

Composition	SnO2/n/i/Pd (all PECVD)	SnO2/n/i/Pd (HWCVD i-layer)
White Response		
Voc (mV)	415	375
Jsc (mA/cm^2)	10.3	8.8
FF	0.66	0.53
Red Response		
Voc (mV)	370	315
Jsc (uA/cm^2)	1250	1190
FF(Red)	0.67	0.64
Blue Response		
Voc (mV)	335	270
Jsc (uA/cm^2)	313	256
FF (Blue)	0.66	0.61
Diode Quality Factor	1.34	1.7
Barrier Height (eV)	1.00	0.94

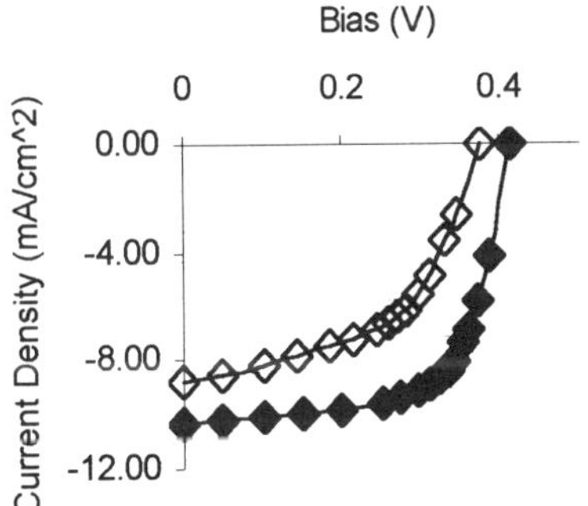

Fig. 3: White J/V curves for HWCVD and PECVD SB devices. (♦ = PECVD i-layer, ◊ = HWCVD i-layer)

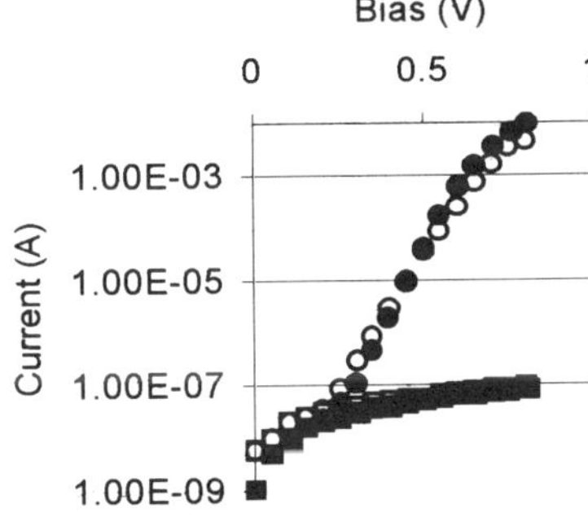

Fig. 4: Dark I/V curves for HWCVD and PECVD SB devices. (Solid symbols are for PECVD i-layer, open symbols are for HWCVD i-layer)

CONCLUSIONS

We have shown that, using our new HWCVD technique, intrinsic, and doped amorphous silicon can be produced and their use in simple SB structures is not too dissimilar from the corresponding PECVD device. The major result here is that the films are reproducible and that in excess of 200um of deposition has been achieved without any discernible change in the "Hot Wire" source. This has a potential for large-scale development. Preliminary results also indicate that micro-crystalline silicon can also be produced.

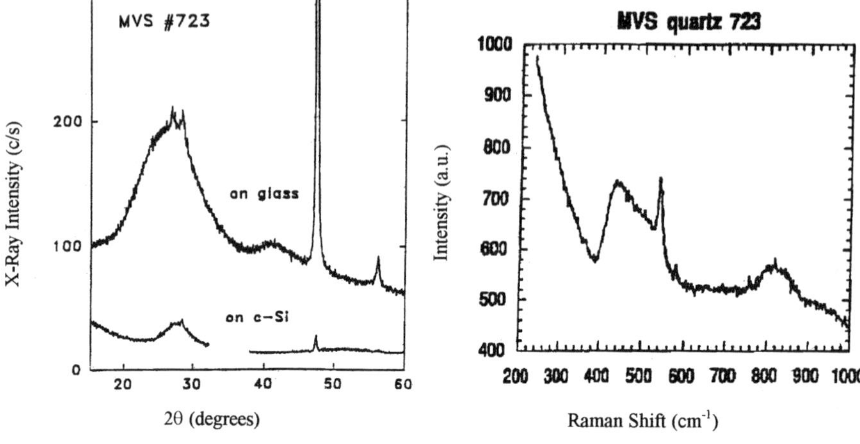

Fig. 5: X-Ray Diffraction Spectrum
of HWCVD μc-Si material

Fig. 6: Raman Spectrum of HWCVD
μc-Si material.

ACKNOWLEDGEMENTS

This work was financially supported, in part by National Renewable Energy Laboratory, Golden, Colorado, under contract #ZAK-7-17619-1. We also thank Don Williamson at the Colorado School of Mines for the XRD data and Craig Taylor at the University of Utah for the Raman data and Ed Valentich for assisting in the preparation of the films.

REFERENCES

1. Matsumura, H., Japn. J. Appl. Phys., **8B**, L1522, 1991.
2. A.Madan, J.Xi and S.Morriosn, patent applied for.
3. J. Puigdollers, J. Bertomeu, J. Cifre, J. Andreu and J. C. Delgado, MRS Proc. **377**, 63, (1995)
4. T. J. McMahon and A. Madan, Appl. Phys. Lett. **57**, 5302, (1985).
5. A. Madan, and M.Shaw, The Physics and Applications of Amorphous Semiconductors (Academic Press, New York, 1988): Plasma Deposition of Amorphous Silicon Based Materials, G. Bruno, P. Capezutto, and A. Madan, eds. Academic Press, 1995.

Structural and optoelectronic properties of amorphous and microcrystalline silicon deposited at low substrate temperatures by RF and HW CVD

P. ALPUIM*, V. CHU*, J. P. CONDE**
*Instituto de Engenharia de Sistemas e Computadores (INESC), Rua Alves Redol, 9, 1000 Lisboa, Portugal, palpuim@eniac.inesc.pt
**Department of Materials Engineering, Instituto Superior Técnico, 1049-001 Lisboa, Portugal

ABSTRACT

The structural and optoelectronic properties of silicon thin films prepared by hot wire chemical vapor deposition and radio frequency plasma enhanced chemical vapor deposition are studied in the range of substrate temperatures (T_{sub}) from 100 °C to 25 °C. The defect density, structure factor and bond angle disorder of amorphous silicon films (a-Si:H) deposited by both techniques are strongly improved by the use of hydrogen dilution. Correlation of these structural properties with important optoelectronic properties, such as photo-to-dark conductivity ratio, is made. Microcrystalline silicon (μc-Si:H) is obtained using HW with a large crystalline fraction for hydrogen dilutions above 85% independently of T_{sub}. The deposition of μc-Si:H by RF requires increasing the hydrogen dilution and shows decreasing crystalline fraction as T_{sub} is decreased. The properties of the low T_{sub} films are compared to those of samples produced at 175 °C and 250 °C in the same reactors.

INTRODUCTION

Hydrogenated amorphous silicon (a-Si:H) is a technologically important material which has been widely applied to devices as diverse as solar cells, flat panel displays and image scanners[1]. Although the specific requirements of the various applications are different, their common need of a semiconductor which is cheap and can be made in large areas has driven the development of the unique properties of a-Si:H. Microcrystalline silicon has been gaining attention in recent years as a possible replacement for a-Si:H in some of these applications. Unlike a-Si:H, it does not suffer light induced degradation (Staebler-Wronski effect) making it a more attractive material for solar cell applications [2]. Also its higher carrier mobility and higher doping efficiency could potentially allow the fabrication of an integrated circuit on glass using thin film transistor technology, making possible integrated driver circuits for flat panel displays.

New applications which require devices to be made on cheap, flexible and unbreakable substrates such as plastic has brought about the need to develop low temperature processing of amorphous and microcrystalline silicon-based devices. Standard deposition temperatures of a-Si:H and μc-Si:H are between 250 -350 °C. In order to expand the range of possible substrates and also to allow for integration of a-Si:H and μc-Si:H with, for example, organic or biological materials, it would be desirable to decrease this deposition temperature to under 150 °C without sacrificing their electronic properties.

In this paper, the structural and optoelectronic properties of a-Si:H and μc-Si:H deposited at 100 °C and 25 °C using two techniques are studied: RF-PECVD and hot-wire CVD. The hot wire technique at standard deposition temperatures (250-350 °C) has been shown to produce high quality a-Si:H and μc-Si:H at deposition rates 10-40 times that of RF-PECVD [3-7]. However there has been no study of this technique applied to low substrate temperatures. At each temperature, the hydrogen dilution is varied and the properties of the material are studied and compared to device quality material deposited at standard deposition temperatures.

Mat. Res. Soc. Symp. Proc. Vol. 557 © 1999 Materials Research Society

EXPERIMENT

A. Film Preparation

The films were deposited by HW-CVD and RF-PECVD in a UHV-quality system with a base pressure $\leq 5 \times 10^{-8}$ Torr. The sample was clamped to the grounded upper electrode which was, except for the room temperature depositions, heated to the temperature of deposition, T_{sub}.

For the HW deposition, a single tungsten filament of 0.5 mm diameter and approximately 7 cm length was placed 5 cm from the substrate and was resistively heated with a DC power supply. The filament temperature was measured with an optical pyrometer ($T_{fil} \sim 2500$ °C) and the pressure was kept constant at 20 mTorr. The details of the HW deposition system are described elsewhere [3]. For the RF deposition, the inter-electrode distance was 3 cm, the density of RF power used was 50 mW/cm^2 (in all but the room temperature depositions where it was 100 mW/cm^2), and the pressure was 100 mTorr. Both in the HW and RF depositions the total gas flow was kept at around 10 sccm, except for the higher dilutions where it was necessary to increase the fluxes so that the SiH$_4$ flux was not less than 0.5 sccm, which was the lower limit for the silane mass flow controller.

B. Film Characterization.

The dark conductivity σ_d was measured between 110 °C and room temperature on coplanar Cr contacts. The activation energy E_a was calculated from $\sigma_d = \sigma_0 \exp\left[-E_a /(k_B T)\right]$. The steady-state photoconductivity σ_{ph} was measured as a function of generation rate using bandpass-filtered illumination to give an approximately uniform carrier generation throughout the thickness of the film. In this work, σ_{ph} refers to the photoconductivity at a carrier generation rate of 10^{21} cm^{-3} s^{-1}.

The constant photocurrent method, CPM, was used to measure the subgap absorption [8]. The deep defect density N_s of the amorphous films was calculated from $N_s=C_{CPM} \times \alpha$ (1.2 eV), with $C_{CPM} =10^{16}$ cm^{-2} [9].

The hydrogen content C_H and the microstructure factor R were determined using infrared spectroscopy. C_H was calculated from the density of silicon atoms, $N_{Si}= 5 \times 10^{22}$ cm^{-3}, and the density of bonded hydrogen atoms, N_H ($C_H = N_H/ N_H+N_{Si}$). N_H was calculated from the integrated absorption coefficient of the Si-H wagging modes located around 630 cm^{-1} [10]. R was calculated from the deconvolution of the stretching band into two peaks, one centered at approximately 2000 cm^{-1} (I_{2000}) and the other centered around 2100 cm^{-1} (I_{2100}), $R = I_{2100}/(I_{2000}+ I_{2100})$ [10].

Raman spectra were measured in the backscattering geometry using a Raman microprobe. The power of the incident beam was set below 50 mW to avoid thermally induced crystallization. For microcrystalline films, the Raman spectrum around the crystalline silicon transverse optical (TO) peak was deconvoluted into their integrated crystalline Gaussian peak, I_c (~ 520 cm^{-1}), amorphous Gaussian peak, I_a (~ 480 cm^{-1}), and intermediate Gaussian peak, I_m (~ 510 cm^{-1}) [11,12]. The crystalline fraction, X_c, was calculated from $X_c = (I_c + I_m)/(I_c + I_m + I_a)$[13]. The crystalline size, d_{Raman}, was calculated from $d_{Raman} = 2\pi\sqrt{(B/\Delta\omega)}$, where $\Delta\omega$ is the shift of the peak for the μc-Si:H sample compared to that of c-Si, and $B=2.0$ cm^{-1} nm^2 [13]. For amorphous films, the bond angle deviation $\Delta\theta$ was related to the full width at half maximum (FWHM) Γ of the TO peak centered at 480 cm^{-1} using the expression: $\Delta\theta= \Gamma/6-2.5$

X-ray diffraction peaks were measured using the Cu $K_{\alpha 1}$ line (λ=1.54056 Å). The samples were measured at grazing incidence (0.5 ° and 1°) using substrate holder rotation (15 rpm). The crystallite size $d_{X\text{-}ray}$ was calculated from the Scherrer formula $d_{X\text{-}ray} = k\lambda /(B\cos\theta_B)$, where

$k\sim0.9$, λ is the wavelength of the X-ray radiation, B is the FWHM of the peaks (in units of 2θ) and θ_B is the angular position of the peak.

RESULTS AND DISCUSSION

Figure 1 shows the effect of hydrogen dilution on the dark conductivity of films deposited at different T_{sub} by RF (a) and HW (b) CVD. One can see a clear difference in behavior: while for RF films, the transition from amorphous-type conductivities ($\sim10^{-11}$-10^{-10} Ω^{-1}cm^{-1}) to microcrystalline-type conductivities (10^{-6}-10^{-4} Ω^{-1}cm^{-1}) occurs abruptly and is substrate temperature dependent, for HW, this transition is gradual and independent of substrate temperature. The critical hydrogen dilution at which the amorphous to microcrystalline transition takes place in RF deposited films increases with decreasing substrate temperature: at 97% for 175 °C, 98% for 100 °C and 99% for 25 °C. This result indicates that the parameter window for depositing μc-Si:H by RF is severely reduced with the reduction of T_{sub}. This is not the case with HW deposited films where all the films, regardless of T_{sub}, have amorphous σ_d below 80% hydrogen dilution, microcrystalline σ_d above 90% hydrogen dilution and intermediate values of σ_d in the continuous transition region in between.

The structure of the films which was inferred from the conductivity results was confirmed by Raman measurements. Figure 2 shows a plot of the dark conductivity as a function of the crystalline fraction, X_c, calculated from Raman spectra. This plot shows that the crystalline fraction scales roughly with the dark conductivity values for both RF and HW films. Amorphous films (i.e. X_C=0) have typical $\sigma_d \sim10^{-12}$- 10^{-10} Ω^{-1}cm^{-1}.

In figure 3 the structure factor R is plotted as a function of hydrogen dilution. A high content of monohydride bonded hydrogen, characterized by a low R, is associated with a low density of electronic traps in a-Si:H. μc-Si:H, although having a much lower hydrogen content than a-Si:H, typically have higher R-values due to the existence of many terminating Si-H$_2$ units in its grain boundaries, which behave like internal

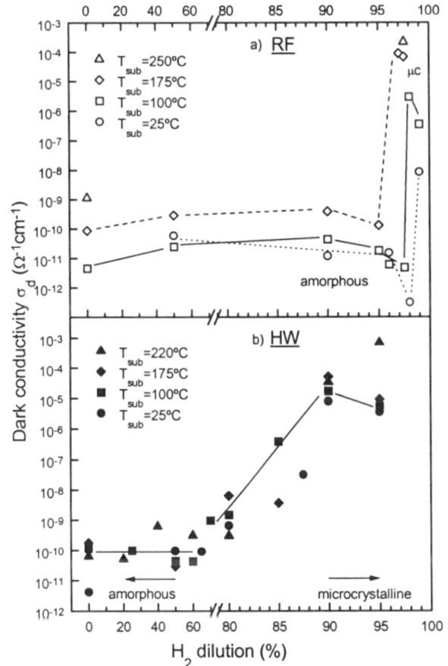

Figure 1. The dark conductivity is plotted as a function of hydrogen dilution for (a) RF and (b) HW deposited films for varying substrate temperatures. The lines are guides to the eye.

surfaces or voids. At 0% H$_2$ dilution, the HW a-Si:H films (fig. 3a) show R values between ~0.25 (T_{sub}= 220 °C) and 0.75 (T_{sub}=25 °C). The R factor reaches a minimum at a H$_2$ dilution of 60-80% where R≤0.1 for T_{sub}= 175 °C and 100 °C and R= 0.27 for T_{sub}= 25 °C. An abrupt increase of R occurs for hydrogen dilution above 80% coinciding with the onset of microcrystallization. In RF films (fig. 3b), with the exception of the films deposited at 25 °C, the R factor has a much smaller dependence on H$_2$ dilution until the onset of microcrystallization, where an abrupt increase is observed. At 25 °C, however, H$_2$ dilution results in a drop of R from ~0.36 at 50% H$_2$ to ~0.06 at 90% H$_2$. The lowest R factors fo RF a-Si:H films deposited at 100 °C and 25 °C are significantly lower than those for HW a-Si:H deposited at the same T_{sub}.

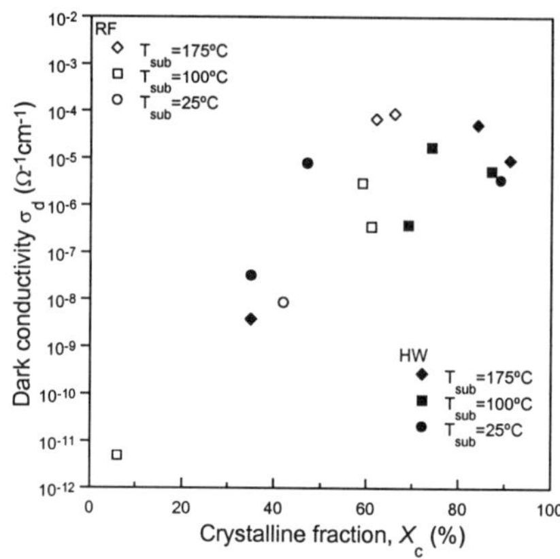

Figure 2. The dark conductivity is plotted as a function of the crystalline fraction calculated from Raman spectra.

Figure 4 shows the absorption coefficient, α, of amorphous films measured by CPM at a photon energy of 1.2 eV. In the amorphous films deposited by RF, the values of $\alpha_{1.2}$ range between 1 cm^{-1} and 10 cm^{-1} ($N_s \sim 10^{16}$-10^{17}cm^{-3}), while in films deposited by HW, $\alpha_{1.2}$ values are mostly between 10 cm^{-1} and 100 cm^{-1} ($N_s \sim 10^{17}$-10^{18}cm^{-3}) for all T_{sub}. For RF films, N_s is independent of H_2 dilution until just before the microcrystalline transition where N_s decreases. For HW, H_2 dilution improves the defect density reaching a minimum around 60% for films deposited at $T_{sub} \geq$ 100 °C. For 25 °C, however, H_2 dilution has no apparent effect on N_s, which remains high.

Figure 5 shows the photosensitivity, σ_{ph}/σ_d, plotted as a function of $\alpha_{1.2}$ (bottom axis) or, equivalently, the defect density in the gap, N_s (top axis). The values of σ_{ph}/σ_d

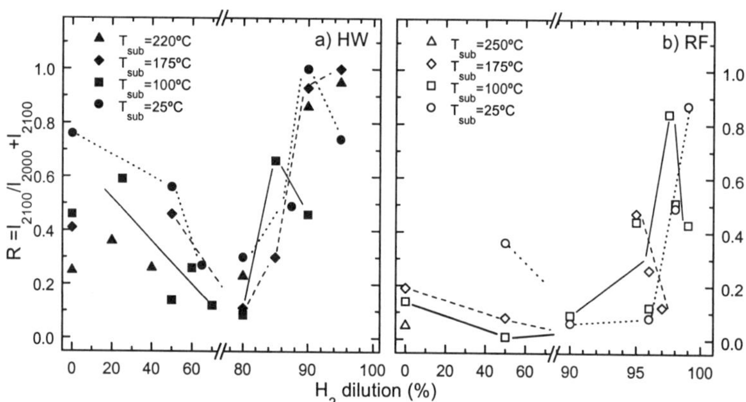

Figure 3. The structure factor, R, from IR spectra is plotted as a function of hydrogen dilution for (a) HW and (b) RF deposited films at different substrate temperatures.

of RF a-Si:H films are, in general, higher than those of films deposited by HW: for all T_{sub} it was possible to obtain films by RF with $\sigma_{ph}/\sigma_d>10^5$ while in HW only the films deposited at $T_{sub}= 220$ °C have $\sigma_{ph}/\sigma_d>10^5$ ($\sim10^6$). At the lower T_{sub}, films by HW have $\sigma_{ph}/\sigma_d<10^5$ which decrease with decreasing T_{sub} ($\sim5\times10^4$ for 175 °C and 100 °C, and $\sim10^3$ for 25 °C).

Figure 6 shows the bond angle disorder ($\Delta\theta$) of amorphous films as a function of hydrogen dilution. At lower hydrogen dilutions, RF films have generally lower $\Delta\theta$ than HW films. At higher hydrogen dilutions, HW films have already become microcrystalline, while RF films are either still amorphous or show very small crystalline fractions according to Raman spectra. The

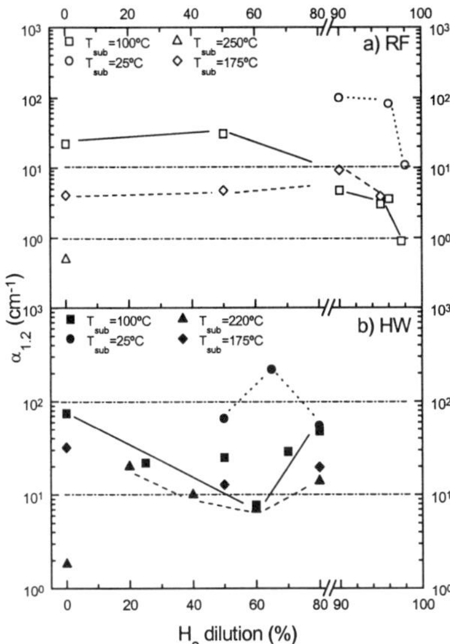

Figure 4. The absorption coefficient of amorphous films measured by CPM at a photon energy of 1.2 eV, $\alpha_{1.2}$, is plotted versus hydrogen dilution for (a) RF and (b) HW deposited films. The lines are guides to the eye.

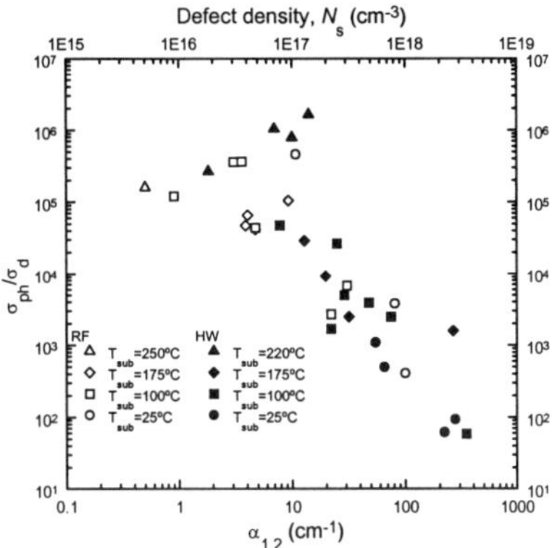

Figure 5. The photosensitivity, σ_{ph}/σ_d, versus the absorption coefficient at 1.2 eV of amorphous films by both RF and HW CVD.

RF films in this region with the lowest $\Delta\theta$ are those which are just before the microcrystalline transition and also show the lowest N_s (fig. 4), highest R (fig. 3) (as well as an increase in hydrogen content) and lowest σ_d (fig. 1) (as well as the highest photosensitivity). It is possible that the nucleation of crystallites in RF is preceded by the formation of a highly porous film which includes small crystallites and a continuous network of high quality amorphous silicon.

CONCLUSIONS

- Amorphous silicon can be prepared by RF-PECVD at low T_{sub} with $\sigma_{ph}/\sigma_d > 10^5$ and $\alpha_{1.2} < 10$ cm^{-2} by using the appropriate H_2 dilution:
  $\Rightarrow$ at 100 ºC: 95-98%
  $\Rightarrow$ at 25 ºC: ~98%
  The electronic properties of a-Si:H by HW

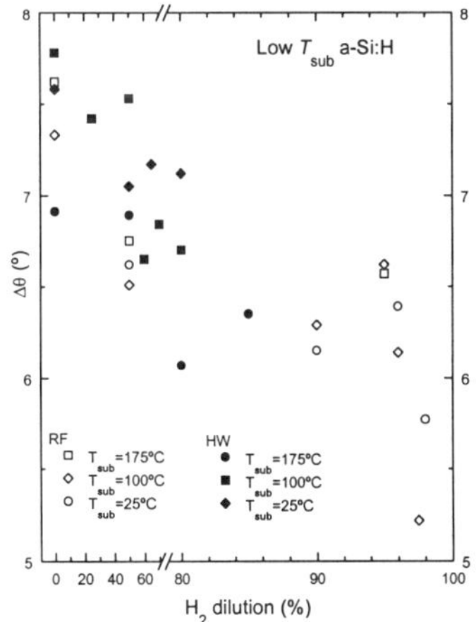

Figure 6. The bond angle disorder, $\Delta\theta$, calculated from FWHM of Raman spectra of amorphous films plotted as a function of hydrogen dilution for different substrate temperatures.

degrade severely with decreasing T_{sub} and is always worse than the best RF films.

- <u>Microcrystalline silicon can be prepared by HW-CVD</u> with crystalline fraction > 80% at all deposition temperatures studied (25 °C $\leq T_{sub} \leq$ 175 °C). For RF, the range of H_2 dilution which can grow μc-Si:H grows smaller with decreasing T_{sub}. At 25 °C, 99% H_2 dilution results in a film with low crystalline fraction and vanishingly small growth rate.
- It is possible that the nucleation of crystallites in RF is preceded by the formation of a highly porous film which includes small crystallites and a continuous network of high quality amorphous silicon.

ACKNOWLEDGMENTS

The authors wish to thank Rui Almeida of INESC/IST for the use of the Raman and FTIR equipment, Olinda Conde of the Faculty of Sciences of the Univ. of Lisbon for help with the XRD. This work was partially supported by FCT through Pluriannual Contracts with UCES/ICEMS (IST) and INESC and by PRAXIS/3/3.1/MMA/1792/95. P. Alpuim thanks the Dept. of Physics at the Univ. of Minho for a leave of absence.

REFERENCES

1. See, for example, R. A. Street, *Hydrogenated Amorphous Silicon* (Cambridge University Press, 1991).
2. J. Meier, P. Torres, R. Platz, S. Dubail, U. Kroll, J. A. Anna Selvas, N. Pellaton Vaucher, Ch. Hof, D. Fischer, H. Keppner, A. Shah, K.-D. Ufert, P. Giannoulès and J. Koehler, *Mat. Res. Soc. Symp. Proc.* **420**, 3 (1996).
3. J. P. Conde, P. Brogueira, V. Chu, *Philosophical Magazine* B **76**, No. 3, 299 (1997).
4. A. H. Mahan, J. Carapella, B.P. Nelson, R.S. Crandall and I. Balberg, *J. Appl. Phys.* **69**, 6728 (1991)
5. R. Zedlitz, F. Kessler and M. Heintze, *J. Non.-Cryst. Solids,* **164-166**, 83 (1993).
6. J. Cifre, J. Bertomeu, J. Puigdollers, M.C. Polo, J. Andreu and A. Lloret, *Appl. Phys.* A **59**, 645 (1994).
7. A.R. Middya, J. Perrin, J. Huc, J.L. Moncel, J.Y. Parey and G. Rose, *Mat. Res. Soc. Symp. Proc.* **377**, 119 (1995).
8. M. Vanacek, J. Kocka, J. Strichlik, Z. Kosicek, O. Stika, A. Triska, *Sol. Energy Mater.* **8**, 411 (1983).
9. N.Wyrsch, F.Finger,T.J.McMahon, M.Vanacek, *J.Non-Cryst. Solids* **137-138**, 347 (1991
10. C.J.Fang, K.J.Gruntz, L.Ley, M.Cardona, F.J.Demond, G. Mueller, S.Kalbitzer, *J.Non-Cryst. Solids* **35-36**, 255 (1980).
11. T. Kaneko, M. Wakagi, K. Onisawa, T. Minemura, *Appl. Phys. Lett* **64**, 1865 (1994)
12. Veprek, F. A. Sarott, Z. Iqbal, *Phys. Rev.B* **36**, 3344 (1987)
13. Y. He, C. Ying, G. Cheng, L. Wang, X. Liu, G. Y. Hu, *J. Appl. Phys.* **75**, 797 (1994).
14. Y. Hiroyama, R. Suzuki, Y. Hirano, F. Sato, T. Motooka, *Jpn. J. Appl. Phys.*, Part 1 **34**, 5515 (1995).

LOW HYDROGEN CONTENT, HIGH QUALITY HYDROGENATED AMORPHOUS SILICON GROWN BY HOT-WIRE CVD

BRENT P. NELSON, RICHARD S. CRANDALL, EUGENE IWANICZKO, A. H. MAHAN, QI WANG, YUEQIN XU, WEI GAO
National Renewable Energy Laboratory, Golden, CO

ABSTRACT

We grow hydrogenated amorphous silicon (a-Si:H) by Hot-Wire Chemical Vapor Deposition (HWCVD). Our early work with this technique has shown that we can grow a-Si:H that is different from typical a-Si:H materials. Specifically, we demonstrated the ability to grow a-Si:H of exceptional quality with very low hydrogen (H) contents (0.01 to 4 at. %). The deposition chambers in which this early work was done have two limitations; they hold only small-area substrates and they are incompatible with a load-lock. In our efforts to scale up to larger area chambers—that have load-lock compatibility—we encountered difficulty in growing high-quality films that also have a low H content. Substrate temperature has a direct effect on the H content of HWCVD grown a-Si:H. We found that making dramatic changes to the other deposition process parameters—at fixed substrate temperature and filament-to-substrate spacing—did not have much effect on the H content of the resulting films in our new chambers. However, these changes did have profound effects on film quality. We can grow high-quality a-Si:H in the new larger area chambers at 4 at. % H. For example, the lowest known stabilized defect density of a-Si:H is approximately 2×10^{16} cm^{-3}, which we have grown in our new chamber at 18 Å/s. Making changes to our original chamber—making it more like our new reactor—did not increase the hydrogen content at a fixed substrate temperature and filament-to-substrate spacing. We continued to grow high quality films with low H content in spite of these changes. An interesting, and very useful, result of these experiments is that the orientation of the filament with respect to silane flow direction had no influence on film quality or the H content of the films. The condition of the filament is much more important to growing quality films than the geometry of the chamber due to tungsten-silicide formation on the filament.

INTRODUCTION

To make a-Si:H with material properties that remain more stable under illumination we lower the H content of the material. We find that films with less than 4 at. % H become significantly more stable than films with higher H content. That is, films with low H content have less change from the initial to the final state, than films with higher H content. Films with a H content less than 2 at. % are the most stable [1]. Typical PECVD-grown a-Si:H films have H on the order of 10 at. %. Most of our work depositing these more stable materials is done in our original HWCVD chamber (Figure 1). We heat the substrates from the back side with a resistive heater (in air) inside the bottom of the heater can. We clamp the substrates to the bottom of the heater can that extends into the reaction chamber. Because of this design, we must take off the heater can to load and unload substrates for each deposition, thus exposing the chamber, heater can, and substrates to air each time.

Because this chamber design is incompatible with a load-lock—and we had initial concerns about nonuniformly heating the substrates—we designed and use a new reaction chamber (Figure 2). This new reactor provides load-lock compatibility. The load-lock is to the left in the side view and the process pumps are to the right. The spacing of the heating elements provide isothermal heating in the growth region of the reactor. We change the filament-to-substrate spacing from 1 to 6 cm by inserting or removing spacers on the filament holder (5 cm shown). Additional information about these reactors is in previous publications [2, 3]. Having a load-lock compatibility allows us to incorporate these materials into solar-cell devices, in which we deposit the doped (p- and n-layers) in a separate chamber, avoiding contamination to the i-layer. This makes it very important for us to grow the high-quality, improved-stability a-Si:H materials (i.e., a-Si:H with < 4 at. % H) in the tube reactor.

Mat. Res. Soc. Symp. Proc. Vol. 557 © 1999 Materials Research Society

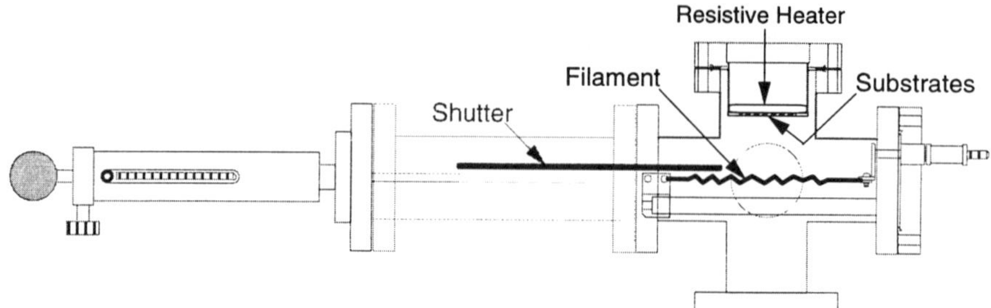

Figure 1: The original HWCVD reaction chamber at NREL consisting of a standard 6-way, vacuum cross with 11.4 cm diameter flanges. In this figure the silane flow is into the page, perpendicular to the filament.

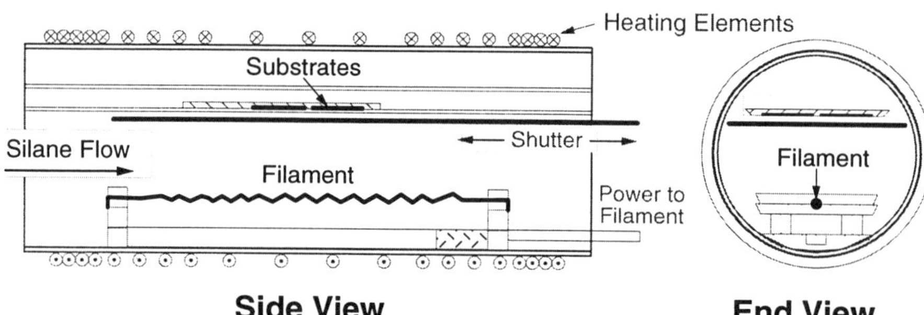

Figure 2: The isothermal HWCVD reaction chamber at NREL consisting of a 10 cm diameter stainless steel tube, 28 cm long. The silane flow is left to right in side view and into the page—parallel to the filament—in the end view.

In both reactor designs we power 0.5 mm diameter, tungsten filaments with currents ranging from 13.5 to 16.5 amps AC. This provides filament temperatures from 1950° to 2250°C, respectively. The powered ends of the filaments receive the voltage necessary to obtain the desired operating current, and the other ends of the filaments are grounded.

The HWCVD a-Si:H materials containing $\geq$ 4 at. % H grown in both chambers are quite similar. The initial material properties are of essentially the same quality and degrade with similar kinetics to comparable stabilized states. However, whenever we grow materials in the tube with H below 4 at. %, we find that the material properties become worse with decreasing H content. This is not true for our a-Si:H grown in the cross reactor, in which the material properties remain exceedingly good down to the detection limit of H by infrared techniques; as well as becoming more stable with decreasing H [1]. The primary way we control the H content of our films is by controlling the substrate temperature. We find that the H content in the a-Si:H we make in our cross reactor decreases monotonically from 20 at. % at substrate temperatures of 180°C, down to $\leq$ 1 at. % at substrate temperatures $\geq$ 400°C [4]. The H content of the a-Si:H we make in our tube reactor also decreases monotonically with substrate temperature, but with a significantly different shape, having about 10 at. % H at substrate temperatures of 200°C, down to $\leq$ 1 at. % H at substrate temperatures $\geq$ 450°C [2]. The influence of substrate temperature on H incorporation of a-Si:H grown in the tube reactor is more similar to our PECVD processes than the HWCVD a-Si:H grown in the cross reactor. The dark conductivity of the films we grow in the tube reactor are very good ($< 10^{-10}$ S/cm) up to substrate temperatures of 400°C. Above 400°C, the dark conductivity of the films rapidly increases with increasing substrate temperature [2] and even begin to develop microcrystalline silicon (μc-Si) inclusions above 550°C. Films made in the cross reactor maintain good (low) dark conductivity with substrate temperatures above 400°C, and do not show similar trends of growing inferior material with increasing substrate temperature.

In this paper, we report on what we have learned about the difficulty of growing low-H ($\leq$ 4 at. %) a-Si:H in our tube reactor that is of the same electronic quality as the a-Si:H we grow in our cross reactor. We have taken two different approaches in this investigation. The first is to vary the deposition parameter space in the tube reactor to see if such changes can lower the H in the resulting a-Si:H to below 4 at. % while maintaining good electronic and stability properties. This second approach is to alter the physical design and geometry of our cross reactor to make it more "tubelike," and see which of these factors influence the deposition of quality a-Si:H below 4 at. % H.

EXPERIMENT

To investigate which deposition parameters influence the H of the a-Si:H we grow in our tube reactor, we performed a screening experiment; that is, a three-factor, two-level design of experiment with a midpoint to check for curvature. We have four deposition parameters that we readily vary in our tube reactor depositions: chamber pressure, silane flow rate, filament current (and thus filament temperature), and substrate temperature. Because we already know that substrate temperature strongly influences the H content of the a-Si:H, we eliminate it from our screening experiment and vary the other three parameters set forth in Table 1. Therefore, we fix our substrate temperatures at 380°C for this experiment. In other words, we are looking for other deposition parameters that have a significant influence on the H content at a fixed substrate temperature in the tube reactor. We also fix our filament-to-substrate spacing at 5 cm. In this experiment, we made films under all eight possible high-low combinations (a three-factor, two-level experiment has eight possible combinations, $2^3 = 8$), repeating the midpoint four times.

Table 1: Factor levels for a two-level, three-factor design of experiments to screen the important deposition parameters on H content in a-Si:H grown in our tube reactor.

Deposition Parameter Level	Pressure (mTorr)	Filament Current (amps)	Silane Flow (sccm)
Low Value	11	13.5	3
Midpoint	40	15.0	50
High Value	69	16.5	97

To investigate if differences in the geometric design of the two different types of deposition chambers influence H incorporation in the growing a-Si:H, we did a second screening experiment. This experiment was also a three-factor, two-level design of experiment, except without the midpoint check for curvature (because we cannot achieve a midpoint between having the filament perpendicular or parallel to the direction of gas flow). The three most obvious differences between the cross reactor and the tube reactor are the filament orientation with respect to the gas flow direction, the line-of-sight exposure of the substrate to the filament ends, and the temperature of the chamber walls. The first design of experiments focused on how to lower the H of a-Si:H grown in the tube reactor at a fixed temperature, while the second design of experiments focused on trying to make the cross reactor more tubelike in order to see if geometric factors influence hydrogen incorporation. In other words, will changes to a chamber that deposits "good" material, with low H, cause the materials to either get "worse" or to have higher H? We altered the geometry of the cross reactor between standard and more tubelike configurations as shown in Table 2 below. Having the standard cross-reactor configurations in this series helps ensure that we compare the results of our changes to results from the same set of depositions under our known conditions that grow "good", low-H content materials. We fix the

Table 2: Factor levels for a two-level, three-factor design of experiments to screen chamber-design considerations on H content in a-Si:H grown in our cross reactor.

Chamber Configuration of Cross Reactor	Filament Orientation (wrt: gas flow)	Filament Length (cm)	Chamber Wall Temperature (°C)
standard	perpendicular	16.5	30
changes to tubelike	parallel	6.4	270

silane flow to 20 sccm, the pressure at 10 mTorr, the substrate temperature at 390°C, and the filament-to-substrate spacing at 5 cm. We replicate each of the eight possible high-low combinations to increase accuracy.

The standard chamber configuration for the cross reactor is a 16.5-cm-long filament, oriented perpendicularly to the direction of silane flow, and with unheated chamber walls. The ends of this filament are back in the ports of the six-way cross, such that there is no line-of-sight exposure of the substrates to the ends of the filament. This is important as the physical clamping of the current leads to the filament create an effective heat sink that lowers the temperature of the filament at the connections. We observe that the filament alloys (creating a tungsten silicide) at a much higher rate at the cool ends of the filament than in the middle—hotter—region.

The standard chamber configuration for the tube reactor is also a 16.5-cm-long filament, except oriented in parallel to the direction of silane flow, and the chamber walls are heated to just above the substrate temperature. The only direct change we can make to the cross reactor to make it more tubelike is the filament orientation. The other changes we implement are in the direction of making the cross reactor more tubelike, but not exact replication. We cannot change the cross reactor to expose the substrates to the ends of the filament using a 16.5-cm-long filament. However, we simulate the tube reactor situation by shortening the filament length in the cross reactor. This brings the filament ends closer to the substrate. We wrap heat tape around the outside of the chamber to the cross reactor to emulate the heated walls of the tube reactor. While there is a significant difference between unheated and heated walls of the cross, the chamber wall temperature in the cross reactor is still below that of the tube reactor.

RESULTS AND DISCUSSION

In the screening experiment to find the important deposition parameters on the H content of a-Si:H grown in our tube reactor, we find that even extensive changes in pressure, filament current, and silane flow—at a fixed substrate temperature of 380°C and filament-to-substrate distance of 5 cm—do not produce films with less than 4 at. % H. We illustrate the change in H content with silane flow rate—distinguishing each combination of pressure and filament current—in Figure 3. The films grown under midpoint conditions exhibit large scatter in H content, which only occurs in H content measurements. All other measured film properties have values much closer to each other (e.g., the conductivity ratio values in Figure 4).

In day-to-day operations, we obtain ~4 at. % H content material in the tube reactor (380°C, 50-70 sccm silane, 15 mTorr, and 14-16 amps). In contrast, films we grow at 380°C in the cross reactor have H contents ~2 at. %. Changing the deposition parameters from our standard conditions in the tube reactor either left the resulting H content unchanged or increased it. The scatter in the H content of the samples grown under midpoint conditions is confusing. Because the tube produces an isothermal heating zone, we are not concerned about thermal contact of the substrates to the substrate carrier. The substrates were preheated for the same time, to the same deposition temperature, therefore the actual substrate temperature should have been very similar for all for of these samples. We have yet to determine a plausible explanation for this scatter.

In this experiment, all the deposition parameters have an influence on the deposition rate. The 11 mTorr, 13.5 amps, and 3 sccm combination has the lowest deposition rate of 2.6 Å/s. The 69 mTorr, 16.5 amps, and 97 sccm combination has the highest deposition rate of 56 Å/s. There are interactions between these parameters such that a change in any two factors has a much

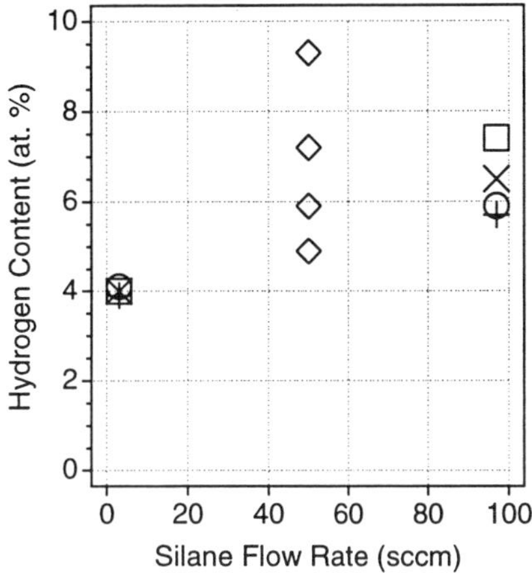

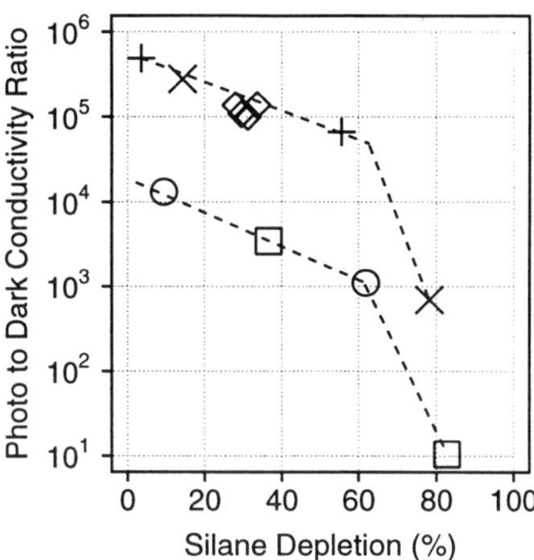

Figure 3: The hydrogen content of a-Si:H films all grown in the tube at 380°C, varying the pressure, filament current, and silane flow rate as set forth in Table 1. The following key is for both Figures 3 and 4.

D = 11 mTorr, 13.5 amps
I = 11 mTorr, 16.5 amps
A = 40 mTorr, 15.0 amps (midpoint)
E = 69 mTorr, 13.5 amps
G = 69 mTorr, 16.5 amps

Figure 4: The ratio of AM1.5 photoconductivity to the dark conductivity as a function of the silane depletion during deposition. The data in this figure are for the same samples as the data in Figure 3. Both an increase in silane flow or a decrease in the film deposition rate decrease silane depletion [5]. That is why films grown at the same silane flow rates (Figure 3) have different silane depletions in this figure.

more dramatic effect on the deposition rate than the additive effect of the changes in the individual factors.

Note that all four of the low-silane flow conditions in Figure 3 are grown in deposition regimes in which the silane is more than 50% depleted, which corresponds to the four points to the far right of Figure 4. We estimate the silane depletion by calculating how much Si goes into a film the thickness of our sample, the area of the inside of a hypothetical tube with a radius equal to the filament-to-substrate distance and a length equal to our filament length. We divide this number by the total number of Si atoms that travel through the chamber during the deposition time, based on the silane flow rate. The two samples grown at over 60% silane depletion have numerous microcrystalline silicon (μc-Si) inclusions.

Because of the decreasing quality of the material with increasing silane depletion, we cannot merely reduce our silane flow below 3 sccm to get the H content below 4 at. %. The dotted lines in Figure 4 illustrate that films we grew at lower pressures—11 and 50 mTorr— follow one trend while the films we grow at higher pressure—69 mTorr—follow a similar, but worse, trend. Both trends drop off sharply at silane depletions > 60%, in which μc-Si inclusions develop.

In the screening experiment to find the important chamber design configurations on the H content of films grown in our cross reactor, we find that we make good materials, with H content below 4 at. %, under all combinations of changes to make the cross reactor more tubelike (Table 2). That is, we grow films with good conductivity ratios ($> 10^5$) with the filament parallel or perpendicular to gas flow, with a 16.5-cm or a 6.4-cm-long filament, and with the chambers walls unheated or heated. In fact, all the films we have grown under these conditions have H ≤ 2 at. %. It is extremely valuable to learn that filament orientation is not important to film quality or H content. However, we still lack the fundamental understanding of the critical deposition factor(s) required for growing high-quality HWCVD a-Si:H with H below 4 at. %.

The thermocouple reading—giving us an indication of the substrate temperature—is the same for those depositions with heated chamber walls as those with unheated chamber walls in this experiment. However, it may be that the actual growing film temperature is higher in the cases where we heat the chamber walls. This may occur because the line-of-sight exposure of the substrate to a heated chamber wall provides a smaller temperature difference for radiation loss. We suspect this is the case, because we find that H is lower in films grown when heating the chamber walls than in the otherwise identical conditions with unheated chamber walls. Because this change was to make the cross reactor more tubelike—by heating the chamber walls—we expect to see the H go up in the heated wall condition, unless the substrate temperature is enough higher to offset any effect from this change. We also see that the ambipolar diffusion length decreases in films grown under the unheated chamber wall conditions compared to the identical conditions when heating the chamber walls; this is another indication our substrate temperature was not the same in these cases. We need to repeat some of the experiments with better control of our substrate temperature.

Finally, we grow a-Si:H in the tube reactor, with a H content of 4 at. %, that has a saturated defect density of 2×10^{16} cm^{-3}. While this value degrades from a lower value, the saturated value is as good as our lowest values from the cross reactor [1]. Many believe this is what is really important to solar cell performance, not the degree of degradation or the H content.

CONCLUSIONS

We find that by using HWCVD in our cross reactor, we can grow a-Si:H materials with unique properties relative to PECVD films and the HWCVD films grown in our tube reactor. We investigated extensive changes in the deposition parameter space for growing films in the tube reactor, but were unable to grow a-Si:H films of high quality containing less than 4 at. % H, as we readily have done in our cross reactors. However, the lowest values of saturated defect density we obtained is 2×10^{16} cm^{-3} in both reactors. This is particularly encouraging as this low saturated defect density was measured on a film with 4 at. % H, grown at 18 Å/s in the tube reactor. Additionally, we made modifications to the chamber configuration of the cross reactor to make it more similar to that of the tube reactor and found that we still could grow high-quality films with less than 4 at. %. More work is necessary to understand how to control the conditions to grow high-quality a-Si:H films containing less than 4 at. % H. Modeling to understand the growth chemistry and kinetics is necessary to be able to control the right factors to grow materials that have promise of increased stability at low defect density. We also are incorporating the lowest saturated defect density materials from both reactors into solar cells.

ACKNOWLEDGMENTS

We gratefully acknowledge helpful discussions with Howard Branz, Jack Thiesen, Bolko von Roedern, Alan Gallagher, and Don Williamson. This work is supported by the U.S. Department of Energy under Contract No. DE-AC36-98-GO10337.

REFERENCES

1. A.H. Mahan and M. Vanecek, *Amer. Inst. of Phy. Conf. Proc.* **234**, 195 (1991).
2. B.P. Nelson, Q. Wang, E. Iwaniczko, A.H. Mahan, and R.S. Crandall, *Mater. Res. Soc. Proc.* **507**, San Francisco, CA, 1998 pp 927-932.
3. Q. Wang, B.P. Nelson, E. Iwaniczko, A.H. Mahan, R.S. Crandall, and J. Benner, 2nd *WCPEC*, Vienna, Austria, 1998.
4. R.S. Crandall, A.H. Mahan, B.P. Nelson, M. Vanecek, I. Balberg, *AIP Conf. Proc.* **268**, Photovoltaic Advanced Research & Development Project, Denver, CO 1992, pp. 81-87.
5. E.C. Molenbroek, A.H. Mahan, A. Gallagher, *Jour. of Appl. Phys.* **82** (4), 1909, 15 August 1997.

Part II

High-Rate Deposition

HIGH-RATE GROWTH OF STABLE a-Si:H

T. TAKAGI, R. HAYASHI, A. PAYNE, W. FUTAKO, T. NISHIMOTO, M. TAKAI,
M. KONDO AND A. MATSUDA
Thin Film Silicon Solar Cells Super Lab., Electrotechnical Laboratory, Ibaraki, JAPAN,
ttakagi@etl.go.jp

ABSTRACT

Correlation between the gas phase species in silane plasma measured by mass spectrometry and the properties of hydrogenated amorphous silicon (a-Si:H) films deposited by plasma enhanced chemical vapour deposition (PECVD) has been investigated. We have especially been interested in the higher-order silane related species in the plasma, whose contribution to the film growth is considered to be the cause of light-induced degradation in the film quality, especially at high growth rate. In this study, we varied excitation frequency, gas pressure and power density to vary the growth rates of a-Si:H films ranging from 2 Å/s to 20 Å/s.

Molecular density ratio of trisilane, representative of higher silane related radicals, to monosilane has shown a clear correspondence to the fill factor after light soaking of Schottky cells fabricated on the resulting films.

INTRODUCTION

A-Si:H film is an important material in the solar cell industry. It is essential to achieve high growth rate and uniform deposition on large area substrates in order to reduce the production cost, but the film properties, initial and more importantly after light-induced degradation, are known to deteriorate with increasing growth rate. The increase in the growth rate is conventionally obtained by increasing the silane partial pressure and the power; however, the amount of gas-phase higher-order silanes is known to increase with the silane pressure, while short lifetime radicals such as Si, SiH and SiH_2 contribute to the film growth under a silane depletion condition [1]. These changes of film-growth precursors in the gas-phase and/or their subsequent interactions on the growth surface are considered to be the cause of the inferior film properties.

It was reported that dihydride bonding configuration ($Si-H_2$) in the a-Si:H network structure is responsible for degradation [2]. The $Si-H_2$ fraction is known to increase with the increase in growth rate. According to the surface diffusion model, the film surface mostly covered with hydrogen and precursors with long diffusion length are the ideal conditions for the growth of films with low dihydride configuration [3]. The causes of $Si-H_2$ are suggested to include the substrate temperature which affects the diffusion process of film precursors, positive ions in the plasma striking the film surface, and the presence of higher-order silane related radicals. These radicals can disturb the surface reaction of film precursors and can be incorporated in the film. As the dihydride modes can be suppressed by increasing the deposition temperature [4], we took notice of higher-order silanes, whose contribution is suggested to be relaxed by increasing the

Mat. Res. Soc. Symp. Proc. Vol. 557 © 1999 Materials Research Society

temperature.

There has been much work on gas-phase analysis using mass spectrometry in order to analyze the ionic and neutral species in the silane rf glow-discharge plasma [1, 5, 6, 7], but most work does not touch upon properties of films deposited from the relevant plasma, especially after light-induced degradation.

In this work, we studied the gas-phase species in pure silane plasmas under various deposition conditions of a-Si:H for growth rates up to 20 Å/s. Our aim was to correlate the gas-phase species with the film properties after light-induced degradation, in order to identify the mechanism of deterioration, and to find the ideal deposition conditions to achieve high rate growth of stable a-Si:H films. We measured the signal intensities of monosilane, disilane and trisilane molecules in the plasma using a mass analysis system and observed the contribution of relevant radicals to film growth. The properties of a-Si:H films after light-induced degradation were monitored by the fill factor of photo current - voltage characteristics in n$^+$ crystalline Si/a-Si:H/Ni Schottky cells after 6 hours of light soaking under 3-sun illumination at 60 °C.

EXPERIMENTAL

Figure 1 shows the experimental setup. A diode-type plasma enhanced chemical vapour deposition (PECVD) system with electrodes of 100 mm in diameter was used. For gas phase diagnosis, a mass spectrometer mounted underneath the anode (grounded electrode) was used. Silane plasma was generated by excitation frequency of 13.56 MHz (RF) and 80 MHz (VHF). The electrode distance was kept constant at 40 mm.

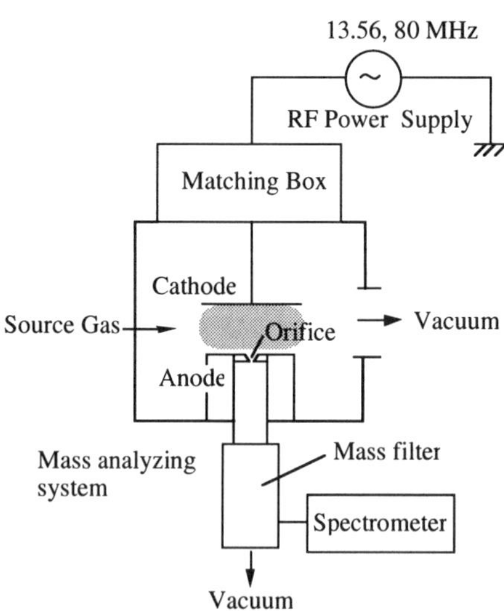

Figure 1 : Schematic diagram of PECVD system with mass spectrometer.

The mass spectrometry measurement was carried out using a differentially evacuated quadrupole mass spectrometer system (ULVAC MSQ-200). The ionic and neutral species, entering the mass-analyzing system through a fine orifice, are dissociated and ionized by electrons with ionization energy of 70 eV. As the density of ionic species in the plasma is five to six orders of magnitude lower than that of the total density of all the species involved in the plasma, the observed signal intensities must be regarded as the fragments of neutral molecules [1]. In this work, the signal intensities of SiH_2^+ (m/e=30), $Si_2H_4^+$ (m/e=60) and $Si_3H_6^+$ (m/e=90), as the most abundant ions in the mass fragmentation, were used to represent monosilane, disilane and trisilane molecules, respectively.

Both plasma diagnosis and film deposition were carried out under various discharge conditions giving a-Si:H growth rates ranging from 2 Å/s to 20 Å/s. The flow rate of silane was kept constant at 30 sccm and the pressure was varied between 20 mTorr and 100 mTorr. The substrate temperature was kept constant at 250 °C and the input power was changed from 2.5 W to 100 W. Lower power density (monitored by external power meter) was enough for 80 MHz than for 13.56 MHz to obtain the same growth rate.

The hydrogen bonding configurations were investigated by FT-IR absorption spectroscopy for the films deposited on intrinsic crystalline silicon wafers. The Si-H and Si-H_2 stretching modes were observed at wavelengths of 2000 cm^{-1} and 2100 cm^{-1}, respectively [4], and the dihydride mode fraction was calculated by [2100 cm^{-1}]/([2000 cm^{-1}]+[2100 cm^{-1}]). Here, [2000 cm^{-1}] and [2100 cm^{-1}] represent the peak heights in FT-IR spectra at wavelengths of 2000 cm^{-1} and 2100 cm^{-1}, respectively, after the background was subtracted. We also calculated the fraction of the areas under these two peaks observed by the integrated Gaussian fit to the peaks at 2100 cm^{-1} and 2000 cm^{-1}. The dihydride fraction obtained from the peak heights and from the integrated areas showed a good correspondence.

Schottky cells were fabricated to monitor the film properties of a-Si:H films. After a-Si:H film with a thickness of 700 ~ 800 nm was deposited on n$^+$ type crystalline silicon wafer, semi-transparent Ni electrodes were evaporated to make the Schottky junction. The thickness of the Ni electrode was 40 Å, and the diameter of the electrodes was 0.5 mm to eliminate the effect of inhomogeneity and pin holes. The Schottky cell performance, initial and after light soaking, was monitored by the fill factor (FF) of photo current - voltage characteristics under AM 1.5 (100 mW/cm^2) light illumination. Light soaking was carried out for 6 hours at 60 °C under 300 mW/cm^2 illumination of Xe lamp through water to cut out infra-red light.

RESULTS AND DISCUSSION

(a) FF of Schottky cells after light soaking

It is well known that the film properties of a-Si:H, initial and after light-induced degradation, deteriorate with increasing growth rate. We monitored the film properties of a-Si:H films by the FF of Schottky cells. Figure 2 shows the relationship between the initial and light-soaked FF and the growth rate of a-Si:H under various growth conditions from pure silane at an excitation frequency of 13.56 MHz. The initial FF shows a slight decrease, and the FF after light soaking shows a clear deterioration with the growth rate. We used the FF after light soaking to characterize the film quality of a-Si:H films.

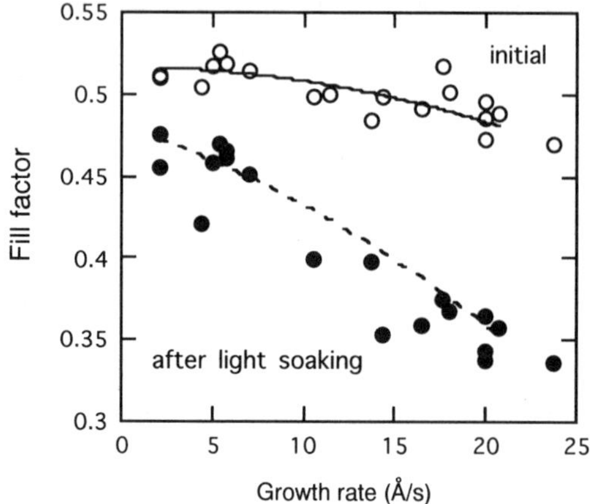

Figure 2 : Fill factor (initial and after light soaking) of Schottky cells as a function of
growth rate from various growth conditions in 13.56 MHz pure SiH$_4$ plasma.

The cause of light-induced degradation is not clear yet, but the correlation between the hydrogen bonding configuration and the degradation properties of solar cells has been pointed out [2]. In our study, we observed the same relationship between the Schottky cell FF after light soaking and the dihydride mode fraction observed from the FT-IR spectra (Figure 3).

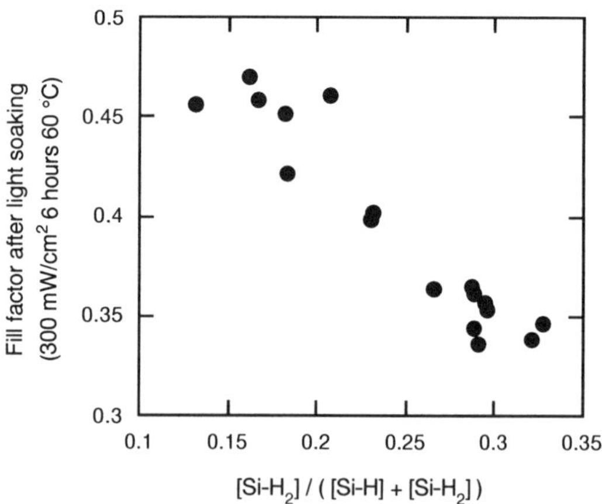

Figure 3 : Relationship between the fill factor of Schottky cells after light soaking
and dihydride fraction of a-Si:H films measured by FT-IR for films
deposited under various conditions in 13.56 MHz pure SiH$_4$ plasma.

The films were prepared in 13.56 MHz pure silane plasma under various growth rates at a constant substrate temperature of 250 °C. As shown in Figure 3, higher FF after light soaking is obtained under lower dihydride mode fraction conditions.

We are considering that the increase of dihydride modes occurs because of the contribution of SiH_3^+ or H^+ ions, or higher-order silane related radicals to the film growth when the substrate temperature is kept constant. As the dihydride modes can be suppressed by increasing the deposition temperature [4], we took notice of higher-order silanes, whose contribution is suggested to be relaxed by increasing the temperature. The surface diffusion model also suggests that relatively large species, such as higher-order silane related radicals, reaching the film surface cannot diffuse long enough [3] . They disturb the surface diffusion of other precursors, such as SiH_3, and as a result a well relaxed film network cannot be produced. It is not clear yet from which size the higher-order silane related radicals affect the film properties.

In our study, we changed the excitation frequency, pressure and power density to alter the film properties under various growth rate conditions. Figure 4 shows the fill factor after light soaking of Schottky cells, made from various deposition conditions including excitation frequencies, as a function of growth rate. The pressures are 20 mTorr and 100 mTorr for 13.56 MHz and 20 mTorr and 50 mTorr for 80 MHz. For each condition, power density was changed to vary the growth rate. It is clearly seen that the Schottky cell performance after light soaking deteriorates with the increase in the growth rate, rather independent of deposition conditions. The FF after light soaking for the 100 mTorr case at 13.56 MHz showed almost the same value for the whole growth rate range. In this case the dihydride mode fraction was high even at low growth rate conditions.

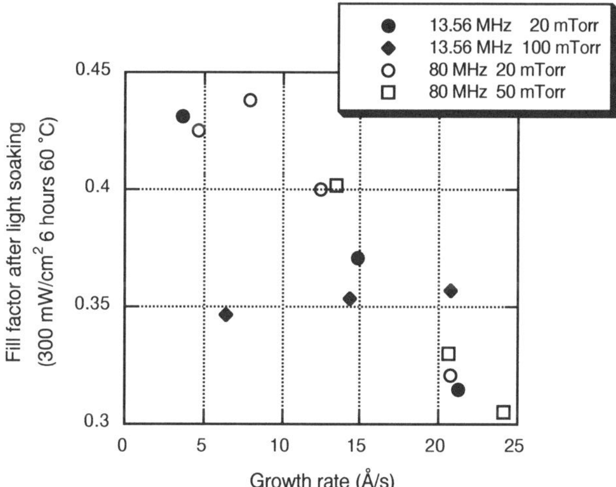

Figure 4 :   Fill factor of Schottky cells after light soaking as a function of
growth rate for films deposited at 20 mTorr and 100 mTorr at
13.56 MHz and 20 mTorr and 50 mTorr at 80 MHz.

To determine whether there is a correlation between the gas phase species in the plasma and the resulting film, we monitored the higher-order silanes using mass spectrometry under the same conditions as for the film depositions.

(b) Mass Spectrometry

A typical mass spectrum observed in our mass spectrometry system is shown in Figure 5. The signal intensities of SiH_2^+ (m/e=30), $Si_2H_4^+$ (m/e=60) and $Si_3H_6^+$ (m/e=90) , as the most abundant ions in the mass fragmentation, were used to record the density of monosilane (SiH_4), disilane (Si_2H_6) and trisilane (Si_3H_8) molecules, respectively. The mass signal intensities related to disilane and trisilane molecules were two and three orders of magnitude lower than that of the monosilane molecule, respectively. In our system, tetrasilane (Si_4H_{10}) related signals were also detected, but they were at the lowest limit of our measurement system, four orders of magnitude lower than the monosilane signal, and are therefore excluded from this work.

Figure 6 shows the monosilane, disilane and trisilane related signal intensities as a function of power. The discharge conditions are pure silane plasma at the pressures of 20 mTorr and 100 mTorr in 13.56 MHz silane plasma. The monosilane related signal intensity decreases with the increase in power, indicating the enhancement of silane dissociation [1]. The disilane and trisilane related signal intensities at high partial pressure (100 mTorr) show a maximum in the low power region, while they increase with power for the low partial pressure case (20 mTorr), approaching the value of the 100 mTorr case in the high power region. We suggest that the decrease in higher-order silanes for the 100 mTorr case over 20 W is related to the depletion of silane molecules as the growth rate started to saturate over this power. For reference, the power needed for the growth rate of 20 Å/s were 60 W for 20 mTorr and 20 W for 100 mTorr.

Mass analysis was also performed for excitation frequency of 80 MHz at 20 mTorr. In this case, the mass intensities were compared under the same growth rate conditions, as the power needed to achieve the same growth rates were different for these frequencies. The disilane intensities were almost the same for both frequencies for the whole growth rate range, while significant difference was seen in trisilane intensities. The trisilane intensities of 80 MHz

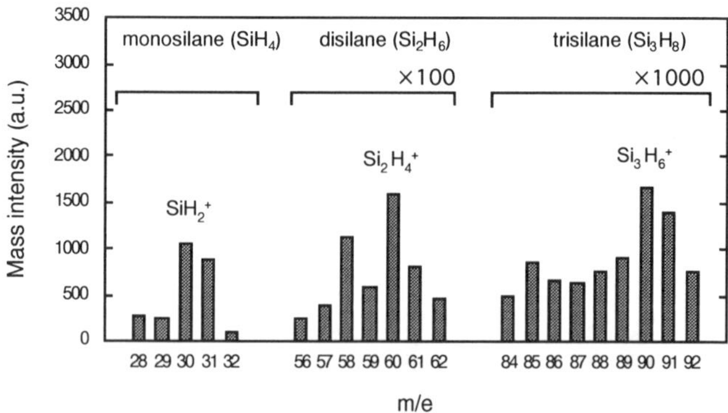

Figure 5 :   A typical mass spectrum observed by mass spectrometry in silane plasma.

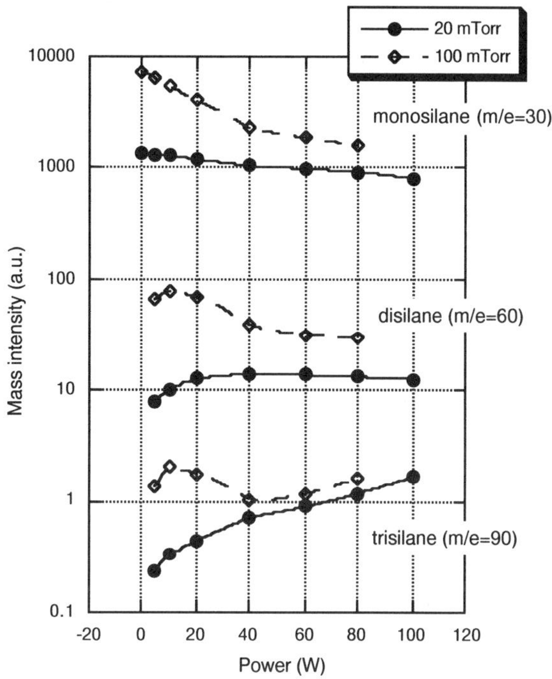

Figure 6 : Mass intensity of monosilane, disilane and trisilane related signals
as a function of power for pure silane at the pressures of
20 mTorr and 100 mTorr in 13.56 MHz pure silane plasma.

case at low growth rate range were comparable to 13.56 MHz case. As the growth rate increased, while the trisilane intensity increased with growth rate in 13.56 MHz case (Figure 6), almost no difference was seen in 80 MHz case. The difference in trisilane intensities for 13.56 MHz and 80 MHz under the same growth rate conditions is ascribed as follows. The electron temperature is lower for higher excitation frequency. This means that at the same growth rate, the number density of higher energy electrons is lower in higher frequency plasma. As the energy needed to dissociate SiH_4 into SiH_2 radicals is higher than for the generation of SiH_3 radicals (the main film precursor), the generation rate of SiH_2 radicals is lower in higher frequency plasma under the same growth rate condition. SiH_2 radicals are considered as the responsible species for the formation of higher order silanes and powders [8]. For example, SiH_2 reacts with SiH_4 to form Si_2H_6, then another SiH_2 reacts with Si_2H_6 to form Si_3H_8, and so on. Although the difference in the generation rate of SiH_2 is small when changing excitation frequencies, as the successive reaction of SiH_2 radicals with higher order silanes goes on, the difference becomes clearer. Therefore, small difference is seen in the steady state density of small size higher silanes, in this case disilane, while as the successive process goes on for the formation of trisilane, tetrasilane and so on, the difference becomes remarkable.

We made the following assumption to estimate the contribution of higher-order silane related radicals to the film growth. The main film growth precursor in SiH_4 plasma, SiH_3, is produced from a single collision of SiH_4 with an electron. Then, the change in the density of SiH_3 in a SiH_4 plasma may be represented as the balance between the generation rate of SiH_3 by the dissociation of SiH_4 and the diffusion (annihilation rate) of SiH_3 ,

$$\frac{d[SiH_3]}{dt} = N_e v_e \sigma_3 [SiH_4] - \frac{[SiH_3]}{\tau_3} \quad . \qquad (1)$$

Here, N_e is the electron density, v_e is the thermal velocity of electron, σ_3 the dissociation cross section of SiH_4 forming SiH_3, τ_3 the characteristic residence time of SiH_3, and $[SiH_4]$ is the molecular density of SiH_4. The density of SiH_3 in the steady state plasma is obtained by setting $d[SiH_3]/dt=0$ to be

$$[SiH_3] = \tau_3 N_e v_e \sigma_3 [SiH_4] \quad . \qquad (2)$$

Similarly, the density of higher-order silane related radicals $[Si_x H_{2x+1}]$ in the steady state plasma is expressed as

$$[Si_x H_{2x+1}] = \tau_{2x+1} N_e v_e \sigma_{2x+1} [Si_x H_{2x+2}] , \qquad (3)$$

where $[Si_x H_{2x+2}]$ is the number density of higher-order silane related molecules.

The contribution ratio of higher-order silane related radicals is expressed from the ratio of (3) to (2) , i.e.,

$$[Si_x H_{2x+1}]/[SiH_3] \propto [Si_x H_{2x+2}]/[SiH_4] , \qquad (4)$$

which is the molecular ratio of relating higher-order silane molecules to the steady state monosilane molecules. For example, the contribution ratio of trisilane related radicals (in case of x=3), such as $Si_3 H_7$, to the film growth, or simply trisilane ratio, can be represented as the ratio of the mass signal intensities related to trisilane molecules to the monosilane molecules ; i.e. [m/e=90] / [m/e=30], where [m/e=90] and [m/e=30] represent the mass signal intensities at mass numbers 90 and 30, respectively.

Figure 7 shows the relationship between the trisilane ratio observed from the mass spectrometry measurement and the growth rate of a-Si:H films from 13.56 MHz and 80 MHz plasma at 20 mTorr. For each condition, the growth rate is increased by increasing the power density. It is shown that the trisilane ratio increases with the increase in growth rate. When compared for the same growth rate, higher trisilane ratios are obtained for 13.56 MHz than for 80 MHz, especially at high growth rate.

Figure 8 shows the relationship between the trisilane ratio and the fill factor of Schottky cells after light soaking. Films for the Schottky cell were fabricated under the same deposition conditions as the mass spectrometry measurement. It is clearly seen that higher FF is obtained for lower trisilane ratio, independent of excitation frequency.

It is still not clear yet from which size the higher-order silane related radicals directly affect the film properties. The trisilane ratio may be only the representative parameters of much higher-order silanes responsible for inferior film properties, however, it can be said that suppressing the trisilane ratio in the gas phase leads to the growth of more stable films.

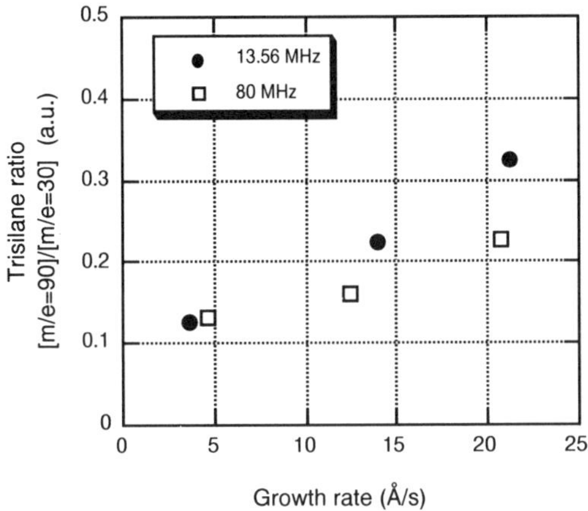

Figure 7 : Trisilane ratios as a function of growth rate of a-Si:H films deposited from 13.56 MHz and 80 MHz pure silane plasma at 20 mTorr.

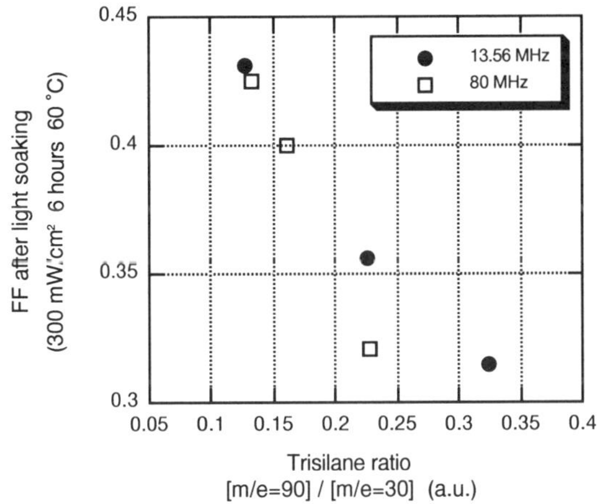

Figure 8 : The relationship between the fill factor of Schottky cells after light soaking and the trisilane ratio obtained by mass analysis from 13.56 MHz and 80 MHz pure silane plasma at 20 mTorr

Further investigation of the gas phase species, searching for the discharge conditions with a low density of higher-order silanes will lead to the observation of much more stable a-Si:H films deposited at a high growth rate.

CONCLUSIONS

We identified the contribution of higher-order silane-related radicals to the film growth as a major cause of the degradation of a-Si:H film properties when increasing the growth rate. It is suggested from the results mentioned here that the suppression of gas-phase higher-order silane-related radicals is a clue for obtaining stable a-Si:H solar cells at high growth rate.

ACKNOWLEDGMENT

This work is partly supported by NEDO under the New Sunshine Project of the Agency of Industrial Science and Technology. We also acknowledge the support of the Science and Technology Agency of Japan.

REFERENCES

1. A. Matsuda and K. Tanaka, Thin Solid Films **92**, 171 (1982).

2. N. Nakamura, T. Takahama, M. Isomura, M. Nishikuni, K. Yoshida, S. Tsuda, S. Nakano, M. Ohnishi and Y. Kuwano, Jpn. J. Appl. Phys. **28**, 1762 (1989).

3. G. Ganguly and A. Matsuda, Phys. Rev. B **47**, 3661 (1992).

4. G. Lucovsky, R. J. Nemanich and J. C. Knights, Phys. Rev. B **19**, 2064 (1978).

5. I. Haller, Appl. Phys. Lett. **37**, 282 (1980).

6. G. Turban, Y. Catherine and B. Grolleau, Thin Solid Films **67**, 309 (1980).

7. C. M. Fortmann, S. Lange, M. Farley and J. O`Dowd, Proc. IEEE Photovoltaic Spec. Conf. 19th, 296 (1987).

8. Y. Watanabe, M. Shiratani, H. Kawasaki, S. Singh, T. Fukuzawa, Y. Ueda and H. Ohkura, J. Vac. Sci. Technol. A 14 (1996) 540.

ANALYSIS OF PLASMA PROPERTIES AND DEPOSITION OF AMORPHOUS SILICON ALLOY SOLAR CELLS USING VERY HIGH FREQUENCY GLOW DISCHARGE

B. YAN[*], J. YANG[*], S. GUHA[*], AND A. GALLAGHER[**]
* United Solar Systems Corp., 1100 West Maple Road, Troy, MI 48084
** JILA, National Institute of Standards and Technology and University of Colorado, Boulder, CO 80309-0440

ABSTRACT

Positive ionic energy distributions in modified very-high-frequency (MVHF) and radio frequency (RF) glow discharges were measured using a retarding field analyzer. The ionic energy distribution for H_2 plasma with 75 MHz excitation at a pressure of 0.1 torr has a peak at 22 eV with a half-width of about 6 eV. However, with 13.56 MHz excitation, the peak appears at 37 eV with a much broader half-width of 18 eV. The introduction of SiH_4 to the plasma shifts the distribution to lower energy. Increasing the pressure not only shifts the distribution to lower energy but also broadens the distribution. In addition, the ionic current intensity to the substrate is about five times higher for MVHF plasma than for RF plasma. In order to study the effect of ion bombardment, the deposition of a-Si alloy solar cells using MVHF was investigated in detail at different pressures and external biases. Lowering the pressure and negatively biasing the substrate increases ion bombardment energy and results in a deterioration of cell performance. It indicates that ion bombardment is not beneficial for making solar cells using MVHF. By optimizing the deposition conditions, a 10.8% initial efficiency of a-Si/a-SiGe/SiGe triple-junction solar cell was achieved at a deposition rate of 0.6 nm/sec.

INTRODUCTION

Initial and stable efficiencies of 15.2% and 13%, respectively, have been achieved for amorphous silicon (a-Si) alloy solar cells [1,2]. These milestones were made at the deposition rate of about 0.1 nm/sec. For industrial production, high deposition rate is desirable to improve the throughput from a given machine. However, the quality of the materials deposited with RF glow discharge normally degrades when the deposition rate is increased, and has been correlated with the increase of the density of microvoids [3]. Consequently, the solar cells made at high deposition rates show poor performance and poor stability. Therefore, development of a new technique to deposit high quality a-Si alloy materials and solar cells is very important for the industrial production of a-Si alloy solar cells.

a-Si alloy made using very-high-frequency (VHF) glow discharge offers an interesting alternative, and single junction cells have been fabricated [4,5] with initial efficiency of about 10%. Previously, we reported that a-Si alloy solar cells made with modified very-high-frequency (MVHF) glow discharge at 0.6 nm/sec show similar initial performance as those made with RF at 0.3 nm/sec. In addition, the stability of 0.6 nm/sec MVHF cells is better than the 0.3 nm/sec RF cells [6]. To further improve the performance of the MVHF solar cells made at high deposition rates, we need to understand the difference between the plasma excited by RF and MVHF, and search for further improvement of the solar cells. For this purpose we have studied the ionic energy distribution of MVHF and RF plasmas as well as the effects of ion bombardment on the performance of a-Si and a-SiGe alloy solar cells.

Mat. Res. Soc. Symp. Proc. Vol. 557 © 1999 Materials Research Society

EXPERIMENTAL

A parallel plate capacitance reaction chamber (MVHF chamber), which was connected to a multi-chamber glow discharge system, was designed to have the capability to couple RF (13.56 MHz) and VHF (75 MHz) frequencies to the plasma. The electrode space between the cathode and anode was 3 cm. The substrate can be grounded, externally biased, or floated, giving us flexibility to study the effect of ion bombardment.

A retarding field analyzer was installed in the plasma chamber to measure the energy distribution of positive ions. The structure of the analyzer is similar to the one used by Ingram and Braithwaite [7]. A hole with a diameter of 4 mm on a 13 μm thick stainless steel substrate allowed positive ions and electrons to flow into the analyzer. A fine grid screen was attached at the front of the hole to prevent the penetration of plasma sheath field to the analyzer. Behind the hole, a second grid screen was negatively biased to block electrons from reaching the collector. Two 5 μm thick Kapton spacers were used to insulate the substrate, the screen, and the collector. The screen voltage, V_s, was set at -40 V for most measurements, which was sufficient to block electrons from reaching the collector. The collector voltage, V_c, (retarding potential) was ramped from -100 V to $+100$ V with a relatively low speed to ensure a steady state collector current, J_c. Obtaining precise ionic energy distribution from measured J_c-V_c characteristic is a very complicated procedure [7]. But as a first order approximation, the derivative of the J_c-V_c characteristic is proportional to the ionic energy distribution. Experimentally, the ion energy distributions of H_2, Ar, SiH_4, and their mixtures were studied as functions of excitation frequency, power density, and pressure. In order to study the effect of ion bombardment, a-Si and a-SiGe alloy component cells were deposited at different pressures and/or with varied external biases. In addition, other deposition conditions such as gas flow rate and substrate temperature were optimized to improve the performance of solar cells.

RESULTS AND DISCUSSION

Ion energy distribution

Figure 1 shows the ionic energy distribution for a H_2 plasma with 75 MHz excitation, where the VHF power was 10 W. At a low pressure (0.1 torr), a sharp peak appears at 22 eV with a half-width of about 6 eV. The cut-off at the high energy side is very steep. The corresponding value could represent the plasma potential. However, at a high pressure (1.0 torr), the peak of the ionic energy distribution shifts to zero and the width becomes significantly broader. Similar measurements were carried out for an RF plasma. Figure 2 shows the results of an RF plasma with the same conditions as in Fig. 1 except the excitation frequency was 13.56 MHz. The peak position is much higher (37 eV) than that shown in Fig.1, and the distribution is also much broader (18 eV of half-width). Similar to the MVHF plasma, there are no high energy ions reaching the substrates at 1.0 torr. In addition, the positive ion current that reached the substrate is about 5 times higher in the MVHF plasma than in the RF plasma for a similar input power. It was also noted that the self-bias on the cathode was higher for the RF plasma than for the MVHF plasma, e.g. at 0.1 torr with 10 W input power, the self-biases were -52 V and -29 V for the RF and MVHF plasmas, respectively.

The peak position of the ionic energy distribution is lower for plasma with a mixture of SiH_4 and H_2 than for plasma with pure H_2. This shift probably indicates that the plasma potential is higher for pure H_2 than for H_2 and SiH_4 mixture. This phenomenon is similar for both RF and MVHF plasmas. Increasing excitation power slightly shifts the energy distribution of MVHF plasma to a higher energy. The shape of the ion energy distribution remains unchanged within

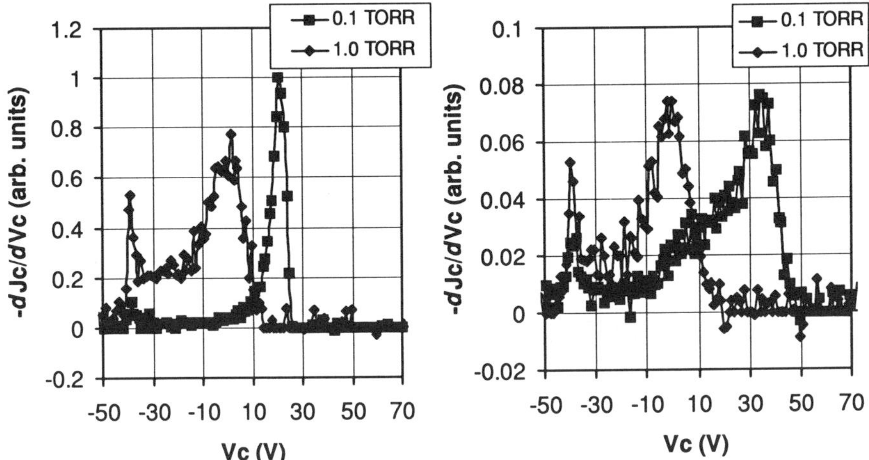

Figure 1. Ionic energy distribution of a H_2 plasma excited by VHF.

Figure 2. Ionic energy distribution of a H_2 plasma excited by RF.

experimental errors. It probably means that the plasma potential is increased by the higher power since the peak-to-peak voltage increases with the increase of power. However, there is no significant change at the cut-off energy for RF plasma. Instead, the shape of the distribution is changed.

From the ion energy distribution measurements we have learned the following. First, positive ion bombardment occurs only at low pressures. For normal solar cell deposition pressures, there is no significant ion bombardment from the plasma itself. Second, at low pressures, the ion bombardment energy is higher in RF plasma than in MVHF plasma. Third, the positive ion flux is higher in MVHF plasma than in RF plasma.

Effect of ion bombardment and optimization of solar cells

In order to find out if ion bombardment is beneficial for making a-Si alloy solar cells with MVHF, we have performed systematic studies of the effects of pressure and external bias on a-Si alloy solar cells. Table I lists the characteristics of a-Si alloy solar cells deposited on stainless steel (SS) substrates. The intrinsic layers were deposited at different pressures. It is noted that pressure significantly influences the deposition rate. The cell made at 0.1 torr is not as good as the cells made at 1.0 torr or 1.5 torr although the deposition rate is only 0.44 nm/sec. This is probably due to the adverse effect of ion bombardment. On the other hand, at pressures higher than 2.0 torr, the open circuit voltage (V_{oc}) and fill factor (FF) drop, which may have resulted from the polymerization in the plasma.

Table II summarizes the performance of a-Si alloy solar cells deposited at 0.1 torr with different external DC biases. The positive and negative biases slightly reduce and enhance the deposition rate, respectively. The FF of the positively biased cells is better than the one deposited on a floating substrate. The negative bias significantly reduces the FF. Physically, a positive bias reduces the energy and flux of the positive ions and a negative bias accelerates the positive ions toward the substrate. Therefore, we may conclude that ion bombardment is not beneficial for making a-Si alloy solar cells using our MVHF with current conditions.

Table I: Characteristics of a-Si alloy solar cells made by MVHF at different pressures. P_{max} is the maximum output power, J_{sc} the short circuit current density, V_{oc} the open circuit voltage, FF, FF_b, and FF_r the fill factor under AM1.5, blue, and red light, respectively, d the thickness of intrinsic layer, P the deposition pressure.

Run #	P_{max} (mW/cm^2)	J_{sc} (mA/cm^2)	V_{oc} (V)	FF	FF_b	FF_r	Rate (nm/s)	d (nm)	P (torr)
7666	6.74	9.75	0.991	0.698	0.767	0.684	0.44	217	0.1
7689	7.67	11.2	0.976	0.702	0.759	0.706	1.4	253	1.0
7664	7.51	10.55	0.990	0.719	0.781	0.706	0.89	214	1.5
7687	6.64	9.91	0.956	0.701	0.770	0.704	0.6	181	2.0
7694	6.67	11.04	0.936	0.646	0.771	0.694	0.3	209	2.5

Table II: Characteristics of a-Si alloy solar cells made at 0.1 torr with different biases.

Run#	P_{max} (mW/cm^2)	J_{sc} (mA/cm^2)	V_{oc} (V)	FF	FF_b	FF_r	Rate (nm/s)	d (nm)	Bias (V)
7716	6.64	9.67	0.989	0.695	0.759	0.681	0.42	204	-5.6*
7717	6.89	9.84	0.998	0.702	0.778	0.689	0.41	200	+30
7718	6.88	9.65	0.999	0.714	0.774	0.689	0.40	198	+60
7738	6.87	9.74	1.000	0.705	0.767	0.679	0.41	200	+90
7739	6.74	9.49	0.997	0.712	0.770	0.700	0.39	193	+120
7740	6.87	9.48	1.005	0.721	0.780	0.695	0.39	193	+150
7741	5.69	9.08	0.992	0.632	0.711	0.671	0.47	228	-60
7742	5.22	8.42	1.005	0.617	0.691	0.635	0.49	241	-120

*Self bias on floating substrate.

The pressure and external bias experiments were also carried out on a-SiGe alloy middle cells. Table III lists the characteristics of a-SiGe alloy middle cells made on SS substrate, where the intrinsic a-SiGe layer was deposited with different external biases. The measurements were made under an AM1.5 solar simulator with a 530 nm cut-on filter. The positive bias does not affect the cell performance, but the negative bias reduces the FF dramatically. It should be pointed out, however, that the negatively biased cells are slightly thicker than the unbiased and positively biased cells. From the above results we may conclude that ion bombardment does not improve the a-SiGe alloy solar cells either.

Table III: Characteristics of a-SiGe alloy middle cells made with external bias.

Run #	P_{max} (mW/cm^2)	J_{sc} (mA/cm^2)	V_{oc} (V)	FF	FF_b	FF_r	d (nm)	V_{bias} (V)
7788	4.06	8.34	0.751	0.648	0.696	0.659	168	-7.2*
7801	3.98	8.71	0.739	0.618	0.679	0.635	163	-6.7*
7802	4.04	8.46	0.747	0.640	0.701	0.644	162	+30
7803	4.00	8.49	0.739	0.638	0.697	0.645	171	+60
7789	3.96	8.48	0.735	0.638	0.686	0.652	167	+120
7804	3.97	8.64	0.736	0.625	0.685	0.632	163	-30
7805	3.78	8.76	0.736	0.587	0.686	0.592	174	-60
7790	3.51	8.39	0.731	0.572	0.690	0.579	172	-120

*Self bias on floating substrate.

In order to achieve high efficiency triple-junction solar cells, we need to optimize the component cells, namely wide band gap a-Si alloy top cells, medium band gap a-SiGe alloy middle cells and low band gap a-SiGe alloy bottom cells. The techniques of the optimization include the improvement of material quality by adjusting the deposition conditions, such as substrate temperature, pressure, and gas flow rates, as well as the band gap engineering, such as profiling the band gap of a-SiGe alloy cells and improving the interfaces between layers. Table IV shows the comparison of MVHF high rate solar cells (0.6 nm/sec) with the RF low rate solar cells (0.1 nm/sec). The performance of the high rate MVHF top cell is slightly lower. However, the performances of a-SiGe alloy middle cells and bottom cells as well as triple-junction cells are still much lower than the corresponding low rate RF cells. There are two factors that limit the further improvement of the MVHF middle and bottom cells. First, the quality of the high rate MVHF a-SiGe alloy materials is still poor. Second, the band gap profiling is not as effective as it is in low rate RF cells. The deposition time is probably too short to allow the actual compositions of Si and Ge to follow the changes of the gas flow rates.

CONCLUSION

The ionic energy distributions were measured for MVHF and RF plasmas. At the same experimental conditions, the ionic energy is lower for MVHF plasma than for RF plasma, however, the ion flux to the substrate is about five times higher for MVHF plasma than for RF plasma. The ionic energy distribution shifts toward low energy with increasing pressure. At ordinary operating pressures for making solar cells in our laboratory, there are no significant high energy ions reaching the substrate, which means that ion bombardment is not the limitation factor of improving the efficiency of high rate solar cells. By reducing the pressure or supplying external bias to enhance the ion bombardment, we found that ion bombardment is not beneficial for making solar cells. Preliminary results show that the MVHF a-SiGe alloy middle and bottom

Table IV: Comparison of the performance of solar cells made using MVHF at 0.6 nm/sec and using RF at 0.1 nm/sec.

		P_{max} (mW/cm^2)	J_{sc} (mA/cm^2)	V_{oc} (V)	FF
Top[a,c]	MVHF	6.21	8.34	0.995	0.749
	RF	6.86	8.68	1.011	0.782
Middle[a,d]	MVHF	4.18	8.35	0.737	0.679
	RF	5.23	9.28	0.791	0.713
Bottom[b,e]	MVHF	3.84	10.0	0.623	0.616
	RF	5.12	12.2	0.631	0.671
Triple[b,c]	MVHF	10.8	7.52	2.191	0.655
	RF	15.2	8.99	2.344	0.722

a. Deposited on a bare stainless steel substrate.
b. Deposited on a textured Ag/ZnO back reflector.
c. Measured under AM1.5 illumination.
d. Measured under AM1.5 with a $\lambda>530$ nm filter.
e. Measured under AM1.5 with a $\lambda>630$ nm filter.

cells are still much poorer than the corresponding RF cells made at low rates, which limits the performance of MVHF triple-junction cells.

ACKNOWLEDGMENTS

The authors are grateful to T. Palmer, D. Wolf, I. Rosenstein, E. Chen, G. Hammond, J. Edens, J. Noch and M. Hopson for experimental assistance and to K. Lord and A. Banerjee for helpful discussion. This work was supported in part by NREL under Subcontract No. ZAK-8-17619-09.

REFERENCES

1. J. Yang, A. Banerjee, and S. Guha, Appl. Phys. Lett. **70**, 2975 (1997).

2. J. Yang, A. Banerjee, K. Lord, and S. Guha, *Proc. of 2nd World Conf. and Exhibition on Photovoltaic Solar Energy Conversion*, Vienna, Austria, (1998), pp.387.

3. S. Guha, J. Yang, S. Jones, Y. Chen, and D. Williamson, Appl. Phys. Lett. **61**, 1444 (1992).

4. H. Chatham and P. K. Bhat, in *Amorphous Silicon Technology-1989*, edited by A. Madan, M.J. Thompson, P. C. Taylor, H. Hamakawa, P.G. LeComber (Mater. Res. Soc. Proc. **149**, Pittsburgh, PA, 1989) pp.447-452.

5. W.G.J.H.M. van Spark, J. Bezermer, R. van der Heijden and van der Weg, in *Amorphous Silicon Technology-1996*, edited by M. Hack, E. A. Schiff, S. Wagner, R. Schropp and A. Matsuda (Mater. Res. Soc. Proc. **420**, Pittsburgh PA, 1996) pp. 21-25.

6. J. Yang, S. Sugiyama, and S. Guha, Mater. Res. Soc. Proc. (1998), in press.

7. S.G. Ingram and N. St. J. Braithwaite, J. Phys. D. **21**, 1496 (1988); J. Appl. Phys. **68**, 5519 (1990).

FAST DEPOSITION OF a-Si:H LAYERS AND SOLAR CELLS IN A LARGE-AREA (40x40 cm^2) VHF-GD REACTOR

U. Kroll[*], D. Fischer[*], J. Meier[*], L. Sansonnens[#], A. Howling[#] and A. Shah[*]

[*] Institut de Microtechnique, Université de Neuchâtel, CH-2000 Neuchâtel, Switzerland

[#] Centre de Recherches en Physique des Plasmas, Ecole Polytechnique Fédérale de Lausanne, CH-1015 Lausanne, Switzerland

ABSTRACT

Large-area deposition of hydrogenated amorphous silicon has been investigated in a single-chamber industrial reactor with electrode dimensions of 40x40 cm^2 in the plasma excitation frequency range of 60 to 120 MHz. The film thickness uniformity, analyzed by a light interferometry technique and a step profiler, has been compared with 2-dimensional interelectrode voltage measurements and calculations. The frequency of 80 MHz has been found to be a good compromise between the gain in deposition rate and the homogeneity requirements necessary for a-Si:H solar cells. Under these conditions and while using hydrogen dilution high deposition rates of 6-7 Å/s with a film uniformity of ±5% over a usable substrate size of 30x30 cm^2 have been obtained.

In the same single-chamber deposition system at 80 MHz, 0.4 µm thick single-junction a-Si:H solar cells with high performance were fabricated in a total process time of 16 minutes applying a continuous deposition process. Spectral response measurements indicate a minor boron contamination of the i-layer. Initial cell efficiencies of 7.1 % could be achieved for such a fast-grown a-Si:H solar cell.

INTRODUCTION

Several studies [1-3] performed in small research reactors have pointed out that a glow discharge at plasma excitation frequencies higher than the standard industrial frequency at 13.56 MHz leads to an increase in the deposition rate. By scaling up the reactor to industrial large-area dimensions, however, this beneficial effect is accompanied by the inconvenience of a less uniform deposition at frequencies in the Very High Frequency (VHF) range. Recently, this has been explained [4-6] by a deterioration in the RF interelectrode voltage uniformity as the quarter of the free-space wavelength becomes comparable to the reactor dimensions in the VHF-range.

In this work, thickness homogeneity for large-area a-Si:H deposition has been compared with voltage distribution measurements and calculations in the VHF-frequency range (60-120 MHz) not fully explored by the former studies [4-6]. Moreover, it was of interest to fabricate a-Si:H solar cells in this single-chamber industrial large-area PECVD reactor and to establish the first proof of concept for the VHF deposition technique under industrial conditions.

EXPERIMENTAL DETAILS

The principal setup for the VHF-reactor including the VHF-electrode is represented in figure 1. The RF-power delivered from a wide range amplifier is coupled via a matching network into the

Mat. Res. Soc. Symp. Proc. Vol. 557 © 1999 Materials Research Society

plasma. The π filter used here allows an impedance matching in the frequency range of 60 to 120 MHz. The 40x40 cm^2 RF-powered electrode is attacked at the center on the back side. Hence, the logarithmic voltage singularity created by the local RF-source connection [4] is placed as far as possible from the actual plasma zone. The process gases are introduced through a showerhead incorporated within the RF-powered electrode. The residual gases are pumped through slits on the top and bottom of the grounded VHF-electrode. The deposition temperature of the a-Si:H layers and cells was 200°C keeping the pressure at 0.1 mbar and the total feed gas flow at 200 sccm. To diminish powder formation the silane was diluted by hydrogen to a silane concentration of 60% during the i-layer deposition. The interelectrode voltage distribution across the electrode surface was measured in the absence of plasma with a passive RF voltage probe (Philips PM 8932).

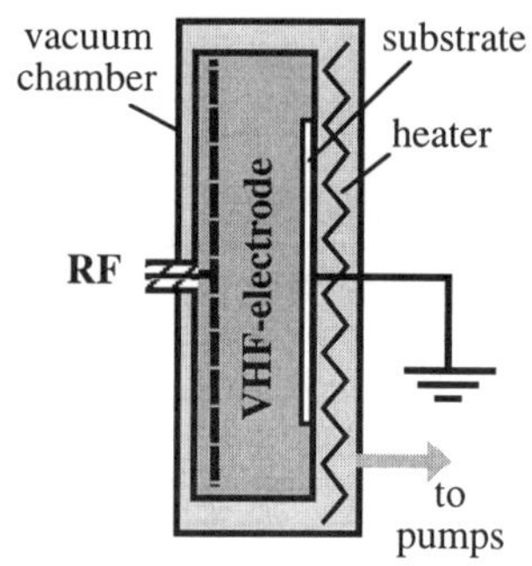

Fig. 1: Schema of the VHF-reactor

The global uniformity of the deposited films was measured by a modified ex-situ interferometry technique as has been discussed in Ref. [7]. Here, instead of the transmitted light the spatial thin-film interference fringes were obtained by monitoring the reflected light. As a rough measure it has been estimated that one fringe (i.e. from one dark region to the next) corresponds to a thickness difference of approximately 80-90 nm.

Independent of the interferometry technique which maps the relative thickness variation, the absolute thickness of the deposited layer was determined at selected spots by a step profiler (Tencor Alpha Step 200).

In this single-chamber reactor p-i-n a-Si:H solar cells were deposited on glass/TCO substrates (Type Asahi U2 based on SnO$_2$:F). The cell deposition was performed at a deposition temperature of 200°C in a continuous process (i.e. there was no plasmastop and no pumpdown between p, i and n layer depositions; these layers were deposited merely by varying the gas composition versus time).

RESULTS AND DISCUSSION

Figure 2 shows the calculated (a) and measured (b) normalized RF voltage distributions across the electrode for 60, 80 and 120 MHz. For a better visibility and due to the symmetry the measured and calculated voltage distribution has been plotted over the full electrode width, but only for half of the depth. The calculation is based on the equations given in Ref.[4]. An excellent agreement of the tendency between the calculation and the voltage measurement was found for all excitation frequencies investigated here. However, the measured voltages at the borders of the electrode drop to slightly lower values than the calculated values.

The voltage distributions determined thereby were compared with a-Si:H films deposited on the bare grounded electrode plate. By directly depositing on the grounded aluminum electrode plate, inhomogeneities caused by the substrate itself, its bad thermal electrical and mechanical [8] contacts etc. can be excluded. The thin-film interferograms in figure 1(c) show the global uniformity of the a-Si:H films deposited at 60, 80 and 120 MHz plasma excitation frequencies having thicknesses between 0.3-0.5 µm in the center of the grounded Al-electrode. The plasma conditions were chosen to limit powder formation, since powder in the plasma zone strongly affects plasma behavior and consequently film homogeneity independently of the voltage distribution [7]. In the 120 MHz case the plasma was not completely powder-free. Therefore, the small amount of powder found at the top side of the electrode close to the exhaust slit is believed to be responsible for the asymmetrical perturbation in the upper half of the plate. Indeed, for a deposition using 20% less RF power, powder formation did not occur and the mentioned asymmetric feature disappeared.

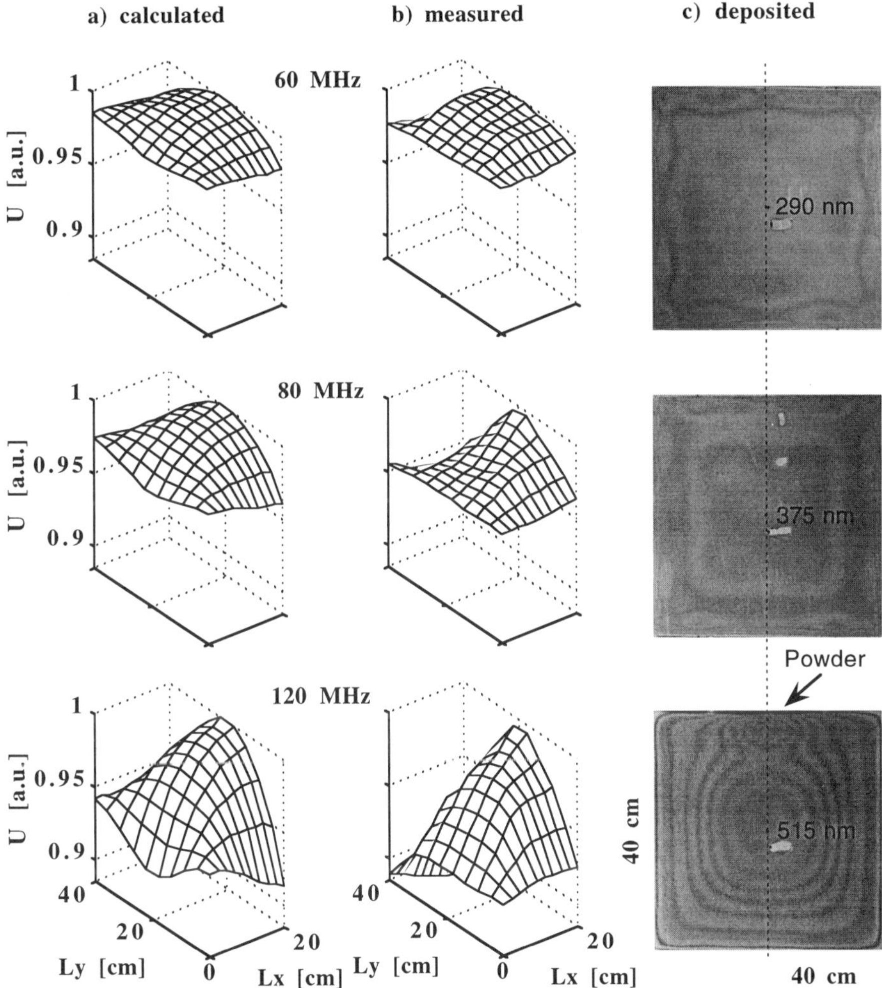

Fig. 2: Calculated (a) and measured (b) normalized interelectrode voltage distribution over the electrode area for different excitation frequencies. Note that for better viewing only half of the electrode depth Lx has been plotted in the graphs;
(c) The interferograms of the a-Si:H layers deposited at the RF power of 100 W on the grounded aluminum electrode plate show the uniformity of the film thickness for each frequency. The numbers represent the absolute film thicknesses in the center of the plate. The vertical dotted line divides the Al-plate corresponding to (a) and (b).

A comparison of the features of the deposited films with the voltage distribution across the electrode indicates that the voltage distribution is the cause of the film inhomogeneity and hence, strongly affects the plasma. In our case other reasons such as source gas depletion, enhanced gas pumping close to the slits etc. may be ignored, since the interferograms are all fourfold symmetric.

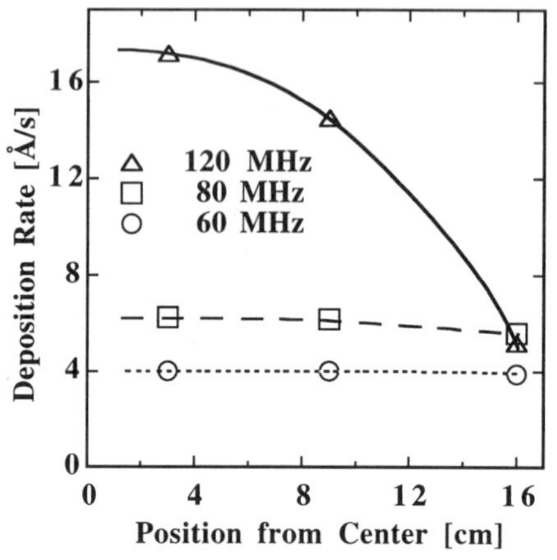

Fig. 3: Deposition rate vs. the position on the aluminum electrode plate for different excitation frequencies.

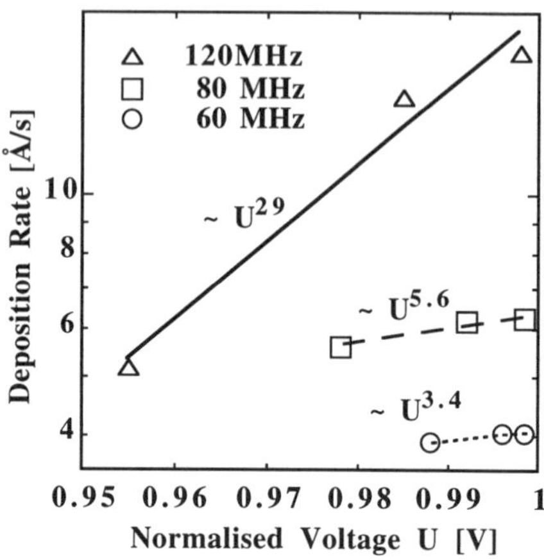

Fig. 4: Logarithm of the deposition rate taken from figure 3 versus the logarithm of the normalized electrode voltage.

Figure 3 shows the deposition rate for selected spots along the indicated dotted line (Fig. 2 (c)). With increasing frequency on one hand an overall increase in the deposition rate has been observed and on the other hand a deterioration of the film uniformity occurs.

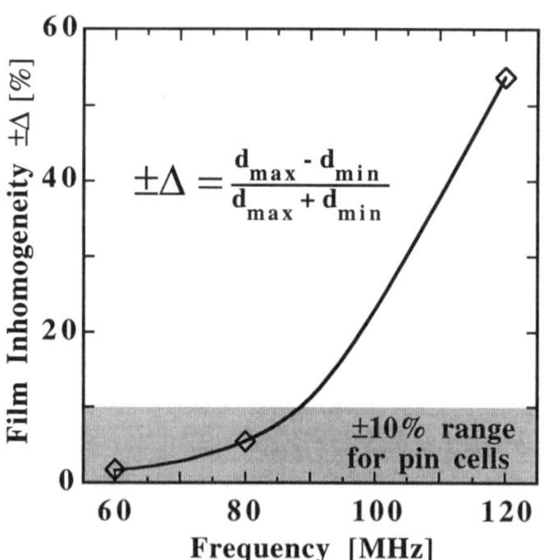

Fig. 5: Film thickness inhomogeneity over a length of 32 cm vs. the plasma excitation frequency.

The power-law relation between the voltage distribution and deposition rate has been represented in figure 4 for the investigated frequencies. In contrast to a formerly found linear relation between the deposition rate and the square of the voltage at 70 MHz [4] assuming an only resistive bulk-plasma wherein the plasma power is dissipated in the present case the exponents are clearly above 2 and moreover, rise with increasing frequency. This needs further investigation: if not experimental, possible physical causes could be inductive plasma currents or circulating currents between the metal electrode and plasma (which would alter the voltage distribution with plasma compared to measurements made in air), or a strong dependence of deposition rate on RF voltage and frequency.

Based on the absolute measured film thicknesses as shown in figure 3 the film inhomogeneity (Fig. 5) has been calculated using the definition shown in the inset of the figure. Assuming necessary homogeneity requirements for single-junction a-Si:H solar cells of maximal ±10 % the plasma excitation frequency of 80 MHz represents a good compromise between the gain in deposition rate and the required film uniformity. At this frequency using hydrogen dilution deposition rates as high as 6-7 Å/s were obtained over a usable substrate size of 30x30 cm^2 with a film uniformity of ±5%.

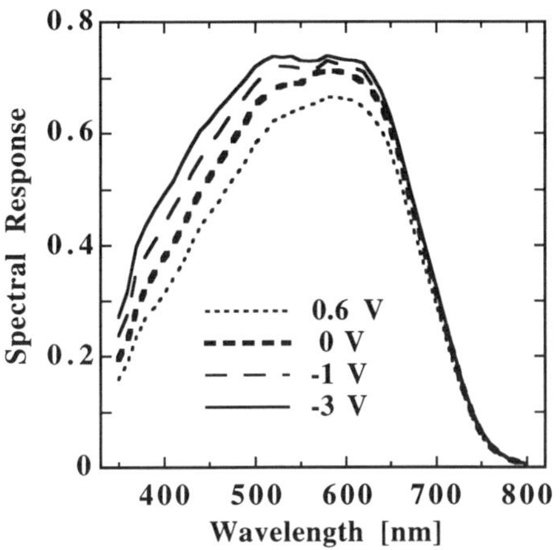

Fig. 6: Spectral response of the pin a-Si:H cell for different bias voltages vs. the wavelength.

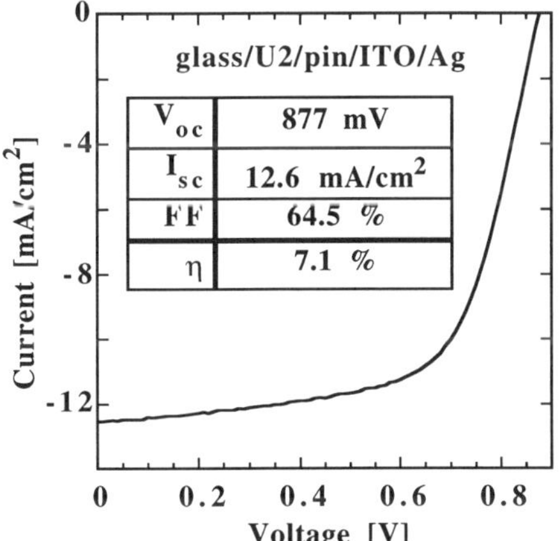

Fig. 7: Current voltage characteristics of a p-i-n a-Si:H cell as deposited under AM 1.5 illumination.

Finally, p-i-n solar cells were deposited on glass/TCO substrates (Ashai type U2) in the described single-chamber reactor. Thereby, after plasma ignition the p, i and n layers were deposited merely by varying the gas composition versus time without interrupting the plasma. Using this continuous process combined with VHF deposition at 80 MHz, 0.4 µm thick single-junction a-Si:H solar cells were fabricated within a total process time of 16 minutes. In comparison, using the standard plasma excitation frequency of 13.56 MHz in a continuous process, typical deposition times for single-junction a-Si:H solar cells in industrial reactors range between 40 min. [9] and 120 min. [10] leading to at least a factor of 2.5 longer process time.

As all layers of the p-i-n solar cell are deposited in the same chamber in a continuous deposition process, control of the boron contamination at the p/i-interface is crucial for cell performance. Boron contamination in the i-layer leads to a reduced electrical field [11] in this layer close to the p-interface and, hence, decreases the p-side blue-light response. Therefore, bias-voltage dependent spectral response measurements (Fig. 6) were carried out to check the cross-contamination behavior of our process in our VHF-reactor. Indeed, a slight enhanced blue light response under reverse voltage-bias can be observed indicating a weak boron contamination of the i-layer at the p-interface, whereas the red light response remains unchanged.

Figure 7 shows the IV-characteristics of a solar cell having an area of 0.25 cm^2. Comparable cell performances have been attained over the whole usable substrate size of 30x30 cm^2. The short-circuit current I_{sc} has been independently determined by the spectral response measurement (Fig. 6) at 0 V. In spite of the slight boron contamination affecting the fill factor and the short circuit current, initial cell efficiencies η of 7.1 % were achieved.

SUMMARY AND CONCLUSIONS

The film uniformity in a single-chamber large-area industrial size reactor with electrode dimensions of 40x40 cm^2 is steadily reduced as the plasma excitation frequency is increased from 60 MHz to 120 MHz. The excellent agreement of the tendency between the measured and calculated 2-dimensional interelectrode voltage distribution and the deposited film uniformity suggests that the voltage inhomogeneity is the main source for the film nonuniformity, and other sources such as gas depletion, residual gas pumping etc. play only a minor role here. However, the quantitative correlation between the voltage distribution and the absolute film uniformity is not yet understood and needs further investigations.

A plasma excitation frequency of 80 MHz was found to be a good compromise between the gain in deposition rate and the film thickness homogeneity as required for a-Si:H solar cells. Under these conditions using hydrogen dilution, high deposition rates of 6-7 Å/s with a film uniformity of ±5% over a usable substrate size of 30x30 cm^2 have been achieved.

Single-junction pin solar cells with initial efficiencies of 7.1 % were fabricated in this single-chamber large-area reactor under industrial production conditions. Using a continuous process (i.e. p, i and n layers deposited merely by varying the gas composition versus time without any interruption of the plasma) in combination with VHF deposition at 80 MHz, 0.4 μm thick single-junction a-Si:H solar cells were deposited within a total time of 16 minutes and showing only a slight boron cross-contamination.

ACKNOWLEDGMENTS

We would like to thank Dr. E. Saurer and J. Meot for their support in the project and Joelle Vuille and Sebastien Dubail for their technical assistance. This work was supported by the Swiss Federal Office of Energy (OFEN) under Research Grant # 19431 and by the EUREKA/Swiss Commission for Technology and Innovation (CTI) under Research Grant # 3522.1/EU 1666.

REFERENCES

1. H. Curtins, N. Wyrsch, M. Favre and A. Shah, Chem. Plasma Processing **7**, 267 (1987).
2. A. Howling, J.-L. Dorier, C. Hollenstein, U. Kroll and F. Finger, J. Vac. Sci. Technol. A **10**, 1080 (1992).
3. M. Heintze, R. Zedlitz and G. H. Bauer, J. Phys. D: Appl. Phys. **26**, 1781 (1993).
4. L. Sansonnens, A. Pletzer, D. Magni, A. A. Howling, C. Hollenstein and J. P. M. Schmitt, Plasma Sources Sci. Technol. **6**, 170 (1997).
5. J. Kuske, U. Stephan, O. Steinke and S. Röhlecke, Proc. of Mat. Res. Soc. Symp. **377** (1995), p. 27.
6. L. Sansonnens, A. A. Howling and C. Hollenstein, Proc. of Mat. Res. Soc. Symp. **507** (1998), p. 541.
7. L. Sansonnens et al., Proc. of 13th EC Photovoltaic Solar Energy Conference 1995, p. 319.
8. H. Meiling, W. G. J. H. van Sark, J. Bezemer, R. E. I. Schropp and W. F. van der Weg, Proc. of 25th PVSC 1996, p. 1153.
9. J. Meot; private communication.
10. R. E. I. Schropp et al., Proc. of 1st WCPEC 1994, p. 626.
11. D. Fischer, Ph. D. thesis, Université de Neuchâtel 1994.

PERFORMANCE OF a-Si-H SOLAR CELLS AT HIGHER GROWTH RATES

G. GANGULY, G. LIN, L.F. CHEN, M. HE, G. WOOD, D. CARLSON, AND R. ARYA
Solarex, A Business Unit of BP/Amoco Solar, 3601 LaGrange Parkway, Toano, VA 23168

ABSTRACT

We have studied the effects of external growth parameters during the deposition of the i-layers of a-Si p-i-n solar cells using dc plasma decomposition of silane-hydrogen mixtures at growth rates of up to 3A/s. The loss of initial performance with increasing growth rate is mainly due to a loss of short-circuit current. The relative degradation of efficiency upon extended light soaking also increases with growth rate, and is mainly due to a decrease in the fill factor. Systematic comparisons of the performance and its degradation with changes in growth conditions reveal that these two components of the total degradation have distinct origins.

INTRODUCTION

Amorphous silicon based solar cell technology has reached the state of commercial large volume production with multi-megawatt factories commissioned by several companies in the US and Japan [1]. This technology promises decrease in cost-per-watt with increasing volume of production. This entails amongst other things, an increase in throughput through the a-Si thin film deposition part of the production process. It has been known for some time that increasing the growth rate of amorphous silicon results in increasing as-grown and light induced defect densities. This is also known to result in the loss of initial and stablized performance of cells using such material. Several techniques have been shown to lessen the loss of performance at higher growth rates like using higher growth temperatures [2], very high frequency [3,4] or microwave decomposition [5]. However, each of these has associated challenges in terms of uniformity and/or reproducibility [6].

Using a bank of knowledge about the plasma gathered through sustained efforts over two decades, preliminary models of the changes in the plasma under conditions required for higher growth rates have highlighted several possible sources of deterioration [7]. The relative importance of these factors can be expected to vary depending on the regime of growth conditions used, including the reactor configuration. In this work we report the effects of changing growth conditions that result in growth rate increases for the i-layer of a a-Si single junction device on its performance and stability.

EXPERIMENT

The single junction cells with the structure *glass/CTO/p/buffer/i/n/ZnO/Al* were fabricated in a double load-lock research reactor using dc plasma decomposition. The remaining parts were held unchanged while the conditions for the i-layer were varied from the standard conditions of pressure (P), hydrogen dilution of silane (D) and plasma power (C). All cells were deposited on Asahi-U type Tin Oxide (SnO_2) coated glass. The cell thicknesses were measured using a profilometer and the deposition rates were calculated after subtracting the combined thickness of the unaltered layers. The nominal i-layer thickness was kept close to 3000A, and any variations are noted. The current voltage (I-V) characteristics were obtained under a calibrated (NREL) solar simulator without masking of the

Mat. Res. Soc. Symp. Proc. Vol. 557 © 1999 Materials Research Society

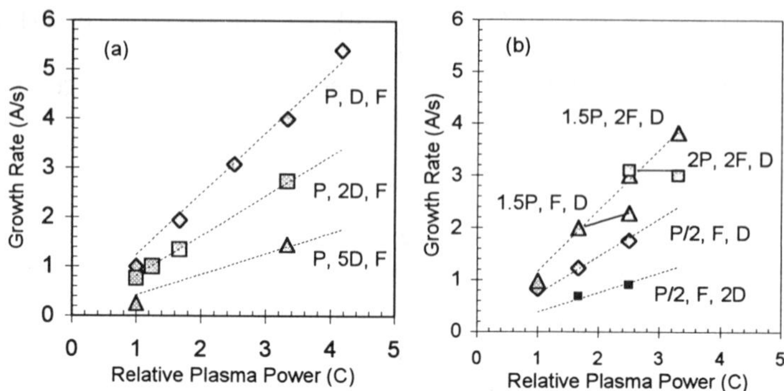

Fig 1 (a) & (b). The growth rate of i-layers plotted as function of the relative plasma current at different relative pressures (P), hydrogen dilutions (D), and flow rates (F). The dashed lines are best fits to data through the origin. The continuous lines indicate saturation due to gas depletion.

0.27 cm^2 devices, but the short circuit current densities (Jsc) were obtained from quantum efficiency measurements.

RESULTS

The growth rate of a-Si:H is expected to vary as the concentration of SiH$_3$ radicals where [7]

$$[SiH_3] = N_{e3} \, v_{e3} \, \sigma_{e3} \, [SiH_4] \, \tau_3 \, / \, (N_{et} \, v_{et} \, \sigma_{et} + \{1/\tau_4\}),$$

where the [ ] denote concentration, Ne, ve and σe are the electron density, velocity and cross section respectively, and τ is the time constant. The subscript 3 denotes SiH$_3$, 4 denotes SiH$_4$ and t denotes total. The data in Fig. 1 are consistent with the expression above in that the dashed lines show that the growth rate increases with plasma power ($N_{e3} \, v_{e3} \, \sigma_{e3}$) when the power level is low ($N_{et} \, v_{et} \, \sigma_{et} \ll \{1/\tau_4\}$). Also, the slopes of the dashed lines decrease when the silane partial pressure ($[SiH_4]$) decreases through lower P or higher D. Finally, we see that when residence times are large at higher P and lower F such that ($N_{et} \, v_{et} \, \sigma_{et} \gg \{1/\tau_4\}$), the growth rates saturate as shown by the continuous lines in Fig. 1(b).

It has been suggested that three potential causes for deterioration of performance with increasing plasma power are higher silane related radicals (Si_nH_{2n+1}, n>1), short-lifetime radicals (SiH_x, x=1,2), and ion-bombardment energies [8]. While discussing the results we will evaluate the possible contributions of these three factors.

The initial efficiencies of single junction devices are shown as function of i-layer growth rates in Fig. 2. It is clear that increasing the growth rate reduces the initial efficiency of the devices. Furthermore, increasing hydrogen dilution and/or decreasing the pressure leads to a faster decrease of the efficiency with increasing growth rate. One common factor between these two variables is that they decrease the partial pressure of silane. This would imply an increase of the contribution to growth of short-lifetime radicals relative to SiH$_3$ and/or an increase of ion-bombardment energies. We find that the decrease in initial efficiency with increasing growth rate is mainly due to decrease of short-circuit current (Jsc). In Fig. 3(a) we

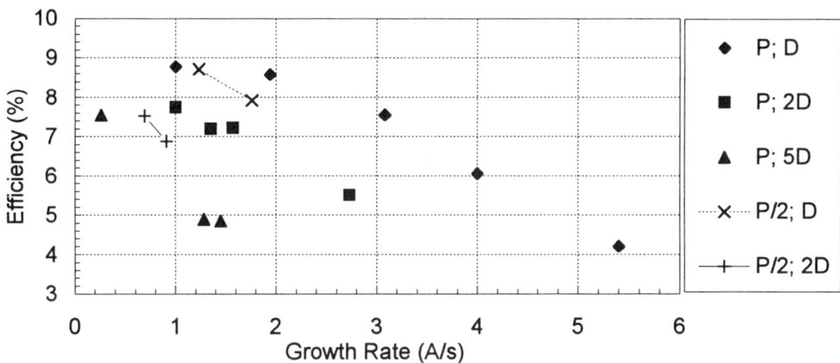

Fig. 2 The efficiency of p-i-n cells with i-layers prepared at different growth rates using a range of pressures (P) and hydrogen dilutions (D).

show the changes of the quantum efficiency with increasing plasma power. The initial decrease appears to be in the blue and is followed by an overall decrease.

In Fig. 3 (b) we have plotted the decrease of the carrier collection (ratio of Jsc to the current when a reverse bias of 1V is applied to the device). It is clear that the current collection drops off with increasing plasma power. Further it would appear from Fig. 3(b) that the drop off is more rapid at higher dilution. This would imply that either increasing contribution of short-lifetime radicals and/or ion-bombardment is responsible for at least part of the decrease in current and hence initial efficiency with increasing plasma power. Fig. 3(b) also shows that higher pressure appears to shift the drop off in current collection to higher growth rates. Increasing pressure increases silane partial pressure and hence leads in a direction of reduced contribution of short-lifetime radicals and/or ion-bombardment energies. Fig. 4 (a) shows that

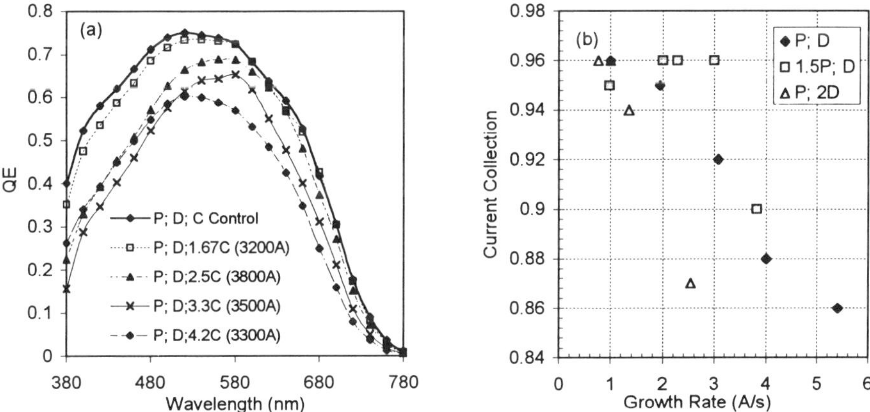

Fig. 3 (a) The external quantum efficiency plots of p-i-n devices with i-layers made using different relative plasma power; (b) the current collection (ratio of current under short circuit and 1V reverse bias) for cells made at different pressures and hydrogen dilutions as function of the i-layer growth rates. The i-layer thicknesses are indicated in parentheses)

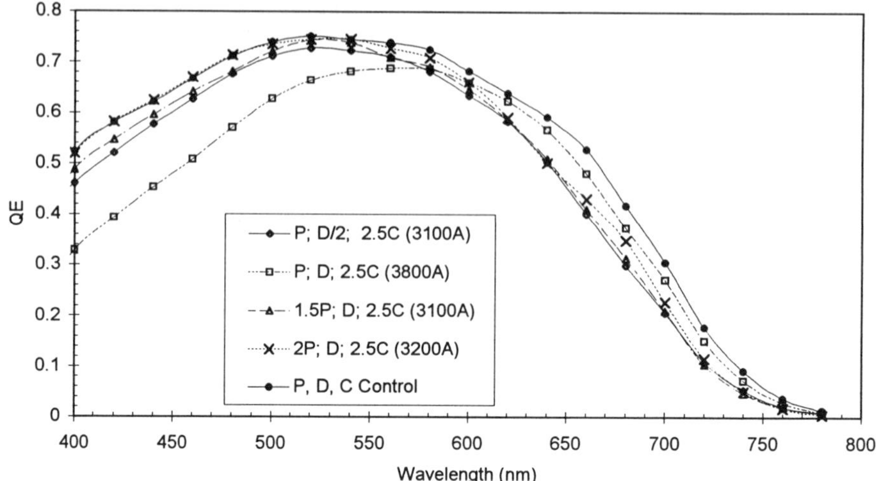

Fig. 4. The external quantum efficiencies of p-i-n devices with i-layers made at different relative pressures and at a lower dilution at growth rates of 3A/s. The data for a cell with control (1A/s) i-layer is also shown for comparison. The i-layer thicknesses are indicated in parentheses.

indeed when the pressure is increased at constant plasma power (yielding a growth rate of about 3A/s), the blue portion of the QE spectrum improves. This figure also shows that increasing the silane partial pressure at constant total pressure by reducing the dilution leads to similar though lesser improvement. Reducing dilution does reduce the contribution of short-lifetime radicals similar to increased total pressure, but it does not reduce the ion-bombardment energies. Combining these pieces of information would suggest that the losses in initial performance with increasing growth rates occur through a combination of increased contribution of short-lifetime radicals to growth and increased ion-bombardment energies, both

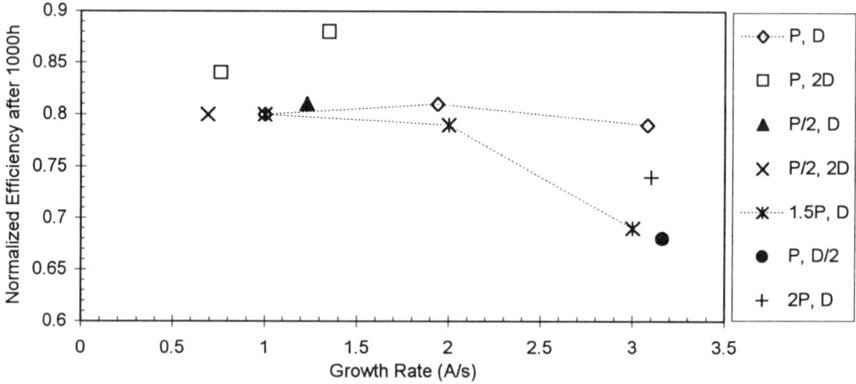

Fig. 5. The relative degradation due to 1000h of ~100mW/cm^2 light soaking of p-i-n cells with i-layers made under different conditions.

of which result in the formation of additional defects during film growth.

The relative degradation of cells made under different conditions is summarized in Fig. 5. One notices that the relative degradation does *not* change significantly under certain conditions from growth rates of 1 to 3 A/s. In the 1 A/s growth rate regime, the relative degradation is sensitive only to dilution, as is well known [9]. In the 3 A/s growth rate regime, the relative degradation becomes sensitive to pressure. This is consistent with an increased contribution of higher silane related radicals at higher growth rates. It occurs because the formation rate of higher silane species is strongly dependent on the plasma power and the discharge pressure [7,8]. The existing inter-silicon bonds in such multi-silicon radicals can be easily imagined to result in poorly interconnected network structure that contributes additional light-induced degradation sites.

In Fig 6, we show the stable efficiencies of cells made at 3 A/s and a control cell made at 1 A/s. We see that increasing the pressure has a significant effect on the stable efficiencies at a growth rate of 3 A/s. However, increasing the silane partial pressure by decreasing the dilution does not have a net effect on the stable efficiency. We see from the figure that at 3 A/s the best cell efficiency is 92% of that at 1 A/s.

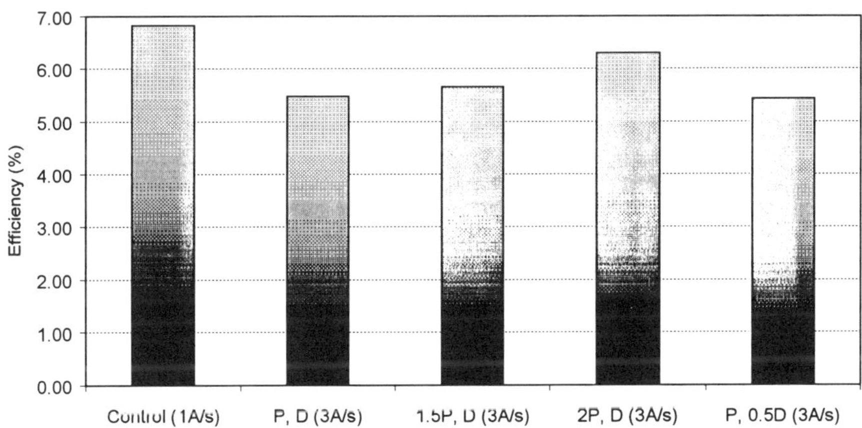

Fig. 6. The stable efficiency of p-i-n cells with i-layers made at a growth rate of 3 A/s using different pressure and hydrogen dilution conditions. A control cell made at 1 A/s is shown for comparison.

CONCLUSIONS

The decrease of initial efficiency of a-Si p-i-n cells with increasing i-layer growth rate is found to be due mainly to a loss of collected current. The loss of current is found to depend on the growth conditions in a way that suggests the short-lifetime radicals and increased ion-bombardment energies both contribute to the formation of poorer material at higher growth rates. On the other hand, the relative light-induced degradation of these cells depend differently on growth conditions which points to higher-silane related radicals being responsible for the additional instability at higher growth rates. The systematic search for optimum growth conditions allows retention of 92% of stable power while tripling the throughput of the rate-limiting i-layer.

ACKNOWLEDGMENTS

This work was partially funded by NREL under Subcontract ZAK-8-17619-02. We would like to thank Prof. C.R Wronski and Prof. R.W Collins and their students at Penn State University for freely sharing many details of their pioneering work. Lively discussions at the NREL/ NCPV Review Meetings have stimulated us to exploring several new approaches.

REFERENCES

1. C.R. Wronski, First WCPEC, Hawaii, Conf. Rec. of the 24th IEEE PVSC, 373 (1994).
2. G. Ganguly and A. Matsuda, Mater. Res. Soc. Symp. Proc. **258**, (MRS, Pittsburg, PA 1992), p. 39.
3. H. Chatham and P.K. Bhat, Mater. Res. Soc. Symp. Proc.**149**, (MRS, Pittsburg, PA 1989), p. 464.
4. K. Prasad, Ph. D. thesis, Institut de Microtechnique de l'Universite de Neuchatel, Neuchatel, 1991.
5. K. Saito, M. Sano, K. Ogawa and I. Kajita, J. Non-Cryst. Solids **164-166**, 689 (1993); K. Saito, M. Sano, J. Matsuyama, M. Hishikawa, K. Ogawa and I. Kajita, Tech. Digest of PVSEC-9, 579 (1996).
6. H. Meiling, W.G.J.H.M. van Sark, J. Bezemer, W.F. van der Weg, J. Appl. Phys. **80**, 3546 (1996).
7. A. Matsuda, Conf. Rec. of the 25th IEEE PVSC, 1029 (1996).
8. G. Ganguly and A. Matsuda, Tech. Digest of PVSEC-9, (1996)
9. R. Arya, A. Catalano, M. Bennett, J. Newton, B. Fieselman, L. Yang, Y.M. Li and R. D'Aiello, Proc. 11th EC PVSEC,199 (1992); M. Bennett, K. Rajan and K. Kritikson, Conf. Rec. of the 23rd IEEE PVSC, 845 (1993).

PREPARATION OF TRIPLE-JUNCTION A-Si:H NIP BASED SOLAR CELLS AT DEPOSITION RATES OF 10 Å/s USING A VERY HIGH FREQUENCY TECHNIQUE

S.J. Jones, X. Deng, T. Liu and M. Izu
Energy Conversion Devices, Inc., Troy, MI 48084

ABSTRACT

In an effort to find an alternative deposition method to the standard low deposition rate 13.56 MHz PECVD technique, the feasibility of using a 70 MHz rf plasma frequency to prepare a-Si:H based i-layer materials at high rates for nip based triple-junction solar cells has been tested. As a prelude to multi-junction cell fabrication, the deposition conditions used to make single-junction a-Si:H and a-SiGe:H cells using this Very High Frequency (VHF) method have been varied to optimize the material quality and the cell efficiencies. It was found that the efficiencies and the light stability for both a-Si:H and a-SiGe:H single-junction cells remain relatively constant as the i-layer deposition rate is varied from 1 to 10 Å/s. Also these stable efficiencies are similar to those for cells made at low deposition rates (1 Å/s) using the standard 13.56 MHz PECVD technique and the same deposition equipment. Using the knowledge obtained in the fabrication of the single-junction devices, a-Si:H/a-SiGe:H/a-SiGe:H triple-junction solar cells have been fabricated with all of the i-layers prepared using the VHF technique and deposition rates near 10 Å/s. Thin doped layers for these devices were prepared using the standard 13.56 MHz rf frequency and deposition rates near 1 Å/s. Pre-light soaked efficiencies of greater than 10% have been obtained for these cells prepared at the high rates. In addition, after 600 hrs. of light soaking under white light conditions, the cell efficiencies degraded by only 10-13%, values similar to the degree of degradation for high efficiency triple-junction cells made by the standard 13.56 MHz method using i-layer deposition rates near 1 Å/s. Thus, use of this VHF method in the production of large area a-Si:H based multi-junction solar modules will allow for higher i-layer deposition rates, higher module throughput and reduced module cost.

INTRODUCTION

The benefit of the a-Si:H-based manufacturing processes over other photovoltaic technologies in terms of low module cost will only be <u>fully</u> achieved when the scale of production is increased significantly over the present rates (5-10 MW/yr.). In order to achieve production above 25MW/yr. in an economically feasible manner, higher deposition rates for the a-Si(Ge):H i-layers over today's 1-3 Å/s rates need to be achieved. Extensive attempts to increase these rates using the standard PECVD process with a 13.56 MHz rf frequency signal has led to enhanced powder and polyhydride formation in the plasma and poorer stable cell efficiencies. At present, the outlook for achieving high deposition rates (5-10 Å/s) calls for the development of an alternative thin film deposition technique.

Of the alternative high deposition techniques tested, the Very High Frequency (VHF) PECVD technique is presently one of the most promising. Several research groups [1-4] have reported higher efficiencies for cells made using frequencies of 70-100 MHz at i-layer deposition rates between 5 and 10 Å/s than for those prepared at the same rates using the rf deposition technique. Also, Chatham et. al. [1] and Shah et. al.[2] found that the deposition rate of a-Si:H films could be increased to 10-15 Å/s without an observable deterioration in the film properties.

In order to incorporate a new high deposition rate technique into ECD's roll-to-roll manufacturing process [5], we have chosen to test the feasibility of using the VHF technique to

Mat. Res. Soc. Symp. Proc. Vol. 557 © 1999 Materials Research Society

prepare small area high efficiency a-Si:H/a-SiGe:H/a-SiGe:H triple-junction solar cells with the ultimate goal of preparing high efficiency large area multi-junction solar modules. The VHF technique is desirable because of the ability to apply the technique to ECD's roll-to-roll process with relatively little hardware changes. Here, we discuss the results from the feasibility study for cells prepared in R&D size deposition systems using high i-layer deposition rates. In this study, we have focused on preparing cells with i-layer deposition rates near 10 Å/s in order attempt to achieve an order of magnitude increase in the deposition rate over the state-of-the-art cells.

EXPERIMENT

The a-Si:H nip/a-SiGe:H nip/a-SiGe:H nip triple-junction solar cell structures were fabricated using a research scale, multi-chamber load locked deposition system. Stainless steel substrates coated with current enhancing Ag/ZnO back reflectors were used as the substrates for preparation of the semiconductor structures. The Ag/ZnO back reflectors were prepared in a roll-to-roll manufacturing machine using a DC sputtering technique. Both the thin doped layers and the thin a-Si:H buffer layers, grown between the VHF deposited a-SiGe:H i-layers and the doped layers, were prepared using the conventional PECVD process in which a 13.56 MHz rf signal is used. To fabricate the a-Si:H and a-SiGe:H i-layers, a fixed VHF frequency of 70 MHz was used. To improve the cell properties, buffer layer and i-layer deposition conditions were altered including substrate temperature, hydrogen dilution, active gas flows, chamber pressure and applied power.

After fabrication of the triple-junction structure, the devices were completed by depositing Indium Tin Oxide (ITO) conductive layers and then Aluminum collection grids. Both the ITO and Al layers were prepared using standard evaporation techniques.

To characterize the cells, standard IV and spectral response (quantum efficiency) measurements were made. For optimization of the a-Si:H cells, standard white AM1.5 light was used to obtain the IV data. For optimization of the a-SiGe:H cells for the green-red light absorbing middle and bottom structures, the AM1.5 light for the IV measurements was filtered using a 530 nm cutoff filter to simulate the absorption due to a top a-Si:H cell for the middle cells and a 630 nm cutoff filter for the bottom cells. During optimization of the component cells, top and middle single-junction cells were fabricated without Ag/ZnO back reflectors while the bottom nip cells were. This was done because in the triple-junction structure, the bottom cell reaps almost all of the benefit of the back reflector layers. To complete light soaking studies, the cells were subjected to 600-1000 hrs. of one sun light with the cell temperature fixed at 50°C. The i-layer thicknesses were determined using capacitance techniques.

RESULTS AND DISCUSSION

Prior to discussing results for the triple-junction cells, data for component a-Si:H and a-SiGe:H cells whose i-layers were prepared at deposition rates of 10 Å/s will be reviewed.

a-Si:H and a-SiGe:H Single-Junction Cells

We have earlier reported results for a-Si:H nip single-junction cells whose i-layers were prepared using the VHF technique at 10 Å/s. The general conclusion from these studies was that by using the VHF technique and careful selection of deposition conditions, the initial and stable cell efficiencies could be made to remain relatively constant with varying i-layer deposition rate up to 10 Å/s for cells prepared without current enhancing Ag/ZnO back reflectors [3]. The cells had i-layer thicknesses near 2300 Å, short circuit currents (J_{sc}) values near 10 mA/cm^2 and stable cell efficiencies near 5.5%. Similar efficiencies were obtained for cells of similar thickness prepared in the same deposition system using the standard rf PECVD technique with a 13.56 MHz frequency and i-layer deposition rates near 1 Å/s. For a-Si:H cells with Ag/ZnO back reflectors, average initial cell efficiencies of 10.3% have been achieved with these cells with i-layer thicknesses of 3000 Å. An IV curve for such a cell is shown in Figure 1. After 100 hrs. of light soaking in white light, these cell efficiencies degrade by 20%, typical degradation percentages for cells of a similar i-layer thickness prepared by the standard rf PECVD technique at i-layer deposition rates of 1 Å/s.

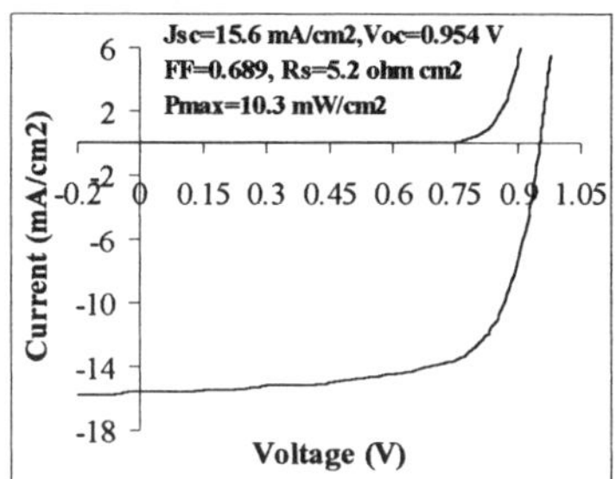

Figure 1. IV plot for a-Si:H cell whose i-layer was prepared at 10 Å/s using VHF technique.

Table I compares data for a-SiGe:H cells prepared with the same deposition equipment using the VHF technique with those made using the standard 13.56 MHz. These cells were also prepared during the same period of time. To judge the potential use of the VHF technique as a method to prepare middle and/or bottom cells for triple-junction structures, we compare in the table IV data obtained using 530 nm filtered AM1.5 light. Again, the 530 nm filter eliminates light typically absorbed by the a-Si:H top cell. Comparing cells with similar J_{sc} values and i-layer thickness, a-SiGe:H cells made by the VHF technique also have similar initial efficiencies to those prepared in the same deposition system using the 13.56 MHz.

The high rate VHF produced a-SiGe:H cells and low rate 13.56MHz produced cells also have similar efficiencies after light soaking as can be seen from the data in Table 2. Cells with similar J_{sc} values have similar stable efficiencies. Some interesting results are the properties for the cell made by the VHF method at a rate of 2 Å/s. The stable efficiencies and FF are actually higher for devices made under these conditions than those made by the 13.56 MHz method at low rates. We have recently improved the efficiencies for the VHF prepared cells and are now achieving higher efficiencies for the a-SiGe:H devices. IV plots for these improved cells prior to light soaking are shown in Figures 2 and 3. Again these data were obtained using AM1.5 light filtered with the 530nm low-bandpass filter.

Table 1.

Data for VHF a-SiGe:H cells prior to light soaking.

Freq. (MHz)	Dep. Rate (Å/s)	i-layer thickness (Å)	J_{sc} (mA/cm^2)	V_{oc} (V)	FF	R_s (Ωcm^2)	Initial P_{max} (mW/cm^2)
13.56	0.95	2400	8.50	0.751	0.552	12.3	3.51
13.56	0.92	2100	7.93	0.752	0.575	11.8	3.42
13.56	6.1	2260	7.43	0.711	0.550	14.4	2.91
70	9.5	2000	8.07	0.760	0.579	10.7	3.55
70	10.2	2150	8.15	0.750	0.571	12.1	3.49

Table 2.

Data for a-SiGe:H cells after light soaking for 600 hrs. under white light conditions.

Freq. (MHz)	Dep.Rate (Å/s)	J_{sc} (mA/cm^2)	V_{oc} (V)	FF	P_{max} (mW/cm^2)
13.56	0.5	6.93	0.703	0.517	2.52
13.56	0.5	7.02	0.704	0.510	2.52
13.56	0.79	6.97	0.703	0.518	2.54
70	2.0	7.06	0.699	0.543	2.68
70	10.0	7.08	0.675	0.527	2.52
70	9.3	6.92	0.686	0.521	2.47

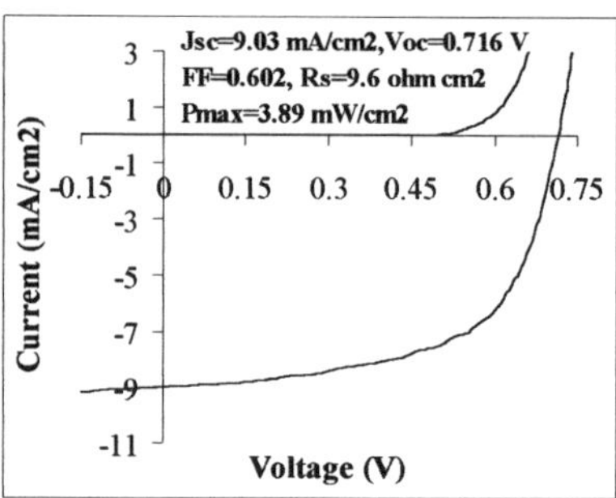

Figure 2. IV curve for a-SiGe:H cell made using VHF technique.

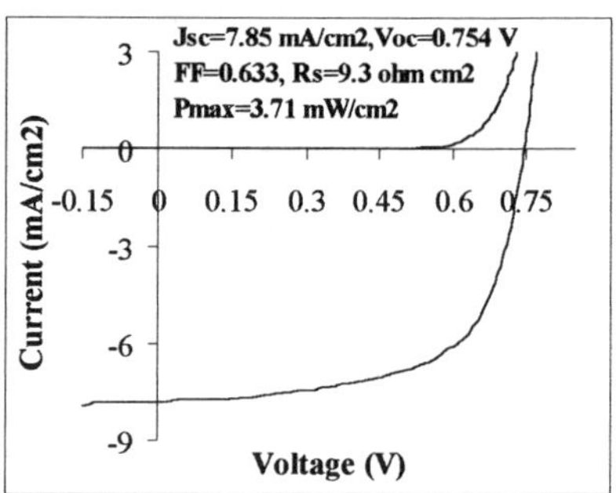

Figure 3. IV curve for a-SiGe:H cell made using VHF technique.

<u>a-Si:H/a-SiGe:H/a-SiGe:H Triple-Junction Cells</u>

Using the optimized conditions obtained from the fabrication of the single-junction structures, a-Si:H/a-SiGe:H/a-SiGe:H triple-junction solar cell devices have been fabricated using the VHF technique to deposit all of the i-layers at rates near 10 Å/s while the doped layers were prepared using the 13.56 MHz technique at rates near 1 Å/s. Using this method, devices with initial active area efficiencies between 9.5 and 10% and total area efficiencies between 9.0 and 9.5% were made. A selection of these cells were light soaked for 600 hrs. under white light conditions and the efficiencies were found to degrade by 10-13%.

However, the uniformity of the depositions across the substrate platform with the optimized conditions initially limited our ability to achieve higher efficiencies for the triple-junction structures. Specifically, under the optimized conditions, there were sharp gradients in thickness, deposition rate and Ge content across the 4" x 4" substrate area. Figure 4 displays the variation in deposition rates across the substrate for the a-Si:H i-layer. The rate was highest near the Gas Inlet, tapering off to the lowest values at the Gas Outlet. This variation in rate was not due to gas depletion since the non-uniform deposits persisted even at very high gas flows. The variations in i-layer thickness and deposition rate led to variations in the J_{sc} and FF across the substrate. The variations in Ge content led to gradients in the V_{oc} across the platform as is shown in Figure 5 for and a-SiGe:H bottom cell. For the triple-junction structure, the variation in i-layer thickness and bandgap made optimization of the current matching between the three component cells difficult.

136

Gas Inlet

10.8		11.9	11.9		11.6		10.0
10.7	11.3	11.4	11.2	11.1	11.1	10.7	9.8
10.0	10.3	10.4	10.3		10.0	9.7	
8.7				9.1	8.9	8.7	8.1
7.5	8.0	8.2	8.2	8.1			
	7.7	7.9		7.8	7.7		6.9
	7.5	7.7		7.6	7.6	7.3	6.8
6.9				7.2	6.7	7.0	6.5

Gas Outlet (Pump Out)

Figure 4. Deposition rate across substrate platform for a-Si:H cell.

Gas Inlet

		525	508		518	548	581
590		536	522		527	548	581
607		563	553	549		570	596
		590		586	588	597	
661	647	639	636	649			
671	658	652	647	657	657		681
678	669			665		676	687
691	681	678	677	679	684	689	697

Gas Outlet (Pump Out)

Figure 5. V_{oc} (mV) across substrate platform for a-SiGe:H cell.

In order to improve the uniformity, a wide range of parameter space was scanned to study each of the parameter's effect on the deposition profile across the substrate. Variation of the chamber pressure was found to have the greatest effect on the deposition profile. Only at very low pressures was it possible to obtain not only uniform profiles but also deposition rates near 10 Å/s. At these low pressures, variation of other deposition parameters was possible to alter the deposition rate and optimize the i-layer quality without altering the deposition profile greatly. Figure 6 displays the uniform deposition rate for the a-Si:H i-layer at the low pressure conditions while Figure 7 shows the improvement in V_{oc} uniformity for the a-SiGe:H bottom cell.

Gas Inlet

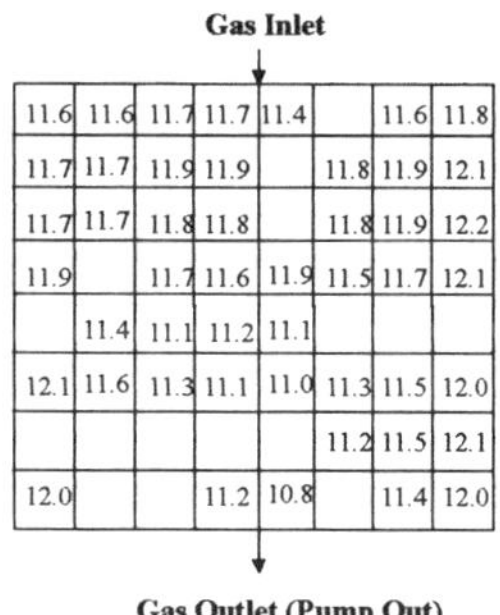

11.6	11.6	11.7	11.7	11.4		11.6	11.8
11.7	11.7	11.9	11.9		11.8	11.9	12.1
11.7	11.7	11.8	11.8		11.8	11.9	12.2
11.9		11.7	11.6	11.9	11.5	11.7	12.1
	11.4	11.1	11.2	11.1			
12.1	11.6	11.3	11.1	11.0	11.3	11.5	12.0
					11.2	11.5	12.1
12.0			11.2	10.8		11.4	12.0

Gas Outlet (Pump Out)

Figure 6. Deposition rate across substrate platform for a-Si:H cell.

Gas Inlet

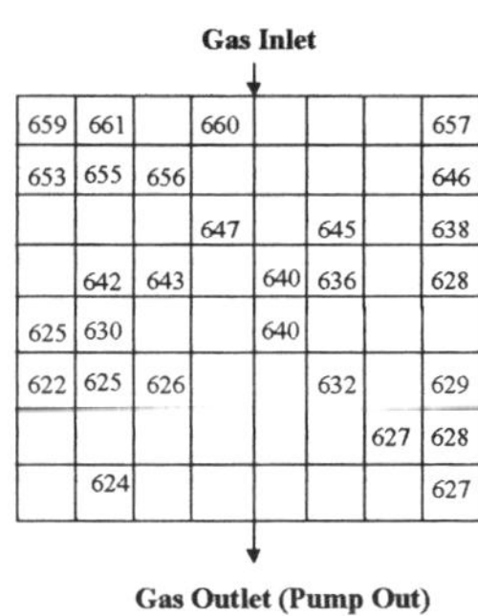

659	661		660				657
653	655	656					646
			647		645		638
	642	643		640	636		628
625	630			640			
622	625	626			632		629
						627	628
	624						627

Gas Outlet (Pump Out)

Figure 7. V_{oc} (mV) across substrate platform for a-SiGe:H cell.

Incorporating the new deposition conditions that gave more uniform film deposits led to the achievement of initial triple-cell active area efficiencies of 10.4 -10.5%. A representative IV curves for such devices are shown in Figures 8 and 9. The current matching for the cells have yet to be fully optimized, thus further improvement in the efficiencies should come with proper matching. While these cells have yet to be sufficiently light soaked, considering the similarity of the i-layer thicknesses for these cells and those of the 9.5-10% cells, it is likely that these cells will degrade by 10-13% after extended light soaking periods. Further optimization of the deposition conditions for these triple-junction devices should lead to higher stable efficiencies.

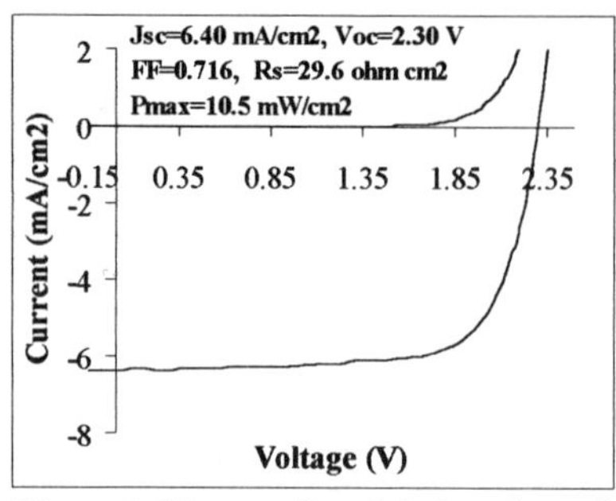

Figure 8. IV curve for triple-junction cell.

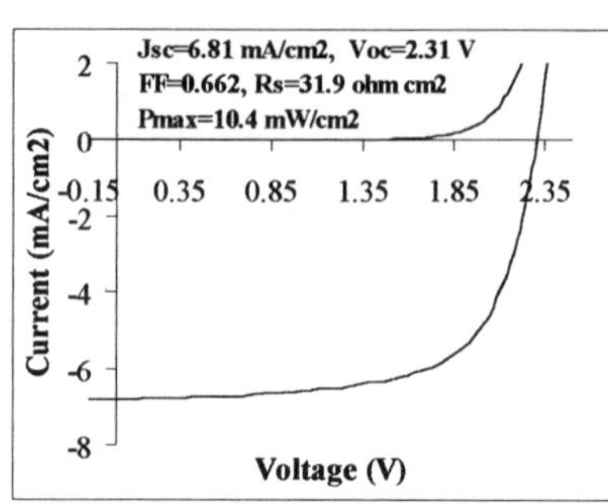

Figure 9. IV curve for triple-junction cell.

DISCUSSION AND CONCLUSIONS

The 10.4 - 10.5% efficiencies obtained for the triple-junction cells are impressive when one considers that 1) all of the i-layers were prepared at deposition rates near 10 Å/s, and 2) that triple-junction cells made in the same deposition system using i-layer deposition rates near 1 Å/s have pre-light soaked efficiencies just above 11%. With further improvement of the current matching of the high rate, VHF produced cells and the baseline efficiencies of our research reactor, higher efficiencies beyond 10.5% should be obtainable.

In terms of the VHF technique for high rate deposition in a large-area production line, proper selection of deposition hardware and the hardware geometry will be critical for obtaining the required deposition uniformity. Without good thin film uniformity, obtaining the current matching needed for high performance modules will not be possible. This does not seem to be an insurmountable problem since some research on cathode hardware for the application of the VHF technology has already been done with some reasonable solutions [6]. Since we have obtained uniform deposits using the VHF technique and low deposition rates (2 Å/s), the non-uniformities we have outlined are primarily due to the high deposition rates we are using. Thus, solutions to this problem should address the general issue of application of high growth rate techniques.

ACKNOWLEDGEMENTS

We would like to thank Dr. H. Fritzsche and Mr. S. R. Ovshinsky for important discussions that helped further advance this work. We would also like to acknowledge B. Viers for her efforts in completing ITO and Al depositions. The work was funded by NREL under the Thin Film Partnership contracts ZAN-4-13318-11 and ZAK-8-17619-18.

REFERENCES

1. H. Chatam and P.K. Bhat, Mat. Res. Soc. Proc. 149, 447 (1989).
2. A. Shah, J. Dutta, N. Wyrsch, K. Prasad, H. Curtins, F. Finger, A. Howling, and Ch. Hollenstein, Mat. Res. Soc. Proc. 258, 15 (1992).
3. S.J. Jones, X. Deng, T. Liu, and M. Izu, Mat. Res. Soc. Proc. 507, 113 (1998).
4. J. Yang, S. Sugiyama, S. Guha, Mat. Res. Soc. Proc. 507, 157 (1998).
5. M. Izu, X. Deng, A. Krisko, K. Whelan, R. Young, H.C. Ovshinsky, K.L. Narasimhan and S.R. Ovshinsky, Proc. 23rd IEEE PV Spec. Conf., p. 919 (1993).
6. L. Sansonnens, A.A. Howling, Ch. Hollenstein, Mat. Res. Soc. Proc. 507, 541 (1998).

HIGH QUALITY a-Si:H FILMS GROWN AT HIGH DEPOSITION RATES

YORAM LUBIANIKER[1], YANYANG TAN[1], J. DAVID COHEN[1] and GAUTAM GANGULY[2,3]
[1]Department of Physics, University of Oregon, Eugene, OR 97403
[2]Electrotechnical Laboratory, Tsukuba-shi, Ibaraki 305, Japan.
[3]Solarex, Toano VA 23168

ABSTRACT

Intrinsic a-Si:H samples were grown with and without hydrogen (H_2) dilution of silane at different growth rates. We find that the dilution leads to a considerable reduction in the defect density, in particular at high growth rates. The defect density is particularly low for samples grown using H_2 dilution conditions at growth rates as high as 10 Å/sec. Using transient photocapacitance measurements we find evidence for a small concentration of *microcrystallites* embedded in the amorphous films. An increase in the microcrystalline fraction correlates with a decrease in the defect density.

INTRODUCTION

Considerable effort has been devoted to obtain hydrogenated amorphous silicon (a-Si:H) films that are grown at high deposition rates but contain a low defect density (before and after light soaking) [1]. Usually, cells deposited by plasma decomposition of pure silane are grown at rates of 3-5 Å/sec. It has been known for some time that hydrogen dilution of silane results in films with better stability [2,3]. A variety of deposition techniques have been utilized to increase the growth rate while maintaining the stabilized cell efficiency [4,5].

In this paper we report measurements of two series of intrinsic a-Si:H films that were deposited with and without hydrogen dilution. The growth rate was changed by altering the radio frequency (rf) power. We show that samples grown from H_2 diluted silane at a relatively high growth rate exhibit a remarkably low defect density. Subband gap absorption spectra reveal that these amorphous materials contain a small fraction of microcrystallites embedded in the amorphous matrix. The correlation between the superior electronic properties and the microcrystalline fraction agrees with recent studies [6-9], and suggests that the most suitable materials for solar cells applications are those that are grown near the amorphous / microcrystalline boundary.

EXPERIMENTAL

Intrinsic a-Si:H films were grown using rf plasma decomposition of silane at a substrate temperature of 250°C. One set of samples was deposited from pure silane at a pressure of 20 mTorr with a gas flow of 30 sccm. The rf power was varied (in the range of 10 to 100 Watts), thus affecting the growth rate. Another set of samples was deposited from hydrogen diluted silane in 4:1 ratio. The silane flow was kept the same as in the first set, so that the total pressure in the chamber was 100 mTorr. In this case the H_2 dilution results in only small decrease of growth rate. For comparison we also grew a non-diluted sample at the higher pressure. The growth conditions are summarized in Table I. More details on the growth system and conditions and more information about these samples are given in ref. 10.

In all cases the samples were co-deposited on p^+ c-Si and quartz, for capacitance and electron spin resonance (ESR) studies, respectively. In addition, a matching set of films, 7000 to 8000Å thick, were grown on n^+ c-Si, and coated with semitransparent (50%) Ni contacts to form Schottky diode solar cells. The fill factors (FF) of these relatively thick devices are expected to be more representative of the bulk material. The FFs were recorded in the annealed state and in the light soaked state (obtained by 6 hours of light soaking using 3 suns at 60°C).

For the capacitance measurements partially transparent Pd dots were evaporated on top of each film. Using admittance spectroscopy [11] we determined the thickness of the films and the Fermi energy position. Defect densities were obtained through the drive-level capacitance profiling (DLCP) method [12]. Subband gap optical absorption spectra were measured using the transient photocapacitance (TPC) technique [13]. In addition, we carried out ESR measurements for the films co-deposited on quartz. All these measurements were first carried out after the samples were annealed in the dark at 220°C for 30 minutes ("State A")). The samples were then light soaked for 100 hours using a red filtered tungsten - halogen lamp with an intensity of 2.2 W/cm^2. During light soaking the samples were immersed in methanol, thus ensuring a low ($\leq$ 65°C) surface temperature. We then repeated all the capacitance measurements in the light degraded state ("state B").

sample	H_2 flow (sccm)	pressure (mTorr)	rf power (Watt)	thickness (μm)	growth rate (Å/sec)	N_d (A) 10^{15} (cm^{-3})	ESR (A) 10^{15} (cm^{-3})	N_d (B) 10^{15} (cm^{-3})
12464	120	100	100	2.1	15.22	1.25	4.6	9.0
12465	none	20	100	2.33	16.2	3.25	28.0	22.5
12466	120	100	60	1.9	10.55	0.55	5.1	4.75
12467	none	20	60	1.6	10.3	1.35	9.2	12.5
12468	120	100	20	1.64	2.5	0.75	-	7.0
12469	none	20	10	1.69	2.8	0.8	2.8	6.1
12451	none	100	20	2.13	14.9	1.45	4.4	9.5

TABLE I. The growth parameters and characteristics of the samples. In all cases the silane flow was 30 sccm. d is the film thickness, and N_d is the defect density as determined from the DLCP method. (A) and (B) symbolize the annealed and degraded states, respectively. The DLCP data was recorded at 360K at a frequency of 11 Hz. Note that unlike previous studies the defect densities given in the table (as well as the figures) are not doubled. This is because of the unusual differences between the DLCP defect densities and the spin densities measured by ESR.

RESULTS

In Fig. 1 we plot the defect densities of the various samples, determined by the DLCP technique, as a function of their growth rate. Closed symbols represent the hydrogen diluted samples whereas the open symbols are for the samples grown from pure silane. One clearly observes that for the non-diluted samples the defect density decreases as the growth rate becomes lower. On the other hand, for the samples grown using hydrogen dilution, the defect density exhibits a minimum at the relatively high growth rate of 10.5 Å/sec. These trends appear both in state A and state B. Furthermore, the hydrogen dilution leads to significantly lower defect densities for the samples grown at the higher rates. One can also see in Fig. 1 that at the highest growth rate an almost identical reduction of the defect density is obtained by increasing the silane pressure instead of dilution with hydrogen.

A very important aspect of the samples in this study is their strikingly low defect densities. Conventional glow discharge a-Si:H exhibits DLCP defect density of roughly 2 x 10^{15} cm^{-3} in state A and about 10 times higher in state B. The present study finds defect densities that are as much as a factor of 4 lower. To verify that this is not due to some anomaly of the DLCP technique we carried out ESR studies for matching films which were co-deposited on quartz. Previous studies have shown [16] that the spin density is typically twice the corresponding DLCP defect density for standard glow discharge a-Si:H samples. As one can see in Table I, the ESR / DLCP ratio for the samples in the current study is considerably higher. A likely source for this difference is discussed below. At this point we simply note that the ESR data also indicates a low defect density at high growth rates, and confirms the trends discussed above. We thus conclude that most of the films under study are indeed of particularly high quality.

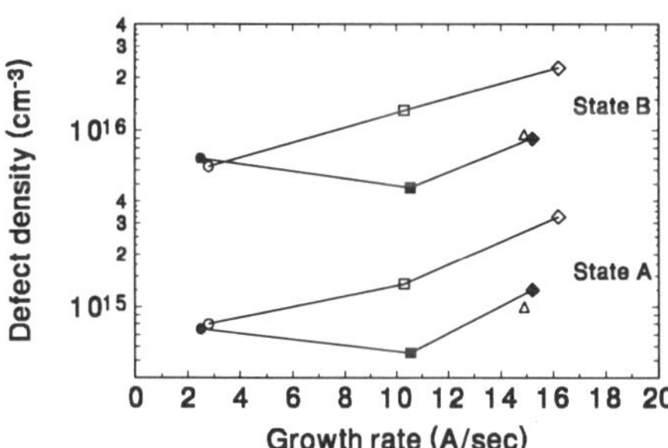

FIG. 1. The defect density as a function of the growth rate. The open symbols represent samples grown from pure silane, and the closed symbols represent samples grown under hydrogen dilution. The triangles represent one sample grown from pure silane under higher pressure.

It has been suggested recently that the most suitable a-Si:H materials for solar cell applications are those that are grown on the verge of microcrystallinity [6-9]. Indeed, the intrinsic layers in the world record [8] cells included a very small fraction of nanometer scale microcrystallites, that was detectable through TEM micrographs [6]. We have recently shown that the TPC technique is very sensitive to the presence of minute amounts (below the detection limit of conventional Raman measurements) of microcrystallites within the amorphous matrix [9,15]. In view

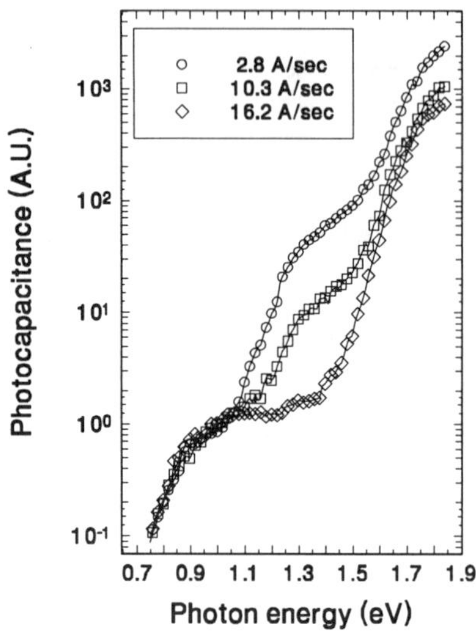

FIG 2: The photocapacitance spectra for the non diluted samples. The symbols match those in Fig. 1. The data is for samples in state B, and was recorded at 380°K.

of the extremely low defect densities displayed in Fig. 1, we have carried out such TPC measurements to determine whether the samples in this study contain microcrystallites.

The results of the TPC measurements for the non-diluted samples are presented in Fig. 2. To compare the spectra we have aligned them in the low energy region (below 1 eV). Such an alignment is helpful in comparing the shapes of the curves despite the differences in their defect densities. Indeed, these curves exhibit a considerable *qualitative* difference in shape at intermediate energies (1.0 - 1.4 eV). The spectra of sample #12465, grown at a high deposition rate, is typical of a-Si:H films. The other two samples (as well as *all* hydrogen diluted samples) exhibit a "shoulder" which is never seen in standard glow discharge a-Si:H films. We have previously shown that this feature is associated with the presence of microcrystallites within the amorphous network [9,15].

An important trend in Fig. 2 is that the magnitude of the 'shoulder' increases as the growth rate decreases for the non-diluted samples. This implies an increase in the microcrystalline inclusion with decreasing growth rate, accompanied by a decrease in the defect density (Fig. 1). All the H-diluted samples exhibited TPC spectra indicating a considerable microcrystalline inclusion. The changes in the microcrystalline fraction among the H_2 diluted samples were much smaller, in accordance with the smaller changes in the defect densities. Sample #12468, which was grown at lower rf power, exhibited the largest microcrystalline feature of the diluted films.

The very low defect densities observed by both DLCP and ESR suggest that solar cells grown under similar conditions would exhibit excellent performance. In Fig. 3 we present the fill factors of such solar cells as a function of the corresponding defect densities of the intrinsic films. We do indeed observe a correlation between a low defect density and a high FF. While the FF values do not appear to be particularly high, this is primarily because the thicknesses (7000 - 8000 Å) are far in excess of those used in standard devices (2000 - 3000Å).

DISCUSSION

Our results show a correlation between low defect densitites and the inclusion of a

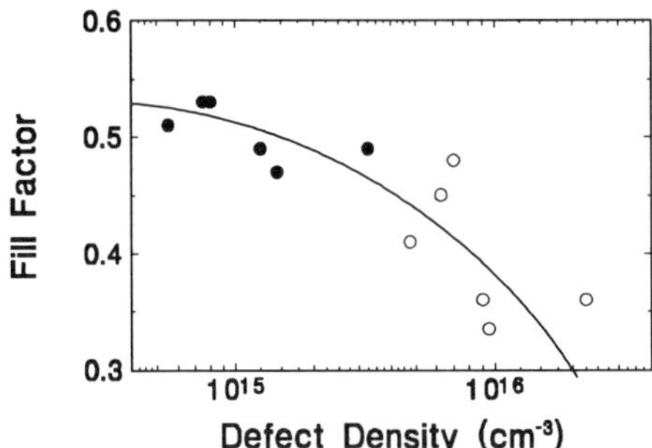

FIG. 3. The Fill Factor of solar cells as a function of the DLCP defect densities in the corresponding intrinsic films. The closed and open symbols represent the annealed and light soaked states, respectively. The line is plotted to guide the eye.

small concentration of microcrystallites. We thus find, in agreement with recent studies [6-9], that growth conditions close to the onset of microcrystalinity yield a-Si:H films with a superior quality. Previous studies have suggested, that the inclusion of a minute concentration of microcrystallites improves the order of the amorphous network, and leads to a low defect density. Of course, it may well be that the growth conditions near the onset of microcrystallinity yield an improved amorphous network and a low defect density, and that the appearance of the microcrystallites is simply a by-product of these growth conditions.

The existence of microcrystallites, revealed by the TPC spectra, may also explain the differences in defect densities obtained from the DLCP and ESR measurements. The DLCP technique measures the bulk exclusively, whereas ESR is also sensitive to surface states. The addition of microcrystallites may introduce additional surface states, and thus give rise to a larger ratio of the ESR / DLCP defect densities. We believe that the lack of a constant ratio between the ESR and DLCP signals may be a consequence of the varying concentration of microcrystallites. Additional causes for these differences are different thickness and effects of the different substrates (c-Si vs. quartz) on the film properties. More detailed measurements are underway to establish the exact relation between the defect densities revealed by the two techniques.

We finally briefly consider the possible effects of growth conditions on the degree of microcrystallinity. The formation of microcrystallites essentially depends on the ratio of H atoms to SiH_3 radicals, with higher values promoting the formation of microcrystallites [16]. Hydrogen atoms, which originate from the dissociation of the silane molecules, collide and react with other silane molecules. This leads to the formation of more SiH_3 radicals accompanied by H_2 molecules [17]. Accordingly, at low silane partial pressures (samples 12647 and 12469) the ratio of H/SiH_3 is high and the formation of microcrystallites is promoted. When the power is increased, H-ion bombardment energies increase and inhibit the growth of microcrystallites (sample 12465) [16]. Increasing the total silane pressure reduces the ion-bombardment energies and again micorcrystallites are formed (sample 12451) [18]. When H_2 is added while maintaining a low silane partial pressure, additional H atoms are formed from the dissociation of H_2 which itself does not react with H atoms, and results in higher H/SiH_3 ratios. This is reflected in the higher microcrystalline inclusion in the diluted samples.

CONCLUSIONS

Intrinsic a-Si:H, deposited under dilution conditions at relatively high growth rates, exhibits low defect density and a corresponding improvement of the Fill Factor in solar cells. The low defect density is intimately related to the inclusion of a small density of microcrystallites within the amorphous matrix. This, and other recent studies, suggest that the most suitable materials for solar cells applications are those grown near the onset of microcrystallinity.

ACKNOWLEDGMENT

The work in Oregon was supported by NREL under subcontract XAF-8-17916-05. The work at ETL was supported under the Sunshine Project by MITI.

REFERENCES

1. G. Ganguly and A. Matsuda, Mat. Res. Soc. Symp. Proc. **258**, 39 (1992).
2. J. Yang, X. Xu and S. Guha, Mat. Res. Soc. Sy,mp. Proc **336**, 687 (1994).
3. R. Arya, A. Catalano, M. Bennet, J. Newton, B. Fiesselman, L. Yang, Y. M. Li and R. D'Aiello, Proc. 11th EC PVSEC, 199 (1992).
4. A.H. Mahan, E. Iwaniczko, B.P. Nelson, R.C. Reedy, R.S. Crandall, S. Guha and J. Yang, Proc. 25th IEEE PVSC, 803 (1996).
5. K. Saito, M. Sano, K. Ogawa and I. Kajita, J. Non-Cryst. Solids **164-166**, 689 (1993).
6. D.V. Tsu, B.S. Chao, S.R. Ovshinsky, S. Guha and J. Yang, Appl. Phys. Lett. **71**, 1317 (1997).
7. J.H. Koh, Y. Lee, H. Fujiwara, C.R. Wronski and R.W. Collins, Appl. Phys. Lett. **73**, 1526 (1998).
8. J. Yang, A. Banerjee and S. Guha, Appl. Phys. Lett. **70**, 2975 (1997).
9. S. Guha, J. Yang, D.L. Williamson, Y. Lubianiker, J.D. Cohen and A.H. Mahan, Appl. Phys. Lett. **74**, 1860 (1999).
10. R. Hayashi, T. Takagi, G. Ganguly, M. Fukawa, M. Kondo and A. Matsuda, Proc. 2nd WCEPVSEC, 925 (1998).
11. J.D. Cohen, in *Semiconductors and Semimetals*, Vol. 21C, edited by J. Pankove (Academic, New York, 1984), p. 9.
12. C.E. Michelson, A.V. Gelatos and J.D. Cohen, Appl. Phys. Lett. **47**, 412 (1985)
13. J.D. Cohen, T. Unold and A.V. Gelatos, J. Non-Cryst. Solids **141**, 142 (1992).
14. T. Unold, J. Hautala and J.D. Cohen, Phys. Rev. B **50**, 16985 (1994).
15. D. Kwon, H. Lee, J.D. Cohen, H-C. Jin and J.R. Abelson, Jour. Non-Cryst. Solids **227-230**, 1040 (1998).
16. A. Matsuda, J. Non-Cryst. Solids **59-60**, 767 (1983).
17. R. Robertson and A. Gallagher, J. Appl. Phys. **59**, 3402 (1986).
18. G. Ganguly, T. Ikeda, K. Kajiwara and A. Matsuda, Mat. Res. Soc. Symp. Proc. **467**, 681 (1997).

VERY WIDE-GAP AND DEVICE-QUALITY a-Si:H FROM HIGHLY H_2 DILUTED SiH_4 PLASMA DECOMPOSED BY HIGH RF POWER

N. TERADA, S. YATA, A. TERAKAWA, S. OKAMOTO, K. WAKISAKA AND S. KIYAMA
New Materials Research Center, Sanyo Electric Co., Ltd.
1-1 Dainichi Higashimachi, Moriguchi, Osaka 570-8502, Japan.

ABSTRACT

The H_2 dilution technique at a high deposition rate (R_D) was investigated by depositing hydrogenated amorphous silicon (a-Si:H) under a high rf power density of 750 mW/cm^2, which is 20 times as large as that of conventional conditions. It was found that the H_2 dilution ratio γ (= [H_2 gas flow rate] / [SiH_4 gas flow rate]) tendency of the film properties, such as the H content (C_H), optical gap (E_{opt}), SiH_2/SiH and photoconductivity (σ_{ph}) of a-Si:H is different for the high rf power (750 mW/cm^2) and the medium rf power (75 mW/cm^2) conditions. Under medium rf power, the C_H, E_{opt} and SiH_2/SiH decrease as γ increases. Under the high rf power, on the contrary, the C_H and E_{opt} monotonously increase while maintaining a low SiH_2/SiH and a high σ_{ph} of 10^{-6} S/cm as γ increases. These results suggest that increasing the rf power enhances the H incorporation reactions due to H_2 dilution. It is thought that a high rf power causes the depletion of SiH_4 and hence the extinction of H radicals, expressed by $SiH_4 + H^* \rightarrow SiH_3^* + H_2$, is suppressed. A high H radical density enhances the incorporation of H into a-Si:H, resulting in very wide-gap a-Si:H with a high C_H. Consequently, very wide-gap a-Si:H with device-quality (E_{opt} of 1.82 eV with an $(\alpha h v)^{1/3}$ plot, corresponding to > 2.1 eV with Tauc's plot, and σ_{ph} of 10^{-6} S/cm) can be obtained at a high R_D of 12 Å/s without carbon alloying.

INTRODUCTION

A stabilized conversion efficiency of 9.5% for a submodule (1200 cm^2) with an a-Si/a-SiGe tandem cell structure has been achieved [1]. To achieve further improvement in the conversion efficiency of solar cells, wide-gap, high-quality a-Si:H, which is suitable for the top i-layer of stacked cells and/or the p-layer, is eagerly awaited [2]. Hydrogenated amorphous silicon carbide (a-SiC:H) is a well-known wide-gap material, but increasing the C content of a-SiC:H to enhance the optical gap causes a decrease in the film quality of a-SiC:H, such as its conductivity, because the incorporation of carbon atoms distorts the silicon network [3]. On the other hand, it was reported that the H plasma treatment and H_2 dilution are useful for preparing wide-gap, high-quality a-Si:H [4], except that the deposition rate decreases (< 1 Å/s).

In this study, a H_2 dilution technique at a high deposition rate was investigated by preparing a-Si:H under the high rf power density of 750 mW/cm^2, which is 20 times as large as that of conventional conditions [4].

EXPERIMENTAL

The a-Si:H films were prepared by using a capacitively coupled rf plasma CVD reactor on Corning #1737 glass substrates and crystalline silicon substrates. The H_2 dilution ratio (γ =[H_2] / [SiH_4]) was changed

Mat. Res. Soc. Symp. Proc. Vol. 557 © 1999 Materials Research Society

Table I. Deposition conditions for α-Si:H

	T_S(°C)	rf power (mW/cm²)	Pressure (Pa)	H_2 dilution ratio
High rf power	180	750	1.7×10^2	0-25
Medium rf power	180	75	1.7×10^2	0-25
Low rf power (Conventional)	80-180	10-30	$1.3\text{-}2.7 \times 10$	0-50

from 0 to 25 at the rf power density of 750 (high rf power) and 75 (medium rf power) mW/cm². The substrate temperature (T_s) and gas pressure (P) were fixed at 180°C and 170 Pa, respectively. The details of the deposition conditions for the high, medium and low rf power (conventional conditions [4]) are given in Table I. The high rf power of 750 mW/cm² is more than 20 times as large as that of conventional conditions.

The hydrogen content (C_H) of the a-Si:H was precisely determined by SIMS measurement. The SiH$_2$/SiH, which is an important factor in the quality of a-Si:H films [5,6], was determined from IR absorption spectra [7,8]. The photoconductivity (σ_{ph}) under AM-1, 100 mW/cm² illumination was measured for a-Si:H films on glass substrates. The optical gap (E_{opt}) was determined from the hν versus $(\alpha h\nu)^{1/3}$ plot, where α and hν denote the optical absorption coefficient and the photon energy, respectively [9]. The optical absorption coefficients were calculated from transmittance (T) and reflectance (R) spectra, and the optical interference effect was suppressed by the T/(1-R) calculation in the analysis.

RESULTS AND DISCUSSION

<u>Opto-electronic properties of a-Si:H film</u>

Figure 1 shows the E_{opt}, B-value and R_D of the a-Si:H as functions of the H_2 dilution ratio (γ) under the rf power density of 750 (high rf power) and 75 (medium rf power) mW/cm². The B-value increases as γ increases under both high and medium rf power. Since the B-value, which is an $(\alpha h\nu)^{1/3}$ gradient of the hν versus $(\alpha h\nu)^{1/3}$ plot, partly reflects on the slopes of the band tails, it seems that a-Si:H films prepared at a high γ have *good* film properties. R_D decreases as γ increases in both cases of rf power. However, R_D is above 10 Å/s at a γ of 25 in the case of the high rf power. The γ tendency of the B-value and R_D is not

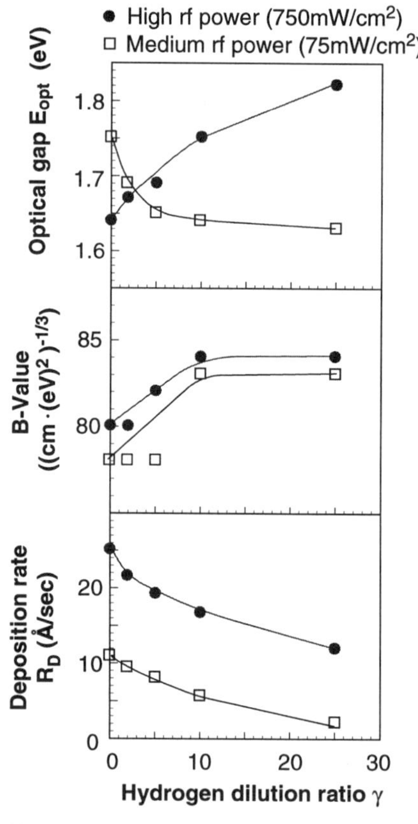

Fig. 1 Optical gap (E_{opt}), B-value and deposition rate (R_D) of α-Si:H films as function of the hydrogen dilution ratio (γ). The solid symbols and open symbols show the data of α-Si:H prepared under the high rf power and medium rf power, respectively.

affected by the rf power. On the contrary, the γ tendency of the E_{opt} depends on the rf power. By increasing γ, E_{opt} monotonously increases under high rf power while it monotonously decreases under medium rf power. It is notable that at a γ of 0, namely 100% SiH_4, the R_D of the high rf power is higher than that of the medium rf power by double, though the E_{opt} of the high rf power is 0.1 eV smaller than that of the medium rf power. This is different from the previously reported [10] relationship that the E_{opt} increases when the R_D increases by changing the rf power under a fixed T_s. Consequently, very wide-gap a-Si:H (E_{opt} of 1.82eV with an $(\alpha h v)^{1/3}$, corresponding to > 2.1 eV with Tauc's plot) was prepared at a relatively high R_D of 12 Å/s under the conditions of the high rf power of 750 mW/cm² and the high γ of 25.

Figure 2 shows the relationship between the E_{opt} and C_H of a-Si:H deposited with various γ. In Fig. 2 the values of γ are shown in parentheses. The shaded area in Fig. 2 indicates the range where data for conventional a-Si:H films (low rf power) [4,10] exist. Under the medium rf power, E_{opt} and C_H decrease as γ increases from 0 to 10, while E_{opt} slightly decreases or saturates and C_H increases as γ increases above 10. Under the high rf power, on the other hand, E_{opt} and C_H monotonously increase as γ increase. The gradient of E_{opt}- C_H for the high rf power is different from that for the low rf power, though that for the medium rf power is similar to that for the low rf power. These results suggest that the network of silicon-hydrogen bonds at the high rf power can be different from that of films prepared under the low or medium rf power.

Figure 3 shows the SiH_2/SiH as a function of C_H. The dark and light shaded areas indicate the range of a-Si:H films from the low rf power (H_2 dilution) [4] and low rf power (100% SH_4) [10], respectively. Un-

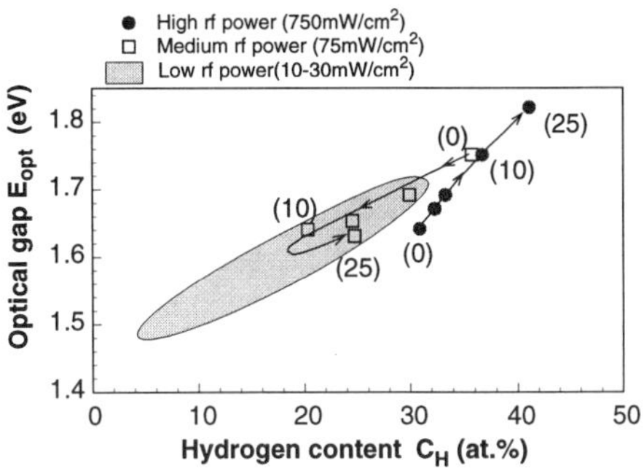

Fig. 2 Optical gap (E_{opt}) of α-Si:H films plotted against hydrogen content (C_H). The shaded area indicates the range where data for α-Si:H films from H_2 diluted SiH_4 under the low rf power and 100% SiH_4 exist [4,10]. The values of the hydrogen dilution ratio (γ) are shown in parentheses. The lines with arrows show the changes which occurred as γ increased.

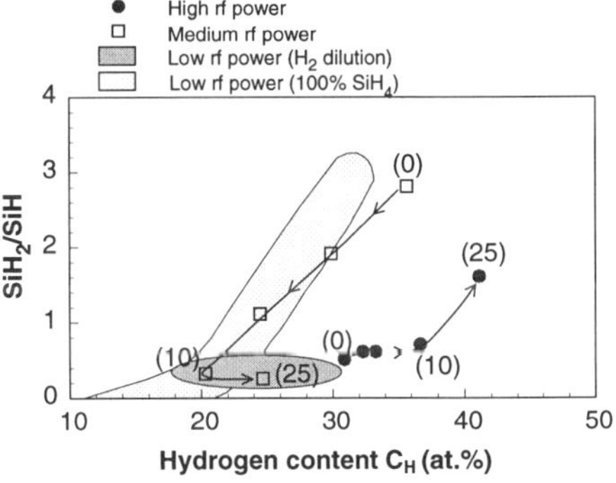

Fig. 3 Hydrogen bond configration (SiH_2/SiH) of a-Si:H films plotted against hydrogen content (C_H). The dark shaded area indicates the range where data for α-Si:H films prepared under the low rf power (H_2 dilution) exist [4]. The light shaded area indicates the range where data for α-Si:H films prepared under the low rf power (100% SiH_4) exist [10].

der the medium rf power, when γ increases, C_H decreases with drastically decreasing SiH_2/SiH at a low γ ≤ 10, and C_H begins to increase while maintaining a low SiH_2/SiH at $\gamma \geq 10$, which is the same tendency as was previously reported for the low rf power with H_2 dilution [4]. On the other hand, with the high rf power, when γ increases, C_H monotonously increases while maintaining a low SiH_2/SiH at within a γ of 10. It is notable that a low SiH_2/SiH of 0.4 and a high C_H of 31% are compatible in a-Si:H film deposited under the high rf power even at $\gamma = 0$ (namely, 100% SiH$_4$) at the same time. The film of $\gamma = 0$ under the high rf power seems to be affected by H$_2$ dilution, because the compatibility of a low SiH$_2$/SiH and a high C$_H$ is a feature of high H$_2$ dilution [4]. Although the SiH$_2$/SiH under the high rf power increases, when γ increases from 10 to 25, that value is considerably lower than that under the low rf power (100% SiH$_4$; light shaded area in Fig. 3).

Figure 4 shows σ_{ph} plotted against E$_{opt}$. The dark and light shaded areas indicate the range of a-Si:H films from the low rf power (H$_2$ dilution) [4] and low rf power (100% SH$_4$) [10], respectively. Under the medium rf power, σ_{ph} monotonously decreases as E$_{opt}$ increases, which is the same tendency as that seen in a-Si:H films from the low rf power (100% SiH$_4$) and a-SiC:H. Under the high rf power, on the other

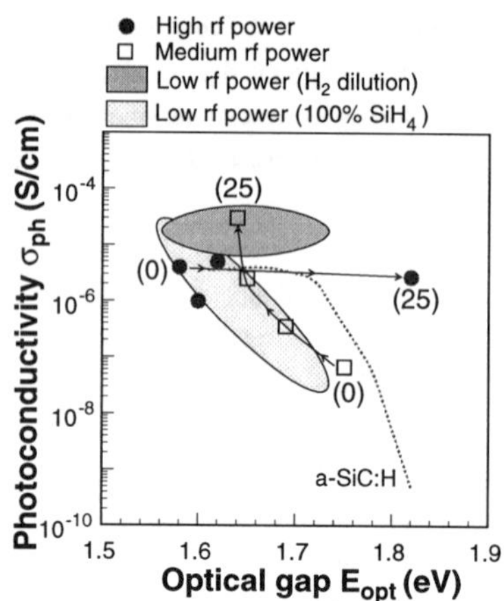

Fig. 4 Photoconductivity (σ_{ph}) of α-Si:H films plotted against optical gap (E$_{opt}$). The dark shaded area indicates the range where data for α-Si:H films prepared under the low rf power (H$_2$ dilution) exist [4]. The light shaded area indicates the range where data for α-Si:H films prepared under the low rf power (100% SiH$_4$) exist [10].

hand, when E$_{opt}$ increases, σ_{ph} maintains a relatively high value of more than 10^{-6} S/cm, which is the same tendency as that seen in the low rf power (H$_2$ dilution). However, the E$_{opt}$ of a-Si:H deposited under the high rf power at a high γ is larger than that of a-Si:H deposited under the low rf power with H$_2$ dilution. The high rf power with high H$_2$ dilution results in a relatively high σ_{ph} of 10^{-6} S/cm at very wide-gap range (E$_{opt}$ of 1.82 eV, corresponding to Tauc's gap of > 2.1 eV), where the σ_{ph} of high rf power is higher than a-SiC:H by 2-3 orders of magnitude and the previous low rf power condition could not prepare the a-Si:H. The wide-gap a-Si:H prepared under the high rf power with high H$_2$ dilution seems to be a good candidate not only for the top i-layer of stacked cells but also for the window layers, such as the p-layer and buffer-layer. Furthermore, the high deposition rate of more than 10 Å/s improves the productivity of a-Si:H solar cells.

Enhancement of the hydrogen dilution effect

The γ dependence of film properties, such as the relationship between the C$_H$ and E$_{opt}$, and the relationship between the C$_H$ and SiH$_2$/SiH, of a-Si:H from the high rf power is quite different from that of a-Si:H from the medium or low rf power. The following results deserve special mention.

(1) At $\gamma = 0$, the E$_{opt}$ decreases while R$_D$ increases as the rf power increases from 75 to 750 mW/cm^2 (See Fig. 1).

(2) Under the high rf power, C$_H$ and E$_{opt}$ monotonously increase as γ increases, while maintaining a low SiH$_2$/SiH (See Figs. 2, 3).

(3) A very low SiH$_2$/SiH is compatible with a high C$_H$ in a-Si:H prepared under the high rf power at $\gamma = 0$, which is an effect of conventional H$_2$ dilution (low rf power) (See Fig. 3).

These results suggest that the effects of H$_2$ dilution are enhanced by high rf power.

Figure 5 shows the SiH$_4$ gas flow rate dependence of R$_D$. Here, γ is fixed at 25. Under the medium rf power, R$_D$ is independent of the SiH$_4$ gas flow rate within the range shown in Fig. 5, which means that the rf power limits the deposition rate. Under the high rf power, on the other hand, R$_D$ increases as the SiH$_4$ gas flow rate increases, which means that the SiH$_4$ gas flow rate limits the deposition rate. Thus, it is believed that SiH$_4$ gas is depleted under high rf power conditions.

Figure 6 shows E$_{opt}$ as a function of the SiH$_4$ gas flow rate. Here, the rf power of 750 mW/cm^2 and the γ of 25 are fixed. E$_{opt}$ monotonously increases when the SiH$_4$ gas flow rate decreases. The depletion of SiH$_4$ gas seems to cause the hydrogen atom incorporation into a-Si:H films, because a decreasing SiH$_4$ gas flow rate corresponds to the further depletion of SiH$_4$ gas. In other words, the depletion of SiH$_4$ gas seems to enhance the effect of H$_2$ dilution.rate corresponds to the further depletion of SiH$_4$ gas. In other words, the depletion of SiH$_4$ gas seems to enhance the effect of H$_2$ dilution.

It is regarded that the effects of H$_2$ dilution are determined by the reaction of H radicals not by H$_2$ molecules. It is known that the H radicals vanish by the following collision with SiH$_4$, expressed by SiH$_4$ + H* $\rightarrow$ SiH$_3$* + H$_2$ [11]. Under high rf power conditions, the depletion of SiH$_4$ gas probably suppresses the extinction of H radical reaction and many more H radicals remain in the plasma. Consequently, the high power rf or the decreasing SiH$_4$ gas flow rate enhances the effect of H$_2$ dilution. The fact that a-Si:H film from the high rf power at $\gamma = 0$ (100% SiH$_4$) appears to be affected by H$_2$ dilution (See Figs. 1, 3) can also be explained by this speculation.

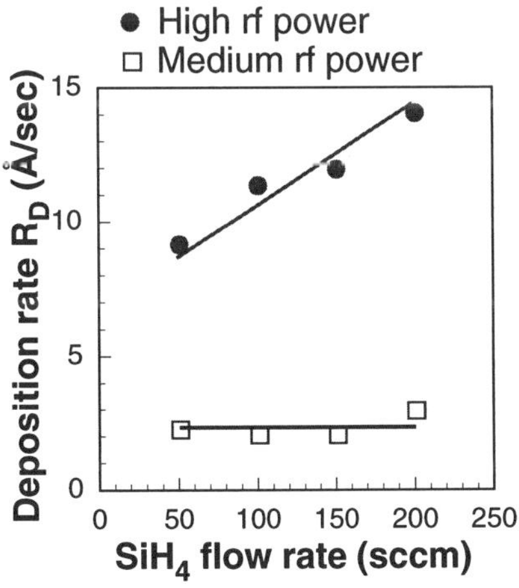

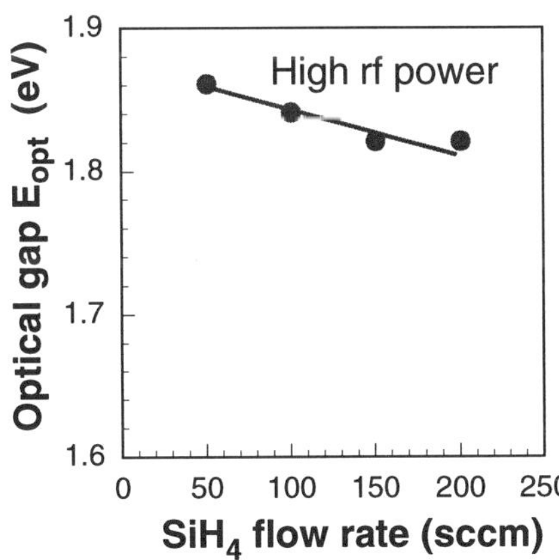

Fig. 5 Deposition rate (R$_D$) of α-Si:H films as function of the hydrogen SiH$_4$ flow rate.

Fig. 6 Optical gap (E$_{opt}$) of α-Si:H films as function of the SiH$_4$ flow rate.

CONCLUSIONS

A H_2 dilution technique at a high deposition rate was investigated by depositing a-Si:H under the high rf power density of 750 mW/cm^2, which is 20 times as large as that of conventional conditions [4].

It was found that the high rf power enhances the H incorporation reactions due to H_2 dilution. It is thought that the high rf power causes SiH_4 depletion and suppresses the extinction of H radicals. A high H radical density enhances the H incorporation into a-Si:H, resulting in very wide-gap a-Si:H with a high C_H. Consequently, very wide-gap a-Si:H with device-quality (E_{opt} of 1.82 eV with an $(\alpha h\nu)^{1/3}$ plot, corresponding to > 2.1 eV with Tauc's plot, and σ_{ph} of 10^{-6} S/cm) can be obtained at the high R_D of 12 Å/s without carbon alloying from highly H_2 diluted SiH_4 plasma decomposed by a high rf power.

ACKNOWLEDGMENT

This work was supported by NEDO as a part of the New Sunshine Program under the Ministry of International Trade and Industry.

REFERENCES

1) Y. Hishikawa, S. Tsuge, N. Nakamura, S. Tsuda, S. Nakano, M. Ohnishi and Y. Kuwano, Mater. Res. Soc. Proc. **192**, 511 (1990).

2) T. Kinoshita, M. Shima, A. Terakawa, M. Isomura, M. Tanaka, S. Kiyama and S. Tsuda, presented at the 14th EU-PVSEC, Barcelona, 1997 (unpublished).

3) W. Beyer, H. Wagner and F. Finger, J. Non-Cryst. Solids **77&78**, 857 (1985).

4) S. Okamoto, Y. Hishikawa and S. Tsuda, Jpn. J. Appl. Phys. **35**, 26 (1996).

5) T. Kamimura, H. Nozaki, N. Sakuma and H. Ito, Jpn. J. Appl. Phys. **27**, L1837 (1998).

6) Y. Hishikawa, S. Tsuge, N. Nakamura, S. Tsuda, S. Nakano and Y. Kuwano, J. Appl. Phys. **69**, 508 (1991).

7) M. Brodsky, M. Cardona and J. Cuomo, Phys. Rev. **B 16**, 3556 (1977).

8) N. Maley and I. Szafranek, Mater. Res. Soc. Symp. Proc. **192**, 663 (1986).

9) Y. Hishikawa, N. Nakamura, S. Tsuda, S. Nakano, Y. Kishi and Y. Kuwano, Jpn. J. Appl. Phys. **30**, 1008 (1991).

10) Y. Hishikawa, S. Tsuda, K. Wakisaka and Y. Kuwano, J. Appl. Phys. **73**, 4227(1993)

11) D. Beeman et. al., Phys. Rev., B32, 874 (1985).

He-DILUTION TO INCREASE DEPOSITION RATE AND FEEDSTOCK UTILIZATION DURING THE GROWTH OF a-Si:H AND a-SiGe:H ALLOYS

A. R. MIDDYA, G. WOOD, G. H. LIN and D. E. CARLSON
Solarex, A Business Unit of BPAmoco Solar, 3601 LaGrange Parkway, Toano, VA 23168.

ABSTRACT

We report on the development of helium diluted a-Si:H and a-SiGe:H solar cells with higher deposition rates and better feedstock utilization than devices made with hydrogen dilution. Both the initial and the stabilized efficiencies of the He-diluted single-junction a-Si:H and a-SiGe:H cells are similar to those of hydrogen-diluted cells with state-of-the-art intrinsic materials. The total fabrication time for tandem cells has been reduced by 17% by using helium dilution without loss in initial and stabilized efficiency.

INTRODUCTION

The correlation between deposition rate (R_d) and stability of a-Si:H of plasma-enhanced chemical vapor deposition (PECVD) is well established; i.e. as R_d increases, the light-induced degradation increases. However it is imperative to increase R_d from 1 A/s to increase throughput in the production line and hence to reduce the cost of a-Si module. Helium (He) dilution technique is one of the several deposition technology have been investigated to increase R_d. Roca i Cabarrocas et. al. has reported that the deposition rate of He-diluted a-Si:H can be increased up to 15 A/s without affecting its initial opto-electronic properties and the defect densities[1]. However, both the initial and the stabilized device characteristics based on these materials were found to be inferior as compare to cells made with intrinsic a-Si:H layers deposited at 1 A/s. Device quality a-SiGe:H alloys by He dilution has been first developed in a particular regime (the γ-regime) of deposition parameter space of RF PECVD[2]. Generally, state-of-the-art a-Si:H has been developed in the parameter space called the "α-regime" of RF PECVD[3]. It has been reported that if the plasma condition is shifted from the "α-regime" to the "γ-regime", the characteristics of He-diluted a-Si:H films are different than what has been reported in the literature; i.e. the density of defect states above Fermi level is significantly lower than that of state-of-the-art a-Si:H [4]. These films contain low bonded H-content (Mostly monohydride H-bonding configuration) and low microvoid fraction ($\leq 0.01\%$). Consequently, the mobility-lifetime product is one to two orders of magnitude higher (independent of Fermi level position) than standard a-Si:H, and most surprisingly, the kinetics of degradation is very fast (the characteristic time for saturation is much lower than state-of-art a-Si:H)[5,6]. These type of materials have been reproduced in a RF PECVD system by Palaiseau group, and because of its complex nanostructure (Si-Si bonding), these materials have been called "polymorph"[7,8]. In this work we report the development of state-of-the-art He-diluted a-Si:H and a-SiGe:H alloys using DC PECVD.

EXPERIMENTAL METHODS

The samples for this study have been deposited on tin oxide coated glass substrates (made by Asahi and AFG) using a DC PECVD system. Both single-junction p-i-n and double junction (tandem) cells (0.27 cm^2 in area) have been fabricated. In the single-junction cells, the p-layer is boron-doped a-SiC:H (~10nm) followed by either a H or He-diluted intrinsic layer and then a phosphorus-doped microcrystalline n-layer. The back contact is formed by sputtering zinc oxide (~100nm) and aluminum (~300nm). The device structure for a-Si:H/a-

Mat. Res. Soc. Symp. Proc. Vol. 557 © 1999 Materials Research Society

SiGe:H tandem cells is glass/TCO/p-i-n/p-i-n/ZnO/Al. The device characteristics of a-SiGe:H cells have been measured under AM1.5 illumination through a red filter (λ>560 nm). Light soaking has been performed both under white light (100 mW/cm^2) and high intensity illumination (60 suns). The temperature of the cells during white light soaking was maintained at 40-50°C by air cooling the sample. The high intensity light soaking was performed using a xenon arc lamp and the temperature of the cells was maintained at 60°C using a flow of nitrogen.

RESULTS AND DISCUSSION

<u>Comparison of deposition rate of H- and He-diluted a-Si:H</u>: A comparison of R_d of H- and He-diluted a-Si:H films deposited under identical condition is shown in Fig. 1. The value of R_d for He-diluted film is higher than that for H-diluted one for any flow rate of silane. We have also compared the variation of R_d with DC power and dilution ratio. The percentage increase of R_d is found to be proportional to flow rate, dilution ratio and inversely proportional to DC power. The deposition rate is a combine effect of gas phase dissociation and surface reaction. The results mentioned above indicates that He-dilution is apparently acting as an extra power to decompose undissociated process gas possibly via excited helium atoms (He*) i.e. improve feedstock utilization.

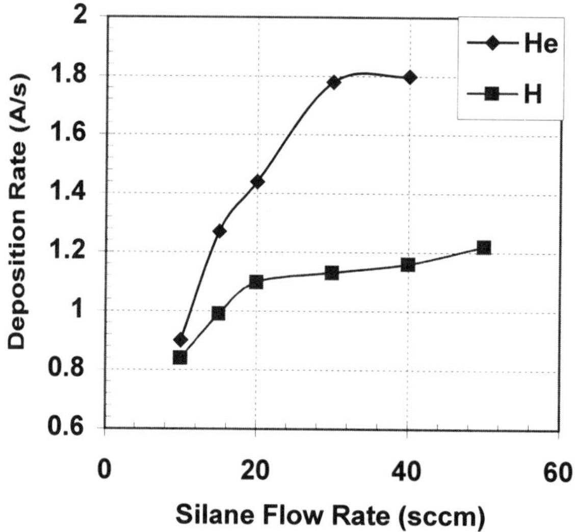

Fig. 1. Variation of deposition rate with silane flow under H- and He-dilution.

<u>II. Comparison of H- and He-diluted a-Si:H and a-SiGe:H cells</u>: Improvements in material properties do not always lead to improvements in device performance. In this study all the experiments have been performed on devices so that the results could be more directly applied to the manufacturing of solar cells. Table I shows a comparison of the device performance and light-induced degradation of H- and He-diluted single-junction a-Si:H cells. The data in Table I represents an average of 10 cells. The first two samples were prepared under identical condition, but in two different runs. For all the samples, only the intrinsic

Table I: Comparison of H- and He-diluted Single-Junction a-Si:H Cells.

Sample # / Diluent	d (A)	R_d (A/s)	V_{OC} (V)	FF	J_{SC}(QE) (mA/cm^2)	Eff. (%)	%Degrd in FF (1000h)	%Degrd. in Eff. (1000h)
C7301-1/H	3000	1.05	0.902	0.712	14.3	9.18	17.2	23.4
C7301-2/H	2990	1.05	0.904	0.714	14.3	9.23	17.0	23.0
C7344-1/He	3120	1.6	0.903	0.711	14.01	9.0	16.0	23.2
C8027-3/He	4800	1.6	0.901	0.68	14.34	8.8	21.0	29.0

layers were different, and the thickness listed in the table is the total thickness of the devices. The estimation of R_d has been done by depositing the intrinsic layers separately. The open-circuit voltage (V_{OC}), fill factor (FF), short-circuit current density (J_{SC}) (calculated from quantum efficiency) and efficiency of H- and He-diluted cells are similar. The spectral

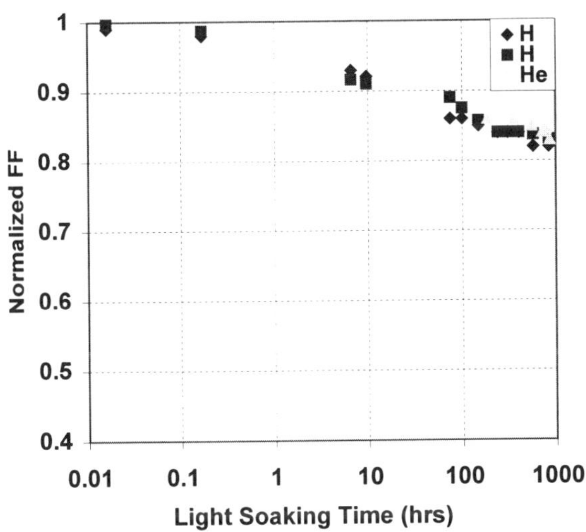

Fig.2a: Variation of FF of H- and He-diluted a-Si:H cells with light soaking time.

response of H- and He-diluted cells is almost identical. The similar value of FF and J_{SC} for comparable thickness indicates that the bandgap and the electronic properties of He-diluted a-Si:H are similar to that of H-diluted a-Si:H which has been observed before in the properties of material (Mobility-lifetime product) [5,6]. The light-degradation of the FF and the efficiency of H- and He-diluted cells after 1000 hours of light soaking are also same around 23-24%. The spectral response of H- and and He-diluted cells after 1000 hours light soaking are similar. The light- degradation in efficiency increases to 29% when the thickness of the He-diluted a-Si:H cells (C8027-3) was increased from 3120 A to 4800 A, which is similar to the degradation observed for H-diluted a-Si:H cells of comparable thickness. Thus, most of

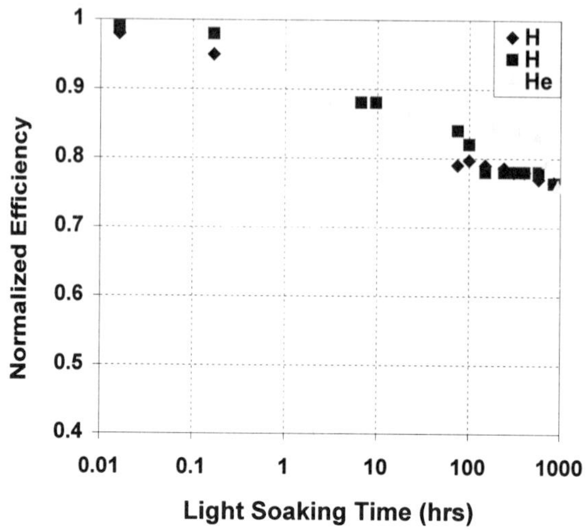

Fig.2b: Variation of efficiency of H-and He-diluted a-S:H cells with light soaking time.

the light-induced degradation is a reflection of bulk properties and not an interface effect. Fig. 2a and Fig. 2b show a comparison of the degradation kinetics of the FF of H- and He-diluted a-Si:H cells with comparable thickness. Each data point in the graph represents the average of 6-7 cells. The initial device characteristics of these cells were shown in Table I. It may be seen from Fig. 1a & Fig. 1b that the kinetics of degradation of FF and efficiency of H- and He-diluted cells is similar. The sudden decrease in efficiency of He-diluted cells after 700-800 hours (After saturation) is also observed in one of the H-diluted cells (C7301-1) which is absent in another H-diluted cells (C7301-2) although the intrinsic layers are identical in both the cases. Thus this slight decrease in efficiency after 700-800 hours is not related to the intrinsic materials.

Table II shows the comparison of the PV characteristics single junction a-SiGe:H cells having intrinsic layer deposited with either H- or He-dilution. The data show the total flow rate of silane and germane could be reduced by 12% while the deposition rate was increased by 20% using He-dilution. The device performance of H- and He-diluted a-SiGe:H cells are similar. The flow rate of GeH_4 could be reduced by as much as 62% while still retaining a cell performance comparable to that of H-diluted cells. The light-induced degradation of these cells under 60 sun illumination has been studied. The efficiency of the most the cells decrease to 2.9-3.0% for 30 minute (a-Si:H cells saturates within this time) exposure both for H- and He-diluted cells despite deposition rate is 20% higher in case of He-diluted cell.

III. <u>Comparison of H- and He-diluted tandem devices</u>: Table III shows a comparison of the PV characteristics of tandem cells made with either H- or He-diluted a-Si:H and a-SiGe:H H intrinsic layers fabricated on either Asahi or AFG tin oxide coated glass substrates. As a result of the higher growth rate of He-diluted a-Si:H and a-SiGe:H alloys, the total process time for He-diluted tandem devices is reduced to 60 minute, which is 17% reduction compared to that of standard H-diluted devices. Simultaneously, the GeH_4 gas utilization was improved by 16.7%. While the values of V_{OC} and FF are comparable, the J_{SC} values are somewhat lower for the He-diluted cells, possibly due to a thinner back junction or lower Ge-content in the a-SiGe:H intrinsic layer thus leading to a lower quantum efficiency at 800nm. All the samples produced with He-dilution exhibited light-induced degradation in the range of 14-17% after 600 hours of light soaking which was independent of the type of substrate. This

Table II: Comparison of H- and He-diluted Single-Junction a-SiGe:H Cells.

Run#	Diln	% Reduction in GeH$_4$ flow	d (A)	R$_d$ (A/s)	V$_{oc}$ (V)	FF	Jsc (QE) (mA/cm^2)	Effn. (%)	Stab-ilized effn. (%)	%Deg in effn. (30 min)
C7135-1	H/12:1	0	1815	1.0	0.65	0.65	17.3	4.07	3.01	26.0
C7055-1	H/12:1	0	2000	1.0	0.62	0.63	17.8	3.95	2.92	26.0
C7198-1	He/13:1	12	1900	1.2	0.69	0.67	16.6	3.9	3.1	20.5
C7205-1	He/13:1	12	1650	1.2	0.66	0.66	16.8	4.0	2.99	25.2
C7209-1	He/13:1	12	-	1.2	0.67	0.66	16.3	4.1	3.06	25.4
C8007-2	He/20:1	62	1750	0.9	0.68	0.68	16.8	3.9	-	-

Table III: Comparison of H- and He-diluted a-Si/a-SiGe:H Tandem Devices.

Diln	V$_{oc}$ (V)	FF	Jsc (mA/cm^2)	J$_{sc1}$ (mA/cm^2)	J$_{sc2}$ (mA/cm^2)	J$_{sc}$ (QE) (mA/cm^2)	Effn (%)	QE (800nm)	%Dgr in effn. (600h)
H	1.54	0.69	8.65	9.20	8.80	18.0	9.64	23.4	-
He	1.53	0.71	8.54	8.38	8.64	17.0	9.23	20.5	16
He	1.48	0.70	8.20	8.54	9.23	17.8	9.23	25.5	14
He[*]	1.50	0.71	8.21	8.27	8.20	16.5	8.90	18.1	17
He[*]	1.49	0.70	8.32	8.04	7.72	15.8	8.50	14.1	14

amount of degradation is similar to that observed in tandem cells produced with H-dilution. Thus, we were able to reduce the total process time of tandem devices by 17% while simultaneously increasing the GeH$_4$ gas utilization by 16.7% using He-dilution, and were able to obtain stabilized efficiencies similar to those H-diluted tandem cells.

IV.Growth Mechanism: In a He-diluted plasma, the electron temperature (T$_e$) of the plasma increases sharply with dilution [9]. For comparable dilutions, T$_e$ is much higher in case of He-dilution than for H-dilution due to its higher ionization energy, which enhances the electron impact dissociation of silane resulting in an improvement in gas utilization and hence higher R$_d$. Because of the high T$_e$, the population of He atoms in the metastable state (He*) in the γ-regime (Plasma sustaining mechanism in DC PECVD) increases. Upon deexcitation, these He* atoms supply engergy (20 eV) and momentum not only for gas decomposition but also for surface reactions. This excess energy can increase the "effective temperature" of growing surface and hence the surface mobility of the radicals thus accelerating the structural relaxation during film growth [10]. Consequently, despite higher deposition rate, Si-Si network structure and H-bonding configuration of He-diluted materials are similar to that of state-of-the-art H-diluted one. Growth of state-of-art a-Si:H and a-SiGe:H alloys in DC PECVD supports the model based on metastable helium induced structural relaxation of a-Si:H *in a continuous plasma* in the γ-regime of RF PECVD[4,5,6]. Modification of the Si-Si network structure of a-Si:H by He*and Ar* treatment of the film surface using the layer-by-layer technique has been demonstrated by I. Shimizu and co-workers[11, 12].

CONCLUSIONS

H- and He-dilution are equivalent in developing device quality materials in DC PECVD, but He dilution has the additional advantage of allowing higher deposition rate. Improved feedstock utilization has also been demonstrated for germane (Which has major effect on the cost of the module) using a He-diluted plasma. The process time for tandem devices was reduced by 17% as a result of the higher deposition rates for He-diluted a-Si:H and a-SiGe:H alloys while maintaining performance at initial and saturated state similar to our standard H-diluted devices. There is a potential that more stable a-Si:H, a-SiGe:H alloys and high quality μc-Si films could be developed by He-dilution technology.

ACKNOWLEDGMENTS

The authors gratefully acknowledge useful discussions with R. R. Arya, L. F. Chen and G. Ganguly. They would also like to acknowledge the support of Dave Bradley (deceased) and K. Rajan during the course of this work. This work was partially funded by NREL under Subcontract ZAK-8-17619-02.

REFERENCES

1. P. Roca i Cabarrocas, J. Non-Cryst. Solids **164-166**, 37 (1993).
2. A. R. Middya, Sukti Hazra and Swati Ray, J. Appl. Phys. **76**, 7578 (1994).
3. J. Perrin, <u>Plasma Deposition of Amorphous-based Materials</u> edited by G. Bruno, P. Capezzuto and A. Madan (Academic Press, 1995)p.216.
4. A. R. Middya, S. Hazra, S. Ray, A. K. Barua and C. Longeaud, J. Non-Cryst. Solids, **198-200**, 1067 (1996).
5. A. R. Middya, Sukti. Hazra, Swati. Ray, C. Longeaud and J. P. Kleider in <u>Amorphous and Microcrystalline Silicon Technology-1997</u> edited by S. Wagner, M. Hack, E. A. Schiff, R. Schropp and I. Shimizu (Mater. Res. Soc. Proc. **467**, Pittsburg, PA, 1997)p.615.
6. A. R. Middya, S. Hazra, S. Ray, C. Longeaud and J. P. Kleider, <u>Conference Record of The Twenty Sixth IEEE Photovoltaic Specialist Conference- 1997</u> (Conf. Rec. 26th IEEE PVSC, Anaheim, CA, 1997)p.655.
7. P. Roca i Cabarrocas, P. St'ahel, S. Hamma and Y. Poissant, in <u>Conference Record of The Twenty Sixth IEEE Photovoltaic Specialist Conference- 1998</u> (Conf. Rec. 27th IEEE PVSC, Vienna, Austria, 1998) (In press).
8. C. Longeaud, J. P. Kleider, M. Gauthier, R. Kaplan, R. Butte, R. Meaudre and P. Roca i Cabarrocas, in <u>Conference Record of The Twenty Sixth IEEE Photovoltaic Specialist Conference- 1998</u> (Conf. Rec. 27th IEEE PVSC, Vienna, Austria, 1998) (In press).
9. H. Nomura, A. Kono and T. Goto, Jpn. J. Appl. Phys. **33**, 4165 (1994).
10. J. Perrin, J. Non-Cryst. Solids **137&138**, 639(1991).
11. H. Shirai, T. Ariyoshi, J. Hanna and I. Shimizu, J. Non-Cryst. Solids **137&138**, 693(1991).
12. W. Futako, K. Fukutani, M. Kannbe, T. Kamiya, C. M. Fortmann and I. Shimizu <u>Conference Record of The Twenty Sixth IEEE Photovoltaic Specialist Conference- 1997</u> (Conf. Rec. 26th IEEE PVSC, Anaheim, CA, 1997).

PROBLEMS OF POWER FEEDING IN LARGE AREA PECVD OF AMORPHOUS SILICON

U. STEPHAN, J. KUSKE, H. GRÜGER[*], A. KOTTWITZ[*]
Forschungs- und Applikationslabor Plasmatechnik GmbH Dresden, D-01217 Dresden, Gostritzer Straße 61-63, F.R. Germany
[*]Semiconductor and Microsystems Technology Laboratory, Dresden University of Technology, D-01062 Dresden, Mommsenstraße 13, F.R. Germany

ABSTRACT

The production of amorphous silicon, e.g. for solar cells, requires large area, high-deposition rate plasma reactors. Increasing the radio frequency from the conventional 13.56MHz up to VHF has demonstrated higher deposition and etch rates and lower particle generation, a reduced ion bombardement and lower breakdown, process and bias voltages.

But otherwise the use of VHF leads to some problems. The non-uniformity of deposition rate increase due to the generation of standing waves (TEM wave) and evanescent waveguide modes (TE waves) at the electrode surface.

Increasing the frequency and/or the deposition area the plasma impedance, the capacitic stray impedance of the RF electrode and other parasitic capacitive impedances decrease. Increasing the frequency and/or the RF power, the phase angle of the discharge and of the impedance at every point at the lines between the RF matching network an the RF electrode tends more and more towards -90°. This results in increasing currents and standing waves with extremly high local current maximas. Increasing resistances of lines and contacts due to the skin effect and loss-caused heating up of the lines the power losses increase extremely, up to 90% and more. In spite of the increasing of the coupled power, the plasma power does not increase. Thermal destructions of the lines due to extreme expansion or melting are possible.

Some solutions to reduce the non-uniformity of the deposition rate like multipower feeding, central backside power feeding, electrode segmentation, use of load impedances, published in former publications, will be discussed in connection with several reactor types (coaxial, large area, long plasma source) in view of the efficiency of power coupling and the practical realization. Solutions to minimize the power losses at the lines will be presented.

INTRODUCTION

Increasing the frequency and the substrate area of the plasma enhanced chemical vapour deposition (PECVD) a higher efficiency of the production of amorphous silicon will result. But on the other hand there are two essential problems:

A) Higher frequencies lead to non-uniformities of the film thickness due to the generation of standing TEM waves and evanescent waveguide modes (TE waves) at the electrode surfaces.

B) The plasma power will be limited due to high power losses on contacts and transmission lines at increasing frequencies and generator powers.

In this paper we discuss and compare some solutions to reduce the non-uniformity of the deposition rate and in this way the film thickness with three reactor types (coaxial and large area reactors, long plasma source). Solutions to minimize the power losses and to increase the efficiency of power-coupling to the plasma also will be presented.

Mat. Res. Soc. Symp. Proc. Vol. 557 © 1999 Materials Research Society

EXPERIMENT

The experiments were carried out on three types of PECVD reactors: a coaxial reactor for producing cylindric substrates (4500cm² average substrate area, 12cm substrate diameter, 20cm diameter of the outer electrode, 120cm substrate length, 13.56 MHz RF); a large area parallel-plate reactor, e.g. for solar cell production (plasma box principle, 60×100cm² double-sided RF-electrode, 27.12 MHz RF) [1]; and a 600mm-long plasma source (780cm² average source area, 60cm maximum substrate width, 35mm electrode/substrate distance, continuous operation principle, 81.36 MHz RF). The RF power is capacitively coupled in all RF electrodes.

RESULTS

A) Standing waves and evanescent waveguide modes

Standing waves (TEM waves) and evanescent waveguide modes (TE waves) cause inhomogeneities of the electric field between the electrodes of large area systems for amorphous silicon deposition. Inhomogeneous voltage distributions, deposition rates and film thicknesses result [2]. Figure 1 shows the sum of TEM and TE waves and the non-uniformity of the deposition rate, calculated and measured at the electrode surface of the large area parallel-plate reactor.

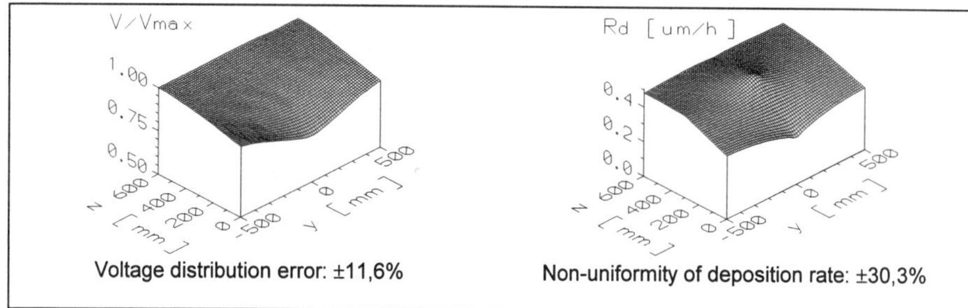

Fig.1: 27.12 MHz RF-singlepower feeding at the point y=z=0: The sum of standing TEM mode and evanescent waveguide modes TE at the electrode surface (left side) and the non-uniformity of the deposition rate R_d (15Pa silane, 10mW/cm²; right side).

Figure 2 shows some solutions to reduce the non-uniformity of the deposition rate, e.g. multipower feeding, central backside power feeding, the use of load impedances and the electrode segmentation. The standing wave and the evanescent waveguide modes were suppressed [2].

Multipower feeding (Fig. 2a,b) can be used for quadratic or rectangular parallel-plate reactor configurations, especially in the case of the impossibility of a central backside feeding at the plasma box principle [1]. The central backside power feeding [3, 4] is suitable for round, quadratic or rectangular parallel-plate reactor configurations (Fig. 2c). Easy access to the backside of the electrode is necessary. High transmission line currents, in comparison to multipower feeding, lead to extreme power losses in large area systems. The use of load impedances (Fig. 2d), especially for coaxial systems but also for quadratic or rectangular parallel-plate reactor configurations, can be realized by load resistances or load inductances. The use of load resistances leads to higher homogenization effects, but the power loss in the impedance (10Ω or less) is 50% or more. Electrode segmentation (Fig. 2e) is the only possibility of homogenization with extremly large electrode areas including one-dimensional extended electrodes. A combination of the discribed methods, e.g. of multipower feeding and load impedances (Fig. 2f), is possible [2].

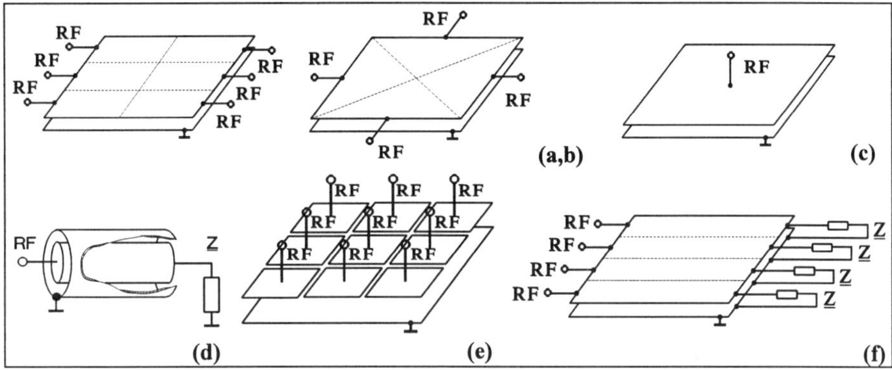

Fig. 2: Solutions to reduce the non-uniformity of deposition rate and film-thickness: Multipower feeding (a and b), Central backside power feeding (c), Load impedances (d), Electrode segmentation (e) and the combination of a and d (f)

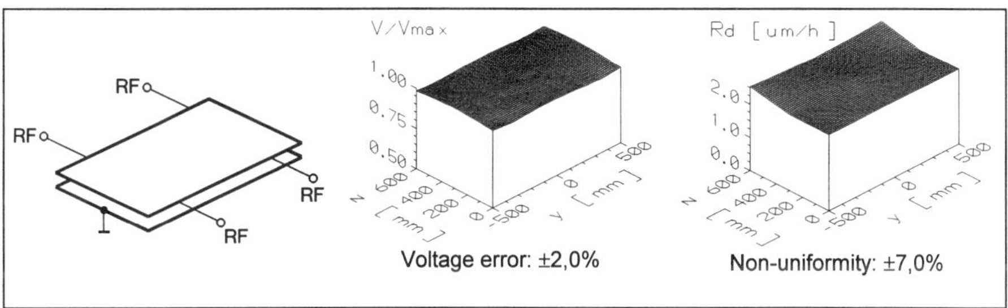

Fig. 3: 27.12 MHz RF-multipower feeding in comparison to Fig. 1· The location of the power feeding points (left side) and the distributions of the voltage V/V_{max} (center) and the deposition rate R_d (10Pa silane, 40mW/cm²; right side)

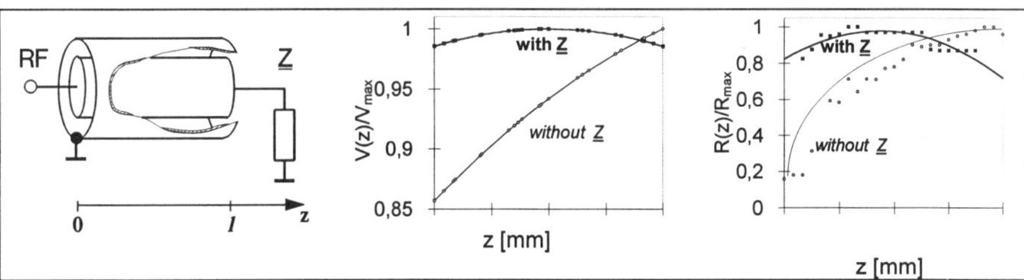

Fig. 4: 13.56 MHz-coaxial system with a load impedance: The location of the power feeding point and the load impedance (left side), the voltage distribution $V(z)/V_{max}$ (center) and the deposition rate R_d (20Pa silane, 16mW/cm²; right side) without and with the load impedance Z

The examples in the Figs. 3-5 represent some results of realized homogenization solutions. Figure 3 shows fourfold-power feeding at the large area parallel-plate ractor like in Fig. 2a. The use of a load impedance at the coaxial reactor is shown in Fig. 4. Figure 5 compares the effect of the electrode segmentation on the voltage and deposition rate distribution in comparison to singlepower feeding. A mechanical separation of the RF electrode is not necessary due to equal magnitudes and phase angles at all power feeding points.

The voltage distribution errors and the non-uniformities of the deposition rate decrease extremely in all three examples.

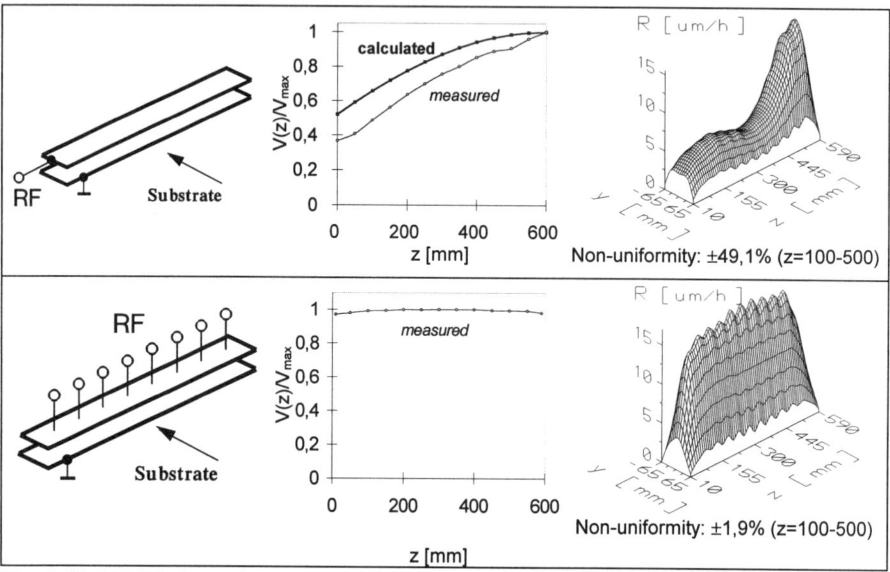

Fig. 5: 81.36 MHz-long plasma source, comparison of singlepower feeding (top) and segmentation (bottom): The location of the power feeding points (left side) and the distribution of the voltage $V(z)/V_{max}$ (center) and the deposition rate R_d (16Pa silane, 13mW/cm²; right side)

B) Limitation of the plasma power

Increasing the deposition power and/or the frequency resulted in the following characteristics:
- The plasma impedance Z_{pl} decreases (Fig. 6a).
- The phase angles φ of the discharge and of the impedance at every point at the transmission lines tends towards -90° (Fig. 6b).
- The impedances of all parasitic capacities, e.g. the RF electrode capacity, decrease extremly.
- The loss-resistances R' of transmission lines and contacts increase due to the skin effect [R'$\propto\sqrt{f/d}$], process- and loss-caused heating of the lines [R'$\propto$(1+const.$\Delta\vartheta$)]. f is the frequency, ϑ the temperature, d is the diameter of the inner conductor of the transmission line.

Extremly low plasma and parasitic reactor impedances $\underline{Z}$ result in very high currents I on the transmission lines. Towards φ=-90° tending phase angles (cos$\varphi\rightarrow$0) together with the necessity of a constant plasma power $P_{pl}=U{\cdot}I{\cdot}\cos\varphi$ also result in extremly high currents (P_{pl}=const. and $\cos\varphi{\to}0$ means: $I{\to}\infty$). These currents in combination with increasing transmission line and contact resistances lead to extreme power losses P_{in}-P_{pl}. The efficiency of the power coupling to the plasma drops down to η=10% and less, see Fig. 7a. In spite of an increase in the coupled power P_{in}, the plasma power P_{pl} does not increase, see Fig. 7b. Short circuits and destructions of the transmission lines due to extreme thermical expansion or melting are possible. A minimization of the power losses is absolutely necessary!

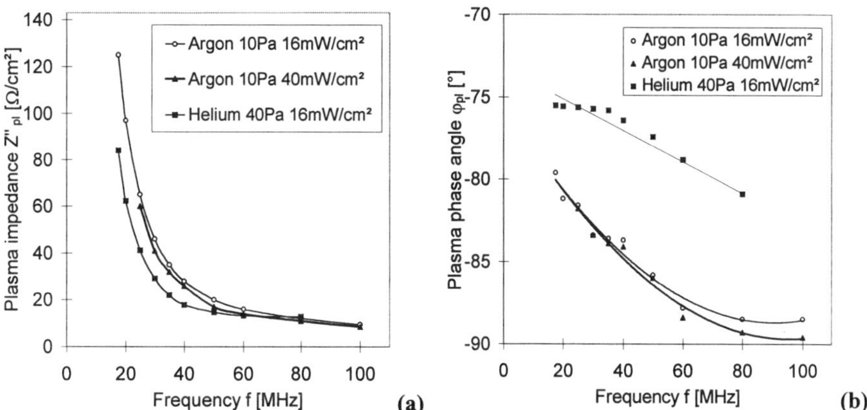

Fig. 6: The influence of increasing frequencies on the impedance Z_{pl} and the phase angle φ_{pl} of the plasma, measured at a parallel-plate reactor (12cm diameter, 32mm electrode distance)

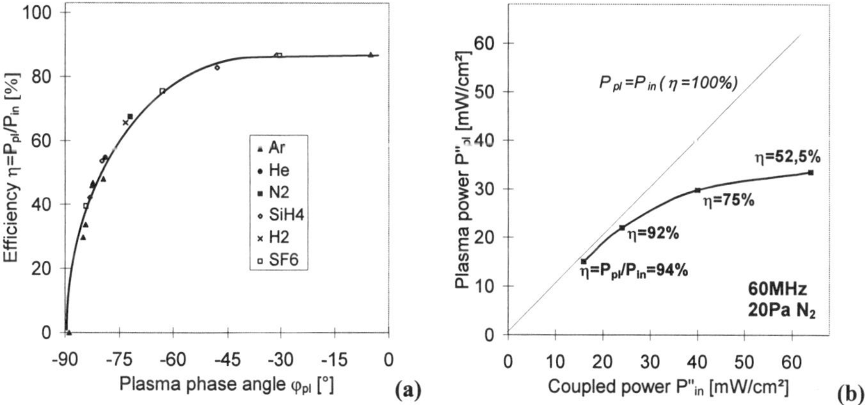

Fig. 7: The influence of the plasma phase angle φ_{pl} on the efficiency η=P_{pl}/P_{in} of the power coupling into the plasma (a); the limitation of the plasma power in spite of increasing powers coupled with the RF generator (b)

Several experiments were carried out at the long plasma source to minimize the currents on the transmission lines. The most advantages results were achieved with impedance-transforming networks (transformers) between the electrode surface and the ends of the transmission lines. The Fig. 8a shows the calculated current I at the transmission lines between the matchbox and the electrode surface without and with the transformer. T11 is one 50Ω-transmission line of 0,8m length, T21-T28 are eight parallel 50Ω or 75Ω-lines between T11 and the electrode segments. Due to the transformers the losses on the transmission lines decrease strongly (see Fig. 8b). Built in transforming networks in the large area parallel-plate reactor also lead to decreasing power losses.

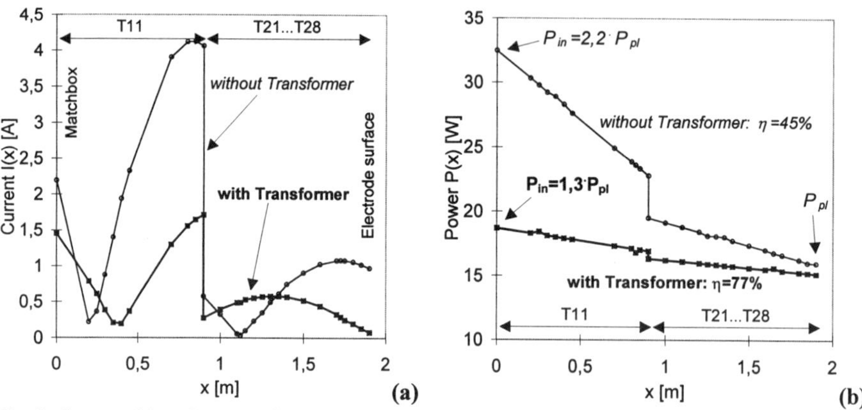

Fig 8: Current (a) and power distribution (b) on the transmission lines for power coupling of the 600mm-long plasma source (40Pa nitrogen, 16mW/cm², 81.36 MHz)

CONCLUSIONS

The homogenization of the voltage and the deposition rate at higher frequencies and deposition areas is theoretically possible and was proven by several experiments. More difficult but also realizable is the minimization of currents and power losses at the transmission lines. This could be seen with calculations and experiments on a long plasma source and a large area system. A doubling of the power coupled to the plasma at a constant generator power is possible. The result is a increasing deposition rate and a increasing lifespan of all transmission lines.

ACKNOWLEDGEMENT

This work was supported by the Bundesministerium für Bildung, Wissenschaft, Forschung und Technologie (F.R. Germany) under contract No. 0329563B.

REFERENCES

[1] A. Bubenzer, J.P.M. Schmitt, Vacuum **41**, pp. 1957-1961 (1990).
[2] J. Kuske, U. Stephan, W. Nowak, S. Röhlecke, A. Kottwitz, MRS Symp. Proc., 467, pp. 591-595 (1997).
[3] J. Kuske, U. Stephan, O. Steinke, S. Röhlecke, MRS Symp. Proc., 377, pp. 27-32 (1995).
[4] L. Sansonnens, A.A. Howling, Ch. Hollenstein, MRS Symp., Amorphous Silicon Technology, 1998, San Francisco, USA

ASSESSMENT OF INTRINSIC-LAYER GROWTH TEMPERATURE TO HIGH-DEPOSITION-RATE a-Si:H n-i-p SOLAR CELLS DEPOSITED BY HOT-WIRE CVD

QI WANG, EUGENE IWANICZKO, YUEQIN XU, BRENT P. NELSON, AND A. H. MAHAN
National Renewable Energy Laboratory (NREL), Golden, CO, USA

ABSTRACT

We report progress in hydrogenated amorphous silicon n-i-p solar cells with the i-layer grown by the hot-wire chemical vapor deposition technique. Early research showed that we grew device-quality materials with low saturated defect density ($2 \times 10^{16}/cm^3$), high initial ambipolar diffusion length (~2000 Å) and low hydrogen content (<1%). One of the major barriers to implementing this material into solar cells is the high substrate temperature required (>400°C). We re-assess the effects of low substrate temperature on the property of the films and the performance of the solar cells as an alternative avenue to solving this problem. We find that the material grown at 300°C can have similar values of saturated defect density and ambipolar diffusion length as the one grown greater than 400°C. We also study the effect of i-layer substrate temperature ranging from 280° to 440°C for n-i-p solar cells. We now consistently grow devices with Fill Factor (FF) greater than 0.66, with the best close to 0.70 at lower substrate temperature. A collaboration with United Solar System, in where they grew the p-layer and top contact, produced devices with initial efficiencies as high as 9.8%. We produce n-i-p solar cells with initial efficiencies as high as 8% when we grow all the hydrogenated amorphous silicon and top contact layers. All these i-layers are grown at deposition rates of 16 to 18 Å/sec. We need to further improve our p-layer and transparent conductor layer to equal the collaborative cell efficiency. We also report light-soaking results of these devices.

INTRODUCTION

Hydrogenated amorphous silicon (a-Si:H) fabricated by hot-wire chemical vapor deposition (HWCVD) has demonstrated its unique structural and electronic properties in the past [1-4]. Among them, the low saturated defect density (N_d^{sat}) ($2 \times 10^{16}/cm^3$), high ambipolar diffusion length (L_{sspg}) (~2000 Å), and the high deposition rate (>10 Å/sec) are the most important factors that apply to solar cells. It means the solar cell made by HWCVD, or at least the intrinsic layer, will have less Staebler-Wronski Effect (SWE) due to the low N_d^{sat}, and better performance due to the high L_{sspg}. Much work has been done to implement this material into solar cells. Unfortunately, we normally grow the material at a substrate temperature (T_{sub}) >400 °C, and the high T_{sub} becomes the major barrier to the success of this implementation. For superstrate cells having a structure of glass/TCO/pin/Ag, the p-layer will change its property due to the high temperature of growing an intrinsic layer on top of it, leading to a poor fill factor [5]. It is well known that n-layers have better resistance to the high processing temperature than p-layers. Therefore, the substrate solar cells having a structure of SS/nip/TCO are the natural choice. However, we still face the difficulty of implementing the material into solar cells at temperatures >400°C [6-9]. Because of this, it is necessary to search for the low substrate temperature materials. We re-assess the effects of low substrate temperature on the property of the films and the performance of the solar cells as an alternative avenue to solving this problem.

In this paper, we focus on searching for the low temperature materials and implementing the materials into the solar cell device. If we succeed, the advantages of the HWCVD technique,

Mat. Res. Soc. Symp. Proc. Vol. 557 © 1999 Materials Research Society

— a simple process, with high deposition rates and good-quality material — may see applications outside the laboratory. Recent achievement of close to 10% initial efficiency [8,9] shows promise.

EXPERIMENTAL DETAILS

We use the two-chamber load-locked T-system to fabricate a-Si:H thin films and hybrid nip solar cells. One chamber is an HWCVD reactor and is only used to grow a-Si:H intrinsic films and solar cell i-layers. The other chamber is a plasma-enhanced (PE) CVD reactor and is only used to fabricate hybrid nip solar cells growing dopant layers such as n- and p-layers. The details of this HWCVD reactor are similar to the one we reported elsewhere [10]. The major differences are that it has a larger cross section of the inductance (5 in. in diameter) and an isotherm tube heater. The substrate is transported to the center of the tube, and a thermocouple is attached to the substrate holder to measure the temperature. This helps us to determine the substrate temperature more accurately. We use a spiral tungsten wire with a diameter of 0.5 mm and place it 5 cm below the substrate. The tungsten filament is heated using an AC current to about 2000°C. A process gas passes by the hot filament, dissociates, and leads to Si/H deposition on the substrate. The PECVD chamber is a standard reactor. With the aid of load-lock system, the substrate can be transported between HWCVD and PECVD chambers without breaking in the air. Therefore, it minimizes the cross contamination in solar cell fabrications.

All the films are grown at the same condition except the substrate temperature. We grow films close to one micrometer on Corning 7059 glass and crystal silicon substrates. The structure of each solar cell is a substrate type of the silver-coated textured stainless steel (SS)/a-Si n/a-Si i/a-Si p/ITO. The top contact is a 700 Å indium tin oxide (ITO) deposited by a reactive thermal-evaporation method. Doped layers are made in the PECVD reactor at a substrate temperature around 200°C, and with a thickness of 400 Å for n-type and 100 Å for p-type. The i-layer thickness is about 2000 Å. We intentionally keep the growth conditions of both doped layers and the thickness of all layers unchanged in all devices. We study a series of five solar cells with various a-Si:H i-layers grown in HWCVD chamber by changing the substrate temperature from 280° to 440°C.

The hydrogen content is determined using the IR spectrometer, and the defect density using the constant photocurrent method (CPM). The thickness of the i-layer is measured using a Tencor Instrument Alpha-Step 200 profilometer. The thickness of each layer in the solar cell is estimated from the deposition rate of each film grown under the same deposition conditions. United Solar provides the silver-coated textured SS. The solar-cell performance is measured using a computer-interfaced JV station under 1-sun illumination (Xenon lamp). We use 1-sun ELH light source to perform the light soaking.

RESULTS AND DISCUSSIONS

One of the advantages of a-Si:H materials made by the HWCVD technique is its low saturated defect density — $2 \times 10^{16}/cm^3$ — we at NREL found in 1991 [1]. This material has brought great attention and great hope for a-Si:H solar cells. It has been believed that such low defect density will lead to a more stable solar cell and increase solar cell performance. Recently many groups [11-14] have grown similar low-defect-density materials using hydrogen dilution by the PECVD technique. Indeed, the solar cells using such materials have shown improved stability against light soaking. On the other hand, we have made steady progress in a-Si:H nip solar cells using the HWCVD technique. In this paper, we will report one of the successful stories of using lower-substrate-temperature intrinsic layer.

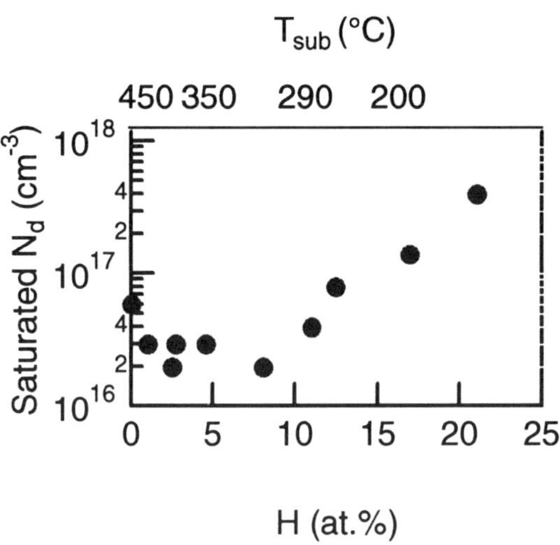

Figure 1. Saturated defect density as function of H content and the substrate temperature.

As an alternative to solving the problem of high-substrate-temperature material incorporated into the solar cell, we study the temperature dependence of the materials [15] and solar cells. We find that the substrate temperature has a profound effect on the material properties such as hydrogen content and optical band gap. For the benefit of a-Si:H solar cell, we focus our attention on the saturated defect density. Figure 1 shows the results of the saturated defect density as a function of hydrogen content and substrate temperature. The defect density is measured using the constant photocurrent method after 1000 hours of 1-sun light soaking. Indeed, the saturated defect density strongly depends on substrate temperature. We find that the materials with a defect density of 2×10^{16}/cm^3 are within 300°-500°C substrate temperature and with 1%-8% hydrogen content, respectively. The material grown at 500°C with close to 0% hydrogen has 6×10^{16}/cm^3 defects. It is worth mentioning that this defect density barely changes during the light soaking. For the materials grown at lower than 300°C, the defect density increases dramatically. With this in mind, we find that we can grow low-defect material even at lower temperature, for example 300°C.

The substrate temperature effect on the nip solar cell is shown in Figure 2. All the cells have a structure of SS/nip/ITO and are fabricated at NREL. The SS substrate with textured back reflector coating is provided by United Solar.

In Figure 2a, we show the dependence of the open-circuit voltage (V_{oc}) and FF on the substrate temperature from 280° to 440°C. As the substrate temperature increases, V_{oc} starts to decrease. This can be explained in that V_{oc} decreases as the band gap of the i-layer decreases [16]. As we mentioned above, increasing substrate temperatures of the material will decrease its band gap. Qualitatively, the decrease of V_{oc} scales with the band gap of the i-layer, which decreases with increasing substrate temperature. We made devices with FF >0.6 from 280° to 400°C. At 440°C the FF drops to 0.5. We believe that high process temperatures damage the material grown underneath and also the interface. We find that such critical temperature with

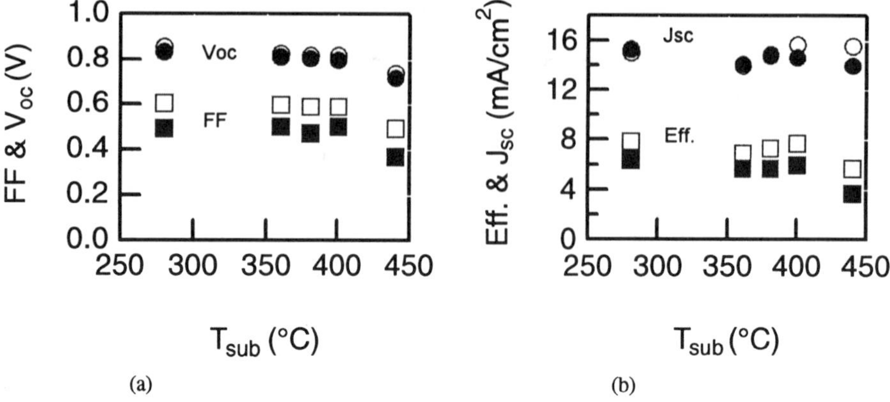

Figure 2. Effects of i-layer growth temperature on the solar cell performance and light-soaking results. Open symbols are the initial values. Solid symbols are the results after 1-sun light soaking for 900 hours.

minimum damage to the solar cell performance is around 400°C. A general trend is that FF declines with the increase of the substrate temperature. We still face the problem of implementing materials grown at greater than 400°C into our solar cell.

In Figure 2b, we show the dependence of the short-circuit current (J_{sc}) and solar cell efficiency (Eff.) on the substrate temperature from 280° to 440°C. J_{sc} increases with both decreased i-layer band gap and increased substrate temperature. However, this increase is only slight and not enough to compensate the loss of V_{oc}. The efficiency of the solar cell — which relates to the product of FF, J_{sc} and V_{oc} — still declines with the increase in substrate temperature.

Despite this, we have achieved initial solar cell efficiencies >8% using in-house deposition capacities, with an i-layer deposition of about 18 Å/sec. The solar cell has J_{sc} = 15.2 mA/cm^2, V_{oc} = 0.855 V, and FF = 0.623. We will address the issue of how to improve the in-house solar cell later.

We also show light-soaking results in both Figures 2a and 2b. Both V_{oc} and J_{sc} show less degradation than FF. The degradation of the efficiency is mostly coming from the degradation of FF. The best efficiency of a degraded solar cell is 6.4%. We note that all the solar cells experience almost the same degree of degradation regardless of substrate temperatures. We question whether this degradation may come from the other sources, such as both dopant layers and interfaces. We believe that the intrinsic layer is more stable than the other layers, based on studying the materials during the light soaking from materials study.

Nevertheless, we are still able to make an initial solar cell with close to 10% efficiency with an i-layer deposition of about 18 Å/sec and when grown <400°C. Figure 3 shows a solar cell made in the collaboration with United Solar. We deposit n- and i-layers at NREL, ship it to the United Solar, and they finish up with the p-layer and ITO top contact. Also they perform the solar cell measurement and light soaking. The best initial efficiency with this collaboration is

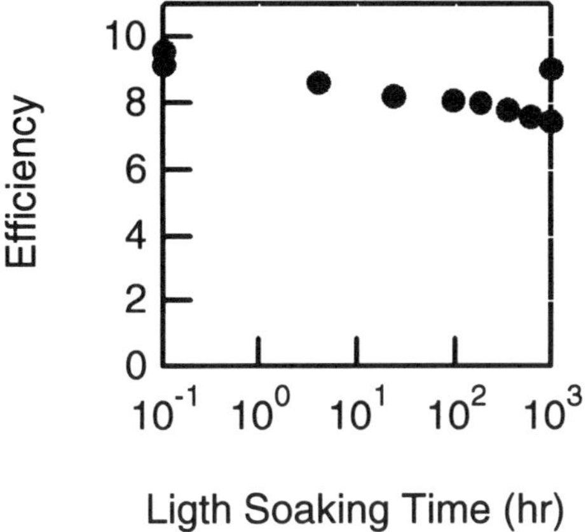

Figure 3. Degradation of the United Solar/NREL solar cell with an HWCVD i-layer at a deposition rate of 18 Å/sec.

9.81% with J_{sc} =16.31 mA/cm^2, V_{oc} = 0.921 V, and FF = 0.652. In comparing with our in-house cell performance, the biggest gain is the V_{oc}. It increases from 0.85 to 0.92 V with a proprietary p-layer. It also shows a significant increase of J_{sc}.

The degradation of the solar cell is composed of two parts. The first part is related to SWE. Our best efficiency after 1000 hours light soaking is 7.7% with a degradation of 14%. The degradation is compatible with the solar cell in which the i-layer material with the similar defect density is grown by the PECVD technique. It is remarkable for an i-layer grown at 18 Å/sec. The cell does recover after annealing. We plot this point at 1000 hours light-soaking time. In second part, there is 4%-5% degradation even without shining the light. We show it at initial light-soaking time for simplicity. Certainly it is not related to the SWE, and it cannot be recovered by annealing. We are investigating this issue. We find some of the solar cells have this effect and others do not. We hope this initial degradation will naturally go away as we improve our process control.

We have achieved high-device-quality intrinsic materials. For solar cell applications, one of the many important issues is how to make all component layers work together. We knew from previous work that there is a fundamental incompatibility between HW intrinsic layers and PECVD dopant layers. Therefore, improving the interfaces becomes a critical issue. To improve the solar cell further, we also need to improve dopant layers — especially the p-layer and other components.

CONCLUSIONS

In studying the substrate temperature dependence of both i-layer materials and nip solar cells, we find that the substrate temperature has a profound effect on the materials and solar cell performance. We find that for i-layers grown at 300°C the saturated defect density of the materials can be as low as 3 x $10^{16}/cm^3$. In-house, we are able to make solar cells with initial 8% efficiencies using lower substrate temperature materials. In collaboration with United Solar, we achieve initial results close to 10%.

ACKNOWLEDGEMENTS

The authors enjoy the fruitful collaboration with United Solar Inc. and specially thank J. Yang and S. Guha for their helps. We are also grateful to H. Branz and R. S. Crandall for critical discussions. This work is supported by DOE subcontract DE-AC-98-G010337.

REFERENCES

1. A.H. Mahan, J. Carapella, B.P. Nelson, R.S. Crandall, and I. Balberg, *J. Appl. Phys.*, **69**, p. 6728 (1991).
2. Y. Wu, J.T. Stephen, D.X. Han, J.M. Rutland, R.S. Crandall, and A.H. Mahan, *Phys. Rev. Lett.*,**77**, p. 2049 (1996).
3. A.H. Mahan, D.L. Williamson, and T.E. Furtak, *Mat. Res. Soc. Proc.*, Vol. **467**, p. 657 (1997).
4. X. Liu, B.E. White Jr., R.O. Pohl, E. Iwaniczko, K.M. Jones, A.H. Mahan, B.P. Nelson, R.S. Crandall, and S. Veprek, *Phys. Rev. Lett.*, **78**, p. 4418 (1997).
5. B.P. Nelson, E. Iwaniczko, R.E.I. Schropp, A.H. Mahan, E.C. Molenbroek, S. Salamon, and R.S. Crandall. *12th PVSC*, pp. 679-682 (1994).
6. P. Papadopulos, A. Scholz, S. Bauer, B. Schroeder, and H. Oechsner, *J. Non-Cryst. Sol.*, 164-166, p. 87 (1993).
7. A.H. Mahan, E. Iwaniczko, B.P. Nelson, R.C. Reedy Jr. T. Unold, R.S. Crandall, S.Guha, and J. Yang, *AIP Conf. Proc.*, **394**, p. 27 (1996).
8. S.Bauer, W. Herst, B. Schroeder, and H. Oechsner, *Proc. 26th IEEE PV Spec. Conf.*, p. 719 (1997).
9. A.H. Mahan, R.C. Reedy Jr., E. Iwaniczko, Qi Wang, B.P. Nelson,Y. Xu, A. C. Gallagher, H.M. Branz, R.S. Crandall, J. Yang, and S. Guha. *Mat. Res. Soc. Proc.*, V**507**, p.119 (1998).
10. M. Vanecek, A.H. Mahan, B.P. Nelson, and R.S. Crandall, *Proc. 11th European PV Solar Energy Conf.*, p. 96. (1993).
11. J. Yang, X. Xu, and S. Guha, *Mat. Res. Soc. Proc.* Vol **336**, p. 687 (1994).
12. L. Yang and L.-F. Chen, *Mat. Res. Soc. Proc.* Vol **336**, p. 669 (1994).
13. P. Siamchai and M. Konagai, *Appl. Phys. Lett.*, **67**, p. 3468 (1995).
14. B. Rech, S. Wieder, F. Siebke, C. Beneking, and H. Wagner. *Mat. Res. Soc. Proc.*, Vol **420**, p. 33 (1996).
15. Daxing Han, Tamihiro Gotoh, Motoi Nishio, Tomonari Sakamoto, Shuichi Nonomura, Shoji Nitta, Qi Wang, E. Iwaniczko, Harv Mahan. *National Center for Photovoltaic Program Review*,**462**, p. 260 (1998).
16. Qi Wang, Homer Antoniadis, E. A. Schiff, and S. Guha, *Mat. Res. Soc. Proc.*,Vol **258**, p. 881 (1992).

Part III

Recrystallization, Amorphization and Porous Silicon

A NOVEL EXCIMER LASER CRYSTALLIZATION METHOD OF POLY-SI THIN FILM BY GRID LINE ELECTRON BEAM IRRADIATION

C-M PARK, M-C LEE, J-H JEON, AND M-K HAN
School of EE, Seoul National University, Kwanak-ku, Seoul 151-742, Korea,
e-mail: mkh@emlab.snu.ac.kr, phone: +82-2-880-7992, fax: +82-2-875-7372

ABSTRACT

Excimer laser annealing technique is proposed to increase the grain size and controlling the microstructure of polycrystalline silicon (poly-Si) thin film. Our method is based on the lateral grain growth during laser annealing. Our specific grid ion beam irradiation method was designed to maximize the lateral growth effect and arrange the location of grain boundaries. We observed well-arranged poly-Si grains up to micrometer order by transmission electron microscopy (TEM).

I. INTRODUCTION

Due to their large field effect mobility and high current driving capability, polycrystalline silicon thin film transistors (poly-Si TFT's) have been extensively investigated for the application in active matrix liquid crystal displays (AMLCD's) [1]. In the fabrication of poly-Si TFT's on glass substrate, the most important process is the recrystallization of amorphous silicon (a-Si) deposited at low substrate temperature. Among various recrystallization methods, excimer laser annealing is considered as most promising because it provides good quality of poly-Si film and low thermal budget on glass substrate [2]. It is well known that melting and solidification occurs during laser irradiation on a-Si film [3]. This is the main reason for the low in-grain defect density of excimer laser annealed poly-Si, in contrast to solid phase crystallized (SPC) poly-Si which does not undergo phase transformation from liquid to solid during the recrystallization. However, the grain size of laser annealed poly-Si film varies sensitively with laser energy density, showing a variety of device performance [4]. In the excimer laser recrystallization, there exists some critical energy level which maximizes the grain size, and this energy level is the so-called SLG (super lateral growth) regime [5]. This energy level may be considered favorable for the best electrical performance, however does not ensure the reproducibility because the pulse-to-pulse energy of excimer laser fluctuates rather significantly. At higher energy level over the SLG regime, a-Si melts fully to the bottom and the homogeneous nucleation, which leads to abrupt reduction of grain size, becomes dominant. Under such full melt energy condition, further increase in the energy density would hardly affect the grain size so that some energy fluctuation may ensure the nearly identical size of grain.

The purpose of this paper is to introduce an excimer laser annealing technique to improve both the grain size and the microstructure of poly-Si film. Our method is carried out at the full melt energy condition immune to the laser energy fluctuation. Furthermore, in order to increase the grain size, our method employs a specifically designed masking window, as illustrated in figure 1.

Our masking window has the ordered array of square patterns and we implanted silicon ions through our masking window (Fig. 1(b)). The implanted ions may form mechanical defects in a-Si layer. XeCl excimer laser beam is irradiated on the a-Si layer (Fig. 1(c)). The mechanical defects from implanted ions act as nucleation seed when the crystallization of poly-Si film is being performed. This implantation step is employed to increase the grain size of poly-Si through the lateral grain growth effect. It is well known that the artificial mechanical defects from ion implantation contribute to the crystallization of poly-Si film [6]. Recently, J. S. Im and R. K. Singh have reported about considerably enlarged poly-Si grains by employing a masking window during the laser annealing [7].

Mat. Res. Soc. Symp. Proc. Vol. 557 © 1999 Materials Research Society

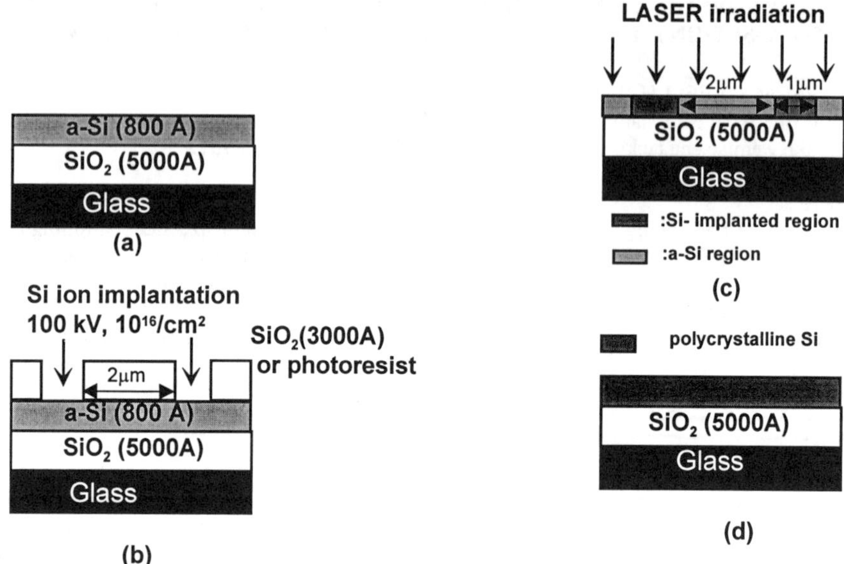

Fig. 1. Schematic diagram of our laser annealing method. The excimer laser irradiates a-Si film, which is implanted by silicon ions through the specifically designed masking window.

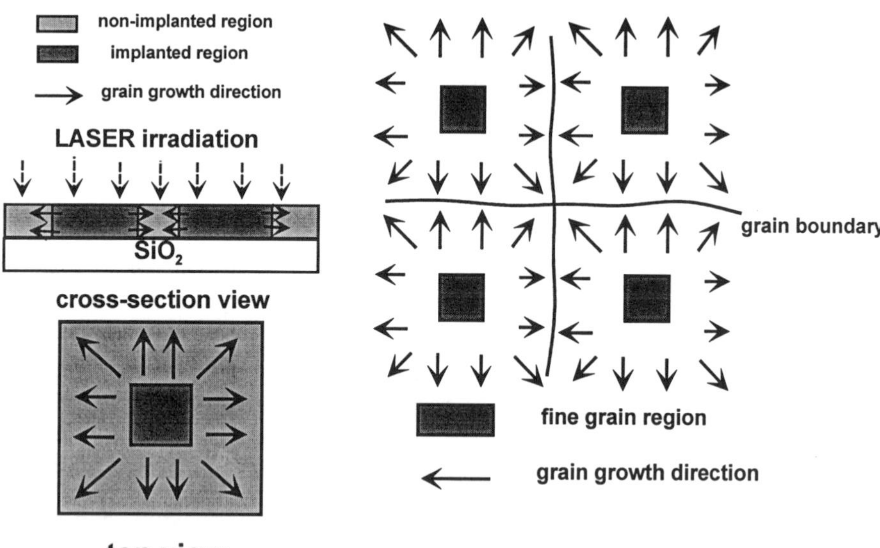

Fig. 2. The illustration of the growth mechanism in our method. Implanted silicon ions act as nucleation seeds, and then the grains solidify in lateral direction.

The enlargement of grain size originated from the lateral growth phenomenon. They verified that when liquid and solid regions have a common boundary due to the partial melting, the liquid region nucleates at the liquid/solid boundary region and the solidifying surface propagates in the perpendicular direction to the boundary, so that the grains grow laterally. In our experiment, we artificially induced mechanical defects by ion implantation and the masking window was specifically designed to maximize the effect of the lateral growth phenomenon. The figure 2 also illustrates the growth mechanism during laser irradiation in this method. The defects in the implanted region may nucleate the a-Si film faster than the other a-Si region, where is not implanted (Fig. 2 (a), (b)). The solidification of a-Si layer proceed in lateral direction, so that the grains may be enlarged until two grains, grown from both the edges, encounter in the middle of implanted region (Fig. 2(c)). Our experimental results show that the distance between making patterns can be expanded up to a few micrometer orders.

II. EXPERIMENTAL RESULTS

We investigated the microstructure of the recrystallized poly-Si films by transmission electron microscopy (TEM). A 800 Å thick a-Si films were deposited with plasma enhanced chemical vapor deposition (PECVD) at 250°C on the glass substrate (Fig. 1(a)). Dehydrogenation was carried out at 400°C for 2hours. After photolithographic step, the a-Si filme was implanted by Si+ ions (100Kev, 10^{16}ions/cm^2) through our masking window (Fig. 1(b)). The XeCl excimer laser(λ=308nm) was irradiated on a-Si film. We employed the single shot of the XeCl excimer laser. The energy density of the laser was 300mJ/cm^2. In our system, this energy level melts 800Å thick a-Si layer fully to the bottom of layer. The substrate temperature was maintained at 200°C.

Figure 3 shows the plane view TEM (transmission electron microscopy) image of conventionally recrystallized poly-Si film by XeCl excimer laser. The fine grains are observed. The plane TEM images of the poly-Si film, which is crystallized by proposed method, are presented in Fig. 4. In the low resolution TEM image (Fig 4(a)), we observed the overall feature of selectively crystallized poly-Si film. The fine grain regions of poly-Si film are allocated in ordered array, which is determined by implantation windows (Fig. 4(b)).

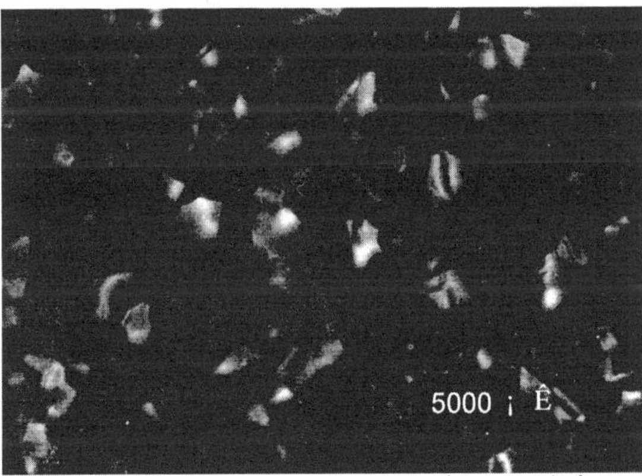

Fig. 3. The plane view TEM (transmission electron microscopy) image of conventionally recrystallized poly-Si film by XeCl excimer laser.

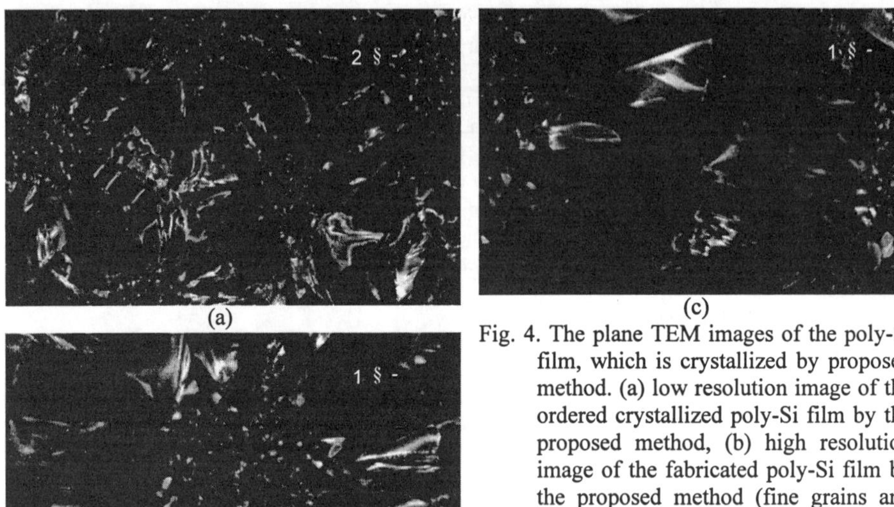

(a)

(c)

(b)

Fig. 4. The plane TEM images of the poly-Si film, which is crystallized by proposed method. (a) low resolution image of the ordered crystallized poly-Si film by the proposed method, (b) high resolution image of the fabricated poly-Si film by the proposed method (fine grains and large grains), (c) large grains meet in the mid-region between the fine grain region.

Fig. 5 The border of fine grain and large poly-Si grain region.

The large and clear poly-Si regions are located around the fine grain region. The structure of the large grain poly-Si indicates that the grain grew along the lateral direction. The shape of fine grain region is rather circular. This is because the angle of square pattern was smoothened during the photolithography and etch step. Most of all, the several large and elongated poly-Si grains are outstanding. They spread out from the residual fine grain regions, which looks like a sunflower. The elongated poly-Si grains grown from both the opposite sides encounter in the mid-region and the length of each grain is nearly 1 μm, which is half the distance between implantation window (Fig. 4(c)). This result indicates that our specific implantation window method is very successful to obtain the lateral growth of grains and the ordered arrangement of grain boundaries. Further increase of grain size is also possible through enlarging the distance of masking patterns and optimizing laser energy density as long as the lateral growth reaches.

Figure 5 shows border of fine grain and large poly-Si grain region. This figure shows from where the grain growth starts during laser crystallization. We can clearly see that the large poly-Si grains grew from the fine poly-Si region edge. Consequently the small grains which originated from the vertical growth were obtained. From these results, we can conclude that lateral growth takes place from the mechanical defects by silicon ion implantation.

IV. CONCLUSION

We proposed a novel excimer laser annealing technique in order to increase the grain size and controlling the microstructure of polycrystalline silicon (poly-Si) thin film. Our method is based on the lateral grain growth during laser annealing. Our specific grid ion beam irradiation method was designed to maximize the lateral growth effect and arrange the location of grain boundaries. We observed well-arranged poly-Si grains up to micrometer order by transmission electron microscopy (TEM).

REFERENCE
1) R. E. Proano, R. S. Misage and D. G. Ast, IEEE Trans. Electron Devices **ED-36**, 1915 (1989).
2) T. Serikawa, S. Shirai, A. Okamoto and S. Suyama, IEEE Trans. Electron Devices **ED-36**, 1929 (1989).
3) H. Watanabe, H. Miki, S. Sugai, K. Kawasaki and T. Kioka, Jpn. J. Appl. Phys. **33**, 4491 (1994).
4) N. Kodama, H. Tanabe, K. Sera, K. Hamada, S. Saitoh, F. Okumura and K. Ikeda, Solid State Devices and Materials, 431, (1993)
5) H. J. Kim and J. S. Im, Appl. Phys. Lett. **63**, 1969 (1993).
6) T. Yamaoka, K. Oyoshi, T. Tagami, Y. Arima, K. Yamashita and S. Tanaka, Appl. Phys. Lett. **57**, 1970 (1990).
7) H. J. Song and J. S. Im, Appl. Phys. Lett. **68**, 3165 (1996).

LOW-TEMPERATURE PREPARATION OF POLY-SI THIN-FILMS HAVING GIANT GRAINS

Wen-Chang Yeh and Masakiyo Matsumura
Department of Physical Electronics, Tokyo Institute of Technology,
2-12-1 O-okayama, Meguro-ku, Tokyo 152-8550, Japan
Email: wyeh@silicon.pe.titech.ac.jp   URL: http://silicon.pe.titech.ac.jp

ABSTRACT

A low heat-capacitance substrate has been introduced to lateral grain-growth for Si thin-films. Grains as large as 13μm could be fabricated by irradiation of a single shot of an excimer-laser light pulse with intensity gradient. The film has an orientation in the (110) direction, different from (111) for the conventional excimer-laser-annealed film, and is applicable to a seed layer for solid-phase epitaxial-growth aiming at solar cell application.

INTRODUCTION

Although solar cells made from Si bulk crystal are characterized fairly well, their production cost is estimated very difficult to be lowered satisfactorily for worldwide use. Thin-film solar cells based on poly-Si or a-Si are very attractive since they can offer dramatic improvement from a cost-performance viewpoint. It is, however, clarified experimentally [1,2] and theoretically [3,4] that recombination of photo-generated carriers at the grain boundary should be minimized for high-performance poly-Si solar cells. Grains should be much more than the active-layer thickness to enable for carriers to travel through the active layer without recombination. Since Si thin-films should be more than several μm for high conversion efficiency, the grain size should be more than a few 10μm. Many methods have been studied so far for preparation of poly-Si films aiming at solar cell application, and a two-step process, i.e., formation of a Si seed layer on non-Si substrate and epitaxial growth on it, seems the most attractive at present. Since the grain size in the epitaxial layer equals to that in the seed layer, the formation of the giant-grain Si seed layer is a key technology for poly-Si solar cells.

We have proposed a new structure and process [5] for poly-Si solar cells where the seed layer is grown by lateral growth initiated by excimer-laser irradiation [6,7]. The lateral growth is enhanced dramatically by introducing the intensity gradient to the excimer laser light, and the growth time is elongated by introduction of a porous SiO_2 film as an underlayer. In this paper, we present detailed experimental results on the formation of the giant-grain seed layer.

EXPECTED EFFECTS OF POROUS SUBSTRATE

Schematic is shown in Fig.1 for lateral-growth mechanism triggered by excimer-laser light irradiation. Due to the light intensity gradient along the film surface, thermal energy stored in the molten Si thin-film changes along the surface. The lateral growth then starts from the edge of the completely molten region (shown as A) using the unmolten Si as the seed just at the finish of laser light irradiation, and stops when the region (shown as B) irradiated with the

Mat. Res. Soc. Symp. Proc. Vol. 557 © 1999 Materials Research Society

highest intensity losses its excess thermal energy Q_{ex} per unit area and solidifies completely. Since heat flows vertically to the substrate by diffusion, Q_{ex} should be written approximately as

$$Q_{ex} = C_{sub}(T_c - T_{sub})\sqrt{\pi D_{sub} t}, \qquad (1)$$

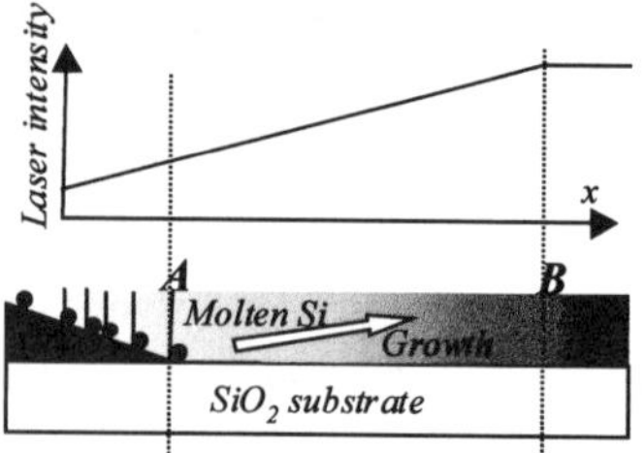

Fig.1 Schematic view of lateral growth mechanism

where C_{sub}, T_c, T_{sub} and D_{sub} are the heat capacitance per unit volume of the substrate, the critical temperature for crystallization, the substrate temperature and the diffusion constant of the substrate, respectively, and t is the time for complete solidification after the laser light irradiation, that is, the lateral growth time. Since the lateral-growth velocity v is on the order of 10m/s [8] at the most, the maximum lateral-growth length L is proportional to t. From these relationships, L is concluded proportional to C_{sub}^{-2}. Since heat capacitance per unit volume of matter is predominantly determined by number density of atoms, it can be reduced dramatically only by making many pores inside the substrate.

Guided by these considerations, we have proposed recently a new poly-Si thin-film solar-cell structure shown in Fig.2 [5]. The active layer is a Si epi-layer grown on a thin Si seed layer. Since the grain size in the active layer equals to that in the seed layer, the gain size of the seed layer plays a critical role for solar cell efficiency. Combination of the phase-modulated excimer-laser annealing (PMELA) method [7] and a porous SiO_2 substrate with porosity P is very effective to grow giant grains for the seed layer. There are also the

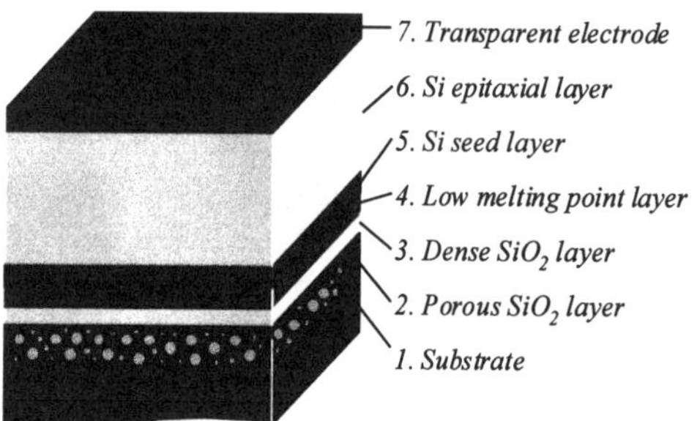

Fig.2 Proposed solar cell structure

low melting-point layer and the dense SiO_2 thin layer. The low melting-point layer plays another important role for enlargement of grain size by reduction of T_c and resulting enlargement of v, but is not discussed in this paper. The dense SiO_2 thin layer is of a flat and high-quality layer, which is very effective for burying small pores on the porous SiO_2 surface and for making a barrier against mixing effects between the Si seed layer and the porous SiO_2 substrate during PMELA. Since this dense SiO_2 layer with thickness d_{ox} acts to store a part of Q_{ex}, thermal equilibrium equation for the structure shown in Fig.2 should be modified as

$$Q_{ex} = C_{ox}(T_c - T_{sub})\left\{d_{ox} + (1-P)(-d_{ox} + \sqrt{\pi D_{sub} t})\right\}, \qquad (2)$$

for the case $d_{ox} \ll \sqrt{\pi D_{sub}t}$. Where C_{ox} is the heat capacitance per unit volume of bulk SiO_2 and we have taken into consider the fact that the heat capacitance per unit volume of porous SiO_2 is roughly given by $C_{ox}(1-P)$. Since thermal conductivity is roughly proportional to $(1-P^{2/3})$ for porous SiO_2 under the assumption of no infrared radiation [9], D_{sub} is given by $D_{ox}(1-P^{2/3})/(1-P)$

where D_{ox} is a diffusion constant of bulk SiO_2. Then L becomes

$$L = L_0 \left\{ 1 - d_{ox} P / \sqrt{\pi D_{ox} t_0} \right\}^2 / (1 - P)(1 - P^{2/3}), \qquad (3)$$

where L_0 and t_0 are the lateral-growth length and time for the non-porous substrate (P=0) .

SAMPLE PREPARATION

The 3μm-thick porous SiO_2 film was deposited by a spin coating technique on 1.2μm-thick thermal oxide, by adding triphenylsilanol to the commercially available SOG (Spin-on-glass) source. Then phenyl groups were evaporated from the film by curing at about 700°C for 30min. The porous SiO_2 films are now studying extensively for the ultra-low dielectric-constant insulating thin-films aiming at ULSIs [10], and it is said that P can be increased to more than 0.9 without difficulty. Mechanical strength, however, decreases quickly with increasing P. Thus we restricted P to less than 0.75. Figure 3 shows a SEM photograph for the porous SiO_2 with P=0.5. Pores were as small as 100nm in diameter.

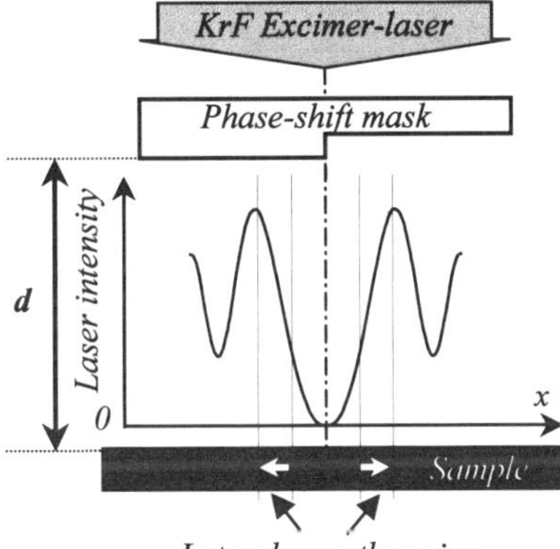

Fig.3 SEM image of porous silica in cross section   Fig.4 Schematic view of PMELA

After the 50nm-thick PECVD SiO_2 layer was deposited on the porous SiO_2 layer at 350°C as the dense SiO_2 thin layer, the 200nm-thick Si layer was CVD-deposited from disilane at 500°C. Then the sample was crystallized by a single shot of KrF excimer laser light pulse through a phase shift mask with a phase retardation of 180°, as schematically shown in Fig.4. Then mutual interference effects modulates the light intensity on the sample surface, that is, light intensity on the center line is zero and spacing between primary (left and right) interference patterns is proportional to square of distance d between the sample and the phase shift mask. Since lateral growth occurs within each interference pattern, L for a given P value seems proportional to square root of d if the vt product is sufficiently large. During the crystallization the sample was heated at 500°C. The average energy density of an incident laser light pulse was

550-600mJ/cm^2 which was the maximum permissible density without peeling-off of the film, and the largest grain size could be obtained in this condition. The film was smooth and had no morphology under as-crystallized conditions. SEM observations were done after grain boundary was clarified by Secco etching.

EXPERIMENTAL RESULTS

Figure 5 shows the surface SEM images for the crystallized Si film on the SiO$_2$ substrate with P=0.75 (a) and P=0 (b), respectively. D was fixed at 5mm. Lateral crystallization starting from low intensity region toward high intensity region could be confirmed. The grain size was 9μm in length along the intensity gradient which was 3 times as large as that with P=0.

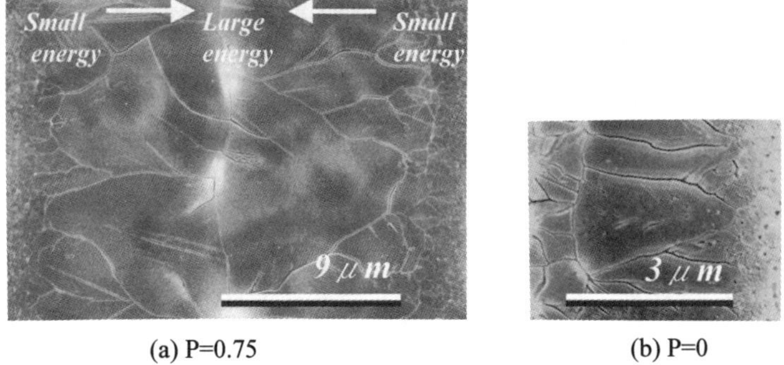

(a) P=0.75   (b) P=0

Fig.5. Surface SEM image of crystallized Si film after Secco-etching

Grain sizes with P=0.75 are shown in Fig.6 as a function of d. It increased at first with increasing d with square root of d, and saturated at 13μm for d more than 5mm.

Figure 7 shows the P dependence of the saturated grain size. Grain size increased with P, and reached 13μm at P=0.75 which was 4 times as big as that at P=0. Solid curves in the figure are the lateral-growth length L estimated from Eq.(3) for several t_0 values. L_0 was fixed at 3μm.

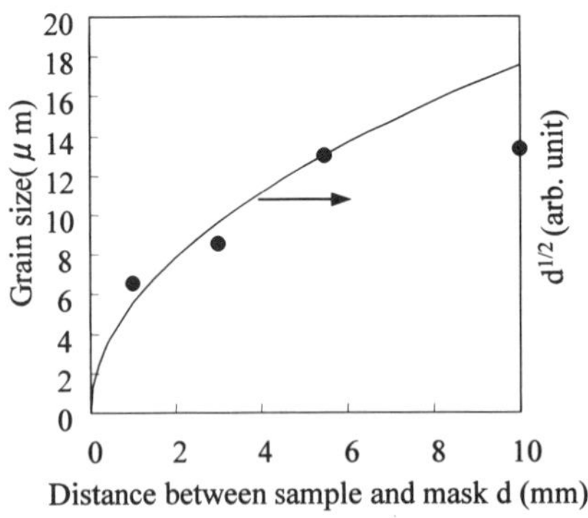

Fig.6. Grain sizes as a function of distance d

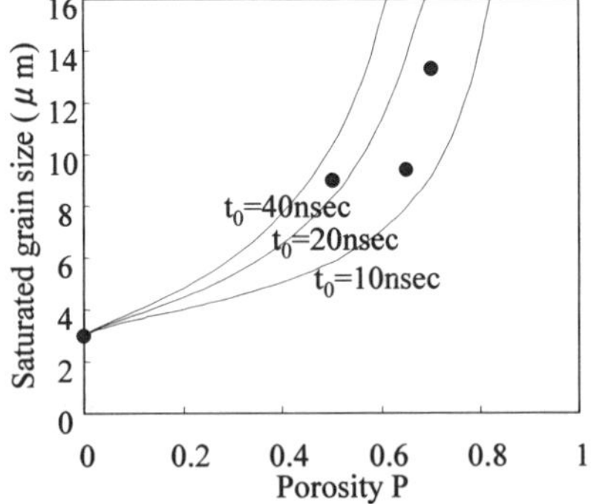

Fig.7. Porosity dependence of saturated grain size

Experimental data coincided with curves for t_0=10ns~40ns. Thus, we expect that grain size will be increased to more than 100μm at P=0.9. Elongation of t_0 is very effective for enlargement of the grain size. This condition can be achieved by increasing Q_{ex} or decreasing T_c. Q_{ex} can be increased by thickening the Si film or inserting high heat capacitance layer. T_c can be reduced by introducing the low melting-point layer as shown in Fig.2.

Figure 8 shows the X-ray diffraction spectrum of the giant-grain Si film grown on a quartz substrate. Two peaks corresponding to (111) and (110) faces were observed at 28.5° and 47.5°, respectively. The (100) peak was not observed at near 69°.

After the measurement, the sample was etched by a solution consisting of ethylenediamine, pyrocatechol and water (EPW) in order to investigate the crystal orientation. Since the etching rate along <110> orientation is 30 times as fast as that along the <111> orientation, (110)-faced grains are selectively eliminated by the EPW etching. Inserts in Fig.8 show zooms of (111) peak and (110) peak before and after the EPW etching. The selective etching between (111) and (110) grains were confirmed.

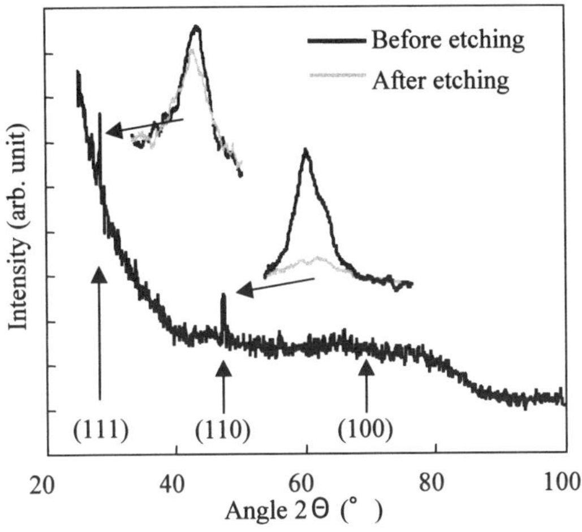

Fig.8. X-ray diffraction spectrum

SEM image of the sample after EPW etching is shown in Fig.9. Insert is the intensity distribution of laser light having irradiated to the sample and the lateral growth region is emphasized with a bold line. Since the lateral growth region was caved in, we can identify that the laterally grown grains were of (110) face. The small grains were subsequently concluded to

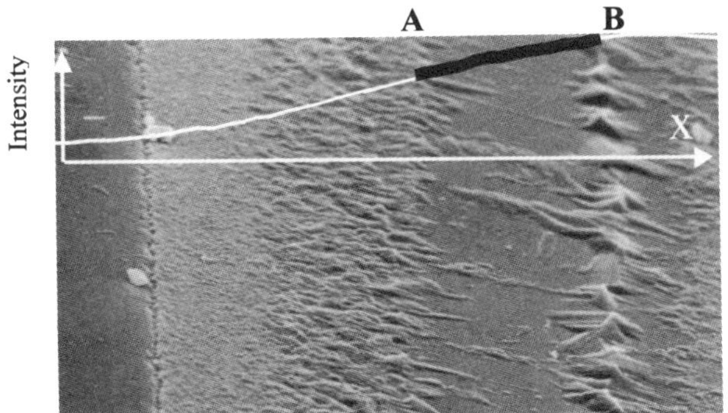

Fig.9 Surface SEM image of the sample after EPW etching

be (111)-faced, consistent with the results for the conventional (uniformly) laser-annealed film [11,12].

For solar cell application, the crystal orientation of the seed layer is very important since it influences easiness of Si epitaxy on it. For example, the film is difficult to grow on the (111)-faced seed layer at low-temperatures [13]. Since epitaxy is much easier for the (110) face compared with the (111) face, we can conclude that the giant grains formed by lateral growth is applicable to the seed layer for epitaxial growth of the active layer in solar cells.

CONCLUSION

Grains as large as 13µm could be fabricated using lateral growth enhanced by combining the advanced excimer-laser annealing method and the low heat-capacitance underlayer. The crystal is of (110)-oriented for the laterally grown region.

REFERENCES

1. S. Kumari, N. K. Arora and G. C. Jain: Solar En. Mat., 5 (1981) 383.
2. K. M. Koliwad and T. Daud: 14th IEEE Photovoltaic Specialists Conf., (1980) 1204.
3. A. K. Ghosh, C. Fishman and T. Feng: J. Appl. Phys., 51 (1980) 446.
4. H. C. Card and E. S. Yang: IEEE Trans. Electron Devices, 24 (1977) 397.
5. W.-C. Yeh and M. Matsumura: Jpn. J. Appl. Phys., 38 (1999) L110.
6. K. Ishikawa, M. Ozawa, C. Oh and M. Matsumura: Jpn. J. Appl. Phys., 37 (1998) 731.
7. C. Oh and M. Matsumura: Jpn. J. Appl. Phys., 37 (1998)5474.
8. J. Narayan and C. W. White: Appl. Phys. Lett., 44 (1984) 35.
9. H. W. Russell: American Ceramic Soc., (1934) 1.
10. C. Jin, J.D. Luttmer, D.M. Smith and T.A. Ramos: MRS Bulletin, 22 (1997) 39.
11. K. Yamamoto, A. Nakashima, T. Suzuki and M. Yoshimi: Jpn. J. Appl. Phys., 3 (1994) L1751.
12. H. Kuriyama, T. Nohda, Y. Aya, T. Kuwahara, K. Wakisaka, S. Kiyama and S. Tsuda: Jpn. J. Appl. Phys., 33 (1994) 5657.
13. R. Drosd and J. Washiburn: J. Appl. Phys., 53 (1982) 397.

THE ROLE OF VACANCIES AND DOPANTS IN SI SOLID-PHASE EPITAXIAL CRYSTALLIZATION

C. M. CHEN*, S. RASSIGA**, T. GESSMANN**, M. P. PETKOV**, M. H. WEBER**, K. G. LYNN** AND H. A. ATWATER*
*Thomas J. Watson Laboratory of Applied Physics, California Institute of Technology, Pasadena, CA 91125
**Dept of Physics, Washington State University, Pullman, WA 99164

ABSTRACT

The role and interaction of vacancies and dopants in the crystallization of amorphous Si (a-Si) by solid-phase epitaxy (SPE) was investigated. To this end, we studied: (i) the solid-phase epitaxy rate measured by time-resolved reflectivity (TRR), (ii) the dopant and carrier concentrations measured by secondary ion mass spectrometry (SIMS) and spreading resistance (SR) analysis, and (iii) the vacancy concentration measured by positron annihilation spectroscopy (PAS). Phosphorus was implanted into a-Si on Si (001), which was previously amorphized by $^{29}Si^+$ implantation, to create a nonuniform P doping profile. Phosphorus doped samples compensated with a similar boron profile were also studied. Samples were vacuum annealed for various times so that the amorphous-crystal interface was stopped at various depths providing frozen frames of the SPE process. These samples were then studied with PAS to investigate the vacancy population and to identify the impurity-defect complexes. Using this method, we have observed a population of phosphorus-vacancy complexes in the epitaxial layer.

INTRODUCTION

Understanding the mechanisms of dopant-enhanced solid-phase Si crystallization is important in developing an optimum process for low-temperature solid-phase growth of large-grained polycrystalline silicon (poly-Si) thin films on glass substrates for future thin film poly-Si photovoltaic applications. It is well known that dopants like P and B strongly influence the solid-phase crystallization [1], but despite extensive study, the mechanism for dopant-enhanced solid-phase epitaxy (SPE) at high doping concentrations ($\geq 10^{18}/cm^3$) is not well understood.

Kinetic models for the dopant enhancement of solid-phase crystallization have been proposed [2]. In these models, both neutral and charged defects (e.g., dangling bonds, vacancies) could contribute to the amorphous to crystalline silicon conversion. With the addition of dopants, the Fermi level shifts, increasing the population of a certain –but as yet unspecified-- charged defect, and thereby enhancing the growth rate. However, the SPE rate has not been correlated with direct measurements of the concentration of any point defects that could enhance the SPE rate. By using PAS [3] to measure to the vacancy population at various annealing steps, it was possible to probe the relationship between vacancy population and doping effects on SPE.

EXPERIMENT

The samples were designed to have a surface a-Si layer atop a c-Si layer to study SPE, with a doping concentration varying with depth. We studied P-doped Si, P-doped Si compensated with B, and undoped Si for comparison. Float-zone (FZ) Si wafers (<100>, p-doped, 200-300 Ω-cm) were amorphized by ion implantation of ^{29}Si at L-N$_2$ temperatures. Two implantations, with energies of 70 keV (dose, $2x10^{15}$ at/cm^2) and 200 keV (dose, $6x10^{15}$ at/cm^2),

Mat. Res. Soc. Symp. Proc. Vol. 557 © 1999 Materials Research Society

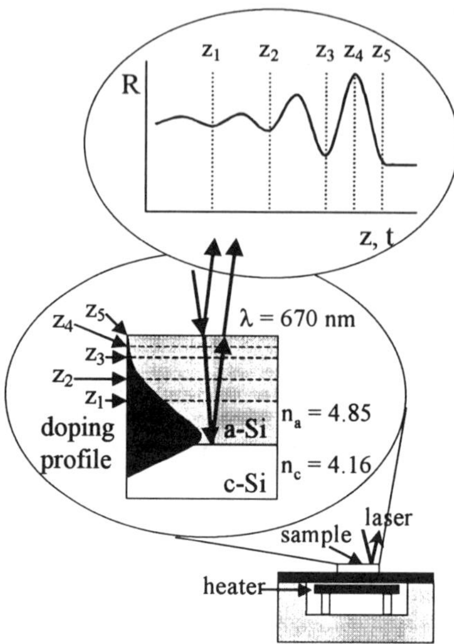

Figure 1. The experimental set-up of the anneal. Samples were vacuum annealed on a hot stage at 600°C, with a 670 nm diode laser incident to monitor the growth rate with TRR. The interface was halted at depths corresponding to extrema in the reflectivity data $(z_1, z_2, \ldots)$.

were used to amorphize the surface; ^{29}Si was used to prevent CO and N_2 contamination during implantation. Cross-sectional TEM analysis showed that the amorphized layer thickness was 346 nm. After the amorphization step, the P-doped samples were implanted with P at an energy of 200 keV, and with a dose of 1.3×10^{14} at/cm^2; (peak concentration, 7×10^{18} at/cm^3). The compensated samples were implanted with both P (above specifications) and B at an energy of 72 keV and with a dose of 1.2×10^{14} at/cm^2 (peak concentration, 7×10^{18} at/cm^3). TRIM simulations predicted that the peak of the doping profile would occurr approximately at a depth of 300 nm, so the a-Si/c-Si interface would encounter a decreasing doping concentration as crystal growth proceeded.

The samples were vacuum annealed ($\sim 10^{-6}$Torr) on a hot stage at 600°C, and growth was monitored by TRR, using a diode laser (λ=670 nm) (see Figure 1). The movement of the a-Si/c-Si interface could be monitored via oscillations in reflectivity as the reflections from the surface and the moving interface constructively and destructively interfered. The period of the oscillations corresponded to a movement of the interface of 69 nm. As schematically shown in Figure 1, the interface was stopped at different depths, each depth corresponding to a different doping concentration. These samples were then dipped into HF to etch the native oxide, and were analyzed with positron annihilation spectroscopy (PAS) to study vacancy population and to probe the vacancy-impurity complexes.

In PAS, incident positrons bombard the sample, thermalize and annihilate with electrons in the sample to produce two gamma rays with energies near 511 keV, which corresponds to the rest mass of an electron. Annihilating electrons with nonzero momentum will shift the gamma ray energies away from 511 keV, so annihilations from valence electrons, which have low momentum, produce gamma rays with little shift from the peak at 511 keV. Also, the lifetime of the positrons is inversely proportional to the electron density, so open volume defects trap positrons. The primary parameter extracted from the energy distribution of the annihilation gamma radiation is the S parameter, which is the ratio of the counts centered at the peak maximum to the whole peak. The S parameter reflects the number of annihilations from valence electrons, which is the most probable annihilation case for positrons in an open volume defect. The S parameter data normalized to FZ Si wafers gives the ratio of open volume defects present in the sample versus clean Si.

Using two synchronized detectors, the momentum of the annihilated electron can be resolved. By taking the ratio of the S parameter vs. momentum data for different samples, characteristic peaks are seen, since core electrons of different elements have different momentum. The location of these peaks can be predicted by theoretical calculations, though amplitude can vary by a factor of two. [4, 5]

Fully regrown samples were analyzed with SIMS and spreading resistance to determine the total doping and carrier concentration profiles, respectively. Cross-sectional transmission electron microscopy (TEM) was also done to study the interface morphology. Growth rates were determined from TRR signals.

RESULTS

From the TRR data from fully recrystallized samples (see Figure 2), the velocity versus interface depth information was calculated (see Figure 3). The P-doped sample had the fastest growth rate, and the compensated sample showed some enhancement. A growth retardation was observed at the surface, as seen in Figure 2, by the long tail at the end of the anneal. An additional gradual retardation of the growth rate was observed in all the samples, as seen in Figure 3. The interface velocity slowed from 0.17 nm/s, 0.47 nm/s, and 0.29 nm/s for undoped, P-doped, and P&B-doped Si respectively, to 0.05 nm/s, 0.13 nm/s, and 0.06 nm/s at the surface.

From cross-sectional TEM, the a-Si/c-Si interface is at a depth of 346 nm after implantation and before annealing, for all three samples. There are a number of dislocations extending from the end of range damage from ion implantation, to the surface. From TEM images of partially recrystallized samples, the interface has a roughness around 10 nm.

Figure 4 shows the SIMS and SR measurements made on fully recrystallized P-doped and P&B-doped samples. Levels of O and C impurities were also monitored by SIMS. The peak concentrations of P and n type carriers coincide in the P-doped Si case, suggesting that the P implant was fully activated. In the P&B-doped sample, the carrier concentration is 2 orders of magnitude less than the dopant profile of P, but not to intrinsic levels, which suggests that the B was not fully activated. This is consistent with the enhancement in the growth rate for the P&B-doped samples, as compared to the undoped sample. End of range damage from ion implantation can be seen as a

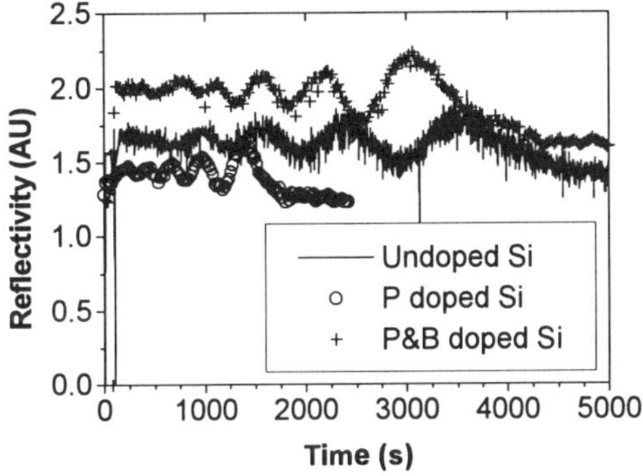

Figure 2. TRR data for undoped, P-doped, and P&B-doped Si samples, annealed at 600°C. The oscillations mark the interface movement as the reflected light alternates between constructive and destructive interference.

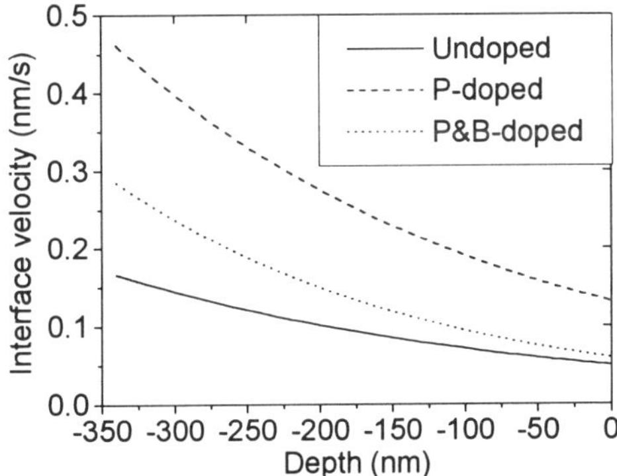

Figure 3. Growth velocity versus interface depth, for the undoped, P-doped, P&B-doped Si samples, annealed at 600°C.

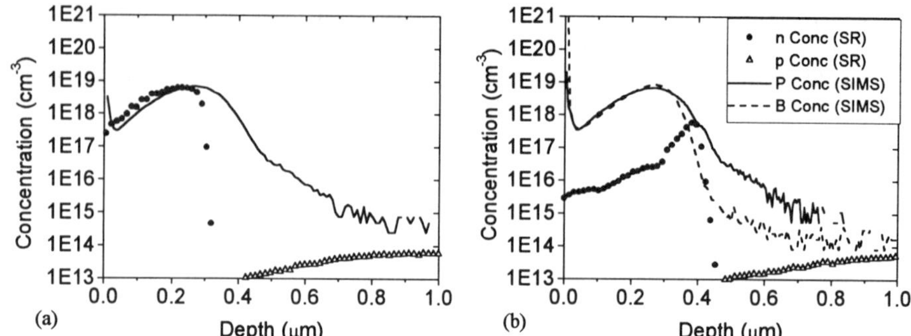

Figure 4. Spreading resistance (SR) and SIMS data for the fully crystallized (a) P-doped sample (P dose, 1.3×10^{14} at/cm^2; peak concentration, 7×10^{18} at/cm^3) and (b) P&B-doped sample, (P specifications same as above; B dose, 1.2×10^{14} at/cm^2; peak concentration, 7×10^{18} at/cm^3), annealed at 600°C.

drop off of the carrier concentration. We see elevated O concentrations at the surface, which correspond well with TRIM simulations of recoil implanted O from a native oxide layer of 20 nm, by implantation of ^{29}Si with the same doses and energies used to amorphize our samples. This elevated concentration of O could be the cause of the retardation of growth at the surface; concentrations of O at 0.5 at % have been observed to retard Si SPE growth by a factor of 0.1 at 550°C [1].

From the positron annihilation spectroscopy data (see Figure 5), several observations can be made. The general reduction of the normalized S (S_N) parameter after the initial anneal implies that some of the defects were annealed out. The advance of the interface can be seen in the depth profiles at each interval of the anneal. In the P-doped samples, the shoulder of the curve, seen advancing toward the surface with each anneal, has a shallower slope, possibly due to an increased trapping of positrons, from an increase of negatively charged vacancies from phosphorus-vacancy complexes (P-v), which have been observed in a-Si [6]. The S_N parameter also dipped below one at the surface of all fully crystallized samples, suggesting a contribution from impurities. This is consistent with the recoil implanted O at the surface forming vacancy-oxygen (v-O) complexes [7]. The P-doped sample has a more prominent dip at the surface, which is attributed to an additional phosphorus related effect, though the mechanism is not understood.

The two detector measurements show a peak in both the P-doped and P&B-doped samples that corresponds to a P atom in the vacancy neighbor shell, as seen in theoretical calculations. This signal suggests that there are P atoms next to vacancies, and that there are more P-v complexes than would be formed by a random distribution of P atoms and vacancies. The P-doped sample has a larger signal than the P&B-doped sample, which is not yet understood [5].

CONCLUSIONS

The interface velocity versus depth was calculated, and a slow retardation was observed for all samples, as well as an additional retardation at the surface, which could be due to recoil-implanted oxygen, which is seen in the SIMS data. The phosphorus was fully activated, but the boron was not, as seen by comparing the spreading resistance and SIMS data. TEM images show that the interface roughness is approximately 10 nm, and there are dislocations extending

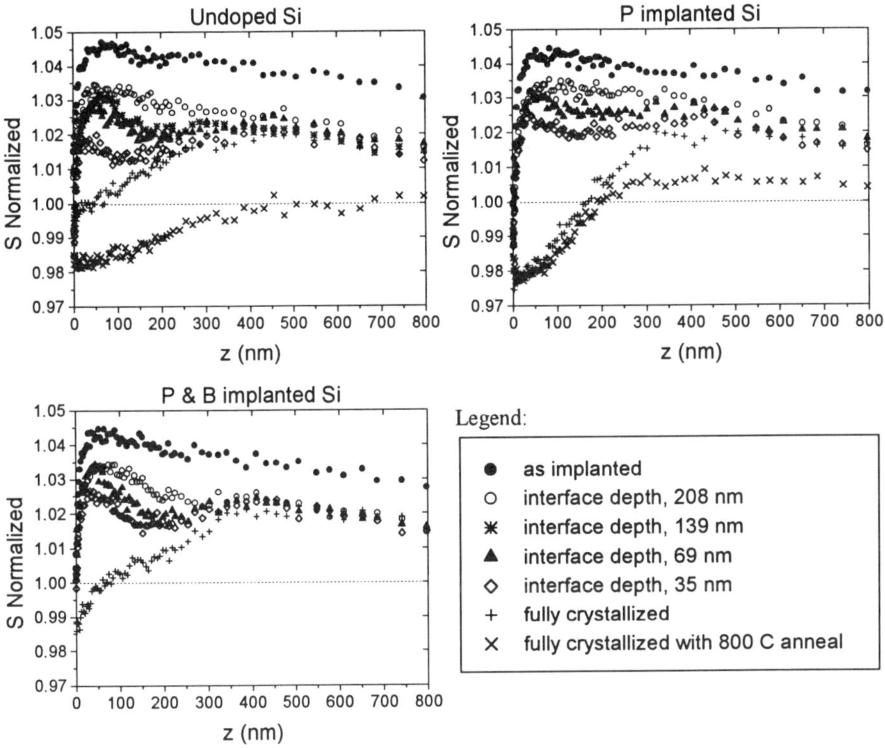

Figure 5. Positron annihilation spectroscopy (PAS) data for undoped, P-doped, and P&B-doped silicon samples. The different curves for each sample include an as-implanted sample (interface at 346 nm), annealing steps with the interface at depths given, a fully crystallized sample, and a fully crystallized sample with an additional 800°C anneal, to anneal out defects in the end-of-range damage region from the ion implantation.

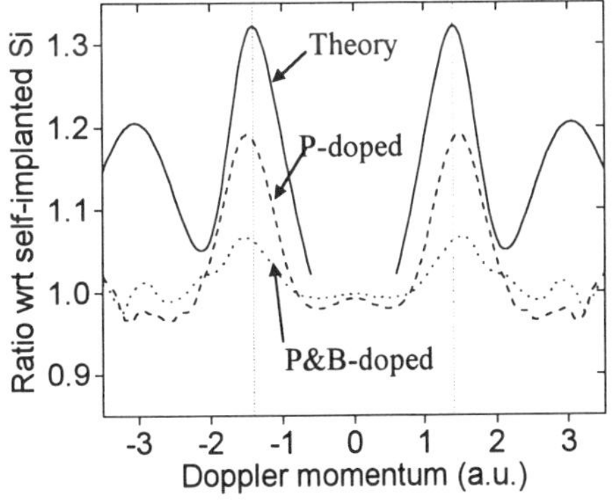

Figure 6. Momentum resolved PAS data for P-doped and P&B-doped Si samples, both divided by the undoped Si data. The resulting data exhibit peaks that correlate with theoretical calculations of P/Si, which suggests there are P-v complexes.

from the end of range damage. The PAS data suggest that there are O-v complexes in all samples, P-v complexes in the P-doped samples, and vacancy clusters in the end of range region. Further work using positron lifetime measurements and depth profiles will be done to identify the type of open volume defects we have, and to correlate the defect profile with the velocity and dopant concentration data.

ACKNOWLEDGMENTS

This work is supported by the United State Department of Energy, Basic Energy Sciences, Lawrence Livermore National Laboratory and the National Renewable Energy Laboratory.

REFERENCES

1. G. L. Olson and J. A. Roth, *Handbook of Crystal Growth 3*, edited by D. T. J. Hurle (Elsevier, Amsterdam, 1994), pp. 257-312.
2. G. Q. Lu, E. Nygren, and M. J. Aziz, J. Appl. Physics **70**, 5323 (1991).
3. C. Szeles and K. G. Lynn, *Encyclopedia of Applied Physics, Vol. 14* (VCH Publishers,1996), pp. 607-632.
4. V. J. Ghosh, M. Alatalo, P. Asoka-Kumar, K. G. Lynn, and A. C. Kruseman, *Appl. Surf. Sci.* **116**, 278 (1997).
5. M. Alatalo, P. Asoka-Kumar, V. J. Ghosh, B. Nielsen, K. G. Lynn, A. C. Kruseman, A. Van Veen, and M. J. Puska, *J. Phys. Chem. Solids* **59**, 55 (1998).
6. M. P. Petkov, M. H. Weber, K. G. Lynn, R. S. Crandall, V. J. Ghosh, *Phys. Rev. Lett.* **82**, (10 May 1999).
7. J. Xu, E. G. Roth, and O. W. Holland, A. P. Mills, R. Suzuki, Appl. Phys. Lett. **74**, 997 (1999).

ESR MEASUREMENTS OF a-Si:H AND a-Si$_{0.5}$Ge$_{0.5}$:H FILMS UNDER SOLID-PHASE CRYSTALLIZATION

I. H. YUN, O. H. ROH AND J.-K. LEE
Department of Physics, Chonbuk National University, Chonju, KOREA

ABSTRACT

We have investigated the solid-phase crystallization of a-Si$_{1-x}$Ge$_x$:H (x=0 and 0.5) films by using electron spin resonance and x-ray diffraction. The films were deposited on Corning 1737 glass in a plasma-enhanced chemical vapor deposition system using SiH$_4$ and GeH$_4$ gases. The films were then annealed to be crystallized at 600°C. It was observed that, for the a-Si:H film, both the spin density and the g-value first increased with annealing time, and then rapidly decreased as the film was crystallized. For the a-Si$_{0.5}$Ge$_{0.5}$:H film, the Ge dangling bond spin density increased from 3 x 10^{18} cm^{-3} to 2 x 10^{19} cm^{-3} for the first stage of annealing and then decreased to 3 x 10^{17} cm^{-3} after being crystallized; the Si dangling bond spin density just increased to about 2 x 10^{17} cm^{-3} and remained nearly constant for further annealing. It is thought that exodiffusion of hydrogen resulted in the increase of spin density in the beginning, and then some portions of amorphous components were converted into the crystalline phase by further annealing. And the keen correlation between the dependence of x-ray peak intensity and the dependence of Ge dangling bond spin density on the annealing time suggests that the Ge dangling bonds rather than Si dangling bonds play an important role in the crystallization of the Si$_{0.5}$Ge$_{0.5}$ film.

INTRODUCTION

Polycrystalline silicon (poly-Si) is an important material for thin film transistor (TFT) applications. Solid-phase crystallization (SPC) of amorphous silicon (a-Si) thin films has been suggested to be a promising approach to obtain these poly-Si thin films [1,2], as opposed to depositing them directly. It has also been proposed that poly-Si$_{1-x}$Ge$_x$ film could be used for TFT applications due to its advantage of lower thermal budget and also for its potential to increase field effect mobility [3].

The crystallization kinetics of a-Si layers deposited by plasma enhanced chemical vapor deposition (PECVD) has been the subject of many studies. It has been suggested that the initial structural disorder of the film plays an important role in the SPC of a-Si films [2,4]. While the SPC of a-Si films has been studied by many researchers, systematic studies on the crystallization of a-Si$_{1-x}$Ge$_x$ alloy films are relatively few. No clear results have been reported yet on the relationship between the defect spin density of the Si$_{1-x}$Ge$_x$ alloy films and the crystallization fraction of the film under SPC.

The purpose of this article is to investigate the solid phase crystallization mechanism of amorphous Si$_{1-x}$Ge$_x$ alloy films in relation to the variation of defect spin densities, and to compare the results with those of pure Si films. In our study, a clear correlation between the variation of Ge dangling bond spin density and the progress of the crystallization of the Si$_{0.5}$Ge$_{0.5}$ film was observed. We suggest that the Ge dangling bonds rather than Si dangling bonds play an important role in the crystallization mechanism of the Si$_{0.5}$Ge$_{0.5}$ film.

Mat. Res. Soc. Symp. Proc. Vol. 557 © 1999 Materials Research Society

EXPERIMENT

The films were deposited on Corning 1737 glass in a PECVD system using SiH_4 and GeH_4 gases. Before deposition, the system was evacuated to 2×10^{-5} Torr by a turbomolecular pump, and during deposition, a mechanical rotary pump was used. The substrate temperature was 200° C and the r.f. power was 3W. The films were then annealed to be crystallized at 600°C in a N_2 atmosphere. The film thickness was measured by the Alpha-Step 200 (Tencor Instruments). Some deposition conditions are given in Table I.

The crystallinity of the films has been investigated by x-ray diffraction (XRD) (Rigaku DMAX/IIIA) using $CuK\alpha$ radiation. The sample size (10mm x 10mm) for the XRD measurements was fixed. The XRD measurements were carried out for a 2θ range of 10°- 80°.

Random and channeled Rutherford backscattering spectrometry (NEC Model: 3SDH) was employed to measure the Ge composition, x. The measurements were made using 2.2 MeV $^4He^+$ ion incident beams.

The spin densities of the films were measured by electron spin resonance (ESR) measurements (Bruker EMX-8) in the x-band at room temperature. The magnetic field modulation frequency was 100 kHz. Average unpaired spin densities were obtained by comparison of the double integrated magnetic field spectrum with that of weak pitch standard under unsaturated conditions.

Table I. PECVD deposition conditions for the films.

	Temperature	Pressure	Gas Flow	Thickness
a-Si:H	200°C	0.6 Torr	SiH_4 6sccm	3.3 μm
a-$Si_{0.5}Ge_{0.5}$:H	200°C	0.6 Torr	SiH_4 3sccm, GeH_4 3sccm	2.8 μm

RESULTS AND DISCUSSION

When the a-Si:H film was deposited directly by PECVD, the x-ray result showed a weak and broad peak at around $2\theta=26.5°$, indicating that the film was deposited in an amorphous state. However, we noted that a weak peak occurred at around $2\theta=28°$ for the 4h-annealed Si film indicating that a crystalline phase had been formed. The crystallized film showed (111), (220), and (311) peaks at around $2\theta = 28°$, 47°, and 55°, respectively, showing a strong preferential orientation of (111). The diffracted x-ray intensities are indicative of the crystallinity of the film [5,6], and we may well investigate the progress of the crystallization of the film by measuring the x-ray (111) peak intensities. Figure 1 shows the variation of XRD (111) peak intensity of the Si film with annealing time. Our x-ray results show that the solid-phase crystallization started between roughly one and ten hours after annealing of the pure Si film. The film required more than 50 h for complete transformation after the incubation time. It has been reported that the incubation time and transformation time depend on the deposition conditions of the film such as hydrogen content and substrate temperature, and our results agree generally with some previously published works [1,5,7,8].

In Fig. 2, the values of the spin density N_s of the Si film estimated from each ESR spectrum are shown. It was observed that the spin density N_s increased from 2×10^{17} cm^{-3} (as-deposited state) to 1.5×10^{19} cm^{-3} after 4 hours of annealing, and then decreased to 3×10^{18} cm^{-3} with

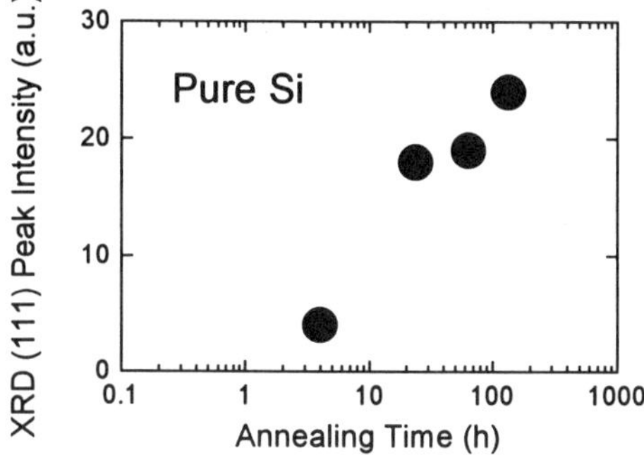

Fig. 1. The variation of x-ray (111) peak intensity of the Si film annealed at 600°C with annealing time.

further annealing. Our results agree with some previously published works [6,9,10]. And, as can be seen in Fig. 1, the film keeps amorphous over the range of annealing time where the N_s value increases. It is presumed that hydrogen diffuses out from the Si film in the beginning stage of annealing, resulting in the increase of spin density. The hydrogen evolves from Si above 400°C. The hydrogen evolution from a Si film was also noticed by Lee et al. [7] who observed that the Raman spectra changed after 1 h of annealing at 600°C, which was shorter than the incubation time. The rapid decrease in N_s values with a further increase in annealing time is associated with the crystallization of the films. In poly-Si, it is known that the ESR signal mainly arises from Si dangling bonds within the region of a grain boundary. An increase of the average grain size results in a reduction of the grain boundary volume and a decrease of the spin density. The value of g reflects the local atomic structure around the Si defect. The g-factor was also found first to increase with the annealing time, and then to decrease after the film was crystallized (Fig. 3).

We noted that the $Si_{0.5}Ge_{0.5}$ film was deposited in an amorphous state since the x-ray results did not show any peaks. However, weak SiGe (111), (220), and (311) peaks occurred for the $Si_{0.5}Ge_{0.5}$ film after only 5 minutes of annealing, indicating that a crystalline phase has been formed during that time. Figure 4 shows the variation of the SiGe (111) peak intensity of the $Si_{0.5}Ge_{0.5}$ film with annealing time. The x-ray results shows that the $Si_{0.5}Ge_{0.5}$ film started to crystallize after about 5 minutes of annealing, and the crystallization proceeded roughly between 0.1 h and 4 h. We note that the time for crystallization is significantly reduced compared to pure Si. Our results agree with some of the previous published works [5,11] which show that both the incubation time and the full crystallization time are significantly reduced in $Si_{1-x}Ge_x$ alloys as compared to pure Si. It may be inferred that the activation energy of the incubation time for the crystallization of a-Ge is lower than that of a-Si [11].

For the as-deposited $a-Si_{0.5}Ge_{0.5}$:H film, we were able to obtain only a Ge dangling bond signal from the ESR spectrum. But as the film was annealed for a few minutes, a Si dangling bond signal also appeared and the profiles of the ESR spectra changed with increasing annealing

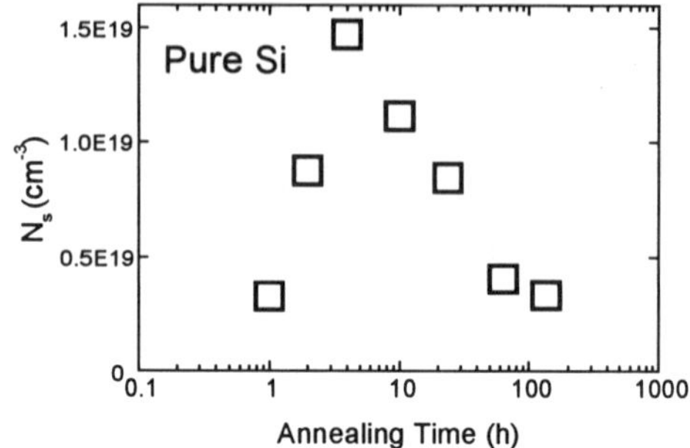

Fig. 2. The variation of spin density N_s of the Si film as a function of annealing time.

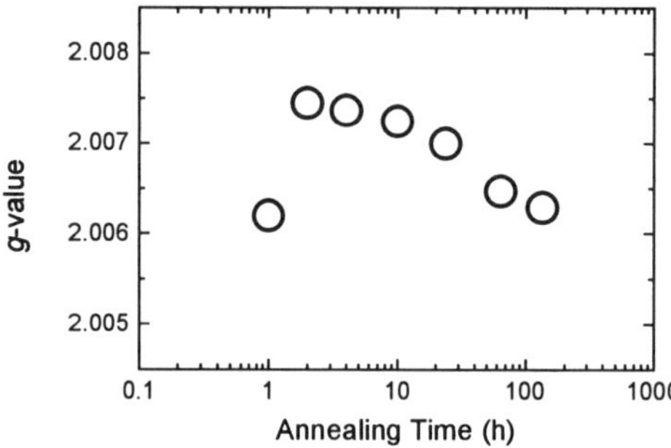

Fig. 3 The variation of g-values as a function of annealing time.

time. Deconvolution of the ESR spectrum of the $Si_{0.5}Ge_{0.5}$ film into a Si dangling bond signal and a Ge dangling bond signal has been attempted. The approach was to model the lineshape of each defect with a Gaussian whose width and center are permitted to vary with alloy concentration to accommodate differing backbonding configurations. Figure 5 shows the variation of spin densities of the $Si_{0.5}Ge_{0.5}$ film as a function of annealing time, where Si-db represents the spin density of a Si dangling bond, and Ge-db indicates that of a Ge dangling bond. In Fig. 5, it is seen that the Ge dangling bond spin density increased from 3×10^{18} cm^{-3} (as-deposited state) to about 2×10^{19} cm^{-3} for the first stage of annealing and then decreased to about 3×10^{17} cm^{-3} after being crystallized; the Si dangling bond spin density increased to about 2×10^{17} cm^{-3} and remained nearly constant for further annealing.

First, it is thought that H diffuses out at the beginning stage of the annealing as in the Si films. But the preferential attachment of hydrogen to Si over Ge occurs here. It may be inferred that a large fraction of the hydrogen evolves from a-Ge:H near 150°C, and the evolution from a-Si:H occurs above 400°C [12]. The rapid decrease in the spin density of Ge dangling bonds with further annealing is associated with the crystallization of the film. The constant spin density of the Si dangling bonds may imply that the grain boundary defects are quite stable.

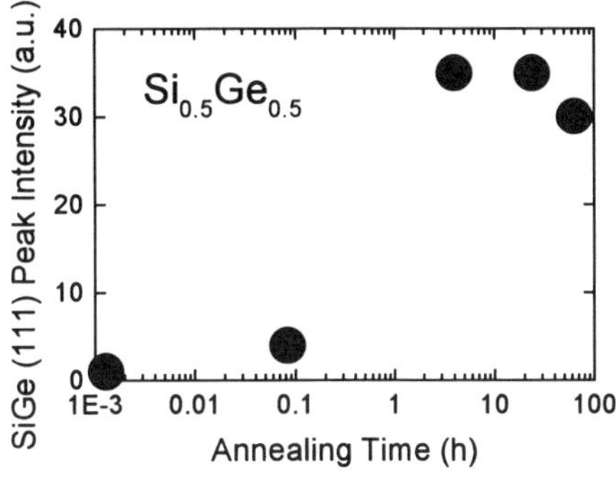

Fig. 4. The variation of x-ray (111) peak intensity as a function of annealing time for the $Si_{0.5}Ge_{0.5}$ film annealed at 600°C.

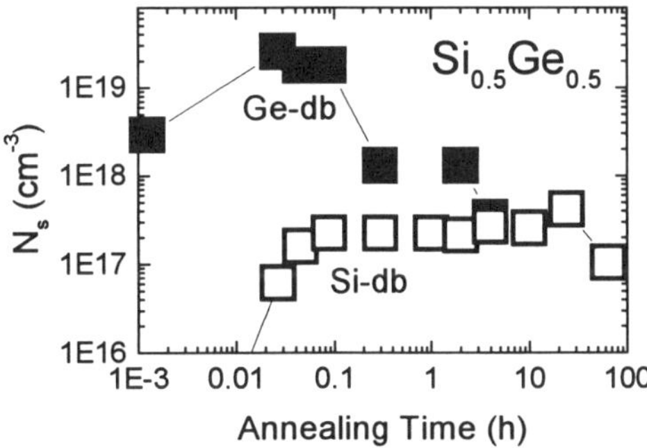

Fig. 5 The variation of defect spin densities of the $Si_{0.5}Ge_{0.5}$ film with annealing time. Si-db indicates the spin density of Si dangling bonds; Ge-db indicates that of Ge dangling bonds.

The variation of Ge dangling bond spin density clearly shows the incubation stage and the crystallization stage. This is consistent with the XRD results shown in Fig. 4, where the $Si_{0.5}Ge_{0.5}$ film started to crystallize within about 5 minutes of annealing and the crystallization proceeded roughly between 0.1 h and 4 h. The clear correlation between the phase variation presumed by the change of Ge dangling bond spin density and that manifested by XRD results may suggest that the Ge dangling bonds rather than the Si dangling bonds play an important role in the crystallization mechanism of the $Si_{0.5}Ge_{0.5}$ film.

CONCLUSIONS

We have presented ESR results on a-Si:H and a-$Si_{0.5}Ge_{0.5}$:H films under solid-phase crystallization. When the a-Si film was annealed at 600°C, both the g and N_s values first increased with annealing time, and then decreased after the film was crystallized. For the $Si_{0.5}Ge_{0.5}$ film, it was observed that the values of the spin density for the Ge dangling bonds first increased with increasing annealing time, and then decreased; the Si dangling bond spin density increased rather slightly and then remained constant.

It is presumed that hydrogen diffuses out at the beginning stage of annealing, resulting in the increased spin density; the decrease in the spin density at longer annealing times is associated with the crystallization of the films. For the $Si_{0.5}Ge_{0.5}$ film, the Ge dangling bonds rather than the Si dangling bonds seem to play the dominant role in the crystallization of the $Si_{0.5}Ge_{0.5}$ film.

ACKNOWLEDGMENTS

The authors wish to acknowledge the financial support of the Korea Research Foundation made in the program year of 1997.

REFERENCES

1. R. Kingi, Y. Wang, S. Fonash, O. Awadelkarim, and Y.-M. Li in *Flat Panel Display Materials II*, edited by M. K. Hatalis, J. Kanicki, C. J. Summers, and F. Funada (Mater. Res. Soc. Proc. 424, Pittsburgh, PA, 1996) pp. 249-254.

2. K. Nakazawa and K. Tanaka, J. Appl. Phys. **68**, 1029 (1990).

3. T.-J. King, K. C. Saraswat, J. R. Pfiester, IEEE Electron Dev. Lett. **12**, 584 (1991).

4. J.-K. Lee, J. of Korean Vac. Sci. & Tech. **2**, 49 (1998).

5. C. W. Hwang, M. K. Ryu, K. B. Kim, S. C. Lee, C. S. Kim, J. Appl. Phys. **77**, 3042 (1995).

6. S. Hasegawa, S. Watanabe, T. Inokuma, Y. Kurata, J. Appl. Phys. **77**, 1938 (1995).

7. J. N. Lee, B. J. Lee, D. G. Moon, B. T. Ahn, Jpn. J. Appl. Phys. **36**, 6862 (1997).

8. R. Kingi, Y. Wang, S. J. Fonash, O. Awadelkarim, J. Mehlhaff, and H. Hovagimian, in *Flat Panel Display Materials II*, edited by M. K. Hatalis, J. Kanicki, C. J. Summers, and F. Funada (Mater. Res. Soc. Proc. 424, Pittsburgh, PA, 1996) pp. 237-241.

9. T. Aoyama, Y. mochizuki, G. Kawachi, S. Oikawa, K. Miyata, Japanese J. of Appl. Phys. **30**, L84 (1990).

10. T. Aoyama, G. Kawachi, N. Konishi, T. Suzuki, Y. Okajima, and K. Miyata, J. Electrochem. Soc. **136**, 1169 (1989).

11. F. Edelman, Y. Komem, M. Bendayan and R. Beserman, J. Appl. Phys. **72**, 5153 (1992).

12. W. Paul, D. K. Paul, B. von Roedern, J. Blake, S. Oguz, Phys. Rev. Lett. **46**, 1016 (1981).

ELECTROCHEMICAL TAILORING AND OPTICAL INVESTIGATION OF ADVANCED REFRACTIVE INDEX PROFILES IN POROUS SILICON LAYERS

S. ZANGOOIE, R. JANSSON, H. ARWIN[+]
Laboratory of Applied Optics, Department of Physics and Measurement Technology
Linköping University, SE-581 83 Linköping, Sweden

ABSTRACT

Porosity depth profiles with exponential or sinusoidal shape were fabricated electrochemically in crystalline silicon using time-variable current densities and studied employing variable angle spectroscopic ellipsometry. Since volume porosity in porous silicon depends on the current density, it was possible to electrochemically tailor porosity depth profiles, which in a first approximation resembled the time modulation of the applied current. Optical characterization of the samples were realized using multilayer optical models and the Bruggeman effective medium approximation allowing variations of the index of refraction according to the applied current density profiles. The analysis also revealed deviations from desired profiles in terms of in-depth inhomogeneities.

INTRODUCTION

Due to room-temperature photo- [1] and electroluminescence [2] properties of porous silicon (PS), this material has received extensive attention from the research community. In addition to the application possibilities within silicon based integrated optoelectronics, further implementations in disciplines such as optics [3-5], and gas- and biosensor technologies [6-8] have been reported and are continuously under development. The optimism concerning applications of the material is further boosted by the simplicity of PS fabrication, which often is realized electrochemically in ethanoic solutions of hydrofluoric acid (HF). The morphology and microstructure of PS is mainly governed by a number of etching parameters such as HF concentration, current density, temperature, illumination, crystal orientation, silicon type and doping concentration [9,10]. By choosing those parameters properly, it is in principle possible to fabricate a PS layer of desired thickness, porosity, pore size distribution and internal surface area. Good control over the etching process parameters is essential for fabrication of optimal and useful microstructures and morphologies in many applications. One important way of control is time modulation of the current density. As a result, complex structures such as Bragg reflectors [3], Fabry-Pérot filters [4] and optical wave-guides [11] can be manufactured. These structures are basically fabricated by changing the current density in a periodic and well-controlled manner. As a result, periods with sublayers of different porosity and thereby indices of refraction can be fabricated. A computer-controlled continuous variation of the current density, demanding a more complex software as well as hardware, opens up new possibilities for tailoring more advanced structures with interesting application opportunities in sight.

In this paper, we report on fabrication of non-conventional refractive index profiles in PS with, for example, exponential or sinusoidal shape. State of the art variable angle spectroscopic ellipsometry is employed to non-destructively characterize microstructural properties of the PS samples. This technique has previously been demonstrated to be an effective tool for both in-depth investigations of PS morphologies and compositional depth variations in adsorption due to vapor and protein exposure [12,13]. Non-idealities in terms of deviations from expected mathematically defined patterns are resolved and discussed.

✢ Corresponding author: han@ifm.liu.se

Mat. Res. Soc. Symp. Proc. Vol. 557 © 1999 Materials Research Society

EXPERIMENTAL AND THEORETICAL DETAILS

PS samples were prepared from p-type 0.010-0.020 Ω cm silicon wafers having (111) crystal orientation from Okmetic ÓY (Finland). Anodization was performed in a mixture of HF, water and ethanol (volume ratio 1.6:2.4:6). The currents were predefined according to the desired porosity profile, and realized with a specially designed computer controlled current source. Prior to etching, the samples were placed in the etching solution for 1 minute to remove the native oxide. The counter electrode was a platinum wire positioned about 2 cm from the silicon wafer. After anodization, the samples were placed in ethanol for 10 minutes to remove remaining HF and finally blown dry with nitrogen gas.

Optical measurements were done with a variable angle spectroscopic ellipsometer of the rotating analyzer type (J. A. Woollam Co., USA) [14]. Analysis of the optical data was done with the WVASE32 software package. The spectroscopic measurements were carried out at room temperature in air at six angles of incidence in the range 50°-75° (accuracy $\pm$ 0.005°) with an interval of 5° in the 400-1700 nm wavelength range. Variable angle spectroscopic ellipsometry is based on measurements of the change in the polarization state of a linear polarized light beam reflected at a sample surface at oblique incidence. The measured quantity in ellipsometry is the complex reflectance ratio $\rho=tan(\psi)\cdot exp(i\Delta)$ where ψ and Δ are the ellipsometric parameters. From these parameters information can be obtained about the optical properties of the material under study in terms of complex refractive index, $N=n+ik$ or alternatively the dielectric function $\varepsilon=N^2$. Examples of microstructural parameters that can be quantified from measured ψ and Δ values are layer thickness, composition and porosity. For quantification the sample must be described with an optical model containing the relevant parameters. The Levenberg-Marquardt multivariate regression algorithm is then employed to find model parameters that minimize the difference between experimental data ψ^{exp} and Δ^{exp}, and model data ψ^{mod} and Δ^{mod}. This fitting procedure is based on minimization of the mean squared error (MSE) :

$$MSE = \frac{1}{2N-M} \sum_{i=1}^{N} \left[\left(\frac{\psi_i^{mod} - \psi_i^{exp}}{\sigma_{\psi,i}^{exp}} \right)^2 + \left(\frac{\Delta_i^{mod} - \Delta_i^{exp}}{\sigma_{\Delta,i}^{exp}} \right)^2 \right] \qquad (1)$$

where N is the number of measured (ψ,Δ) pairs, M is the number of variable parameters in the model, and the $\sigma's$ are the standard deviations in the experimental data points. The MSE value, the 90% confidence intervals in the model parameters and a parameter correlation table serve as the basis for estimating the quality of the match between the data calculated from the model and the experimental data as well as for judging the credibility of the model employed.

The Bruggeman effective medium approximation (EMA) was used to calculate the effective optical properties (ε_{eff}) of sublayers in the models according to the following expressions:

$$\sum_{i=1}^{m} f_i \frac{\varepsilon_i - \varepsilon_{eff}}{\varepsilon_i + 2\varepsilon_{eff}} = 0 \qquad (2)$$

$$\sum_{i=1}^{m} f_i = 1 \qquad (3)$$

where m is the number of constituents, and f_i and ε_i are the volume fraction and the complex dielectric function of constituent i, respectively. The volume fraction, f_i, is expressed in percentage below. This EMA has previously been employed for characterization of different types of PS [3,4,12].

RESULTS AND DISCUSSION

Figure 1a illustrates experimental and model-generated ψ and Δ spectra for a PS sample with a volume porosity depth profile obtained using a periodic variation in the current density according to an exponential equation. Here, only the fit for one angle of incidence (70°) is shown for clarity. The current i versus time t, repeated in each period, can be expressed as:

$$i(t) = C_1 + C_2 \exp(t) \tag{4}$$

where C_1 and C_2 are constants which are used to define the initial and final magnitudes as well as the general shape of the current density. The results presented in Fig. 1 were obtained on a sample for which C_1 and C_2 were chosen so that the current varied between 9 and 300 mA. Thus, considering a total sample area of 3 cm^2, current densities between approximately 3 and 100 mA/cm^2 were obtained. The etching time for each period was 5 s and a total number of five periods was used. In Fig. 1b, the modeled depth profile of the multilayer sample in terms of the real part of the index of refraction as well as the volume porosity is presented. The modeling procedure is described below.

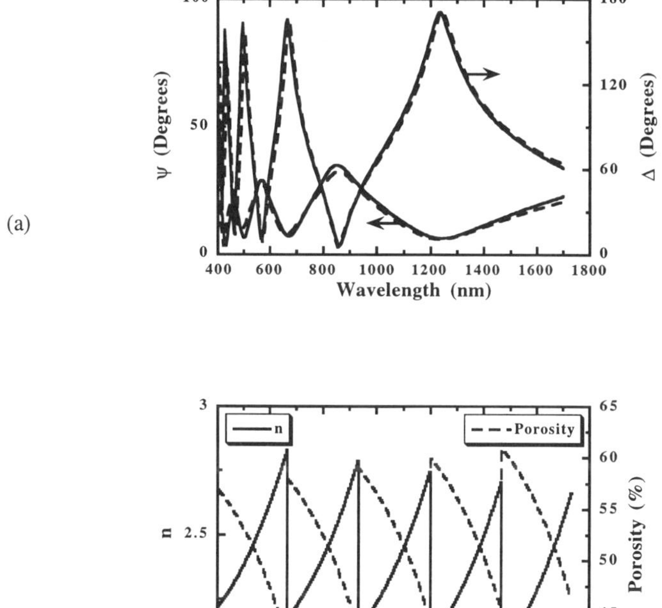

(a)

(b)

Figure 1. (a) Experimental (---) and modeled (—) ψ and Δ spectra at an angle of incidence of 70°. (b) Depth profile of the real part (n) of the refractive index at wavelength of 500 nm (—) and volume porosity (---). The imaginary part of the refractive index shows large similarity to n and is not shown here.

Optical characterization of the sample is realized by defining a mathematical model for the depth variations of the volume porosity and consequently the index of refraction using an equation similar to eq. (4). The volume porosity is changed gradually using 50 slices in each period, where the porosity is varied slightly from slice to slice. Further improvements are made by introducing additional terms in order to model non-idealities. One example is the tilt of the index profile, which represents a possible long-range depth inhomogeneity. The expression used for the depth variation of the void volume, $f(d)$, in each sublayer $l \in [0,4]$, can thus be expressed as:

$$f(d) = A + B \cdot (\frac{d}{D} + l) + C \cdot \exp(\frac{d}{D} \cdot E) \qquad (5)$$

where d is a depth parameter which varies between 0 at the bottom to D at the top of each sublayer. D is the thickness of each sublayer and is a fitting parameter. Other fitting parameters in this model are A, B, C and E. Hence, the total number of variable parameters in this model is five which, considering the complexity of the material and the amount of available data, is a low number. It has to be noted that, in order to be able to introduce the tilt as simply as possible, the fitting parameters in the five separate sublayers were coupled to each other. The fitting parameters were therefore identical for the five sublayers. The only difference between the sublayers were the second term in eq. (5), where $l=0$ for the sublayer next to the silicon substrate and $l=4$ for the sublayer next to the ambient. Using an exponential inhomogeneous term did not improve the fit. Table I contains the evaluated parameters and their 90% confidence intervals.

Table I. Result of fitting to the model as given by eq. (5). (MSE = 24).

A	72.9 ± 0.3
B	1.08 ± 0.03
C	-16.1 ± 0.4
D (nm)	66.6 ± 0.1
E	0.66 ± 0.01

The modeling suggests existence of a significant depth inhomogeneity, where the volume porosity decreases towards the silicon substrate. Such inhomogeneities have previously been reported and discussed [3,15,16]. It has to be noted that a linear inhomogeneity consideration may not fully represent the complex nature of this phenomenon. However, considering more complicated inhomogeneities using, e.g., polynomial or exponential relations would increase the number of fitting parameters and thereby parameter correlation without decreasing the MSE significantly. We therefore believe that the assumption of a linear deviation, at least as a first approximation, is a good representation for the observed inhomogeneity.

In Fig. 2, depth profiles for two other PS structures are shown. Figure 2a represents the depth profile of a PS layer with a periodic sinc structure (five periods), whereas Fig. 2b illustrates a profile fabricated using a sinusoidal current density. Also here a linear term is included in the modeling. It is obvious that it is possible to fabricate porous structures that to a good degree resemble the applied current density. Thus, it should be possible, by proper modifications of the current density, to fabricate structures that more ideally follow the desired patterns. For instance, by increasing the current density with time it should be possible to compensate for the decrease in volume porosity closer to the silicon substrate. This is due to that a larger current density result in larger volume porosity in PS [9]. However, such experiments have not yet been performed, since it was considered to be of a more laborative character and not within the frames of this work.

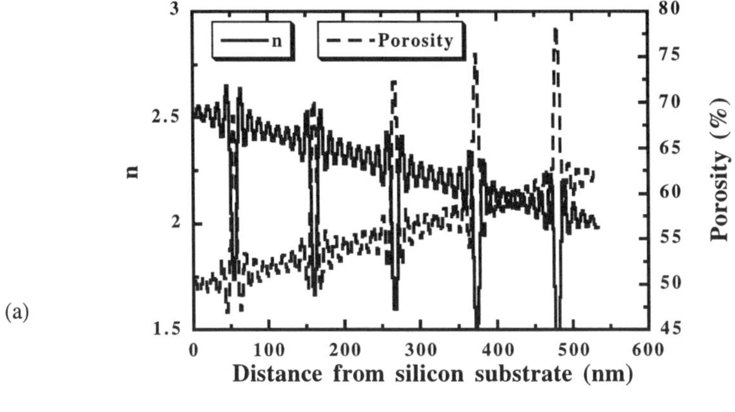

(a)

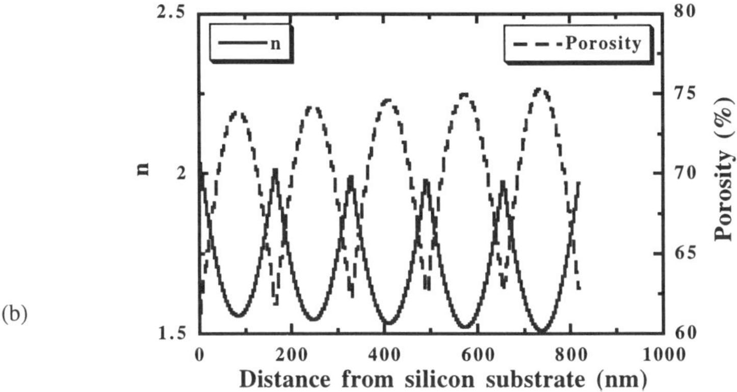

(b)

Figure 2. (a) Depth profile of a PS sample with a sinc structure (MSE=35) and (b) with a sinusoidal structure (MSE=46). Refractive indices shown are at a wavelength of 500 nm.

CONCLUSIONS

It was shown that by controlled time variation of the current density, it is possible to fabricate porosity profiles in PS, which to a good degree resemble a desired structure. Three different profiles based on sinusoidal and exponential mathematical relations were demonstrated here. It should be noted that variation of the current density is the most obvious and simplest approach for microstructural tailoring in PS. Linear deviations from the ideal models were found. However, it is suggested that by proper variations of the current density, in addition to the current profile for the target microstructure, such deviations can be compensated for.

These types of tailored structures may find applications in disciplines such as optoelectronics and sensor technology since the profiling capability provides interesting optical and microstructural characteristics different from the properties found in conventional single layer or multilayer PS structures with the refractive index of the sublayers being constant. In addition, since larger currents result in formation of larger pores, it should be possible to fabricate structures with different pore size distributions at different positions relative to the silicon substrate. Possibility of such designs may be of importance for the gas sensitivity of the material as well as future applications within biotechnology.

ACKNOWLEDGEMENTS

Financial support was provided by the Swedish Research Council for Engineering Sciences.

REFERENCES

1. L.T. Canham, Appl. Phys. Lett. **57**, 1406 (1990).
2. A. Halimaoui, C. Oules, G. Bomchil, A. Bsiesy, F. Gaspard, R. Herino, M. Ligeon, and F. Muller, Appl. Phys. Lett. **59**, 304 (1991).
3. S. Zangooie, R. Jansson, and H. Arwin, J. Vac. Sci. Technol. A **16**, 2901 (1998).
4. S. Zangooie, R. Jansson, and H. Arwin, J. Appl. Phys., In press.
5. U. Grüning and V. Lehmann, *Photonic Band Gap Materials*, Kluwer Academic Publishers, Netherlands, 1996, p. 453.
6. S. Zangooie, R. Bjorklund, and H. Arwin, Sens. Actuators B **43**, 168 (1997).
7. T. Laurell, J. Drott, L. Rosengren, and K. Lindström, Sens. Actuators B **31**, 161 (1996).
8. A. Janshoff, K.S. Dancil, C. Steinem, D.P. Greiner, V.S.-Y. Lin, C. Gurtner, K. Motesharei, M.J. Sailor, and M.R. Ghadiri, J. Am. Chem. Soc. **120**, 12108 (1998).
9. R.L. Smith and S.D. Collins, J. Appl. Phys. **71**, R1 (1992).
10. S. Zangooie, R. Jansson, and H. Arwin, Appl. Surf. Sci. **136**, 123 (1998).
11. A. Loni, L.T. Canham, M.G. Berger, R. Arens-Fischer, H. Müunder, H. Lüth, H.F. Arrand, and T.M. Benson, Thin Solid Films **276**, 143 (1996).
12. S. Zangooie, R. Bjorklund, and H. Arwin, J. Electrochem. Soc. **144**, 4027 (1997).
13. S. Zangooie, R. Bjorklund, and H. Arwin, Thin Solid Films **313-314**, 827 (1998).
14. R.M.A. Azzam and N.M. Bashara, *Ellipsometry and Polarized Light*, North-Holland, New York, 1997.
15. S. Billat, M. Thönissen, R. Arens-Fischer, M.G. Berger, M. Krüger, and H. Lüth, Thin Solid Films **297**, 22 (1997).
16. M. Thönissen, M.G. Berger, S. Billat, R. Arens-Fischer, M. Krüger, H. Lüth, W. Theiβ, S. Hillbrich, P. Grosse, G. Lorendel, and U. Frotscher, Thin Solid Films **297**, 92 (1997).

A STEM STUDY OF P AND Ge SEGREGATION TO GRAIN BOUNDARIES IN $Si_{1-x}Ge_x$ THIN FILMS

W. Qin *, D. G. Ast ** and T. I. Kamins ***
* Institute of Microelectronics, 11 Science Park Road, Singapore 117685
** Department of Materials Science and Engineering, Cornell University, Ithaca, NY 14853
*** Hewlett-Packard Laboratories, 3500 Deer Creek Road, Palo Alto, CA 94303

ABSTRACT

The segregation of phosphorus to grain boundaries in phosphorus implanted $Si_{0.87}Ge_{0.13}$ films, deposited by chemical vapor deposition (CVD), was directly observed by scanning transmission electron microscopy (STEM) with energy dispersive x-ray (EDX) microanalysis. The segregation was determined to be a thermal equilibrium process by measuring and comparing the average phosphorus concentrations at the grain boundaries in $Si_{0.87}Ge_{0.13}$ films subjected to 700, 750 or 800 °C annealing, following the implantation and 1000 °C annealing processes. The measured segregation energy was 0.28 eV/atom. No Ge segregation was found at grain boundaries in phosphorus implanted $Si_{0.87}Ge_{0.13}$ films by STEM x–ray microanalysis. Neither was evidence shown by STEM microanalysis that Ge segregated to grain boundaries in intrinsic $Si_{1-x}Ge_x$ films with x = 0.02, 0.13 and 0.31. Secondary ion mass spectrometry (SIMS) analysis showed that these intrinsic $Si_{1-x}Ge_x$ films contained 10^{19} to 4×10^{19}/cm^3 H, depending on the deposition temperature.

INTRODUCTION

It is well known that phosphorus atoms like to segregate into the grain boundaries in polysilicon because of the huge difference in the heat of sublimation between Si and P [1]. Electrical measurements showed that the resistivity values and carrier concentrations in P doped polysilicon varied with the annealing temperature [2] and these phenomenon were interpreted as the thermal activation of P atoms segregating to the grain boundaries. The accumulation of P atoms in grain boundaries in the polysilicon was also directly detected by the STEM x-ray microanalysis [3]. In recent years, polycrystalline $Si_{1-x}Ge_x$ has been successfully used for manufacturing ultra-shallow junctions in short channel length MOS transistors [4], for making low resistance contact [5] and for replacing polysilicon as the gate material [6]. Studying the segregation phenomena in polycrystalline $Si_{1-x}Ge_x$ would be helpful to control device electrical properties and manufacturing process. However, such research has not been reported.

Assuming that the system is a regular solid solution in which the free energy change on segregation is only due to the change in entropy, when the solute concentration is much less than 1, the segregation can be described as [3]:

$$X_{GB} = \frac{X_{BK}\exp(\frac{\Delta F}{kT})}{1 + X_{BK}\exp(\frac{\Delta F}{kT})} \tag{1}$$

where X_{GB} and X_{BK} are the equilibrium atom fraction of the segregated component at the interface and in the bulk, respectively. ΔF is the surface segregation energy.

Mat. Res. Soc. Symp. Proc. Vol. 557 © 1999 Materials Research Society

The above expression is also based on the assumption that the density of grain boundary states is negligible compared to the lattice sites in the bulk of the grains. To more precisely describe the segregation in a finite grain size system, the density of grain boundary sites, N_t, and the total number of sites in the system, N_{Si}, should be introduced [2]:

$$X_{GB} = \frac{X_{BK}\exp(\frac{\Delta F}{kT})}{1+(\frac{N_t}{N_{Si}})\exp(\frac{\Delta F}{kT})} \tag{2}$$

where the term N_t / N_{Si} can be related to the average grain size, L, by the expression $1.62\times10^{-7}/L$ for Si, if the grains are assumed to be cubes with length of dimension L in cm.

EXPERIMENT

Poly-$Si_{1-x}Ge_x$ films with x=0.02, 0.13 and 0.31, respectively, were grown by CVD to a thickness about 300 nm on a poly-Si seed layer using SiH_2Cl_2 and GeH_4 gases, in the temperatures range from 600 to 790 °C, depending on the Ge content. The poly-Si seed layer is about 60 nm thick and was grown by CVD on a SiO_2 layer thermally on a Si substrate. The Ge concentration and film thickness were measured by RBS. Part of the $Si_{0.87}Ge_{0.13}$ wafer was then implanted with P at a dose of 10^{16} /cm^2 with zero degree tilt and 110 KeV implantation energy. After implantation, the wafer was capped with 100 nm PECVD oxide to prevent the P from out-diffusing, followed by furnace annealing at 1000 °C in N_2 for 1 hour to stabilize the grain structure. The wafer was then broken into three pieces and annealed in N_2, at 800 °C for 10 hrs, 750 °C for 24 hrs and 700 °C for 48 hrs, respectively. The PECVD oxide was removed after the annealing process. Planar TEM samples were prepared for three intrinsic $Si_{1-x}Ge_x$ samples (with x=0.02, 0.13 and 0.31), and for three P implanted $Si_{0.87}Ge_{0.13}$ samples (annealed at 700, 750 and 800 °C). Sample preparation was performed by dimpling from the backside of the sample, followed by ion milling until part of the sample was electron beam transparent. Then the P implanted $Si_{0.87}Ge_{0.13}$ samples with different annealing temperatures and the three intrinsic $Si_{1-x}Ge_x$ samples with x=0.02, 0.13 and 0.31, respectively, were analyzed by STEM equipped with EDX detector, with an electron beam probe diameter of 1 nm. Grain boundary sites were selected using the STEM images and grain boundary spectrums were obtained by placing the probe on the grain boundary. During the acquisition time, the STEM image was frequently checked to make certain that the probe was still accurately positioned on the grain boundary. After data were acquired at the grain boundary, the probe was then moved to either side of the boundary, and bulk x-ray spectrums were acquired for comparison. Ge, P and Si concentrations were then calculated. The excess concentration of the element at the grain boundary was determined by comparison of the spectrums obtained at the grain boundary and in the bulk. The diffraction-contrast images of the TEM samples were recorded in the conventional microscope, to evaluate the defects and average grain size. SIMS depth profile analysis was performed at Evans East Co. on three intrinsic $Si_{1-x}Ge_x$ samples with x=0.02, 0.13 and 0.31, respectively.

EXPERIMENTAL RESULTS AND DISCUSSIONS

In all three intrinsic $Si_{1-x}Ge_x$ samples, no Ge accumulation was detected at the grain boundaries by STEM x-ray microanalysis, which leads to the conclusion that no Ge segregated at the grain boundaries in our intrinsic $Si_{1-x}Ge_x$ samples. The SIMS data showed that all intrinsic samples contained H distributed uniformly in depth. The concentration of H ranges from 10^{19} to $4\times10^{19}/cm^3$, with a higher H concentration for samples deposited at a lower temperature (higher Ge fraction).

STEM x-ray microanalysis results showed that P segregated to grain boundaries in all P implanted $Si_{0.87}Ge_{0.13}$ samples with different annealing temperatures, while the Ge concentration did not appear to be substantially different at a grain boundary and away from the boundary site. Fig. 1 shows two x-ray spectrums obtained at a grain boundary and a bulk site, respectively, in the P implanted $Si_{0.87}Ge_{0.13}$ sample annealed at 700 °C. The peak height of Si in the boundary is slightly lower than that of a bulk site, and the P accumulation could clearly be seen by the P peak at the grain boundary site while no P peak is seen in the spectrum from the bulk. The peak height of Ge is about the same for grain boundary site and bulk site. It was found that the amount of accumulated P at the grain boundary was different at different grain boundaries and some grain boundaries do not show higher concentration of P than that in the bulk. This implies that the configuration of the grain boundary structure may play an important role in determining the level of segregation, as the density of dangling bonds to accommodate the segregated atoms depends on the disorder of the grain boundary, which is determined by the atomic structure of the grain boundary. Thermodynamically, the increase of the disorder at the grain boundary would lead to a

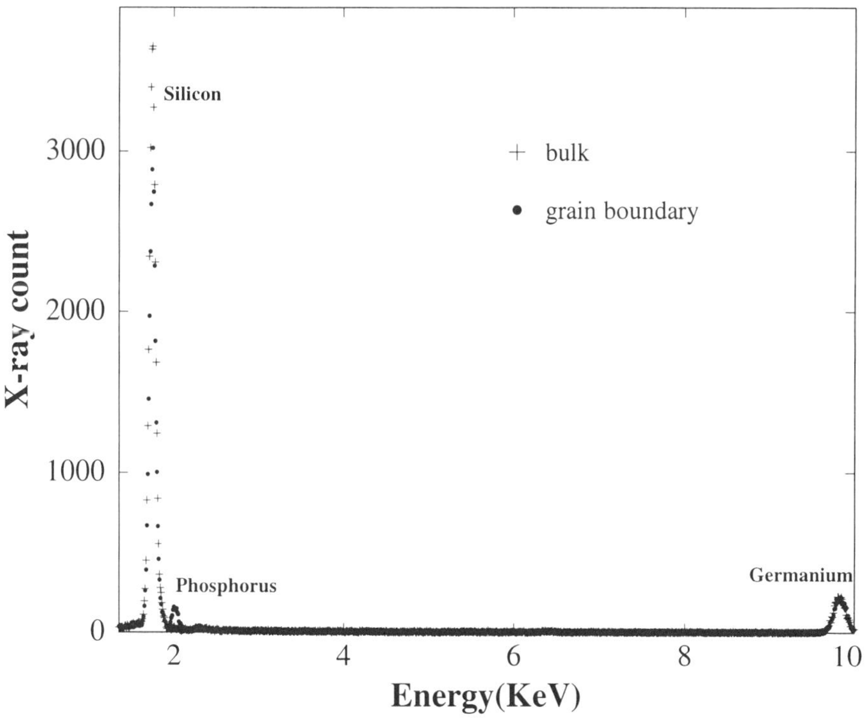

Figure 1: X-ray spectrums obtained at the grain boundary and away from the grain boundary in P implanted $Si_{0.87}Ge_{0.13}$, annealed at 700 °C.

higher free energy difference between grain boundary and bulk and, therefore, favor more P atom segregation. X-ray spectrums obtained in twin boundaries in all samples investigated do not reveal excess P atoms compared to that in the bulk. This could be explained by the high degree of order of twin boundaries and absence of dangling bonds in the twin boundaries. From the conventional TEM studies, no contrast caused by clusters of P or a high density of dislocation-loop type defects, which are usually generated by the precipitation of excess P in the film, was observed, as shown in Fig. 2. This indicates that most of the P atoms either stay in the bulk of the grains or segregate to the grain boundaries.

EDX analysis of an intrinsic $Si_{0.87}Ge_{0.13}$ film showed that the Ge concentration in the film was about 10.3%, which is about 21% lower than the result of RBS. Table 1 lists the average grain size, average Ge and Si concentrations, average and the maximum observed volume concentrations of P and X_{GB}, at the grain boundaries in $Si_{0.87}Ge_{0.13}$ thin films with various annealing temperatures. The concentration of doping in the matrix, X_{BK}, by implantation is 0.57 at.%. The table shows that the amount of segregated P atoms increases as the annealing temperature decreases, indicating an equilibrium segregation process. Fig. 3 is a plot of average P concentration at the grain boundaries versus the annealing temperature. Here, we suppose that the grain boundary consists of one monolayer indexed as (100) plane, which has $6.8 \times 10^{14}/cm^2$ atomic density. It is assumed that this monolayer plane can be saturated by segregated atoms. Therefore, $C_{GB} = 1$ in Fig. 3 corresponds to an atomic density of $6.8 \times 10^{14}/cm^2$. The experimental data is

Figure 2: Planar TEM image of P implanted $Si_{0.87}Ge_{0.13}$ thin film annealed at 700 °C.

fitted with Eqs. (1) and (2), where the average grain size is taken as 0.22 μm. Though Eq. (2) uses one more parameter, the grain size, than Eq. (1) does, the two fitted curves almost overlapped over the entire range of annealing temperatures. The fitted segregation energy obtained by fitting Eq. (1) is 0.282 eV/atom, and it is 0.286 eV/atom by fitting Eq. (2).

Table 1: Grain size (μm), average and maximum concentration (at. %) of P, average Ge and Si concentration (at. %) at grain boundaries with different annealing temperatures.

	700 °C	750 °C	800 °C
Grain size	0.21	0.22	0.24
P (maximum)	3.44	3.06	2.55
P (average)	1.87	1.67	1.54
Ge	11.4	10.4	10.1
Si	86.7	87.9	88.4

Rose et al. [3] measured the P segregation energy of 0.32 eV/atom in polysilicon via STEM x-ray microanalysis with probe size of 10 nm. Their films were doped by diffusion with doping concentration 0.6 at.%, which is close to the doping level in our sample, and annealed in the temperature range from 650 to 800 °C. Mandurah et al. [2] measured the P segregation energy of 0.44 eV/atom by resistivity and Hall measurement on implanted samples with 0.04 at.% doping level annealed in the temperature range from 800 – 1000 °C. As the segregation level largely depends on the density of grain boundary states, which is determined by the film growing process, doping level and process of thermal treatment, slight differences in the value of segregation energy could be expected for different films and different measurement techniques, though more work is needed to decide if these differences are meaningful. Ge appears to compete with P, since from the experiment, the Ge concentration at the boundary weakly, but systematically increases with decreasing temperature. However this small difference of concentration is still within the measurement error of EDX. Therefore, we cannot conclude that Ge segregates.

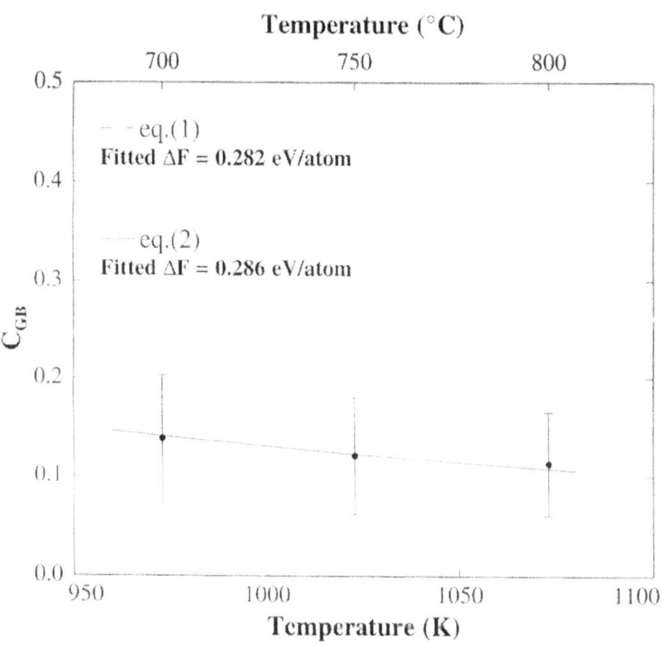

Figure 3: Theoretical fits, based on thermodynamic theory, to the experimentally measured P segregation in $Si_{0.87}Ge_{0.13}$. $C_{GB}=1$ corresponds to the density of (100) plane, i.e. $6.8\times10^{14}/cm^2$.

The Si-Ge binary equilibrium diagram predicts that Ge atoms would segregate in the grain boundaries in polycrystalline SiGe alloy, according to the criteria to determine the segregated component in a two-component solid solution [7]. The same conclusion could also be reached by comparing the difference of heat of sublimation between the Ge and Si atom. However, our intrinsic $Si_{1-x}Ge_x$ films are not a pure two-component system. During growth, H is mobile and probably diffuses quickly along grain boundaries, as they are the fast diffusion paths in the polycrystal. Grützmacher et al. [8] presented the evidence that Ge segregation to a Si/SiGe interface was prevented by H introduced during atmospheric-pressure CVD. It was reported that H could also passivate dangling bonds and other defects at grain boundaries [1]. The unsaturated grain boundary density in polysilicon was around 10^{12} to $10^{13}/cm^2$, depending on the grain boundary structure [1]. The grain size in our intrinsic $Si_{0.87}Ge_{0.13}$ film was about 63 nm measured by TEM. To passivate the expected number of electrically active states at the grain boundaries requires about $10^{18}/cm^3$ H, which is an order of magnitude lower than the H concentration measured in our film. Taking the boundary as a (100) plane, the grain boundary density is around $6.8\times10^{14}/cm^2$, which corresponds to a volume density of about $10^{20}/cm^3$. If some of the broken bonds at the grain boundary are terminated by Si-Si, Si-Ge and Ge-Ge bonds, $1-4\times10^{19}/cm^3$ H may be sufficient to locally saturate the rest of the boundary states. Thermodynamically, the saturation of the dangling bonds by H reduces the grain boundary energy, particularly the strain energy due to the lattice mismatch of Si and Ge. On the other hand, Ge and Si have similar chemical properties, like four-fold coordinated bonds. Chemically, Ge may not be as likely to segregate to grain boundaries as elements with different chemical properties from Si, such as H and P. Therefore the suppression of Ge segregation by H could be expected.

In our P implanted $Si_{0.87}Ge_{0.13}$ film, the P may play the same role as H does in the intrinsic SiGe film to saturate the grain boundary. This could lead to the reduction of the grain boundary energy, and therefore prevent the Ge segregation. Ge, Si and P have similar covalent radii, but the difference of heat of sublimation between the Si and P atom is larger than that between the Si and Ge atom. Segregation of P will therefore make the free energy of the system lower than will Ge segregation.

CONCLUSIONS

We have shown that Ge does not accumulate in grain boundaries in CVD grown intrinsic $Si_{1-x}Ge_x$ films with Ge fractions = 0.02, 0.13 or 0.31. H, detected by SIMS depth profile, is postulated to saturate the grain boundary to prevent Ge segregation. We have shown that P segregated to the grain boundaries in $Si_{0.87}Ge_{0.13}$ films in the annealing temperature range from 700 to 800 °C and the measured segregation energy was about 0.28 eV/atom. No Ge segregation in the grain boundary in P implanted $Si_{0.87}Ge_{0.13}$ films was observed. It is suggested that the reduction of grain boundary energy by P segregation is responsible for this phenomenon.

REFERENCES

[1] T. I. Kamins, *Polycrystalline Silicon for Integrated Circuit Applications*, Kluwer Academic Publisher, Massachusetts, 1988, p.113, p.186 and p.177

[2] M. M. Mandurah, K. C. Saraswat, C. R. Helms and T. I. Kamins, J. Appl. Phys., **51**, 5755 (1980)

[3] J. H. Rose and R. Gronsky, Appl. Phys. Lett., **41**, 993 (1982)

[4] T. Uchino, T. Shiba, K. Ohnishi, A. Miyauchi, M. Nakata, Y. Inoue and T. Suzuki, International Electron Devices Meeting 1997, p. 479

[5] Y. S. Chieh, J. P. Krusius, D. Green and M. Ozturk, IEEE Elec. Dev. Lett., **17**, 360 (1996)

[6] T. J. King, J. R. Pfiester and K. C. Saraswat, IEEE Elec. Dev. Lett., **12**, 533 (1991)

[7] J. J. Burton and E. S. Machlin, Phys. Rev. Lett., **37**, 1433 (1976)

[8] D. A. Grützmacher, T. O. Sedgewick, A. Power, M. Tejwani, S. S. Iyer, J. Cotte and Cardone, Appl. Phys. Lett., **63**, 2531, (1993)

EXPERIMENTAL VERIFICATION OF A RANDOM MEDIUM MODEL FOR THE OPTICAL BEHAVIOUR OF ULTRATHIN CRYSTALLINE SILICON LAYERS GROWN ON POROUS SILICON

L. STALMANS[*], A.A. ABOUELSAOOD[$], M.Y. GHANNAM[+], J. POORTMANS[*],
H. BENDER[*], M. CAYMAX[*] and J. NIJS[*,§]
[*]IMEC vzw, Kapeldreef 75, 3001, Leuven, Belgium, stalmans@imec.be
[+]ECE Dept., College of Engineering and Petroleum, Kuwait University, Kuwait
[$]Physics and Mathematics Dept., Faculty of Engineering, Cairo University, Egypt
[§]Also Professor at KULeuven, Faculty of Engineering

ABSTRACT

The random medium model for porous silicon (PS) is applied to a porous silicon layer to be used as a back reflector in a thin film silicon solar cell realized in an ultrathin epitaxially grown Si layer on PS. Three-layer-structures (epi/PS/Si) have been fabricated by RPCVD of 150-1000 nm epitaxial silicon layers on the porous surface of silicon wafers. The light reflection has been measured in the 300-1000 nm wavelength range. An excellent agreement is found between the experimentally measured reflectance curves and those calculated using the random medium model. The analysis shows that the epitaxial growth has led to an appreciable reduction in the porosity in the intermediate PS layer from about 60% to 20-30%.

INTRODUCTION

In a previous work [1] it has been shown that a front surface porous silicon (PS) layer acts as a perfect light diffuser for the transmitted light, which is an important requirement for light confinement. Recently PS has been introduced in crystalline Si (c-Si) thin-film silicon solar cell technology as a means to realize thin c-Si films by epi-lift-off from a reusable high-quality Si substrate [2]. Another thin-film approach consists of using the PS layer as a backside diffuse reflector. Preliminar experimental results have recently been reported independently in [3,4]. Light trapping is of utmost importance in this extreme case of ultrathin (2-3 μm) epitaxially grown silicon layer on PS. Without any intermediate structure between the c-Si thin film and the silicon substrate, a large part of light is lost due to absorption in the highly doped substrate. The porous layer however, acts as an optical barrier because of its different optical properties, which are linked to its porosity.

Recently, we have proposed a semiquantitative model for the optical properties of low-porosity silicon based on electromagnetic wave propagation in a random medium [5]. Both the coherent (specular) and fluctuating (diffuse) field components are considered and both light absorption as well as scattering (using Mie's theory) are taken into account for the attenuation of the coherent field when propagating through the medium. This model differs from the effective medium models for heterogeneous media in which light scattering is ignored. The purpose of the present study is to check the validity of this model by comparing its predictions to results of recent experiments in which the reflectance of epitaxial silicon thin films grown on porous silicon layers has been measured as a function of the wavelength of the incident light.

EXPERIMENT

Samples with the structure sketched in Fig.1 have been prepared and analyzed, both structurally and optically. The porous layers are prepared by conventional electrochemical anodization in an HF-based electrolyte. The anodization time is adjusted to obtain a layer thickness in the range of 150-450 nm. Highly-doped (0.01 Ω.cm) (100) p$^+$-Si substrates are used

Mat. Res. Soc. Symp. Proc. Vol. 557 © 1999 Materials Research Society

for a twofold purpose. Firstly no back contact is required for anodization. The presence of metal would be incompatible with the subsequent thermal chemical vapor deposition (CVD) process. Secondly, porous etching of a highly doped material results in "columnar" meso-porous silicon which is the preferred type of porous silicon for the subsequent CVD-growth from the viewpoint of epitaxial quality of the deposited silicon layer. The samples can be divided into two groups according to the initial porosity of the PS layer before the epitaxial growth (about 20% and 60%). This initial porosity depends on both the electrolyte composition and the applied anodic current density.

Crystalline silicon layers of thickness in the range of 150-1500 nm are epitaxially grown on porous silicon in an Epsilon-One epitaxial reactor manufactured by ASM Epitaxy. It consists of a horizontal, lamp-heated quartz chamber with a SiC-coated graphite susceptor. For this application, various deposition temperatures have been used in the range 700-850 °C. The CVD-reactor is operated at a reduced pressure (RPCVD), typically 40 Torr. SiH_2Cl_2 is used as a silicon source gas, which is diluted in a flow of H_2 as carrier gas. The time between the porous silicon formation and the loading into the CVD chamber system is kept as short as possible to avoid ageing.

The structural analysis of the epitaxial silicon growth and the resulting morphology of both the porous and the epitaxial layers reveal that the PS remains largely intact, while the epi-layer quality strongly depends on the initial porosity and the CVD growth temperature [3]. The thickness of the porous and epitaxial layers has been determined from XTEM analysis. The porosity and thickness of the porous layers have been estimated by means of spectroscopic ellipsometry (SE) using the Sopra regression software package Winelli 3.01 that makes a best fit to the data on the basis of the Bruggeman effective medium model. The total reflectance of the samples (normal incidence) has been measured in the wavelength range 300-1000 nm by means of an integrating sphere.

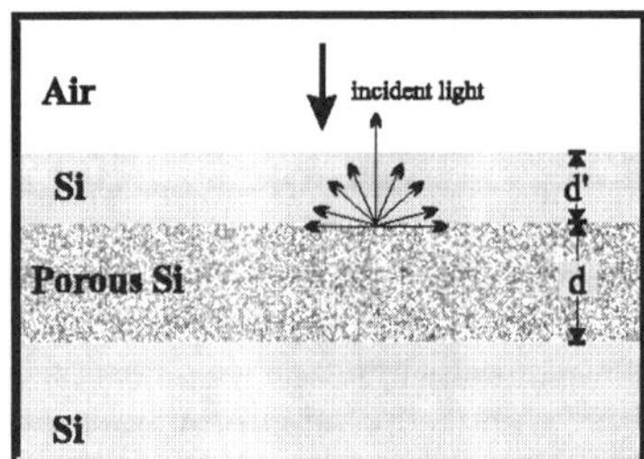

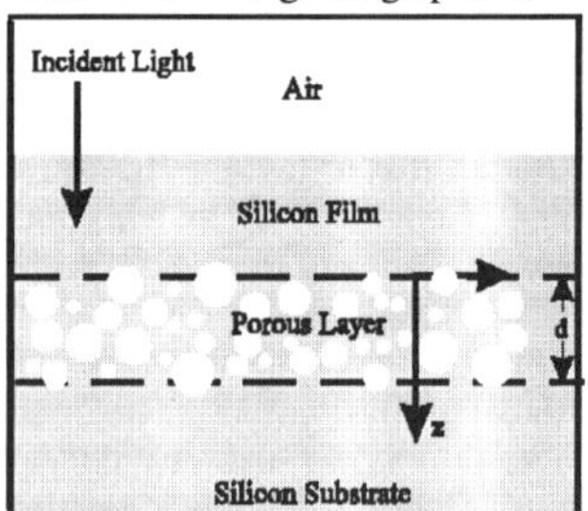

Fig.1: A PS layer of thickness d used as a diffusing back-reflector to an epitaxial silicon film of thickness d' grown on it.

Fig.2: Random medium representation of a PS layer as a silicon host medium containing spherical cavities.

TOTAL REFLECTANCE ANALYSIS

Calculations based on the random medium model [5] show that the diffuse transmittance of a porous silicon layer is considerably higher than its diffuse reflectance at any given wavelength, and that almost all diffuse reflected light is absorbed in the epi-layer and very little of it is retransmitted through the upper surface of the samples. It is therefore reasonable to identify the total reflectance of a sample with its specular component. The main task then is to fit the total reflectance data by means of the model for each sample. The basic free parameters in the model are:

1. the effective thickness d of the porous layer which is the thickness of the effective layer in which the centers of the spherical cavities are randomly distributed. This thickness is somewhat smaller than the geometrical thickness of the porous layer as measured by XTEM since cavities with centers on or near the boundary of the effective porous layer protrude outside it as we can see from Fig.2.

2. the effective thickness d' of the epitaxial silicon layer defined as the distance between its upper surface and the upper boundary of the effective porous layer.

3. the effective porosity γ of the PS layer defined as the ratio between the total volume of the cavities and the total volume of the effective PS layer. Initial values for these parameters are taken from XTEM and ellipsometry analysis keeping in mind the distinction between the geometrical and effective values.

4. the size distribution parameter ν and the average radius r_m of the pores. Standard distributions will be considered with $\nu = 0$ (uniform), 2 (half Lorentzian), power law or half Gaussian when ν tends to infinity.

A 1% relative change in the parameters d and d' and a 1% absolute change in γ makes the calculated results deviate drastically form the experimental data, which implies that they are determined from the fit with a high degree of confidence. The results are less sensitive to the other parameters r_m and ν, which are determined with an accuracy of 20% relative only.

DATA FITTING RESULTS

Samples with low initial porosity

The first group includes those samples made from low-porosity layers. The initial porosity of about 20% is derived from SE-analysis according to the Bruggeman model ($\sigma = 0.016$). The SE-data and fit of the initial porous layer are shown in Fig.3. Fig.4 shows the random medium model fit to the reflectance of a sample with a 445 nm epi-layer over a 150 nm PS layer according to the XTEM analysis. The optimum fit is obtained with d = 153 nm, d' = 407 nm and $\gamma = 0.23$. The fit is almost perfect, although there is a slight difference between the calculated reflectance and the experimental data at the longest wavelengths. Free carrier absorption in the highly doped substrate has been considered but did not result in an improved fitting since it appears to be significant only at wavelengths beyond 1100 nm which is outside the range of our fittings. The additional light reflection at the backside of the finite-thickness (675 μm) substrate has been incorporated in a rather crude way, but also this can only affect the fit beyond 1000 nm since the absorption depth for this wavelength light is 150 μm (or less in the case of a highly doped substrate).

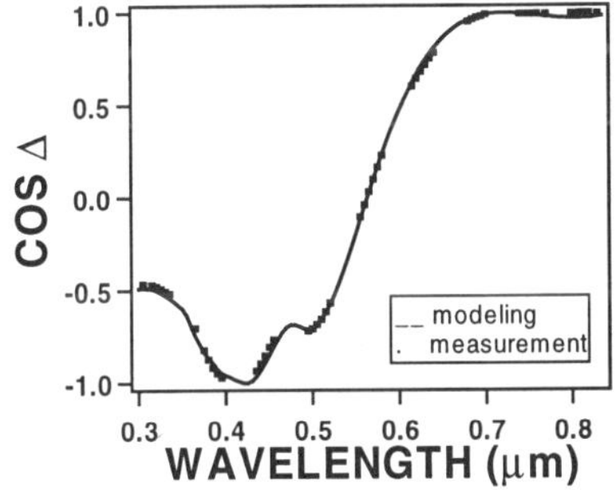

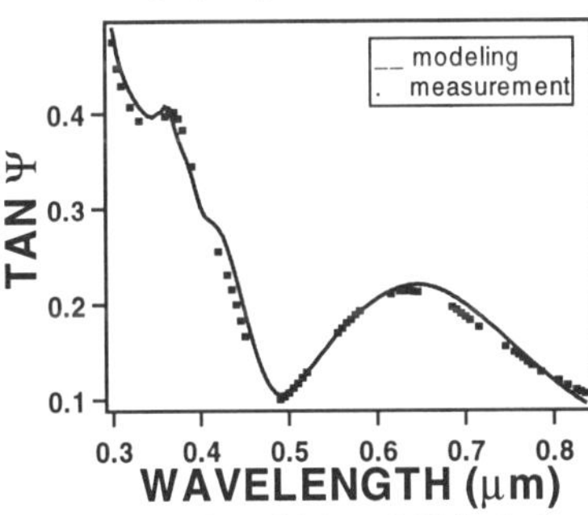

Fig.3: SE-data and fit ($\sigma=0.016$) of the initial porous layer for which $\gamma=0.20$ is derived.

Fig.5 shows a fitting to the same experimental data by means of Bruggeman's model. This model gives almost the same values for d and d', but a slightly lower value of 0.21 for the porosity. The calculated results show more pronounced oscillations in the short wavelength reflectance compared with the experimental results, indicating that the Bruggeman model has underestimated the imaginary part (extinction coefficient) of the difference in dielectric constant between porous silicon and crystalline silicon in this wavelength range, which is due to the fact that the Bruggeman model does not take into account scattering induced absorption. Figure 6 shows the random medium model fit to the reflectance data of another sample in the first group with an epi-layer of thickness 800 nm grown over a 143 nm porous layer according to the XTEM analysis. The model gives the effective thickness d'=788 nm and d = 142 nm, and the value of 0.17 for the effective porosity. The best pore distribution function for the samples in the first group corresponds to $r_m = 30$ nm and $v = 3$.

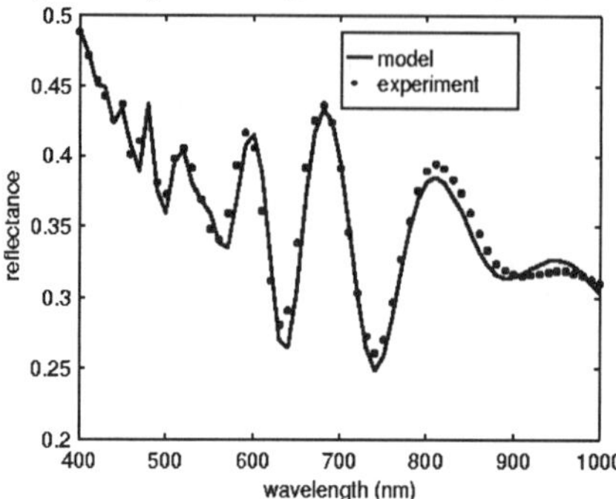

Fig.4: Random model fit to the reflectance of a sample with a 445nm epi-layer over 150nm PS according to XTEM analysis, obtained with d=153nm, d'=407nm and γ=0.23.

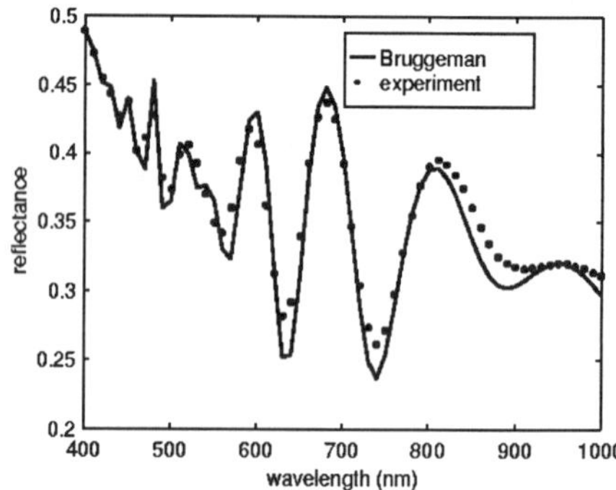

Fig.5: A fitting to the same experimental data as Fig.3 but using Bruggeman's model. A lower value of γ=0.21 is used.

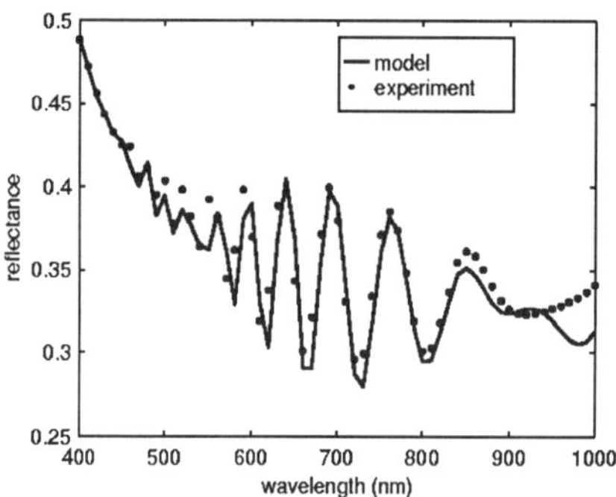

Fig.6: Random model fit to the reflectance of a sample with a 800nm epi-layer over 143nm PS according to XTEM analysis, obtained with d=142nm, d'=788nm and γ=0.17.

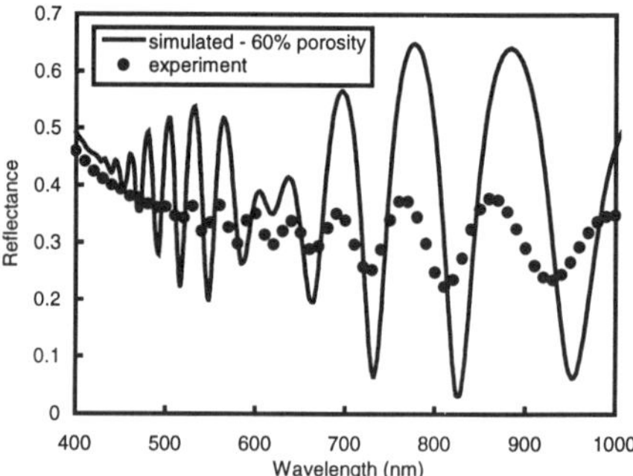

Fig.7: Reflectance of a sample with a 675nm epi-layer over 280nm PS with an initial porosity of 60% and the simulated specular reflectance assuming a γ=0.6 after epi-growth.

Samples with high initial porosity

The second group of samples is made from high-porosity PS layers for which an initial porosity of about 60% is derived from SE-analysis using the Bruggeman model (σ= 0.091). Fig.7 shows the reflectance data for a sample with a 675 nm epi-layer on top of 280 nm PS layer according to XTEM analysis. A comparison is made to the calculated specular reflectance of an epi/PS/Si layer structure in which the PS layer is assumed to maintain its 60% porosity and using the layer thickness as derived from XTEM analysis. Besides the fact that the locations of the extrema of the reflectance occur at similar wavelength values, two specific properties are to be mentioned:

1. The moderate oscillations in the measured reflectance suggest that during the epitaxial growth process, the porosity has been brought down to a value not very different from those of the samples made from low porosity PS layers. A decreased porosity (15-20%) of the buried PS is derived from SE-analysis as well. An independent quantification of the porosity is not possible using XTEM due to the small dimensions of the pore and crystallite sizes. On the other hand preliminar results of Rutherford Back Scattering analysis (RBS) on porous samples reveal that the presence of a porous surface layer gives rise to an increase of the channeled signal. Asssuming that the porous layer corresponds to this region of enhanced channeled signal allows the calculation of the Si-counts within this area and the ratio to the Si-counts that are expected normally for an equally thick Si layer results in a value for the porosity. In this way a 60% porosity is derived for the high-porosity PS, which agrees well with the value obtained from SE-analysis. Subsequently RBS analysis of a sample with a buried PS layer has been performed and similarly a porosity has been calculated from the region with an enhanced channeled signal. In this case the Si-counts approach the value that is expected for bulk Si, which implies a strongly decreased porosity. Hence also some experimental evidence is presently available to support the picture of a significantly decreased porosity when depositing a c-Si film on top of high-porosity PS.

2. Furthermore, different from the case of samples with a low initial porosity and different from the calculated specular reflectance behaviour, the amplitude of the reflectance oscillations does not become small at certain wavelengths (outside the strong absorption region) that correspond to destructive interference between the light waves reflected at the two PS layer surfaces. This means that the reflectance of one of the two surfaces is appreciably smaller than that of the other. At first sight, the more obvious possibility is that the porous silicon still has a high porosity, hence a significant amount of light that enters the layer is diffused on its round-trip to and back from the back surface, and is eventually absorbed in the silicon epi-layer according to the light trapping scenario. If this picture was true, the upper surface of the PS layer would contribute much more to the specular reflectance of the sample than the PS layer back surface. This effect becomes increasingly important for thicker porous layers because of the stronger light diffusing behaviour [5]. However, the fact that the apparent absence of significant destructive interference between the light waves reflected by the two surfaces of the PS layer does not seem to depend on its thickness and the indication of a generally lowered porosity (see above) are not in favour of this first picture. The other possibility is that the PS layer may have a low porosity, but no sharp upper boundary: it may be that the porosity increases gradually from the epi-layer over a certain distance before we finally reach a genuine porous layer. In that case, the PS layer back surface would be the more efficient specular reflector because of the higher porosity. An observation in favour of this picture is the fact that the locations of the extrema of the reflectance seem to be governed by the sum of the thickness of the epi-layer and the PS layer, which means that the specular reflection from the lower surface of the PS layer is the dominant one.

We are led accordingly to accept the picture of a porosity which is decreased after the epitaxial growth, the presence of a transition layer of effective thickness d" (to be treated as an additional fitting parameter) between the PS layer of thickness d and the epi-layer of thickness d'. We further assume that the porosity changes linearly from its value γ in the porous layer to zero at the other side of the transition layer. The value of d" is determined with a relative uncertainty of 10% and the value of d" virtually equals zero for the samples with a low intial porosity. This assumption has actually been used to make a fitting (Fig. 8) to the data of Fig.7, leading to d = 170 nm, d' = 630 nm, d" = 50 nm and $\gamma = 0.25$, with $r_m = 30$ nm and $v = 3$ as in the previous examples. Almost the same values for the latter two parameters are obtained while fitting the sample in Fig.9 giving d = 155 nm, d' = 1215 nm, d" = 60 nm and $\gamma = 0.21$. These values should be compared with the XTEM values for the PS layer thickness which equals 235 nm and for the epi-layer thickness which equals 1350 nm. The less good data fit for these samples with a high initial porosity could be attributed to the assumption of a linear porosity change in the transition layer. It can be mentioned that the inclusion of a transition layer (roughness) in between the epi-layer and the PS layer also results in an accurate fitting of the SE-data for this epi-on-PS sample. ($\sigma = 0.060$).

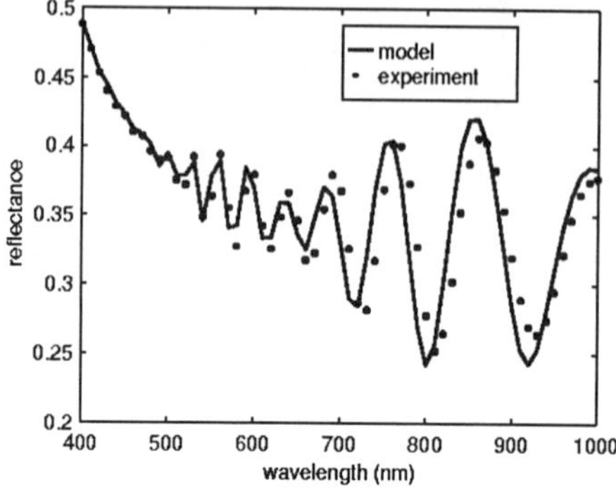

Fig.8: Random model fit to the reflectance of a sample with a 675nm epi-layer over 280nm PS according to XTEM analysis, obtained with d=170nm, d'=630nm,d"=50nm and γ=0.25.

Fig.9: Random model fit to the reflectance of a sample with a 1350nm epi-layer over 235nm PS according to XTEM analysis, obtained with d=155nm, d'=1215nm, d"=60nm and γ=0.21.

CONCLUSIONS

The total reflectance of epi/PS/Si-structures is measured and modeled in the 300-1000nm wavelength range. For some samples the initial porosity before the epitaxial growth was low (20%), while for some others it was relatively high (60%). These data are fitted by means of a random medium model, treating the thickness of the PS and of the epi-layers and the porosity as fitting parameters. An excellent agreement is found between the experimentally measured reflectance curves and those calculated using the proposed model. The values of the layer thickness used to get such perfect fitting correspond, within a reasonable experimental error, to those obtained independently by experimental methods. Such agreement provides an experimental verification that the random-medium approach can be applied to porous silicon and certainly validates the use of the proposed model in the low porosity case. The analysis shows that the epitaxial growth results in an appreciable porosity decrease from 60% to 20-30% for the initially high-porosity layers and the creation of a transistion layer between the porous material below and the crystalline material above. No such drastic change in porosity has been observed for samples made from initially low-porosity layers.

ACKNOWLEDGEMENTS

The authors are indebted to Dr. S. Jin (IMEC) for performing the XTEM-analysis, and to B. Brijs (IMEC) for RBS analysis and interpretation. The research of M. Ghannam is supported by Kuwait University research grant EE-108. L. Stalmans is supported by a fellowship from IWT (Flemish Institute for Promotion of Scientific-technological research in the Industry, Belgium).

REFERENCES

[1] L. Stalmans, J. Poortmans, H. Bender, M. Caymax, K. said, E. Vazsonyi, J. Nijs and R. Mertens, *Progress in Photovoltaics:Research and applications* **6**, p. 233-246 (1998).
[2] R. Brendel, *Proceedings 14th European PV solar energy conference*, p. 1354-1357 (1997).
[3] L. Stalmans, J. Poortmans, M. Caymax, H. Bender, S. Jin, J. Nijs, and R. Mertens, *Proceedings 2nd World conference on PV Solar Energy Conversion*, p. 124-127 (1998).
[4] J. Zettner, H.V. Campe, M. Thoenissen, J. Ackermann, Th. Hierl, R. Brendel and M. Schulz, *Proceedings 2nd World conference on PV Solar Energy Conversion*, p. 1766-1769 (1998).
[5] A.A. Abouelsaood, M.Y. Ghannam, and J.F. Nijs, *Proceedings 2nd World conference on PV Solar Energy Conversion*, p. 176-179 (1998).

POSSIBILITY OF QUASI-SINGLE-CRYSTALLINE
SEMICONDUCTOR FILMS

T. Noguchi, S. Usui, D. P. Gosain and Y. Ikeda*

Frontier Science Laboratories, *Personal AV Company, Sony Corporation

2-1-1, Shinsakuragaoka, Hodogaya-ku, Yokohama-shi, 240-0036, JAPAN,

noguchit@src.sony.co.jp

ABSTRACT

The existence of a novel tetrahedral semiconductor Quasi-Single-Crystalline (QSC) phase is posited. In the QSC semiconductor phase, the films consist of grains with a diamond structure of tetrahedral elements, such as Si, Ge and C, and the grains have a preferred orientation, such as <111> or <100> normal to the film. The lattices perpendicular to the grain boundaries are quasi-matched with neighboring grains. The grains in the films form a regular array, and are distributed more uniformly than in conventional poly-crystalline semiconductor films. Because of the small-angle grain boundaries, a tetrahedral QSC semiconductor such as QSC Si films are expected to have extremely low energy barriers at the grain boundaries.

1. STATE-OF-THE ART CRYSTALLIZATION OF SI FILMS

Research on TFTs (Thin Film Transistor) has been undertaken for application to SRAM (Static Random Access Memory) and LCD (Liquid Crystal Display). [1, 2] In order to enhance electronic properties such as carrier mobility in TFTs (Thin Film Transistors), the grain size should be enlarged, and various approaches to this have been tried using SPC (Solid Phase Crystallization) or/and ELA (Excimer Laser Annealing). [3, 4, 5, 6] Prefererred orientation of the {110} face has been reported using SPC after the channeling technique of Si+ ion implantation [7]. The silicon thin film consisting of silicon poly-crystal grains whose selected orientation is approximately the <111> direction can be formed by repeatedly irradiating an excimer laser on an amorphous layer. [8] On the other hand, controlling the location of the seeds of crystallization, as shown in Fig.1, as well as control of grain size have been reported using SPC. [9, 10]. By using ELA for seed formation, large and single-crystal-like disk-shaped grains of a preferable orientation of the {111} face or hexagonal grains have been reported. [10, 11, 12] In order to realize SOI (Silicon On Insulator) or system LSIs on glass or on other substrates, some other promising seed-controlling techniques for enlarging of grain size have been reported using SPC in a low-temperature process. [13, 14] By using ELA, extreme enlargement of grains laterally has

Mat. Res. Soc. Symp. Proc. Vol. 557 © 1999 Materials Research Society

been reported as well as improving the crystallinity, which affects the device characteristics. [15] More precise control of single-crystalline large grains with a uniform distribution and with the preferred orientation is required.

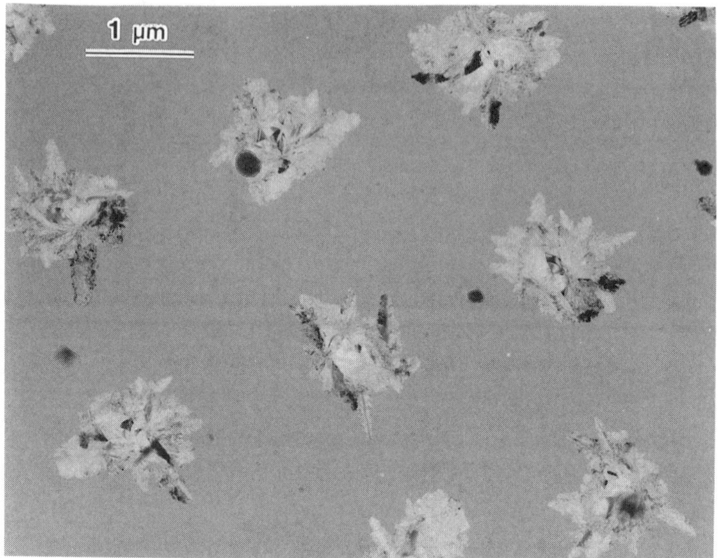

Fig.1 TEM micrograph of location-controlled poly-crystalline Si grains obtained after SPC.

2. QSC AS A NEW CRYSTALLINE PHASE FOR TETRAHEDRAL SILICON FILMS

We posit a novel semiconductor Quasi-Single-Crystalline (QSC) phase between single-crystalline and poly-crystalline phase, and we propose the existence of the QSC for tetrahedral silicon films. For the QSC semiconductor phase, the films consist of single-crystalline grains with a diamond structure of tetrahedral elements, such as Si, Ge and C, and the grains have a preferred orientation, such as <111> or <100> normal to the film. The lattices perpendicular to the grain boundaries are quasi-matched with neighboring grains. The grains in the films form a regular array, and are distributed more uniformly than in conventional poly-crystalline semiconductor films.

In the case of QSC films with <100> orientation, the grains form a square-based shape, as shown in Fig.2. The Si films are made of substantially single-crystalline grains which are preferentially {100} face oriented, have a square-based shape, and are closely aligned in rows and

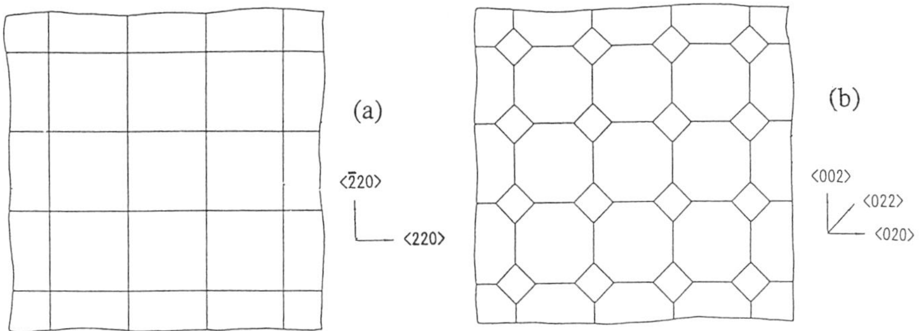

Fig.2 Schematic figure of array for QSC with <100> orientation.

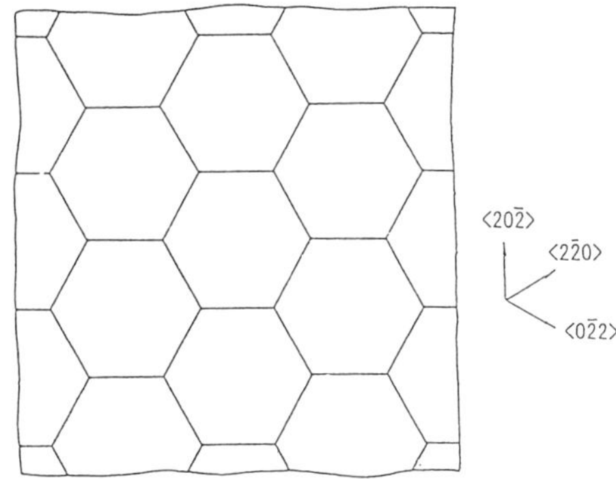

Fig.3 Schematic figure of array for QSC with <111> orientation.

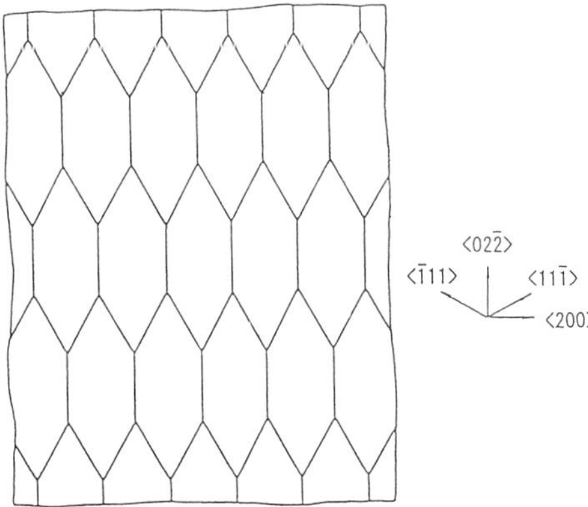

Fig.4 Schematic figure of array for QSC with <110> orientation.

columns. The four side surfaces of each crystal grain are {002} oriented surfaces. Adjacent crystal grains are lattice-matched at least in a part of their grain boundaries. On the other hand, in the case of QSC films with <110> or <111> orientation, the grains have a hexagonal shape, as shown in Fig.3 and Fig.4. Single-crystalline grains, which are preferentially <111> oriented, are aligned in two dimensions in an equilateral turtle shell pattern. The six side surfaces of each crystal grain are {202} oriented. Size of the crystal grains for QSC is larger and more uniform than that of conventional polycrystalline Si films. A thickness between 10nm and 100nm, typically 40nm, and the grain size between 100nm and 10µm, can be controlled for the required application. Because it is made up of substantially single-crystalline grains oriented preferentially, and adjacent crystal grains are lattice matched at least in a part of their grain boundaries, a tetrahedral QSC semiconductor such as QSC Si film is expected to have extremely low energy barriers for conducting carriers.

In order to realize QSC silicon films, nucleation of the grains can be achieved using by SPC or ELA or other technique at fixed sites. [7, 9, 10, 12] Subsequently, lateral grain growth must be controlled from the fixed sites by similar crystallization technique such as SPC or ELA. [9, 10, 11, 14, 15] Otherwise, the preferred grain orientation is expected to be realized by surface-energy-driven grain growth in an extremely high temperature near the liquid-solid interface. [16, 17] Some promising results for <111> oriented disk-shaped or corresponding to hexagonal-shaped grains have been realized using SPC subsequently after ELA, as shown in Fig.5. [10, 11] Also, square based single crystal-like grains with <100> preferred orientation is indicated experimentally in an article of specification [18], as shown in Fig.6.

The QSC Si film formed on a SiO_2 under-layer is expected to have better electronic characteristics than conventional polycrystalline Si films. Therefore, a good-quality SOI structure equivalent to bulk single-crystalline Si can be realized using these three types of QSC silicon films. When the QSC semiconductor films with highly preferred orientation and uniform grain size is realized, innovative devices based on SOI-liked TFTs can be possible by fabricating the channel either in each single crystalline grain or even across the grain boundaries.

3. CONCLUSION

We posit a novel crystalline phase between known single-crystalline and poly-crystalline phase for tetrahedral semiconductor. Because of the small-angle grain boundaries, the novel tetrahedral QSC semiconductor films are expected to have extremely low energy barriers at the grain boundaries and are expected to exhibit electronic properties superior to conventional poly-crystalline silicon films and should be suitable for application to innovative devices such as high-performance TFTs on insulating substrates equivalent to existing SOI MOS transistors in bulk Si.

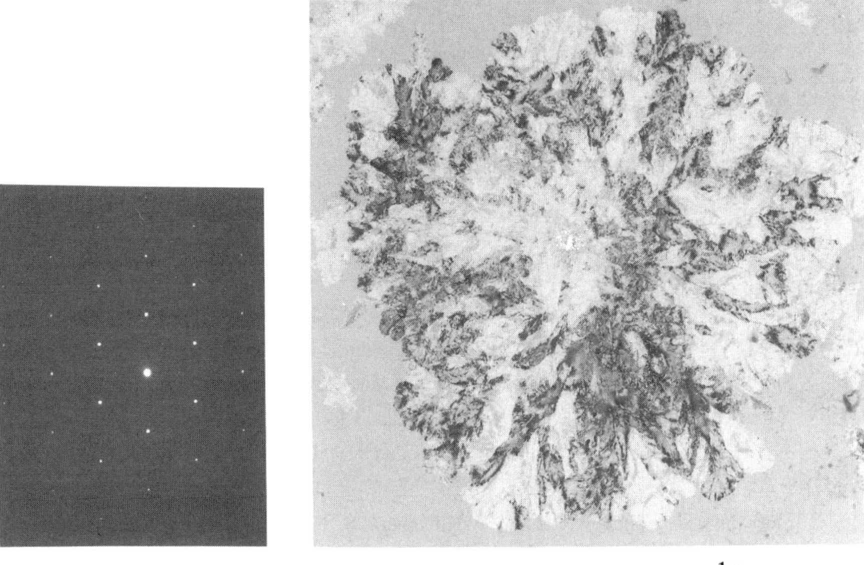

Fig.5 TEM micrograph of a disk-shaped or hexagonal-shaped
Si grain with <111> orientation.

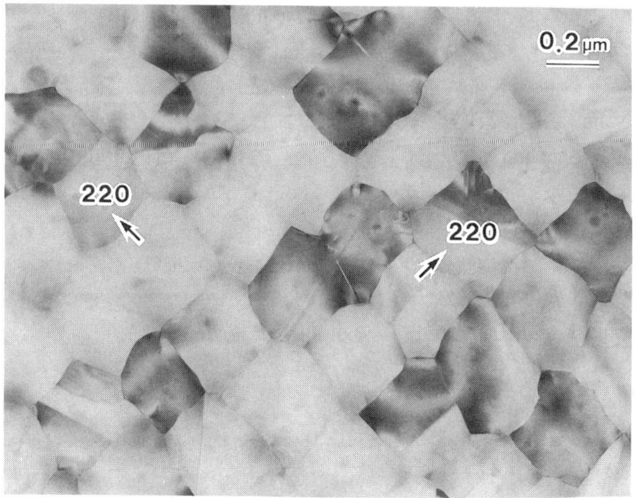

Fig.6 TEM micrograph of arrayed Si grains
with <100> preferred orientation.

4. ACKNOWLEDGMENTS

We would like to thank Dr. Y. Mori in Sony Research Center and Mr. S. Hayashida in Sony Semiconductor Company for their encouragement.

REFERENCES

[1] T. Katoh, IEEE Trans. Elec. Dev. 35, 923(1988).

[2] S. Morozumi, K. Oguchi, S. Yazawa, T. Kodaira, H. Ohshima and T. Manao, SID 83 Digest 156(1983).

[3] T. Noguchi, K. Tajima and Y. Morita, Proc. of Mat. Res. Soc. 146, 35(1985).

[4] N. Yamauchi, J-J. Hajar and R. Reif, IEEE Trans. Electron. Devices 38, 55(1991).

[5] S. Usui, T. Sameshima and M. Hara, Optoelectronics-Devices and Technologies 4, 235(1989).

[6] H. Kuriyama, T. Nohda, S. Ishida, T. Kuwahara, S. Noguchi, S. Kiyama, S. Tsuda and S. Nakano, Jpn. J. Appl. Phys. 32, 6190(1993).

[7] K. T-Y. Kung, R. B. Iverson and R. Reif, Appl. Phys. Lett., 46, 683(1985).

[8] B-H. Jung, C.-J. Youn, C.-W. Hwang, B. S. Bae, J.-H. Sohn, Y. H. Sun and S. Chen, Abstract of AM-LCD 95, 117(1995).

[9] H. Kumomi and T. Yonehara, Extended abstracts of IC SSDM 26(1993).

[10] T. Noguchi and Y. Ikeda, Proc. of Sony Research Forum 200(1992).

[11] Y. Ikeda and T. Noguchi, Proc. of 44th. Symp. on Semiconductors and Integrated Circuits Tech., 187(1993).

[12] J. H. Kim and J. Y. Lee, Thin Solid Films 292, 313(1997).

[13] T. Asano, T. Aoto and Y. Okada, Jpn. J. Appl. Phys. 36, 1415(1997).

[14] J. Jang, J. Young Oh, S. K. Kim, Y. J. Choi, S. Y. Yoon and C. O. Kim, Nature, 395, 481(1998).

[15] M. A. Crowder, P. G. Carey, P. M. Smith, R. S. Sposili, H. S. Cho and J. S. Im, IEEE Ele. Dev. Lett. 19, 306(1998).

[16] C. V. Thompson and H. I. Smith, J. Appl. Phys., 58, 763(1985).

[17] M. W. Geis, H. I. Smith, B-Y. Tsauer, J. C. C. Fan, D. J. Silversmith and R. W. Mountain, J. Electrochem. Soc., 2812(1982).

[18] T. Noguchi, Y. Kanaya, M. Kunii, Y. Ikeda and S. Usui, Japanese Patent Application No. J10-041234, Opened in (1998).

LATERAL SOLID PHASE CRYSTALLIZATION OF AMORPHOUS SILICON UNDER HIGH PRESSURE

Seung-Mahn Lee and Rajiv K. Singh
Department of Materials Science and Engineering,
University of Florida, Gainesville, FL 32611-6400

ABSTRACT

We have investigated a novel surface-seeded crystallization technique at low processing temperatures ($\leq$ 550°C) and high pressures (10MPa~25MPa) using polished polycrystalline diamond seeds. By controlling the high pressure, the nucleation and growth of silicon can be controlled to obtain improved quality silicon films on amorphous substrates at low temperatures. Depending on the annealing temperature and applied pressure, the orientation of crystallized silicon thin films varies as seen by x-ray diffraction and transmission electron microscopy results. In addition, crystallization of amorphous silicon thin films has effect on their roughness.

INTRODUCTION

In recent years, there has been a constant research effort in the area of silicon on insulator technology (SOI). The reason for this is its inherent advantages, such as low leakage current, perfect isolation, low parasitic capacitance, and high radiation immunity [1,2]. Especially, crystalline-silicon (c-Si) thin-film transistors (TFTs) on a transparent substrate are very attractive technology for active-matrix liquid-crystal-display (AMLCD) technology [3-5]. Most widely used approaches for this technology are annealing, SIMOX (separation by implantation of oxygen) [6], and BESOI (bond and etch back SOI) [7,8]. Annealing is very suitable for AMLCD. Due to low temperature processing (below 600°C), an inexpensive glass can be used compared to polycrystalline films deposited by low pressure chemical vapor deposition (LPCVD) [9], while SIMOX and BESOI processes are carried out at high temperatures. The low processing temperatures can minimize the nucleation from the a-Si/SiO$_2$ interface and impurity profiles in the a-Si films are not redistributed [10].

Many of the electrical properties of polycrystalline silicon depend on its grain size and defect density (e.g. microtwins in (111) orientation). However, major drawbacks of thermal annealing are the long processing time, high defect density and small grain size.

It is reported that solid phase epitaxy growth rate strongly depends on the intrinsic residual strain in the amorphous silicon thin films induced by hydrostatic pressure and doping [11]. Ishiwara *et* al. [2] have reported non-hydrostatic ultra high pressure (>1GPa) effect to grow the lateral solid phase epitaxy using amorphous silicon films deposited on SiO$_2$/Si structure with seeding area. The method we have described in this letter is based on the use of top polished diamond seed which is pressed onto the amorphous silicon thin films using high pressure (10~25MPa) at low processing temperatures ($\leq$550°C). We call this as surface-seeded crystallization technique because crystallization is achieved from the surface layer. By controlling the applied pressure, the nucleation and growth of silicon can be controlled to obtain improved quality silicon films on amorphous substrates at low temperatures. In a previous paper, we have reported the novel surface-seeded crystallization technique at low pressure and low temperature using roughed single crystal silicon seeds [10]. In this study, using polished polycrystalline diamond seeds at high pressure, selective nucleation and growth of silicon films were obtained at low temperature.

Mat. Res. Soc. Symp. Proc. Vol. 557 © 1999 Materials Research Society

EXPERIMENT

1000Å amorphous silicon films were deposited on 6000Å SiO$_2$/Si substrate using LPCVD (low pressure CVD). Deposition conditions were: deposition temperature ~540°C, silane flow rate ~120sccm, and deposition rate ~20Å/min. To eliminate problems of film and seed breakage associated with the roughed silicon seed and problems caused by bonding between the amorphous silicon film and the flat silicon seeds at high pressure, polished diamond (RMS~3nm) was used as the top seed material because of its high hardness, thermal conductivity, and chemical inertness.

Figure 1 shows the schematic diagram of the surface-seed crystallization method. In this method, the a-Si/SiO$_2$/Si structure with top diamond seed was annealed at low processing temperature ($\leq$ 550°C) for time ranging from 1 to 5 hrs and for pressure ranging from 10 MPa to 25MPa in a hot pressure furnace in <10^{-3} Torr argon ambient. During annealing, temperature was controlled at ±1°C using a thyristor-controlled power supply. This technique is followed by HF treatment to remove the native oxide silicon layer on a-Si film.

X-ray diffraction (XRD), transmission electron microscopy (TEM), and atomic force microscopy (AFM) were used to characterize the crystallinity and microstructure of the silicon films obtained by this technique.

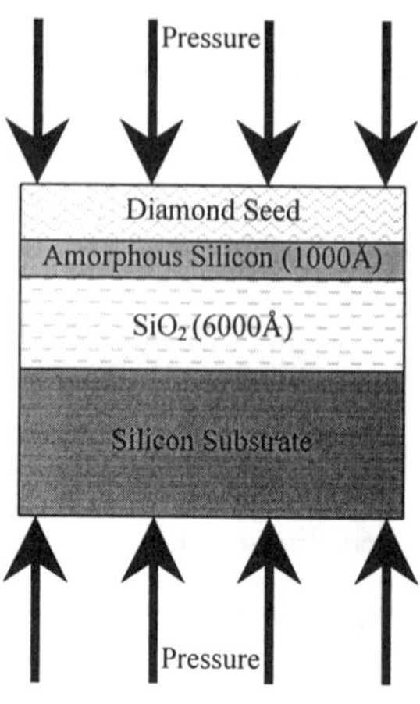

Figure 1. Schematic diagram of Surface-seeded crystallization method

RESULTS AND DISCUSSION

Crystal structures and crystallinity of the films were analyzed using XRD. Figure 2 shows x-ray diffraction patterns of amorphous silicon film and silicon films annealed for 1hr by the surface-seeded crystallization in high pressure as a function of annealing temperatures and pressure. Comparing Fig. 2(b) to Fig. 2(c), there is nothing crystallized when annealed at 10MPa and 500°C for 1 hr, whereas the Si films crystallized by the surface-seeded crystallization show (110) Si peak (indicated by (220) peak) when annealed at 25MPa and 500°C for 1 hr. (111) and (110) Si peaks appear at even relatively low pressure (10Mpa) above 525°C as shown in Fig. 2(d). Moreover, (331) peak can also be seen at 550°C as seen Fig. 2(e). From these XRD patterns, it is observed that high pressure (~25MPa) at relatively low temperatures can induce (110) orientation crystallization in

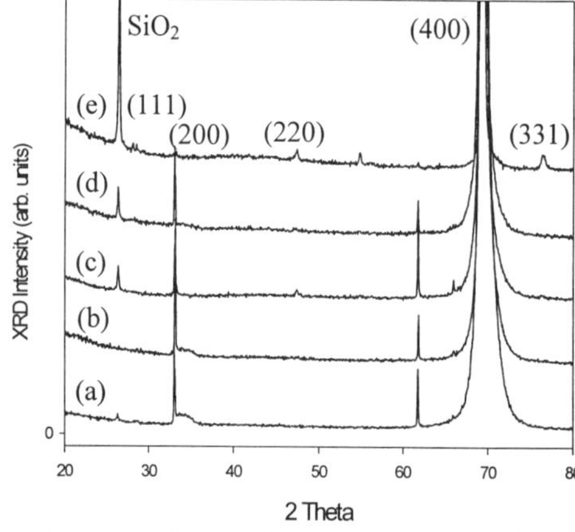

Figure 2. X-ray diffraction spectrum from 1000Å amorphous silicon film and after annealing with top diamond seed at (b) 500°C, 10MPa for 1hr, (c) 500°C, 25MPa for 1hr, (d) 525°C, 10MPa for 1hr, and (e) 550°C, 20MPa for 1hr

silicon rather than (111) orientation. As annealing temperature increases using same pressure, (111), (110), and (331) Si peaks are generated. In Fig. 2(e), the a-Si film was transited to polycrystalline silicon film. It is understood that (110) oriented grains are dominant in high-pressure region, and the nucleation and growth of (111) and (331) grains are more dependent on the temperature than the orientation of (110) grains. Similar observation has been reported by Ishiwara et al. [2] and this XRD analysis also agrees with the result of Sakai et al.[12] and Lee and Rha [13]. High pressure can modify the total strain on the surface of the a-Si film by adding extrinsic strain so that the lateral solid phase growth rate increases. From the XRD results, it is clear that (110) orientation is a preferred orientation in high strain region caused by pressure and the selective nucleation can be achieved at low temperature (500°C) and high pressure (25MPa) by surface-seeded technique.

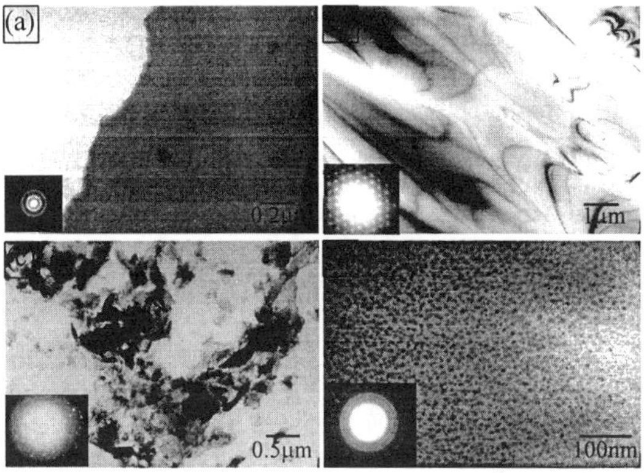

Figure 3. Transmission electron micrographs of films: (a) amorphous silicon film, (b) (110) orientation in 25MPa at 500°C for 1hr, (c) (110) & (111) grains in 10MPa at 525°C for 1hr, and (d) (111) grains in 10MPa at 525°C for 1hr

To investigate the microstructures of the crystallized Si films, plan view TEM was used. Figure 3 shows transmission electron micrographs of a-Si film and crystallized Si films using the surface-seeded crystallization at various temperatures and pressures. In Fig. 3(b), it is observed that almost single crystallized (110) silicon film can be seen in silicon film annealed at 10MPa and 500°C for 1hr which agrees with the XRD results. However, much more area, where amorphous silicon films were not crystallized, was also seen in the same film. It might be the effect of the roughness of the polished diamond seed. Though the RMS roughness of the Diamond seed is ~3nm, the local roughness on some areas is much higher than the average due to the presence of bumps (image Z range ~50nm). So, higher pressure (> 25 MPa) can be concentrated on the interface between amorphous silicon film and the bumps of the top diamond seed. In the area of higher pressure, it is considered that crystallized (110) orientation silicon films can be nucleated and grown due to the highly strained surface. On the other hand, relatively lower pressure can not give the amorphous films enough driving force to nucleate (110)

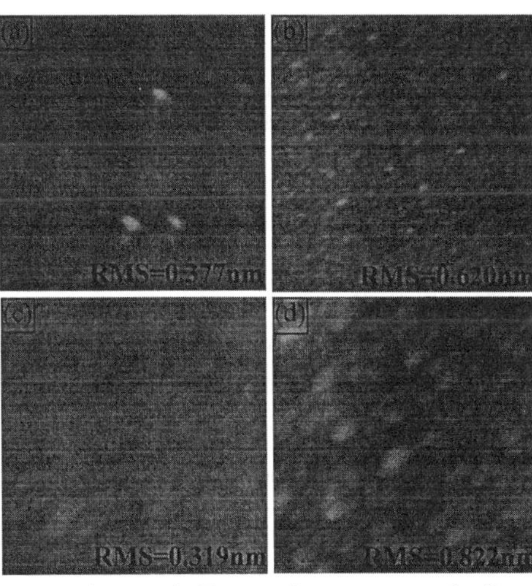

Figure 4. Atomic Force microscopy graphs for a silicon thin films (size: 2.5 x 2.5 μm). (a) a-Si, annealed (b) at 500°C and 25MPa for 1 hr, (c) at 525°C and 10MPa for 1 hr, and (d) at 550°C and 20MPa for 1 hr

grains. Figures 3(c) and (d) show the silicon film annealed at 10MPa and 525°C for 1hr. As seen in selected area diffraction pattern, Fig. 3(c) shows (110) and (111) grains together in areas which were in contact with seeds and it can be seen from Fig. 3(d) that (111) grains begin to nucleate on the silicon film. From these TEM results, it is known that the annealing temperature has an important role in nucleating (111) grains, because it is found that very fine (111) grains (<10nm) are observed at all over the Si film annealed at 525°C and 10MPa for 1hr in transmission electron micrograph.

AFM was conducted to investigate the effect of crystallization on the roughness. Figures 4(b) and 4(d) show that the roughness increases in the silicon thin films annealed at 500°C and 25MPa for 1 hr and at 550°C and 20MPa for 1 hr respectively, while Fig. 4(c) shows that the roughness does not change much in the silicon thin film at 525°C and 10MPa for 1 hr. This is considered to be the effect of crystallization. As mentioned earlier, amorphous silicon films can be crystallized partially due to concentrated higher pressure (Fig. 4(b)) and as temperature increases, (111) grains start coarsening, which is anticipated (Fig. 4(d)). This might be the reason why the roughness increases. In contrast, very fine (111) grains are spread on silicon thin film and (110) grains are of sub-micron size at 525°C and 10MPa for 1 hr so that these small grains might not have much effect on roughness due to their sizes. With this reason, it is seen that the roughness in Fig. 4(c) does not change as much as due to the other conditions.

CONCLUSION

A novel surface-seeded crystallization technique for lateral solid phase crystallization on a-Si/SiO$_2$/Si systems has been developed. The experimental results show that depending on the annealing temperature and uni-axial pressure, the orientation of crystallized silicon thin films varies. When annealed at 500°C and 25MPa for 1hr, only one orientation, (110) grain, was detected in XRD and TEM results. It is clear that (110) orientation is a preferred orientation in high strain region caused by higher pressure (>25MPa), and selective nucleation of (110) grains can be achieved by surface-seeded crystallization technique. The annealing temperature has been found to play an important role in nucleation of (111) grains, as very fine (111) grains (<10nm) begin to appear at all over the Si film annealed at 525°C and 10MPa for 1hr. The crystallization of a-Si films can affect the surface roughness. The roughness increases in the case of partially crystallized film or coarse grain. Very fine grains do not have an effect on the variation of the roughness.

ACKNOWLEDGMENT

The authors wish to thank K. G. Cho for useful discussion and technical assistance.

REFERENCE

1. N. G. Einspruch, and G. Gildenblat, Advanced MOS device physics, Academic Press, San Diego (1991)
2. H. Ishiwara, H. Wakabayashi, K. Miyazaki, K. Fukao, and A. Sawaoka, Jpn. J. Appl. Phys. 32, 308 (1993)
3. W.C. Omara, "Liquid Crystal Displays" in Manufacturing Science and Technology, Van Nostrand and Reinhold, NY (1993)
4. G. Fortunato, Thin Solid Films 296, 82 (1997)
5. O. S. Panwar, R. A. Moore, S. H. Raza, H.C. Gamble, and B. M. Armstrong, Thin Solid Film 237, 255 (1994)

6. O. W. Holland, D. S. Zou, and D. K. Thomas, Appl. Phys. Lett. 63 ,896 (1993)
7. W. P. Maszara, G. Goetz, A. Caviglia, and J. B. McKitterick, J. Appl. Phys. 64(10), 4943 (1988)
8. A. Benitez, J. Esteve, and J. Bausells, Sensors and Actuators A 50, 99 (1995)
9. S.D. Brotherton, D.J. McCulloch, J.B. Clegg, J.P. Gowers, IEEE Trans. Elect. Devs. 40, 407 (1993)
10. R. K. Singh, S.-M. Jung, S.-M. Lee, and R. E. Hummel, J. Electrochem. Soc. 145(11), 3963 (1998)
11. G.-Q. Lu, E. Nygren, and M. J. Aziz, J. Appl. Phys. 70,5323 (1991)
12. A. Sakai, H. Ono, K. Ishida, T. Niino, and T. Tatsumi, Jpn. Appl. Phys. Lett. 30, L941 (1991)
13. E. G. Lee and S. K. Rha, J. Mat. Sci. 28, 6279 (1993)

SPECTROSCOPIC INVESTIGATIONS OF CRYSTALLINITY AND ELECTRONIC-STRUCTURAL TRANSITIONS DUE TO SOLID PHASE CRYSTALLIZATION OF AMORPHOUS $Si_{1-x}Ge_x$

S. YAMAGUCHI, N. SUGII, K. NAKAGAWA, and M. MIYAO
Central Research Lab., Hitachi Ltd., Tokyo 185-8601, Japan, yamaguci@crl.hitachi.co.jp

ABSTRACT

Solid phase crystallization of $Si_{1-x}Ge_x$ have been investigated by means of change in crystallinity and crystallization temperature using the ellipsometric spectroscopy (2-5 eV). Dispersion analysis on the spectra revealed that the $Si_{1-x}Ge_x$ alloys show both abrupt increase and rapid saturation of the crystallinities according to the structural transitions. The saturated values of crystallinities did not reach those of single crystals in all samples and rapidly decreased with increase of Ge concentration. Ion-implanted amorphous $Si_{1-x}Ge_x$ exhibited much lower activation energies of re-crystallization than those of the deposited amorphous samples. Furthermore, their crystallization temperature were strongly depend on the species of implanted ions such as As^+ and BF_2^+. It was found that the crystallization velocity of the As^+-implanted $Si_{1-x}Ge_x$ was estimated as 4 times faster than that of the BF_2^+-implanted $Si_{1-x}Ge_x$. Itinerant property of Ge in Si may give rise to the decrease of the crystallinities as well as the decrease of the activation energy of solid phase crystallization which may be strongly affected by the catalytic effects of the implanted impurities.

INTRODUCTION

The rapid progress of SiGe-based heterostructure enables the introduction of band engineering into Si large scale integration (LSI) fields. This feature has been effectively used for the formation of new high-speed devices such as modulation-doped field effect transistors [1,2] and heterojunction bipolar transistors [3]. It has been shown, however, that Ge segregation to the epitaxial surface during crystal growth inhibits the formation of abrupt heterointerfaces [4]. Re-crystallization of ion-implanted SiGe also has the problem related to the itinerant property of Ge in which diffusion of doped layer destroy the band profile so that the band discontinuity is insufficient to utilize high carrier concentration. To avoid such a thermodynamic diffusion and intermixing between Si and Ge, it is strongly requested the rapid progress of low-temperature growth process such as solid phase crystallization (SPC) techniques.

The present research describes the crystallinity and electronic-structural transitions due to SPC by mean of an optical investigation such as ellipsometric spectroscopy. The conventional methods to measure those phenomena are Raman measurement and Rutherford Back Scattering (RBS). However, the Raman uses far-IR region where the penetration depth of incident light is far above the thickness of target layers and obtained spectra may be affected by large contribution from the substrates. The RBS is also not a very accurate technique to measure thin layer. Since the optical response of SiGe is quite sensitive to both composition and atomic-level alignment, the present optical spectroscopy can be sensitive probe for characterization of the averaged lattice-randomness near the sample surface.

Mat. Res. Soc. Symp. Proc. Vol. 557 © 1999 Materials Research Society

EXPERIMENT

$Si_{1-x}Ge_x$ films were grown on p-type (100) Si substrates by a solid source molecular beam epitaxy (MBE) system. Epitaxial layer of $Si_{1-x}Ge_x$ (hereafter referred as epi SiGe) were grown to a thickness of 500nm at the substrate temperature of 600°C. Deposited amorphous layer of $Si_{1-x}Ge_x$ (depo SiGe) were formed (500nm thickness) by lowering the substrate temperature below 250°C. The surface concentration of Ge atoms was measured using the X-ray photoelectron spectroscopy. Ion-implanted amorphous $Si_{1-x}Ge_x$ (impla SiGe) were prepared by ion-implantations into the above mentioned epi SiGe with As^+ (100keV dose=1E15 cm^{-2}) and BF_2^+ (70keV dose=1E15 cm^{-2}). Annealing were performed at 400-800°C for 15 min in electrical furnace with dry N_2 atmosphere. Ellipsometric spectra were carried out using a Sopra ES4G spectrometer with an incident angle of 75 degree. A focused light has a diameter of 3x10 mm^2 on the sample surface. All measurements were performed at room temperature in the spectral range from 2.0 to 5.0 eV.

RESULTS

Figure 1 shows imaginary part (ε_2) of the dielectric tensor obtained from the ellipsometric measurements for (a) the epi SiGe and (b) the depo SiGe, respectively. The spectra of the epi SiGe exhibit characteristic features which are similar to those reported for the single crystals [5]. The spectra consists of three interband transitions which are

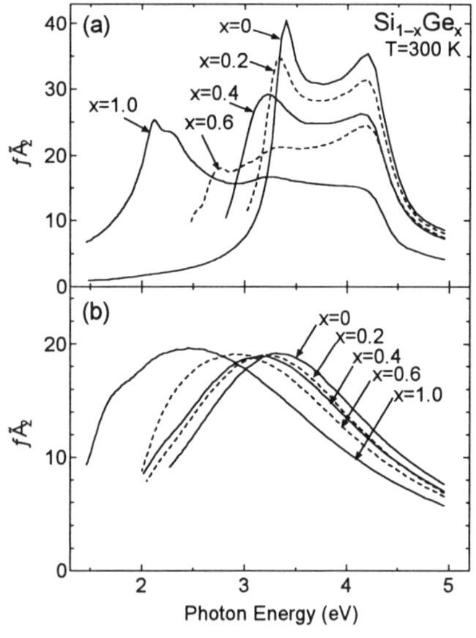

Figure 1. Imaginary part of the dielectric tensor elements (ε_2) obtained from the ellipsometric measurements for (a) the epitaxially grown $Si_{1-x}Ge_x$ and (b) the deposited amorphous $Si_{1-x}Ge_x$ samples, respectively.

clearly resolved for the x=1.0 spectrum at photon energies 2.1, 3.1, and 4.2 eV. These transitions have been assigned, from low to high energies, as E_1, E_0', and E_2 [5]. Although the spectra of lower x also consist of the same transitions, their spectral shapes are not well resolved. This is because the band gaps are strongly dependent on x so that both E_1 and E_0' nearly coexist in the main peak at 3.4 eV in the x=0 spectrum. The spectra of (b) the depo SiGe exhibit extreme broadening of the spectra due to increase of randomness in atomic alignment. Each spectra, however, have small anisotropy against photon energy implying local-lattice alignment of $Si_{1-x}Ge_x$ alloy still exist and small contributions from above mentioned three transitions exist.

Figure 2 shows temperature dependence of ε_2 spectrum for the depo $Si_{0.8}Ge_{0.2}$. The result shows abrupt spectral change at temperature from 600°C to 610°C, and contrastive stability above 610°C The final state of the spectrum does not reach the one for the epi $Si_{0.8}Ge_{0.2}$ (c-$Si_{0.8}Ge_{0.2}$). Penetration depth of the incident light is estimated from absorption coefficient [6] as 10-20nm for Si at the photon energy corresponding to the E_2 transition peak (4.2 eV). With such a shallow observation depth, a front surface of SPC may pass through the observed area in short period and, therefore, the spectral change is sufficiently rapid compared to the temperature interval of 10°C.

To quantify the crystallinities, we attempted dispersion analysis on the ε_2 spectra using the Lorentzian equation such as

$$\varepsilon = \varepsilon_\infty + \sum_j \frac{S_j \omega_j^2}{\omega_j^2 - \omega^2 + i\Gamma_j \omega} \qquad (1)$$

where S_j is the oscillator strength, ω_j the oscillation frequency, Γ_j the attenuation coefficient, j the index of the j-th oscillator, ω the photon frequency, and ε_∞ the dielectric element at infinity. The calculated results are shown in Fig. 3. Atomic displacements in samples attenuate the optical mode oscillation so that the degree of randomness is probed by the Γ_j. Since the E_2 transition is well resolved from the other transitions as shown in Fig. 3, we define crystallinity C of samples using the Γ of the E_2 transition such as

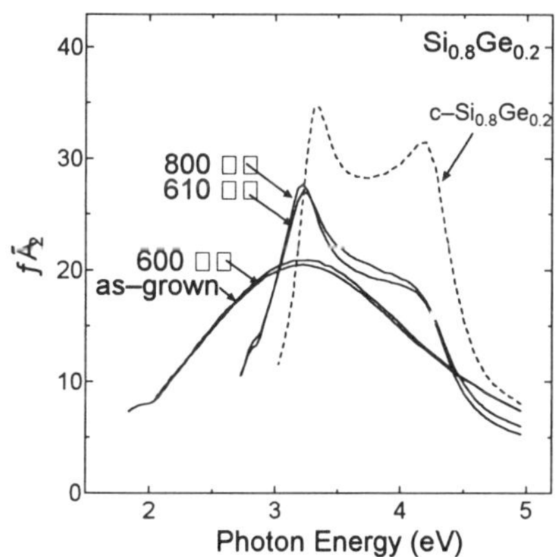

Figure 2. Anneal-temperature dependence of ε_2 spectrum for deposited amorphous $Si_{0.8}Ge_{0.2}$ with annealing time for 15min. Dotted line is the spectrum for the epitaxially grown single crystal c-$Si_{0.8}Ge_{0.2}$ as a reference.

$$C \equiv \frac{\Gamma_a - \Gamma_x}{\Gamma_a - \Gamma_c} \qquad (2)$$

where Γ_a is the attenuation coefficient of E_2 transition for the depo SiGe, Γ_c the one for the epi SiGe, and Γ_x the one for the target sample. With use of this definition, the C is normalized between 1 and 0 which are corresponding to the crystallinity of single crystal (=1) and deposited amorphous (=0), respectively.

Figure 4 shows the results of C for various x. Each result shows abrupt increase of C at crystallization temperature (T_C), in which SPC has arrived at the surface of samples, and rapid saturation of C above T_C. The saturated values of C as well as T_C decrease with increase of x.

The results are illustrated in Figs. 5 and 6. The T_C decreases monotonically with x from 710°C (x=0) to 470°C (x=1.0). It should be noted that these temperatures are much higher than those estimated from the previously reported crystallization velocity [7]. This may be cause by temperature profile of annealing time (shortage of effective time), but their relative change is essential since the annealing condition is carefully kept as constant. The rapid decrease of T_C as - 240°C from x=0 to x=1.0 is, therefore, caused by the role of Ge such as small bond energy in Si-Ge or equivalently the intermixing between Ge and Si. The C shows rapid decrease from 0.88 to 0.55 with increase of x as shown in Fig. 6. This phenomenon can also be interpreted as a result of the increase of intermixing because itinerant property of Ge in Si may enhance atomic displacements so that the randomness of lattice is increased.

In addition to these phenomena, we investigated the SPC of ion-implanted samples.

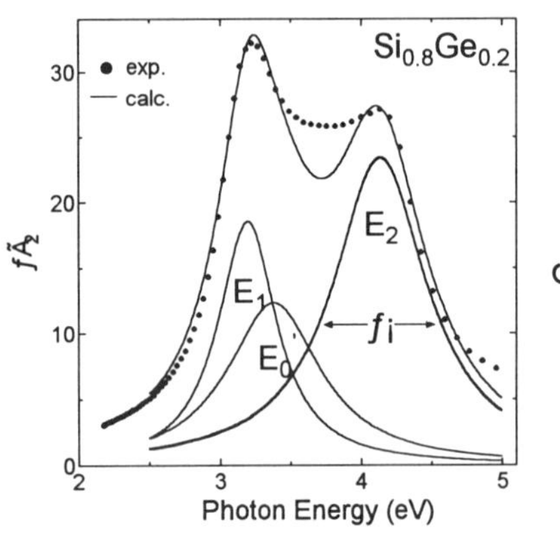

Figure 3. Measured (dots) and calculated (line) ε_2 spectra for depo amorphous Si$_{0.8}$Ge$_{0.2}$ after annealed at 800°C for 15min. The calculated spectrum consists of three components of interband transitions such as E_1, E_0', and E_2 peaks.

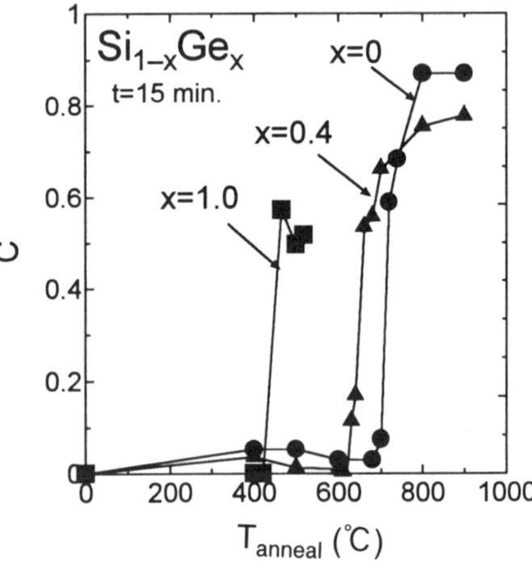

Figure 4. Crystallinities C for depo amorphous Si$_{1-x}$Ge$_x$ with change of anneal-temperature.

Figures 7 and 8 show T_C and C for implanted SiGe which are obtained by the same procedure as that for deposited SiGe. The spectral transitions due to SPC are also similar to those of depo SiGe where the abrupt increase and rapid saturation are found in the spectral shapes. However, we must note that there exist large discrepancies in SPC phenomena between depo

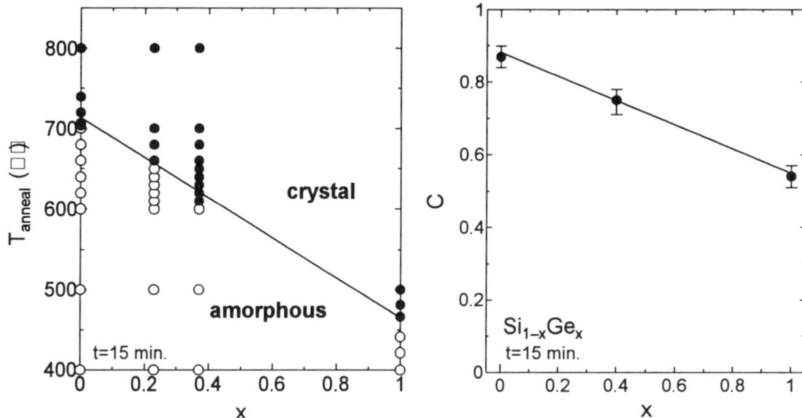

Figure 5. Temperature dependence of crystallinity such as amorphous (open circles) and crystal (close circles) obtained by the spectral shape of ε_2. A solid line indicates estimated crystallization temperature T$_C$.

Figure 6. Crystallinity C of SPC grown Si$_{1-x}$Ge$_x$ averaged above the crystallization temperature T$_C$.

SiGe as shown in Fig. 7. Furthermore, large difference in T$_C$ is found between the As$^+$-implanted and BF$_2^+$-implanted SiGe. Note that estimated projected ranges (R$_P$) for both ions are almost the same as 60nm and the Dose rates are kept as constant (1E15 cm^{-2}). Vacancy densities by our simulation predict As$^+$-implanted samples has more vacancies than BF$_2^+$-implanted ones, however according to our experimental results, broadness of as-implanted spectra are found to be almost the same for these two species. Therefore the drastic decrease of the crystallization temperature has been found in the As$^+$-implanted SiGe compared to the BF$_2^+$-implanted SiGe.

Since the SPC velocity of SiGe is known to be a simple activation type phenomena, we can estimate the difference in SPC velocity among non-impla, As$^+$, and BF$_2^+$ samples by referring to the previously reported velocities [7]. We found that the SPC velocity of As$^+$-implanted SiGe is about 4 times faster than that of BF$_2^+$-implanted SiGe. This is surprisingly large difference and implies that SPC properties are strongly affected by the catalytic effects of the implanted impurities as well as the role of itinerant Ge in Si.

CONCLUSIONS

The SPC phenomena of Si$_{1-x}$Ge$_x$ have been investigated by the ellipsometric spectroscopy (2-5 eV). The crystallinity of SPC grown Si$_{1-x}$Ge$_x$ have been found to decrease with increase of Ge concentration. Ion-implanted amorphous Si$_{1-x}$Ge$_x$ exhibited similar SPC phenomena, however much lower activation energies of SPC than those for the deposited amorphous Si$_{1-x}$Ge$_x$. Furthermore, large discrepancy in SPC velocity has been found such that the As$^+$-implanted Si$_{1-x}$Ge$_x$ has 4 times faster value than that of the BF$_2^+$-implanted Si$_{1-x}$Ge$_x$.

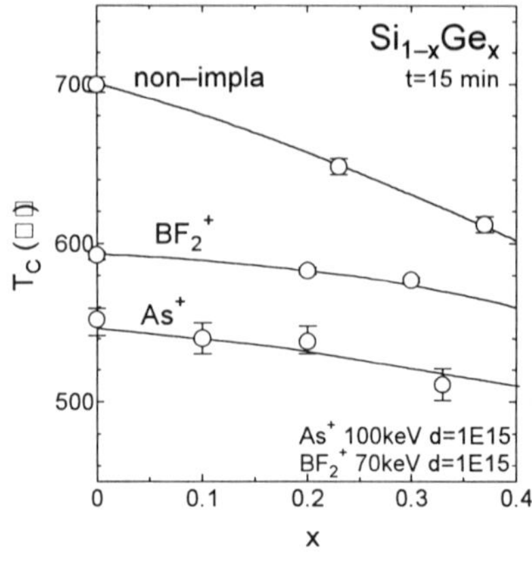
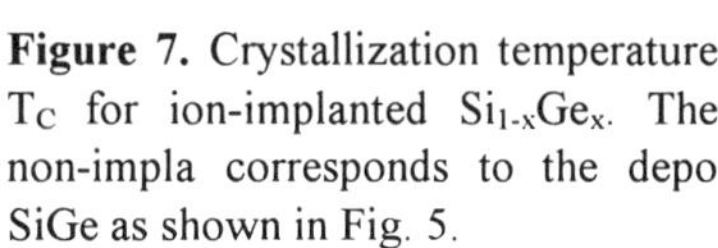

Figure 7. Crystallization temperature T$_C$ for ion-implanted Si$_{1-x}$Ge$_x$. The non-impla corresponds to the depo SiGe as shown in Fig. 5.

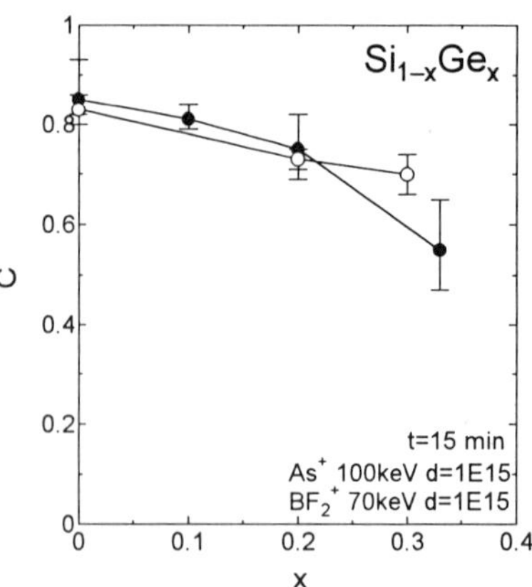

Figure 8. Crystallinity C of SPC grown Si$_{1-x}$Ge$_x$ averaged above the crystallization temperature T$_C$ for As$^+$-implanted (filled circles) and BF$_2$$^+$-implanted (open circles)

Itinerant property of Ge in Si may give rise to the decrease of the crystallinities as well as the decrease of the activation energy of SPC which is also affected by the catalytic effects of the implanted impurities.

REFERENCES

1. G. Abstreiter, H. Brugge, and T. Wolf, Phys. Rev. Lett. **54**, p.2441 (1985).
2. K. Ismail, B. S. Meyerson, and P. J. Wang, Appl. Phys. Lett. **58**, p.2117 (1991).
3. M. Miyao and K. Nakagawa, Jpn. J. Appl. Phys. **33**, p.3791 (1994).
4. S. S. Iyer, J. C. Tsang, M. C. Copel, P. R. Pukite, and R. M. Tromp, Appl. Phys. Lett. **54**, p.219 (1989).
5. J. Humlicek, M. Garriga, M. I. Alonso, and M. Cardona, J. Appl. Phys. **65**, p. 2827 (1989).
6. S. M. Sze, *in Physics of Semiconductor Devices* (Wiley, New York, 1981), p.42.
7. P. Kringhoj and R. G. Elliman, Phys. Rev. Lett. **73**, p.858 (1994).

STRUCTURAL PROPERTIES OF AMORPHOUS SILICON PRODUCED BY ELECTRON IRRADIATION

J. Yamasaki and S. Takeda
Department of Physics, Graduate School of Science, Osaka University-Toyonaka, 1-16,
Machikaneyama-cho, Toyonaka, Osaka 560-0043, Japan

ABSTRACT

The structural properties of the amorphous Si (*a*-Si), which was created from crystalline silicon by 2 MeV electron irradiation at low temperatures about 25 K, are examined in detail by means of transmission electron microscopy and transmission electron diffraction. The peak positions in the radial distribution function (RDF) of the *a*-Si correspond well to those of *a*-Si fabricated by other techniques. The electron-irradiation-induced *a*-Si returns to crystalline Si after annealing at 550 °C.

INTRODUCTION

Amorphous silicon (*a*-Si) has been long studied from both fundamental and technological points of view [1] . We have recently found a new route of amorphization in Si: Crystalline silicon (*c*-Si) is rendered into *a*-Si by MeV electron irradiation at low temperature [2]. A MeV electron beam can be focused on an area smaller than 0.2 μ m in diameter, can be scanned and can penetrate about 0.5 μ m into *c*-Si, so the finding may be applied to fabricating artificial photonic crystals and quantum dots of the two Si structures with different dielectric and electronic transport properties. In this work, we examine both the structural properties and the thermal stability of the *a*-Si, in order to obtain the basic data for applications.

EXPERIMENTAL PROCEDURES

The samples were cut from a Czochralski(CZ) -grown, P-doped Si {110} wafer. The samples were thinned in a solution of HNO_3 and HF, and then irradiated by 2 MeV electrons in a high voltage transmission electron microscope (HVTEM), which was equipped with a liquid-helium cooling stage. The irradiation temperature was 28 ± 4 K. The electron beam was set parallel to the <110> zone axis and its diameter on a specimen surface was about 1 μm. Using a Faraday cage installed in the microscope, we estimated the electron flux on the central area of the beam on a specimen surface, 300 nm in diameter.

After electron irradiation, the samples were transferred to a 200 keV transmission electron microscope (TEM), and post-irradiation examined at 25 °C (room temperature), 500 °C and 550 °C.

ELECTRON-IRRADIATION-INDUCED AMORPHIZATION IN Si

Figure 1 shows an image taken from a series of *in-situ* HVTEM observation of the amorphization process at 28 ± 4 K. The electron energy was 2 MeV. The central area, which appears to be white in Fig. 1, received the dose of 8.4 x 10^{22} cm^{-2} and was fully amorphized. In fact, the transmission electron diffraction shown in Fig. 1 exhibits the halo rings due to *a*-Si. The details about the amorphization in Si by electron irradiation are reported elsewhere [2].

Mat. Res. Soc. Symp. Proc. Vol. 557 © 1999 Materials Research Society

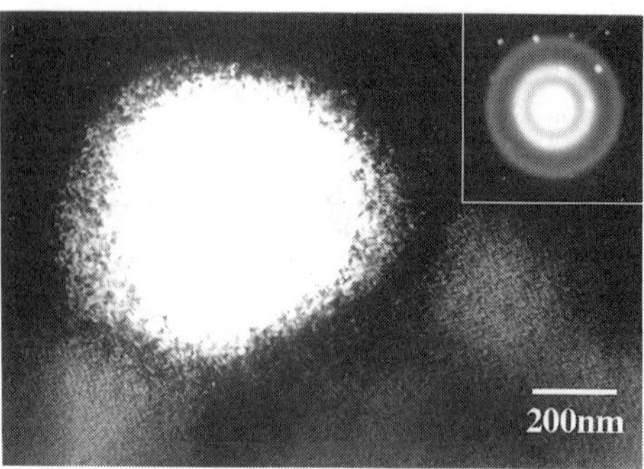

Fig. 1: Amorphized Si by 2 MeV electron irradiation at 28±4 K. HVTEM image and electron diffraction were recorded at 28±4 K.

Figure 2 summarizes the post-irradiation TEM observations of the irradiated areas which received the different electron doses. The irradiation flux was 5.8 x 10^{20} cm^{-2}s^{-1}. The insets are the corresponding electron diffraction patterns, in which the selected area aperture covered the central area of about 250 nm in diameter.

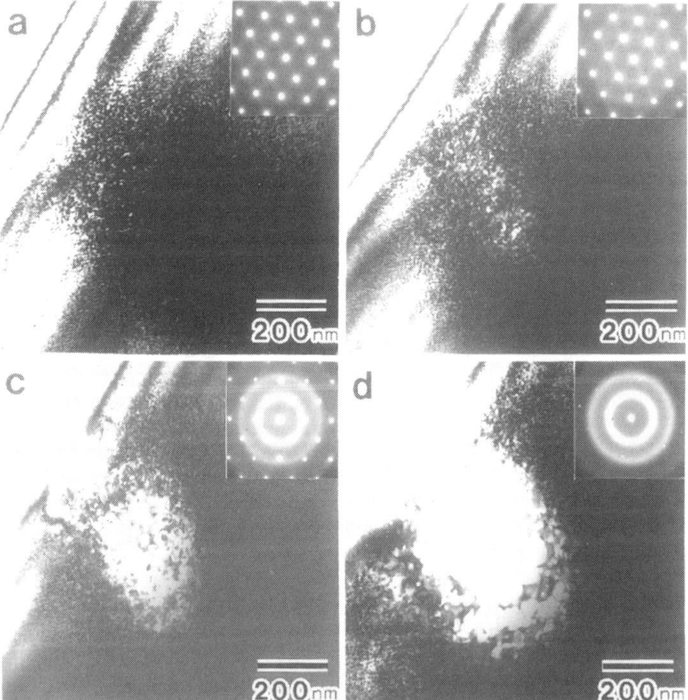

Fig. 2: Electron irradiated areas which received the doses of (a) 1.7 x 10^{22} cm^{-2}, (b) 3.5 x 10^{22} cm^{-2}, (c) 7.0 x 10^{22} cm^{-2} and (d) 9.9 x 10^{22} cm^{-2}.

As shown in the diffraction patterns in Fig. 2(a) and (b), the crystallinity seems to be retained in the early stage of irradiation, even though numerous defects were introduced in the irradiated areas. Clearly, the areas which received the doses of 7.0 x 10^{22} cm^{-2} (Fig. 2(c)) and 9.9 x 10^{22} cm^{-2} (Fig. 2(d)) give rise to halo rings in the diffraction patterns, and correspondingly, in the images, the central areas appear bright with many fragmented black regions, which are attributed to c-Si islands. The diffraction spots from c-Si almost disappear at the dose of 9.9 x 10^{22} cm^{-2} and the bright area is more extended as seen in Fig. 2(d). The central area of about 200 nm in diameter was fully amorphized.

RADIAL DISTRIBUTION FUNCTIONS OF a-Si

Figure 3 shows the radial intensity profiles of electron diffraction patterns of a non-irradiated area and the irradiated areas shown in Fig. 2(b), (c), and (d). With the increase of dose, the intensities of the peaks due to c-Si (at 3.2, 3.7, 5.2, 6.1 nm^{-1} etc.) gradually decrease, and the halo rings appear. Assuming that the background intensity due to inelastically scattered electrons is identical in both irradiated and non-irradiated areas of the same foil thickness, we converted the halo intensity profiles in Fig. 3 to the radial distribution functions (RDF) using a standard method [3]. The maximum scattering wavenumber for Fourier transformation is 108.4 nm^{-1}. As shown in Fig. 4, the peaks are located in the RDF (the dose: 7.0 x 10^{22} cm^{-2}) at 0.235 nm, 0.382 nm, 0.588 nm and 0.755 nm, and in the RDF (the dose: 9.9 x 10^{22} cm^{-2}) at 0.238 nm, 0.378 nm, 0.581 nm and 0.753 nm. All of them agree with the peak potions of the RDF obtained in evaporated pure a-Si [4], i. e. 0.234 nm, 0.384 nm, 0.584 nm and 0.768 nm within the accuracy of 2 %. A few small peaks which appear in the range r < 0.2 nm are artifacts, since the heights and/or locations of the peaks are changed largely with changing of fitting parameters and the maximum scattering wavenumber for Fourier transformation. Therefore, we conclude that, in terms of pair-wise atomic correlation, the structures of the a-Si induced by electron irradiation are similar to those of a-Si created by other existing techniques. In addition, RDFs do not change whenever halo rings are detected in diffraction patterns. This suggests that tiny but fully amorphized regions are produced at an early stage.

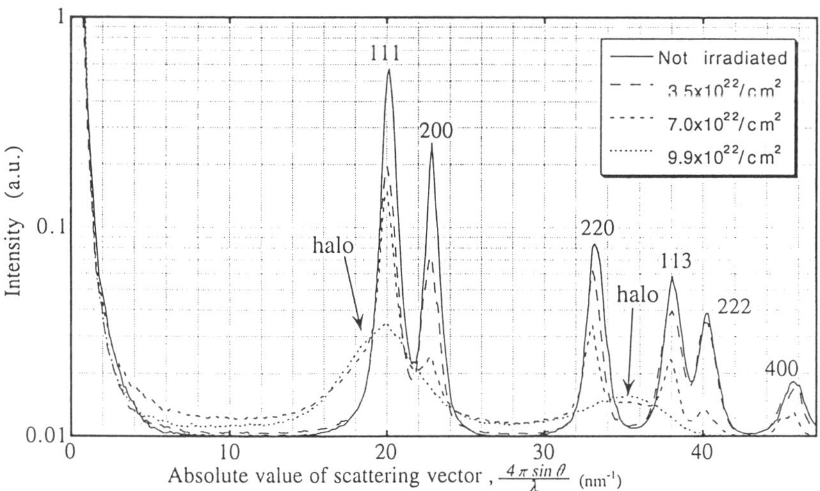

Fig. 3: Radial intensity distribution of electron diffraction of a non-irradiated area and the areas shown in Fig. 2(b) to (d).

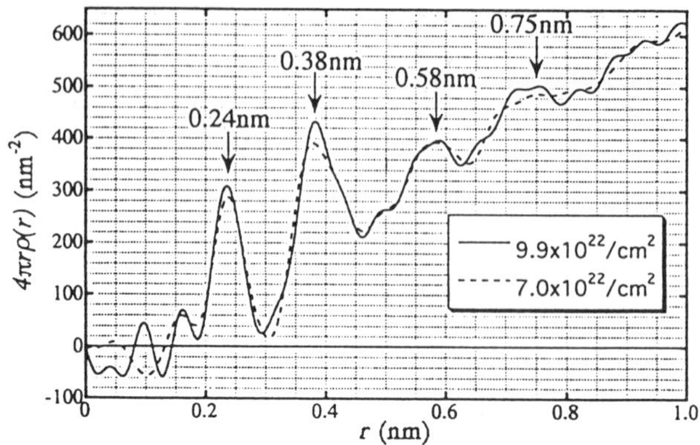

Fig. 4: Radial distribution functions of *a*-Si from the intensity patterns shown in Fig. 2(c) and (d).

STRUCTURES OF THE TRANSLATIONAL REGIONS BETWEEN *a*-Si AND *c*-Si

The translational regions between *a*-Si and *c*-Si extend over about 50 nm as shown in Fig. 1, and a partially amorphized area (for instance, Fig. 2(c)) exhibits the fragmented black TEM contrast. Figure 5 shows a more magnified image of a translational region.   The image was observed at room temperature using a 200 keV TEM.   In the translational area, we observed localized amorphized areas in the damaged crystal and many fragmented *c*-Si areas embedded in *a*-Si.   This result suggests that the amorphization by electron irradiation occurs in a heterogeneous manner.

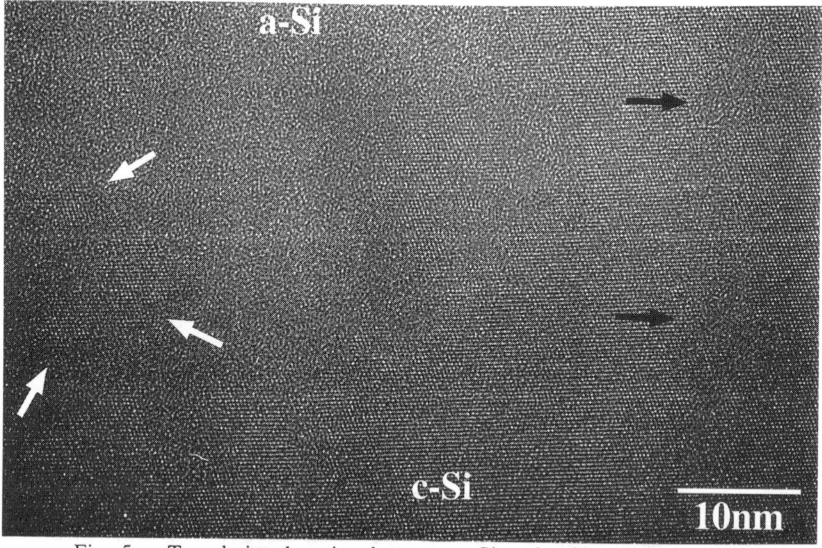

Fig. 5 : Translational region between *a*-Si and *c*-Si.   Post-irradiation HRTEM observation.   The black and white arrows indicate the localized *a*-Si and the fragmented *c*-Si, respectively.

RECRYSTALLIZING OF *a*-Si

Figure 6 shows an *in-situ* observation of the recrystallizing process of the *a*-Si. Before annealing (Fig. 6(a)), halo rings appear in the diffraction pattern and, as mentioned above, the *a-c* interface is rough.   After annealing at 500 °C for 120 minutes (Fig. 6(b)),   the recrystallizing occurs slightly due to solid phase epitaxial (SPE) growth, and the interface roughness is slightly reduced. The velocity of SPE was roughly measured to be 4.5 nm/hour, which corresponds to the SPE velocity measured at 470 °C in *a*-Si created by other methods [5].   In the diffraction pattern, any apparent change cannot be detected. The annealing temperature was then increased from 500 °C to 550 °C, and after annealing for 35 minutes at 550 °C (Fig. 6(c)), the SPE recrystallizing progressed further at the *a-c* interface. The velocity of SPE was roughly measured to be 46 nm/hour, which corresponds to the velocity at about 510 °C of the *a*-Si created by other methods [5]. The recrystallizing in the electron-irradiation-induced *a*-Si is slightly slower than that of *a*-Si produced by the other methods. In addition to the SPE recrystallizing, we observed that several *c*-Si grains nucleated not at the *a-c* interface but inside the *a*-Si region.   In the corresponding diffraction pattern, the Debye-Scherrer rings came up in addition to the halo rings (Fig. 6(c)), and after annealing for 43 minutes at 550 °C (Fig. 6(d)), the amorphous phase was completely converted to poly-crystalline Si. It is likely that the *a*-Si shown in Fig. 6(a) includes fragmented tiny *c*-Si regions, as seen in Fig. 5, and they act as the nucleation centers for poly-crystallization. This annealing behavior is similar to that of the *a*-Si created by indentation [6], and differs from that of the *a*-Si created by ion-implantation, which crystallizes into single crystalline silicon by SPE [5].

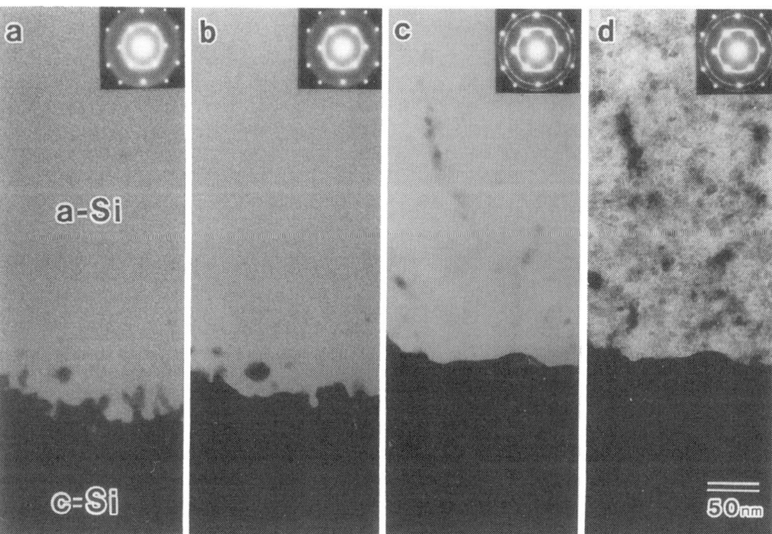

Fig. 6: *In-situ* TEM observation of the recrystallizing process of *a*-Si.   (a) As irradiated, (b) after annealing at 500 °C for 120 minutes, (c) after annealing at 550 °C for 35 minutes and (d) after annealing at 550 °C for 43 minutes.

CONCLUSIONS

We have examined the structures of the amorphized Si induced by 2 MeV electron irradiation at about 25 K. The peak positions in the RDFs derived from the almost fully and partly amorphized regions agree with those of a-Si created by the other methods. The structures of the a-Si induced by electron irradiation do not significantly differ from that of the a-Si produced by other existing techniques.

The a-c interface is rough and extends over several nm. This means that the amorphization by electron irradiation occurs in a heterogeneous manner.

The a-Si changes into poly-crystalline Si after annealing at 550 °C. This suggests that the tiny c-Si regions which remain in the a-Si act as nucleation centers for poly-crystallization.

ACKNOWLEDGMENTS

This work was partially supported by the Ministry of Education, Science, Sports and Culture, Grant-in-Aid for Scientific Research (A)(2) No. 10305006, 1998.

REFERENCES

1. For instance, T. Motooka, S. Harada and M. Ishimaru, Phys. Rev. Lett. **78**, 2980 (1997).
2. S. Takeda and J. Yamasaki, submitted for publication (1999).
3. For instance, S.S. Nandra and P.J. Grundy, J. Phys. 7, 207 (1977).
4. S. Kugler, G. Molnar, G. Peto, E. Zsoldos, L. Rosta, A. Menelle and R. Bellissent, Phys. Rev. B**40**, 8030 (1989).
5. J.S. Custer, Michael O. Thompson, D.J. Eaglesham, D.C. Jacobson, and J.M. Poate, J. Mater. Res. **8**, 820 (1993).
6. D.R. Clarke, M.C. Kroll, P.D. Kirchner, R.F. Cook, and B.J. Hockey, Phys. Rev. Lett. **60**, 2156 (1988).

Part IV

Ordering and Hydrogen

AMORPHOUS SILICON ALLOY MATERIALS AND SOLAR CELLS NEAR THE THRESHOLD OF MICROCRYSTALLINITY

J. YANG AND S. GUHA
United Solar Systems Corp., 1100 West Maple Road, Troy, MI 48084

ABSTRACT

One of the most effective techniques used to obtain high quality amorphous silicon alloys is the use of hydrogen dilution during film growth. The resultant material exhibits a more ordered microstructure and gives rise to high efficiency solar cells. As the hydrogen dilution increases, however, a threshold is reached, beyond which microcrystallites begin to form rapidly. In this paper, we review some of the interesting features associated with the thin film materials obtained from various hydrogen dilutions. They include the observation of linear-like objects in the TEM micrograph, a shift of the principal Si TO band in the Raman spectrum, a sharp, low temperature peak in the H_2 evolution spectrum, a shift of the wagging mode in the IR spectrum, and a narrowing of the Si (111) peak in the X-ray diffraction pattern. These spectroscopic tools have allowed us to optimize deposition conditions to near the threshold of microcrystallinity and obtain desired high quality materials. Incorporation of the improved materials into device configuration has significantly enhanced the solar cell performance. Using a spectral-splitting, triple-junction configuration, the spectral response of a typical high efficiency device spans from below 350 nm to beyond 950 nm with a peak quantum efficiency exceeding 90%; the triple stack generates a photocurrent of 27 mA/cm^2. This paper describes the effect of the improved materials on various solar cell structures, including a 13% active-area, stable triple-junction device.

INTRODUCTION

Over the last two decades, photovoltaic (PV) technology using amorphous silicon (a-Si) alloy materials has advanced significantly to a stage where large-scale manufacturing is taking place worldwide [1]. Most of the products to date are aimed at terrestrial applications such as battery charging and building-integrated photovoltaics. On the other hand, due to the tremendous growth in the field of telecommunication, intensive efforts are underway to qualify the lightweight inexpensive thin-film solar cells for space applications [2]. Today, state-of-the-art a-Si alloy triple-junction modules on thin, flexible substrates are being seriously evaluated in orbit such as those from United Solar Systems Corp. (United Solar) on board the Mir space station.

Despite the substantial progress made in recent years, the biggest challenge for both terrestrial and extraterrestrial applications today is achieving even higher conversion efficiencies. In the United States, a National Thin-film Partnership Team including members from the PV industry, universities, and national laboratories was organized in 1992 by the National Renewable Energy Laboratory (NREL) to coordinate and address various aspects of the thin-film materials and devices [3]. A broad spectrum of activities, ranging from fundamental material studies to novel multijunction designs, has been addressed. In the last several years, the most important finding from these activities for improving stabilized cell efficiencies using plasma-enhanced chemical vapor deposition (PECVD) is probably the realization of the benefit

Mat. Res. Soc. Symp. Proc. Vol. 557 © 1999 Materials Research Society

of using hydrogen dilution during film growth [4]. Although the first report on the effect of hydrogen dilution was made in 1981 by Guha et al. [5], it was not until the 1990's that improvement on solar cell performance using this approach became available in the literature. These cells exhibit higher efficiencies and better light stability.

A recent study by Tsu et al. [6] further revealed that a-Si alloy materials grown using appropriate hydrogen dilution result in more-ordered microstructures in that intermediate-range orders have been observed. As the hydrogen dilution ratio is increased, there exists a transitional region beyond which microcrystallites begin to form rapidly. The best quality a-Si alloy solar cells are obtained when one deposits the material just below the threshold of the amorphous to microcrystalline transition. In this paper, we review some interesting features associated with the materials near the threshold and discuss the corresponding solar cell performance.

EXPERIMENTAL RESULTS AND DISCUSSION

<u>Near Threshold Material Characteristics</u>

The first experimental observation of improved stability against light soaking by using hydrogen dilution during film growth was reported by Guha at al. in 1981 [5]. They found that films obtained using 10% silane (SiH_4) and 90% hydrogen (H_2) exhibited better stability than those grown using 100% SiH_4.

In 1994, we reported that a-Si alloy solar cells whose intrinsic layers were made with high hydrogen dilution displayed better stability than those made with low dilution at a same deposition temperature of 300 °C [4]. In order to understand the difference in the two types of materials, we measured film properties such as infrared (IR) absorption, optical bandgap, hydrogen evolution, and small-angle X-ray scattering (SAXS). From the IR absorption spectra and optical bandgap measurements, we found that both types of materials contain ~10% of bonded hydrogen, and their optical bandgaps around 1.72 eV are also similar. On the other hand, the hydrogen evolution spectra for the two types of materials are distinctly different, as shown in Fig. 1. The low dilution sample exhibited a broad evolution peak near 500 °C (Fig. 1(a)), while the high dilution sample displayed a sharp, narrow peak near 400 °C

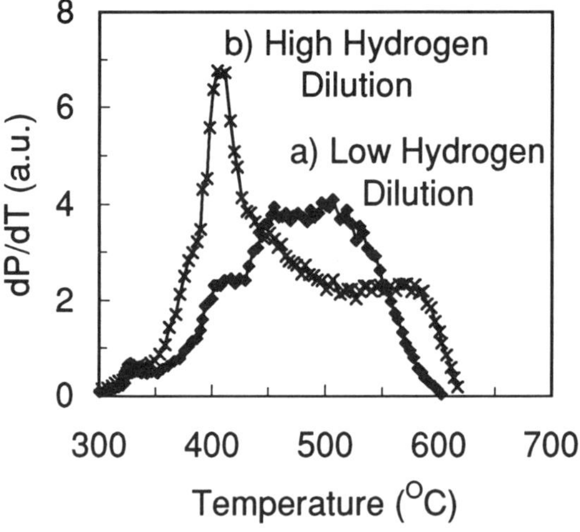

Figure 1. Hydrogen evolution spectra for (a) the low dilution and (b) the high dilution samples.

(Fig. 1(b)) [7]. This pronounced difference suggests that the microstructure of the two types of films may be quite different. Based on the IR analyses, it was pointed out that the low temperature evolution peak was not associated with dihydride or polyhydride structures. In addition, SAXS data showed that high hydrogen dilution samples contain oriented microstructures parallel to the film growth direction [8,9]. Based on hydrogen evolution and SAXS data, it appears that the high dilution samples may have a heterogeneous structure.

Furthermore, we have made a-Si alloy solar cells using a gas mixture of SiD_4 and D_2 and compared the cell performance to those obtained from SiH_4 and H_2. We found that, using similar deposition conditions and intrinsic layer thicknesses, the deuterated cells show better stability against light soaking than the hydrogenated counterparts [10]. What is also interesting is that the deuterium evolution spectrum also showed a sharp, low temperature peak at an even lower temperature of 300-350 °C, displayed in Fig. 2.

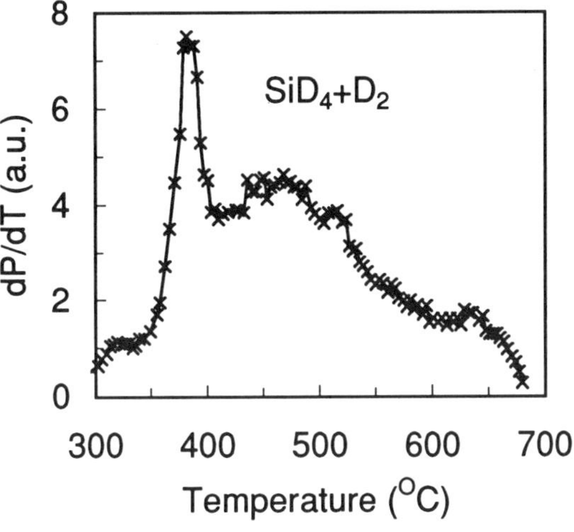

Figure 2. Deuterium evolution spectrum from the deuterated sample.

From the above experimental results, it appears that the deposition conditions used for achieving more stable solar cells also lead to materials having a low temperature peak in the evolution spectra. A better stability seems to be correlated with an improved microstructure.

In order to further investigate the correlation between enhanced stability and the effect of hydrogen dilution, we collaborated with Tsu et al. [6] and carried out a systematic study on samples prepared with low, medium, and high hydrogen dilutions. We used high resolution transmission electron microscopy (TEM) and Raman scattering measurements to look for qualitative and quantitative differences in the three types of films. Figure 3 shows the Raman spectra for a-Si alloy films of ~500 Å thick deposited on 7059 Corning glass. The low, medium, and high dilution films are shown in Fig. 3(b), 3(c), and 3(d), respectively. Also shown in Fig. 3(a) is a standard a-Si alloy film prepared with no hydrogen dilution. The following two observations are made. First of all, there is a shift in the Si TO peak position, indicated by the dashed line, from no dilution to high dilution. Secondly, a sharp band appeared at 516.5 cm^{-1} for the high dilution sample (Fig. 3(d)), and is attributed to microcrystalline inclusions. From this study, a clear trend emerges in that as the hydrogen dilution increases, it promotes the formation of microcrystallites in the amorphous matrix.

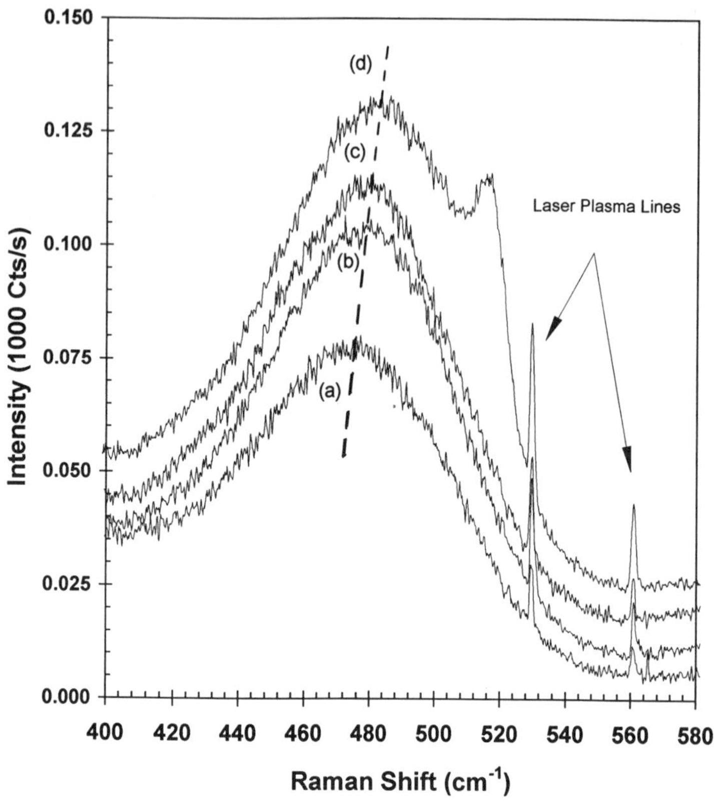

Figure 3. Raman spectra versus hydrogen dilution of four samples: (a) no dilution, (b) low dilution, (c) medium dilution, and (d) high dilution.

The films were subsequently lifted off the 7059 glass substrates and their microstructures analyzed by TEM. For the high dilution sample, distinct microcrystallites were indeed observed embedded in the featureless amorphous matrix even under low magnification. These spherical microcrystallites have sizes ranging from 2 to 10 nm in diameter. Under high magnification, TEM micrograph revealed that the high dilution sample exhibited long parallel lines separated by 0.31 nm, corresponding to the d spacing of Si (111) planes. In addition, one-dimensional linear-like objects that appear to have some degree of order along the length were observed. These objects have widths of 2-3 nm and lengths of up to a few 10's of nm. In fact, a careful examination of the other three samples showed that they all contain, to a lesser degree, these linear-like objects. This is true even for the standard sample prepared with no hydrogen dilution. Based on TEM and Raman data, we conclude that the intermediate range order of different degrees indeed exists in a-Si alloy films. The volume fraction occupied by microcrystallites within the amorphous material increases with increasing hydrogen dilution.

Recently, we have conducted a series of X-ray diffraction (XRD) studies on a-Si alloys prepared from various hydrogen dilutions to see if XRD signals can be correlated with hydrogen dilution ratios [11]. We found that the full-width-at-half maximum (FWHM) intensity of the first X-ray diffraction peak is narrower in high hydrogen diluted samples than those in low or no hydrogen dilution samples. This confirms the improved medium range order for high dilution samples. A detailed account is given by Williamson in this symposium [12].

Very recently, in collaboration with Mahan et al., we carried out systematic IR and H_2 evolution studies for hydrogen diluted samples. A careful analysis of the IR spectra revealed that the peak frequency of the Si-H wagging mode shifts downward from ~640 cm^{-1} to 620 cm^{-1} for high H_2 dilution. The 620 cm^{-1} frequency has been interpreted as being due to H bonded to the surfaces of very small, highly eccentric Si microcrystallites. These crystallites catalyze the crystallization of the remainder of the amorphous matrix upon annealing and resulting in the observed low temperature peak in the H_2 evolution spectrum. A more detailed description of this work is given in a separate paper in this symposium [13].

Kamei et al. [14] recently investigated effects of embedded crystallites in hydrogenated amorphous silicon on light-induced metastable dangling bond defect creation. The crystalline-like nanostructures were obtained by using hydrogen dilution during film growth, confirming the observation by Tsu et al. [6]. They also found that inclusion of a small-volume fraction of crystallites into the amorphous matrix significantly suppresses defect creation against moderate light illumination. They propose that the mechanism of improved stability is related to the trapping/recombination of photocarriers by embedded crystallites.

From the above discussion, it should be clear that appropriate hydrogen dilution can create intermediate range order and improve material properties. However, dilution beyond certain critical value often results in excess microcrystallite inclusions and degrades solar cell performance. In fact, a-Si alloy *p i n* cells with excess hydrogen dilution give rise to low open-circuit voltages and poor cell performance. We should also point out that the threshold condition is system dependent. In one system, a dilution ratio of ten to one may be the optimum, while in a different system, one may need a very different ratio. Other deposition parameters, such as pressure, temperature, electrode spacing, chamber geometry, and gas flows, all affect the growth kinetics. In fact, conditions too close to the threshold may not be easily reproducible. They may be too sensitive to small variations in the deposition conditions, as shown by a recent work using real-time spectroscopic ellipsometry [15].

Near Threshold Solar Cell Performance

From the above discussion, one should recognize the importance of using appropriate hydrogen dilution in the PECVD process to optimize the deposition conditions so that high quality materials can be obtained near the amorphous to microcrystalline transition. We have followed this approach and studied four different types of solar cell structures shown in Fig. 4. These structures use stainless steel substrates and textured silver/zinc oxide as back reflectors. Figure 4(a) shows the basic *p i n* structure most commonly used to evaluate the quality of the a-Si alloy materials. To achieve high efficiency in this structure, not only does one need high quality intrinsic material such as those discussed above, but also other key requirements such as light trapping, transparent conductive oxide, and interface properties. The thickness of the intrinsic layer is also very important. If the intrinsic layer is too thick, cells will suffer large degradation due to the Staebler-Wronski effect (SWE) [16]. If the intrinsic layer is too thin, it does not absorb sufficient photons, thus resulting in low efficiencies. An optimum thickness

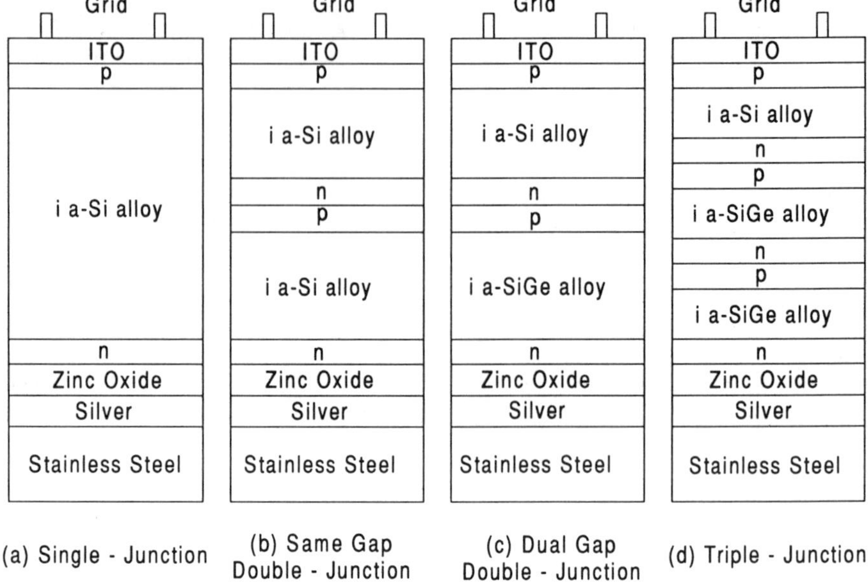

Figure 4.   Schematic diagram of four solar cell structures: (a) single, (b) same bandgap double, (c) dual bandgap double, and (d) triple junctions.

of ~2200 **Å** was used to achieve the highest stabilized active-area efficiency of 9.3%. The initial and stable J-V characteristics are listed in Table I. It should be pointed out that the degradation due to SWE for this structure is only 12.3%, a value much lower than the typical 50% degradation when the quality of the material was poor or the thickness of the intrinsic layer was too thick. Using hydrogen dilution during deposition, one generally observes less degradation and an earlier saturation [17,18].

For the same bandgap double-junction structure shown in Fig. 4(b), one takes advantage of the thin top cell. The thin top cell gives rise to a high internal field [19], thus less degradation. The top cell also absorbs the high energy photons and allows only low energy photons to reach the bottom cell. From Table I, we see that the initial fill factor of the same bandgap structure is higher than the single-junction structure; the stabilized fill factor is also higher. The stabilized efficiency of 10.1% is better than the 9.3% obtained from the single-junction structure.

In order to further increase the cell efficiency, one resorts to narrow bandgap materials such as a-SiGe alloys to broaden the spectral response. The carrier transport properties in the a-SiGe alloy materials, however, are often worse than those in the a-Si alloys. Using the hydrogen dilution approach, we have been able to improve the performance of the a-SiGe alloy solar cells substantially. Hydrogen dilution also allows for a lower deposition temperature without deteriorating the fill factor of the device [20]. Figure 5 shows the J-V characteristic

Table I.   Highest stable cell efficiencies (active area ~0.25 cm^2) at United Solar for different junction configurations.

		J_{sc} (mA/cm^2)	V_{oc} (V)	FF	η (%)	Deg. (%)
a-Si:H	Initial	14.65	0.992	0.730	10.6	
single junction	Stable	14.36	0.965	0.672	9.3	12.3
a-Si:H/a-Si:H	Initial	7.9	1.89	0.76	11.4	
same gap	Stable	7.9	1.83	0.70	10.1	11.4
double junction						
a-Si:H/a-SiGe:H	Initial	11.04	1.762	0.738	14.4	
dual gap	Stable	10.68	1.713	0.676	12.4	13.9
double junction						
a-Si:H/a-SiGe:H/	Initial	8.57	2.357	0.723	14.6	
a-SiGe:H	Stable	8.27	2.294	0.684	13.0	11.0
triple junction						

of an improved a-SiGe alloy device having an average germanium concentration of ~20% in the intrinsic layer. The intrinsic layer has also incorporated a bandgap profiling design [21] to improve the cell performance. The 11.6% initial active-area efficiency is the highest reported to date for a single-junction a-SiGe alloy solar cell.

Using the improved a-SiGe alloy in the bottom cell, we then proceeded to make dual bandgap double-junction solar cells using the structure shown in Fig. 4(c). In this structure, the top cell needs to be thicker than that of the same gap structure in order to match the higher photocurrent generated by the a-SiGe alloy bottom cell.

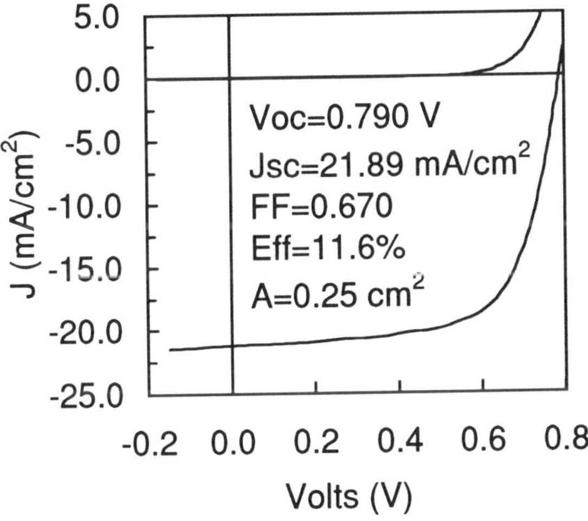

Figure 5.   Initial J-V characteristic of an a-SiGe alloy *p i n* solar cell.

The initial and stable characteristics of the best dual gap double-junction cell is also shown in Table I. Comparing the dual gap and same gap structures, it is obvious that the main advantage of the dual gap structure is the substantially higher current density, which overcomes the lower open-circuit voltage and fill factor. In fact, the dual gap structure shows a more than 20% gain in the stabilized efficiency than the same gap structure. The degradation of the dual gap device, on the other hand, is higher than the same gap counterpart. This is mainly due to the thicker top cell and the somewhat poorer a-SiGe alloy material in the bottom cell.

To further enhance the efficiency and broaden the spectral response, we next discuss the triple-junction configuration shown in Fig. 4(d). The structure incorporates a low bandgap a-SiGe alloy in the bottom cell. In this configuration, the bandgap and thickness of the top and middle cells are dictated by the quality of the bottom cell. If one can successfully incorporate a higher concentration of germanium into the bottom cell and further extend the spectral response into the long wavelength region, one may split the spectral response into blue, green, and red regions, corresponding to the top, middle, and bottom cells, respectively. The goal of this design is, of course, that the gain in current can overcome the loss in V_{oc} and FF. We have again used the hydrogen dilution method and successfully achieved 14.6% initial and 13.0% stable conversion efficiencies for a 0.25 cm^2 active-area device in the triple-junction configuration. Their J-V characteristics are also listed in Table I. Figure 6 shows the J-V characteristic of the 14.6% triple-junction cell; the quantum efficiency versus wavelength plot for this device is shown in Fig. 7.

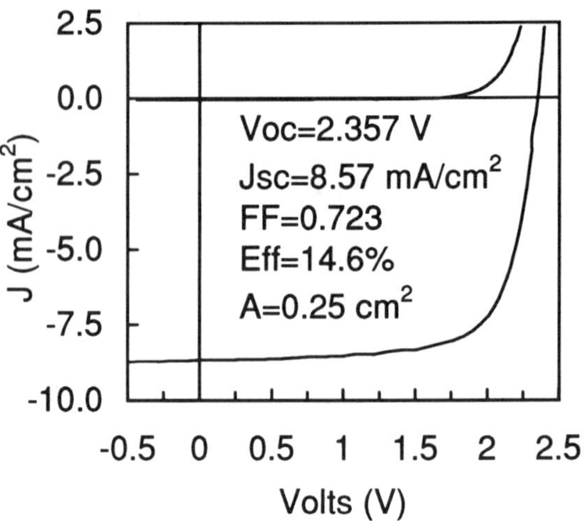

Figure 6. Initial J-V characteristic of the 14.6% triple-junction cell.

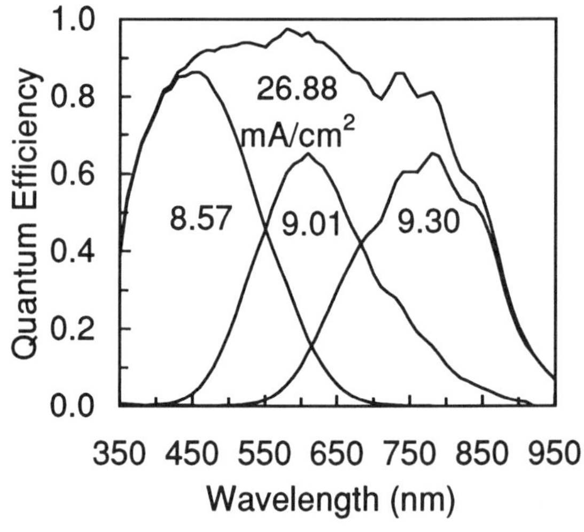

Figure 7. Quantum efficiency of the device shown in Fig. 6.

It should be pointed out that the stabilized efficiency for each structure in Table I represents the highest value reported to date in the literature. Hydrogen dilution indeed played a major role in achieving this result. We should also point out, however, that high quality material is a necessary but not sufficient condition for making high efficiency solar cells. Other key factors, such as the design of the device structure, the appropriate back reflector, the thickness and the optical bandgap of the intrinsic layer, the bandgap profiling of the a-SiGe alloys, the interface layers, the tunnel junctions, the component cell current matching, and the top conductive oxide, all play an important role in achieving the high efficiency cells [22].

Based on the experimental data of Table I, it is easy to conclude that the triple-junction structure gives rise to the highest initial and stable efficiencies. The quantum efficiency curve in Fig. 7 clearly shows the spectral-splitting feature of the triple junction design. The current mismatch observed among the three component cells is designed to limit the triple-junction current by the top cell which usually has the best fill factor. Any improvement in the fill factor of the middle or bottom cell can allow us to further reduce the current mismatch and improve the cell efficiency. Our continuous effort in optimizing the device design and deposition conditions has led us to achieve a 15.2% initial active-area

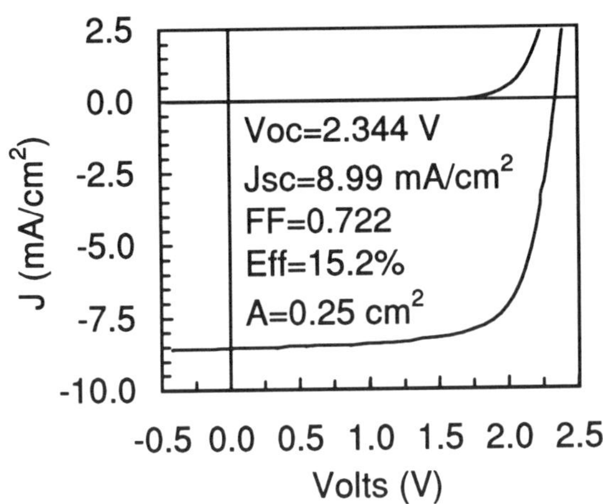

Figure 8. Initial J-V characteristic of the 15.2% triple-junction cell.

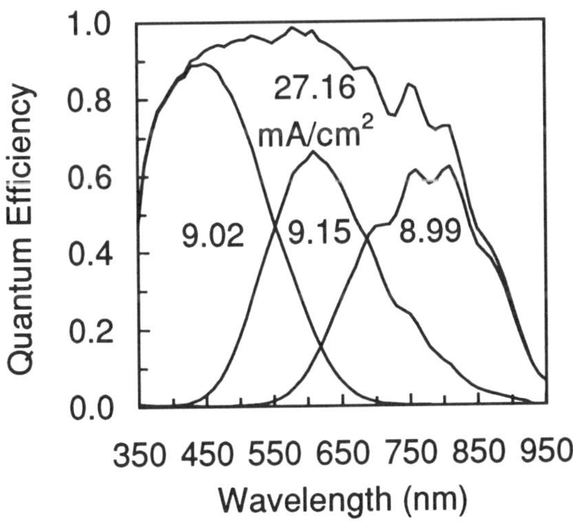

Figure 9. Quantum efficiency of the device shown in Fig. 8.

efficiency. The J-V characteristic is shown in Fig. 8, and the
quantum efficiency plot is shown
in Fig. 9. From Fig. 9, one can
see that the spectral response
spans from below 350 nm to
beyond 950 nm with the peak
quantum efficiency exceeding
90% in the green region. The
triple stack generates a total
photocurrent of 27 mA/cm^2. The
current mismatch among the
component cells has been
virtually eliminated.

Based on the above
experimental results and
discussion, it is clear that
hydrogen dilution improves the
microstructure and enhances the
cell performance. The optimum
condition is very close to the
amorphous to microcrystalline
transition. The challenge is to
approach the threshold without
exceeding the critical limit. In
order to find out the cell
characteristics at optimum
conditions, we have evaluated a
series of a-Si alloy top cells on
stainless steel substrates by
adjusting the hydrogen dilution
condition near the threshold.
The requirement is that the J_{sc}
value must be around 8 mA/cm^2
for top cell applications. Figure
10 shows the best cell obtained
to date in terms of high fill
factor and high open-circuit
voltage. The cell exhibits FF =
0.785, V_{oc} = 1.010 volt, and J_{sc} =
8.25 mA/cm^2. As soon as the
dilution ratio exceeds a certain
critical value, a sudden drop in
FF and V_{oc} occurs, signifying
that the threshold has been
exceeded.

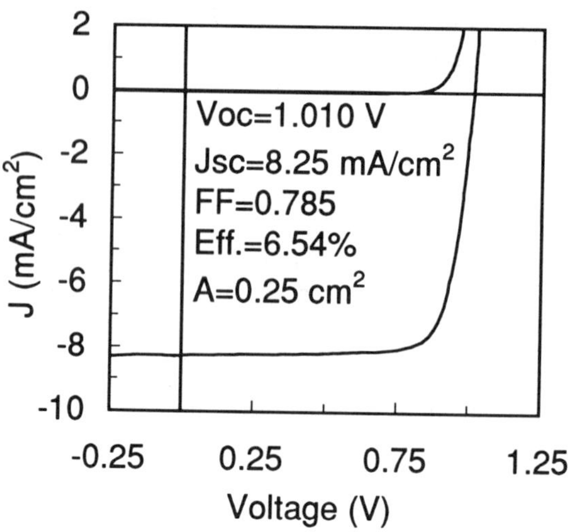

Figure 10. Initial J-V characteristic of an a-Si alloy top cell deposited at ~1 Å/s.

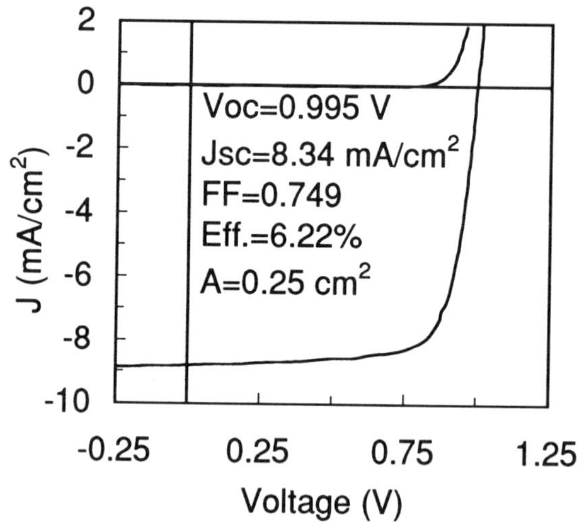

Figure 11. Initial J-V characteristic of an a-Si alloy top cell deposited at ~8 Å/s by MVHF.

Hydrogen dilution in general lowers the deposition rate. The high efficiency cells described above have a typical deposition rate of ~1 Å/s. Can one obtain high quality materials and solar cells using high deposition rates? This is of particular interest concerning the throughput in large volume productions. We have used a modified very high frequency (MVHF) approach and studied the hydrogen dilution effect on high-rate-deposited top cells [23]. Figure 11 shows the best cell obtained to date using MVHF with hydrogen dilution at a deposition rate of 8 Å/s. The J-V characteristic is very close to that of the 1 Å/s cell shown in Fig. 10. High performance a-Si alloy cells can indeed be achieved at high rates using an appropriate hydrogen dilution. More work is needed to study the a-SiGe alloy using high deposition rates.

SUMMARY

Using hydrogen dilution during film growth in the PECVD process, we have shown that the microstructure of the material and solar cell performance are improved. Interesting features associated with hydrogen diluted films near the amorphous to microcrystalline transition are reviewed. They include the observation of linear-like objects in the TEM micrograph, a shift of the principal Si TO band in the Raman spectrum, a low temperature H_2 evolution peak, a shift in the wagging mode frequency in the IR spectrum, and a narrowing of the Si (111) peak in the X-ray diffraction pattern.

World record high efficiency solar cells in four different device structures have been achieved by optimizing hydrogen dilution conditions near the threshold of the amorphous to microcrystalline transition. More work is needed to understand these hydrogen diluted materials to improve the efficiencies further.

ACKNOWLEDGEMENT

We wish to thank D. Wolf for carrying out the H_2 evolution experiments and V. Trudeau for preparation of the manuscript. We also thank E. Chen, J. Edens, G. Hammond, M. Hopson, J. Noch, and T. Palmer for their assistance in the sample preparations and measurements. Stimulating discussions with S. R. Ovshinsky, H. Mahan, D. Williamson, D. Cohen, A. Banerjee, B. Yan, and K. Lord are acknowledged. This work is supported in part by NREL under subcontract no. ZAK-8-17619-09.

REFERENCES

1. J. Rannels, *Proc. of 2nd World Conf. and Exhibition on Photovoltaic Solar Energy Conversion*, Vienna, Austria (1998), p. LXXXVII.

2. S. Guha, J. Yang, A. Banerjee, T. Glatfelter, G.J. Vendura, Jr., A. Garcia, and M. Kruer, *Proc. of 2nd World Conf. and Exhibition on Photovoltaic Solar Energy Conversion*, Vienna, Austria (1998) p. 3609.

3. B. von Roedern, K. Zweibel, E. Schiff, J.D. Cohen, S. Wagner, S.S. Hegedus, and T. Peterson, AIP Conference Proceedings **394**, 3 (1997).

4. J. Yang, X. Xu, and S. Guha, Mater. Res. Soc. Symp. Proc. **336**, 687 (1994).

5.   S. Guha, K.L. Narasimhan, and S.M. Pietruszko, J. Appl. Phys. **52**, 859 (1981).

6.   D.V. Tsu, B.S. Chao, S.R. Ovshinsky, S. Guha, and J. Yang, Appl. Phys. Lett. **71**, 1317 (1997).

7.   X. Xu, J. Yang, and S. Guha, J. Non-Crys. Solids **198-200**, 96 (1996).

8.   S.J. Jones, Y. Chen, D.L. Williamson, X. Xu, J. Yang, and S. Guha, Mater. Res. Soc. Symp. Proc. **297**, 815 (1993).

9.   D.L. Williamson, Mater. Res. Soc. Symp. Proc. **377**, 251 (1995).

10.   S. Sugiyama, J. Yang, and S. Guha, Appl. Phys. Lett. **70**, 378 (1997).

11.   S. Guha, J. Yang, D.L. Williamson, Y. Lubianiker, J.D. Cohen, and A.H. Mahan, Appl. Phys. Lett. (1999) (to be published).

12.   D.L. Williamson, this symposium.

13.   A.H. Mahan, J. Yang, and S. Guha, this symposium.

14.   T. Kamei, P. Stradius, and A. Matsuda, Appl. Phys. Lett. **74**, 1707 (1999).

15.   J.H.Koh, Y. Lee, H. Fujiwara, C.R. Wronski, and R.W. Collins, Appl. Phys. Lett. **73**, 1526 (1998).

16.   D.L. Staebler and C.R. Wronski, Appl. Phys. Lett. **31**, 292 (1977).

17.   J. Yang, A. Banerjee, T. Glatfelter, K. Hoffman, X. Xu, and S. Guha, 1st World Conf. on Photovoltaic Energy Conversion, Waikoloa, Hawaii, Dec. 5-9, 1994, p. 380.

18.   L. Yang and L. Chen, Mater. Res. Soc. Symp. Proc. **336**, 669 (1994).

19.   L. Jiang, Q. Wang, E.A. Schiff, S. Guha, J. Yang, and X. Deng, Appl. Phys. Lett. **69**, 3063 (1996).

20.   M. Shima, A. Terakawa, M. Isomura, M. Tanaka, S. Kiyama, and S. Tsuda, Appl. Phys. Lett. **71**, 84 (1997).

21.   S. Guha, J. Yang, A. Pawlikiewicz, T. Glatfelter, R. Ross, and S.R. Ovshinsky, Appl. Phys. Lett. **54**, 2330 (1989).

22.   J. Yang, A. Banerjee, and S. Guha, Appl. Phys. Lett. **70**, 2975 (1997).

23.   J. Yang, S. Sugiyama, and S. Guha, Mater. Res. Soc. Symp. Proc. **507**, 157 (1998).

MEDIUM-RANGE ORDER IN a-Si:H BELOW AND ABOVE THE ONSET OF MICROCRYSTALLINITY

D.L. WILLIAMSON
Department of Physics, Colorado School of Mines, Golden, CO 80401

ABSTRACT

Medium range order (MRO) and the formation of microcrystallites in a-Si:H prepared by plasma-enhanced chemical vapor deposition (PECVD) and hot-wire chemical vapor deposition (HWCVD) have been probed by systematic x-ray diffraction studies with films as thin as those used in solar cells. Effects of substrate temperature, hydrogen dilution, film thickness, and type of substrate have been examined. High-hydrogen-diluted films of 0.5 µm thickness, using optimized deposition parameters for solar cell efficiency and stability, are found to be partially microcrystalline (µc) if deposited directly on stainless steel (SS) substrates but are fully amorphous provided a thin (20 nm) n-layer of a-Si:H or µc-Si:H is first deposited on the SS. The latter predeposition does not prevent partially microcrystallinity if the films are grown thicker (1.5 to 2.5 µm) and this is consistent with a recently proposed phase diagram of thickness versus hydrogen dilution. Analysis of the first (lowest angle) scattering peak of the a-Si:H phase demonstrates that its width, directly related to MRO, is reduced by heavier hydrogen dilution in PECVD growth or by increased substrate temperature in HWCVD growth. The narrowest width of fully amorphous material correlates with better solar cell stability and this is not likely related to bonded hydrogen content since it is quite different in the optimized PECVD and HWCVD a-Si:H. A wide range of MRO apparently exists in the residual amorphous phase of the mixed a/µc material.

INTRODUCTION

There is a growing body of experimental evidence that the best device-quality a-Si:H materials have some type of improved, or at least modified, structural order as determined or inferred by a variety of techniques [1-21]. All of these studies can be divided into two deposition methods being used to reach these new levels of order: (i) high hydrogen dilution of silane or disilane in the plasma-enhanced chemical vapor deposition (PECVD) method [1-12] or (ii) elevated substrate temperatures in the hot-wire chemical vapor deposition (HWCVD) method [13-21]. These two types of films are dramatically different in hydrogen content, ranging from 7 to 16 at.% for the PECVD H_2-diluted films and from less than 1 at.% to about 4 at.% for the HWCVD high-substrate-temperature films. It is also interesting that the deposition rates of the HWCVD process (~1 nm/s) are significantly faster than those from the PECVD process ($\leq$0.3 nm/s), for fabrication of films of comparable opto-electronic quality. In the above studies, structural characteristics have been inferred from NMR (hydrogen distributions), Raman (short-range order), IR (hydrogen bonding configurations), spectroscopic ellipsometry (film nucleation, growth, and microcrystallinity), AFM (surface topography), internal friction, Urbach energy, hydrogen evolution, residual stress, and mass density. Several of the studies have used the more direct structural methods of transmission electron microscopy (TEM) [4,5,10,11,21] and x-ray diffraction (XRD) [7,10,12,18]. High-resolution TEM has provided direct evidence of ordered regions on the nanometer scale within the amorphous matrix of the high-hydrogen-diluted PECVD material [4,5,10] and cross-sectional TEM of HWCVD material shows the onset of crystallinity near the substrate with increasing film thickness [21]. In addition to providing

Mat. Res. Soc. Symp. Proc. Vol. 557 © 1999 Materials Research Society

direct information on the degree of microcrystallinity [7,10,12] and its sensitivity to the nature of the substrate [12], XRD has recently been used to show direct evidence of improved medium-range order in both the elevated-substrate-temperature HWCVD material [18] and in the high-hydrogen-dilution PECVD films [12]. This approach involves a careful examination of the width of the lowest angle x-ray scattering peak from the a-Si:H matrix. Here we will provide more details and results of such studies, with a more complete comparison of data from the PECVD and HWCVD materials, after first providing a brief review of the concept of medium-range order, particularly as detected by XRD.

MEDIUM-RANGE ORDER

Medium-range order (MRO) is a term used to discuss the structural ordering of amorphous systems on a length scale larger than nearest neighbor distances (short-range order) up to an ill-defined upper limit around 2 to 5 nm. MRO has received a great deal of attention recently with respect to amorphous covalent glasses. Most of the focus has been on the oxide and chalcogenide glasses due to the "pre-peak" or "first sharp diffraction peak (FSDP)" or "first scattering peak (FSP)" in the x-ray and neutron diffraction patterns as described in review articles [22-25] that reveal the considerable controversy on this topic. This feature occurs at unusually low scattering angles or momentum transfer, $q \approx 10$ to 20 nm^{-1}, $q = (4\pi\sin\theta)/\lambda$, where 2θ is the scattering angle and λ is the radiation wavelength. There has been some modeling research [26-28] directed specifically to understanding the FSP of a-Si and a-Si:H which occurs near the upper end of the q-range associated with MRO (q=19.5 nm^{-1}, or at $2\theta\approx27.6^{o}$ when λ=0.154 nm, i.e. Cu-K$_\alpha$ x-rays). It is interesting to note that this peak has little to do with the (111) Bragg diffraction peak of c-Si despite its similar location at $q = 20.0$ nm^{-1} [28] or with diffraction associated with the well-defined nearest neighbor distance in a-Si since, via the Bragg law, this nearest neighbor distance would produce a scattering peak at $q \approx 2\pi/0.235$ nm $= 26.7$ nm^{-1}. The latter q is actually near the *minimum* between the first and second scattering peaks of a-Si (and a-Si:H) [26-28].

Calculations based on various a-Si structural models [26-28] clearly demonstrate that the FSP has significant contributions from MRO, while the higher q peaks have little sensitivity. Most recently, Uhlherr and Elliott [28] have shown that weak oscillations in the real-space pair distribution function g(r) of a-Si extending from r = 1.0 to 3.3 nm contribute about half the intensity of the FSP. The partial pair distribution functions based on next-nearest neighbor separation seems to play the dominant role leading to this extended range order [28]. These studies make it clear that an experimental focus on the FSP of a-Si:H should lead to information on the MRO and, indeed, this was recognized several years ago [29] and applied to comparisons of a-Si:H prepared by sputter-deposition and PECVD. A simple, experimental, quantitative measure of the correlation length, L, over which the atomic density fluctuations contribute to this FSP can be obtained from its width, W (full-width-at-half-maximum intensity in radians), via the Scherrer equation [30], $L= 0.9\lambda/W\cos\theta$, or via a nearly equivalent expression [25,31], $L = 2\pi/\Delta q$, where Δq is the width of the peak in nm^{-1}. The Scherrer equation is most often used to estimate the average crystallite size of microcrystalline (μc) Si.

Another x-ray scattering technique used to provide detailed structural information on a-Si:H and related thin-film materials is small-angle x-ray scattering (SAXS) occurring at very low q [32]. However, this information is more appropriately characterized as medium- or extended-range *disorder*, associated with inhomogeneities in the films such as nanovoids and columnar-like growth features [32]. Studies of both the high-hydrogen-diluted PECVD and elevated-temperature HWCVD materials of interest here demonstrated undetectable levels of inhomogeneities on the nanoscale (void volume fractions less than 0.01%), but evidence of larger-scale features (> 20 nm), probably associated with residual columnar-like structure or surface roughness [2,32].

Recently, a new TEM method has been developed to examine MRO in amorphous materials [33] and evidence has been found for changes in MRO in a-Si:H induced by light soaking [34].

EXPERIMENTAL METHODS

Numerous PECVD and HWCVD a-Si:H films have been prepared for XRD on stainless steel (SS) substrates, both because of the weak interference of the SS with the FSP in the XRD data and because this is a common substrate for the best solar cells made from these materials [2,35,36]. Several series of PECVD undoped (intrinsic - i) films were made to explore effects of hydrogen dilution (none, low, high), film thickness (0.5 μm to 2.5 μm), and substrate nature (bare stainless steel - i/SS, amorphous n-layer-coated stainless steel - i/a-n/SS, and microcrystalline n-layer-coated stainless steel - i/μc-n/SS). The n-layers were typically about 20 nm thick. For XRD the films were deposited on 2.5 cm x 2.5 cm stainless steel (SS) substrates providing adequate area for good diffraction signals. The level of hydrogen dilution in the "high-hydrogen-dilution" i-layers studied here are similar to those used in recent world-record-efficiency solar cells [35]. Some of the films have been further processed to examine solar-cell open circuit voltages, V_{oc}, and defect concentrations via drive-level capacitance profiling and these results have been recently presented in correlation with XRD results [12]. Three series of HWCVD films have been prepared [13] in two different deposition systems and with different filament currents, typically to a thickness of 1-2 μm. All three series were made with substrate temperature as the deposition variable. XRD and Raman experiments with one series have been reported earlier [18].

The XRD measurements were carried out in the symmetric Bragg-Brentano geometry operating with Cu-K$_\alpha$ radiation as selected by a graphite monochromator in the scattered beam. The instrumental linewidth in the region of the FSP of a-Si:H is $W_i = 0.12^\circ$ based on measurements using a NIST LaB$_6$ reference powder. The width used in the Scherrer equation (given above) is typically $W = (W_m{}^2 - W_i{}^2)^{1/2}$, where W_m is the measured width from the sample. Since the absorption length of this 0.154 nm radiation is about 70 μm in Si, most of the scattering signal originates from the SS substrate. However, this signal is strongly localized in the bcc/fcc diffraction peaks and some secondary phase peaks, none of which interfere with the FSP (located near $2\theta = 27.6^\circ$) from a-Si:H nor with the (111), (220), or (311) c-Si diffraction peaks. Long counting times were used to achieve good signal-to-noise and the bare SS substrate signal was subtracted from each XRD pattern (with appropriate absorption corrections). Since the SS apparently has different degrees of polycrystalline texture for the various samples, this would not typically remove the strongest bcc peaks at $2\theta = 44^\circ$ and 65°. These were then removed by a smoothing procedure in order to examine more carefully the a-Si:H and μc-Si:H peaks. The full-width-at-half-maximum intensity (W_m) of the FSP of a-Si:H, as well as the fractional area (integrated intensity) of the μc-Si:H (111)+(220)+(311) peaks compared to that of the total area from 15° to 65° were extracted via a least-square fitting routine based on a superposition of Voigt line-shape functions. This function is symmetric but has two width parameters used to adjust the shape. Typically the second a-Si:H peak centered near $2\theta = 51^\circ$ was included in the fit due to the slight overlap of its low-angle tail with the FSP. To account for some slight asymmetry in this second peak, a third peak was included at a fixed angle near 45° in all fits of fully amorphous films. An example of the data and this type of fit is shown in Fig. 1 for a fully amorphous film. Also shown is a fit to a partially microcrystalline film where the sharper (111), (220), and (311) peaks of the μc phase are indicated. The residual amorphous phase FSP remains clear enough to track its width in the partially microcrystalline material.

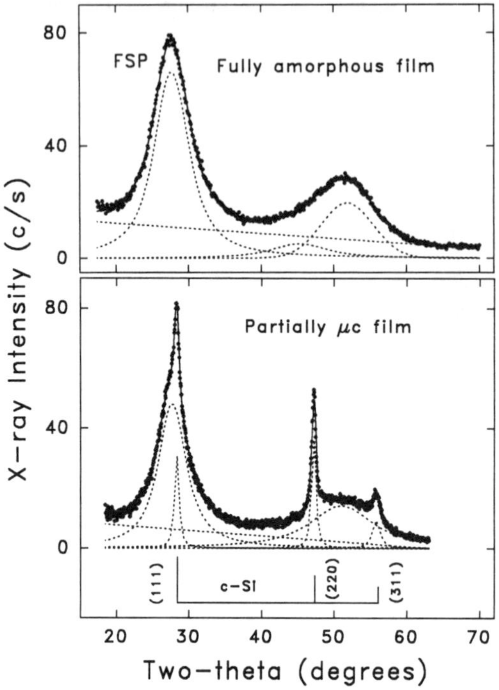

FIG. 1. Example least-square fits of X-ray diffraction patterns from the two type of films studied here, completely amorphous-Si:H (with the first scattering peak, FSP, indicated) and mixed amorphous/microcrystalline. Superpositions of Voight line-shape functions and a linear background were used as indicated by the dashed curves to generate the solid line fit through the data. The stick diagram for randomly-oriented c-Si is shown with the lowest 3 Bragg peaks.

RESULTS AND DISCUSSION

PECVD Material

Figure 2 demonstrates the substrate sensitivity of the high-H_2-diluted PECVD films to the formation of partially μc material. These films were all made under nominally identical conditions and grown to a common thickness of 0.5 μm. One can see that the presence of the thin intermediate a-n or μc-n layer on the SS prevents the partial microcrystallinity, at least to the sensitivity level of the XRD which is estimated to be about 2 to 4 vol.% for the data shown. The latter estimate is based on the result that the fractional areas of the μc peaks for the two samples shown are 10% and 15% so that 2 to 4% should still be detectable within the counting statistics obtained. Duplicate runs are shown to demonstrate reproducibility of the effect. Sensitivity of microcrystallite formation to the type and morphology of the substrate has been demonstrated by others [9,37-41]. Somewhat similar to the present result, deposition on a pre-deposited a-Si:H layer apparently hinders microcrystallinity [9,37,40], although increase in dilution [9], longer hydrogen plasma treatment [37], or higher plasma frequency [40] leads to initiation of microcrystallinity even on this self-similar substrate.

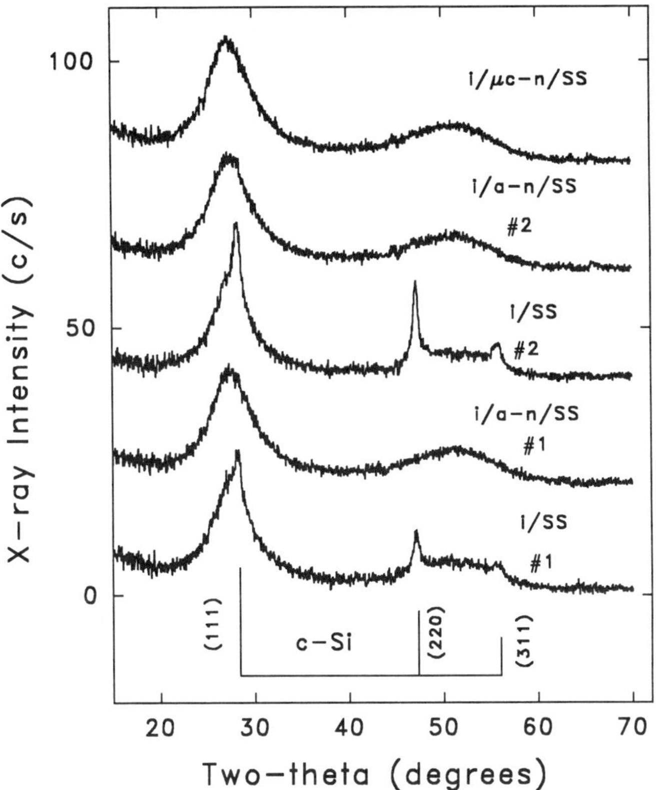

FIG. 2. X-ray diffraction patterns from a series of i-layers on SS, a-n coated SS, and μc-n coated SS. All i-layers were grown under identical high-hydrogen-dilution PECVD conditions and all are 0.5 μm thick. Duplicate runs are indicated.

A series of films made with increasing thickness under nominally identical high-hydrogen-dilution conditions on the a-n coated SS shows a transition to partially μc material as demonstrated in Fig. 3. Also shown is a thick film made at low dilution which remains fully amorphous. Thus, between 0.5 and 1.5 μm the intrinsic a-Si:H becomes partially microcrystalline under the high dilution condition. From the widths of the μc peaks, the grain sizes via the Scherrer formula given in the experimental section are 6 to 22 nm, depending on orientation. The thicker 2.5 μm film develops a strong preferred orientation of the (220) planes parallel to the film surface as indicated by its much stronger relative intensity compared to the random c-Si intensities shown by the stick diagram in Fig. 2. These grains also tend to be at the large end of the size range. The transition to partially μc-Si:H is accompanied by large drops in the open circuit voltage, V_{oc}, for devices made by completion of a p-i-n structure on co-deposited films [12]. In addition, defect levels as determined by capacitance measurements [12], tend to decrease with distance from the substrate, suggesting improved order with thickness. Supporting the transition thickness shown in Fig. 3, photocapacitance spectra show the onset of microcrystallinity between 1.0 and 1.3 µm for films made under the same high-dilution conditions used for the XRD studies [12].

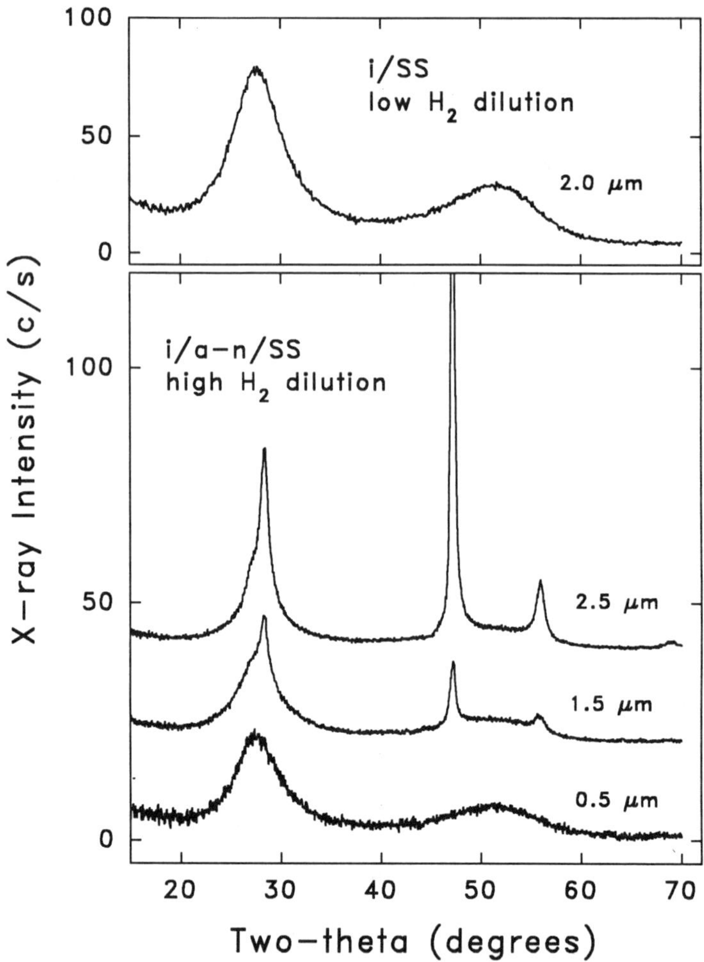

FIG. 3. X-ray diffraction patterns from series of i-layers grown under identical high-hydrogen-dilution PECVD conditions on a-n coated SS grown with increasing thickness as indicated. The top pattern shows no evidence for microcrystallinity under the low-hydrogen-dilution condition. The intensities of the 1.5 and 2.5 µm films have been divided by 3 and 5, respectively, for better comparison to the 0.5 µm film.

Although others have examined thickness effects on the structure of a/μc-Si:H [6,9, 21,39,40], the work by Koh et al [6,9] is particularly relevant to the present results since hydrogen dilution levels and substrate effects were examined in connection with the onset of microcrystallinity versus film thickness. A thickness versus hydrogen dilution "phase diagram" was proposed [9] in which a phase boundary exists between amorphous and microcrystalline material. For their deposition system, this phase boundary was explored via spectroscopic ellipsometry below about 200 nm down to 10 nm for H_2/SiH_4 gas ratios up to 80. This phase diagram led to the development of a two-step process for solar cell fabrication whereby an initial growth of the i-layer was done at high dilution to begin this layer close to but below the onset of μc-Si:H formation, followed by a reduction of the dilution ratio to finish the layer to remain in the fully amorphous state, but again close to the onset. The region of the phase diagram near the phase boundary, but below the onset, was termed the "proto-crystalline" regime [9]. Improved device behavior was demonstrated with this two-step procedure [9]. Although the present results are far from the phase boundary in the 10 nm thickness regime, it is clear that such a boundary exists between 0.5 and 1.3 μm for the high-dilution condition on the a-n/SS substrate and it is shifted to less than 0.5 μm on the bare SS.

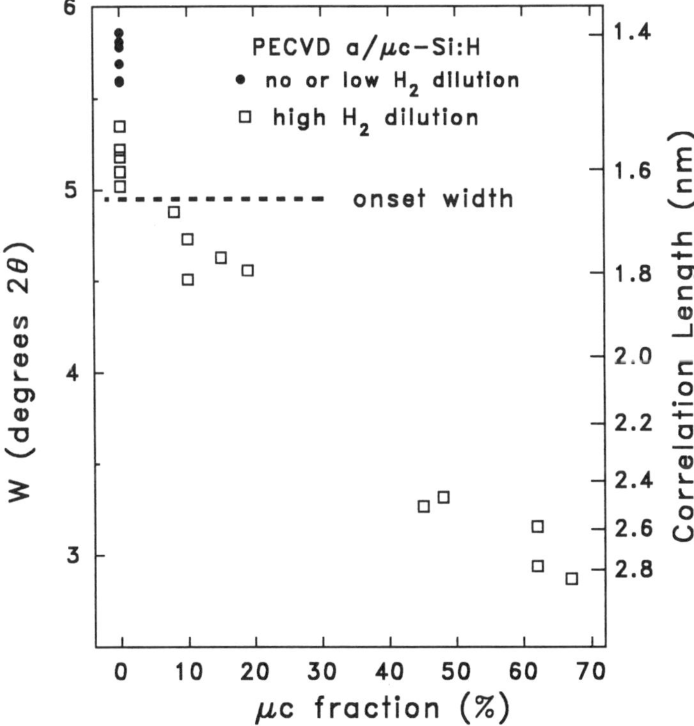

FIG. 4. Summary of XRD width values for the FSP of PECVD a-Si:H. Right-hand scale is based on Scherrer equation. The μc fraction is estimated from the fitted XRD integrated intensities of the (111)+(220)+(311) Bragg peaks as described in the experimental section.

Evidence for improved structural order is presented in Fig. 4 in exactly the same regime suggested as "proto-crystalline" in nature [9]. The width of the FSP is displayed as a function of the amount of microcrystalline material in each PECVD sample as determined by the fractional area of the (111)+(220)+(311) diffraction peaks. Note that there are several samples shown that are fully amorphous but they fall into two groups, a broader-width band above 5.5° associated with the low- and no-hydrogen-diluted material and a narrower-width band between 5.0° and 5.4° generated by the high-hydrogen-diluted films. All films with some μc fraction then fall below a rather well-defined "onset width" value of $W = 4.95 \pm 0.1^{\circ}$ as indicated by the dashed line in Fig. 4. The width of the FSP decreases systematically with increasing microcrystallinity. Using the Scherrer equation to estimate changes in correlation length associated with the MRO, the right hand scale shows that the correlation length increases by about a factor of two from 1.4 to 2.8 nm for the range of W's observed. The improvement in MRO associated with the fully amorphous material covers a rather modest range of just over 0.2 nm, or about a 15% change. However, this improvement seems to be correlated with improvement in device properties as will be discussed after the HWCVD data are included.

A comparison with earlier PECVD material is possible from the XRD study of Dixmier et al [27,29] which correlated the Urbach energy of a-Si:H prepared by both PECVD and sputtering with the width of the FSP. A plot was made based on the Scherrer equation and it was observed that the PECVD material was grouped at the longer correlation end of the trend, with Urbach energies approaching 50 meV. It is interesting that the longest length corresponds to $W = 5.8^{\circ}$ (for $\lambda = 0.154$ nm) which is in aggreement with the present results in Fig. 4 for the low- or no-dilution materials. An Urbach energy for the present high-dilution PECVD material was measured to be 46 meV [12]. The sputtered material showed Urbach energies up to about 170 meV and shorter correlation lengths corresponding to W's up to 8.9°, much larger than the PECVD material.

A look at the position of the FSP is warranted based on earlier evidence of a sensitivity to network relaxation [27]. The peak remains located in the narrow range $27.6 \pm 0.1^{\circ}$ for all the fully amorphous films and the group of partially μc films with fractions below 20% (Fig. 4) but increases slightly to the range 27.9 ± 0.1 for the group with higher fractions (45 - 67%).

HWCVD Material

Evidence of improved MRO in HWCVD a-Si:H was obtained by variation of the substrate temperature, T_s, during deposition [18]. The width of the FSP was shown to decrease systematically with increasing T_s. The same samples were studied by Raman spectroscopy and no evidence was found for improvements in the SRO based on the width of the Si-Si TO mode [18]. The XRD data are reproduced in Fig. 5 along with new data from HWCVD films made with a different system ("tube" system vs the earlier "cross" system). The solid curve guides the eye through the earlier data and dashed lines show the trends for two series of HWCVD films made with two different filament currents (14 and 16 A). There seems to be a slight difference in the trends for the two different HWCVD deposition systems, but both show improvements in MRO with increasing T_s. All the HWCVD material was made without any hydrogen dilution and all of the films remained fully amorphous. However, similar to the PECVD trend, using hydrogen dilution leads to microcrystalline material [42] and there is recent evidence for a thickness-induced transition to the partial μc state [21], however at a thickness of only about 70 nm for the conditions used.

A detailed comparison with the PECVD film results is now made by adding these data to Fig. 5 along the vertical line at $T_s = 300^{\circ}C$ where all these films were made. The V_{oc}'s obtained [12] are also shown for the three groups of PECVD samples (fully amorphous, low μc fraction, and high μc fraction - as indicated in Fig. 4). All the values for the fully amorphous films remain uniformly high (0.9 to 1.00 V). One can see that the low- and no-hydrogen-dilution widths are the same as the HWCVD film widths for a similar substrate temperature but the high-dilution widths clearly fall below the HWCVD trends. The reader is reminded that the higher-temperature HWCVD material is that which is of interest for device applications due to evidence of improved light stability [36,43]. Also, the high-hydrogen-dilution PECVD material is being used to produce the best and most stable solar cell efficiencies to date [2,35]. Figure 5 therefore suggests a band of FSP width (shaded region) that is associated with the best device material made by *either* PECVD or HWCVD methods. This band represents the best MRO possible, as determined by XRD, before the onset of microcrystallinity and the associated reductions in V_{oc} [12] as indicated in Fig. 5.

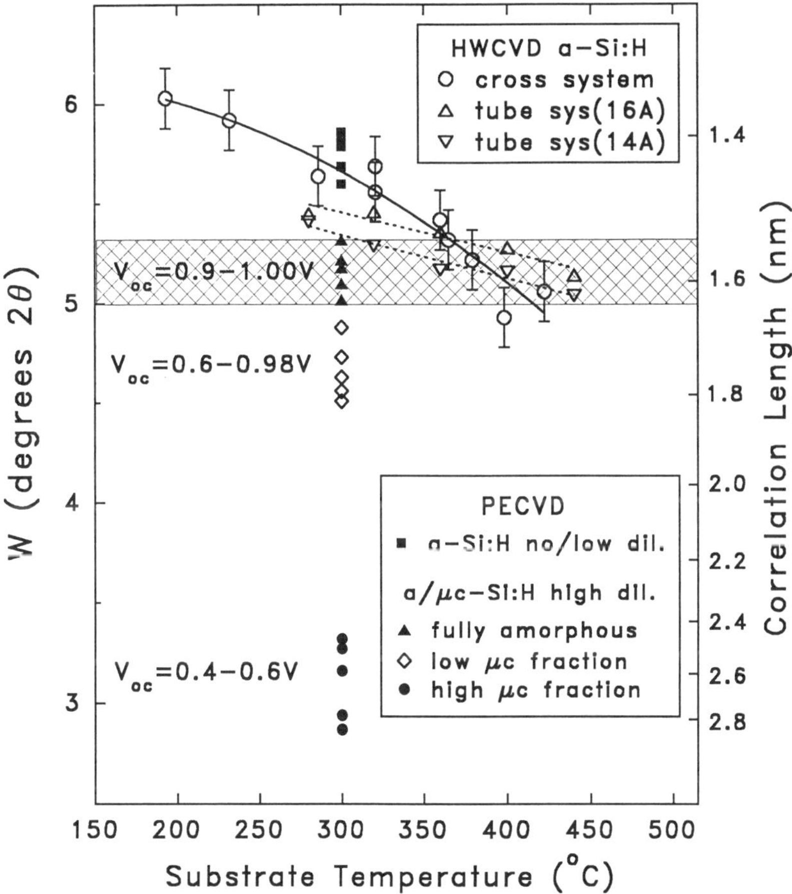

FIG. 5. Comparison of XRD FSP widths for HWCVD and PECVD i-layer films. The HWCVD films were made at various substrate temperatures while the PECVD films were all made at 300°C. The open circuit voltages shown were all obtained from co-deposited PECVD films (via completion of a p-i-n dot solar cells [12]).

SUMMARY

Focus on the first x-ray scattering peak of a-Si:H made by recent state-of-the-art methods has provided direct evidence of improved MRO in the materials presently being used as i-layers in the best solar cells fabricated by each method. These improved materials lie just below the boundary of the transition to the microcrystalline state, consistent with the concept of a proto-crystalline regime [9]. Although PECVD and HWCVD deposition conditions have been developed that result in very similar levels of MRO as determined by XRD, the two materials remain very different in their bonded H content and apparently in their H distributions as detected by NMR [1,16]. There are no HRTEM studies yet that have identified nanoscale ordered regions in the HWCVD material, in contrast to such observations for the PECVD films. Such material containing nanometer-sized ordered features has recently shown evidence of faster crystallization under annealing [10]. This may be a sensitive method for the detection of such features. It would be interesting to probe via cross-sectional TEM the depth distribution and morphology of the microcrystallites that develop in the high-hydrogen-dilution PECVD films as the thickness is increased. Such recent studies of HWCVD films show cone-shaped crystals that appear to extend from near the substrate-film interface to the surface for a film 74 nm in thickness. It would be surprising if this occurs in the samples studied here which can show no evidence of microcrystallinity until thicknesses in excess of 1000 nm are reached.

An important implication of the observed thickness- and substrate-dependent changes in MRO, microcrystalline fraction, and defect density of the material prepared near the a-Si:H/μc-Si:H boundary is that films made for various property measurements requiring differing thicknesses or substrates will likely give conflicting results.

Perhaps modeling studies along the lines recently described [24,25,28] could examine which type of changes in MRO can induce the width changes in the FSP documented here.

Acknowledgements

This work is based on films supplied by groups at United Solar Systems Corporation and the National Renewable Energy Laboratory (NREL). Discussions and input from S. Guha, J. Yang, H. Mahan, Y. Lubianiker, and D. Cohen have been valuable. The work was supported by a subcontract from NREL (#XAK-8-17619-31).

References
1. K-C. Hsu and H-L. Hwang, Appl. Phys. Lett. **61**, 2075 (1992).
2. X. Xu, J. Yang, and S. Guha, J. Non-Cryst. Solids **198-200**, 60 (1996).
3. U. Kroll, J. Meier, A. Shah, S. Mikhailov, and J. Weber, J. Appl. Phys. **80**, 4971 (1996).
4. D.V. Tsu, B.S. Chao, S.R. Ovshinsky, S. Guha, and J. Yang, Appl. Phys. Lett. **71**, 1317 (1997).
5. P. Roca i Cabarrocas, S. Hamma, S.N. Sharma, G. Viera, E. Bertran, and J. Costa, J. Non-Cryst. Solids **227-230**, 871 (1998).
6. J. Koh, H Fujiwara, R.W. Collins, Y. Lee, and C.R. Wronski, J. Non-Cryst. Solids **227-230**, 73 (1998).
7. U. Kroll, J. Meier, P. Torres, J. Pohl, and A. Shah, J. Non-Cryst. Solids **227-230**, 68 (1998).
8. S. Sheng, X. Liao, G. Kong, and H. Han, Appl. Phys. Lett. **73**, 336 (1998).
9. J.H. Koh, Y.H. Lee, H. Fujiwara, C.R. Wronski, and R.W. Collins, Appl. Phys. Lett. **73**, 1526 (1998).
10. E. Bertran, S.N. Sharma, G. Viera, J. Costa, P. St'ahel, and P. Roca i Cabarrocas, J. Mater. Res. **13**, 2476 (1998).
11. T. Kamei, P. Stradins, and A. Matsuda, Appl. Phys. Lett. **74**, 1707 (1999).
12. S. Guha, J. Yang, D.L. Williamson, Y. Lubianiker, J.D. Cohen, and A.H. Mahan, Appl.

Phys. Lett. **74**, 1860 (1999).

13. A.H. Mahan, J. Carapella, B.P. Nelson, and R.S. Crandall, J. Appl. Phys. **69**, 6728 (1991).

14. P. Papadopulos, A. Scholz, S. Bauer, B. Schröder, and H. Oechsner, J. Non-Cryst. Solids **164-166**, 87 (1993).

15. M. Heintze, R. Zedlitz, H.N. Wanka, and M.B. Schubert, J. Appl. Phys. **79**, 2699 (1996).

16. Y. Wu, J.T. Stephen, D.X. Han, J.M. Rutland, R.S. Crandall, and A.H. Mahan, Phys. Rev. Lett. **77**, 2049 (1996).

17. Xiao Liu, B.E. White, Jr., R.O. Pohl, E. Iwanizcko, K.M. Jones, A.H. Mahan, B.N. Nelson, R.S. Crandall, and S. Veprek, Phys. Rev. Lett. **78**, 4418 (1997).

18. A.H. Mahan, D.L. Williamson, and T.E. Furtak, Mat. Res. Soc. Symp. Proc. **467**, 657 (1998).

19. Z. Remes, M. Vanecek, P. Torres, U. Kroll, A.H. Mahan, and R.S. Crandall, J. Non-Cryst. Solids **227-230**, 876 (1998).

20. S. Bauer, B. Schröder, and H. Oechsner, J. Non-Cryst. Solids **227-230**, 34 (1998).

21. A.M. Brockhoff, E.H.C. Ullersma, H. Meiling, F.H.P.M. Habraken, and W.F. van der Weg, Appl. Phys. Lett. **73**, 3244 (1998).

22. S.C. Moss and D.L. Price, in *Physics of Disordered Materials*, edited by D. Adler, H. Fritzsche, and S.R. Ovshinsky (Plenum, New York, 1985) p.77.

23. S.R. Elliott, Nature **354**, 445 (1991); J. Phys.:Condens. Matter **4**, 7661 (1992).

24. L. Cervinka, J. Non-Cryst. Solids **232-234**, 1 (1998).

25. K. Tanaka, Jpn. J. Appl. Phys. **37**, 1747 (1998).

26. J. Bletry, Phil. Mag. B **62**, 469 (1990).

27. J. Dixmier, J. Phys. I France **2**, 1011 (1992).

28. A. Uhlherr and S.R. Elliott, Phil. Mag. B **71**, 611 (1995).

29. M. Essamet, B. Hepp, N. Proust, and J. Dixmier, J. Non-Cryst. Solids **97&98**, 191 (1987).

30. B.E. Warren, *X-ray Diffraction* (Addison-Wesley, Reading MA, 1969).

31. A.P. Sokolov, A. Kisliuk, M. Soltwisch, and D. Quitmann, Phys. Rev. Lett. **69**, 1540 (1992).

32. D.L. Williamson, Mat. Res. Soc. Symp. Proc. **377**, 251 (1995).

33. M.M.J. Treacy and J.M. Gibson, Acta Cryst. **A52**, 212 (1996).

34. J.M. Gibson, M.M.J. Treacy, P.M. Voyles, H-C. Jin, and J.R. Abelson, Appl. Phys. Lett. **73**, 3093 (1998).

35. J. Yang, A. Banerjee, and S. Guha, Appl. Phys. Lett. **70**, 2975 (1997).

36. A.H. Mahan, R.C. Reedy Jr., E. Iwaniczko, Q. Wang, B.P. Nelson, Y. Xu, A.C. Gallagher, H.M. Branz, R.S. Crandall, J. Yang, and S. Guha, Mat. Res. Soc. Symp. Proc. **507**, 119 (1998).

37. P. Roca i Cabarrocas, N. Layadi, T. Heitz, B. Drevillon, and I. Solomon, Appl. Phys. Lett. **66**, 3609 (1995).

38. H. Shirai, Jpn. J. Appl. Phys. **34**, 450 (1995).

39. R. Ossikovski and B. Drevillon, Phys. Rev. B **54**, 10530 (1996).

40. M. Tozlov, F. Finger, R. Carius, and P. Hapke, J. Appl. Phys. **81**, 7376 (1997).

41. L.L. Smith, E. Srinivasan, and G.N. Parsons, J.Appl. Phys. **82**, 6041 (1997).

42. A.H. Mahan, M. Vanacek, A. Poruba, V. Vorlicek, R.S. Crandall, and D.L. Williamson, Mat. Res. Soc. Symp. Proc. **507**, 825 (1998).

43. A.H. Mahan and M. Vanacek, AIP Conf. Proc. **234**, 211 (1991).

44. S. Guha, Current Opinion in Solid State and Materials Science **2**, 425 (1997).

45. M Shima, M. Isomura, E. Maruyama, S. Okamoto, H. Haku, K. Wakisaka, S. Kiyama, and S. Tsuda, Jpn. J. Appl. Phys. **37**, 6322 (1998).

46. R. Platz, S. Wagner, C. Hof, A. Shah, S. Wieder, and B. Rech, J. Appl. Phys. **84**, 3949 (1998).

KINETICS OF LIGHT-INDUCED CHANGES IN P-I-N CELLS WITH PROTOCRYSTALLINE Si:H

R.J. Koval, J. Koh, Z. Lu, Y. Lee[*], L. Jiao, R.W. Collins, and C.R. Wronski
The Center For Thin Film Devices, The Pennsylvania State University,
University Park, PA 16802, USA, crwece@engr.psu.edu

ABSTRACT

Studies have been carried out on the thickness dependent transition between the amorphous and microcrystalline phases in intrinsic Si:H materials (i-layers) and its effect on p-i-n solar cell performance [1]. P(a-SiC:H)-i(a-Si:H)-n(μcSi:H) cell structures were deposited with the intrinsic Si:H layer thickness and the flow ratio of hydrogen to silane, R=[H$_2$]/[SiH$_4$], guided by an evolutionary phase diagram obtained from real-time spectroscopic ellipsometry. The thickness range over which the fill factors are controlled by the bulk was established and their characteristics investigated with different protocrystalline i-layer materials (i.e., materials prepared near the amorphous to microcrystalline boundary but on the amorphous side). Insights into the properties of these materials and the effects of the transition to the microcrystalline phase were obtained from the systematic changes in the initial fill factors, their light-induced changes, and their degraded steady states for cells with i-layers of different thickness and H$_2$ dilution.

INTRODUCTION

Significant progress has been made in improving the performance and stability of a-Si:H based p-i-n and n-i-p solar cells using intrinsic layers prepared with hydrogen dilution of silane [2-5]. It is found that both materials and solar cells prepared with H$_2$ dilution of SiH$_4$ exhibit degradation kinetics distinctly different from their undiluted counterparts. They not only exhibit less degradation under AM1.5 illumination (100mW/cm^2), but they also reach a degraded steady state within approximately 100 hours [5]. These are distinguishing characteristics of what are called protocrystalline materials—a term used to describe a-Si:H grown close to the microcrystalline phase boundary [1,6,7]. However, there are many unanswered questions not only about the growth of these materials but also about the nature and properties of the amorphous phase, and in particular about the material in the optically deduced transition region between the amorphous and microcrystalline regimes.

The optical properties and microstructural evolution of the transition region can be characterized using real-time spectroscopic ellipsometry (RTSE) which can guide systematic studies on further evaluating the basic material properties as well as their effects on the stabilized performance of solar cells [8]. However, it is very difficult to characterize the basic electronic properties of such a transition region even from quite sophisticated thin film measurements due to the limitations of the experimental techniques with respect to depth profiling [9]. Since the protocrystalline materials by their very nature evolve with thickness, the operation of the coplanar structures used in such thin film measurements is such that the transition material is probed in parallel with the amorphous phase rather than in series. Since in p-i-n (n-i-p) solar cells two such constituent materials are effectively in series, the solar cell can be used to obtain insights not only about the "sharpness" of these transitions but also about changes which are on the order of 10^{17}cm^{-3} and cannot be detected with optical techniques such as RTSE. However, since p-i-n (n-i-p) cell structures are complex, it is very important to establish that the cell

Mat. Res. Soc. Symp. Proc. Vol. 557 © 1999 Materials Research Society

characteristics being studied, particularly in the degraded steady state (i.e., "end-of-life"), are determined by the bulk intrinsic layer and not by the p-layer and/or the p/i interface region.

A detailed study was carried out on a variety of a-Si:H p-i-n cells whose structures were guided by an evolutionary phase diagram obtained via RTSE. These cells of different thicknesses were fabricated from two types of p contacts and i-layer materials, in which the i-layers were prepared using H_2 dilution ratios, $R=[H_2]/[SiH_4]$, ranging from 0 to 20. The effects of protocrystalline material and its phase transition to microcrystallinity in the i-layers of the p-i-n cell structures were characterized by studying the solar cell characteristics from their annealed states to their AM1.5 degraded steady states.

EXPERIMENTAL DETAILS

The a-SiC:H (p)/a-Si:H (i)/μc-Si:H (n) structures were deposited in a Tek-Vak MPS 4000 LS multi-chamber rf PECVD system at $T_s=200°C$ on specular SnO_2. Two different a-SiC:H p-layers (p_1 and p_2) were deposited from the same feedstock gases [SiH_4, $B(CH_3)_3$, CH_4], with the only difference being their relative doping ratios. p_1 used a doping ratio of $D=[B(CH_3)_3]/[SiH_4]=.01$ and p_2 used a doping ratio $D=.005$. Two types of intrinsic layers (i_1 and i_2) were incorporated into the structures using the same deposition conditions but with different electrode spacings (1.91 cm and 2.54 cm, respectively) and hence different deposition rates. The majority of the experiments were carried out with the i_2 intrinsic materials deposited at a rate of 0.36Å/s for R=10. The cells were fabricated with 250Å p-type a-SiC:H layers, 350Å n-type μc-Si:H layers, and 1000Å Cr contacts under deposition conditions that have been reported earlier [5].

The light I-V characteristics of the p-i-n solar cells at 25°C were obtained with AM1.5 illumination from an Oriel model 6258 solar simulator. The light-induced degradation kinetics were obtained under illumination by the solar simulator and by ELH tungsten-halogen lamps, with IR and KG2 filters, calibrated for AM1.5. The cells were illuminated until a degraded steady state was reached at 25°C. Similar results were obtained using the solar simulator and the ELH lamps.

RESULTS AND DISCUSSION

RTSE was used to investigate the structural evolution of films deposited with different H_2 dilution ratios R on a-Si:H (R=0) substrates and under conditions similar to those used in the fabrication of the p-i-n solar cells. The deduced structural evolution identifies the regimes of film thickness and H_2-dilution ratio within which the a-Si:H and the μc-Si:H phases are obtained. The resulting deposition phase diagram was then used to guide the fabrication of the i-layers for the different cell structures in this study.

Figure 1 shows the surface roughness layer thickness (d_s) versus bulk layer thickness (d_b) for three Si films prepared with R=15, 30, and 80. The initial roughness value at the first

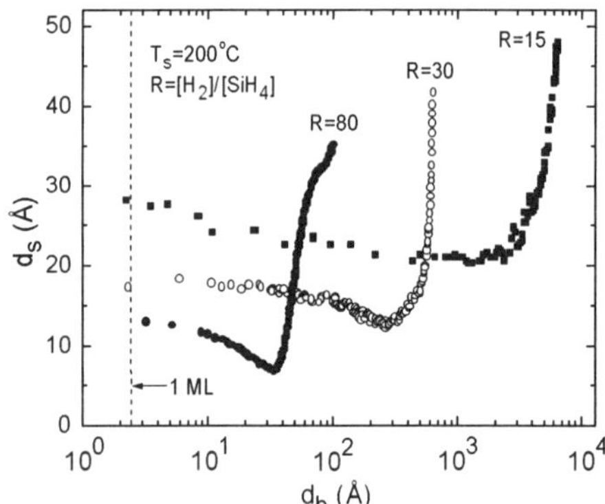

Figure 1. Surface roughness layer thickness (d_s) versus bulk layer thickness (d_b) obtained from RTSE measurements during the growth of three Si:H films prepared with $R=[H_2]/[SiH_4]$ of 15, 30, and 80. The underlying substrate is a-Si:H prepared with R=0. The vertical line at $d_b=2.5$Å corresponds to a monolayer thickness.

monolayer of bulk film growth (d_b=2.5Å, vertical broken line) is controlled by the roughness of the underlying R=0 a-Si:H substrate film. The key features in Figure 1 are the abrupt increases in surface roughness layer thickness which signify the amorphous to microcrystalline transition for the different R values. These results give rise to an evolutionary phase diagram shown in Figure 2. This phase boundary separates the deposition of a-Si:H (left) and μc-Si:H (right) thin films as a function of layer thickness d_b and H_2 dilution ratio R on a-Si:H (R=0) substrates. Even though the intrinsic layers of the p-i-n cells used in this study were deposited in a different processing chamber and onto different substrates (p-type a-SiC:H), these results can be used as a guide based on the correleations described later in this section. Although studies reported below

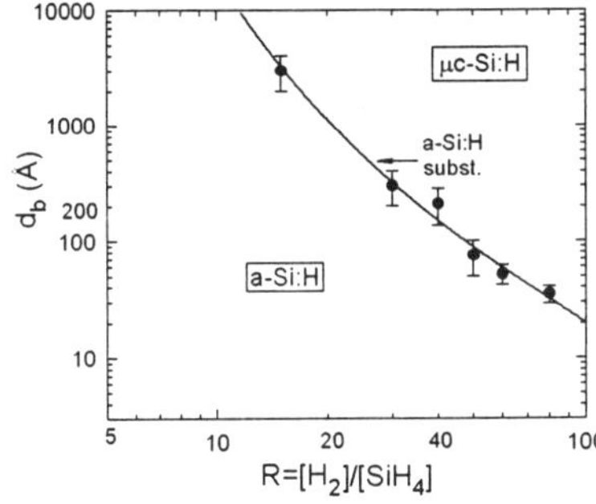

Figure 2. Evolutionary phase boundary (solid line) separating the deposition of a-Si:H (left) and μc-Si:H (right) as a function of layer thickness d_b and H_2 dilution ratio R on an a-Si:H (R=0) substrate.

suggest that for R≥15 the phase boundary is the same for a-SiC:H p-layers and a-Si:H (R=0) film substrates, additional studies need to be performed at lower R values.

The information about the protocrystalline materials and the onset of the phase transition was obtained primarily from the fill factors of the different cell structures. As a result, studies were carried out first to establish the thickness range of the i-layers over which the fill factors in the degraded steady state were determined solely by the bulk. As a first step, the dependence of fill factor on the i-layer thickness in the annealed state was investigated for R=10 cells with the same p-layers and p/i interfaces. The results obtained with p_2-i_2 (R=10)-n structures are shown in Figure 3. There is a systematic increase in the annealed state fill factors from 0.67 at 5500Å to 0.75 at 1800Å after which there is no further increase. Similarly, the fill factors in

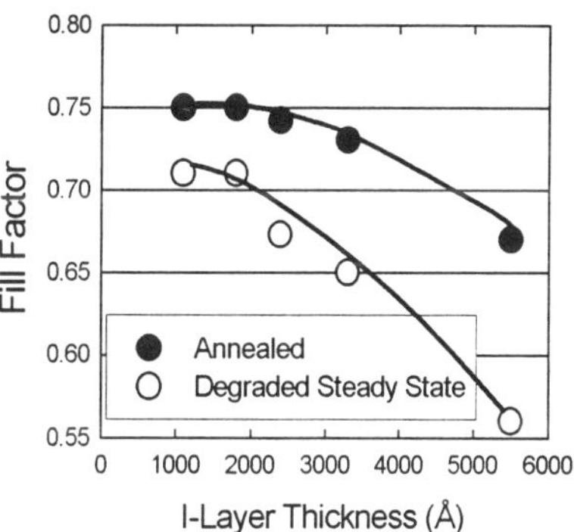

Figure 3. Annealed and degraded steady state fill factors of R=10 p_2-i_2-n cells versus cell thickness.

the degraded steady state increase from 0.56 at 5500Å to 0.71 at 1800Å. These results indicate that above a thickness of 1800Å the fill factors are not limited by the contacts even in the annealed state. It is pointed out here that in this region excellent fill factors are obtained in the degraded steady state, for example 0.71 for 1800Å and 0.65 for 3300Å thick i-layers.

Even though these results indicate bulk domination of the fill factors, it is necessary to ensure that there are no effects on the degraded steady state fill factor due to possible light induced changes in the material near the p/i interfaces [10]. The absence of such effects is indicated by the stability of the open circuit voltages to prolonged illumination. This is illustrated in Figure 4 where the open circuit voltages are shown as a function of AM1.5 illumination time for different thickness p_2-i_2-n cells with an R=10 i-layer. Since V_{oc} is very sensitive to the p/i interface region [11], the results in Fig. 4 lead to the conclusion that there is no effect of the p/i interface on the stabilized fill factors.

The stability of the V_{oc} under AM1.5 illumination indicates that there are not major increases in the defect density in the i-layer near the p/i interface region, a fact that results in fill factors dominated by the bulk i-layer in the degraded steady state. The absence of such an increase in defect density at the p/i interface as well as a contribution of the p-layer to the degraded steady state of the fill factor was further established by comparing the degradation kinetics of cells with different p-layer contacts. Even though the initial fill factors were up to 0.04 higher for cells having p_2 layers and thicknesses from 1800 to 3300Å (compared to cells with p_1 layers) the fill factors were the same in the degraded steady state. This is illustrated in Fig. 5 with the results obtained on two p-i_1-n solar cells with a 3300Å thick R=10 i-layers.

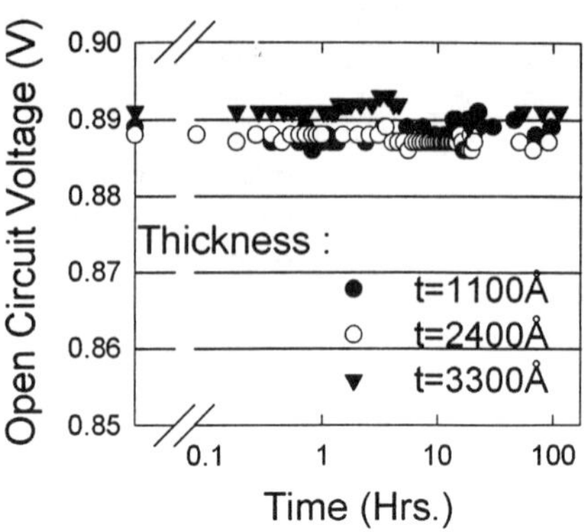

Figure 4. AM1.5 25°C degradation of the V_{oc} of p_2-i_2 (R=10)-n cells of different thickness.

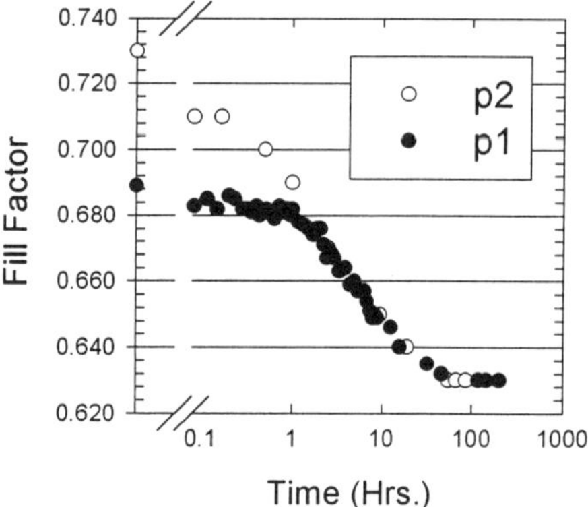

Figure 5. AM1.5 25°C degradation of p-i_1(R=10)-n cells having different p-layers.

To assess what improvements in cell performance and stability occur with increasing hydrogen dilution and the corresponding changes in the protocrystalline intrinsic materials, p-i-n cells were studied in which the 3300Å i-layer was deposited with R=0 to R=20. These cells exhibit improved AM1.5 stabilized fill factors as R increases from 0 to 10 and then remain the same up to R=20. This is illustrated in Figure 6 where the degradation kinetics are shown from the annealed to the degraded steady states for p_1-i_1-n cells with R=5, R=10 and R=20 intrinsic layers. Figure 6 also illustrates that the fill factors in the annealed state are constant below R=10 and then decrease significantly for R>10. The observed decreases in the initial fill factor for R>10 are not related to any degradation of the p/i interface region since the insertion of a 200Å R=20 p/i interface region in a cell with R=10 bulk i-layers leads to improved open circuit voltages in both the initial and degraded steady state [11]. The same trend is also present in the p-i-n cells with i_2 intrinsic layers. Based on the deposition phase diagram, however, such decreases in annealed state fill factors can be attributed to the phase transitions in the i-layers.

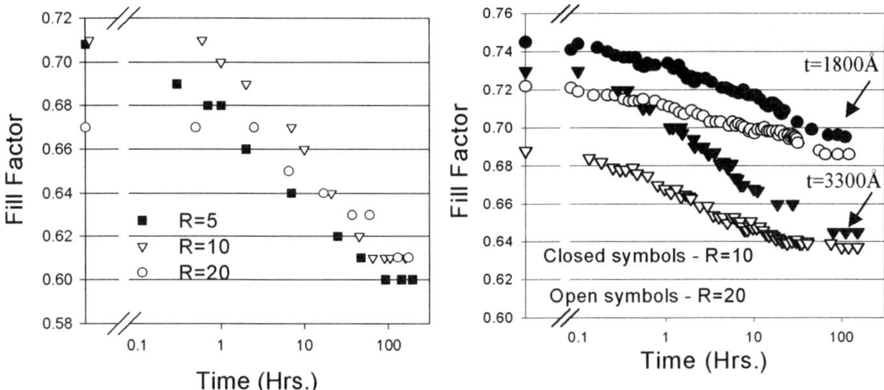

Figure 6. AM1.5 25°C degradation of 3300Å
p_1-i_1-n cells with R=5, 10, 20.

Figure 7. AM1.5 25°C degradation of 1800Å
and 3300Å p_2-i_2-n cells with R=10 and 20.

Specifically, Fig. 2 shows that the phase boundary to microcrystallinity for R=15 and R=20 occurs at thicknesses of about 3000Å and 1000Å, respectively.

To further assess the nature of protocrystalline materials (i.e., materials prepared just on the amorphous side of the boundary), the onset of the phase transition, and their effects on solar cell characteristics, cells with R ranging from 10 to 20 and i-layer thicknesses $\leq$ 3300Å were investigated. It is found that as the thickness of the i-layers and hence the contribution from the material in the transition region decreases, the values of the annealed state fill factors of the R=15 and R=20 cells approach the values for the R=10 cells. This is illustrated in Fig. 7 where results are shown for cells with 1800Å and 3300Å thick i_2 layers having R=10 and 20. Because the fill factors in cells less than 1800Å are limited by the contacts, the initial and degraded steady state fill factors obtained with 1100Å thick i-layers with R=10 and R=20 were the same. Because of this, the changes that accompany the transition with decreasing thickness across the boundary in Fig. 2 into the amorphous regime could not be evaluated for R=20. Such a transition with decreasing thickness could be studied, however, with R=15 i-layers because a thickness less than 3000Å should be amorphous according to Figure 2. In this case R=10 and 15 cells with 2400Å and 1800Å thick i-layers gave identical fill factors both in the annealed and degraded states and their light induced changes were identical as well. These results clearly demonstrate the absence of deleterious effects on the annealed state of the materials associated with the phase transition range. The results also indicate that in this region of thickness the properties of the protocrystalline materials fabricated with R=10 and R=15 are indistinguishable. Thus, there appears to be no significant benefit in increasing R from 10 to 15 while remaining in the amorphous regime (thicknesses less than 3300Å). This is consistent with optical studies that have shown that the largest improvement in ordering occurs in the protocrystalline regime of larger R (~40) and thinner layers (~200Å) [1].

CONCLUSIONS

A study has been carried out on p-i-n solar cells with protocrystalline intrinsic materials which was guided by an evolutionary phase diagram obtained from real-time optics. New information was obtained about the nature of the protocrystalline materials and the transitions to the microcrystalline phase. In this study, a range of cell thicknesses was identified for which the fill factors could be directly related to the bulk i-layer material. From results on fill factors in the annealed and AM1.5 degraded steady states, the presence and the effects of the transition region from the amorphous to the microcrystalline phase were evaluated. The results clearly show that the i-layer materials with R=15 and 20 deposited to a thickness that spans the amorphous to microcrystalline transition have inferior properties in the annealed state but exhibit higher stability than materials that remain amorphous throughout the entire thickness. By maintaining the amorphous phase throughout the entire thickness for R=15 intrinsic layers it was found that their contributions to cell performance and stability were indistinguishable from R=10 materials. Additional studies must be carried out to further characterize the protocrystalline materials and the nature of the phase transitions on the nanoscale and its effects on defect densities less than 10^{17} cm^{-3}.

ACKNOWLEDGMENTS

The authors wish to acknowledge the many helpful discussions with Dr. G. Ganguly. This research was supported by the National Renewable Energy Laboratory (NREL) under subcontract # XAF-8-17619-22, the Electric Power Research Institute (EPRI) Thin Film Solar Cell Program, and the New Energy Development Organization (NEDO) International Joint Research Grant.

REFERENCES

1.   J. Koh, Y. Lee, H. Fujiwara, C.R. Wronski, R.W. Collins, Appl. Phys. Lett. **73**, 1526 (1998).

2.   J. Yang, X. Xu, and S. Guha, Mater. Res. Soc. Symp. Proc. **336**, 687 (1994).

3.   M. Bennett, K. Rajan, and K. Kritikson, Conf. Record of 23rd IEEE PVSC, (IEEE, New York, 1994), p. 845.

4.   L. Yang and L.-F. Chen, Mater. Res. Soc. Symp. Proc. **336**, 669 (1994).

5.   Y. Lee, L. Jiao, H. Liu, Z. Lu, R.W. Collins, and C.R. Wronski, Conf. Record of the 25th IEEE PVSC, (IEEE, New York, 1996), p.1165.

6.   Y. Lu, S. Kim, M. Gunes, Y. Lee, C.R. Wronski, R.W. Collins, Mater. Res. Soc. Symp. Proc. **336**, 595 (1994).

7.   D.V. Tsu, B.S. Chao, S.R. Ovshinsky, S. Guha, and J. Yang, Appl. Phys. Lett. **71**, 1317 (1997).

8.   Joohyun Koh, Y. Lu, S. Kim, J. S. Burnham, C. R. Wronski, and R. W. Collins, Appl. Phys. Lett. **67**, 2669 (1995).

9.   L. Jiao, Ph.D. Thesis, The Pennsylvania State University, 1998.

10.   R. Platz, D. Fischer, C. Hof, S. Dubail, J. Meier, U. Kroll, and A. Shah, Mater. Res. Soc. Symp. Proc. **420**, 51 (1996) .

11.   Y. Lee, A.S. Ferlauto, Z. Lu, J. Koh, H. Fujiwara, R.W. Collins, and C.R. Wronski, 2nd World Conference and Exhibition on Photovoltaic Solar Energy Conversion, (Arte Stampa, Daverio [VA], 1998), p. 940.

* Currently at Silicon Storage Technology, Inc., 1171 Sonora Court, Sunnyvale, CA 94086.

STRUCTURAL CHANGES IN a-Si:H FILMS DEPOSITED ON THE EDGE OF CRYSTALLINITY

A.H. MAHAN, J. YANG*, S. GUHA*, AND D.L. WILLIAMSON**
National Renewable Energy Laboratory, Golden CO 80401
United Solar Systems Corp., Troy, MI 48084
Colorado School of Mines, Golden, CO 80401

ABSTRACT

Using infrared, H evolution and x-ray diffraction (XRD), the structure of high H dilution, glow discharge deposited a-Si:H films 'on the edge of crystallinity' is examined. From the Si-H wag mode peak frequency and the XRD results, we postulate the existence of very small Si crystallites contained within the as-grown amorphous matrix, with the vast majority of the bonded H located on these crystallite surfaces. Upon annealing at ramp rates of 8-15°C/min, a H evolution peak at ~400°C appears, and film crystallization is observed at temperatures as low as 500°C, both of which are far below those observed for a-Si:H films grown without H dilution using similar rates. While the crystallite volume fraction is too small to be detected by XRD in the as-grown films, these crystallites enable the crystallization of the remainder of the amorphous matrix upon moderate annealing, thus explaining the existence of the low temperature H evolution peak.

INTRODUCTION

Hydrogenated amorphous silicon (a-Si:H) films, deposited by the glow discharge (GD) process using silane highly diluted in hydrogen, have become increasingly important both from a technological as well as from a scientific point of view. Since the first observation that H dilution led to an improved stability (1), this process has been used extensively in the production of a-Si:H films and solar cells. It has been well established not only (a) that increasing H dilution leads eventually to the deposition of microcrystalline silicon (μc-Si) (2), but also (b) that the improved stability of the a-Si:H films as well as solar cells, deposited using H dilution, has been obtained using films deposited 'on the edge of crystallinity' (3,4). However, several experimental factors such as thickness and substrate effects (5,6), suggest that it is not easy to identify and investigate this material in a cohesive (consistent) fashion.

Nevertheless, new information is starting to emerge regarding the properties of thcsc 'on the edge' materials. A succession of papers has examined properties of 'on the edge' a-Si:H films deposited by the same group using high H dilution. Xu et. al. (7) first reported the existence of a low temperature H evolution peak and an anisotropic (columnar-like) structure effect as measured by SAXS. Tsu et. al. (4) then reported, as seen by TEM, the existence of large linear-like objects; these were interpreted, with the help of a shifted Raman peak, to be intermediate in order between the continuous random network (amorphous) and crystalline phases. These (randomly oriented) objects appeared even for films deposited using undiluted disilane, and the number of these objects increased at high H dilution. Finally, Guha et. al. (8) showed that the width of the first x-ray diffraction (XRD) peak narrowed at high H dilution, and that this narrowing was suggestive of an improvement in medium range order.

In this work we examine, from a different perspective, 'on the edge' films deposited by the same group using infrared (IR) spectroscopy, XRD, H evolution, and (partial) annealing. We find that, even though the XRD shows no evidence of crystallinity and the peak frequency of the IR active stretch mode remains at ~2000 cm-1, indicating monohydride bonding associated with device quality a-Si:H, the peak frequency of the Si-H wag mode shifts downward from ~640 cm-1 to 620 cm-1 at high H dilution. This latter frequency is identified as being due to H bonded on crystalline Si surfaces (9). This shift in peak position suggests that, even though the matrix is still almost completely amorphous, the vast majority of the H is bonded on the surfaces of Si crystallites. We show that the only way to simultaneously satisfy the IR (H content) and XRD (sensitivity) results is for these crystallites to be very small and to be randomly oriented. Under

Mat. Res. Soc. Symp. Proc. Vol. 557 ©1999 Materials Research Society

these conditions, we estimate that the crystallite volume fraction (vf) can be ~ 5-10%. Although this vf is still too small for the grain boundary precursors to serve as H percolation (evolution) pathways, thus explaining the existence of the low temperature H evolution peak, the crystallites serve as nucleation sites for further crystallization upon sample annealing. That is, as there is no nucleation barrier to overcome upon sample heating, such samples can be expected to crystallize at annealing temperatures far lower than samples which are entirely amorphous (10). Indeed, crystallinity is detected some 200°C lower in 'on the edge' samples as compared to fully amorphous samples comparably annealed (11), thus (indirectly) confirming the existence of these small Si crystallites.

EXPERIMENT

An extensive series of intrinsic films were deposited by the glow discharge (GD) process at varying levels of H dilution, all at a deposition temperature of 300°C, onto c-Si substrates. Three different H dilutions were used; 'no dilution', 'low dilution', and 'high dilution'. The 'high dilution' films used in solar cells demonstrate high efficiency and improved stability (3). Sample thicknesses ranged between 0.5-1.05μm. IR measurements were performed using a dual beam Perkin-Elmer 580-B Spectrometer, with an instrumental resolution of 2.3 cm-1. Transmission data, after appropriate baseline subtraction, were transformed into absorption peak profiles using the analysis of Brodsky et. al. (12). H evolution was performed using ramp rates between ~6-14°C/min, with the rates adjusted such that the total time in the H evolution furnace for each (piece of) film was ~ 1hr. For step-wise evolution experiments on films which were later examined for crystallinity, the samples were promptly removed from the H evolution apparatus when the ramp temperature reached the appropriate value; these temperatures were 400, 500, and 700°C respectively.. The XRD measurements were carried out with a Siemens D-500 Diffractometer operating with Cu-Kα radiation in the Bragg-Brentano geometry. Long counting times were used to achieve good signal-to-noise ratios.

RESULTS

Fig. 1 shows IR spectra for (a) the Si-H stretch and (b) wag modes for the 'no dilution' and 'high dilution' films; both film thicknesses were ~1.0 μm. The 'high H dilution' film was seen by SEM to lift off from its substrate while remaining intact structurally. While there exists some difference in the stretch mode shape, the peak frequencies are both centered around 2000 cm-1, indicating predominantly monohydride (Si-H) bonding, and thus state-of-the-art device quality. From Fig. 1b, the film H contents (CH) do not change appreciably with dilution; that is, the CH's are 9.6, and 8.2 at.% respectively for the 'no dilution' and 'high dilution' films. The wag mode full width at half maximums (FWHM) for the films are also almost identical (100 cm-1), a value which is typical of device quality a-Si:H films containing similar CH. On the other hand, the wag mode peak frequency is seen to shift from 635-640 cm-1 for the 'no dilution' film to ~620 cm-1 for the 'high dilution' film. While the former peak frequency is typical for 'standard' a-Si:H (13), the latter lies distinctly outside this frequency range.

We consider it significant that the wag frequency of H bonded in the monohydride (Si-H) configuration on both the c-Si (100) and (111) surfaces is identically at ~620 cm-1 (9). While H bonded in other (SiH2, (SiH2)n) configurations on c-Si can show a wide variety of peak frequencies (from 620 cm-1 to as high as 675 cm-1) (14), for purely Si-H monohydride bonding only one wag mode frequency is observed. Hence, combining these observations with the present measurements, we assert that the IR frequency shift for the 'high dilution' film indicates the presence of crystallites contained within the amorphous matrix, with the large majority of the H bonded on these crystallite surfaces.

This is to be contrasted with XRD data. This is seen in Fig. 2, where the XRD intensity for the 'no dilution' and 'high dilution' sample is plotted vs. the scattering angle 2θ from 15°-60°. Also noted are the angular positions where the first three peaks in powdered (randomly

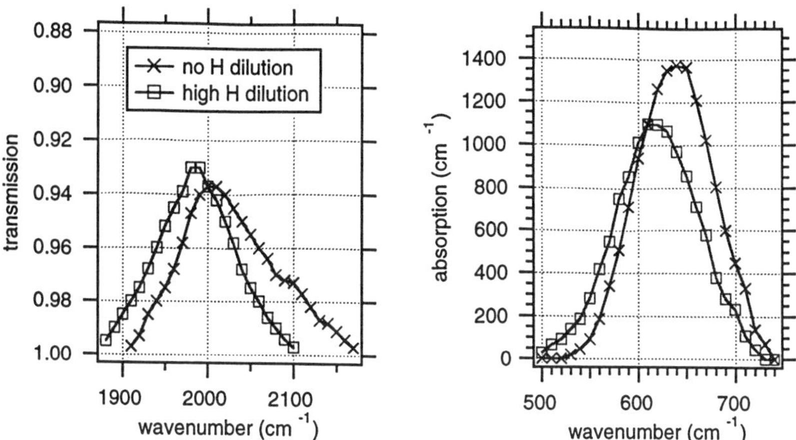

Figure 1. IR stretch mode peak position (a) and wag mode absorption spectra (b) for the 'no dilution' and the 'high dilution' a-Si:H samples.

oriented) c-Si (the (111), (220), and (311) crystal orientations) would occur. The breaks in the curves correspond to second order diffraction effects from the c-Si substrate, which are omitted. As can be seen, the only features observed in the spectrum are the strong (first amorphous) peak occurring at roughly 28.5°, and a less intense (second amorphous) peak occurring at ~51°. The substrate reference is included for comparison. Based on these results, the film appears to be fully amorphous.

The problem, then, is to reconcile these two sets of measurements. That is, we must, for the 'high dilution' film, understand how to incorporate enough crystallinity into the sample to enable enough H bonding on the surfaces of these crystallites to account for the majority (> 5 at.%) of the bonded H (thus satisfying the wag mode peak frequency shift), while at the same time not generating any detectable crystalline features in the XRD spectrum. We address this by considering crystallite size. From the Scherrer formula and the XRD sensitivity, we estimate that large c-Si crystallites (>50Å), yielding sharp diffraction peaks, can easily be distinguished from any (broad) amorphous diffraction signal with a sensitivity of 1-2% in vf. On the other hand, the first XRD peak FWHM of the 'high dilution' a-Si:H film is ~5° (8), so if the dimension of any crystallites contained within the film were reduced from the >50Å size previously discussed to ~15-20Å, the XRD FWHM of these crystallites now becomes comparable to the width of the amorphous XRD signal. The only method of distinguishing these (smaller) crystallites from the (first XRD peak) amorphous feature now becomes the very small difference in peak positions, and crystallite detection is now more difficult than before, enabling an upper 'adjustment' in the vf limit. Going further with this argument, the existence of even smaller crystallites, on the order of ~10Å in diameter, which produce even broader XRD features, would be even more difficult to detect.

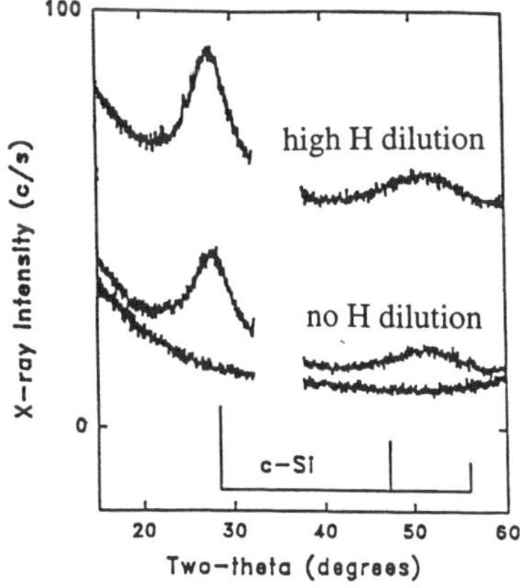

Figure 2. X-ray intensity vs. 2θ (°)

271

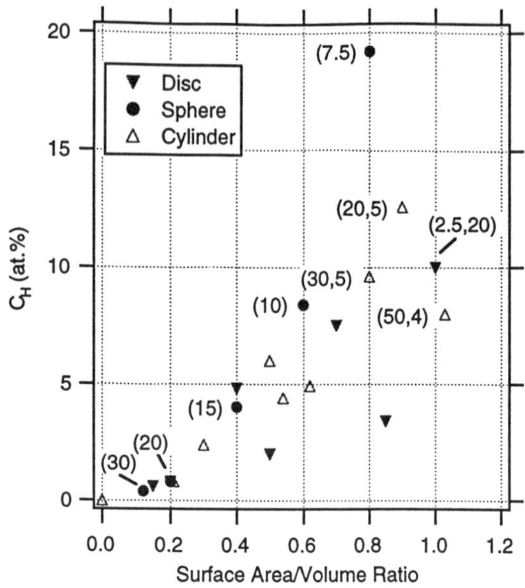

Figure 3. CH vs. crystallite (surface area/volume) ratio.

Thus, size arguments alone can enable the existence of a considerably larger real crystallite volume fraction than that given by our initial estimate (1-2%). In Fig. 3 we summarize numerically the arguments made in the preceding paragraph. In making this figure, we have assumed an upward sliding scale in vf vs. decreasing crystallite size, with a sensitivity limit of 5% for crystallite sizes of 15-20Å (yielding XRD FWHM's of ~5°), and a limit of 8% for even smaller crystallites having sizes of 10-15Å (yielding XRD FWHM's of >9°). Using Aw = 2.3e19/cm3 (15) and assuming a H coverage density of 1e15/cm2 on c-Si surfaces (16), we plot the total amount of H on these crystallites vs. the crystallite (surface area/volume) ratio. We consider spheres, discs, and cylinders, with representative dimensions in Å (length, diameter) noted in the figure; for crystallites with a large eccentricity, we assume that they are randomly oriented in the film (7) and take the average of the three dimensions to be the size detected by XRD (17). We note some scatter in the data. This is ascribed to different surface area/volume ratios for crystallites of different shapes as well as the fact that crystallites having one large dimension weight the averaged dimension as seen by XRD to larger values; hence, in our model, a lower CH is thus observed due to the higher XRD sensitivity. However, the overall trend is easily seen. That is, large crystallites have small (surface area/volume) ratios, and we are unable to put enough bonded H on the surfaces to satisfy the IR results, while for small crystallites with large (surface area/volume) ratios, we exceed our 5 at.% H limit for all three shapes. Thus, we are able to satisfy the XRD and the IR results simultaneously by considering the inclusion of small crystallites within an amorphous matrix

Finally, given the presence of small crystallites in our 'high dilution' a-Si:H film (up to a vf = 5-10%), we now examine in this context the H evolution results (6). That is, we need to explain the existence of a H evolution peak which occurs at a temperature ~100°C lower than that observed in 'standard' a-Si:H films of similar thickness (18). To do this, we partially anneal pieces of our films up to 400°C, 500°C, and 700°C respectively in our H evolution furnace, and then examine them for crystallinity. While the 'high H dilution' film remains amorphous after the 400°C anneal, crystallinity is observed after the 500°C anneal, and after the 700°C anneal the film appears to be almost fully crystallized. From the IR, 40% (75%) the bonded H is lost after

film appears to be almost fully crystallized. From the IR, 40% (75%) of the bonded H is lost after the 400°C (500°C) anneal, while no bonded H remains after the 700°C anneal. This is to be contrasted with results for 'standard' a-Si:H, which lose almost no bonded H at low (400°C) anneal temperatures and crystallize only at very high temperatures (>750°C) (11). In addition, low CH films deposited by the hot wire (HW) technique, which exhibit as similar narrowing of the first XRD peak (19), also show no crystallinity when annealed in a similar fashion.

DISCUSSION

While acknowledging the strong substrate dependence on film structural properties, as well as the sensitivity of film structure to small changes in film deposition conditions for films in this "on the edge" region, we nevertheless comment on how the present results add to the structural picture emerging for these films. In particular, they add to and extend the results of Tsu et al. (7). In that publication, TEM and Raman measurements showed the existence of "somewhat ordered linear-like objects", which were randomly oriented, and which made up from Raman about 10-15% of the total Raman signal. Assuming that this percentage relates to volume fraction, the similarity of those results to that of the present work, particularly the magnitude of the vf, is highly suggestive. In particular, we suggest that the "objects" discussed by Tsu et al. (7) may in fact be crystallites, that those objects observed by TEM must be those at the very extreme (upper) end of a Gaussian size distribution, and that most of these crystallites must be very small in size. Further, we show how the presence of crystallites in an a-Si:H matrix explains the existence of the low temperature H evolution peak seen by Xu et al. (6). It is well known, for a low temperature H evolution peak to occur, that some sort of connective path to the surface (microvoids, grain boundaries) is needed. By both accounts (Mahan, Tsu), the vf of the present as grown films (5-10%, 10-15%) is not large enough for percolation to occur unless these objects are highly oriented; yet, in both cases this preferential orientation is not seen to occur. On the other hand, if small crystallites already exist in the as grown material, there is no nucleation barrier to overcome upon low temperature sample annealing (10). Therefore, the existing crystallites can grow rapidly in size with moderate annealing, enabling H to effuse out of the film along the suddenly accentuated grain boundaries, and thus contribute to the low temperature H evolution peak. However, these are not grain boundaries in the traditional sense, as they exhibit only one of the two IR signatures commonly observed (the wag mode shift). That is, traditional grain boundaries in μc-Si show a shift in both the stretch and wag mode peak frequencies (2); as mentioned previously, the peak frequency of the stretch mode of the present "high dilution" sample remains at 2000 cm-1, which is not indicative of H bonded on c-Si surfaces. We note, however, that the present results are in fact not in contradiction to the existing literature, as Rath et al. (20) have shown that very compact grain boundaries (compact in the sense that the stretch mode peak position is still at 2000 cm-1) can in fact occur in μc-Si.

Finally, we comment on the better lattice ordering, postulated to occur by Tsu et al. (7) and recently observed by XRD by Guha et al. (8) for the "high dilution" films. Both publications agree that a more ordered structure is produced using high H dilution. We note, however, that a more ordered structure has been previously been obtained, as observed by XRD, with low CH films deposited by the HW technique (19). In those films, NMR measurements (21) showed that H clustering in the HW films was so enhanced, and the CH so low, that there existed large spatial regions without appreciable H; from XRD, these regions were observed to be better ordered. We postulate that the same thing is happening in these high H dilution GD films, but for a different reason. What may be needed deposition-wise to produce the narrower XRD FWHM is a method to produce a H spatial inhomogeneity throughout an a-Si:H film; what results are regions containing minimal H with better ordering. In the HW films, on the one hand, higher H clustering and lower CH produced these better ordered (devoid of H) regions, while for the present high H dilution films, H preferentially bonded on the surfaces of the small crystallites serves the same function.

CONCLUSIONS

We have examined GD a-Si:H films, deposited "on the edge of crystallinity" using high H dilution, by IR, H evolution and XRD. The Si-H wag mode peak position, along with the XRD results, are interpreted as evidence for the existence of very small Si crystallites contained within the as-grown amorphous matrix, with the vast majority of the bonded H located on these crystallite surfaces. Upon annealing, a low temperature H evolution peak appears, and film crystallization is observed at temperatures as low as 500°C, which is far below that observed for a-Si:H films deposited without H dilution. While the crystallite vf is too small to be detected by XRD in the as-grown films, the crystallites catalyze the crystallization of the remainder of the amorphous lattice after annealing, thus providing a connectivity path to the surface and explaining the existence of the low temperature H evolution peak.

ACKNOWLEDGEMENTS

The authors thank F. Hassoon and M. al Jassim for SEM measurements, W. Beyer for many stimulating discussions and ideas relating to this work, and L. Gedvillas and J. D. Webb for IR measurement support. The research at NREL, United Solar, and the Colorado School of Mines was supported by the U. S. Department of Energy under subcontracts DE-AC36-98-GO10337, ZAK-8-17619-09, and XAF-8-17619-05 respectively.

REFERENCES

1. S. Guha, K. L. Narasimhan, and S. M. Pieruszko, J. Appl. Phys. 52 (1981) 859.
2. U. Kroll, J. Meier, A. Shah, S. Mikhailov, and J. Weber, J. Appl. Phys. 80 (1996) 4971.
3. J. Yang, A. Banerjee, and S. Guha, Appl. Phys. Lett. 70 (1997) 2975.
4. D. V. Tsu, B. S. Chao, S. R. Ovshinsky, S. Guha, and J. Yang, Appl. Phys. Lett. 71 (1997) 1317.
5. P. Roca i Cabarrocas, N. Layadi, T. Heitz, B.Drevillon, and I. Solomon, Appl .Phys. Lett. 66 (1995) 3609.
6. J. Koh, Y. Lee, H. Fujiwara, C. R. Wronski, and R. W. Collins, Appl. Phys. Lett. 73 (1998) 1526.
7. X. Xu, J. Yang, and S. Guha, J. non-Cryst. Sol. 198-200 (1996) 60.
8. S. Guha, J. Yang, D. L. Williamson, Y. Lubianiker, J. D. Cohen, and A. H. Mahan, Appl. Phys. Lett. 74, 1999, in press.
9. H. Wagner and W. Beyer, Solid State Comm. 48 (1983) 585.
10. R. B. Iverson and R. Reif, J. Appl. Phys. 62 (1987) 1675.
11. W. Beyer, private communication.
12. M. H. Brodsky, M. Cardona, and J. J. Cuomo, Phys. Rev. B 16 (1977) 3556.
13. W. Beyer, in <u>Semiconductors and Semimetals</u>, N. Nickel, Ed. (Academic Pr., 1999), p.165.
14. J. A. Schaefer, T. Balster, V. Polyakov, U. Rossow, S. Sloboshanin, U. Starke, and F. S. Tautz, MRS Proc. 513 (1998) 3.
15. A. A. Langford, M. L. Fleet, B. P. Nelson, W. A. Lanford, and N. Maley, Phys. Rev. B45 (1992) 13367.
16. R. J. Culbertson, L. C. Feldman, P. J. Silverman, and R. Haight, J. Vac.Sci. Tech. 20 (1982) 868.
17. B. E. Warren, <u>X-Ray Diffraction</u>, Dover Publications (New York, 1990), pg. 256.
18. W. Beyer, in <u>Tetrahedrally Bonded Amorphous Semiconductors</u>, D. Adler and H. Fritzsche, Eds. (Plenum, 1985), pg. 129.
19. A. H. Mahan, D. L. Williamson, and T. E. Furtak, MRS Symp. Proc. 467 (1997) 657.
20. J. K. Rath, A. Barbon, and R. E. I. Schropp, J. non-Cryst. Solids 227-230, 1277 (1998).
21. Y. Wu, J. T. Stephen, D. X. Han, J. M. Rutland, R. S. Crandall, and A. H. Mahan, Phys. Rev. Lett. 77 (1996) 2049.

THEORY OF HYDROGEN INTERACTIONS WITH AMORPHOUS SILICON

CHRIS G. VAN DE WALLE* and BLAIR TUTTLE**
*Xerox Palo Alto Research Center, 3333 Coyote Hill Road, Palo Alto, California 94304;
vandewalle@parc.xerox.com
**Department of Physics, Computational Electronics Group, Beckman Institute, University
of Illinois, Urbana, IL 61801

ABSTRACT

We present an overview of recent results for hydrogen interactions with amorphous silicon (a-Si), based on first-principles calculations. We review the current understanding regarding molecular hydrogen, and show that H_2 molecules are far less inert than previously assumed. We then discuss results for motion of hydrogen through the material, as relating to diffusion and defect formation. We present a microscopic mechanism for hydrogen-hydrogen exchange, and examine the metastable $\equiv SiH_2$ complex formed during the exchange process. We also discuss the enhanced stability of Si-D compared to Si-H bonds, which may provide a means of suppressing light-induced defect generation.

INTRODUCTION

We have recently witnessed a resurgence of interest in the properties of hydrogen in semiconductors in general, and in amorphous silicon in particular. Several exciting developments have spurred this renewed interest: focusing on a-Si, they include new information about Si-H bond dissociation, hydrogen diffusion, and new models for light-induced defect generation; the potential for more stable defect passivation by using deuterium instead of hydrogen; and progress in the ability to experimentally observe interstitial H_2 molecules. First-principles computations have significantly contributed to our understanding of hydrogen-related phenomena. In this paper we will focus on recent work closely connected to the issues outlined above. The computational results have all been obtained using a state-of-the-art first-principles approach based on density-functional theory, *ab initio* pseudopotentials, and a supercell geometry.

In the first part of the paper, we will focus on hydrogen molecules. We will review experimental observations of interstitial H_2 molecules in crystalline and amorphous semiconductors, and describe the theoretical framework for understanding the physics of incorporation of a strongly bound molecule in a semiconducting environment.

The second part of the paper will deal with hydrogen motion, as occurs in diffusion and in light-induced defect generation. We first discuss an exchange process between trapped and interstitial H that plays a significant role in diffusion. We have determined a low-energy pathway for exchange which involves an intermediate, metastable $\equiv SiH_2$ complex with both H atoms strongly bound to the Si atom. The energy barrier for the exchange is less than 0.2 eV, consistent with observations of hydrogen-deuterium exchange in a-Si:H(D) films. We also discuss potential implications of the $\equiv SiH_2$ complex for metastability and defect generation. On the issue of stability of Si-H bonds, finally, we discuss the dissociation path and the connection to vibrational properties. We then show how these insights into the microscopic mechanisms immediately explain the enhanced stability of Si-D bonds.

275

Mat. Res. Soc. Symp. Proc. Vol. 557 © 1999 Materials Research Society

METHODS

We have performed comprehensive and systematic calculations for hydrogen interactions with silicon using a state-of-the-art first-principles approach based on density-functional theory in the local-density approximation [1]. We employ a plane-wave basis set and a supercell geometry, with *ab initio* pseudopotentials for the semiconductor host atoms [2, 3]. Relaxation of host atoms was always included, and 32-atom supercells were typically used. This approach has produced reliable results for bulk properties of many materials, as well as properties of surfaces, interfaces, impurities, and defects. More details about the application of the method to the study of hydrogen can be found in Refs. [4], [5], and [6]. We estimate the uncertainty on the energies quoted here to be ± 0.1 eV.

HYDROGEN MOLECULES

It has been known for some time that H_2 molecules are one of the more stable forms of hydrogen in many semiconductors. This knowledge was based on computational studies (see, e.g., Ref. [4, 5, 7]) as well as on interpretation of experimental data. Direct observation of H_2 molecules proved very challenging, however, because of sensitivity problems in techniques such as NMR (nuclear magnetic resonance) and vibrational spectroscopy. Recently, however, great progress has been made in this area.

A thorough understanding of the incorporation of H_2 in the lattice is essential for the many technologically important processes that involve hydrogen: passivation of defects at the Si/SiO_2 interface; the "smartcut" process for producing silicon-on-insulator structures [8]; passivation and generation of defects in amorhous silicon; etc. In amorphous silicon, it has long been known that much more hydrogen is incorporated than is strictly needed for defect passivation. Work by Norberg *et al.* [9] suggests that a significant fraction of this hydrogen could be in the form of interstitial molecules. For many of these processes, it is essential to understand how H_2 interacts with existing defects or contributes to the formation of new defects; one aspect of such interactions is also addressed in the section on **Hydrogen Motion**.

Experimental observation of interstitial H_2 in crystalline semiconductors

Vibrational spectroscopy provides an excellent tool for observing hydrogen-related centers in semiconductors, but the lack of a dipole moment in the symmetric H_2 molecule makes it difficult to observe with infrared (IR) spectroscopy. Even if the interaction with its environment induces a dipole moment in the molecule, it is expected to be very weak. Raman spectroscopy, on the other hand, is not hampered by the inversion symmetry of the molecule.

Recently, impressive progress has been made in the experimental observation of H_2 molecules in crystalline semiconductors. Hydrogen molecules have been detected with Raman spectroscopy in GaAs [10] and Si [11]. In both cases, the vibrational frequency of the stretch mode was found to be significantly lower than the frequency in H_2 gas. More recently, the vibrational modes of interstitial H_2 in Si have also been observed with IR spectroscopy [12, 13]; at 10 K, a value of 3618 cm^{-1} was found, in excellent agreement with the Raman results. In Si, care has to be taken to distinguish between truly interstitial H_2 molecules, and H_2 in small voids or near platelets; the latter exhibit a vibrational frequency close to H_2 in vacuum [14].

First-principles calculations of interstitial hydrogen molecules

We have performed first-principles computational studies of interstitial H_2 in a number of different semiconductors: Si, GaAs, InAs, and GaN (in the zincblende structure) [15]. These investigations show that incorporation of H_2 in an interstitial position results in a lowering of the binding energy, an increase in the bond length, and a lowering of the vibrational frequency. These effects can be attributed to the immersion of the molecule in a low-density electron gas near the interstitial site. Indeed, the decrease in binding energy and corresponding lowering of the vibrational frequency correlate with the charge density near the interstitial site. Our calculated lowering of the frequency for H_2 in GaAs and in Si agrees well with the experimental values [10, 11], and with other recent calculations by Hourahine *et al.* [16] and by Okamoto *et al.* [17].

For each of the semiconductors, we placed H_2 in various interstitial configurations, in a number of different orientations. For the lowest-energy configurations, a series of calculations was subsequently carried out for different bond lengths in order to obtain a potential energy curve for determination of the vibrational frequencies, including both harmonic (ω) and anharmonic ($\Delta\omega$) terms. The vibrational frequencies turned out to be remarkably insensitive to the orientation of the molecule. We also calculated the difference between the energy of interstitial H_2 in the semiconductor and the energy of H_2 in vacuum; for c-Si, this energy difference ΔE is 0.8 eV per molecule.

Our systematic investigation of H_2 in different semiconductors allowed us to extract trends and investigate the underlying mechanisms. The vibrational frequency decreases along with the lattice constant a of the semiconductor – up to a point. For large a there is little direct interaction between the H_2 molecule and the host atoms; the relaxation of the host atoms is small and its effect on the vibrational frequency is negligible. The interaction increases as the lattice constant goes down, resulting in lower values of the vibrational frequency ω, larger values of the anharmonicity, $\Delta\omega$, and higher values of ΔE. Repulsion between H atoms and host atoms (which would lead to a stiffening of the force constant and an increase in frequency) does not seem to play any significant role, until the lattice constant becomes smaller than that of silicon. In fact, immersion of the molecule in the semiconductor charge density leads to a weakening of the bond. This trend obviously breaks down for semiconductors with a very small lattice constant, such as GaN. In this case strong repulsion occurs between the molecule and the host atoms; in essence, there is not enough room in the interstitial cage to accommodate the molecule, resulting in a large value of ΔE and an increased value of ω.

These insights allow us to predict that in amorphous silicon, the vibrational frequencies corresponding to interstitial H_2 would probably correspond to a broad band, reflecting the slightly different environments at the different interstitial sites. The width of this band would probably make detection by vibrational spectroscopy very difficult.

Diffusion of H_2

We also performed calculations for diffusion of interstitial H_2 in c-Si. We found that the saddle point occurs at the hexagonal interstitial site, with a migration barrier of 0.95 eV. An alternative diffusion mechanism for H_2 diffusion consists of dissociation of the molecule, followed by atomic diffusion. Results obtained in Ref. [5] indicate that dissociation of H_2 into two neutral interstitial hydrogen atoms costs 1.74 eV. Dissociation into an H^+-H^- pair would cost 1.34 eV. It is conceivable that the dissociation energy could be even lower, for instance, if additional energy can be gained from conversion of one of the charge states, or maybe through a catalytic interaction with an impurity.

Still, the migration barrier of 0.95 eV for H_2 to diffuse as a molecule seems low enough to form a viable channel for diffusion under many circumstances. This value agrees well with the observed diffusion of interstitial H_2 reported in Ref. [12]. Markevich and Suezawa [18] also derived an activation energy for diffusion of a hydrogen-related species responsible for the formation of O-H complexes in c-Si; their value was 0.78 ± 0.05 eV. Based on other evidence, they argued that this species was interstitial H_2 molecules.

We also mention that during the Raman experiments of Leitch *et al.* [11] a decrease of the interstitial H_2 signal was observed during the measurement, indicating that the H_2 molecules are either diffusing or dissociating [19]. Since this effect is observed even at 4 K, it must be attributed to the interaction of (above band-gap) light with the molecule, either directly or through the generation of carriers.

Hydrogen molecules in amorphous silicon

Much of what we discussed above for H_2 in crystalline silicon is likely to apply to amorphous silicon as well. The presence of H_2 molecules in a-Si had been considered previously, but mainly in the context of molecular hydrogen trapped in voids or microbubbles [20, 21]. Device-quality hydrogenated amorphous silicon typically contains up to 15% hydrogen – a concentration that is much larger than the amount of hydrogen needed to passivate defects. The configuration in which this massive amount of hydrogen is incorporated has been debated for a long time.

Norberg and coworkers [9] have recently performed deuteron and proton nuclear magnetic resonance (NMR) measurements on high-quality plasma-enhanced CVD a-Si films, showing that a significant fraction of the hydrogen in these samples (2% or more) is not involved in Si-H bonds. On the basis of their measurements they conclude that nearly all of this non-bonded hydrogen is present as isolated H_2 molecules, located in centers of atomic dimensions, perhaps in the analogue of T_d-sites in crystalline silicon. These molecules are *not* the small population of densely packed molecules in the occasional microvoids. Norberg has also found that the photoelectronic quality of the a-Si films increases as the fraction of non-bonded hydrogen increases. The non-bonded hydrogen also appears to be in the vicinity of light-induced defects, suggesting the molecular hydrogen may play a role in Staebler-Wronski degradation [22]. We will return to this issue in the section on **Hydrogen Motion**.

One may wonder whether molecular hydrogen could play any role in structural rearrangements or electronic defect formation. In this context, we point out the following experimental result for crystalline silicon: Heating boron-doped c-Si in H_2 gas at temperatures exceeding 900°C allows hydrogen to diffuse into the material. When the sample is quenched to room temperature, some of the hydrogen is found to passivate boron acceptors, but the majority of hydrogen atoms (about 70%) are found in a different configuration, which was recently identified to be H_2 molecules [23]. Annealing of the sample at 175°C results in the dissociation of the H_2 molecules and formation of additional H-B complexes. The B acceptors probably play a role in the H_2 dissociation; still, these findings indicate that H_2 molecules can dissociate at modest temperatures. Estreicher *et al.* [24] recently pointed out that interaction of H_2 with vacancies or self-interstitials also leads to dissociation of the molecule, with a substantial gain in energy.

Regarding diffusion, both experiment and theory now point towards an activation energy for H_2 diffusion in c-Si of less than 1.0 eV. This makes molecular hydrogen far more mobile than previously assumed. This result presumably applies also to a-Si, although it is hard to predict how the disorder in the network will affect the diffusion.

HYDROGEN MOTION

Interstitial hydrogen can diffuse through crystalline silicon with an activation energy of about 0.5 eV [4, 25]. Hydrogen interacts strongly with other impurities as well as with defects in the crystal. The strongest of these interactions is with silicon dangling bonds, where Si-H bonds are formed with bond strengths up to 3.6 eV [5, 6], similar to those in silane. Silicon dangling bonds thus form deep traps for hydrogen. In a-Si:H hydrogen diffusion is understood to occur by the dissociation of Si-H bonds, producing interstitial hydrogen; the latter diffuses rapidly along interstitial sites until trapped at a dangling bond or other defect site. The measured activation energy for hydrogen diffusion in a-Si:H is about 1.5 eV [26].

Exchange of deeply trapped and interstitial hydrogen

Hydrogen exchange between deeply trapped and mobile states plays an important role in the diffusion process [27, 28]. If such exchange takes place by first dissociating a Si-H bond and subsequently placing another H at the dangling bond, the activation energy would be high. Experimentally, however, the exchange is known to proceed very efficiently, based on observations of deuterium replacement of hydrogen [27, 29, 30, 31]. Unraveling the microscopic mechanisms by which a neutral interstitial hydrogen can exchange with a deeply trapped hydrogen was a challenge we tackled with first-principles calculations.

Our calculations focused on hydrogen trapped at an isolated Si-H bond in bulk crystalline silicon [32]. To isolate a Si-H bond, we create a small void in the silicon crystal following the procedure outline in Ref. [33]. The local environment in this model, including the open void above the Si-H bond, is similar to the environment of deeply trapped, isolated Si-H bonds in explicit simulations of a-Si:H [6]. Recent *ab initio* total energy calculations also indicate that the bond strengths of isolated Si-H bonds in a-Si:H are similar to those in our c-Si model [6]. In addition, the interactions governing the energetics of the exchange process seem to be fairly localized in nature (as illustrated by the insensitivity to relaxations beyond second-nearest neighbors); we therefore suggest our results apply to an amorphous environment as well.

The main result is that H-H exchange can proceed with an energy barrier of less than 0.2 eV. The first part in the process consists of an interstitial H atom approaching the Si-H bond, resulting in a hydrogen in a bond-center (BC) site next to the Si-H bond. The H-H exchange then proceeds via an intermediate, metastable state, in which *both* H atoms are equally bonded to the Si atom, a configuration which we label $\equiv$SiH$_2$. In this configuration the two atoms can easily rotate; the H atom that was originally deeply bound can then jump to a BC position and diffuse away, completing the exchange.

Figure 1 displays the exchange process schematically. Note that Figure 1 includes neither all the atoms of the supercell nor all the atoms relaxed in our simulations. We use the following notation: hydrogen at a bond-center site is labeled H-BC; for the isolated dangling bond we use DB, and if it is passivated by hydrogen we use H-DB or Si-H, interchangeably; for hydrogen in a BC site next to a DB site, we use (H-BC,H-DB); finally, if H-BC is far from a DB site we use (H-BC)+(H-DB).

Figure 1(a) is a schematic of the fully relaxed (H-BC,H-DB) complex which is the starting point for the exchange. The energy of the (H-BC,H-DB) complex is 0.15 eV higher than the energy of (H-BC)+(H-DB). This modest increase in energy does not constitute much of a barrier for an interstitial H atom to approach the Si-H bond.

The exchange of the H-BC with the H-DB involves a concerted motion of the two H atoms, along with relaxations of all the Si atoms. First, we move the H atom from BC towards H-DB, fixing this H at a number of positions along a path labeled "Path I" in Fig. 1(a). Path I leads to an intermediate $\equiv$SiH$_2$ configuration, as shown in Fig. 1(b). The energy barrier along Path I is $\sim$0.06 eV. Our calculations for this configuration established that $\equiv$SiH$_2$ is actually a metastable configuration (i.e., a local minimum in the energy surface).

The energy of the $\equiv$SiH$_2$ complex is 0.01 eV lower than the (H-BC,H-DB) complex. An investigation of the electronic structure reveals that along path I the highest occupied eigenlevel descends from near the conduction-band edge to near mid-gap, at which point the eigenstate is localized on the silicon and both hydrogens. We found that the H atoms in the $\equiv$SiH$_2$ complex can "rotate" with relative ease around the [111] direction of the original Si-H bond, as sketched in Fig. 1(b). We calculated the energy barrier for the re-orientation of the $\equiv$SiH$_2$ complex to be smaller than 0.04 eV. To complete the exchange, the original H-DB moves into a BC site along a path which, by symmetry, is the reverse of path I, with the same energy barrier.

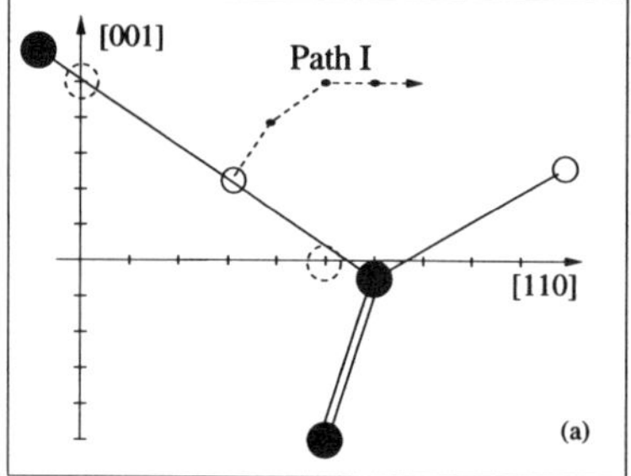

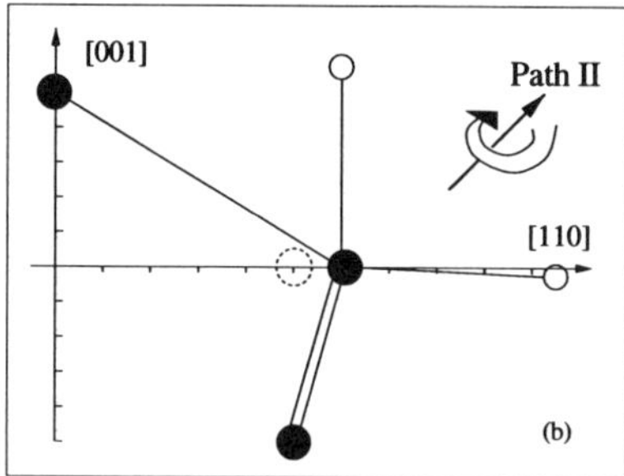

Figure 1: Schematic illustration of the hydrogen-hydrogen exchange process. A dangling bond at atom Si2 passivated by a H atom (H-DB). The small open circle represents a hydrogen atom; the large filled circles represent silicon atoms. The solid lines represent bonds in the plane of the page [the $(1\bar{1}0)$ plane]; the double lines indicate bonds to the Si3 and Si4 atoms, which lie in front of, resp. behind, the plane of the page. (a) The bond-centered H atom moves by the path labeled I towards the dangling-bond region, resulting in a metastable $\equiv$SiH$_2$ complex. The dotted circles represent the initial position of the silicon atom. The solid circles show the position of the Si atoms in the (H-BC,H-DB) complex. (b) In the $\equiv$SiH$_2$ complex the two H atoms can "rotate" around the [111] direction, as schematically illustrated by Path II. To complete the exchange, the originally deeply bound H atom moves to a new BC position, along a path that is the equivalent of Path I.

Considering the full exchange process, we find that the energy barriers along Paths I and II in Fig. 1 are both smaller than 0.1 eV. Since the migration barrier for interstitial H is about 0.5 eV [4, 25], the barriers along paths I and II can easily be overcome at the modest temperatures at which interstitial H is mobile. The activation energy of the exchange process is therefore dominated by the energy cost of 0.15 eV needed to place the interstitial (transport-level) hydrogen in a (H-BC,H-DB) state.

In the course of our investigations we explored a number of possible paths and exchange mechanisms other than the one reported in Fig. 1. For instance, we examined the possibility that as the H-BC moved towards the dangling-bond region, the H-DB would simultaneously move to one of the two neighboring BC sites. We found that such a process would encounter energy barriers over 0.8 eV. We conclude that having at least one H to passivate the dangling bond at any time is essential to a low-energy exchange. Only mechanisms which consistently maintain at least one H passivating the dangling bond will be competitive with those reported in Fig. 1.

It should be noted that our calculations apply to *neutral* interstitial H exchanging with a deeply bound H. Interstitial hydrogen can occur in different charge states [4, 5]. In the positive charge state H still prefers the BC site, and we expect only minor modifications to the exchange path described above. In the negative charge state, H prefers to sit in open interstitial positions, and the exchange mechanism (if present) may be quite different.

Exchange: experiments

Detailed experimental information about the exchange process in amorphous silicon has been obtained in hydrogen-deuterium tracer diffusion experiments in a-Si:H(D) [29, 30]. To explain the data, Branz *et al.* found it essential to take deep trap levels into account, in addition to shallow traps and a transport level; in addition, it was found that (1) there is significant exchange between the transport level and the deeply trapped hydrogen and (2) the energy barriers for H-H(D) exchange are much smaller than the barriers for long-range diffusion (1.5 eV). Kemp and Branz [30] proposed that H-D (or H-H) exchange occurs via an intermediate state with two H atoms bonded to a Si atom, i.e., where the Si atom is over-coordinated. They also noted that the energy of this intermediate state has to be nearly degenerate with the energy of an isolated Si-H plus a transport-level hydrogen. The properties of the $\equiv SiH_2$ emerging from our calculations are entirely consistent with the mediatory complex suggested by Kemp and Branz [30]. We find an energy difference of 0.14 eV between $\equiv SiH_2$ and (H-BC)+(H-DB). The consistency of our present results with the experimentally based constraints for hydrogen exchange suggests that we have captured important features of the exchange mechanisms for H in a-Si:H.

The $\equiv SiH_2$ complex and light-induced defect generation

In this study, we have examined the $\equiv SiH_2$ complex which involves a five-fold coordinated silicon atom. Overcoordination defects are known to be present in bulk silicon; they have also been suggested to be an important intrinsic defect in *amorphous* silicon [6, 34]. The present complex differs from over-coordinated defects found in bulk silicon. In the $\equiv SiH_2$ complex, the underlying silicon defect is a 3-fold coordinated atom. The overcoordinated silicon complex occurs when two hydrogen atoms are placed in the dangling-bond region. A similarity between the 5-fold coordination defect studied in the context of a-Si and the $\equiv SiH_2$ complex is that both result in a deep level which is rather localized on several atoms [6, 34, 35].

The $\equiv\mathrm{SiH_2}$ in the c-Si model discussed above is a locally stable configuration, but the barrier to go to the (H-BC,H-DB) configuration is less than 0.1 eV high; the $\equiv\mathrm{SiH_2}$ would therefore not be stable for a long time. It is conceivable that certain sites in a-Si would provide a slightly greater stability of this complex, due to increased flexibility of the surrounding network.

One might speculate that the $\equiv\mathrm{SiH_2}$ complex will be a precursor for generation of $\mathrm{H_2}$, leaving a dangling bond behind. In fact, our calculations indicate that such a configuration, in which the $\mathrm{H_2}$ is still close to the dangling bond, is close in energy to the $\equiv\mathrm{SiH_2}$ complex (within 0.1 eV). A full determination of the barrier between the two configurations was beyond the scope of our study; however, preliminary investigations suggested the barrier was about 1.0 eV high [36].

We suggest that the $\equiv\mathrm{SiH_2}$ complex might play a role in defect formation. Since the complex is electrically active, light or free carriers could enhance the dissociation of the complex which may lead to formation of a dangling bond and an $\mathrm{H_2}$ molecule. This $\mathrm{H_2}$ molecule may diffuse away (again potentially assisted by light), leaving behind a dangling bond, i.e., an electrically active defect.

It is well accepted that hydrogen plays a role in light-induced defect formation. Explanations of this effect need to include a mechanism for generating mobile hydrogen (usually attributed to excitation from Si-H bonds due to photo-induced carriers), as well as a mechanism for trapping the hydrogen in the final state. Recently, Branz [37] proposed a model in which two mobile H atoms collide, forming an immobile complex containing two Si-H bonds. Biswas and Pan [38] proposed specific configurations for this final configuration based on molecular dynamics simulations. Here we speculate that $\mathrm{H_2}$ molecules could also be candidates for the final state. Instead of requiring two mobile hydrogens to come together, a single mobile hydrogen (which has left a dangling bond behind) could approach an Si-H bond, forming the $\equiv\mathrm{SiH_2}$ complex. Photo-generated carriers could promote the reaction to form $\mathrm{H_2}$ plus a second dangling bond. The $\mathrm{H_2}$ may then diffuse away from the dangling bond. Alternatively, two mobile H atoms may come together (as in the Branz model) and form an $\mathrm{H_2}$ molecule.

Thermal annealing of light-induced defects proceeds with an activation energy of about 1.1 eV [22]. Recent experiments in c-Si (see Ref. [23], discussed above) indicate that annealing at 175°C dissociates most of the $\mathrm{H_2}$ present in a quenched sample; this anneal temperature would be consistent with an activation energy of 1.1 eV.

While this suggestion for $\mathrm{H_2}$ involvement in light-induced defect generation is not nearly as developed as the comprehensive model proposed by Branz [37], we feel that the information that has recently become available about the behavior of $\mathrm{H_2}$ in silicon should prompt a careful examination of the role $\mathrm{H_2}$ may play in defect generation.

Dissocation mechanisms of Si-H bonds; H/D isotope effect

A few years ago it was discovered that Si-D bonds behave very differently from Si-H under electronic excitation: Si-D were found to be orders of magnitude harder to break. This giant isotope effect was first observed for Si-H bonds on Si surfaces [39, 40], and quickly applied to passivation of defects at the $\mathrm{Si/SiO_2}$ interface in MOS transistors [41]. Since hydrogenated amorphous Si suffers from carrier- and light-induced degradation, it should be expected that the observed enhanced stability of Si-D as compared to Si-H would also apply to Si-D bonds in a-Si. Experimental observations of the enhanced stability of deuterated

a-Si under light exposure were recently reported by Wei *et al.* [42] and by Sugiyama *et al.* [43]. Replacing hydrogen with deuterium has also been reported to greatly reduce the PL degradation of porous silicon [44].

These observations may seem surprising, because H and D are entirely equivalent from an electronic point of view: indeed, the static electronic structure of the Si-H and Si-D bonds is identical. The difference must therefore be attributed to dynamics. We have proposed a mechanism which provides a natural explanation for the difference in dissocation rates [45]. The dissociation of Si-H bonds has been proposed to proceed via multiple-vibrational excitation by tunneling electrons (at least in the low-voltage regime) [46]. The extent to which vibrational energy can be stored in the bond depends on the lifetime, i.e., on the rate at which energy is lost by coupling to phonons. Because the lifetime of H on Si is long [47, 48], efficient vibrational excitation is expected. The question then is: why would Si-D behave qualitatively differently?

The path followed by the hydrogen (or deuterium) atom during the breaking of a Si-H (Si-D) bond turns out to play a crucial role in the dissociation mechanism. It was often implicitly assumed that dissociation would proceed by moving the H atom away from the Si along the direction of the Si-H bond away from the Si atom; however, this is unlikely to be the most favorable path, for two reasons: (a) the initial rise in energy in that direction is high, as indicated by the high vibrational frequency (around 2100 cm^{-1}) for the Si-H stretch mode; (b) this path eventually leads to a position of the H atom in the interstitial channel, which is not the lowest energy site for H in the neutral or positive charge state (in c-Si). Both of these arguments actually favor a different path in which the H atom stays at approximately constant distance from the Si atom to which it is bound: (a) the barrier in that direction is much lower, as indicated by the vibrational frequency (around 650 cm^{-1}) for the Si-H bending mode; (b) this path leads to H positions closer to the Si atom, which are more favorable for H^0 and H$^+$ in c-Si. A detailed examination of these dissociation paths was recently presented in Ref. [49].

The vibrational lifetime is thus mostly controlled by the Si-H *bending* modes, as discussed above. The vibrational frequency of the bending mode for Si-H is around 650 cm^{-1}, and the estimated frequency for Si-D is around 460 cm^{-1}. The latter frequency turns out to be very close to the frequency of bulk TO phonon states at the X point (463 cm^{-1}) [50]. We therefore expect the coupling of the Si-D bending mode to the Si bulk phonons to result in an efficient channel for deexcitation. While it is quite possible to reach a highly excited vibrational state in the case of Si-H, this will be more difficult for Si-D. These qualititative differences between H and D have recently been confirmed in tight-binding molecular dynamics studies by Biswas *et al.* [51]. Deuterium should therefore be much more resistant to STM-induced desorption and hot-electron induced dissociation, due to the relaxation of energy through the bending mode.

We also point out that displacements along the "bond-bending path" cause energy levels to be introduced into the band gap (near the valence band and near the conduction band), enabling the complex to capture carriers; after changing charge state there is virtually no barrier to further dissociation. The barrier for dissociation can therefore be significantly reduced when carriers are present, such as in a scanning tunneling microscope (STM) desorption experiment; at an Si/SiO$_2$ interface, during device operation of the MOS transistor; or in a-Si, in the form of injected or light-induced carriers.

CONCLUSIONS

In summary, we have discussed a number of areas in which first-principles calculations have recently provided new insights into microscopic mechanisms relevant for hydrogenated amorphous silicon. We pointed out that theory as well as experiments on crystalline silicon indicate that H_2 molecules may play a more important role than previously thought, because they diffuse and dissociate more easily than had been assumed. With relevance for hydrogen diffusion, we have determined a low-energy exchange mechanism between interstitial and deeply bound hydrogen, which requires an activation energy of only 0.15 eV. The microscopic mechanism involves an intermediate $\equiv SiH_2$ complex. Finally, we have discussed dissociation of the Si-H bond and our explanation for the enhanced stability of Si-D.

ACKNOWLEDGMENTS

We gratefully acknowledge stimulating interactions and collaborations with J. Adams, C. Herring, W. Jackson, N. Johnson, N. Nickel, J. Neugebauer, S. Pantelides, R. Street, and N. Troullier. B. T. would like to acknowledge funding from the Xerox Foundation, NSF (Grant No. NSF-DMR-91-58584PYI), DOE (Grant No. DEFG 02-96-ER45439) and the Office of Naval Research.

References

[1] P. Hohenberg and W. Kohn, Phys. Rev. **136**, B864 (1964); W. Kohn and L. J. Sham, *ibid.* **140**, A1133 (1965).

[2] D. R. Hamann, M. Schlüter, and C. Chiang, Phys. Rev. Lett. **43**, 1494 (1979).

[3] N. Troullier and J. L. Martins, Phys. Rev B **43**, 1993 (1991); N. Troullier and J. L. Martins, Phys. Rev B **43**, 8861 (1991).

[4] C. G. Van de Walle, P. J. H. Denteneer, Y. Bar-Yam, and S. T. Pantelides, Phys. Rev. B **39**, 10791 (1989).

[5] C. G. Van de Walle, Phys. Rev. B **49**, 4579 (1994).

[6] B. Tuttle and J. Adams, Phys. Rev. B **57** 12859 (1998).

[7] For a review, see S. K. Estreicher, Mat. Sci. Engr. Reports **14**, 319 (1995).

[8] M. Bruel, Electron. Lett. **31**, 1201 (1995).

[9] R. E. Norberg, D. J. Leopold, and P. A. Fedders, J. Non-Crystalline Solids, **227-230**, 124 (1998).

[10] J. Vetterhöffer, J. Wagner, and J. Weber, Phys. Rev. Lett. **77**, 5409 (1996).

[11] A. W. R. Leitch, V. Alex, and J. Weber, Phys. Rev. Lett. **81**, 421 (1998).

[12] R. E. Pritchard, M. J. Ashwin, J. H. Tucker, R. C. Newman, E. C. Lightowlers, M. J. Binns, S. A. McQuaid, and R. Falster, Phys. Rev. B **56**, 13 118 (1997).

[13] R. E. Pritchard, M. J. Ashwin, J. H. Tucker, and R. C. Newman Phys. Rev. B **57**, R15 048 (1998).

[14] A. W. R. Leitch, V. Alex, and J. Weber, Solid State Commun. **105**, 215 (1998).

[15] Chris G. Van de Walle, Phys. Rev. Lett. **80**, 2177 (1998).

[16] B. Hourahine, R. Jones, S. Öberg, R. C. Newman, P. R. Briddon, and E. Roduner, Phys. Rev. B **57**, 12 666 (1998).

[17] Y. Okamoto, M. Saito, and A. Oshiyama, Phys. Rev. B **56**, 10 016 (1997).

[18] V. P. Markevich and M. Suezawa, J. Appl. Phys. **83**, 2988 (1998).

[19] J. Weber, private communication.

[20] Y. J. Chabal and C. K. N. Patel, Phys. Rev. Lett. **53**, 1771 (1984).

[21] Y. J. Chabal and C. K. N. Patel, Rev. Mod. Phys. **59**, 835 (1987).

[22] D. L. Staebler and C. R. Wronski, Appl. Phys. Lett. **31**, 292 (1977).

[23] R. E. Pritchard, J. H. Tucker, R. C. Newman, and E. C. Lightowlers, Semicond. Sci. Technol. **14**, 77 (1999).

[24] S. K. Estreicher, J. L. Hastings and P.A. Fedders, Phys. Rev. B **57**, R12 663 (1998).

[25] A. Van Wieringen and N. Warmoltz, Physica **22**, 849 (1956).

[26] R. A. Street, C. C. Tsai. J. Kakalios, and W. B. Jackson, Philos. Mag. B **56**, 305 (1987).

[27] W. Jackson, J. Non-Crystalline Solids **164-166**, 263 (1993).

[28] C. G. Van de Walle and R. A. Street, Phys. Rev. B **51**, 10 615 (1995).

[29] H. M. Branz, S. E. Asher, B. P. Nelson and M. Kemp, J. Non-Cryst. Solids **164-166**, 269 (1993).

[30] M. Kemp and H. M. Branz, Phys. Rev. B **52**, 13946 (1995).

[31] R. A. Street, in *Amorphous Silicon Semiconductors – Pure and Hydrogenated*, edited by A. Madan, M. Thompson, D. Adler, and Y. Hamakawa, MRS Symposia Proceedings **95** (Materials Research Society, Pittsburgh, 1987), p. 13.

[32] B. Tuttle, C.G. Van de Walle, and J.B. Adams, Phys. Rev. B **59**, 5493 (1999).

[33] C.G. Van de Walle and R. A. Street, Phys. Rev. B **49**, 14766 (1994).

[34] S. T. Pantelides, Solid State Comm. **84** 221 (1992).

[35] R. Biswas, C.Z. Wang, C.T. Chan, K.M. Ho and C.M. Soukuolis, Phys. Rev. Lett. **63** 1491 (1989).

[36] B. Tuttle (unpublished).

[37] H. M. Branz, Solid State Commun. **105**, 387 (1998).

[38] R. Biswas and B. C. Pan, Appl. Phys. Lett. **72**, 371 (1998).

[39] J. W. Lyding, T. C. Shen, J. S. Hubacek, J. R. Tucker, and G. C. Abeln, Appl. Phys. Lett. **64**, 2010 (1994).

[40] Ph. Avouris, R. E. Walkup, A. R. Rossi, T.-C. Shen, G. C. Abeln, J. R. Tucker, and J. W. Lyding, Chem. Phys. Lett. **257**, 148 (1996).

[41] J. W. Lyding, K. Hess, and I. C. Kizilyalli, Appl. Phys. Lett. **68**, 2526 (1996).

[42] J.-H. Wei, M.-S. Sun, and S.-C. Lee, Appl. Phys. Lett. **71**, 1498 (1997).

[43] S. Sugiyama, J. Yang, and S. Guha, Appl. Phys. Lett. **70**, 378 (1997).

[44] T. Matsumoto, Y. Masumoto, and N. Koshida, Mat. Res. Soc. Symp. Proc. **452**, 449 (1997).

[45] C. G. Van de Walle and W. B. Jackson, Appl. Phys. Lett. **69**, 2441 (1996).

[46] T.-C. Shen, C. Wang, G. C. Abeln, J. R. Tucker, J. W. Lyding, Ph. Avouris, and R. E. Walkup, Science **268**, 1590 (1995).

[47] P. Guyot-Sionnest, P. Dumas, Y. J. Chabal, and G. S. Higashi, Phys. Rev. Lett. **64**, 2156 (1990).

[48] P. Guyot-Sionnest, P. H. Lin, and E. M. Miller, J. Chem. Phys. **102**, 4269 (1995).

[49] B. Tuttle and C. G. Van de Walle, Phys. Rev. B (May 15, 1999, in press).

[50] O. Madelung, editor, *Data in Science and Technology: Semiconductors*, (Springer-Verlag, Berlin, 1991).

[51] R. Biswas, Y.-P. Li, and B. C. Pan, Appl. Phys. Lett. **72**, 3500 (1998).

INTERSTITIAL TRAPPED HYDROGEN MOLECULES IN PECVD AMORPHOUS SILICON

R. BORZI, F. Mascarenhas, P. A. Fedders, D. J. Leopold, R. E. Norberg
Department of Physics, Washington University, One Brookings Drive,
St. Louis, MO 63130, ren@howdy.wustl.edu
P. Wickboldt, W. Paul
Harvard University, Division of Applied Science, Cambridge, MA 02138

ABSTRACT

New NMR measurements show that interstitial T site-trapped molecular hydrogen can amount to more than one third of the contained hydrogen in high quality PECVD amorphous silicon. Microvoid-contained dense molecular hydrogen is negligible in these good films. Experiments on a sequence of hydrogenated and/or deuterated a-Si films have characterized individually-trapped molecular HD and D_2 in films deposited from SiD_4, and from SiH_4+D_2. The T site-trapped molecular hydrogen fraction observed here is larger than previously reported because of recent efforts to measure very slowly relaxing molecular components and the employment of radiofrequency pulse sequences to detect ortho-D_2 with nuclear spin $I=2$. The population of interstitially trapped molecular hydrogen increases with increasing photovoltaic quality over a range of an order of magnitude in photoresponse product $\eta\mu\tau$. Above 200 K, hopping transport of molecular hydrogen among the amorphous equivalent of interstitial T sites occurs with an activation energy near 50 meV.

INTRODUCTION

Recent deuteron magnetic resonance (DMR) measurements have examined high quality amorphous silicon films prepared by plasma enhanced chemical vapor deposition (PECVD). The measurements have demonstrated the presence of populations of trapped molecular hydrogen significantly larger than those reported earlier [1]. The increase arises from the observation of previously undetected slowly-relaxing nuclear spin $I=2$ orthodeuterium populations.

EXPERIMENT

DMR at 30 MHz has been used between 1.2 and 300 K to examine high quality a-Si:D and a-Si:H,D films prepared by PECVD from SiD_4 and from SiH_4+D_2. Photoresponse products $\eta\mu\tau$ and other photovoltaic parameters were measured on intact films [2]. The new DMR measurements have sought to optimize observation of ortho-D_2 nuclear spin $I=2$ signals by employing very long magnetization recovery intervals (up to more than 10^4 sec) and radiofrequency pulse rotation sequences near 90_x-τ-45_y. 90_x-τ-26_y DMR pulse sequences have been used to detect molecule-specific multiple echoes [3] which occur only for molecular ortho-D_2.

RESULTS

Figure 1 shows 90_x-τ-90_y DMR solid echo spectra for high quality a-Si:D,H at 4.2 K and 32 K at three very different magnetization recovery delays (0.4, 50, and 10,000 sec).

Mat. Res. Soc. Symp. Proc. Vol. 557 © 1999 Materials Research Society

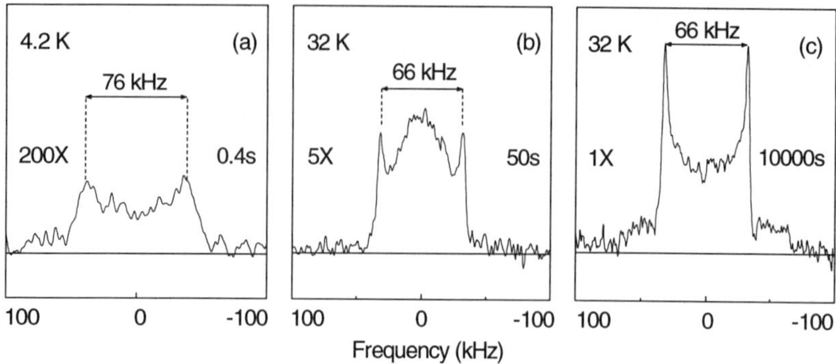

Figure 1. DMR 90_x-τ-90_y solid echo resonance line shapes at three pulse delays in deuterated a-Si.

The ordinate scales, as indicated, are different for the three spectra. Figure 1a shows, at 0.4 sec, the fast relaxing 76 kHz DMR doublet arising from a small population of para-D_2 molecules physisorbed on internal surfaces, which produce large crystal fields. Figure 1b shows a DMR spectrum at an intermediate 50 sec recovery time. Here a slowly-relaxing 66 kHz SiD doublet begins to appear along with a molecular deuterium 33 kHz FWHM gaussian broad central (BC) feature arising primarily from D_2 and HD molecules singly-trapped at the amorphous equivalent of interstitial T sites in the a-Si. At 50 sec the BC gaussian component is nearly fully relaxed. Figure 1c shows, at 10,000 sec, the fully relaxed DMR spectrum with a much-increased 66 kHz doublet SiD component and including the BC feature of Fig. 1b. This 10,000 sec spectrum nevertheless remains incomplete, since the 90_x-τ-90_y solid echo sequence suppresses DMR signals from nuclear spin $I=2$ ortho-D_2 molecules.

Figure 2 shows an example of this 90_x-τ-90_y $I=2$ signal suppression in data taken at 85 K and a recovery time of 1000 sec in a high quality perdeuterated a-Si:D film (H579). The upper panel shows the DMR response to a standard 90_x-τ-90_y solid echo pulse sequence. The lower panel shows the response to a 90_x-τ-45_y pulse sequence designed to improve the spin $I=2$ ortho-D_2 DMR signal. Here the plotted signal amplitude has been doubled to bring the spin $I=1$ 66 kHz SiD doublet component up to the same amplitude as for the solid echo sequence in the upper panel. A central $I=2$ component now has appeared and arises principally from the anticipated 9.5 kHz ortho-D_2 signal from molecules physisorbed on internal surfaces. This central ortho-D_2 signal is not at its full intensity since the anticipated ortho-D_2 spin lattice relaxation time T_1 here may exceed 1000 sec and since the rf pulse sequence is not set at 90_x-τ-26_y.

Figure 3 summarizes temperature variations of DMR resonance linewidths for several trapped molecular hydrogen species in high quality films. The 76 kHz component from rapidly-relaxing $I=1$ para-D_2 molecules on internal surfaces (Fig. 1a and open circles in Fig. 3) does not narrow appreciably from 1.5 to about 30 K. Above 30 K the para-D_2 line narrows as T^{-1}. The power law narrowing probably reflects line shape changes arising from phonon-induced molecular Δm_J transitions within the $J = 1$ molecular rotational state [4].

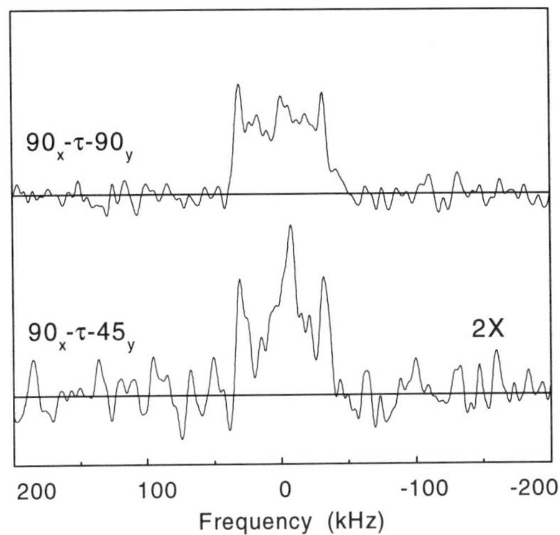

Figure 2. Changing from a DMR 90$_x$-τ-90$_y$ solid echo to a 90$_x$-τ-45$_y$ sequence brings up a central ortho-D$_2$ DMR molecular signal component in deuterated a-Si.

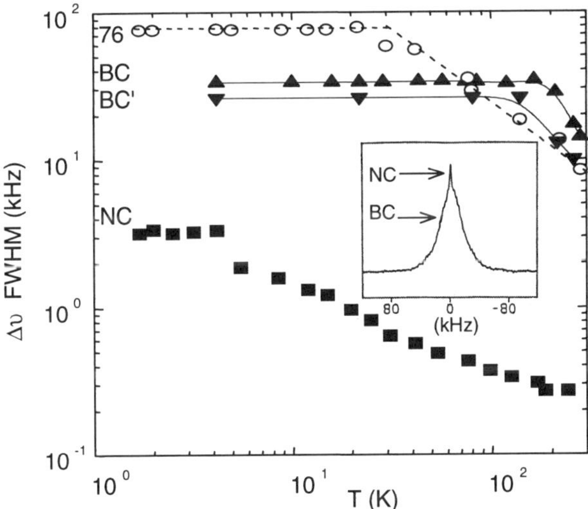

Figure 3. Four molecular D$_2$ and HD components in a-Si show different temperature variations of resonance line widths.

The broad central (BC) 33 and 25 kHz lines from T-site-trapped HD and para-D$_2$ (inset and solid triangles in Fig. 3) do not narrow below 100 K. The 33 kHz BC line from HD narrows above 160 K and the 25 kHz BC line from para-D$_2$ narrows above 120 K.

Both narrowings are exponential with reciprocal temperature and reflect intersite hopping transport of the trapped molecules with an activation energy near 50 meV [1].

At the lower left of Fig. 3 the solid squares show the line width of the very small narrow central (NC) population of D_2 and HD trapped in microvoids. This locally mobile population shows motional narrowing near 30 K and exhibits a power law behavior up to about 80 K. In lower quality films this microvoid molecular population becomes much larger. Examples of the BC and NC DMR lines in a high quality a-Si:D,H film are shown in the inset in Fig. 3.

Some relevant properties of four PECVD a-Si films are listed in Table I. The films were deposited on 230°C substrates. Three were deposited on a powered electrode and one on an unpowered electrode. Details on gas flow rates and photoresponse products have been given elsewhere [5]. The total H and D contents in atomic percent have been determined by NMR and by gas evolution. The final column lists the DMR-determined molecular hydrogen fraction in percent of the contained hydrogen. These molecular percentages are plotted for the three powered electrode films as a function of photoresponse produce $\eta\mu\tau$ in Fig. 4. These fractions are somewhat larger than previously reported, because of efforts to properly detect populations of ortho-deuterium molecules.

Table I

Photoresponse products and hydrogen content for PECVD amorphous silicon films.

Film	Type	Starting Gas	$\eta\mu\tau$ $(\mathrm{cm^2V^{-1}})$	$n(\mathrm{H})$ at.%	$n(\mathrm{D})$ at.%	Molecular Fraction %
541 P	a-Si:H,D	SiH_4+D_2	1.6×10^{-5}	5.2	4.0	40.9
579 P	a-Si:D	SiD_4	4.8×10^{-6}	—	8.0	36.5
760 P	a-Si:H,D	SiH_4+D_2	1.5×10^{-6}	9.6	14.7	29.4
760 U	a-Si:H,D	SiH_4+D_2	9.0×10^{-6}	8.0	9.3	25.7

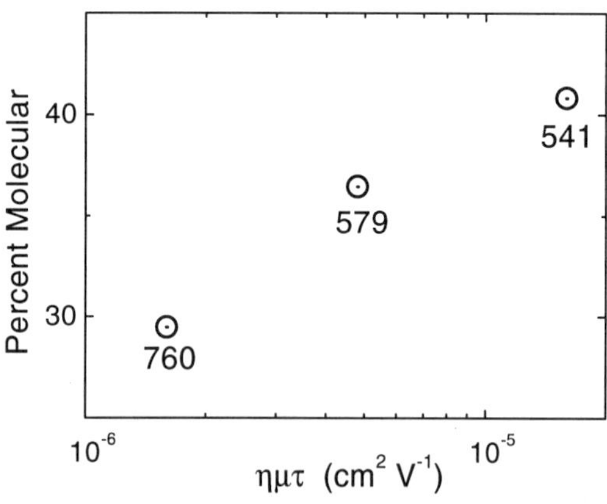

Figure 4. DMR-determined molecular percentage of contained deuterium as a function of $\eta\mu\tau$ for three powered electrode a-Si films.

CONCLUSION

High quality PECVD amorphous silicon typically contains more than one third of its hydrogen as molecular hydrogen. Most of that molecular hydrogen is interstitially trapped in the amorphous equivalent of T sites. This trapped molecular hydrogen gives rise to the less-clustered proton narrow 3 kHz NMR line [1].

ACKNOWLEDGMENTS

This work was supported in part by a Sirovich Fellowship and by the National Science Foundation under DMR 93-05344.

REFERENCES

1. R. E. Norberg, D. J. Leopold, and P. A. Fedders, Jour. Non-Cryst. Sol. **227–230**, 124 (1998).
2. W. A. Turner, S. J. Jones, D. Pang, B. F. Bateman, J. H. Chen, Y. M. Li, F. C. Marques, A. E. Wetsel, P. Wickboldt, W. Paul, J. Bodart, R. E. Norberg, I. ElZawawi, and M. L. Theye, J. Appl. Phys. **67**, 12 (1990).
3. M. P. Volz, P. Santos-Filho, M. S. Conradi, P. A. Fedders, R. E. Norberg, W. Turner, and W. Paul, Phys. Rev. Letters **63**, 2582 (1989).
4. P. A. Fedders, Phys. Rev. **B 34**, 7489 (1986).
5. R. E. Norberg, P. A. Fedders, and D. J. Leopold, Mat. Res. Soc. Symp. Proc. Vol. **420**, 475 (1996).

NMR STUDY OF ORTHO-MOLECULAR HYDROGEN IN HYDROGENATED AMORPHOUS SILICON

TINING SU*, P.C. TAYLOR*, SHENLIN CHEN*, R.S. CRANDALL** AND A.H. MAHAN**
*University of Utah, Department of Physics, Salt Lake City, Utah
**National Renewable Energy Laboratory, Golden, Colorado.

ABSTRACT

A Jeener-Broekaert three-pulse sequence is used to investigate ortho-molecular hydrogen (o-H_2) in device quality amorphous silicon films prepared by plasma enhanced chemical vapor deposition (PECVD) and hot wire CVD (HWCVD). For the PECVD sample, the concentration of hydrogen molecules is $\sim 11\%$ of the total hydrogen concentration, one order of magnitude larger than that inferred from spin-lattice relaxation time measurements ($\sim 1\%$). Hence, most of the hydrogen molecules do not serve as effective relaxation centers. For HWCVD samples with ~ 3 to 4% hydrogen and very low void densities, the concentrations of hydrogen molecules are $\sim 1\%$ of the total hydrogen concentration. In these samples, spin-lattice relaxation measurements for bonded hydrogen indicate that the concentration of hydrogen molecules that contribute to spin-lattice relaxation is at most 0.1% of the total hydrogen concentration. Spin-lattice relaxation time (T_1) measurements of ortho-molecular hydrogen indicate two very different T_1's. The longer T_1 is ~ 0.6 s, possibly due to an electric quadrupole-quadrupole (EQQ) interaction between o-H_2 molecules and, the shorter T_1 is ~ 3 ms, very close to that calculated for a two-phonon Raman process for rotating o-H_2.

INTRODUCTION

It is well known that molecular hydrogen (H_2) exists in PECVD-prepared hydrogenated amorphous silicon (a-Si:H) [2,5], but the concentrations and distributions of H_2 are controversial [1,2,11]. In this paper we report a new way to measure the fraction of hydrogen that exists as H_2 in a-Si:H.

There are two forms of H_2, ortho-hydrogen (o-H_2) and para-hydorgen (p-H_2). Because p-H_2 has zero nuclear spin, nuclear magnetic resonance (NMR) is sensitive only to o-H_2. At room temperature the ratio of o-H_2 to p-H_2 is approximately 3:1, and on typical laboratory time scales, the ration is frozen-in on cooling the samples. Ortho-hydrogen has total nuclear spin I = 1 and rotational angular momentum J = 1. In an anisotropic crystal field, the three fold degeneracy of the $m_J = 0, \pm 1$ states is lifted. The upper energy level is two-fold degenerate ($m_J = \pm 1$), and the lower level is non-degenerate ($m_J = 0$). At low temperature, the probability that o-H_2 occupies the $m_J = 0$ state is 1, i.e., the o-H_2 molecules are locally ordered with respect to their local crystal fields. The resultant NMR lineshape is a powder-averaged Pake doublet with splitting between the two divergences $\delta v = 173$ kHz, for an axially symmetric crystal field [1,12,13].

For solid H_2, the transition temperature for producing the Pake doublet is $T_c = 1.6$ K. This transition temperature is controlled by the electric quadrupole-quadrupole interaction between o-H_2 molecules [13]; however, in a-Si:H T_c can be as high as 20 K for unannealed samples [1], due to an additional electric field gradient (EFG) from the Si lattice, which interacts with the quadrupole moment of o-H_2. This latter effect (EFG effect) is the major contribution to T_c in a-Si:H since T_c is much higher than in solid H_2.

Mat. Res. Soc. Symp. Proc. Vol. 557 © 1999 Materials Research Society

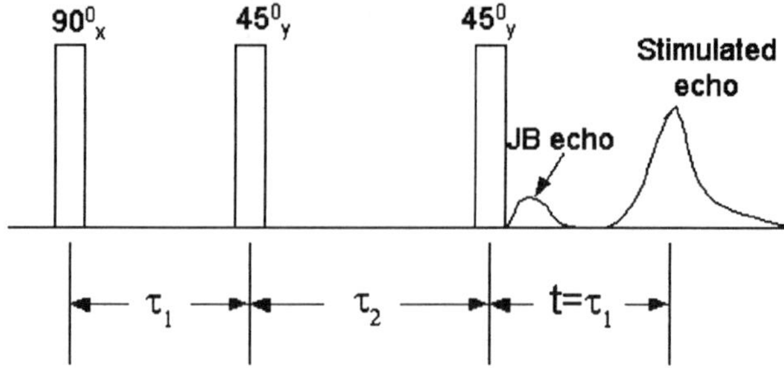

Fig. 1. Jeener-Broekaert pulse sequence. The stimulated echo forms at $t = \tau_1$ after the third pulse.

Since the Pake doublet is very broad, the conventional measurement of the free induction decay (FID) following a 90° pulse suffers severely from the recovery time of the equipment. Also, the low concentration of hydrogen molecules makes it difficult to distinguish this broad line from that due to bonded hydrogen. For these reasons, we used a Jeener-Broekaert three-pulse sequence. Figure 1 shows a schematic diagram of the Jeener-Broekaert 3-pulse sequence (JB sequence). Immediately after the third pulse, a JB echo forms. At $t = \tau_1$ after the third pulse, a stimulated echo forms. The amplitude of the stimulated echo can be expressed as [3]:

$$f(\tau_1, \tau_2) \propto e^{-2\tau_1/T_2} e^{-\tau_2/T_{1d}} \quad .$$

(1)

By measuring the amplitude of each component at different values of τ_1 and τ_2, one can extrapolate to the initial intensities of each component as τ_1 and τ_2 approach zero. At $\tau_1 = \tau_2 = 0$ the relative intensities yield the fraction each component contributes to the total hydrogen signal. Also, since o-H_2 has much larger values of T_2 and T_{1d} than bonded H [3], it is possible to emphasize the Pake doublet at large values of τ_1 and τ_2, and therefore one can obtain more accurate measurements of the Pake doublet component.

EXPERIMENTAL DETAILS

All experiments were done at a static field $H_0 \sim 2$ T. The concentration measurements were performed at $T = 8$ K. The T_1 measurements of o-H_2 were performed at $T = 40$ K, which is the temperature at which the T_1 minimum of bonded hydrogen occurs. Details of the NMR apparatus and measurement procedures are available elsewhere [14].

RESULTS AND DISCUSSION

It is well known [2,4,5] that NMR of bonded hydrogen in a-Si:H yields two inhomogeneously broadened lines--a narrow (~ 4 kHz) Lorentzian component due to hydrogen that is essentially randomly distributed, and a broad (~ 20 kHz) Gaussian component that is due to hydrogen clustered together in specific regions of the sample. In our JB stimualted-echo experiments three components are observed--the two mentioned above and a third component due to o-H_2. A representative JB

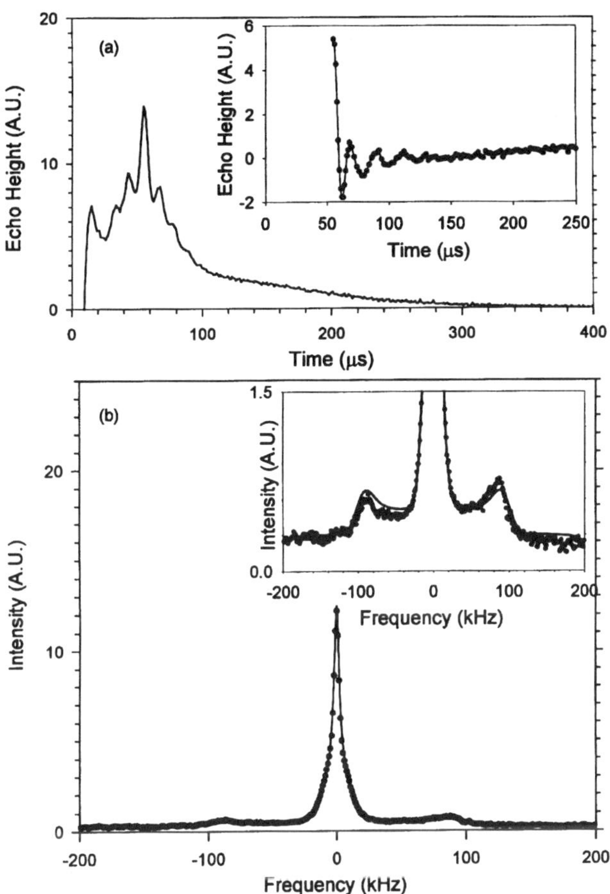

Fig. 2. The stimulated echo in the time domain and the corresponding spectrum in the frequency domain, which is obtained by Fourier Transformation. (a) Stimulated echo after the JB pulse sequence with $\tau_1 = 60$ μs, $\tau_2 = 9$ ms. The inset is the o-H_2 signal after subtracting the Gaussian and Lorentzian components (due to bonded hydrogen) from the echo. The line is an aid to the eye. (b) The Fourier-transformed spectrum of the echo in (a). The solid circles are the experimental data. The solid line is a fit to the data assuming a Pake-doublet lineshape. The inset shows the Pake doublet enlarged. The splitting of the two divergences in the Pake doublet is 176 ± 5 kHz.

stimulated echo at 8 K is shown in Fig. 2a for the PECVD sample. The inset to Fig. 2a shows the o-H_2 component after the two contributions due to bonded hydrogen have been subtracted. Figure 2b shows the Fourier transform of the data in Fig. 2a. The Pake doublet is most easily recognized as the two peaks at approximately $\pm$ 90 kHz in the inset.

Figure 3a shows the τ_2 dependence of the intensities of these three components in the PECVD sample. The intensities of the two bonded-hydrogen components (4 kHz Lorentzian and 20 kHz Gaussian) are best fitted with two T_{1d}'s while the o-H_2 has a single T_{1d} which is much longer than either of the other two components. This large value for T_{1d} is probably due to the very broad linewidth of o-H_2 which cannot be relaxed by the fluctuations in dipolar fields caused by the other two bonded-hydrogen components. Additionally, the low concentration of o-H_2 causes these molecules to relax very slowly.

Figure 3b shows the τ_1 dependence of the three components. The total intensity at $\tau_1 = \tau_2 = 0$ has been normalized to one to show the fraction of each component directly. All three components have similar large values for $T_2 \sim 400$ μs. In addition, the 20 kHz Gaussian line also has a short T_2 ~ 60 μs, and the o-H_2 has a short $T_2 \sim 70$-100 μs. The most interesting result is the very large signal fraction from o-H_2, which is $\sim$10 % of the total hydrogen signal. In this same film, the o-H_2 fraction obtained from the T_1 minimum of bonded hydrogen is $\sim$ 1 % of the total hydrogen. There is a difference of about one order of magnitude.

Similar results are obtained for three HWCVD samples. The total hydrogen concentrations in these samples were between 1 and 5 at. %. These samples have very low void densities, and internal friction measurements indicate that these films are more ordered than typical PECVD samples of a-Si:H. In the HWCVD samples the o-H_2 fractions are about 1 % of the total hydrogen concentration while the T_1 of

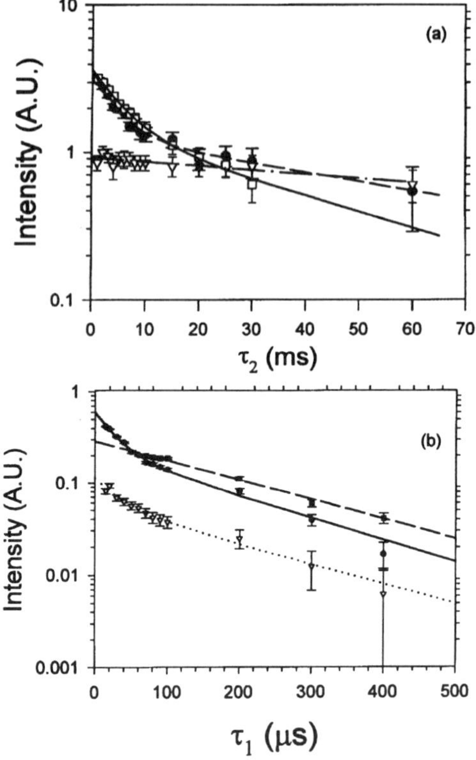

Fig. 3. Fitting of the signal intensities as functions of τ_1 and τ_2. Open squares, solid circles and open triangles represent the Lorentzian, Gaussian and o-H_2 components, respectively. The dashed line, solid line and dotted line represent the corresponding fits. (a) T_{1d} dependence of the signal. The short values of T_{1d} for the Gaussian and Lorentzian components are 4.2 and 5.8 ms, respectively. The long values of T_{1d} are 69 and 40 ms, respectively. The T_{1d} of o-H_2 is 153 ms. (b) T_2 dependence of the signal intensities for $\tau_2 = 0$. The short values of T_2 for Gaussian, and o-H_2 components are 59 and 66 μs, respectively. The long values of T_2 are 368 and 411 μs, respectively. T_2 for the Lorentzian component is 408 μs.

bounded hydrogen at (T $\approx$ 40 K) yields an upper bound for the o-H_2 fraction of at most 0.1 % of the total hydrogen. In fact, the T_1 of bonded hydrogen is almost independent of temperature from 8 K to 300 K, indicating that a different spin-lattice relaxation mechanism probably determines T_1 in these HWCVD samples.

Figure 4 shows the spectrum of one HWCVD sample and one PECVD sample at T = 8 K as measured by the JB stimulated-echo sequence. At T = 8 K, there is still strong o-H_2 motion in the HWCVD sample as is evident from the much less prominent Pake doublet and larger unresolved component of $\sim$ 60 kHz full width that appears in the HWCVD spectrum. This behavior indicates a much smaller T_c in HWCVD samples. Since T_c is mainly determined by the EFG of the Si lattice, this lower value of T_c is evidence that in a-Si:H made by HWCVD the underlying Si structure is much more ordered. Greater order results in a smaller EFG due to smaller distortions of the Si lattice. This conclusion is consistent with other experimental results [10] obtained on similar samples.

The large discrepancy between the o-H_2 concentrations as measured by the JB sequence and the T_1 minimum of bonded hydrogen suggests that only a small portion of the o-H_2 molecules participate in the spin-lattice relaxation of the bonded hydrogen. Two different o-H_2 environments have been proposed [8], which should lead to two different values of T_1 for o-H_2 in a-Si:H. In one environment for the o-H_2, relaxation is due to Raman phonon processes as

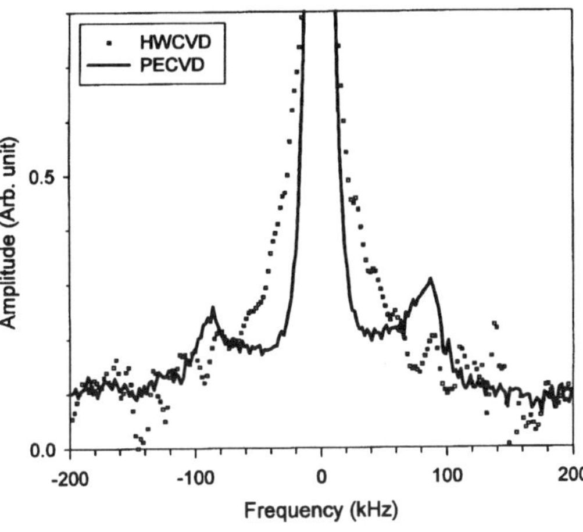

Fig. 4. The hydrogen NMR spectrum of one HWCVD sample and one PECVD sample at $T = 8$ K as obtained from a JB pulse sequence. Strong motional narrowing can be seen from the narrowing of the Pake doublet to a central line about 70 kHz wide in the HWCVD sample.

typically assumed in the early NMR studies. In the second o-H_2 environment [8], the relaxation is due to an EQQ interaction between o-H_2 molecules. This interaction is essentially temperature independent and leads to a much longer value for T_1. To examine these two environments further, the T_1 of o-H_2 was measured at the T_1 minimum of bonded hydrogen ($T = 40$ K) over a time scale from 1 ms to 2 s. These data were taken on the PECVD sample. The results, as shown in Fig. 5, show two very different T_1's. The larger value of T_1 is about 0.6 s, which is of the same order of magnitude as the value in solid hydrogen (~ 0.3 s) caused by the EQQ interaction. The smaller value of T_1 is about 3 ms, which is very close to the value calculated from the model due to Fedders [7]. These data are direct evidence that there are two very different o-H_2 environments in a-Si.H.

CONCLUSION

By using a Jeener-Broekaert sequence, the o-H_2 concentrations in a-Si:H prepared by both PECVD and HWCVD are one order of magnitude larger than previously inferred from spin-lattice relaxation time measurements on bonded hydrogen. Measurements of T_1 for o-H_2 reveal two distinct T_1's which may be related to two different local environments for the o-H_2--one dominated by the EQQ interaction and one by two-phonon Raman processes. The much lower freeze-in temperature, T_c, in HWCVD samples indicates that these samples are much more ordered than typical PECVD samples.

ACKNOWLEDGMENTS

The research was supported in part by NREL under subcontract number XAK-8-17619-13 and by NSF under grant number DMR-9704946.

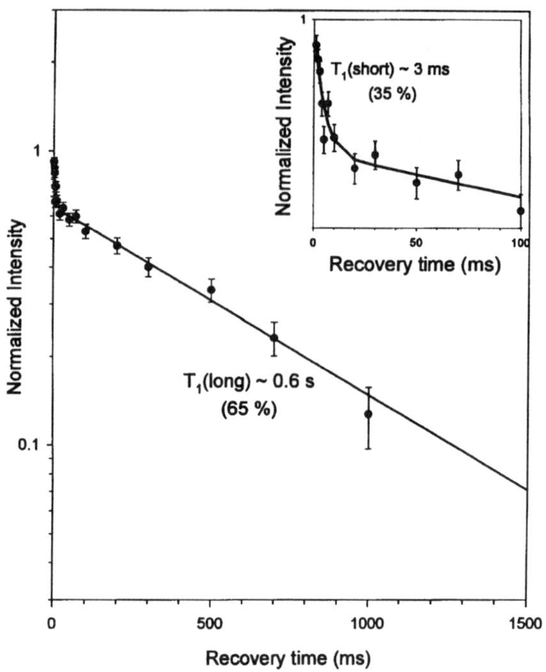

Fig. 5. The o-H$_2$ signal intensity as a function of recovery time after a saturating comb of pulses. The signal is normalized to I (t = ∞). The short T$_1$ is about 3 ms, and the longer T$_1$ is about 0.6 s. The signal fractions of these two components are 35 % and 65 %, respectively.

REFERENCES

1. J. B. Boyce, and M. Stutzmann, Phys. Rev. Lett. **54**, 562 (1985).
2. W. E. Carlos and P. C. Taylor, Phys. Rev. B **25**, 1435 (1982).
3. P. H. Chan, Ph.D. Thesis, Washington University, September, 1993 (unpublished).
4. M. S. Conradi, K. Luszczynski, and R. E. Norberg, Phys. Rev. B **20**, 2594 (1979).
5. M. S. Conradi and R. E. Norberg, Phys. Rev. B **24** 2285 (1981).
6. D. A. Drabold and P. A. Fedders, Phys. Rev. B **39**, 6325 (1989).
7. P. A. Fedders, Phys. Rev. B, **20**, 2588 (1979).
8. P. A. Fedders, R. Fisch, and R. E. Norberg, Phys. Rev. B **31**, 6887 (1985).
9. J. Jeener and P. Broekaert, Phys. Rev., **157**, 232 (1967).
10. Xiao Liu, B. E. White, Jr., and R. O. Pohl, Phys. Rev. Lett. **78,** 4418 (1997).
11. R. E. Norberg, D. J. Leopold, P. A. Fedders, J. Non-cryst. Solids, **227-230**, 124 (1998).
12. F. Reif and E. M. Purcell, Phys. Rev. **91**, 631 (1953).
13. I. F. Silvera, Rev. Mod. Phys. **52**, 393 (1980).
14. P. Hari, P.C. Taylor and R.A. Street, J. Non-Cryst. Solids **198-200**, 736 (1996), and references contained therein.

COMPARATIVE STUDY OF HYDROGEN DIFFUSION IN HOT-WIRE AND GLOW-DISCHARGE-DEPOSITED a-Si:H

J. SHINAR,[1] R. SHINAR,[2,3] K. E. JUNGE,[1] E. IWANICZKO,[4] A. H. MAHAN,[4] R. S. CRANDALL,[4] AND H. M. BRANZ[4]
[1]Ames Laboratory - USDOE and Department of Physics and Astronomy, Iowa State University, Ames, IA 50011
[2]Microelectronics Research Center, Iowa State University, Ames, IA 50011
[3]Microanalytical Instrumentation Center, Iowa State University, Ames, IA 50011
[4]National Renewable Energy Laboratory, Golden, CO 80401

ABSTRACT

Long-range atomic H motion in hot-wire deposited (HW) a-Si:H is compared directly to that in glow-discharge deposited (GD) a-Si:H by monitoring the deuterium secondary ion mass spectrometry (DSIMS) profiles in [GD a-Si:H]/[GD a-Si:(H,D)]/[HW a-Si:H] multilayers vs annealing temperature and time. While the profiles in the GD layer are in excellent agreement with complementary error-function behavior and previous studies, the profiles in the HW layer suggest that the multiple-trapping motion of the H and D atoms is much slower, possibly due to an interface layer of defects. However, an exponential "tail" of D atoms extends deep into the HW layer, probably due to a long diffusion length of mobile D atoms, consistent with the established release times of H and D from the GD layer and H loss typical during growth of HW films. The results are also discussed in terms of the H exchange model and compared to previous NMR studies of HW a-Si:H, which suggest that most of the hydrogen in the HW layer is concentrated in H-rich clusters dispersed in a network of very low H content.

INTRODUCTION

Various studies of HW a-Si:H [1 – 3] indicate that it is a promising photovoltaic material which is quite different from a-Si:H prepared by any other method. Specifically, its H content of ~2 to ~3 at.% is much lower than that of device-quality GD films [1 - 3], its ^{1}H NMR spectrum suggests that most of the H atoms are concentrated in clusters [4], and it exhibits an internal friction which is anomalously low for amorphous materials [5]. These properties suggest a material in which the isolated H-rich regions are surrounded by a Si network which is significantly more ordered than that of GD films. The close relation between the metastable defect dynamics (i.e., the Staebler-Wronski effect) and H dynamics in a-Si:H suggested by numerous studies [6 – 9] warrant a study of hydrogen motion in this material. This paper describes a preliminary deuterium secondary ion mass spectrometry (DSIMS) study of long-range H diffusion in this material.

Previous DSIMS studies of H diffusion in a-Si:H showed that the H diffusion constant is generally time-dependent [10 - 15], usually decreasing with time t as $D(t) = D_{00}(\omega t)^{-\alpha}$ where the dispersion parameter $0 \le \alpha \le 1$. $D(t)$ is determined experimentally by fitting the DSIMS profiles of annealed [a-Si:H]/[a-Si:(H,D]/[a-Si:H] multilayers to a complementary error function

$$c(x,t) = n_0 \mathrm{erfc}\{x/[2\sqrt{(\langle r^2(t)\rangle)}]\} \equiv n_0 \mathrm{erfc}\{z\} \tag{1}$$

where

$$\langle r^2(t)\rangle \equiv \int_0^t D(\tau)d\tau = [D_{00}(\omega t)^{1-\alpha}]/[\omega(1-\alpha)] \equiv r_0^2 t^{1-\alpha} \tag{2}$$

Mat. Res. Soc. Symp. Proc. Vol. 557 © 1999 Materials Research Society

is the mean square displacement of the diffusing atoms during the annealing time t. The behavior of $D(t)$ was first explained within a multiple-trapping mechanism [11]. However, subsequent work demonstrated that α increases sharply with increasing nanovoid and microvoid content C_{nV} of the films [12,14]; a sufficiently high C_{nV} suppressed the long-range motion of H and D atoms [12]. The overall picture that emerged from these and other studies [15 - 17] suggested that in addition to C_{nV}, α is largely determined by dynamical Si network relaxation processes. The previous studies also showed that $D(t)$ decreases sharply, by a factor $> 10^4$, as the H content decreases from ~17 at.% to ~2 at.% [15].

At the microscopic level, H and D motion through the Si network is believed to proceed via bond-center transport states followed by trapping at a Si-bonding site. In models proposed by Kemp and Branz [18], the trapping event likely involves an exchange between a mobile and a trapped H or D atom; it is manifest in an exponential tail of the D profile during the early stages of deuterium diffusion into the initially undeuterated layer.

This paper shows that the multiple-trapping motion of the H and D atoms in HW a-Si:H is much slower than in GD films, consistent with the low H content of the HW films. However, it also reveals the existence of an exponential tail of D atoms, which extends deep into the HW layer and is probably due to a long diffusion length λ of the mobile D atoms, consistent with the established release times of deuterium from the GD layer [18] and H loss typical during growth of HW films [1- 3].

EXPERIMENTAL PROCEDURE

The [2000 Å GD a-Si:H]/[2000 Å GD a-Si:(H,D)]/[4300 Å HW a-Si:H] multilayer samples #1 and #2 were deposited on stainless steel; sample #3 was deposited on a Si wafer. In all of the multilayers, the bottom HW layer was deposited at 360°C; the middle and top GD layers were deposited at 200°C. In addition, in Samples #2 and #3, the HW layer was annealed at 200°C in an H plasma for 5 and 15 min, respectively, prior to deposition of the GD layers. It was noted that considerable hydrogen evolved out of the HW layer during the growth at 360°C. The samples were annealed in evacuated quartz tubes and DSIMS measurements were performed on a Perkin-Elmer PHI 6300 system as described elsewhere [12,14,15].

RESULTS AND DISCUSSION

The DSIMS profiles of all of the samples annealed under the same conditions were qualitatively similar. Figures 1 and 2 show DSIMS profiles of Samples #2 and #3 annealed for various periods at 300°C, and 360°C. The open circles are the experimental results; the lines which extend into the middle deuterium-source layer are the best fit of Eq. (1), i.e., of $n_0\mathrm{erfc}\{z\}$ $+ C_0$, where C_0 is the background D level, over the regions shown. The solid lines in the tails of the DSIMS profiles deep in the HW layers are the best fits to an exponential tail $n_1\exp[-(x - x_1)/\lambda] + C_1$, where C_1 is the background D level in those regions.

As clearly seen, the profiles in the GD layers are in excellent agreement with a complementary error function over the whole measured range. In particular, the exponential tails of the profiles far from the deuterium source layer are generally very small or unobservable.

Figure 3 compares the values of $<r^2(t)>$ in GD films reported previously [10,14,15] with the values obtained by the present fitting procedure for the GD and HW layers in this work. As clearly seen, the present results on the GD layers are consistent with the previous reports, as the error in all of the values is probably ±20%. However, we note that some of previous measurements did demonstrate an exponential tail in GD films in the early-time regime [7].

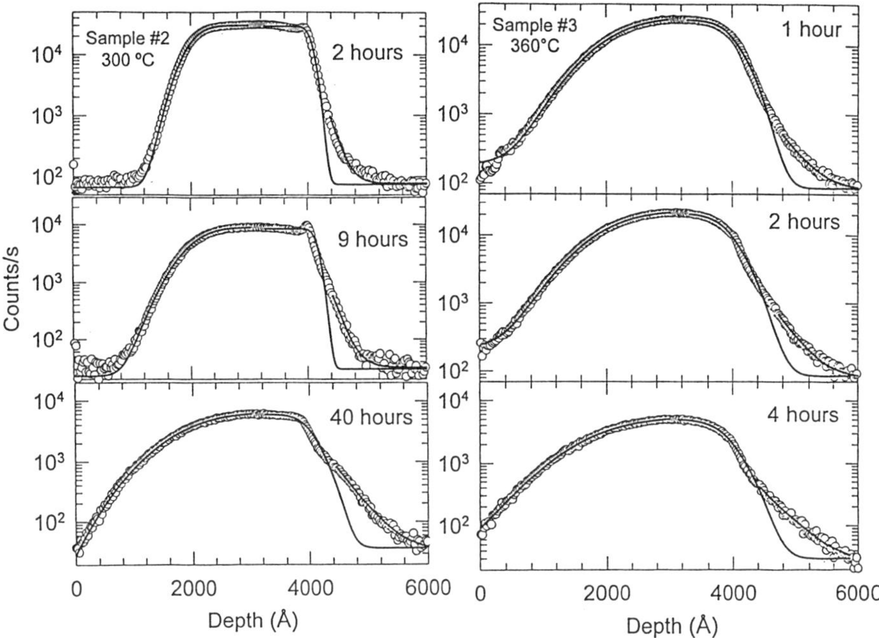

Figure 1. DSIMS profiles of Sample #2 annealed as shown. The lines from the deuterated source layer are the best fit of $n_0\mathrm{erfc}\{z\} + C_0$; those in the tail of the profile in the (bottom) HW layer are the best fit of $n_1\exp[-(x-x_1)/\lambda] + C_1$.

Figure 2. DSIMS profiles of Sample #3 annealed as shown. The lines are the best fits of $n_0\mathrm{erfc}\{z\} + C_0$ and $n_1\exp[-(x-x_1)/\lambda] + C_1$ as explained in the caption of Figure 1.

Figs. 1 – 3 clearly show that the values of $\langle r^2(t)\rangle$ are generally considerably smaller in the HW layers than in the GD layers. Since the DSIMS profiles the trapped deuterium, the results indicate that the penetration of the D atoms into the trapping sites of the (bottom) HW layers is much slower than into those of the (top) GD layers. At the same time, however, the exponential tail of the D atoms which is prominent above the $\mathrm{erfc}\{z\}$ profile in the HW layer but absent in the GD layers demonstrates that the mean diffusion distance λ between trapping events is much higher in the HW layers. The values of λ obtained from the fits shown in Figures 1 and 2 are summarized in Table I. They range from ~150 Å following annealing for $t = 2$ hours at 300°C to ~400 Å after $t = 4$ hours at 360°C. At a given temperature they generally increase with increasing t, suggesting that the release times $\tau \sim 2$ hours at 300°C and 360°C [18]. They also suggest that the true diffusion distance $\lambda_0 \le \lambda$ ranges is ~150 and ~250 Å at these respective temperatures.

The slow penetration and diffusion of the H and D atoms which undergo multiple trapping probably results from two properties of the samples: (i) a large trap density at the GD/HW inter-

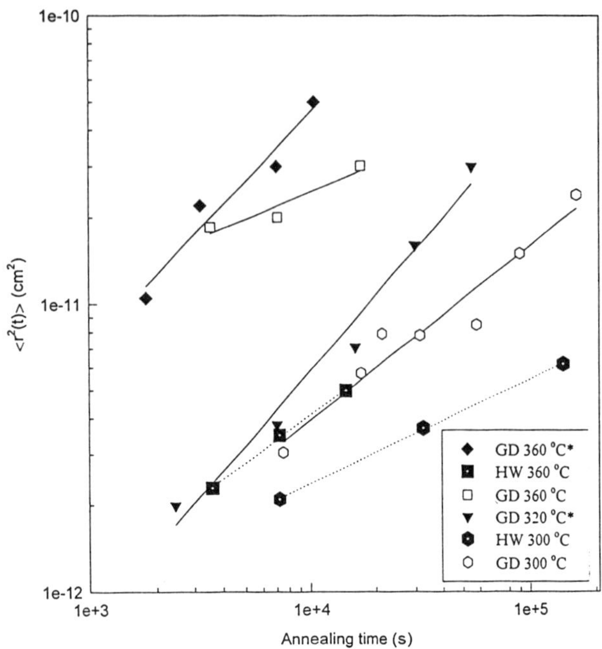

Figure 3. $<r^2(t)>$ of D profiles in GD and HW films vs annealing time t at 300°C and 360°C, obtained by fitting Eq. (1) to the measured profiles. *The data points marked ▼ at 320°C and ◆ at 360°C are previous results [14,15].

Table I. The values of λ obtained by fitting $n_1\exp[-(x-x_1)/\lambda] + C_1$ to the tails of the D profiles in the HW layers shown in Figures 1 and 2.

Sample #2 $T = 300°C$		Sample #3 $T = 360°C$	
Anneal time t (hours)	λ (Å)	**Anneal time t (hours)**	λ (Å)
2	150	1	280
9	180	2	260
40	300	4	400

face, possibly due to, among others, the large H density [19], and consequent frequent exchange events in that region [18]. (ii) The low H density in the bulk of the HW layer also implies that detrapping of a D atom in that region via exchange [18] would be infrequent relative to its occurrence in the GD layer. Thus, the exchange mechanism also accounts for the sharp decrease in the multiple-trapping motion with decreasing H content [15].

The observed evolution of H out of the HW films during growth [1-3] and following annealing [19] is also consistent with the scenario in which the multiple-trapping motion of the H and D atoms is slow, but since λ is long, they may evolve out of the HW layer much like their evolution from crystalline Si (c-Si). We note that since the microvoid and nanovoid content of HW films studied by small angle x-ray scattering is low [20], the scenario of H and D atoms diffusing a short distance to the nearest void and evolving out of the film at that void's surface is improbable.

Finally, the scenario of rapid diffusion of H and D atoms in the Si network, reminiscent of H diffusion in c-Si, concomitant with slow diffusion of the multiply-trapped D atoms, strongly supports the suggested structure of HW a-Si:H as a relatively ordered and almost H-free Si network which contains small, isolated H-rich domains [4,5]. Within this suggested structure of HW material, the H-rich domains are precisely the H and D trapping sites, and the trapping event most likely proceeds via an exchange mechanism.

SUMMARY

In summary, H motion in HW a-Si:H was compared directly with that in GD a-Si:H by monitoring the deuterium SIMS (DSIMS) profiles in [GD a-Si:H]/[GD a-Si:(H,D)]/[HW a-Si:H] multilayers vs annealing temperature and time. The profiles in the (top) GD layer followed a complementary error function closely, with diffusion lengths $<r^2(t)>$ which were in agreement with previous studies. In the HW layer, the values of $<r^2(t)>$ were considerably smaller and an exponential tail indicative of a long diffusion length of mobile D atoms was prominent. The results suggest that the penetration and diffusion of H and D atoms which undergo multiple trapping is slow, but the diffusion of mobile H and D atoms is rapid and reminiscent of their behavior in crystalline Si. This scenario is consistent with the exchange mechanism of Kemp and Branz, in which the low H content of the HW films and the concentration of the H atoms in isolated H-rich domains results in infrequent exchange events, and the suggested structure of HW films, in which the Si network surrounding the H-rich domains is relatively well-ordered.

ACKNOWLEDGEMENTS

Ames Laboratory is operated by Iowa State University for the US Department of Energy under Contract W-7405-Eng-82. This work was supported by the Director for Energy Research, Office of Basic Energy Sciences. The work at NREL was supported by the DOE under Contract DE-AC36-98-GO10337.

REFERENCES

1. A. H. Mahan, J. Carapella, B. P. Nelson, R. S. Crandall, and I. Balberg, J. Appl. Phys. **69**, 6728 (1991).

2. A. H. Mahan and J. Vanacek, in *Amorphous Silicon Materials and Solar Cells*, edited by B. L. Stafford, AIP Conf. Proc. **234**, 195 (AIP Press, NY, 1991).

3. D. Kwon, J. D. Cohen, B. P. Nelson, and E. Iwaniczko, in *Amorphous Silicon Technology – 1995*, edited by M. Hack, E. A. Schiff, A. Madan, M. Powell, and A. Matsuda, Mat. Res. Soc. Symp. Proc. **377**, 301 (1995).

4. Y. Wu, J. T. Stephen, D. X. Han, J. M. Rutland, R. S. Crandall, and A. H. Mahan, Phys. Rev. Lett. **77**, 2049 (1996).

5. X. Liu, B. E. White, Jr., R. O. Pohl, E. Iwanizcko, K. M. Jones, A. H. Mahan, B. N. Nelson, R. S. Crandall, and S. Veprek, Phys. Rev. Lett. **78**, 4418 (1997).

6. P. V. Santos, N. M. Johnson, and R. A. Street, Phys. Rev. Lett. **67**, 2686 (1991).

7. H. M. Branz, S. E. Asher, and B. P. Nelson, Phys. Rev. B **47**, 7061 (1993).

8. R. Biswas, Q. Li, B. C. Pan, and Y. Yoon, Phys. Rev. B **57**, 2253 (1998); R. Biswas and B. C. Pan, Appl. Phys. Lett. **72**, 371 (1998); R. Biswas and Y.-P. Li, Phys. Rev. Lett. **82**, 2512 (1999).

9. H. M. Branz, Sol. St. Comm. **105**, 387 (1998).

10. R. A. Street, C. C. Tsai, J. Kakalios, and W. B. Jackson, Phil. Mag. B **56**, 305 (1987).

11. J. Kakalios, R. A. Street, and W. B. Jackson, Phys. Rev. Lett. **59**, 1037 (1987).

12. J. Shinar, R. Shinar, S. Mitra, and J.-Y. Kim, Phys. Rev. Lett. **62**, 2001 (1989).

13. X.-M. Tang, J. Weber, Y. Baer, and F. Finger, Phys. Rev. B **41**, 7945 (1990); *ibid.* **42**, 7277 (1990).

14. J. Shinar, R. Shinar, X.-L. Wu, S. Mitra, and R. F. Girvan, Phys. Rev. B **43**, 1631 (1991).

15. R. Shinar, J. Shinar, H. Jia, and X.-L. Wu, Phys. Rev. B **47**, 9361 (1993).

16. V. Halpern, Phys. Rev. Lett. **67**, 611 (1991).

17. J. Shinar, H. Jia, R. Shinar, Y. Chen, and D. L. Williamson, Phys. Rev. B **50**, 7358 (1994).

18. M. Kemp and H. M. Branz, Phys Rev. B **47**, 7067 (1993); *ibid.* **52**, 13946 (1995).

19. R. Shinar, J. Shinar, E. Iwaniczko, A. H. Mahan, R. S. Crandall, and H. M. Branz, unpublished results.

20. D. L. Williamson, in *Amorphous Silicon Technology – 1995*, edited by M. Hack, E. A. Schiff, A. Madan, M. Powell, and A. Matsuda, Mat. Res. Soc. Symp. Proc. **377**, 251 (1995)..

HYDROGEN TRANSPORT IN PHOSPHORUS AND BORON DOPED POLYCRYSTALLINE SILICON

N. H. NICKEL AND I. KAISER
Hahn-Meitner-Institut Berlin, Rudower Chaussee 5, D-12489 Berlin, Germany.

ABSTRACT

Hydrogen diffusion in phosphorus and boron doped polycrystalline silicon was investigated by deuterium diffusion experiments. The presence of dopants enhances hydrogen diffusion. The effective diffusion coefficient D_{eff} is thermally activated and the activation energy varies between 0.1 and 0.4 eV. This is accompanied by a variation of the diffusion prefactor by 12 orders of magnitude. Using the theoretical diffusion prefactor the actual energy $\overline{E}_A$ was calculated from D_{eff}. $\overline{E}_A$ also depends strongly on the Fermi energy and exhibits a similar dependence as the formation energies of H^+ and H^- in single crystal silicon.

1. INTRODUCTION

Polycrystalline silicon (poly-Si) is becoming very attractive as an active layer for a variety of applications such as solar cells and thin-film transistors. Although there are many methods to produce poly-Si solid-state and laser crystallization of amorphous silicon are becoming very attractive methods, recently. However, these methods have in common that the resulting poly-Si is not suitable for device applications since the crystallization process causes the out-diffusion of hydrogen leading to high defect concentrations.

In order to improve the electrical properties grain-boundary defects have to be passivated. Commonly, this is achieved by exposing the samples to a hydrogen plasma at elevated temperatures.[1] The major drawback of this method is the fact that defect passivation advances from the surface and is governed by diffusion.[2]

Hydrogen diffusion in disordered silicon has been studied in great detail over the past decade. Although most of the research has been devoted to H diffusion in amorphous silicon[3,4], recently an increasing effort is being made to understand H migration in poly-Si.[5] Independent of the degree of disorder (e.g.: amorphous or polycrystalline) it was found that at high hydrogen concentrations the diffusion is dispersive depending on the hydrogenation time and in both material systems hydrogen diffusion can be described in terms of a two-level model.[6,7]

In this paper we investigate the influence of the Fermi energy on the H transport properties in solid-state crystallized poly-Si. Hydrogen diffusion was studied as a function of the phosphorus and boron concentrations, the substrate temperature, and the hydrogen concentration. We find that the diffusion properties of H are strongly influenced by the Fermi energy and the hydrogen concentration.

2. EXPERIMENTS

Polycrystalline silicon samples were fabricated by solid-state crystallization of a-Si:H at 600 °C. Doping of the samples was achieved by multiple implantations with phosphorus or boron.

Mat. Res. Soc. Symp. Proc. Vol. 557 © 1999 Materials Research Society

The dopants were activated in a furnace anneal at 900 °C. According to transmission-electron microscopy micrographs the specimens reveal an average grain size of 1500 Å.

The hydrogenation or deuteration sequence consisted of the following steps. First the native oxide was removed with dilute HF to avoid a barrier to hydrogen incorporation. Then the specimens were exposed to monatomic deuterium produced in a remote plasma passivation system. Deuterium concentration depth profiles were obtained from secondary-ion-mass spectrometry (SIMS) measurements.

In the following sections the terms hydrogenation and deuteration will be used interchangeably since no difference in the diffusion properties of the two species was found.

3. PHOSPHORUS DOPED POLYCRYSTALLINE SILICON

The influence of the substrate temperature on hydrogen diffusion in poly-Si doped with a phosphorus concentration of 10^{16} cm^{-3} is shown in Fig. 1. The samples were exposed to monatomic deuterium at the indicated temperatures for 5 min. With increasing substrate temperature the deuterium surface concentration increases and deuterium diffuses deeper into the bulk. At higher temperatures ($T \geq 206$ °C) the deuterium concentration decays according to a complementary error function (erfc). The dashed curves represent fits to the convolution of the SIMS depth-resolution and an erfc.[7] At a deuterium concentration of 6×10^{18} - 1×10^{19} cm^{-3} the data deviate from the fit and exhibit an exponential decrease with a characteristic slope of $x_0 \approx 24$ nm. This is consistent with previous observations in undoped poly-Si[5] and a-Si:H.[6,7]

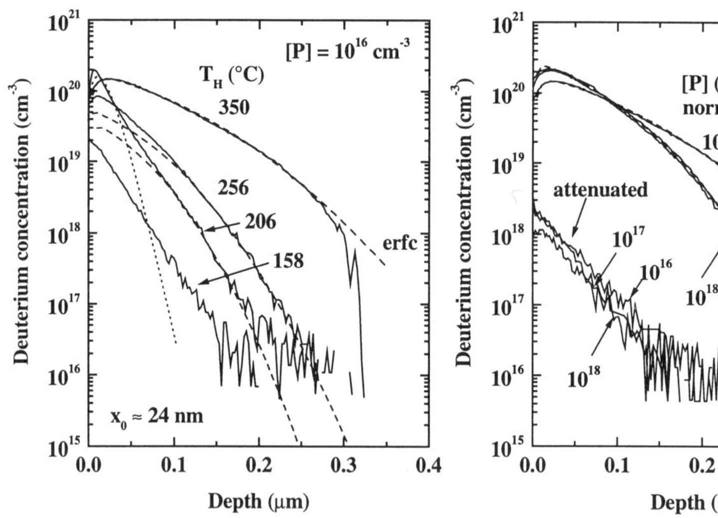

Fig. 1. D concentration depth profiles in P doped poly-Si. The specimens were exposed to monatomic D for 5 min at the indicated temperatures. The dashed lines are fits to an erfc and the dotted lines indicate exponential decays with a characteristic slope x_0 of ≈24 nm.

Fig. 2. Influence of the P and deuterium concentrations on deuterium diffusion. The samples were exposed to monatomic deuterium for 5 min at 350 °C. The curves labeled "attenuated" represent D depth profiles after an attenuated plasma exposure.

The influence of the phosphorus concentration on deuterium diffusion is shown by the solid curves labeled "normal" in Fig. 2. The phosphorus concentration was varied between 10^{16} and 10^{18} cm^{-3}. The specimens were deuterated for 5 min at 350 °C. A significant change in the deuterium depth profiles is observed as the phosphorus concentration increases from 10^{16} to 10^{17} cm^{-3}. The deuterium surface concentration increases while the D concentration in the bulk decreases with increasing phosphorus concentration. A further increase of the phosphorus concentration to 10^{18} cm^{-3} does not cause a change in the D depth profile. A similar but more pronounced influence of the P concentration on deuterium migration in c-Si was observed previously. It was suggested that as the P concentration increases the surface concentration of the dominating diffusion species decreases since more and more hydrogen interacts with phosphorus and forms complexes.[8]

The influence of the deuterium concentration on the diffusion properties was investigated by attenuating the D flux. This results in a decrease of the D surface concentration by 2 orders of magnitude (Fig. 2). In addition, the depth profiles exhibit an exponential decay with a characteristic slope of $\approx$32 nm which is independent of the phosphorus concentration. A similar behavior is observed at lower deuteration temperatures.

The effective diffusion coefficient D_{eff} was obtained by fitting the data to the convolution of the SIMS depth resolution and a complementary error function. In Fig. 3a D_{eff} is plotted as a function of the reciprocal temperature. The open and solid symbols denote D_{eff} for high and low deuterium concentration diffusion, respectively. For comparison the temperature dependence of D_{eff} in undoped poly-Si is plotted by the diamonds. Compared with undoped poly-Si D_{eff} for high D concentration diffusion increases by roughly one order of magnitude due to phosphorus doping. This result shows that phosphorus doping of poly-Si enhances hydrogen diffusion. D_{eff} is thermally activated with an apparent activation energy varying between 0.26 and 0.35 eV. For low concentration D diffusion the effective diffusion coefficient is smaller compared with D_{eff}

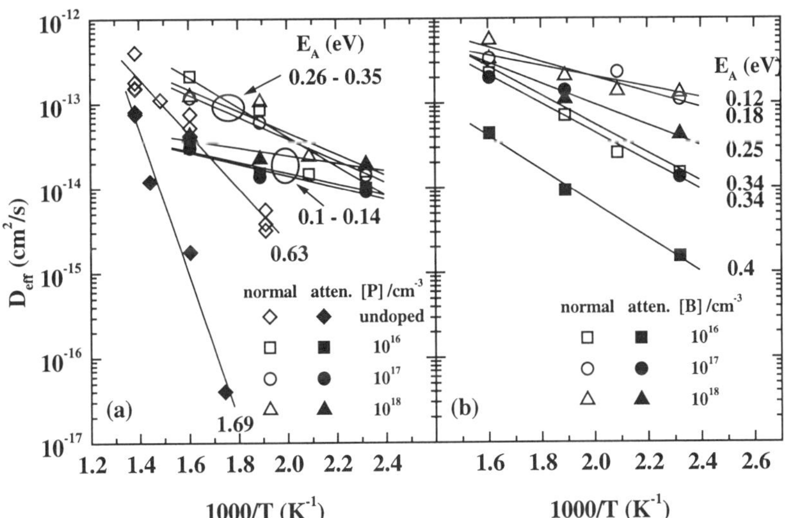

Fig. 3. Effective diffusion coefficient as a function of the reciprocal temperature for H diffusion in (a) undoped and phosphorus doped poly-Si and (b) in boron doped poly-Si. The open (solid) symbols denote high (low) concentration diffusion.

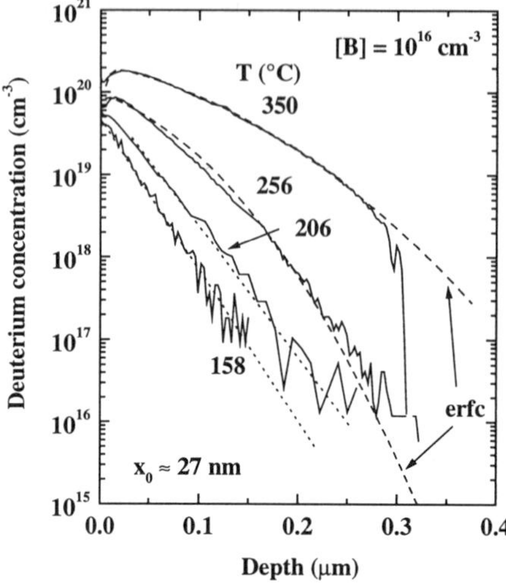 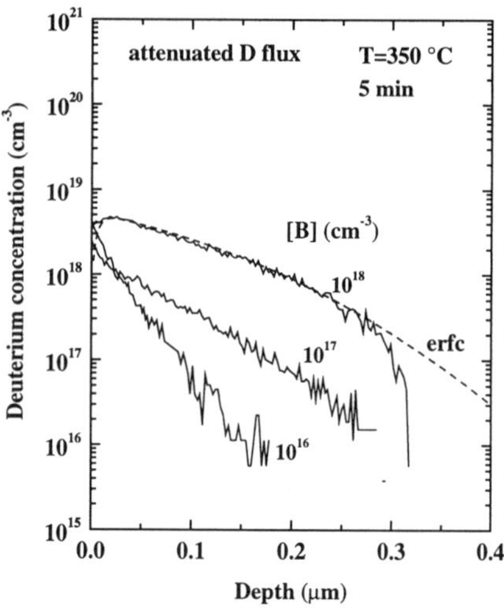

Fig. 4. D concentration depth profiles in boron doped poly-Si. The samples were exposed to monatomic D for 5 min at the indicated temperatures. The dashed lines are fits to an erfc and the dotted lines indicate exponential decays with a slope x_0 of $\approx$27 nm.

Fig. 5. Deuterium depth profiles after an exposure to an attenuated D plasma for 5 min at 350 °C. The samples were doped with the indicated boron concentrations. The dashed line is a fit to an erfc.

for high concentration diffusion. However, in contrast to undoped poly-Si (diamonds in Fig. 3a) E_A decreases to values between 0.1 and 0.14 eV.

4. BORON DOPED POLYCRYSTALLINE SILICON

Deuterium concentration depth profiles in boron doped poly-Si are plotted in Fig. 4. At substrate temperatures $T \leq 206$ °C the profiles decay exponentially with depth with a characteristic slope of $\approx$27 nm. The value of the slope is similar to results obtained for low D concentration diffusion in undoped[5] and phosphorus doped poly-Si. At higher passivation temperatures the D concentration profiles decay according to an erfc. This behavior is similar to deuterium diffusion in phosphorus doped poly-Si. However, deuterium diffusion in boron doped poly-Si is enhanced compared with D diffusion in undoped[5] *and* phosphorus doped material (Figs. 1 and 4).

The results obtained from low deuterium concentration diffusion experiments in boron doped poly-Si are shown in Fig. 5. Attenuation of the D flux results in a decrease of the surface concentration by about two orders of magnitude. However, an exponential decay of the deuterium concentration is observed only for the sample with the lowest boron concentration. At boron concentrations $\geq 10^{17}$ cm^{-3} the depth profiles decay according to an erfc. With increasing boron concentration the effective diffusion coefficient increases. This shows that low concentration deuterium diffusion in boron doped poly-Si is significantly enhanced compared with undoped and phosphorus doped specimens.

The temperature dependence of the effective diffusion coefficient is shown in Fig. 3b. In response to an increase of the deuterium and boron concentrations the apparent activation energy decreases from 0.4 eV to 0.12 eV. In contrast to phosphorus doped poly-Si (Fig. 3a) the activation energy for low D concentration diffusion is always larger than E_A for high D concentration diffusion for a given phosphorus concentration.

5. DISCUSSION

In the preceding sections we showed that the apparent activation energy of the effective diffusion coefficient varies over a range of 1.6 eV due to a change of the Fermi energy. The effective diffusion coefficient can be written as

$$D_{eff} = D_0 \exp\left(-\frac{E_A}{kT}\right) , \qquad (1)$$

where D_0 is the diffusion prefactor, T is the temperature, and k is the Boltzmann constant. According to Eq. (1) the diffusion prefactor should be similar for all data shown in Fig. 3. This, however, is not the case. Extrapolating the data in Fig. 3 to $1/T = 0$ K^{-1} reveals that D_0 varies between 10^{-1} and 10^{-13} cm^2/s. These widely varying values for the activation energy and the diffusion prefactor do not provide insight into the microscopic mechanism governing hydrogen migration. Theoretically, the diffusion prefactor should be independent of the deuterium and dopant concentrations. The theoretical diffusion prefactor is given by $D_{0,T} = 1/6\,\nu a^2$, where ν is the attempt frequency and a is the mean free path. With reasonable assumptions of $\nu \approx 10^{12}$ Hz and $a \approx 0.3$ nm the theoretical prefactor amounts to $D_{0,T} \approx 10^{-3}$ cm^2/s. With this prefactor the actual energy of the diffusion process at a given temperature was calculated using the relation

$$\overline{E}_A = -kT \ln\left(\frac{D_{eff}}{D_{0,T}}\right) . \qquad (2)$$

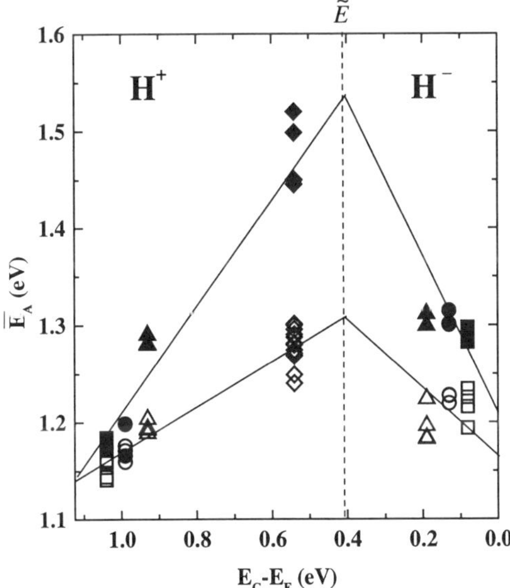

Fig. 6. $\overline{E}_A$ at 350 °C versus the Fermi energy for high (open symbols) and low (solid symbols) deuterium concentration. At $E_C-E_F = \tilde{E}$ the charge state of monatomic H changes from H$^+$ to H$^-$.

In the following we will focus on the values of $\overline{E}_A$ at 350 °C. In Fig. 6 $\overline{E}_A$ is plotted as a function of the Fermi energy. For high D concentration $\overline{E}_A$ first increases from ≈ 1.15 to 1.28 eV and then decreases to ≈ 1.2 eV as the Fermi energy moves from the valence-band edge to the conduction-band edge. A similar but more pronounced change of $\overline{E}_A$ is observed for low D concentration (solid symbols). It is important to note that the dependence of $\overline{E}_A$ on the Fermi energy is similar to that of the formation energy for H^+ and H^- in c-Si.[9] Based on this behavior we suggest a model for trap-limited H diffusion, in which $\overline{E}_A$ is the sum of the migration barrier for interstitial H, and the difference between the formation energies for a hydrogen interstitial (H^+ or H^-, depending on the Fermi energy), and the formation energy of H at a deep trap.[10]

6. SUMMARY

In summary, we have shown that H diffusion in poly-Si depends strongly on the Fermi energy. Hydrogen transport in doped poly-Si is enhanced. The effective diffusion coefficient is thermally activated and varies between 0.1 and 1.69 eV in response to a change of the D concentration and the Fermi energy. This is accompanied by a variation of the diffusion prefactor over 12 orders of magnitude. Using the theoretical diffusion prefactor the energy $\overline{E}_A$ necessary to yield the observed diffusion coefficient was calculated. $\overline{E}_A$ also depends on the Fermi energy and reveals a dependence similar to that of the formation energy for H^+ and H^- in c-Si. Our data support a model for trapping controlled H transport in which $\overline{E}_A$ is the sum of the migration barrier for interstitial H, and the difference between the formation energies for a hydrogen interstitial (H^+ or H^-, depending on E_F), and the formation energy of H at a deep trap.[10]

ACKNOWLEDMENTS

This work was supported by the BMBF under contract number 03299773.

REFERENCES

[1] T. J. Kamins and P. J. Marcoux, IEEE Electron Device Lett. **EDL-1**, 159 (1980).

[2] N. H. Nickel, N. M. Johnson, and W. B. Jackson, Appl. Phys. Lett. **62**, 3285 (1993).

[3] P. V. Santos, N. M. Johnson, R. A. Street, M. Hack, R. Thompson, and C. C. Tsai, Phys. Rev. B **47**, 10244 (1993), and references therein.

[4] H. M. Branz, S. E. Asher, and B. P. Nelson, Phys. Rev. B **47**, 7061 (1993).

[5] N. H. Nickel, W. B. Jackson, and J. Walker, Phys. Rev. B **53**, 7750 (1996).

[6] W. B. Jackson and C. C. Tsai, Phys. Rev. B **45**, 6564 (1992).

[7] P. V. Santos and W. B. Jackson, Phys. Rev. B **46**, 4595 (1992).

[8] C. Herring and N. M. Johnson, in *Hydrogen in Semiconductors*, eds. J. I. Pankove and N. M. Johnson (Academic Press, San Diego, 1991) vol. 34, Chap. 10.

[9] C. G. Van de Walle, P. J. Denteneer, Y. Bar-Yam, and S. T. Pantelides, Phys. Rev. B **39**, 10791 (1989).

[10] N. H. Nickel, I. Kaiser, and C. G. Van de Walle, submitted for publication.

ANISOTROPY IN HYDROGENATED AMORPHOUS SILICON FILMS AS OBSERVED USING POLARIZED FTIR-ATR SPECTROSCOPY

J. D. WEBB, L. M. GEDVILAS, R. S. CRANDALL, E. IWANICZKO, B. P. NELSON, A. H. MAHAN, R. REEDY, R. J. MATSON
National Renewable Energy Laboratory, Golden, CO

ABSTRACT

We used polarized attenuated total reflection (ATR) measurements together with Fourier transform infrared (FTIR) spectroscopy to investigate the vibrational spectra of hydrogenated amorphous silicon (a-SiH$_x$) films 0.5-1.0 microns in thickness. We deposited the films using hot-wire or plasma-enhanced chemical vapor deposition methods (HWCVD or PECVD, respectively) on crystalline silicon and cadmium telluride substrates. Our ATR technique gave a spectral range from 2100-400 cm^{-1}, although the Si-H wagging mode absorption band at 640 cm^{-1} was somewhat distorted in the a-SiH$_x$/Si samples by impurity and lattice absorption in the silicon ATR substrates. We report the identification of a Si-O-C impurity band with maximum intensity at 1240-1230 cm^{-1}. The assignment of this band to a Si-O-C vibration is supported by secondary-ion mass spectrometry (SIMS) measurements. Our polarized FTIR-ATR spectra of HWCVD and PECVD a-SiH$_x$ films on <111> Si ATR substrates show that the impurity dipoles are oriented strongly parallel to the film growth direction. The wagging mode absorbance band is more intense in the film plane. This trend is less pronounced for the Si-H stretching vibrations. These observations are consistent with some degree of anisotropy or medium-range order in the films. The anisotropy in the Si-H bands may be related to residual stress in the films. Our scanning electron microscopy (SEM) analyses of the samples offer additional evidence of bulk structural anisotropy in the a-SiH$_x$/Si films. However, the Si-O-C impurity band was not observed in the polarized ATR-FTIR spectra of the a-SiH$_x$/CdTe samples, thus indicating that the Si substrates influence formation of the impurity in the a-SiH$_x$/Si films.

INTRODUCTION

Infrared (IR) spectroscopy and hydrogen secondary-ion mass spectrometry (SIMS) are the two most accessible methods for determining the hydrogen content of hydrogenated amorphous silicon (a-SiH$_x$) films [1,2]. For IR transmission spectroscopy, the a-SiH$_x$ films are commonly deposited on crystalline silicon substrates. However, in IR transmission, the electric field vector of the transmitted radiation is aligned predominantly with the sample plane.

Attenuated total reflection (ATR) infrared spectroscopy involves the use of an IR - transparent waveguide of refractive index n_1, in contact with an IR-absorbing sample of refractive index n_2, where typically $n_1 > n_2$. If infrared light is directed into the waveguide (substrate) at an incidence angle (θ) greater than its critical angle, total internal reflection takes place, except at wavelengths where the evanescent standing wave extending into the sample interacts with the IR-active dipoles and is attenuated [3]. The ATR-IR spectrum of a sample resembles its IR absorbance spectrum, except for the wavelength dependence in the ATR-absorbance band intensities induced by the wavelength-dependent penetration depth of the evanescent wave into the rarer sample medium [3]. However, in cases where n_2 nearly equals or slightly exceeds n_1, as with a-SiH$_x$ films on Si or CdTe substrates in air, total internal reflection takes place at the sample outer surface (Fig. 1), rather than at the substrate/sample interface. In

Mat. Res. Soc. Symp. Proc. Vol. 557 ©1999 Materials Research Society

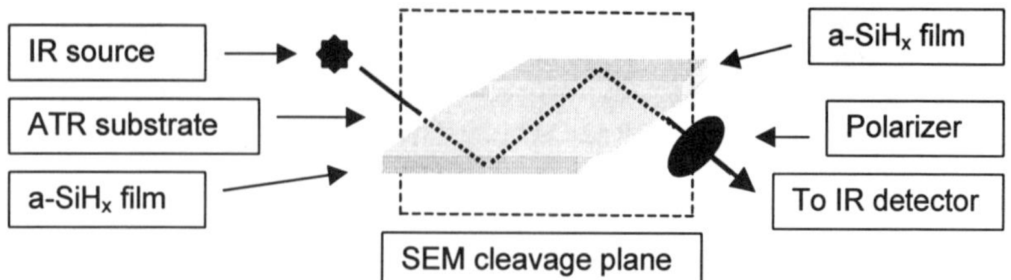

Fig. 1. Schematic drawing of the attenuated total reflection configuration for infrared spectroscopy of a-SiH$_x$ films. The cleavage plane for SEM measurements on the <111> Si ATR substrates is also shown.

these cases there is no wavelength dependence of the ATR-absorbance band intensities [3], although dispersion in n$_2$ may have some effect on the ATR-IR absorbance band shapes of α-SiH$_x$ films on Si substrates [4]. Unlike transmission spectroscopy, ATR spectroscopy allows polarization of the incident light in the film growth direction, as well as in the film plane (Fig.1).

An early work pioneering the application of ATR spectroscopy to the analysis of a-SiH$_x$ films [5] described the measurement of the first overtone absorbance of the Si-H stretching vibration near 3950 cm^{-1}, as well as the absorbance from 3900-5400 cm^{-1} corresponding to high-pressure molecular hydrogen in microvoids in the films. To observe these weak absorbance bands, an optical path of 6.3 cm through the Si ATR substrates was used. However, at such long path lengths, impurity, lattice, and free-carrier absorption in the Si substrates limited their transparency to frequencies higher than about 1500 cm^{-1} [5,6]. In this work, we used short-path-length (0.7-1.4 cm) Si and CdTe ATR substrates to extend the range of our spectroscopic measurements to frequencies as low as 400 cm^{-1}.

EXPERIMENT

Thin films of hydrogenated amorphous silicon were deposited on both of the <111> faces of crystalline silicon ATR substrates and on one face of polycrystalline cadmium telluride ATR substrates, as shown in Fig. 1. These substrates were blown clean with dry nitrogen before film deposition, but were otherwise used as received from the manufacturer (Harrick Scientific Co.), as is our typical practice in preparing a-SiH$_x$ films on Si substrates for IR analysis of H content. The gaps in film coverage of the substrate (Fig. 1) result from the clamp used to secure the substrates to the heater can in the deposition system [1]. The clamp also shielded the 45° faces from the film deposition reaction. The ATR substrates were of the 45° single-pass parallelepiped (SPP) type [4], and were either 10 x 5 x 1 mm or 5 x 5 x 1 mm in size. The <111> orientation of Si and other cubic semiconductors is commonly used for ATR substrates, since it facilitates producing them in the 45° SPP geometry. We used hot-wire chemical vapor deposition (HWCVD) [1] and plasma-enhanced chemical vapor deposition (PECVD, or glow-discharge) [2] methods to deposit the a-SiH$_x$ films. We prepared other HWCVD a-SiH$_x$ films on <100> crystalline Si wafers. These samples were analyzed both by clamping them to an uncoated <111> Si ATR substrate and by IR transmission. A Harrick Model 4XV 4x beam condenser equipped with a Perkin-Elmer wire-grid polarizer was operated with average incidence angle set to 45 ± 15 degrees to the film plane to collect the FTIR-ATR spectra. The wide range of actual incidence angles is caused by the ~F1 beam condenser optics. The condenser optics minimized the effects of the nominally odd number of IR reflections in the 5 x 5 x 1 mm, 45° SPP ATR substrates [7].

The infrared spectra were collected using either a Nicolet Magna 550 FTIR spectrophotometer or a Nicolet 510 FTIR spectrophotometer with deuterated tryglycine sulfate detector. The polarized FTIR-ATR spectra (I) of the films deposited on ATR substrates were taken by averaging $10^3 - 10^4$ scans of the samples at 8 cm^{-1} resolution, and were converted to absorbance units by taking the ratio $A=-\log_{10}(I/I_0)$, where I_0 is the unratioed spectrum of the uncoated substrate taken at similar polarization and resolution. The polarized ATR spectra of the films deposited on Si wafer substrates were taken similarly, except that the samples were clamped to an uncoated 45°, 10 x 5 x 1 mm, SPP ATR crystal for analysis. Transmittance FTIR spectra (100I/I$_0$, or %T) were recorded by collecting 128 scans of the samples (I) and substrates (I$_0$) at 4 cm^{-1} resolution and computing the ratio. The transmittance spectra were converted to absorbance units for comparison to the polarized ATR-absorbance spectra or were converted to absorbance coefficient spectra (α_v) for calculation of hydrogen content [1,2].

After collection of their infrared spectra, some of the samples were subjected to SIMS depth profiling for Si, H, O, C, and/or N. The SIMS depth profiles of the samples were collected using a Cameca IMs-3f system with a Cs$^+$ primary-ion source, and detection of negative secondary ions. The diameter of the ion etch pit was ~100 μm at the sample surfaces. Certain of the samples were then cleaved and imaged in cross-section using scanning electron microscopy (SEM). Sample films deposited on the <111 > faces of Si ATR substrates were cleaved through the <111> tilt plane, as indicated in Fig. 1. The SEM images were collected with a JEOL JSM field-emission scanning electron microscope with an effective spatial resolution of ~1 nm. Table 1 gives a summary of the analyses done on each sample.

Table 1. Deposition parameters and analytical methods for a-SiH$_x$ film samples. All films were deposited on both sides of 5x5x1 mm Si ATR substrates, unless otherwise noted.

Sample Number	Deposition Method, Substrate Type	$T_{substrate}$ (K)	100*[H]/[H+Si], Si=5E22 cm^{-3}	Thickness (μm)	Analysis Methods
H473	GD	470.5	10.0** (11.5)	(0.8)	ATR, %T, SIMS, SEM
H497	HW 10x5x1 CdTe, 1-side	541.5	9.3**	1.5 †	ATR, SEM
H498	GD 10x5x1CdTe, 1-side	470.5	10.0** (12.3)	(0.9)	ATR, %T, SIMS, SEM
T896	HW	541.5	9.8** (9.1)	(0.5)	ATR, %T, SIMS
H496	HW, 1-side stepwise deposition	541.5	9.8** (7.6) 14.0*	(0.7)	ATR, %T, SIMS
HW64	HW, 1-side Si wafer	598.0	12.4*	[0.88]	Contact ATR, SEM
HW76	HW, 1-side Si wafer	613.0	8.4*	[0.55]	Contact ATR, SEM
HW58	HW, 1-side Si wafer	648.0	1.7*	[1.50]	Contact ATR, SEM
H350i	HW	669.3	2.2** (2.9)	(0.9)	ATR, %T, SIMS
H472	HW, 1-side bilayer deposition	690.6 (bulk), 541.5 (15 nm)	1.2, 9.3** (0.3 - 0.6, bulk, 15 nm)	(0.9)	ATR, %T, SIMS, SEM
H351I	HW	719.0	0** (0.6)	(4.7)	ATR, SIMS
			*FTIR ( ) SIMS	[ ] profiler ( ) SIMS	
			**estimated from T_{sub}	† SEM	

RESULTS

Figure. 2 shows the polarized ATR-FTIR spectrum of PECVD sample #H473. Our ATR technique enabled detection of the Si-H_x fundamental stretching band from 2100 - 2000 cm^{-1} and a previously unreported Si-O-C impurity band with maximum at 1240 - 1230 cm^{-1}, visible only with p-polarization (parallel to the film growth direction). The shape of the wagging mode fundamental band at 640 cm^{-1} was distorted by the strong substitutional carbon absorbance at 607 cm^{-1} and by the phonon absorbance at 619 cm^{-1} in the Si substrates. However, a higher intensity of the 640 cm^{-1} band in s-polarization (parallel to the film plane) is evident, indicating some enhancement of the wagging modes in the film plane. The stretching mode band from 2100 – 2000 cm^{-1} shows little or no polarization dependence (indicating little or no orientation of the Si-H dipoles), and is about five times more intense than the same band measured in transmission mode because of enhancement by multiple reflections. The polarized ATR-FTIR spectra of the other a-SiH$_x$/Si samples qualitatively resemble those shown in Fig. 2.

We considered assigning the impurity band to the first overtone of the fundamental Si-H wagging vibration at 640 cm^{-1}, but its intensity is too great, even exceeding that of the supposed parent band in the case of sample # H472, and bears no relationship to the hydrogen content of the samples. However, the intensity of the impurity band does have a strong positive correlation with both the average oxygen (Fig. 3) and the average carbon content of the a-SiH$_x$ films on Si substrates, integrated from the film surface through the interfacial region to the substrate, as measured by SIMS. We were able to fit the polarization-dependent components of this band with the sum of two Gaussian distributions centered at 1228 and 1165 cm^{-1}, respectively (Fig. 4). These frequencies are consistent with the C-O and Si-O stretching vibrations in a Si-O-C impurity phase, which is highly oriented in the film growth direction. The strong orientation in the film growth direction explains why the impurity band has not been observed previously using transmission IR spectroscopy, in which the

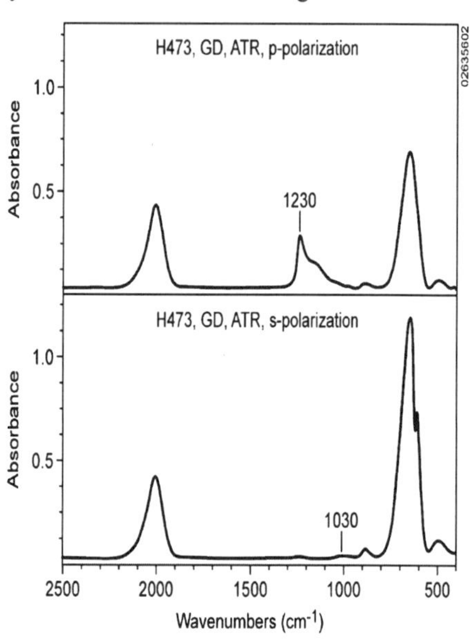

Fig. 2. Polarized FTIR-ATR spectra of a-SiH$_x$/Si sample #H473.

electric field of the incident radiation is oriented in the film plane. The impurity bands at 1228 and 1165 cm^{-1} resemble the bands at 1204 and 1108 cm^{-1} in the absorbance spectrum of trimethoxysilane [8]. There is some indication of the impurity bands in the contact FTIR-ATR spectra (not shown) of samples HW58, HW64, and HW76 (Table 1), but these spectra are noisy because of poor contact between the rigid samples and the equally rigid ATR substrates, and because of an unfavorable optical mismatch [3]. The intensity of a shoulder at 1030 cm^{-1} on the band structure in Fig. 4 is invariant with polarization (Fig. 2), and is consistent with a SiO$_x$ impurity phase. Fig. 5 shows that the intensity of the Si-O-C impurity band is linearly dependent on the deposition time during a stepwise HWCVD deposition of sample #H496, for which a film thickness of 0.9 µm was measured at the end of the single-sided deposition.

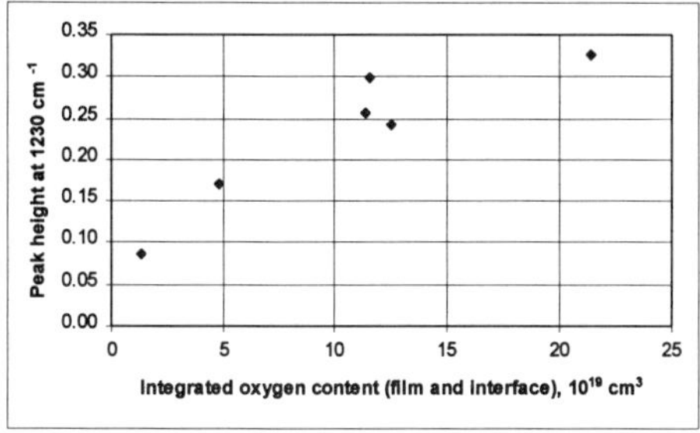

Fig. 3. Relationship between FTIR/ATR band height at 1230 cm^{-1} and integrated oxygen content of a-SiH$_x$/Si film samples determined from SIMS depth profiles.

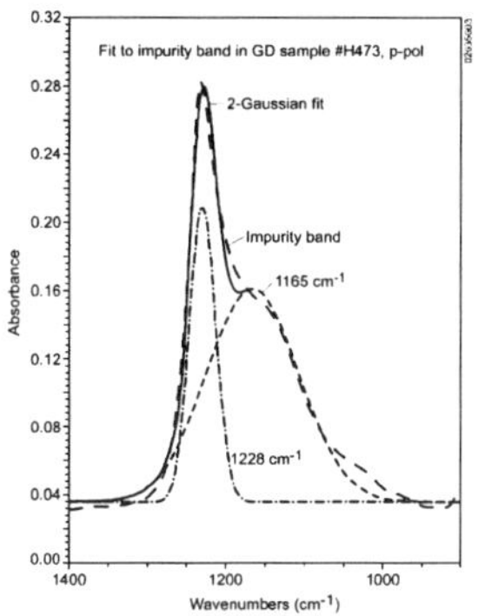

Fig. 4. Two-Gaussian fit to the Si-O-C impurity band observed in the FTIR-ATR spectrum of a-SiH$_x$/Si sample #H473, with ν_1 = 1228 cm^{-1}, $\Delta\nu_1$ = 39 cm^{-1}, ν_2 = 1165 cm^{-1}, and $\Delta\nu_2$ = 137 cm^{-1}.

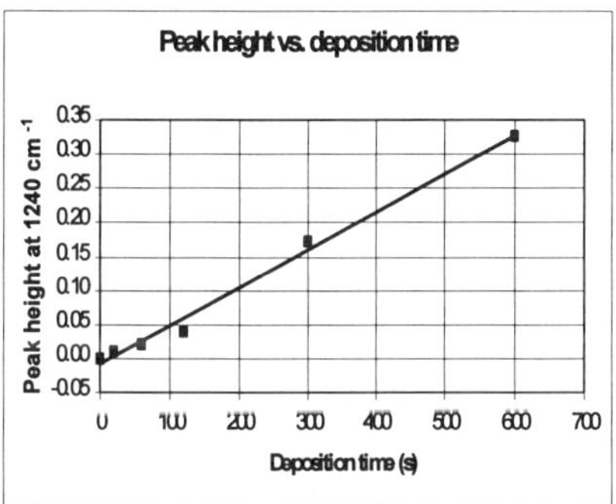

Fig. 5. Intensity of the Si-O-C impurity band at 1240 cm^{-1} for p-polarization as a function of deposition time in step-deposited HWCVD sample #H496.

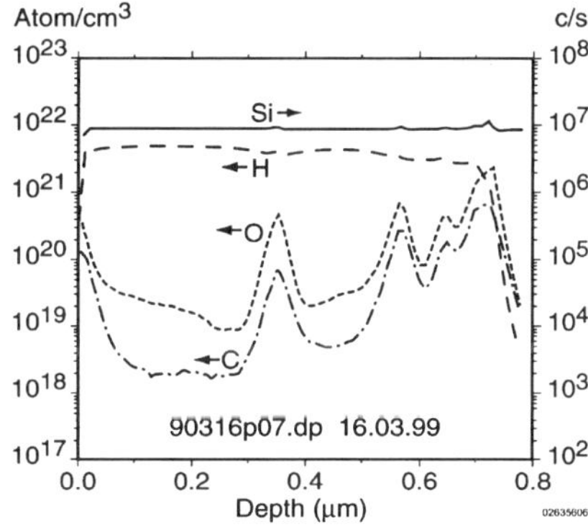

Fig. 6. SIMS Si, H, O, and C depth profile of layer-deposited a-SiH$_x$/Si sample #H496.

A SIMS depth profile through sample #H496 (Fig. 6) shows high levels of C and O at the film surface and at the film/substrate interface. These features are common to all the samples analyzed via SIMS. Also evident in Fig. 6 are accumulations of C and O at the individual deposition interfaces. These contaminants apparently accumulated when the sample was removed from the evacuated deposition chamber for FTIR-ATR spectroscopic measurements. It is likely that the C and O impurities introduced to the sample surface after each deposition reacted with the silane to yield the oriented Si-O-C impurity phase in each subsequent deposition step during growth of this sample. Consistent with earlier results [9], the SIMS analyses show that there is little penetration of H into the crystalline Si substrates.

An interesting variation in the FTIR data is the complete absence of the Si-O-C impurity band in the polarized ATR spectra (not shown) of the a-SiH$_x$ films deposited on polycrystalline CdTe ATR substrates (H497 and H498, Table 1). Although carbon and oxygen levels in these samples were similar to those observed in some of the a-SiH$_x$/Si samples, it may also be necessary to have a surface layer rich in SiO$_x$ to serve as a precursor to the Si-O-C impurity. The polarized ATR-FTIR spectra of these samples are otherwise similar to those in Figure 2, except that they show the wagging mode absorbance bandshape at 640 cm^{-1} without substrate absorption artifacts. The polarized ATR-FTIR spectra of samples H497 and H498 also indicate some enhancement of the wagging absorbance in the film plane. In the SEM image of HWCVD sample #H497 (not shown), which was deposited on CdTe, the a-SiH$_x$ film appears amorphous in

areas of poor adhesion and structured in areas where it makes good contact with the substrate. The SEM images (not shown) of PECVD a-SiH$_x$/CdTe sample #H498 reveal no evidence of microstructure, but also show poor film adhesion, perhaps from residual stress.

The SEM micrographs of the cleaved cross-sections of the HWCVD a-SiH$_x$ films on silicon substrates show a distinct columnar microstructure and some areas of poor adhesion. The microstructure is especially evident in the SEM image of sample #H472 (Fig. 7), which incorporated a 15-nm a-SiH$_x$ interfacial layer nominally 9.3 % in H. However, the SEM micrographs of PECVD a-SiH$_x$/Si sample #H473 (not shown) indicate that the columnar microstructure is discontinuous in this a-SiH$_x$ film.

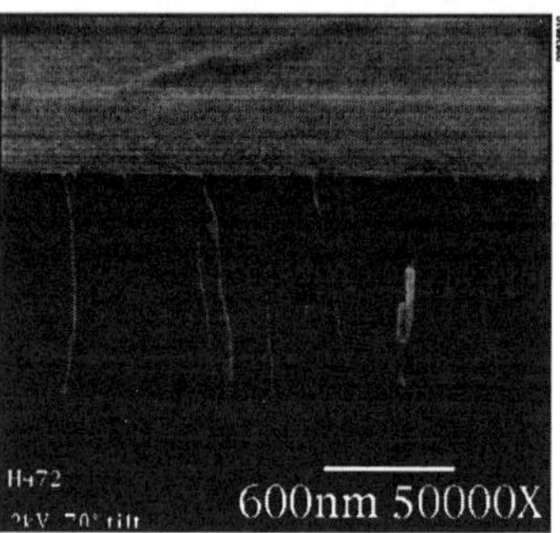

Fig. 7. Cross-sectional SEM micrograph of cleaved a-SiH$_x$/Si sample #H472.

CONCLUSIONS

Using polarized FTIR-ATR spectroscopy and SIMS depth profiling, we have demonstrated the existence of an impurity phase, likely to contain Si-O-C bonds, which is highly oriented in the film growth direction in a-SiH$_x$/Si films. The impurity phase is not observed in a-SiH$_x$/CdTe films. We have also shown that the Si-H wagging absorbance is enhanced in the film plane in a-SiH$_x$ films on <111> Si substrates and on polycrystalline CdTe substrates, perhaps because of residual stress. Our SEM measurements support the FTIR-ATR measurements of structural anisotropy in the a-SiH$_x$/Si films. The anisotropy is influenced more by the substrates, and their surface contaminants, than by the film deposition method. The anisotropy in the wagging absorbance at 640 cm^{-1} may explain some of the scatter in hydrogen content calibration measurements made using transmission-mode FTIR spectroscopy of a-SiH$_x$/Si samples [1].

REFERENCES

1. B. P. Nelson, Y. Xu, A. Mason, R. Reedy, J. Webb, L. Gedvilas, and W. A. Lanford, J. Non-Cryst. Solids (1999), in press.
2. A. A. Langford, M. L. Fleet, B. P. Nelson, W. A. Lanford, and N. Maley, Phys Rev. B **45**, 23, pp. 13367-13377 (1992).
3. N. J. Harrick, *Internal Reflection Spectroscopy* (Interscience Publishers, New York, 1967), pp. 13-66.
4. *Ibid.* [3], pp. 67-88.
5. Y. J. Chabal and C. K. N. Patel, Phys. Rev. Lett. **53**, 2, pp. 210-213 (1984).
6. *Ibid.* [3], p. 144.
7. *Ibid.* [3], p. 130.
8. Nicolet Instrument Corp., Madison, WI 53711. The complete Nicolet digital library contains the mid-IR spectra of roughly 50,000 compounds (mostly organic chemicals).
9. B. Sopori, M. I. Symko, R. Reedy, K. M. Jones, and R. J. Matson (Proc. 26th IEEE PVSC, Anaheim, CA, 1997), pp. 25-30.
10. *Ibid.* [3], p. 31.

CHARACTERIZATION OF HYDROGEN
IN HYDROGENATED NANO-CRYSTALLINE SILICON

K. YAMAMOTO*, T. ITOH*, K. USHIKOSHI*, S. NONOMURA**, S. NITTA*
*Department of Electrical Engineering, Gifu University, 1-1 Yanaido, Gifu 501-1193, Japan
**Environment and Renewable Energy Systems, Graduate School of Engineering,
 Gifu University, 1-1 Yanaido, Gifu 501-1193, Japan

ABSTRACT

The characterization of hydrogen in hydrogenated nano-crystalline silicon (nc-Si:H) films with various crystallinity has been studied using FTIR absorption spectroscopy and gas effusion spectroscopy. In gas effusion spectrum of a nc-Si:H, three evolution peaks of H_2 are found near 400, 500 and 600 °C. From the deconvolution of the gas effusion spectra, the hydrogen concentration in amorphous region dose not depend on the volume fraction of crystalline. The surface density of hydrogen at the interface and in the grain boundaries of nano-crystalline grains decreases with the volume fraction of crystalline. The properties of hydrogen in nc-Si:H films are discussed with the results of FTIR absorption spectroscopy.

INTRODUCTION

Hydrogenated nano-crystalline silicon (nc-Si:H) films can be prepared by PECVD method using high dilution of SiH_4 in H_2. The nc-Si:H films are more stable by light irradiation than a-Si:H films. The hydrogen concentration in nc-Si:H films is almost same as that in a-Si:H films. In the growth and the properties of nc-Si:H, the role of hydrogen is very important. The properties of hydrogen in nc-Si:H films, however, have not been understood in detail yet. In this report, nc-Si:H films with various crystallinity were prepared at RF-power range between 9 and 76 W. The thermal evolution and the bonding states of hydrogen in nc-Si:H films have been studied by using gas effusion spectroscopy and FTIR absorption spectroscopy. From these results, the properties of hydrogen in nc-Si:H films are discussed.

EXPERIMENT

The samples were prepared on sapphire and c-Si substrates by PECVD at 13.56 MHz using SiH_4 and H_2. The total gas pressure was 4 Torr with the ratio $R=H_2/SiH_4$ of 40. The substrate temperature was ~250 °C and the RF-power was varied from 9 to 76 W. The film thickness was from 0.9 to 1.4 μm. A hydrogenated amorphous silicon (a-Si:H) film was prepared by the same method at RF-power of 70 W with R=9 to compare to nc-Si:H films. The film thickness was ~0.9 μm.

The crystallinity of the sample was investigated using Raman spectroscopy and X-ray diffraction (XRD). Raman spectra were measured using He-Ne laser at a wavelength of 633 nm.

Mat. Res. Soc. Symp. Proc. Vol. 557 © 1999 Materials Research Society

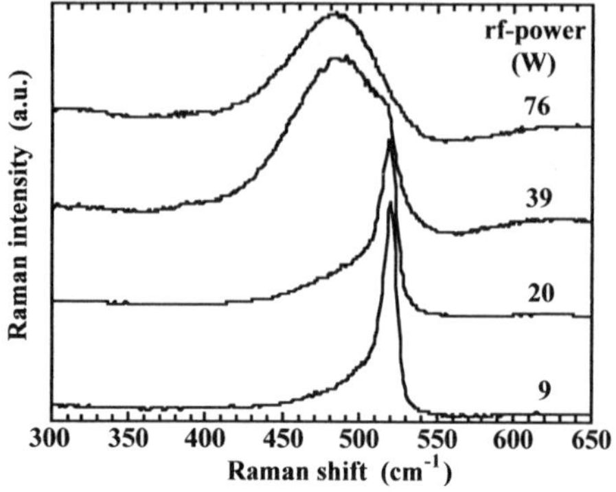

Fig. 1. Raman spectra of the samples prepared at various RF-power.

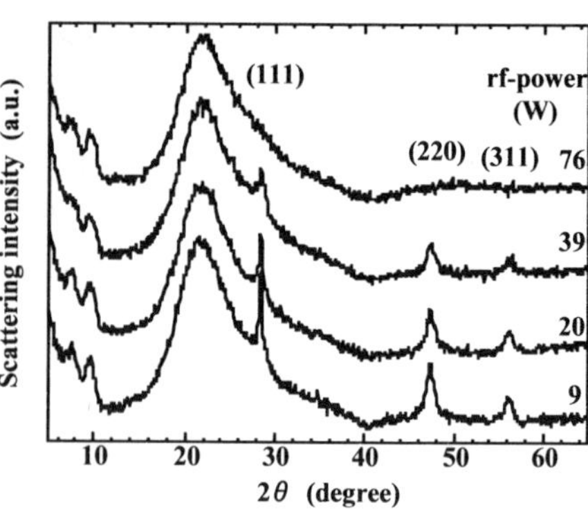

Fig. 2. XRD patterns of the samples prepared at various RF-power.

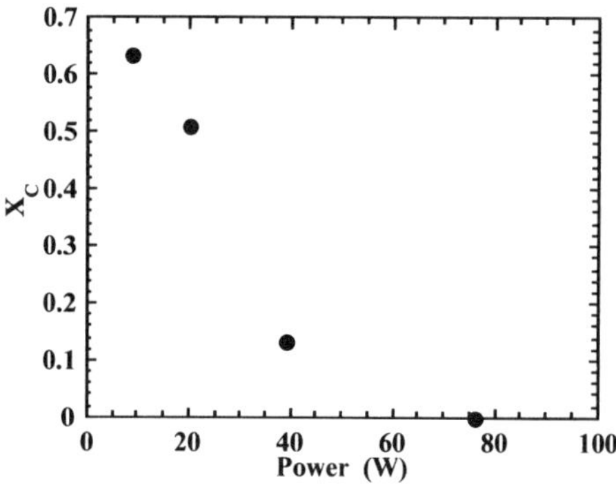

Fig. 3 (a). RF-power dependence of volume fraction of crystalline in the samples.

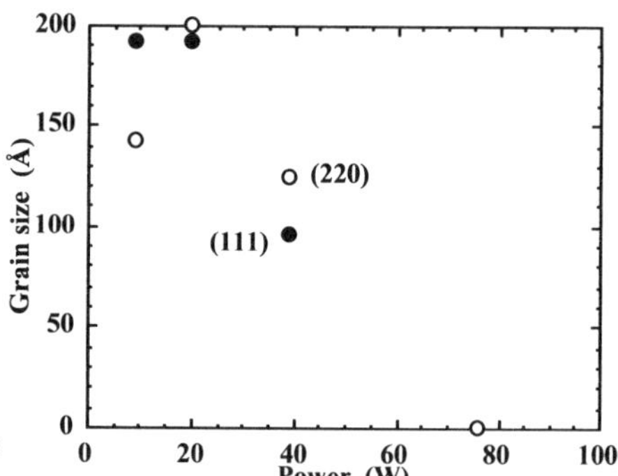

Fig. 3 (b). RF-power dependence of grain size in the samples.

The volume fraction of crystalline X_c was estimated as the ratio $I_c/(I_c+I_a)$, where I_a and I_c corresponds to the integrated intensities of Raman peaks near 488 and 520 cm^{-1}, respectively. The grain size of the samples were determined from XRD using Scherrer formula. Infrared (IR) absorption spectra were measured by using a furrier-transform infrared spectrometer JES-FE1X. Gas effusion spectra were measured at 5 °C/min as described elsewhere [1].

RESULTS AND DISCUSSION

Figures 1 and 2 show Raman spectra and XRD patterns of samples prepared at various RF-power. In the Raman spectra, the peak intensity near 520 cm^{-1} corresponding to c-Si increases and that near 480 cm^{-1} corresponding to a-Si decreases with RF-power. In the XRD

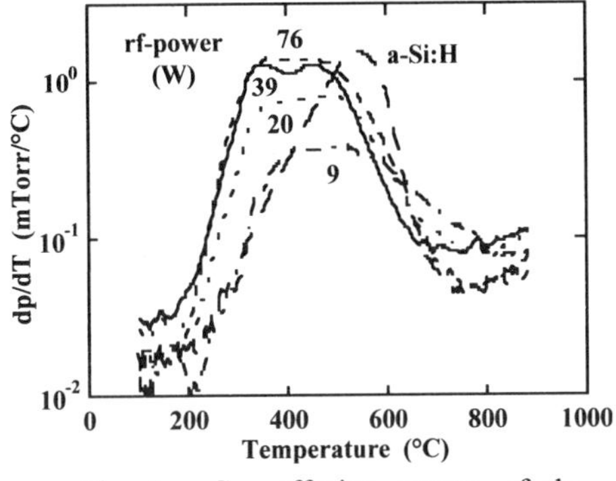

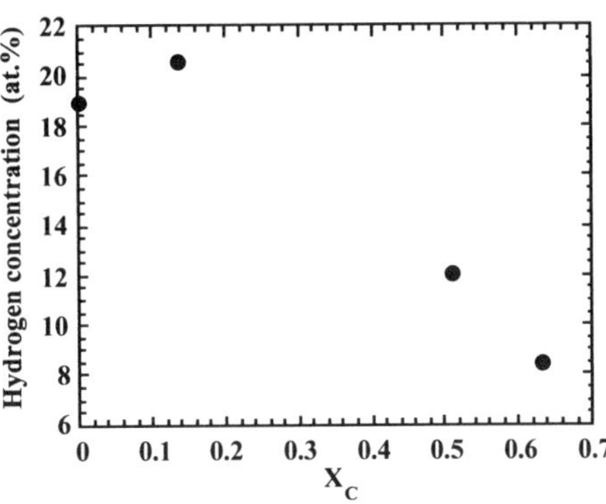

Fig. 4. Gas effusion spectra of the samples prepared at various RF-power. Spectrum of a-Si:H film for comparison.

Fig. 5. Dependence of hydrogen concentration in the samples on volume fraction of crystalline.

patterns, the (111), (220) and (311) peaks are found and all peak intensities decrease with RF-power. The sample prepared at RF-power of 76 W is amorphous from the results of Raman spectroscopy and XRD. Figure 3 (a) and (b) show the RF-power dependence of volume fraction of crystalline X_c and grain size, respectively. From these results, the volume fraction of crystalline and the grain size in the samples decrease with RF-power.

Figure 4 shows the gas effusion spectra of the samples in fig. 1 and 2. In gas effusion spectra of nano-crystalline samples, three main peaks of H_2 evolution were found near 400, 500 and 600 °C. This result agrees with previously published data [2]. Comparing previous reports [3, 4], the origins of these H_2 evolutions would be assigned as follows;the H_2 evolutions near 400 and 500 °C are the evolutions from the dihydride and the monohydride at the interface and in the grain boundaries of nano-crystalline grains, respectively. The H_2 evolution near 600 °C is the evolution from Si-H bonds in amorphous region. In gas effusion spectrum of the amorphous sample prepared at rf-power of 76 W, three main evolution peaks of H_2 are also found near 400, 500 and 600 °C. In that of the a-Si:H film prepared with R=9, however, only one main peak of H_2 evolution is found near 600 °C. These results could suggest that the amorphous sample prepared with high dilution of SiH_4 in H_2 has nano-particles like seed crystals. The hydrogen concentration in the sample is determined from gas effusion spectrum. Figure 5 shows the dependence of the hydrogen concentration in the samples on the volume fraction of crystalline X_c. In this figure, the hydrogen concentration decreases with the volume fraction of crystalline. The hydrogen concentration in the samples with high volume fraction crystalline is almost same as that in device quality a-Si:H films. Three components of the H_2 evolution near 400, 500 and 600 °C are deconvoluted in the gas effusion spectra. Using the component of H_2 evolution near 600 °C and the volume fraction of amorphous, 1-X_c, in the samples, the hydrogen concentration in amorphous region is estimated. Figure 6 shows the dependence of the hydrogen concentration in amorphous region on the volume fraction of crystalline X_c. In this figure, the hydrogen concentration in amorphous region dose not depend on the volume fraction of crystalline. The

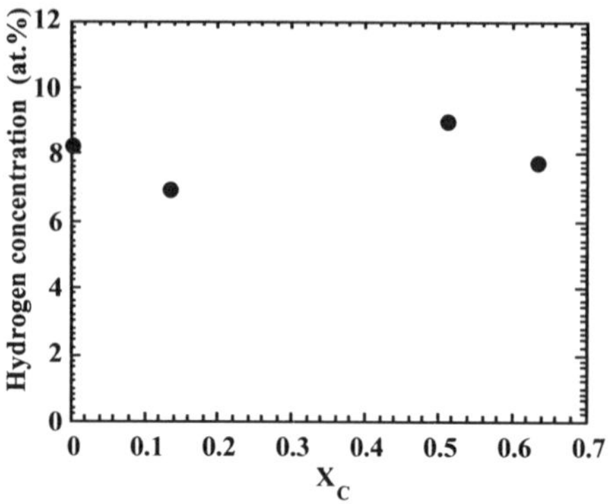
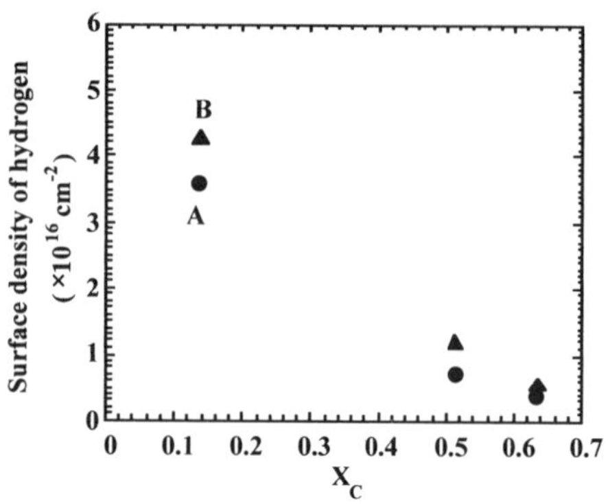

Fig. 6. Dependence of hydrogen concentration in amorphous region on volume fraction of crystalline.

Fig. 7. Dependence of surface density of hydrogen at surface and in grain boundaries of nano-crystalline grains on volume fraction of crystalline. Plots A and B from the components near 400 and 500 °C in fig. 4, respectively.

estimated hydrogen density in amorphous region is almost same as that in device quality a-Si:H films and that in the sample with the highest volume fraction of crystalline. This result suggests that the interface and the grain boundaries of nano-crystalline grains could contain a large quantity of hydrogen. Therefore the quantity of hydrogen at the interface and in the grain boundaries of nano-crystalline grains would affect the hydrogen concentration in nc-Si:H films. Using the components of hydrogen evolution near 400 and 500 °C, the volume fraction of crystalline X_c and the grain size in the samples, the surface density of hydrogen at the interface and in the grain boundaries of nano-crystalline grains is obtained. Here it is supposed that the grain shape is a ball and its radius is the averaged grain size determined from (111) and (220) XRD peaks. Figure 7 shows the dependence of the surface density of hydrogen at the interface and in the grain boundaries of nano-crystalline grains on the volume fraction of crystalline X_c. In this figure, the surface density of hydrogen at the interface and in the grain boundaries of nano-crystalline grains would be depend on the volume fraction of crystalline in the samples. This result shows that the quantity of hydrogen around the nano-crystalline grains would be related to the growth of nc-Si:H.

Figure 8 shows the IR absorption spectra of the samples prepared at various RF-power. In fig. 8 (a), the peaks near 2000 and 2100 cm^{-1} are found in the spectra of the samples. In the spectrum of the a-Si:H film prepared with R=9, the peak near 2000 cm^{-1} is found. The peak intensity near 2000 cm^{-1} decreases and that near 2100 cm^{-1} increases with the volume fraction of crystalline. Therefore the peaks near 2000 and 2100 cm^{-1} would be related to hydrogen in amorphous region and hydrogen at the surface and in the grain boundaries of nano-crystalline grains. These results agree with previous report [2]. In fig. 8 (b), the peaks near 840 and 880

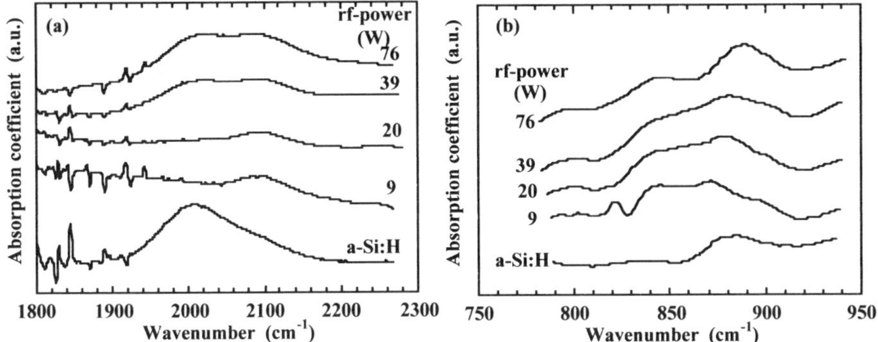

Fig. 8.   IR absorption spectra of the samples prepared at various RF-power.   (a) Wavenumber range between 1800 and 2300 cm^{-1}, and (b) wavenumber range between 750 and 950 cm^{-1}. Spectra of a-Si:H film for comparison.

cm^{-1} are found in the spectra of the samples.   In the spectrum of a-Si:H film prepared with R=9, the peak near 840 cm^{-1} is not found. The peak intensity near 880 cm^{-1} decreased with the volume fraction of crystalline.   In the IR absorption spectrum of the sample prepared at rf-power of 20 W, the peak near 840 cm^{-1} disappears by the annealing at 500 °C.   The IR absorption peak near 840 cm^{-1} was assigned as $(Si-H_2)_n$ wagging mode [5].   Therefore these results would indicate that the samples have the $(Si-H_2)_n$ chain structures.   Comparing with the results of gas effusion spectroscopy, the $(Si-H_2)_n$ chain structures could be in the grain boundaries of nano-crystalline grains.   The existence of the $(Si-H_2)_n$ chain structures in the grain boundaries of nano-crystalline grains would explained that the nc-Si:H films show high hydrogen concentration.

SUMMARY

The volume fraction of crystalline and the grain size in the nc-Si:H films decrease with rf-power.   The hydrogen concentration determined from gas effusion spectra decreases with the volume fraction of crystalline.   The analysis of gas effusion spectra shows as following results; the hydrogen concentration in amorphous region dose not depend on the volume fraction of crystalline.   The surface density of hydrogen at the interface and in the grain boundaries of nano-crystalline grains decreases with the volume fraction of crystalline.   These results would suggest that the interface and the grain boundaries of nano-crystalline grains contain a large quantity of hydrogen and the quantity of hydrogen around the nano-crystalline grains is related to the growth of nc-Si:H.   The IR absorption spectra could indicate that the nc-Si:H films have the $(Si-H_2)_n$ chain structures.   The $(Si-H_2)_n$ chain structures could be in the grain boundaries of nano-crystalline grains.   The existence of the $(Si-H_2)_n$ chain structures in the grain boundaries of nano-crystalline grains would explained that the nc-Si:H films show high hydrogen concentration.

ACKNOWLEDGMENT

This work was supported by the foundation of 'Research for the Future' by the Japan Society for the Promotion of Science.

REFERENCES

1. T. Itoh, S. Nitta, S.L. Wang, and P.C. Taylor, J. Appl. Phys. **80**, p.6028-6031 (1996).

2. W. Beyer, P. Hapke, and U. Zastrow in *Amorphous silicon Technology-1997*, edited by S. Wagner, M. Hack, E.A. Schiff, R. Schropp, and I. Shimizu (Mater. Res. Soc. Proc. **467**, Pittsburgh, PA 1997), p.343-348.

3. K. Sinniah, M.G. Sherman, L.B. Lewis, W.H. Weinberg, J.T. Yates, Jr., and K.C. Janda, Phys. Rev. Letters **62**, p.567-570 (1989).

4. W. Beyer, Physica B **170**, p.105-114 (1991).

5. S. Nitta, M. Kawai, M. Sakaida, I. Murase, and A. Hatano in *Tetrahedrally-Bonded Amorphous Semiconductors*, edited by D. Adler and H. Fritzsche (Plenum Press, New York, NY 1985), p.501-511.

STRUCTURAL ORIGIN OF BULK MOLECULAR HYDROGEN IN HYDROGENATED AMORPHOUS SILICON

Xiao Liu *+, R.O. Pohl *, R.S. Crandall **
* Department of Physics, Cornell University, Ithaca, NY 14853, pohl@msc.cornell.edu
+ present address: SFA, Inc., Largo, MD 20774
** National Renewable Energy Laboratory, Golden, CO 80401

ABSTRACT

The elastic anomaly observed previously at the triple point of bulk molecular hydrogen in hydrogenated amorphous silicon films prepared by hot-wire chemical-vapor deposition has also been observed in deuterated films at the triple point of D_2. The origin of this anomaly has now been traced to bubbles formed at the crystalline-amorphous interface. An upper limit of the pressure in these bubbles at their formation temperature, 440°C, has been estimated to be 11 MPa, and is suggested to be a measure of the bonding strength between film and substrate at that temperature. Bubble formation after heat treatment at 400°C has also been observed in films prepared by plasma-enhanced chemical-vapor deposition. The internal friction anomalies resemble those observed previously in cold-worked hydrogenated iron where they have been interpreted through plastic deformation of solid hydrogen in voids.

INTRODUCTION

We have recently presented evidence [1] for the existence of bulk molecular hydrogen in certain hydrogenated amorphous silicon films (a-Si:H) prepared by hot-wire chemical-vapor deposition (HWCVD) [2]. The location as well as its origin had to be left as open questions, and will be answered in this investigation. We will also explore the physical nature of the elastic anomalies observed in these films.

EXPERIMENTAL

Internal friction and transverse sound velocity of the amorphous silicon films were measured on the antisymmetric mode (~ 5500 Hz) of double-paddle oscillators fabricated out of high purity crystalline silicon wafers with extremely small background internal friction (~ 2×10^{-8}). The technique was described in [1] and will not be repeated here. All measurements were performed at a strain-amplitude $\varepsilon_m = 1\times10^{-6}$, with the exception of one shown in Fig. 5. The HWCVD a-Si:H film used in this work was one of those reported in [1] (H128, 2 at.% H), where the film preparation also was described. The HWCVD a-Si:D film (H375) was prepared under the same conditions as H128, in an attempt to produce a similar dopant concentration and elastic anomaly. A film of PECVD a-Si:H (T485) was also prepared at NREL, with deposition temperature and rate of 230°C and 1.3 Å/s, respectively. This resulted in an optimal device-quality PECVD film with ~ 9 at.% H. Hydrogen concentrations were measured both by infrared absorption and thermal evolution which resulted in similar values to within 20%, and the values given here are averages. The deuterium concentration in H375 has not been determined, but is expected to be close to 2 at.%.

RESULTS

Fig. 1 compares internal friction and sound velocity for the HWCVD a-Si:D film (H375) with measurements on HWCVD a-Si:H (H128) reported previously [1]. In the a-Si:D, the elastic anomalies are very similar, except that they occur at 18.7K, the triple point of bulk D_2 [3]. This result provides convincing evidence that the abrupt change of internal friction and sound velocity are closely related to the triple points, i.e. to the onset of freezing in the bulk liquids. We mention only in passing that below ~ 2K the internal friction of the deuterated film is even smaller than that of the a-Si:H film, 4×10^{-7} vs. 8×10^{-7}. We had previously reported that with the appropriate hydrogen concentration (~ 1 at.%) in HWCVD films an anomalously small internal friction was observed in this temperature range, ~ 5×10^{-7} [4]. As seen in Fig. 1(a), deuteration shows the

Mat. Res. Soc. Symp. Proc. Vol. 557 © 1999 Materials Research Society

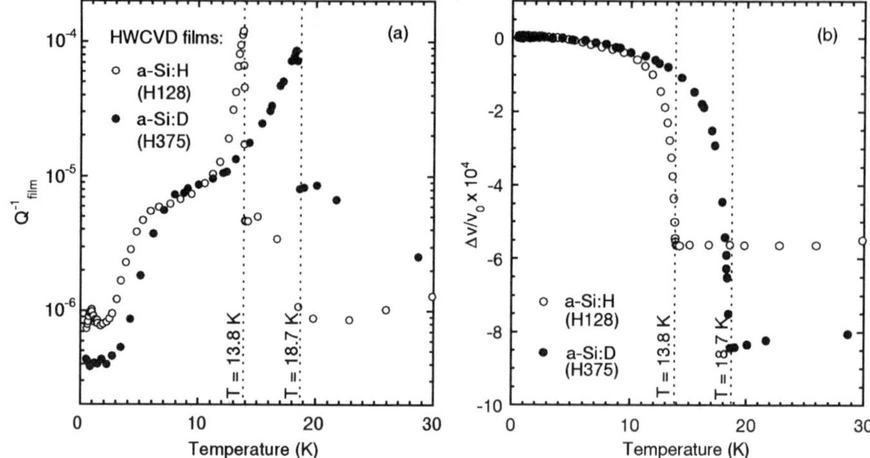

Fig. 1 Internal friction, (a), and relative variation of sound velocity, (b), of a HWCVD a-Si:H film (sample H128, 2 at.% H, 4.5 μm thick, as deposited, the same data as in [1]), compared with those of a HWCVD a-Si:D film (H375, deuterium concentration not known, 1.8 μm thick, as deposited). The dotted lines mark the respective triple points [3]. The reference value v_0 indicates the sound velocity at the lowest temperature of measurement. Measuring technique and evaluations as described in [1].

same striking effect which sets these a-Si films apart from all other amorphous solids studied to date, which have internal frictions typically three orders of magnitude larger. In fact, the extremely small internal friction of these a-Si:H and a-Si:D films enables the elastic anomaly to be studied in this work.

Inspection of the a-Si:H film measured in Fig. 1 under an optical microscope revealed a number of large bubbles, as shown in Fig. 2(a). Note that the picture shows the entire width of the neck of the double-paddle oscillator, and half of its full length. The cross-sectional profile of the bubbles has been measured with a stylus profilometer. An example is shown in Fig. 2(b), with the measurements along the white line in Fig. 2(a). The bubbles are typically 10 μm high and 100 μm wide. They were then removed by carefully scraping with fine sandpaper, see Fig. 2(c), which also shows some unavoidable scratch marks. Stylus measurement, Fig. 2(d), shows that the bubbles extended to the base of the a-Si:H film, which was 4.5 μm thick in this case. Removal of the bubbles led to the complete disappearance of the elastic anomaly near and below the triple point, Fig. 3. Here, the elastic measurements of the paddle carrying the film are shown, in order to allow a comparison with bare paddle data (the solid curves in both Fig. 3(a) and 3(b)). Below ~ 3K, the internal friction of the hydrogenated film even with the bubbles is very small, as mentioned above. Removal of the bubbles actually increases the internal friction in this temperature range, which is probably the effect of the scratches (Fig. 2(c)). The important point here is that the pronounced structures of internal friction and of sound velocity at and below 13.8K, which extend as low as ~ 1K, has been clearly identified through this experiment with the bubbles shown in Fig. 2.

In Fig. 4 it is shown that the same elastic anomalies can also occur in PECVD a-Si:H films, taking T485 as an example. Although absent in the as-deposited film, the anomalies arise after an anneal at 400°C for 15 minutes. No bubbles were observed before the annealing. However, bubbles of ~ 25 μm in diameter were visible under the optical microscope after the annealing. Finally it is seen in Fig. 5 that the internal friction is strongly dependent on the strain amplitude, ε_m, much in contrast to amorphous solids. An increase of ε_m by a factor of five increases the internal friction by more than one order of magnitude near the triple point. Strain-amplitude dependence was observed even at the smallest ε_m.

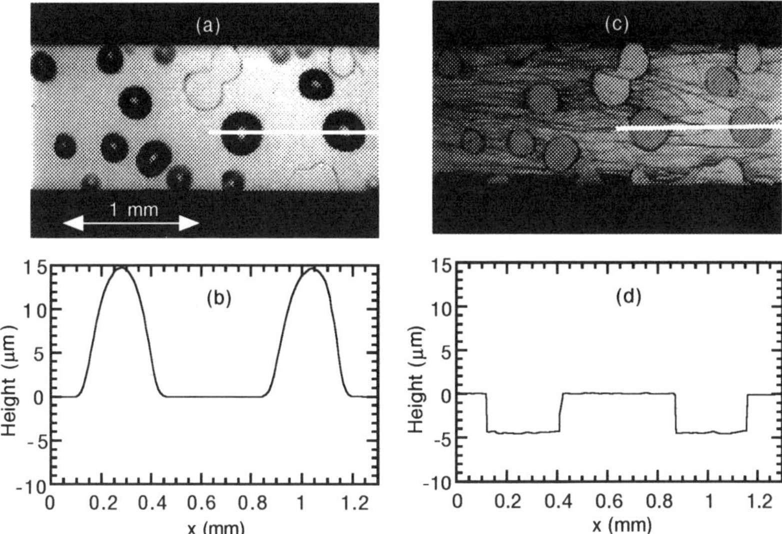

Fig. 2(a) and 2(c): Optical micrographs of one-half the length of the neck of sample H128 (see Fig. 1), before and after removal of the bubbles (note the scratch marks in (c)). Fig. 2(b) and 2(d): Cross-sectional profile measured along the white line indicated in Figs. 2(a) and 2(c), before and after removal of the bubbles. Note the different scales for height and width. To determine the bubble volume, the area under the curves in Fig. 2(b) was used (above the horizontal line at 0 μm).

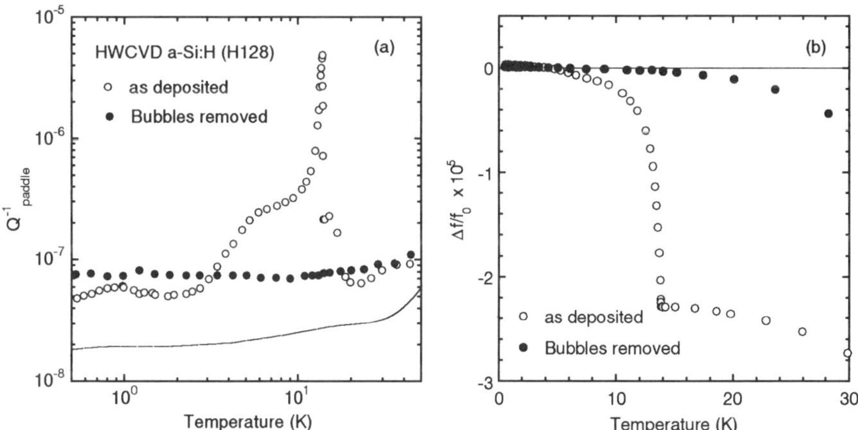

Fig. 3 Internal friction, (a), and relative change of resonance frequency, (b), of the paddle carrying a 4.5 μm thick HWCVD a-Si:H film (H128, Fig. 1), before and after removal of the bubbles. From the difference between the raw data and the background (solid lines), the internal friction and sound velocity of the films can be calculated, as described in [1], and shown in the other Figs. in this paper.

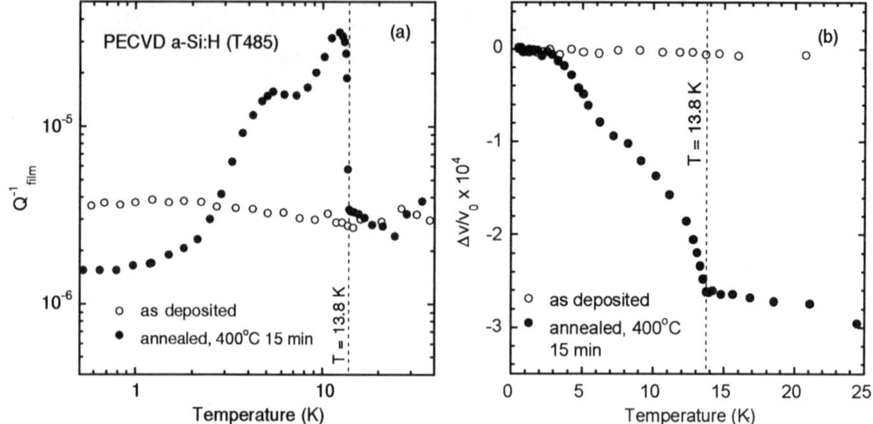

Fig. 4 Internal friction, (a), and relative variation of sound velocity, (b), of a PECVD a-Si:H film (sample T485, ~ 9 at.% H, 0.92 μm thick), as deposited and after a 400°C anneal for 15 minutes in a vacuum of 5×10^{-7} Torr.

DISCUSSION

Our measurements on the PECVD film have shown that heat treatment leads to the accumulation of bulk molecular hydrogen. We conclude that the bubbles observed in the HWCVD films, Fig. 2, are also caused by excess hydrogen leaving the amorphous structure due to the high deposition temperature. Some of it collects at the interface to the substrate, and if its pressure exceeds the bonding strength, a separation of the a-Si film from the substrate will occur.

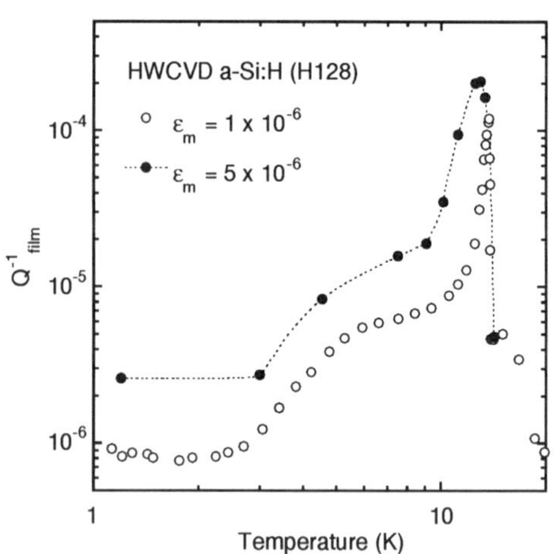

Fig. 5 Internal friction of the HWCVD a-Si:H film H128 (Fig. 1) at two different strain amplitudes. The strain amplitude dependence was observed even at the smallest strain amplitude used.

We can estimate an upper limit of this bonding strength by assuming that the total hydrogen content, which is 2 at.% for H128, collects in the bubbles in molecular form. The total volume of the bubbles on the neck of the oscillator is 1.3×10^{-2} mm^3 as measured by stylus, see Fig. 2, while the volume of the a-Si film is 3.0×10^{-2} mm^3. This results in 1.5×10^{18} Si atoms and 1.5×10^{16} H$_2$ molecules in the film. From this we calculate that at low temperatures 4.5% volume of the bubbles in this film will be filled with solid H$_2$, which has a mass density of 0.09 g·cm^{-3} [5]. Furthermore, a pressure of 11 MPa in the bubbles at the deposition temperature, 440°C, can be calculated using the ideal gas law, with an error estimated to be ± 30%. Since we have ignored all the hydrogen that remains bonded to the silicon, the pressure, and thus the bonding strength between film and substrate is expected to be smaller than this value. Using the same assumption, i.e. that most of the

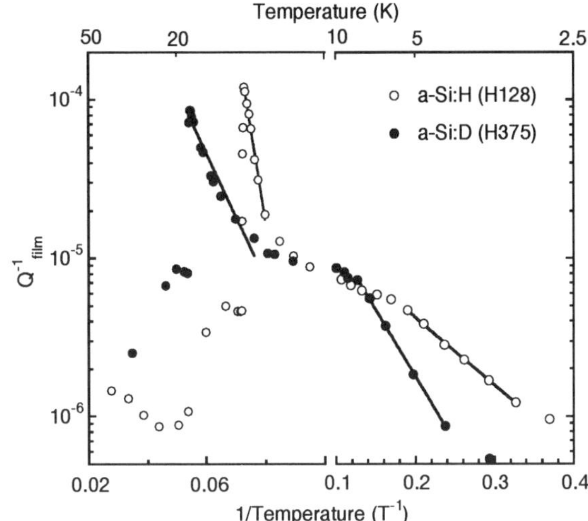

Fig. 6 Arrhenius plots of the internal friction of the films shown in Fig. 1. Note the different temperature scales close to the triple points, and below 10K. The solid straight lines were used to determine activation energies discussed in the text.

hydrogen collects in the bubbles, we can also estimate a lower limit of the sensitivity of our technique for the detection of bulk H_2. Inspecting Fig. 3(a), we believe that we can detect an anomaly at the triple point with a peak height as small as 1×10^{-8}, 200 times smaller than the one produced by 2 at.% H in H128. Thus, 100 ppm at. H in bulk molecular form should be detectable.

The solidification of the liquid in the bubbles is expected to increase the stiffness of the neck of the oscillator, and hence also the resonance frequency of the antisymmetric mode of the paddle. We have tried to estimate the frequency change $\Delta f = f - f_0$ of the antisymmtric mode, where f_0 and f are the frequencies before and after the molten H_2 freezes, using the simplified picture that all of the H_2 is spread uniformly over the entire neck forming a layer of thickness $t_{layer} = 82$ nm. From [6]:

$$\frac{\Delta f}{f_0} = \frac{3}{2}\frac{t_{layer} G_{layer}}{t_{sub} G_{sub}}, \tag{1}$$

Where t and G are thickness and shear modulus of solid H_2 layer and substrate composite (including crystalline Si substrate and the a-Si film), respectively. Neglecting the contribution from the a-Si film, we have $t_{sub} = 300$ μm, and $G_{sub} = 62$ GPa which is the shear modulus of the crystalline Si in the <110> orientation of the neck. As G_{layer}, we use the shear modulus of bulk hydrogen, 0.11 GPa [5]. From these values, the estimated frequency change is $\Delta f / f_0 = 7.4\times10^{-7}$, over thirty times smaller than the change observed in Fig. 3(b) which is $\Delta f / f_0 = 2.4\times10^{-5}$. The reason why the H_2 in the bubbles has such a large effect requires further study.

The abrupt increase of the internal friction as the liquid is cooled below the triple point is believed to be caused by plastic deformation. Golter and Roshchupkin [7] have observed a very similar internal friction anomaly in heavily cold-worked iron containing voids, which had been subsequently loaded electrolytically with hydrogen. From the very rapid decrease of the internal friction just below the triple point ($T > 10$K), and from the slower drop-off around 5K, which can be expressed as $Q^{-1} \propto \exp(-U/T)$, they determined two different activation energies, $U_2 = 195$K, and $U_1 = 10$K, respectively. The authors suggested that these energies might be associated with self-diffusion of H_2 molecules, as expected in grain boundary motion, and for dislocation motion across Peierls barriers, respectively. From our measurements, we similarly obtain the activation energies 270K and 10K for H_2, and 90K and 20K for D_2, as determined from the slopes of the straight lines in Fig. 6. The strain-amplitude dependence observed in our experiments, Fig. 5, may make this agreement somewhat fortuitous, but on the other hand, this strain-amplitude dependence is also what is known for the motion of atoms or dislocations under stress. Details on the plasticity of bulk solid H_2 are contained in the Section 8.3 of Ref. [5].

CONCLUSIONS

Through measurements of internal friction, the presence of bulk molecular hydrogen in macroscopic voids can be detected with a sensitivity corresponding to an average concentration in the a-Si film of less than 100 at. ppm. The freezing of the liquid H_2 has been found to have an unexpectedly large influence on the effective shear modulus of the film. The technique used here offers a promising way for the study of solid H_2 near its melting point, and because of its sensitivity also for the study of H_2 in very small voids. Finally, the formation of the H_2 bubbles can be used to estimate the strength with which the a-Si film is bonded to the substrate.

ACKNOWLEDGMENTS

We thank Eugene Iwanizcko, Wei Gao, Yuechin Xu, and Brent Nelson for sample preparation. Work was supported by the Semiconductor Research Corporation under Grant No. 95/Sc/069, the National Science Foundation under Grant No. DMR-970972, the NREL FIRST Program under Contract No. DE–AC02–83CH10093, and the NREL Grant No. RAD-8-18668. Additional support was received from the National Nanofabrication Facility at Cornell University, NSF Grant No. ECS-9319005.

REFERENCES

1. Xiao Liu, E. Iwaniczko, R.O. Pohl, R.S. Crandall, in *Amorphous and Microcrystalline Silicon Technology—1998*, edited by R. Schropp, H.M. Branz, M. Hack, I. Shimizu, and S. Wagner (Mat. Res. Soc. Symp. Proc. **507**, Pittsburgh, PA, 1998), pp. 595-600.

2. A.H. Mahan, J. Carapella, B.P. Nelson, R.S. Crandall, I. Balberg, J. Appl. Phys. **69**, 6728 (1991).

3. R.B. Scott, *Cryogenic Engineering*, (Van Nostrand, Princeton, NJ, 1959), p. 291.

4. Xiao Liu, B.E. White, Jr, R.O. Pohl, E. Iwaniczko, K.M. Jones, A.H. Mahan, B.N. Nelson, R.S. Crandall, S. Veprek, Phys. Rev. Lett. **78**, 4418 (1997).

5. V.G. Manzhelii and Y.A. Freiman, *Physics of Cryocrystals*, (American Institute of Physics, Woodbury, NY, 1997), p. 125.

6. B.E. White and R.O. Pohl in *Thin Films: Stresses and Mechanical Properties V*, edited by S.P. Baker, C. A. Ross, P. H. Townsend, C. A. Volkert, P. Borgesen (Mat. Res. Soc. Sym. Proc. **356**, Pittsburgh, PA, 1995), pp. 567-572.

7. A.E. Golter and A.M. Roshchupkin, Sov. J .Low Temp. Phys. **8**, 622 (1982); ibid., **13**, 105 (1987).

MICROSTRUCTURE AND HYDROGEN DYNAMICS IN a-Si$_{1-x}$C$_x$:H

R. SHINAR,[1,2] J. SHINAR,[3] D. L. WILLIAMSON,[4] S. MITRA,[5] H. KAVAK,[1,6] and V. L. DALAL[1]
[1]Microelectronics Research Center, Iowa State University, Ames, IA 50011
[2]Microanalytical Instrumentation Center, Iowa State University, Ames, IA 50011
[3]Ames Laboratory - USDOE and Department of Physics and Astronomy,
Iowa State University, Ames, IA 50011
[4]Department of Physics, Colorado School of Mines, Golden, CO 80401
[5]Department of Engineering Physics, University of Tulsa, Tulsa, OK 74104
[6]Physics Department, Cukurova University, l01330 Adana, Turkey

ABSTRACT

Small angle x-ray scattering (SAXS), IR spectroscopy, and deuterium secondary ion mass spectrometry (DSIMS) were used to study the microstructure and hydrogen dynamics of undoped and boron-doped rf-sputter-deposited (RFS) and electron cyclotron resonance (ECR)-deposited hydrogenated amorphous silicon carbides (a-Si$_{1-x}$C$_x$:H) with $x \leq 19$ at.%. The SAXS measurements indicated residual columnar-like features and roughly spherical nanovoids of total content $C_{nV} \leq 1.0$ vol.%. The growth of C_{nV} with annealing was due largely to an increase in the average nanovoid radius. It was noticeably smaller than in RFS a-Si:H films. The IR spectra demonstrated H transfer by annealing from mostly bulk-like Si-H groups to C-bonds. The H diffusion and its temperature dependence in undoped films resembled those of a-Si:H and were consistent with the SAXS and IR data. Suppression of long-range motion of most of the H atoms, consistent with increased C_{nV}, was observed in B-doped ECR films. However, a small fraction of the H atoms appeared to undergo fast diffusion, reminiscent of the fast diffusion in doped a-Si:H. The results are consistent with impeded relaxation processes of the Si network, caused by the presence of C atoms, and H trapping at C-H bonds.

INTRODUCTION

In previous studies, SAXS measurements provided a quantitative measure of the distribution of void shapes and sizes, their preferred orientation, and overall content C_{nV} in a-Si:H [1-3]. One study combined SAXS and IR measurements of the Si-H bond content and configuration [2]. It demonstrated a strong correlation between the decreasing Si-bonded H content $C_{\text{Si-H}}$ and the growth of C_{nV} during annealing at temperatures $T \leq 430°C$.

Studies of DSIMS showed that the H diffusion constant $D(t)$ usually decreases with time t as $D(t) = D_{00}(\omega t)^{-\alpha}$, where the dispersion parameter $0 \leq \alpha \leq 1$ [4–8]. $D(t)$ is determined experimentally by fitting the DSIMS profiles to a complementary error function [6]. The observed behavior of $D(t)$ and decrease of α with increasing T were first attributed to a multiple-trapping mechanism [5]. However, subsequent work showed that at $T \geq 350°C$ α increases with increasing T [8]. Other work demonstrated that α increases sharply with increasing C_{nV} of the films [6,7]; a sufficiently high C_{nV} suppressed the long-range motion of H atoms [6]. The overall picture that emerged suggested that in addition to C_{nV}, α is largely determined by dynamical Si network relaxation processes that affect the distribution of H site energies. Finally, the study mentioned above [2] confirmed the relation between H diffusion and structural relaxation processes; it showed that C_{nV} in undoped RFS a-Si:H films sharply increases at $T \geq 350°C$, consistent with the observed increase in α.

This paper compares SAXS, IR, and DSIMS studies of microstructure evolution and H dynamics in undoped and B-doped a-Si$_{1-x}$C$_x$:H to those reported for a-Si:H.

EXPERIMENT

Bilayer RFS a-Si$_{1-x}$C$_x$:(H,D)/a-Si$_{1-x}$C$_x$:H films were fabricated by rf sputtering a Si target [6,8] in an Ar, H$_2$, and CH$_4$ plasma. Undoped and boron doped bilayer ECR films were fabricated using a system described elsewhere [9]. SAXS measurements were conducted on the RFS films codeposited on Al foil; the samples were annealed in a tube furnace under flowing high purity He.

For IR and DSIMS measurements, films deposited on Si wafers were annealed in evacuated pyrex tubes. $C_{\text{Si-H}}$ (with an estimated error of ±15%) was determined from the 640 cm^{-1} IR wagging mode [10,11]. Initial values are given in Table I. Since the 3000 cm^{-1} C-H stretch band was unobservable, the C-bonded H content $C_{\text{C-H}}$ could not be determined directly. The Si-CH$_n$ and C-H wagging modes at ~770 and ~1000 cm^{-1}, respectively, provide only a qualitative measure of $C_{\text{C-H}}$, due to overlap with other modes. The 2000-2150 cm^{-1} Si-H stretch bands provide insight into the nature of the Si-H bonding configuration. The 2000 cm^{-1} band is the vibration frequency of a Si-H embedded in the a-Si network [10,11]. The 2080–2150 cm^{-1} band is associated with the vibration of Si-H bonds on internal surfaces of voids, Si-H bonds in SiH$_n$ configurations, or H bonded to Si-O or Si-C.

The C content x was determined by Auger electron spectroscopy (AES) or electron probe microanalysis (EPMA). In the ECR samples $x \approx 14$ at.%, consistent with the measured energy gap of ~2 eV. The boron levels were determined using SIMS.

Table I. The carbon content x, boron level C_B, and the initial Si-bonded H-content $C_{\text{Si-H}}$ of the RFS- and ECR-deposited films.

Sample	x (at.%) (a)From EPMA (b)From AES	$C_{\text{Si:H}}$ (at.%)	Sample	C_B (at.%)	$C_{\text{Si:H}}$ (at.%)
RFS1	2.2 (b)		ECR1	0	12
RFS2	2.0(a), 3.0(b)		ECR2	0.01	12
RFS3	1.6(a)	12.3	ECR3	0.2	12
RFS4	3.1(a)	14.0	ECR4	0.6	12
RFS5	5.7(a)	12.5	ECR5	0	6
RFS6	2.9(a)				
RFS7	6.0(a), 6.2(b)	16.8			
RFS8	7.1(a), 7.4(b)	9.0			
RFS9	19.0(a)				

RESULTS AND DISCUSSION

a. SAXS and the Microstructural Dynamics of the RFS Films.

The SAXS intensity $I(q)$, where the scattering wavevector $q = (4\pi/\lambda)\sin\theta$ ($\lambda = 1.54$ Å is the x-ray wavelength and 2θ is the scattering angle), of the RFS films increased steeply with decreasing q at low q (see Figure 1). In this region, the scattering curve $I_L(q)$ results generally from structural features which are large relative to the instrumental resolution of ~20 nm. The existence or absence of preferred orientation of the structural features may be obtained from tilting SAXS measurements, in which $I(q)$ is measured at different angles of the film with respect to the incident x-ray beam [1]. Figure 2 shows the effects of the tilt angle on the SAXS of RFS3 and RFS4, which were also observed in other samples. The large tilt-angle dependence at $q \leq 0.6$ nm^{-1}, which corresponds to the large features, indicates that these features are highly elongated and have a preferred orientation. We suggest that they are due to residual columnar-like domains [1,2] and are defined by lower density regions at their boundaries.

In addition to $I_L(q)$, all RFS samples showed a clear q-independent diffuse scattering region I_D and a nanostructural contribution $I_N(q)$ [1]. $I_N(q)$ corresponds to the broad "shoulder" in the high-q region. The nanovoids average radius $R = 0.55$ nm and C_{nV} of 0.5-1.0 vol.% were calculated from this region.

Following a sequence of annealing steps up to 420 °C, the integrated SAXS intensity Q_{SAXS} of samples RFS2–RFS7 increased monotonically with annealing temperature or time. In all cases Q_{SAXS} increased by $\leq 100\%$ from the initial to the most annealed state. In comparison, Q_{SAXS} of RFS a-Si:H, including films with pronounced columnar microstructures, increased by more than 200% following similar annealing steps [2]. Thus, the weaker increase in C_{nV} in the carbides (see Table II), which cannot be attributed to the residual columnar morphology, is suspected to be due to a slower evolution of the microstructure, associated with the significantly stronger Si-C and C-H bonds, in comparison to Si-Si and Si-H bonds. However, the dynamics of the microstructure was largely independent of x for $1.6 \leq x \leq 19$ at.%.

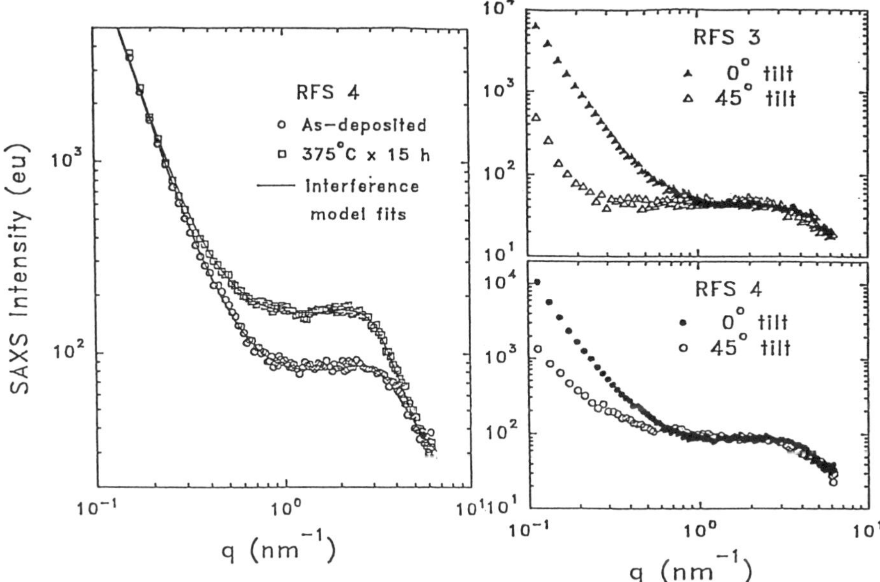

Figure 1. SAXS with interference model fits of as-deposited and annealed sample RFS4.

Figure 2. The effect of the tilt angle on the SAXS.

The analysis of the SAXS indicated that the large-scale features did not change significantly upon annealing (see Fig. 1). However, I_D decreased by 25%–40%, probably due to the loss of bonded H [3]. The analysis of $I_N(q)$ indicated that the voids' radius increased to $R = 0.65$ nm. This increase would account for a 65% increase in C_{nV}. Hence, it appears that as in RFS a-Si:H [2], the expansion of the average void size is the primary process affecting C_{nV} rather than void generation. Additionally, the SAXS data were consistent with little or no change in the spatial distribution of the nanovoids during annealing.

In general, it appears that species derived from methane in the plasma impede dynamical processes occurring during growth (observed for a-Si:H films), which eliminate columnar microstructure. Moreover, the strength of the C-H bond and consequent incorporation of CH_n groups in the growing film probably promote the formation of nanovoids.

b. IR Measurements of H Evolution and Bonding Configurations in RFS and ECR Films.

IR spectra were measured before and following sequential annealing for 1 hr at 300, 350, and 375°C, followed by additional annealing for a total of 3 hrs and 8 hrs at 375°C. The evolution of the spectra was qualitatively similar in all films. The decrease of C_{Si-H} following the final annealing step, C_{nV} before and following annealing, and x are summarized in Table II. While it is difficult to correlate the decrease of C_{Si-H} with x, it does appear to correlate with C_{nV}. This correlation is plausible since the average distance from a Si-H group to the nearest void surface, where the H may recombine to H_2, decreases with increasing C_{nV}.

Table II. Fractional reduction in C_{Si-H} ($\Delta C_{Si-H}/C_{Si-H}$) following annealing.

Sample	Initial C_{nV} (vol.%)	C_{nV} (vol.%)	$\Delta C_{Si-H}/C_{Si-H}$ (%)	x (at.%)
RFS3	0.5	1.4	20	1.6
RFS5	0.7	1.4	32	5.7
RFS7	0.9	1.7	34	6.0
RFS4	0.9	1.9	50	3.0

The bulk Si-H stretch band at 2000 cm^{-1} weakened relative to the (overlapping) ~2100 cm^{-1} band. This is consistent with detrapping of H atoms from Si-H bonds embedded in the Si network and diffusion to deeper H trapping sites, vibrating at ~2100 cm^{-1}. There is also a shoulder at ~750 cm^{-1} and peaks at ~910 and ~1000 cm^{-1}, which are usually associated with vibrations of Si-CH_n, Si-O, and C-H, respectively. The intensities of the Si-CH_n and the C-H modes increased relative to the Si-H wagging at 640 cm^{-1}. This behavior is consistent with the net transfer of hydrogen from shallower Si- to deeper C-bonded trap sites.

c. H Diffusion in RFS a-$Si_{1-x}C_x$:H, $x \le 3$ at.% and ECR a-$Si_{0.86}C_{0.14}$:H.

Figure 3 shows typical example of a "smeared" DSIMS depth profile of annealed sample RFS1 and the complementary error function fit.

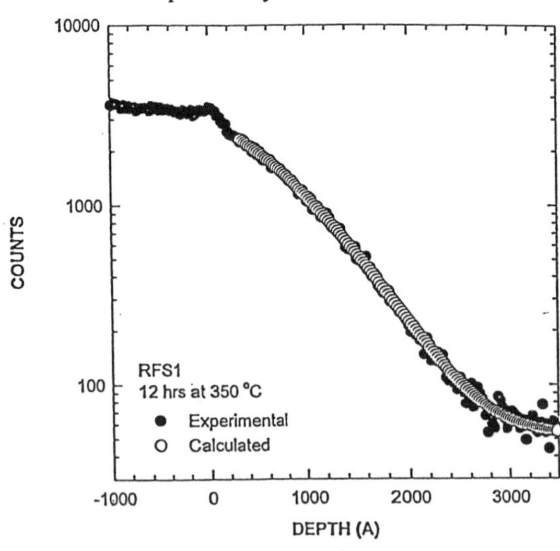

Figure 3. DSIMS profile of sample RFS1 following annealing for 12 hours at 350°C. The open circles are the fitted complementary error function.

The values of α, summarized in Table III, indicate that within each sample, the temperature dependence of α was too weak to be determined. This situation is in contrast to that of a-Si:H of low-to-moderate C_{nv}, where α typically decreases with increasing T at $T \leq 350°C$, but increases strongly at higher T, regardless of the deposition procedure [7,8].

Table III. Values of the dispersion parameter for diffusion α in RFS1-RFS3.

Sample RFS1		Sample RFS2		Sample RFS3	
T (°C)	α	T (°C)	α	T (°C)	α
350	0.63	350	0	375	0
380	0.47	380	0.20	425	0.09
420	0.52	420	0.07	476	0.30

As in RFS films, D diffusion in undoped ECR1 behaves as a power law function of the annealing time t. The values of α were 0.3, 0.3, and 0.1 at 350, 400, and 450°C, respectively. These values are small and suggest a relatively low C_{nv}. While α does appear to decrease with increasing temperature, the decrease is very moderate.

For a diffusion length $L = 1000$ Å, we obtained activation energies E_a of 1.7, 1.4, 0.7, and 1.0 eV in RFS1, RFS2, RFS3, and ECR1, respectively. The first two values are similar to those found for a-Si:H [3] and indicate that incorporation of up to 3 at.% C does not change H diffusion significantly. This conclusion can be readily understood within a scenario in which C atoms are either bonded tightly to other atoms in the network, or are present as saturated CH_n groups at the surfaces of nanovoids. Such carbon would probably interact less than Si with the hopping H. The nature of the anomalously low E_a value found for RFS3 and ECR1 is not clear.

The DSIMS together with the SAXS results of the RFS films demonstrated that a residual columnar morphology and C_{nv}, which increased to ≤ 2 vol.% by annealing, are not sufficient to suppress the long range D motion. Since α was much less than 1, it is suspected that boundaries of the residual columnar-like features, that are largely covered by saturated CH_n groups, are not as efficient H-traps as the surfaces of spherical voids.

Several SIMS depth profiles of B-doped sample ECR3, annealed at $T \leq 450°C$ and normalized to the maximal D level, are shown in Figure 4. The profiles clearly show that the diffusion of most of the H and D atoms is suppressed, even at $T = 450°C$. However, the "tail" of the profiles in the undeuterated layer indicates rapid diffusion of a small fraction of the D atoms.

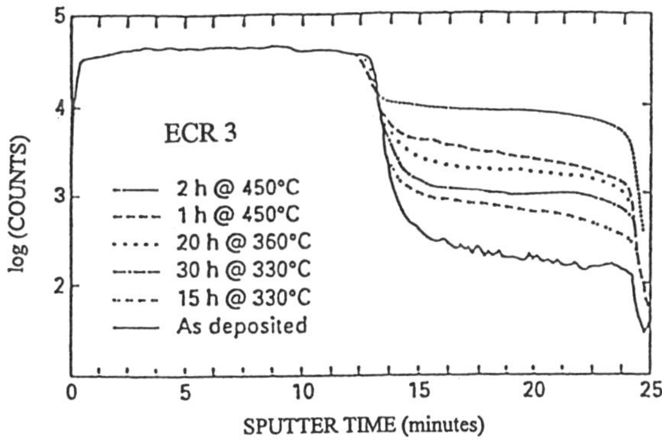

Figure 4. DSIMS depth profiles of sample ECR3 normalized to the maximum D level.

These discontinuous profiles may indicate trap-controlled diffusion with weak exchange [12]. The growth of the deuterium level in the initially undeuterated layer with annealing time at

360°C was faster for ECR3 (C_B ~0.2 at.%) than in ECR4 (C_B ~0.6 at.%). This behavior is in stark contrast to the dramatic B-induced enhancement of H diffusion in a-Si:H [5], but consistent with the introduction of H deep trapping in B-induced complexes and nanovoids.

SUMMARY

SAXS, IR, and DSIMS studies of microstructure and H dynamics in RFS a-Si$_{1-x}$C$_x$:H (1.6 ≤ x ≤ 19 at.%) and ECR a-Si$_{0.86}$C$_{0.14}$:H were described. The SAXS of the RFS films indicated a residual columnar morphology and a nanovoid content $0.5 \leq C_{nV} \leq 1.0$ vol.%. The larger-scale residual columnar microstructure was unaffected by annealing of up to 4 hours at 420°C, but C_{nV} increased by ~100%, mostly due to a ~20% increase in the average nanovoid radius. This increase was smaller than that observed in RFS a-Si:H, suggesting that the C atoms incorporated in the Si network weakened network relaxation processes. The IR measurements indicated a reduction in C_{Si-H}. This reduction appeared to be correlated with C_{nV}, as expected, since the distance from any given Si-H bond to the nearest surface decreases with increasing C_{nV}. A stronger reduction was observed in the bulk Si-H bond (2000 cm^{-1}) in comparison to the 2080-2150 cm^{-1} bands, characteristic of surface Si-H, C-Si-H, and O-Si-H groups. This and the net transfer of H from Si-H to C-H bonds are consistent with the migration of Si-bonded H from bulk to deeper trapping sites and the greater strength of the C-H bond relative to the Si-H bond.

Long-range H motion, similar in nature to that of a-Si:H, was observed in undoped ECR and RFS films with x ≤ 3 at.%, in spite of the residual columnar morphology and C_{nV} which reached ~2 vol.% following annealing. However, the dependence of α on the annealing temperature was weaker than that observed in a-Si:H, also consistent with weaker network relaxation dynamics. The DSIMS profiles of the B-doped ECR films indicated that most of the H and D atoms were deeply trapped, possibly at surfaces of B-induced voids and complexes. A small fraction of the H and D atoms exhibited a fast diffusion, reminiscent of the doping-induced enhanced H diffusion in a-Si:H.

ACKNOWLEDGEMENTS

DLW acknowledges the support of NREL through subcontract No. RAN-4-13318-04. Ames Laboratory is operated by Iowa State University for the US Department of Energy under Contract W-7405-Eng-82. This work was supported by the Director for Energy Research, Office of Basic Energy Sciences.

REFERENCES

1. D. L. Williamson, Mat. Res. Soc. Symp. Proc. **377**, 251 (1995).
2. J. Shinar, H. Jia, R. Shinar, Y. Chen, and D. L. Williamson, Phys. Rev. B **50**, 7358 (1994).
3. S. Acco, D. L. Williamson, P. A. Stolk, F. W. Saris, M. J. van den Boogaard, W. C. Sinke, W. F. wan der Weg, S. Roorda, and P. C. Zalm, Phys. Rev. B **53**, 4415 (1996).
4. R. A. Street, C. C. Tsai, J. Kakalios, and W. B. Jackson, Phil. Mag. B **56**, 305 (1987).
5. J. Kakalios, R. A. Street, and W. B. Jackson, Phys. Rev. Lett. **59**, 1037 (1987).
6. J. Shinar, R. Shinar, S. Mitra, and J.-Y. Kim, Phys. Rev. Lett. **62**, 2001 (1989).
7. X.-M. Tang, J. Weber, Y. Baer, and F. Finger, Phys. Rev. B **41**, 7945 (1990).
8. R. Shinar, J. Shinar, H. Jia, and X.-L. Wu, Phys. Rev. B **47**, 9361 (1993).
9. R. D. Knox et al., J. Vac. Sci. Tech. A **11**, 1896 (1993).
10. M. L. Albers, J. Shinar, and H. R. Shanks, J. Appl. Phys. **64**, 1859 (1988).
11. M. Cardona, Phys. Stat. Sol. B **118**, 463 (1983).
12. M. Kemp and H. M. Branz, Phys. Rev. B **52**, 13946 (1995).

Part V

Metastability

PHOTOINDUCED EXPANSION IN HYDROGENATED AMORPHOUS SILICON

S. NONOMURA*, T. GOTOH*[1], M. NISHIO*, T. SAKAMOTO*, M KONDO**,
A. MATSUDA** and S. NITTA*
*Department of Electrical Engineering, Gifu University, 1-1 Yanagido, Gifu 501-1193, Japan
**Thin Film Silicon Solar Cells Superlab, ETL, Ibaraki, Japan

ABSTRACT

The structural aspect of photodegradation effect in hydrogenated amorphous silicon has been investigated by the use of the simple and sensitive detection technique, the laser optical-lever bending method, for a small expansion or extraction in thin films. The volume change induced by the thermal expansion due to the photothermal effect and the residual expansion was observed in hydrogenated amorphous silicon prepared by PECVD. The latter residual expansion was persistent after light soaking and was recovered by thermal annealing at 200 °C. The time dependence of the volume expansion with light soaking shows the same time dependence of photoinduced defect density. The photoinduced volume changes normalized by the initial volume are the order of 10^{-6}~10^{-5}, which values are two orders smaller than chalcogenide glasses such as a-As_2S_3. The normalized volume change of a-Si:H with the different sample preparation conditions of PECVD such as the hydrogen dilution ratio r (r = SiH_4/H_2) and substrate temperature is shown. Also it is demonstrated that the photoinduced expansion is observed in hydrogenated amorphous silicon prepared by photo CVD and hot-wire CVD methods. The spatial extent related to a photoinduced defect creation in a-Si:H is estimated.

INTRODUCTION

The light induced decrease of photoconductivity by Steabler-Wronski [1] is a first finding of photoinduced phenomenon in hydrogenated amorphous silicon (a-Si:H). It was elucidated that the phenomenon is caused by the defect creation [2,3]. Many models [4] have tried to interpret the mechanism of the light induced defect creation, but the conclusive consensus is not obtained even in the present stage. It is well known that chalcogenide glasses show the large fractional thickness change due to the band gap light soaking. The estimated thickness change is 3.8×10^{-3} in a-As_2S_3 [5], and the structural change is detectable by XRD. On the other hand, no direct evidence of structural change in a-Si:H has been reported until now. Recently, several groups have reported experimental evidences of macroscopic structural change using the polarized electroabsorption method [6] and the laser ablation time of flight mass spectroscopy [7], after proposing the possibility of structural change [8] by

---

1. Present address; Laboratory of Photoelectronic Materials, Department of Applied Physics, Hokkaido University,

Mat. Res. Soc. Symp. Proc. Vol. 557 ©1999 Materials Research Society

Fritzsche. These obtained changes are indirect and very subtle as compared to those of chalcogenide glasses. In generally, a-Si:H has a high restricted network and the small number of dangling bond's creation (the order of 10^{16}) by light soaking. These are the reasons being difficult to detect the related volume change in a-Si:H by the current methods.

The sensitive bending technique combined with optical-lever method [9,10] has been developed for the detection of a infinitesimal volume change in semiconducting thin films. Normally, a thin film is deposited on a substrate and the sample's structure is same to that of bimorph. If we use the very thin substrate, an infinitely small expansion becomes detectable, because an infinitesimal volume change converts a bending of a sample. The change in the bending angle of a sample is detected as a laser beam deflection. This hybridized method has been utilized in the photothermal bending spectroscopy (PBS) developed by us [9,11]. This PBS technique is a kind of photothermal spectroscopy for a measurement of the sub gap absorption spectrum of a-Si:H, because an a-Si:H is thermally expanded due to a nonradiative recombination of a electron and a hole excited by light absorption. It's spectral range covers from ~3 eV to ~0.25 eV. We have applied this method in order to study a photostructural change accompanied with the photoinduced defect creation.

In this paper, we demonstrate the more direct experimental evidence of the photoinduced microscopic structural change, the photoinduced expansion, in a-Si:H using the laser optical-lever bending method. It is pointed out that the defect creation by the light soaking is accompanied with the expansion of the lattice. The volume changes are studied for a-Si:H films deposited at various deposition conditions, a substrate temperature and a hydrogen dilution ratio of SiH_4. The results of a-Si:H and μc-Si:H prepared by a PECVD, a photo CVD and a hot-wire CVD are also demonstrated. We estimate the spatial extent in a random network around a induced defect by a light soaking.

EXPERIMENT

Samples used in this study are hydrogenated amorphous silicon (a-Si:H) and microcrystalline silicon (μc-Si) thin films with the thickness of 500 ~ 1000 nm. The a-Si:H films are undoped one prepared by a plasma enhanced chemical vapor deposition (PECVD), a photo CVD and a hot-wire CVD [12]. Th pure silane was used for the preparation of a-Si:H films with changing the substrate temperature from 100 to550 °C. And the a-Si:H films deposited at 250 °C were employed for the study related to the hydrogen dilution ratio ($H_2/SiH_4 = r = 0$~39). The tentative substrate is the polished quartz with the thickness of 50 ~100 μm. The sample's width and length are 2 mm and 20 mm, respectively.

The utilized highly sensitive detection technique developed by us is the laser optical-lever bending method, which has DC mode and AC mode. The measurement system is constructed by two parts of amplifying technique. One is an amplification using the bimorph structure of a sample which is formed by a semiconductor film deposited on a thin substrate. The laser optical-lever method is the second stage amplifier. The one edge of a sample (a-

Si:H on quartz substrate) is held on the sample holder, and the other edge is free. The expansion or contraction of a-Si:H is converted to the displacement of the sample's free edge. We applied the super invar as a metal material of the sample holder, because this metal has a extremely small expansion coefficient, 0.4×10^{-6} K^{-1}, similar to those of glasses. The use of the super invar prevents the displacement of the sample holder itself by a room temperature drift.

Figure 1 shows the schematic diagram of the measurement system for the

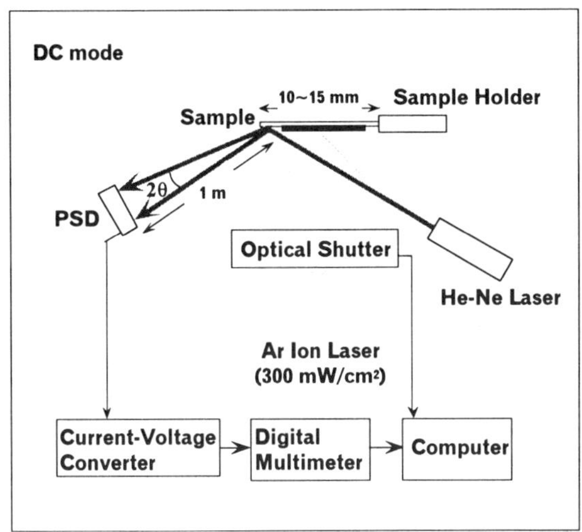

Fig.1. A schematic diagram of the DC mode laser optical-lever bending method in order to detect the photoinduced structural change.

DC mode laser optical-lever bending method [10]. The multi-mode Ar* laser (λ=488 and 514 nm) was used for making the photodegradation in a-Si:H. The light power was about 300 mW/cm^2, and the light irradiation area on the sample was 2x10 mm^2. The intermittent light irradiation was carried out for avoiding the temperature rise of the sample by a light soaking. The light-on and -off periods are both 5 minutes, which are controlled by a mechanical shutter. The He-Ne laser of λ = 632.8 nm was employed as a probe beam for detecting the displacement of the sample's free edge resulting from the volume change. The probe beam is reflected at the sample's free edge and fed into the position sensitive detector (PSD). The free edge as shown in Fig.1 is the smooth surface of quartz substrate without deposition of a-Si:H in order to prevent the dispersion of the reflected He-Ne laser light. The deflected light is detected by PSD and amplified by a DC voltage amplifier after the current-voltage conversion circuit.

The internal stress σ resulting from an a-Si:H film and a thin substrate was obtained from a bending angle θ using Stoney's equation [13],

$$\sigma = E_s d_s^2 \tan(\theta/2)/3(1-v_s)d_f a \quad \dots\dots\dots\dots\dots\dots\dots\dots\dots\dots\dots\dots(1)$$

where, E_s is Young's modulus, v is Poisson's ratio, θ is the bending angle and a is the sample's length. The subscript s and f mean the substrate and the film, respectively. This equation is correct in the condition of $E_s d_s \gg E_f d_f$, because the value values of E_f, E_s, n_f and n_s used in this study were 1.3x1011 Pa, 7.31x1010 Pa, 0.28 and 0.17, respectively. The photoinduced change of the internal stress $\Delta\sigma$ gives a photoinduced change of strain $\varepsilon = \Delta a/a$ by

$$\varepsilon = \Delta a/a = (1 - v_f)\Delta\sigma/E_f \;...(2)$$

The photoinduced volume change $\Delta V/V$ is calculated from the strain ε assuming the homogeneous structure of a film in three dimensions. The obtained detection limit of $\Delta V/V$ was 2×10^{-7} in our detection system. This high sensitivity means that our method has much potential being applicable to the detection of thin film's small volume changes by a thermal expansion, an infinitesimal structural change, etc.

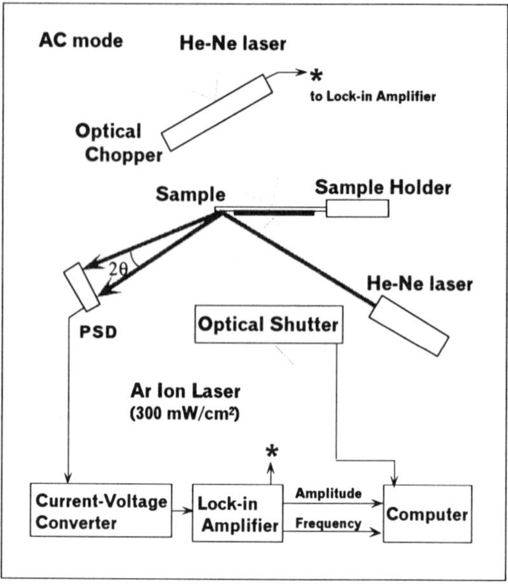

Fig.2. A schematic diagram of the AC mode laser optical-lever bending method. The chopping frequncy is from 100 Hz to 2 kHz.

Figure 2 shows the experimental set up for the AC mode laser optical-lever bending method [9]. In this technique, the change of bending angle at the sample's free edge is also monitored by a laser optical-lever method. The small vibration in a sample is caused by another He-Ne laser light used as a modulated excitation beam. The tentative chopping frequency is from ~3 to 2 kHz. Sweeping the chopping frequency, the maximum signal intensity of sample's bending is detected by lock-in amplifier at resonant frequency defined by the sample's size, the thickness, Young's modulus and so on. This resonant frequency is measured by the computer controlled lock-in amplifier (Stanford Research Systems, SR830) and the spectrum before and after light soaking are compared. The information of a volume change is derived from the shift of resonant frequency.

RESULTS

The displacement of the sample's free edge by the intermittent light irradiation is shown in the Fig.3 (a). The light-on and -off periods are 5 minutes, respectively. The displacement signal increases abruptly by the light exposure of 300 mW/cm², which causes a heat expansion induced due to a nonradiative recombination of excited electrons and holes. This means the production of a compressive stress in a-Si:H. After cutting off the light exposure, the signal decreases and returns back to the almost same value of the initial displacement signal within ten seconds. But the additional increase of the displacement signal is observed at 3 hours. The displacement signals of the light-off state are redrawn in Fig.3 (b). The induced residual expansion increases gradually with the light exposure and keeps constant for 2 hours during the light-off (at 7 ~ 9 hours). If the residual volume expansion is caused by the thermal expansion of the sample and sample holder, the signal will decrease during this light-off period because of the releasing of the thermal energy to the air. The keeping at the constant value means that the residual expansion is a persistent phenomenon, but not by a thermal effect. The same increase of the residual expansion (2nd) with light exposure was also observed after the thermal annealing at 200 °C. Our observed photoinduced expansion is a persistent and the phenomenon is recovered by thermal annealing. The photoinduced volume expansion is a reversible phenomenon by thermal annealing at 200 °C. We have to emphasize that these experimentally obtained

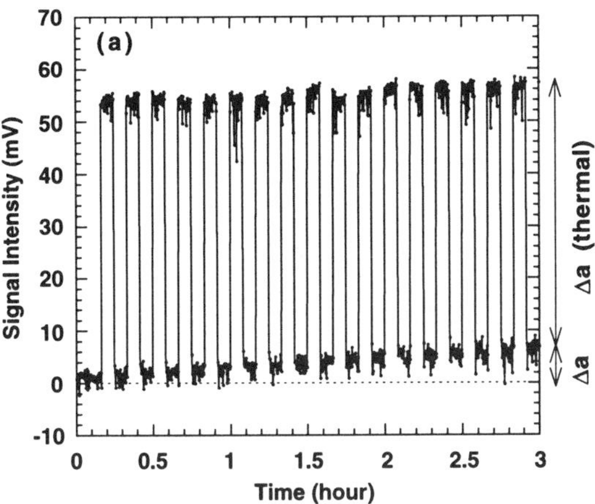

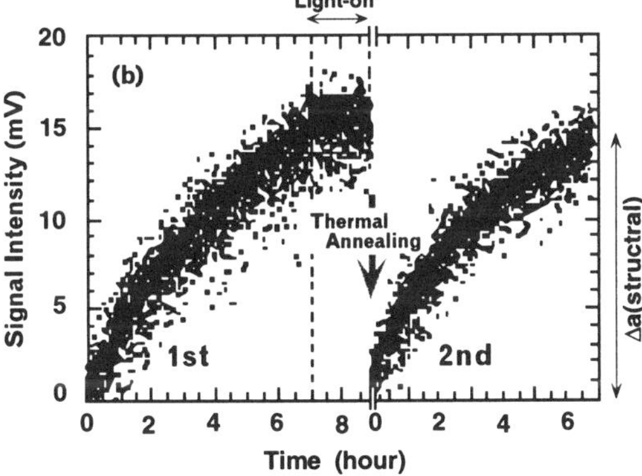

Fig.3 The changes of the DC mode laser optical-lever bending signal are shown. (a) shows the signal with Ar ion laser light-on and -off. The light-on and -off time are 5 minutes, rrespectively. (b) shows the signal change (1st) with the light-off state, which are redrawn from Fig.3(a). The intermittent light exposure is done at 0~7 hours. At the following 7~9 hours, the light is not irradiated, and the sample is thermal annealed at 200 °C. The signal change (2nd) is remeasured after the thermal annealing.

results are similar to those of the photoinduced defect creation in a-Si:H. Figure 4 shows the value of the photoinduced volume expansion $\Delta V/V$ converted from the displacement signal. The obtained $\Delta V/V$ of a-Si:H at various deposition conditions are less than $\sim 3 \times 10^{-5}$. The defect densities N_d estimated from a constant photocurrent method are also plotted by the open squares in Fig.4. One can see the good coincidence between the change of $\Delta V/V$ and N_d with light exposure time.

We also applied AC mode laser optical-lever bending method in order to confirm the volume expansion by the light soaking. Fig. 5 shows the frequency dependence of the displacement signal detected at the sample's free edge. The frequency means the chopping frequency of He-Ne laser light as an excitation beam. The spectrum of a-Si:H before the light soaking is denoted by open circles. A maximum value is seen at 674 Hz. This resonant frequency is determined by the thickness, geometry and Young's modulus in the a-Si:H film and the substrate. After the light soaking for 600 min, the resonant frequency shifts to higher frequency of 679 Hz. Here, the light power for the photodegradation was 500 mW/cm^2. The shift to the higher frequency means the increase of the sample's hardness. In generally, the experimentally obtained large shift of the resonant frequency can not be explained by the change of Young's modulus. Figure 6 is a photograph of the sample and the sample holder. It is seen that the sample bends in the direction of the sample's length. The sample has a convex shape, and the internal stress of a-Si:H is compressive. We can easily suppose that the photoinduced volume expansion also enhances the bending in the direction of sample's width. This enhanced bending increases virtually sample's hardness. This interpretation and experimental

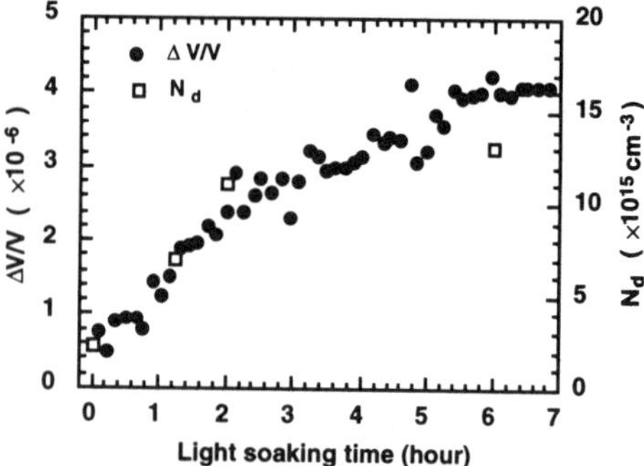

Fig.4 The comparison of the light exposure time dependences of the photoinduced volume expansion ΔV/V and the photoinduced defect density.

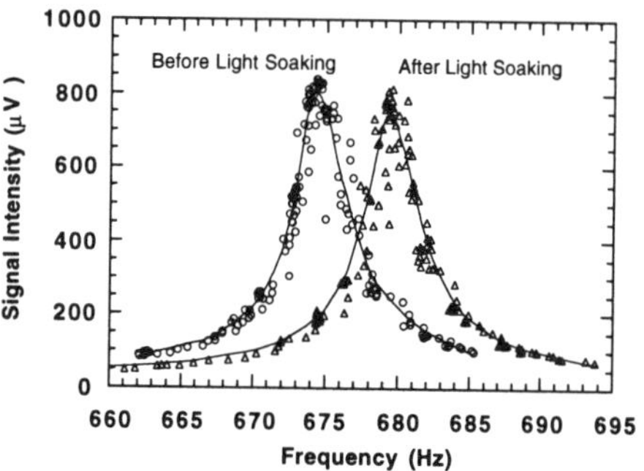

Fig.5 The chopping frequency dependence of the displacement signal in a-Si:H. The signals before and after ligth soaking are shown by open circles and open triangles. The light sorce for the photodegradation was Xe lamp. The light power was 500 mW/cm^2 cutting the IR light by a optical filter.

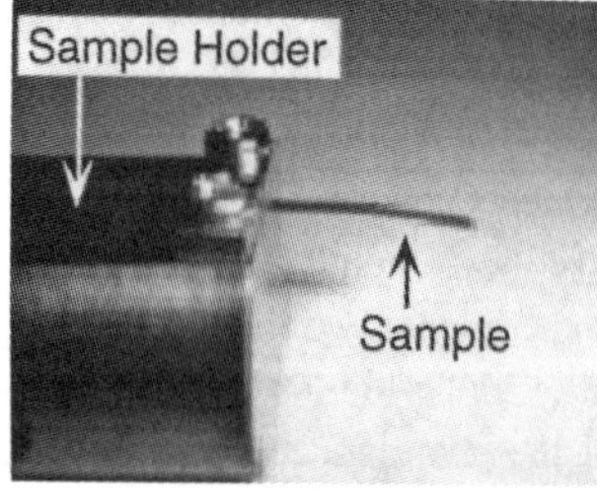

Fig.6 The photgraph of the sample and sample holder. The sample bends convexly. The surface of a-Si:H is an upper side.

results support the photoinduced volume expansion detected by the DC mode laser optical-lever bending method.

The undopoed a-Si:H was prepared by a PECVD, a photo CVD and a hot-wire CVD [12] for the study of a photoinduced volume change $\Delta V/V$. Figure 7 shows the $\Delta V/V$ of a-Si:H deposited at different deposition temperatures from 100 °C to 550 °C. The $\Delta V/V$ has a maximum value around 300 °C and becomes smaller at higher deposition temperature than 300 °C. This tendency coincide with the smaller photodegradation at higher deposition temperature, because of the defect creation by an effusion of hydrogen atoms from an a-Si:H network. The reduction of $\Delta V/V$ is observed at the lower deposition temperature from 250 °C to 100 °C. It is inconsistent with the creation of defect states in a-Si:H prepared at room temperature. But the observable $\Delta V/V$ is a macroscopic quantity. In generally, the structure of a-Si:H prepared at lower temperature has much microvoid and defect states. So, the detection of $\Delta V/V$ might be difficult, because the induced stress related to the defect creation will relax by the flexible network of a-Si:H with much microvoid and much defect density.

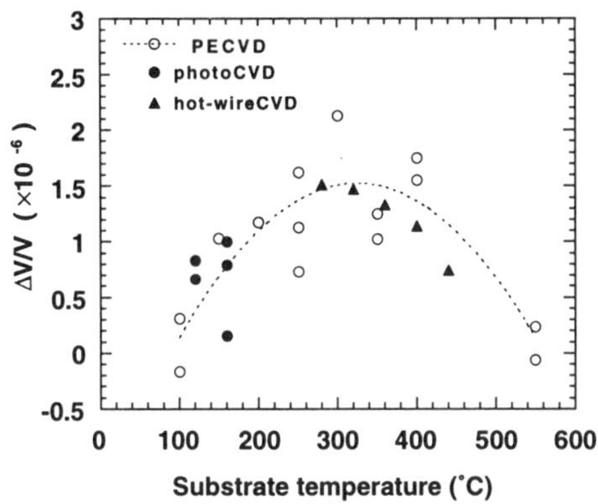

Fig.7 The substrate temperature dependence of photoinduced volume change. The open circles, closed circles and colsed triagles show the $\Delta V/V$ of a-Si:H prepared by a PECVD, a photo CVD and a hot-wire CVD.

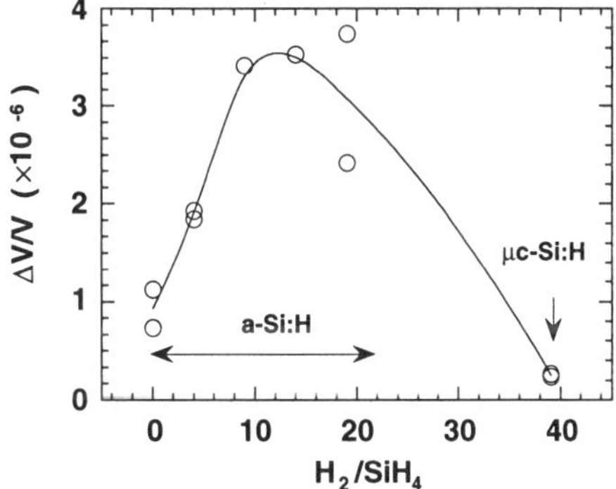

Fig.8 The change of photoinduced expansion in a-Si:H prepared by the diffrent hydrogen dilution ratio. The μc-Si:H with low defect density is obtained at r=39.

The photoinduced volume changes of a-Si:H prepared by a photo CVD and a hot-wire CVD are also shown in Fig.7. The decreases of the photoconductivity by light soaking are observed in all a-Si:H films. We emphasize that the photoinduced volume expansion is in accompany of the photoinduced defect creation in a-Si:H.

The hydrogen dilution ratio is one of the important deposition parameters. The high quality a-Si:H films are obtained at r~10. Figure 8 shows the $\Delta V/V$ of a-Si:H deposited at different hydrogen dilution ratio from 0 to 39. The $\Delta V/V$ becomes larger with the increase of the hydrogen dilution ratio and has a maximum value at around r~10. The μc-Si:H with the defect density below 10^{16} cm^{-3} is obtained at r=39 in our PECVD system. It is interesting that the μc-Si:H has negligibly small value of $\Delta V/V=2.5\times10^{-7}$. The value is closely same to the detection limit of our measurement system. This result agrees with the missing of the photodegradation effect in our μc-Si:H film prepared by same deposition conditions. The two contradicting results for the photodegradation have been reported in a μc-Si:H. Nickel et al. [14] observed the photoinduced defect creation in hydrogen passivated μc-Si:H prepared at ~600 °C. But Miere et al. [15] reported no change of photoconductivity by light irradiation in a μc-Si:H deposited at ~200 °C. Recently, Kondo et al. [16] confirmed both the increase of defect states and the formation of oxide layer by light irradiation in the μc-Si:H with the defect density of 5×10^{15} cm^{-3}. They interpreted the increase of defect states by the band bending model resulting from the formation of a oxide layer, because the covering of μc-Si:H surface by a thin a-Si:H (60 Å) prevents the surface oxidation and the photoinduced defect creation. Our obtained negligible small volume change in μc-Si:H supports that a high quality μc-Si:H film has no photoinduced defect creation, and means that an amorphous network is one of the essential factors for the revealing of photoinduced phenomena including the photoinduced defect creation.

DISCUSSIONS & CONCLUSIONS

We detected the volume change of a-Si:H induced by light soaking. The typical values of $\Delta V/V$ and the defect density were 4.2×10^{-6} and 2.16×10^{16} cm^{-3}, respectively. The volume change per defect is roughly estimated to be 9.7 molecular volume assuming a diamond structure for the Si. The volume change is well studied in alkali halide crystals [17] and c-SiO$_2$ [18]. The volume change and the optical absorption accompanied with the creation of E_1' center is observed and decays after cutting off the irradiation. The reported volume change per E_1' center is 3.8~4.7 molecular volumes in c-SiO$_2$ during the irradiation of. x-ray or electron pulses. On the other hand, chalcogenide glasses show the large volume change, which is caused by the lattice reconstruction being detectable by XRD. In a-As$_2$S$_3$, the volume expansion is 1.26×10^{-2}. This significant volume change is understood in terms of the network flexibility due to its low coordination number. The smaller volume change of a-Si:H originates from the higher coordination number than those of chalcogenide glasses. But the volume change much larger than unit molecular volume suggest that the defect is created by the microscopic structural change over the medium range scale, not by the simple bond breaking. Though an a-Si:H has a rigid network than chalcogenide glasses, the average coordination number must be decreased by the incorporation of hydrogen-silicon bonds in order to keep an amorphous network. The microscopic area with higher flexibly in a-Si:H network decrease an average energy states around the defect, and stabilizes the photoinduced defect.

In conclusion, the photoinduced volume expansion is persistent and reversible phenomenon. The kinetics of the photoinduced volume expansion are identical to those of the photoinduced defect creation in a-Si:H prepared by a PECVD, a photo-CVD and a hot-wire CVD. The photoinduced volume expansion reveals in accompanied with the defect creation by the light soaking. The negligible small volume change is observed in the annealed a-Si:H at 550 °C and μc-Si:H which shows no photodegradation. The estimated volume change per defect suggests that the microscopic structural change occures over a medium range scale.

ACKNOWLEDGMENTS

We greatly thank to Prof. K. Morigaki, Prof. H. Okamoto, Prof. M. Konagai, Prof. K. Shimakawa, Prof. Ke. Tanaka, Prof. C.M. Fortmann and Prof. Y. Hamakawa for helpful discussion and suggestion.

This work is partially founded by NEDO as part of the New Sunshine Program of MITI and Grant-in Aid for Science Research (C), the Japan Securities Scholar Ship Foundation and the program, "Research for the Future" of the Japan Society for the Promotion of Science.

REFERENCES

1. D.L. Staebler and C.R. Wronski, Appl. Phys. Lett. **31**, p.292 (1977).

2. H. Dersch et al., Phys. Stat. Sol. **(b) 105**, p.265 (1981).

3. M. Stuzmann et al., Phys. Rev. **B 34**, p.63 (1986).

4. M. Stuzmann et al., Ouyoubutsuri (in Japanese) **60**, p.1004 (1991).

5. H. Hamanaka, K. Tanaka and S. Iijima, Solid State Commun. **19**, p.499 (1976).

6. K. Shimizu, T. Shiba, T. Tabuchi and H. Okamoto, Jpn. J. Appl. Phys. **36**, p.29 (1997).

7. T. Kato, et al., Technical Digest Int. PVSEC-9, Miyazaki, C-V-3, (1996), p.561-562.

8. H. Fritzsche, Solid State Commun. **94**, p.953 (1995).

9. T. Gotoh, S. Nonomura, S. Hirata, N. Masui and S. Nitta, Solar Energy Materials & Solar Cells **49**, p.13-18 (1997).

10 T. Gotoh, S. Nonomura, M. Nishio, S. Nitta, M. Kondo and A. Matsuda, Appl. Phys.Lett.**72**, p.2978 (1998).

11. T. Gotoh, S. Nonomura, S. Hirata and S. Nitta, Progress in Natural Science **6**, p.S-34 (1996).

12. D. Han, T. Gotoh, M. Nishio, T. Sakamoto, S. Nonomura, S. Nitta, Q. Wang and E. Iwaniczko, Materials Research Society Symp. Proceedings **505**, p.445 (1998).

13. G.G. Stoney, Proc. Roy. Soc. **A82**, p.172 (1909).

14. N.H. Nickel, W.B. Jackson and N.M. Johnson, Phys. Rev. Lett. **71**, p.2733 (1993).

15. J. Meire, R. Fluckiger, H. Keppner, A. Sah, Appl. Phys. Lett. **65**, p. 860 (1994).

16. M. Kondo, T. Nishimiya, K. Saito, A. Matsuda, J. Non-Cryst. Solids, **227-230**, p.1031 (1998).

SLOW DEGRADATION OF HYDROGENATED AMORPHOUS SILICON PHOTOCONDUCTIVITY UNDER PULSED ILLUMINATION

STEPHAN HECK AND HOWARD M. BRANZ
National Renewable Energy Laboratory, Golden, CO, 80401

ABSTRACT

We degraded hydrogenated amorphous silicon (a-Si:H) using red light pulses of 32 microseconds to 2 milliseconds. These metastable photoconductivity degradations were compared to degradation with continuous light of the same intensity and same exposure time. For a given integrated exposure time we observe higher degraded photoconductivities (by up to 40 %) as we shorten the pulses. For example, to obtain the same amount of degradation with 120 microsecond pulses as with continuous illumination, the integrated sample exposure time must be doubled. Experiments were conducted to exclude thermal effects. Our result cannot be explained with a simple recombination-driven degradation mechanism, because electron and hole populations rise and fall to steady-state values in only a few microseconds. We conclude that carrier recombination creates a more proximate precursor to metastable degradation of a-Si:H. When illumination begins, this precursor's density rises more slowly than the density of photocarriers and therefore degradation is delayed.

INTRODUCTION

Ever since Staebler and Wronski (SW) discovered that light exposure degrades the photoconductivity of hydrogenated amorphous silicon (a-Si:H) [1], light- and carrier-induced property changes of a Si:H have been a fascinating scientific puzzle. The SW effect is also of great technological interest, because of its deleterious effect on a-Si:H solar cells and thin film transistors. However, despite great efforts made since the discovery of the SW effect in 1977, a generally accepted model remains elusive [2, 3].

In this paper, we present the first report of slow photoconductivity degradation under pulsed illumination in a-Si:H. We degrade our samples with continuous (CW) and pulsed light of same intensity and compare the resulting photoconductivities (σ). We find less σ degradation during pulsed than during CW degradation. The pulse times used in our experiment are much longer than the rise and decay times [4] of σ (a few microseconds), therefore photoexcited carriers alone cannot be responsible for the SW effect.

Our experiments show that light induced degradation of a-Si:H involves a second timescale of order 0.5 ms. We identify this timescale with the rise-time of a more proximate precursor to degradation that is excited by the photocarriers. For example, in models based on mobile hydrogen (H) as a precursor for metastable defect formation, this 0.5 ms timescale would be the time to reach a steady-state population of mobile H. During illumination with pulses shorter than

Mat. Res. Soc. Symp. Proc. Vol. 557 © 1999 Materials Research Society

this mobile H rise time, the peak mobile H concentration is lower than during continuous illumination, and defect creation is suppressed.

EXPERIMENTAL

We use two device-quality intrinsic a-Si:H films on Corning 7059 glass for the experiments. The first, 0.6 μm thick, was grown in our laboratory by plasma-enhanced chemical vapor deposition (PECVD) at 230 °C on predeposited chromium contacts. The second was used only to confirm results obtained on our own sample. This second film was grown at the University of Chicago, also by PECVD at 230 °C. It is 1.1 μm thick and has post-evaporated chromium contacts.

To degrade the samples, we use two different light sources which give similar results: a 150 W Xe arc lamp filtered through a 665±30 nm (1.88 eV) bandpass filter or a laser diode at 653±2 nm (1.90 eV). Our sample has a bandgap of 1.9 eV with an absorption coefficient of $1.0{\times}10^4$ cm^{-1} at 665 nm and $1.3{\times}10^4$ cm^{-1} at 653 nm, so the red light efficiently creates a nearly uniform distribution of electron hole pairs. The light passes through a slit aperture and is focused on the sample. The arc lamp produces an intensity at the sample of 0.15 W/cm^2; the laser diode, 0.1 Wcm^{-2}. After degrading the sample either with CW or pulsed illumination, we attenuate the same light source by 10^{-3} with a neutral density filter and measure the low light σ. In all cases, the dark conductivity is at least three orders of magnitude smaller than σ.

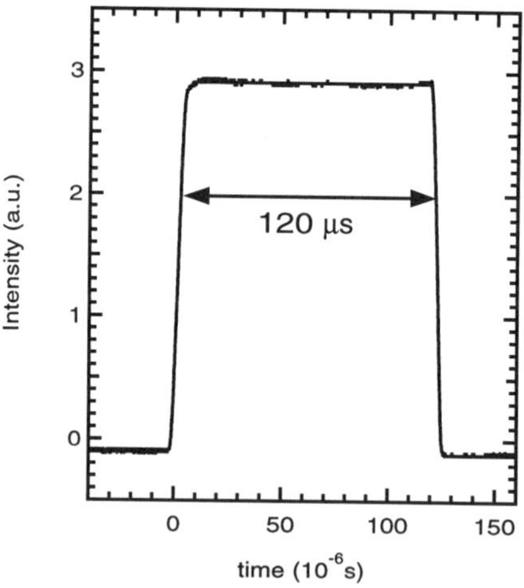

Figure 1: Illumination intensity for a 120 μs pulse as measured with a photodiode and oscilloscope. The rise time is 7 μs.

For the pulsed degradations, we mount a chopper wheel comprised of two identical chopper blades adjacent to the slit aperture. The relative angle of the two blades can be adjusted; this enables us to control the duty cycle of the pulsed illumination between about 1 and 50 %. We measure the pulse length, t_p, by monitoring either the response of σ with a current-to-voltage amplifier or the signal from a silicon photodiode. Fig. 1 shows the pulse shape for a 120 μs pulse. We define t_p as the full width at two thirds of the maximum. With this measurement, we also determine the dark time between pulses, t_d. We define the exposure time, $t_e = N_p \times t_p$, where N_p is the number of pulses.

We avoid any movement of the sample between experiments by integrating a cartridge heater into the sample holder. This enables us to anneal and to perform experiments at elevated temperature. The temperature, T, is monitored by a thermocouple pressed with a clamp on a piece of 7059 glass close to the sample. The sample compartment is continuously flushed with N_2 gas to provide cooling and avoid oxidation of the sample surface.

The sample was first pre-annealed for 22 h at 190 °C. Then, prior to each subsequent exposure, the sample is re-annealed at 190 °C for 20 min (or for 2 h during certain experiments). The long pre-anneal stabilizes σ against irreversible changes during the anneal/degradation cycles. Still, irreversible changes due to annealing cause 5 % run-to-run variations in σ — our largest source of scatter in the measurements. We usually alternate several times between pulsed and CW illuminations to ensure reproducibility of the results

We made a careful study of the film temperature rise in our system to ensure that T did not increase significantly, even during CW illumination. We measured the low light intensity open circuit voltage (V_{OC}) of a 0.5 μm-thick solar cell illuminated in our apparatus. The fast and slow changes in V_{OC} are monitored by an oscilloscope and a voltmeter respectively. Isothermal calibration reveals a voltage decrease of 2.8 mV/°C of this solar cell at the measurement intensity we used. We found that the temperature rise after 15 minutes CW exposure is less than 5 °C. The T rise with pulsed illumination is obviously even smaller. In separate experiments we found almost no dependence of σ or its CW degradation upon T, between 6 and 36 °C.

RESULTS

Figure 2 shows the result of a degradation experiment with $t_p = 120$ μs and $t_d = 3.9$ ms. σ degrades about 25 % less with pulsed than with CW illumination in the time range studied. Another formulation of this result is that comparable degradation of the sample requires <u>twice</u> as much exposure time with pulsed light than with CW.

In Figure 3 we show data for several of the many pulse times we studied. We show only one (typical) set for each pulse length and three for CW degradation. The 580 μs degradation differs little from the CW illumination. As the pulse lengths are shortened, we get progressively less degradation. A more detailed analysis [5] reveals a characteristic time scale of about 0.5 ms.

We probe the dark time dependence with experiments at a constant $t_p = 0.48$ ms with $0.5 < t_d < 80$ ms. For $t_d > 3$ ms, pulsed illumination is much less effective than CW illumination

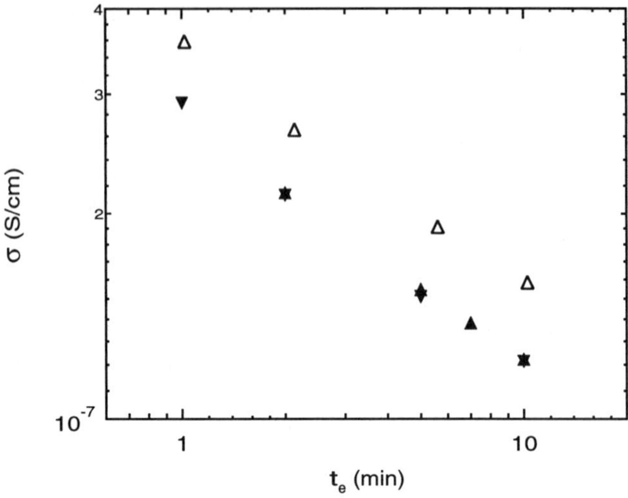

Figure 2: Degradation of photoconductivity σ with continuous and pulsed illumination of pulse length $t_p = 120$ μs.

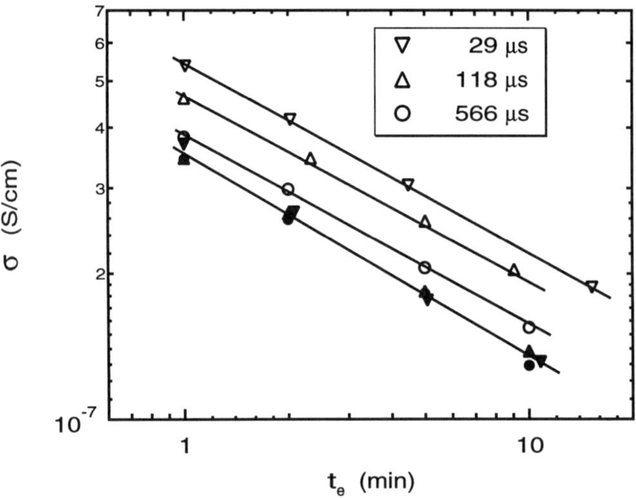

Figure 3: Photoconductivity degradation at various pulse lengths. See key for pulse length. Solid symbols are CW degradation. Lines are guides to the eye.

in causing photoconductivity degradation. The difference between pulsed and CW illumination gradually shrinks for $t_d < 3$ ms, disappearing almost completely at $t_d = 0.5$ ms.

Fig. 4 shows preliminary results for experiments at 60 °C. For comparison, the data for room temperature is included. Here $t_p = 118$ µs and $t_d = 9.9$ ms. We observe a similar decrease in pulsed degradation at both temperatures.

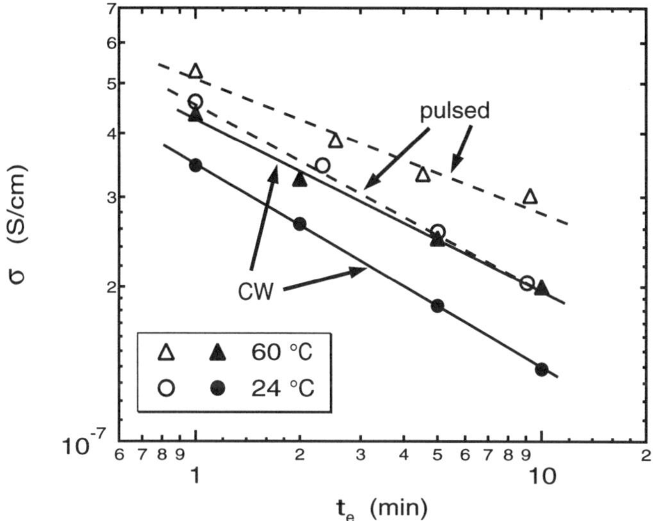

Figure 4 : Pairs of CW and pulsed degradation experiments with $t_p = 118$ µs and
$t_d = 9.9$ ms at 24 and 60 °C. Filled symbols are CW degradations and
hollow symbols are pulsed degradation

DISCUSSION

Our results cannot be explained by a greater temperature rise during CW compared to pulsed illumination. First, an internal measure of film temperature reveals less than 5 °C rise during CW illumination. Second, Fig. 4 shows that a rise in temperature causes <u>less</u> degradation, not more, in our sample.

Reduced degradation for pulsed illumination cannot be explained by a simple carrier-driven metastability model. With a rise and fall time of σ in the order of 3 µs, [4] σ follows closely the rise and fall of the light pulses. If electron-hole pairs alone were responsible for the degradation, then there should be no difference between pulsed and CW degradation.

We therefore suggest a more proximate precursor to degradation than electron-hole recombination. This precursor is excited by recombination, but has a much slower rise and decay time compared to the carrier density. We identify the characteristic times derived from our experiments as the rise (0.5 ms) and decay (3 ms) of this precursor at room temperature. If the light pulses are shorter than about 0.5 ms and the time between pulses is longer than about 3 ms, then the precursor cannot reach its steady state density and degradation is suppressed.

A slow, intermediate precursor to degradation was proposed in the hydrogen collision model of the SW effect [6]. This precursor is mobile hydrogen, whose rise time is determined by its annihilation rate at dangling-bond defects. We will discuss this and other models of intermediate precursors in detail elsewhere [5].

CONCLUSIONS

We observe reduced degradation for pulsed illumination, if the pulses used are shorter enough (< 0.5 ms) and if the time between pulses is long enough(> 3 ms). A carrier-recombination model alone, cannot account for our Staebler Wronski metastability results. Therefore, we suggest a more proximate precursor to metastable degradation than electron-hole recombination. This precursor has a rise time of about 0.5 ms under illumination and a decay time of about 3 ms in the dark. Though the precursor is created by electron-hole recombination, a microscopic description of the Staebler Wronski effect must include this precursor.

ACKNOWLEDGMENTS

We thank Panos Tzanetakis for suggesting this class of experiment. We acknowledge the help and suggestions of Hellmut Fritzsche, Walter Fuhs, Scott Ward, Richard Crandall and Qi Wang, and thank Brent Nelson and H. Fritzsche for providing samples. This research was supported by the U.S. DOE under contract DE-AC36-98G010337.

REFERENCES

1. D. L. Staebler and C. R. Wronski, Appl. Phys. Lett. **31**, 292 (1977).
2. H. Fritzsche, in *Proceedings of the Amorphous and Microcrystalline Silicon Technology-1997*, edited by E. A. Schiff, M. Hack, S. Wagner, R. Schropp, and I. Shimizu (Materials Research Society, Pittsburgh, San Francisco, 1997), p. 19 .
3. M. Stutzmann, in *Proceedings of the Amorphous and Microcrystalline Silicon Technology-1997*, edited by E. A. Schiff, M. Hack, S. Wagner, R. Schropp, and I. Shimizu (Materials Research Society, Pittsburgh, San Francisco, 1997), p. 37 .
4. M. Hoheisel and W. Fuhs, Phil. Mag. B **57**, 411 (1988).
5. S. Heck and H. M. Branz, to be published.
6. H. M. Branz, Phys. Rev. B **59**, 5498 (1999).

LOW TEMPERATURE KINETICS FOR THE GROWTH AND DECAY OF BAND-TAIL CARRIERS AND DANGLING BONDS IN HYDROGENATED AMORPHOUS SILICON

NIKO SCHULTZ AND P.C. TAYLOR
Department of Physics, University of Utah, Salt Lake City, Utah 84112

ABSTRACT

In hydrogenated amorphous silicon (a-Si:H), the kinetics of the light induced production of silicon dangling bonds and long-lived band-tail electrons and holes has been measured at temperatures between 65 and 340 K using light induced electron spin resonance (LESR). Below about 150 K the measurement of Si dangling bonds is masked by the accumulation of long-lived band-tail carriers. The kinetics of the growth and decay of these long-lived, trapped band-tail carriers consists of very fast components ($\tau < $ ms) and very long components ($\tau > $ h). Optical quenching of these long-lived carriers is not efficient at quenching energies of 0.6 eV. After removal of these long-lived band tail carriers by annealing at about 250 K we find that the total production of silicon dangling bonds at 65 K after 10 h of illumination is about a factor of five less than at 340 K. The dangling bond production resulting from 10 h of illumination is well fit to an underlying mechanism that, if thermally activated, exhibits an activation energy of approximately 10 meV.

INTRODUCTION

The most common metastability in the electronic properties of hydrogenated amorphous silicon (a-Si:H) is the Staebler-Wronski (S-W) effect [1], which is a decrease in both the photo- and dark conductivities after excitation with light of energy close to the optical bandgap. Most models that attempt to explain this effect attribute the decreases in conductivity to a metastable increase in the density of silicon dangling bonds. For this reason it is important to understand the kinetics for the production of silicon dangling bonds. Photoconductivity experiments [2,3] have recently shown that the S-W effect is only weakly dependent on temperature between 4 K and 300 K. To explain this very weak temperature dependence a photo-enhanced diffusion process for hydrogen in a-Si:H has been proposed [4]. In this paper we show that the production of silicon dangling bonds, as measured by electron spin resonance (ESR), is much less efficient at 65 K than at 300 K and that the ESR measurements below about 100 K are complicated by the presence of long-lived, optically excited carriers trapped in localized band-tail states. At temperatures below about 100 K, previous measurements have shown that the lifetimes of optically excited carriers vary by several orders of magnitude [5,6]. Some carriers recombine rapidly ($<$1ms) [5], however, a subset of photoexcited carriers can have lifetimes in excess of several hours [6]. Some of these long-lived carriers can be induced to recombine (i.e., the densities can be bleached) by irradiation of the sample with low energy, infrared light. Presumably, the infrared light re-excites the trapped band-tail carriers into extended states, and therefore this light promotes diffusion and recombination of electrons and holes. This infrared bleaching or quenching mechanism is inferred from infrared light induced changes in photoluminescence (PL) [7], light induced ESR (LESR) [7], and photoconductivity (PC) [2,8,9]. However, although infrared quenching is known and has been frequently applied, little is known about its effectiveness in removing all long-lived, band-tail carriers as measured by LESR. Such measurements are crucial because the long-lived carrier densities can easily exceed the density of the silicon dangling bonds [10,11]. It is very difficult to distinguish the dangling bonds from the long-lived electrons trapped in conduction band-tail states by ESR measurements. The ESR response

Mat. Res. Soc. Symp. Proc. Vol. 557 © 1999 Materials Research Society

attributed to dangling bonds occurs at a g-value of approximately 2.0055, while the signals attributed to long-lived band-tail electrons and holes occur at g-values of approximately 2.004 and 2.01, respectively. Since the line widths for all three resonances are several Gauss, it is difficult to resolve these three resonances, especially the signals due to dangling bonds and band-tail electrons.

EXPERIMENTAL DETAILS

Using the standard detection scheme for ESR, which is to detect in-phase at the modulation frequency of the magnetic field, the sensitivity to detect dark ESR and LESR in a-Si:H at low temperatures is greatly reduced due to microwave saturation. This saturation is caused by long spin-lattice relaxation times. For this reason, we employ a technique, which has recently been applied to a-Si:H by Yan et al. [12], that at least partially eliminates this problem. In this "rapid passage" technique, the signal is detected at twice the modulation frequency (second-harmonic detection), and the resulting spectrum resembles the absorption signal. At temperatures below about 100 K, the effective signal-to-noise ratios for ESR and LESR in a-Si:H can be improved by up to four orders of magnitude over those obtained using the standard detection technique. This technique therefore allows detailed kinetic experiments to be performed even at small ESR spin densities. Details are available elsewhere [12]. The present LESR results provide direct information on both the microscopic origins of the optically induced defects and the kinetics for their production and decay. The a-Si:H sample, which was employed to study the photoexcitation and bleaching of electrons and holes trapped in localized band-tail states, consisted of a-Si:H flakes confined in a quartz sample tube. The sample volume was approximately 1.5×10^{-3} cm^{-3}, and the dark ESR spin density due to residual silicon dangling bonds was approximately 2×10^{16} cm^{-3}. A HeNe laser (approximately 3 mW/cm^2 at 6328 Å at the sample) was employed as the excitation source, and a glowbar with long wavelength pass filters was used for the infrared bleaching of the carriers (approximately 0.5mW/cm^2 at the sample with a 2000 nm long pass filter). For the experiments designed to measure the production of silicon dangling bonds, a film of a-Si:H on a quartz substrate was employed. This approximately 2 μm thick sample had a dark ESR spin density of less than 2×10^{16} cm^{-3}. Since light induced defect production is a very inefficient process, for these experiments, an Ar-ion laser (approximately 0.5 W/cm^2 in multi-line mode at the sample) was employed as the excitation source. Under these excitation intensities sample heating becomes an important issue. Therefore, the temperatures were measured on the reverse side of the sample while optical excitation took place on the front surface. Although helium gas was permanently passing over the photoexcited sample surface, the sample temperature increased by about 10 K to 20 K during the optical excitation. After every low-temperature degradation measurement the sample was annealed at 175 C for 30 minutes.

RESULTS

We first examine the kinetics and efficiency of infrared bleaching of long-lived photoexcited carriers. We set the magnetic field to the resonance field (approximately 3390 G) and monitored the peak of the resonance over time. The sample was cooled down in the dark. At t = 0 the ESR signal is solely due to the dark spin density and the concentration of long-lived, photoexcited carriers is zero, as shown in Fig. 1. First, we illuminate the sample for approximately 20 minutes with low intensity light. As shown in Fig. 1, for this illumination intensity, the LESR spin density initially rises rapidly and is followed by a slow component that increases for several hours at the light intensities employed in these experiments. After about 20 minutes of illumination we turn off the excitation and keep the sample in the dark. The sharp features in the spectrum between about 10 and 25 minutes are artifacts due to thermal instabilities during the measurement. After an initial rapid recombination

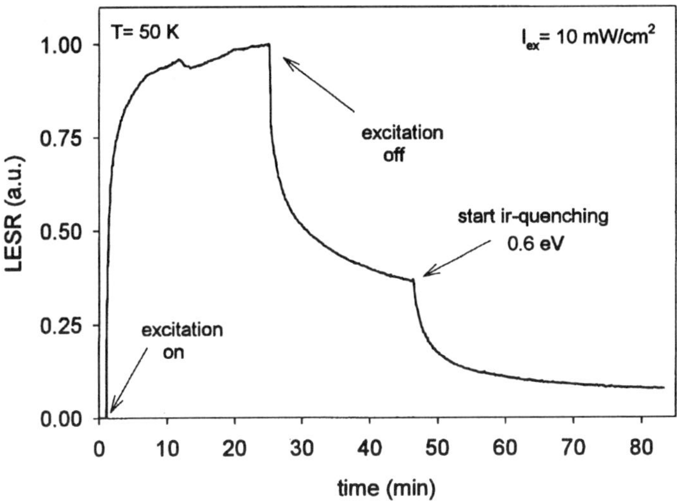

Fig. 1. Infrared quenching of residual LESR. The data up to approximately 45 min show the rise and decay of the LESR signal after the excitation source has been turned on and off, respectively. Infrared quenching is initiated after approximately 45 min with an infrared cut-off energy of 0.6 eV.

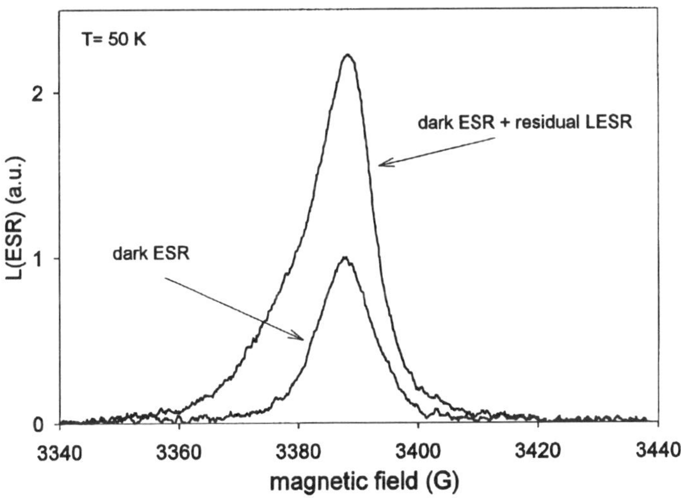

Fig. 2. ESR signal measured at t=0 min and t=85 min in Fig. 1. The dark ESR signal was taken immediately after cooling the sample down in the dark and is entirely due to silicon dangling bonds. The spin increase observed in the signal taken at 85 min over the dark ESR signal indicates a significant residual LESR carrier concentration that is not quenched by the ir-light.

of short-lived carriers, a very slow decay of the ESR signal continues due to long-lived carriers. This residual LESR signal remains even after several hours. We then initialize the photoquenching with low energy infrared light using a 0.6 eV high-energy cut-off filter. After about 45 min of ir-quenching, the residual LESR spin density is greatly reduced; however, a significant fraction of long-lived carriers remains. This situation is more clearly demonstrated in Fig. 2, which shows the ESR spectra taken immediately after the sample was cooled down in the dark, i.e., at t = 0 in Fig. 1, and immediately after the quenching, i.e., at t = 85 min. Clearly, the ESR signal taken after 45 min of ir-quenching (t = 85 min in Fig. 1) shows a significant increase over the dark ESR signal, indicating a residual LESR spin density that is not affected by the ir-quenching light. We performed several photoexcitation-and-quenching cycles as shown in Fig. 1 with quenching energies that were cut off at different energies [14]. We find that photoquenching of long-lived carriers becomes observable for photoquenching energies above about 0.35 eV, indicating that the most shallowly trapped, long-lived carriers (presumably the electrons) are trapped about 0.35 eV below the band edges. However, under all quenching light energies we find that a significant fraction of deeply trapped carriers is unaffected by the photoquenching. Light of approximately 1 eV either induces LESR or bleaches existing LESR, depending on the preexisting density of long-lived carriers.

Next, we examine the kinetics of the light induced production of silicon dangling bonds between 65 K and 340 K. From the measurements just described at low excitation intensities, we know that infrared quenching does not remove all the long-lived carriers. Therefore, to assure that the ESR signal is only due to dangling bonds, we annealed the sample for several minutes at 250 K before every low temperature ESR measurement at the irradiation temperature. Figure 3 shows two representative growth curves of silicon dangling bonds produced at 65 K and 340 K. Note the linear scales. After a total of 600 min of illumination the sample irradiated at 340 K shows a density that is about a factor of five greater than the sample irradiated at 65 K. Although the illumination time was not long enough, the upper curve seems to flatten at long times. This curve also is consistent with the commonly found sublinear dependence of the degradation on illumination time [13]. At 65 K a dependence on illumination time that is closer to linear is found.

Figure 4 shows the light induced dangling bond densities after 10 h of illumination at four temperatures between 65 and 340 K. It follows from Fig. 4 that the production rate of silicon dangling bonds decreases with decreasing temperature and drops by about a factor of five between 300 K and 65 K. If this process is assumed to be thermally activated, then the activation energy is approximately 0.01 eV.

CONCLUSION

It is known that, at low temperatures, infrared irradiation of a sample that has been previously excited with visible light induces a large fraction of the long-lived carried to recombine [9, 11]. However, the present LESR results suggest that a significant fraction of deeply trapped carriers is not affected by the quenching light on a laboratory time scale (t < 10^4 s). This result is in contrast to previous measurements [9-11] that suggested that photobleaching with energies of about 0.6 eV completely removes the long-lived carriers. The optical removal of the deeply trapped carriers is generally a very inefficient process, because the absorption coefficient, associated with these carriers drops exponentially as the photoquenching energy is reduced below the band gap energy. One reason why not all long-lived carriers are affected by the photoquenching light is that, as trapped carriers are removed through optical re-excitation and subsequent recombination, the probability for excitation of additional carriers is reduced due to the lower density of the remaining photoexcited carriers. We find that only thermal annealing at temperatures above about 150 K removes all the long-lived carriers. The difference between the thermal excitation energy (kT ~ 12 meV at 150 K) and the

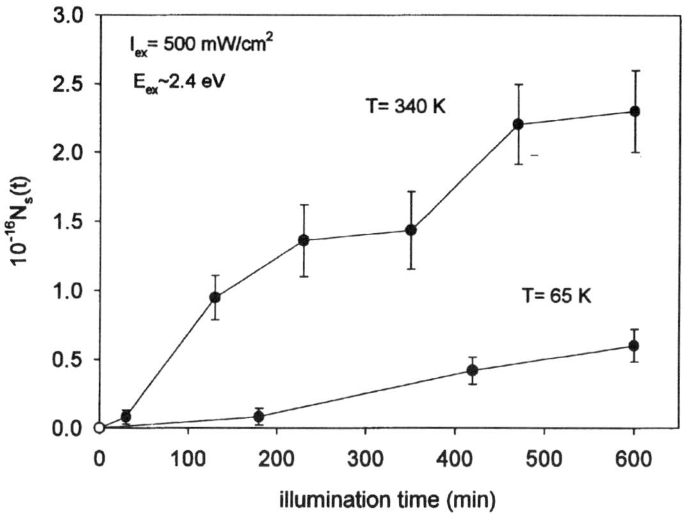

Fig. 3. Light induced silicon dangling bond production at 65 K and 340 K. ESR measurements were performed at the illumination temperature. At 65 K the sample was annealed at 250 K before each measurement.

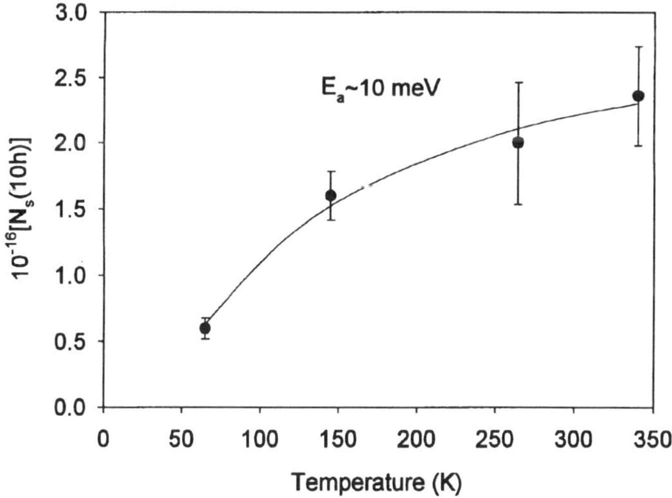

Fig. 4. Light induced increase in spin density $N_s(t)$ after t=10 h of illumination at four temperatures. If a thermally activated process is assumed, an activation energy of approximately 10 meV is found, as indicated by the solid line.

optical energy (> 350 meV) to re-excite the long-lived trapped carriers is very large. This apparent discrepancy is easily explained by consideration of the optical transition matrix elements. Since the probability for optical transitions depends exponentially on the overlap of the wave functions for the initial and final states, the probability for transitions between two localized states in the band tails is very low. Significant photobleaching becomes important only when the carriers are excited into extended states. However, only much smaller thermal energy is necessary for trapped carriers to tunnel to nearby localized states and either recombine or diffuse further. The hopping of trapped band-tail carriers into states of higher energy increases the localization radius and the overlap with other localized states so that recombination becomes more likely.

The production of silicon dangling bonds in a-Si:H after 10 h of high intensity illumination is about a factor of five less efficient at 65 K than at 300 K. This result appears to be in contrast to previous results obtained from photoconductivity experiments [2,3], which show that the light induced production of defects at 4.2 K is about 70% of its value at 300 K. If the observed temperature dependence is assumed to be a thermally activated process, then the activation energy is about 10 meV. This relatively small activation energy, has little effect on the efficiency for dangling bond production at room temperature, but the kinetics of the Si dangling bond production at low temperatures is greatly reduced, and any theoretical model must take this fact into account.

ACKNOWLEDGMENTS

We thank Z. V. Vardeny for helpful discussions and W. Wingert for help in sample preparation. This study was supported by NREL under subcontract number XAK-8-17619-13 and NSF under grant number DMR-9704946.

REFERENCES

1.   D. L. Staebler and C. R. Wronski, Appl. Phys. Lett. **31**, 292 (1977).
2.   P. Stradins and H. Fritzsche, Phil. Mag. **B69**, 121 (1994).
3.   P. Stradins and H. Fritzsche, J. Non-Cryst. Solids **198-200**, 432 (1996).
4.   H. Branz, Solid State Commun. **105**, 387 (1998).
5.   T. Umeda, S. Yamasaki, J. Isoya, A. Matsuda, K. Tanaka, Phys. Rev. Lett. 77, 4600 (1996).
6.   R. A. Street and D. K. Biegelsen, Solid State Commun. **44**, 501 (1982).
7.   R. Carius and W. Fuhs, AIP Conf. Proc. **120**, 125 (1984).
8.   P. Persans, Phil. Mag. **B46**, 435 (1982).
9.   M. Q. Tran, P. Stradins, and H. Fritzsche, MRS Symp. Proc. **336**, 431 (1994).
10.   H. Fritzsche, S. Heck, and P. Stradins, J. Non-Cryst. Solids **198-200**, 153 (1996).
11.   F. Boulitrop, AIP Conf. Proc. **120**, 178 (1984).
12.   B. Yan and P. C. Taylor, MRS Symp. Proc. **507**, 805 (1998).
13.   M. Stutzmann, W. B. Jackson, and C. C. Tsai, Phys. Rev. **B32**, 23 (1984).
14.   To be published elsewhere.

STRUCTURAL CHANGES IN a-Si:H STUDIED BY X-RAY PHOTOEMISSION SPECTROSCOPY

SHURAN SHENG*, EDWARD SACHER*, ARTHUR YELON*, HOWARD M. BRANZ, AND DENIS P. MASSON*****
* Groupe de Recherche en Physique et Technologie des Couches Minces & Département de Génie Physique et de Génie des Matériaux, École Polytechnique de Montréal, Montréal (Québec) H3C 3A7 Canada
** National Renewable Energy Laboratory, Golden, CO 80401, USA
*** Nortel Technology, Ottawa (Ontario) K1Y 4H7 Canada

ABSTRACT

X-ray irradiation-induced structural changes in undoped a-Si:H have been investigated in detail by X-ray photoemission spectroscopy (XPS). The Si2s and the Si2p peaks were found to shift simultaneously to lower bonding energies, by the same amount, with X-ray irradiation. The shifts are near saturation, at about 0.1 eV, after one hour of irradiation at the intensity used; they can be reversed almost completely, seemingly with an activation energy lower than that for the metastable changes in electronic properties (Staebler-Wronski effect). The present results suggest that essentially the whole Si network structure is affected by the X-ray irradiation.

INTRODUCTION

Owing to its importance in technological applications, the light-induced degradation of hydrogenated amorphous silicon (a-Si:H) [Staebler-Wronski effect (SWE)] has drawn much attention since its discovery [1]. In addition, the effect involves interesting physics in a metastable system. With thorough and careful investigations of the effect, it has been recognized that any simple model which associates it only with changes in individual bonds can never elucidate the full complexity of the SWE phenomenology, since light-induced metastable defect creation is not the only manifestation of the SWE. Indeed, there is experimental evidence of metastable structural changes over the whole or, at least, a major part of the Si network [2-6]. Among these experimental observations, Masson *et al.* [4] used X-ray photoemission spectroscopy (XPS) to study the structural changes in undoped a-Si:H and observed a reversible shift of about 0.1 eV in the Si2p peak to lower bonding energy following 24 hour exposure to white light of 0.1 W/cm^2 intensity, but without a corresponding shift in the Si2s peak. Density functional calculations indicated, however, that both the Si2p and the Si2s core levels should be displaced simultaneously. The absence of the corresponding shift in the Si2s peak is difficult to understand because both peaks would have been expected to shift simultaneously in the same direction and by the same amount.

In almost all XPS measurements of a-Si:H, including those in which light-induced changes in the XPS spectra were investigated, the possible effect of the X-rays on the experimental results has not been considered. In their review of XPS analysis of the SiO$_2$/Si system, Iwata and Ishizaka [7] pointed out that, in XPS, in which a surface is irradiated by X-rays, this irradiation may affect the electronic state of the sample. Thus, what one measures may not be the state without X-ray irradiation. Indeed, shortly after the discovery of the SWE, it was found that such irradiation could also produce metastable changes in both the optoelectronic and structural properties of a-Si:H [8]. Consequently, the observed shift in the Si2p peak [4] may be attributed to both X-ray irradiation-induced and light-induced metastable effects, as well as

Mat. Res. Soc. Symp. Proc. Vol. 557 © 1999 Materials Research Society

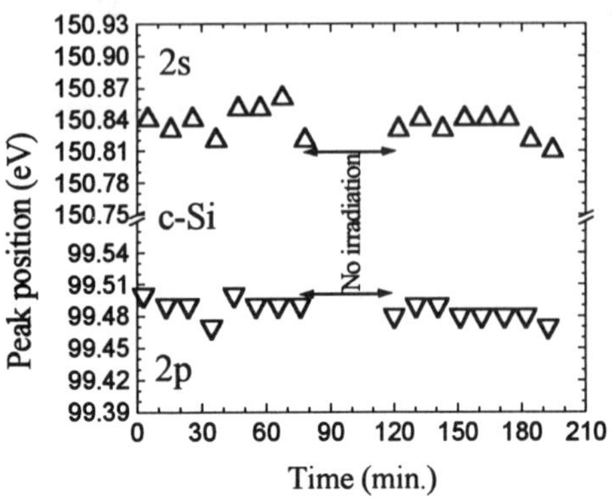

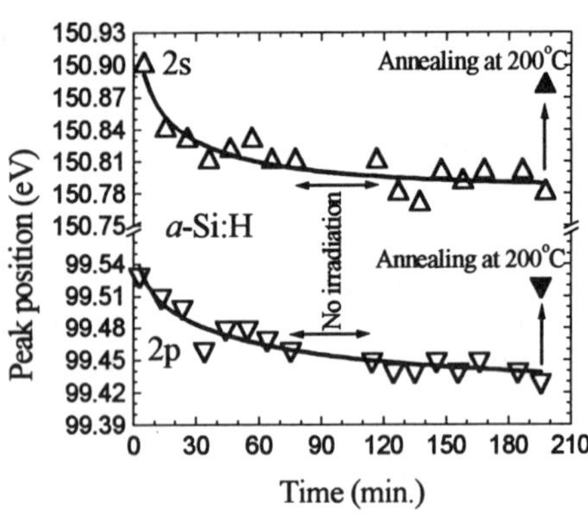

Fig.1 X-ray irradiation time dependences of the 2s and 2p peak positions of a c-Si wafer.

Fig.2 X-ray irradiation time dependences of the 2s and 2p peak positions of an a-Si:H film. The curves are a fit of the data to eq. (1).

charging effect. The difference in shifts between the Si2s and Si2p peaks might therefore be due to differences in measurement conditions. The present work reports detailed investigations of the X-ray irradiation-induced structural changes in undoped a-Si:H as measured by XPS. A further study of light-induced structural changes in a-Si:H will be reported elsewhere [9].

EXPERIMENTAL

Undoped device-quality a-Si:H films were deposited at NREL on crystalline silicon (c-Si) substrates at 230°C from pure silane by the rf plasma-enhanced chemical vapor deposition (PECVD) technique. The rf power density was 60 mW/cm^2, the silane flow rate was 50 sccm and the working pressure was 0.6 Torr. The thicknesses of the samples were about 0.9 μm.

XPS spectra of these and of c-Si samples were taken in a Vacuum Generators dual chamber ultrahigh vacuum (UHV) system. It consists of an ESCALAB MK II electron spectrometer, an analysis chamber for taking the XPS spectra, operating under a vacuum of better than 5×10^{-10} Torr, and a preparation chamber for sample introduction and treatment. The X-ray source was non-monochromated Mg K_α at 1253.6 eV (15 kV, 20 mA), and was also used for the irradiation of the samples. In order to maximize the contribution from the bulk signal, we chose a photoelectron take-off angle of 0° (perpendicular to the sample surface).

The irradiation-induced structural changes in the samples were expected to be quite small. Therefore, in order to trace such subtle changes, special attention needed to be paid to the stability of the instrument and of sample surface potentials during the measurements. For this purpose, our samples were clamped onto Cu supports by thin Au rings. This kept the sample surface well grounded, avoiding changes due to band bending. The position of the Au4f peaks from the retaining ring was used for calibration so as to eliminate shifts or drifts due to experimental artifacts. During all of the XPS measurements, fluctuations in the Au4f peak positions were ± 0.01 eV or less.

Samples were unavoidably subjected to atmospheric surface contamination. In order to remove these contaminants, they were, first, cleaned successively in trichloroethane, acetone and 2-propanol together with the sample holders and the Au ring, to remove organic contaminants, before rinsing in 18 MΩ deionized water, and were then gently sputtered, for about 30 minutes, by 500 eV Ar ions in the UHV preparation chamber. The sputtering rate was about 20 Å/min. After sputtering, the samples were annealed at 200°C for an hour in the same chamber. Possible surface damage caused by the Ar ion bombardment was eliminated by this annealing procedure. Any components of the Si2s and Si2p peaks due to surface oxide layers were undetectable after the sputter-anneal cycle. The amount of impurities (C, O, *etc.*) after cleaning and annealing was at or below the XPS C1s and O1s detection levels. A trace amount of embedded Ar introduced during the sputtering could be observed in the XPS spectra, but was too weak to be used for calibration or to verify the stabilities of the measurements. The evolution of the XPS spectra of the samples with X-ray irradiation time was recorded continuously, and each XPS spectrum took about 2 minutes to accumulate. For both *a*-Si:H and c-Si, one sample was investigated in detail, while another was measured briefly to ensure that the behavior initially observed was reproducible.

RESULTS AND DISCUSSION

The positions of the 2s and 2p peaks of a c-Si sample as a function of X-ray irradiation time are shown in Fig.1. We observe that, within experimental accuracy, they do not show any observable shift with X-ray irradiation time, as reported by Iwata and Ishizaka [7]. This indicates that c-Si shows neither structural changes observable by XPS nor charging effects. While the *a*-Si:H samples are far more resistive, we do not expect charging in this case either. This is because samples were irradiated by X-ray photons during the XPS measurements. Under irradiation, the resistivity of *a*-Si:H would be reduced considerably and comparable with that of c-Si.

Figure 2 shows the changes in peak positions of the 2s and 2p spectra of an *a*-Si:H film with irradiation time. It can be seen that both peaks shift to lower binding energy with increasing irradiation time, almost saturating after irradiation for about one hour. Further, the separation between the 2s and 2p peaks does not show any observable change, indicating that both peaks shift in the same direction and at the same rate. Before a conclusion is drawn concerning the structural changes in *a*-Si:H due to the X-ray irradiation, let us consider whether the phenomenon observed here could be caused by some artifacts, such as charging due to photoelectron emission. Iwata and Ishizaka [7] systematically studied the SiO_2/Si system by XPS, and found an irradiation time dependence for the 2p spectra. However, an obvious difference from our results is that the Si2p peak shifted in the opposite direction, to higher binding energy. They attributed this to positive oxide charges induced by X-rays, that is, to a charging effect, because this time dependence could no longer be observed when the oxide layers were removed. For our *a*-Si:H samples, the surface oxide layers were removed during the sputter-anneal cycle. The observed shifts here are, therefore, very likely related to structural changes in the films. In order to verify this, we turned off the X-ray source for about half an hour, and again turned it on to continue the measurements. We found that the positions of both peaks almost matched the values at the end of the previous irradiation, and that further shifts to lower bonding energy with irradiation time were very small. This is strong evidence that the shifts observed here are not caused by the charging effect; otherwise, changes in peak positions would not survive when the X-ray irradiation is turned off after the first extended irradiation.

Those changes, shown in Fig.2, can be described by a stretched exponential rule, which can be expressed as:

$$P(t) = P_s - (P_s - P_o) \cdot \exp[-(t/\tau)^\beta], \qquad (1)$$

where P represents the peak position, P_s is the saturation value, P_o the initial value (at $t = 0$), τ the time constant, and β the dispersive factor. As it is impossible for us to obtain the position of the

2s or the 2p peak before irradiation by means of the XPS measurements, we use the first measured value for P_o. For the data shown in Fig.2, we find the saturated $P_s - P_o \approx -0.1$ eV, $\tau = 11.1$ min., and $\beta = 0.45$ for both peaks.

As pointed out above, the difference in shifts between the 2s and 2p peaks reported by Masson *et al.* [4] might be attributed to differences in measurement conditions. From our present results, we see that it may be caused by different X-ray irradiation conditions, particularly, different irradiation times. A subsequent publication on light-induced structural changes in *a*-Si:H [9] will show that both 2s and 2p peaks shift by an equal energy with light-soaking.

It is very important to verify whether the observed X-ray irradiation-induced peak shifts in both peaks are reversible. For this reason, we first annealed the samples at 200°C for an hour before recording the XPS spectra under the same conditions. As shown in Fig.2, the irradiation-induced shifts in both peaks were largely reversed. We know that the SWE changes in electronic properties can be annealed, usually at elevated temperatures (≥ 150°C) in a few hours. Thus, it is of interest to see how the X-ray-induced structural changes can be annealed. We consequently annealed the samples at 110°C for up to 7 hours. It is seen from Fig.3 that, with increasing annealing time, the irradiation-induced shifts in both peaks are gradually reversed. After a 7 hour anneal at 110°C, the shifts are almost completely reversed. Based on these two annealing experiments, we may roughly estimate the annealing activation energy of XPS changes by means of an Arrhenius plot of $\ln(t_A)$ versus T_A^{-1}, where t_A is the annealing time required to reverse the XPS shifts, and T_A the annealing temperature. It is estimated to be about 0.4 eV, much lower than that for the SWE changes in electronic properties (~1.1 eV), suggesting that the X-ray-induced effect is metastable, but appears to has a lower annealing energy. Further work is needed on this point.

It is well known that, while the sampling depth of *a*-Si:H by XPS is of the order of tens of Angstroms (>50 Å) [10], the path length of the X-ray photons is of the order of micrometers (1-10 µm). Therefore, we believe that X-ray photons cause structural changes over the whole Si network by interacting with Si network atoms, not merely at the surface of *a*-Si:H, detectable by XPS. The shifts observed in both 2s and 2p peaks may be due mainly to subtle changes in the bulk structure of *a*-Si:H films. Furthermore, we know that the XPS detection limit is about 10^{19} cm^{-3} [7]. Consequently, the X-ray-induced structural changes observed here indicate that a large number of atoms are actually involved in the process. As reported by Masson *et al.* [4], some long-range structural rearrangements of the amorphous network must take place. It is important to note that the entire 2s and 2p peaks shift to lower binding energy with increasing X-ray irradiation time rather than broadening or showing a shoulder, in agreement with the observations for light-induced effects [4,9]. This also strongly indicates that the microscopic environment of almost all Si atoms is affected by the X-ray irradiation.

From the results given here, we see that the reversible structural changes in *a*-Si:H films show saturation after the X-ray irradiation for as little as about one hour, and are largely reversible, following a stretched exponential rule similar to that observed for many other metastable changes in *a*-Si:H films, but with a lower activation energy. Pontuschka *et al.* [8] investigated X-ray irradiation-induced paramagnetism in doped and undoped *a*-Si:H films and found that the intensities of most electron spin resonance (ESR) responses appeared to saturate at irradiation times less than 50 minutes. However, their experiments were performed at 77 K. Further work is needed to clarify the correlation between X-ray irradiation-induced structural changes and defect creation to determine whether they are both caused by the same effect, or if one is the primary effect and the other, an auxiliary process.

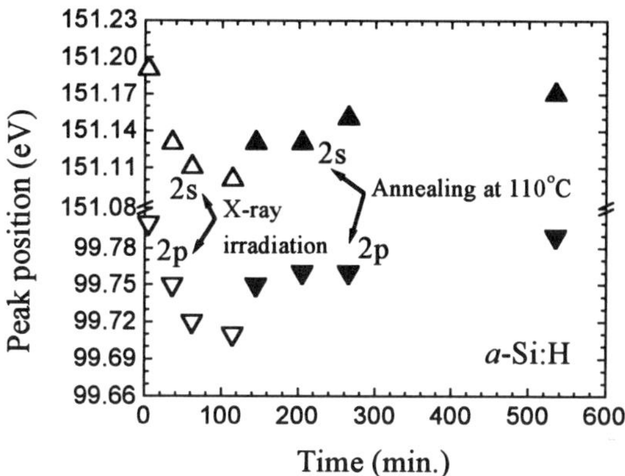

Fig.3 Annealing of the X-ray irradiation-induced shifts in both the 2s and 2p peaks of an *a*-Si:H film.

CONCLUSIONS

We have found that both Si2s and Si2p peaks in *a*-Si:H films shift simultaneously to lower binding energies by the same amount and at the same rate, as a function of X-ray irradiation time, and nearly reach saturation shortly after the outset of irradiation. The saturation change is about 0.1 eV for the samples measured, and seemingly can be annealed with a lower activation energy than the electronic properties. As we shall show elsewhere [9], these changes are similar to those induced by light. The whole Si network structure is apparently affected by the X-ray irradiation. The present results are, we believe, helpful for the understanding of the full complexity of the SWE.

ACKNOWLEDGMENTS

The authors wish to thank the Natural Sciences and Engineering Research Council of Canada and the U.S. Department of Energy for financial support. We are also grateful to S. Poulin for help and advice in obtaining and analyzing the XPS data, and to B.P. Nelson and Y. Xu for preparing the samples.

REFERENCES

1. D.L. Staebler and C.R. Wronski, Appl. Phys. Lett. **31**, 292 (1977).
2. R.E. Norberg, P.A. Fedders, J. Bodart, R. Corey, W. Paul, W. Turner and S. Jones, J. Non-Cryst. Solids **137&138**, 71 (1991).
3. H. Fritzsche, Solid State Commun. **94**, 953 (1995); in *Amorphous and Microcrystalline Silicon Technology*, edited by S. Wagner, M. Hack, E.A. Schiff, R. Schropp, I. Shimizu, (Mat. Res. Soc. Symp. Proc. **467**, San Francisco, CA, 1997), pp.19-30.
4. D.P. Masson, A. Ouhlal, and A. Yelon, J. Non-Cryst. Solids **190**, 151 (1995).
5. G.L. Kong, D.L. Zhang, G.Z. Yue, and X.B. Liao, Phys. Rev. Lett. **79**, 4210 (1997).

6. G.Z. Yue, G.L. Kong, D.L. Zhang, Z.X. Ma, S.R. Sheng, and X.B. Liao, Phys. Rev. B **57**, 2387 (1998).
7. S. Iwata and A. Ishizaka, J. Appl. Phys. **79**, 6653 (1996).
8. W.M. Pontuschka, W.W. Carlos, P.C. Taylor and R.W. Griffith, Phys. Rev. B **25**, 4362 (1982).
9. S.R. Sheng, E. Sacher and A. Yelon, (unpublished).
10. Z.H. Lu, E. Sacher and A. Yelon, Philos. Mag. B **58**, 385 (1988).

STABILITY OF AMORPHOUS SILICON THIN FILM TRANSISTORS

R.B. WEHRSPOHN, S.C. DEANE, I.D. FRENCH, J. HEWETT, M.J. POWELL
Philips Research Laboratories, Redhill, Surrey, RH1 5HA, United Kingdom

ABSTRACT

Dangling bond defects are created during positive bias stress of amorphous silicon thin film transistors and there is an energy barrier between 0.9 and 1 eV for this process. We have studied how this energy barrier depends on the material parameters of the amorphous silicon, namely hydrogen content, hydrogen bonding, Urbach energy and intrinsic, deposition induced stress. We observe no dependence on the hydrogen content or hydrogen bonding type, but we do observe a clear dependence on the Urbach energy and the intrinsic stress. These measurements support a localized model for defect creation involving Si-Si bond breaking and the switching of a neighboring H atom to stabilize the broken bond. These results suggest that stable amorphous silicon TFTs can be obtained at low deposition temperatures by control of the deposition-induced, intrinsic stress.

INTRODUCTION

Amorphous silicon thin film transistors (a-Si:H TFT) are widely used for active-matrix addressing in AMLCDs and image sensors [1]. Mobility has a direct influence on the display performance, particularly in high-resolution displays or sensors. Stability influences the lifetime of the device, and is in particular an important issue for low-temperature deposition of amorphous silicon on polymer substrates. Moreover, high-stability TFTs would make integrated row and column drivers possible.

A-Si:H TFTs exhibit threshold voltage shifts after prolonged applied bias to the gate electrode. Two mechanisms can lead to a threshold voltage shift, namely charge trapping in the gate dielectric and defect-state creation in the amorphous silicon. While it is clear that both mechanisms can occur under certain conditions, we always find that for high-quality silicon-nitride gate dielectric, state creation in the a-Si:H layer near the interface dominates at low positive bias [2] which is the technologically most important region. Recent quasi-static capacitance measurements on our samples independently carried out by Kleider et al. [3] confirm these results. Moreover, we observe that charge trapping is characterized by a weak temperature dependence, while state creation is a thermally activated process [4].

In a recent report [5], we have analyzed the effect of the deposition conditions on the mobility of the devices, focussing on the two different plasma regimes: the sheath heating or α-regime, and the plasma-bulk Joule heating or γ'-regime. In the present paper, we extend this analysis to the stability of these a-Si:H TFTs.

SAMPLE PREPARATION

The depositions were made in an ANELVA IL V-9104 PECVD reactor with a RF frequency of 13.56 MHz. The material properties of the samples have been studied in addition to those in Ref.5 by exo-diffusion and stress measurements. Exo-diffusion has been carried out on 1 μm a-Si:H flakes with a heating ramp of 6 K/min. The response of the system used in the dynamic mode was calibrated using similar samples to those measured, with known hydrogen content previously determined by static mode exo-diffusion and cross-checked by elastic recoil detection (ERD). The total stress of the a-Si:H thin films has been determined by the change of angle of reflection of a laser beam as it is scanned across the sample. For room-temperature measurements a 400nm thick a-Si:H film is deposited on Schott D263, whereas for the temperature dependent measurement the films were deposited on 6" silicon wafers (100). The heating ramp was 0.5 K/min and the maximum temperature was set to about 20°C below deposition temperature. Transfer characteristics and bias stress experiments have been carried

Mat. Res. Soc. Symp. Proc. Vol. 557 © 1999 Materials Research Society

out on n-type, bottom gate, back channel etch a-Si:H TFTs with a silicon-nitride gate insulator (d_{ins} = 300 nm, ε_{ins} = 6.4), an intrinsic a-Si:H layer of about 160nm and an n^+ layer of about 50 nm. The deposition recipes for the a-Si:N:H and the n^+ a-Si:H layer were the same for all TFTs, except that the deposition temperature was adjusted to be the same as that of the amorphous silicon. The mobility was measured at 30°C in the saturated regime (V_{SD} = 20V) from long-channel TFTs in order to ensure that the results were not limited by the contacts. The threshold voltage shifts were determined in the linear regime (V_{SD} = 0.25 V) for different stressing times and at different temperatures. The stress voltage has been fixed to be 1 MV/cm over the initial threshold voltage.

Experimentally, the most stable TFTs with reasonable mobility have been reported at a deposition temperature around 300°C in the γ'-regime for PECVD deposition [5,6]. However, for lower temperatures, the situation is more complicated as shown in Figures 1(a) and 1(b). Here, we have plotted for a deposition temperature of 240°C the results for the different deposition conditions (pressure and dilution) as a function of the growth rate, which distinguished between the two regimes [5,7]. Whereas the mobility is about constant in the α-regime and strongly decreasing in the γ'-regime with deposition rate, the stability is increasing strongly in the α-

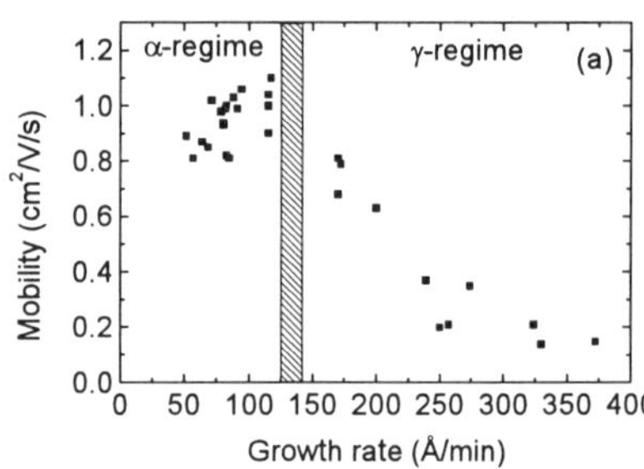
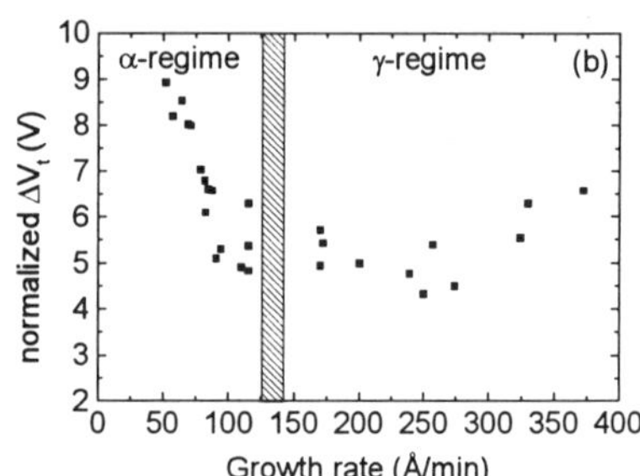

Fig. 1: (a) Mobility as a function of growth rate at 240°C deposition temperature for different hydrogen dilutions and pressures. (b) The normalized threshold voltage shifts of the same TFTs for 1MV/cm bias stress over threshold at 60°C for t = 1000s. The dashed area is the transition region between the two regimes.

regime with growth rate, constant at the beginning of the γ'-regime and again decreasing for very high deposition rates. This trend is generally true for all low-T depositions down to ~ 150°C, the lower limit of our deposition machine. In this paper, we analyze the stability of a-Si:H TFTs and make predictions for deposition conditions for high-stability, high mobility a-Si:H TFTs deposited at low temperature.

STABILITY MODEL

Metastable defects are created during bias stress according to the relationship [2]

$$\Delta N_D \approx \frac{\varepsilon_0 \varepsilon_{ins}}{q d_{ins}} \Delta V_t \qquad (1)$$

where ΔN_D is the number of created defects and ΔV_t the threshold voltage shift.

Since the barrier for defect annealing is much larger than for defect creation under electron accumulation, one might consider only the forward reaction rate, i.e., the excitation over a single barrier. However, experimentally, the kinetics of defect creation in a-Si:H TFTs strongly deviate from a single barrier rate equation and have been much better described by a stretched-exponential behaviour [8] which obeys the following differential equation:

$$\frac{d\Delta N_D}{dt} = -\frac{dN_{BT}}{dt} \propto [N_{BT}(t)]^\alpha \, \frac{t^{\beta-1}}{t_0^\beta} \qquad (2)$$

where N_{BT} is the bandtail carrier density, $\beta = kT/kT_0$, $t_0 = \nu^{-1}\exp(E_A / kT_0)$ and $\alpha = 1$ for a stretched exponential. One way to obtain this differential equation is the replacement of a single barrier by an exponential distribution of barriers $\{D(E) \propto \exp[(E\text{-}E_A)/kT_0]\}$. Two different microscopic models have been shown to give such an exponential distribution: (a) Emission of hydrogen out of a silicon-hydrogen bond to the hydrogen mobility edge, followed by *dispersive* long-range hydrogen diffusion and final trapping of the hydrogen in a silicon-silicon bond creating a SiHD defect [9]. This is a singly hydrogenated Si-Si bond, electrically indistinguishable from an isolated Si dangling bond. (b) Local reaction of a silicon-silicon bond with nearby hydrogen [10] (for example Si-Si + SiHHSi -> 2SiHD). This rearrangement will involve the energy to break a silicon-silicon bond, which is known to be exponentially distributed in energy.

However, recently several groups have observed a significant deviation of this stretched exponential dependence [6,9,11,12]. We have shown in a recent article [11], that this error arises from the underestimated bandtail carrier dependence of defect creation. Experimentally, the bandtail carrier dependence can best be determined by constant current stress (N_{BT} = const.) for which the defect creation kinetics have a power-law time dependence [Eq.(2)]. By plotting the threshold voltage shift ΔV_t as a function of the gate bias over initial threshold [V_{bias} - $V_t(0)$], α can be determined. We observe a small variation of α with the mobility, varying from α =1.6 for a mobility of $\mu = 0.2$ cm^2/V/s to about 1.9 for $\mu = 1.3$ cm^2/V/s. These values are in line with other groups [6,9,13]. Thus, Eq. (2) has to be modified by setting $\alpha \sim 1.6...1.9$ and an analytic solution yields

$$\Delta N_D = [N_{BT}(t=0) - N_D(t=0)]\left\{1 - \left[1 + \left(\frac{t}{t_0}\right)^\beta\right]^{-1/\varepsilon}\right\} \qquad (3)$$

with $\varepsilon = \alpha$-1. The experimentally observed super-linear bandtail carrier dependence is not yet understood completely.

Following Stutzmann et al. [14], we can further simplify Eq. (3). Since we assume a distribution of energy barriers $D(E)$, which could account for the observed kinetic behavior, to a first-order approximation after a time t at a temperature kT, all possible defect creation sites with $E \le kT \ln(\nu t)$ will have converted into defects. A thermalization energy can therefore be defined by $E_{th} = kT \ln(\nu t)$ where ν is the attempt-to-escape frequency. Applying this thermalization energy concept to Eq. (3) and replacing N_D by ΔV_t [Eq.(1)], Eq. (3) yields:

$$\Delta V_t(E_{th}) = [V_{bias} - V_t(0)]\left[1 - \frac{1}{\{\exp[(E_{th} - E_A)/kT_0] + 1\}^{1/\varepsilon}}\right] \qquad (4)$$

with V_{bias} is the applied gate bias, $V_t(0)$ the initial threshold voltage. Excellent fits are obtained by this "stretched hyperbola" [Eq. (4)] for the degradation kinetics of all our TFT samples for different times and temperatures [11] as shown in Fig.2(a). The attempt-to-escape frequency has been estimated to be $\nu = 10^{10}$ Hz by the optimal overlap of the degradation kinetics for different temperatures and it seems to be a unique value for all a-Si:H TFTs. Even more insight is given by the derivative of Eq. (4), which represents the probability distribution of defect creation barriers. The maximum E_{max} and the FWHM of the probability distributions shown in Fig. 2(b) are related to the parameter pair E_A and kT_0 by $E_{max} = E_A - kT_0\ln(1/\varepsilon)$ and FWHM $\approx 3kT_0$. We have observed that E_A varies slightly with the gate bias, thus in the present study we have fixed the stress voltage to 1 MV/cm over threshold which corresponds to a gate bias of 30V for a 300 nm thick silicon nitride layer. Thus the stability of an a-Si:H TFT can now be described by these two parameters. Note, that E_A is the dominant parameter in determining the threshold voltage shifts [Eq. (4)] under typical conditions.

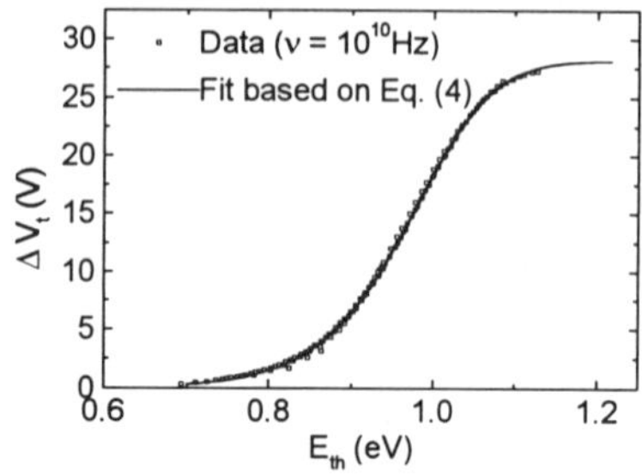
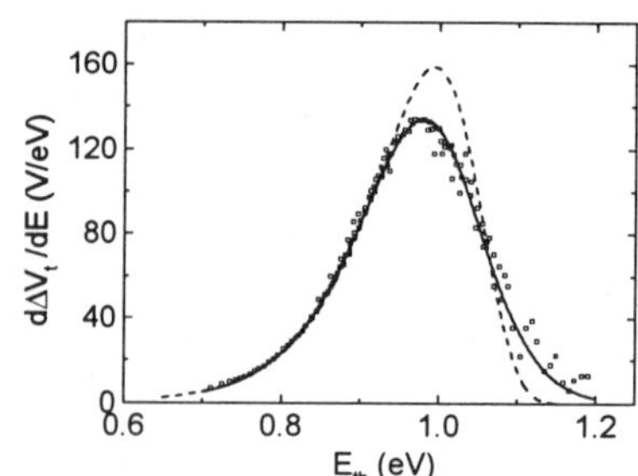

Fig.2: (a) Threshold voltage shift for different times (10-10⁶s) and temperatures (60-110° *unified as a function of the thermalization energy E_{th}. The solid line is a fit based on Eq. (4). (Derivative of the data and the fit. The dashed line is a fit based on a stretched exponential.*

RESULTS AND DISCUSSION

In the following, we analyze how the parameter pair E_A and kT_0 depend on the bulk a-Si:H properties. The TFTs have been deposited at different temperatures (between 240 and 330°C) and under different deposition conditions covering the whole $\alpha-\gamma'$ regime. Typically, we observe an increase of kT_0 in the γ'-regime with increasing growth rate (from about 65 meV up to 80 meV) whereas kT_0 is nearly constant in the α-regime. Conversely, the fit-parameter E_A is nearly constant in the γ'-regime and decreases with decreasing growth rate in the α-regime. First, based on the models described above, we analyze the kT_0 dependence on the Urbach energy. One possibility to get an estimate of the Urbach energy near the interface is to measure the field-effect mobility as shown by Sherman et al. [15]. Figure 3 shows that the fit parameter kT_0 clearly correlates with the field effect mobility μ. We have cross-checked for two TFTs the bulk Urbach energy by CPM. For a 0.35 cm²/V/s TFT, it is 55 meV and for a 0.7 cm²/V/s, it is 50 meV. The values of kT_0 and the Urbach energy differ by a the amount of 15 meV, which might be due to increased disorder at the interface.

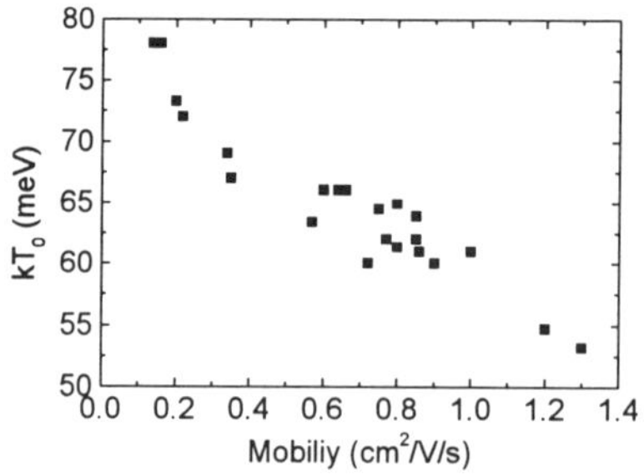

Fig.3: Fit parameter kT_0 vs. mobility for TFTs deposited between 240 and 330°C.

Second, we have analyzed the E_A dependence on the silicon-hydrogen and silicon-silicon bond strength in order to distinguish between the two defect creation models proposed above.

We have analyzed the hydrogen content and bonding of a-Si:H deposited at different temperatures in α- and γ'-regime by FT-IR [5] and static and dynamic exo-diffusion. Films in the α-regime deposited at temperatures lower than 300°C exhibit a single Si-H stretching mode at 2000 cm⁻¹ , which shows that hydrogen is mainly bonded as mono-hydride. Films deposited in the γ'-regime show an additional absorption in the 2090 cm⁻¹ region, typically attributed to di-hydrides or clustered hydrogen. Films deposited at or above 300°C have only a single 2000 cm⁻¹ peak similar to the α-regime material. Static exo-diffusion measurements show that films deposited over a temperature range from 150 to 270°C in the α-regime have almost constant hydrogen content of 10 at.% +/- 1% decreasing to 8% at 330°C. Film deposited in the γ'-regime have a increasing hydrogen content from 8% to 19% with decreasing deposition temperature T_{dep}

from 330 to 150°C. This increase is mainly related to an increase in SiH_2 bonding (2090 cm^{-1} vibration) which increases the disorder and therefore kT_0. The parameter E_A exhibits a correlation with the growth rate in the α-regime, whereas E_A is nearly constant in the γ'-regime. This trend anti-correlates with the hydrogen content. The hydrogen content changes significantly with T_{dep} in the γ'-regime whereas no change is observable in the α-regime. Moreover, dynamic exo-diffusion spectra do not show any difference for samples in the α-regime as compared to high-T samples [16] even though the stability changes significantly. Therefore, we conclude that the stability parameter E_A is not correlated to the hydrogen bonding or content. The threshold voltage shift under typical stress conditions does increase slightly with the hydrogen content, i.e. growth rate in the γ'-regime (Fig.1b) but this is due to a change in kT_0 , i.e., the Urbach energy, and not E_A [Eq.(4)] !

If the silicon-silicon bond strength plays an important role in the instability, then the weakening of the silicon-silicon bonds by intrinsic stress should influence the stability of the device. Stutzmann has already put this idea forward more than 10 years ago [17]. However, he tried to increase the internal stress by applying external mechanical stress (bending the substrate). This idea has been discussed controversially and other groups have shown contrary results [18]. We try to avoid this by analyzing the deposition-induced intrinsic stress. The total stress of thin films can be determined by the curvature of the substrate [19]. The intrinsic stress can then be determined by subtracting the change in curvature due to thermal mismatch from the total stress

$$\sigma_i = \sigma_{tot} - \left[Y_F / \left(1 - v_F\right)\right]\left(\alpha_S - \alpha_F\right)\left(T_{dep} - T_{meas}\right) \qquad (5)$$

where Y_F and v_F are the Young's modulus and the Poisson ratio of the thin a-Si:H film, α_S and α_F the thermal expansion coefficient of the substrate and the a-Si:H film, respectively and T_{dep} and T_{meas} the deposition and the measurement temperature, respectively. The α_F for the Schott D263 substrate is 7.2×10^{-6} K^{-1}. For a-Si:H, we have determined α_F by temperature-dependent stress measurements. The values are around $\sim 4 \times 10^{-6}$ K^{-1} for a-Si:H with a hydrogen content of 10-20%. For $Y_F/(1-v_F)$, we have taken the crystalline silicon value (181 GPa).

Figure 4 shows the the parameter E_A versus the intrinsic, deposition-induced stress. For the compressive stress region, an almost linear relationship is observable. Typically, the most unstable transistors have an E_A of about 0.92 eV and very high compressive stress, whereas the low compressive stress samples and slightly tensile ones have a much higher E_A, with a maximum at 1.03 eV for about 200MPa stress. Notice that the change in E_A from the most stable to the most instable is only about 100 meV. In the tensile region, there is no clear trend of the stability with the stress.

From a practical point of view, the maximum of the ratio E_A / kT_0 determines the deposition regime for most stable TFTs. For a-Si:H deposited at $T < 300$°C, this is the case at the transition of the α-γ' regimes. For deposition temperatures higher than 300°C, the process window is much larger, since the Urbach energy is constant in whole γ'-regime. This might explain why first reports claim that transistors deposited around 300°C are most stable [6].

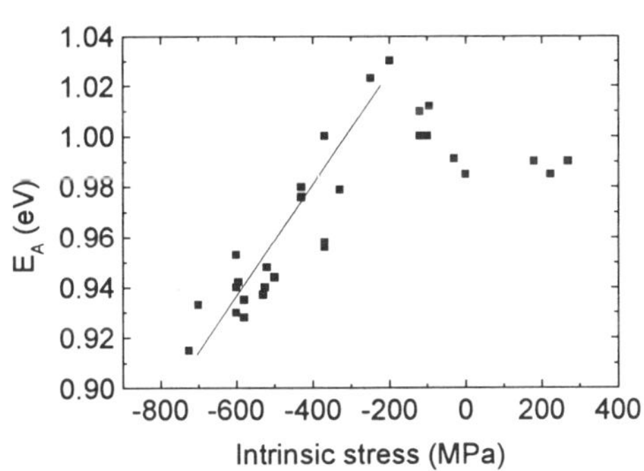

Fig.4: E_A as a function of the intrinsic stress for TFTs deposited between 240 and 330°C.

Our results have important implications for the TFT performance: a small Urbach energy does not only increase the mobility but also increases the stability of the TFT device slightly. However, the more significant parameter E_A has only a clear-cut relation to the deposition-induced compressive stress. Since stress arises mainly from ion-bombardment, a controlled-ion energy process like PECVD-IDECR [20] or ECWR [21] would be highly preferable. In such a process the ion-energy could be adjusted to a range where sufficient energy is available to improve the material density by ion-induced surface diffusion of the growth species [7] but to prevent compressive stress. This would be typically in an energy-range of $E_{ion} = 10\text{-}20$ eV. A similar effect could be also achieved by increasing the plasma-frequency to frequencies of about 50-70 MHz as shown by Dutta et al. [22]. Our results imply that highly stable and good mobility amorphous silicon TFTs should be obtained not only at high temperature (~ 300°C) but also with an ion-energy-controlled process at any deposition temperature.

CONCLUSION

In conclusion, we have analyzed the amorphous silicon bulk parameter dependence of threshold voltage shifts in TFTs. Based on improved description of the defect creation kinetics (stretched hyperbola) and the thermalization energy concept, we have obtained excellent fits with two physically meaningful parameters: kT_0, the slope of the exponential distribution of barrier states, and E_A, the typical energy barrier for defect creation under a given level of electron accumulation. We observe a linear relationship of kT_0 on the Urbach energy and of E_A on the intrinsic stress, i.e., the silicon-silicon bond strength. Contrarily, we cannot observe any correlation of E_A to hydrogen content or hydrogen bonding, i.e., the silicon-hydrogen bond strength, determined by exo-diffusion. We therefore believe that the mechanism for defect creation under bias-stress is a localized reaction involving the breaking of a silicon-silicon bond and the switching of nearby hydrogen in order to stabilize it. From a technological point, our results imply that for highly stable and good mobility TFTs, a dense amorphous silicon layer with low deposition-induced compressive stress and low Urbach energy is necessary.

REFERENCES

[1] M.J. Powell, IEEE Trans Electron Dev. **36**, 2753 (1989)
[2] C. van Berkel and M.J. Powell, Appl. Phys. Lett. **51**, 1094 (1987).
[3] J.P. Kleider and F.Dayoub, Phys. Rev. B **58**, 10401 (1998).
[4] M.J. Powell, C. van Berkel, and J.R. Hughes, Appl. Phys. Lett. **54**, 1323 (1989).
[5] I.D. French, S.C. Deane, D.T. Murley, J. Hewett, I.G. Gale and M.J. Powell, MRS Symp. Proc. **467**, 875 (1997)
[6] Y. Kaneko, A. Sasano and T. Tsukada, J. Appl. Phys. **69**, 7301 (1991).
[7] J. Perrin, in *Plasma Deposition of Amorphous Silicon-Based Materials*, ed. by G. Bruno, P. Capezuto and A.Madan (Academic press, NY, 1994).
[8] W.B. Jackson, J.M. Marshall, and M.D. Moyer, Phys. Rev. B **39**, 1164 (1989).
[9] W. B. Jackson, Phys. Rev. B **41**, 1059 (1990).
[10] R. S. Crandall, Phys. Rev. B **43**, 4057 (1991).
[11] S.C. Deane, R.B. Wehrspohn and M.J. Powell, Phys. Rev. B **58**, 12625 (1998).
[12] F.R. Libsch and J. Kanicki, Appl. Phys. Lett. **62**, 1286 (1993).
[13] Y. Fujimoto, IBM J. Res. Develop. **36**, 76 (1992).
[14] M. Stutzmann, W.B. Jackson, and C.C. Tsai, Phys. Rev. B **32**, 23 (1985).
[15] S. Sherman, S. Wagner, and R.A. Gottscho, Appl. Phys. Lett. **69**, 3242 (1996).
[16] R.B. Wehrspohn, S.C. Deane, I.D. French, M.J. Powell and R. Brüggemann, to be published.
[17] M. Stutzmann, Appl. Phys. Lett. **47**, 21 (1985).
[18] A. Ghaith, Phil. Mag. Lett. **55**, 197 (1987) and references therein.
[19] D.L. Smith, *Thin-Film Deposition*, McGraw-Hill,New York,1995, p.196.
[20] P. Bulkin, N. Betrand, and B. Drevillon, Thin Solid Films **296**, 66 (1997).
[21] H. Oechsner, in *Plasma Processing of Semiconductors*, ed. by P.F. Williams (Kluwer, Netherlands, 1997), p. 157
[22] J. Dutta, U. Kroll, P. Chabloz, A. Shah, A.A. Howling, J.-L. Dorier, and Ch. Hollenstein, J. Appl. Phys. **72**, 3220 (1992).

HYDROGEN FLIP MODEL FOR METASTABLE STRUCTURAL CHANGES IN AMORPHOUS SILICON

R. BISWAS* AND Y.-P. LI**
*Department of Physics and Astronomy, Microelectronics Research Center and Ames Laboratory-USDOE, Iowa State University, Ames, Iowa 50011
**Department of Physics, University of Science and Technology of China, Hefei 230026, People's Republic of China

ABSTRACT

We propose a new metastable defect associated with hydrogen atoms in amorphous silicon. A higher energy metastable state is formed when H is flipped to the backside of the Si-H bond at monohydride sites. This defect is described by a double–well potential energy and occurs in addition to metastable dangling bonds. The dipole moment of this "H-flip" defect is larger and increases the infrared absorption. This defect accounts for large structural changes observed on light soaking including larger absorption and volume dilation.

The Staebler-Wronski (SW) effect [1] or light-induced degradation of hydrogenated amorphous silicon (a-Si:H), remains a leading area of study. It is well established that light-soaking generates metastable dangling bonds with midgap electronic states at densities of 10^{16}–10^{17} cm^{-3}. Hydrogen has been suggested to play a key role in degradation [2], since annealing of metastable defects occurs at temperatures (> 180 $^{\circ}$C) where H diffusion becomes important. Recently several new experiments characterizing the SW effect have demonstrated additional large changes of the amorphous network [3-8] that can not be accounted for by the low density of dangling bond (D^0) defects. Infrared absorption [4] and x-ray photoemission [5] find changes involving $\approx 10^{19}$ cm^{-3} sites on light soaking. Nuclear magnetic resonance [6], $1/f$ noise statistics [7], compressive stress, and photodilation [8] all indicate metastable changes to the network that are much large than the small fraction of sites (10^{-6}–10^{-5}) at which the dangling bond D^0 defects are formed. The origin of these large structural changes is unclear. Recently, mobile H has been suggested [9] to cause such changes. We propose a simple "H-flip" model of rearrangements of bonded H atoms that accounts naturally for many of these new observations. These metastable H rearrangements may be closely related to the anomalous low temperature properties [10] universally found in amorphous solids that indicate the existence of a broad distribution of low energy excitations.

Our calculations are based on 240-atom or 60-atom a-Si:H cells where all the H are in monohydride (Si-H) groups. This cell had been very successful in modeling a-Si:H [11,12] with the tight-binding molecular dynamics model.

Our basic finding is that H atoms in monohydride configurations in a-Si:H surprisingly have *two* bonding configuration with differing energies. The lower energy configuration is the normal tetrahedral bonding, with a H-Si$_1$ bond length of 1.48 Å and a distribution of H-Si$_1$–Si bond angles around tetrahedral (Fig.1). In the higher energy configuration (Fig.1), H is bonded to the same Si$_1$ atom but in its backplane. Effectively, the H atom is flipped to the backside of the H-Si bond. The Si$_1$ also moves about 0.5 Å into the backplane, resulting in a very similar H-Si$_1$ bond length. The movement of the Si$_1$ into the backplane helps the defect to form. In the H-flip configuration the back bonds of the Si$_1$ atom become more planar, with the H-Si$_1$-Si bond angles deviating away from tetrahedral and toward 90°. The Si displacements at this

Mat. Res. Soc. Symp. Proc. Vol. 557 © 1999 Materials Research Society

to the floppy mode eigenvectors previously found for dangling bonds [13] in *a*-Si.

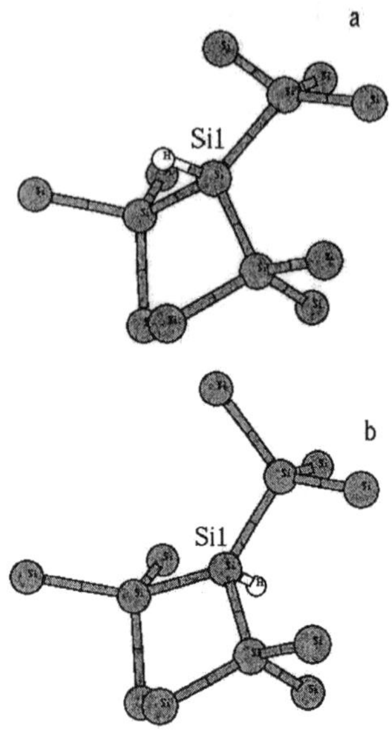

Fig. 1. a) Normal bonding configuration for a H-atom at a silicon site (Si1) in the a-Si:H network. For clarity only a portion of the network upto the second neighbor shell of Si1 is shown.b) The higher energy H-flip defect where the H is at the backplane of Si1. The Si1 also moves ~0.5 Å.

All the H atoms in our model display these two locally stable energy states. The flipped configuration is a metastable local minimum calculated to be between 0.18 and 0.7 eV higher than the original configuration. The energy of the H-flip defect scales well with the average bond angle on the Si site. Monohydride sites that are distorted toward planar are easier to convert to the H-flip defect and have a lower defect energy. We have checked these tight-binding results by performing *ab initio* calculations and found very good agreement. Using the *ab initio* Car-Parrinello method with plane wave cutoffs of up to 18 Ry, we also found locally stable configurations with very similar geometries as the tight-binding results for two of the H-sites we examined. These *ab initio* defect energies were within 0.05-0.12 eV of the tight-binding results.

A substantial energy barrier exists between the original (Fig. 1a) and metastable H-flip defect (Fig. 1b) configurations. We examined a few simple transition paths between the two configurations, and found lower barriers (~1.8-2 eV) when H passes between the Si-Si$_1$ back bonds [analogous to a bond-centered BC site in *c*-Si], accompanied by relaxation of the Si-atoms (Fig. 2). Finding optimized transition paths in this complex network requires recent activated techniques [14] for exploring the global energy surface. We may adopt the experimental activation energy for H-diffusion of ~1.5 eV for the estimated energy barrier.

This value is also supported by *ab initio* calculations of the BC state in *c*-Si [15]. Each H-atom then resides in a double-well potential energy surface (Fig. 2) with the energy wells (Figs. 1a and 1b) differing by an energy (ΔE) and separated by an energy barrier (E_b) close to 1.5 eV We expect a uniform distribution of energies (ΔE) would emerge in calculations for large *a*-Si:H models. Our calculations are consistent with this distribution. The energy surface of the amorphous material then has multiple local minima connected by H-motions. This is among the first identifications of a simple double well potential configuration in amorphous semiconductors.

With molecular dynamics we found that a diffusing unbonded interstitial-D (D_i), approaching the back plane of a normal Si-H bond, can break the SiH bond and form a new distorted Si-D bond where the configuration is flipped. The first H is released as an interstitial. The D and H effectively exchange, through

$$D_i + \text{Si-H} \rightarrow \text{Si-D} + H_i , \qquad (1)$$

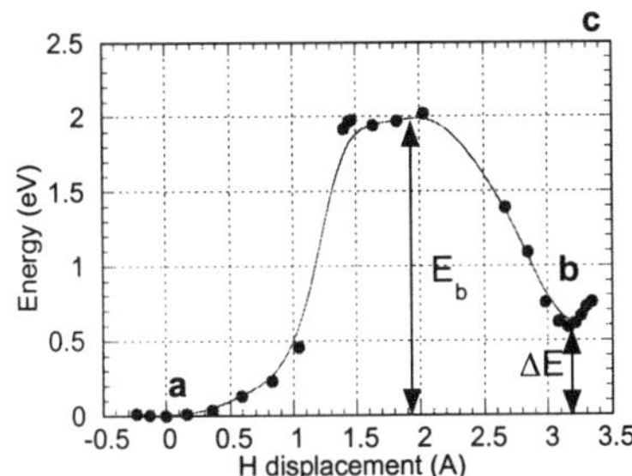

Fig. 2. A calculated double-well potential energy surface connecting configurations in Fig. 1 (a and b). The H approximately moves in an arc through the Si1-Si back bond.The network was relaxed for each H-position.The line is a smooth fit to the calculated energies (circles).

suggesting that the H-flip defect can be created by motion of mobile H/D during light-soaking. When H-diffusion occurs at higher temperature, the defect can be flipped back to its original configuration by a diffusing H-atom, through the reverse of (1). This process may be an important component of H-D exchange during H-diffusion [16].

This new H-flip defect accounts for several recent observations of light-induced changes, including the unexplained increase of infrared (ir) absorption [4]. The integrated ir absorbance I is proportional to the H-content N_H through $N_H = AI$. The absorption coefficient A is inversely proportional to the square of the dynamic effective charge e^*_d of the oscillator mode (i.e. A $\propto e^{*2}_d$). The oscillator strength is dominated [17] by the dynamic effective charge e^*_d, rather than the static charge.

At the H-flip defect the back-bonds of the Si_1 are weakened resulting in a charge transfer to the H, slightly increasing the static effective charge e^* of the H, from the normal bonding state. In the Si-H stretch mode, there is a much larger variation of the effective charge of the defect (Fig. 1b) than the normal configuration (Fig.1a) (Fig. 3). We calculated the dynamic dipole moment $p(l) = e^*(l) \times l$ as a function of the Si-H bond length (l) (Fig. 3 for a typical case), and extract the effective charge e^*_d from the slope dp/dl. At every H-flip defect, we find the dynamic effective charge e^*_d increases substantially (by a factor of 1.2-2.2) (Fig. 4) depending on the local geometry. Our interpretation of the ir measurement [4] is that the number of Si-H bonds remains constant unlike earlier suggestions [4]. Instead, a fraction of the Si-H bonds change their bonding environment when the H-flip defects are created, and these defect sites have a substantially larger dynamic charge e^*_d that increases the ir-oscillator strength. This can also explain enhanced Si-H oscillator strengths found after ion-bombardment of a-Si:H [18]. We do not find appreciable Si-H frequency shifts between the normal and defect states.

A density of H-flip defects can be created by either thermal effects or light-soaking. To deconvolute these two contributions, we used rate equations for thermally-induced and light-induced generation and annealing in the double-well system. The strength of the light-induced terms is proportional to the light-intensity [9]. Computational details will be published elsewhere. We estimate that light-excitation can generate a saturated macroscopic fraction (~0.5-1 %) of H-flip defects. For a typical H-content of 10 % this predicts a macroscopic density of H-flip defects ~10^{19} cm^{-3} - much larger than the small density of metastable dangling bonds (10^{16}-10^{17}cm^{-3}). When 0.5-1 % of bonded H increase their dynamical effective charge e^*_d by a large factor (~1.2-2.2, Table 1), the increase in ir absorption of the Si-H stretch is (0.75 % - 1.5 %), very similar to the 1% increase observed experimentally [4].

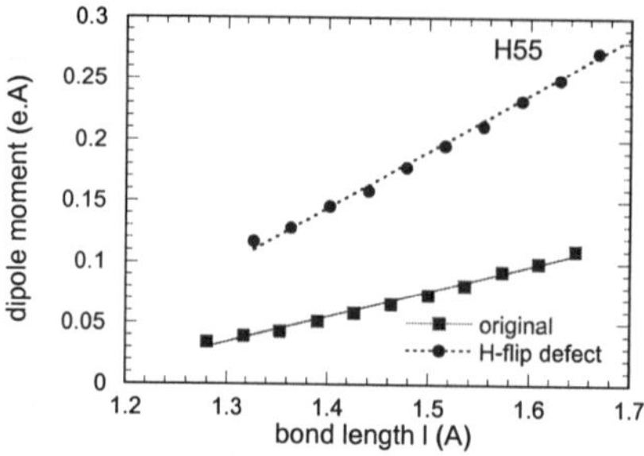

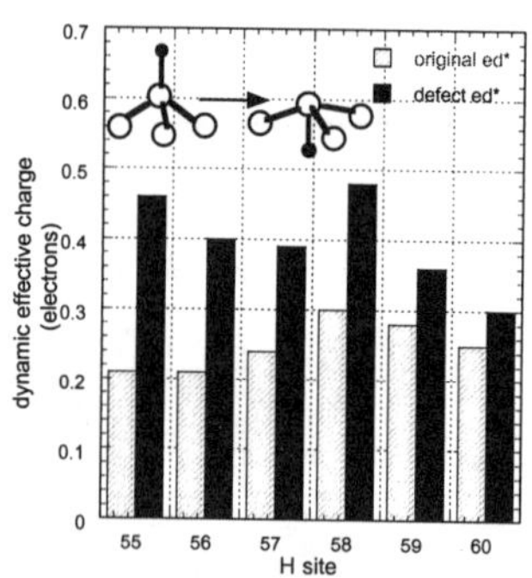

Fig. 3. The dynamic dipole moment (p=e*l) for the SiH stretch mode as a function of the bond length l. Results for a representative H-flip defect are compared with the original bonding configuration. The dynamic effective charge e_d* is the slope.

Fig. 4. The dynamic effective charge e_d* of the SiH stretch for each of the H-flip defects in our model, compared with the initial H-configuration. The IR-absorption is proportional to the square of the effective charge. The inset is a schematic picture of the structural change.

TABLE I. Energy of H-flip defect ΔE, their dynamic effective charge e^*_d before (original) and after formation of defect, and the volume changes induced by the defects.

H	ΔE eV	e^*_d Original	e^*_d Defect	Volume Change %
55	0.69	0.21	0.46	0.613
56	0.56	0.21	0.40	0.088
57	0.48	0.24	0.39	0.516
58	0.60	0.30	0.48	0.074
59	0.18	0.38	0.36	0.000
60	0.28	0.25	0.30	0.306

There is a substantial local motion of 3 - 3.5 Å of the H that have converted to the H-flip defect (Fig. 1). The H-flip defect significantly distorts the local atomic structure around it, through displacements of 0.1-0.2 Å for several neighboring silicon atoms. We unexpectedly find other H-atoms in the vicinity of this H-flip defect are displaced by 0.1-0.4 Å, involving a reorientation of their H-Si bonds. There is a large local motion of ~1 % of H by 3-4 Å, and additional smaller local motions of several nearby H by ≤ 0.4.Å. A similar conclusion of enhanced local motion of a macroscopic number of H was reached from NMR measurements of a-Si:H [6] after light-soaking. In our model, the local H-flip defects may be generated by diffusing H, hence one expects a natural correlation between the local H-motion and long-range H-motion as found earlier [6]. The local motion of H between the normal configuration and the H-flip defect, together with atomic distortions in its vicinity, can also generate conductance fluctuations in current-carrying microchannels and be connected to the random telegraph noise observed [7] in a-Si:H. H-motion has been suggested [7] to cause these fluctuations.

The H-flip defect induces a small volume expansion of the network (Table 1). Since the Si-Si back bonds at this defect site are strained, the network preferentially expands to relieve the bond-angle strain. In our computational 60 atom cell with one H-flip defect, we find (Table 1) volume expansions between 0 % - 0.6 % or an average expansion of 0.26 %. Scaling the density of H-flip defects to the expected experimental case of ~1 % of the H-content, we predict volume expansions less than 4×10^{-4} or averaging $(1-2) \times 10^{-4}$. This compares very favorably with the measured increase in compressive stress [8], although is slightly larger than optical-lever based measurements[19] of the light-induced expansion of the film.

We find the H-flip defect is a high energy local minima in crystalline silicon and consequently less important in c-Si. We calculate the H-flip defect for H at Si(111) surface to be $\Delta E \cong 0.9$ eV above the initial bonding state. At a hydrogenated vacancy we find an even higher energy of $\Delta E \cong 1.3$ eV. The much higher defect energies in c-Si are because the lattice cannot deform easily to relax the strain, as the amorphous network can. The high energies of the defect imply that the annealing barrier (E_b -ΔE) is only 0.6-0.2 eV in these crystalline systems. Hence at room temperature such defects would easily self-anneal.

We find double-well potentials associated with the H-atoms. Quantitatively relating the H displacements at our defect to the observed tunneling states in a-Si:H, whose density can be controlled by varying the H-content of hot-wire a-Si:H [10], requires further work.

In conclusion, we have found a new metastable defect in amorphous silicon, involving the flipping of bonded H atoms. This defect has a double-well potential energy surface involving simple atomic displacements and a close connection to two-level systems. This defect occurs in addition to the formation of metastable dangling bonds. Hence dangling bonds alone can not describe degradation effects, which may be important in coorelating device degrataion with material degradation. The dipole moment of the Si-H bond increases accounting naturally for increased ir absorption changes and recently observed metastable changes of the structure. We expect this defect to occur in other amorphous Group IV materials, a-SiGe:H alloys, or glasses containing H or low coordinated atoms.

ACKNOWLEDGEMENTS

This research was supported by the Electric Power Research Institute. We acknowledge support from the Department of Energy through the Center for the Synthesis and Processing of Advanced Materials. Ames Laboratory is operated by the U.S. Department of Energy by Iowa State University under Contract No. W-7405-Eng-82. We thank B. C. Pan, Q. Li, H. Branz, Z. Y. Lu and C. Z. Wang for valuable discussions and computational programs.

REFERENCES

[1] D. L. Staebler, and C. R. Wronski, Appl. Phys. Lett. **31**, 292 (1977).
[2] M. Stutzmann, W. B. Jackson, and C. C. Tsai, Phys. Rev. B **32**, 23 (1985).
[3] H. Fritzsche, Solid State Commun **94**, 953 (1995).
[4] Z. Yiping, Z. Dianlin, K. Guanglin, P. Guangqin, and L. Xianbo, Phys. Rev. Lett. **78**, 558 (1995).

[5] D. P. Masson, A. Ouhlal, A. Yelon, J. Non-Cryst. Sol. **190**, 151 (1995).

[6] P. Hari, P.C. Taylor, and R. A. Street, Mat. Res. Soc. Symp. Proc. **337**, 329 (1994); J. Non-Cryst. Sol. **198-200**, 52 (1996).

[7] C.E. Parman, N.E. Israeloff, and J. Kakalios, Phys. Rev. Lett. **69**, 1097 (1992); J. Fan and J. Kakalios, Phil. Mag. B **69**, 595 (1994).

[8] D. Han, T. Gotoh, M. Nishio, T. Sakamoto, S. Nonomura, S. Nitta, Q. Wang, E. Iwanickzko, Mat. Res. Soc. Symp. Proc. **505**, 445 (1998).

[9] H. Branz, Solid State Comm. **105**, 387 (1998).

[10] X.Liu, B.E. White, R.O. Pohl, E. Iwanizcko, K.M. Jones, A.H. Mahan, B.N.Nelson, R.S. Crandall, and S. Veprek, Phys. Rev. Lett. **78**, 4418 (1997).

[11] R. Biswas, and B.C. Pan, Appl. Phys. Lett. **72**, 371 (1998).

[12] R. Biswas, B.C. Pan, Q. Li, and Y. Yoon, Phys. Rev. B **57**, 2253 (1998).

[13] R. Biswas *et al*, Phys. Rev. Lett. **60**, 2280 (1988).

[14] G. Barkema and N. Mousseau, Phys. Rev. Lett. **77**, 4358 (1996).

[15] C. VandeWalle and R.A. Street, Phys. Rev. B **49**, 14766 (1994).

[16] M. Kemp and H. Branz, Phys. Rev. B **52**, 13946 (1996).

[17] A.A. Langford, M.L. Fleet, B.P. Nelson, W.A. Lanford, and N. Maley, Phys. Rev. B **45**, 13367 (1992).

[18] S. Oguz, D.A. Anderson, W. Paul, and H.J. Stein, Phys. Rev. B **22**, 880, 1980.

[19] T. Gotoh, S. Nonomura, M. Nishio, S. Nitta, M. Kondo, and A. Matsuda, Appl. Phys. Lett. **72**, 2978 (1998).

HYDROGEN DIFFUSION IN THE HYDROGEN COLLISION MODEL OF AMORPHOUS SILICON METASTABILITY

HOWARD M. BRANZ
National Renewable Energy Laboratory, Golden, CO 80401, hbranz@nrel.gov

ABSTRACT

The trap-controlled model of H diffusion that underpins the H collision model of amorphous silicon metastability provides new insight into thermal and light-enhanced H diffusion in a-Si:H. The longstanding puzzle of the linear DB-dependence of diffusion coefficient with doping is resolved. Expressions for the light-induced H diffusion coefficient are derived with and without D-for-H exchange reactions. It is shown that D-for-H exchange does not affect the long-time diffusion coefficient that is measured under illumination.

INTRODUCTION

Recent publications [1-4] demonstrate that the "hydrogen collision" model can explain qualitatively and quantitatively the main experimental observations of light-induced degradation (the Staebler-Wronski effect [5]) in a-Si:H. For example, the model explains the creation kinetics for production of metastable threefold-coordinated Si dangling bond defects (DB's) by continuous illumination [1-3], pulsed lasers [2, 3], and electron beams [4].

In this paper, I explore the relationship of the H collision model to H diffusion in a-Si:H. The "hydrogen collision" kinetic models [1-4] rely upon the assumption that DB defects dominate the retrapping of mobile H. Mobile H which are excited from Si-H bonds normally annihilate at DB's to reform Si-H bonds, thereby contributing to H diffusion. Also, the parameters of the H collision model can be bounded by light-enhanced D tracer diffusion experiments [6]. However, interpretation of these experiments is complicated by the existence of a D-for-H exhange mechanism in a-Si:H [7, 8]. To test metastability models involving long-range H motion, it is vital to understand how exchange reactions impact light-induced D tracer diffusions measurements.

HYDROGEN REACTIONS AND RATES

In the H collision model, the first step of light-induced or thermal diffusion of H is emission of mobile H (H_m) from Si-H bonds leaving behind an isolated DB:

Mat. Res. Soc. Symp. Proc. Vol. 557 © 1999 Materials Research Society

$$Si\text{-}H \dashrightarrow DB + H_m \quad . \tag{1}$$

This simplified model neglects the observed difference between clustered and isolated H enviroments.

The H_m thermal emission rate (Reaction 1) is given (in $cm^{-3}\text{-}s^{-1}$) by

$$R_{th} = N_H \; v_{th}^{0} \exp(-E_H/kT) \tag{2}$$

where N_H is the immobile Si-H density v_{th}^{0} is a thermal emission prefactor and E_H is the energy of mobile H emission from a Si-H bond. Under illumination, the emission rate of mobile H from a Si-H bond (Reaction 1) is $v_{ill}(G,N_{db})$. This rate is a function of the DB density (N_{db}) and the steady-state volume generation and recombination rate of electron-hole pairs (G). The total H_m emission rate is

$$R_{ill} = N_H \; v_{ill}(G, N_{db}). \tag{3}$$

N_H can be taken as constant because the mobile H density is negligible at all times compared to the density of Si-H bonds.

This mobile H diffuses rapidly through a-Si:H once it is formed [9]. Normally, H_m retraps to a DB by the reverse of Reaction (1), though not necessarily to the same DB from which the H_m was excited. The annihilation rate of mobile H to DBs by the reverse of Reaction (1) is

$$R_{db} = k_{db} N_m N_{db} \quad . \tag{4}$$

Here, N_{db} is the dangling bond density, N_m is the mobile H density and k_{db} is a rate constant (in $cm^3\text{-}s^{-1}$).

Isolated DBs created by Reaction (1) become metastable only when

$$H_m + H_m \dashrightarrow M(Si\text{-}H)_2 \; , \tag{5}$$

a rare side-reaction in which two mobile H collide and associate to form $M(Si\text{-}H)_2$, the metastable two-H complex. However, due its rarity, Reaction (5) is normally irrelevant to H diffusion.

THERMAL DIFFUSION

In this Section, I derive a useful expression for D_H and apply it to the longstanding puzzle of the doping dependence of D_H in a-Si:H. The well-known kinetic formula for diffusion [10] is

$$D_H = v_H \lambda_t^2 = R_H \lambda_t^2 / N_H. \tag{6}$$

Here, $v_H = R_H/N_H$ is the emission rate per Si-H of mobile H and λ_t is the mean distance H_m travels before retrapping. Because a mobile H is retrapped only when it meets a DB, it takes roughly N_{Si}/N_{db} random steps before retrapping. For a random walk in three dimensions, it can be shown [10] that

$$\lambda_t = (6aN_{db})^{-1/2}. \tag{7}$$

For thermal diffusion, substitution of $v_H = R_{th}/N_H$ from Eqn. (2) and λ_t from Eqn. (7) into Eqn. (6) yields

$$D_H = (v_{th}^0/6aN_{db}) \exp(-E_H/kT). \tag{8}$$

Previously, Street et al. [11] considered the possibility that $D_H \propto 1/N_{db}$, as in Eqn. (8). However, they rejected this inverse dependence based upon an apparent discrepancy with their doping-dependent diffusion data. These authors observed [11] that as the dopant density is varied in n- and p-type a-Si:H, $D_H(240°C)$ *increases* roughly linearly with N_{db}, an observation that has remained unexplained. In this section, I obtain their result using the present model of H diffusion and the observed thermal equilibrium DB densities.

The formation energy (F) of charged DBs in a-Si:H depends upon the electronic Fermi energy (E_f) because of charge exchange with the Fermi sea [12]. Pierz et al. [13] observed that

$$N_{db} \propto \exp^{E_f/kT_e} \tag{9}$$

in both n-type and p-type material, showing the defect density does, in fact, equilibrate with the electronic Fermi energy. Here, T_e is an effective equilibration temperature of about 350°C in n-type and 200°C in p-type samples [14]. Eqn. (9) implies that the formation energy (F) of a charged DB has the expected forms [12] F_0-E_f (n-type) and F_0+E_f (p-type), where F_0 is a reference energy. Earlier publications [11, 15] explained the reduction of the H diffusion activation energy in doped a-Si:H by assuming that H emission from Si-H (Reaction 1) is a step in H diffusion and noting the E_f-dependance of F.

Mobile H is well-described as a complex of a Si-H bond and an accompanying DB [16]. Emission of a mobile H into transport (Eqn. 1) therefore requires the formation of two DBs, one at the original Si-H site and one which moves with H_m. These defects are charged in doped a-Si:H.

Thus, the formation energy of H_m is roughly the formation energy of these two DBs (the number of Si-H bonds is conserved).

Assuming that mobile H emission is controlled by the DB formation energies rather than any energy barrier, the thermal emission rate is

$$v_{th} \propto \exp^{2E_f/kT}. \tag{10}$$

The factor of 2 arises because two charged DBs are formed. Combining Eqns. (9) and (10) yields,

$$v_{th} \propto N_{db}^{2} \tag{11}$$

for $T \approx T_e$, as during H diffusion measurements in a-Si:H. Substituting $v_H = v_{th}$ and λ_t from (7) into Eqn. (6),

$$D_H \propto N_{db} . \tag{12}$$

Eqn. (12) describes the results of Street et al. [11] for films with varying E_f.

EXCHANGE IN LIGHT-INDUCED DIFFUSION

Light-induced H diffusion measurements are used to test models of light-induced metastability in a-Si:H and to bound the model parameters [6]. However, these measurements are complicated by the D-for-H exchange reactions that dominate over direct capture and emission processes during thermal D tracer diffusion [7]. For example, mobile H exchanges with bonded D according to

$$H_m + Si\text{-}D \rightarrow D_m + Si\text{-}H, \tag{13}$$

where D_m denotes mobile D. Kemp and Branz [8] showed that the long-time diffusion profiles and D_H are unchanged by introduction of exchange, but early-time profiles depend upon the rate of exchange. Here, I demonstrate that similar conclusions hold for light-induced D tracer diffusion experiments.

I first derive D_H for light-induced diffusion in the absence of exchange. For light induced diffusion, substitution of $v_H = v_{ill}$, and λ_t from Eqn. (7) into Eqn. (6) yields

$$D_H = v_{ill}(G, N_{db})/6aN_{db}. \tag{14}$$

I next consider the basic exchange model in which H_m-for-D (Reaction 13) and D_m-for-H exchange rate constants are equal and exchange dominates over direct D emission and retrapping. Further, the mobile and trapped H concentrations are always much greater than the corresponding D tracer concentrations. There is a small probability, $x \ll 1$, of exchange when a D atom meets an H atom --- but a near unity probability of trapping when H or D meets a D. The exchange rate of Reaction (13) is therefore

$$R_x = x R_{db} = x k_{db} N_m N_{db} , \qquad (15)$$

where the rightmost form is obtained by substitution of R_{db} from Eqn. (4). The measured D tracer diffusion coefficient generalizes from Eqn. (6) to

$$D_D = v_D \lambda_x^2 . \qquad (16)$$

Here v_D is the exchange emission rate per D atom and and λ_x is the exchange retrapping distance of mobile D atoms. Exchange does not affect the mobile H concentration. Therefore, I equate $R_{ill} = R_{db}$ at steady-state and obtain $N_m = v_{ill} N_H / k_{db} N_{db}$ by substitution from Eqns. (3) and (4). Substituting N_m into Eqn. (15), the emission rate of mobile D from Si-D by exchange with H_m is

$$v_D = x k_{db} N_m = x N_H v_{ill} / N_{db} . \qquad (17)$$

The effective density of retrapping sites for D is $x N_H$, because mobile D exchanges with H from any Si-H with probability, x. By analogy with Eqn. (7),

$$\lambda_x = (6 a x N_H)^{-1/2} . \qquad (18)$$

Substituting Eqns. (17) and (18) into Eqn. (16), the exchange diffusion coefficient of D is

$$D_D = v_{ill} (G, N_{db}) / 6 a N_{db} , \qquad (19)$$

identical to the Eqn. (14) result for D_H. As in thermal diffusion [7, 8], exchange increases v_D and decreases the retrapping distance, but leaves the diffusion coefficient unchanged. Thus, the early-time regime of the light-induced tracer diffusion experiments [6] probes exchange rates, while the long-time regime measures H diffusion.

CONCLUSIONS

In the "H collision" model of metastability, mobile H which is excited from Si-H bonds anihillates by retrapping at DB defects. The longstanding puzzle of the doping-dependence of D_H is understood by considering the effects of E_f and N_{db} upon v_H and λ_t. To understand D tracer diffusion measurements in a-Si:H, it is essential to consider the impact of exchange. However, long-range light-induced diffusion coefficients (like the thermal diffusion coefficients) are unaffected by exchange.

ACKNOWLEDGEMENTS

I thank Prof. Panos Tzanetakis for providing an excellent environment for this work at the Univ. of Crete Physics Dept. The research was largely supported by the U.S. DOE under contract DE-AC36-83CH10093. The Fulbright Foundation and the Foundation of Research and Technology Hellas in Greece supplied additional financial support.

REFERENCES

1. H. M. Branz, Solid State Commun. **105/6**, 387 (1998).
2. H. M. Branz, in *Amorphous and Microcrystalline Silicon Technology-1998*, <u>507</u>, edited by S. Wagner, M. Hack, H. M. Branz, R. Schropp, and I. Shimizu (Materials Research Society, Pittsburgh, 1998), p. 709.
3. H. M. Branz, Phys. Rev. B **59**, 5498 (1999).
4. A. Yelon, H. Fritzsche, and H. M. Branz, unpublished .
5. D. L. Staebler and C. R. Wronski, Appl. Phys. Lett. **31**, 292 (1977).
6. H. M. Branz, S. Asher, H. Gleskova, and S. Wagner, Phys. Rev. B **59**, 5513 (1999).
7. H. M. Branz, S. E. Asher, Y. Xu, and M. Kemp, in *Amorphous Silicon Technology-1995*, <u>377</u>, edited by M. Hack, E. A. Schiff, A. Madan, M. Powell, and A. Matsuda (Materials Research Society, Pittsburgh, 1995), p. 331.
8. M. Kemp and H. M. Branz, Physical Review B **52**, 13946 (1995).
9. H. M. Branz, S. E. Asher, and B. P. Nelson, Phys. Rev. B **47**, 7061 (1993).
10. M. Kemp and H. M. Branz, Phys. Rev. B **47**, 7067 (1993).
11. R. A. Street, C. C. Tsai, J. Kakalios , and W. B. Jackson, Phil. Mag. B **56**, 305 (1987).
12. H. M. Branz, Phys. Rev. B **39**, 5107 (1989).
13. K. Pierz, W. Fuhs, and H. Mell, Phil Mag B **63**, 123 (1991).
14. K. Pierz, W. Fuhs, and H. Mell, J. Non-Cryst. Solids **114**, 651 (1989).
15. G. Muller, in *Amorphous Silicon Technology*, edited by A. Madan, M. J. Thompson, P. C. Taylor, P. G. LeComber, and Y. Hamakawa (Materials Research Society, Pittsburgh, 1988), p. 321.
16. R. Biswas, Q. Li, B. C. Pan, and Y. Yoon, Phys. Rev. B **57**, 2253 (1998).

STRUCTURAL CHANGES AND HYDROGEN MOTION IN A-SI:H OBSERVED BY PROTON NMR

Jonathan Baugh*, Daxing Han*, Qi Wang** and Yue Wu*
*Department of Physics & Astronomy, University of North Carolina,
Chapel Hill, NC 27599-3255, yuewu@physics.unc.edu
**National Renewable Energy Laboratory, 1617 Cole Blvd, Golden, CO 80401

ABSTRACT

Proton nuclear magnetic resonance (NMR) is applied to investigate hydrogen dynamics and microstructures in a-Si:H. In addition to the generic broad and narrow lines observed in all a-Si:H, an additional narrow line (about 1 kHz wide) is observed as the temperature is raised above RT. This narrow line is shifted to the up-field by about 4 ppm with respect to the generic narrow line of a few kHz in width. Below 150°C, the change of the proton spectrum with temperature is reversible; the hydrogen associated with this additional narrow line is shown to originate from hydrogen originally associated with the broad line. The spin-lattice relaxation time T_1 of this up-field shifted narrow line is about 0.4 s. This short T_1, along with its small linewidth, suggests that this line is associated with molecular hydrogen, possibly trapped in sites of atomic dimensions.

INTRODUCTION

It is widely believed that hydrogen is involved in the generation of metastable defects upon light soaking [1], an effect generally referred to as the Staebler-Wronski effect (SWE). After over 20 years of intensive investigation, the microscopic mechanism of the SWE remains unclear [1, 2]. Microscopic models of SWE often invoke hydrogen in explaining the metastability of the light-induced defects [3-5]. This idea is further supported by the observation of light-induced hydrogen diffusion in a-Si:H [6, 7]. In addition, thermal equilibrium defect generation has been observed above an equilibrium temperature T_{eq} with an activation energy of about 0.3 eV [4, 8, 9]. Below T_{eq}, which is around 200°C in intrinsic a-Si:H [9], the system is out of thermal equilibrium with a frozen-in defect concentration as a result of the slow kinetics below T_{eq} attributed to hydrogen glass transition [10]. This idea is supported by observations which establish a close relationship between hydrogen diffusion and thermal equilibrium defect generation [10].

Hydrogen diffusion measures hydrogen displacement over a macroscopic length scale and the concept of a hydrogen glass describes the average behavior of hydrogen kinetics. However, proton NMR studies clearly show that hydrogen in device quality a-Si:H exists either as isolated Si-H in the dilute phase, which gives rise to a narrow peak of a few kHz, or as Si-H clusters which give rise to a broad peak of over 25 kHz [11, 12]. It is not yet clear what role such hydrogen microstructure plays in the metastability of a-Si:H although some models of SWE incorporate explicitly such hydrogen microstructure in the theory [4]. Obviously, it would be very informative for understanding the SWE to investigate hydrogen dynamics of both the dilute and clustered Si-H around T_{eq} using NMR. In addition to the dilute and clustered phases, a few percent of the incorporated hydrogen in device quality a-Si:H are identified by proton NMR as molecular hydrogen, attributed to H_2 trapped in microvoids [13, 14]. However, a systematic deuterium NMR study shows that in addition to molecular hydrogen trapped in microvoids, a much larger fraction (typically over 10 percent) of incorporated hydrogen is in fact molecular hydrogen trapped in centers of atomic dimensions;

Mat. Res. Soc. Symp. Proc. Vol. 557 © 1999 Materials Research Society

it is suggested that the narrow proton line might correspond mostly to such trapped H_2 [15]. Furthermore, such molecular hydrogen is suggested to be related to SWE [15]. Clearly, there is a need to examine the proton spectra in more detail and to carry out a systematic proton NMR study around T_{eq}. Motivated by these considerations, we began a systematic proton NMR study of a-Si:H above room temperature. Some preliminary results are reported here.

EXPERIMENTAL

Proton NMR is carried out on a pulsed NMR spectrometer operating at a radio frequency (rf) of 200 MHz in an external magnetic field of 4.7 Tesla. The NMR probe is a home-made high temperature probe capable of carrying out experiments up to 1000°C. The temperature is achieved by placing the rf coil in a resistive furnace. The molybdenum heating wire is wound in such a way to minimize the heating current-induced magnetic field. Steady flow of nitrogen gas through the interior of the furnace prevents oxidation of the furnace. In the present experiments, high spectral resolution is desirable. Therefore, in order to completely remove the effect caused by the heating current, the current power supply is controlled by the pulse programmer of the spectrometer through a TTL pulse, which turns on or off the heating current in about 50 ms. The heating current is off during the NMR data acquisition time of about 16 ms and is turned back on immediately thereafter. Homogeneous and stable temperature of the sample can be achieved by this method over long time periods. The employed 90° rf pulse is 1.5 μs. All a-Si:H films were deposited on Al foil substrates which were removed by HCl etching and the samples were vacuum-sealed in quartz tubes. Although many samples have been investigated, the discussion here will refer mainly to results obtained on one hot-wire (HW) a-Si:H made by the NREL group and one glow-discharge (GD) a-Si:H made by Solarex Inc. The HW sample was deposited at a substrate temperature T_s of 300°C and a filament current of 14 A according to procedures described elsewhere [16]. The GD a-Si:H was deposited at $T_s = 200°C$ by standard dc GD decomposition of pure silane [17]. For comparison, a spectrum is shown of an ac GD sample deposited at $T_s = 200°C$ provided by the Beijing group [18].

RESULTS

Figure 1 shows the room temperature (RT) spectrum of the Solarex GD sample. The overall features are the same as reported before, namely, a narrow peak and a broad peak. The narrow peak contains about 13 percent of the total hydrogen content. However, a closer look at the spectrum reveals some fine features. First of all, the broad peak cannot be fit with a single Gaussian line as reported earlier for other samples [11, 12]. As shown in Fig. 2 (a), a good fit of the broad peak can be achieved with two Gaussian lines, one with a full width at half height (FWHH) of 43 kHz and the other with a FWHH of 60 kHz. However, the fit is far from unique, especially the width and the intensity of the broad Gaussian line can be varied considerably. This broad Gaussian line is not evident in the Beijing GD sample as shown also in Fig. 1. There, the broad peak can be fit with one Gaussian line of FWHH=31 KHz. It is not clear whether there are two distinct hydrogen environments, which lead to the two different broad Gaussian lines, or the difference simply arises from differences in dipolar interactions, which lead to different lineshapes for the broad peak. It is interesting to note that the broad peak of the best quality HW a-Si:H is a

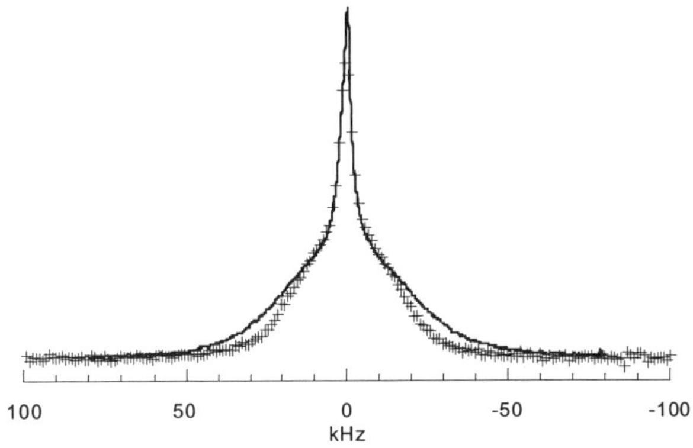

Fig.1: RT proton NMR spectra of the Solarex GD a-Si:H (—) and the Beijing
GD a-Si:H (+).

Gaussian line of over 50 kHz in FWHH [19]. There seems to be a correlation between the
lineshape of the broad peak and the microstructure of a-Si:H. This issue needs further
investigation. Figure 2 (b) shows the details of the narrow peak. The peak is asymmetric and
can be fit with two Lorentzian lines, one with a FWHH of 4.5 kHz and the other with a FWHH
of 1.6 kHz.

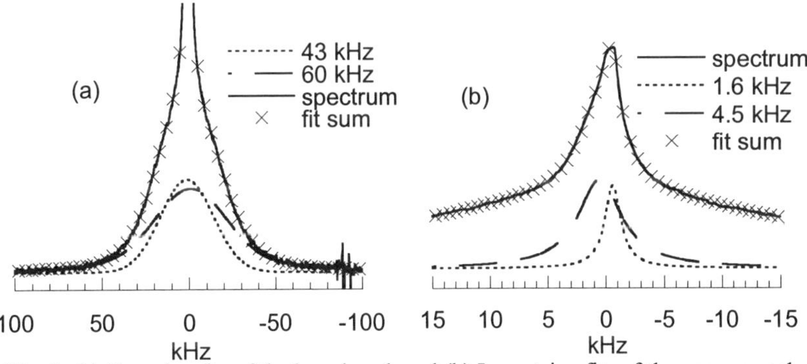

Fig. 2: (a) Gaussian fits of the broad peak and (b) Lorentzian fits of the narrow peak of
the Solarex GD a-Si:H. The FWHH of the fits are indicated in the figure. The fit sum
is the total sum of all four fits.

The 1.6 kHz line, which possesses about 4 percent of the total hydrogen content, is shifted up-
field by about 4 ppm with respect to the 4.5 kHz line. The 4.5 kHz peak is the generic narrow
peak reported earlier and is attributed to isolated Si-H bonds. It was reported previously that
the proton spectrum under magic angle spinning shows two resonance peaks of nearly equal
intensity separated by 0.6 ppm; the up-field shifted peak is attributed to molecular hydrogen
trapped in microvoids [20]. One of the characteristics of the molecular hydrogen peak is its
short spin-lattice relaxation time T_1 [13]. The result of T_1 measurements for the 4.5 and 1.6
kHz lines is shown in Fig. 3. Indeed, the 1.6 kHz line has a short T_1 of 0.45 s whereas the 4.5

kHz line has a typical T_1 of 2.2 s for the Si-H peak. The short T_1 value indicates that the 1.6 kHz line could be associated with molecular hydrogen. It is interesting to note, however, that this peak is shifted significantly up-field with respect to the 4.5 kHz line. Within experimental error the proton chemical shift of molecular hydrogen gas filled in a sample tube is identical to that of the 4.5 kHz line. Samples (e.g., annealed above 500°C) containing a substantial amount of molecular hydrogen trapped in microvoids also give a molecular hydrogen peak at the same shift as that of the 4.5 kHz line. Although paramagnetic centers could cause such a up-field shift, no direct experimental evidence for such a mechanism is available at this stage of the investigation and further studies are required.

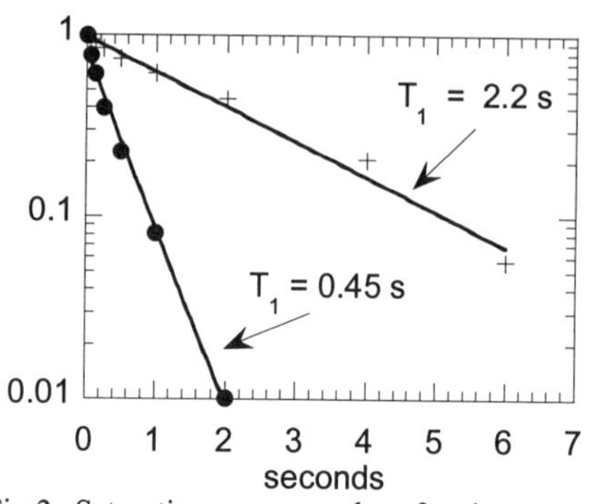

Fig.3: Saturation recovery data for the 1.6 kHz line ($\bullet$) and the 4.5 kHz line (+). The straight lines are exponential fits and the resulting T_1 are indicated in the figure.

Fig.4: Proton spectra of the Solarex GD a-Si:H measured at RT and at 150°C.

Figure 4 shows the temperature dependence of the spectrum for the Solarex GD a-Si:H. The details are shown in Fig.5. Figure 5 (a) shows that the wings far away from the center lose

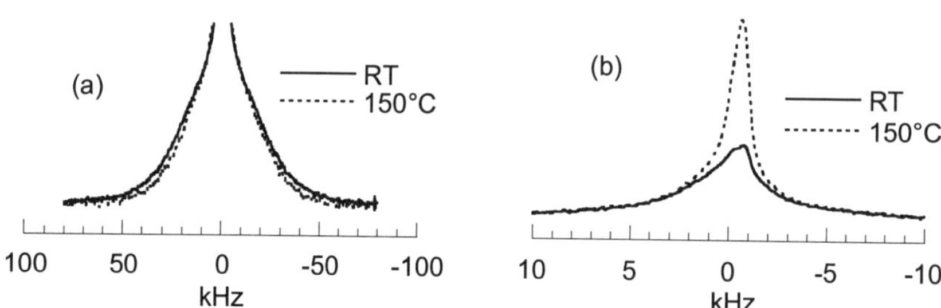

Fig.5: Proton spectra of the Solarex GD a-Si:H measured at RT and at 150°C. (a) The vertical scale is expanded to show the intensity decrease of the wings. (b) Zoomed around the center of the spectra to show the intensity increase of the 1.6 kHz peak.

more intensity as temperature increases whereas the increase of the narrow peak intensity is exclusively due to the increase of the 1.6 kHz line. This behavior is not the same as expected from motional narrowing of the broad peak. It is consistent with the scenario where hydrogen contributing to part of the broad line is converted to mobile molecular hydrogen as temperature increases. The change is completely reversible below 150°C. Figure 6 shows the RT spectra of the HW a-Si:H taken before and after heating at 247°C for a few hours. Surprisingly, irreversible changes occur even though the sample is deposited at T_s=300°C. The narrow line increases in intensity whereas the broad line loses intensity. The details of the narrow peak changes are shown in Fig.7.

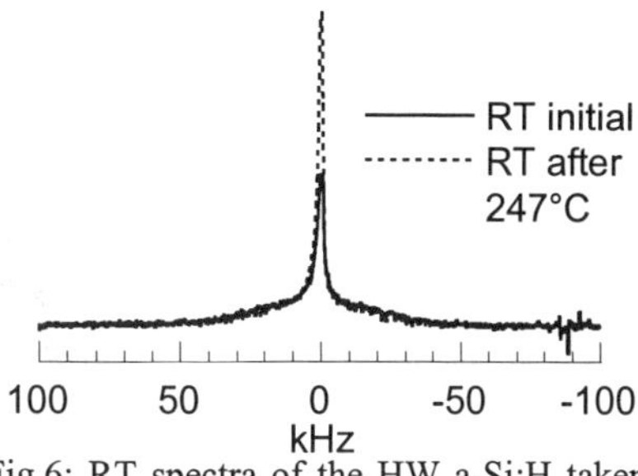

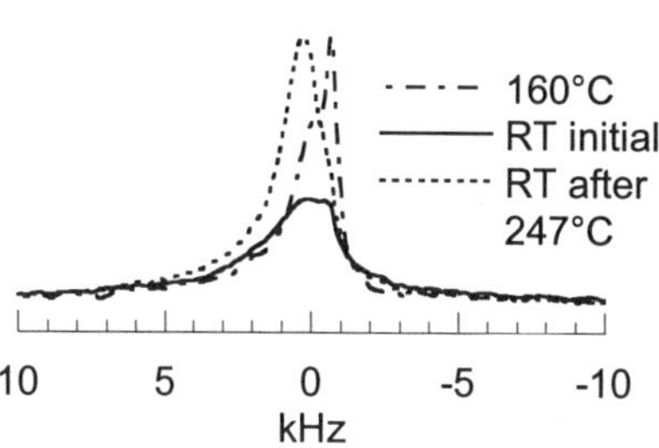

Fig.6: RT spectra of the HW a-Si:H taken before and after heating at 247°C for a few hours.

Fig.7: RT spectra shown in Fig.6 expanded about the center frequency. A spectrum measured at 160°C is also shown.

It is interesting to note that the sharp up-field shifted line discussed previously is also visible for HW a-Si:H above RT. This is demonstrated in Fig.7 where a spectrum taken at 160°C is shown. Although irreversible changes occur upon heating at 247°C, the up-field shifted line remains reversible.

What is the mechanism for the temperature dependence of the intensity of the up-field shifted narrow line? A systematic deuterium NMR study suggests that a large fraction of the hydrogen in a-Si:H is molecular hydrogen trapped at centers of atomic dimensions [15]; the properties of the broad central deuterium peak associated with such molecular hydrogen are not the same as those of H_2 molecules trapped in microvoids. Motional narrowing of the broad central peak starts above 200 K whereas motional narrowing is complete above 50 K for H_2 trapped in microvoids [14, 15]. So far, the relationship between this broad central deuterium peak and the proton spectrum is not quite clear. It was suggested that the narrow peak of the proton spectrum ariscs mainly fiom H_2 trapped in centers of atomic dimensions [21]. Since the intensity and the linewidth of the 4.5 kHz line shown in Fig.2 (b) and 5(b) are independent of the temperature below 200°C, we believe that this line is not due to H_2 and is associated with isolated Si-H as interpreted in the earlier work [11, 12]. Instead, the up-field shifted line reported in this work could correspond to the broad central deuterium line. If this is the case, a large fraction of H_2 trapped at centers of atomic dimensions still contributes to the broad peak of the proton spectrum at RT. One possible scenario for the temperature dependence of the proton spectrum reported here is: the trapped H_2 phase is converted thermally to a relatively mobile H_2 phase. Motional narrowing is another possibility. However, the change of the linewidth seems to be very abrupt which is not consistent with motional narrowing. Further studies are needed to clarify this issue.

CONCLUSIONS

A narrow up-field shifted line is identified in the proton spectrum of a-Si:H. This line increases in intensity as temperature increases above RT. The properties of this line suggest that it is associated with H_2 molecules that are different from molecular hydrogen trapped in microvoids. Possibly, these H_2 molecules are trapped in centers of atomic dimensions as suggested previously based on a deuterium NMR study.

ACKNOWLEDGMENTS

This work is supported by NREL under sub-subcontract XAK-8-17619-11 and thin film PV partnership, by North Carolina Space Grant Consortium Fellowship, NGT5-40011, and by NSF under Contract No. DMR-9802101.

REFERENCES

1. H. Fritzsche, in Amorphous and Microcrystalline Silicon Technology, edited by S. Wagner, M. Hack, E. A. Schiff, R. Schropp, I. Shimizu, (Mater. Res. Soc. Proc. **467**, San Francisco, CA, 1997), pp. 19-30.
2. M. Stutzmann, in Amorphous and Microcrystalline Silicon Technology, edited by S. Wagner, M. Hack, E. A. Schiff, R. Schropp, I. Shimizu, (Mater. Res. Soc. Proc. **467**, San Francisco, CA, 1997), pp. 37-48.
3. M. Stutzmann, W. B. Jackson, and C. C. Tsai, Phys. Rev. B **32**, 23 (1985).
4. S. Zafar and E. A. Schiff, Phys. Rev. Lett. **66**, 1493 (1991).
5. H. M. Branz, Phys. Rev. B **59**, 5498 (1999).
6. P. V. Santos, N. M. Johnson, and R. A. Street, Phys. Rev. Lett. **67**, 2686 (1991).
7. H. M. Branz, S. E. Asher, and B. P. Nelson, Phys. Rev. B **47**, 7061 (1993).
8. R. A. Street, J. Kakalios, and T. M. Hayes, Phys. Rev. B 34, 3030 (1986).
9. Z. E. Smith and S. Wagner, in Amorphous Silicon and Related Materials, edited by H. Fritzsche (World Scientific Co. Singapore, 1989), pp. 409-460.
10. J. Kakalios and W. B. Jackson, in Amorphous Silicon and Related Materials, edited by H. Fritzsche (World Scientific Co. Singapore, 1989), pp. 207-245.
11. J. A. Reimer, R. W. Vaughan, and J. C. Knights, Phys. Rev. B **24**, 3360 (1981).
12. W. E. Carlos and P. C. Taylor, Phys. Rev. B **26**, 3605 (1982).
13. M. S. Conradi and R. E. Norberg, Phys. Rev. B **24**, 2285 (1981).
14. J. B. Boyce and M. Stutzmann, Phys. Rev. Lett. **54**, 562 (1985).
15. R. E. Norberg, P. A. Fedders, and D. J. Leopold, in Amorphous Silicon Technology, edited by M. Hack, E. A. Schiff, M. Powell, A. Matsuda, and A. Madan (Mater. Res. Soc. Proc. **420**, San Francisco, CA, 1996), pp. 475-483.
16. A. H. Mahan, J. Carapella, B. P. Nelson, R. S. Crandall, and I. Balberg, J. Appl. Phys. **69**, 6728 (1991).
17. L. Yang and L. Chen, Appl. Phys. Lett. **63**, 400 (1993).
18. Y. Zhao, D. Zhang, G. Kong, G. Pan, and X. Liao, Phys. Rev. Lett. **74**, 558 (1995).
19. Y. Wu, J. T. Stephen, D. Han, J. M. Rutland, R. A. Crandall, and A. H. Mahan, Phys. Rev. Lett. **77**, 2049 (1996).
20. B. Lamotte, Phys. Rev. Lett. **53**, 576 (1984).
21. R. E. Norberg, D. J. Leopold, and P. A. Fedders, J. Non-Cryst. Solids **227-230**, 124 (1998).

LIGHT-INDUCED INCREASE IN TWO-LEVEL TUNNELING STATES IN HYDROGENATED AMORPHOUS SILICON

Xiao Liu *+, R.O. Pohl *, R.S. Crandall **
* Department of Physics, Cornell University, Ithaca, NY 14853, pohl@msc.cornell.edu
+ present address: SFA Inc., Largo, MD 20774
** National Renewable Energy Laboratory, Golden, CO 80401

ABSTRACT

We observe an increase of the low-temperature internal friction of hydrogenated amorphous silicon prepared by both hot-wire and plasma-enhanced chemical-vapor deposition after extended light-soaking at room temperature. This increase, and the associated change in sound velocity, can be explained by an increase of the density of two-level tunneling states, which serves as a measure of the lattice disorder. The amount of increase in internal friction is remarkably similar in both types of films although the amount and the microstructure of hydrogen are very different. Experiments conducted on a sample prepared by hot-wire chemical-vapor deposition show that this change anneals out gradually at room temperature in about 70 days. Possible relation of the light-induced changes in the low-temperature elastic properties to the Staebler-Wronski effect is discussed.

INTRODUCTION

The first observation of light-induced changes in hydrogenated amorphous silicon, a-Si:H, were made in 1977 by Staebler and Wronski [1]. They measured large decreases in photo and dark conductivity, owing to intense illumination with band-gap light, that could be reversed by annealing near 170°C. This discovery started an investigation of what has become a central problem in a-Si:H, namely, its metastability. Dersch et al. [2] found, using electron spin resonance, that these changes were associated with a metastable increase in neutral dangling-bond defects, D^0. Light [1], single charge injection [3], charge extraction [4], or quenching from elevated temperature [5], all produce these metastable changes. Consequently there has been intense attempts to understand this effect, now named the Staebler-Wronski Effect, SWE, after its discoverers. With over 20 years of research there is still no universally acceptable model for its cause.

In the past few years there have been reports of large scale structural change that appears to affect a few % of the Si-H bonds as well as the Si environment. Since the SWE results in an increase of the density of dangling bonds to a few 10^{17} cm^{-3} at most, it seems unlikely that these few dangling bonds are responsible for the observed large scale change. Using X-ray photoemission spectroscopy, Masson et al. [6] observed a reversible shift of about 0.1 eV of the Si 2p peak to lower binding energy on light-soaking a sample of a-Si:H. Since this change requires a rearrangement of a few % of the available Si atoms it is unlikely to be caused by the amorphous lattice restricted to the immediate surroundings of the only 10^{17} cm^{-3} defects responsible for the SWE. Thus the SWE appears to be accompanied by long-range structural rearrangements of the amorphous network.

Changes in the hydrogen microstructure were observed by several groups. Using infrared absorption, Zhao et al. [7] observed a significant increase in the strength of the Si-H vibration upon light-soaking that they interpreted as an increase in the number of Si-H bonds. This effect correlated well with the SWE and could be reversed by annealing. However, it is hard to see how the number of Si-H bonds could increase reversibly on light-soaking. It seems more likely that structural changes in the entire lattice affect the optical absorption strength. Other effects on the H environment have been observed using nuclear magnetic resonance, NMR. Kernan et al. [8] found light-induced rearrangements in a-Si:(H,D) films containing a 50-50 mixture of H and D during and after illumination. These reversible changes were probed by measuring the deuteron NMR. Hari et al. [9] reported that light-soaking affected the local motion of H in P doped a-Si:H. They performed H^1 NMR dipolar echo measurements which showed that the dipolar spin-lattice relaxation time, T_{1D}, decreases to approximately 4 ms from 11 ms owing to light-soaking with white light. After thermal anneal, T_{1D} returns to its pre-irradiation value. All of the above experiments showing that the majority of the H are affected, point to large scale effects that may or may not be connected to the much smaller increase in the density of dangling bonds.

Fan et al. [10] measured the $1/f$-noise power spectra of coplanar current fluctuations in P-doped a-Si:H before and after light-soaking. They interpreted the non-Gaussian vs. Gaussian noise spectral density as due to an increase in the disorder at the mobility edge in the light-soaked state.

In this paper we will demonstrate another aspect of light-induced increase of lattice disorder by presenting the results of the low-temperature internal friction measurements on light-soaked a-Si:H. We find a dramatic change in the low-temperature elastic properties of a-Si:H that anneals away on the time scale of the usual SWE.

EXPERIMENTAL DETAILS

Measurements of internal friction and of sound velocity were performed by using the double-paddle oscillator technique, described in detail in ref. [11]. The oscillators were fabricated out of high purity, undoped silicon wafers. The overall dimension of the oscillators was about 28 mm high, 20 mm wide, and 0.3 mm thick. Torsional oscillation of the antisymmetric mode at its resonance was excited capacitively with a frequency of $\sim$ 5500 Hz. The antisymmetric mode of a bare paddle oscillator has an exceptionally small internal friction ($\sim 2\times10^{-8}$), at low temperatures, which is reproducible for different oscillators within $\pm$ 10%. The results of internal friction and of sound velocity presented in this work were obtained exclusively using this mode to maximize the detection sensitivity.

Deposition of a thin film onto the double-paddle oscillator will change the internal friction of the paddle, Q_{paddle}^{-1}. From the increase above the bare paddle internal friction, Q_{sub}^{-1}, the internal friction of the thin film *itself*, Q_{film}^{-1}, can be calculated through [11]

$$Q_{film}^{-1} = \frac{G_{sub}t_{sub}}{3G_{film}t_{film}}\left(Q_{paddle}^{-1} - Q_{sub}^{-1}\right),\tag{1}$$

where G and t refer to the shear moduli and thicknesses of the paddle (substrate) and the film, respectively. For crystalline silicon twisting around a <110> orientation, $G_{sub} = 6.2\times10^{11}$ dyne/cm^2. For the a-Si:H films, G_{film} is unknown. Since these films have the similar mass densities to that of a-Si prepared by ion-implantation [12,13], we assume their shear moduli to be equal, which is $G_{film} = 5.54\times10^{11}$ dyne/cm^2, based on sound velocity [14] and mass density measurements [13] of the latter.

The relative variation of sound velocity of a thin film can be measured as follows. If $(\Delta f / f_0)_{sub}$ and $(\Delta f / f_0)$ are the relative variations in the resonant frequency of a paddle oscillator before and after the deposition of a thin film, respectively, with f_0 being the frequency at some reference temperature T_0, the relative variation of sound velocity of the thin film is given by:

$$\left(\frac{v_0}{v}\right)_{film} = \frac{G_{sub}t_{sub}}{3G_{film}t_{film}}\left[\left(\frac{\Delta f}{f_0}\right) - \left(\frac{\Delta f}{f_0}\right)_{sub}\right].\tag{2}$$

Several a-Si:H films were light-soaked in this work. All of them were prepared at the National Renewable Energy Laboratory. The results of two particular films will be presented below. One, H203, was prepared by hot-wire chemical-vapor deposition (HWCVD), the other, T485, by plasma-enhanced chemical-vapor deposition (PECVD). H203 was deposited at a rate of 10.0 Å/sec onto double-paddle oscillator held at 440°C, resulting in a 3600 nm thick film with 2 at.% H. For T485, the parameters were 1.3 Å/sec, 230°C, 915 nm, and 9 at.% H. The hydrogen concentration was determined by both the infrared absorption as bonded H and by hydrogen effusion as total H including H$_2$, which resulted in similar values to within 20%.

RESULTS

Two PECVD and two HWCVD samples were light-soaked for 7 days at room temperature with an ELH set at roughly 1 AM1 equivalent onto the films. The paddles were adequately ventilated during light-soaking to ensure their temperature did not exceed 50°C. One of the PECVD samples, T485, had low enough damping so that changes in internal friction could be observed. Both HWCVD samples had extremely small damping to show clear effects of light-soaking.

Fig. 1 is a good example of the changes in both internal friction and sound velocity produced by light–soaking. The virtually featureless data of Q_{film}^{-1} in Fig. 1(a) is the so-called "plateau" region of two-level tunneling states, TLS, characteristic of amorphous solids, the value of which is proportional to the density of TLS. The known universality of amorphous solids is that all amorphous solids have an internal friction plateau in the range between 1.5×10^{-4} and 1.5×10^{-3} below ~ 10K (called "glassy range"), with the tetrahedrally bonded a-Si and a-Ge being exceptions [15]. As shown in Fig. 1(a), the plateau value for T485 is over one order of magnitude lower than the lower edge of the "glassy range". For H203 to be presented below, the plateau value is even over two orders of magnitude lower in the as-deposited state. The variability of the number density of TLS in tetrahedrally bonded amorphous solids already indicates that the amorphous structure in these materials is not unique. It depends on the preparation methods and post-treatments. This has to do with the rigid four-fold coordinated bonding which effectively reduces the lattice disorder under certain conditions [15]. It has recently been proved by low temperature internal friction measurements of e-beam a-Si films under magnetic field up to 6 T [16] that the plateau observed in a-Si films is indeed caused by the atomic tunneling states, and is not associated with electronic excitations like dangling bonds. The data in Fig. 1(a) show that the TLS double in density after light-soaking. Unfortunately, we do not know if these defects produced by light can be annealed away because the sample was inadvertently used for another study, and subsequently was not suitable for studying the annealing. However, from the annealing behavior of light-soaked HWCVD sample to be presented below we expect the same behavior to occur also in the PECVD sample.

Change in Q_{film}^{-1} is also reflected in change of the transverse sound velocity, v, shown in Fig. 1(b). An empirical relation has been established [17] in which the plateau value of Q_{film}^{-1} is

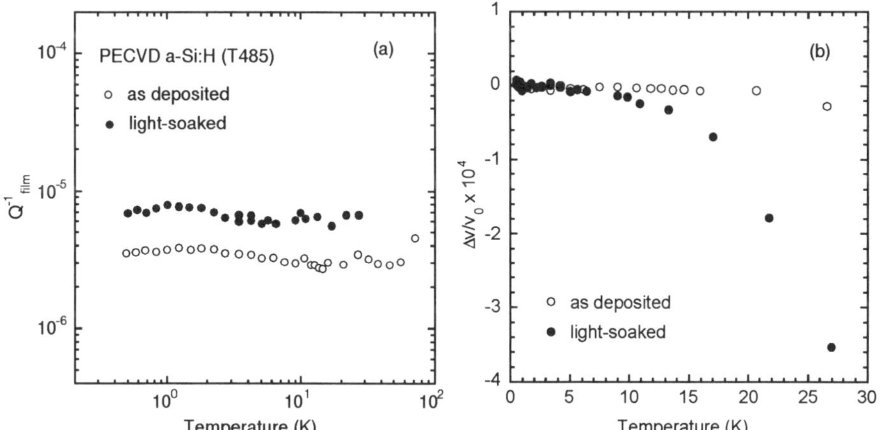

Fig. 1 Internal friction, (a), and relative variation of sound velocity, (b), of a PECVD a-Si:H film (T485, 9 at.% H, 915 nm thick) in the as deposited and light-soaked states (7 days at room temperature).

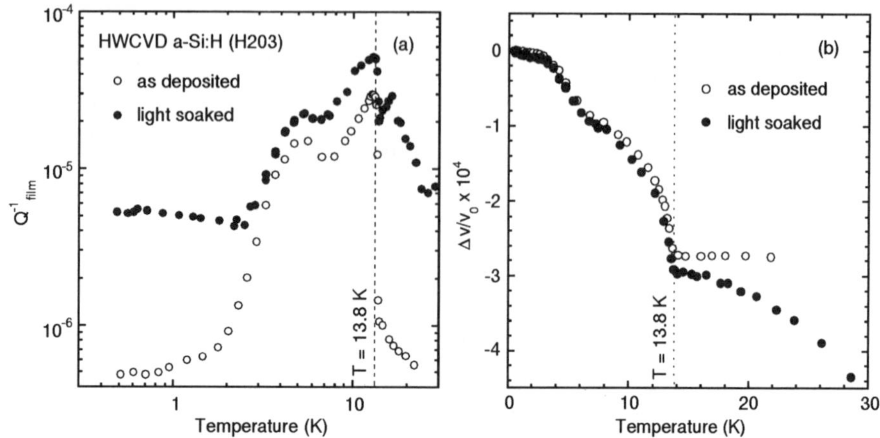

Fig. 2 Internal friction (a), and relative variation of sound velocity, (b), of a HWCVD a-Si:H film (H203, 2 at.% H, 3600 nm thick) in the as deposited and light-soaked states (7 days at room temperature).

proportional to the slope of the linear variation of $\Delta v / v_0$ in temperature. In comparison with the as-deposited state, the slope of $\Delta v / v_0$ increases significantly after light-soaking. However, the slopes for this sample and for the HWCVD one to be shown below increase more than expected and are not at all linear in temperature.

Fig. 2 shows how light-soaking affects the internal friction and sound velocity of the HWCVD sample, H203. The Q_{film}^{-1} data in Fig. 2(a) exhibit two distinct regions: a plateau region below ~ 2K and perhaps also a little above ~ 13.8K, and a dramatic elastic anomaly culminating at the triple point of hydrogen, 13.8K. This anomaly, which is discussed elsewhere [18], is caused by the freezing of bulk H_2 contained in bubbles located at the interface between the a-Si:H film and the crystalline Si substrate of the paddle oscillator. The damping in the plateau region below 2K as well as above 13.8K, where the internal friction is least influenced by the bulk H_2, increases by a factor of 10 on light-soaking, much larger than that in the PECVD sample. However, an interesting observation is that light-soaking raises the Q_{film}^{-1} to a similar plateau value ($5 \times 10^{-6} \sim 8 \times 10^{-6}$) in both types of films although the amount of Si-H bonds, their distributions, and microstrucutres are very different. Note that Q_{film}^{-1} also increases noticeably in the region of the elastic anomaly although the background Q_{film}^{-1} is much higher there, because of the solid H_2 [18]. Fig. 2(b) shows light-induced changes in $\Delta v / v_0$ that are similar to those for PECVD material, Fig. 1(b), at temperatures above 13.8K. Below that, $\Delta v / v_0$ increases rapidly with decreasing temperature

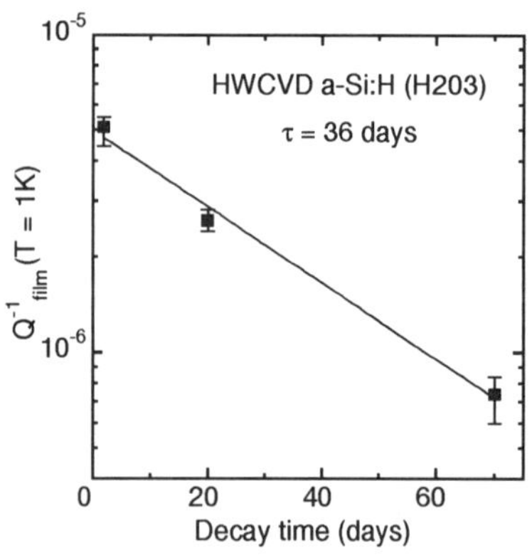

Fig. 3 Logarithmic internal friction of HWCVD a-Si:H film, H203, measured at T = 1K as a function of room-temperature annealing time following light-soaking. The error bars indicate the variation of the internal friction plateaus in temperature.

because of the elastic anomaly, consistent with the abrupt change in Q_{film}^{-1}. We also mention that $\Delta v / v_0$ changes with light-soaking even in the region of the elastic anomaly although not as much pronounced as those above 13.8K, see Fig. 2(b). This is also consistent with the internal friction change at the anomaly shown in Fig. 2(a).

Furthermore, we found that the light-induced effect in H203 annealed slowly at room temperature over a period of 3 months. Fig. 3 shows the decay of Q_{film}^{-1}. Following this anneal the sample was again light-soaked for 7 days at room temperature. Instead of a factor of 10 increase of Q_{film}^{-1} in the plateau region as it did after the first light-soaking, it only increased by a factor of two (not shown), which is the same as that in PECVD film, T485, after light-soaking. At this point, it is not clear why the second light-soaking resulted in a smaller increase in TLS than the first one, even though the internal friction and the sound velocity had annealed back at room temperature to their original values. Presumably, there is an irreversible structural change along with the reversible change associated with the initial light-soaking in HWCVD films if only room temperature annealing takes place. In the future we will study if this is reproducible on other samples, and will also carry out annealing at different temperatures. Perhaps a high temperature anneal is required to return the sample to its original state.

DISCUSSION

The results presented above show that light-soaking produces reversible changes in the amorphous structure which can be described in terms of TLS. This change enhances the TLS and reduces the stiffness of a-Si:H as exhibited by the increase in the plateau of Q_{film}^{-1} and the decrease in $\Delta v / v_0$.

Both PECVD and HWCVD samples show an increase of internal friction to about the same value following light-soaking. This indicates that similar densities of the TLS are produced in both types of films. One PECVD sample did not show any change on light-soaking, which had an initial Q_{film}^{-1} already above those of T485 and H203 after light-soaking. Thus there may be a limiting value to which the internal friction plateau can be increased, just as in the case of the original SWE. In terms of TLS, this limit is about 10 times smaller than the internal friction plateau observed in e-beam a-Si — highest among a-Si [15]. This means that the light-soaking effect can be easily masked by other factors that also induce structural disorder. It is tempting to speculate that structural rearrangement of dangling bonds is the source of the TLS in a-Si. However, when all of the hydrogen was driven out of the HWCVD samples, the internal friction increased only to 2×10^{-6} [15], even though the density of dangling bonds is known to increase to nearly 10^{19} cm^{-3}. On the contrary, we expect that light-soaking increases the density of dangling bond to less than 10^{17} cm^{-3}, however, it increases Q_{film}^{-1} to 6×10^{-6}. Therefore dangling bonds can not be a direct source of the TLS and we must look for another explanation. Nevertheless, first principles molecular-dynamics simulations by Fedders et al. [19] show large scale changes in the lattice accompanying a change of charge state of a defect such as a dangling bond. Since a dangling bond changes its charge many times under illumination (it is the dominant recombination center) it stands to reason that this could be the driving force for large scale lattice rearrangements. These arrangements then might produce the increase in the TLS that we observe as a measure of macroscopic disorder, although a microscopic picture of TLS is still missing.

The fact that the TLS annealed out in a few months seems to indicate that the observed effect might not be associated with the SWE. Typically, the SWE anneals out at temperatures around 150°C. However, this is just to speed up the annealing. All metastable effects have annealing activation energies, E_a, and attempt-to-escape frequencies that obey a Meyer-Neldel-Relation [20]. Measured E_a extend from as low as 0.4 eV to as high as 1.8 eV. We can use this relation and the measured TLS annealing time constant of 36 days at room temperature to estimate an E_a of 0.79 eV, a reasonable value. We also performed a light-soaking experiment on a HWCVD a-Si:H film after driving out all of the hydrogen in the film. No light-soaking effect was found, suggesting that the change of the low-temperature elastic properties reported here may indeed be related to the existence of hydrogen.

SUMMARY

In this paper we have described the first experiments demonstrating that light-soaking also produces changes of internal friction and sound velocity in a-Si:H. These changes show that the density of the TLS, which serves here as a measure of structural disorder, increases with light-soaking, similar to the other structural changes that have been observed. These changes are metastable and can be annealed out in a few months at room temperature.

ACKNOWLEDGMENTS

We thank Eugene Iwanizcko, Wei Gao, and Yuechin Xu for sample preparations and Ann Dillon for making H effusion measurements. Work was supported by the Semiconductor Research Corporation under Grant No. 95/Sc/069, the National Science Foundation under Grant No. DMR-970972, the NREL FIRST Program under Contract No. DE–AC02–83CH10093, and the NREL Grant No. RAD-8-18668. Additional support was received from the National Nanofabrication Facility at Cornell University, NSF Grant No. ECS-9319005.

REFERENCES

1. D.L. Staebler and C.R. Wronski, Appl. Phys. Lett. **31**, 292 (1977).
2. H. Dersch, J. Stuke, and J. Beichler, Appl. Phys. Lett. **38**, 456 (1981).
3. R.S. Crandall, Phys. Rev. B **24**, 7457 (1981).
4. D.V. Lang, J.D. Cohen, and J.P. Harbison, Phys. Rev. Lett. **48**, 421 (1982).
5. J. Kakalios and R.A. Street, Phys. Rev. B **34**, 6014 (1986).
6. D.P. Masson, A. Ouhlal, and A. Yelon, J. Non-Cryst Solids **190**, 151 (1995).
7. Y.P. Zhao, D.L. Zhang, G.L. Kong, G.Q. Pan, and X.B. Liao, Phys. Rev. Lett. **74**, 558 (1995).
8. M. J. Kernan, R. L. Corey, P. A. Fedders, D. J. Leopold, R. E. Norberg, W. A. Turner, and W. Paul, in *Amorphous Silicon Technology - 1995*, edited by M. Hack, E.A. Schiff, A. Madan, M. Powell, and A. Matsuda (Mater. Res. Soc. Sym. Proc. **377**, Pittsburgh, PA, 1995), pp. 395-399.
9. P. Hari, P.C. Taylor, and R.A. Street, in *Amorphous Silicon Technology – 1994*, edited by E.A. Schiff, M. Hack, A. Madam, M. Powell, and A. Matsuda (Mater. Res. Soc. Sym. Proc., Pittsburgh, PA, 1994), 329.
10. J. Fan and J. Kakalios, Phys. Rev. B **47**, 10903 (1993).
11. B.E. White, Jr., and R.O. Pohl, in *Thin Films: Stresses and Mechanical properties V*, edited by S.P. Baker, C.A. Ross, P.H. Townsend, C.A. Volkert, P. Borgesen, (Mat. Res. Soc. Sym. Proc. **356**, Pittsburgh, PA 1995), pp. 567-572.
12. Z. Remes, M. Vanecek, A.H. Mahan, and R.S. Crandall, Phys. Rev. B **56**, R12710 (1997).
13. D.L. Williamson, S. Roorda, M. Chicoine, R. Tabti, P.A. Stolk, S. Acco, and F. W. Saris, Appl. Phys. Lett. **67**, 226 (1995).
14. R. Vacher, H. Sussner, and M. Schmidt, Solid State Commun. **34**, 279 (1980).
15. Xiao Liu, B.E. White, Jr, R.O. Pohl, E. Iwaniczko, K.M. Jones, A.H. Mahan, B.N. Nelson, R.S. Crandall, S. Veprek, Phys. Rev. Lett. **78**, 4418 (1997); Xiao Liu and R.O. Pohl, Phys. Rev. B **58**, 9067 (1998).
16. T.H. Metcalf, et al, (to be published).
17. B.E. White, Jr., and R. O. Pohl, Z. Phys. B **100**, 401 (1996).
18. Xiao Liu, E. Iwanizcko, R.O. Pohl, R.S. Crandall, *Amorphous and Microcrystalline Silicon Technology—1998*, edited by R. Schropp, H.M. Branz, M. Hack, I. Shimizu, and S. Wagner (Mat. Res. Soc. Symp. Proc. **507**, Pittsburgh, PA, 1998), pp. 595-600; Xiao Liu, R.O. Pohl, R.S. Crandall, (this Symposium).
19. P.A. Fedders, Y. Fu, and D.A. Drabold, Phys. Rev. Lett. **68**, 1888 (1992).
20. R.S. Crandall, Phys. Rev. B **43**, 4057 (1991).

A COMPARISON OF THE DEGRADATION AND ANNEALING KINETICS IN AMORPHOUS SILICON AND AMORPHOUS SILICON-GERMANIUM SOLAR CELLS

D. E. CARLSON*, L. F. CHEN*, G. GANGULY*, G. LIN*, A. R. MIDDYA*, R. S. CRANDALL** AND R. REEDY**
*Solarex, A Business Unit of BP/Amoco Solar, 3601 LaGrange Parkway, Toano, VA 23168
**National Renewable Energy Laboratory, 1617 Cole Boulevard, Golden, CO 80401

ABSTRACT

The degradation and annealing kinetics of both a-Si:H and a-SiGe:H single-junction solar cells were investigated under varying conditions. In every case, the kinetics associated with degradation and annealing were slower for a-SiGe:H cells than for a-Si:H cells. Since deuterium diffusion studies indicate that the hydrogen in our a-SiGe:H films diffuses more slowly than that in the a-Si:H films, hydrogen motion may play a role in determining both the degradation and annealing kinetics of the devices.

INTRODUCTION

All amorphous silicon based solar cells exhibit varying degrees of light-induced degradation [1] depending on the device structure and deposition conditions. Single-junction a-Si:H cells typically degrade about 20 to 30% before reaching a steady state while a-Si:H / a-SiGe:H tandem cells typically degrade about 10 to 18%. However, there is still considerable debate about the microscopic origin of the light-induced degradation and whether hydrogen motion plays a role.

In the current study, we have examined the degradation of a-Si:H and a-SiGe:H single-junction cells and also the annealing kinetics of these cells under a variety of illumination and bias conditions. In an attempt to elucidate the role of hydrogen in the degradation and annealing kinetics, we also performed some experiments to examine the diffusion of hydrogen in these materials.

EXPERIMENT

The a-Si:H and a-SiGe:H cells in this study were all p-i-n diodes (0.27 cm^2 in area) deposited on Asahi (type U) tin oxide coated glass using DC PECVD. The p-layers consisted of about 10 nm of boron-doped a-SiC:H, and the n-layers were phosphorus-doped a-Si:H (~ 200 nm thick). A thin hydrogen-rich buffer layer (~ 10 nm thick) was deposited between the p- and i-layers. The i-layers were deposited using 10:1 hydrogen dilution, and the germane to silane ratio was ~ 0.28 during the deposition of the a-SiGe i-layers. The back contacts were formed by sputter depositing ~ 100 nm of zinc oxide and then ~ 300 nm of aluminum.

Accelerated light soaking was performed using a xenon arc lamp to illuminate the cells with high intensity white light (up to 60 suns), and the temperature was controlled by adjusting the flow of nitrogen gas that was directed onto the cell. The temperature of cells exposed to intense illumination was measured using a thin foil (0.0005") thermocouple that was cemented to the back contact of the cell with a high thermal conductivity, electrically insulating paste.

Mat. Res. Soc. Symp. Proc. Vol. 557 © 1999 Materials Research Society

RESULTS

We first looked at the relative rate of light-induced degradation in both a-Si:H and a-SiGe:H single-junction cells when exposed to intense illumination (60 suns at 60°C) under open-circuit conditions. As shown in Fig. 1, the normalized conversion efficiency fell more slowly over the first 20 minutes for an a-SiGe:H cell as compared to an a-Si:H cell of comparable thickness ($\sim 150 - 170$ nm). While the a-Si:H cell tended to reach a steady state after about an hour of illumination, the a-SiGe:H cell did not exhibit saturation. When the thickness of the i-layer in the a-Si:H cell was increased from 150 to 300 nm, the cell degraded even more rapidly and took longer to saturate.

We exposed some of the cells to 60 suns illumination at 60°C for times up to 73 hours. An a-Si:H cell with a 300 nm thick i-layer saturated at a normalized efficiency of ~ 0.6 after about 10 hours of illumination. However, the a-SiGe:H with a 170 nm thick i-layer exhibited a normalized efficiency of 0.48 after 73 hours and still showed no evidence of saturation.

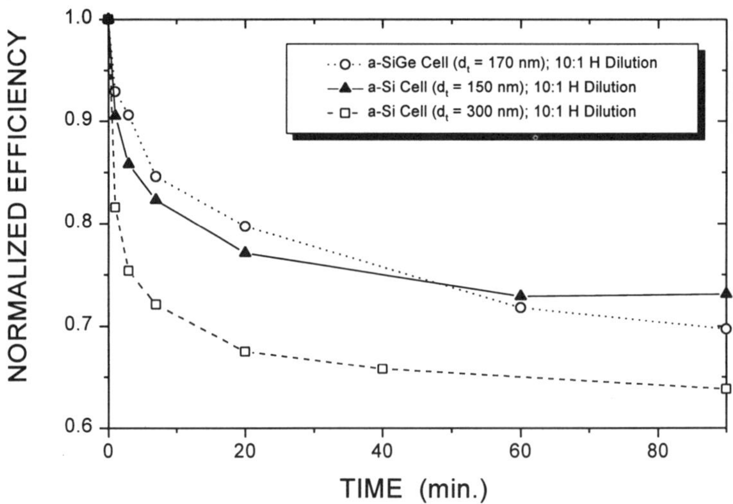

Fig. 1. The normalized conversion efficiency of a-Si:H and a-SiGe:H cells as a function of time exposed to 60 suns illumination at 60 °C under open circuit.

Several experiments were performed to investigate the effects of strong reverse biases on the recovery of light-soaked a-SiGe:H single-junction cells [2]. In the first set of experiments, we looked at the effects of a strong reverse bias on the annealing of degraded a-Si:H and a-SiGe:H cells in the dark at 100^0C. The cells were initially degraded by exposing them to 50 suns for 30 minutes at 60^0C under open-circuit conditions. The cells were then annealed in the dark at 100^0C under open-circuit conditions and at a reverse bias of - 4 V. The a-Si:H cell was ~ 290 nm thick while the a-SiGe:H cell was ~ 240 nm thick. As shown in Fig. 2, a reverse bias accelerated the annealing process in both a-Si:H and a-SiGe:H cells, but the recovery process both at open circuit and at - 4 V proceeds more slowly in the a-SiGe:H cells. Terakawa et al.[3] also observed slower degradation for a-SiGe:H cells than for a-Si:H cells made using RF PECVD, but similar open-circuit annealing rates for both types of cells.

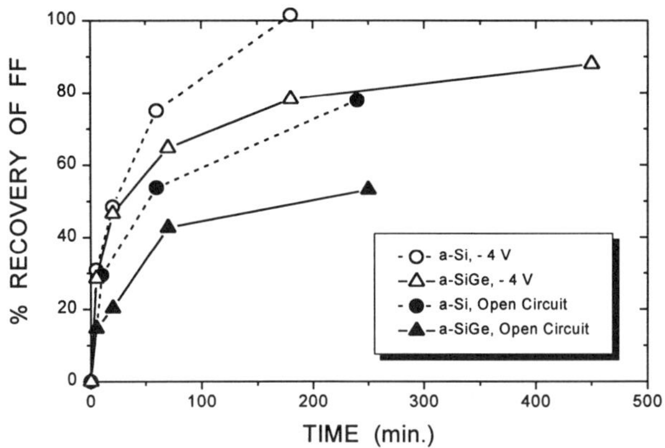

Fig. 2. The recovery of the fill factor of degraded a-Si:H and a-SiGe:H cells as a function of annealing time at 100^0C in the dark for different bias conditions.

In another series of experiments, a-Si:H and a-SiGe:H cells were degraded by exposure to 60 suns for 30 minutes at 60^0C under open-circuit conditions. They were then exposed to 60 suns for 200 minutes while under a reverse bias of - 4 V at 100^0C. As shown in Fig. 3, both the a-Si:H and a-SiGe:H cells exhibit a significant recovery of the light-induced degradation when they are exposed to both intense illumination and strong reverse biases [4]. As in the case of annealing in the dark, the recovery process appears to be somewhat slower for the a-SiGe:H cells. As also shown in Fig. 3, the recovery process is even slower for an a-Si:H cell made in a pure silane discharge. The a-Si:H cells were ~ 350 nm thick while the a-SiGe:H cell was ~ 170 nm thick. Since earlier work [4, 5] showed that the recovery process speeds up as the electric field increases, we would expect the recovery of the a-SiGe:H cell to be even slower if the thickness of the cells were all equal.

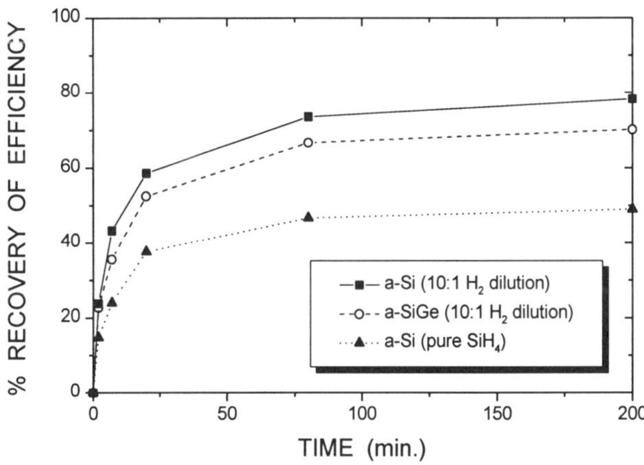

Fig. 3. The recovery of the efficiency of degraded a-Si:H and a-SiGe:H cells as a function of time at 100^0C while exposed to 60 suns at $-$ 4 V.

We also examined the effects of reverse bias annealing on a-Si:H and a-SiGe:H cells. Both a-Si:H and a-SiGe:H p-i-n cells were made in the same deposition system using the same deposition conditions for the p-layer, the thin buffer layer and the n-layer. These cells were annealed at 170°C for 30 minutes after the ZnO/Al back contacts were deposited, and the cells were then annealed at 120°C using different bias voltage conditions. The i-layer of the a-Si:H cell was ~ 390 nm thick while that of the a-SiGe:H cell was ~ 140 nm thick. A reverse bias of –4V was applied to the a-SiGe:H cell and a reverse bias of –6V was applied to the a-Si:H cell. Thus, the internal electric field was stronger by a factor of ~ 1.8 in the a-SiGe:H cell.

As shown in Fig. 4, despite the stronger field, the reverse bias annealing effect was significantly slower in the a-SiGe:H cell as compared to the a-Si:H cell. This result implies that the reverse bias enhancement can not be due solely to the activation of boron sites in the p-layer and the buffer layer since these layers were similar in the two cells. Thus, the i-layer of the p-i-n cells must also be playing a role in the reverse bias enhancement effect.

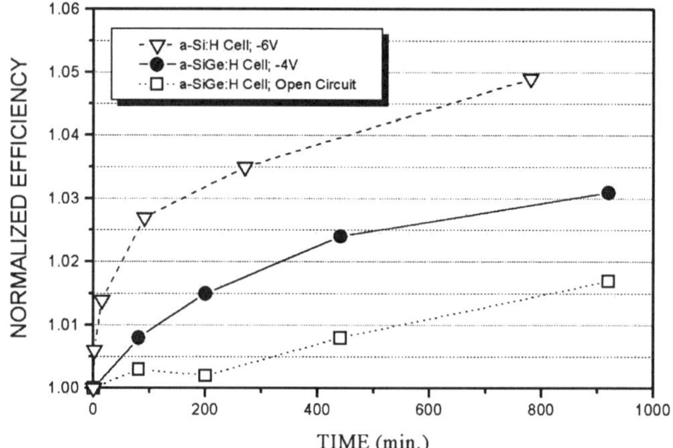

Fig.4. Normalized efficiency vs. annealing time at 120°C for a-Si:H and a-SiGe:H p-i-n cells made in the same deposition system.

We also looked at the behavior of the spectral response with annealing time. Fig. 5 shows the change in spectral response after annealing for 204 hours at 120°C for an a-SiGe:H cell subjected to a reverse bias of – 4V as compared to an adjacent cell annealed under open-circuit conditions. The data in Fig. 5 shows that the short-wavelength response of the biased a-SiGe:H cell has decreased more than that of the unbiased cell (V_{OC} of the biased cell increased from 0.654 to 0.668 V and the fill factor increased from 0.666 to 0.684). We have observed similar reductions in the short-wavelength response of a-Si:H cells after long reverse-bias treatments. After prolonged reverse bias treatments (up to 800 hours), all the biased a-SiGe:H cells start to exhibit excessive leakage in reverse bias, and the photovoltaic performance decreased.

One possible explanation is that some hydrogen ion motion is occurring during the reverse bias treatments, and this hydrogen motion leads to the creation of new defect states that eventually degrade the junction. Hydrogen ion motion has been observed for reverse bias treatments under intense illumination at elevated temperatures [6].

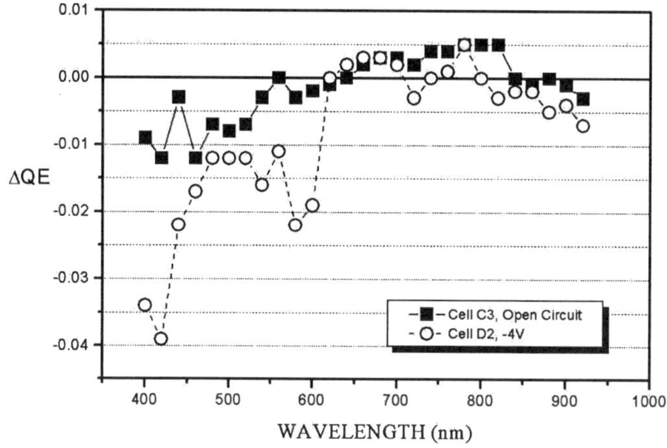

Fig. 5. Difference in spectral response for a-SiGe:H cells annealed for 204 hours at 120°C under open-circuit and reverse-biased conditions.

Since hydrogen motion may be involved in the light-induced degradation and annealing of a-Si:H and a-SiGe:H cells, we performed some experiments to examine the diffusion of deuterium in both a-Si:H and a-SiGe:H films. We fabricated sandwich structures of a-Si:H/a-Si:D:H/a-Si:H and a-SiGe:H/a-SiGe:D:H/a-SiGe:H using 10:1 deuterium dilution for the middle layers in both structures. Both sets of samples were annealed for various times at temperatures ranging from 250 to 350 °C. The deuterium profiles were then measured at NREL using SIMS (secondary ion mass spectrometry), and the SIMS profiles are shown in Figs. 6.

As is apparent from a comparison of the profiles in Figs. 6, the extent of the deuterium diffusion is significantly less for similar heat treatments in all cases for the a-SiGe:H films as compared to the a-Si:H films. This is contrary to the observations of Beyer et. al. [7] who

(a)   (b)

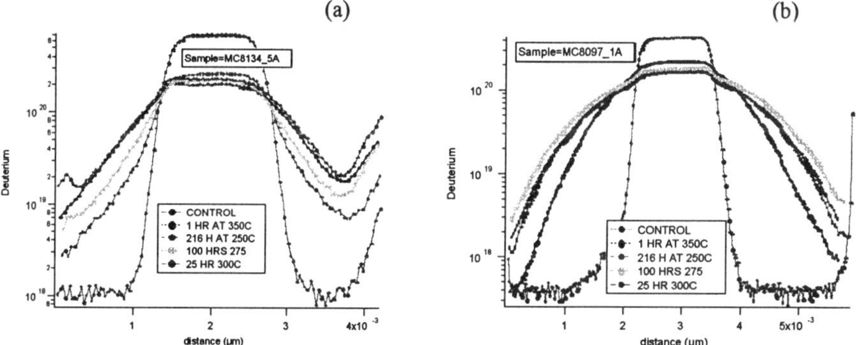

Fig. 6. SIMS profiles of deuterium in (a) an a-SiGe:H sandwich structure and in (b) an a-Si:H sandwich structure after various heat treatments.

obtained evidence for faster hydrogen diffusion in a-SiGe:H than in a-Si:H for films made using both RF and DC PECVD.

An analysis of the data in Fig. 6 indicates that the diffusion coefficients of deuterium in a-Si:H and in a-SiGe:H are comparable to the earlier published data [8] with activation energies are in the range of 1.4 to 1.6 eV. The more limited extent of the deuterium diffusion in a-SiGe:H as compared to that in a-Si:H may be related to differences in the pre-exponential factor (i.e. the jump frequency or jump distance) which may be caused by differences in the microstructure.

CONCLUSIONS

We find that our a-SiGe:H single-junction cells degrade more slowly under illumination and also anneal more slowly than a-Si:H cells under a variety of bias and illumination conditions. Degraded a-SiGe:H cells recovered more slowly than a-Si:H cells when subjected to a strong reverse bias either in the dark or when exposed to intense illumination (up to 60 suns). In addition, we found that when as-deposited a-SiGe:H and a-Si:H cells were exposed to a strong reverse bias at elevated temperatures (reverse bias annealing), the a-SiGe cells exhibited a much smaller and slower improvement in performance than the a-Si:H cells.

Thus, in every case, the kinetics associated with degradation and annealing are slower for a-SiGe:H cells than for a-Si:H cells. Since the deuterium diffusion studies indicate that the hydrogen in a-SiGe:H diffuses more slowly than that in a-Si:H, hydrogen motion may play a role in determining both the degradation and annealing kinetics of the devices. Therefore, the current investigation provides support for models which assume that hydrogen motion is associated with the light-induced degradation and annealing of a-Si:H based alloys [5, 9, 10].

ACKNOWLEDGMENTS

This work was partially funded by NREL under Subcontract ZAK-8-17619-02. The authors would also like to thank Rebecca Lind, George Wood and Mark Dicolli for their assistance in fabricating devices.

REFERENCES

1. D. L. Staebler and C. R. Wronski, Appl. Phys. Lett. **31**, p. 292 (1977).
2. D. E. Carlson and K. Rajan, Proc. of 14th European Photovoltaic Solar Energy Conf., pp. 652-655 (1997).
3. A. Terakawa, M. Isomura and S. Tsuda, Jpn. J. Appl. Phys. **35**, pp. 5612-5617 (1996).
4. D. E. Carlson and K. Rajan, Appl. Phys. Lett. **70**, p. 2168 (1997).
5. D. E. Carlson and K. Rajan, J. Appl. Phys. **83**, pp.1726-1729 (1998).
6. D. E. Carlson and K. Rajan, Appl. Phys. Lett. **69**, p. 1447 (1996).
7. W. Beyer, H. C. Weller and U. Zastrow, J. Non-Cryst. Solids **137&138**, pp. 37-40 (1991).
8. D. E. Carlson and C. W. Magee, Appl. Phys. Lett. **33**, pp. 81-83 (1978).
9. H. Dersch, J. Stuke and J. Beichler, Appl. Phys. Lett. **38**, p. 456 (1980).
10. H. M. Branz, Phys. Rev. B **59**, 5498 (1999).

Part VI

Defects, Band Tails
and Transport

A MOLECULAR DYNAMICS STUDY OF BAND TAILS IN a-Si:H

PETER A. FEDDERS*, D. A. Drabold**
*Department of Physics, Washington University, One Brookings Drive,
St. Louis, MO 63130, paf@wuphys.wustl.edu
**Department of Physics and Astronomy, Ohio University, Athens, Ohio 45701–2979

ABSTRACT

Band tail states are routinely invoked in models of a-Si:H including defect pool models and models of light induced defects. These models describe the band tail states as being localized on a single stretched bond. However to our knowledge, there is no theoretical or experimental work to justify these assumptions. In this work we use *ab initio* calculations to support earlier tight-binding calculations that show that the band tail states are very delocalized — involving large numbers of atoms as the energy is varied from midgap into the tails. Our work also shows that the valence band tail states are statistically associated with *short* bonds (not long bonds) and the conduction-band states with long bonds. We have slightly modified a 512 atom model of a-Si due to Djordjevic *et al.* to produce a large, defect free model of a-Si:H with realistic band tails.

INTRODUCTION

One of the most widely held views of a-Si:H is that valence band tail states are localized on a single stretched bond and that these states are intimately connected (through defect pool models) [1] with the number of dangling bonds. Further, numerous models of the Staebler Wronski Effect of light induced defects [2] are based on the same picture. To our knowledge there is no experimental or theoretical evidence that these valence band tail states are well localized or associated with stretched bonds. In fact, there is some evidence to the contrary. In earlier *ab initio* work we have shown [3] that single or very localized highly strained bonds lead to mid gap states and not to band tail states. More recent work [4] using tight-binding with supercells of thousands of Si atoms has shown that band tails are very delocalized and range over hundreds of atoms when one gets well into the tail. This work was performed using orthogonal tight binding for the energy eigenvalues and the cell itself was built by Djordjevic, Thorpe and Wooten [5] (DTW), using a (classical) Keating potential.

In this paper we investigate band tailing with *ab initio* calculations [6] on fully relaxed supercells of a-Si:H with over 500 atoms and with no topological irregularities leading to a gap state. Although our supercells are considerably smaller than those of Dong and Drabold [4], our results generally confirm that work. However, we find that the details of the distribution of sites in the band tails for the hydrogenated supercells differs substantially from the structure in the unhydrogenated supercell models. In the unhydrogenated supercells we find a very definite wandering string-like structure that is not observed in the hydrogenated supercells. Surprisingly, we also find that average bond lengths weighted by localization yields bond lengths for the valence band tail states that are shorter than average and for the conduction band tail states that are longer than average. The DTW cells are created with no geometrical defects but do contain a few spectral defects (localized states in the gap) due to badly strained bonds with abnormal bond angles. However, selective removal of Si atoms and hydrogenation removes these spectral defects.

Mat. Res. Soc. Symp. Proc. Vol. 557 © 1999 Materials Research Society

SUPERCELL PROPERTIES

We have found that the well known work of Wooten and his collaborators [7] is still the best starting point for modeling a-Si. In this scheme, a simple Keating potential is used to describe interatomic interactions, and four-fold coordination is constrained (which enables the use of the simple potential). As dangling bonds form a very small fraction of the sites in good quality a-Si, this is not only *reasonable*, but also physically *realistic*.

To create the first a-Si model consistent with a first principles Hamiltonian involving of order 512 Si atoms, we simply performed a steepest descent quench of the coordinates provided by DTW, and after some rather minor rearrangements obtained the first structural model employed in this paper. The supercell is denoted by S512H0.

For this paper, we focus primarily on the electronic properties. The electronic density of states for S512H0 is shown in Fig. 1 including 15 states above and below the Fermi level. Each bar on the figure represents an energy eigenvalue and the quantity $Q_2(E)$ is a measure of the localization of the state

$$Q_2(E) = \sum Q(n, E)^2 \tag{1}$$

where $Q(n, E)$ is the localized charge density associated with the eigenvalue E at the site n. For a perfectly uniform extended state $Q_2(E)$ would be 1 and for a perfectly localized state on only one atom it would be $N = 512$, the number of atoms in the supercell. In an amorphous or glassy substance one does not expect $Q_2(E)$ to be unity for any energy eigenvalue because all states have more weight on some sites than others. This does not mean that the states are "localized" in the sense of its charge falling off exponentially in an infinite model, it merely reflects the amorphous nature of the model. The average value of $Q_2(E)$, averaged over all 2048 energy eigenvalues for the cell under discussion, is 1.9 which is about what one would expect.

In supercell S512H0, we note that the two states in the lower and upper regions of the gap, just above and just below the Fermi level, are quite localized. This is universal in all unhydrogenated supercells that we have looked at and is due to badly strained regions of the cell. This type of behavior has been seen in numerous supercells of a-Si and a-C. Such defects are either extremely rare or non existent in real a-Si, and thus to make a really satisfactory model of a-Si they must be removed.

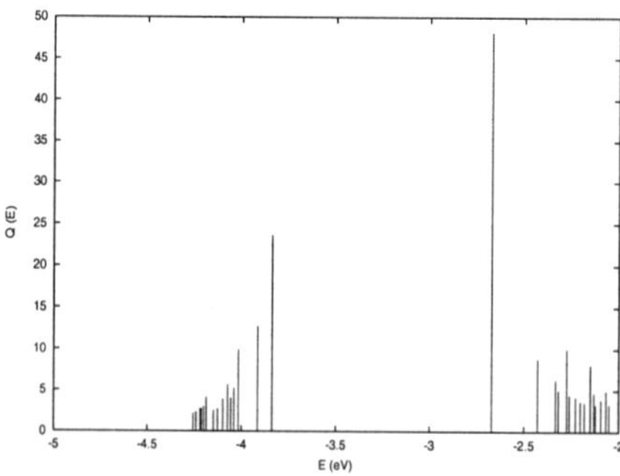

Figure 1: The electronic density of states for supercell S512H0 for the 15 states above and below the gap. The quantity $Q(E)$ is a measure of localization defined in the text.

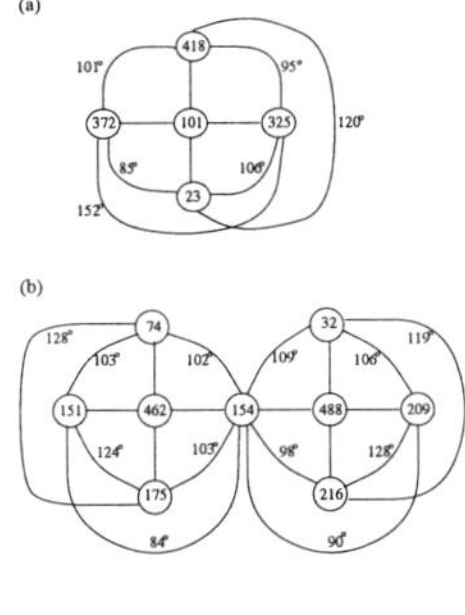

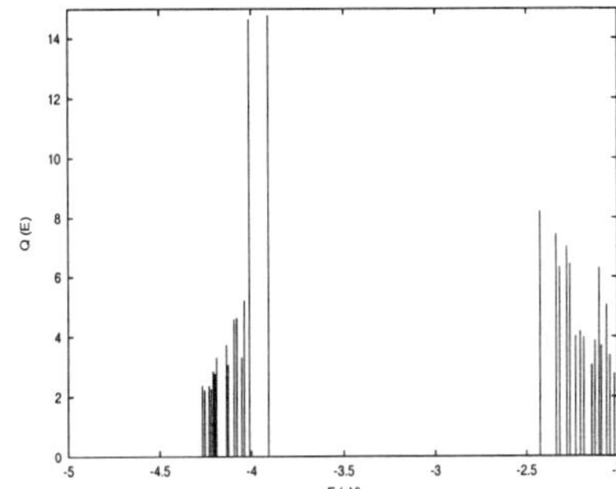

Figure 2: Depiction of the two badly strained regions in supercel S512H0.

Figure 3: The electronic density of states for supercell S509H10.

It is straightforward to make models of a-Si:H starting with a DTW cell while at the same time relieving strain in the network. In particular, one can simply remove atom(s) responsible for deep tail (nearly midgap) states in the H-free model, and hydrogenate the resulting dangling bonds. Supercells hydrogenated with the above method yield a NMR broad line in good agreement with experiments [8]. It is of course a limitation of this approach that the H is always clustered, and it cannot be used to create models with isolated (hydrogenated or not) dangling bonds. For the first model with 512 Si atoms and no H atoms we merely quenched the DTW model since annealing this large a supercell was not feasible.

In detail, to form the hydrogenated model, we removed Si atom 101 and inserted 4 H atoms to tie up the dangling bonds and removed the neighboring Si atoms 462 and 154 and inserted 6 H atoms to tie up the dangling bonds. Figure 2 depicts the strained areas removed. This supercell model is thus referred to as S509H10. The density of states is exhibited in Fig. 3. Note the increased gap. The gap is still too small and this is also universally observed in computer generated supercells.

BAND TAILS IN a-Si AND a-Si:H

First we will consider that average bond length and bond angle associated with band tail states. In practice, in this context, band tail state merely means states near the gap because our supercells are not large enough to identify actual band tails. As noted earlier, a weight or fraction of the charge can be assigned to each atom for a given energy eigenvalue.

To specify an average bond length associated with an energy eigenvalue, e, we sum over all pairs of nearest neighbor atoms weighted by the product of the weights (or charges) of the eigenvalue at the two sites comprising the pair. The results for the average bonds length vs. eigenvalue is given in Fig. 4. Note the abrupt jump at the Fermi level (between occupied and unoccupied states). The average bond length vs eigenvalue number for the hydrogenated supercell S509H10 is virtually identical. Clearly short Si-Si bonds are associated with v-band states and long Si-Si bonds with c-band states. We believe that the average bond lengths is 2.36 Å in both cases. A similar plot for the Si-H bond is presented in Fig. 5. The Si-H bond does not appear to be correlated with anything.

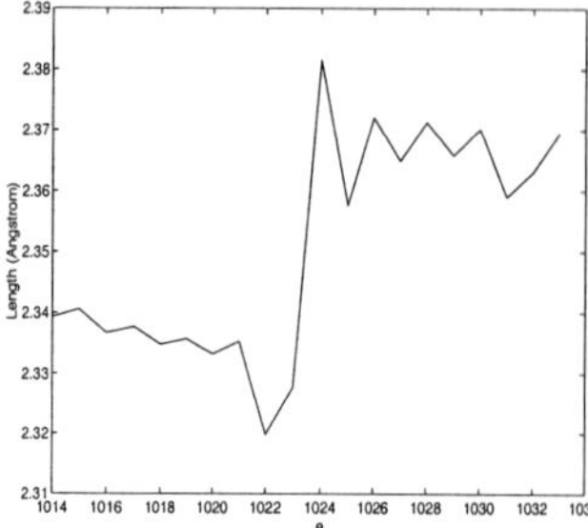

Figure 4: Average Si-Si bond length vs. eigenvalue number for supercell. The Fermi level is between eigenvalues 1023 and 1024.

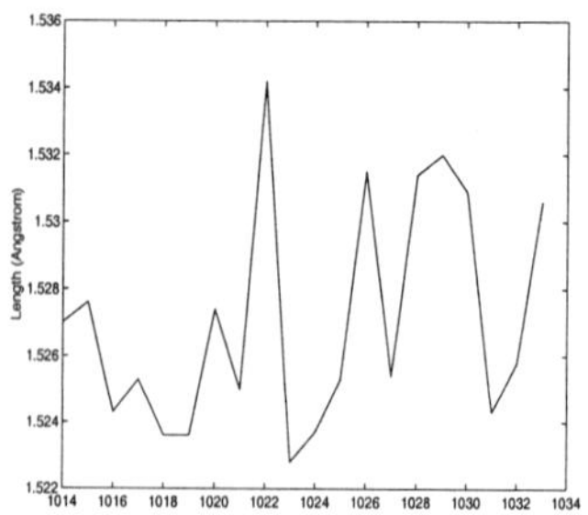

Figure 5: Average Si-H bond length vs. eigenvalue number for supercell S509H10.

Now the bond angles. We have tried various type of weightings here and the results appear to be robust. The results are presented in Fig. 6 where the weighting is the sums of the weights of the three atoms making up each triad. Interestingly enough, hydrogenation has very little effect on the average bond lengths and the Si-H bond lengths do not correlate in the same way that Si-Si bond lengths do. However, as we shall see, H does have a large effect of the topology of the strained bonds.

These results are rather startling, at least to us. Thus we have checked the results in a couple of different ways. First we calculated the same averages for a model of 71 atoms (61 Si and 10 H) that was created in an entirely different way, was thoroughly annealed, and has been shown to be absolutely stable at reasonable temperatures. The results were qualitatively the same and different in detail only slightly. We also checked those results with a local orbital code with 9 Si orbitals and 4 H orbitals. Thus we believe the results to be independent of supercell model as long as one has relaxed the supercell with a good *ab initio* code.

DISCUSSION AND CONCLUSION

The extension of the bandtails observed here and elsewhere is a point of some importance and we emphasize that it should strongly affect DPMs, and the modeling of transport in amorphous Si.

We have also detected some statistical correlations between bond lengths and energy eigenvalues; the valence tails being associated with short bonds and the conduction tails with long bonds. In some sense the conduction band result can easily be "explained". As atoms are pushed further apart, the conduction band will drop since it is made of anti-bonding states. However, then one might expect the valence band to drop as the atoms get closer together and this is not observed. The answer lies along different directions. If one takes a crystal and distorts it slightly, both the conduction and valence bands will tend to move toward the center of the gap. The reason is simply that symmetries are broken and degenerate states split. To first order, the center of gravity of the split does not move as some states move up and some move down. Thus the gap will tend to close up. The same effect holds with a-Si — at least with the Bethe-lattice. With a perfect Bethe lattice,

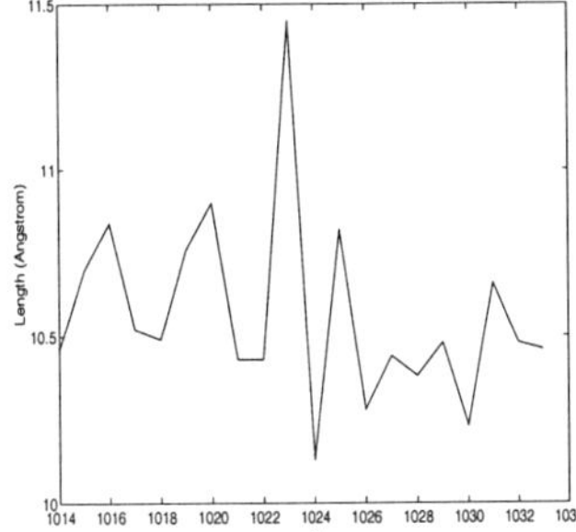
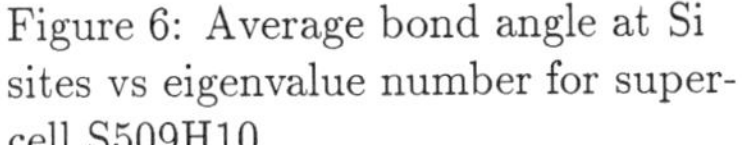

Figure 6: Average bond angle at Si sites vs eigenvalue number for supercell S509H10.

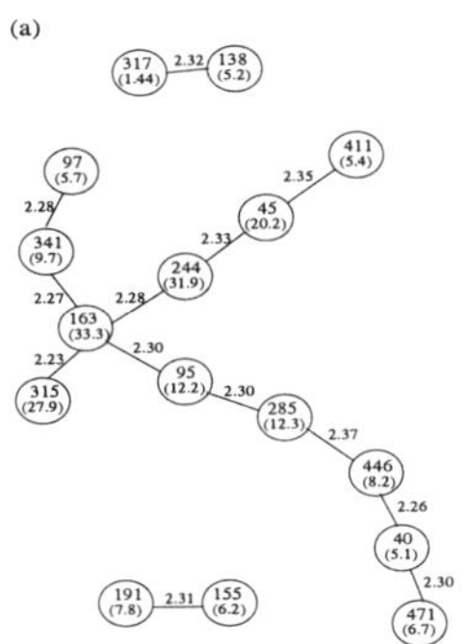

Figure 7: Example of the connectivity of atoms with a high localization for a band tail state in supercell S512H0.

any local perturbations in bond length or bond angle will tend to move the conduction band down and the valence band up. We believe that an analogue of this is happening in our model. However, we would have expected more of a bond angle effect than a bond length effect. We must be careful to distinguish two things being correlated with one thing causing another. Here the bond lengths and band tails are correlated but not necessarily causally.

Finally we consider the topology and localization of the strained bonds. As far as the localization vs position in the gap, we can only say that our results are consistent with the earlier study [4] and those results are much more detailed because the supercell was much larger. However, we find that the topology of the strained bonds is greatly affected by hydrogenation. These results are rather hard to present so we resort to crude maps of sketches of the topology of the sites that are most localized for each eigenvalue. An example in Figs. 7 is for a typical eigenvalue for the unhydrogenated supercell model S512H0. In this figure all atoms are included with values of $Q(e, n)$ greater than the stated amount and the bond lengths are noted. One can see that the sites where the state is most localized (and the shortest bonds) for a sort of string or random walk as noted earlier. All or almost all of the sites with the largest weighting are nearest neighbors! This structure was also found by Dong and Drabold [4].

This effect is absent for the hydrogenated model S509H10. In fact, for the then states nearest at the gap, there was at most one pair per state that were nearest or next-nearest neighbors. Table I shows the site numbers of the most localized sites and their distances from other such localized sites. The site appear to be almost scattered at random! Thus, the topology of the band tail states and strains are profoundly altered by the addition of a very small amount of H.

The defect pool model is usually derived (or justified) on rather detailed kinetic arguments and/or rather detailed models on the formation of dangling bond defects from the band tail. We believe that our work casts an enormous amount of doubt on the theoretical underpinning of these models. However, that is not to say that we believe that the defect

Table I

Atom 1	Atom 2	Separation (Å)	Atom 1	Atom 2	Separation (Å)
133	140	13.5079	324	389	12.8045
133	324	14.0299	324	279	10.0070
133	338	10.8199	324	326	15.0473
133	389	11.1224	324	257	6.3679
133	279	3.8445	338	389	10.8146
133	326	17.3294	338	279	5.2526
133	257	15.5986	338	326	5.6931
140	324	5.2071	338	257	7.1921
140	338	9.5261	389	279	10.7911
140	389	7.0415	389	326	16.4000
140	279	8.7139	389	257	20.1649
140	326	7.3633	279	326	10.6819
140	257	10.0923	279	257	13.7813
324	338	7.4569	326	257	3.8955

pool is nonsense. In fact, one does not need the detailed models alluded to above in order to obtain the defect pool model. The basic laws can be assumed much as the rule for tight binding can be assumed. While orthogonal tight binding is based on unphysical assumptions orthogonality, overlap, and lack of three center terms are wrong, tight binding works quite well for some things; especially things that are close to situation at the fitting parameters came from. We view the defect pool model to be in a similar situation.

ACKNOWLEDGMENTS

DAD acknowledges the support of NSF under grant DMR 96-18789. We thank Dr. Jianjun Dong for many helpful discussions.

REFERENCES

1. See, for example, R. A. Street, Phys. Rev. **B 43**, 2454 (1991); R. A. Street and K. Winer, Phys. Rev. **B 40**, 6236 (1989); M. J. Powell and S. C. Deane, Phys. Rev, **B 10815**, (1993); P. C. Taylor, in *Semiconductors and Semimetals*, edited by J. I. Pankove (Academic, Orlando, 1984), Vol. 21C, page 99, and references therein.
2. There are a vast number of models and references on this subject. For a recent summary see H. Fritzsche, Mat. Res. Soc. Symp. Proc. **467**, 19 (1997), and references therein.
3. P. A. Fedders, D. A. Drabold, and Stefan Klemm, Phys. Rev. **B 45**, 4048 (1992); P. A. Fedders and D. A. Drabold, Phys. Rev. **B 47**, 13277 (1993).
4. Jianjun Dong and D. A. Drabold, Phys. Rev. Lett. **80** 1928 (1998).
5. B. R. Djordjevic, M. F. Thorpe, and F. Wooten, Phys. Rev. **B 52**, 5685 (1995).
6. A. A. Demkov, J. Ortega, O. F. Sankey, and M. P. Grumbach, Phys. Rev. **B 52**, 51618 (1995).
7. F. Wooten and D. Weaire, in *Solid State Physics*, edited by H. Ehrenreich and D. Turnbull (Academic, New York, 1984), Vol. 40, p2.
8. P. A. Fedders and D. A. Drabold, Phys. Rev. **B 47**, 13277 (1993).

EFFECTS OF CHLORINE ON DOPANT ACTIVATION IN a-Si:H

ADAM M. PAYNE AND SIGURD WAGNER
Department of Electrical Engineering, Princeton University, Princeton, New Jersey 08544

ABSTRACT

The incorporation of chlorine has a significant effect on the dark conductivity of doped and undoped hydrogenated amorphous silicon (a-Si:H). The dark conductivity of a-Si:H films deposited from dichlorosilane ($SiCl_2H_2$) and SiH_4, and doped with diborane, increases by as much as a factor of 100 over the usual a-Si:H,B films deposited without $SiCl_2H_2$. The effect is observed at gas phase concentrations of diborane ranging from 0.006% to 0.5%, and for both DC and RF plasma depositions, although it is more noticeable for the DC discharge. An increase in dark conductivity is also observed in B doped a-Si,C:H films deposited with dichlorosilane, albeit coupled with a change in the Tauc gap. Chlorine reduces the conductivity of undoped and phosphorus doped a-Si:H films. Undoped a-Si:H films deposited from $SiCl_2H_2$ and SiH_4 have a dark conductivity of $\sim 1 \cdot 10^{-12}$ S·cm^{-1}, which is an order of magnitude lower than films deposited from pure SiH_4. We discuss several alternatives for the mechanism of chlorine enhanced or reduced dopant activation. We have made solar cells using chlorinated p-type a-SiC:H films as the p-layers.

INTRODUCTION

The exploration of a-Si:H has been accompanied by experiments with structural modifiers, in particular the alloying elements Ge and C, and with alternative bond terminators, in particular F [1] and Cl [2]. Chlorinated silane gases have been used to deposit a-Si:H,Cl with a long-term goal of improving the quality of the material for application in solar cells or thin-film transistors [3, 4]. Mixtures of dichlorosilane (DCS or $SiCl_2H_2$) with silane (SiH_4) enabled a rapid deposition rate and thus provided further motivation for exploring the deposition from chlorinated source gases [5, 6]. A series of solar cells was deposited using a DCS/SiH_4 mixture for the i-layer in an effort to decrease i-layer deposition time [7]. The cell structure of this previous work is shown in Fig. 1. The characteristics of these cells are given in Table 1. Note that all of the cells with Cl-containing layers adjacent to the n-layer had V_{oc} of ~ 0.66V, whereas those cells which had no Cl adjacent to the n-layer had V_{oc} ~ 0.76V. A variety of other cells were deposited with chlorine-free regions between the i-layer and the n-layer, all of which confirmed this effect of Cl on V_{oc}.

Table 1. Cells with 50nm Cycled SiH_4 and 100nm DCS Layer

i-layer	Efficiency (%)	FF	V_{oc} (V)	J_{sc} (mA/cm^2)
all DCS	1.35	.39	.661	5.21
p/i interface	0.75	.30	.645	3.88
central i-layer	1.50	.39	.647	6.00
i/n interface	2.25	.47	.765	6.26
all silane	4.42	.59	.766	9.84

Mat. Res. Soc. Symp. Proc. Vol. 557 © 1999 Materials Research Society

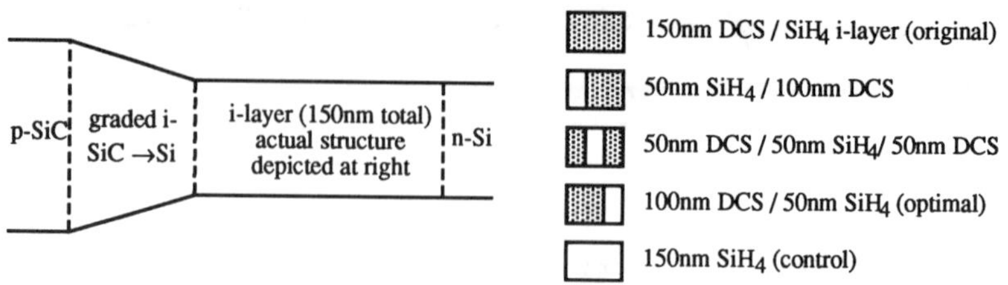

Fig. 1 shows the cell structure on the left with a depiction of the i-layer on the right. The five cells grown are detailed in Ref. [7].

EXPERIMENTS
Film Growth

To investigate this loss of V_{oc} we grew two phosphorus doped films, one from a source gas having a small amount of DCS and one without DCS. These films were grown under identical deposition conditions except for the addition of dichlorosilane ($T_{substrate}$ = 300°C, deposition pressure 500 mTorr (62 Pa), DC excitation, power density ~75-100mW/cm^2, 50 or 47 sccm SiH_4, 0 or 3 sccm $SiCl_2H_2$, and 1% PH_3 in H_2 at 6 sccm). Adding the DCS tends to lower the applied voltage, therefore lowering the power slightly, but the deposition rates were unchanged at ~4 - 5 Å/s. Co-planar thermally evaporated aluminum electrodes 7.5 mm long with a gap of 0.5 mm were used to determine the electrical conductivity and its thermal activation energy in the range between room temperature and 110°C. Optical transmission spectra were evaluated to determine the film thickness and absorption coefficient, from which the Tauc gap was calculated. From these measurements we observed that Cl reduces the n-type conductivity of a-Si:H from $2 \cdot 10^{-3}$ S/cm to $8 \cdot 10^{-6}$ S/cm. This confirms a previous result for phosphorus-doped films deposited from H_2 and $SiCl_4$ [8]. Other groups depositing from dichlorosilane did not observe this effect, probably because their gas phase concentration of PH_3 was significantly higher than ours, and because their films are microcrystalline rather than amorphous [4, 9, 10]. We hypothesize that in the solar cells described above, when the Cl in the i-layer comes in contact with the n-layer, the presence of the Cl reduces the n-layer conductivity and thus decreases V_{oc}.

In addition to the phosphorus-doped films, two series (13.56 MHz RF and DC) of boron-doped films were also deposited under the same conditions, except for the use of 1% B_2H_6 in H_2 (varied from 0.3sccm to 50sccm) instead of PH_3 and a substrate temperature of 260°C. Fig. 2 shows the variation in dark conductivity for the DC and RF deposited films as a function of the gas phase concentration of diborane. The corresponding activation energies and Tauc gaps are shown in Fig. 3. From these Fig.s we see that the chlorinated films have higher conductivity except at the highest gas phase concentration of diborane, where the conductivities are the same. We further note that the conductivity enhancement with Cl is more noticeable for the DC deposited films.

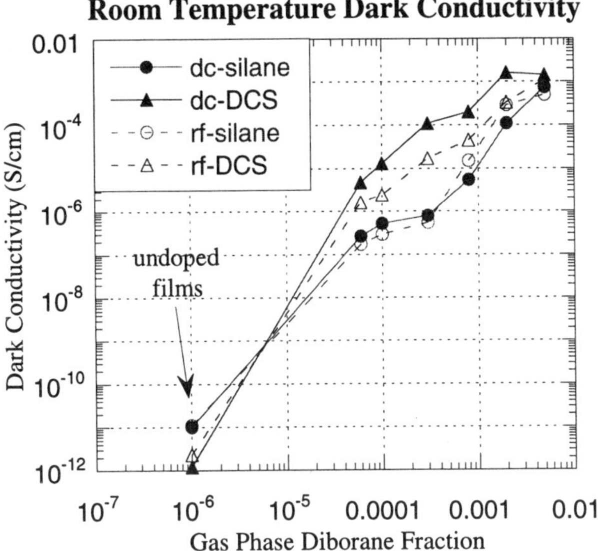

Fig. 2. Room temperature dark conductivity for four sets of films (dc-silane, dc-DCS, rf-silane, rf-DCS) with varying boron concentration.

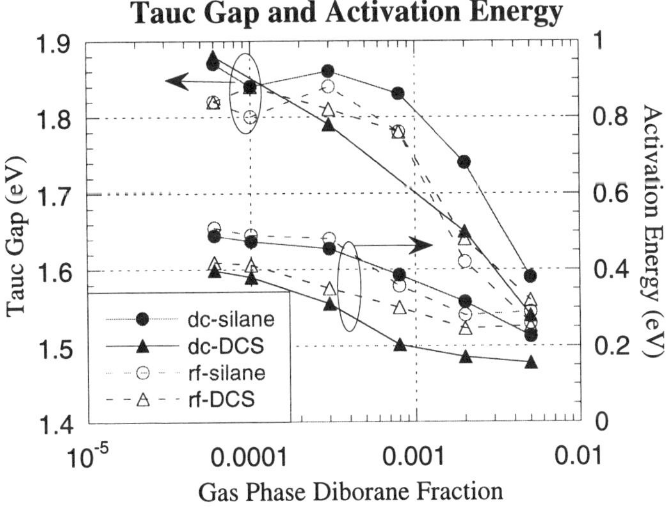

Fig. 3. Tauc gap and activation energy for four sets of films (dc-silane, dc-DCS, rf-silane, rf-DCS) with varying boron concentration.

Film Properties and Discussion

One possible explanation for the observed increase in conductivity is that the band gap of the material changes when chlorine is added. However, the Tauc gaps of these films are not changed in any systematic way by chlorine. The decrease in Tauc gap at the highest boron incorporation rates has been observed previously, but is unrelated to the chlorine content of the films [11]. Another hypothesis is increased boron incorporation when Cl is also present, and a third possible explanation is that Cl enhances the activation of boron in a Si:H. Increased B incorporation is ruled out on the basis of the Tauc gaps as shown in Fig. 3; however, secondary ion mass spectroscopy (SIMS) was performed for confirmation. Table 2 shows the results of the SIMS spectra taken at Evans East. The boron incorporation is enhanced by approximately 30% when Cl is also present for the DC deposited films; however, a 10% decrease in boron incorporation is observed for the chlorinated RF film. A 30% increase in [B] should cause only a 30% increase in conductivity which would barely be detected with our conductivity measurement, but cannot explain the factor of 10 enhancement actually observed. Therefore, we conclude that Cl *enhances the doping efficiency of boron* in a-Si:H.

Table 2. Impurity Concentrations in Films (in 10^{19} cm^{-3})

Film	[H]	[O]	[C]	[Cl]	[B]
p-Si 0.3[†]	200	*	*	*	4.6
p-Si 1.5	200	*	*	0.004	17.0
p-Si 4.5	200	*	0.20	0.020	31.0
p-DCS 0.3	200	3.0	0.20	30.0	5.8
p-DCS 1.5	250	2.5	0.20	30.0	22.0
p-DCS 4.5	250	2.0	0.30	40.0	41.0
i-Si	200	*	*	*	*
i-DCS	200	2.0	0.30	20.0	*
n-Si	300	*	*	*	*
n-DCS	200	2.0	0.10	20.0	*
p-Si 1.5 rf	300	*	*	0.005	18.0
p-DCS 1.5 rf	200	*	*	20.0	16.0
Background	--	1.0	0.04	0.002	0.01

†The numbers represent the B_2H_6 flow rate in sccm.

* Indicates a concentration equal to the detection limit of SIMS (listed in the last row).

Only the group at Bari has published results on boron doped films deposited using chlorinated silane [8]. When looking closely at their data, and noting that they deposited films using $SiCl_4$ heavily diluted with hydrogen (1 $SiCl_4$: 20 H_2), we conjecture that they observed microcrystalline behavior in their highly conductive B-doped a-Si:H,Cl films. This is supported by the unusually large optical gaps they report (1.95eV or higher) [12]. Here, we report the new effect of Cl acting to enhance the conductivity of B-doped *a-Si:H*. It has already been determined that Cl acts as a weak p-type dopant, based both on the decrease in conductivity of intrinsic a-Si:H,Cl [13] (an effect we confirm) and on thermoelectric power measurements [4]. Because the concentration of Cl is roughly the same for all concentrations of boron (Table 2), Cl, by itself, cannot account for the conductivity increase.

It is known that hydrogen passivates boron in B-doped crystalline silicon [14]. Recent calculations [15] indicate that a similar effect is responsible for the low doping efficiency of boron in a-Si:H. These calculations also indicate that hydrogen preferentially diffuses to sites were it passivates boron in the a-Si:H matrix. If Cl prevents H from such diffusion, then we could expect more B to be electronically active. Unfortunately, for this hypothesis, the films would need 10 times more Cl atoms than they currently possess to passivate all the H. (See Table 2) Another possible explanation is that the boron and chlorine preferentially form a cluster. Another possible explanation is that the boron and chlorine preferentially form a cluster. We conjecture that such a cluster is a doping site unaffected by the passivating presence of nearby hydrogen.

Table 3. Properties of a-SiC:H,Cl films

Film	σ_{dark} (S/cm)	E_{act} (meV)	E_{Tauc} (eV)
dc p-SiC:H	$1.2 \cdot 10^{-5}$	366	1.87
dc p-SiC:H, Cl	$1.4 \cdot 10^{-4}$	290	1.80
rf p-SiC:H	$2.2 \cdot 10^{-6}$	404	1.80
rf p-SiC:H, Cl	$2.3 \cdot 10^{-4}$	214	1.72
dc i-SiC:H	$7.5 \cdot 10^{-13}$	967	1.90
dc i-SiC:H,Cl	$6.4 \cdot 10^{-14}$	984	1.90

<u>Solar Cell Results</u>

Films of B-doped a-SiC:H were also deposited to identify any effect of the enhanced conductivity of the p-type window layer on solar cell performance. We made the a-SiC:H films by DC and by RF excitation. Deposition conditions were 260°C, 500 mT, DC power density ~90 mW/cm^2 (RF power density ~30mW/cm^2), 4.5 sccm CH_4, 41 (40) sccm SiH_4, 0 (3) sccm $SiCl_2H_2$, and 5 sccm 1% B_2H_6 in H_2. The dark conductivity and Tauc gap of these films were measured in the same fashion as those above and are listed in Table 3 which also includes two intrinsic films of a-SiC:H. In these intrinsic films the Cl does not alter the Tauc gap, but in the DC boron-doped films Cl appears to decrease the Tauc gap by 0.07eV. This Tauc gap variation could account for the difference in conductivity of the DC deposited films, but the conductivity difference of the RF deposited films is too great for this explanation alone. Solar cells were fabricated using these DC deposited B-doped a-SiC:H films as the p-layers. The cells were deposited on Solarex textured SnO$_2$ coated glass substrates with a structure of 13 nm p-layer (a-SiC:H) / 8 nm buffer layer / 1500 nm i-layer (a-Si:H) / 18 nm n-layer (a-Si:H) / Al (thermally evaporated) back contact. The two best devices for the DC deposited p-layers are shown in Table 4. The RF deposited p-layers resulted in very poor solar cells, probably because the recipe was not optimized for the Solarex substrates. Because the chlorinated p-layer has a higher conductivity we might expect the open circuit voltage of this cell to increase; however, the improvement in cell efficiency occurs because of the increase in the short circuit current density despite the chlorinated p-layer having a narrower Tauc gap. We conjecture that using DCS diminishes the reduction of the TCO surface which occurs in a H plasma. This would explain the increased J_{sc} when using a chlorinated p-layer.

Table 4. Solar Cell Results

Cell Type	Efficiency (%)	Fill Factor	V_{oc} (V)	J_{sc} (mA/cm^2)
Standard	7.1	0.66	0.783	13.7
Cl p-layer	7.8	0.67	0.778	14.9

CONCLUSION

The conductivity in B-doped a-Si:H,Cl is enhanced over B-doped a-Si:H. This increase occurs from a gas phase dopant range of 60 ppm to 0.5% for both DC and RF plasma excitation techniques. The conductivity increase is also observed for a-SiC:H, albeit there it is coupled with a reduction in Tauc gap. Band gap changes and enhanced incorporation of boron were ruled out as causes for this effect in a-Si:H. A cluster of B-Cl not passivated by nearby H is proposed as the origin of the enhanced conductivity. Solar cells were fabricated utilizing chlorinated a-SiC:H p-layers. The solar cells showed improved efficiency through increased short circuit current density, the origin of which is under investigation.

ACKNOWLEDGEMENTS

This work is supported by the Electric Power Research Institute, the Princeton Plasma Physics Laboratory, and the Princeton Environmental Institute. We thank Robert Stabinsky and Solarex for providing us with textured tin oxide substrates.

REFERENCES

1. R. Ovshinsky and A. Madan, Nature **276**, p 482 (1978).

2. D. E. Carlson and C. Wronski in *Topics in Applied Physics*, Volume **36**, edited by M. H. Brodsky, pp. 310 -- 311 (1979).

3.. J. N. Bullock and S. Wagner, Mater. Res. Soc. Symp. Proc. Volume **336**, pp. 97 -- 102 (1994).

4. Keun Soo Lee, Jong Hyun Choi, Sung Ki Kim, Hong Bin Jeon, and Jin Jang, Applied Physics Letters **69** (16), p. 2403 (1996).

5. Masami Nakata and Sigurd Wagner, Applied Physics Letters **65** (15), pp. 1940 -- 42 (1994).

6. A.M. Payne and S. Wagner, Mater. Res. Soc. Symp. Proc. Volume **420**, pp. 883 -- 888 (1996).

7. Adam Payne, Brian Crone, and S. Wagner, Mater. Res. Soc. Symp. Proc. Volume **467**, p. 789 (1997)

8. Vincenzo Augelli, Teresa Ligonzo, Roberto Murri, and Luigi Schiavulli, Thin Solid Films **125**, pp. 9 -- 16 (1985).

9. Sung Ki Kim, Byeong Yeon Moon, Jae Seong Byun, Hong Bin Jeon, and Jin Jang, Journal of Applied Physics **69** (8), p. 1131 (1996).

10. Jin Jang, personal communication, August 1997.

11. Satoshi Yamasaki, Akihisa Matsuda, and Kazunobu Tanaka, Japanese Journal of Applied Physics **21, part 2 Letters** (12), pp. L789 -- L791 (1982).

12. Giovanni Bruno, Pio Capezzuta, Francesco Cramarossa, Vincenzo Babarossa, Vincenzo Augelli, and Roberto Murri, Journal of Non-Crystalline Solids **59 & 60**, pp. 815 -- 18 (1983).

13. A.E. Delahoy and R. W. Griffith, Journal of Applied Physics **52** (10), pp. 6337 -- 46 (1981).

14. J.I. Pankove, D. E. Carlson, J. E.Berkeyheiser, and R.O. Wance, Physical Review Letters **51** (24), pp. 2224 -- 25 (1983).

15. Peter A. Fedders and D. A. Drabold, Journal of Non-Crystalline Solids **227 -- 230,** Pp. 376 -- 379 (1998).

CURRENT NOISE MEASUREMENTS OF SURFACE DEFECT STATES IN AMORPHOUS SILICON

P. W. WEST and J. KAKALIOS
University of Minnesota, School of Physics and Astronomy, Minneapolis,
MN 55455 USA

ABSTRACT

Measurements of conductance fluctuations in coplanar hydrogenated amorphous silicon (a-Si:H) are reported as a function of surface etching treatments. The noise power spectrum displays a broadened Lorentzian peak, associated with surface damage by CF_4 reactive ion etching (RIE), whereas surface etches using ion milling or wet chemicals remove the Lorentzian spectral feature and only a 1/f spectral form for frequency f is observed. The Lorentzian spectral feature can be explained by trapping-detrapping from surface states induced by the RIE etch, which cause fluctuations in the depletion width of the space charge region near the film surface. The thermally activated Lorentzian corner frequency is a measure of the degree of band bending and the Fermi energy at the thin film surface.

INTRODUCTION

The properties of surfaces and interfaces, as well as the near surface region, of hydrogenated amorphous silicon (a-Si:H) thin films play an important role in determining electronic transport properties and device performance. The near surface typically contains a higher density of defect states, creating a space-charge region and band bending that extends into the bulk of the material. Moreover, the first and last few hundred Ångstroms of plasma-deposited a-Si:H are known to have differing structural and electronic properties [1] than the bulk of a 1 μm thick film. The surface region is also known to be sensitive to the presence and diffusion of impurities, surface preparation, process details, and the nature of an adjacent interface. Conductance fluctuations in a number of different semiconductor materials and devices may originate from the thin film surface, the bulk of the thin film, and the metal/semiconductor interface. Trapping and emission of charge carriers into and out of gap state defects are known to yield measurable fluctuations in the conductance of semiconductors [2], and the density, spatial location, and energy location of these defect states may be determined from noise measurements. In this paper measurements of the current noise in a-Si:H are described which reflects the occupancy of surface defect states after certain etching treatments. The noise provides a non-invasive method for assessing surface properties of thin film a-Si:H and is highly sensitive to changes in the surface space charge and the Fermi energy at the thin film surface.

SAMPLE PREPARATION AND EXPERIMENTAL DETAILS

All samples studied here were synthesized at the University of Minnesota using plasma-enhanced chemical vapor deposition of silane (SiH_4) under standard deposition parameters known to produce high-electronic-quality films. The substrate temperature during deposition was 250°C and the rf power was 3 watts for an electrode area of ~50 cm^2, corresponding to a deposition rate of ~2 Ångstroms per second. All films were doped by dynamically mixing phosphine (PH_3) with silane for a gas-phase doping level of either 4×10^{-4} or 10^{-3} [PH_3]/[SiH_4]. One micron of n-type a-Si:H was deposited onto Corning 7059 glass substrates followed by 500 Å of an n+ (10^{-2} [PH_3]/[SiH_4]) layer for a reduction of the metal/a-Si:H Schottky barrier. Chromium (500 Ångstroms) followed by Gold (3000 Ångstroms on top of the Chromium) metal pads were evaporated onto select areas, forming the electrodes in a standard coplanar configuration. The n+ layer was then removed from the region between the metal pads using either a reactive ion etch, an ion mill, or a wet chemical etch. Reactive ion etching of the surface was performed using a Plasmatherm Dual chamber reactive ion etcher. The etching time ranged from 60-140 seconds at an etch rate of (10 Ångstroms/sec), using a gas flow of 25 sccm CF_4 and 1 sccm O_2 at a rf power

Mat. Res. Soc. Symp. Proc. Vol. 557 © 1999 Materials Research Society

of 100 Watts. Ion milling of the surface was performed in a Technics Ion Mill using Argon and an acceleration voltage of 100 Volts, giving an etch rate of ~13 Angstroms/min. Typical etch times for the ion mill were greater than 30 min., since it was necessary to remove the 500 Angstroms of n+, or alternately to insure that all effects of the reactive ion etch were removed from the surface. Wet chemical etching with a mixture of 98% HNO_3 and 2% HF near room temperature for 3-25 seconds was also used. This was a very fast uncontrolled etch and was not used extensively but was employed as an alternative etching method as described below. After removal of the n+ layer, the samples were patterned into a Van der Pauw four-probe structure using standard photolithographic techniques, giving an effective sample volume of ~10^{-6} cm^3.

For coplanar conductance fluctuation measurements, the samples are placed inside an integrated circuit package, and gold thread is wire bonded from the gold contact pad to the IC package pins. The IC package resides under vacuum in thermal contact with a temperature controlled copper block in the measurement chamber. The sample is annealed at 430 K for one hour to remove any surface adsorbates which might influence the conduction [3] and to remove any effects of prior light exposure [4]. By using the diagonal contacts of the four-probe structure, measurements are performed in a two-probe configuration commonly used for low level noise studies of high impedance films [5,6]. An Ithaco 564 low-noise current preamplifier is used to amplify the current fluctuations, while a HP 3561 spectrum analyzer digitizes and calculates the power spectral density from the amplified current fluctuations. The third and fourth contacts of the four-probe structure are utilized to insure that noise from the metal contact/a-Si:H region is not present. By contact detection methods outlined elsewhere [7], all noise results reported were verified to be free of contact noise.

RESULTS

The current noise from these a-Si:H films typically shows a power spectral density which varies as $f^{-\gamma}$ for frequency f, with $\gamma \sim 0.9$-1.2. However, when the top surface of the a-Si:H is reactive ion etched, the noise power spectrum changes dramatically, as illustrated in Figure 1, developing a sharp "Lorentzian" feature ($S(f) \sim 4\tau/(1+(2\pi f\tau)^2)$) in addition to the usual 1/f noise component. Shown in Fig. 1 for comparison is a power law 1/f as well as a Lorentzian frequency dependence. The frequency dependence of the spectral density clearly deviates from a pure 1/f power law, but is not described by a pure Lorentzian spectrum either. It is difficult to determine from Figure 1 whether the spectral feature near 100 Hz is broadened, or whether a 1/f component in addition to a pure Lorentzian accounts for the broadened appearance. It is this feature in the noise spectral density that is the focus of this paper, which appears only after a RIE treatment of the thin film surface.

This feature is identified as arising from the near surface region, as it is highly sensitive to the surface etching preparation. Shown in Figure 2 is the resulting noise power spectrum for the film in fig. 1 after the top 200 Angstroms of the RIE etched film are removed by ion milling. The Lorentzian spectral feature is completely absent and only a 1/f power spectrum is observed. An additional RIE etch of the surface restores the spectral feature as in fig. 1, and several cycles of surface treatments have been performed to verify that the Lorentzian component is reversibly associated with the RIE etch. The spectral feature can also be removed by wet chemical etching the RIE prepared surface. While there is some variation in the exact location and magnitude of the spectral feature for different samples and different etching runs, the phenomenon is highly reproducible, with the feature observed in seven different films synthesized in different deposition runs and RIE treated for varying times. For example, if the reactive ion etch is performed for 60-90 seconds with no other variations in the etching process, the lifetime τ of the spectral feature as well as the noise magnitude may typically vary by a factor of two at a given temperature between etching runs.

The spectral density of the RIE treated films varies with temperature, as shown in fig. 3, with the peak of the spectral feature shifting to higher frequencies as the temperature is increased. Figure 4 plots the log of the lifetime τ of the spectral feature, obtained by fitting the spectral density to a Lorentzian frequency dependence, as a function of inverse temperature. The spectral feature can usually be observed over a bandwidth of 1-1000 Hz and a temperature range of 290-360 K, limited by background noise at high frequencies and thermal limitations of the experimental setup at low frequencies. The activation energy of the lifetime is E_a=0.49 eV with a pre-exponential factor

Figure 1: Log-log plot of the noise power spectrum against frequency of an n-type a-Si:H film after the top surface was reactive ion etched. Shown for comparison is a power law 1/f form as well as a Lorentzian form $(S(f)\sim 4\tau/(1+(2\pi f\tau)^2))$. The frequency dependence of the power spectrum clearly deviates from a purely 1/f power law, but also appears not as sharp as a pure Lorentzian spectrum.

Figure 2: The noise power spectrum of a RIE etched film before and after the top 200 Ångstroms are removed by ion milling. While the 1/f component of the noise appears to remain essentially the same, the spectral peak is completely absent from the spectrum after the ion mill etch is performed. An additional RIE etch of the surface restores the spectral feature.

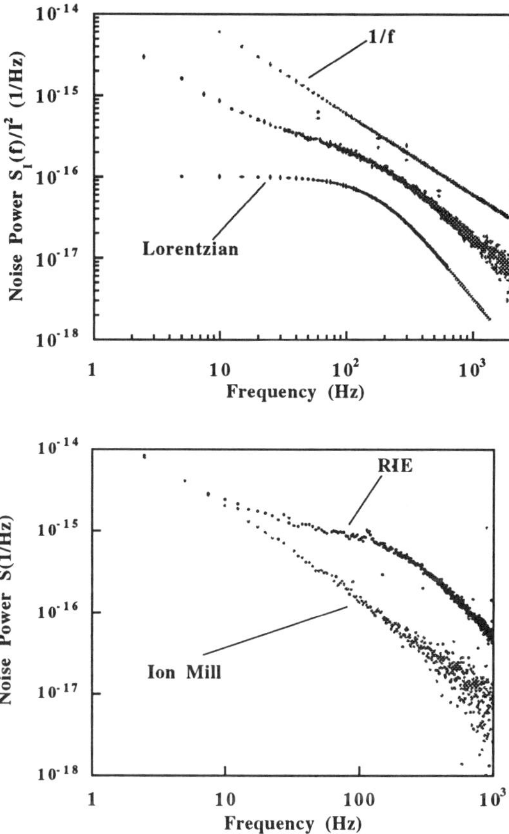

of $\tau_0=3\times10^{-10}$ sec. The temperature dependence of the power spectra has been measured for several different 4×10^{-4} and 10^{-3} [PH$_3$]/[SiH$_4$] n-type doped films with RIE etching times of 60-90 seconds, and the measured activation energies are always in the range $E_a=0.52\pm0.05$ eV and $\tau_0=4\times10^{-10}$ sec to within a factor of 5.

DISCUSSION AND ANALYSIS

It can be safely concluded that the "Lorentzian" spectral feature in fig. 1 originates from the surface of the thin film, given its sensitivity to details of the surface preparation. RIE is known to cause electronic and structural damage to semiconductor surfaces, creating defects at the surface or near surface region through impinging ions, electrons, or high energy photons [8-10]. The heaviest damage is generally confined to the top 100 Å of the film, and surface defects may arise from the etching induced disorder or the etching species, for example, CF$_4$ etching is known to create carbon related electronic defects. The RIE-induced surface defects affect the film's conductance through changes in the band bending at the a-Si:H surface. As indicated in the sketch in fig. 5, the space charge layer arises from negatively charged defect states at the semiconductor surface, and is balanced by an equal amount of positive charge extending into the film. Additionally, the actual state of the surface may be more complex, with the density and type of defect states varying throughout the band bending region. The space charge is responsible for band bending and reduces the conductivity near the surface. If a voltage is applied in a coplanar geometry across the semiconductor thin film, the width of the space charge region affects the measured conductance by defining the thickness of the conducting region. Conductance fluctuations occur when a charge carrier is trapped or emitted from a defect state near this surface region, causing a change in the charge state of the defect and fluctuations in the width and depth of the surface space charge region. Only defects near the Fermi energy contribute to the conductance fluctuations, even in materials with a continuous distribution of gap states, such as a-Si:H. Defects located more than a few kT above or below the Fermi energy are either always filled or empty, and do not significantly contribute to the conductance fluctuations [11].

As a defect alternates between being occupied and unoccupied due to charge trapping and emission (random two state switching), the resulting current-current autocorrelation function is given by $<I(t)I(0)> \sim \exp[-t/\tau]$ with a corresponding power spectra that has a Lorentzian frequency dependence $(S(f) \sim 4\tau/(1+(2\pi f\tau)^2))$. The characteristic frequency of the Lorentzian is determined by $1/\tau = 1/\tau_c + 1/\tau_e$, where $1/\tau_e$ is the emission rate from the defect and $1/\tau_c$ is the capture rate. If the defect resides at or near the Fermi Energy, then $\tau_c=\tau_e$ and the defect spends an equal amount of time occupied and unoccupied. While an electron may escape from the defect by tunneling, the noise reported here is thermally activated and only emission of a charge carrier from a defect state by thermal activation to the mobility edge is considered. Since the a-Si:H films are n-type doped, the relevant trapping mechanism is of electrons to and from the conduction band. The lifetime τ_e of an electron in a trap is thermally activated,

$$1/\tau_e = f_o \exp[-E_a/kT] \qquad\qquad (1)$$

where f_o is the attempt-to-escape frequency, and the activation energy E_a is defined by the energy separation between the defect level and the conduction band edge. In a material with an ensemble of traps, each with differing activation energies E_a and corresponding lifetimes τ, the observed spectral density is the superposition of the individual Lorentzian power spectra. A "broad" distribution of lifetimes τ leads to the effective total power spectrum having a 1/f frequency dependence.

In crystalline semiconductors with discrete gap state defect levels, a Lorentzian power spectra of the current noise may arise from states crossing the Fermi level in the band bending region [12]. In a-Si:H the continuous distribution of states across the Fermi energy throughout the band bending region should lead to a 1/f noise spectrum reflecting the superposition of Lorentzia with a distribution of characteristic frequencies. However, the relatively sharp spectral feature in figures 1 and 2 suggests that the noise originates from surface defects, where the highest density of defect states likely exists and where all noise creating defects will have similar activation energies (labeled E_1 in Figure 5). The activation energy of the spectral feature's lifetime (fig. 4) is a measure of the band bending at the thin film surface, and will differ from the bulk conductivity activation energy. For example, in a film that has had a 90 second RIE treatment, the activation energy of the spectral feature (Fig. 4) is a measure of E_1 and is 0.49 eV, while the conductivity activation energy (labeled E_2 in Fig. 5) is $E = 0.13$ eV for this film. Reactive ion etching appears to induce substantial band bending so that the electron lifetimes for these surface states (Eq. 1) reside in the observable bandwidth. Further support for this interpretation is that if a small (~ 25 Å) damaged layer created by the RIE process is removed by ion milling, the corner frequency of the spectrum shifts to higher frequencies, consistent with a decrease in the band bending. As further material is removed by additional ion milling, the spectral feature continues to move to frequencies beyond the observable frequency bandwidth, and only a 1/f spectral density remains.

Previous work on coplanar a-Si:H conductance fluctuations have observed non-Gaussian noise properties believed to be induced by fluctuations in the long range disorder [5,13,14]. In this study the noise has a purely Gaussian statistical nature which is consistent with the noise mechanism of trapping-detrapping from electronic defect states. Details of the difference in noise properties are discussed elsewhere [15,16]. A similar spectral feature as in fig. 1 has been reported in a study of undoped a-Si:H in a sandwich structure by Verleg and coworkers [17].

Figure 3: Log-log plot of the frequency dependence of the noise power spectrum, at 308 K and 350 K, for a a-Si:H film having a 60 second RIE surface etch.

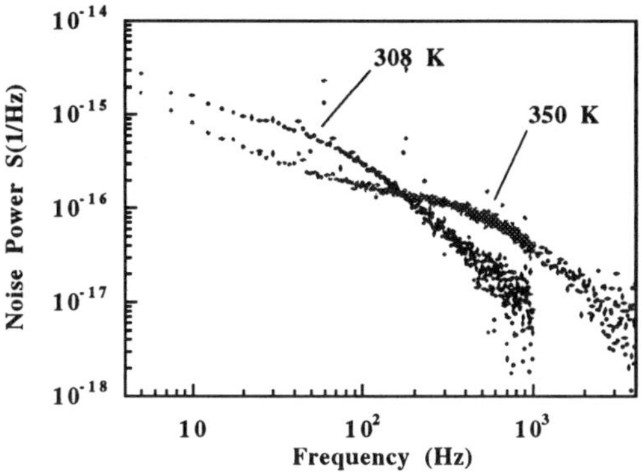

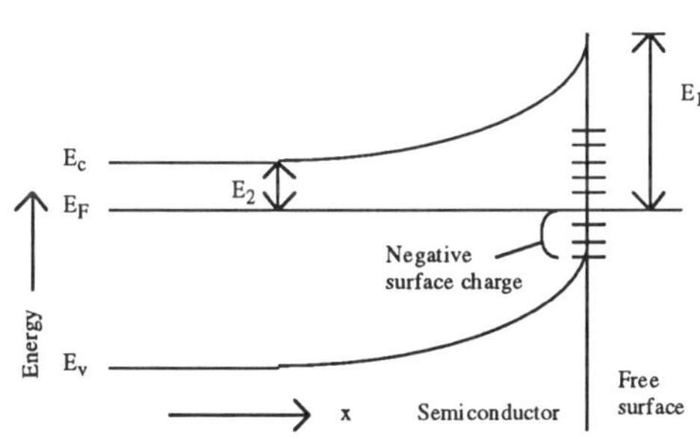

Figure 4: Arhennius plot of the lifetime τ of the spectral feature. The lifetime is thermally activated with an activation energy of E_a=0.49 eV and a pre-exponential factor of τ_0=3x10^{-10} sec.

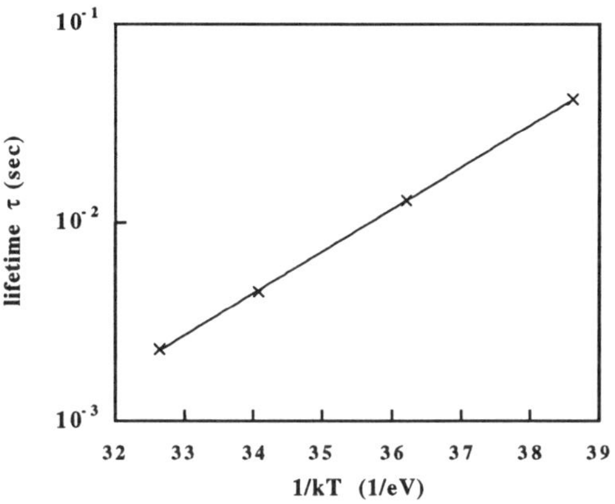

Figure 5: A sketch of the energy band bending at the a-Si:H free surface. A space charge layer is present due to the high density of negatively charged defect states at the semiconductor surface. Charge trapping and emission at the surface and near surface region cause fluctuations in the space charge and depletion width of the band bending region, resulting in a fluctuation of the measured conductivity.

419

They observed a broad thermally activated spectral feature in the noise power spectra and attribute it to generation (emission) and recombination (trapping) of charge carriers between the bands and Fermi level traps, although fluctuations in the number of free charge carriers are believed to be responsible for the conductance fluctuations. The noise in the transverse structure is also believed to be a bulk material effect, with the spectral feature broadened by band bending near the metal interface [17]. Similar Lorentzian features in the noise spectral density have also been reported for crystalline silicon MOSFET devices [18]. For certain gate voltages, trapping and emission from defect states at the oxide/semiconductor interface can create a Lorentzian spectral feature in the channel noise whose corner frequency depends on the energy difference between the Fermi energy at the interface and the conduction or valence band. These authors also cite number fluctuations as giving rise to the spectral feature, rather than space charge effects as described here. Although trapping at surface defects near the Fermi energy can induce fluctuations in the density of free electrons, the recombination rate in a-Si:H leads to number fluctuations having a characteristic frequency of MHz [17,19], far outside of our observable bandwidth. Fluctuations in the space charge due to the surface state defects are not affected by this recombination rate, and we consequently attribute the Lorentzian spectral feature to fluctuations in band bending at the top surface.

This work was supported in part by NSF DMR-9424277 and the University of Minnesota Graduate School Grant-In-Aid Program.

REFERENCES

1.   H. Fritzsche, in Semiconductors and Semimetals, edited by J.I. Pankove, (Academic Press, Inc., New York, 1984), vol. 21, part C
2.   M.B. Weissman, Rev. Mod. Phys 60, 537 (1988)
3.   M. Tanielian, Philosophical Magazine B 45, 435 (1982)
4.   D.L. Staebler and C.R. Wronski, J. Appl. Phys. 51, 3262 (1980)
5.   K.M. Abkemeier, Phys. Rev. B 55, 7005 (1997)
6.   R.E. Johanson, D. Scansen, S.O. Kasap, Phil. Mag. B 73, 707 (1996)
7.   P.W. West, accepted for publication in Rev. Sci. Inst.
8.   G. Jager-Waldau, H.-U. Habermeier, G. Zwicker, and E. Bucher, J. Elect. Mat. 23, 363 (1994)
9.   G.S. Oehrlein, Mater. Sci. Eng. B 4, 441 (1989)
10.   D.J. DiMaria, L.M. Ephrath, and D.R. Young, J. Appl. Phys. 50, 4015 (1979)
11.   K. Lee, K. Amberiadis, and A. Van Der Ziel, Sol.-State El. 25, 999 (1982)
12.   T.G.M. Kleinpennig, S. Jarrix, and G. Lecoy, J. Appl. Phys. 78, 2883 (1995)
13.   C.E. Parman, N.E. Israeloff, and J. Kakalios, Phys. Rev. B 47, 12 578 (1993)
14.   G.Khera, J. Kakalios, Phys. Rev. B 56, 1918 (1997)
15.   P.W. West, submitted for publication to J. Appl. Phys.
16.   P.W. West, Ph.D. thesis, University of Minnesota, 1999
17.   P. Verleg, Ph.D. thesis, Universiteit Utrecht, Netherlands, 1997
18.   B.K. Jones and G.P. Taylor, Sol.-State El. 35, 1285 (1992)
19.   A.D. van Rheenen, G. Bosman, and R. Zijlstra, Sol.-State El. 30, 259 (1987)

PHOTOCONDUCTIVITY TRANSIENT RESPONSE FROM THE STEADY STATE IN AMORPHOUS SEMICONDUCTORS

C. MAIN*, S. REYNOLDS*, J.- H. ZOLLONDZ* and R. BRÜGGEMANN**
*School of Science and Engineering, University of Abertay Dundee, Dundee DD1 1HG, Scotland, UK, c.main@tay.ac.uk
**F.B. Physik, Carl von Ossietzky Universität Oldenburg, D-26111 Oldenburg, Germany

ABSTRACT

We present analysis, computer modelling and experimental measurements of the photoconductive decay which occurs on cessation of illumination, in amorphous semiconductors. We explore the processes of relaxation of the excess carrier distributions, and examine the relative rôles of re-trapping and recombination in a model case of an exponential trapping state profile, with monomolecular recombination. A variety of possible decay behaviour is revealed. We examine several plausible intuitive explanations of the decay process, including (a) the assumption that the rate limiting step in the decay process is the thermal release of trapped carriers from the vicinity of the quasi - Fermi level, and (b) multiple re-trapping at the quasi - Fermi level prior to recombination. Actual decay rates, however are often much faster than that predicted by these assumptions, and the generation rate dependencies do not follow the relation expected. These explanations are shown in detail to be largely erroneous. Results of experimental measurements of the decay from steady state and TPC in films of a-Si$_{1-x}$C$_x$:H are presented. While these appear initially to be at variance with the predictions of the present work, we demonstrate that the observations can be reconciled fully with theory.

INTRODUCTION

Recent publications[1-4], have reported on the decay of photocurrent from the steady state in amorphous semiconductors, after steady illumination has been switched off. Analysis of this situation has been the subject of numerous works dating back at least to the classic work of Ryvkin[5] and Rose[6] in the 1950s and 60s. More commonly, studies on amorphous semiconductors have concentrated on the photocurrent response to *impulse* optical excitation, i.e. the transient photocurrent experiment (TPC). In this experiment, an excess carrier density ΔN is created by a short pulse of uniformly absorbed light in a semiconductor sample with coplanar ohmic contacts, and the subsequent decay of the photocurrent is followed. Under these conditions, with the *proviso* that recombination is not significant, the instantaneous photocurrent is also a measure of the instantaneous mobility of the *ensemble* of ΔN excess carriers.

In the switch-off experiment, in contrast to TPC, conditions just prior to cessation of illumination are dominated by *recombination* - *viz* the steady state excess electron density is given by $dn = G\tau_R$ where τ_R is the recombination lifetime. This lifetime may be a constant, in the case of monomolecular recombination, or may be dependent on the generation rate G. We might be led to assume that recombination should play an important part in the switch-off photocurrent decay *from the outset*, in contrast to the TPC experiment. However, since high densities of traps exist in amorphous semiconductors, it is not expected that the initial decay time τ_i of photocurrent should reflect *directly* the recombination time for free electrons, since excess carriers will interact with traps, slowing down the response. All of the above implies that interpretation of switch-off decay could be much more difficult than for TPC.

Mat. Res. Soc. Symp. Proc. Vol. 557 © 1999 Materials Research Society

The first of the "intuitive" models we present asserts that the initial decay of the whole excess is determined by the rate at which electrons trapped in states at the quasi-Fermi energy are released thermally to the conduction band prior to recombining. The rate limiting step, and hence the observed decay time, is then just the thermal release time τ_e of electrons at the quasi Fermi energy, E_{Fn}, *viz:*

$$\tau_e = \upsilon^{-1}\exp\left(E_{Fn}/kT\right) = \tau_i .$$ (1)

The second "intuitive" model contends that for any free electron, *multiple trapping* occurs in states close to the quasi-Fermi energy prior to the electron finally recombining. In this case, for every free electron recombining, there will be renewal from the reservoir of trapped electrons, most of which are close to E_{Fn}. The observed initial decay time is then given by

$$\tau_m = \frac{\tau_e + \tau_t}{\tau_t} \times \tau_R \approx \frac{n_t}{n} \times \tau_R \approx \tau_i ,$$ (2)

where n and n_t are the free and trapped electron densities, and τ_t ($\ll \tau_e$) is the trapping time into states close to the quasi Fermi energy.

It is evident that while the above arguments may seem plausible, they do not agree in detail. We will demonstrate the inadequacies of each approach below, by numerical modelling which also provides a framework of insight for more appropriate approximate analytical treatments, results of which are presented below. Following this, we then direct attention to measurements of the photocurrent decay from steady state in thin film specimens of the amorphous alloy a-Si$_{1-x}$C$_x$:H, to apply the insights gained from simulation.

COMPUTER MODELLING: RESULTS AND DISCUSSION

The fate of both free and trapped charge following switch-off or an impulse excitation, may be visualised using a numerical solution scheme has been developed by the authors[7] In this section, we present the results of a computer simulation of the decay from steady state in a deceptively simple system consisting of an exponential band tail and a fixed recombination time for free electrons. These are used to support simple analytical expressions for the decay under different conditions. The parameters employed are listed below.

These parameters thus give

$$\alpha = \frac{T}{T_c} = 0.69 .$$

Figure 1 shows the decay curves of free electron density after steady volume generation rate G ranging

Band edge density of states	$g(E_C)$	4×10^{21} cm^{-3} eV^{-1}
Characteristic temp. - band tail	T_C	290 K
Attempt-to-escape frequency	υ	1×10^{12} s^{-1}
Free electron recombination time	τ_R	6.9×10^{-8} s
Temperature	T	200 K

from 10^{14} to 10^{22} cm^{-3}s^{-1}. It is clear that there are several well-defined regions in which power-law decay occurs. These are classified below, and the indices are compared with the approximate analytical expressions obtained in this work.

		index	analytical index
Low excitation ($G < 10^{16}$ cm^{-3}s^{-1})	short times	- 0.7	$-\alpha$ = - 0.69
	long times	- 1.7	$-(1+\alpha)$ = - 1.69
High excitation ($G > 10^{20}$ cm^{-3}s^{-1})	short times	- 3.0	$-1/(1-\alpha)$ = - 3.22
	long times	- 1.7	$-(1+\alpha)$ = - 1.69

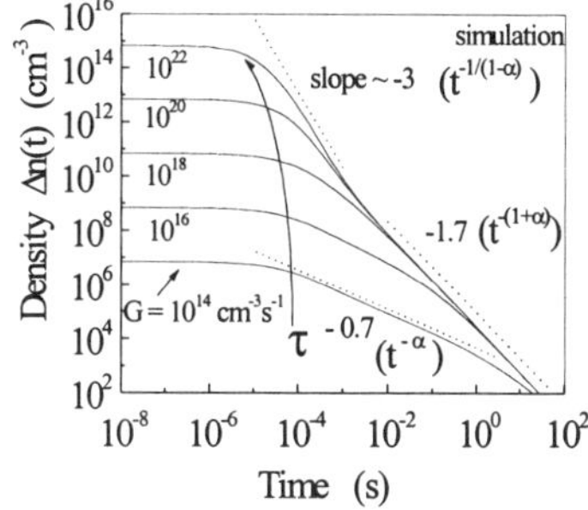

Figure 1. Simulated decay of excess electron density following switch-off of steady generation. Recombination time 7×10^{-8}s.

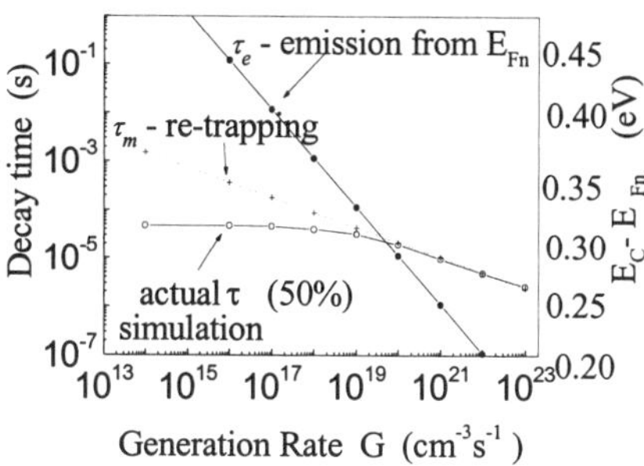

Figure 2. 50% decay time vs Generation rate, for computed decays and for two intuitive models.

In figure 2 we plot on a log -log scale the relation between the time to decay to 50% of the steady state density $\tau_{1/2}$, vs the prior steady generation rate G. With the computed curve we superimpose curves obtained using the two intuitive models described in section 1. We also display on the right ordinate axis of the figure an energy scale corresponding to the time scale on the left, using the relation $E(t) = kT\ln(\upsilon t)$. The results from the simulation show that at low excitation, the decay time is independent of generation rate, and that for G above about 10^{19} cm^{-3}s^{-1}, the decay time follows the relation $\tau_{1/2} \propto G^{-\beta}$, with $\beta = 0.28$. In this work it was found by analysis that under the conditions prevailing the index $\beta = 1-\alpha$, giving a value of 0.31, close to that found by simulation. We note that the value predicted by earlier analyses, of $\beta = \gamma = 0.59$, does not hold.

It is also clear from fig 2 that the 'intuitive' models for the decay are not in agreement with the simulation. The model which assumes that emission from states at the quasi-Fermi energy controls the decay is invalid over the whole range of G, while the simple multi-trapping model is in agreement at high excitation, but incorrectly predicts a variation of decay time with G at low excitation. It is significant that the three curves intersect at a single point, with an associated energy of about 0.3 eV. This "critical" energy E_{crit} corresponds to the level in the DOS for which the probability of free electron trapping into deeper states is equal to the probability of recombination, i.e

$$\int_{-\infty}^{E_{crit}} C_n g(E)dE = 1/\tau_R ,\qquad(3)$$

where C_n is the capture coefficient of the tail states for free electrons, giving, approximately,

$$E_{crit} \approx kT_c \ln\left(v\,\tau_R/\alpha\right). \tag{4}$$

Thus in any given decay, electrons already trapped below E_{crit} will upon release, recombine rather than be re-trapped below E_{crit}. Of course, many trapping and release events will occur for any such released electron, in shallow states prior to eventual recombination, but these transitions are not rate limiting.

Figures 3 and 4 illustrate the consequences of this observation, and help explain the various forms of the decay. Figure 3 shows the time - evolution of the trapped charge distribution during decay after a high steady generation rate $G = 10^{22}$ cm^{-3}s^{-1} which places the quasi-Fermi energy E_{Fn} (in this case equal to the trapped charge quasi-Fermi energy) *above* the critical energy E_{crit}. During the first part of the decay, while the peak of the excess distribution lies above E_{crit},

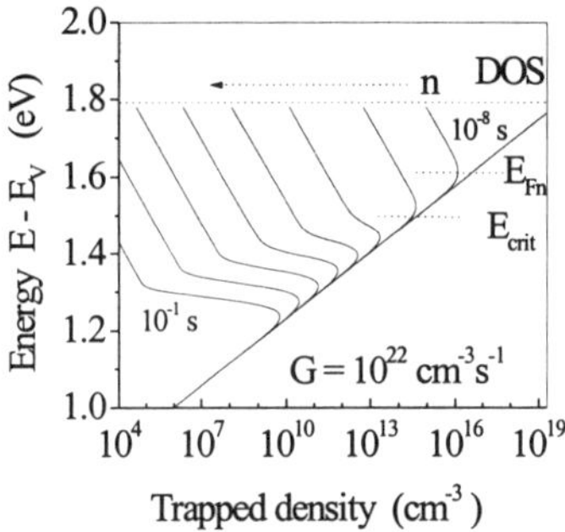

Figure 3. Trapped electron distributions at various times, from 10^{-8}s to 10^{-1}s after switch-off. Generation rate G=10^{22} cm^{-3}s^{-1}

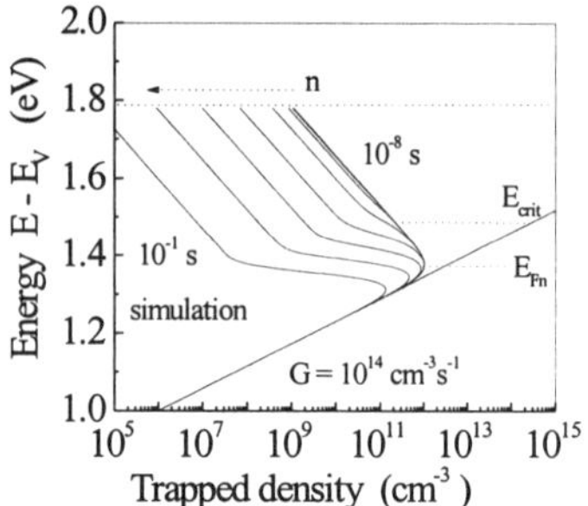

Figure 4. Trapped electron distributions at various times, from 10^{-8}s to 10^{-1}s after switch-off. Generation rate G=10^{14} cm^{-3}s^{-1}

the decay thus takes place under conditions in which re-trapping is more likely than recombination. The trapped charge packet moves downward without change of form as it is uniformly depleted by recombination loss. Under these conditions, the intuitive 'multi-trapping' concept is actually valid, and the effective instantaneous decay time is determined by Eq. (2) as described in section 1 above, where the trapped charge density n_t is essentially the integrated charge close to the peak of the distribution. The steepness of the decay in this time - region, i.e. $i(t) \propto t^{-1/(1-\alpha)}$ is a consequence of the variation of the 'multiplying' factor n_t/n during the decay.

When the peak of the trapped charge distribution reaches E_{crit}, a qualitative change occurs in the progression of the decay. Now the decay is simply rate limited by emission of trapped charge from states at energy $E_d(t) = kT\ln(vt)$. This causes the 'kink' in the shape of the distribution, a feature which also occurs in the post-transit or post- recombination portions of TOF and TPC respectively, and results in the same power-law decay as seen in these experiments, $i(t) \propto t^{-(1+\alpha)}$.

Figure 4 shows the time-evolution of the trapped charge distribution during decay after a low steady generation rate $G = 10^{14}$ cm^{-3}s^{-1} which places the quasi-Fermi energy E_{Fn} *below* the critical energy E_{crit}. During the first part of the decay, charge is lost uniformly by states down to E_{crit} but *no deeper*. The initial decay time is thus set by the emission time from E_{crit}, and *not* E_{Fn}, accounting for both the faster than expected decay (since E_{crit} lies above E_{Fn}), and the generation rate independence of the decay time in this regime. The decay assumes a shallow power law form $i(t) \propto t^{-\alpha}$ as charge is stripped out in this way from a trapped distribution which is increasing in density with depth, until the peak of the charge distribution at E_{fn} is reached. At this point, the familiar post-recombination form is observed, and the decay reverts to the form $i(t) \propto t^{-(1+\alpha)}$.

EXPERIMENTAL RESULTS AND DISCUSSION

In this section we report on experimental photocurrent decay measurements on a-Si$_{1-x}$C$_x$:H films produced by PECVD deposition at IPE Stuttgart, and fitted with coplanar electrodes. Film thickness was 1 μm, and the gap dimensions were typically 0.5 mm wide and 1 cm long. The material's optical (Tauc) gap is 1.87 eV . Uniform excitation was obtained using a high output red LED emitting at 660 nm (Stanley H3K). Constant current drive was used, employing a dc supply with suitable series resistor. Rapid 'turn-off' was ensured for a wide range of feed resistance by using an FET switch shunting the LED. In this way, the RC time constant for the LED drive current switch-off was kept low, independent of the feed resistance value.

Firstly, from the pre switch-off steady state photocurrent, the steady state power law index γ is found to be 0.9 at room temperature, indicating an approximately constant free electron lifetime[6]. Figure 5 shows a typical decay from steady state, at 290K, for an incident photon flux of 1.4×10^{17} cm^{-2}s^{-1}, giving generation rate G ~ 10^{21} cm^{-3}s^{-1}. The decay time for this excitation is seen to be about 10^{-6} s. After a shallow decay section at short times, the long time decay for this excitation, and for a range of values down to ~ 10^{18} cm^{-3}s^{-1} is of power law form, with index $\beta = -$ 0.68. This is remarkably shallow. In all cases studied above by simulation and analysis, the expected long-time decay in the presence of recombination should be steep - at least t^{-1} in form. We superimpose on the figure, TPC data measured on the same sample. It can be seen that the long time decay is very similar in form to that of the switch-off case. It is usually accepted that in the case of dispersive transport (as observed here) the pre-recombination section of TPC will exhibit a power law slope of less than unity and a post-recombination slope of greater than unity. Since the decay from steady state is supposedly controlled by recombination from the outset, then the similarity between the curves is problematic.

One explanation for this apparently inconsistent behaviour makes use of the concepts developed above, but with an unexpected 'twist'. We must invoke a *negative effective*

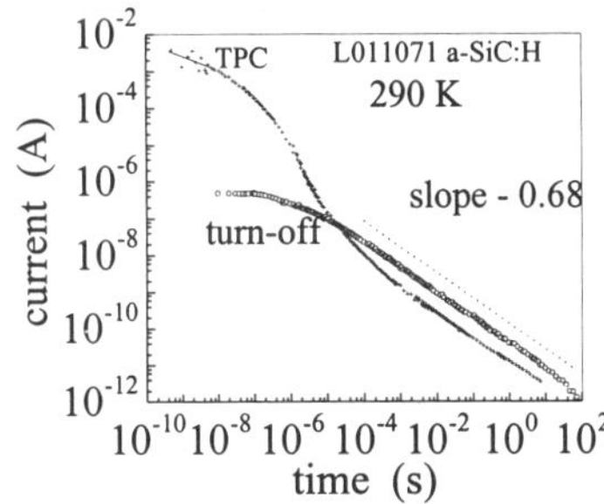

Figure 5. Experimental switch-off and TPC decays in a-Si$_{1-x}$C$_x$

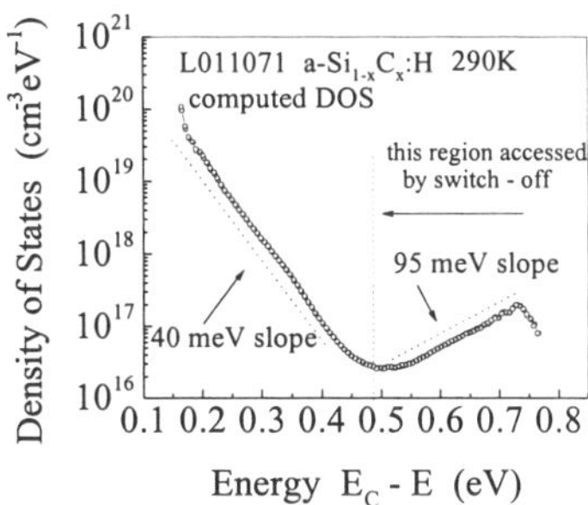

Figure 6. DOS computed for a-Si$_{1-x}$C$_x$ from TPC data using Fourier Transform method.

temperature to characterise the states in which the excess electrons reside - i.e. the density of states involved must *rise* toward the middle of the gap. We have shown that a Fourier transform technique can be used to compute the DOS profile from TPC[8]. Figure 6 shows the DOS obtained in this way for the silicon carbide film. There is an exponential conduction band tail, with characteristic energy 40 meV. At energies deeper than 0.47 eV, the DOS *rises* again in a fashion which may be described, in this case, with a single *negative* characteristic energy of about -95 meV (or a deep 'T_{CD}' = -1100K). This deeper part of the DOS is probably associated with dangling bond defects. Applying the appropriate result from the analysis of decay under monomolecular conditions, we obtain, for the long-time decay, a value of dispersion parameter $\alpha = T/T_{CD} = -0.27$, with index $\beta = -(1 + \alpha) = -0.73$, close to the observed slope of -0.68.

CONCLUSIONS

We have shown that even in relatively simple model cases, the decay from steady state photoconductivity in amorphous semiconductors can assume a variety of forms which are determined by the relative probabilities of free carrier *capture* by traps and recombination centres. Computer modelling illustrates clearly the internal processes of such photodecay. Plausible 'intuitive' models for the initial rate of decay have been shown to be erroneous or limited in application. Measurements of the switch off decay and transient photocurrents in a-$Si_{1-x}C_x$:H may be explained in terms of the above, by invoking a negative effective characteristic energy for the distribution of the localised states.

ACKNOWLEDGEMENTS

The authors acknowledge support from EPSRC Research Grant GR/M16696. We thank H-D Mohring, IPE Stuttgart, for sample preparation, the British Council and DAAD for support under the British - German ARC scheme, the University of Abertay Dundee for a postgraduate studentship.

REFERENCES

[1]   G.J.Adriaenssens, S.D.Baranovskii, W.Fuhs, J.Jansen and Ö.Öktü, Phys. Rev B51, 9661, 1995.

[2]   R.Brüggemann, Solid State Comms. 101, 199, 1997.

[3]   H.Cordes, G.H.Bauer and R.Brüggemann, Phys. Rev. B58, 16160, 1998.

[4]   P.Popovic, E.Bassanese, F.Smole, J.Furlan, S.Grebner and R.Schwarz, J.Appl. Phys. 82, 4504, 1997.

[5]   S.M.Ryvkin, *Photoelectric Effects in Semiconductors* (Consultants Bureau New York 1964), ch6.

[6]   A.Rose, RCA Rev. 12, 362, 1951.

[7]   C.Main, J.Berkin and A.Merazga, in *New Physical Problems in Electronic Materials,* ed M.Borissov,N.Kirov,J.M.Marshall, and A.Vavrek, (World Scientific Press, Singapore, 1991) pp 55-86.

[8]   C.Main, in *Amorphous and Microcrystalline Silicon Technology-1997,* ed. M. Hack, E.A. Schiff, S. Wagner, A. Matsuda and R. Schropp, MRS Symp. Proc., Vol 467, (MRS, Pittsburgh 1997) Ch.143, pp.167-178.

AN EXPERIMENTAL EVALUATION OF MODULATED PHOTOCURRENT SPECTROSCOPY AS A DENSITY OF STATES PROBE

S. REYNOLDS[*], C. MAIN[*], D.P. WEBB[†], M.J. ROSE[#]
*School of Science and Engineering, University of Abertay Dundee, Dundee DD1 1HG, UK,
s.reynolds@tay.ac.uk
†Department of Electronic Engineering, City University of Hong Kong, Tat Chee Avenue,
Kowloon, Hong Kong.
#Department of APEME, University of Dundee, Dundee DD1 4HN, UK.

ABSTRACT

Modulated and Fourier-transformed transient photocurrent (MPC and TPC-FT) spectroscopies have been evaluated through a study of the density and capture properties of localised states in as-prepared and light-soaked PECVD a-Si:H samples over a range of temperatures and optical excitations. Both techniques return a conduction band tail state characteristic energy of approximately 22 meV. However, defect state spectra differ in detail and are strongly influenced by dc optical excitation. A feature correlating with the quasi-Fermi level position is observed, but the capture coefficient implied (of order 10^{-6} cm^3 s^{-1}) is some two orders greater than that calculated from thermal activation of emission frequencies. Such a value would suggest an implausibly low absolute density of defects. Possible explanations are briefly discussed and additional investigations proposed.

INTRODUCTION

Modulated Photocurrent (MPC) Spectroscopy is now an established method for mapping transport parameters in disordered semiconductors. Developments in interpretation of MPC data [1] have led to its widespread use as a density of states (DOS) probe, and as a technique for measuring defect-state capture coefficients [2]. Its comparative experimental and analytical simplicity, combined with the ability to monitor the kinetics of light-induced defects *in-situ* has contributed to the popularity of the technique.

A 'direct' MPC experiment involves measurement of the complex photocurrent response $I(\omega)$ to a sinusoidal optical excitation $G(\omega)$, and is frequently carried out using an LED light source and phase-sensitive detector. However, it has been shown [3] that there are advantages to be gained in terms of improved sensitivity and increased energy range by carrying out a transient photocurrent (TPC) experiment and subsequently Fourier-transforming the I *vs. t* data to obtain $I(\omega)$. Henceforth this will be referred to as the 'TPC-FT' method.

Computer simulation of both transient and direct modulated photocurrent responses of multiple-trapping systems has led to a deeper understanding of the benefits and limitations of the direct MPC and TPC-FT approaches [4,5]. In this paper our interests lie in the application of these methods to the experimental investigation of a-Si:H samples. Specifically, we will compare the DOS profiles obtained using each method to identify features that appear consistently in both and to highlight those areas where discrepancies arise.

Direct MPC measurements unavoidably involve the application of a dc optical excitation which, in general, will give rise to quasi Fermi-level (QFL) shifts that invalidate the DOS analysis at deeper energies. However, the transition between the emission-limited regime, where DOS spectroscopy is possible, and the capture-limited regime, where the occupancy of states governs the response, may be identified experimentally and used to deduce a value for the

Mat. Res. Soc. Symp. Proc. Vol. 557 © 1999 Materials Research Society

capture coefficient of defect states. 'Tuning' the QFL by varying the dc excitation in principle permits contributions from different species of defect to be resolved [6]. Here, we present results obtained using this procedure, and also from TPC-FT experiments where dc optical bias has been applied, in order to assess whether these methods are compatible. Capture coefficients have also been deduced from thermal activation measurements on features in the DOS, allowing a check on self-consistency to be made.

OVERVIEW OF THEORY

A summary of the results of the frequency-domain DOS analysis, extended to include the capture-limited response is given below for the case where electron transport predominates and hole transitions may be neglected.

In the spectroscopic régime, *i.e.* within the largely unoccupied states above the quasi-Fermi level E_{Fn}, the trap density N_{ti} at energy $E_\omega = kT\ln(v/\omega)$ may be obtained in the following approximate analytic form to within a resolution kT:

$$N_{ti} \approx \frac{2}{C_{ni}\pi\,kT}(B_\omega - \omega)$$

(1)

where

$$B_\omega = \frac{G_\omega e\mu FA'\sin(\phi(\omega))}{|I(\omega)|},$$

(2)

μ is the free carrier mobility, G_ω the generation rate amplitude, A' the conduction cross-section and F the electric field strength. B_ω is obtained from the MPC or TPC-FT measurements. In most cases of interest $B_\omega \gg \omega$.

In the capture-limited régime, where states are largely filled, equation 1 no longer reflects the DOS. Assuming a constant capture coefficient C_{ni},

$$B_\omega = \left(\frac{N_{ti}C_{ni}\pi kT}{2} + \omega\right)\left(1 - f_d(E_\omega)\right),$$

(3)

where f_d is the Fermi-Dirac function in which E_F is replaced by E_d, the energy below which capture-limited behaviour dominates. A plot of B_ω vs. E_ω is thus expected to depart from the true DOS (ignoring resolution limitations) at the onset of capture-limited behaviour. If the DOS at E_d is not too steep a function of energy this transition should be identifiable by a change to a kT slope associated with f_d.

Provided the hole capture rate is small compared with the electron capture rate, E_d lies close to the electron quasi-Fermi level E_{Fn} and the associated emission frequency

$$\omega_d = C_n N_c \exp(-(E_c - E_d)/kT) = 2C_n n_0,$$

(4)

enabling the capture cross-section of states at E_d to be measured.

EXPERIMENTAL

Measurements were carried out on undoped PECVD a:Si-H films of thickness 0.7 μm deposited on Corning 7059 glass substrates. Coplanar interdigitated Al contacts, of effective length 50 cm and separation 60 μm allowed a satisfactory signal-to-noise ratio to be achieved even at low excitations. The sample was mounted in a custom-built cryostat allowing measurements to be made between 120 K and 400 K. An applied electric field of 1.7×10^4 V cm^{-1} was used throughout.

The sample used in most of this work (designated ASi549) was found, in the 'as-prepared' state, to have a room temperature conductivity $\sigma = 9 \times 10^{-10}$ S cm^{-1} and conductivity activation energy $\Delta E_\sigma = 0.74$ eV. Assuming a free electron mobility $\mu_0 = 10$ cm^2 (Vs)$^{-1}$ and effective density of states $N_c = 10^{20}$ cm^{-3} leads to a free carrier density $n_0 = 6 \times 10^8$ cm^{-3} and a Fermi level position of 0.67 eV (*i.e.* slightly n-type). A reduction in conductivity of approximately one order was observed following light-soaking for 1 hour under AM1 conditions.

For MPC measurements, sinusoidal excitation was provided by a Stanley H3000 red LED (660 nm) driven via a series resistor by a Thurlby-Thandar TG1304 function generator. An EG&G 7602 lock-in amplifier was used as detector, covering a frequency range of 10 mHz to 250 kHz. Custom-built preamplifiers and rigorous calibration [7] were employed to optimise signal to noise performance and to minimise the effects of instrumental phase shifts.

The TPC system comprises a Laser Science VSL337 N$_2$ - pumped 5 ns pulse dye laser source (640 nm) and Tektronix TDS380 oscilloscope as transient recorder. By suitable adjustment of settings a time range of typically 1 ns to 10 s could be covered. Appropriate improvement in signal-to-noise ratio was achieved by repetitive averaging of 10 to 100 data sets. Optical bias could be applied by means of an LED. Thorough screening was employed to minimise crosstalk between sources and inputs.

RESULTS AND DISCUSSION

Comparison of MPC and TPC-FT DOS Spectroscopies

Representative DOS spectra obtained using direct MPC and TPC-FT methods for an as-prepared sample are shown in Figure 1. As the frequency range of direct MPC is less than that of TPC-FT due to instrumental limitations, experiments at three different temperatures have been carried out and pieced together to enable a wider energy coverage. Deviation from spectroscopic behaviour is clearly visible in the MPC plots.

There is good agreement between the (assumed exponential) tail state energy, of approximately 22 meV, which is particularly steep. Superficially the defect region appears similar also, with a prominent feature occurring at approximately 0.58 eV. However there are differences in detail, which we shall return to below.

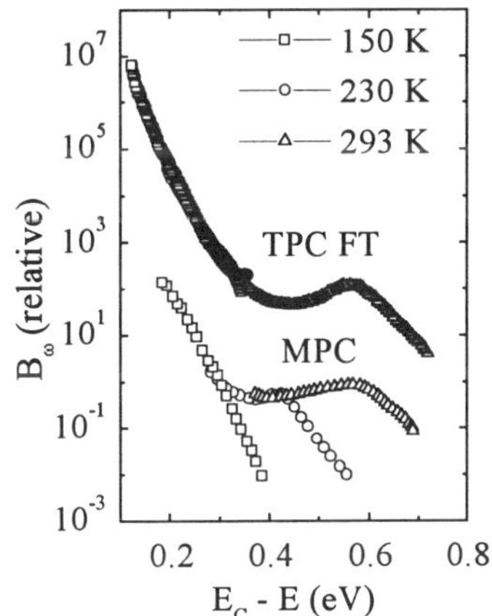

Figure 1. DOS plots: direct MPC and TPC-FT methods.

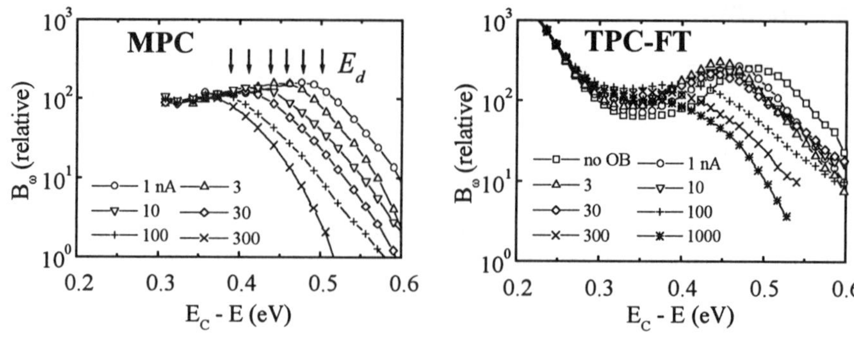

Figure 2. Direct MPC and TPC-FT plots at 250 K versus dc photocurrent. Energy scale assumes $\nu = 10^{12}$ s^{-1}.

Effect of Optical Bias

The effect of optical bias on the MPC and TPC-FT spectra is shown in figure 2. These data were taken following light-soaking, although similar results have been obtained from as-prepared material. The TPC-FT data were taken using a low pulse density (3×10^{13} cm^{-3}) to avoid problems associated with non-linear behaviour [8].

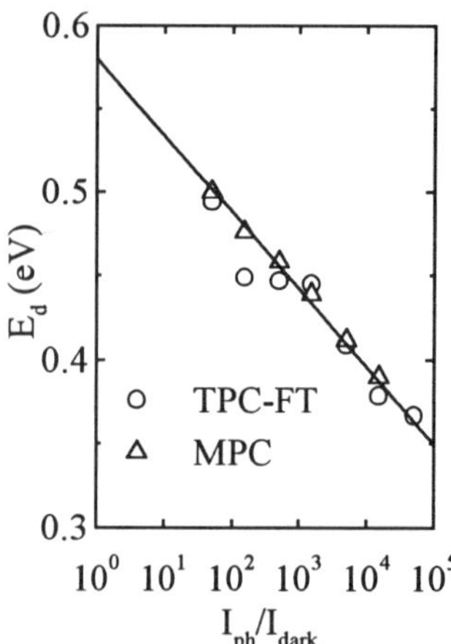

Figure 3. E_d vs. ratio of photocurrent to dark current (data from figure 2).

The presence of an optical bias-dependent feature is immediately apparent in both sets of data, although there are discrepancies in detailed behaviour. The turn-over feature energy E_d is plotted as a function of photocurrent in figure 3, revealing a move to shallower energy in sympathy with E_{Fn} (line of slope kT). However, the intercept at $I_p = I_{dark}$ at 0.58 eV is 0.10–0.15 eV shallower than the equilibrium Fermi level position. This suggests either that E_d should not be considered equal to E_{Fn} due to an incomplete theory, or that the energy scale is incorrect. Rescaling to fit requires the attempt-to-escape frequency to be increased from the value of 10^{12} s^{-1} assumed here to about 10^{14} s^{-1} and, taking $N_c = 10^{20}$ cm^{-3}, a resultant increase in C_n from 10^{-8} cm^3 s^{-1} to 10^{-6} cm^3 s^{-1}.

Temperature-dependence of TPC-FT data

The temperature-dependence of B_ω vs. ω between 155 K and 330 K was investigated for a light-soaked sample using the TPC-FT method. A variation in temperature is predicted to have two effects on B_ω vs. ω data. Firstly, as can be seen from equation 1, B_ω is predicted to scale as $C_n T$ for a given constant trap density. Empirically, it is found that at the peak of the defect 'bump' $B_\omega \propto T^2$ which suggests a linear dependence of C_n on T for these states. Following adjustment of each data set to compensate for this, the energy scale was then calibrated by calculating the attempt-to-escape frequency required to achieve optimal overlap o*

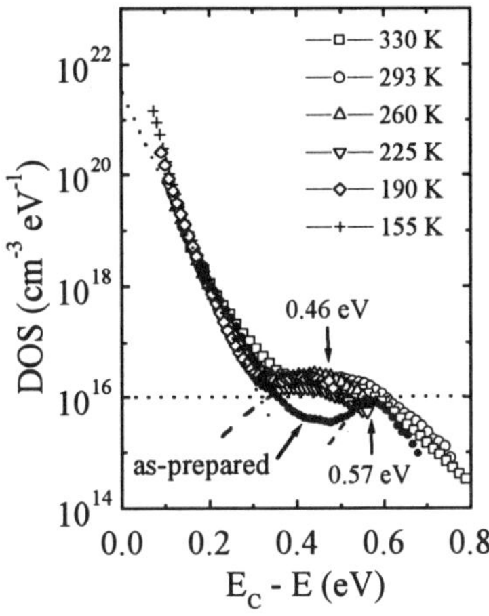

Figure 4.
DOS distributions *vs*. T.

the N_{ti} *vs*. ω curves. A best-fit value of $\nu = 2.5 \times 10^{12}$ s^{-1} ($C_n = 2.5 \times 10^{-8}$ cm^3 eV^{-1}) was obtained for tail states (allowing for poorer resolution at higher temperatures). A slightly lower but less precise value was obtained based on the defect feature. The scaled DOS plots are shown in figure 4, with data obtained from an as-prepared sample at 293 K included for comparison. The DOS magnitude scale has been fixed by extrapolation of the conduction band tail to an estimate of 4×10^{21} cm^{-3} eV^{-1} at the band edge.

Within this framework the magnitude of the peak of the defect bump centred at 0.46 eV is of order 3×10^{16} cm^{-3} eV^{-1}, quite low in comparison with literature values but in reasonable agreement with the value inferred from DOS *vs*. pulse density measurements made on the same material [10]. However, if the approximately two orders larger defect capture coefficient suggested by the optical bias results is adopted the density would be only 3×10^{14} cm^{-3} eV^{-1}, which is implausibly small.

Possible explanations for discrepancy in capture coefficient values

Both approaches we have employed in measuring the capture coefficient assume a density of states that is constant relative to E_c and is independent of temperature and optical bias. The validity of these assumptions has recently been questioned, as discussed briefly below.

The 'defect relaxation' model allows the DOS to accommodate the prevailing photocarrier density by moving to shallower energies, and it has been proposed [9] that this effect rather than solely a shift in E_{Fn} is responsible for the behaviour shown in figure 2. However, we have previously reported measurements [10] on as-prepared samples which show that E_d is influenced by photocarrier densities well below the thermal density, suggesting a more subtle process may be involved. A larger than anticipated hole capture rate could have such an effect.

The 'random temporal fluctuations' model asserts that the density of states is subject to random fluctuations in energy of a *1/f* character. One consequence of this is a reduction in the observed emission times for deep states particularly at low temperatures, since as they are on average occupied for longer, they have a greater opportunity to release prematurely due to a large amplitude fluctuation. However, at least superficially, this model [11] might be expected to affect the results from the thermal activation and optical bias experiments to a similar extent.

CONCLUSIONS

Direct MPC and Fourier-transformed TPC (TPC-FT) methods can be used with success to investigate localised state distributions in PECVD a-Si:H. By varying the experimental temperature it is possible to widen the energy range accessible using direct MPC, allowing a comparison to be made with TPC-FT over both the conduction band tail state and defect state energy ranges. A good agreement is obtained for conduction band tail states which, in this study, were found to fit an exponential distribution of characteristic energy 22 meV. Over the defect state energy range the agreement is less satisfactory. In order to investigate further the role of the dc excitation term inherent in direct MPC, a comparison with TPC-FT plus applied dc optical bias has been carried out. The general behaviour observed is similar in both cases, in that a feature that shifts in energy and is linked to the position of the electron quasi-Fermi level can be identified. This behaviour is observed in both as-prepared and light-soaked material. However, for this feature to coincide with the QFL calculated from the dc photoconductivity, an attempt-to-escape frequency of order 10^{14} s^{-1} (corresponding capture coefficient 10^{-6} cm^3 s^{-1}) must be assumed. This conflicts with a value of approximately 10^{12} s^{-1} obtained here from thermal activation analysis of TPC-FT data with no optical bias applied, and would also result in a very low value for the absolute density of states.

While the defect relaxation model may predict this discrepancy, it is unclear how it could account for the fact that dc photocarrier densities substantially below thermal densities are seen to influence the MPC spectrum. It is proposed that future work will include a more thorough experimental study of the effect of optical bias on the MPC spectrum in the tail state region, and further computer simulations of this system to determine whether the inclusion of hole transitions might account for the behaviour observed.

ACKNOWLEDGMENTS

The authors would like to thank R. Brüggemann for valuable discussions during the course of this work, and EPSRC for support under grant GR/M 16696.

REFERENCES

1. R. Brüggemann, C. Main, J. Berkin and S. Reynolds, Phil. Mag. B **62**, 29 (1990).
2. K. Hattori, Y. Niwano, H. Okamoto, and Y. Hamakawa, J. Non-Cryst. Sol. **137-138**, 363 (1991).
3. C. Main, R. Brüggemann, D.P. Webb and S. Reynolds, Sol. St. Commun. **83**, 401 (1992).
4. C. Main, D.P. Webb and S. Reynolds, in *Electronic, Optoelectronic and Magnetic Thin Films,* edited by J.M. Marshall, N. Kirov and A. Vavrek, (John Wiley-Research Studies Press, New York 1995), pp. 12-25.
5. C. Main in *Amorphous and Microcrystalline Silicon Technology*, edited by M. Hack, E.A. Schiff, S. Wagner, A. Matsuda and R. Schropp (Mater. Res. Soc. Proc. **467**, Pittsburgh, PA 1997), pp. 167-178.
6. K. Hattori, H. Okamoto, and Y. Hamakawa, J. Non-Cryst. Sol. **198-200**, 288 (1996).
7. D.P. Webb, PhD thesis, University of Abertay Dundee, UK, 1994.
8. D.P. Webb, C. Main, S. Reynolds, R. Brüggemann, Y.C. Chan and Y.W. Lam, J. Non-Cryst. Sol. **227-230**, 211 (1998).
9. J.D. Cohen and D. Kwon, J. Non-Cryst. Sol. **227-230**, 348 (1998).
10. S. Reynolds, C. Main, D.P. Webb and M.J. Rose, 10th International School on Condensed Matter Physics (Varna, Bulgaria, 1-4 Sept. 1998), in press.
11. V.I. Arkhipov and G.J. Adriaenssens, J. Non-Cryst. Sol. **227-230**, 166 (1998).

MODELLING OF DRIFT MOBILITY EXPERIMENTS ON a-Si:H

WEN CHAO CHEN*, LOUIS-ANDRÉ HAMEL
Groupe de recherche en physique et technologie des couches minces (GCM),
Département de Physique, Université de Montréal C.P. 6128,
Succursale centre-ville, Montréal, Québec H3C 3J7, Canada

MATHIEU KEMP
Zoology Department, Duke University, Durham, NC 27708-0325 USA

ARTHUR YELON
Groupe de recherche en physique et technologie des couches minces (GCM),

Département de Génie Physique et Génie des Matériaux, École Polytechnique de Montréal
C.P. 6079, Succursale centre-ville, Montréal, Québec H3C 3A7, Canada

ABSTRACT

We present recent results of a study of the behavior of electronic carriers in a-Si:H, using the model of multiple trapping (MT) in an exponential density of states. In previous publications, using Monte Carlo simulations, we showed that the standard low field MT model gives reasonable agreement with experiment particularly if the Meyer-Neldel effect is included in the model. We report here on the results of including two other effects. First, we have included a simple model of field assisted detrapping, to take account of the effect of high fields. We obtain very good agreement with the results of measurements on both electrons and holes, from a number of laboratories. In addition, we show here that the validity of an effective temperature approach can be checked easily by comparison with experiment. Second, we have presented a simple model of rapid relaxation of trapped carriers. This model offers the possibility of removing the apparent inconsistency between these measurements, and other experiments.

INTRODUCTION

Analyses of time-of-flight (TOF) experiments using approximate analytical solutions are very imprecise for the prediction of the dispersion parameters α and the microscopic mobility. Thus, we have modelled the experimental drift mobility for a-Si:H by Monte Carlo (MC) simulations[1,2] for the multiple trapping (MT) model with an exponential density of states under low electric field conditions. The calculations have shown clearly that the standard MT model[1] can explain most of the low field time-of-flight (TOF) experiments by using a mobility $\mu_0 = 0.5\,\mathrm{cm^2\,V^{-1}s^{-1}}$, as is usually found, for holes, but $\mu_0 = 4\,\mathrm{cm^2 V^{-1}s^{-1}}$ for electrons, in contrast to the value of $13\,\mathrm{cm^2 V^{-1}s^{-1}}$ which is usually reported.[1,3,4] We also found[2] that the agreement between model and experiment, especially at the highest and lowest temperatures, is improved by including the Meyer-Neldel effect (MNE) in the MC simulations. However, when the electric field is increased up to $10^5\,\mathrm{V/cm}$, we have found that this model underestimated the experimental drift mobilities for high electric field TOF[5-7]. We have therefore proposed[8] a model of field assisted detrapping, the diagonal jump model (DJM).[8] In DJM, the carrier emission rate is maximum along a diagonal path to the mobility edge, which depends upon position under the effect of the field. The jump combines the effect of temperature and electric field.[8] The effective temperature, T_{eff}, can be defined from the electron distribution. The DJM gives a particular, simple analytical form for T_{eff}. However, as we shall demonstrate below, the validity of the effective temperature concept can be verified, and the value of the localization length can be determined without evoking a particular form for T_{eff}, directly from experimental data.

We have also recently proposed[9] a simple model of rapid relaxation of trapped carriers. We find that this model can give agreement with TOF data equivalent to that of the standard MT model, but with a higher value of μ_0. The two models cannot be distinguished by drift

Mat. Res. Soc. Symp. Proc. Vol. 557 © 1999 Materials Research Society

mobility measurements. However, the relaxation model offers the possibility of removing the apparent inconsistency between these measurements, measurements of photoluminescence, and the absence of geminate recombination in a-Si:H. In both of the new models, MNE has not been taken into account, for simplicity, but it can be readily included.

TRANSPORT MODELS

Diagonal jump model (DJM)

In this section, we describe the DJM for high field electronic transport which can be found in details in Ref. [8]. We assume that the capture rate remains the same as in the standard MT model which, is usually written as:

$$K_c^0(E) = \frac{\nu_0}{kT} \exp(-E/kT_0),$$ (1)

where k is the Boltzmann constant, T the temperature, T_0 the characteristic temperature of the bandtail states, ν_0 the attempt-to-escape frequency and E the energy depth from the mobility edge. The basic idea of our model is that all the carriers are assumed to be released along the diagonal path for which the release rate is maximum. This means that the easiest way of detrapping to the mobility edge, which depends upon position is by thermally assisted tunneling, rather than only by absorbing energy E, or only by elastically tunneling through a distance E/eF. We assume that the result is not very different from what we would get if we took account of all of the possible paths both for trapping and detrapping. We also assume detailed balance between the escape and release mechanisms, this gives a release rate of[8]:

$$K_r^F(E) = \nu_0 \exp(-E/kT_{eff})$$ (2)

where T_{eff} is the effective temperature, defined as

$$T_{eff} = \sqrt{T^2 + (eaF/2k)^2)}.$$ (3)

In Eq. (3), e is the elementary charge, F is the electric field and a is the localization length[10]. We note that the localization length is a very important parameter to be determined from experiments. We also note that when at low field $eaF/2 << kT$, the DJM reduces to the standard MT model.

We have carried out[8] MC simulations in which a is a simulation parameter, keeping the parameters of our earlier studies, ν_0, T_0, and μ_0, fixed. The carrier energy distributions thus obtained were in fact Fermi-Dirac corresponding to the temperature T_{eff} (above the demarcation energy). The calculation yields reasonable agreement with experiments for electrons[6,11,12] and for holes[13], with values of a between 8 and 12 Å. However, as we shall now show, it is possible to verify the existence of an effective temperature, to obtain an analytic form and to determine a, directly from experiment, without any simulations.

The T_{eff} model

If we simply assume that there exists a $T_{eff}(T, F)$, that is, Eqs. (1) and (2) are correct, then the Fermi function of trapped carriers is determined by T_{eff} and by E_f^*, the quasi-Fermi level. It is then easy to demonstrate that the number of mobile carriers is given by

$$n = g_0 kT \exp(-E_f^*/kT_{eff})$$ (4)

where g_0 is the effective density of states at the mobility edge, and

$$E_f^* = E_f^*(T_{eff}, t) \tag{5}$$

where t is the time after the carriers are generated, but less than the transit time t_T. It is then obvious from Eq. (4) that for a pair of temperatures and fields, T_1, F_1 and T_2, F_2, if the currents obey

$$\frac{I_1(t)}{T_1 F_1} = \frac{I_2(t)}{T_2 F_2} \tag{6}$$

then

$$T_{eff}^1 = T_{eff}^2 . \tag{7}$$

If Eq. (6) is satisfied for all $t < t_T$, then T_{eff} is a well defined quantity. In this case, T_{eff} must be a monotonically increasing function of both T and F that satisfies the conditions $T_{eff}(T, F = 0) = T$ and $T_{eff}(T = 0, F) = eaF/2k$. A suitable form for T_{eff} is a generalization of Eq. (3) in which the square law is replaced by an x-power law with $x \geq 1$.

With these assumptions, it may be observed that the localization length a can be extracted from experimental data without any reference to a specific model. In Fig. 1, we show a curve (T,F) of constant T_{eff}. Such curves of constant T_{eff} can be obtained experimentally from Eqs. (6) and (7). For measurements at high T and low field (region 1), $T_{eff} \approx T_1$ and for measurements at low T and high field (region 2), $T_{eff} \approx eaF_2/2k$, irrespective of a specific model. By finding two such regions corresponding to the same T_{eff}, according to Eq. (6) the localization length is immediately found

$$a = 2kT_1/eF_2 . \tag{8}$$

Other measurements in intermediate regions (region 3) would provide the exact shape of the function $T_{eff}(T, F)$. In particular, the power law may be tested and the value of the power x may be extracted.

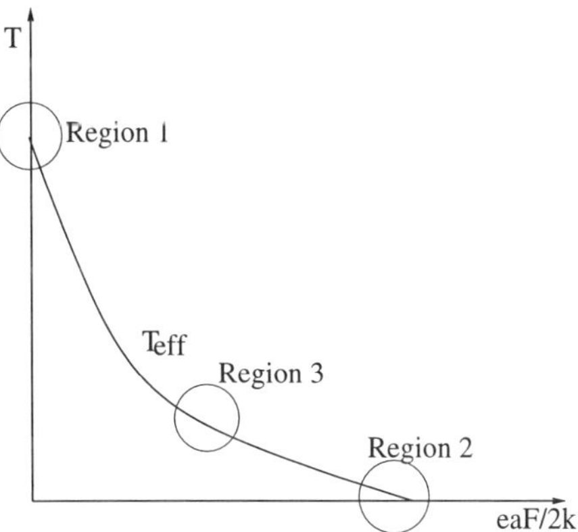

Fig. 1. Region of interest for the effective temperature model. T_{eff} is a monotonic function which satisfied the conditions $T_{eff}(T = 0, F) = eaF/2k$ and $T_{eff}(T, F = 0) = T$.

In table 1, we show the calculation of the localization length from the experimental data of Nebel et al.[7] for two different values of T_{eff}. T_{eff} are extracted from Fig. 4 of Ref. [7] by using the relation

$$\mu_d(T_{eff}, F = 0) = \mu_d(T, F) . \tag{9}$$

By finding two sets of temperatures and fields corresponding to the same value of T_{eff} as discussed above, we obtained $T_1 = T_{eff}$ at low field ($F_1 \approx 0$) and a value of F_2 at very low temperature ($T_2 \approx 0$). The localization length is calculated using Eq. (8). In both cases, we found $a \approx 8$ Å which is not very far from the value obtained previously when a was an adjustable parameter in the MC simulations[8].

Fig. 2 shows the analysis of electron drift mobility measurements of Gu et al.[6] for three temperatures, as function of effective temperature using DJM. In this case, we assume the validity of Eq. (3) for the effective temperature model. The temperatures and electric fields correspond to region 3 of Fig. 1. We note that the drift mobility, μ_d, is calculated from the transient current, $I(t)$, at 30 ns, far from the transit time, and is given by[6]

$$\mu_d = I(t)L/eF , \tag{10}$$

where L is the sample thickness. We find that Eq. (3) fits the data and that the best value of the localization length is $a = 8$ Å. This value is consistent with the values of Ref. [8].

Rapid relaxation model (RRM)

In all of our previous MC simulations, with or without the Meyer-Neldel effect or for high field effects, the TOF experimental results were best described with a relatively low value of the electron microscopic mobility, μ_0, of $4\,\mathrm{cm^2V^{-1}s^{-1}}$. Such a low value of μ_0 appears to be incompatible with the absence of geminate recombination[14].

The rapid relaxation of trapped carriers has been proposed by Kemp[15] to explain the time and spectral characteristics of low temperature photoluminescence (PL) in a-Si:H. In this model, once trapped, the electrons relax in energy logarithmically in time, until they recombine or escape and move away from holes. This relaxation is due to the response of the local atoms to a change in the electric charge of the trap.

We incorporated rapid ralaxation in our MC simulations[9] to determine to what extent this mechanism might affect TOF measurements. In our rapid relaxation model (RRM), trapped carriers relax logarithmically in time until they are released to the extended states or until the stabilized polaron energy is reached. A carrier which is emitted and retrapped is again initially in an unrelaxed trap. The assumption of logarithmic relaxation is plausible[9], but another form would probably yield similar results. In contrast, the limiting minimum energy, which was not neccessary for the PL model[15] is essential here. Otherwise, some carriers will never be collected. We have found[9] that the dispersion parameters α_1 and α_2, before and after the transit time, are unaffected by rapid relaxation. On the other hand, the release rate of relaxed carriers is reduced, leading to lower drift mobilities. The microscopic mobility μ_0 must thus be increased significantly (up to 2 orders of magnitude) in order to yield the experimental drift mobilities. Once μ_0 is readjusted, the drift mobilities of the RRM are identical to those from there obtained from the standard MT model, making the RRM impossible to prove or disprove from TOF data. The RRM, with higher values of μ_0, is then compatible with TOF measurements and, with appropriate values of the relaxation rate and of the microscopic mobility, can be made also compatible with PL measurements and explain the absence of geminate recombination. With the RRM, a wider class of phenomena can thus be explained by a single model.

CONCLUSIONS

We have shown that if the effective temperature concept is correct, the localization length a can be found from measurements at different T and F corresponding to the same T_{eff} independently of a specific model. We found $a \approx 8$ Å for electrons by using two different methods of analysis. This result is consistent with our previous work using MC simulation[8].

Table 1. Calculation of localization length from data in regions of interest 1 and 2 of the T_{eff} model. Experimental data are taken from Nebel et al.[7].

$T_{eff}(K)$	$T_1(K)$	F_1 (V/cm)	$T_2(K)$	F_2 (V/cm)	a (Å)
196	196	≈ 0	≈ 0	4.0×10^5	8.44
204	204	≈ 0	≈ 0	4.25×10^5	8.27

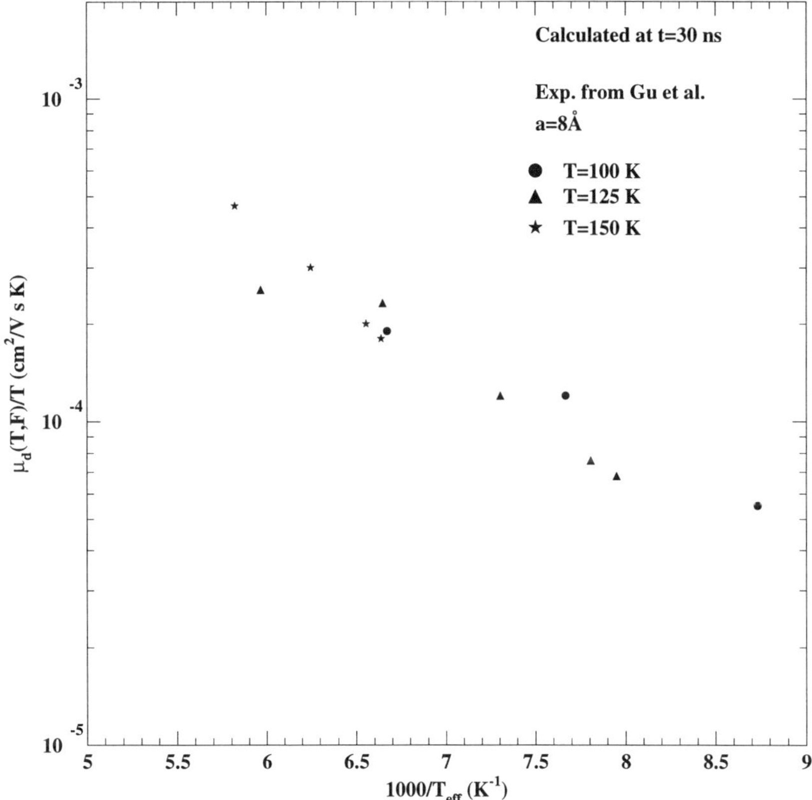

Fig. 2.   Analysis of electrons drift mobility as a function of effective temperature using DJM. Experimental data are taken from Gu et al.[6] for three different values of temperatures (T). The localization length, a, is found to be 8 Å .

We note that it is also very interesting to measure the hole drift mobility at a given time t and make a similar analysis for holes in order to determine a. We also proposed a rapid relaxation model (RRM) of trapped carriers and found that RRM predictions are equivalent to those of the standard MT model, except that a higher value of μ_0 is required. RRM and standard MT cannot be distinguished by drift mobility measurements but RRM is the first model to be consistent at the same time with the TOF experiments, PL measurements and the absence of geminate recombination.

* Presently at ANIQ R&D Inc., Université de Montréal C.P. 6128, Succursale centre-ville, Montréal, Québec H3C 3J7, Canada

REFERENCES

1. W.C. Chen, L.A. Hamel, in: Amorphous Silicon Technology – 1996, ed. M. Hack, E.A. Schiff, S. Wagner, R. Schropp and A. Matsuda (Materials Research Society, Pittsburgh, PA, 1996) p. 759.
2. W.C. Chen, L.A. Hamel, A. Yelon, J. Non-Cryst. Solids **220**, 254 (1997).
3. T. Tiedje, in Semiconductors and Semimetals, edited by J. Pankove (Academic Press, New York, 1984) Vol. 21C, p.207.
4. T. Tiedje, in Physics of Hydrogenated Amorphous Silicon, edited by J. D. Joannopoulos and G. Lucovsky (Springer, New York, 1984), Vol. 2, p. 261.
5. H. Antoniadis, E.A. Schiff, Phys. Rev B **43**, 13957 (1991).
6. Qing Gu, E.A Schiff, J.B. Chévrier and B. Equer, Phys. Rev. B **52**, 5695 (1995).
7. C.E. Nebel, R.A. Street, N.M. Johnson, J. Kocka, Phys. Rev. B **46**, 6789 (1992).
8. W.C. Chen, L.A. Hamel, A. Yelon, accepted for publication in J. Non-Cryst. Solids..
9. W.C. Chen, M. Kemp, L.A. Hamel, A. Yelon, to be published.
10. B.I. Shklovskii, E.I. Levin, H. Fritzsche and S.D. Baranovskii, in Advances in Disordered Semiconductors, Vol 3, ed. H. Fritzsche (World Scientific, 1990), p.161.
11. C.E. Nebel, G.H. Bauer, Phil. Mag. B **59**, 463 (1989).
12. G. Juška, K. Arlauskas, J. Kocka, J. Non-Cryst. Solids, **189-200** 202 (1996).
13. C.E. Nebel, R.A. Street, Phil. Mag. B **67**, 407 (1993).
14. M. Silver, D. Adler, in Optical Effects in Amorphous Semiconductors, AIP Conf. Proc. No. 120, ed. by P.C. Taylor and G. Bishop (Snowbird, Utah, 1984) p.197.
15. M. Kemp, in Amorphous Silicon Technology - 1995, ed. by M. Hack, E.A. Schiff, A. Madan, M. Powell and A. Matsuda (Materials Research Society, Pittsburgh, 1995) p.157.

TWO CARRIER SENSITIZATION AS A SPECTROSCOPIC TOOL FOR a-Si:H

L.F. FONSECA*, S.Z. WEISZ*, R. RAPAPORT** AND I. BALBERG**
*Department of Physics, University of Puerto Rico San Juan 00931, PR
**The Racah Institute of Physics, The Hebrew University, Jerusalem 91904, Israel

ABSTRACT

In a recent letter we have reported the first observation of the phenomenon of minority carrier-lifetime sensitization in hydrogenated amorphous silicon (a-Si:H). We find now that combining the study of this phenomenon with the study of the well-known phenomenon of majority carrier lifetime sensitization, in this material, can provide direct information on its density of states (DOS) distribution. This finding is important in view of the limitations associated with other methods designed for the same purpose. We have carried out then an experimental study of the effect of light soaking on the phototransport in a-Si:H. We found that the increase of the dangling bond concentration with light soaking affects the sensitization and thermal quenching of the majority carriers lifetime. Using computer simulations, we further show that the details of the observations associated with the sensitization effect yield semiquantitative information on the concentration and character of the recombination centers in a-Si:H.

INTRODUCTION

The phenomenon of thermal quenching of the photoconductivity in intrinsic a-Si:H is known for many years [1]. The interpretation of this phenomenon is two-fold. First, it requires that the capture coefficients of the electrons is much smaller for the bandtail states (BTS) than for the dangling bonds (DB), and second, it requires that with increasing temperature (T) the number of BTS available for recombination, P_v, decreases until at some T it equals the concentration of the dangling bonds N_D. While this basic picture is in accord with the results of numerical simulations [2,3], we do not know of a direct experimental proof that establishes this model and that ties this phenomenon with the phenomenon of sensitization. The latter phenomenon is that the addition of excess recombination centers can prolong the lifetime of the charge carriers in a given photoconductor [4]. In this paper we study this effect by applying the tool of light soaking (LS) in order to change the concentration of desensitizing recombination centers in a-Si:H. We do that here in the "reverse mode" to the one used classically [4], i.e. by keeping the total concentration of the sensitizing centers nearly constant and changing the concentration of the desensitizing recombination centers.

After establishing experimentally the above basic picture of the sensitization, we have turned to numerical simulations that enabled us to evaluate the relation between the manifestations of the phenomenon of the sensitization and various parameters of the concentration and capture coefficients of the recombination centers in the mobility gap of a-Si:H. As we show below, these features give further semiquantitative spectroscopic information on the concentration and character of the recombination centers in the material.

Mat. Res. Soc. Symp. Proc. Vol. 557 © 1999 Materials Research Society

EXPERIMENTAL STUDIES

The samples used in the present experimental study were device quality a-Si:H films which were deposited [5] on a Corning 7059 glass and had a thickness of 1.3μm. The temperature of the substrate at deposition was 250°C. Two planar contacts of NiCr were predeposited on the substrate in a configuration that we have described previously [6]. The photoconductivity σ_{ph} was measured as a function of the (HeNe laser) illumination intensity and T, in the T range of 80 $\leq$ T $\leq$ 300K. The latter yielded a carrier generation rate, G, of $5 \times 10^{18} \leq G \leq 5 \times 10^{20} cm^{-3}s^{-1}$. The value of the mobility-lifetime product $(\mu\tau)_e$ of the electrons (the majority carriers) was derived from the relation $(\mu\tau)_e = \sigma_{ph}/qG$ where q is the electronic charge. It was thus implicitly assumed that the carrier generation quantum efficiency is unity and the absorption of the light is uniform. To find the light intensity exponent, γ_e, at a given T, the best fit of the data to a linear relation between log σ_{ph} and log G has been used. The importance of the value of γ_e is, as will be discussed below, that it reflects semiquantitatively some basic features of the recombination centers [4]. In particular, when $\gamma_e > 1$, the T dependence of γ_e yields information on their energy position, their concentration and their capture coefficient. In the present study we have measured the T dependence of $(\mu\tau)_e$ and γ_e, before and after LS of the a-Si:H film. This was done in order to find out what spectroscopic information can be derived from these dependences. In the present study the light soaking was carried out under a HeNe illumination of 100 mW/cm^2 for 10 hours.

The results of our measurements are presented in Fig. 1. In this figure we show the T dependence of $(\mu\tau)_e$ before and after LS. It is clearly seen that before LS there is a clear change in the sign of the slope of the $\sigma_{ph}(T)$ dependence around 150°K. We call the temperature of this change the T of thermal quenching, T_{tq}. After LS the $(\mu\tau)_e$ drops, as expected, and the T_{tq} shifts to a somewhat lower T. This is in agreement with previous experimental results and numerical simulations [2] of the effect of the increase of the DB concentration on the $(\mu\tau)_e$ dependence on T. The sensitization, as mentioned above, is manifested by a $\gamma_e > 1$ value [4]. In the present case we find that, by comparison of the above behavior with the $\gamma_e(T)$ dependence given in the inset of Fig. 1, the sensitization and the thermal quenching are two aspects of the same phenomenon. This is since we see that there is a clear downward shift in the $\gamma_e(T)$ peak as well as a decrease in its amplitude around T_{tq}.with the increase of LS

The above features of the $(\mu\tau)_e$ and γ_e dependences on T are simply explained along the lines of the

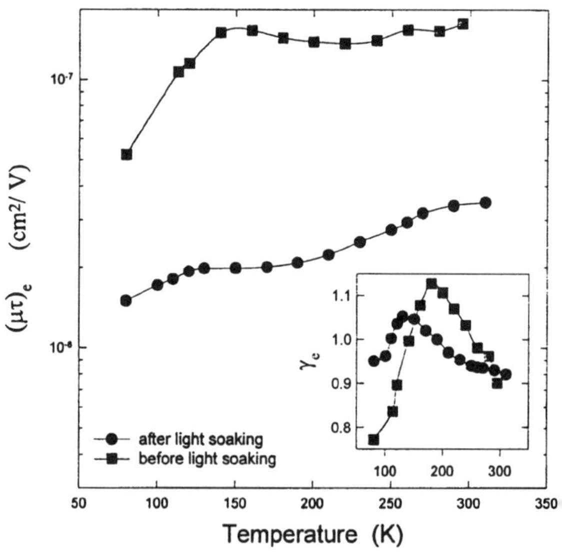

Figure 1. The measured T dependence of the mobility-lifetime product and the corresponding light intensity exponent for a typical sample of intrinsic a-Si:H.

general Rose model [4] as follows. The fact that the $(\mu\tau)_e$ values, before LS, are much closer to each other at low T indicates that the recombination is not dominated there by the recombination centers that are induced by the LS. This is in contrast with the situation at the higher T. Following the well-known observation that LS increases the concentration of the dangling bonds N_D but does not cause an increase in the concentration of the valence band tail (VBT) states [7,8] it is expected that the latter states are responsible for the recombination at low temperatuares. It thus appears that at the higher T the recombination is controlled by the dangling bonds (DB), while at the lower T it is controlled by the VBT states. We consider the VBT states rather than the conduction band tail (CBT) states since the lowering of T causes the shift of the quasi-Fermi level for holes towards the valence band edge thus increasing the concentration of centers available for electron recombination. The analysis of the above data in view of the classical picture of sensitization suggests then that as T is lowered there is a sensitization of the DB states by the VBT states, and that around T_{tq} the conditions for sensitization [4] are fulfilled. As (following the LS) there are more desensitizing DB the conditions for sensitization require more sensitization centers i.e., in our case, the decrease of T. In all of this we must assume that the electron capture coefficient of the BT states is considerably smaller than that of all relevant (known as the D^o and D^+) DB. Hence, all the above results are consistent with our suggested explanation of the T_{tq} and with the observed $\gamma_e > 1$ values. The question arises then whether the data shown in Fig. 1 can yield information beyond the above general picture. To consider this question we will discuss, in the next section, the T dependence of γ_e. This dependence is used since, as can be appreciated from Fig. 1, $\gamma_e(T)$ has more conspicuous features than the $(\mu\tau)_e$ dependence. For example, as far as we know the shape of γ_e has not been considered before. We will find out then the significance of the features in the of $\gamma_e(T)$ dependence by using numerical simulations.

NUMERICAL SIMULATIONS

Let us start by establishing the basic sensitization and thermal quenching picture described above. In Fig. 2(a) we show the T dependence of $(\mu\tau)_e$ for the standard model of a-Si:H with parameters similar [9,10] to those used previously by Tran [2], in his "B1" model, but with four values of N_D. For the highest N_D, we have, below room temperature (RT), almost a T independent $(\mu\tau)_e$, while for the lowest N_D we have a strongly dependent $(\mu\tau)_e$. (The behavior above RT has to do with the thermal excitation of the carriers and will be discussed elsewhere). The above results can be simply interpreted by the fact that the quasi-Fermi levels for electrons and holes E_{Fn} and E_{Fp} embrace all the DB levels in the entire T range so that the concentration of DB centers available for electrons recombination, N_D', does not change in the T range under study. In contrast, the concentration of the VBT states available for electron recombination, $P_v(T)$, is sensitive to the shift of E_{Fp} because of its "movement" through the exponential distribution of the density of states (DOS) in this tail. Hence, when we eliminate the DB (smallest N_D) we get (considering the power-law T-dependence of $P_v(T)$ in an exponential tail [4]) that the recombination is dominated by $P_v(T)$. This interpretation leads naturally to the expected behavior of $(\mu\tau)_e$ for the intermediate values of N_D. For these values there is a T ($\approx T_{tq}$) below which the recombination via the available DB states $N_D'C_{Dn}$ equals the recombination via the VBT states $P_v'C_{vn}$. Here N_D' is the concentration of "hole" occupied DB, C_{Dn} is their capture coefficient, $P_v'(T)$ is the available concentration of holes in the VBT states and C_{vn} is their capture coefficient. When T is lowered so that $P_v'C_{vn} > N_D'C_{Dn}$ the sensitization is completed. The latter is of course true only if $C_{vn} \ll C_{Dn}$ as discussed by Rose [4]. Correspondingly, we expect the transition from a recombination via the DB to a recombination via the VBT yielding the effect of the thermal quenching, as is clearly confirmed by the numerical results seen for the intermediate N_D values. We further see that T_{tq} shifts towards low T with increasing N_D. Again, this is in complete agreement with the above picture since the above condition will be fulfilled then for larger $P_v(T)$ values, i.e. at lower T. Hence, the above picture is consistent with both the experimental results and the numerical simulations.

All the above becomes even much more conspicuous when we consider the T dependence of γ_e. This dependence is shown in Fig. 2(b). When there is no thermal quenching the γ_e values are smaller than unity while when thermal quenching is present a very distinct peak with a $\gamma_e > 1$ value is observed around T_{tq}. This is the well-known manifestation of the sensitization effect [4]. The clear shift of the position of the γ_e peak with increasing N_D to a lower T, as in Fig. 1 establishes the association of this effect with the thermal quenching. We further see that with the increase of N_D the value of γ_e decreases and

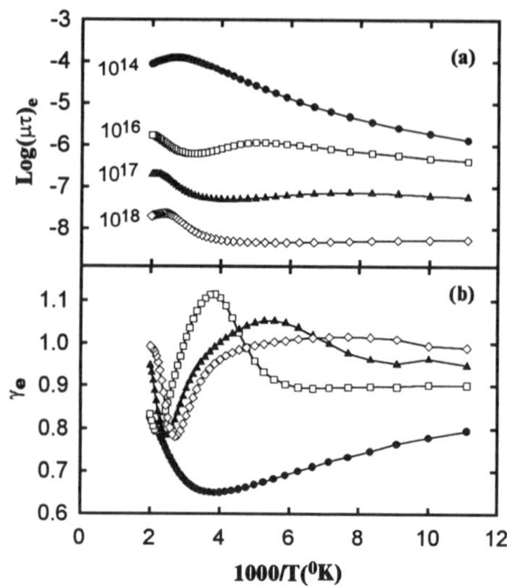

Figure 2. The computed T-dependence of the mobility-lifetime product (a) and the light intensity exponent (b) for four values of the dangling bond concentration (in cm^{-3})

its width, at its peak, increases. Again, this is consistent with the experimental observation shown in Fig. 1, and the above interpretation, since a "wider" section of the VBT (and thus a wider T-range) is required in order to dominate the recombination process. Hence, the basic picture described in the context of Fig. 1 is confirmed by the simulations. On the other hand, the details of the T dependences of $(\mu\tau)_e$ and γ_e can convey many more details regarding the recombination centers in a-Si:H. In the present case the lower and wider the $\gamma_e(T)$ peak the larger the ratio between N_D and the concentration of the VBT states. In the rest of this section we will derive such semiquantitative information from the $\gamma_e(T)$ dependences around T_{tq} showing that one can follow the changes in the electronic structure in a series of samples by just measuring the above easily measured phototransport properties.

Our first check was then to find out the behavior, when the bandtails are eliminated, in order to evaluate the role of the bantails. We see in Fig. 3 that this elimination causes the absence of sensitization and that above about RT $\gamma_e(T)$ is determined by the thermal excitation of carriers from the DB. The fact that we have found a similar result in the presence of the CBT without the presence of the VBT establishes the conclusion discussed above that the VBT states are the sensitizing centers of the DB. Hence, for reasonable parameters which are consistent with all we know about the CBT states, these states are not the sensitizing centers of the DB. We also conclude from the simulations that an increase in the peak value (to $\gamma_e \geq 1$) without a significant shift in T_{tq} indicates a change in $P_v(T)$ without a change in N_D.

To check the effect of the first condition for sensitization, i.e that C_{Dn}/C_{vn} will be large, we carried out a simulation with two values of this ratio. Indeed, as seen in Fig. 4, if this ratio is 3 the value of γ_e does not exceed 1 while if this ratio is 30 it does exceed the value of 1. We note then that since the capture coefficients are not expected to change significantly with preparation conditions, the above interpretation of the variation in γ_e values when a series of samples is studied, is valid.

The value of $P_v(T)$ is also determined by the energy separation of the dark Fermi level E_F and the valence band edge E_v. As to be expected, the smaller the E_F-E_v the deeper the E_{Fp} for a given illumination intensity and T, and thus the larger the $P_v(T)$. Indeed, the comparison made in Fig. 5 of the γ_e values for E_F-E_v = 0.9 eV and E_F-E_v = 1.3 eV reveals this point. In the former we can get

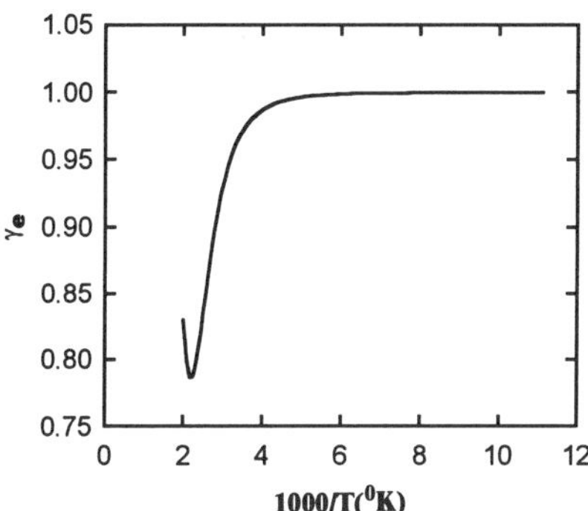

Figure. 3. The computed T-dependence of the γ_e for dangling bonds in the case of negligible concentration of bandtail states.

sensitization with γ_e > 1 while in the latter γ_e < 1. This is important since it indicates, as was found in many experimental studies [11-13], that in doped (n-type) samples no sensitization or thermal quenching will be observed. Since, most of the recombination center parameters are fixed in a-Si:H, we can conclude that the absence of sensitization and thermal quenching is usually a signature of a large E_F-E_v, i.e. of a doped or contaminated sample. Whether this is the reason for the absence of sensitization can be distinguished from the reasons mentioned above, by finding (from the T dependence of the dark conductivity in a corresponding set of samples) whether a change in E_F-E_v took place

From all the above we can further conclude that the appearance of a γ_e > 1 limits the parameter space to an E_F located just above the midgap, to an N_D concentration of about 10^{16}cm^{-3} and to a VBT, of a width of about 0.05 eV and a DOS at the valence band edge of about 10^{21}eV$^-$

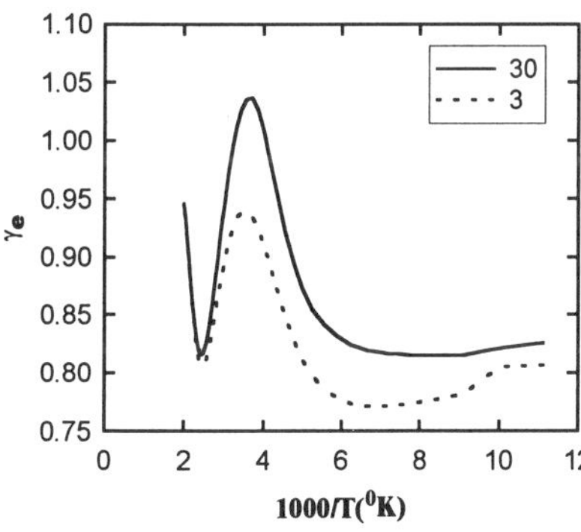

Figure 4. The computed temperature dependence of the light intensity exponent for two values of the C_{Dn}/C_{vn}.

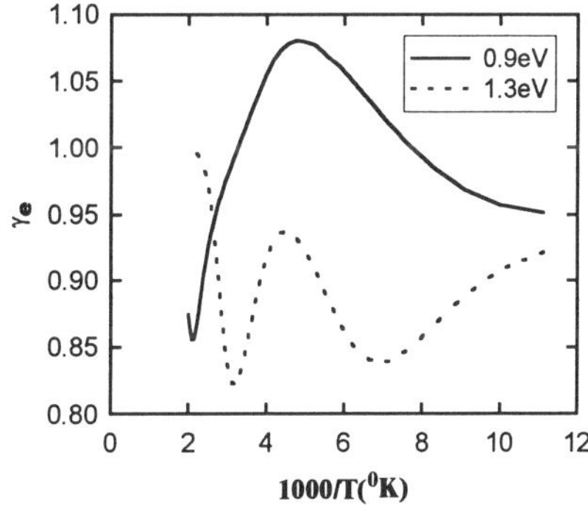

Figure 5. The computed T-dependence of the light intensity exponent for two positions of the Fermi level.

^{1}cm^{-3}. These values are consistent with results derived from very different measurements [2] and thus shows that combining the computer simulation with the experimental features associated with the sensitization can indeed narrow the parameter space of the recombination centers, thus yielding quantitative spectroscopic information on these parameters.

CONCLUSIONS

We have shown that the thermal quenching and sensitization of the electrons' lifetime in a-Si:H follows the Rose model of two types of recombination centers. The VBT states are the sensitizing centers of the DB. The examination of the T dependence of the light intensity exponent in the vicinity of the thermal quenching T reveals then non-trivial information on the recombination centers. This is useful in particular when a series of materials is considered as follows; A shift of the γ_e peak towards low T indicates an increase of the concentration of the DB. A decrease and broadening of the γ_e peak without a T shift indicates a reduction in the concentration of the VBT states, provided there is no significant change in the (easily measured) activation energy of the dark conductivity. Hence, given a series of samples, the simple measurements of the T dependence of the conductivity and photoconductivity can yield semiquantitative information on the recombination centers in a-Si:H. It will be also shown that combining such data with that of the ambipolar diffusion length can yield a more quantitative information.

ACKNOWLEDGMENTS

The authors are indebted to Dr. A. Catalano for the samples used in this study. The present work was supported by US ARO grant No. DAAHO4-96-1-0405.

REFERENCES

1. P.E. Vanier, A.E. Delahoy and R.W. Griffith, J. Appl. Phys. **52**, 5235 (1981).
2. M.Q. Tran, Philos. Mag. B **872**, 35 (1995).
3. R. Bruggemann, Photoconductive Properties, in *Future Directions in Thin Film Science and Technology*, Proc. 9th ISCMS, edited by J.M. Marshall and N. Kirov, (World Scientific, Singapore, 1997), p. 80.
4. A. Rose, *Concepts in Photoconductivity and Allied Problems* (Wiley Interscience, NY, 1963). For a recent manifestation of the sensitization effect see, I. Balberg and R. Naidis, Phys. Rev. B57, R6783 (1998). A semiquantitative theory of this manifestation in a-Si:H has been developed recently: R. Rapaport, M.Sc. Thesis, The Hebrew University, Jerusalem.
5. Y.-M. Li, B.F. Fieselmann and A. Catalano, Proc. of XXII PVSC (IEEE, NY, 1991), p. 123.
6. I. Balberg and Y. Lubianiker, Phys. Rev. B **48**, 8709 (1993).
7. M. Stutzmann, Philos. Mag. B **60**, 531 (1989).
8. W. Graf, K. Leihkamm, M. Wolf, J. Ristein and L. Ley, Phys. Rev. B **53**, 4522 (1996).
9. Y. Lubianiker, I. Balberg and L. Fonseca, Phys. Rev. B **55**, R15997 (1997).
10. R. Rapaport, Y. Lubianiker, I. Balberg and L. Fonseca, Appl. Phys. Lett. **72**, 103 (1998)
11. M. Hoheisel and W. Fuhs, Philos. Mag. B **57**, 411 (1988).
12. B. Cleve and P. Thomas, Mater. Res. Soc. Symp. Proc. **192**, 317 (1990).
13. R. Naidis and I. Balberg, unpublished.

STEP RESPONSE OF a-Si:H PHOTODIODES

M. Mulato, M. Ramón and S. Wagner
Department of Electrical Engineering, Princeton University, Princeton-NJ, 08544-5263, USA

ABSTRACT

We correlate the dark current-transients of a-Si:H p-i-n photodiodes with solar cell performance and quantum efficiency. For devices with solar cell efficiency varying from 7.2 to 2.2 % subjected to ± 1 V and a sampling frequency of 21 kHz: i) the reverse-bias charging time is shorter, with the initial charging time being reduced by ~ 1.5 µs, while the time for final current stabilization is reduced by ~ 5 µs.; ii) the forward-bias time for the onset of space charge limited current dominated regime is ~ 1 µs faster, the final current limit is achieved also in shorter time (~ 9 µs), and the individual final limit current is diminished by ~ 8 % due to carriers trapping, while the relative final limit current between different diodes is diminished by ~ 50 %.

INTRODUCTION

Hydrogenated amorphous silicon (a-Si:H) photodiodes are finding increasing applications in photosensor arrays. Two of the main p-i-n a-Si:H photodiodes characterization techniques are current *versus* voltage in the dark and under illumination, and current-response as a function of the excitation photon wavelength, *i.e.* the quantum efficiency. From these measurements the quality of the i-layer film and also the quality of the doped layers can be inferred [1-3]. Given that the step-response characterization is faster than these techniques, it could find important industrial applications. Also, the diodes current-response to voltage steps becomes important for such photosensor arrays, because this current sets the sensitivity at a given frame rate.

Many researchers have already presented several theories and computer simulations [4-9] and also experimental results [4,7-12] related to this subject. There are also works related to the transient response of other device [13] and silicon structures, such as microcrystalline [14,15] and porous [15,16] materials.

To the best of our knowledge, no correlation of the a-Si:H photodetectors step-response data to solar cell performance and quantum efficiency has been presented. Therefore, we deposited hydrogenated amorphous silicon single layers and p-i-n photodiodes, and independently characterized the single layers and the photodiodes to correlate these data to the devices' pulsed step-response.

EXPERIMENTAL

All single films and p-i-n photodiodes were deposited from dc-excited glow discharge in a three-chamber PECVD system Solarex S900. Single layers were deposited onto Corning 7059 glass and c-Si substrates at 250°C. Specific deposition conditions and thin film typical characterization parameters can be found in ref. [3].

The photodiodes were produced by the deposition of a 16 nm-thick p^+-layer on top of glass (9 inches square) previously covered with a textured transparent conducting oxide (TCO). In sequence, a 550 nm-thick i-layer was deposited, followed by a 21 nm–thick n^+-layer. Evaporated chromium (100 nm-thick, 0.25 cm^2) was used as the back metal contact. The front

Mat. Res. Soc. Symp. Proc. Vol. 557 ©1999 Materials Research Society

contact was achieved by scratching through the film and by attaching a gold wire with silver paste to the TCO. No patterning or structuring procedure was used.

We first characterize the diodes as solar cells using: i) current-voltage (I-V) in the dark and under AM1.5, 100mW/cm^2 illumination; and ii) quantum efficiency (QE) measurements. Next, the step-responses of the photodiodes were measured by connecting the diodes in series with a probe resistor (150Ω) and a square-wave ($\pm$ 1V) pulse generator operated at 21kHz. The time response of the photodiodes was analyzed by measuring the voltage drop across the probe resistor with the use of a digitizing oscilloscope (HP 54201A). All the measurements were performed at room temperature.

RESULTS

The I-V results for three different diodes under light illumination are shown in figure 1. Diode A corresponds to the standard optimized diode [3]. It is a diode sitting exactly in the center of the substrate. Diodes B and C correspond to devices sitting radially toward and close to the edge of the substrate. Because of edge effects in the glow discharge plasma, film thickness and quality vary from the center to the edge of the substrate. The diodes close to the edge have poorer quality. We used them to simulate poor layers' materials. The solar cell efficiencies calculated from the data in figure 1 are 7.2, 3.1 and 2.2 % for diodes A, B and C, respectively.

Figure 2 shows the experimental quantum efficiencies for the three diodes.

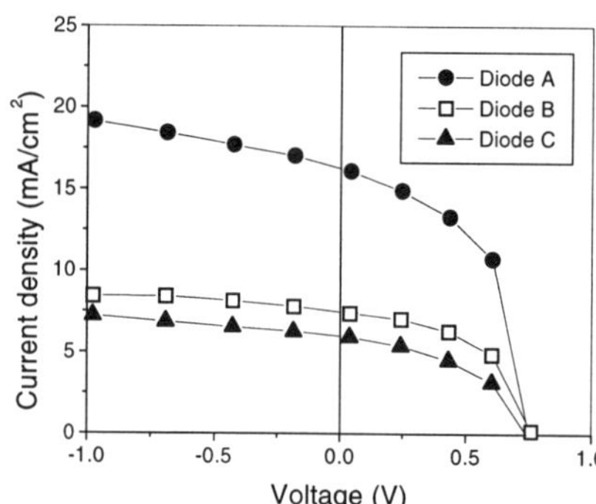

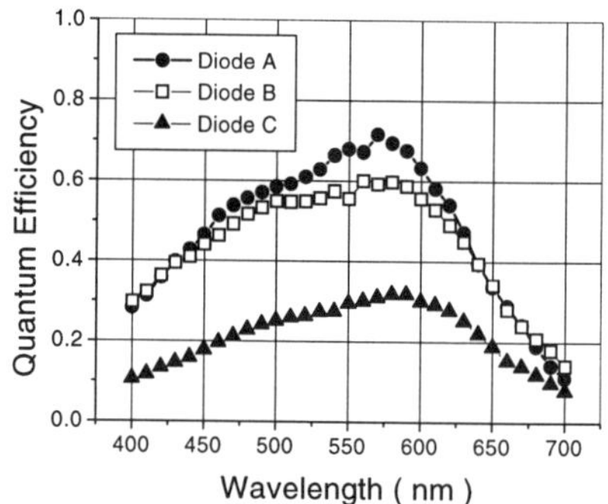

Figure 1 – I-V response of the diodes under light illumination. From top to bottom, the curves represent diodes A, B and C, respectively, which have solar cell efficiencies of 7.2, 3.1 and 2.2 %.

Figure 2 – Current response as a function of the excitation photon wavelength, i.e. the quantum efficiency of the three diodes.

Figure 3 shows the detailed characteristics of the a-Si:H step-response profile (top curve for forward, and bottom curve for reverse biasing). After pulsing from reverse to forward bias (top curve) the current inverts immediately. Then, after a first decrease in the order of ft1 microseconds, follows a signal recovery at ft2 microseconds. A detailed inset on the top-right side of the figure indicates that - in reality - after the second local maximum is achieved at ft2, there is a slow decrease to the final limit value. After pulsing from forward to reverse bias (bottom curve) the current also changes sign quickly. Then its magnitude decays and approaches the final limit value.

In figure 3, in the forward case, ft1 is the time difference between the edge of the applied pulse and the initial current drop. ft2 it the time difference between the edge of the applied pulse and the intersection of two imaginary lines: the first corresponding to the slope of the recovery current, and the second corresponding to the second current local maximum (maximum current value seem in the top-right side inset) [9]. ft3 is the time difference between the edge of the applied pulse bias and the time when the current is only 0.5 % higher than the final half-period limit value. Fmm is the relative difference between the second current maximum and the final current value. fi is the difference between the final current values of two different photodiodes.

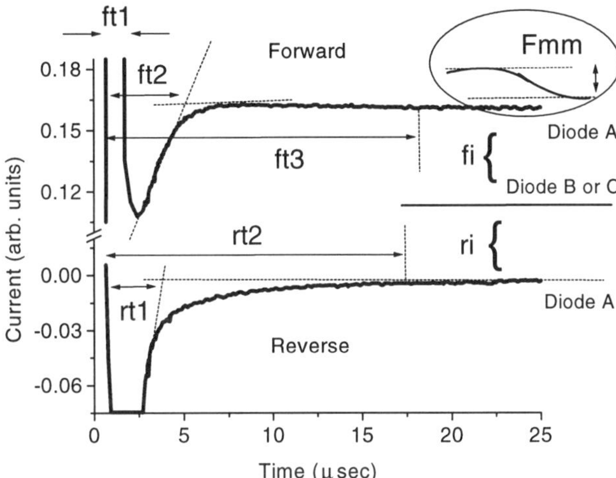

Figure 3 – Detailed step-response of Diode A. Top: after pulse from reverse to forward, and bottom: after pulse from forward to reverse biasing. The sampling frequency is 21kHz. The shown parameters are explained in the text.

In the reverse case, rt1 is the time difference between the edge of the applied pulse bias and the time defined by the intersection of two imaginary lines: the first corresponding to the slope of the decreasing current, and the second corresponding to the final reverse current value. rt2 represents the time difference between the edge of the applied pulse bias and the time when the current is only 10 % higher than the final half-period limit value. ri is the relative difference between the final currents of two different photodiodes. Because the current y-axis in figure 3 is plotted in linear scale, it is not clearly seen that the final reverse current is always different than zero.

DISCUSSION

In the case of the forward pulse, the rapid increase in current (within ns) is followed by a space-charge-limited-current (SCLC) decrease (faster than 1.5 μs, as indicated by ft1). This happens because transport is dominated by electrons, since the effective electron drift mobility is two orders of magnitude higher than the holes. This is illustrated in fig. 4(a). In consequence, an increasing number of electrons reach their back contact, diminishing the potential barrier for holes emission, and increasing hole injection. This is illustrated in fig 4(b). The transition from the SCLC dominated regime to recombination current (RC) dominated regime occurs (as represented by ft2) when enough holes have reached their back contact. This is illustrated in fig.

4(c). After that, a steady-state is achieved (as represented by ft3). The relative difference between the second current local maximum and the final steady state (represented by Fmm) is probably due to an increase in trapped charges, which achieve equilibrium at ft3. This is illustrated in fig. 4(d).

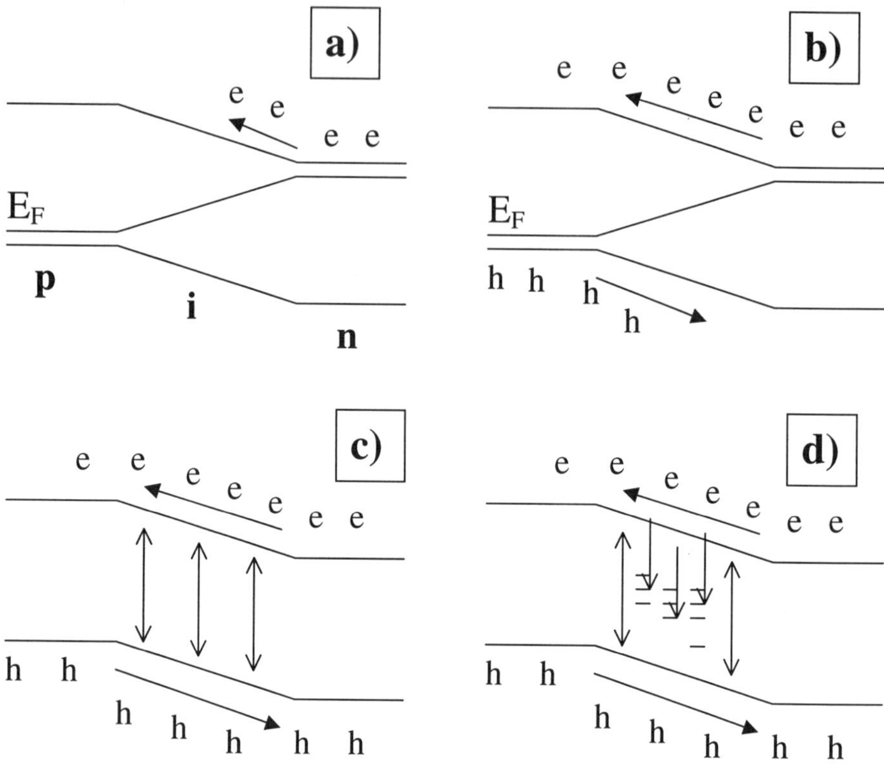

Figure 4 – Energy band diagrams illustrating the different transport mechanisms during the transient forward biasing. E_F stands for the Fermi level, while e and h stand for electrons and holes respectively. a) SCLC dominated regime (less than 1 μs); b) enough electrons have reached their back contact reducing the barrier for holes injection (between 1.5 and 3 μs); c) recombination current starts to dominate (~ 3 μs); and d) trapping of the carriers (also starts above 3 μs, but reaches equilibrium at ~ 20 μs).

The step-response parameters defined in fig. 3 are plotted in figure 5 *versus* solar cell efficiency for the three previously discussed diodes. The error bar associated to each data point is smaller than their printed size.

In the case of pulling to reverse biasing, trapping leads to the charging of the device. The poorer the quality of the material, the higher its capacity for trapping charges, and the shorter the time for this charging (represented by rt1), and the shorter the time to reach its final current value (represented by rt2). Note that this final current value is determined by the experimental sampling frequency, among other effects. It should not be confused with the steady-state thermal generated current, which is know to be reached at longer times [17]. Devices with lower solar cell efficiencies are the ones with poorer i-layer properties, resulting from higher sub-gap defect densities. Decreasing the device quality leads to a decrease in the time to: i) the onset of SCLC (ft1); ii) the end of the SCLC dominated period (ft2); iii) the reverse-bias trapping and charging of the device (rt1); and iv) reaching a "steady-state" regime (rt2, ft3). Also, poorer i-layer quality leads to an increased difference in the final current of different devices (ri, fi).

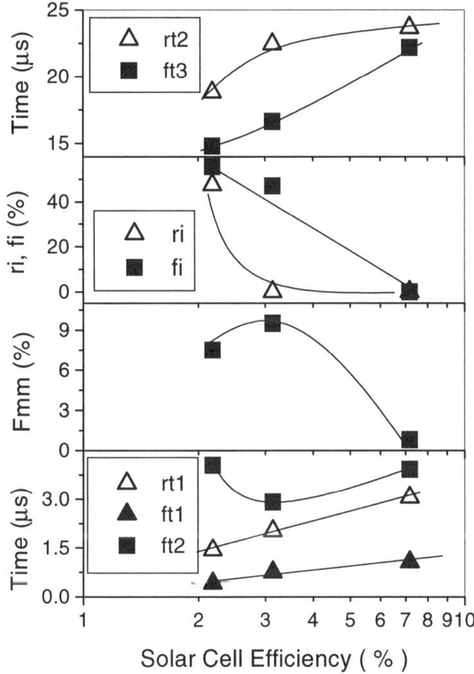

Figure 5 – Step response parameters defined in figure 3 plotted as a function of the solar cell efficiency of the selected photodiodes. The lines are guides to the eyes. The error bars are much smaller than the size of the printed data.

Our results compare well with previous reported data about the double injection transients of p-i-n diodes [4]. The authors conclude that the same transport mechanisms dominate in individual time ranges, with the time for the onset of current recovery for forward-biasing (the change from SCLC to RC dominated regime) being a function of pulse repetition rate, illumination conditions, degradation state of the diode, applied bias, etc. According to their experimental and simulation results this time can vary from μs to ms. They also conclude that degraded metastable defects close to the n^+ contact have a more dominant effect than those close to the p^+ contact. This can be interpreted according to our Fig. 4. Once that electrons injection occur from the n^+ side, defects siting close to this interface will have an important contribution for the trapping of these carriers.

Our experimental data are also in accordance with the ones from ref. [9]. The authors conclude that the product of ft2 by the final limit current for forward-biasing is a constant, for each material, regardless of the applied voltage.

They provide values for 2 and 10 μm-thick devices. Considering our 0.55 μm-thick diode A, we observe a linear dependence of the above product as a function of the optimized-device thickness [18]. Decreasing the device quality (diodes B and C) decreases the value of the above product. Their main conclusion is that ft2 is related to the transit time of the holes, in agreement to our suggestion of the maximum of the RC dominated regime.

CONCLUSIONS

We have studied the transient step-response of a-Si:H photodiodes. For the first time the transient parameters were correlated to solar cell efficiency. We have qualitatively identified the dominant transport mechanisms, and we have also quantified their onset and duration times for a $\pm 1V$ step-voltage biasing. Decreasing the device quality leads to a decrease in the time to both the onset and the end of the SCLC dominated regime. For reverse biasing, worse devices have shorter trapping times, and also achieve limit current values in shorter periods. Their final forward-biasing steady state regime is reached faster. The poorer i-layer quality leads to an increased difference in the final current of different devices.

A computer program linking the experimental data, and device parameters and single layer properties would lead to faster characterizations, what could also be a useful industrial tool for monitoring the quality of the produced devices.

ACKNOWLEDGEMENTS

This work was supported by The Electric Power Research Institute and Intel Corporation. We thank A. M. Payne for experimental help, and H. Gleskova for experimental help and a critical reading of the manuscript.

REFERENCES

1. R. A. Street, Hydrogenated Amorphous Silicon, Cambridge University Press, Cambridge, 1991, p.383.
2. ref 1, p. 367.
3. A. M. Payne, PhD Thesis, Department of Electrical Engineering, Princeton University,1998.
4. M. Hack and R. A. Street, J. Appl. Phys. **72**(6), 2331-2339 (1992).
5. P. Popovic, E. Bassanese, J. Furlan, F. Smole, and I. Skubic, J. Non-Cryst. Solids **191**, 184-192 (1995).
6. M. Hack and J. G. Shaw, Mat. Res. Soc. Symp. Proc. Vol. **219**, 315-320 (1991).
7. R. Carius, F. Becker, R. Brüggemann, and H. Wagner, J. Non-Cryst. Solids **198-200**, 246-250 (1996).
8. F. Lemmi and N.M. Johnson, Appl. Phys. Lett. **74**, 251-253 (1999).
9. B. Yan, G.J. Adriaenssens, A. Eliat, and D. Han, J. Non-Cryst. Solids **190**, 85-94 (1995).
10. G. Schmid, N. Bernhard, and M. Schubert, J. Non-Cryst. Solids **198-200**, 206-209 (1996).
11. D. Han, K. Wang, and M. Silver, J. Non-Cryst. Solids **164-166**, 339-342 (1993).
12. B. Yan, A. Eliat, and G. J. Adriaenssens, Appl. Phys. Lett. **65**(18), 2338-2340 (1994).
13. R.I. Hornsey, K. Aflatooni, and A. Nathan, Appl. Phys. Lett. **70**, 3260-3262 (1997).
14. M. Vieira, Appl. Phys. Lett. **70**, 220-222 (1997).
15. L.P. Kazakova and E.A. Lebedev, Semiconductors **32**, 169-173 (1998).
16. L.P. Kazakova, A.A. Lebedev, and E.A. Lebedev, Semiconductors **31**, 517-518 (1997).
17. R.A. Street, Appl. Phys. Lett. **57**, 1334-1336 (1990). See also ref. [1] p.370.
18. Ifinal x ft2 (C/cm^2)= -4.47 x 10^{-7} + 1.348 x 10^{-6} * d, where d is the film thickness in μm. For our diodes Ifinal x ft2 range from 1.7 x 10^{-8} down to 7 x 10^{-9} C/cm^2, from diode A to C, respectively.

VIBRATIONAL PROPERTIES OF a-Si:H FILMS CONTAINING VOIDS: EXPERIMENT AND MODELING

N. BARRIQUAND*(barri@ramansco.ups-tlse.fr),V. PAILLARD*, P. ROCA I
CABARROCAS**,G. LANDA*, M. DJAFARI-ROUHANI*,¶
*Laboratoire de Physique des Solides, Université Paul Sabatier, 118 route de Narbonne, 31062
Toulouse cedex-4, France, ESA5477 associée au CNRS
**Laboratoire de Physique des Interfaces et des Couches Minces, UMR 7647 associée au CNRS,
Ecole Polytechnique, F-91128 Palaiseau Cedex, France
¶Laboratoire d'Analyse et d'Architecture des Systèmes-CNRS, 7 avenue du Colonel Roche, 31077
Toulouse cedex, France

ABSTRACT

In this paper, we present results about the vibrational properties of hydrogenated amorphous
silicon films. We expect to explain the slight differences observed in the Raman spectra using
atomic-scale modeling. In particular, we focuse on the correlation of our results to the density of
samples. This should give quantitative structural information which could be correlated to both
macroscopic properties and elaboration conditions.

INTRODUCTION

Improving the efficiency of a-Si:H-based solar cells is an actual technological matter. In order
to optimize both elaboration conditions and electronic properties, the aim is still the structural
characterization of a-Si:H films which are known to contain voids and/or clusters, both of
nanometric size. Since these small domains cannot be easily evidenced by conventional structural
characterization technique such as X-Ray diffraction or TEM, vibrational properties can be used
to obtain a structural information. Nevertheless, Raman spectra generally exhibit only slight
changes, that are difficult to attribute unambiguously to a specific structure.
In the following, we show Raman scattering experiments of different a-Si:H films obtained by
plasma-enhanced chemical vapor deposition (PECVD) and calculations of the density of
vibrational states (DVS.) The calculations are carried out on highly disordered silicon with
different structures, modeled through the size of voids and their distribution in the system.
Therefore, we expect a quantitative analysis of the experimental Raman spectra.

EXPERIMENT AND MODELING

Experiments

1 micron-thick a-Si:H films are obtained by PECVD from a silane / argon mixture. As
explained elsewhere [1,2], the strong dilution of silane conducts to materials with improved
electronic transport properties. It has been proposed that the structure of such materials is
different from conventional a-Si:H films, and may contain small ordered domains formed by
silicon clusters deposited from the plasma [1,2]. However, these clusters are very difficult to
evidence experimentally by TEM or X-ray diffraction since their diameter should be less than 2
nm. During elaboration, pressure (60 mTorr), SiH_4/Ar ratio and RF power are kept constant. The
variable parameter is the substrate temperature (T_{sub}) ranging from 100 °C to 250 °C.

Mat. Res. Soc. Symp. Proc. Vol. 557 © 1999 Materials Research Society

The electronic properties of the films are measured by dark conductivity, defect density and Urbach tail. The Raman scattering experiments were performed using a triple monochromator spectrometer (Coderg T800), coupled with a photomultiplier followed by a photon counting device. The 514.5 nm excitation wavelength was provided by an ionized argon laser. The power on the sample, focused through a cylindrical lens in a 1 mm^2 line, was limited to 50 mW. To prevent both air scattering and temperature increase, the samples were kept in helium gas (1 atm.) during the spectrum acquisition.

Modeling

In order to build highly-disordered silicon, we use a supercell approach with periodic boundary conditions (PBC) coupled with a continuous random network (CRN) [3]. An 8% vacancy ratio was introduced, in agreement with some other works [4,5]. Detailed effects of this ratio have been previously studied [6]. Since CRN models give an higher density for amorphous silicon than for crytalline silicon [4], some voids are introduced by choosing a central atom and removing their neighbors inside a given sphere, in order to lower the density of our amorphous system below the c-Si one. The second step is the relaxation of these systems. Anharmonic potential must be used to describe the inter-atomic forces in these kind of system [6,7] due to the high disorder. The last step is the diagonalization of the dynamical matrix (D). In order to smooth the DVS, we have calculated D with different wavevectors in the supercell Brillouin zone (SBZ). The DVS is sufficiently well-defined, with a sampling of $d\omega = 6cm^{-1}$, with ten wavevectors randomly distributed over the SBZ.

To introduce anharmonicity in the system, the usual way is to use the Stillinger-Weber' potential [8]. But it is a cutoff potential and during the relaxation phase, the number of interaction taken into account increases. Indeed, we allow a variation of the supercell volume, which decreases. Then, to overcome this fact, we use the potential defined by Rücker *et al.* [10], which is a variation of the Keating's potential i.e. the number of interaction for one atom is limited to its first and second neighbors (in term of chemical bonds.) This potential is well-checked for group-IV alloys with chemical disorder and the results obtained here are in good agreement with experimental ones.

In the following discussion, we need the inverse participation ratio (IPR) [4,11] to analyze the variation on DVS at low-frquency. The IPR gives an indication about the number of atoms participating to one given mode. The IPR runs from 1, if only one atom participates to the mode, to 1/N if all atoms have the same vibrational amplitude.

RESULTS

Fig.1 shows the Raman spectra recorded on different samples, grown in fixed operating conditions, except the deposition temperature T_{sub}. The Raman spectra do not show any evidence of small crystallites (no sharp phonon peak), but they exhibit rather strong differences. With the decrease of T_{sub}, we observe both a decrease of the optical mode wavenumber and an evident deviation from Debye's law, which is more distinct on the anti-Stokes part of the spectra. This is certainly due to less multiple-order Raman processes compared to the Stokes part. In order to link these changes to the structure of the samples, we carried out calculations with one void of different sizes in a CRN model, and with several voids. This should allow to identify the influence of the size as well as the concentration of voids.

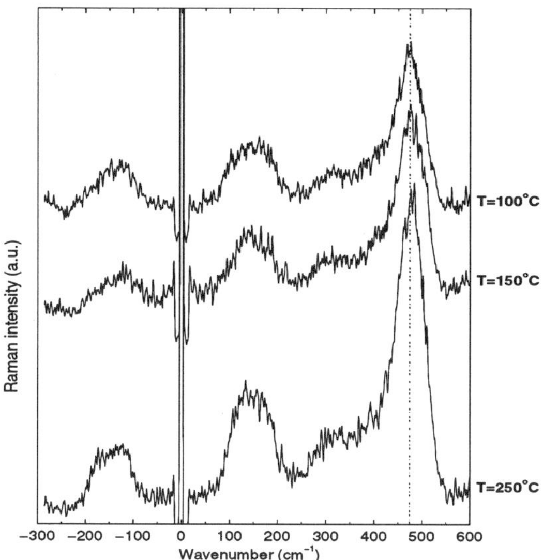

Fig.1. Raman spectra of a-Si:H films versus temperature depositions.

The calculated DVS for different structures are shown in the left part of Fig.2.We observe an obvious correlation between the optical-mode-shift and the film density, whatever the distribution of the voids. There is however one exception for the system with sixteen voids of five atoms (the so-called 16x5 system) which shows the same features as an homogenous system in our original supercell of 432 atoms and where PBC effects are strong.

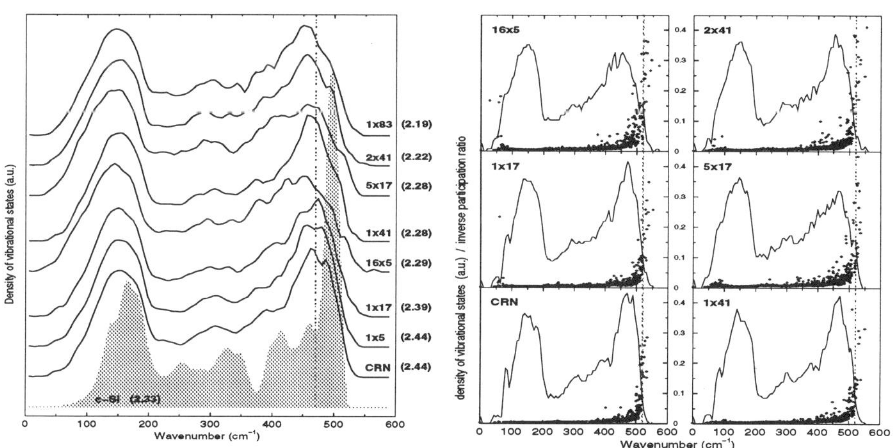

Fig.2. The left part shows Density of vibrational states (DVS) with different sizes and distributions of voids in an initial supercell of 432 atoms with a continuous random network (8%.) 2x41 means that the system contains 2 voids of 41 atoms each. The value under brackets is the mass density (in g.cm^{-3}.) The right figure gives the inverse participation ratio and DVS calculated in the center of the Brillouin's zone.

In agreement with experimental Raman spectra, we obtained a decrease of the frequency of the optical peak with the density of sample.This decrease runs from the CRN to the multiple voids systems (in the 1x83 system, the void is in strong interaction with itself due to its size which is near than dimension of the supercell.)

In order to analyze the variation from the Debye's law, we have calculated the IPR shown in the right part of Fig.2 (calculations are carried out at the center of the SBZ.) We clearly show the correlation between the high IPR-optical modes and the strain in the samples. Indeed, such modes are localized on the compressed bonds and only one or two atoms participate to these modes. We also see that, from the CRN to systems with voids, the modes with high IPR appear at low frequency. For these acoustic modes we cannot clearly associate the low-frequency modes with high IPR to a special structure. Indeed, this seems to depend on the distribution of the voids, i.e. the interaction between them. In fact, these acoustic modes are similar to breathing modes of the void when the supercell only contains one void with a diameter which is smaller than the size of supercell. Therefore, there is no strong interaction between the void and itself. For the structures which contains more than two voids, we obtain breathing modes and / or modes where the participating atoms are localized between voids. Further work is in progress to characterize the exact influence of both size and concentration of voids.

Finally, the Raman spectrum of sample elaborated at the highest temperature corresponds rather well to the DVS of the dense CRN or to systems containing few small voids. As T_{sub} decreases, the Raman spectra correspond to the system which contain large voids and / or high concentration of them. This is in agreement with the electronic transport properties shown in Table I. Indeed, it has been published that the highest semiconducting properties are caused by a low concentrations of small voids[12].

T_{sub}	dark conductivity (S.cm)	defect rate (cm^{-3})	Urbach tail (meV)
250	5×10^{-10}	1×10^{16}	57
150	4×10^{-12}	1×10^{16}	57
100	5×10^{-13}	2×10^{17}	97

Tab. I : Electrical properties of a-Si:H films. The film quality of sample decreases with the substrate temperature T_{sub} during elaboration.

CONCLUSION

The present results are the first step of a systematic quantitative correlation between Raman spectra, the film structure and the electronic properties. The experiments show a decrease of the optical-mode frequency with T_{sub}. which corresponds to a better organization of the sample when the deposition temperature increases. Similar agreement has been obtained by atomic-scale modeling. We link the evolution of the optical-mode frequency as a function of the density which depends on the void rate. The dependency of low-frequency modes with the void-distribution will be specifically analyzed in further work. In particular, the influence of the hydrogen-ratio and the nature of hydrogen bonds will be included in the modeling.

ACKNOWLEDGMENTS

This work was supported by the CNRS and the ADEME (Agence de l'Environnement et de la Maîtrise de l'Energie) under the Ecodev program.

REFERENCES

1. P. Roca i Cabarrocas, P. St'ahel, S. Hamma and Y. Poissant, Proc. of the 2nd World Conf. photovoltaic Solar Energy Conversion, Vienna, Austria, 1998, thin Fiml Cells and Technol., p. 680-683.
2. P. Roca i Cabarrocas, S. Hamma, S.N. Sharma, G. Viera, E. Bertran and J. Costa, *J. of Non-cryst. Sol.*, **871**, p. 227-230 (1998).
3. F. Wooten and D. Weaire, *Sol. State Phys.*, p. 1,32 (1987).
4. A. Chehaidar, M. Djafari Rouhani and A. Zwick, *J. of Non-cryst. Sol.*, **192**, p.238-242 (1995).
5. M.G. Duffy, D.S. Boudreaux and D.E. Polk, *J. of Non-cryst. Sol.*, **15**, p.435-454 (1974).
6. F. Finkemeier and W. von Niessen, *Phys. Rev. B*, **58**, p. 4473,4483 (1998).
7. J.L. Feldman, P.B. Allen and S.R. Bickham, *Phys. Rev. B*, **59**, p. 3551,3559 (1999).
8. F.H. Stillinger and T.A. Weber, *Phys. Rev. B*, **31**, p. 5262,5271 (1985).
9. P.N. Keating, *Phys. Rev. B*, **145**, p. 637,645 (1966).
10. H. Rücker and M. Methfessel, *Phys. Rev. B*, **52**, p. 11059,11072 (1995).
11. J.L. Feldman, M.D. Kluge, P.B. Allen and F. Wooten, *Phys. Rev. B*, **48**, p.12589,12601 (1993).
12. X. Liu, B.E. White, R.O. Pohl, E. Iwanizcko, K.M. Jones, A.H. Mahan, B.N. Nelson R.S. Crandall and S. Veprek, *Phys. Rev. Lett.*, **78**, p.4418,4421 (1997).

INFRARED ELECTROABSORPTION SPECTRA IN AMORPHOUS SILICON SOLAR CELLS

J. H. LYOU[a] AND ERIC A. SCHIFF
Department of Physics, Syracuse University, Syracuse, NY 13244-1130

STEVEN S. HEGEDUS
Institute of Energy Conversion, University of Delaware, Newark, DE 19716-3820

S. GUHA AND J. YANG
United Solar Systems Corporation, Troy, Michigan 48084

ABSTRACT

We report measurements of the infrared spectrum detected by modulating the reverse-bias voltage across amorphous silicon *pin* solar cells and Schottky barrier diodes. We find a band with a peak energy of 0.8 eV. The existence of this band has not, to our knowledge, been reported previously. The strength of the infrared band depends linearly upon applied bias, as opposed to the quadratic dependence for interband electroabsorption in amorphous silicon.

The band's peak energy agrees fairly well with the known optical transition energies for dangling bond defects, but the linear dependence on bias and the magnitude of the signal are surprising if interpreted using an analogy to interband electroabsorption. A model based on absorption by defects near the *n/i* interface of the diodes accounts well for the infrared spectrum.

INTRODUCTION

Electroabsorption is the change in optical absorption due the application of a voltage bias across a sample. It is thus a form of modulation spectroscopy, and can offer insights not readily available from more direct optical measurements. Electroabsorption was first developed to study the band structure of crystalline solids. In amorphous semiconductors electroabsorption measurements corresponding to interband optical transitions indicate that electric fields warp the wavefunctions of nearly localized electronic states near mobility edges [1], the effect is quadratic in the applied electric field. Investigators have explored the relationship between electroabsorption and the mobilities that govern photocarrier transport in the bands [2] as well as the possibility of light-soaking effects on electronic states at the bandedges [3]

There has been relatively little work done on electroabsorption corresponding to defect-related optical properties. The defect-related analog of the interband effect just described would be several orders of magnitude weaker, since the absorption coefficient for transitions involving mid-band gap states is far lower than the interband absorption, and would presumably share its quadratic dependence upon electric field. While this type of electroabsorption has not been observed, defect-related electromodulation was reported by Eggert and Paul [4] in forward-biased a-Si:H diodes. In this case, electric-field induced warping of electronic states is no longer responsible for the electromodulation, and the effect is not quadratic in electric field. Changes in defect occupancy in the intrinsic layer due to injection of space-charge by forward bias lead to the readily observable changes in infrared optical properties. The measurements are closely related to sub-bandgap *photomodulation* measurements, which have been extensively studied in the last twenty years [5]

---

[a] Permanent address: Department of Physics, Korea University, Chungnam 399-800, South Korea.

Mat. Res. Soc. Symp. Proc. Vol. 557 © 1999 Materials Research Society

In the present work we present measurements of sub-bandgap optical properties in *reverse biased* diodes; injection currents are negligible, and hence the Eggert & Paul mechanism is not applicable directly. We nonetheless find a readily measurable infrared electromodulation band with a prominent maximum near 0.8 eV in several samples. Based primarily on the fact that this band depends linearly upon the electric-field across the device, we attribute it to changes in infrared absorption due to depletion of bandtail states near the interface of the strongly doped, n^+ layer and the intrinsic layer. The linearity of the signal derives from the linear dependence on reverse bias of the space-charge stored near the two interfaces.

This electromodulation band has several interesting features. First, its principal feature at about 0.8 eVis fairly narrow (roughly 0.1 eV), which we believe to be a direct measurement of the width of the D^- (negatively charged dangling bond) band in n-type a-Si:H. This electromodulation feature is quite similar to the photoluminescence spectrum for n-type a-Si:H. The narrowness can be understood using the proposed model. The final states involved in electromodulation are near the Fermi energy of the n^+ material; they are not spread throughout the conduction band of the material as for most other forms of sub-bandgap spectroscopy. We also observe a broader spectral feature at 1.1 eV, which we have not identified.

The present paper is a preliminary account of the infrared measurements. If the proposed interpretation is correct, then infrared electromodulation spectra should be an interesting diagnostic of interfaces in semiconductor diodes. Additionally, the infrared spectrum is an interesting complement to the known interband electroabsorption spectrum and the infrared photomodulation spectra. We hope to exploit these features in subsequent work.

SAMPLES AND MEASUREMENT DETAILS

Two types of sample have been used in the experiments. The first type consisted of Schottky diodes prepared at the Institute of Energy Conversion at the University of Delaware; the diodes were deposited in the sequence glass/SnO$_2$/a-Si:H:P/a-Si:H/ITO. The diodes had an absorber layer thickness of 0.50 µm and an area of 0.40 cm^2. The second type of sample consisted of *nip* solar cells prepared at United Solar Systems Corp.; these cells were deposited in the sequence stainless-steel/Ag/a-Si:H:P/a-Si:H/µc-Si:H:B/ITO. The cells had an absorber layer thickness of 0.225 µm and an area of 0.25 cm^2.

Electromodulation spectra (EA) were measured by applying a sinusoidal bias potential across a sample and measuring the corresponding modulation of the transmission or reflection of an optical beam. The beam itself came from a small illuminator/monochromator with 1 nm resolution. The beam intensity was detected using either a Si or an InGaAs photodetector. We typically used a 10 kHz modulation frequency. We employed a computer to record the electroabsorption spectrum; the spectra reported here are the average of many such spectra taken as rapidly as possible. In general the actual spectrum reported is for the modulation ratio of the optical intensity; the actual amplitude of the modulated detector photocurrent has been divided by the unmodulated photocurrent. This procedure yields a relative modulation spectrum δT/T or δR/R. Some additional details may be found in prior publications from our research group [6,7].

We comment briefly on the modulated reflection spectra on the *nip* solar cells prepared on stainless steel substrates. For longer wavelengths, most of the incident light passes through the top interfaces and propagates to the rear of the sample, where it is reflected by the silver "back reflector." For this reason the "modulated reflection" spectra are sensitive to bulk electroabsorption processes as well as to true "electroreflectance" effects.

EXPERIMENTAL RESULTS AND DOPED LAYER ABSORPTION MODEL

In Fig. 1 we present measurements of the modulated reflectance $\delta R/R$ for a *nip* cell from United Solar, along with a cartoon illustrating the proposed origins of the two components of the spectrum. Both the infrared band (peaking near 0.8 eV) and the visible band (peaking near 1.8 eV) are affected by interference fringes; these effects are much reduced in the measurements shown subsequently.

We emphasize the very different dependences of the two bands upon the DC value of the reverse bias voltage. The behavior of the visible band is fairly well known: it results from interband electroabsorption, and follows the approximate form:

$$\delta R\big/_R \approx 4\alpha''(h\nu)\delta E(V - V_{bi}) \tag{1}$$

where δE is the modulation field, V is the applied DC bias, V_{bi} is the built-in potential across the diode. $\alpha''(h\nu)$ is the electroabsorption coefficient of the *intrinsic* layer of the diode, which depends upon optical energy and polarization. In writing this expression, we assume that only illumination which is reflected from the back surface of the solar cell is detectable; we neglect several refinements involving interference fringes and electroabsorption in the doped layers.

The strength of the infrared band is essentially independent of the applied DC potential across the cell. Although not shown here, it is proportional to the amplitude of the reverse bias voltage δV across the cell. It is also important to note that the cells act as a nearly perfect capacitor under reverse bias at the illumination levels of the experiment.

The only model which we have found to account for these measurements invokes the optical properties of the space-charge which flows into the doped layers of the diode as the bias voltage

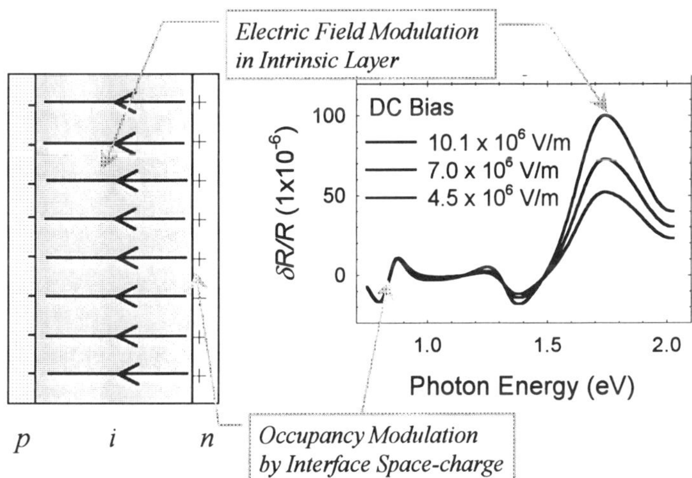

Fig. 1: To the right of the figure we illustrate measurements of the electroreflectance modulation $\delta R/R$ resulting from a reverse bias modulation $\delta V = 1.0$ V (rms). The different currents represent differing DC reverse bias levels as indicated. The annotation and cartoons indicate the well-established model for interband electroabsorption (peak at 1.8 eV) as well as the proposed model for the infrared spectrum.

is modulated. In particular consider the consequences of *depleting* the *n*-type layer in the diode. The resulting positive surface charge density modulation $\delta\sigma = \varepsilon\varepsilon_0(\delta E)$ in the *n*-layer corresponds to removing electrons from levels lying at Fermi energy 0.1-0.2 eV below the conduction band edge. The depletion should have two effects. This depletion will *bleach* any optical absorption originally associated with these electrons, and it will *induce* optical absorption by opening up levels for optical transitions originating from electrons in deep-levels or the valence band. It is worth commenting that such transitions presumably involve localized initial and final states.

We do not provide a more quantitative analysis here, noting only that one would expect the strength of the induced optical absorption to depend strongly on the extent of doping in the phosphorus doped layer. The model that the infrared effect originates with the *n*-type layer is based on the comparable spectrum we found in the Schottky diode samples from the Institute of Energy Conversion (IEC). We present corresponding measurements in Fig. 2. This sample was measured using transmittance, and did not exhibit the pronounced interference effects found in the *nip* solar cell from United Solar Systems Corp.. Additionally, the Schottky diode has no doped *p*-type layer. Since the spectrum is comparable in location and magnitude to that measured in the cell from United Solar Systems Corp., we infer that the *n*-type amorphous layers present in both diodes account for the spectrum.

DISCUSSION

In Fig. 3 we illustrate the level diagram for *n*-type a-Si:H which we invoke to account for the infrared electromodulation spectrum. As is evident from the figure, the 0.8 eV peak position (cf. Fig. 2) is consistent with the identification with an optical transition of an electron between the D^- state of a dangling bond defect (in phosphorus-doped a-Si:H) and a state near the conduction band edge; previous electrical and luminescence measurements have fairly consistently placed the (0/-) level of this defect at 0.9 eV below the conduction bandedge[8]. The

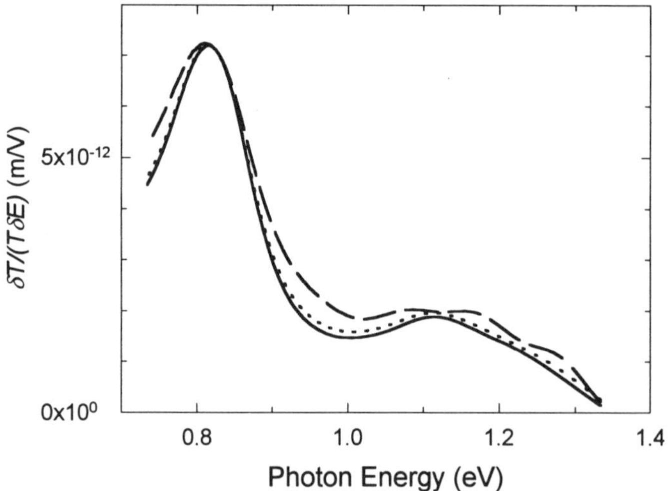

Fig. 2: Infrared electromodulation spectra for a Schottky diode structure from the Institute for Energy Conversion; from lowest to highest, the spectra were measured at 3.6, 4.5, and 15×10^6 V/m. The diode incorporates an a-Si:H:P n^+ layer. The model described in the text explains the spectrum in terms of induced defect absorption in this doped layer.

detailed mechanism is reminiscent of the mechanism proposed previously for the infrared luminescence in doped a-Si:H, which involves the radiative capture of a bandtail electron by a dangling bond:

$$B^- + D^0 \rightarrow B^0 + D^- + h\nu,$$

where B denotes a bandtail state and D denotes a dangling bond. It may seem sensible to presume that the infrared absorption is just the reverse transition invoked to account for the infrared photoluminescence.

While we have not yet completed a more quantitative analysis, we wish to note that this model may be an inadequate explanation for the infrared electromodulation spectrum. Implicit in the argument above is the idea that the bandtail and defect states are uncorrelated states which happen to occur in close proximity. If this view is the correct one, and if depletion near the Fermi level is sufficient to induce a defect-related spectrum, then we would anticipate a far larger induced spectrum corresponding to optical transitions from valence-band states to the conduction bandtail. No such signal is found in our measurements.

We therefore note an alternative: the infrared spectrum may reflect an internal transition of a neutral defect-dopant complex $D^- P_4^+ + h\nu \rightarrow D^0 P_4^0$. In this view, depletion of the n-layer would neutralize complexes which were originally negatively charged $D^- P_4^0$, thereby inducing the observed infrared spectrum.

We have not yet found a plausible identification for the relatively weak absorption near 1.1

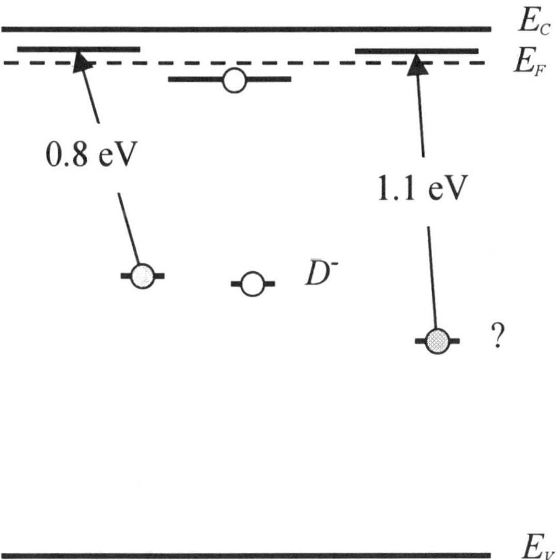

Fig. 3: Level diagram indicating the proposed identification of the 0.8 eV infrared electromodulation band as a transition of an electron from a negatively-charged dangling bond (D^-) to a bandtail state. One difficulty with this model is noted in the text, along with an alternative interpreation of this diagram invoking dangling bond - dopant complexes.

eV in Fig. 2. It is possible that   this absorption band belongs to a previously unidentified interfacial defect or complex; we hope that experimental work on additional samples will clarify its origin.

ACKNOWLEDGMENTS

This work has been supported through the Thin Film Photovoltaics Partnership of the National Renewable Energy Laboratory. The authors thank Nikos Kopidakis for discussions.

REFERENCES

1. S. Al Jalali and G. Weiser, J. Non-Crystalline Solids **41**, 1 (1980); G. Weiser, U. Dersch, and P. Thomas, Phil. Mag. **B 57**, 721 (1988).
2. H. Okamoto, K. Hattori, and Y. Hamakawa, *J. Non-Cryst. Solids* **137&138**, 627 (1991) and **164-166**, 893 (1993).
3. L. Jiang, Q. Wang, E. A. Schiff, S. Guha, and J. Yang, *Appl. Phys. Lett.* **72**, 1060 (1998).
4. J. R. Eggert and W. Paul, *Phys. Rev. B* **35**, 7993 (1987).
5. L. Chen, J. Tauc, J.-K. Lee, and E. A. Schiff, *Phys. Rev. B* **43**, 11694 (1991) and references therein.
6. Q. Wang, E. A. Schiff, and S. S. Hegedus, in Amorphous Silicon Technology-1994, edited by E. A. Schiff, M. Hack, A. Madan, M. Powell, and A Matsuda (Materials Research Society, Pittsburgh, 1994), Vol. **336**, p. 365.
7. Lin Jiang, Qi Wang, E. A. Schiff, S. Guha, J. Yang, Xunming Deng, *Appl. Phys. Lett.* **69**, 3063 (1996).
8. R. A. Street, *Hydrogenated Amorphous Silicon* (Cambridge University Press, Cambridge, 1991), p. 130.

CAPACITANCE SPECTROSCOPY OF DEFECTS IN a-SI:H/c-SI HETEROSTRUCTURES

M. RÖSCH, T. UNOLD, R. POINTMAYER, G.H. BAUER
Fachbereich Physik, Universität Oldenburg, 26111 Oldenburg, F.R.Germany

ABSTRACT

We investigate defects at the interface in heterodiodes of hydrogenated amorphous silicon and monocrystalline silicon by frequency and temperature dependent capacitance measurements. The interpretation of the experimental results is supported by numerical simulations of capacitance experiments via transient calculations of defect charging and decharging in the diodes. A defined variation of waver surface treatments prior to amorphous silicon deposition shows a clear correlation of interface defects determined by capacitance measurements with current-voltage characteristics.

INTRODUCTION

Heterodiodes of hydrogenated amorphous silicon (a-Si:H) and crystalline silicon (c-Si) are of interest for photodetectors and for solar cells, which have been shown to yield efficiencies up to 20% in this device structure [1]. The treatment of the wafer surface with wet and plasma etching prior to the deposition of the amorphous silicon films has been reported to be a very important step in the diode preparation [2]. Also, the introduction of a few nanometer thick intrinsic a-Si:H layer between the c-Si and the doped a-Si:H has been shown to increase the device efficiency [3]. In several studies it has been concluded that these surface treatments influence the defect density at the heterojunction interface [4,5]. Defects at the heterojunction interface have been studied by capacitance measurements in doped a-Si:H/c-Si diodes [4,5] and also in intrinsic a-Si:H/c-Si Schottky structures [6].The interface defects may substantially affect the current transport across the heterojunction and therefore the device quality.The aim of this work is the study of the defect density at the heterojunction interface and their effect on the transport properties of the diodes. We will show that low frequency capacitance measurements are a sensitive tool to study such interface defects. By systematically varying wafer surface cleaning conditions and comparing capacitance measurements, numerical simulation and IV curves, we will demonstrate how interface defects influence the capacitance and the transport properties of the heterodiodes.

EXPERIMENTS

Heterodiodes were prepared by a-Si:H deposition on p-doped 1Ωcm float zone silicon wafers (100) with aluminium back contacts. A series of 5 samples with different surface treatments was prepared for the defined variation of surface quality, and interface defect density. For all samples the oxide overlayers of the wafers were removed by wet etching with buffered HF. For some diodes a H-plasma was applied to the wafer surface, and on some samples a 5nm thick intrinsic a-Si:H layer was deposited onto the wafer by using a standard DC glow-discharge system (see Table I). Finally on all samples a highly n-doped (1% PH_3/SiH_4) a-Si:H film was deposited. Aluminium was evaporated as a semitransparent front contact. Frequency and temperature dependent capacitance measurements (100Hz-1MHz) were performed using a HP-

Mat. Res. Soc. Symp. Proc. Vol. 557 © 1999 Materials Research Society

4192A impedance bridge and a liquid nitrogen cryostat. Current-voltage measurements were performed in the dark and under illumination using a electrometer and filtered tungsten-halogen light.

Table I: Heterodiodes prepared for this study

sample	H-plasma	(i)-a-Si:H layer	$(n)^+$-a-Si:H
1		5 nm	60 nm
0	-	-	100 nm
H	Yes	-	100 nm
I	-	5 nm	95 nm
HI	Yes	5 nm	95 nm

CAPACITANCE MEASUREMENTS

In figure 1 the temperature and frequency dependent capacitance is shown for sample 1. At temperatures below 150 K a capacitance step of 0.23 nF is observed. The temperature at which the step occurs shifts to higher temperatures for higher measurement frequencies. At higher temperatures the high-frequency capacitance increases little, whereas for low frequencies the increase is much bigger.

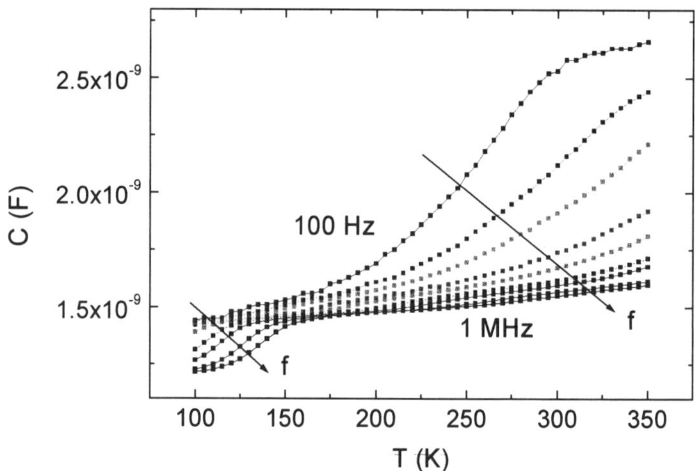

Fig 1 Temperature dependent capacitance of sample 1 at different frequencies.

The step below 150 K is due to the low conductivity of a-Si:H at these temperatures, which causes the dielectric relaxation frequency to be smaller than the voltage modulation frequency. The value of 0.23nF corresponds well to the geometrical capacitance calculated for the 65 nm thick a-Si:H layer. The big increase of the low-frequency capacitance can be explained with the charging and decharging of defects. At a given defect energy, E_d, the defects contribute only at modulation frequencies lower than the thermal emission frequency ν_{emiss}

$$v_{emiss} = v_0 \exp\left(-\frac{E_d}{kT}\right) \qquad (1)$$

where v_0 is the attempt-to-escape frequency, typically 10^{12}-10^{13} s^{-1}. That is, if the defect states are located deep enough, they affect the capacitance much more at low frequencies at a given temperature than at high frequencies. Also, because the capacitance increases over a broad temperature range, it cannot be caused by the activation of one discrete defect level, but must be caused by a broad distribution of defects in the gap. But where in the diodes are the defects located? Since the defect density in the float-zone wafer is very low, we don't expect these defects to contribute significantly to this increase in the low-frequency capacitance. Consequently, either the defects in the n$^+$-a-Si:H, for which we estimate the defect density at approximately $N_d = 10^{19}$ cm^{-3} from optical absorption measurements, or defects at the heterojunction interface must contribute to the capacitance. In order to understand which of these two possibilites is correct, we have carried out numerical simulation which are reported in the next section.

SIMULATION OF CAPACITANCE

Model

Our numerical simulation contains the Poisson equation, the continuity equation, and current transport including drift, diffusion, and thermionic emission over the barriers in the valence and the conduction band. Subband gap states are introduced by assuming exponential band tails and a quasi-continuous distribution of amphoteric dangling bond states. Emission and capture of free charge carriers from bands into defect states is taken into account. The simulation program is based on a program previously used for calculations of transient phenomena in a-Si:H pin diodes and films [7],[8]. The program has been adapted to heterostructures by including thermionic emission currents over barriers.

For the simulation of capacitance we calculate a stationary solution at a given bias voltage and then switch the applied potential ($\Delta V = 50$ mV) so that emission or capture of charge carriers can be observed. The time dependent calculation starts with times smaller than the dielectric relaxation time in the bulk materials. Then, the transient current, $I(t)$, is calculated up to times, when a stationary value of $I(t)$ is reached. The imaginary part of the admittance, which is interpreted as the product of the capacitance and the frequency ω, is obtained by the ratio of the complex Laplace transform of $I(t)$ and $V(t)$.

Comparison of simulation with experiments

In order to distinguish between the influence of defects in the amorphous silicon and defects at the interface two different structures have been modeled. For both structures the (n$^+$)-a-Si contains a gaussian defect band ($N_d=10^{19}$ cm^{-3}) around midgap and tail states with characteristic tail energies of 130 meV and 200 meV at the conduction and the valence band, respectively. For the c-Si 10^{12} cm^{-3} defects located at midgap have been assumed in order to yield a realistic diffusion length of several hundred μm. In one of the structures a 1 nm thick defect layer at the crystalline side of the heterojunction between a-Si:H and c-Si has been included with bandtails and a gaussian defect peak at midgap qualitatively similar to those in a-Si:H.

In fig. 2 the measured frequency-dependent capacitance is compared with the calculated capacitance for heterostructures with and without the extra defect layer. It can be seen that the simulated capacitance of the structure without a interface defect layer is nearly independent of

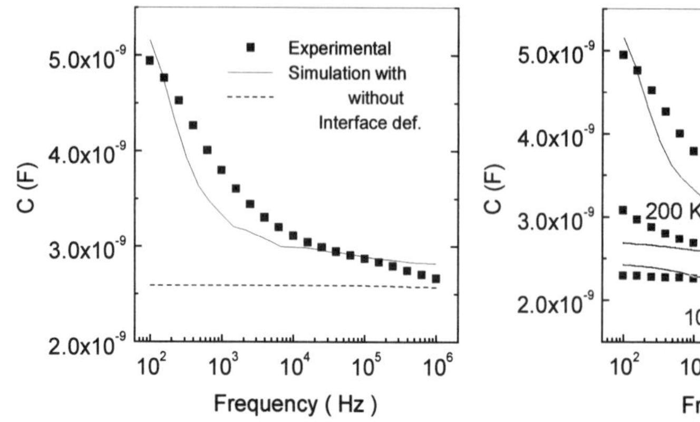

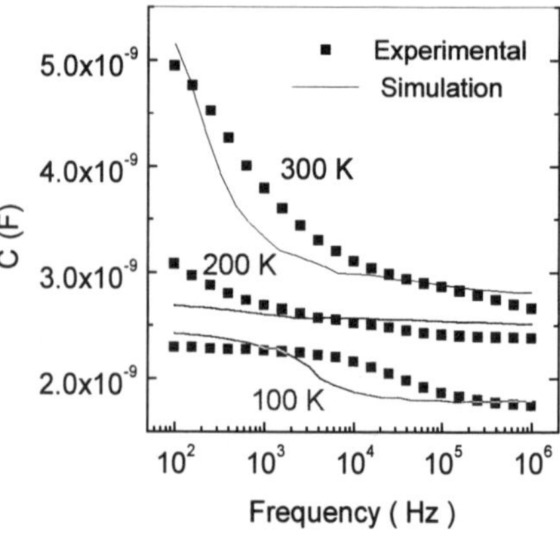

Fig 2: Comparison of experimental capacitance with simulation with and without interface defects at 300 K

Fig 3: Comparison of experimental capacitance with simulation at different temperatures

frequency whereas for the other structure the simulated capacitance increases for low frequencies in good agreement with the experimental data. This clearly indicates that not the defects in the n-doped a-Si:H but those at the heterojunction interface are responsible for the increase in capacitance at low frequencies.

The frequency-independence of the capacitance for the first structure can be understood if we consider that the position of the Fermi-level in the (n)-a-Si:H is quite shallow (E_c-E_f = 0.25 eV at 300 K) and that therefore the charge in deep defects is nearly unchanged by applying a small-signal voltage. If there is an additional defect layer the situation changes. This is because the position of the Fermi level is substantually shifted at the a-Si:H/c-Si interface and therefore allows for a variation of the occupation of defects. Because of the dependence of the emission rate on the energetic depth of the defects (equation 1), the contribution of deep defects to the capacitance signal increases at low measurement frequencies. We also find that higher magnitudes of the interface defects densities in the simulation lead to higher low frequency capacitance values. Good agreement of measured and calculated data for different temperatures (fig. 3) was reached with an interface defect distribution that consists of bandtails with characteristic energies of 0.1 eV at E_v and E_c and a gaussian defect peak at 0.45 eV above E_v corresponding to a total surface defect density of $5 \cdot 10^{12} cm^{-2}$.

CAPACITANCE AND I-V CHARACTERISTICS

In fig. 4 capacitance vs. frequency measurements for samples with different surface treatments are shown. As can be seen, the sample without special treatment (sample 0) has the largest capacitance values at low frequencies indicating that in this sample the interface defect density is the highest. The samples additionally etched with a H-plasma (samples H, and HI) exhibit lower capacitance values. The smallest low-frequency capacitance, and hence the lowest interface defect density, is found for sample I, which contains an intrinsic a-Si:H layer and has seen no H-plasma etching. This might be due to a passivation of the surface defects by the deposition of intrinsic a-Si:H. We also conclude that the H-plasma has a cleaning effect on the wafer surface, because we find less interface defects for sample H than for sample 0. However,

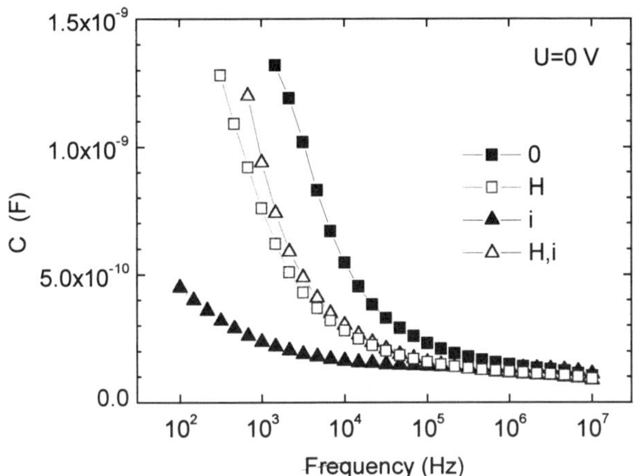

Fig 4: Frequency dependent capacitance at 300 K for samples with different wafer treatment

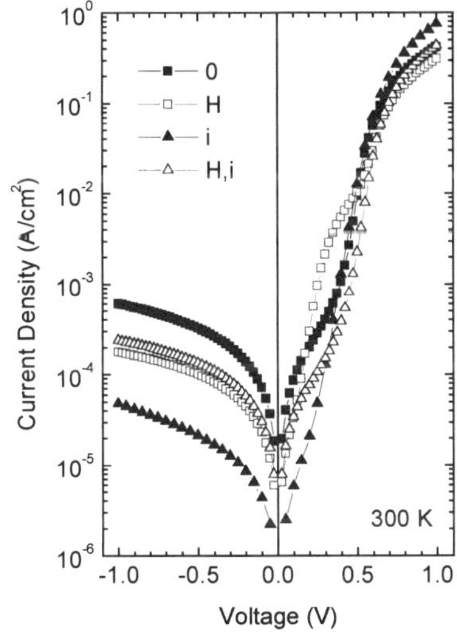

Fig. 5: Dark I-V curves for samples with different wafer treatment (300 K)

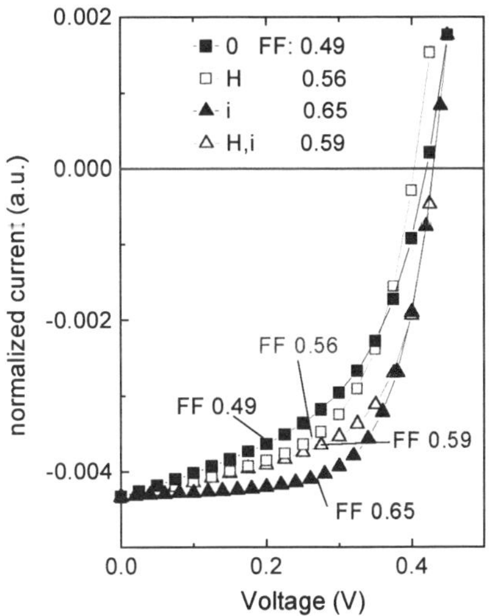

Fig. 6: I-V curves under illumination for samples with different wafer treatment (300 K)

it is also likely that the hydrogen plasma introduces some damage to the wafer, possibly through ion bombardment, since sample HI has more interface defects than sample I.

The influence of interface defects on the dark I-V curves is shown in figure 5. It can be seen that the samples with higher interface defect densities exhibit higher reverse currents as well as higher forward currents at low bias voltages. This shows that the charge transport in the heterodiodes is significantly influenced by the interface defects, e.g. by hopping in subgap states as suggested by a multi tunneling capture emission model [9]. Interface states have also a detrimental influence on I-V curves under illumination. As can be seen in figure 6, the fill factors decrease from 0.65 to 0.49 with increasing interface defect densities for the different diodes.

CONCLUSION

We have shown that defect states at the (n)-a-Si:H/(p)-c-Si heterojunction interface can be characterized by temperature and frequency-dependent capacitance measurements. The absolute values of the defect densities are obtained from comparison with numerical simulation. Qualitatively, the defect densities deduced from the capacitance measurements of a series of several differently prepared diodes have been shown to correlate with the transport properties found for the same samples.

ACKNOWLEDGEMENTS
The authors acknowledge helpful discussions with Dr. Rudi Brüggemann and financial support from the Universitäts-Gesellschaft Oldenburg e.V.

REFERENCES

[1] T. Sawada, N. Terada, S. Tsuge, T. Baba, T. Takahama, K. Wakisaka, S. Tsuga, S. Nakano: Proc. of the 1st World Conf. On Photovoltaic Energy Conversion (IEEE, Piscataway, 1994), p. 1219
[2] B. Jagannathan, W.A. Anderson, Solar Energy Materials and Solar Cells **44**, p.165- 176 (1996)
[3] M.Tanaka, M. Taguchi, T. Matsuyama, T. Sawada, S. Tsuda, S. Nakano, H. Hanafusa, Y. Kuwano, Jpn. J. Appl. Phys. **31**, p. 3518-3522, (1992)
[4] D. Caputo, G. de Cesare, A. Nascetti, F. Palma, R. De Rosa, M. Tucci, Proc. of the 2nd World Conf. On Photovoltaic Energy Conversion (Vienna,1998), in press
[5] B. Jagannathan, W.A. Anderson, J.Appl. Phys. **82**, p. 1930 (1997)
[6] J.M. Essick, J.D. Cohen, Appl. Phys. Lett. **55**, p. 1232-1234 (1989)
[7] R. Brüggemann, Solid State Phenomena **44-46**, Part 1, p.505 (1995)
[8] R. Brüggemann, C. Main, G.H. Bauer, MRS Symposium Proceedings **258** (MRS, Pittsburg), p. 729 (1992)[9] H.Matsuura, T. Okuno, H. Okushi, K. Tanaka, J.Appl. Phys. **55**, p. 1012-1019 (1984)

OPTICAL TRANSITIONS IN LIGHT-EMITTING NANOCRYSTALLINE SILICON THIN FILMS

T. TOYAMA, Y. KOTANI, A. SHIMODE, S. ABO, H. OKAMOTO
Department of Physical Science, Graduate School of Engineering Science, Osaka University
Toyonaka, Osaka, 560-8531 Japan

ABSTRACT

Optical transitions in nanocrystalline Si (nc-Si) thin films with different mean crystal sizes ranging from < 2 nm to ~3 nm have been studied by electroreflectance (ER) spectroscopy. At 293 K, ER signals are observed at 1.20–1.37 eV to be corresponding to fundamental gap in bulk crystalline Si. With a decrease in the mean crystal sizes of nc-Si, the transition energy of the fundamental gap is increased and the ER signal is intensified. The bandgap widening would be due to quantum confinement (QC) in nc-Si, and the increased signal indicates appearance of direct transition nature. The ER signals are also observed at 2.2 eV and at E_1 (E_0') direct gap of 3.1–3.4 eV, while photoluminescence (PL) peak energies are located at 1.65–1.75 eV and at 2.3 eV. With the reduced mean crystal size, the 1.7-eV PL peak energy is also increased, suggesting that QC is also responsible for the increased PL peak energy.

INTRODUCTION

Nanocrystalline Si (nc-Si) is one of the promising materials for optoelectronic devices [1]. We have investigated thin-film light emitting diodes (TFLEDs) using plasma-deposited p-type nc-Si as a light emitting active layer [2], because the thin-film devices have many advantages (e.g. low cost fabrication on a large-area substrate) which have been already realized on amorphous Si (a-Si:H) devices [3]. In our previous reports, we have fabricated the nc-Si TFLED with a structure of SnO$_2$/p-type nc-Si/Al, and studied electroluminescence (EL) properties of the TFLEDs in conjunction with photoluminescence (PL) properties of the nc-Si thin films [2]. For further improvement of the luminescent efficiency from material aspects, characterization of the electronic energy structure of nc-Si should be required. In general, optical transitions measured by spectroscopic techniques provide a rich source of information on the electronic energy structure of semiconductors. Regarding porous Si which is the most intensively studied nc-Si, optical transitions have been mainly studied by optical transmittance or photoluminescence excitation (PLE) spectroscopy [4]. The obtained transmittance and PLE spectra, however, show featureless broadwide tails. Therefore it should be introduced an experimental approach with high resolution as well as high sensitivity for extraction of structures from the broadwide lineshape. Electroreflectance (ER) spectroscopy is an appropriate tool because the obtained lineshape is related to the derivative of the unperturbed dielectric function, thereby the structure-less background is largely reduced and sharp peaks should appear at the interband critical points. In addition, spectral structures with respect to confined systems (e.g. excitons, subband states in quantum wells) can be extracted by use of the ER spectroscopy [5].

In this article, we present experimental results of the electroreflectance (ER) measurements performed on the nc-Si thin films [6]. We discuss ER features observed at 1.2–1.4 eV, 2.2 eV, and 3.1–3.4 eV with a particular emphasis on the ER feature at 1.2–1.4 eV in comparison with data on crystal sizes and PL peaks of the nc-Si thin films.

Mat. Res. Soc. Symp. Proc. Vol. 557 © 1999 Materials Research Society

EXPERIMENT

The *p*-type *nc*-Si thin films were deposited on SnO_2-coated glass substrates by rf plasma chemical vapor deposition (p-CVD), then anodized in HF aqueous solution [2]. The crystal size was changed mainly by the deposition conditions (rf power and gas mixture ratio) and slightly decreased by the anodizing current density. The optical transmittance of the samples used here was found to be largely increased near 2 eV. The deposition temperature of 180°C and pressure of 133 Pa were maintained during the deposition. A mixture of SiH_4, B_2H_6 and H_2 gases was used as a source gas. The fluoride acid was diluted by de-ionized water into 2.4 vol.%. The typical anodizing current density and time were 13 mA/cm^2 and 1 min, respectively. The thicknesses of the *nc*-Si thin films were roughly estimated as 0.7–1.4 µm after anodization. Finally, 80-nm thick indium tin oxide (ITO) was evaporated at a substrate temperature of 200°C in O_2 atmosphere on the *nc*-Si thin film as a transparent front electrode.

Experimental configuration of the ER measurement was a standard setup [4]. Monochromatic light from tungsten-halogen lamp or xenon lamp was applied on a sample, then the reflection light was detected by a Si or InGaAs photodiode. A dc bias voltage, V_{dc} (2.5–4 V) and a modulation ac voltage, ΔV_{pp} (5–8 V, $\Delta V_{pp} = 2V_{dc}$) were applied. We set the ITO electrode on the *nc*-Si layer as the ground. A modulation frequency was set as 500 Hz. The ER signal, $\Delta R/R$, was obtained from the reflectance component, R, and the modulation component, ΔR. For the PL measurements, Ar$^+$-ion laser light (488 nm) was applied through the ITO layer onto the *nc*-Si film. The detail experimental setup for PL measurements has been described in our previous report [9]. Raman scattering measurements were performed by micro-Raman system (JEOL SYSTEM-1000) including Ar$^+$-ion excitation laser (514.5 nm).

RESULTS AND DISCUSSION

Mean Crystal Size

Before showing ER results, let us explain the mean crystal size of nanocrystals, L_0, deduced from the Raman lineshapes at TO mode. In Fig. 1, in associated with experimental results on the *nc*-Si thin films, downshift in center of Raman scattering frequencies from that of bulk crystalline Si (*c*-Si), $\Delta\omega$, are plotted against the full width at half maximum (FWHM) of the Raman spectrum, Γ_{TO}. Also plotted in Fig. 1 are experimental data on porous Si [7]. Broadening of the linewidth and downshift of the peak frequency of *nc*-Si are both theoretically correlated with the crystal size in terms of the phonon confinement model [8]. The influence of the change in inter-atomic distance is not taken into account. As shown in Fig. 1, the theoretical line is fitted to the experimental data both on the *nc*-Si thin films and on porous Si in the range of Γ_{TO} of 20–60 cm^{-1} which is corresponding to L_0 of 4–2 nm. By contrast, the Ra-

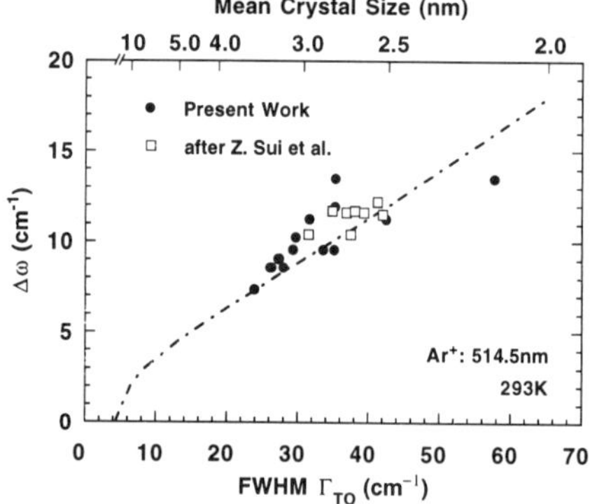

FIG. 1. Downshift in center of Raman scattering frequencies from *c*-Si ($\Delta\omega$) plotted against FWHM of Raman spectrum (Γ_{TO}) for *nc*-Si thin films (●). Open squares are data on porous Si after Z. Sui *et al.* in Ref. 7. Dashed-dotted line is obtained from a theoretical model as a function of mean crystal size of nanocrystals represented at upper horizontal axis.

man lineshape with $\Gamma_{TO} > 60$ cm^{-1} cannot be expressed by the model mainly due to appearance of amorphous component at low frequency, so that $L_0 < 2$ nm estimated by this way should involve large inaccuracy.

ER Spectrum

Figure 2 shows the 293-K ER spectra of the *nc*-Si thin films with different crystal sizes. The mean crystal sizes, L_0, are estimated to be < 2 nm (a), 2–3 nm (b), and ~3 nm (c), respectively. All of the three ER spectra show ER features at 1.2–1.4 eV, i.e., just above the fundamental gap of *c*-Si (1.12 eV at 293 K), and those at 3.1–3.4 eV corresponding to the direct E_1 (and/or E_0') gap in *c*-Si [9]. The E_1 transitions have been found in porous Si and p-CVD as-deposited *nc*-Si thin films by means of spectroscopic ellipsometry or electrolyte ER spectroscopy [10].

The ER features at 1.2–1.4 eV depend distinctly on L_0 as found in Fig. 2. With a decrease in L_0, the ER transition energy increases, and the ER signal is largely intensified. The relationship between the reflectance modulation and the corresponding modulation of dielectric function is expressed by [4]:

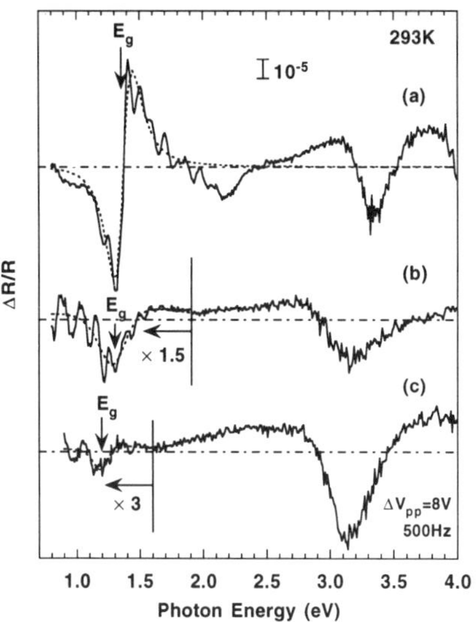

FIG. 2. ER spectra of *nc*-Si thin films with different mean crystal sizes [(a); < 2 nm, (b); 2–3 nm, (c); ~3 nm]. ER spectra are measured at 293 K. Dotted lines are least-square fitting results to Eq. (2). Allows exhibit fundamental gaps, E_g.

$$\frac{\Delta R}{R} = \alpha \Delta\varepsilon_1 + \beta \Delta\varepsilon_2, \tag{1a}$$

with the Seraphin coefficients α and β defined by

$$\alpha = \frac{\partial \ln R}{\partial \varepsilon_1}, \quad \beta = \frac{\partial \ln R}{\partial \varepsilon_2}, \tag{1b}$$

where ε_1, ε_2 denote real and imaginary parts of the dielectric function, respectively, and $\Delta\varepsilon_1$, $\Delta\varepsilon_2$ denote their modulation components. If the dielectric function is approximated by a Lorentzian lineshape, the ER signal, $\Delta R/R$, can be expressed by the following equation [4]:

$$\frac{\Delta R}{R} = \mathrm{Re}\left[Ce^{i\theta}\left(E - E_g + i\Gamma\right)^{-m}\right], \tag{2}$$

where E denotes the photon energy of the probe light, E_g the transition energy, Γ the broadening factor, C and θ the amplitude and phase factors depending only weakly on E. The exponent $m = 2$ is used for the confined system [4]. As found in Fig. 2, the theoretical lines are well fitted to the experimental results. Thereby E_g is determined as 1.37 eV (a), 1.30 eV (b), and 1.20 eV (c), respec-

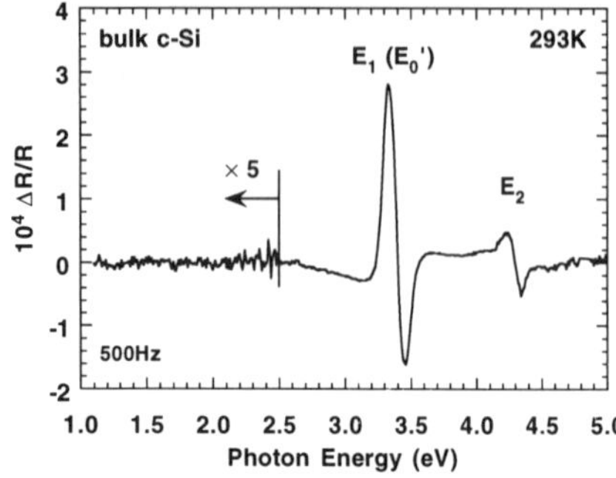
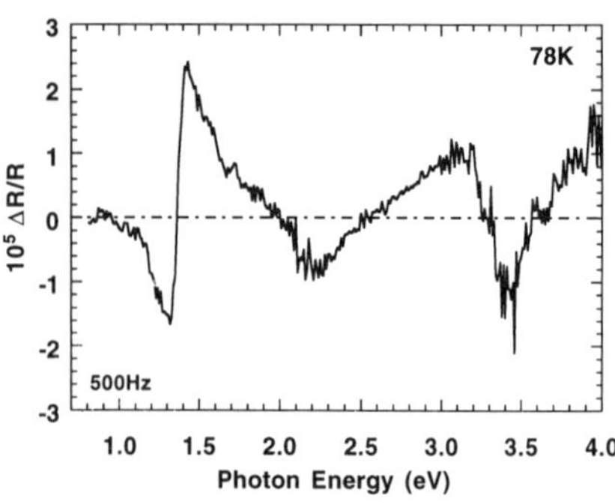

FIG. 3. ER spectra of bulk *c*-Si at 293 K. ER features appear at direct gaps; E_1 (E_0') gap at 3.4 eV and E_2 gap at 4.2 eV.

FIG. 4. ER spectrum of *nc*-Si thin film used in Fig. 2 (a) taken at 78 K.

tively. The blueshift of E_g against the decreased L_0, although it is relatively smaller than that of the optical bandgap determined by the optical transmittance or PLE spectra [4], leads us to an important conclusion that the ER features arise from the optical transitions at the fundamental gap in *nc*-Si, and the bandgap widening would be actually due to quantum confinement (QC). In addition, the increased amplitude of the ER feature with the reduced L_0 strongly indicates that the transition nature changes from indirect to direct-assisted probably owing to mixing of the wave functions at different k-points which might be induced by the break of translational symmetry arisen from the finite size [11]. Such an intense ER feature as found in Fig. 2 has never been observed at the fundamental gap in *c*-Si. For the confirmation, we have measured the ER spectrum of bulk *c*-Si employing a *p*-type c-Si/ITO Schottky-like diode as shown in Fig. 3. No particular ER response is found at the range of 1–2.5 eV including indirect transitions such as Γ-X (the fundamental gap of *c*-Si) and Γ-L transitions. By contrast, two ER features corresponding to direct gaps, i.e., E_1 (E_0') gap at 3.4 eV and E_2 gap at 4.2 eV, are clearly observed.

From the fitting results, Γ of 0.1–0.2 eV is obtained, and the obtained Γ seems quite huge. As shown in Fig. 4, the huge Γ was observed in the ER spectrum taken at 78 K, so that thermal broadening would be excluded from the dominant mechanism of the broadening. The quite huge broadening factor is likely to be caused by the crystal size, L, distribution in the *nc*-Si thin films. The presence of the L-distribution in the *nc*-Si thin film agrees with observation of the cross-sectional views of the *nc*-Si thin films by the transmittance electron microscope (TEM) measurement. Since the single-electron energies depend strongly on L, the L-distribution introduces a pronounced inhomogeneous broadening of the observed spectrum. This has been theoretically identified for the optical absorption spectrum of CdS nanocrystals [12]. Applying contribution of the L-distribution to E_g, E_g should be described as a function L. E_g deduced from the ER spectrum would represent $E_g(L_0)$, i.e., the fundamental gap of the nanocrystals with a size of L_0.

Dependence of Optical Transition Energy on Crystal Size

In the following section, we discuss the relation between the PL peak and the fundamental gap. In Fig. 5, PL spectra of the samples used in Fig. 2 are displayed. The PL spectrum of the sample (a) consists of two Gaussian components with peak energies of 1.75 eV and 2.28 eV. In the

PL spectra of the samples (b) and (c), the higher energy band is not clearly found, thus the PL spectra are fitted by a Gaussian component with peak energies of 1.69 eV (b) and 1.65 eV (c), respectively. The 1.7-eV PL peak energy, E_{PL}, is increased with the decreased L_0 as we have reported [2]. Moreover, as mentioned above, E_g is also increased with the decreased L_0. In Fig. 6, both E_{PL} and E_g are plotted against Γ_{TO} (or L_0). The energy difference, E_{PL} - E_g, is almost unchanged (~0.4 eV) or slightly decreased with a decrease in L_0.

The PL peaks being ~0.4 eV larger than E_g may be explained by a following model in terms of the L-distribution [13]. The PL lineshape is described as a function of the oscillator strength, f_{osc} ($\propto L^{-\beta}$), the upshift of E_g due to QC, ΔE_g ($\propto L^{-x}$, $x = 1.4$–2), the volume of the single nanocrystal, V ($\propto L^n$, $n = 2$ for column or $n = 3$ for sphere), the mean crystal size, L_0, and the variance of the L-distribution, σ_L (now assuming the L-distribution as a Gaussian function [13]). If f_{osc} depends strongly on L, say $\beta = 6$, and remaining parameters are set as typical values obtained from porous Si, $E_{PL} > E_g$ is concluded. Even though the

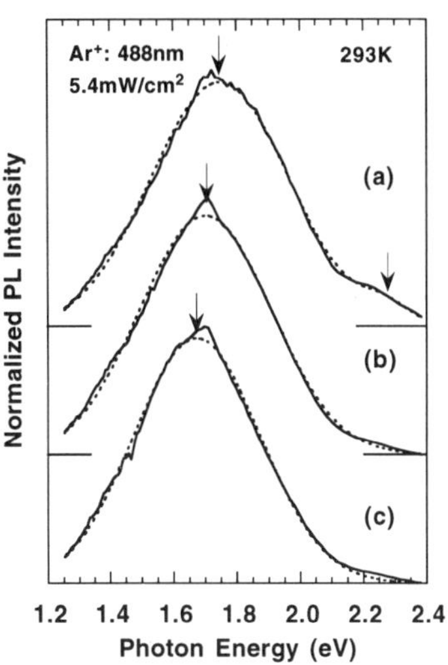

FIG. 5. PL spectra of samples used in Fig. 2 taken at 293 K (solid lines) excited by Ar$^+$ ion laser ($\lambda = 488$ nm, $P_{in} = 5.4$ mW/cm^2). Dashed lines are fitting results to two (a) or one ((b), (c)) Gaussian component(s) of which peak energies are exhibited by arrows.

emission energy is almost equal to the bandgap in a single nanocrystal, the mean value, E_{PL}, of the integrated emission from the whole nanocrystals with various crystal sizes could be different from the mean value, E_g, of their bandgaps in the above condition. Consequently, the relation of $E_{PL} > E_g$ found in Fig. 6 implies that PL of the nc-Si thin film is essentially attributed to QC with f_{osc} which depends strongly on L. The energy difference, E_{PL} - E_g, however, seems too large to be explained by this model alone, so that the additional emission mechanisms, i.e., recombination via states which are insensitive for the electric-field-modulation, might be responsible for spectral parts being much larger than E_g in the broadwide PL.

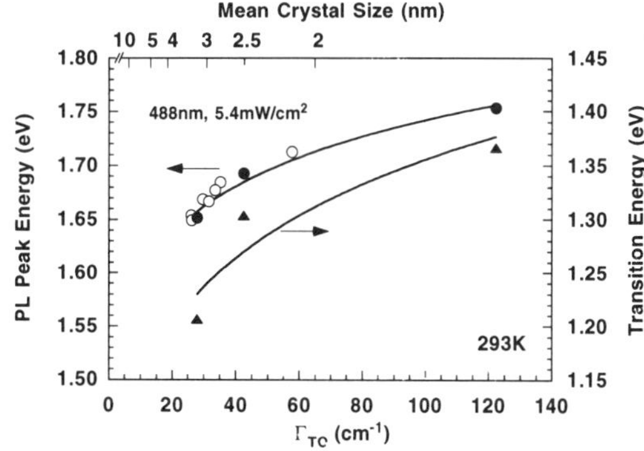

FIG. 6. PL peak energies (circles) and transition energies (triangles) as a function of FWHM of Raman TO band spectrum (Γ_{TO}). Upper horizontal axis is mean crystal size of nanocrystals as shown in Fig. 1. Solid symbols represent data obtained from samples (a), (b) and (c) in Figs. 2 and 5. The lines are a guide for the eye.

We briefly describe a weak feature found at 2.2 eV in the ER spectrum in Fig. 2 (a). Two mechanisms would be responsible for the 2.2-eV transition. The first mechanism is that the transition would be attributed to amorphous region surrounding nanocrystals because we have observed the amorphous component in the Raman spectrum of the sample (a) as described before. Employing the diode structure as used here for electroabsorption (EA) measurement of a-Si:H, the EA signal appears near optical absorption edge due to reflection from the back-side contact through the a-Si:H layer [14]. The alternative mechanism is Γ-L transition in Si nanocrystals, if the transition nature changes from indirect to direct-assisted as discussed for the fundamental gap. The 2.2-eV transition would correspond to the PL small peak at 2.28 eV shown in Fig. 5, so that the 2.3-eV PL band would be interband emission.

CONCLUSION

We have studied optical transitions in nc-Si by means of ER spectroscopy in conjunction with crystal sizes and PL peaks. The optical transitions of nc-Si are found at 1.20–1.37 eV, 2.2 eV and 3.1–3.4 eV. When the mean crystal sizes are decreased, upshift of the fundamental gap of 1.20–1.37 eV and an increase in the amplitude of the ER signal occur, strongly indicating the bandgap widening due to QC as well as appearance of direct transitions probably due to mixing of the wave functions. Optical transitions corresponding to dominant 1.7-eV PL band has not been observed in the ER spectra, while the PL peak being located above the fundamental gap by ~0.4 eV is shifted toward blue when the mean crystal size is decreased. A likely emission mechanism for the 1.7-eV PL band has been demonstrated in terms of a QC model including the crystal size distribution.

REFERENCES

1. D. J. Lockwood, *Light Emission in Silicon From Physics to Devices, Semiconductors and Semimetals*, vol. 49, edited by R. K. Willardson and E. R. Weber (Academic Press, New York, 1997) and references there in.
2. T. Toyama, Y. Kotani, A. Shimode, K. Shimizu, and H. Okamoto, Mater. Res. Soc. Symp. Proc. **507**, 243 (1999), T. Toyama, Y. Kotani, H. Okamoto, and H. Kida, Appl. Phys. Lett. **72** 1489 (1998).
3. R.A. Street, *Hydrogenated Amorphous Silicon* (Cambridge University Press, Cambridge, 1991).
4. V. Lehmann and U. Gösele, Appl. Phys. Lett. **58**, 856 (1991), Y. Kanemitsu, Phys. Rev. B **48**, 12357 (1993), A. Kux and M. Ben Chorin, Phys. Rev. B **51**, 17535 (1995).
5. P. Y. Yu and M. Cardona, *Fundamentals of Semiconductors*, (Springer, Berlin, 1996), F. H. Pollak, in *Optical Properties of Semiconductors*, *Handbook on Semiconductors Completely Revised Edition*, edited by T. S. Moss, (Elsevier Science, Netherlands, 1994) vol. 2, p. 527, D. E. Aspnes, Surf. Sci. **37**, 418 (1973).
6. T. Toyama, Y. Kotani, A. Shimode, and H. Okamoto, to appear in Appl. Phys. Lett.
7. Z. Sui, P. P. Leong, I. P. Herman, G. S. Higashi, and H. Temikin, Appl. Phys. Lett. **60**, 2086 (1992).
8. H. Richter, Z. P. Wang and L. Ley, Solid State Commun. **39**, 625 (1981), I. H. Campbell and P. M. Fauchet, Solid State Commun. **58**, 739 (1986).
9. P. Lautenschlager, P. B. Allen, and M. Cardona, Phys. Rev. B **31**, 2163 (1985) and references there in.
10. C. Pickering, in *Porous Silicon*, edited by Z. C. Feng and R. Tsu (World Science, Singapore, 1994) p. 3, A. Daunois and D. E. Aspnes, Phys. Rev. B **18**, 1824 (1978), V. I. Gavarilenko, J. Humlicek, N. I. Klyui, and V. G. Litovchenko, phys. stat. sol. (b) **155**, 723 (1989).
11. L. Brus, J. Phys. Chem. **98**, 3575 (1994), M. S. Hybertsen, in *Porous Silicon Science and Technology*, edited by J.-C. Vial and J. Derrien (Springer, Berlin, 1995) p. 67.
12. H. Haug and S. W. Koch, *Quantum Theory of the Optical and Electronic Properties of Semiconductors* (World Scientific, Singapore, 1990) p. 332.
13. V. Ranjan, V. A. Singh, and G. C. John, Phys. Rev. B **58**, 1158 (1998).
14. Y. Hamakawa, in *Hydrogenated Amorphous Silicon, Semiconductors and Semimetals*, vol. 21B, edited by J. I. Pankove (Academic Press, New York, 1984) p. 141.

COLLECTION EFFICIENCIES GREATER THAN UNITY BY ELECTRON OR HOLE GATING IN a-Si:H p-i-n DIODES

J.-H. ZOLLONDZ, C. MAIN, S. REYNOLDS
School of Science and Engineering, University of Abertay Dundee, Dundee DD1 1HG, UK,
phrhz@tay.ac.uk

ABSTRACT

We report measured electron and hole gating in thick a-Si:H (3.5 μm) p-i-n diodes under reverse bias conditions. Previous publications have shown very high collection efficiency values for electron gating (p-side bias, n-side probe) of up to 50 (i.e. 5000%) for measured and simulated data and predictions of up to 400 (i.e. 40000%) from simulations. Reversing the usual sides of illumination for (electron) gating a situation can be created where, by n-side bias and p-side probe illumination, holes can be gated to travel through the sample to be collected at the contact. Even though the holes have much lower mobility, by this process we can still obtain collection efficiencies greater than unity. This measurement is more difficult because of unwanted illumination by stray bias beam photons on the more sensitive p-side, caused by reflections within the apparatus. Simulation of this situation corroborates qualitatively the measured data. A wide ranging study of the gating phenomenon in relation to different incident wavelengths and photon fluxes for bias and probe beam is reported. We present comparisons of electron and hole gating by measurement and simulation and explain the phenomenon for both electron and hole gating in terms of field changes near to the incident bias interface.

INTRODUCTION

Collection efficiencies or quantum efficiencies (QE) greater than unity have been reported by a number of groups [1-6], including our own. Electron photogating is the explanation for such a phenomenon. An external 'gain' results from internal field changes produced when a low level probe beam is applied in the presence of a much larger bias beam (< 1:100).

Hou and Fonash [1], and Rubinelli [2], explained the electron photogating phenomenon as follows. Strongly absorbed bias light creates a low-field region in the front of the i-layer, near the p-i interface, while increasing the field in the remainder of the i-layer. Additional illumination with a weakly absorbed probe beam generates holes throughout the sample. Trapping of these holes results in a field increase in the front low field region. This increase of the internal electric field in a region of high absorption of the bias beam results in a higher collection of the photocarriers generated by the bias beam.

In contrast to much earlier work we use a combination of reverse bias and steady state measurement. We illuminate first by the bias beam only and take a steady state measurement after the current is settled. Then we add the probe beam and measure the current again after it is settled. We observed long settling times (of the order of seconds) for both measurements, especially for the high collection efficiency situations. In this case, a steady state measurement is more appropriate than using a chopped beam and lock-in technique which gives lower collection efficiency values.

Mat. Res. Soc. Symp. Proc. Vol. 557 © 1999 Materials Research Society

EXPERIMENTAL SETUP AND DEFINITIONS

In this work we measured and simulated the photocurrent response of thick (3.5μm) a-Si:H p-i-n diodes under various illumination situations with a two beam steady state experiment. As bias beam source we used high power LEDs, which can reach photon fluxes up to 3×10^{16} cm^{-2} s^{-1} with typical FWHM of 10 – 20 nm. The probe beam was produced by a halogen lamp, *via* a double monochromator. The bias wavelength was 450 nm and the photon flux was of the order of 1×10^{16} cm^{-2} s^{-1} in the cases where it was required to be constant. For the probe beam we used a wavelength of 622 nm and a photon flux of the order of 1×10^{12} cm^{-2} s^{-1} when the probe beam was required to be constant. A steady state measurement technique was chosen, because the response time of the sample was very slow (of the order of seconds).

The probe collection efficiency is determined as

$$QE = \frac{j_{ph}(\Phi_p + \Phi_b) - j_{ph}(\Phi_b)}{e\Phi_p} \qquad (1)$$

where $j_{ph}(\Phi_p + \Phi_b)$ is the photocurrent under probe and bias illumination with the photon flux Φ_p and Φ_b respectively. Note that the maximum value for QE should not exceed 1 if all and only probe light generated carriers are collected.

EXPERIMENTAL AND SIMULATED RESULTS

The measurements on degradation effects are made on a 3.5 μm p-i-n a-Si:H sample with 200Å p- and n-layers. The optical transmission of the Cr-contacts in the visible range of the spectrum was measured as ~30%. All other measurements were undertaken on a 3.5 μm thick a-Si:H p-i-n diode with ITO contacts with typical transmissions of 90% and comparable doped layer thicknesses.

Electron gating

We use the term electron gating for the case where electrons, if released at the field

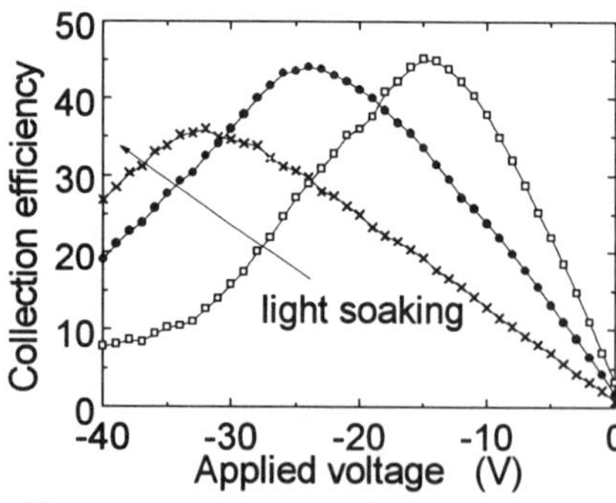

Figure 1: Experimentally determined collection efficiencies *vs* applied reverse bias, after successive AM1 light soak exposures.

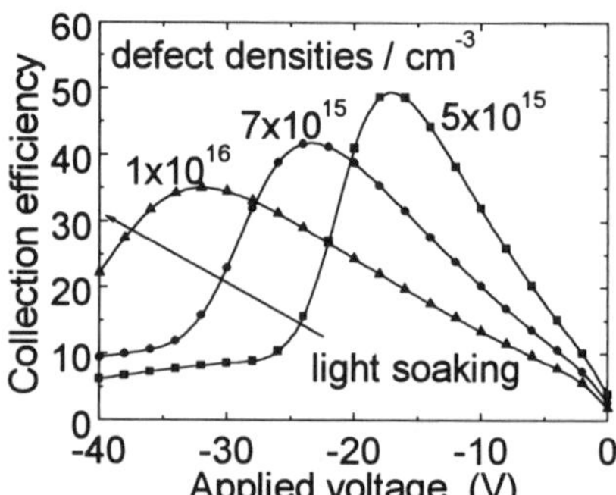

Figure 2: Numerically modelled QE *vs* reverse bias; illumination conditions equivalent to those used experimentally (Fig. 1).

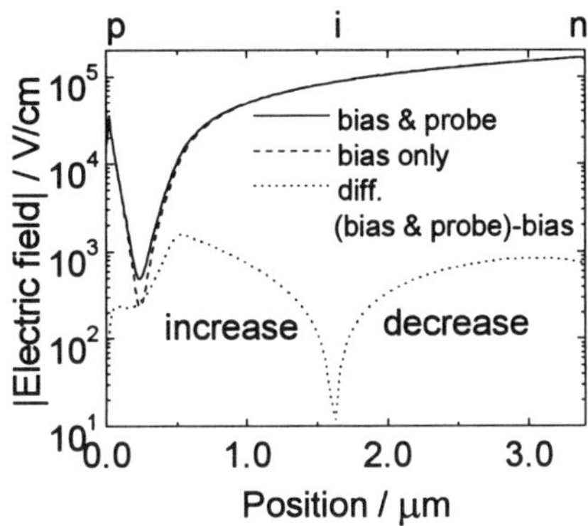

Figure 3: The effect of the addition of probe illumination on the internal electric field at -28 V reverse voltage. One of the curves is the *change* in field on application of probe illumination.

minimum near the p-i interface must travel to the n-contact to be collected (reverse bias). This is the case where the strongly absorbed bias beam is incident through the p-side contact while the probe beam is incident through the n-side.

Figures 1 and 2 show measured and simulated data of collection efficiency values *vs* applied bias voltage after different light soaking times [4] for electron gating. For this experiment and simulation the wavelength of the bias beam and probe beam were 430 nm and 622 nm respectively. The peak observed in the collection efficiency response shifts to higher voltages for longer light soaking times, and hence higher defect densities. Comparison with simulated data shows that even a small change in the defect densities $\approx 2 \times 10^{15}$ cm^{-3} gives a big shift in the position of the maximum (~10 V shift) and a clear decrease of the maximum (~10%). This measurement is thus a sensitive indicator of defect density variation as a p-i-n device is progressively degraded.

To explain the phenomenon, we make use of Figure 3 which shows the internal electric field distribution. The dashed line represents the electric field at −28 V, with bias illumination only. Such a field minimum occurs for illumination with a steep absorption profile. Additional n-side probe illumination (solid line) results in a field increase near the p-i interface while a decrease over the rest of the sample is found. The maximum relative field change is in the region of the field minimum. Because generation of the charge carriers by the bias beam is high in this region, more are now collected. Because the photon flux of the bias beam is higher than the probe flux by orders of magnitude, collection efficiency values of up to 140 (14000%) can be measured and explained. Collection efficiencies higher than unity can be alternatively viewed as resulting from changes in recombination rates [6].

Figure 4 shows *measured* collection efficiencies for varying photon fluxes of the bias beam. For bias beam photon flux values of 1×10^{12} cm^{-2}s^{-1} the collection efficiency values never exceed 1. Even for 1×10^{14} cm^{-2}s^{-1} the collection efficiency values only exceed 100% for low reverse voltages and by a small margin. Increasing the flux, i.e. increasing the difference between probe and bias beam, results in high collection efficiency values, up to 140 (14000%)

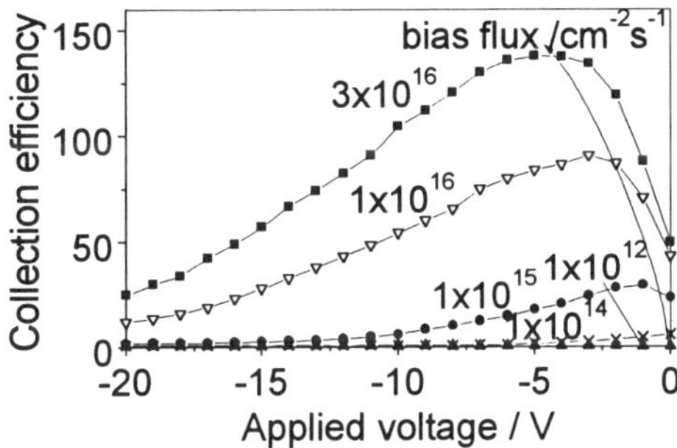

Figure 4: Experimentally determined collection efficiencies *vs* applied reverse bias. The bias beam flux is varied while the probe beam flux is held constant at 1×10^{12} cm^{-2} s^{-1}.

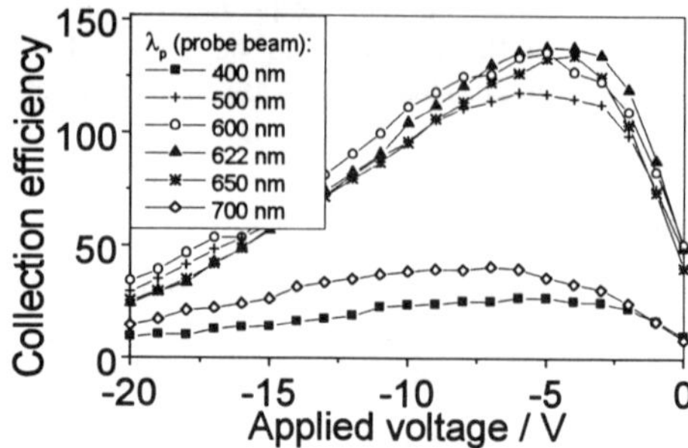

Figure 5: QE *vs* reverse bias for varied probe wavelengths ($\Phi_p = 1\times10^{12}$ cm^{-2} s^{-1}) and blue bias illumination conditions ($\lambda_b = 470$ nm, $\Phi_b = 3\times10^{16}$ cm^{-2} s^{-1}).

for the maximum bias photon flux of 3 $\times$ 10^{16} cm^{-2}s^{-1}.

Note that these very high values were taken with a steady state measurement, in contrast to lower values obtained by ac methods by other groups.

Results of variation of the probe *wavelength* while keeping the bias wavelength constant at 450 nm and constant photon fluxes of 3 $\times$ 10^{16} cm^{-2}s^{-1} for bias and 1 $\times$ 10^{12} cm^{-2}s^{-1} for probe illumination are shown in Figure 5. Different probe wavelengths produce different changes in the field minimum formed by the bias beam. However the result of this modification of the internal electric field in the region of the field minimum seems to be quite similar for probe beam wavelengths between 500 nm and 650 nm. For 400 nm and 700 nm, collection is far lower, but still higher than unity. For a probe beam of 700 nm the collection efficiency is lower because the absorption depth $1/\alpha$ is approximately of the order of the sample thickness so that the change in the field minimum is not as big as for lower wavelengths. If the incident probe wavelength is 400 nm, which is highly absorbed, this results in losses in the n-doped layer and in a secondary field change far from the "main" field minimum. A lower collection of gated carriers results.

Variation of the bias beam wavelength results in a very different picture. In Fig. 6 the measured data are shown. Here the probe wavelength was kept at 622 nm and a photon flux of 1 $\times$ 10^{12} cm^{-2}s^{-1} while the bias photon flux was 1 $\times$ 10^{16} cm^{-2}s^{-1} for each wavelength used. As long as a significant field minimum can be expected (from 470 nm – 525 nm wavelength) a high gating value is measured. Decreasing maxima for decreasing wavelength are found because the field minimum widens and shifts deeper in the i-layer for illumination at lower wavelength as the absorption is less for these wavelengths. This results in a relatively lower release of gated photo carriers. For wavelengths of 590 nm and 612 nm the minimum widens further, until for 660 nm there is no field minimum evident. Over this range, collection of gated photo carriers decreases to zero.

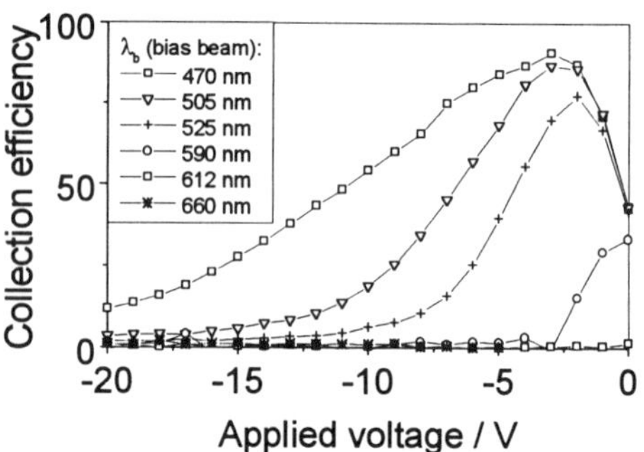

Figure 6: QE *vs* reverse bias for varied bias wavelengths ($\Phi_b = 1\times10^{16}$ cm^{-2} s^{-1}) and red probe illumination conditions ($\lambda_p = 622$ nm, $\Phi_p = 1\times10^{12}$ cm^{-2} s^{-1}).

Hole gating

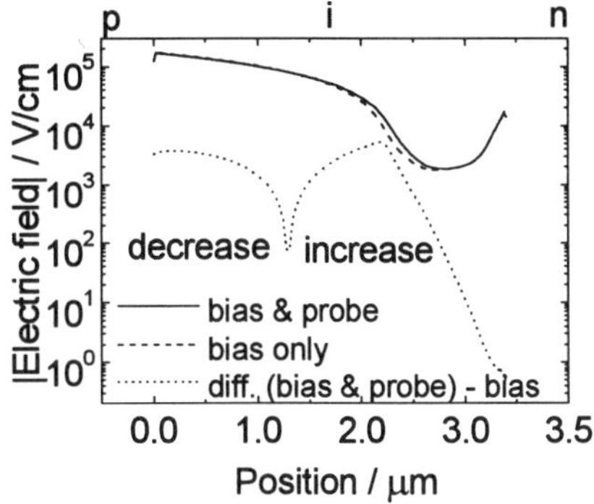

Figure 7: QE *vs* reverse bias obtained by numerical modelling for red probe and blue bias illumination conditions equivalent (reversed) to those in Fig.3.

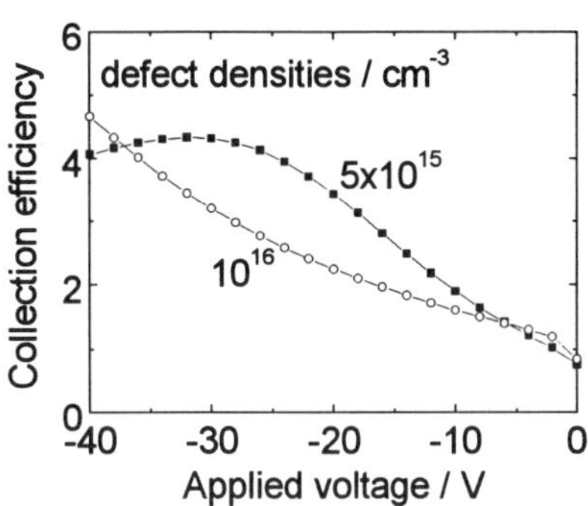

Figure 8: QE *vs* reverse bias obtained by numerical modelling for red probe and blue bias illumination conditions equivalent to those in Fig.2.

We now examine the case of inverting the experiment for electron gating (with p-sided bias and n-sided probe illumination) to n-sided bias and p-sided probe illumination, which results in a field minimum next to the n-layer. Similarly to the electron gating situation an increase in the internal electric field minimum is the reason for a gating effect. Figure 7 shows that this minimum is broader than its analogue for electron gating and deeper in the i-layer for the same simulation parameters as for Fig. 2 but with reversed illumination sides for bias and probe beams.

We call this process *hole gating* because under reverse bias voltage when photocarriers from the region of this minimum are released, holes have a longer way to be collected at the p-contact, just as electrons drift through nearly the full i-layer in electron gating.

Figure 8 shows simulated collection efficiencies for the hole gating experiment. This results in a maximum at high reverse voltages. The results look similar to those of electron gating for a fairly degraded sample but the QE values are far lower. The simulated data do match the measured data in form (see Fig. 9) for the same input parameters as for Fig. 2, but we could not fit the simulated results quantitatively to get the same peak values as measured. The reason for that could be that variation of the defect density with a uniform profile does not reflect the real defect density distribution in the sample. Degradation for a sample of 3.5 μm will happen in the layers near the doped areas.

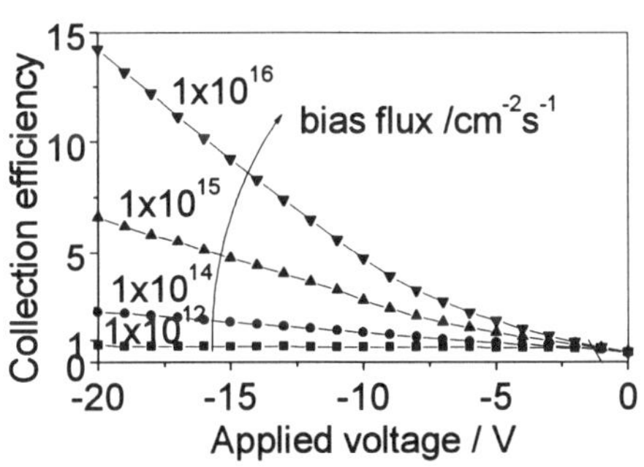

Figure 9: QE *vs* reverse bias obtained by numerical modelling for red probe and blue bias illumination conditions equivalent to those in Fig. 3.

Figure 9 shows similar results as in Fig. 4: collection is far higher for higher bias photon fluxes while keeping the probe photon flux at 1×10^{12} cm^{-2}s^{-1}. For a low ratio between bias and probe photon flux a small effect is observed. For a probe flux equal to the bias photon flux the collection efficiency values never exceed 1, i.e. no gating is observed.

In Fig. 10 hole gating collection efficiency values *vs* reverse bias voltage are shown. As in Fig. 6 the probe wavelength was kept at 622 nm with a photon flux of 1×10^{12} cm^{-2}s^{-1} while the bias photon flux was 1×10^{16} cm^{-2}s^{-1} and the bias wavelength was varied.

This variation of the bias wavelength results in a wide spread in the response for the collection efficiency values: lowest collection efficiencies for a red bias beam up to highest values for a blue bias beam. Again the gating effect is observed for those wavelengths that form a clear field minimum. For 660 nm bias beam wavelength, where no electric field minimum can be expected, the collection efficiency never exceeds unity.

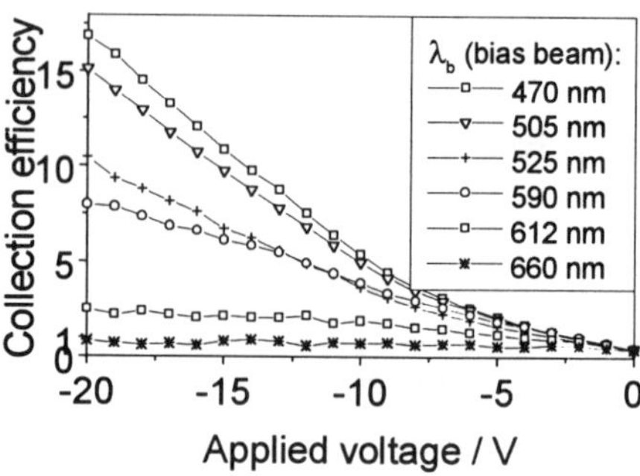

Figure 10: QE *vs* reverse bias for varied bias wavelengths (Φ_b = 1×10^{16} cm^{-2} s^{-1}) and red probe illumination conditions (λ_p = 622 nm, Φ_p = 1×10^{12} cm^{-2} s^{-1}, p-side).

CONCLUSIONS

Under a reverse bias voltage situation and bias beam illumination, probe beam collection efficiencies greater that unity for electrons or holes can be achieved. This photogating effect can reach gain values up to 140 for electron gating in thick a-Si:H p-i-n samples, which are the highest measured values so far to our knowledge. We achieved this by taking long relaxation times into account, using a steady state technique instead of the previously reported lock-in methods. The photogating effect is very sensitive to variations in defect density and provides a high resolution diagnostic tool to investigate degradation or quality of a p-i-n diode. Variation of the wavelength of the probe and bias beam give an understanding of field dependent processes involved in the gating effect.

ACKNOWLEDGMENTS

The authors acknowledge funding from the British Council and DAAD via an ARC collaboration. One of the authors (J.-H. Zollondz) is grateful to the University of Abertay Dundee for the provision of a research studentship.

REFERENCES

1. J. Hou and S. J. Fonash, Appl. Phys. Lett. **61**, 186 (1992).
2. A. Rubinelli, J. Appl. Phys. **75**, 998 (1994).
3. P. Chatterjee, J. Appl. Phys. **75**, 1093 (1994).
4. J.-H. Zollondz, W. Gao, R. Brüggemann, C. Main, and G. H. Bauer (Mater. Res. Soc. Proc. **420**, Pittsburgh, PA 1996) p. 251-256.
5. R. Brüggemann, J.-H. Zollondz, C. Main, and W. Gao (Mater. Res. Soc. Proc. **467**, Pittsburgh, PA 1997) p. 759-764.
6. C. Main, J.-H. Zollondz, S.Reynolds, W. Gao, R. Brüggemann, and M. J. Rose, J. Appl. Phys. **85**, p. 296-301 (1999).

Part VII

Heterogeneous Materials
and Devices

MICROCRYSTALLINE SILICON - RELATION OF TRANSPORT PROPERTIES AND MICROSTRUCTURE

J. KOČKA, A. FEJFAR, V. VORLÍČEK, H. STUCHLÍKOVÁ, J. STUCHLÍK
Institute of Physics, Academy of Sciences of the Czech Republic, Cukrovarnická 10,
162 53 Prague 6, Czech Republic
phone: +420-2-24311137, fax: +420-2-3123184, e-mail: kocka@fzu.cz

ABSTRACT

Understanding of transport in hydrogenated microcrystalline silicon (μc-Si:H) is difficult due to its complicated microstructure (grains, grain boundaries, amorphous tissue). μc-Si:H layers often exhibit preferential orientation leading to transport anisotropy. Furthermore, specific μc-Si:H growth features lead to the thickness dependence of the structure and properties.

μc-Si:H incubation layer was studied by AFM with conductive cantilever measuring simultaneously morphology and local conductivity maps with submicron resolution. Clear identification of Si crystallites (with size of few tens of nanometers) is demonstrated. The crystalline fraction at the surface may be easily evaluated.

For the charge collection in solar cells we need to study transport perpendicular to the substrate. Measurement of frequency spectra of A.C. conductivity is introduced as a new tool which can exclude the influence of contact barriers in sandwich geometry and can be used for finding the "true" conductivity perpendicular to the substrate. Using this technique transport anisotropy in some μc-Si:H samples was clearly demonstrated.

Finally, it is shown how the transport properties change with growing μc-Si:H thickness and how these changes correlate with the structure observed by AFM.

INTRODUCTION

Recently demonstrated high efficiency and stability of solar cells based on hydrogenated microcrystalline silicon (μc-Si:H) [1,2] led to a concentrated research effort in many laboratories. As a result we know a lot about the structure and growth of μc-Si:H and its optical properties. Although the improvement of the solar cells proceeds mainly by trial and error, efficiencies as high as 8.5% were reported for single μc-Si:H junction or 12% for "micromorph" cell, i.e. in tandem of μc-Si:H cell and a cell based on amorphous hydrogenated silicon (a-Si:H) [3,4].

On the other hand, the real understanding of electronic transport in μc-Si:H remains elusive. One reason is the complicated μc-Si:H microstructure which includes grains, grain boundaries, remaining amorphous tissue and possibly also void fraction. The material is not only inhomogeneous, it also often exhibits preferential orientation and texture and thus anisotropic transport may be expected. Further complication is due to the complex μc-Si:H growth process which leads to the thickness dependence of the structure. The other reason is that μc-Si:H represents rather a rich class of the materials and thus widely spread values of important parameters like for example mobility were reported [5-7].

The standard measurement methods often yield only effective values of the transport parameters and thus we need new measurement methods which would be able to select and clarify the underlying phenomena, e.g. differences in local conductivity or transport anisotropy.

Mat. Res. Soc. Symp. Proc. Vol. 557 © 1999 Materials Research Society

EXPERIMENTAL RESULTS AND THEIR DISCUSSION

Characterization of µc-Si:H initial growth by local transport studies

For understanding the transport in µc-Si:H we need to know how the charge carriers behave in the material constituents, that is the silicon crystallites, grain boundaries, amorphous tissue or near the voids. That means we need a technique capable of measuring selectively the properties of the films with resolution better than the typical size of these features. The necessary resolution can be easily achieved by using a sharp probe of either scanning tunnelling microscope (STM) or atomic force microscope (AFM). The previous studies using standard AFM have been used to study the µc-Si:H surface morphology and in some cases a correlation between grain size and transport properties of µc-Si:H could be observed [8].

Recently we have started to use the conductive cantilever in contact mode AFM for measurements of local conductivity [9,10]. The main advantage is that we obtain simultaneously the image of sample surface morphology and the local conductivity map. Surface morphology is

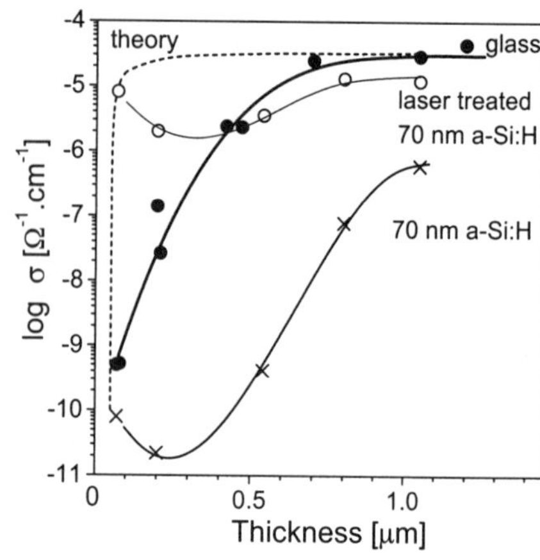

Figure 1 Dark conductivity of µc-Si:H for films with different thicknesses deposited on glass (●), on glass coated with thin initial a-Si:H layer (d=70 nm) (X) or on the the same a-Si:H treated by a single excimer laser pulse (O). The dashed curve shows results calculated from a simple model of 50 nm amorphous incubation layer with $\sigma = 10^{-10}$ $\Omega^{-1}cm^{-1}$ on which µc-Si:H with $\sigma = 3 \cdot 10^{-5}$ $\Omega^{-1}cm^{-1}$ grows.

deduced from the deflection of the cantilever. External D.C. bias is applied to the cantilever and the induced D.C. current flowing between the cantilever and the sample bottom electrode is used to obtain the local current image. The contact mode of the AFM ensures good electrical contact between the tip of the cantilever and the sample surface. This condition is necessary for excluding the contact (tunnelling) barrier which would otherwise limit the current and mask the properties of the underlying sample.

We have used this technique to study the µc-Si:H in the initial stages of growth. We have intentionally selected growth conditions (see below) near the amorphous/crystalline transition so that the thickness of so called incubation layer reached almost 0.5 µm. This is documented in Fig. 1 by relatively slow increase of D.C. room temperature conductivity with increasing µc-Si:H thickness. Due to the technological importance of a-Si:H/µc-Si:H interface we have prepared also a sample with a 70 nm thin a-Si:H sublayer (using low power 13.56 MHz PECVD with 8% dilution of SiH_4 in hydrogen) on which the growth continued under conditions readjusted for µc-Si:H growth by changing silane dilution in hydrogen to 4.5%. Dramatic delay of nucleation and increase of the thickness necessary for achievement of conductivity typical for µc-Si:H is evident from Fig. 1. After the deposition, the samples have been transferred into the measurement chamber without breaking vacuum (p ≤ 1x10⁻⁷ mbar). We have used Omicron UHV AFM with conductive cantilever made of highly doped silicon coated with platinum to obtain the morphology and current images. The image of the surface morphology has been rendered by scanning the profile of constant force [9]. A D.C. voltage -3 V was applied to the cantilever and the current flowing through the sample to the grounded nickel-chromium bottom electrode was registered.

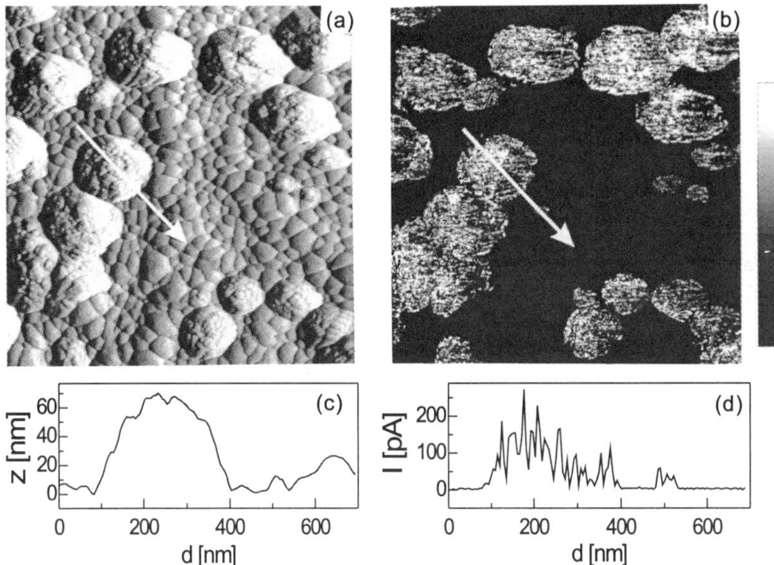

Figure 2 (a) Normal force image of surface morphology (1500 x 1500 nm) of the µc-Si:H layer grown on 70 nm a-Si:H (total thickness about 500 nm) obtained in AFM constant force mode.
(b) Map of simultaneously measured local current flowing between the conductive cantilever and the bottom NiCr electrode. Microcrystalline areas are clearly resolved against the dark (less conductive) amorphous background. Even very small crystallites may be identified in the current map.
(c) Profiles of the sample surface height and (d) local current along the line shown by the arrows. Note the large irregularities of the current observed on top of the crystallite. The reliability of the crystalline area identification can be illustrated by comparison of both signals at positions around d=650 nm. Even though a peak is observed in sample height, the current remains low, indicating an amorphous region. On the other hand, the current increase on a small bump at d=500 nm clearly identifies small crystalline grain.

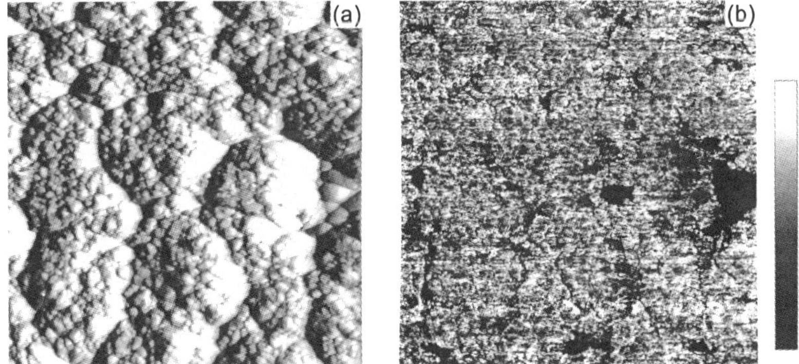

Figure 3 Surface morphology (a) and map of local current (b) of the µc-Si:H layer grown on laser treated a-Si:H initial layer. The images were obtained in the same way and on the sample grown in the same deposition as in Fig. 2, only in this case the underlying 70 nm thick initial a-Si:H layer was treated by single excimer laser pulse. Note much higher crystallinity at the surface even though some residual amorphous areas still remain (e.g. dark region on the right).

The resulting AFM surface morphology and current map of about 500 nm thick μc-Si:H grown on the amorphous initial layer are shown in Fig. 2. Even at this thickness the transition to fully microcrystalline layer is still not complete due to the deposition conditions and to a-Si:H as a substrate. In the surface morphology (Fig. 2a) we can resolve roughly circular protrusions above the smoother and relatively flat rest of the surface. We have assumed that these protrusions correspond to the surface of the silicon crystallites surrounded by the amorphous matrix. This assumption was clearly proved by the current map (Fig. 2b) which shows that the local current is much higher on the protrusions than on the rest of the sample, in agreement with expected higher conductivity of μc-Si:H in comparison with a-Si:H.

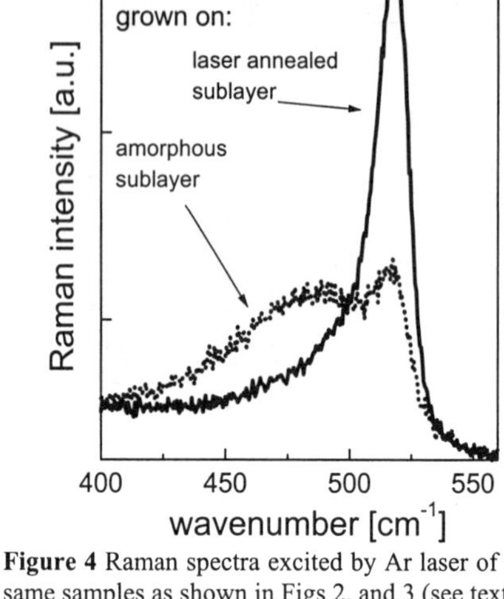

Figure 4 Raman spectra excited by Ar laser of the same samples as shown in Figs 2. and 3 (see text).

Notice that the parallel measurement of AFM morphology and current map allows clear identification not only of the large crystallites but also small crystalline grains which are indistinguishable in the AFM surface morphology (see Fig. 2a and b). It also revealed the presence of inner grain boundaries in the large grains of the sample from Fig. 2. The sizes of these inner grains relate well to the size deduced from the X-ray diffraction, commonly within range 20 to 50 nm.

The incubation layer is detrimental to the solar cell efficiency and thus various techniques were used to shorten this transition. We have applied the combined AFM measurement to study the effect of the laser annealing of the initial layer. Part of the initial a-Si:H sublayer of the samples deposited as described above was treated by a single pulse of ArF excimer laser with wavelength 193 nm and energy density 90 mJ/cm^2, i.e. just above the a-Si:H melting threshold. Resulting sample exhibits substantially different surface morphology and also local current image (Fig. 3). In this case we observe very high fraction of crystallites and negligible amorphous tissue.

High contrast between crystalline and amorphous phase in the current image (due to the large difference of their conductivities) allows us to find the crystallinity of the films at the surface as the ratio of the conductive part to total sample area. The corresponding values of crystallinity for films were 40% (sample on a-Si:H sublayer, Fig. 2) and 95 % (sample on laser treated a-Si:H sublayer, Fig. 3). In order to verify these data we have measured the Raman spectra from the surface of these samples using Ar laser excitation with short penetration depth (see Fig. 4). The crystallinity deduced simply from the ratio of the magnitude of Raman intensity at 520 and 480 cm^{-1} reached 56% and 88% respectively, in qualitative agreement with the values found from the local current image.

The versatility of the combined AFM/local conductivity method was further illustrated by imaging the a-Si:H layer laterally structured by laser pulse split into two beams forming a stripe-like interference pattern. Where such a periodic profile exceeds a-Si:H melting threshold, the material is crystallized. This results in stripes of microcrystalline silicon separated by a-Si:H [10]. Again we could obtain high contrast between the crystalline and amorphous parts of the sample.

The combined AFM measurement thus give us a tool which is able to study the conductivity on single crystalline grains. However, further research is needed to understand the influence of the underlying sample structure on the resulting images and the resolution limit of this method as well as the role of relatively high electrical field concentrated near the sharp tip of the cantilever.

Anisotropy of transport in μc-Si:H

Depending on the deposition conditions, the growth of μc-Si:H leads to the preferential crystallographic orientation of the film. The silicon crystallites quite often have a prolonged shape, with the main axis oriented perpendicular to the substrate. For such samples an anisotropy of transport properties may be expected, with obvious consequences for the solar cells function. However, the experimental methods capable of characterizing this anisotropy have to overcome many difficulties. For instance we cannot directly compare the dark conductivities measured parallel to the substrate (with coplanar electrodes) with the values obtained from the measurement with sandwich electrodes (perpendicular to the substrate). In the latter case the measurement result is decided by depleted regions at the barriers formed near the contacts and/or incubation layer.

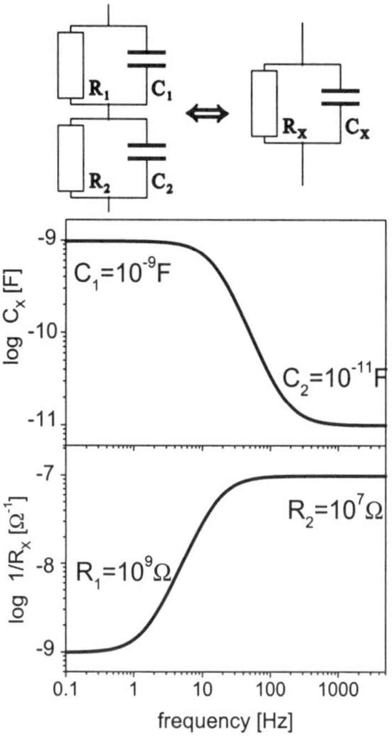

We have recently demonstrated how the frequency spectra of A.C. conductivity can be used to exclude the depleted regions at the contacts from the results and the "true" value of conductivity σ ($\perp$) of the material perpendicular to substrate can be obtained. Using this approach we were able to demonstrate for the first time the anisotropy of transport in μc-Si:H [8].

The principle of the determination of the "true" conductivity from the A.C. spectra can be illustrated simply by considering the sample to be composed of a series combination of the R_1 and C_1, representing contact region(s) and of the R_2 and C_2, representing remaining volume of the film (see Fig. 5). The measurement of A.C. impedance then yields effective values of resistivity R_x and capacity C_x of equivalent circuit. The simulated frequency spectra of A.C. impedance in Fig. 5 illustrate how the contact regions dominate at low frequencies. At high frequencies the observed values get close to the volume properties.

Typical frequency spectra measured for a 8.4 μm thick μc-Si:H sample at different temperatures are shown in Fig. 6. We can see that the "true" conductivity of the volume of the sample can be found from the plateau of the conductivity at frequencies where also the capacitance comes close to the geometrical value ($\varepsilon_{eff.} = C \cdot d/S \approx 11$). Thus we can obtain the value of the conductivity perpendicular to the substrate. The measurements are limited by other factors such as for example the resistance of the metal (semitransparent) contacts to the sample. Even a small value

Figure 5 Equivalent circuit of the sample composed of the series combination of the depleted contact region(s) (R_1 and C_1) and the remaining sample volume (R_2 and C_2). The simulated A.C. frequency spectra illustrate how the contact regions dominate at low frequencies. At high frequencies the observed values get close to the volume properties.

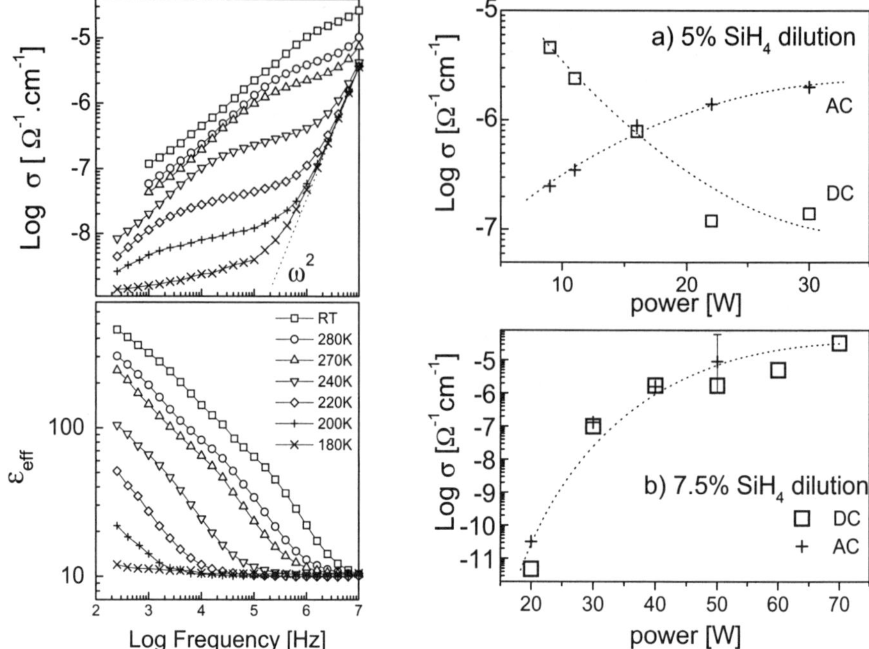

Figure 6 A.C. spectra of conductance (a) and effective permittivity (b) of the 8.4µ thick µc-Si:H layer at different temperatures. The true conductance of the sample perpendicular to the substrate can be found from the plateau at medium frequencies where the effective permittivity ($\varepsilon_{\text{eff}} = C_{\text{meas}} \cdot d/S$) reaches the correct geometrical value (~11). The activation energy and prefactor of the conductance can be evaluated from the shift of the plateau with temperature.

Figure 7 Comparison of the perpendicular A.C. "plateau" conductivity (+) and D.C. coplanar conductivity ($\square$) for 5% (a) and 7.5% silane dilution series (b) of µc-Si:H layers prepared in IMT Neuchatel [8] as a function of discharge power (see text). These results demonstrated for the first time anisotropy of conductivity in µc-Si:H.

of the contact resistance (of the order of few tens of ohms) leads to an apparent increase of the conductivity ($\approx \omega^2$) at higher frequencies [11]. High ε_{eff} even in MHz range (see Fig. 6) indicates that due to relatively high conductivity of µc-Si:H any applied electric field will be quickly redistributed and concentrated to the contact-related region. This may lead to violation of basic assumption used in Time-of-Flight (TOF) and makes the measurement of drift mobilities in µc-Si:H very difficult.

We have first tested the possibility to compare parallel D.C. conductivity - σ ($\parallel$) with the perpendicular $\sigma(\perp)$ evaluated from the A.C. data on two series of µc-Si:H samples prepared in Neuchatel [8] at 5% and 7.5% SiH₄ dilution in hydrogen. The results [8] are displayed in Fig. 7. For 7.5 % series both conductivities practically coincide. For 5% dilution series there are great differences between the conductivities in both directions, exhibiting opposite trends with changing power to the discharge. Although comparable values of σ ($\parallel$) and σ ($\perp$) at room temperature do not guarantee isotropy of transport (see below), their difference is a strong indication of anisotropy. Independent support for the existence of the anisotropy of the transport properties of µc-Si:H was

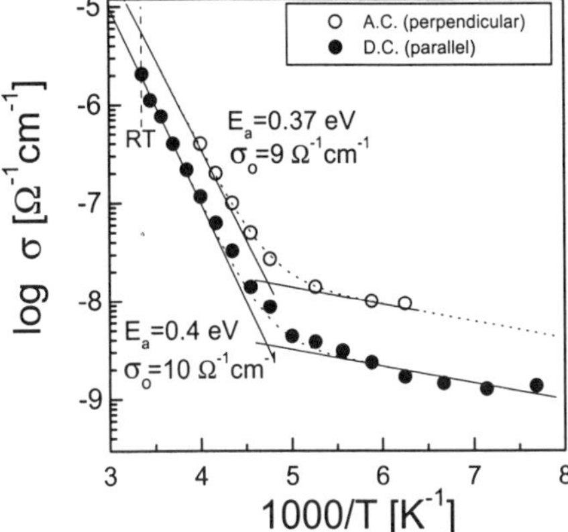

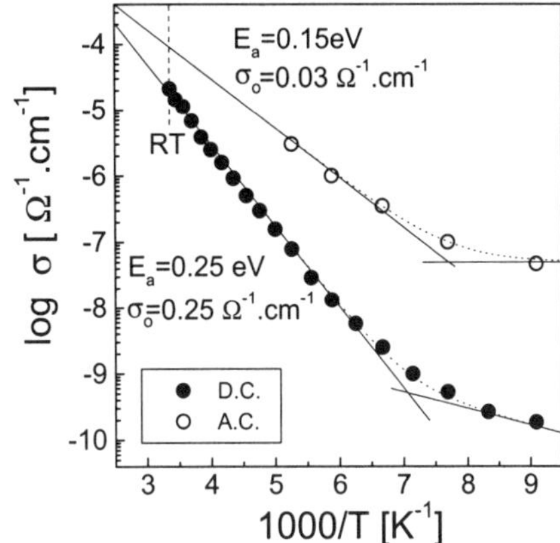

Figure 8 Arrhenius plot of D.C. conductivity (parallel to the substrate) and A.C. conductivity (perpendicular to the substrate) measured on the 2.75 μm sample from 5% dilution series μc-Si:H prepared at 19W. This sample lies close to the crossing of the curves in Fig. 7a and thus isotropic conductivity is expected. The activation energies and prefactors of conductivities parallel and perpendicular to the substrate agree within the experimental uncertainty, confirming the same transport mechanism in both directions.

Figure 9 Arrhenius plot of D.C. conductivity $\sigma(\parallel)$ (parallel to the substrate) and A.C. conductivity $\sigma(\perp)$ (perpendicular to the substrate) measured on the μc-Si:H sample with 2.2 μm thickness prepared in Prague. Although the values of conductivities are almost the same at room temperature, the activation energies and prefactors are different, indicating different transport mechanisms.

demonstrated by comparing values of diffusion length obtained by Steady-State Photocarrier Grating (SSPG) and by Surface Photovoltage (SPV) methods [12]. Recently detailed studies of diffusion length by both SSPG and SPV on 5% SiH_4 dilution series [13] again clearly indicated anisotropy of transport in μc-Si:H.

The temperature dependence of the conductivity $\sigma(\perp)$ found from the A.C. conductivity plateau can be used to measure also the activation energy (E_a) and often overlooked conductivity prefactor (σ_0). The importance of these values becomes clear from the Arrhenius plots of μc-Si:H conductivity in figures 8 and 9. Both of the samples exhibited comparable room temperature values of conductivity perpendicular (from A.C.) and parallel (from D.C.) to the substrate. However, only the sample in the Fig. 8 is truly isotropic, exhibiting the same activation energy and prefactor of conductivities in both directions, while a substantial difference revealed in Fig. 9 indicates that in this sample the transport is not isotropic!

The knowledge of E_a and σ_0 is important for yet another reason. Recently we have shown [10] that the apparent paradox (conductivity increasing and charge carrier mobility decreasing with deposition power for 7.5% dilution μc-Si:H series) can be explained by the shift of the transport path which is evident by decrease of both E_a and σ_0. Such a shift of the transport path from above the mobility edge to the tail state region would naturally explain the experimentally observed decrease of V_{oc} with increase of VHF power [8] because the shifting transport path indicates higher number of states in the tails. This leads to lower splitting of the quasi-Fermi levels and to lower V_{oc}.

Thickness dependence of the transport properties

Are properties of μc-Si:H for thicknesses larger than the thickness of incubation layer constant or is the thickness dependence of the properties another reason why the results of transport measurements in different laboratories and/or on different samples are often different? The room temperature (coplanar) dark conductivity measured on μc-Si:H with $d \lesssim 1$ μm indicated (see Fig. 1 and "■" in Fig. 10) that the incubation layer was completed, conductivity saturated and almost full crystallinity reached. However, AFM topography of the series of films prepared under the same conditions with the thickness of film changing from 0.6 to 4.8 μm (Fig.11) indicates pronounced changes of the morphology even for $d > 1$ μm. As the growth continues, larger and larger crystallites prevail and the average size of the crystallite changes from approx. 50 nm to ≥ 200 nm.

The corresponding changes are observed not only in conductivity (see Fig. 10) but also in the other transport data measured on the same samples. Fig. 12 shows the Arrhenius plots of the coplanar conductivities of the μc-Si:H layers with thicknesses increasing from 0.6 to 4.7 μm. The evaluated activation energies E_a and prefactors σ_0 are summarized in Fig. 13 and show the same sharp initial drop of σ_0 and E_a which was observed for 7.5 % SiH_4 dilution series with increasing deposition power [8]. The inset in Fig. 13 schematically illustrates the downward shift of the transport path responsible for this behaviour. The gradual increase of conductivity with thickness > 1μm (Fig. 10) is mainly due to the increasing prefactor σ_0 (Fig. 13) which is probably related to the decreasing number of grain boundaries, stimulated by the increase of grain size (see Fig. 11).

Finally, in Fig. 14 we compare the diffusion lengths measured on the same samples by SSPG measurements, i.e. coplanar $L_{diff.}(\parallel)$, and by the SPV measurements, i.e. perpendicular $L_{diff.}(\perp)$. In both cases the diffusion length increases with the μc-Si:H sample thickness. However, the diffusion lengths perpendicular to the substrate are considerably larger than along the substrate and this difference increases with the sample thickness, again confirming the μc-Si:H anisotropy.

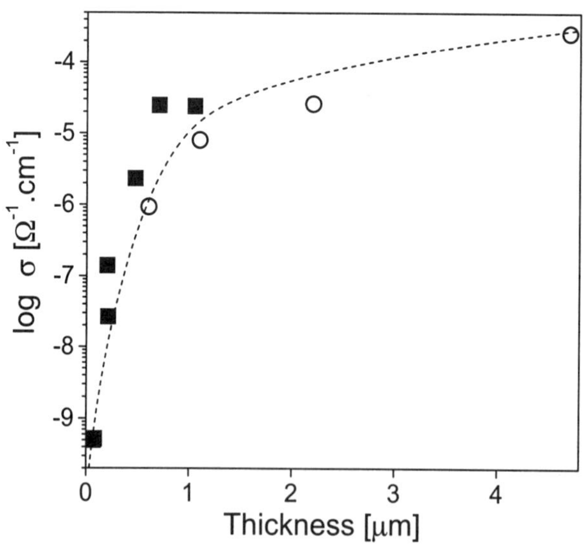

Figure 10 D.C. dark (coplanar) conductivity measured at room temperature versus thickness of μc-Si:H samples. Samples labelled "■" were prepared with thickness $d \lesssim 1$μm for the study of the incubation layer (see Fig 1). Samples labelled "O" were prepared under the same conditions and used for study of surface morphology and transport (see Figs. 11-14).

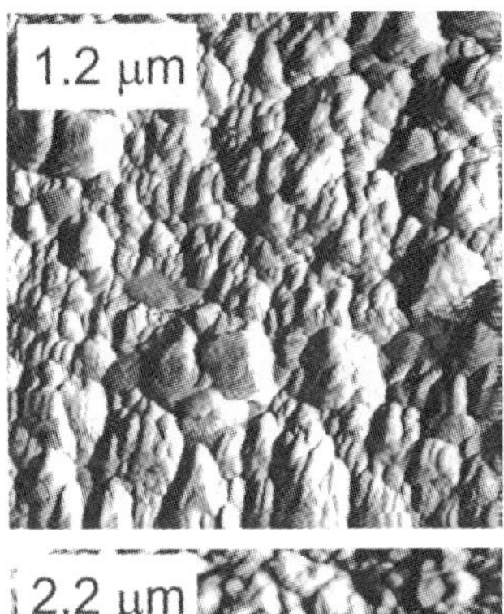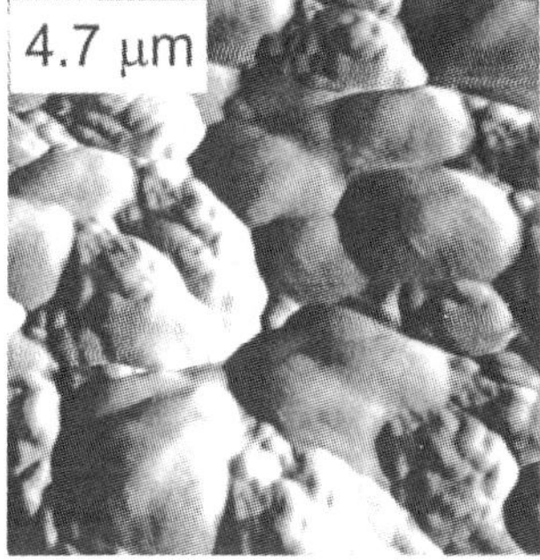

Figure 11 AFM morphology of μc-Si:H samples deposited at the same deposition conditions but with different thicknesses (series labelled by "O" in Fig. 10). The field of view is 1×1 μm. Normal force images are shown.

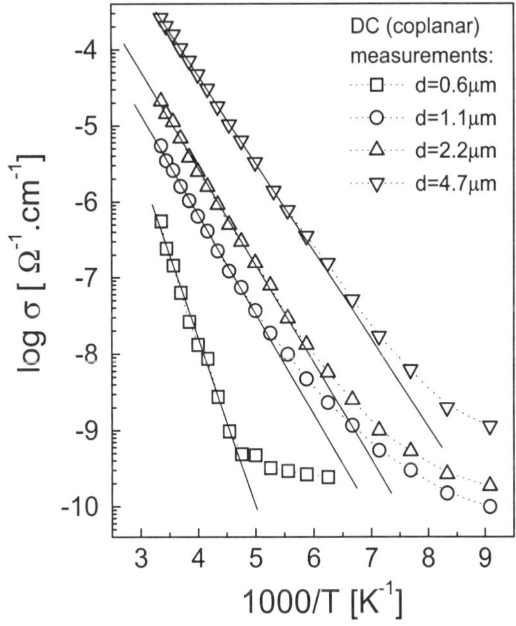

Figure 12 Arrhenius plot of D.C. conductivity (parallel to the substrate) measured on the μc-Si:H sample series with different thicknesses (see Fig. 10 and 11).

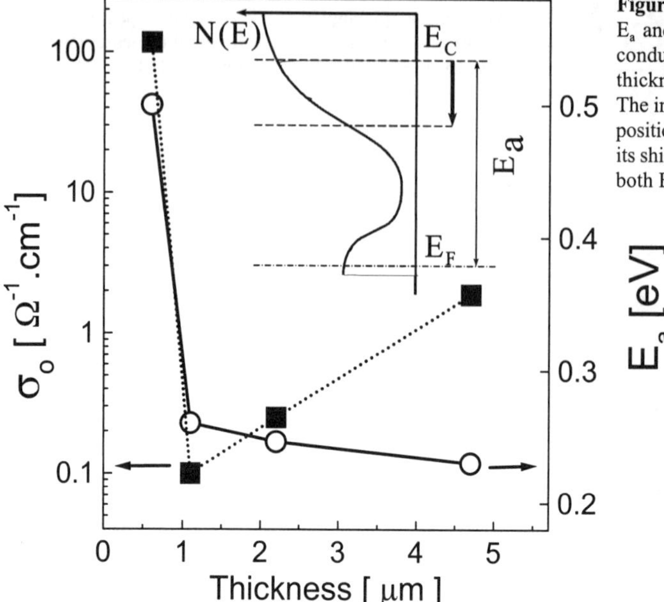

Figure 13 The activation energy E_a and prefactor σ_0 of the D.C. conductivity of the μc-Si:H thickness series (see Fig. 12). The inset illustrates the probable position of the transport path and its shift, indicated by decrease of both E_a and σ_0 .

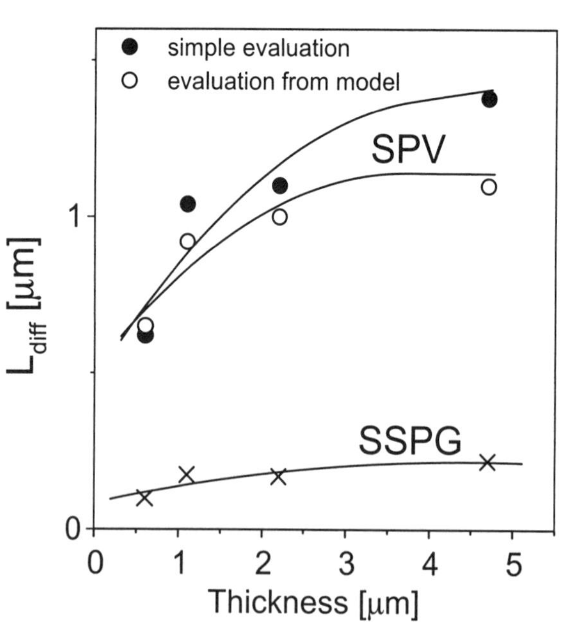

Figure 14 The diffusion lengths measured parallel to the substrate (by steady state photo-carrier grating, SSPG) and perpendicular to the substrate (by surface photovoltage, SPV) for the μc-Si:H thickness series. The SPV results are obtained either by standard method or from a model including the depleted regions at the surface.

SUMMARY

We have demonstrated the usefulness of the AFM combined with local conductivity measurement for identification of crystallites at the surface of μc-Si:H samples. We have used this approach to study the initial stages of μc-Si:H growth and shown how the incubation layer, detrimental to solar cells, can be reduced by treating the initial layer with single excimer laser pulse.

We have used A.C. impedance frequency spectra to evaluate conductivity perpendicular to the substrate $\sigma(\perp)$ and compared it with conductivity parallel to the substrate $\sigma(\parallel)$ found from D.C. measurements. The conductivity anisotropy was clearly shown for selected μc-Si:H samples. The existence of anisotropy is further supported by comparison of diffusion lengths found from SSPG and SPV.

We have further shown the importance of comparing the activation energy (E_a) and prefactor (σ_0) of conductivities in both directions. Their changes indicate change of transport path, which directly influences the function of solar cells (for example V_{oc}). Transport can be anisotropic even if $\sigma(\parallel) \doteq \sigma(\perp)$ coincide at room temperature but the prefactors (σ_0) and activation energies (E_0) are not the same.

Structure and transport properties of μc-Si:H are changing with the sample thickness. These changes continue even at thicknesses well above completing the transition to fully crystalline growth.

ACKNOWLEDGEMENTS

The authors would like to thank the group of prof. A. Shah from IMT, Neuchatel for providing some of the μc-Si:H samples and for fruitful discussions. This work was supported by projects GAAV A1010809, GAČR 202/98/0669, by NEST project of EC and by NEDO.

REFERENCES

[1] J. Meier, R. Flückiger, H. Keppner, A. Shah, Appl. Phys. Lett. 65 (1994) 860.

[2] J. Meier, S. Dubail, J. Cuperus, U. Kroll, R. Platz, P. Torres, J.A. Anna Selvan, P. Pernet, N. Beck, N. Pellaton Vaucher, Ch. Hof, D. Fischer, H. Keppner and A. Shah, J. Non-Crystal. Solids **227-230** (1998) 1250.

[3] J. Meier, S. Dubail, L. Feitknecht, Y. Ziegler, P. Torres, C. Hof, U. Kroll, D. Fischer, J. Cuperus, H. Keppner, A. Shah, Proc. 2nd World Conf. Photovoltaic Solar Energy Conversion, Vienna, July 1998, p. 375.

[4] K. Saito, M. Sano, K. Matuda, T. Kondo, T. Nishimoto, K. Ogawa, I. Kajita, Proc. 2nd World Conf. Photovoltaic Solar Energy Conversion, Vienna, July 1998, p. 351.

[5] N. Wyrsch, M. Goerlitzer, N. Beck, J. Meier and A. Shah, in *Amorphous Silicon Technology - 1996*, edited by M. Hack, E.A. Schiff, S. Wagner, R. Schropp, A. Matsuda (Mater. Res. Soc. Proc. **420**, Pittsburg, PA 1996), pp. 801-806.

[6] N. Beck, P. Torres, J. Fric, Z. Remeš, A. Poruba, Ha Stuchlíková, A. Fejfar, N. Wyrsch, M. Vaněček, J. Kočka, A. Shah, in *Advances in Microcrystalline and Nanocrystalline Semiconductors - 1996* edited by R. W. Collins, P. M. Fauchet, I. Shimizu, J.-C. Vial, T. Shimada, A. P. Alivisatos (Mater. Res. Soc. Proc. **452**, Pittsburg, PA 1997) p. 761-766.

[7] A. Fejfar, N. Beck, H. Stuchlíková, N. Wyrsch, P. Torres, J. Meier, A. Shah, J. Kočka: J. Non-Crystal. Solids, 227-230 (1998) 1006-1010.

[8]   J. Kočka, A. Fejfar, H. Stuchlíková, B. Rezek, A. Poruba, M. Vaněček, P. Torres, J. Meier, N. Wyrsch, A. Shah and A. Matsuda, Proc. 2nd World Conf. Photovoltaic Solar Energy Conversion, Vienna, July 1998, p. 785.

[9]   B. Rezek, J. Stuchlík, A. Fejfar, J. Kočka, C.E. Nebel and M. Stutzmann, phys. stat. sol. (a) 170, R1, 1998.

[10]   B. Rezek, J. Stuchlík, A. Fejfar, J. Kočka, Appl. Phys. Lett. 74 (1999) 1475.

[11]   R.A. Street, G. Davies, A.D. Yoffe, J. Non-Crystal. Solids 5(1971) 276.

[12]   J. Kočka et al., NEST project workshops, Prague and Julich 1998.

[13]   N. Wyrsch, P. Torres, M. Goerlitzer E. Vallat, U. Kroll, A. Shah, A. Poruba and M. Vaněček, in *Polycrystalline Semiconductors V - Bulk Materials, Thin Films and Devices* edited by J.H. Werner, H.P. Strunk, H.W. Schock, (Solid State Phenomena series, Scitech Publ., Uettikon am See, Switzerland 1999), to appear.

ELECTRONIC TRANSITIONS IN MIXED PHASE CRYSTALLINE/AMORPHOUS SILICON IN THE LOW CRYSTALLINE FRACTION REGIME

J. David COHEN*, Daewon KWON*, Chih-Chiang CHEN*, Hyun-Chul JIN[+], Eric HOLLAR[+],
Ian ROBERTSON[+], and John R. ABELSON[+]
*Department of Physics, University of Oregon, Eugene, OR 97403
[+]Coordinated Science Laboratory and Department of Materials Science and Engineering,
University of Illinois, Urbana, IL 61801

ABSTRACT

Amorphous silicon films were prepared by dc reactive magnetron sputtering under
conditions approaching the phase transition to microcrystallinity. Using TEM imaging these
films were found to contain clusters of 5 to 50 nm sized Si crystallites embedded in an
amorphous silicon matrix. Photocapacitance and transient photocurrent sub-band-gap optical
spectra of this material appear to consist of a superposition of a spectrum typical of amorphous
silicon together with an optical transition, with a threshold near 1.1eV, that exhibits a very large
optical cross section. This transition arises from valence band electrons being optically inserted
into empty levels lying within the amorphous silicon mobility gap. Using modulated
photocurrent methods we have determined that these states also dominate the electron deep
trapping in this material. We argue that these states arise from defects at the crystalline-
amorphous boundary.

INTRODUCTION

Amorphous silicon films produced just below the amorphous/crystalline phase boundary
have recently been receiving a great deal of attention since these appear to be the most stable
with respect to light-induced degradation in photovoltaic devices [1 ,2]. However, it is also
known that, once the μc-Si volume fraction becomes appreciable, the performance of such
devices suffers considerably [1,3]. In this paper we report results from a study of samples
produced by dc sputtering under hydrogen or deuterium dilution. Using TEM we have confirmed
that these samples contain clusters of Si microcrystallites embedded in a matrix that is
predominantly high quality amorphous silicon. We have characterized the electronic properties
of these samples using photocapacitance, transient junction photocurrent, and modulated
photocurrent methods. Our results reveal a unique optical transition associated with the
embedded crystallites which behaves like a defect level within the amorphous silicon host but
with an unusually large optical cross section. We believe this defect may have a considerable
impact on the performance of devices employing such mixed phase materials.

SAMPLE PREPARATION

Hydrogenated and deuterated amorphous silicon films were deposited at 100Å/min by dc
reactive magnetron sputtering of a 5"×12" planar Si target in an Ar + H_2 or Ar + D_2 plasma. We
employed a substrate temperature of 250°C, an Ar partial pressure of 1.5 mTorr, and varied the
H_2 or D_2 from 0.3 to 0.8 mTorr. Films grown under similar conditions using slightly lower
hydrogen pressure have been found to have electronic properties fully equivalent to a-Si:H
deposited by plasma CVD [4], and we previously have demonstrated that the transition to fully
μc-Si growth occurs at a hydrogen partial pressure of 4 mTorr [5], and roughly 2 mTorr for
deuterium [6].

To carry out the photocapacitance and photocurrent studies described below we evaporated
semitransparent Pd contacts onto each film; however, we reverse biased the junction at the p$^+$ c-
Si substrate to probe the films in a region roughly 3000 to 6000Å from the substrate interface.
Samples were studied in "state A" (annealed at 470K in the dark), and in a light soaked "state B",
produced using red-filtered tungsten-halogen light at an intensity of 300mW/cm^2 for 150 hours.

Mat. Res. Soc. Symp. Proc. Vol. 557 © 1999 Materials Research Society

FIG. 1. Bright-field TEM image of sample 1. The diffraction pattern from the region circled indicates that the dark regions in the micrograph consist of fine grained crystalline silicon.

EXPERIMENTAL METHODS

The presence and morphology of the μc-Si component of these films was obtained using plan-view TEM. An example is shown in the bright-field image of Fig. 1. The selected area diffraction pattern from the region indicated by the circle confirms that the darker regions in the bright-field image consist of an agglomerate of small crystallites. We also carried out Raman spectroscopy on these films which confirmed a small μc-Si component for our samples and allowed us to estimate the microcrystalline volume fraction to be at or below 5% [7].

We established the energy distribution of defects in these samples by recording the sub-band-gap spectra using transient photocapacitance (TPC) and transient junction photocurrent (TPI) spectroscopy. As we have discussed previously [8,9], the TPC method detects the optically induced charge change within the depletion region while the TPI method detects the charge motion. Therefore, a transition between a gap state to the majority (conduction) band will *increase* the junction capacitance while those to the minority band will *decrease* the capacitance. Thus, the TPC spectra reflect the difference *between* these two types of transitions. In contrast, the sign of the current arising from either type of transition is the same, so that the TPI spectra reflect the *sum*.

The actual TPC (or TPI) spectra are obtained by averaging the capacitance (or current) transient signal during a time window following an applied voltage filling pulse, with and without the presence of the sub-band-gap light. The difference of the averaged signal between light and dark, normalized to the photon flux, is then plotted as a function of photon energy to provide the photocapacitance (or the junction photocurrent) spectrum. Examples of TPC and TPI spectra for one of our samples is shown in Fig. 2. These spectra will be discussed in detail in the next Section.

An additional method we employed in this study was the modulated photocurrent (MPC) method. The MPC method is used to disclose the thermal energy distribution of deep defects that act a majority carrier traps [10,11]. It involves measuring the amplitude and phase of the photocurrent response to a weak oscillating light source. The experimental details of the manner in which we have implemented this techniques have been given elsewhere [12]. Example spectra obtained for one of our samples are presented in the next Section.

RESULTS

Figure 2 displays TPC and TPI spectra for one of our deuterated samples in state B at several intensities of sub-band-gap light. We have overlapped the two types of spectra in the lowest energy regime (below 0.9eV) since, at optical energies less than one half the gap, only majority carrier transitions are possible and the current and capacitance signals should be strictly proportional. Moreover, in this energy regime the TPI and TPC obey a strict linear relation with light intensity; hence, these normalized spectra appear independent of intensity.

However, while the spectral dependence below 1.0eV appears consistent with most previously reported spectra for a-Si:H [9] indicating transitions from the dominant ("dangling bond") deep defect band, these spectra reveal a very different behavior compared to standard glow discharge a-Si:H samples in the intermediate energy regime, between 1.0 to 1.5eV. In the TPI spectra we observe an intermediate energy "shoulder" with a threshold close to the c-Si

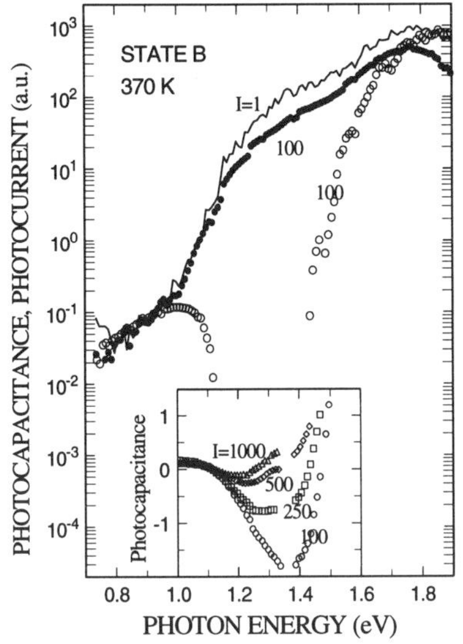

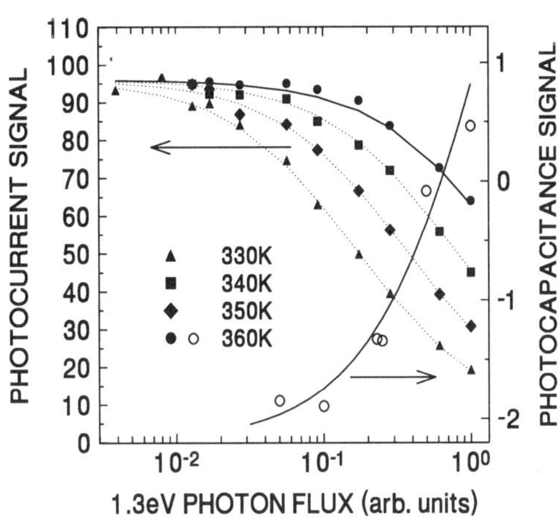

FIG. 2. Photocapacitance (open symbols) and photocurrent spectra (lines and solid symbols) of sample 1 in its light soaked state. The results at several different light intensities are shown. The inset shows the intensity dependence of the photocapacitance in more detail on a *linear* scale.

FIG. 3. The intensity dependence of the normalized photocapacitance and photocurrent spectra at 1.3eV for state B of sample 1. The highest light flux was 4.5×10^{15} cm^{-2}s^{-1}. The solid lines are fits to the 360K data based upon the set of optical transitions shown in the inset of Fig. 6. The dotted lines are fits to the photocurrent spectra at other temperatures obtained only allowing the parameter γ to vary.

bandgap which exhibits a marked deviation from strict linearity. This feature appears in the entire series of films and its magnitude is roughly proportional to the integrated crystalline Raman component for these samples [7]. Thus, we believe it is associated with the crystallites contained within these films. In the photocapacitance spectra this feature has a negative sign, implying a dominance of a minority band transitions. Like its TPI counterpart, this TPC spectral feature displays a marked non-linearity with light intensity.

The normalized photocapacitance and photocurrent signals at 1.3eV (the energy where the feature is most prominent) are displayed over a much wider range of light intensities in Fig. 3. The photocurrent signal dependence is also shown for a series of lower temperatures. Unfortunately, at temperatures below 360K the photocapacitance signals are difficult to measure due to the much longer dielectric response times. All of the intensity dependences can be fit quite well using a simple set of transitions which are described in the next Section.

Finally, in Fig. 4 we display MPC spectra for this same sample in State B at the same set of measurement temperatures used for the TPI spectral measurements. At each measurement temperature the MPC spectra display a peak for a particular modulation frequency. These peak frequencies, displayed in an Arrhenius plot in Fig. 5, indicate a thermally activated electron detrapping process with an activation energy near 0.76eV. While standard glow discharge a-Si:H samples exhibit similar MPC peaks, the activation energy is generally about 0.1eV smaller, and the distribution is considerably narrower than those indicated in Fig. 4. [12]

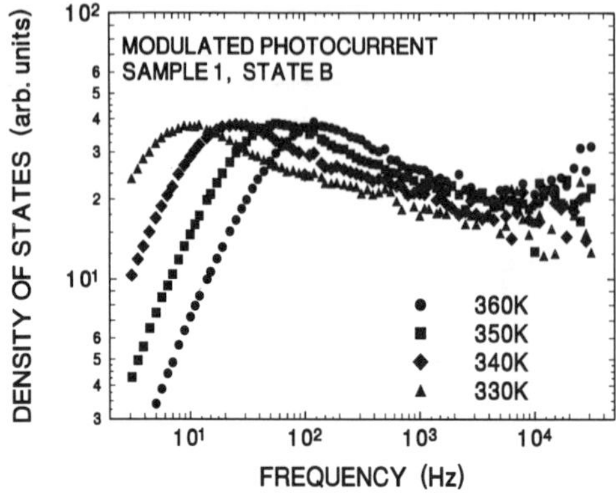

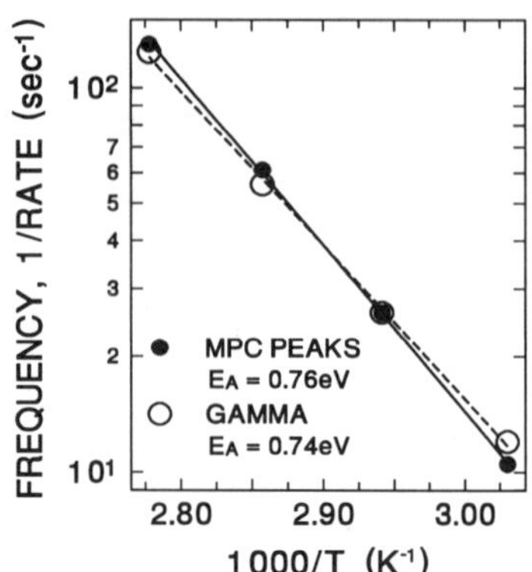

FIG. 4. Modulated photocurrent spectra for sample 1 in state B. The four temperatures used match those employed for the transient photocurrent measurements displayed in Fig. 3.

FIG. 5. Arrhenius plot of peak frequencies of MPC spectra from Fig. 4 (solid circles) and of the parameter γ obtained from fitting the intensity dependence of the transient photocurrent data of Fig. 3.

INTERPRETATION

To better understand the characteristics of the unusual intermediate energy optical transition in Fig. 2, we display in Fig. 6 the actual capacitance transients, with and without the presence of the 1.3eV light. One clearly observes that there are two distinct types of optical transitions, a fast component which *decreases* the capacitance as soon as the light is turned on, and a component that *increases* the capacitance much more slowly. When the light is turned off, the faster component of the capacitance change is reversed within a few milliseconds. This behavior suggests the set of transitions shown in the inset involving two distinct gap states. We have labeled these states D and X.

We next demonstrate that this proposed set of transitions can account in detail for the intensity dependence of the TPI and TPC signals displayed in Fig. 3. First, we note that our spectra are recorded during a time window extending from 0.3s to 0.7s following the onset of illumination. This means that steady-state conditions (between optical filling and thermal emptying) have been achieved for the fast component (X), while the optical emptying of the slow component (D) is still in its initial stage.

We assume that level D is occupied in thermal equilibrium so that the sub-band-gap light excites electrons from D and into the conduction band at the rate $\alpha\Phi$, where Φ denotes the optical flux. Since this rate is slow relative to the time delay, τ, before the experimental window, this implies the *loss* of $\alpha\Phi\tau N_D$ electrons per unit volume from these states (relative to the sample in the dark). Level X is *unoccupied* in equilibrium and the light excites valence band electrons into these states at a rate $\beta\Phi$. However, these electrons are lost by thermal excitation to the conduction band at a rate γ. Under steady state conditions, the electronic occupation fraction of level X will be given by: $f = \beta\Phi/(\beta\Phi+\gamma)$. This corresponds to the *addition* of fN_X electrons per unit volume into the gap. Thus, the TPC signal will be proportional to the total charge change normalized to the flux, or:

FIG. 6. Capacitance transient data for sample 1 at two different light intensities using 1.3eV light. Note the two types of optical excitation processes that occur during the light-on region: a fast process (labeled $\beta\Phi$) that *decreases* the junction capacitance, and a slow process (labeled $\alpha\Phi$) that *increases* the capacitance. When the light is turned off, a fast thermal process (labeled γ) reverses the change caused by the fast optical process. The inset indicates our proposed model for the types of processes which account for the observed changes in capacitance.

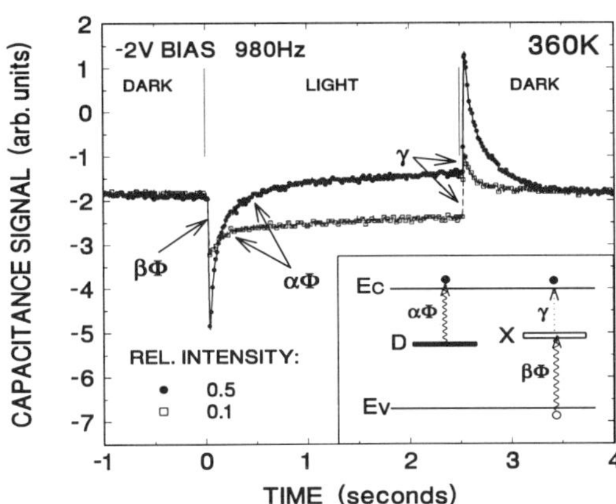

$$S_{TPC} \;\propto\; \alpha\tau N_D - \frac{\beta}{\beta\Phi+\gamma}N_X \; . \tag{1}$$

The corresponding TPI signal due to these transitions states will be proportional to the density of electrons inserted into the conduction band per unit time plus a fraction, ξ, of the number of holes (limited by their lower mobility). One has, therefore:

$$S_{TPI} \;\propto\; \alpha N_D + \frac{\beta\gamma}{\beta\Phi+\gamma}N_X(1+\xi) \; . \tag{2}$$

The parameters τ and ξ are known independently, and we have normalized both signals to the same value at low optical energies (which sets the constants of proportionality). This leaves only four parameters α, β, γ and the ratio N_X/N_D, to fit the TPC plus TPI intensity dependence.

These fits are displayed along with the data in Fig. 3. A single set of the four parameters were used except for the TPI data at the lower temperatures, for which we varied only the value of γ, the thermal escape rate for the electrons from X. The variation of the γ values with T is plotted along with the activation energy of the MPC spectral peak in Fig. 5, and indicates a nearly identical activation energy of (0.74eV *vs.* 0.76eV for the MPC peaks).

DISCUSSION

The simple model described above is able to quite successfully account for the intensity dependence of both the TPC and TPI spectra. We believe that the D level arises from bulk defects in the a-Si:D host material including, perhaps, an excess of such defects in the vicinity of the microcrystallites. To support this we note that the value of the fitting parameter α is quite consistent with the optical cross section determined for deep defects in a-Si:H (of 1 to 3×10^{-16} cm^2 [8]).

The optical cross section for X, on the other hand, is much too high for a normal point defect. This is indicated, first of all, by the raw capacitance transient data of Fig. 6 which indicates a rate faster than $(10\text{ms})^{-1}$ at a photon flux near 3×10^{15} cm^{-2} s^{-1}. This implies an optical cross section of at least 10^{-14} cm^2. Moreover, we know that this transition can be saturated at relatively high temperatures while competing with a thermal process having an activation energy 0.74eV. Indeed, the close match of activation energies between the peak of the MPC band and the thermal escape rate from the X defect (Fig. 5) suggests that they are the same.

Since we know the prefactor for this MPC peak (1×10^{13} s^{-1}), this implies an optical cross section of 4×10^{-14} cm^2.

Given that the optical threshold for X lies very near the c-Si bandgap energy, we hypothesize that X is populated via interband transitions in the embedded crystallites. The electrons introduced into the conduction band of the crystallites might then be trapped at defects near the crystallite/amorphous silicon host interface. This would help account for the unusually large effective optical cross section. Surprisingly, even with this proposed mechanism the observed cross section is still at least an order of magnitude larger than would be predicted from the optical absorption of bulk crystalline silicon at 1.3eV assuming 100-500Å diameter crystallites. Therefore, additional factors, such as quantum size effects, must also be playing a role to enhance the absorption in the embedded crystallites.

The proposed mechanism for the optical population of the X defect does not depend in any obvious manner on the band offset energies between the amorphous Si and the crystallites. However, one might also speculate that if the majority of the offset were at the conduction band edge, then the bottom of the c-Si conduction band would be nearly iso-energetic with the defect and give rise to a closely coupled system. This might also help to account for the magnitude of optical cross section that is observed. This last hypothesis disagrees with recent studies which indicate that the major part of this offset occurs at the valence band [13 ,14]. However, the present case of embedded small crystallites may exhibit fundamentally different interfacial energies than those of the large planar interfaces of those other studies.

Finally, we suggest that the defect that we have discovered may help contribute to the known significant degradation of performance of amorphous silicon devices fabricated from material near this mixed phase boundary once the microcrystalline fraction becomes significant [3]. Additional studies will hopefully help clarify the role this defect may have with respect to photo-transport processes in amorphous silicon.

ACKNOWLEDGMENTS

We thank Hao Lee and Pete Sercel for carrying out Raman measurements on several of these samples and acknowledge useful discussions with Yoram Lubianiker. This work was supported by NSF Grant DMR-9624002 at Oregon and, at Illinois, by the Joint Services Electronics Project of the Office of Naval Research, and by DOE contract DEFG02-91-ER45349.

REFERENCES

1. D.V. Tsu, B.S. Chao, S.R. Ovshinsky, S. Guha, and J. Yang, Appl. Phys. Lett. **71**, 1317 (1997).
2. J.H. Koh, Y. Lee, H. Fujiwara, C.R. Wronski, and R.W. Collins, Appl. Phys. Lett. **73**, 1526 (1998).
3. S. Guha, J. Yang, D.L. Williamson, Y. Lubianiker, J.D. Cohen, and A.H. Mahan, Appl. Phys. Lett. **74**, 1860 (1999)..
4. M. Pinarbasi, M. J. Kushner, and J. R. Abelson, J. Appl. Phys. **68**, 2255 (1990).
5. G. Feng, M. Katiyar, Y. H. Yang, J. R. Abelson, and N. Maley, Mat. Res. Soc. Symp. Proc. **258**, 179 (1992).
6. J.E. Gerbi and J.R. Abelson, Mat. Res. Soc. Symp. Proc. **507**, 429 (1998).
7. D. Kwon, H. Lee, J.D. Cohen, H.-C. Jin, and J.R. Abelson, J. Non-Cryst. Solids **227-230**, 1040 (1998).
8. A.V. Gelatos, J.D. Cohen, and J.P. Harbison, J. Non-Cryst. Solids **77&78**, 291 (1985).
9. A.V. Gelatos, K.K. Mahavadi, J.D. Cohen, and J.P. Harbison, Appl. Phys. Lett. **53**, 403 (1988).
10. H. Oheda, Philos. Mag. B**52**, 857 (1985).
11. G. Schumm and G.H. Bauer, Phys. Rev. B**39**, 5311 (1989).
12. F. Zhong and J.D. Cohen, , Mat. Res. Soc. Symp. Proc. **258**, 813 (1992).
13. R.C. Fang and L. Ley, Phys. Rev. B**40**, 3818, (1989).
14. J.M. Essick and J.D. Cohen, Appl. Phys. Lett. **55**, 1232 (1989).

POLYMORPHOUS SILICON: TRANSPORT PROPERTIES AND SOLAR CELL APPLICATIONS

C. Longeaud*, J. P. Kleider*, M. Gauthier*, R. Brüggemann*, Y. Poissant[**], and P. Roca i Cabarrocas**
* Laboratoire de Génie Electrique de Paris (UMR 8507 CNRS), Supélec, Universités Paris VI et XI, Plateau de Moulon, 91192 Gif sur Yvette Cedex, France, longeaud@lgep.supelec.fr
** Laboratoire de Physique des Interfaces et Couches Minces (UMR 7647 CNRS), Ecole Polytechnique, 91128 Palaiseau, France

ABSTRACT

Transport properties of hydrogenated polymorphous silicon layers (pm-Si:H) deposited at 150 °C under various pressures in the range 80-293 Pa in sandwich (Schottky and p-i-n diodes) and coplanar structures have been compared to those of hydrogenated amorphous silicon (a-Si:H) samples deposited at the same temperature in standard conditions. The layers have been studied as-deposited, annealed and after light-soaking. With increasing pressure up to 240 Pa: i) the density of states above the Fermi level decreases as determined by means of the modulated photocurrent technique, ii) the mobility-lifetime products of electrons and holes measured by means of steady-state photoconductivity and photocarrier grating techniques both increase. The highest values for the diffusion length of minority carriers exceed 200 nm. Capacitance measurements as a function of frequency and temperature show that the density of states at the Fermi level is lower in the pm-Si:H than in the a-Si:H films. After light-soaking the diffusion length of minority carriers in a-Si:H is reduced by a factor oftwo whereas it is less reduced or not affected in the pm-Si:H layers. Solar cells including this new material present an excellent stability.

INTRODUCTION

Exploring the space of parameters in radio frequency powered plasma enhanced chemical vapour deposition (RF-PECVD) we have shown that it is possible to deposit a new type of material exhibiting enhanced transport properties at least as far as the majority carriers are concerned [1, 2]. In particular we have found that, deposited at a substrate temperature of 250 °C under high dilution of silane into hydrogen, helium or argon with a high pressure of the gas mixture and at high RF power, films of this new material present very high values of the mobility lifetime product for electrons ($\mu\tau_e$) compared to films of hydrogenated amorphous silicon (a-Si:H) deposited under standard conditions (no dilution of silane, low pressure and low RF power) [3]. Moreover, the behaviour of this new material under light-soaking makes it very promising for photovoltaic applications [4]. We believe that these peculiar properties are due to the microstructure of the material made of ordered nanoparticles of silicon embedded into an amorphous matrix as shown by high resolution transmission electron microscopy [5]. These nanoparticles are produced in the plasma and deposited onto the film since the glow discharge occurs in a regime close to powder formation [3,5]. Because of the presence of nano-sized ordered domains within the amorphous phase we have named this material polymorphous silicon (pm-Si:H).

While most of the materials studies have focused on layers deposited at 250 °C, the solar cells are usually prepared at 150 °C for example to avoid the reduction of the SnO_2 substrate by the hydrogen of the discharge as well as to decrease the diffusion of impurities from the substrate into the intrinsic layer. Therefore we focus here on pm-Si:H films prepared at a substrate temperature of 150 °C and show that this material is indeed an excellent candidate for photovoltaic applications.

Mat. Res. Soc. Symp. Proc. Vol. 557 © 1999 Materials Research Society

SAMPLES AND EXPERIMENTS

The samples studied were deposited in a radio frequency powered (13.56 MHz) PECVD system [6] at a substrate temperature T_s of 150 °C. For the polymorphous films, the silane was highly diluted into hydrogen (3% of SiH_4 in 97 % of H_2) and the RF power P_{rf} was of the order of 110 mW/cm^2. The total pressure P in the chamber was varied between 80 and 293 Pa. The transport properties of these samples were compared to those of an a-Si:H sample prepared under standard conditions (T_s=150 °C, P_{rf} =5 mWcm^{-2}, P= 5 Pa). For both types of materials different types of samples were prepared: simple films deposited onto 7059 Corning glass, Schottky diodes and p-i-n diodes, the i-layer being made of either pm-Si:H or a-Si:H.

Two parallel ohmic aluminum electrodes 2 mm apart were thermally evaporated on the films deposited onto glass in order to perform the following experiments: conductivity and steady-state photoconductivity (SSPC) at different light intensities, modulated photocurrent (MPC) and steady-state photocarrier grating (SSPG). From these experiments we determined respectively the value of the dark conductivity and of the activation energy, the mobility-lifetime ($\mu\tau_e$) product of the majority carriers, here assumed to be the electrons, the shape of the density of states above the Fermi level, hereafter referred to as the MPC-DOS, and the mobility-lifetime ($\mu\tau_h$) product for the minority carriers (i.e. the holes).

On the Schottky diodes we performed capacitance measurements as a function of frequency and temperature to determine the density of states at the Fermi level. We also performed time of flight measurements to determine the drift mobilities of the carriers.

Finally, the I-V characteristics of the solar cells were measured at different light intensities to obtain the effective $\mu\tau$ of carriers in the intrinsic layer of the pm-Si:H cells by means of the variable irradiance method (VIM) [7].

The samples were studied in their as-deposited, annealed and light-soaked states. The light soaking was achieved under white light with a flux of the order of 400 mW/cm^2 at a temperature of 80 °C during 70 hours for the films and 50 °C during 24 hours for the pin diodes. For the films the kinetics of the light soaking process was monitored by the evolution of the dark and photo conductivity measured at 20°C.

RESULTS

For all the studied samples the room temperature dark conductivities and activation energies range between 3×10^{-12} and 6×10^{-11} S.cm^{-1} and 0.7 and 0.9 eV, respectively. In Fig. 1 we present the evolution with pressure of the mobility-lifetime product for electrons and holes measured under red light (λ=670 nm) with a flux of 10^{17} cm^{-2}s^{-1}. These products were measured before and after light-soaking. The comparison with the results obtained on a standard a-Si:H sample shows that in the as-deposited state the $\mu\tau$ products for holes and electrons are of the same order of magnitude or slightly lower than those measured on the standard sample. This is in contrast to what we obtain at a lower flux of 10^{14} cm^{-2}s^{-1} where the $\mu\tau$ products for electrons for the pm-Si:H material are higher than for the a-Si:H layer (see Figure 2). It can also be seen that as the pressure increases up to 238 Pa both $\mu\tau_e$ and $\mu\tau_h$ increase simultaneously underlining a continuous improvement of the material quality. Finally it appears that in the range 200-260 Pa there is an optimum for the transport parameters and that in this range of pressure there is only a small influence of the light-soaking process on $\mu\tau_h$. As a consequence, after light-soaking, the $\mu\tau_h$ of the sample deposited at 238 Pa is three times higher than for the a-Si:H sample. Moreover, the diffusion length of the minority carriers in the light-soaked state is higher than 0.2 μm for the pm-Si:H sample deposited at 238 Pa.

We present in Figure 2 the evolution with pressure of the value of the MPC-DOS measured at 0.5 eV below the conduction band edge for samples in the as-deposited state. Also presented are the evolution of $\mu\tau_e$ measured under a dc flux of 10^{14} cm^{-2}s^{-1}. Clearly, there is a good correlation between these two quantities: the lower the MPC-DOS the higher the $\mu\tau_e$. Note that the $\mu\tau_e$

products were obtained here on as-deposited samples without any annealing. An annealing of the samples at the deposition temperature of 150 °C during 1-2 hours leads to an increase of these products by a factor of five to ten for the pm-Si:H material and not for the a-Si:H sample.

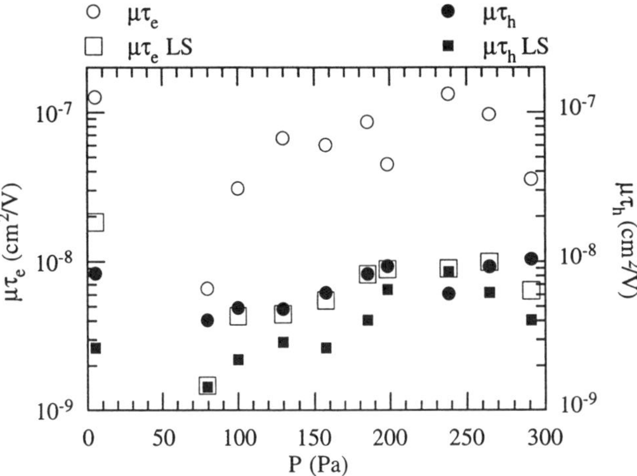

Figure 1 : Evolution vs pressure of the $\mu\tau$ products of electrons and holes, measured with a dc flux of 10^{17} cm^{-2}s^{-1} before and after light-soaking (LS). Also shown are the $\mu\tau$ products of electrons and holes for a standard a-Si:H sample prepared at the same temperature at P = 5 Pa.

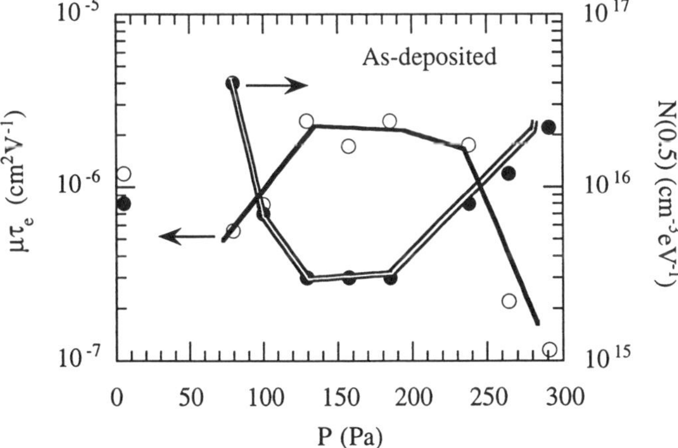

Figure 2 : Evolution vs pressure of the MPC-DOS measured at 0.5 eV below the conduction band edge for samples in the as-deposited state. Also presented are the $\mu\tau_e$ obtained at $F_{dc}=10^{14}$ cm^{-2}s^{-1}. These values can be compared to those obtained on a standard sample shown at P=5 Pa. Lines are guides to the eyes.

To understand this increase we have performed capacitance measurements versus temperature and frequency $C(T,\omega)$ on a Schottky diode, the layer of which was deposited at 238 Pa, since it is around this pressure that the pm-Si:H material seems to present the best transport parameters (see Fig. 1). The value of the DOS at the Fermi level we found was lower than 10^{15} cm^{-3}eV^{-1}, a value which is among the lowest ever found in intrinsic a-Si:H samples. This lowering of the Fermi level DOS in pm-Si layers compared to a-Si:H layers can be considered as the main reason of the high values of the $\mu\tau_e$ products. Indeed, time-of-flight measurements have shown that the drift mobility was of the order of 1.5 cm^{2}V^{-1}s^{-1} comparable to the value found in standard a-Si:H. The increase of the $\mu\tau_e$ products in pm-Si:H material prepared at 150 °C is thus mainly due to a lowering of the DOS instead of an increase of the mobility as already suggested for the pm-Si:H prepared at 250 °C [2].

We present in Figure 3 the evolution with pressure of the value of the MPC-DOS measured at 0.5 eV below the conduction band edge for samples in the light-soaked state. Also presented are the evolution of $\mu\tau_e$ measured under a dc flux of 10^{14} cm^{-2}s^{-1}. Again, there is a good correlation between these two quantities: the lower the MPC-DOS the higher the $\mu\tau_e$. It can also be seen that the lower values of MPC-DOS in the light-soaked state are obtained for the samples deposited in the pressure range 200-260 Pa.

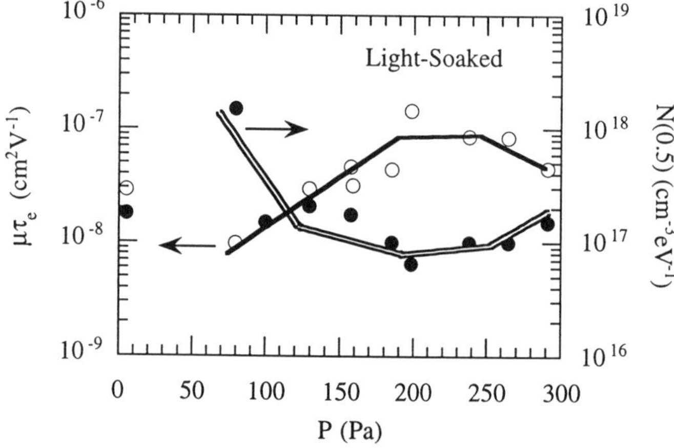

Figure 3 : Evolution vs pressure of the MPC-DOS measured at 0.5 eV below the conduction band edge for samples in the light-soaked state. Also presented are the $\mu\tau_e$ obtained at $F_{dc}=10^{14}$ cm^{-2}s^{-1}. These values can be compared to those obtained on a standard sample shown at P=5 Pa. Lines are guides to the eyes.

Considering the results presented in Figs. 1 - 3 one expects the best solar cells to be produced in the pressure range of 200 - 260 Pa. This is confirmed by the results presented in Table I where we report the material and solar cells parameters for a series of p-i-n solar cells deposited at 150 °C and having an i-layer thickness of the order of 0.4 µm. The p-layer for this series of cells consists of an a-SiC:H alloy obtained from the dissociation of a gas mixture consisting of 6 sccm of silane, 5 sccm of methane, and 60 sccm of hydrogen. An undoped a-SiC:H buffer layer was inserted between this p-layer and the intrinsic layer.

As shown in Table I the effective $\mu\tau$ product, deduced from the VIM measurements, presents a maximum at 210 Pa in agreement with the fill factor of the solar cell. We note that the values of fill factor and efficiencies reported in this table are lower than those previously reported for pm-

Si:H solar cells [4]. This is attributed to a non-optimization of the solar cells presented here since our purpose was mainly to correlate the material transport properties to the solar cell parameters.

Table I: I(V) characteristics and $(\mu\tau)_{eff}$ of pm-Si:H cells deposited at different pressures.

Pressure (Pa)	$(\mu\tau)_{eff}$ (cm^2V^{-1})	FF	V_{oc} (V)	I_{sc} (mA/cm^2)	Eff. (%)
132	1.31×10^{-8}	0.64	0.79	8.8	4.3
158	1.53×10^{-8}	0.66	0.82	9.9	5.3
210	3.71×10^{-8}	0.67	0.85	10.9	6.2
238	1.44×10^{-8}	0.58	0.81	10.9	5.2
263	1.42×10^{-8}	0.62	0.82	11.2	5.6

Turning now to the stability of the solar cells, Fig. 4 shows the spectral response of a solar cell produced at 238 Pa. Indeed, at this pressure we expect a stable efficiency (see Figure 1) since the hole diffusion length is not altered by the light-soaking process. As shown in Figure 4, the spectral response of the solar cell does not change after 24 hours white-light soaking under 4 AM1 at 50 °C except slightly at short λ. As indicated above, the solar cells deposited at high pressure have not been optimized. Therefore the absolute values of the spectral response are low with respect to those obtained in solar cells with pm-Si:H layers deposited at lower pressure and presented elsewhere [4].

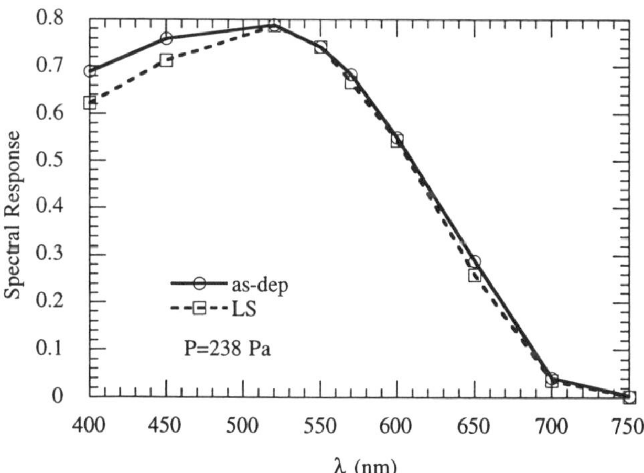

Figure 4 : Spectral Response before (as-dep.) and after 24 h light-soaking (LS) of a pm-Si:H cell deposited at 238 Pa. The cell thickness is 0.42 µm. Lines are guides to the eyes.

The importance of the optimization on the absolute efficiency of the solar cells is demonstrated by the results reported in Table II. These results have been obtained with a p-i-n solar cell in which we replaced the p(a-SiC:H) layer by a p(μc-Si:H) layer, the intrinsic layer being deposited at 238 Pa. As compared to the results of Table I we obtain higher values of the fill factor and open circuit voltage. Morevover the short circuit current of the solar cell remains almost constant, in agreement with the spectral response results (Fig. 4). The higher value of Jsc is due to the smaller absorption in the p(μc-Si:H) layer. As previously reported [4], we can see in Table II that although the FF of the solar cell decreases upon light-soaking, the efficiency remains constant because of the increase in open circuit voltage.

Table II: I(V) characteristics of a pm-Si:H cell in its as-deposited state and after 24 hours light-soaking at 50 °C under 4 AM1 conditions.

State	FF	J_{sc} (mA/cm^2)	V_{oc} (V)	Efficiency (%)
As-deposited	0.63	13.4	0.89	7.5
Light-soaked	0.60	13.0	0.94	7.3

CONCLUSION

We have studied the transport properties of both minority and majority carriers as well as stability of polymorphous silicon films produced at different pressures at 150 °C. We observe a clear pressure range (200 $\leq$ P $\leq$ 260 Pa) where the $\mu\tau$ products for electrons and holes are maximum, corresponding to a minimum in the defect density. Moreover, for the best samples, the $\mu\tau$ product of holes does not show any degradation. The optimum in the material properties has been correlated to the solar cell parameters, both in terms of efficiency and stability.

ACKNOWLEDGEMENTS

This work was supported by the Centre National de la Recherche Scientifique (CNRS) and the Agence de l'Environnement et de la Maitrise de l'Energie (A.D.E.M.E.) within the framework of their joint photovoltaic program (ECODEV).

REFERENCES

1. A. R. Middya, S. Hazra, S. Ray, A. K. Barua and C. Longeaud, J. Non-Cryst. Solids **198-200**, 1067 (1996).

2. C. Longeaud, J.P. Kleider, P. Roca i Cabarrocas, S. Hamma, R. Meaudre and M. Meaudre, J. Non-Cryst. Solids **227-230**, 96 (1998).

3. P. Roca i Cabarrocas, in *Amorphous and Microcrystalline Silicon Technology*, edited by R. Schropp, H. M. Branz, M. Hack, I. Shimizu, and S. Wagner (Mat. Res. Soc.. Proc. **507**, San Fransisco, CA, 1998)

4. P. Roca i Cabarrocas, P. St'ahel and S. Hamma, Proc. of the 2nd World Conference on Photovoltaic Solar Energy Conversion, edited by J. Schmid, H.A. Ossenbrink, P. Helm, H. Ehmann and E.D. Dunlop, Vienne, Austria, 355 (1998).

5. P. Roca i Cabarrocas, S. Hamma, S. N. Sharma, G. Viera, E. Bertran and J. Costa, J. Non-Cryst. Solids **227-230**, 871 (1998).

6. P. Roca i Cabarrocas, J. B. Chevrier, J. Huc, A. Lloret, J. Y. Parey and J. P. M. Schmitt, J. Vac. Sci. Technol. **A 9**, 2331 (1991).

7. J. Merten, J.M. Asensi, C. Voz, A.V. Shah, R. Platz, and J. Andreu, IEEE Trans. on Electron Devices **45**, 423 (1998).

A NOVEL P-TYPE NANOCRYSTALLINE SI BUFFER AT THE P/I INTERFACE OF A-SI SOLAR CELLS FOR HIGH STABILIZED EFFICIENCY

Chang Hyun LEE, Koeng Su LIM
Department of Electrical Engineering, Korea Advanced Institute of Science and Technology, 373-1 Kusong-dong, Yusong-gu, Taejon 305-701, Korea, chlee@pretty.kaist.ac.kr

ABSTRACT

We investigated the properties of a novel p-type nanocrystalline Si (p-nc-Si) prepared onto p-a-SiC and the effect of using the buffer with an energy bandgap over 1.9 eV at the p/i interface on the performance of p/i/n type amorphous silicon based solar cells. At the initial growth stage of the p-nc-Si onto p-a-SiC, Si nanocrystallites are proved to be formed in amorphous matrix within the thickness of less than $100\,\text{Å}$. The open circuit voltage and the blue response of the cell were improved significantly by inserting the film at the p/i interface as a buffer as compared with the bufferless cell. We found from a numerical simulation using the Gummel-Sharfetter method that the buffering effect of the p-nc-Si is originated from the reduction of highly defective region with a short life-time in the vicinity of the p/i interface.

INTRODUCTION

The abrupt bandgap discontinuity and mismatch originated from the p-a-SiC (E_g = 2.0 eV)/i-a-Si (E_g = 1.72 eV) heterojunction result in the formation of thin several nanometer-thick high defect zone with a short carrier life time at the interface and thereby the carrier recombination loss increases severely in amorphous silicon solar cells. To solve this problem, the carbon-alloy graded or constant bandgap a-SiC buffer layer inserted at the p-a-SiC and i-a-Si has been developed [1-3]. The buffer layer improves the open circuit voltage (V_{oc}) and the short wavelength response. A high conversion efficiency of 12 % has been achieved using the buffer layer [4].

In this paper, we report on the properties of a novel p-type ultrathin nanocrystalline Si (p-nc-Si) buffer layer that was developed in this study and the effects of inserting the buffer at the p-a-SiC/i-a-Si interface on the cell performance of p/i/n type amorphous silicon (a-Si) based solar cells.

EXPERIMENTAL

We used a three separated-chamber mercury-sensitized photochemical vapor deposition (photo-CVD) apparatus for fabricating p-nc-Si films and a-Si based solar cells. P-nc-Si:H films were prepared from SiH_4, H_2 and B_2H_6. The reactant gases were passed through a mercury pot maintained at $20\,°C$ to the p-chamber. The hydrogen dilution ratio (H_2/SiH_4) was 30. The substrate temperature was maintained at $250\,°C$. Fabricated solar cells have a simple structure of glass/SnO_2/p-a-SiC/p-nc-Si/i-a-Si/n-μc-Si/Al. The cell area is 0.09 cm^2. The p-nc-Si film was prepared on Corning 7059 glass. The dark and photo-conductivity measurements were performed on the sample with a parallel electrode configuration. The optical properties of the film were characterized by a spectroscopic ellipsometer. The absorption coeffcients were deduced by fitting the dielectric functions of a-Si and c-Si to the measured psi and delta from Bruggeman effective medium theory [5].

Mat. Res. Soc. Symp. Proc. Vol. 557 © 1999 Materials Research Society

RESULTS

<u>Preparation of p-type nanocrystalline Si film</u>

Figure 1 shows the dark and photo conductivities of the p-nc-Si:H films prepared onto glass and 35 Å-thick p-a-SiC as a function of film thickness. With decreasing the thickness of the p-nc-Si film, the dark conductivity is reduced exponentially because electrons transport through a larger fraction of amorphous tissue layer with a high photosensitive and resistive characteristics among the crystallites.

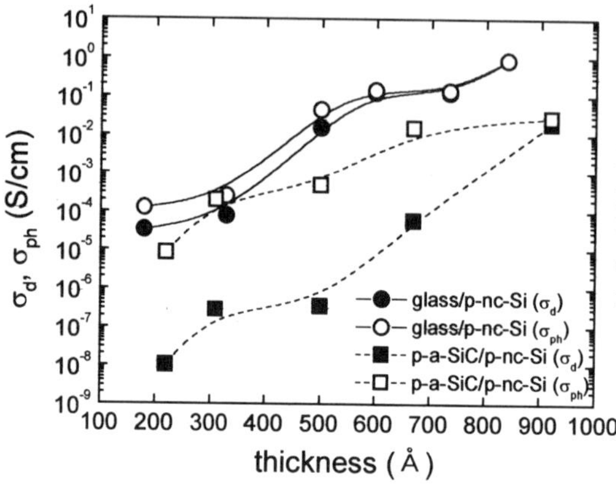

Fig.1 Dark conductivity (σ_d) and photo conductivity (σ_d) of p-nc-Si film grown on glass (● σ_d, ○ σ_{ph}) and on p-a-SiC(■ σ_d, □ σ_{ph}) as a function of film thickness.

When the electrical conductivity of the film deposited onto p-a-SiC is compared with that of the film onto glass, it can be known that the presence of an initial p-a-SiC layer hinders the formation of crystallites and enhances the formation of a void-free denser amorphous matrix than that of the p-nc-Si film prepared onto glass. Thereby the p-nc-Si film has higher photosensitivity and lower dark conductivity when it is deposited onto p-a-SiC. It has been known that crystallization process proceeds after a hydrogen-rich porous amorphous layer, which is called an incubation layer, is formed within ~100 Å thickness when microcrystalline Si is grown onto a-Si substrate [6]. According to the fact, it is not certain if thin ~100 Å-thick p-μc-Si layer, as used as a window layer in solar cell, has microcrystalline phase or fully

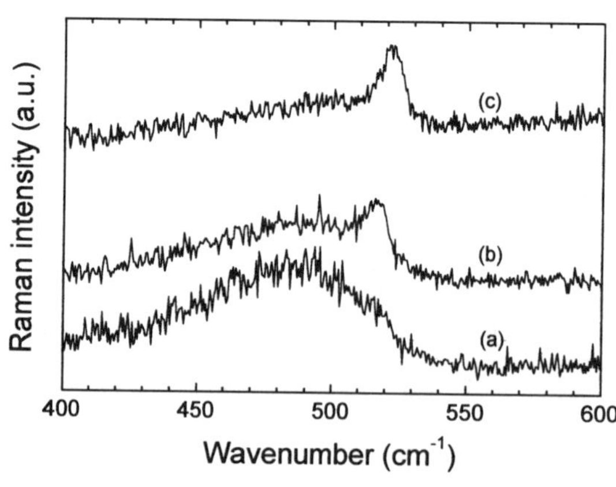

Fig. 2 Raman spectra of the p-nc-Si films grown on p-a-SiC with a structure of
(a) (52 Å-thick p-a-SiC/80 Å-thick p-nc-Si)$_5$ multilayer
(b) (17 Å-thick p-a-SiC/61 Å-thick p-nc-Si)$_{10}$ multilayer
(c) (35 Å-thick p-a-SiC/910 Å-thick p-nc-Si)$_1$.

amorphous phase. To investigate the structural properties of the ultrathin p-nc-Si grown onto p-a-SiC, the multilayers of p-a-SiC and p-nc-Si were prepared as shown in Fig. 2. The existence of nanometer-size Si crystallites in the p-nc-Si film grown onto p-a-SiC at an initial growth regime below a thickness of ~100 Å is confirmed by the 520 cm^{-1} crystal peak from Raman spectrum of the (52 Å-thick p-a-SiC/80 Å-thick p-nc-Si)$_5$ multilayer and the (17 Å-thick p-a-SiC/61 Å-thick p-nc-Si)$_{10}$ multilayer as displayed in Fig. 2. The larger crystal peak intensity in the (17 Å-thick p-a-SiC/61 Å-thick p-nc-Si)$_{10}$ multilayer than that in the (52 Å-thick p-a-SiC/80 Å-thick p-nc-Si)$_5$ multilayer is attributed to the reduced amorphous peak originating from a thinner p-a-SiC. Therefore, the incubation layer formed at the interface between the p-a-SiC and the p-nc-Si, is demonstrated to consist of nanocrystal seeds and amorphous phase. In the FTIR spectrum of the p-nc-Si as shown in Fig. 3, we find the dominant peak indicating the Si-H bond at 610 cm^{-1}. Absorption at 2090 cm^{-1} due to SiH$_2$ bond is not found in the p-nc-Si films prepared on p-a-SiC. A spectroscopic ellipsometry analysis was performed for the film structure (c-Si/SiO$_2$/35 Å-thick p-a-SiC/p-nc-Si). With increasing film thickness, the absorption coefficient decreases due to larger crystal volume fraction in Fig. 4. The tauc's optical bandgap of the film with the thickness of 220 Å was 1.9 eV.

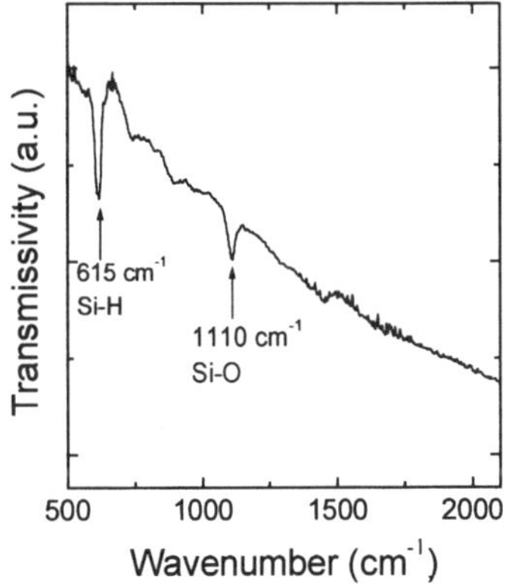

Fig. 3 FTIR spectrum of the 310 Å-thick p-nc-Si prepared onto c-Si/35 Å-thick p-a-SiC.

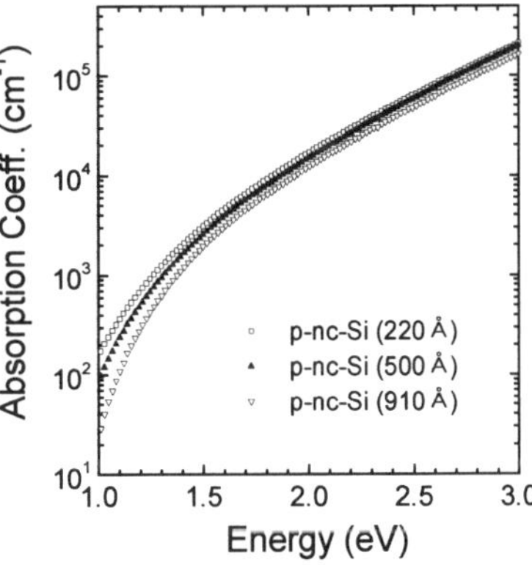

Fig. 4 Absorption coefficient (cm^{-1}) of the p-nc-Si films prepared onto c-Si/35 Å-thick p-a-SiC with the thickness of (a) 220 Å (b) 500 Å (c) 910 Å.

<u>P-type nanocrystalline Si buffered solar cells</u>

In order to study the influence of insertion of the ultrathin p-nc-Si layer at the p/i interface on the cell performance, we fabricated a-Si solar cells with a simple structure of glass/SnO$_2$/p-a-SiC/p-nc-Si/i-a-Si/n-μc-Si/Al. Figure 5 shows the effect of boron doping ratio in ultrathin p-nc-Si on the cell performance. The V_{oc} increases with the boron doping ratio. The increase in the V_{oc} with higher boron addition is attributed to the enhancement of i-layer internal field by shifting fermi-level of the p-nc-Si to its valence band. The fact implies that the higher boron-doped p-nc-Si buffer acts as a p-layer as well as a buffer. The J_{sc} of the cell with p-nc-Si, doped with boron of 2140 ppm is higher than that of the cell with undoped p-nc-Si, originating from the enhancement of the short wavelength response, as shown in Fig. 6. But as boron doping ratio increases above 2140 ppm, the bandgap of p-nc-Si shrinks and the

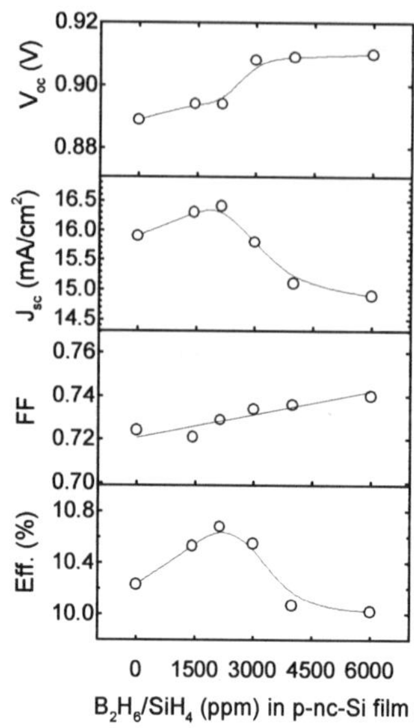

Fig. 5 The dependence of a-Si solar cell characteristics on the B_2H_6/SiH_4 doping ratio (ppm) of ~50 Å-thick p-nc-Si layer inserted at the p/i interface.

absorption loss in the p-nc-Si increases, and thus the J_{sc} decreases again. The FF increases with increasing the boron doping ratio owing to its higher conductivity. As compared with the bufferless cell, photo-carrier collection efficiency in the short wavelength and the V_{oc} are improved by introducing p-nc-Si, doped with 2140 ppm boron. Such a buffering role of the p-nc-Si is thought to reduce a severe carrier recombination loss at the interface of the p-a-SiC and i-a-Si. A moderate bandgap (1.9 eV) p-nc-Si buffer between p-a-SiC (2.0 eV) and i-a-Si (1.72 eV) is thought to relieve a stress originating from an abrupt heterojunction between the p-a-SiC and i-a-Si and thereby prohibit the formation of highly defective region with a short carrier life time at the p/i interface. The cell efficiency increased from 9.6 % (V_{oc} = 0.882 V, J_{sc} = 14.7 mA/cm^2, FF = 0.742) for the bufferless cell to 10.7 % (V_{oc} = 0.894 V, J_{sc} = 16.4 mA/cm^2, FF = 0.729) for the cell with the p-nc-Si buffer. By applying ZnO/Ag/Al back reflector onto n-layer of the cell with structure of glass/SnO$_2$/p-amorphous diamond-like carbon/p-a-SiC/p-nc-Si/i-a-Si/n-µc-Si, we could obtain an initial conversion efficiency of 12.0 % (V_{oc} = 0.900V, J_{sc} = 18.3 mA/cm^2, FF = 0.727), the cell having a final conversion efficiency of 9.5 % after 36 hour under 1 sun light exposure.

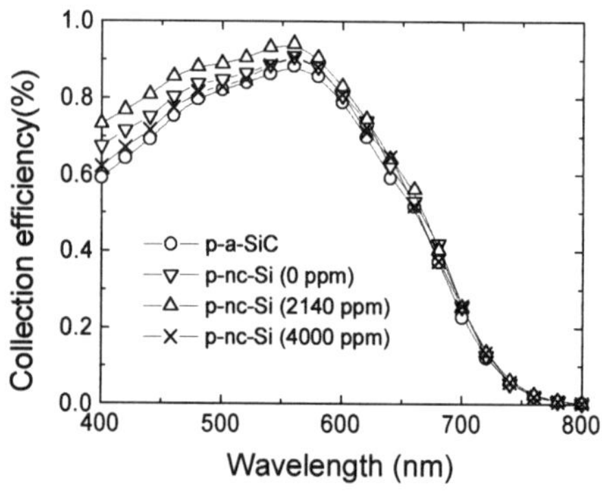

Fig. 6 Spectral responses of the cell with p-a-SiC (○), he cell with p-a-SiC/p-nc-Si (B_2H_6/SiH_4 = 0 ppm) (▽), the cell with p-a-SiC/p-nc-Si (B_2H_6/SiH_4 = 2140 ppm) (△) and the cell with p-a-SiC/p-nc-Si (B_2H_6/SiH_4 = 4000 ppm) (×).

Numerical analysis of p-type nanocrystalline Si buffered solar cells

We analyzed the buffering effect of the p-nc-Si inserted at the interface of the p-a-SiC and i-a-Si through a numerical simulation using Gummel-Sharfetter method [7]. In the analysis, we have chosen the N_A = 5×10^{18} cm^{-3}, the mobility gap = 1.9 eV and the electron affinity = 3.90 eV as a modeling parameter of the p-nc-Si film. The simulated band structure of the cell with the buffer is shown in Fig. 7. All the parameters of each p, i, n layer used in our analysis are listed in table I. The p/i interface region to cause a recombination loss at the heterojunction of p-a-SiC and i-a-Si was modeled as a 50 Å-thick section with a shorter carrier life time (2×10^{-13} sec) in the bufferless cell [8]. Also, such a region having the same properties as those in the bufferless cell was postulated to exist at the interface of the p-nc-Si and i-a-Si, as shown in Fig 7. Assuming that the interfacial layer next to the p-nc-Si have the same thickness and properties as that in the bufferless cell, a positive change in the characteristics of the p-nc-Si buffer such as a mobility gap, a density of acceptor and D-state was found from the simulation not to cause an enhancement of the blue response by inserting the buffer. On the other hand, the shrinkage of the interfacial defective layer next to the buffer was found to enhance the blue response and the V_{oc} but to enhance the FF slightly as shown in Fig. 8.

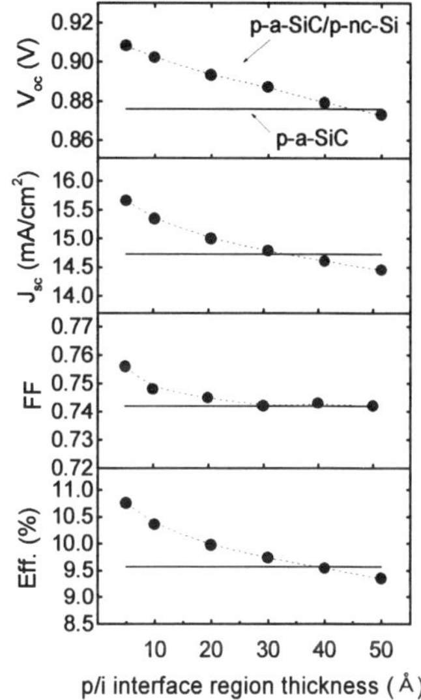

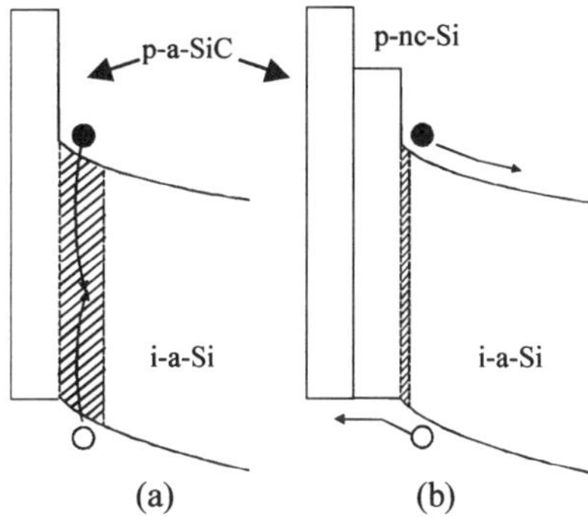

Fig. 7 Schematic band diagrams of the cell. The shadowed regions indicate a highly defective region near the p/i interface. (a) A cell with p-a-SiC, (b) A cell with p-a-SiC/p-nc-Si.

Fig. 8 Simulated cell performance as a function of the thickness of a highly defective region near the interface of the p-nc-Si and i-a-Si in the cell with p-a-SiC (80 Å)/p-nc-Si (60 Å). A solid line indicates the bufferless cell with a 50 Å-thick interface layer.

Table I: Simulated model parameters of p-i-n cells used in the analysis

parameter	p-a-SiC	p-nc-Si	p/i interface	i-a-Si	n-μc-Si
Thickness (Å)	80	60	0 ~ 50	5000	250
Electron affinity (eV)	3.86	3.90	4.0	4.0	3.96
Mobility gap (eV)	2.0	1.90	1.72	1.72	1.90
μ_n (cm^2/Vs)	10	10	10	10	10
μ_p (cm^2/Vs)	1	1	1	1	1
N_A, N_D (cm^{-3})	4×10^{18}	7×10^{18}			5×10^{18}
D-state conc. (cm^{-3})	1×10^{18}	3×10^{18}	5×10^{15}	5×10^{15}	1×10^{18}

In the microstructural view, when an initial interfacial i-a-Si film grows onto the p-a-SiC, it appears to be a highly defective interfacial region, resulting in the poor V_{oc} and a severe recombination loss of a-Si solar cells. However, growing onto the p-nc-Si film which consists of nanocrystallites and an amorphous matrix with Si-H bond configuration, a negative heterojunction problem being present at the junction of p-a-SiC and i-a-Si is excluded and thereby the formation of a highly defective interface is not carried out.

CONCLUSIONS

In order to improve the conversion efficiency of a-Si based solar cells, we have developed for the first time the new p-nc-Si film as an alternative buffer to the conventional a-SiC buffer. By inserting the buffer at the p/i interface, the V_{oc} and the J_{sc} were improved significantly due to the reduced recombination loss near the p/i interface. With increasing the boron doping ratio in the p-nc-Si, the V_{oc} and the FF become higher due to the increased doping efficiency but the J_{sc} decreases due to the increased absorption loss. The cell efficiency increased from 9.6 % (V_{oc} = 0.882 V, J_{sc} = 14.7 mA/cm^2, FF = 0.742) for the bufferless cell to 10.7 % (V_{oc} = 0.894 V, J_{sc} = 16.4 mA/cm^2, FF = 0.729) for the cell with the p-nc-Si buffer. From the results of the numerical simulation, it is suggested that p-nc-Si buffer acts as p-layer to build a higher internal field and hinders the formation of a defective p/i interface layer at the junction between the p-nc-Si and i-a-Si.

ACKNOWLEDGEMENTS

The work was supported by the Korea Electric Power Research Institute (KEPRI).

REFERENCES

1. K. S. Lim, M. Konagai, and K. Takahashi, J. Appl. Phys. 56 (1984) 538

2. R. R. Arya, A, Catalano, and R. S. Oswald, Appl. Phys. Lett. 49 (1986) 1089

3. W. Y. Kim, H. Tasaki, M. Konagai, and K. Takahashi, J. Appl. Phys. 61 (1987) 3071

4. P. Chaudhuri, Swati Ray, A. K. Batabyal, A. K. Barua, Solar Energy Materials and Solar Cells 36 (1994) 45

5. D. E. Aspnes and J. B. Teeten, Phys. Rev. B20 (1979) 3292

6. P. Roca i Cabarrocas, N. Layadi, T. Heitz, B. Drevillon, and I. Solomon, Appl. Phys. Lett. 66 (1995) 360

7. D. L. Scharfetter and H. K. Gummel, IEEE Trans. Electron Devices ED-16 (1969) 64

8. H. Tasaki, W. Y. Kim, M. Hallerdt, M. Konagai, and K. Takahashi, J. Appl. Phys. 63 (1988) 550

HIGH RATES AND VERY LOW TEMPERATURE FABRICATION OF POLYCRYSTALLINE SILICON FROM FLUORINATED SOURCE GAS AND THEIR TRANSPORT PROPERTIES

T. Kamiya, K. Nakahata, K. Ro, C. M. Fortmann and I. Shimizu

The Graduate School, Tokyo Institute of Technology, 4259 Nagatsuta, Midori-ku, Yokohama, JAPAN

ABSTRACT

Low temperature (50-300°C) growth of polycrystalline silicon (*poly*-Si) by very high frequency (100MHz) glow-discharge plasma enhanced CVD using SiF_4 and H_2 mixtures was studied. The *poly*-Si microstructure was strongly affected by the SiF_4/H_2 gas flow ratio. For example, either (220) or (400) preferentially oriented films were prepared by appropriate SiF_4/H_2 ratio selection. The addition of small SiH_4 flows to the SiF_4/H_2 mixtures could be used to increase the growth rate while the SiF_4/H_2 continued to control the film structures such as preferential orientation. Highly crystalline films were grown at a growth rate of 0.52nm/s using $SiF_4/H_2/SiH_4$ flow rates of 30/90/2.0sccm (respectively). However, at higher SiH_4 flows amorphous films were deposited. Under the small SiF_4/H_2 ratio condition, highly crystallized *poly*-Si was grown at temperatures as low as 50°C. N/i/Pt Schottky diode solar cells were prepared using these *poly*-Si for both the n- and the i-layers. These solar cells exhibited good performance; for example, open circuit voltages over 0.32V. N-i-p solar cell results are very promising with 6.2% of conversion efficiency being achieved in the initial trials.

INTRODUCTION

Previously Ishihara *et al.* [1] reported that by using of SiF_4 and H_2 gas mixtures and a remote-type microwave plasma (RT-MW) polycrystalline silicon (*poly*-Si) thin films could be grown on glass substrates at temperatures below 400°C. These *poly*-Si films showed excellent electronic transport with a high photoconductivity and high electron mobility ($\sim20cm^2/Vs$ measured at room temperature). This excellent transport was attributed to well passivated dangling bonds and large grain sizes (>100nm).[2,3] Despite the low temperature growth, high growth rate (>1nm/s) was attained. Therefore, a favorable chemical reaction among hydrogen, halogen and the growth surface was suggested for the SiF_4/H_2 reaction system. In addition by selecting deposition parameters, in particular, the SiF_4/H_2 ratio [R], preferentially oriented films could be deposited. For example, (220) preferentially oriented films were grown with [R]<60/10sccm, while the (400) preferentially oriented films were deposited at [R] ~90/10sccm.[4]

Capacitive coupled parallel electrode reactors are easier to scale to industrial size than the RT-MW used in the previous studies. For example capacitive coupled parallel electrode reactors has

Mat. Res. Soc. Symp. Proc. Vol. 557 © 1999 Materials Research Society
"

been successfully applied to the production of *a*-Si:H solar cells. In this study we attempted to deposit *poly*-Si films from the fluorinated sources using a VHF (100MHz) capacitive coupled parallel electrode reactor. In particular the low temperature, high rate growth deposition of preferentially oriented *poly*-Si films was investigated. Transport properties were also studied using photo and dark conductivity measurements as well as Hall analysis. The photovoltaic performances were evaluated using Schottky diodes, n/i/Pt solar cell device structures.

EXPERIMENTS

A capacitive coupled VHF (100MHz) PECVD apparatus was used to fabricate *poly*-Si thin films from a mixture of SiF_4 and H_2. Small amounts of SiH_4 ($\leq$2sccm) were added to the gaseous mixture of SiF_4 and H_2 in order to enhance the decomposition of SiF_4. The gas flow rate ratio SiF_4/H_2 (=[R]), reaction pressure and substrate temperature (Ts °C) were systematically varied to optimize film structure and properties. Corning 7059 glass substrates were used for all cases.

Film structure was studied using X-ray diffractometer (XRD) analysis. The crystal fraction was estimated from the area ratio of the sharp crystalline TO Raman feature to that of the broad feature associated with the amorphous phase. Electronic transport was examined using dark and photo (560nm monochromatic illumination at 1×10^{15}/s·cm^2 photon flux) conductivity analysis for the undoped cases. Electron mobility of phosphorous doped films was evaluated using Hall analysis.

N/i/Pt Schottky diodes were used to evaluate i-layer quality. All n-layers were deposited on Ga doped ZnO coated glass substrates at 300°C with PH_3 doping ($[PH_3]/[SiH_4]$=5000ppm), and with $SiF_4/H_2/SiH_4$ flow rates of 30/90/0.25sccm (respectively) and a pressure of 200mTorr. The i-layer deposition conditions were the same as those of the n-layer except for growth temperature (200°C) and SiH_4 flow rate that was varied from 1 to 2sccm. Schottky barrier solar cell devices were prepared by the deposition of 1.5 or 3.0mm in diameters of semi-transparent (transmittance was 25-35%) Pt contacts onto the i-layer. The photovoltaic performance of these solar cells was measured under 100mW/cm^2 white light illumination incident on the semi-transparent Pt contact.

RESULTS AND DISCUSSIONS

Figure 1 shows the film structure of *poly*-Si deposited at Ts=360°C, 400mTorr pressure, 20W VHF power, and with a [R] value of 60/10sccm. The SiH_4 flow rate was varied. The addition of SiH_4 resulted in an increased growth rate (increasing from 0.04 to 0.42nm/s as SiH_4 flow rate was increased from 0 to 0.75sccm). Interestingly, no discernible changes in crystal content or preferential orientation occurred. These *poly*-Si films exhibited narrow XRD peaks consistent with (220) preferential orientation. This peak is much narrower than that typical found in microcrystalline silicon (*mc*-Si:H) prepared from H_2 diluted SiH_4 (H_2/SiH_4) (for example, see ref.[5]). In

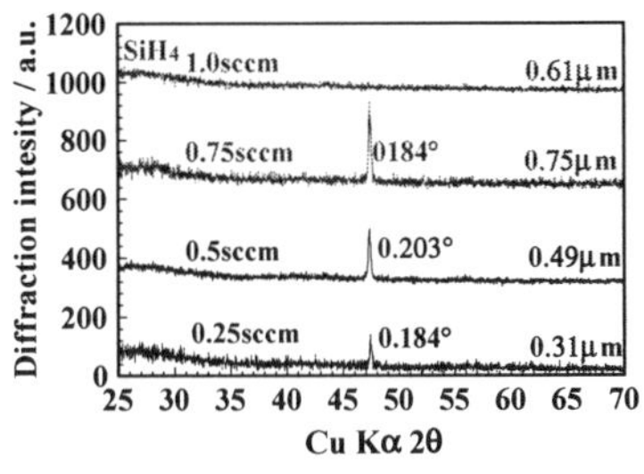
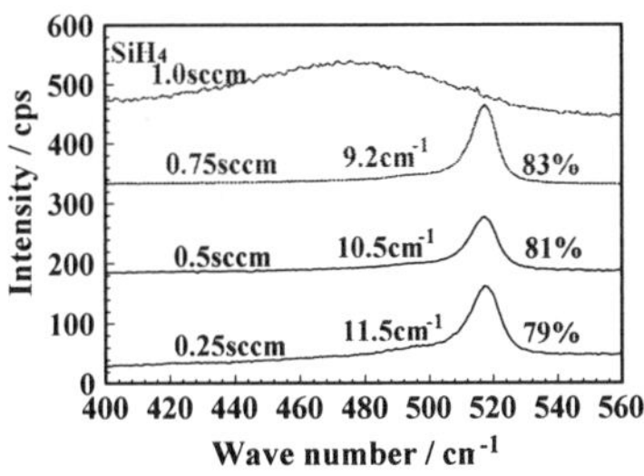

Fig. 1 XRD profiles (left) and Raman spectra (right) of *poly*-Si as a function of SiH₄ flow rate.

addition, we should note that strong (220) orientation structure is easily obtained from SiF₄/H₂ gas mixtures over very wide temperature and gas flow ranges. However, when SiH₄ flow rates larger than 1.0sccm were used amorphous films were deposited (Figs.1 and 2). This structure difference possibly originates from the deposition precursors and/or the radicals in the plasma. The SiF₄/H₂ plasma generates SiH_nF_m (n+m<4) deposition precursors and other fluorine related radicals such as F. While SiH_n (n=1,2,3) radicals are the major deposition precursors in the H₂/SiH₄ systems. The above results suggest that fluorine related precursors play an important role in the large size grain growth seen in this study. Two possible explanations for the growth rate increase resulting from the addition of SiH₄ are: (1) enhanced decomposition of SiF₄ molecules due to an increase in hydrogen radicals caused by the decomposition of SiH₄ and (2) the SiH₄ may scavenge atomic F thereby reducing the etch rate at the growth surface. In any case the addition of small amounts of SiH₄ compared with the major source SiF₄ is essential for increased growth rate with control of microstructure. Also, excessive SiH₄ flows cause SiH_n to become the major deposition precursors, resulting in microstructure changes including *a*-Si:H growth.

Figure 3 summarizes the relationship between crystal fraction of *poly*-Si, growth temperature

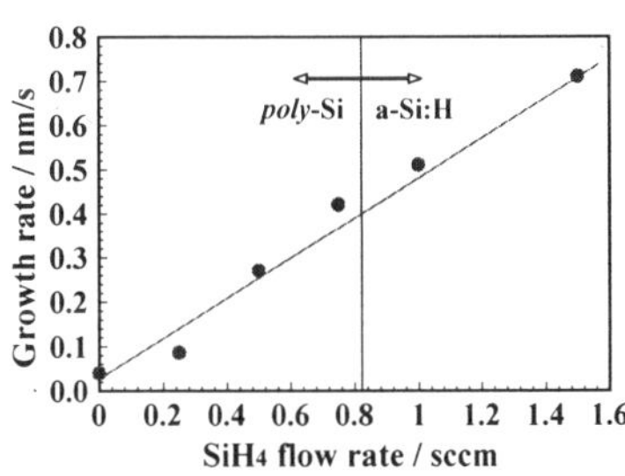

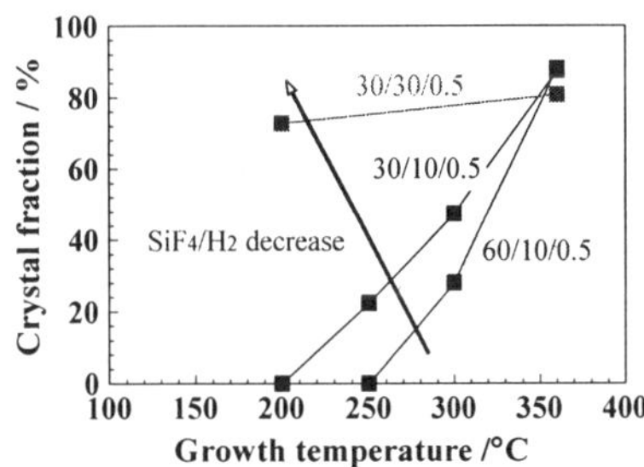

Fig. 2 Growth rate as a function of SiH₄ flow rate.

Fig. 3 Crystal fraction of *poly*-Si as functions of [R] and growth temperature. Numbers in the figure indicates SiF₄/H₂/SiH₄ gas flow rates.

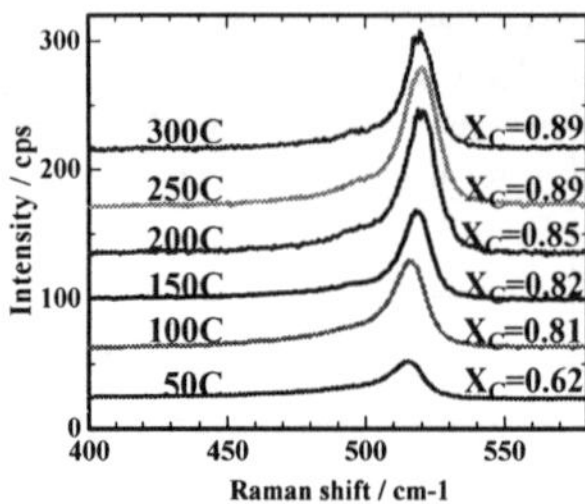

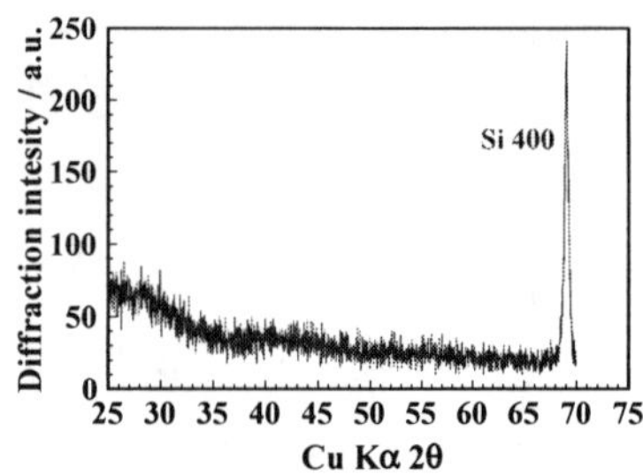

Fig.4 Raman spectra of *poly*-Si as a function of growth temperature.

Fig.5 XRD profile of *poly*-Si grown at SiF$_4$/H$_2$/SiH$_4$=60/2.5/0sccm.

and [R] ratio. When [R] ratio was large the crystal fraction largely decreased with decreasing growth temperature. While, at lower [R] ratio the decrease in the crystal fraction at low temperatures was suppressed. Therefore, we chose a low [R] value, 30/40/0.25sccm, for the preparation of *poly*-Si at low temperatures. Figure 4 shows Raman spectra of *poly*-Si fabricated at 200mTorr using the 30/40/0.25sccm value as a function of substrate temperature. Similar film structures were obtained when the substrate temperatures ranged between 150°C and 300°C. In particular the crystalline fraction and the Raman TO band peak position were almost the same for films grown within this temperature range. However, at substrate temperatures less than 100°C the TO band peak position shifts to lower energy; but, even at 50°C, relatively high crystalline fractions 62% were obtained. The XRD profiles are consistent with (220) preferential orientation at temperatures greater than 150°C; while, films grown at 50°C had random orientation and with the XRD (220) peak width increased to 0.89°. The shift in the TO band to lower energy and the wide XRD (220) peak indicate that the grain size becomes very small at 50°C substrate temperatures. The temperature dependence of dark conductivity for the 150°C T$_S$ *poly*-Si case is consistent with a mid-gap Fermi level. The photoconductivity activation energy was only 0.15eV. The room temperature conductivity values were 2×10^{-7}S/cm and 5×10^{-5}S/cm in the dark and under light illumination, respectively.

At larger [R] ratios, ~60/2.5sccm, (400) oriented films were deposited (Fig.5). As was the case with the RT-MW deposition case (220) and (400) oriented structure was controlled by appropriate [R] ratio selection. While, (220) preferential oriented films were grown under small [R] ratio, (400) preferential oriented films were prepared under larger [R] ratio. This relationship between the [R] ratio and film structure is remarkably similar to that obtained by the RT-MW. Hall measurements were performed with these (220) and (400) oriented films which were doped with P concentrations ([PH$_3$]/[SiH$_4$]) ranging from 5ppm to 0.5% for (220) and from 50ppm to 4% for (400) orientation condition. Relatively large electron mobilities of 5.8cm^2/Vs and 8.4cm^2/Vs were obtained for 1μm thick (220) and (400) oriented films having ~10^{18}/cm^3 carrier density, respectively. Surprisingly, the mobility activation energy decreases from 38meV

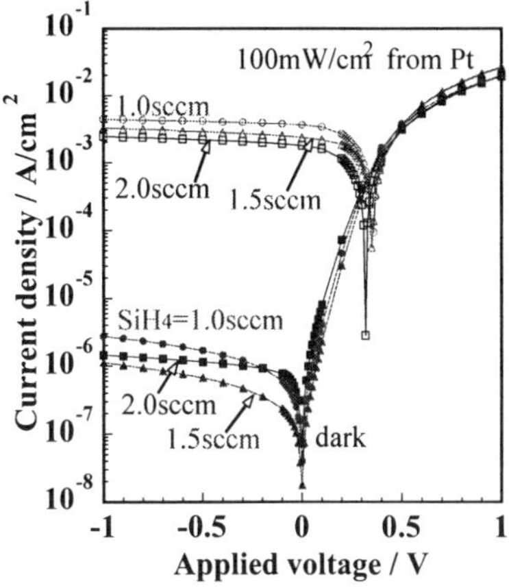

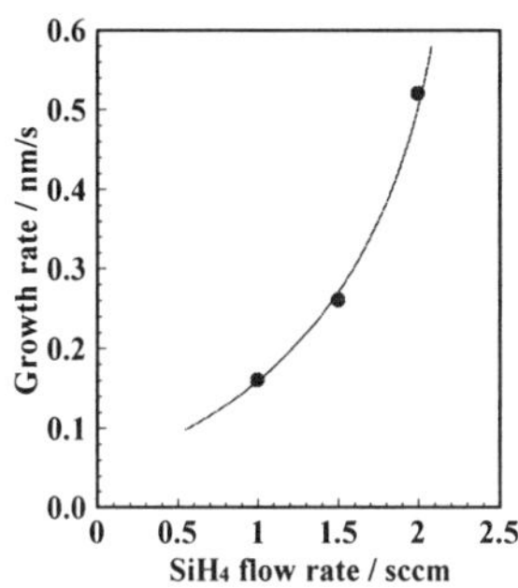

Fig. 6 I-V characteristics of n/i/Pt Schottky diodes as a function of SiH₄ flow

Fig. 7 Growth rate of i-layer as a function of SiH₄ flow rate.

Table 1. I-V parameters and crystal fraction (Xc) of i-layer of n/i/Pt Schottky diodes as a function of SiH₄ flow rate for i-layer growth.

SiH_4 (sccm)	V_{oc}(V)	J_{sc} (mA/cm²)	F.F.(%)	Eff(%)	X_c
1.0	0.36	3.65	41.0	0.53	0.89
1.5	0.35	2.36	46.1	0.38	0.84
2.0	0.32	1.79	40.1	0.23	0.78

to 20meV as carrier density increases from 10^{16} to 2×10^{19}/cm³. This result is not consistent with the Seto deep trap model [6], suggesting that the carrier transport in the *poly*-Si may be limited by a shallow trap located at the grain boundaries (the details will be discussed in elsewhere).

Figure 6 and table 1 shows I-V characteristics of n/i/Pt Schottky diodes as a function of SiH₄ flow rate for i-layer deposition. The i-layers of these devices were prepared at a higher H₂ flow rate ([R]=30/90sccm) so that the crystal content could be maintained at higher SiH₄ flow rates and growth rates. The open circuit voltage Voc exceeds 0.32V for all cases. A non-linear growth rate increase is observed with increasing SiH₄ flow rate. A relatively high rate of 0.52nm/s was obtained at SiH₄ flow rate of 2.0sccm (Fig.7). The crystal fraction of this film was 78% and it was still (220) preferential oriented. A fill factor of 46% was obtained with 1.5sccm SiH₄ i-layer. These I-V characteristics demonstrates that growth rate can be increased without evident degradation of device performance by optimizing the various flow rates. However, some Voc decrease can be seen with increasing SiH₄ flow rate. A small number of n-i-p solar cells were prepared using improved light trapping techniques (including a textured ZnO/Ag substrate) as a function of i-layer growth temperature. One n-i-p solar cell had an energy conversion efficiency of 6.2% (Voc=0.416V, short circuit current Jsc=24.2mA/cm², fill factor=61.8%; i-layer was deposited with SiH$_4$ flow of 1.0sccm at 185°C). Further optimization is expected to further improve film quality, device performance and increase growth rate.

CONCLUSION

In conclusion, polycrystalline silicon thin films were fabricated on glass substrates using SiF$_4$/H$_2$/SiH$_4$ gas mixtures in a 100MHz VHF glow-discharge CVD system. The SiF$_4$/H$_2$ gas flow ratio is the major factor controlling preferential orientation and crystalline fraction. By optimizing the SiF$_4$/H$_2$ ratio and the SiH$_4$ flow rate, high crystalline fraction (>80%) *poly*-Si are obtained at temperatures greater than 100°C and relatively high crystallinity (63%) microcrystalline silicon were also grown at temperatures as low as 50°C. The addition of small SiH$_4$ flows can increase the growth rate without influencing the film structure. Large electron mobilities of 5.8cm^2/Vs and 8.4cm^2/Vs were obtained for P doped (220) and (400) oriented 1μm thick *poly*-Si films, respectively. However at larger SiH$_4$ flows amorphous films are deposited. A high growth rate of 0.52nm/s was obtained by optimizing the various flow rates. N/i/Pt Schottky diodes exhibited good photovoltaic performance with relatively large open circuit voltages over 0.32V for this type of solar cell. A n-i-p solar cell achieved an energy conversion efficiency of 6.2%. Further improvement in solar cell performance is expected from more optimized growth condition and solar cell designs.

ACKNOWLEDGEMENTS

We thank PVTEC for partial support of this work. This study was also supported in part by a Grant-in-Aid for Scientific Research on Priority Area, "The Chemistry of Inter-element Linkage" (No.09239104) of The Ministry of Education, Science and Culture (No.09305002) and also by the New Sunshine Project of MITI.

REFERENCES

1. S. Ishihara, D. He, T. Akasaka, Y. Araki and I. Shimizu, Mat. Res. Soc. Symp. Proc., **297**, 79 (1993).
2. D. He, N. Okada and I. Shimizu, Solar Energy Materials and Solar Cells **34**, 271 (1994).
3. M. Nakata and I. Shimizu, Mat. Res. Soc. Symp. Proc. **283**, 591 (1993).
4. T. Kamiya, K. Nakahata, A. Miida, C. M. Fortmann and I. Shimizu: Thin Solid Films **337**, 18 (1999).
5. T. Nishimiya, M. Kondo and A. Matsuda, Mat. Res. Soc. Symp. Proc., **467**, 397 (1997).
6. J. Y.W. Seto, J. Appl. Phys., **46**, 5247 (1975).

ATOMIC FORCE MICROSCOPY STUDY OF INITIAL NUCLEATION IN THE DEPOSITION OF μc-Si:H

P. Brogueira[*], V. Chu[**] and J.P. Conde[***]
[*]Dept. of Physics, Instituto Superior Técnico, Lisbon, Portugal
[**]Instituto de Engenharia de Sistemas e Computadores, Lisbon, Portugal.
[***]Dept. of Materials Engineering, Instituto Superior Técnico, Lisbon, Portugal

ABSTRACT

The initial stages of microcrystalline silicon growth of n^+ doped films prepared by rf plasma enhanced chemical vapor deposition (PECVD) and of intrinsic films prepared by hot-wire chemical vapor deposition (HW-CVD) are studied using atomic force microscopy, Raman spectroscopy and parallel dark conductivity measurements. The effect of the use of a plasma hydrogen treatment, of chamber conditioning prior to this treatment, of the type of substrate (glass or c-Si) used and the effects of a seed layer on the film properties are discussed.

INTRODUCTION

Thin-film transistors, TFT, made from hydrogenated amorphous silicon, a-Si:H, are used in matrix-addressed arrays such as flat-panel displays and image scanners.[1] Since microcrystalline silicon, μc-Si:H, has higher bulk electrical conductivity than a-Si:H[2,3] it is a natural candidate to replace the a-Si:H as the TFT active layer. The channel of a bottom-gate μc-Si:H TFT is formed within 20 Å of the interface with the dielectric and so the performance of the device is determined in the initial nucleation phase. In this region the material may still be either amorphous or fine-grained due to incipient nucleation of the crystalline grains. Doped hydrogenated microcrystalline silicon is also used as contact layers in TFTs and in solar cells.[4] Current limitations due to poor contacts often degrade the performance of TFTs. In this paper, we present a study of the nucleation of n^+-μc-Si:H prepared by rf glow discharge (PECVD) and of intrinsic μc-Si:H prepared by Hot Wire (HW-CVD) on glass and c-Si exposed to different surface pre-treatments.

EXPERIMENTAL PROCEDURES

Sample preparation

n^+-μc-Si:H thin films were prepared by PECVD in a parallel plate reactor (13.56 MHz). Intrinsic μc-Si:H material was deposited by HW-CVD using a single tungsten filament, 0.5 mm in diameter, placed 3 cm from the substrate. Silane (SiH_4) and hydrogen (H_2) were used as reaction gases. For the preparation of n-type material, 2% phosphine (PH_3) diluted in H_2 was added. The hydrogen dilution (%H_2) is defined as $\%H_2 = FR(H_2)/(FR(H_2)+FR(SiH_4))$, where $FR(H_2)$ and $FR(SiH_4)$ are the gas flow rates of hydrogen and silane, respectively. For each deposition method (PECVD and HW-CVD) two series of samples with varying thicknesses (between 50 and 2000 Å) were prepared: one with exposure of the substrate to a H_2 plasma pre-treatment prior to film deposition and the other without H_2 plasma pre-treatment. The experimental conditions used for the H_2 plasma pre-treatment were: p~0.1 Torr, rf power of 150 mW/cm^2, substrate temperature of T_{sub}=250 °C and FR(H_2)=15 sccm for 5 min. Preceding the H_2 plasma treatment, a 1 hour sacrificial deposition of a-Si:H (~1 μm film near the electrode) was performed using p~0.1 Torr, rf power of 100 mW/cm^2, FR(SiH_4)=10 sccm and T_{sub}=250 °C.

Mat. Res. Soc. Symp. Proc. Vol. 557 © 1999 Materials Research Society

PECVD n^+-μc-Si:H samples were deposited using p~0.1 Torr, rf power of 300 mW/cm^2, FR(SiH$_4$)=1 sccm, %H$_2$=97.5%, %PH$_3$=1% and T$_{sub}$=250 °C. Together with the sacrificial deposition and H$_2$ plasma pre-treatment, these experimental conditions were found to optimize the conductivity of the 500 Å n^+-μc-Si:H layers used as contact layers for TFTs.[5] HW-CVD μc-Si:H samples were prepared using p~20 mTorr, T$_{fil}$~2500 °C, FR(SiH$_4$)=5 sccm, %H$_2$=90% and T$_{sub}$=300 °C.

Sample characterization

Films were deposited on two different substrates simultaneously: (1) Corning 7059 glass for parallel transport measurements, atomic force microscopy (AFM), Raman scattering measurements and x-ray diffraction; (2) double-side polished <100> Si for AFM. For parallel conductivity measurements, thermally evaporated parallel Cr contacts, 6 mm long and 1 mm apart were used. A Multimode with a Nanoscope IIIa controller from Digital Instruments was used for AFM. The Multimode is mounted on an optical table for vibration isolation. The imaging was performed in *tapping mode* using commercial TappingMode Etched Silicon Probes and a 10×10 μm^2 scanner. All measurements were performed on 200×200 nm^2 and 1×1 μm^2 areas under ambient conditions. The root mean square (rms) roughness of the sample surface was calculated from $R_q = \sqrt{\sum_i (z_i - \bar{z})^2 / N}$ where $\bar{z}$ is the average height of the scanned area and z_i is the height value of each point. Raman spectra were measured in backscattering geometry using a Spex micro-Raman spectrometer. For each sample, the power of the laser was kept below the level that would thermally induce crystallization (~6×10^4 W/cm^{-2}). The crystalline fraction was calculated from the deconvolution of the Raman spectrum[6-10] and the crystalline size, d_{Raman}, was calculated from the transverse optical peak shift.[11] X-ray diffraction peaks were measured with a Siemens D-5000 x-ray diffractometer using the Cu Kα_1 line.

RESULTS AND DISCUSSION

n-doped silicon by PECVD

Figure 1 shows the room-temperature σ_d as a function of the thickness, for n^+ type doped samples deposited by PECVD, with and without H$_2$ plasma pre-treatment. For depositions without H$_2$ plasma pre-treatment an eleven orders of magnitude increase in σ_d from ~10^{-11} Ω^{-1}cm^{-1} to ~1 Ω^{-1}cm^{-1} is observed as the thickness increases from 50 to 2000 Å. At the same time the activation energy for σ_d, E_{a,σ_d}, decreases from approximately 0.85 eV to ≤ 0.1 eV, typical of doped microcrystalline silicon. For samples having a H$_2$ plasma pre-treatment, the increase in σ_d is steeper and occurs at a lower thickness. σ_d increases from ~10^{-10} Ω^{-1}cm^{-1} to ~0.1-1 Ω^{-1}cm^{-1} for films with thicknesses increasing from 50 to ~500 Å. For thicker films, σ_d remains at ~1 Ω^{-1}cm^{-1}. E_{a,σ_d} decreases from approximately 0.8 eV to ≤0.1 eV when the thickness increases from 50 to ≥500 Å. As the inset shows, the Raman spectrum for the 1000 Å thick sample prepared with pre-treatment is typical of an amorphous-to-microcrystalline transition material. A crystalline fraction of X$_c$~15 % was calculated. For the series deposited without H$_2$ plasma pre-treatment a microcrystalline character is distinguishable by Raman only for the 2000 Å thick sample (X$_c$~20 %). Amorphous-like spectra are obtained in both series with samples thinner than, respectively, 500 and 2000 Å. The high σ_d values, when compared to the low X$_c$ observed, suggest that the microcrystalline layer is located only on the surface of the sample. Preceding the hydrogen plasma pre-treatment by a 1 hour sacrificial deposition of μc-Si:H film (p~0.1 Torr, rf power of 300 mW/cm^2, FR(SiH$_4$)=1 sccm, %H$_2$=97.5% and T$_{sub}$=250 °C) results in a 500 Å thick amorphous n-type doped film. The ability of the atomic hydrogen to selectively remove a-Si:H covered electrode surfaces, in contrast to μc-Si:H covered electrode surfaces to promote initial nucleation on the substrate could be responsible for this result.[12]

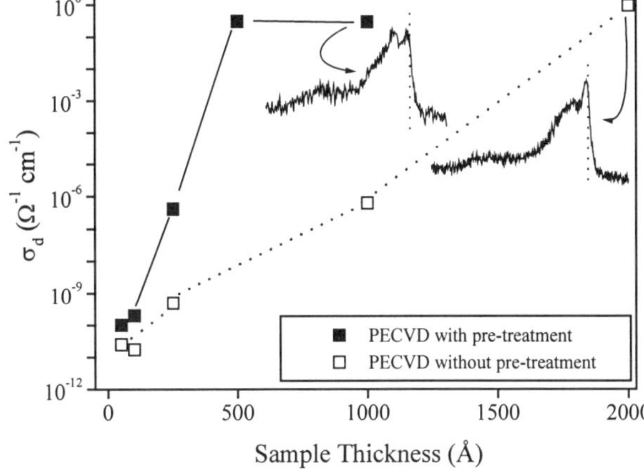

Figure 1: The dark conductivity, σ_d, for n^+ type doped samples deposited by PECVD, with and without H_2 plasma pre-treatment is plotted as a function of the film thickness. The Raman spectra of the amorphous-to-microcrystalline transition samples for each series are plotted in the inset. The vertical dashed lines indicate the position of the TO phonon peak for c-Si (520 cm^{-1}). The other lines are guides to the eye.

Figure 2 shows the AFM micrographs for n^+ type doped samples deposited by PECVD with H_2 plasma pre-treatment and varying thickness: i) 100 Å, ii) 250 Å and iii) 1000 Å. The surface morphology of the 100 Å thick sample exhibits the presence of individual nuclei with an approximately random distribution (confirmed in a scan performed on 1×1 µm^2 area) with an average diameter of ~15 nm. The topography of the 250 Å sample shows an increase in the grain size to 18-28 nm while keeping a roughly similar nuclei density. The presence of nuclei was not revealed in AFM surface measurements on samples with similar thicknesses prepared without H_2 plasma pre-treatment. For the 1000 Å thick sample, individual or clustered "hills" with lateral dimensions in the range, respectively, of 15-25 nm and of 30-40 nm are observed. R_q for these samples were: i) 0.35 nm, ii) 0.63 nm and iii) 4.1 nm. The samples deposited on double-side polished <100> Si substrates show improved nucleation and R_q values of: i) 0.41 nm, ii) 0.73 nm and iii) 1.99 nm. Similar trends in R_q were previously observed by Smith $et\ al.$[13]

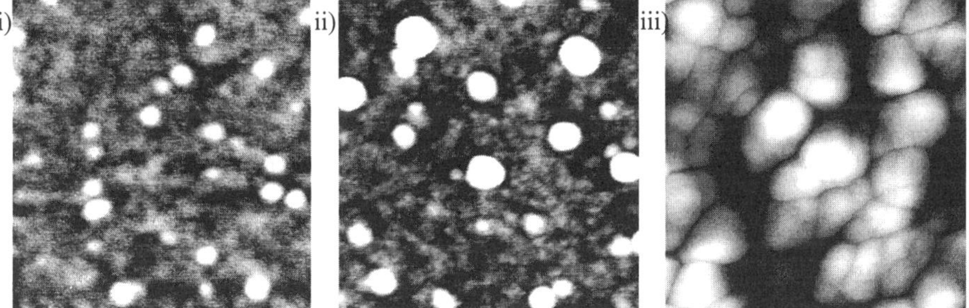

Figure 2: Top-view AFM surface imaging using tapping mode of n^+ type doped samples prepared by PECVD with H_2 plasma pre-treatment and varying thickness: i) 100 Å , ii) 250 Å and iii) 1000 Å. Scanned areas: 200×200 nm^2. Height color scale: i) and ii) 2 nm ; iii) 25 nm.

R_q calculated from AFM measurements on a 200×200 nm^2 area of the bare Corning glass substrates were typically 0.15 nm and of the silicon wafers used as substrates were 0.14 nm. However, the surface of the glass shows broader "hills" than that of the silicon, which is characterized by a "spiky" surface. The values for R_q, which are equivalent to an integral of the power density spectra over the entire surface, will give similar values as long as, on average, the distance between measurement point and the height averages are similar, independently of local fluctuations. The surface topography of the glass substrates immediately after the H_2 plasma pre-treatment, maintains R_q~0.16 nm, but shows a power spectral density approaching that of c-Si.

Intrinsic silicon by HW-CVD

Figure 3 shows the room-temperature σ_d as a function of the thickness, for samples deposited by HW-CVD, with and without H_2 plasma pre-treatment. For depositions without pre-treatment, σ_d increases from ~8×10^{-11} $\Omega^{-1}cm^{-1}$ for the 50 Å thick sample to ~2×10^{-6} $\Omega^{-1}cm^{-1}$ for the 1000 Å thick sample. At the same time E_{a,σ_d}, decreases from ~0.85 eV to ~0.6 eV. For samples with pre-treatment σ_d increases from ~10^{-10} $\Omega^{-1}cm^{-1}$ for the 50 Å thick sample to ~8×10^{-6} $\Omega^{-1}cm^{-1}$ for the 1000 Å thick sample and E_{a,σ_d}, decreases from ~0.9 eV to ~0.5 eV. The similar transport properties of the samples in both series suggest a little influence of the H_2 plasma pre-treatment on the properties of μc-Si:H HW-CVD samples The peaks related to diffraction at the (111), (220) and (311) planes of c-Si are clearly resolved in the spectrum shown in the inset of figure 3 for the 1000 Å thick sample prepared with pre-treatment.

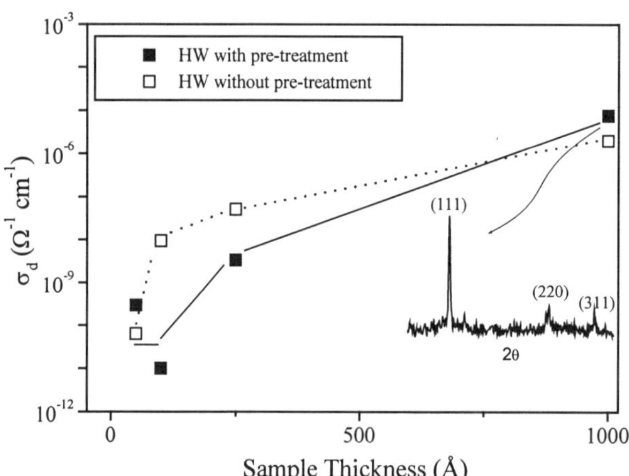

Figure 3: The dark conductivity, σ_d, for samples deposited by HW-CVD, with and without H_2 plasma pre-treatment are plotted as a function of the film thickness. The x-ray spectrum of the 1000 Å thick sample prepared with pre-treatment is plotted in the inset. The lines are guides to the eye.

Figure 4 shows the AFM micrographs (top) and Raman spectra (bottom) for intrinsic samples deposited by HW-CVD without H_2 plasma pre-treatment and varying thickness: i) 100 Å, ii) 250 Å and iii) 1000 Å. In contrast to the PECVD results shown in figure 2, after 100 Å of film deposition, the nuclei on the glass substrates appear to have almost completely covered the surface. The Raman spectrum is typical of an amorphous-to-microcrystalline transition material with a crystalline fraction X_c~9 % and a grain size d_{Raman}~5.6 nm. As the films become thicker and their crystalline character as measured by Raman grows, a rougher grain structure consisting of round domains with 10-30 diameter nm is revealed. X_c~14%, d_{Raman}~6 nm and X_c~47%, d_{Raman}~9.8 nm were calculated, respectively, for the 250 and 1000 Å thick films. R_q for the three samples were: i) 1.06 nm, ii) 1.27 nm and iii) 1.32 nm. AFM measurements performed in a 1×1 μm^2 area results in very similar surface topographies for these films. The AFM micrographs obtained for the samples prepared by HW-CVD with and without H_2 plasma pre-treatment show vanishingly small differences.

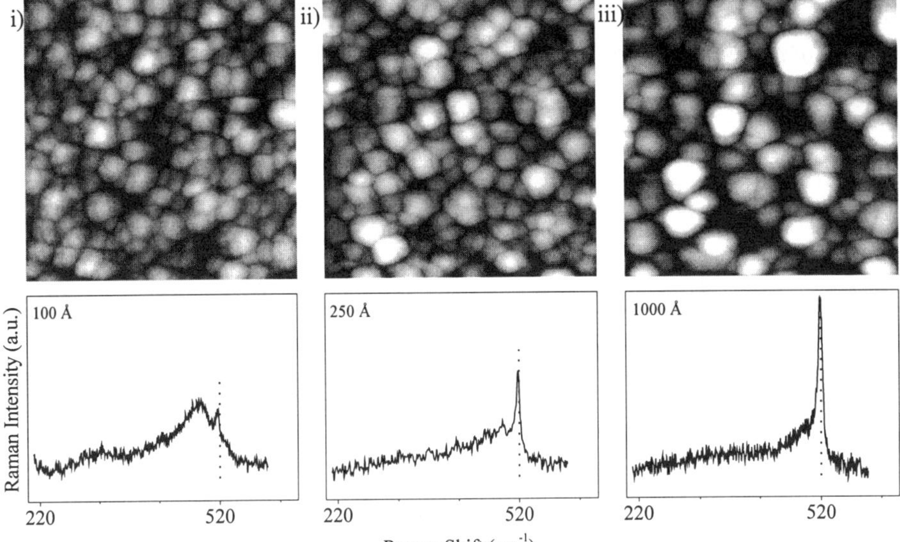

Figure 4: Top-view AFM surface imaging using tapping mode (top) and Raman spectra (bottom), of intrinsic samples prepared by HW-CVD with pre-treatment and varying thickness: i) 100 Å , ii)250 Å and iii)1000 Å. Scanned areas: 200×200 nm^2. Height color scale: 10 nm. The dashed line (520 cm^{-1}) at the Raman spectra indicates the position of the TO phonon peak for c-Si.

Use of seed layers

In order to promote nucleation and increase the size of the grains observed by AFM while keeping a relatively high growth rate, a double-layer deposition was tested. 250 Å thick films were prepared in two steps: 1) a 50 Å layer was deposited to create nucleation seeds using the deposition conditions previously optimized for the preparation of bulk microcrystalline material; 2) subsequent deposition of the remaining 200 Å using lower values of hydrogen dilution.

Figure 5 shows the AFM micrographs for double-layered, 250 Å thick, intrinsic samples deposited by HW-CVD without H$_2$ plasma pre-treatment. The initial 50 Å layer was deposited with 90 %H$_2$. The final 200 Å layer was deposited using: i) 90 %H$_2$ and ii) 87.5 %H$_2$. A sample with a higher compactness and exhibiting features with crystal-like facets was obtained when the hydrogen dilution of the second layer was slightly decreased (ii) but no significant increase of the grain size was observed. Simultaneously σ_d increased by more than one order of magnitude

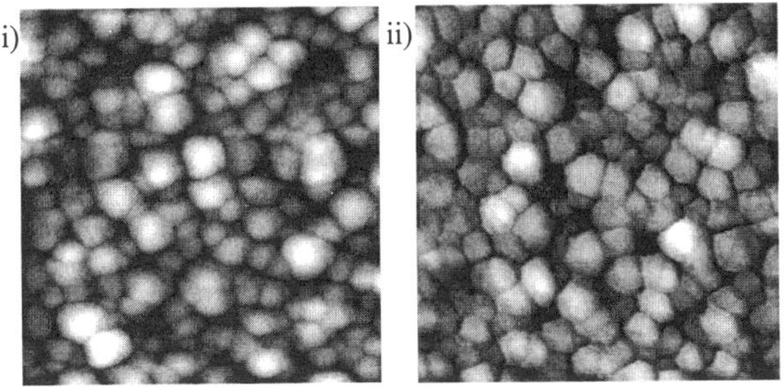

Figure 5: Surface topography of 250 Å thick double-layer samples, prepared by HW-CVD without pre-treatment. A 50 Å layer for both samples were deposited using 90 % H$_2$ dilutuion. For the 200 Å the following hydrogen dilution were used: i) 90 % and ii) 87.5%. Scanned areas: 200×200 nm^2. Height color scale: 10 nm

and E_{a,σ_d} decreased to 0.55 eV. The film in figure 6.ii), appears less rough than the film shown in figure 6.i). R_q for these samples were: i) 1.27 nm and ii) 1.04 nm. The surface topography analysis of the 50 Å seed layer already exhibits crystalline-like features and is characterized by $R_q\sim0.62$ nm.

CONCLUSIONS

The surface morphology of the samples varied significantly depending on the deposition technique used, the substrate material and whether or not it was exposed to a H_2 plasma treatment. The glass surface after exposure to the H_2 plasma pre-treatment maintains the R_q but becomes more "spiky" and similar to that of c-Si. n$^+$-μc-Si:H doped films as thin as 500 Å could be deposited by PECVD, if preceded by an adequate hydrogen plasma pre-treatment of the substrate. AFM imaging of 100 Å-thick layers reveals the presence of nuclei randomly distributed on the surface of samples prepared with H_2 plasma pre-treatment in contrast with similar measurements on samples deposited without the pre-treatment where no nuclei are observed. μc-Si:H intrinsic films as thin as 100 Å could be deposited by HW-CVD. In this case the H_2 plasma pre-treatment is not crucial for the nucleation. Decreasing the hydrogen dilution after a first layer of nucleation resulted in a more compact surface exhibiting features with crystal-like facets, although no grain size increase was observed. The grain size seems to be self-limited by the growth process: independently of the use of HW-CVD or PECVD as deposition method, with or without hydrogen plasma pre-treatment, and for different values of hydrogen dilutions, all microcrystalline samples were characterized by grain sizes inferior to 25-30 nm in diameter

ACKNOWLEDGEMENTS

The authors acknowledge Luisa Paramés and Olinda Conde of the Faculty of Science of the University of Lisbon for help with the X-ray measurements. One of the authors (P.B.) gratefully acknowledges FLAD for a travel grant. This work was supported by the FCT through Pluriannual Contratcs with UCES/ICEMS (IST) and INESC and by project PRAXIS/3./3.1/MMA/1775/95.

REFERENCES

1. R. A. Street, *Hydrogenated Amorphous Silicon* (Cambridge University Press, Cambridge, 1991).
2. M. Konuma, H. curtins, F. A. Sarott and S. Veprek, Phil. Mag. B **55**, 377 (1987).
3. A. Matsuda, J. Non-Cryst. Solids **59-60**, 767 (1983).
4. M. Yoshimi, W. Ma, T. Horiuchi, C.C. Lim, S.C. De, K. Hahari, H. Okamoto and Y. Mamakawa, Mater. Res. Soc. Symp. Proc. **258**, 845 (1992).
5. J.P. Conde, H. Silva and V. Chu, Mater. Res. Soc. Symp. Proc., to be published (1998).
6. J.S. Lanin, *Semiconductors and Semimetals, Part B* (Academic, New York, 1984), Vol 21.
7. T. Kaneko, M. Wakagi, K. Onisawa and T. Minemura, Appl. Phys. Lett. **64**, 1865 (1994).
8. S. Veprek, F. A. Sarrot and Z. Iqbal, Phys. Rev. B **36**, 3344 (1987).
9. R. Tsu, J. G. Hernandez, S. S. Chao, S. C. Lee and K. Tanaka, Appl. Phys. Lett. **40**, 534 (1982).
10. H. Kakinuma, M. Mohri and T. Tsuruoka, J. Appl. Phys. **70**, 7374 (1991).
11. Y. He, C. Yin, G. Cheng, L, Wang, X.Liu and G.Y. Hu, J. Appl. Phys. **75**, 797 (1994).
12. K. Saitoh, M. Kondo, m. Fukawa, T. Nishimiya, A. Matsuda, W. Futako and I. Shimizu, Appl. Phys. Lett. **71**, 3403 (1997).
13. L.L. Smith, E. Srinivasan and G.N. Parsons, J. Appl. Phys. **82**, 6041 (1997).

STRUCTURE AND OPTOELECTRONIC PROPERTIES AS A FUNCTION OF HYDROGEN DILUTION OF MICRO-CRYSTALLINE SILICON FILMS PREPARED BY HOT WIRE CHEMICAL VAPOR DEPOSITION

GUOZHEN YUE*, JING LIN*, QI WANG** and DAXING HAN*

*Department of Physics & Astronomy, University of North Carolina at Chapel Hill, Chapel Hill, NC 27599-3255, USA.
**National Renewable Energy Laboratory, Golden, CO 80401, USA.

ABSTRACT

Films prepared by hot wire CVD using H dilution ratio, $R=H_2/SiH_4$, from 1 to 20 were studied by X-ray, Raman, PL, and conductivity measurements. We found that (a) when the dilution ratio reached R=3, the structure transition from amorphous to microcrystalline growth occured; meanwhile, PL spectrum showed a dual-peak at 1.3 and 1.0 eV; (b) the total intensity, band width, and peak position of the low energy PL band decreased with increasing H dilution; (c) both the Raman and PL measured from the transparent substrate side showed that initial growth tends to be amorphous and a portion of μc-Si was formed when $R \geq 5$; and (d) the conductivity activation energy first decreased from 0.68 to 0.15 eV when the film transition from a- to μc-Si; then increased slightly with increasing μc-Si fraction. The results demonstrate that the variation of the H-dilution ratio has significant effects on both the film structures and the optoelectric properties.

INSTRUCTION

Microcrystalline silicon (μc-Si:H) has been used in thin-film device technology, such as contact materials in MOS devices or as an active layer in thin film transistors (TFTs). Many methods have been used to grow the μc-Si[1, 2]. However, a one-step low temperature and high growing rate process is desirable for device applications. The hot wire chemical vapor deposition (HWCVD) can achieve high growth rate with reasonable material quality. In this method, μc-Si can be formed by varying growth parameters, such as the filament temperature (T_{fil}), gas pressure (P), substrate temperature (T_S), and H dilution ratio ($R=H_2/SiH_4$) [3-6]. When one has a proper filament and substrate temperature, H-dilution is expected to be a simple and promising method for producing high quality μc-Si film. However, the effect of H dilution on the film microstructure is somewhat in contradiction. Brogueira et al. [3] reported that μc-Si:H growth was achieved without H dilution at T_{fil} =1500 °C, T_S=220 °C, and P $\leq$ 0.1 Torr. Conde et al. [4] indicated that at T_S=220 °C, H dilution does not significantly change the a-Si:H film properties at 4<R<9, while μc-Si can be obtained using R>9. Diel et al. [5] demonstrated that at T_S=230 °C, an alteration of the H-dilution ratio 2.5 <R<25 does not affect the crystallite size and the optical absorption edge remains the same. On the other hand, Wanka et al. [6] showed at high substrate temperature T_S=400 °C the grain size and the crystalline quality are strongly dependent on H-dilution ratio.

In this work, we report systematical studies of H-dilution effect on microstructure and optoelectric properties of μc-Si:H films prepared at low T_S (240 °C) by using X-ray, Raman, PL spectroscopies and conductivity temperature dependence.

EXPERIMENTAL

The films were deposited at a substrate temperature of 240 ºC on Corning 7059 glass . We used a spiral tungsten filament with a diameter of 0.5 mm and place 5 cm below the substrate. The filament was heated to ~2000 ºC by an AC current of 16 A. We changed the hydrogen to silane ratio from 1 to 20 while keeping the chamber pressure the same at 30 mT. The Raman spectra were measured by using a Spectra-Physics model 2017 with a Dilor XY spectrometer equipped with a LN_2-cooled CCD camera. The data were taken at room temperature using 514.5

nm Ar^+ laser line. The laser beam power on the sample was below the level which would thermally induce crystallization. A 100 mW/cm^2 514.5 nm Ar^+ laser beam was used for the PL excitation. The sample was mounted on a cold stage with a temperature range from 80 K to 300 K. The PL spectra were analyzed using a grating monochromator and detected by a LN_2-cooled Ge detector. The optical system was calibrated by a standard lamp.

RESULTS

Structure Properties:

Table I   Sample structure and parameters

Sample No.	H_2/SiH_4 (R)	XRD	Grain size, L (nm) (111), (220)		Thickness (μm)	Dep. Rate ($\overset{\circ}{A}$/s)	Dark-, Photo- conductivity (ohm·cm)	
1	1	a-Si	~,	~	4.1	17.2	1.0e-9,	2.0e-5
2	2	a-Si	~,	~	3.0	12.5	1.6e-9,	7.6e-5
3	3	a-Si+μc	10.5,	13.2	1.7	7.1	4.0e-7,	9.3e-5
4	4	μc	7.1,	26.9	2.4	8.0	1.2e-6,	9.8e-5
5	5	μc	10.7,	35.1	2.5	7.0	5.2e-4,	1.0e-3
6	10	μc·	18.7,	32.1	1.4	3.9	2.6e-6,	3.4e-5
7	20	μc	28.2,	10.7	1.1	2.3	1.6e-6,	1.9e-5

Table I lists the sample structure and parameters. The X-ray diffraction data show that the threshold for structure transition from a- to μc-Si growth was 3; and the crystalline structure orientation changed gradually from (220) to (111) direction as the H-dilution ratio increasing from 3 to 20. The average crystalline size was determined from the full width at half-maximum(FWHM) of (220) and(111) diffraction lines by Scherre's formula [7]. As shown in Table I, the grain size is larger than 10 nm and depends on H-dilution ratio R.

The Raman spectra for the films were presented elsewhere[8]. A typical broad 480 cm^{-1} peak characterizes a-Si:H phase, and a sharp 520 cm^{-1} peak appears when R$\geq$3. The results are in agreement with the results of X-ray diffraction. The Raman spectrum was decomposed into three components: the crystalline component, the amorphous component, and a intermediate component associated with the bond dilation at grain boundaries (g.b.)[9]. The crystalline and g.b. volume fractions, X_c, and X_{gb} were estimated from $X_c=(I_c+I_{gb})/(I_c+I_{gb}+yI_a)$ and $X_{gb}=I_{gb}/(I_c+I_{gb}+yI_a)$ where I_c, I_a, and I_{gb} are integrated intensities of the crystalline, amorphous, and intermediate peaks, respectively, and y is the ratio of the cross section for amorphous to crystalline phase which varies with the grain size L (see in Table I), $y(L)=0.1+exp[-(L/250)]$.[10] Fig. 1 shows the crystalline fraction, X_c and the grain boundary fraction X_{gb} as a function of the H-dilution ratio. One can see that the X_c increases monotonically from 68% to 92% as the H-dilution ratio is increased; meanwhile, the g.b. fraction X_{gb} decreases from 48% to 21%, that indicated an improvement of the crystalline quality.

Optoelectrics Properties:

The total intensity of the PL decreased by three orders of magnitude when the film structure changed from a- to μc-Si. This is simply because of the high recombination rate in amorphous phase due to a large number of trapped carriers in the localized tail states and lose of momentum conservation; in contrast, the carrier recombination rate is low in c-Si phase. We believe that the low energy luminescence is from the two-dimensional grain boundary region. Fig. 2 shows the normalized photoluminescence spectra at 80 K from the same group of samples as in Fig. 1. The open circles are the measured data and the solid lines are the fitting curves. The distinction between the spectra are apparent for the films with varied H-dilution. For the R=1 and 2 films, the PL spectra show typical amorphous characteristic, but the peak energy is 0.14 eV higher and the FWHM is 0.07 eV narrower in the R=2 than that of the R=1 film. This implies the

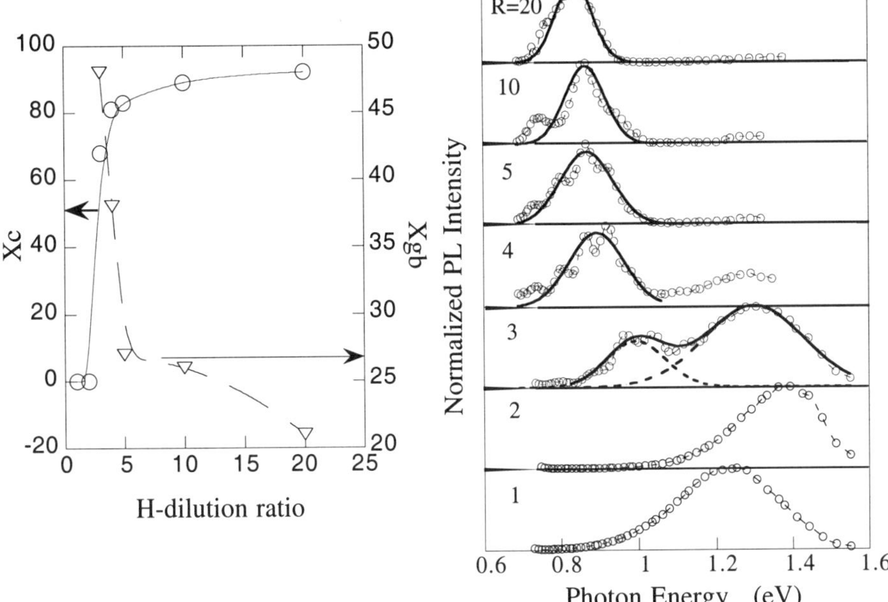

Fig. 1 The crystalline fraction Xc (o) and the grain boundary fraction Xgb (∇) as a function of H-dilution ratio R.

Fig. 2 Normalized PL spectra at 80 K. The open circles are experimental data, the solid lines are fitting curves.

widen of the optical gap and narrowness of the band tails by H-dilution effect in a-Si:H [11]. Further increasing H-dilution ratio to R=3, the PL peak energy decreased. This can be understood by a reduction of the optical gap due to formation of dense material with lower H content. For the R=3 film, in addition to the 1.3 eV a-Si PL band, a distinctive luminescent band appeared at 1.0 eV when the material contains the mixed μc-Si and a-Si phases as proved by both X-ray and Raman data. As the μc-Si volume fraction increases with increasing H-dilution, the a-Si luminescence fades away and the low energy PL dominates gradually. Clearly, the low energy PL band is related to the μc-Si phase. As the hydrogen-dilution ratio R increases from 3 to 20, several interesting features of the low energy PL were observed as shown in Fig. 3: (a) the PL intensity decreases about a factor of two orders of magnitude; (b) the PL peak energy red-shifts from 1.0 to 0.84 eV; (c) the PL band width decreases from 0.177 to 0.117 eV. All of these observations can be explained by radiative tail-to-tail recombination in two types of grain boundaries that are between the isolated grains and the a-Si matrix, and between the c-Si grains. The model was described elsewhere for details.[8]

In order to further confirm that the PL is originated from tail-to-tail but not from defect-to-tail transition, we measured the PL intensity temperature dependence. Fig.4 shows that PL intensity for R=4 film in the temperature range of 80~280 K. The PL intensity decreases with increasing temperature following the relationship of $I=I_o\exp(T/T_o)$ in the range of 80~200 K. This is consistent with the carrier thermalization in exponential band tail states. Thus, it rules out the possibility of attributing the low energy PL to defect recombination from a-Si phase[12].

Photo- and dark-conductivity are also measured. One set of room temperature results are listed in Table I. It is clear that the samples have a high ratio of photo- to dark conductivity when R<3. As the film structure changes from a- to μc-phase, the material changes from a high- to low-sensitive photoconductor. Fig. 5 shows the room temperature conductivity and the

527

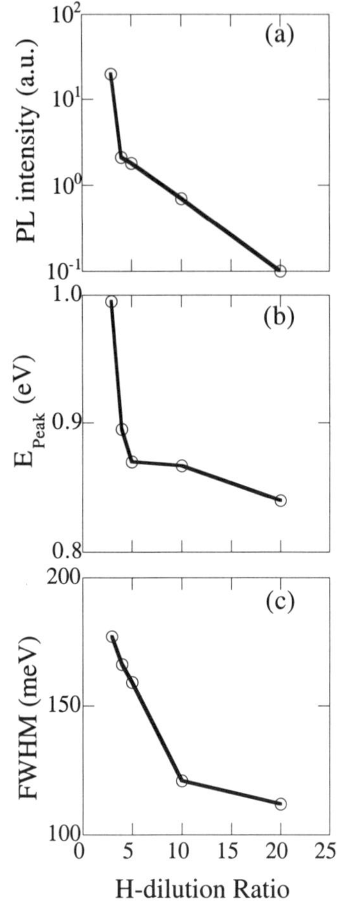

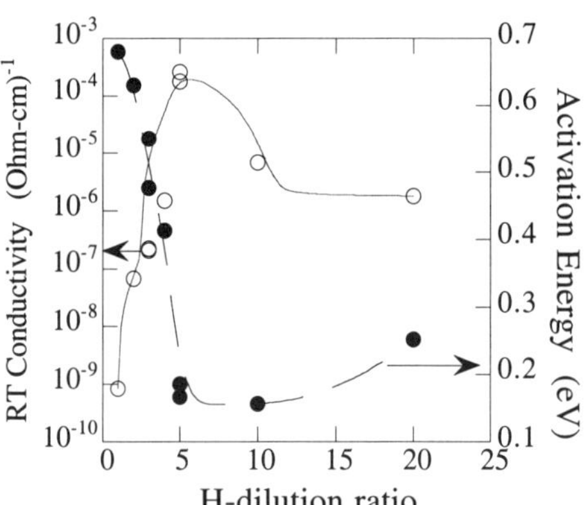

Fig.4 PL intensity as a function of temperature for the R=4 sample.

Fig.3 The low energy PL features as a function of H-dilution ratio. (a) PL intensity, (b) peak energy, and (c) band width.

Fig. 5 Conductivity and activation energy as a function of H-dilution ratio.

activation energy (Ea) as a function of H-dilution ratio. We obtain the Ea by an exponential fitting. The Ea value first decreased from 0.68 to 0.15 eV when the film transition from a- to μc-Si; then Ea increased slightly with increasing μc-Si fraction. The data of dark conductivity (averaged from several measurements) show an opposite temperature dependent behavior compared to Ea.

Substrate effect on the film growth

The above Raman and PL were taken by 514.5 nm laser beam illumination from the top surface. So, the results give information in the top surface layer. Whereas, the structure of the initial growth interface is crucial for the operation of the thin film devices. In order to investigate the effect of H-dilution on the interface layers, Raman and PL were performed with the laser

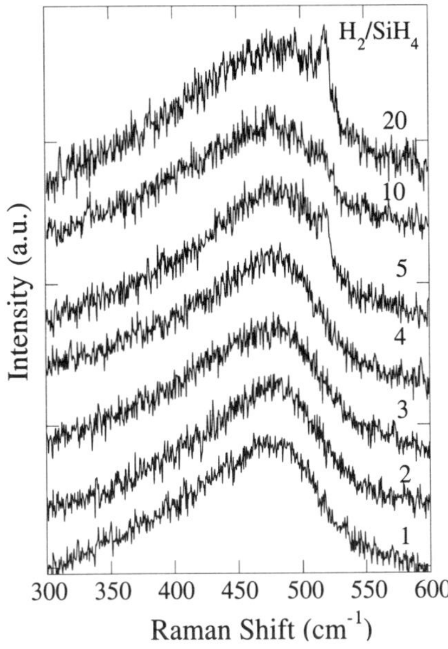

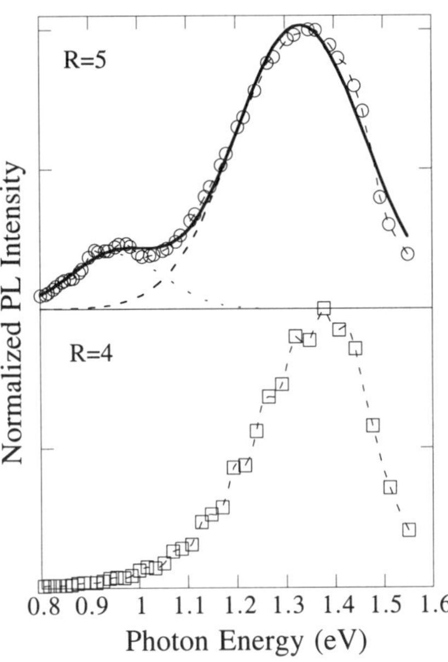

Fig. 6 Raman spectra from near substrate interface layer as a function of H-dilution.

Fig. 7 PL spectra of the R=4 and 5 films, the illumination is from the substrate side.

beam incident through the transparent substrate. The Raman results are shown in Fig. 6. It shows, in the depth of about 80 nm (the absorption depth of the 514.5 nm light in a-Si:H), no crystalline phase for the 1<R< 4 films. Unlike the crystallization occurring at R=3 H-dilution ratio for the top surface film, the crystalline peak appeared from the R=5 film near the substrate interface. Up to the R=20 film, although 92% crystalline volume fraction was observed from the top surface Raman spectrum, the amorphous phase domination remained in the 80 nm interface layer. Fig. 7 shows corresponding PL results of the R=4 and 5 films. We observed the double-peak structure appearing when the H-dilution ratio was R=5, which is in agreement with the Raman results. Both Raman and PL indicate the transition from a- to μc-Si starting at R=5.

SUMMARY AND DISCUSSIONS

X-ray, Raman and PL spectra as well as conductivity measurements demonstrated that the variation of H-dilution ratio has significant effects on both the film structure and the optoelectronic property.

(a) Both grain size and crystalline volume fraction increase with increasing the H-dilution ratio, meanwhile, the decrease of grain boundary volume fraction indicated an improvement of the crystalline quality. The largest grains size 35 nm is achieved at H-dilution ratio 5. The highest crystalline volume fractions 92% is achieved at H-dilution ratio 20.

(b) Corresponding to the structure change from a- to μc-Si, we observed a gradual change of the electronic density of states characterized by PL spectroscopy.

In the literature, there appears three typical explanations about PL spectra in μc-Si. Anderson et al.[12] described that the high density of defect states in the amorphous phase is responsible for the low energy PL (peaked at 0.95 eV with a band width of 0.18 eV). Bhat et al.[13] suggested that the low energy PL band in μc-Si films results from radiative recombination at defect centers in crystalline environment. Recently, Savchouk et al.[14] interpreted that the PL in polycrstalline-Si films originates from tail-to-tail transition in grain boundary.

In this study, we observed interesting features of the PL spectra in the μc-Si phase, such as the energy red-shift, the intensity decrease, and the narrowness of the band-width with the increasing of H-dilution. We described that the low energy PL originated from tail-to-tail recombination in grain boundary regions. When the bonding configuration in the grain boundary becomes more ordered with increasing H-dilution ratio, the optical band gap shrinks and the band tails narrows down, that results in a peak energy red-shift and band width narrowness. In additiona to better ordering of the grain boundary, the grain-boundary volume fraction decreases with increasing of the H-dilution ratio. These two factors result a rapid decrease in the PL efficiency. The PL temperature dependence shown in Fig. 4 further confirmed the exponential band tail nature in the grain boundary region, and ruled out the possibility of attributing the low energy PL to defect luminescence from a-Si or c-Si phase.

(c) The Raman and PL results from the 80 nm interface layer suggested that the amorphous matrix is formed first and an increase in H-dilution up to R=5 can improve the amorphous quality and form a portion of μc-Si. The changes of both microstructure and electronic properties have not saturated yet up to R=20 H-dilution.

(d) The conductivity activation energy is 0.68 eV for a-Si film, then decreased to 0.15 eV when the film structure changed to μc-Si. A slight increase of Ea is found with increasing μc-Si fraction. This may indicate the transport mechanism starting to change from defect-dominant to more close to crystalline silicon nature.

ACKNOWLEDGMENTS

The work is supported by NREL sub-subcontract XAK-8-17619-11 and thin film PV partnership. Yue is partially supported by NSF-Int-9604915. Han is partially supported by CGP Fund, NSF-Int-9802430. Lin is supported by Chinese Educational Fellowship. Wang is supported by DOE subcontract DE-AC02-83CH10093. The X-ray data were furnished by Prof. William at school of Mine. The Raman spectra were measured by Dr. Lorentzen in the Department of Physics and Astronomy at UNC.

REFERENCES

1. M. Luysberg, P. Hapke, R. Carius and F. Finger, Phil. Mag. **A75**, 31 (1997).
2. F. Wang, D. Wolfe, and G. Lucovsky in *Amorphous and Microcrystalline Silicon Technology*, edited by S. Wagner, M. Hack, E. A. Schiff, R. Schropp and I. Shimizu (Mat. Res. Soc. Symp. Proc. **467**, 1997) pp. 433-438; References therein.
3. P. Brogueira, V. Chu and J.P. Conde in *ibid*, edited by M. Hack, E.A. Schiff, A. Madan, M. Powell and A. Matsuda (Mat. Res. Soc. Symp. Proc. **377**, 1995) pp.57-62.
4. J.P. Conde, P. Brogueira, R. Castanha and V. Chu, in *Amorphous Silicon Technology*, edited by M. Hack, E.A. Schiff, S. Wagner, R. Schropp and A. Matsuda (Mat. Res. Soc. Symp. Proc. **420**, 1996), pp.357-362.
5. F. Diehl, W. Herbst, B. Schroder and H. Oechsner in *ibid*, edited by S. Wagner, M. Hack, E.A. Schiff, R. Schropp and I. Shimizu (Mat. Res. Soc. Symp. Proc. **467**, 1997) pp.451-456.
6. H.N. Wanka, R. Zedlitz, M. Heintze and M.B. Schubert in *ibid*, edited by M. Hack, E.A. Schiff, S. Wagner, and R. Schropp (Mat. Res. Soc. Symp. Proc. **420**, 1996) pp.295-300.
7. H.P. Klug and L.E. Alexander, *X-ray Diffraction Procedure* (Wiley, New York, 1974).
8. Guozhen Yue, J.D. Lorentzen, Jing Lin, Daxing Han and Qi Wang, submitted to APL.
9. S. Veprek, F.A. Sarott, and Z. Iqbal, Phys. Rev. **B36**, 3344 (1987).
10. E. Bustarret, M.A. Hachicha, and M. Brunel, Appl. Phys. Lett. **52**, 1675 (1988).
11. Daxing Han, Keda Wang and Liyou Yang, J. Appl. Phys. **80**, 2475 (1996).
12. D.A. Anderson, G. Moddel, R.W. Collins and W. Paul, Solid State Commun. **31**, 677 (1979).
13. P.K. Bhat, G. Diprose, T.M. Searle, I.G. Austin, P.G. LeComber and W.E. Spear, Physica **117B&118B**, 917 (1983).
14. A.U. Savchouk, S. Ostapenko, G. Nowak, J. Lagowski, and L. Jastrzebski, Appl. Phys. Lett. **67**, 82 (1995).

HYDROGEN DILUTION EFFECT ON THE CRYSTALLINITY OF SILICON FILMS GROW BY HOT WIRE CELL METHOD

M. ICHIKAWA, J. TAKESHITA, A. YAMADA AND M. KONAGAI
Department of Electrical and Electronic Engineering, Tokyo Institute of Technology,
2-12-1 O-okayama, Meguro-ku, Tokyo, Japan

ABSTRACT

A new process, the Hot Wire Cell method, was developed and successfully used to grow polycrystalline silicon thin films at a low temperature and high growth rate. In the Hot Wire Cell method, reactant gases are decomposed as a result of reacting with a heated tungsten filament placed near to a substrate and polycrystalline silicon films can be deposited at a growth rate of 1.2nm/s without hydrogen dilution and 0.9nm/s with the use hydrogen dilution. The film crystallinity changed from amorphous to polycrystalline due to the addition of hydrogen, thus hydrogen dilution was effective for improving film crystallinity. Furthermore, we obtained (220) oriented polycrystalline silicon thin films with a 90% crystal fraction by the use of hydrogen dilution. These results showed that the Hot Wire Cell method is promising for the deposition of device-grade polycrystalline silicon films for photovoltaic applications.

INTRODUCTION

The Hot Wire Cell method is a new and very promising process for depositing polycrystalline silicon thin films for photovoltaic applications. In this process, a heated tungsten filament is used to induce the catalytic and/or pyrolytic dissociation of reactant gases such as SiH_4 and H_2[1] and produce atomic hydrogen as a reaction by-product, which then reacts with silane molecules to produce SiH_3 radicals. In this work, since the filament is positioned perpendicular to the substrate holder, the reactant gas is efficiently decomposed during its passage through the filament thus enhancing its decomposition rate. In the Hot Wire Cell method, atomic hydrogen is also produced as a result of the decomposition of the silane. We have previously obtained polycrystalline silicon thin films without hydrogen dilution and demonstrated that polycrystalline silicon films could be deposited at low temperatures of 175-400°C at a relatively high growth rate of 1.0nm/s[2]. In this study, we present our experimental results on the effect of filament temperature and hydrogen dilution on the growth rate and structural properties of polycrystalline silicon films.

EXPERIMENT

A schematic diagram of the deposition chamber incorporating the Hot Wire Cell is shown in Fig. 1. The Hot Wire Cell consists of gas inlet and a tungsten filament where the axis of the gas inlet into the chamber was perpendicular to the plane of the substrate holder. A tungsten wire with a diameter of 0.3mm was used as the filament. The filament, coiled with a diameter of 4mm and length of 15mm, was arranged to be parallel to the gas inlet and gas flow and the distance between the filament and the substrate was about 6cm. The temperature of the filament was measured using an optical pyrometer through a window of the chamber. The filament was heated by a D.C. power supply and the power kept constant throughout the experiment by P.I.D. control. Thin films were deposited on Corning 7059glass substrate using silane or hydrogen diluted silane and in the case of hydrogen dilution, H_2 gas was also passed through the heated filament.

Mat. Res. Soc. Symp. Proc. Vol. 557 © 1999 Materials Research Society

The temperature of the substrate was measured by a calibrated thermocouple attached to the substrate holder where the calibration of the substrate temperature was carried out using a second thermocouple attached to the substrate. The gas pressure was controlled by adjusting a main valve positioned between the chamber and the mechanical booster pump.

The morphological and structural properties of the deposited films were characterized by Scanning Electron Microscopy (SEM), X-ray diffraction (XRD) and Raman spectroscopy. The Raman spectra were deconvoluted into their integrated crystalline (I_c ~520cm^{-1}), amorphous (I_a ~480cm^{-1}) and intermediate (I_m ~510cm^{-1}) phonon peak intensities. The crystalline fraction, X_c, was calculated from the following relationship[3]:

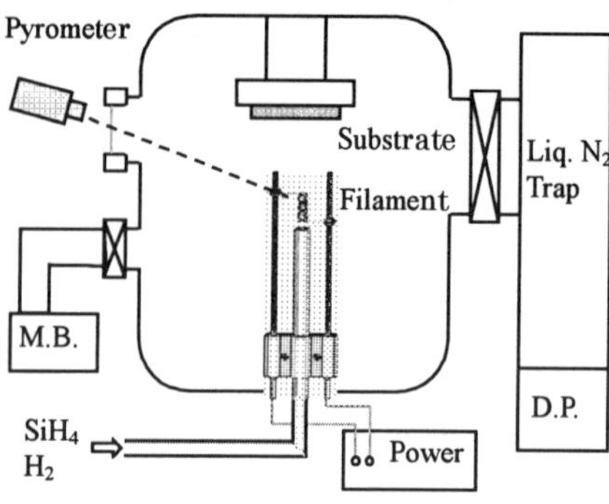

Fig. 1. Schematic of the deposition chamber

$$X_c = (I_c + I_m) / (I_c + I_m + I_a). \qquad (1)$$

The thickness of the films was measured using a profilometer.

RESULTS

<u>Effect of using the filament</u>

The results of deposition carried out without hydrogen dilution showed that the crystalline fraction of the films depended critically on the temperature of the filament[2]. When the filament temperature was low (T_{fil}<1800°C), the films were amorphous in nature and if the filament temperature was increased, polycrystalline silicon films were obtained even at a low substrate temperature of 175°C. These results indicate the efficient generation of atomic hydrogen by the tungsten filament. In order to investigate the effect of the dimensions of the filament on the properties of the films, we deposited films using a filament having a shorter length of 8mm. Figure 2 shows the growth rate as a function of the filament temperature for both the 15mm and 8mm filaments. We measured the uniformity of the film thickness and found that it was almost the same within the substrate area (2 x 2cm). For deposition using the 15mm filament length, a polycrystalline silicon film was obtained with a relatively high growth rate of 1.2nm/s at a filament temperature of 1800°C. However, for

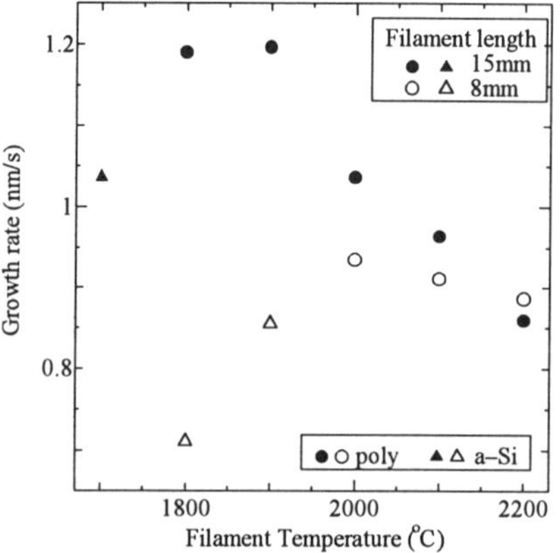

Fig. 2. Dependence of the growth rate on the filament temperature (T_{sub}= 325°C).

deposition with the 8mm filament, the resulting films were found to be amorphous at filament temperatures of 1800°C and 1900°C, but at a filament temperature of 2000°C, polycrystalline silicon films were obtained. These results suggest that the efficient dissociation of reactant gases with the evolution of atomic hydrogen, is required to crystallize silicon films at low filament temperatures.

Figure 3 shows the dependence of the dark and photo conductivity of the films on the filament temperature for the 15mm and 8mm filaments. The amorphous films showed typical characteristics with a low dark conductivity that varied between 4×10^{-10}S/cm and 1×10^{-8}S/cm and a photosensitivity (σ_{ph}/σ_d) of 10^4. However, for polycrystalline films, the dark conductivity

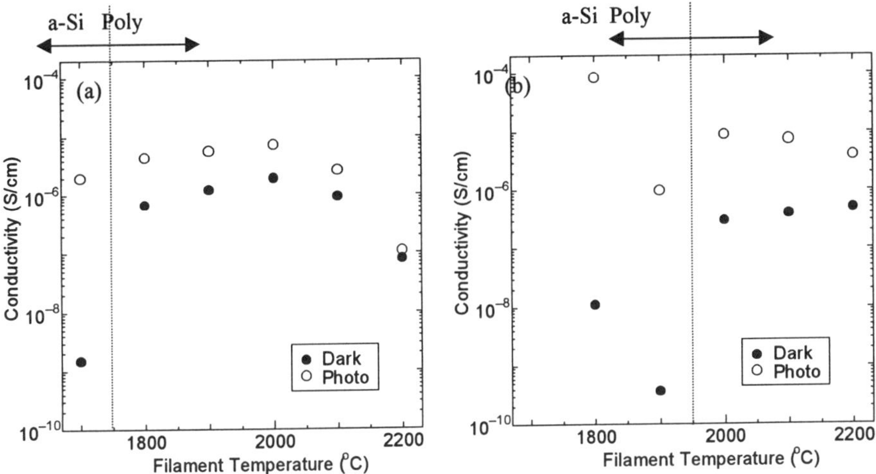

Fig. 3. Dependence of the dark and photo conductivity of the films on the filament temperature at the filament length of 15mm (a) and 8mm (b).

remained approximately constant at 10^{-6}S/cm for the 15mm filament and 4×10^{-7}S/cm for 8mm filament. The maximum photosensitivity was obtained for films deposited using the 8mm filament at a filament temperature of 2000°C ($\sigma_{ph}/\sigma_d \sim 30$).

<u>Hydrogen dilution</u>

In order to study the effect of hydrogen dilution, we deposited films under two deposition conditions. First, we deposited films at a substrate temperature of 210°C and filament temperature of 1800°C with hydrogen dilution. It was found that when the reactant gas was not diluted with hydrogen then amorphous silicon films were obtained. The total pressure of the chamber was kept constant at 0.1Torr for all H_2 flow rates. Figure 4 shows the dependence of the deposition rate on the H_2 flow rate. It can be seen that the deposition rate decreases from 1.2nm/s without hydrogen dilution to 0.7nm/s when a H_2 flow rate of 60sccm was used. The film crystallinity changed from amorphous to polycrystalline as a result of hydrogen dilution. Figure 5 shows the Raman spectra of the films as a function of the H_2 flow rate. For the films deposited at H_2 flow rates of less than 15sccm, the spectrum showed only a broad peak with a low-intensity centered around 480cm^{-1} which is a characteristic of an amorphous phase. When the flow rate was increased to the 23sccm, the peak was observed at 520cm^{-1} which is a characteristic of a

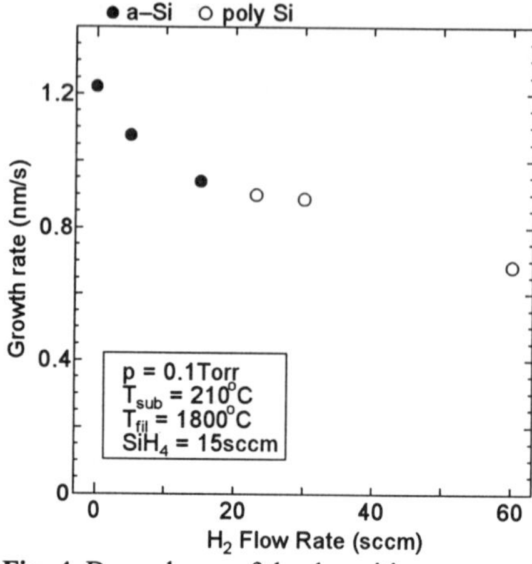

Fig. 4. Dependence of the deposition rate on the H_2 flow rate.

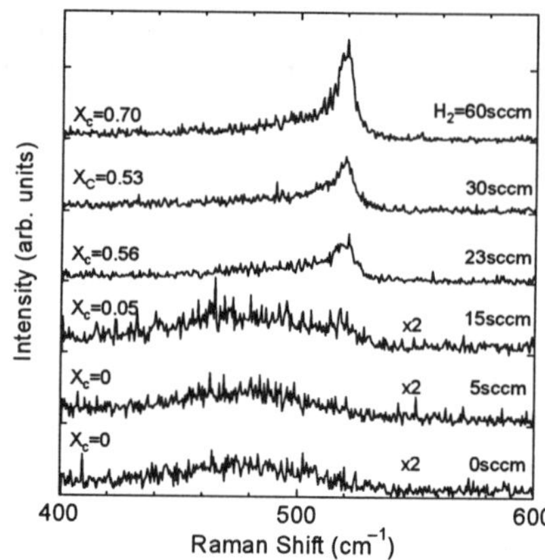

Fig. 5. Dependence of the Raman spectra on the H_2 flow rate.

crystalline phase, and the crystalline fraction was calculated to be 56%. Further, when the H_2 flow rate was 60sccm, only a narrow peak centered at 520cm^{-1} was observed and the crystalline fraction of the film was about 70%. Thus it was found that the crystal structure of silicon film was sensitive to hydrogen dilution.

In a second set of experiments, we deposited films at a substrate temperature of 400°C and filament temperature of 2000°C with hydrogen dilution. When the reactant gas was not diluted with hydrogen, we obtained polycrystalline silicon films. Figure 6 shows the deposition rate as a function of H_2 flow rate. The growth rate decreased monotonically from approximately 0.9nm/s without hydrogen dilution to 0.75nm/s at a H_2 flow rate of 15sccm. When the H_2 flow rate was increased from 15sccm to 30sccm, the growth rate was almost constant at 0.75nm/s. Figure 7 shows the Raman spectra as a function of the H_2 flow rate. It can be seen that all samples exhibit a peak centered around 520cm^{-1} and that the crystalline fraction of the film was increased from 74% without hydrogen dilution to 91% at the H_2 flow rate of 30sccm. Figure 8 shows the XRD pattern as a function of H_2 flow rate. The polycrystalline silicon films were mainly composed of grains oriented in the (111), (220) and (311) planes, which are perpendicular to the substrate surface. Without hydrogen dilution, the X-ray diffraction pattern indicated that the films were randomly orientated. As the H_2 flow rate was increased from 0 to 30sccm, the intensity of the (220) diffraction peak increased indicating that the film had a very strong (220) preferential orientation at a H_2 flow rate of 30sccm. These results suggest that the crystallinity and orientation of the films depends on the amount of atomic hydrogen present during deposition. When the reactant gas was diluted with

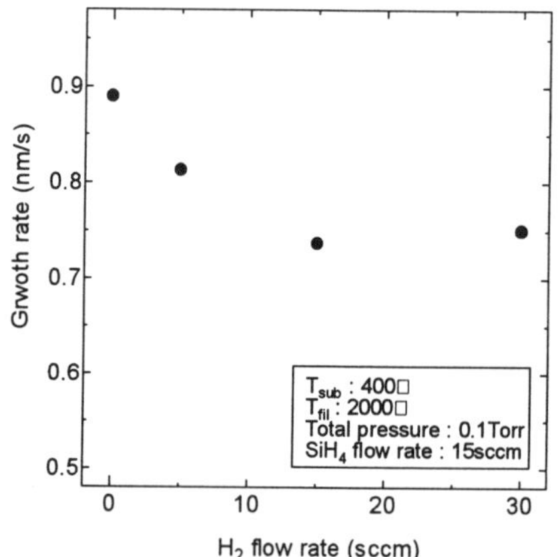

Fig. 6. Dependence of the deposition rate on the H_2 flow rate.

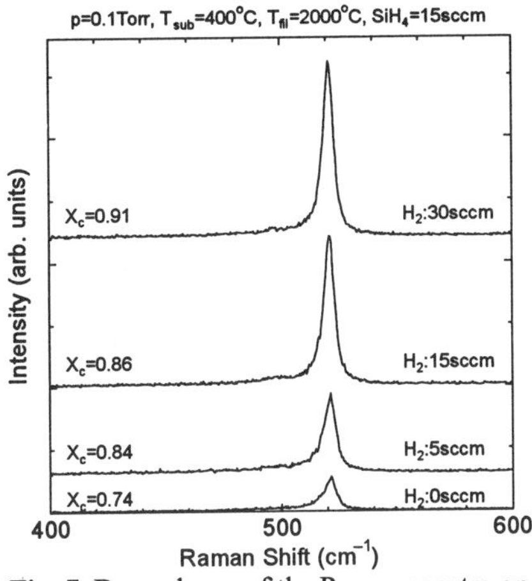

Fig. 7. Dependence of the Raman spectra on the H_2 flow rate.

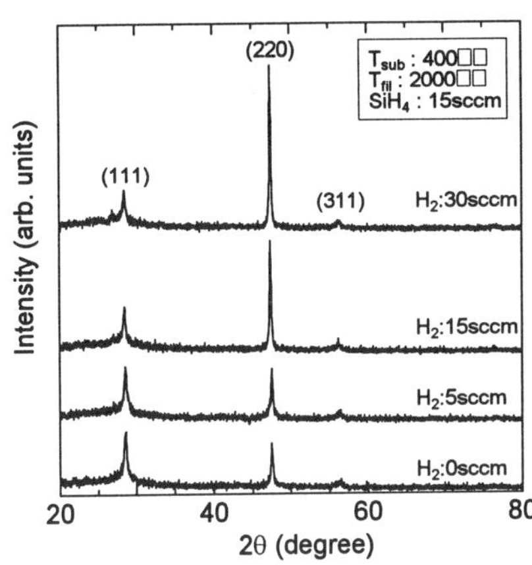

Fig. 8. Dependence of XRD pattern on the H_2 flow rate.+

hydrogen, the hydrogen molecules were dissociated into atomic hydrogen by the hot tungsten filament and the atomic hydrogen thus generated may improve the quality of the polycrystalline films.

Figure 9 shows the dark and photo conductivity of the films as a function of H_2 flow rate.

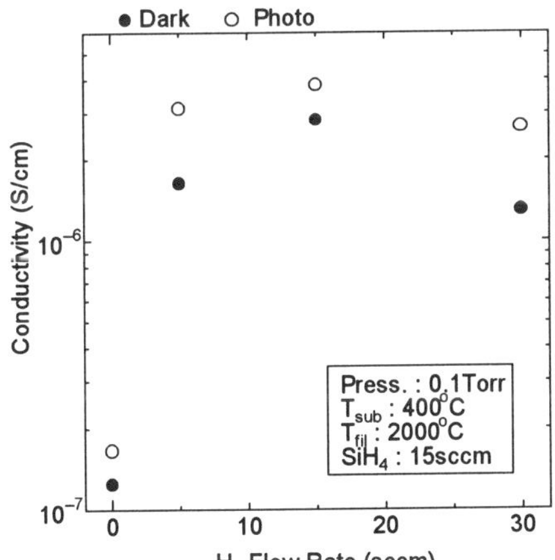

Fig. 9. Dependence of the dark and photo conductivity on the H_2 flow rate.

Without hydrogen dilution the low dark conductivity was 1.3×10^{-7} S/cm and the photo-conductivity also had a low value of 1.7×10^{-7} S/cm. However, with hydrogen dilution, the conductivity increased to the order of 10^{-6} S/cm, but the photosensitivity was still low ($\sigma_{ph}/\sigma_d \sim 2$).

CONCLUSION

Polycrystalline films were deposited at a the growth rate of 1.2nm/s on glass substrates by the Hot Wire Cell method without hydrogen dilution. The crystalline structure of the silicon films changed from amorphous to polycrystalline depending on the efficiency of the gas decomposition at the tungsten filament. Hydrogen dilution was also effective in enhancing the deposition of the high quality polycrystalline silicon films by the Hot Wire Cell method. When the reactant gas was diluted by hydrogen, the film crystallinity improved with increasing hydrogen gas flow rate. The polycrystalline silicon film with a crystalline fraction of over 90% was obtained having a strongly preferential orientation in the (220) plane with hydrogen dilution. The results of this study showed that the Hot Wire Cell method is very promising for depositing high quality polycrystalline silicon films.

ACKNOWLEDGMENTS

This study was supported in part by NEDO as a part of the New Sunshine Program under the Ministry of International Trade and Industry.

REFERENCES

1. H. Matsumura, Jpn. J. Appl. Phys. **37**, 3175 (1998)
2. M. Ichikawa, J. Takeshita, A. Yamada and M. Konagai, Jpn. J. Appl. Phys. **38**, L24 (1999).
3. T. Kaneko, M. Wakagi, K. Onisawa and T. Minemura, Appl. Phys. Lett. **64**, 1865 (1994)

STRESS IN HYDROGENATED MICROCRYSTALLINE SILICON THIN FILMS

D. PEIRÓ, C. VOZ, J. BERTOMEU, J. ANDREU, E. MARTÍNEZ AND J. ESTEVE
Department of Applied Physics and Optics, Universitat de Barcelona, Av. Diagonal 647, 08028-Barcelona, Catalunya, Spain, jandreu@fao.ub.es.

ABSTRACT

Hydrogenated microcrystalline silicon films have been obtained by hot-wire chemical vapor deposition (HWCVD) in a silane and hydrogen mixture at low pressure ($<5\times10^{-2}$ mbar). The structure of the samples and the residual stress were characterised by X- ray diffraction (XRD). Raman spectroscopy was used to estimate the volume fraction of the crystalline phase, which is in the range of 86 % to 98%. The stress values range between 150 and -140 MPa. The mechanical properties were studied by nanoindentation. Unlike monocrystalline wafers, there is no evidence of abrupt changes in the force-penetration plot, which have been attributed to a pressure-induced phase transition. The hardness was 12.5 GPa for the best samples, which is close to that obtained for silicon wafers.

INTRODUCTION

Hydrogenated microcrystalline silicon (μc-Si:H) has attracted much attention because of its applications in photovoltaic conversion and large-area electronics [1,2]. This material can benefit both from the available amorphous silicon technology and from the stability and good electronic properties of crystalline silicon. Several deposition techniques have been used to obtain μc-Si:H [1,3]. Another technique that could be useful in large-area μc-Si:H deposition is the known as catalytic-CVD or hot-wire CVD (HWCVD). This technique allows deposition of μc-Si:H films with good structural and optoelectrical properties at low substrate temperatures [2,4-6].

In order to achieve large-area devices on cheap substrates, low temperature processes are required. Moreover, good mechanical properties of the films on such substrates must be ensured (i.e. good adhesion and low stress). Although the structural, optical and electrical properties of μc-Si:H films have been extensively studied, much less attention has been paid to the mechanical properties. In addition, the study of these properties presents a clear interest for the application of this material to mechanical sensors (displacement, pressure, ...).

In this paper, we study the mechanical properties of μc-Si:H films deposited by HWCVD in the low temperature (LT) and low pressure (LP) range. Special deposition conditions were chosen to study the influence of pressure, temperature and hydrogen dilution. The structure of the samples was also analysed. The results show that the films deposited onto glass substrates at LT and LP are not stressed. The hardness of our best films is close to that of silicon wafers.

EXPERIMENTAL

Five μc-Si:H films were deposited under different conditions onto Corning 7059 glass and Si (111) substrates in a HWCVD multichamber reactor equipped with a load-lock chamber and base pressures below 10^{-8} mbar. A gas mixture of silane and hydrogen was introduced into the deposition chamber with inverted geometry, designed so as to minimize the formation of pinholes due to powder formation. The dissociation of the gases was obtained by means of a hot tungsten wire (0.5 mm diameter, basket-shaped) at 1740°C, placed in front of the gas inlet 4 cm below the substrates.

Mat. Res. Soc. Symp. Proc. Vol. 557 © 1999 Materials Research Society

The thickness of the films was measured with a stylus profiler. The main technological parameters of the samples are summarised in Table I.

The structure of the samples deposited onto glass was studied by Raman Spectroscopy using a 514 nm laser radiation and an energy density of 10 W/cm^2. The crystalline fraction in the films was obtained from the deconvolution of the amorphous and the crystalline contribution in the Raman spectra [7]. X-ray diffraction in the Bragg-Brentano geometry was used to analyse preferential orientations in samples deposited onto glass substrates. The crystallite grain size of the films was calculated from the (111) peaks. By using the Scherrer formula residual stress was characterised in a Philips MRD diffractometer, by analysing the (111) diffraction peak in the range of 2θ between 26.4° and 30.4° for Ψ-tilts from 0 to 80°, for both Φ=0° and 180°. These peaks were fitted with a Pseudo-Voigt function giving 16 lattice spacings. The stress values were calculated assuming that Young's modulus and Poisson's ratio are 160 GPa and 0.17, respectively [8].

Microhardness values were evaluated by the dynamical microindentation technique. Measurements were performed in a nanoindentation system manufactured by MicroMaterials Ltd. This system uses a Berkovich diamond head, which performs shallow indentations on the sample surface, with load and depth resolutions better than 1 μN and 0.1 nm respectively, both in the loading and unloading cycles. These measurements were performed for samples deposited onto Si(111) wafers. To eliminate the influence of the substrate on the hardness measurements, one sample deposited onto glass was also measured and no differences were found at low loads. Adhesion of the films to the substrate was evaluated by microscratch measurements. They were performed with the same microindentation system. In this case the machine uses a spherical diamond indenter with a tip radius of 50 μm, and applies increasing loads.

Table I. Main technological parameters of the samples used in this study

Sample	P (mbar)	T_S (°C)	$f_{H2}/(f_{H2}+f_{SiH4})$ (%)	t(μm)
981019	7.0×10^{-3}	150	95	1.2
981124	1.0×10^{-2}	150	95	0.9
990216	3.5×10^{-2}	150	95	2.1
990209	7.0×10^{-3}	300	95	1.6
990108	7.0×10^{-3}	150	90	1.1

RESULTS AND DISCUSSION

The deposition conditions were chosen to determine the influence of the process pressure, substrate temperature and hydrogen dilution on the mechanical properties. We selected sample 981019 as a reference sample obtained at low pressure, low temperature and high hydrogen dilution (Table I). Samples 981124 and 990216 were deposited at increased process pressures, while the effect of the substrate temperature was studied in sample 990209. Finally, sample

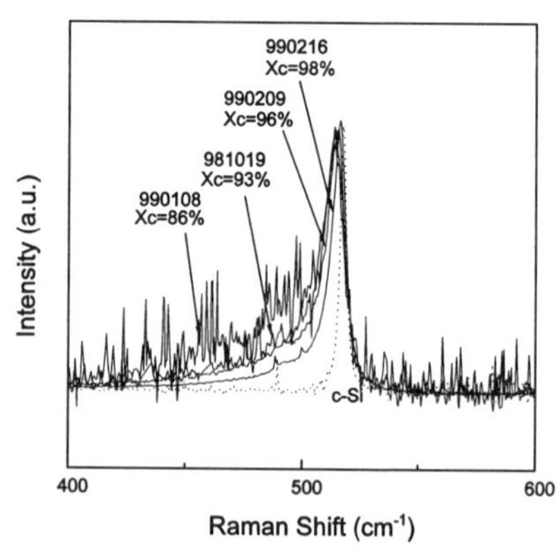

Figure 1. Raman spectra for 4 of the samples used in this study and for a c-Si reference.

990108 was obtained at a lower hydrogen dilution.

The Raman spectra of the films are presented in Figure 1. The calculated crystalline fractions are presented in Table II. These results indicate that higher hydrogen dilutions, higher pressures and higher temperatures favor the crystallinity of the films. The peaks related to the TO-phonon mode are shifted respect to that of monocrystalline silicon. The values of this displacement are also presented in Table II.

The diffraction spectra of the samples are presented in Figure 2. All the samples showed a clear (111) preferential orientation. The crystallite sizes, as determined from the

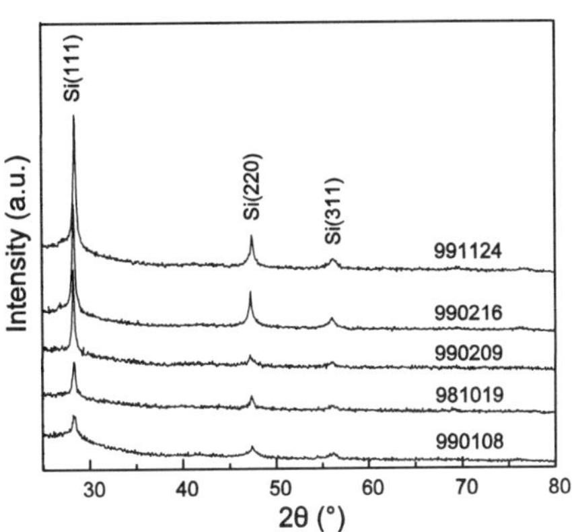

Figure 2. X-ray diffractograms obtained in the θ/2θ mode for all the samples.

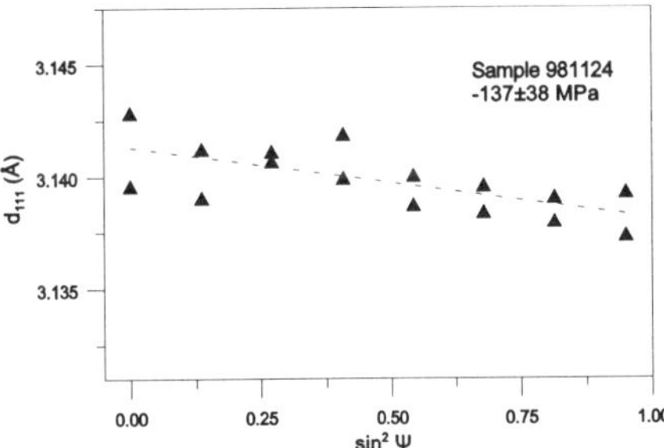

Figure 3. Lattice spacings obtained from the Si(111) diffraction peak as a function of the $\sin^2\Psi$ for the sample 981124. The negative slope indicates a compressive stress.

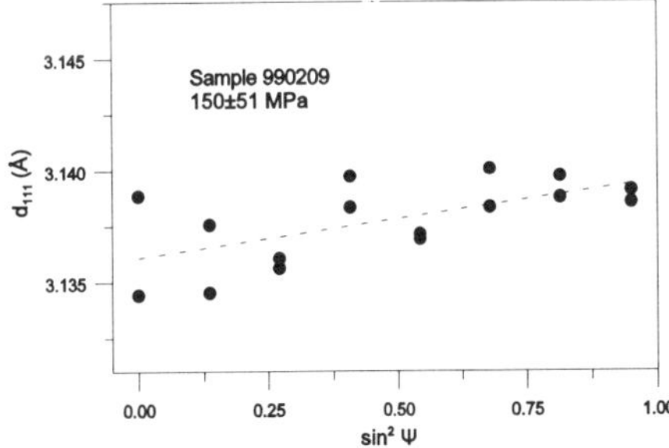

Figure 4. Lattice spacings obtained from the Si(111) diffraction peak as a function of the $\sin^2\Psi$ plot for the sample 990209. The positive slope indicates a tensile stress.

analysis of the (111) peak, ranged between 12 and 39 nm. These values (Table II) are correlated with the observed displacement of the Raman peaks. The samples with largest crystallites presented a lower displacement. This suggests that the displacement is mainly due to crystal size effects. The possible contribution of stress to this displacement is discussed below.

The (111) peak was chosen to analyse the stress of the films, as detailed above, since it is the most intense. The $\sin^2\Psi$ plots (d_{111} vs. $\sin^2\Psi$) for the samples with higher stress values are shown in Figures 3 (compressive stress) and 4 (tensile stress). The stress values for the whole set of samples are presented in Table II. These values are lower than those obtained in a previous study with samples deposited at higher pressures and higher substrate temperatures [9]. Samples 981019, 990216 and 990108 can be considered as unstressed since the error bar is larger than the stress measured. The tensile stress presented by sample 990209 could be the explanation for the relatively high shift in the position of the TO-phonon mode with respect to that of sample 990216, which presents similar crystallinity.

The hardness values were obtained from the load curves (Figure 5) in the unload cycle.

Table II. Summary of the structural and mechanical properties of the films: crystalline fraction (X_c), shift of the TO-phonon peak, grain sizes (g.s.), (111) lattice spacing, stress and hardness.

Sample	X_c (%)	Shift (cm^{-1})	g.s. (nm)	d_{111}(Å)	Stress (MPa)	Hardness (GPa)
981019	93	2.44	26	3.152	99 ± 119	12.1+0.2
981124	not meas.	not meas.	39	3.156	-137 ± 38	12.1+0.1
990216	98	1.92	31	3.151	-21 ± 34	10.0+0.2
990209	96	2.6	30	3.134	150 ± 51	12.5+0.2
990108	86	2.98	12	3.152	64 ± 99	11.0+0.3

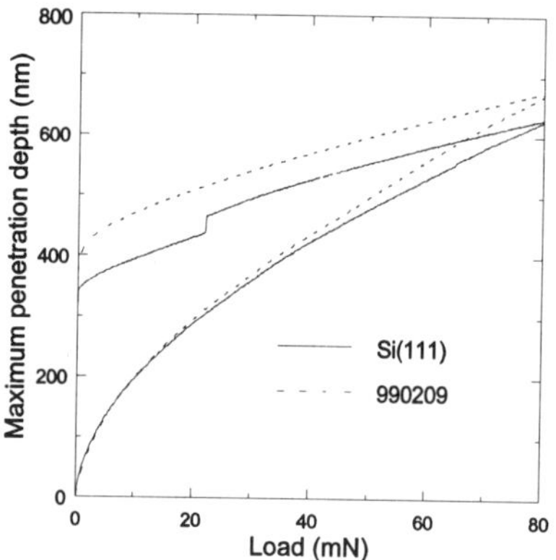

Figure 5. Load curves used for the determination of the film hardness. The solid curve corresponds to a c-Si(111) reference, whereas the dotted curve was obtained for sample 990209.

Figure 6. Indentation print for samples 981019 (a) and 990209 (b) as observed by optical microscopy.

These values were calculated from the ratio between the applied load and the indentation print area [10]. Two typical indentation prints observed by optical microscopy in the samples are presented in Figure 6. Microcracks arose for all the samples, and film detachment was occasionally observed around the indentation print.

The load curves present the maximum penetration depth as a function of the applied load. A maximum load of 80 mN was applied to all the samples. The shape of the unload cycle depends on this maximum applied load, and for high maximum penetration depths, the influence of substrate is present. To avoid this, several load curves were performed for each sample with different maximum penetration depths. From each curve, a different value of hardness was obtained. These values are represented as a function of the maximum penetration depth, giving the behavior observed in Figure 7. The hardness values presented in Table II were obtained from the initial part of the hardness vs. maximum penetration depth plot (Figure 7) by only considering the initial values (for depths lower than 200 nm). The values obtained for samples on c-Si and Corning glass

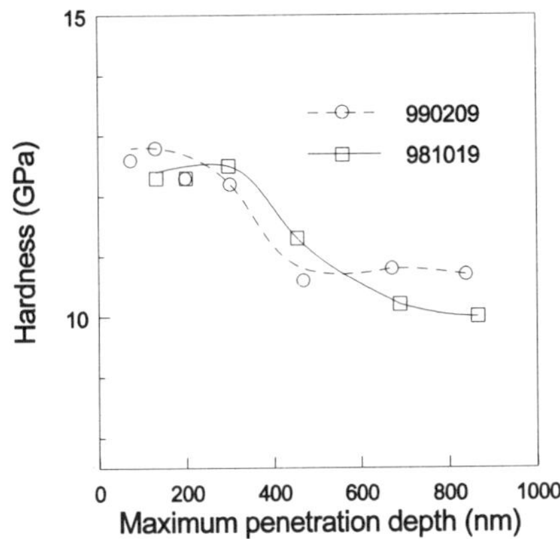

Figure 7. Hardness of samples 990209 and 981019 as a function of the maximum penetration depth. The hardness values in Table II were deduced from the values at low maximum penetration depths.

the load required to delaminate the film) was in the range of 300-500 mN for all the samples. In Figure 8, a typical view of a scratched sample is presented. A common feature of all the samples is a very low friction coefficient before the failure of the film. This is responsible for the absence of prints on the surface before the failure. Microscratch tests were performed on films grown in the same run on c-Si and substrates were the same at low loads. These values were close to that of monocrystalline silicon for all the samples, except 990216. This sample was deposited at higher pressure, at which gas phase radical formation from Si and H atoms could arise, giving a less dense structure in the film, with some voids. This voidy structure could also explain the absence of stress in this film. Another feature common to all the films is the absence of the step that appears in the unload cycle of the Si(111). This step is usually attributed to a phase transition in the silicon structure [11].

Microscratch tests indicate that the films showed good adhesion. The critical load (i.e.

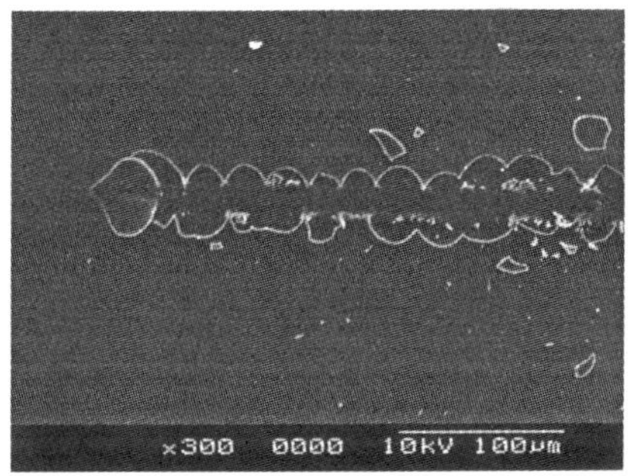

Figure 8. Scanning electron microscropy image of the scratch test on sample 981019.

Corning glass, and no significant differences in the print nor the critical loads were found between them. In some cases, a slightly higher critical load was found on c-Si substrates, indicating better film adhesion.

CONCLUSIONS

The structural and mechanical properties of several samples deposited by HWCVD have been studied. The crystalline fraction was higher than 85% for all the samples. Crystallinity increased with hydrogen dilution, substrate temperature and process pressure. All the samples showed a rather low stress, ranging from tensile to compressive depending on the deposition conditions. The sample with highest amorphous fraction showed lowest hardness. The samples with higher crystalline fraction presented a hardness values very near to that of monocrystalline silicon, except for the sample deposited at the highest pressure, which, surprisingly, presented a lower hardness. Good adhesion to the substrate was deduced from microscratch measurements for all the samples.

The results on mechanical properties of the µc-Si:H films deposited at low temperatures and low pressures by HWCVD show that this material is suitable to be deposited onto large scale glass substrates.

ACKNOWLEDGEMENTS

The authors thank the Scientific-Technical Services of the Universitat de Barcelona, for Raman and XRD measurements. This work was supported by the EC under contract JOR3-CT97-0126 and the CICYT of the Spanish Government (MAT97-1714-CE and MAT98-1305-CE).

REFERENCES

1. P. Torres, J. Meier, R. Flückiger, U. Kroll, J.A. Anna Selvan, H. Keppner and A. Shah, Appl. Phys. Lett. **69**, p. 1373 (1996).
2. H. Meiling, A.M. Brockhoff, J.K. Rath and R.E.I. Schropp, J. Non-Cryst. Solids **227-230**, p. 1202 (1998).
3. K. Yamamoto, M. Yoshimi, T. Suzuki, Y. Tawada, Y. Okamoto and A. Nakajima, Proc. 2nd World Conference on Photovoltaic Solar Energy Conversion, Vol. II, 1999, p. 1284.
4. H. Matsumura, Jpn. J. Appl. Phys. **30**, p. L1522 (1991).
5. J. Cifre, J. Bertomeu, J. Puigdollers, M.C. Polo, J. Andreu and A. Lloret, Appl. Phys. A **59**, p. 645 (1994).
6. A.R. Middya, A. Lloret, J. Perrin, J. Huc, J.L.Moncel, J.Y. Parey and G.Rose in Amorphous Silicon Technology, edited by M. Hack, E.A. Schiff, S. Wagner, A. Matsuda and R. Schropp (Mat. Res. Soc. Proc. **377**, San Francisco, CA, 1995) pp. 119-124.
7. M.H. Brodsky, M. Cardona and J.J. Cuomo, Phys. Rev. B **16**, p. 3556 (1977).
8. T.I. Kamins, Sensors and Actuators **A21-23**, p. 817 (1990).
9. D. Peiró, J. Bertomeu, F. Arrando and J. Andreu, Mater. Lett. **30**, p. 239 (1997).
10. W. C. Oliver and G. M. Pharr, J. Mater. Res. **7**, p. 1564 (1992).
11. E. R. Weppelmann, J. S. Field and M. V. Swain, J. Mater. Res. **8**, p. 830 (1993).

CHARGE TRANSPORT IN THE TRANSITION FROM HYDROGENATED AMORPHOUS SILICON TO MICROCRYSTALLINE SILICON

A. Kattwinkel [*], R. Braunstein [*], G. Sun [*], Qi Wang [**]
[*]Department of Physics and Astronomy, University of California, Los Angeles, CA 90024
[**] National Renewable Energy Laboratory, 1617 Cole Boulevard, Golden, CO 80401-3393

ABSTRACT

The electronic transport properties of a series of samples prepared by hot-wire chemical vapor deposition with a transition from a-Si:H to μc-Si:H were measured applying the photoconductive frequency mixing technique. We found both improved stability against light-soaking and different values for the photomixing electron lifetime and mobility close to the onset of microcrystallinity as compared to the amorphous state. In particular, the mobility-lifetime product of charge carriers in some of the μc-Si:H samples turns out to lie about two orders of magnitude higher than that of a-Si:H films. The mobility on the other hand, is shown rather to decrease in the transition to the μc-region. Additional measurements of the range and the depth of long range potential fluctuations yield a possible explanation for our results in that grain boundaries may serve as scattering centers and barriers against recombination.

INTRODUCTION

Microcrystalline silicon (μc – Si:H) has been the object of studies for many years to understand its electronic, optical and structural properties. However, recently there has been a renewed interest in incorporating μc – Si:H in different solar cell structures since such devices have shown no degradation after prolonged light soaking [1,2]. It has been shown that a-Si:H films grown under hydrogen dilution close to the onset of microcrystallinity exhibit a higher degree of stability against light soaking compared to a-Si:H. In hot-wire assisted   chemical vapor deposition (HW-CVD) [3-7], the decomposition of silane and hydrogen gas mixture allows one to prepare material with a transition from amorphous to microcrystalline growth by variation of the silane to hydrogen ratio. By varying the grain size, an enhanced absorption of microcrystalline compared to crystalline silicon has been observed [8]. The (HW-CVD) method allows a continuous preparation of material with a smooth transition from amorphous hydrogenated silicon to microcrystalline silicon. The photoconductivity in microcrystalline silicon has been studied as a function of Fermi-level to determine the mobility-lifetime product ($\mu\tau$) [9] with the observation that ($\mu\tau$) increased significantly by shifting the Fermi level from the mid-gap towards the conduction or valence band; this was attributed to the increase in the majority carrier lifetime due to a change in the thermal occupation of defect centers by the shift of the Fermi level. It is of interest to separately determine the mobility and lifetime in the transition from the amorphous to the microcrystalline state;   dc photoconductivity measurements determine the $\mu\tau$ product. The independent   determination of μ and τ was accomplished by employing the photoconductive frequency mixing technique [10-15]. By observing the increase in the drift mobility as a function of electric field, the range and the depth of the long-range potential fluctuations [13] as a function of hydrogen content and hence the change in these quantities in the transition from amorphous to microcrystalline states were determined.

Mat. Res. Soc. Symp. Proc. Vol. 557 © 1999 Materials Research Society

EXPERIMENTAL

A series of samples was deposited on 1" x 1" Corning 7059 glass at a substrate temperature of about 240° C using the HWCVD technique. The details of the HWCVD reactor are reported elsewhere [16]. A tungsten filament was used with a diameter of 0.5 mm and 6 inches in length and placed 5 cm below the substrate. The filament was heated by an ac current to about 2000° C. The substrate was heated using a resistive heater and monitored using a thermocouple meter. All the samples were grown at the same pressure of 30mTorr while varying the hydrogen to silane ratio (RH) from 1 to 20. Table I summarizes the sample codes, film thicknesses, and deposition rates. The thickness of the i-layer was measured with a Tencor Instrument Alfa-step 200 profilmeter. The deposition rate was calculated from the film thickness and deposition time.

Table I: Summary of the sample characterization results

Sample ID	H/SiH$_4$	XRD Structure	(μc) av. Grain size(nm)	Sample Thickness (Å)	H-content (%)	Deposition rate (Å /s)
T516	1	a-Si		41,400	13.3	17.25
T517	5	μc Si	12	21,350	5.2	5.93
T518	10	μc Si	13	14,300	4.7	3.97
T519	20	μc Si	14	11,200	3.9	2.33
T528	5	μc Si	18	25,400	3.1	7.05
T529	4	μc Si	12	24,000	4.2	8.00
T530	3	a-Si+μc Si	9	16,400	4.4	7.80
T531	2	a-Si		25,000	8.9	10.41
T532	3	a-Si+μc Si	11	17,000	4.0	7.08
T534	2	a-Si		30,000	10.9	12.50

The charge transport parameters were determined by the photoconductivity frequency mixing technique [10-15]. This technique is based on the separate detection of the dc and microwave part of the photocurrent, which is the result of heterodyning two monochromatic laser beams incident on a photoconductive sample to which a dc electric field is applied. These two parts of the photocurrent allow a separate determination of the drift mobility (μ_d) and the photomixing lifetime (τ) corresponding to the dominant photogenerated carriers. In the present work the longitudal modes of a helium-neon laser at an angular frequency of 1.58 GHz were employed. The range and the depth of the long range potential fluctuations were determined by the electric field dependence of the mobility [13].

RESULTS

Results of photomixing measurements

In a-Si:H, the electronic transport depends tightly on the hydrogen content, so the photoconductivity is usually plotted against the amount of H-atoms introduced into the samples. As to be seen in Table I, by increasing the dilution ratio one drives the film into the microcrystalline state. At the same time, the hydrogen content drops. We decided to plot the transport properties versus the hydrogen content so that both the behavior of amorphous silicon with changes in the hydrogen content can be kept track of and a clear separation of the different states is provided. Figure 1 shows that with decrease of the H-content, the photoconductivity is more or less constant as long as the hydrogen concentration is over about 5% to 6%. However, it shows a sudden increase when the H-concentration drops below 5%, i.e. the microcrystalline regime is entered.

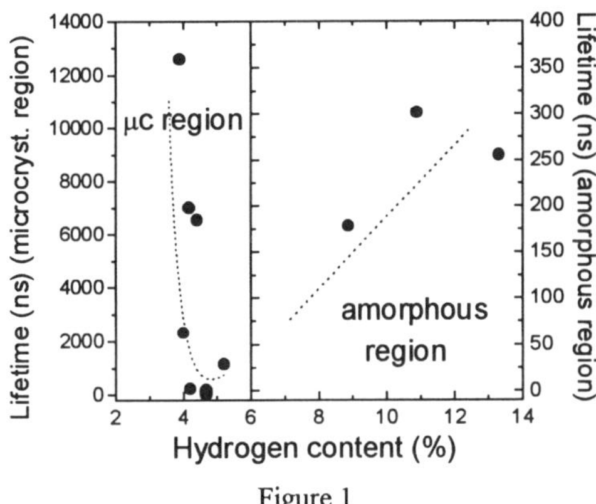

Figure 1

Figure 2

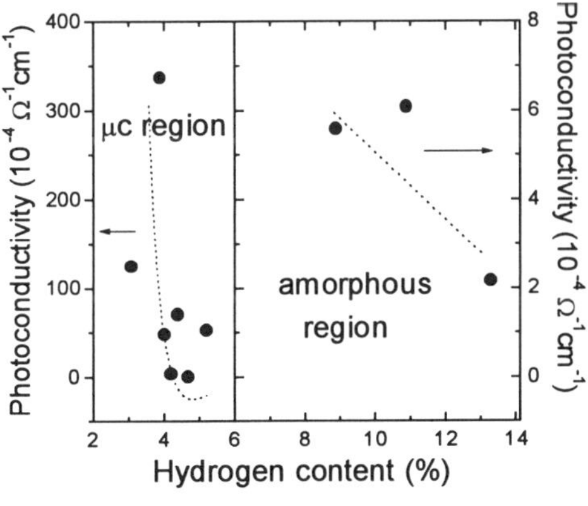

Figure 3

Figures 1-3:

Lifetime, mobility, and photoconductivity of all samples are plotted against their respective hydrogen content with amorphous and microcrystalline regions separated.
In the case of lifetime and photoconductivity, different scales were used for the different regions.

To investigate what quantity this increase is due to, we determined the photomixing lifetime τ and mobility μ for each sample. The results are shown in Figures 2 and 3. Both τ and μ are affected by the transition. Similarly to the behavior of the photoconductivity, the lifetime exhibits a sudden increase up to two orders of magnitude when the hydrogen content drops to values less than 5% (Figure 3). The mobility, however, shows somewhat opposite behavior (Figure 2). In the amorphous region, the mobility *increases* with decreasing H-concentration. This is surprising since one usually finds the mobility to increase when the hydrogen content grows. For H-contents under 5%, the mobility *drops* to values not detectable anymore. However, except for the samples T518 and T528, μ only drops to about 20% of it's maximum value, so the photoconductivity, being straight proportional to τ and μ, still increases dramatically.

Long-Range Potential Fluctuations

Long-range Potential Fluctuations (LRPF) [17] are thought to reflect the density and arrangement of *charged* defects in amorphous silicon. To determine the LRPF within the samples, we performed field-dependent measurements of the mobility and lifetime [13]. The results are shown in Figures 4 and 5. Note that the range increases to values of around 50nm in the μc – regime, which might reflect the occurrence of grain boundaries. Also the depth of the potential fluctuations increases. These results lead to a possible explanation for both the drop of the mobility and the dramatic increase of the lifetime in the case of microcrystalline silicon. Grain boundaries might serve as scattering centers as well as barriers against recombination.

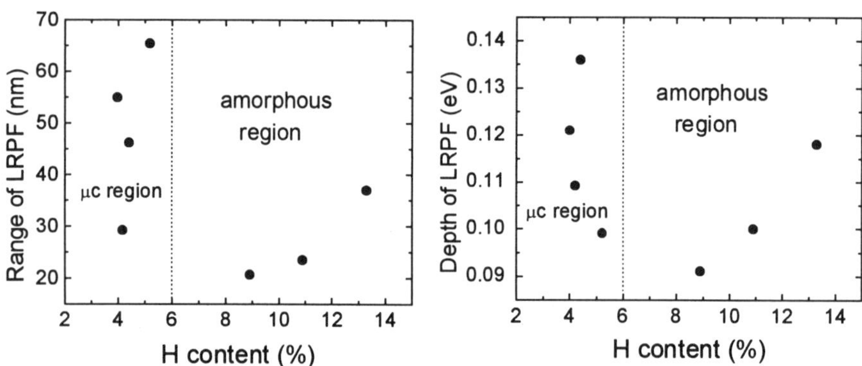

Figures 4 and 5: Range and depth of long-range potential fluctuations obtained from the field-dependence of the mobility. The x-axis position of each sample is associated with it's respective hydrogen content.

<u>**Light-induced Decay Measurements**</u>

This section covers measurements of the light-induced degradation (Staebler-Wronski-effect) for samples in the amorphous state, mixed state (a-Si + μc-Si) and the microcrystalline state (μc-Si). Figure 6 shows the normalized photoconductivity for the three amorphous samples, which exhibit behavior as expected from former results. In each case the samples were first annealed for 1 hour at a temperature of 150 °C.

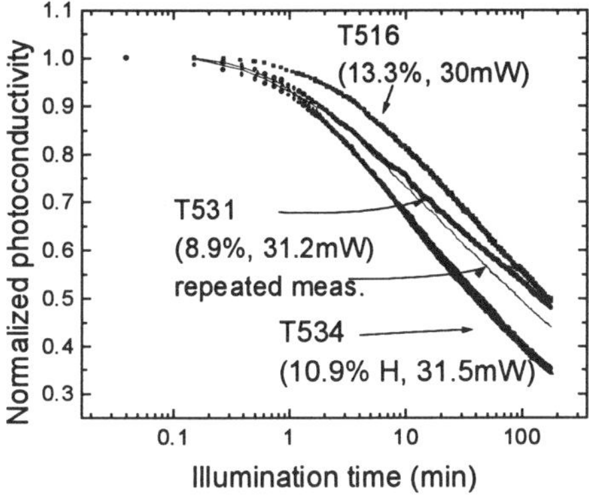

Figure 6:
Normalized photoconductivity vs. illumination duration for T516, T531 and T534 (amorphous state). The light intensities are given in the Figure.

Since in the case of microcrystalline silicon, we found both high dark and photocurrent, we had to restrict the number of points in order to prevent those samples from heating. Also, we could not measure lifetime and mobility at high fields. This results in an increased error and so a certain scattering of both lifetime and mobility. Nonetheless Figure 7 shows increased stability against light-induced defects for samples in both the mixed and microcrystalline phase. The increase of the photocurrent in the case of the samples T529 and T530 is probably due to a slight heating.

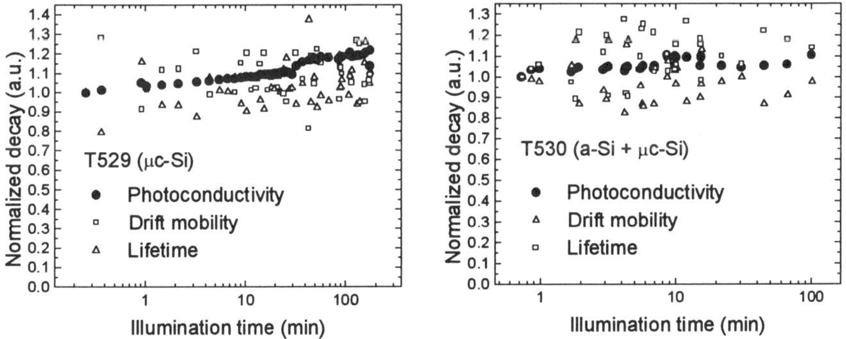

Figure 7: Photoconductivity, mobility and lifetime vs. illumination duration for samples in the microcrystalline and mixed state.

CONCLUSIONS

Summarizing the above results, we found both increased stability and a huge increase of the photomixing lifetime resulting in an increase of the photoresponse for samples entering the microcrystalline regime. The increase in lifetime and decrease in mobility as the microcrystalline regime is entered indicates that grain boundaries can serve as scattering centers as well as barriers against recombination. In order to further illucidate the mechanisms responsible for the dramatic change in lifetime and mobility as one makes the transition from the amorphous to the microcrystalline state, it would be useful to perform similar measurements on samples prepared by different methods which result in different microstructures.

ACKNOWLEDGEMENTS

This work was supplied by the National Renewable Energy Laboratory subcontract XAK-8-17619-24. A. Kattwinkel gratefully acknowledges the support of the Berliner-Ungewitter Stiftung.

REFERENCES

1. J. Meier, P. Torres, R. Platz, S.Dubail, U. Kroll, J.A. Anna Selvan, N. Pellton Vaucher, Ch. Hof, D. Fischer, H. Keppner, A. Sha, K. D. Lifert, P. Giannulés, J. Koehler, MRS Symp. Proc. **420**, p. 3 (1996).
2. D.V. Tsu, B.S.Cho, S.R.Ovshinsky, S. Guha and J. Yang, Appl Phys. Lett. **71**, p. 1317 (1997).
3. H. Wiesmann, A. K. Ghosh, T. McMahon, and M. Strongin, J. Appl. Phys. **50**, p. 3752 (1979).
4. H. Matsumera. Jpn. J Appl. Phys., Part 2, **25**, p. L949 (1986).
5. J. Cifre, J.Bertomeu, J. Puigdollers, M.C. Polo, J. Andreu, and A. Lioret, Appl. Phys. A: Solids Surf. **59**, p. 645 (1994).
6. A. R. Middya, J. Guillet, J. Perrin, A. Lioret, and J.E.Boirree in *Procedings of the 13th European Photovoltaic Solar Energy Conference, Nice* 1995 edited by W. Freiesleben, W. Palz, H.A. Ossenbrink and P. Helm (H.S. Stephens & Associates, Bedford, U. K., 1995). p. 3.
7. M. Heintze, R..Zedlitz, H.N.Wanka, and M. B.Schubert, J. Appl. Phys. **79**, p. 2699 (1996).
8. F. Diehl, W. Herbst, B. Schröder, H. Oechsner, MRS Symp. Proc. **467**, p. 451 (1997).
9. R. Brüggemann, C Main Phys. Rev. D **57**, p. R15080 (1998).
10. E.R.Geissinger, R. Braunstein, S.Dong, and R. Martin, J. Appl .Physics **69**, p. 1469 (1991).
11. Y. Tang, R. Braunstein, and B. von Roedern, Appl. Phys. Lett. **63, p.** 2393 (1992).
12. Y.Tang, R. Braunstein, B. von Roedern, F.R. Shapiro. Mater. Res. Soc. Symposium Proc. **297,** p. 407 (1993).
13. Y. Tang and R. Braunstein, Appl. Phys. Lett. **66**, p. 721 (1995).
14. Y. Tang and R. Braunstein, J. Appl. Phys.**79**, p. 850 (1996).
15. Y. Tang, S. Dong, R. Braunstein, and B. von Roedern, Appl. Phys. Lett. **68**, p. 640 (1996).
16. M. Vaneck, A. H. Mahan, B. P. Nelson, R. S. Crandall, *Proc. 11th European Photovoltaic Solar Engergy Conference* edited by L. Guimares, W. Palz, C. Dereyff, H. Kiess, and P. Helm (Harvard Acad. Publ., Switzerland, 1993), p. 96.
17. H. M. Branz and M. Silver Phys. Rev. **B 42**, p. 7420 (1990).

ANISOTROPIC TRANSPORT IN MICROCRYSTALLINE P-I-N DEVICES

[1]M. VIEIRA, [2]A. FANTONI, [1]M. FERNANDES, [1]A. MAÇARICO, [1]J. MARTINS, [1]P. LOURO AND
[3]R. SCHWARZ
[1]Electronics and Communications Department, ISEL, Lisboa, Portugal
[2]FCT/UNL, Monte da Caparica, Portugal
[3] Physics Department, IST, Lisboa, Portugal

ABSTRACT

Entirely μc-Si:H p-i-n structures presenting an enhanced sensitivity to the near infrared region and a positive spectral response under forward bias higher than the open circuit voltage are analysed under different external voltage bias and illumination conditions.

A two phase model to explain the transport properties is proposed using as input parameters the measured experimental data. The results suggest that the transport is preferentially concentrated inside the crystalline grains. The conduction within the amorphous regions is poor. The percolation path is different for electrons and holes and is determined by the local fields at the boundaries. These local fields are independent of the externally applied condition, and they can be related to the persistence of the small photocurrent observed when a bias voltage higher than the open circuit voltage is applied.

INTRODUCTION

The first successful deposition of microcrystalline silicon (μc-Si) films has been realized in the late sixties [1]. Since then, μc-Si has been a subject of interest due to its high electrical conductivity and absorption coefficient, making this material a good candidate for future photovoltaic applications such as solar cells, thin film transistors, and large-area optoelectronic sensors [2, 3].

The understanding of μc-Si:H based devices is still in an early stage and some questions about its properties are actually a subject of investigation and are under discussion [4, 5]. The μc-Si:H is a two-phase material. Its composition can be considered as grains of crystalline silicon imbedded in an a-Si:H tissue. The structural properties of μc-Si:H are strongly dependent on the deposition process used. Evidence of different film morphology has been found in samples deposited with different processes [6, 7].

In this study we have used the closed-chamber chemical vapor deposition method [7] to produce thin entirely μc-Si:H p-i-n structures resulting in wide spectral range device with an enhanced sensitivity to the visible and near infrared regions and a positive spectral response under forward bias. In order to gain insight into the transport mechanism we have correlated the spectral responsivity with a detailed simulation analysis. An anisotropic model, based on the band discontinuities near the grain boundaries, is presented and supported by numerical simulation.

EXPERIMENTAL DETAILS AND DEVICE CHARACTERIZATION

All layers of the analyzed p-i-n structures have been prepared by Closed-Chamber Chemical Vapor Deposition. The μc-Si films have a high crystalline fraction (>80%) and can be efficiently doped both <n> and <p> type. A conic-like growth for the crystallites is assumed with an amorphous-to-crystalline transient sublayer in the initial stages of μc-Si film growth [8]. For the characterization of the individual layers, optoelectronic measurements

Mat. Res. Soc. Symp. Proc. Vol. 557 © 1999 Materials Research Society

have been applied. Optical gaps similar to amorphous silicon (1.7-1.8 eV) but higher subgap absorption (α(1.2 eV)=20 cm^{-1}) were inferred.

The device consists of a ZnO covered glass substrate, followed by a p$^+$ μc-Si:H/i μc-Si:H/n$^+$ μc -SiH (μc-p-i-n) structure and Al top contact. The current-voltage characteristics in dark or under different illumination conditions were recorded at room temperature, using a tungsten-halogen lamp of about 40 mWcm^{-2}. An open circuit voltage (V_{OC}) of the order of 0.4 V and a short circuit current of about 10 mAcm^{-2} were measured. The spectral characterization was obtained by measuring the photocurrent, I_{ph}, in the range of 300 nm to 1100 nm using a 25 cm focal length monochromator and different applied voltage (in the range of -1 V to 1 V).

The light was incident either through the glass (<p> side) or through the rear part of the photodiode (<n> side). All the measurements were carried out by using lock-in technique to suppress the dark current.

RESULTS AND DISCUSSION

<u>The 2-phase model</u>

Microcrystalline silicon has its own macroscopic characteristics that are different from each of the single-phase ones. As shape and size of the crystalline grains can be highly variable and their distribution within the amorphous matrix is different a proper description of the structural properties of this material can be a quite complicated task.

The problem of the material anisotropy can be approached applying the heterostructure transport equations on a two-dimensional rectangular domain [9]. The ASCA simulation program has been adopted for the analysis of the μc-Si:H devices. Details about the simulator and the set of the material parameters used can be found elsewhere [10, 11]. The μc-Si:H intrinsic layer is defined by crystalline grains, whose size is 0.3 μm, imbedded in an a-Si:H matrix and separated by a distance of 0.1 μm. The doped layers are standard <p> and <n> type amorphous silicon (to avoid band mismatch at the interfaces) whose thickness is fixed at 0.1 μm. The simulation takes into account the transverse dimension (perpendicular to the junction plane, denoted as x-direction) and one of the lateral dimensions (parallel to the junction plane, denoted as y-direction). Fig. 1 depicts the photo-generation rate profile (a) and the electric charge distribution (b) under AM1.5 illumination, with the light incoming from the <p> side.

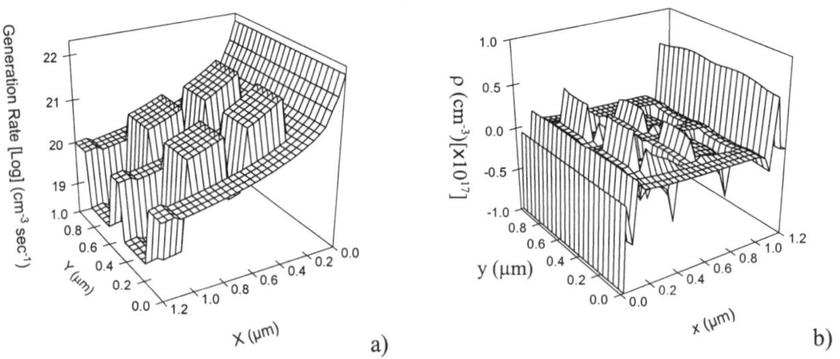

Fig. 1   (a)The photo-generation rate profile and (b) the electric charge distribution under AM1.5 illumination and short circuit condition.

As shown in Fig. 1a the photo-generation process is mainly localised inside the grains, where the red and infrared radiation are better absorbed. Here, the lower density of localised states determines a higher concentration of free carriers. Due to the high gradient of concentration, the photo-generated carriers tend to diffuse out towards the amorphous regions. Most of them are trapped at the borders leading to a space charge concentration at the grain borders as depicted in Fig.1b. It is important to notice that the space charge is always negative at the crystalline border while at the amorphous one it is positive towards the p-i interface and negative towards the i-n one. This localised space charge leads to an anisotropic distribution of the transverse (E^x) and the lateral (E^y) electric fields as reported in Fig. 2.

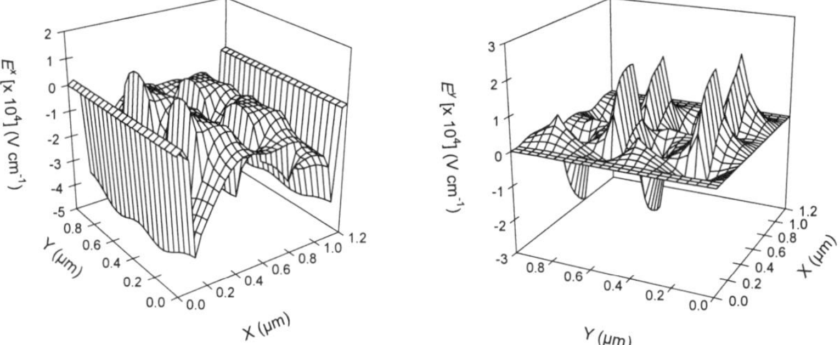

Fig. 2 Transverse (E^x) and lateral (E^y) electric field distributions under AM1.5 illumination and short circuit condition.

Simulated profiles show that E^x is dominated by the junction field, whose peaks at the p-i and the i-n interfaces are about 4×10^4 V cm^{-1}. High peaks of about the same intensity can be observed at the grain boundaries. Also E^Y profile reveal local peaks of the same order but directed towards the bulk of the grains. As no junction field appears in the lateral direction, its intensity only depends on the grain size.

The transverse built-in electric field together with the non-uniform field distribution leads to preferential transport paths in x-direction, depending on the grain distribution and size. Fig. 3 reports the transverse current components for the hole (J_p^x) and for the electron (J_n^x), receptively.

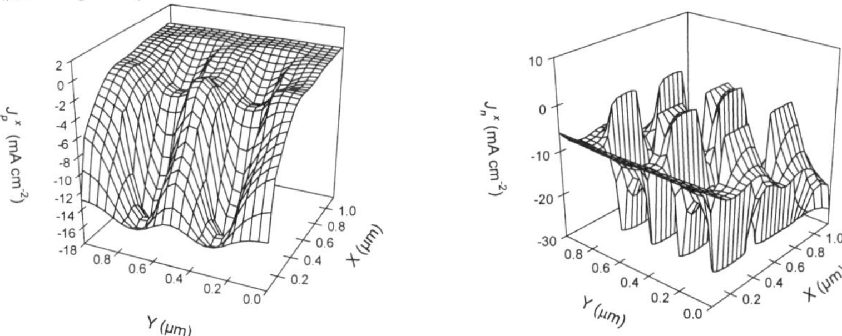

Fig. 3 Transverse holes (J_p^x) and electrons (J_n^x) current density profiles within a μc-Si:H p-i-n structure under AM1.5 illumination and short circuit condition.

The transport of holes is preferentially made in the bulk of the grains along the x-direction following the junction built-in electric field. Transverse hole current is only possible if the intensity of the junction field overcomes the local one. Opposite behaviour is observed for electrons, the junction field pushes them toward the collecting contacts, and the superposition of the local fields makes them laterally percolating through the crystalline grains, close to the borders. Fig. 4 reports the potential profile distribution in open circuit condition under: a) AM1.5 spectrum; b) under red, green and blue light illumination.

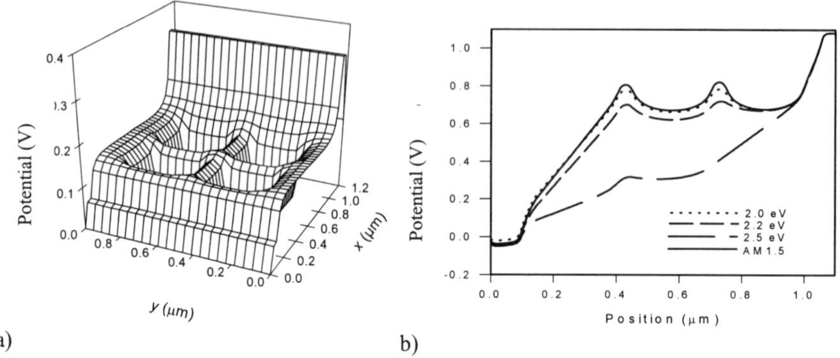

Fig. 4   Potential profile under: a) AM1.5 spectrum and open circuit condition; b) Under red (2.0 eV), green (2.2 eV), and blue (2.5 eV) light illuminations and short circuit condition.

It can be well seen how the local electric fields result in a lumped potential distribution acting the grains as regions of low potential inside the i-layer. The potential profile depends on the irradiation wavelength. As the photon energy increases the photogeneration rate inside the grains decrease due to the low absorption of the crystalline material. The local electric fields peaks are minimized and the potential distribution smoothes hiding the heterogeneity ascribed to the microcrystalline nature. The local peaks do not change significantly their profile with the applied bias. Even when the built-in junction field is almost null (open circuit condition) a difference of potential between the amorphous and the crystalline regions is present. So, percolation of carriers between grains is possible and a small photocurrent can be expected when an external bias higher than V_{OC} is applied as shown in Fig.5 and reported in previous publications [12, 13].

<u>Anisotropy in spectral response</u>

The conic-like growth of the crystallites inside the amorphous phase [14] leads to a p-i-n structure where the crystalline fraction is not uniform and increases from the p-i to the n-i interfaces. The measured enhanced absorption below 1.8 eV and the reduced absorption above 2 eV can only be understood as average values and are dependent on the grain size and their relative position inside the device. In order to check this anisotropy we have illuminated the device from both sides and correlated the experimental spectral behavior with the simulated results.

In Fig.5a and Fig.5b the responsivity curves normalized to their maximum are displayed for different applied voltages with the light incoming from the p-side and n-side, respectively. As shown the spectral response depends on the illuminated side. Under front illumination (Fig.. 5a) the spectral response is extended beyond 1000 nm and has a maximum near 500 nm. If the light is incident through the rear part of the diode (Fig. 5b) the infrared response is increased and the maximum is shifted towards the near-infrared with a rejection of the blue spectrum.

The heterogeneous nature of the device together with the results from the simulation can help to explain the spectral behavior:

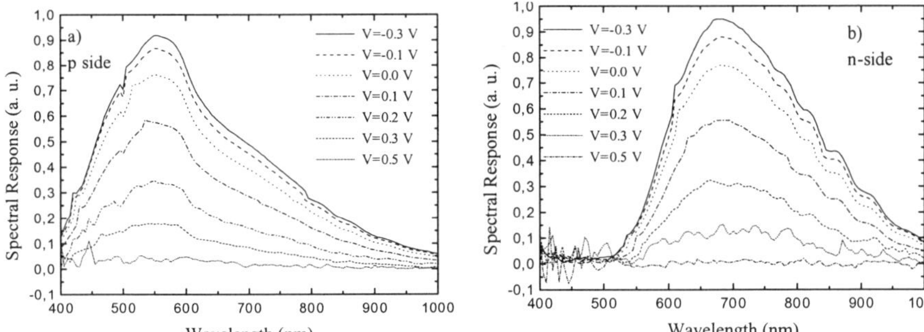

Fig.5   Measured spectral photoresponse curves under different bias voltages under(a) p-side illumination, and (b) n-side illumination.

Fig. 5 outlines very clearly the potential dependence (see Fig. 3) on the wavelength. In the visible/near infrared range the carriers mainly generated inside the crystallites (low potential regions) flow along the columnar grains in the x direction. Only a small fraction of them can drift inside the amorphous matrix after passing the grain boundaries where they are mainly trapped or recombine. The enhanced red/infrared sensitivity even for forward bias higher than V_{OC} is ascribed to the carriers collected through a diffusion-limited process inside the grains. With the increase of the photon energy the amount of carriers generated inside the amorphous phase increases and decreases inside the grains, smoothing the potential profile. The collection is then a field driven process resulting in a spectral response dependent on the forward applied bias and on the incoming light side.

If the incoming light on the <p> side is blue the photons are strongly absorbed at the front part of the p-i-n structure (mainly amorphous). The photo carriers drift inside the amorphous matrix and are collected dependent on the applied bias. The transport is drift-limited and so a transition from the primary to the secondary photocurrent is expected near the V_{OC}. Under <n> side illumination the high energy photons are weakly absorbed near the rear part of the p-i-n device, mainly crystalline. The junction field decreases near the <n> layer and the carriers flow inside the crystallites where they mainly recombine (holes) or are trapped (electrons) at the grain borders. Thus, no spectral response is observed in the blue range since the carriers can not be extracted.

CONCLUSIONS

Thin entirely µc-Si:H p-i-n structures were analyzed resulting in wide spectral range photodiodes with an enhanced sensitivity to the near infrared region and a positive spectral response under forward bias.

A heterojunction model based on the presence of additional local electric fields near the grain boundaries is presented to explain the internal electric behaviour of a µc-Si:H p-i-n photodiode. The transport mechanism is mainly concentrated inside the crystalline grains.

Conduction within the amorphous region contributes to the transport only by allowing a percolation of the carriers through the crystalline grains. The percolation path is different for holes and electrons and it is determined by the distribution of the local fields at the grain boundaries. These local fields are not dependent on the externally applied conditions, and they can be related to the persistence of a small photocurrent observed in μc-Si:H p-i-n when a bias higher than the open circuit voltage is applied.

With the increase of the photon energy the amount of carriers in the amorphous matrix increases and the potential profile smoothes. The transport is then drift limited and will proceed along both the amorphous and the crystalline phases.

ACKNOWLEDGEMENT

The authors would like to thank S. Koynov and A. Maçarico for the device deposition and to R. Martins for the discussion during the ASCA simulator development. This work has been financially supported by Fundação Luso-Americana para o Desenvolvimento, by the Portuguese program PRAXIS XXI and by INCO-COPERNIUS PL9781104.

REFERENCES

1. S. Veprek and V. Marecek, Solid State Electronics **11** (1968) 683.
2. J. Meier, S. Dubail, R. Fluckiger, D. Fischer, H. Keppner, and A. Shah, Proc. 1st WCPEC (1994) 1237.
3. R. Weisfield, and C.C. Tsai, Mat. Res. Soc. Symp. Proc., **192** (1990) 423.
4. N. Wyrsch, P. Torres, J. Meier and A. Shah, Journal of Non-Crys. Solids, **227** (1997) 1272.
5. J. Zimmer, H. Stiebig, and H. Wagner, Mat. Res. Soc. Symp. Proc. (S. Francisco, April, 1988) in print.
6. S.Veprek, Mat. Res. Soc. Symp. Proc. **164** (1990) 39.
7. S. Koynov, R. Schwarz, T. Fisher, S. Grebner, H. Munder, Jap. Jn. of Appl. Physics **33** (1994) 4534.
8. R. Krankenhagen, M. Schmidt, W. Henrion, I. Sieber, S. Koynov, S. Grebner, and R. Schwarz, Solid State Phenomena **47** (1995) 607.
9. S. Lundstrom and R. J. Shuelke, IEEE Transaction on Electron Devices **ED-30** (1983) 1151.
10. A. Fantoni, M. Vieira and R. Martins, in Amorphous Silicon Technology, edited by M. Hack, E. A. Schiff, A. Madan, M. Powell, A. Matsuda, Mat. Res. Soc. Proc. **336** (1994) 711.
11. A. Fantoni, M. Vieira J. Cruz , R. Schwarz and R. Martins, Journal of Physics D: Applied Physics **29** (1996) 3154.
12. M. Fernandes, M. Vieira, A. Maçarico, S. Koynov, A. Fantoni, and R. Schwarz, Mat. Res. Soc. Symp. Proc. (S. Francisco, April, 1988) in printing.
13. J. Meier, S. Dubail, D. Fisher, J. A. Anna Selvan, N. Pellaton Vaucher, R. Platz, Ch. Hof, R. Fluckiger, U. Kroll, Nwyrsch, P. Torres, H. Keppner, A. Shah, K. D. Ufert, 13th European Photovoltaic Solar Energy Conf. Proc. (Nice, France, 1995) 1445.
14. H. N. Liu, Y. L. He, F. Wang, and S. Grebner, J. of Non-Cryst. Solids **164-166** (1993) 1005.

AMORPHOUS SILICON PRECIPITATES IN (100) c-SI FILMS GROWN BY ECRCVD

M. BIRKHOLZ, J. PLATEN, I. SIEBER, W. BOHNE, J. RÖHRICH, W. FUHS

Hahn-Meitner-Institut Berlin, Abteilung Photovoltaik, Rudower Chaussee 5, D - 12489 Berlin

ABSTRACT

Silicon films were grown on (100) n-Si with an electron-cyclotron resonance chemical vapor deposition (ECRCVD) system by decomposition of SiH_4 at 325°C. Structure and composition of thin films were investigated by SEM, Raman spectroscopy, elastic recoil detection analysis (ERDA) and TEM. Excellent epitaxial growth was achieved for some hundred nm thickness. For more than 1 μm thick films, however, SEM revealed the occurrence of conical structures orientated upside-down with their basal plane in the film surface. Depth-profiling of the elemental composition of thin films by means of ERDA showed the hydrogen content C_H to exhibit a pronounced increase with increasing film thickness. Raman spectroscopy evidenced the coexistence of c-Si and a-Si:H by the occurrence of two bands at 520 and 480 cm^{-1}, the ratio of which was found to depend sensitively upon the position of the laser spot on the sample. All experimental results could be consistently explained by assuming the conical precipitates to consist of a-Si:H which was finally proven by coherent electron beam diffraction (CEBD).

INTRODUCTION

Thin-film crystalline silicon is currently investigated and optimized with respect to its suitability for solar cell applications [1, 2]. Various processes like PECVD, ECRCVD, laser crystallization etc. are used for this purpose and different device concepts are being considered which would even include other thin-film materials. In this work we report on epitaxial Si thin films as prepared on CZ-grown Si wafers by virtue of electron cyclotron resonance (ECR) CVD at low temperatures, see also [3]. Our activities aim at the development of a thin film silicon solar cell deposited on glass and on crystalline seed layers. In the following the preparation of nominally undoped films and their structural characterization by scanning electron microscopy (SEM), elastic recoil detection analysis (ERDA), Raman spectroscopy and transmission electron microscopy (TEM) will be presented. It will be shown that the growth of epitaxial film was accompanied by the formation of a-Si:H precipitates after the film's thickness had passed a critical value.

EXPERIMENTS AND RESULTS

Thin Si film depositions were performed in a high-vacuum ECRCVD chamber which is described in detail in [4]. The base pressure before the deposition was less than 10^{-5} Pa. Although the process was performed at 0.66 Pa, which is much lower than in normal PECVD systems, growth rates of 10-12 nm/min were obtained. The frequency of the exciting microwave was set to 2.45 GHz to obtain resonant absorption of the plasma which was enclosed by a magnetic field of 87.5 mT. The microwave power was adjusted to 1000 W and the susceptor temperature

Mat. Res. Soc. Symp. Proc. Vol. 557 © 1999 Materials Research Society

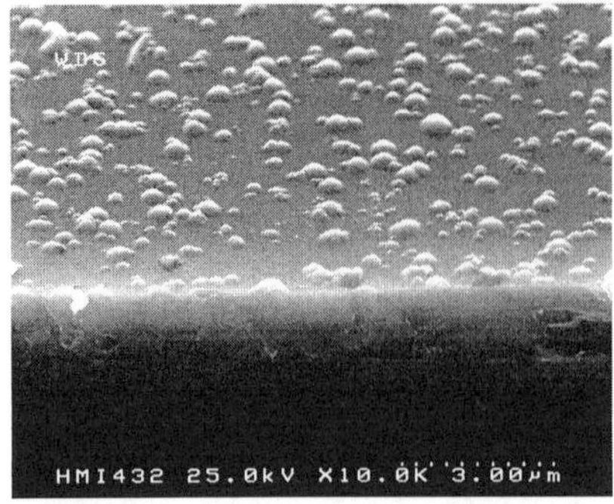 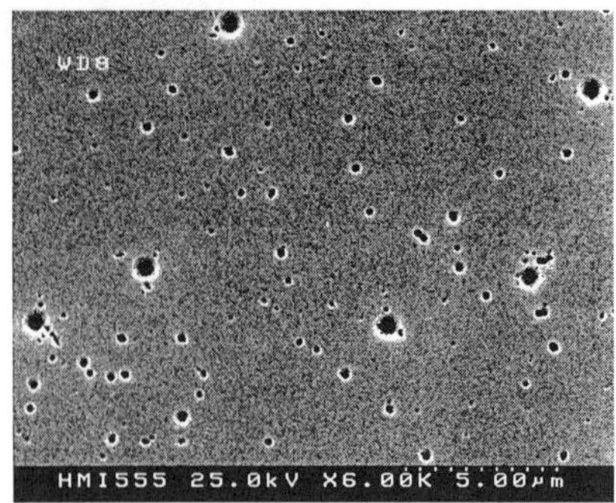

Figure 1

(a) scanning electron micrograph of as-grown homoeptiaxial 1.3 μm thick Si film.
(b) SEM picture of a comparably produced film after Secco etching.

amounted to 325°C. Intentionally undoped films were grown with gas flow rates of 20 sccm SiH_4 and 60 sccm H_2. Prior to the deposition CZ-grown Si (100) wafers were cleaned by a standard RCA procedure and an HF-dip, but plasma cleaning procedures were avoided.

Figure 1(a) displays the SEM view onto the surface of a sample viewed under an angle of 30°. The film surface shows numerous circular structures which are isolated from each other and which yield a significantly different SEM contrast than the embedding matrix. The diameter of the circles is in range of 100 - 900 nm. The thickness of the deposited film amounted to 1.3 μm. Such circular structures have always been observed when the thickness exceeded a critical value of about 500..600 nm. The different SEM contrast implies that such structures account for the occurrence of another phase different from the epitaxial matrix layer. Figure 1(b) displays the scanning electron micrograph of a comparatively produced sample of the same thickness which has been subjected to a Secco etch [5] after deposition. In spite of the precipitates circular holes can be identified within the thin film which suggest the precipitates to exhibit a conical shape. This preferential etching behavior underlies the notion of a second phase precipitate. Our analysis will show that these precipitates consist of amorphous silicon.

In order to perform a concentration profiling of contaminants some films were investigated by elastic recoil detection analysis (ERDA). In the following the results as obtained for a 1.42 μm thick film will be presented. In the ERDA experiment the sample was irradiated by 140 MeV ^{129}Xe ions impinging the sample with an angle of 60° measured normal to the surface. Out-scattered sample atoms were detected under a scattering angle θ of 58.5° and have been measured with respect to their energy E and time-of-flight *TOF* within a fixed flight path. The investigation was carried out at the Ion Beam Facility ISL of the Hahn-Meitner-Institut Berlin and details of the experimental ERDA setup can be found in [6]. The determination of both kinematic parameters E and *TOF* allows - in contrast to the related RBS method – for a mass separation of detected sample atoms and, therefore, for each element occurring in the sample a separate energy spectrum is obtained.

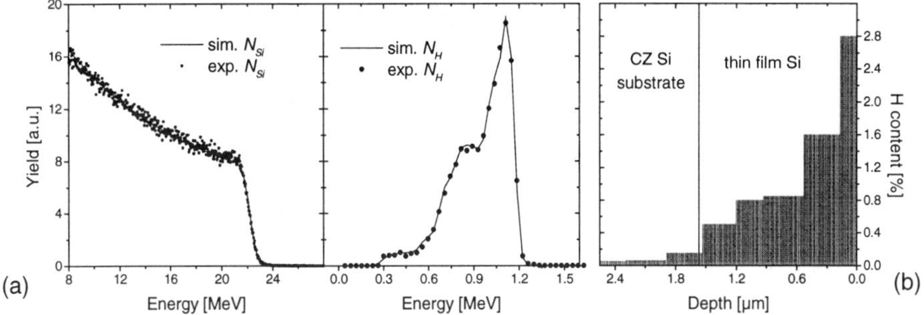

Figure 2

(a) ERDA energy spectra of Si and H of 1.42 µm thick Si film on Si wafer.
(b) Layer model with $C_H(z)$ depth profile as derived from simulation of ERDA spectra.

Energy spectra of Si and H are given in Fig. 2(a). Count rates at high energies account for elemental concentration in layers close to the surface, with the onset energy given by the kinematical conditions, i.e. being dependent on the ejectile mass M. Scattering events from deeper parts of the sample are reduced in energy by an amount with which Xe projectiles and sample atoms have been retarded along their way through the sample. The Si spectrum exhibits an $1/E^2$ behavior typical for an element whose concentration does not depend on depth. No discontinuities which would possibly account for one or the other interface in the homoepitaxial material system can be detected for the Si spectrum. The H spectrum, however, differs drastically from the $1/E^2$ dependency and suggests a hydrogen concentration profile $C_H(z)$ that increases towards the film surface.

A quantitative determination of $C_H(z)$ has been carried out using the simulation code SIMNRA [7]. In order to achieve a satisfying agreement between experiment and simulation we subdivided the sample in eight different layers of varying H content. It has to be emphasized, however, that the modeled step-like H concentration will in fact be an approximation of the probably smooth profile in the sample. The simulated H spectrum is likewise presented within Fig. 2(a). The layer model assumed for the simulation is shown in Fig. 2(b). The result is that C_H decreases from 2.8 % (1.4×10^{21} cm^{-3}) in the upper-most part of the sample to 0.05 % (2.5×10^{19} cm^{-3}) in the substrate wafer. Due to surface roughness, energy straggling and multiple scattering the absolute error in $C_H(z)$ increases with increasing depth z. Values for C_H at the top-most layers, however, are highly reliable as is the observed trend in C_H from some percent down to almost zero near the substrate. We checked moreover for the possibility of degassing hydrogen by the irradiation. No influence of dose in the H spectrum could be detected within the limits of statistical error ($1/\sqrt{N}$) and we therefore conclude that the measured spectra indeed account for an intrinsic property of the sample.

In relating the observed H profile to the film morphology we have to consider that the diameter of the ion beam on the sample is in the range of mm. Therefore, the measured values of C_H account for an integral averaging over both conical precipitates and embedding epitaxial matrix. If the H content is assumed to be approximately constant in both phases, $C_H(z)$ as obtained by ERDA can qualitatively be understood. Just due to geometrical reasons, the integral

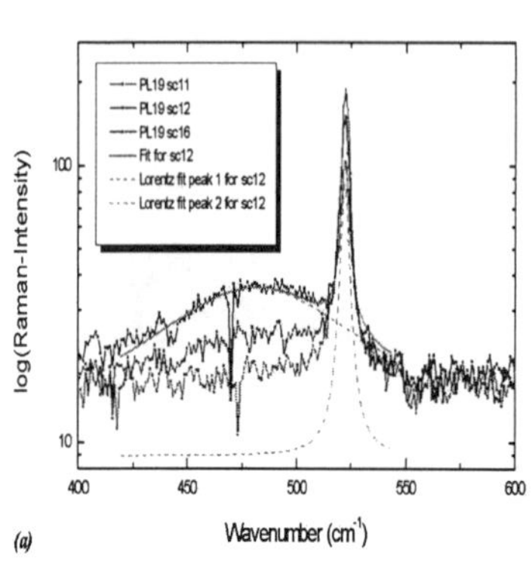 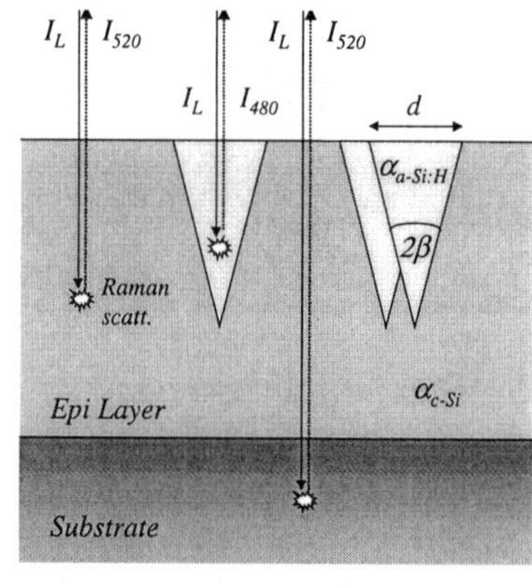

Figure 3

(a) Raman spectra of various spots ($\varnothing \approx 1\,\mu m$) of as-grown homoeptiaxial 1.3 μm thick Si film.
(b) Microstructural model of conical a-Si:H precipitates in epitaxial Si matrix as used for the calculation of Raman spectra.

H content would increase with increasing film thickness. According to the layer model presented in Fig. 2(b) the critical thickness of about 500..600 nm where cone growth is initiated would be associated with a hydrogen content between 0.6 and 0.85 %.

In order to examine the optical properties of precipitates we characterized the films by Raman spectroscopy (DILOR/ISA LabRAM 010). The excitation was performed by a 15 mW HeNe laser ($\lambda = 632.8$ nm). Spectra were taken by use of an objective of 100× magnification restricting the probed surface area to a few μm^2. The intensity of the excitation beam was reduced to avoid crystallization of a possible redundant amorphous phase. Fig. 3(a) displays some typical spectra that were measured at varying positions of the 1.3 μm thick sample. Next to the expected peak at 520 cm⁻¹ accounting for scattering within crystalline silicon, a broad signal at 480 cm⁻¹ showed up which is indicative of a-Si:H. Moreover, a pronounced lateral inhomogenity is realized, since the ratio I_{520}/I_{480} varies for different positions. The spot diameter of the laser beam on the sample of about 1 μm was larger than the typical dimension of the precipitates. Therefore, their individual characterization by Raman spectroscopy was unattainable with the spectrometer's lateral resolution. Solely surface areas with – uncontrolled – varying density of precipitates could be measured.

Nevertheless these spectra strongly suggest that the conical precipitates consist of amorphous silicon containing hydrogen in the range of some percent, a-Si:H, while the embedding matrix is made of crystalline silicon. This hypothesis was tested by assuming a simple model which should bring the variety of measured Raman spectra in accordance with the observed density of cones. Fig. 3(b) displays the optical model of a thin c-Si film to which a-Si:H cones are confined. For reasons of simplicity only absorption and Raman scattering had been included, while refraction and reflection effects at phase boundaries within the film were neglected. It can be shown that the ratio of integrated intensities of both Raman bands is given by

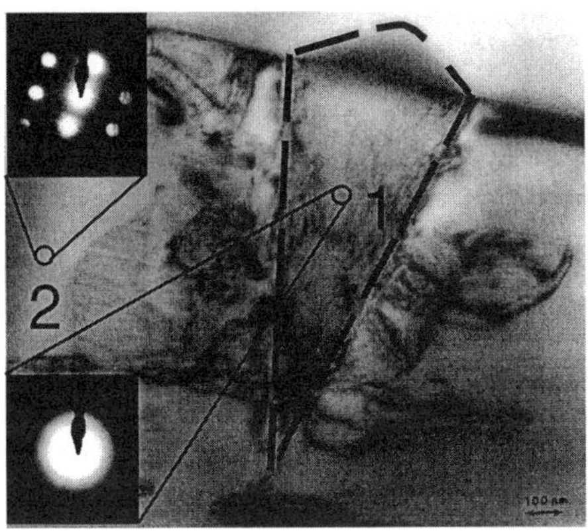

$$\frac{\int I_{520}d\lambda}{\int I_{480}d\lambda} = \frac{\sigma_c\alpha_a}{\sigma_a\alpha_c}\,\frac{\alpha_a^2(2\delta_{ac}-1+e^{-2\delta_{acc}})r_f - 2\delta_a^2\alpha_{ac}^2(1-r_f)}{\alpha_{ac}^2(2\delta_a(\delta_a-1)+1-e^{-2\delta_a})r_f} \tag{1}$$

where σ_c, σ_a account for the Raman scattering cross sections of c-Si and a-Si:H, and α_c, α_a ($\alpha_{ac} = \alpha_a + \alpha_c$) stand for the optical absorption coefficients of both phases at the wavelength of the exciting laser. Furthermore, the optical path lengths and $2\delta_a$ and $2\delta_{ac}$ are derived from the geometrical parameters through $2\delta_a = \alpha_a\,d\,cot\beta$ with 2β being the cone's aperture angle and d the surface diameter. r_f is the share of the illuminated surface area that is covered by the amorphous phase.

This formula may be applied to the measured Raman spectra in order to understand the observed intensity variations of Raman bands at 520 and 480 cm^{-1}. For this purpose the geometrical parameters of the cones have to derived from SEM (see Fig. 1) and inserted in (1). The intensity ratios and their variations are then found to be in agreement with the cone density and their lateral distribution. Since the laser spot size on the sample is larger than the average cone diameter, the Raman spectrometer yields an incidental spectrum depending on the number and size of precipitates actually illuminated. We conclude that Raman spectroscopy supports the model of the cones to be composed of a-Si:H.

TEM investigations, too, support this conclusion. Figure 4 displays a typical cross sectional micrograph of one sample. The conical precipitate can clearly be recognized. The aperture angle 2β was determined from a set of various TEM micrographs of different cones to amount to 20..30°. The cones' axis of rotation was surprisingly identified to deviate by 4-5° from the surface normal. In the vicinity of the cones an extended network of crystal faults can be realized the detailed nature of which remains unclear yet. In some cases the defect network continued from the cone's apex down to the interface. The conical shape of precipitates was revealed to be a reliable approximation, but for larger cones the volume increase with increasing film thickness could be identified of being smaller than for a perfect geometrical cone. Insets in figure 4 display coherent electron beam diffraction (CEBD) patterns that were taken from the cones and the embedding matrix. While the CEBD of the cone simply exhibits broad amorphous-like diffraction rings, a distinctive diffraction spot pattern can be observed for the matrix region.

Accordingly both areas can unequivocally be identified as amorphous and crystalline silicon, respectively.

CONCLUSIONS

We have shown that Si homoepitaxy by ECRCVD at low temperatures is accompanied by the formation of a-Si:H precipitates. The microstructure of this foreign phase within an otherwise epitaxial matrix is that of a cone with a hillock-like protrusion above the film surface. Their diameters within the surface is in the range of 100..900 nm for more than 1 µm thick films. Identification of precipitates as being composed of a-Si:H is corroborated by the depth-dependent H-content determined by ERDA, the lateral inhomogenity on a µm length scale as determined by position-sensitive Raman spectroscopy and was finally proven by coherent electron beam diffraction as performed with a TEM.

Regarding the intended applications of the synthesized film as photovoltaic absorber layer it is evident that any precipitating foreign phase should be avoided. If the growth of the precipitates would be that of a perfect cone their share of the build up volume would increase with the cube of their diameter ($V=d^3 cot\beta/24$) and the c-Si growth would inevitably break down. Although we observe the cone's volume increase to deviate slightly from that of a perfect cone their effect on the growth and electronic properties of the film seems to be deleterious. Moreover, low-temperature grown epitaxial films presented in this work were already of astonishing thickness when compared with other works [8]. The cones appear as the main obstacle for the growth of even thicker c-Si layers and avoiding the cones would enable the growth of as thick homoepitaxial Si layers as limited by deposition time. We conclude that the control of a-Si:H cone growths in thin c-Si films would offer interesting perspectives for application in photovoltaics.

ACKNOWLEDGEMENT

We greatly appreciate the support of G. Keiler, B. Rabe, D. Patzek and G. Röschert for technical assistance. The work was partially supported by the BMBF under contract #329773.

REFERENCES

1. Z. Shi, M.A. Green, Progr. Photovolt. Res. Appl. **6**, 247 (1998).
2. R.B. Bergmann, *Siliziumsolarzellen auf Glassubstrat* (1998).
3. J. Platen, U. Zeimer, B. Selle, K. Kliefoth, S. Brehme, W. Fuhs, presented at the 1999 MRS Spring Meeting, San Francisco, CA, 1999 (to be published).
4. P. Müller, E. Conrad, T.R. Omstead, P. Kember, in 13th EC Photovoltaic Solar Energy Conference, Nice, 1995), p. 1742-1745.
5. S. Wolf, R.N. Tauber, *Silicon processing for the VLSI era* (Lattice Press, Sunset Beach, 1986) , p. 533.
6. W. Bohne, J. Röhrich, G. Röschert, Nucl. Instr. Meth. **B136-138**, 633 (1997).
7. M. Mayer, *SIMNRA User's Guide - Version 3.6* (Max-Planck-Institut für Plasmaphysik, München, 1997).
8. D.J. Eaglesham, H.-J. Gossmann, M. Cerullo, Phys. Rev. Lett. **65**, 1227 (1990).

GROWTH AND PROPERTIES OF MICROCRYSTALLINE GERMANIUM-CARBIDE ALLOYS

JASON HERROLD, VIKRAM L. DALAL
Dept. of Electrical and Computer Engr., Iowa State University, Ames, IA 50011.

ABSTRACT

We report on the preparation and properties of microcrystalline (Ge,C) alloys grown using a remote, reactive H plasma beam deposition technique. The plasma beam was generated using an ECR reactor. The films were grown at low temperatures (300 - 400° C) on glass, stainless steel and c-Si substrates. The optical properties of the films were measured using spectrophotometer and two-beam photoconductivity techniques. The C content was measured using XPS techniques. We find up to 3% C incorporation in the lattice. The degree of crystallinity determined using Raman spectroscopy was very good. X-ray diffraction measurements indicated grain size in the range of a few tens of nm. The best crystallinity was obtained on conducting substrates, indicating the importance of H ion bombardment in promoting crystallinity. We find that the absorption curves for increasing C content remain similar to the curves for c-Ge, but are shifted to higher energies. Thus, the absorption curve is sharper than for c-Si in the same energy range. The defect densities remain low for the range of C content measured. Because of its sharper absorption curve compared with c-Si, the material may be attractive for photovoltaic energy conversion.

INTRODUCTION

Group IV alloys have been actively researched[1] for their potential use in lattice matching or bandgap engineering applications. This has not been the case, though, for the (Ge, C) system, as C is not soluble in Ge under thermal equilibrium at all temperatures and pressures. However, some work has been done in fabricating metastable thin films of $Ge_{1-x}C_x$ by molecular beam epitaxy (MBE)[2-3], and by ultra-high vaccum chemical vapor deposition (UHV-CVD) using germylmethane precursor gasses.[4] Sputtering[5-6] has been used to fabricate amorphous $Ge_{1-x}C_x$ films.

In this paper, we report on the results of using ECR remote plasma deposition for the fabrication of polycrystalline $Ge_{1-x}C_x$ thin films. The goal of the work was to correlate deposition parameters to the material properties of the resulting films. Emphasis was placed on controlling a continuous elevation of the bandgap and contraction of the lattice parameters, while maintaining good crystal quality and grain size.

EXPERIMENTAL

The 7059 Corning glass and polished stainless steel substrates were purchased clean and were then further cleaned immediately prior to deposition by a triple rinse in DI water to remove any particle contamination and subsequent boiling in acetone and methanol to degrease the surface. The [100] oriented p-type Si wafers were cleaned by the standard RCA method, followed by a bath in 50:1 HF, which strips any oxide from the surface and leaves it H terminated. The wafers were then immediately loaded into the vacuum chamber and the H atoms were stripped from the surface by a low energy hydrogen plasma, leaving a bare Si surface for deposition.

The films were prepared using an ECR plasma CVD system, shown schematically in Figure 1. A beam of H ions was produced from the plasma source, approximately 25 cm from the substrate, using H_2 gas. In close proximity to the substrate, the H ions reacted with the process gasses, germane and methane, to produce GeH_3 and CH_3 radicals

Mat. Res. Soc. Symp. Proc. Vol. 557 © **1999 Materials Research Society**

respectively. These radicals are believed to be the dominant species produced due to the heavy H_2 dilution, which tends to suppress the formation of GeH_2 and CH_2.

ECR processing is appropriate for growing this type of material for several reasons. First, reactive ion etching of the film during deposition by the H beam promotes crystalline growth at low substrate temperatures, as non-crystalline material is etched away. Low substrate temperatures are necessary to grow this material so that the Ge and C atoms do not phase separate. Secondly, the ECR plasma is relatively dense at low pressures and low pressures enhance the above etching process and promote faster growth rates. Finally, H passivation of the surface, which is a result of the H_2 dilution, homogenizes the growing surface. This helps to suppress the formation of islands[7] leading to the desired 2-D growth.

RESULTS AND DISCUSSION

For each experiment, a set of parameters similar to those shown in Table 1 was used as a control set. Investigations were then made into how the methane and germane flow rate, substrate temperature, microwave power, chamber pressure, and the type of substrate deposited on affected the material properties of the resulting films. All experiments of deposition parameters were done on amorphous substrates, as a strain free surface was desired.

The first parameter investigated was the flow rate of methane. The bandgap of the films was determined from the absorption data. The dependence of bandgap on methane flow rate is shown in Figure 2. As can be seen, the bandgap initially increases rapidly as the CH_4 flow is increased, but then the values increase only nominally for high CH_4 flow. The crystalline quality of the films was characterized by Raman spectroscopy. All of the films displayed the Ge crystalline peak at 300 cm^{-1}. The full width at half maximum

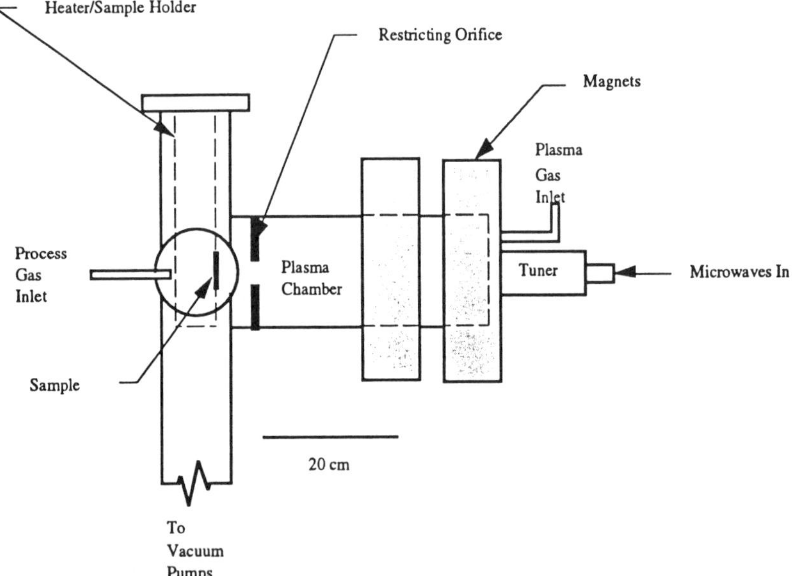

Fig. 1: ECR reactor schematic.

Table 1. Deposition base parameters.

Parameter	Value
Substrate Temperature	350° C
Microwave Power	200 Watts
Chamber Pressure	7 mTorr
GeH_4 Flow Rate	0.5 sccm
H_2 Flow Rate	25 sccm
CH_4 Flow Rate	0 - 22 sccm

(FWHM) value of the Ge peak for the methane flow series is shown in Figure 3. The crystal order of the films was degraded as methane flow increased, but there is not a direct correlation between methane flow and the crystal quality of the films. Activation energy measurements were made on the films, and the values followed a similar trend as the Raman data, with energies ranging from 0.22 to 0.15 eV. The photosensitivities of the films were all close to unity, with dark conductivity values ranging from 10^2 s/cm for pure Ge to 10^0 s/cm range for the $Ge_{1-x}C_x$ samples. The decrease in conductivity was attributed to the incorporation of C. The cubic lattice parameter a was measured by XRD, and the results are shown in Figure 4. As expected, the unit cell became smaller as the flow of methane was increased. As was seen in the bandgap data, the effect of C incorporation was reduced at high methane flow rates. Vegard's Law calculations give approximately 1.5 atomic percent of C estimation for the given shift in a. The XRD spectrums showed evidence of inhomogeneous stress in the films, and this stress obscured any direct measurements of grain size. The XRD data also showed some evidence of a preferred orientation to the [220] plane. XPS measurements were made to directly quantify the amount of C in the films. The atomic percent of C in the films is shown in Figure 5. As can be seen, the values increase to a value of nearly 2 atomic % at moderate methane flow rates and then reduce for high methane flow rates.   Investigations were made into the other deposition parameters using the characterization techniques as above. From these experiments it was found that only GeH_4 flow rates of 0.3 sccm or less produced polycrystalline films. The temperature investigations found that 350° C gave films with the best crystalline quality, as can be seen from the Raman data shown in Figure 6. Films at 400° C had the most C incorporated. It is speculated that lower temperatures did not give sufficient surface diffusional energy for Ge and C to combine in a crystalline structure, and higher temperatures allowed Ge and C to phase separate, which may have reduced the crystal order. The experiments done on microwave power showed that only films of 200 W produced polycrystalline films, while lower powers produced films that were mostly amorphous. The amount of C in the films also decreased as power increased, as shown in Figure 7. Because H etches C very efficiently[8] it is thought that the lower C concentration is due to increased etching at the higher powers. The pressure data was quite complicated, indicating that more than one process in the system is affected by pressure.   The data seemed to generally indicate that 10 to 15 mTorr is best for producing crystalline deposits with up to 3 atomic % substitutional C. Higher pressures gave C concentrations as high as 14 atomic %, but careful examination of the data showed this to be amorphous carbon and not carbon alloyed into the crystalline Ge lattice. The final experiment was to repeat the same deposit, with parameters similar to Table 1, on stainless steel, glass, and Si wafers. The results are summarized in Table 2. As expected, the Si wafer had the most crystalline deposit, and consequently the least amount of C. The steel deposit was more crystalline than the glass deposit, suggesting that the conductivity of the substrate plays a role in the etching effects of the plasma.

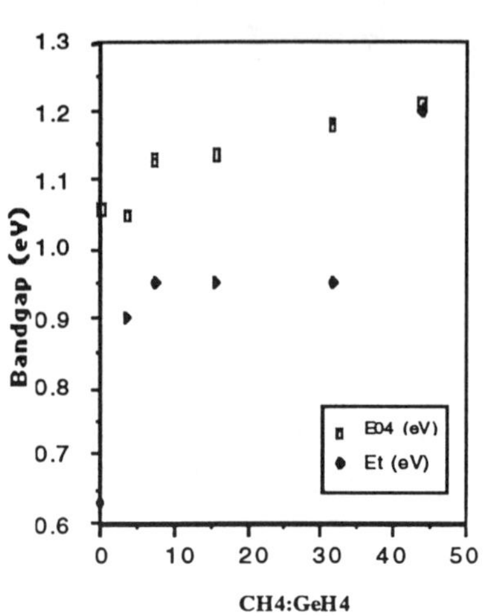

Fig. 2: Bandgap as a function of methane flow rate.

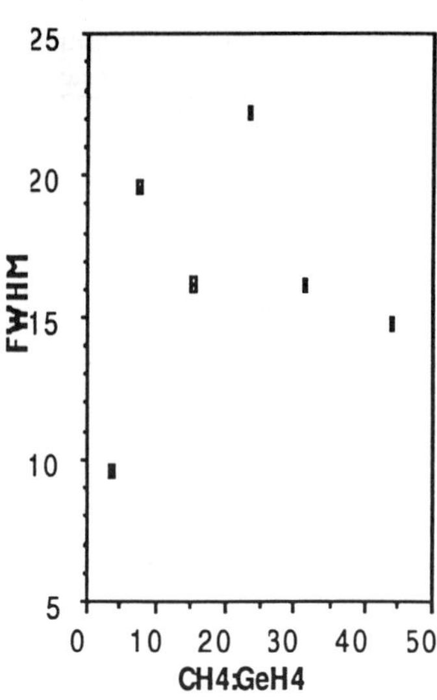

Fig. 3: Raman FWHM as a function of methane flow rate.

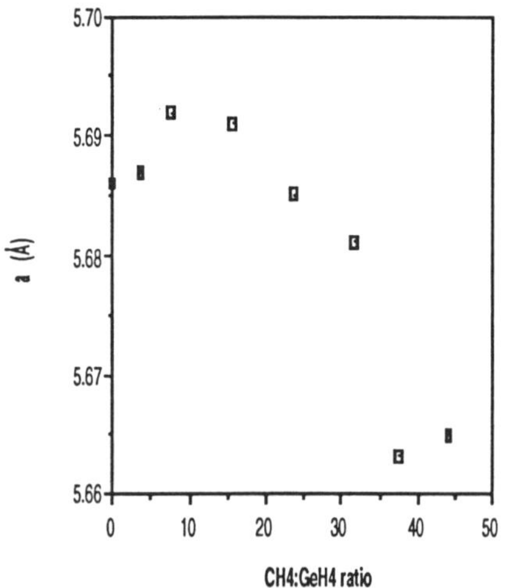

Fig. 4: Cubic lattice parameter a as a function of methane flow rate.

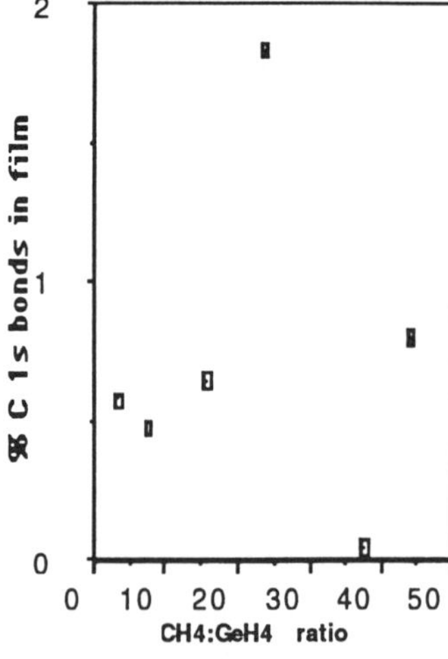

Fig. 5: Atomic % C incorporation as a function of methane flow rate.

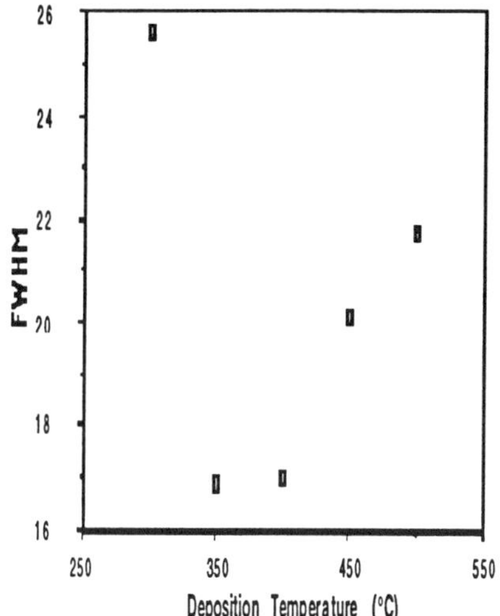

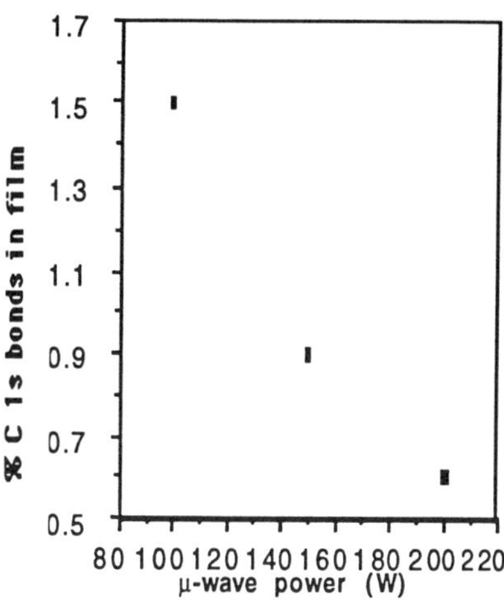

Fig. 6: Crystallinity as a function of temperature.

Fig. 7: Atomic % C incorporation as a function of power.

Table 2: Summary of characterization values for different substrates.

Substrate	Raman Wave # (1/cm)	Raman Width (FWHM)	E04 Bandgap (eV)	% C1s
glass	302.5	19.6	1.11	N/A
steel	302.3	10	N/A	0.98
Si wafer	303	5.7	0.87	0.6

CONCLUSIONS

In summary, investigations were made into how the ECR deposition process parameters affected the properties of polycrystalline $Ge_{1-x}C_x$ thin films. We successfully produced $Ge_{1-x}C_x$ thin films that had optical properties superior to that of Si on both amorphous glass, steel, and crystalline Si substrates. We found that the elevation in bandgap of up to 0.2 eV and the contraction of the lattice of up to 0.03 Å could be continuously controlled by the flow of methane into the reactor. These effects were contributed to the incorporation of C in the Ge lattice, which was as high as 3 atomic % for the films. It was found that GeH_4 flow of 0.3 sccm, 350° to 400° C, 10 to 15 mTorr, and 200 W were the best deposition parameters for depositing this type of material. The type of substrate was shown to be very important to the resulting film, with Si wafers giving more crystalline films with less C incorporated and conductive substrates being conducive to crystalline deposits.

REFERENCES

[1] R.A. Soref, Proc. Inst. Electr. Eng. **81** 1687 (1993).

[2] M. Krishnamurthy, J. Drucker, and A. Challa, J. Appl. Phys. **78** 7070 (1995).

[3] J. Kolodzey, P. A. O'Neil, S. Zhang, B. A. Orner, K.M. Unruh, C. P. Swann, M. M. Waite, and S. Ismat Shah, Appl. Phys. Lett. **67** 1865 (1995).

[4] M. Tod, J. Kouvetakis, and D. Smith, Appl. Phys Lett. **68** 2407 (1996).

[5] Z. Liu, J Zhu, N. Xu, and X Zheng, Jpn. J. Appl. Phys. **36** 3625 (1997).

[6] T. John, J. Bläsing, P.Veit, and T. Drüsedau, MRS Proceedings **472** (1997).

[7] V. Dalal, S. Kaushal, T. Maxson, R. girvan, and A. Boerner, Proc. of Amer. Ass. of Phys. Inst. of Phys. **394** 33 (1997).

[8] B. Deryagin, and D. Fedosev, Growth of Diamond and Graphite From the Vapor Phase, IX. (Nauka, Moscow, USSR 1997) p. 78.

USE OF A GAS JET TECHNIQUE TO PREPARE MICROCRYSTALLINE SILICON BASED SOLAR CELLS AT HIGH I-LAYER DEPOSITION RATES

S.J. Jones, R. Crucet, X. Deng, J. Doehler, R. Kopf, A. Myatt, D.V. Tsu and M. Izu,
Energy Conversion Devices, Inc., Troy, MI 48084

ABSTRACT

Using a Gas Jet thin film deposition technique, microcrystalline silicon (μc-Si) materials were prepared at rates as high as 15-20 Å/s. The technique involves the use of a gas jet flow that is subjected to a high intensity microwave source. The quality of the material has been optimized through the variation of a number of deposition conditions including the substrate temperature, the gas flows, and the applied microwave power. The best films were made using deposition rates near 16 Å/s. These materials have been used as i-layers for red light absorbing, nip single-junction solar cells. Using a 610nm cutoff filter which only allows red light to strike the device, pre-light soaked currents as high as 10 mA/cm^2 and 2.2-2.3% red-light pre-light soaked peak power outputs have been obtained for cells with i-layer thicknesses near 1 micron. This compares with currents of 10-11 mA/cm^2 and 4% initial red-light peak power outputs obtained for high efficiency amorphous silicon germanium alloy (a-SiGe:H) devices. The AM1.5 white light efficiencies for these microcrystalline cells are 5.9-6.0%. While the efficiencies for the a-SiGe:H cells degrade by 15-20% after long term light exposure, the efficiencies for the microcrystalline cells before and after prolonged light exposure are similar, within measurement error. Considering these results, the Gas Jet deposition method is a promising technique for the deposition of μc-Si solar cells due to the ability to achieve reasonable stable efficiencies for cells at i-layer deposition rates (16 Å/s) which make large-scale production economically feasible.

INTRODUCTION

The poor performance of red light absorbing a-SiGe:H cells as compared with the blue-green light absorbing a-Si:H cells has been widely documented[1]. Larger defect densities, weaker hydrogen bonds and heterogeneous microstructures have all been attributed to the poorer performance for the alloy material. With a-SiGe:H alloys commonly used as red light absorbing i-layers for the component cells of high stable efficiency triple-junction devices, the poorer performance limits the achievable efficiencies for the triple-junction modules presently made in production. In addition, the cost of germane gas presently used to make the alloy material is a significant expense in the manufacturing process while silane gas is a low material cost item. Thus, it would be desirable in terms of both stable cell efficiency and module cost to find an alternative low bandgap, red light absorbing material to replace a-SiGe:H as i-layers for bottom cells of the triple-junction structure.

One possible alternative material is microcrystalline Si (μc-Si), which has a bandgap near 1.1 eV. Shah et. al.[2] have shown that high quality μc-Si pin devices with 8.5% efficiencies that absorb a significant amount of the red light and do not degrade with long term light exposure can be made using the Very High Frequency PECVD technique (VHF). However in using this or any standard PECVD technique, the deposition rates of the μc-Si materials are limited to 1-5 Å/s. Because the uc-Si i-layers must be nearly 1 micron thick due to their low light absorption efficiencies as compared with the amorphous materials (amorphous i-layers are typically 1000-2000 Å thick), the low deposition rates make the use of the PECVD techniques economically unsound for the large scale production of microcrystalline materials and solar cells.

Mat. Res. Soc. Symp. Proc. Vol. 557 ©1999 Materials Research Society

Here we report results for μc-Si materials and nip cells made using a new Gas Jet technique[3]. This technique, in which high speed gas is subjected to high intensity microwaves, allows for the preparation of these microcrystalline materials at rates as high as 15-20 Å/s.

EXPERIMENT

The μc-Si materials were prepared in a single chamber system whose vacuum was maintained using a diffusion pump. A microwave source with a fixed 2.54 GHz frequency was used to dissociate the feed gases that included silane, hydrogen and various fluorine-based etching gases. To optimize the material properties, the following parameters were systematically varied; substrate temperature, active gas flows, hydrogen dilution, and the applied microwave power.

To fabricate nip solar cell structures, doped layers were prepared using the standard PECVD process and deposition rates near 1 Å/s. Also, current enhancing Ag/ZnO back reflectors were deposited on the stainless steel substrates prior to fabrication of the nip semiconductor structures. The back reflectors were prepared using a DC sputtering technique. Because of equipment limitations, the doped layers were prepared in a separate deposition system so that the n/i and i/p interfaces were exposed to the atmosphere and are likely oxidized. The authors recognize that this oxidation could limit our cell efficiencies. After fabrication of the nip structure, the devices were completed by depositing Indium Tin Oxide (ITO) conductive layers and then Al collection grids. Both the ITO and Al layers were prepared using standard evaporation techniques.

To characterize the cells, standard current vs. voltage (IV) and spectral response (quantum efficiency) measurements were made. Since our goal is to use the microcrystalline cells as the red light absorbing components of multi-junction cells, IV measurements were completed using not only the direct AM1.5 white light but also the while light filtered with a 610 nm low band pass filter to simulate the absorption due to higher bandgap, blue-green light absorbing top layers. To complete light soaking studies, the cells were subjected to 400-1400 hrs. of one sun light with the cell temperature fixed at 50°C. The i-layer thicknesses were determined using standard capacitance techniques.

To classify the microcrystalline film microstructure, Infrared Absorption Spectroscopy and Raman Spectroscopy measurements were completed to characterize the structural bonding in the films and the degree of microcrystallinity. Also photoconductivity and dark conductivity measurements were completed using AM1.5 light intensities.

RESULTS

Material Properties

Figure 1 displays photocurrent-to-dark current (photo/dark) ratios for Gas Jet single layer films deposited on 7059 glass and prepared with different silane (SiH_4) gas flows. All other deposition conditions were nominally fixed for this set of samples. With the SiH_4 flow decreasing from 30 to 20 sccm there is a sudden drop in the photo/dark ratio indicative of a change in film microstructure. The drop is almost exclusively due to an increase in the dark conductivity which is indicative of the growth of microcrystals in the previously amorphous film. Figure 2 displays Raman data for two of these films made using SiH_4 flows of 20 and 40 sccm. The sharp peaks near 530 and 560 cm^{-1} are Ar Laser lines and are not related to the measured film microstructure. The presence of the peak near 520 cm^{-1} is related to Si microcrystals. The fact that this peak is present in the spectra for the film made with SiH_4=20 sccm and not in the spectra for the SiH_4=40 film demonstrates that the change in the electrical properties with SiH_4 flow is related to an amorphous Silicon-to-microcrystalline Si transition.

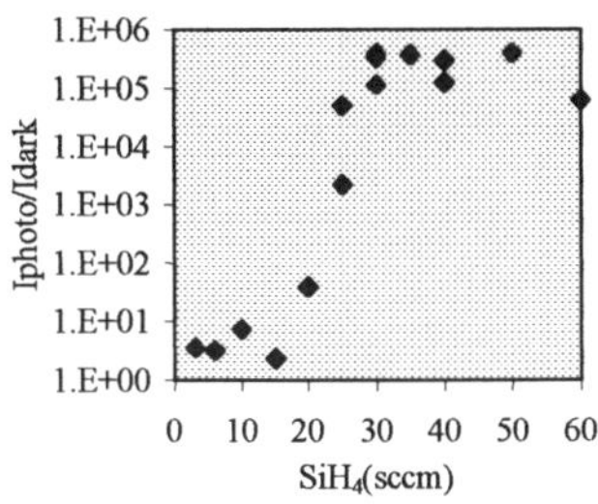

Figure 1. Photo-to-dark current values for films made using different SiH₄ flows.

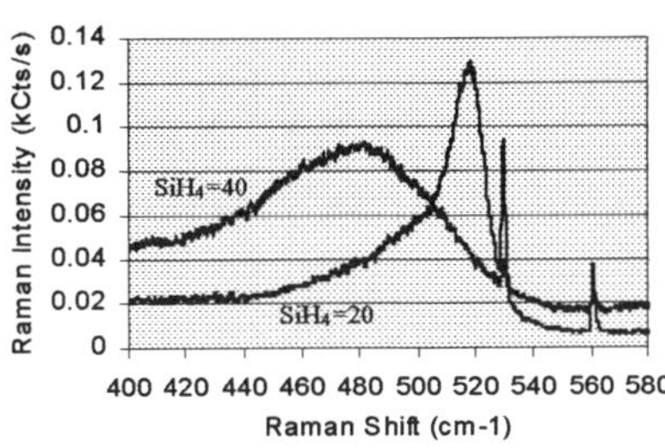

Figure 2. Raman Spectra for films made using different SiH₄ flows.

Figure 3 compares Photothermal Deflection Spectroscopy (PDS) data for films made using SiH₄ flows of 15 and 40 sccm. The enhanced absorption at lower energies for the film made with the lower SiH₄ flow is again consistent with a lower bandgap, red-light absorbing microcrystalline silicon film. Figure 4 demonstrates that as the material becomes μc-Si with the lower SiH₄ flow, the deposition rates decreases. However, when a SiH₄ flow between 15 and 25 sccm is used, the films are microcrystalline and deposition rates between 10 and 20 Å/s are obtained. The degree of microcrystallinity can further be enhanced while maintaining high rates through proper selection of hydrogen dilution, gas flows and applied microwave power.

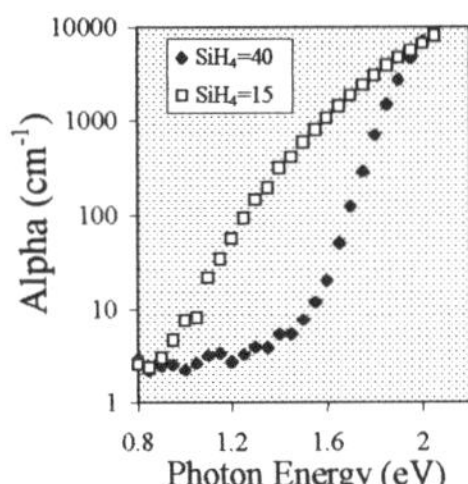

Figure 3. PDS Spectra for films made using different SiH₄ flows.

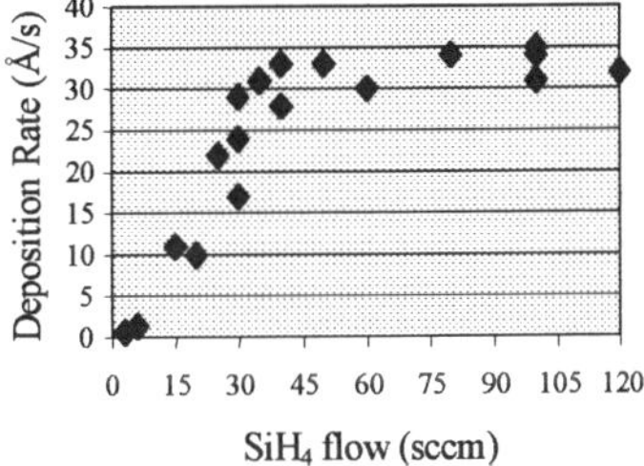

Figure 4. Deposition rates for films made using different SiH₄ flows.

<u>Cells Properties</u>

The nip cells with μc-Si i-layers were fabricated using the procedures outlined in the experimental section. The i-layer thicknesses were between 0.8 and 1.1 microns with an average deposition rate of 16 Å/s. This made the average i-layer preparation time equal to 10.5 min. Figure 5 compares a spectral response curve for a μc-Si device with high quality a-Si:H and a-SiGe:H solar cells. Both of the amorphous i-layers were prepared using the standard PECVD technique and rates near 1 Å/s. The a-Si:H i-layer is much thicker than what is typically used in a triple-junction device while the a-SiGe:H device is representative of a bottom component cell. From the figure, it is clear that the μc-Si cell absorbs a significant amount of red light in the 750-950 nm region of the spectrum, light not collected by the a-Si:H device. Compared with the a-SiGe:H device, the present microcrystalline devices do not absorb as much total red light as the alloy cells do, however they do absorb more light above 900 nm due to their lower bandgap.

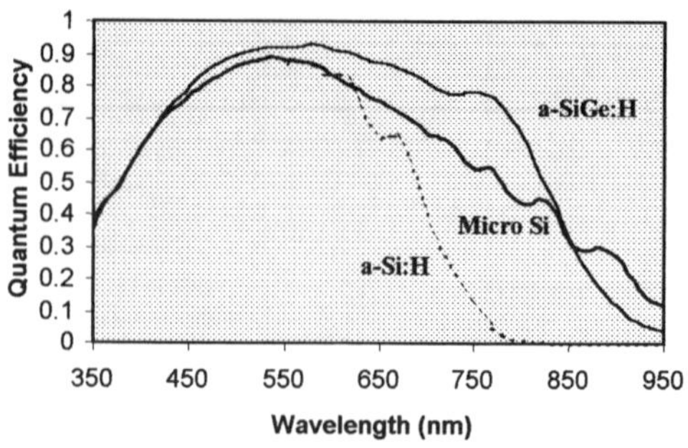

Figure 5. Comparison of quantum efficiency curves for a-Si:H, a-SiGe:H and μc-Si cells.

Table I displays data from IV measurements of some of these microcrystalline devices made in the early part of the program. This IV data was obtained using AM1.5 light that was filtered using a 610nm low bandpass filter. Again, use of this red light truly tests the potential use of the cell as a red-light absorbing bottom cell in a multi-junction solar cell structure. Under these conditions, the microcrystalline cells typically have V_{oc} values of 0.42-0.43 V, J_{sc} values between 8.5 and 10 mA/cm^2 (21-23 mA/cm^2 under white light conditions) and FF of 0.47-0.53. These data can be compared with those listed in the same table for a high quality a-SiGe:H cell typically used as a bottom cell for a triple-junction a-Si:H/a-Si(Ge):H/a-SiGe:H solar cell. This cell was prepared at United Solar using the standard PECVD technique and an i-layer deposition rate near 1 Å/s. From the data it is clear that the a-SiGe:H cells are superior, particularly in terms of pre-light soaked V_{oc} and FF. The smaller V_{oc} for the microcrystalline cells is mostly due to the

Table I

Light soaking completed using unfiltered white light.

IV data taken using AM1.5 light filtered with 610nm cutoff filter (red light).

Cells	Light Soak Time(hr.)	V_{oc} (V)	J_{sc} (mA/cm^2)	FF	R_s (ohmcm^2)	P_{max} (mW /cm^2)	Q.E. 700 nm	Q.E. 900 nm
Gas Jet μc-Si cell 1	0	0.421	9.4	0.523	9.4	2.02	0.63	0.21
	170	0.410	9.3	0.530	9.2	2.01	-	-
	1100	0.412	9.1	0.524	9.6	1.97	-	-
Gas Jet μc-Si cell 2	0	0.420	10.0	0.469	10.5	1.96	0.68	0.22
	170	0.418	9.8	0.472	10.8	1.92	-	-
	1100	0.419	9.7	0.460	12.1	1.87	-	-
Gas Jet μc-Si cell 3	0	0.429	9.1	0.485	11.8	1.89	0.72	0.20
	500	0.425	9.2	0.474	12.4	1.86	-	-
	1400	0.428	9.1	0.464	13.7	1.80	-	-
Gas Jet μc-Si cell 4	0	0.428	8.4	0.534	10.9	1.91	0.62	0.19
	500	0.428	8.4	0.526	11.3	1.90	-	-
	1400	0.430	8.3	0.519	12.3	1.84	-	-
Gas Jet μc-Si cell 5	0	0.426	9.2	0.519	10.1	2.02	-	-
Gas Jet μc-Si cell 6	0	0.428	8.8	0.530	10.1	2.00	-	-
PECVD a-SiGe:H	0	0.615	10.5	0.595	14.1	3.84	0.81	0.12

smaller bandgap for the material. Some improvements in the V_{oc} should come with optimization of the doped layer deposition conditions. The deposition conditions used for doped layers in these devices were optimized for amorphous Si based i-layers but have yet to be optimized for the microcrystalline i-layers. These cells were light soaked under AM1.5 white light conditions for over 1000 hrs to test their stability. The cell efficiencies degrade by less than 5%, values close to the measurement error. This high degree of stability is consistent with μc-Si solar cells and is not found for the amorphous based devices.

Higher red-light peak power output (P_{max}) for the Gas Jet produced microcrystalline cells were obtained when different fluorine-based gases were used in the process. Table II shows IV data for these devices. Higher FF and V_{oc} led to P_{max}=2.2-2.3 mW/cm^2 and larger currents should be obtained for thicker i-layers. Figures 6 and 7 display IV curves for these cells measured under the standard white light conditions demonstrating 5.9-6.0% efficiencies. Using Raman spectroscopy, it was demonstrated that the use of this fluorine-based gas led to enhanced crystallinity, suggesting that the increase in V_{oc} is not related to an increase in the amorphous tissue but to an improvement in the material and/or the i-layer/doped layer interface quality. These cells were light soaked along with a high quality a-SiGe:H cell to directly compare the stability of the two types of cells. Since we are testing the feasibility of using these cells as bottom cells for the triple-junction structure, the light soaking was completed using white light which was filtered using a low bandpass filter to simulate the light absorbed by the top and middle cells. As can be seen from the data in the table, the P_{max} values for the μc-Si devices remain exposure, the P_{max} for the a-SiGe:H cell degrades by 17% demonstrating the enhanced stability for the μc-Si cells.

Table II

Light Soaking Conditions: white light filtered with 570nm low band pass filter, 50°C.
IV data taken using AM1.5 light filtered with 610nm cutoff filter (red light).

Cell Type	Light Soak Time (hr.)	V_{oc} (V)	J_{sc} (mA/cm^2)	FF	R_s (ohmcm^2)	P_{max} (mW/cm^2)	% of Degr.
PECVD a-SiGe:H	0	0.615	10.5	0.595	14.1	3.84	-
	1030	0.595	10.3	0.522	24.4	3.18	17.2
Gas Jet μc-Si cell 1	0	0.462	8.22	0.584	8.9	2.22	-
	1030	0.467	8.16	0.574	10.2	2.19	1.4
Gas Jet μc-Si cell 2	0	0.473	8.52	0.560	9.6	2.26	-
	1030	0.481	8.56	0.550	10.9	2.26	0

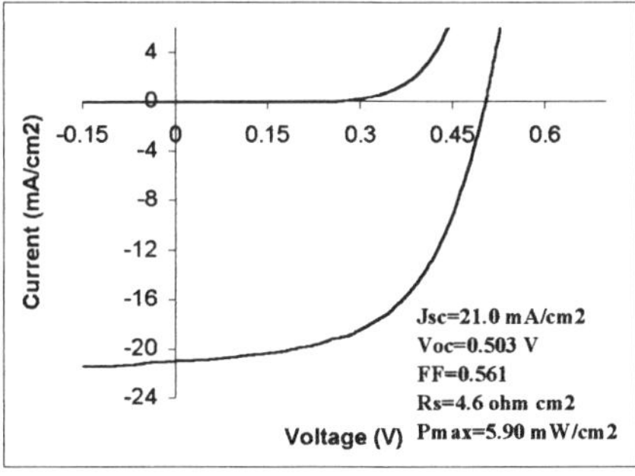

Figure 6. IV plot for μc-Si cell.

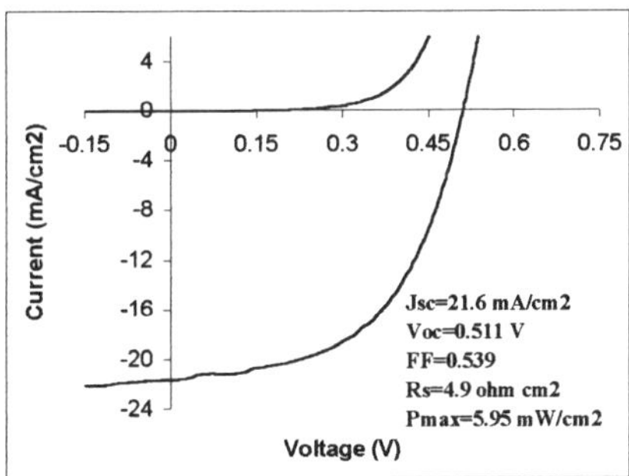

Figure 7. IV plot for μc-Si cell.

DISCUSSION AND CONCLUSIONS

The ability to create μc-Si material at such high deposition rates with the Gas Jet technique is related to the use of microwaves to achieve high decomposition rates for SiH_4 while still allowing for high flows of hydrogen gas for enhanced etching at the film surface. Similar conditions can be obtained using conventional microwave techniques where high deposition rates for μc-Si growth have also been reported[4]. However with the use of the high-speed gas jet with this technique, there is little time for gas reactions after SiH_4 decomposition thus minimizing the chance for polysilane radical formation which may limit material quality and cell performance.

The efficiencies for the cells are respectable however they do not match the 8.5% AM1.5 efficiency for the μc-Si cells[2] made at 1 Å/s using the VHF method nor the light soaked 3.2% red-light efficiency for high quality a-SiGe:H cells (see Table 2). However, expanding the VHF technique to deposition rates of 11 and 14 Å/s, only lower efficiencies of 5.2 and 3.6%, respectively, have been achieved[5]. It should be noted that for the cells reported here, both doped/i-layer interfaced were exposed to air prior to subsequent layer deposition, and the doped layer and the back reflector deposition conditions have yet been optimized specifically for the microcrystalline i-layer. In particular, the microcrystalline material has a more textured surface than the typical amorphous silicon i-layers leading to different requirements for the doped layers and the back reflector layers. Further optimization of the different cell layers and preparation of all semiconductor layers in a load-locked system should lead to higher cell performance.

In terms of other techniques with which μc-Si can be made at high deposition rates, the Hot Wire technique has been used to make these materials at rates of 15-20 Å/s [6]. However, μc-Si cells made using this technique have, to the author's knowledge, yet to achieve AM1.5 white light efficiencies above 4%. However, both the Hot wire and the Gas Jet techniques are in their infant stages in terms of their use for making μc-Si devices and thus further studies are needed to fully assess their potential application in making high rate μc-Si i-layers for solar cells.

In conclusion, it has been demonstrated that the Gas Jet technique can be used to prepare μc-Si i-layers for nip solar cell devices at rates of 16 Å/s. The cells have 2.2% red-light P_{max} (6.0% white-light) that remain stable with long term light exposure. With improvement of the cell performance, this process could be applicable for use in the production of large-area modules with high stable efficiencies due to the ability to make the microcrystalline materials at high deposition rates.

ACKNOWLEDGEMENTS

We would like to thank A. Banerjee, J. Noch, N. Jackett, J. Edens and T. Palmer of United Solar and B. Viers of ECD for their efforts in preparation of doped layers and ITO contacts. This work was partially funded by the Department of Energy under the Grant No. DE-FG02-96ER82162 and by the National Science Foundation under the Award No. DMI-9760805.

REFERENCES

1. For review, W. Luft, Proc. of 20th IEEE Photovoltaics Spec. Conf. 218 (1988).
2. J. Meier, H. Keppner, S. Dubail, U. Kroll, P. Torres, P. Pernet, Y. Ziegler, J.A. Anna Selvan, J. Cuperus, D. Fischer, A. Shah, Mat. Res. Soc. Symp. Proc. 507, 139 (1998).
3. J. Doehler, S. Hudgens, S.R. Ovshinsky, B. Dotter, L. R. Peedin, J.M. Krisko, A. Krisko, United States Patents #4,883,686 and 5,093,149 (1988).
4. J.Schellenberg, R. Mcleod, S. Mejia, H. Card, K. Kao, Appl. Phys. Lett., 48(2), 163 (1986).
5. P.Torres,J.Meier,U.Kroll, N.Beck, H.Keppner, A.Shah, 26th IEEE Photo. Sp. Con.,711(1997).
6. J. Rath, A. van Zutphen, H. Meiling, R. Schropp, Mat. Res. Soc. Proc. 467, 445 (1997).

MICROSTRUCTURAL DEFECTS OF DEVICE QUALITY HOT-WIRE CVD POLY-SILICON FILMS

J.K. RATH, F.D.TICHELAAR* and R.E.I. SCHROPP
Debye Institute, Utrecht University, P.O.Box 80000, 3508 TA Utrecht, the Netherlands
*National centre for HREM, Rotterdamseweg 137, 2628 AL Delft, the Netherlands

ABSTRACT

Two types of poly-Si:H thin films made by Hot Wire CVD have been evaluated with respect to utilisation in solar cells. Poly-Si:H films made at high hydrogen dilution are highly porous and have large interconnected voids. The void density is $25000/\mu m^3$ as determined by XTEM. On the other hand, poly-Si:H layers made at low hydrogen dilution have a compact structure and a much smaller density of voids. In these films, two types of voids exist: globular voids smaller than 15 nm, and elongated voids, often located between columns of large crystals of 150-250 nm wide at the top. The density for the 5 - 15 nm spherical voids is usually ~ $50/\mu m^3$, but larger concentrations often occur locally, up to $1000/\mu m^3$, i.e., 0.05% volume fraction. High oxygen content in the poly-Si films made at high hydrogen dilution is largely due to post deposition intrusion of water vapour through the interconnected voids. *Profiled* layers are made by depositing device quality poly-Si:H layers (low hydrogen dilution) on top of a seed layer (high hydrogen dilution) of high nucleation density. Cells incorporating *profiled* poly-Si:H films as i-layers at a deposition rate of 0.5 nm/s were made on stainless steel substrates in the configuration SS/n-μc-Si:H(PECVD)/i-poly-Si:H(HWCVD)/p-μc-Si:H(PECVD)/ITO. For our n-i-p solar cell with poly-Si i-layer we obtained an efficiency of 4.41% and a FF of 0.607. Due to native surface texture a current density of 19.95 mA/cm^2 is generated in only ~1.22 μm thick i-layer without back reflector.

INTRODUCTION

Some of the requirements of the poly-Si films for the solar cell application are that they should be intrinsic and should have low oxygen incorporation. The actual mechanism of oxygen incorporation into the material is not clear and the extent of oxygen incorporation depends on the type of deposition, growth process and purity of gases [1]. We have been able to achieve, by hot wire chemical vapour deposition (HWCVD), device quality poly-Si:H films with very intrinsic nature (activation energy of 0.54 eV and band gap of 1.1 eV) [2]. These films were made at a low hydrogen dilution of silane gas. However, the level of oxygen incorporation can change with the change of deposition conditions. At high hydrogen dilution the poly-Si contains a large concentration of oxygen [3]. It is the aim of this paper to find a (possible) correlation between microstructural defects and the deposition condition, in particular the hydrogen dilution and investigate the oxygen incorporation mechanism. Moreover, the defect density in the material depends on the microstructural defects and the degree of passivation of internal surfaces of the voids. It is necessary to find a proper deposition regime where microstructural defects can be minimized maintaining good initial crystalline growth. In our earlier paper [3] we had proposed such a deposition scheme, where a *profiled* poly-Si was made integrating two regimes of growth i.e., high and low hydrogen dilution of silane gas. Our preliminary results obtained for actual thin film solar cells on a foreign substrate, i.e., complete n-i-p solar cells using thin n and

Mat. Res. Soc. Symp. Proc. Vol. 557 © 1999 Materials Research Society

p-layers fabricated on a stainless steel substrate, yielded an efficiency of 3.7% and a current density of 23.5 mA/cm^2 with a 1.5 μm thick poly-Si i-layer [3]. In this paper we present our latest solar cell characteristics and investigate the role of impurities and the structure of the cell as the limiting factors in the solar cell efficiencies.

EXPERIMENTAL

Poly-silicon films were deposited on 10 cm x 10 cm Corning 7059 glass and c-Si wafer substrates by HWCVD in one of the chambers of an ultra high vacuum multichamber system (PASTA). Details of the Hot Wire-set up and deposition process have been described elsewhere [4]. The thickness of the films was measured with a Dektak profilometer and with reflection/transmission measurements. Samples were characterised by laser Raman spectroscopy, Fourier transform infrared (FTIR) spectroscopy, hydrogen evolution, secondary ion mass spectrometry (SIMS) and electrical conductivity in the dark and in white light. Bright-field cross sectional transmission electron micrographs (XTEM) were made with a Philips CM30T electron microscope operated at 300kV. High-resolution electron micrographs (HREM) of the material were made to detect crystallinity, especially near the interfaces. N-i-p cells were made on various types of stainless steel (SS) substrates in the configuration SS/n-μc-Si:H(PECVD)/i-poly-Si:H(HWCVD)/p-μc-Si:H(PECVD)/ITO. ITO was deposited by reactive evaporation. Gold grid lines were deposited on the ITO. The cells were characterized by light I-V measurements at AM1.5 illumination by a dual beam solar simulator (WACOM, Japan) and spectral response measurements.

RESULTS

Deposition

The wire temperature (T_w) and chamber pressure (P_r) are at an optimum value of 1800 °C and 0.1 mbar respectively. The substrate temperature is below 500 °C. The structure of the poly-Si films is strongly dependent on hydrogen dilution. Films made at high hydrogen dilution of 99% are named Poly1 layers throughout this paper. They are characterised by poor electronic properties and high defect density of

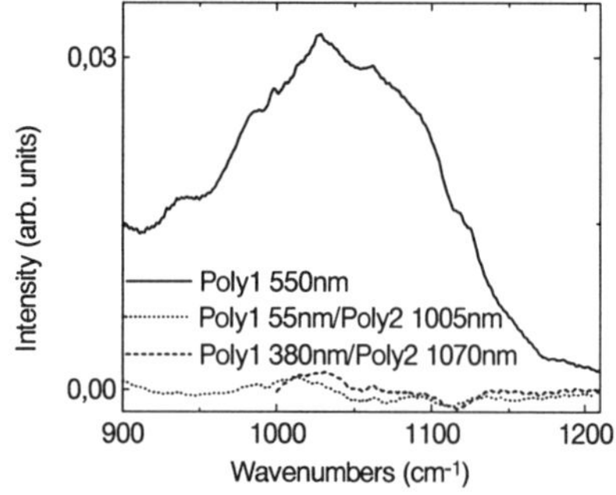

Fig.1: IR spectra of poly-Si films.

2.9×10^{18}/cm^3. The films indeed are quite porous. However, Poly1 layers possess good initial crystalline growth. On the other hand, the films made at low hydrogen dilution of 90% (labeled as Poly2 layers) are of device quality (purely intrinsic nature, a low ESR defect density of 7.8×10^{16}/cm^3). The presence of only 2000 cm^{-1} IR vibration and hydrogen evolution at a high temperature (~600 °C) is evidence for a very compact structure of these films [3]. However, the initial growth of this material is amorphous over about 50 nm [3].

The compactness of the poly-Si films can be best understood by considering the oxygen depth profile in them. For the case of the Poly1 layer it is observed that the oxygen homogeneously penetrates the film at 2×10^{21} cm^{-3} (measured by SIMS) within a few days of exposure to air. In the Poly2 layer, the oxygen profile (even after one year of exposure to air) shows a sharp decrease at a depth of 50 nm from the surface. The oxygen concentration inside the film is nearly constant at about 5×10^{18} cm^{-3}. The bulk value of oxygen is comparable to the

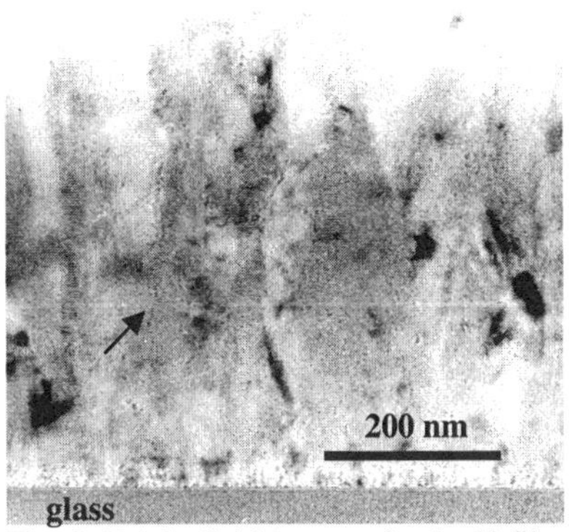
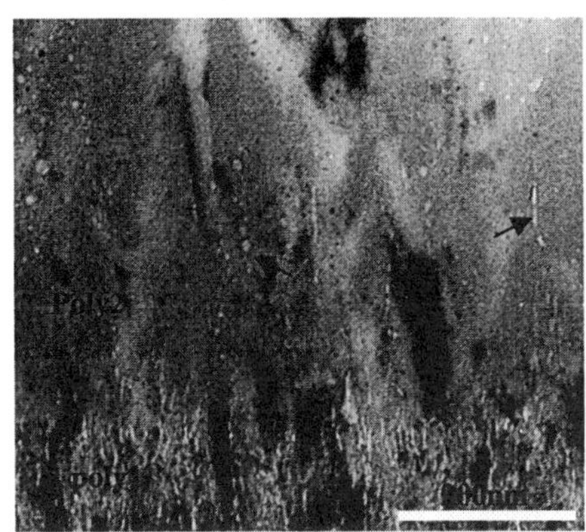

Fig. 2: Bright field image of poly-Si film with the objective lens over focus. Voids show up as areas of dark contrast.

Fig. 3: Part of the Poly1 layer and the Poly2 layer above. Foil thickness 90±10 nm.

oxygen content in c-Si grown by the Czochralski technique [5] and to that of a microcrystalline film deposited by VHF CVD using an extra gas purifier [1]. The IR spectra of Poly1 (Fig.1) shows a strong Si-O band (~1050 cm^{-1}). This suggests that the incorporated oxygen is chemically bonded to silicon, most likely at the grain surfaces. It is also seen that the hydrogen content (measured by SIMS) in Poly1 is much higher than in Poly2. In fact, the ratio of hydrogen to oxygen in Poly1 has been determined to be 2:1. This suggests that the oxygen incorporation in Poly1 is by adsorption of moisture from air. This adsorption is absent in Poly2 film. For a *profiled* films (Poly2 on Poly1) also no Si-O IR absorption could be seen (Fig.1). The difference in oxygen penetration is due to the difference in the structure of these films. The annealing characteristics of this *profiled* poly-Si film show that there is no change of electrical conductivity upon exposure to air. However the Poly1 film showed a metastable increase in conductivity when exposed to air and it is reversible with annealing. It is necessary to study the microstructural defects of the poly-silicon films which will enable us to find the deposition scheme that yields the best device-quality poly-silicon.

Microstructural defects

Voids can be recognised in TEM by comparing BF under- and over-focus (objective lens) images of the same area. Voids will show up (shown with arrows in the figures) dark in over-focus (Fig.2) and light in under-focus (Fig.3) images. Elongated voids occur between columns. In HREM images, these elongated voids between columns may be recognised. It is difficult to be conclusive about the presence of voids from HREM images for the following reasons: i) in very thin foils where voids run through the complete foil in the viewing direction, the voids will be filled with ion milling debris and/or other contaminating material, ii) in thicker areas of the foil the surrounding crystalline matrix will show fringes. Figure 3 shows

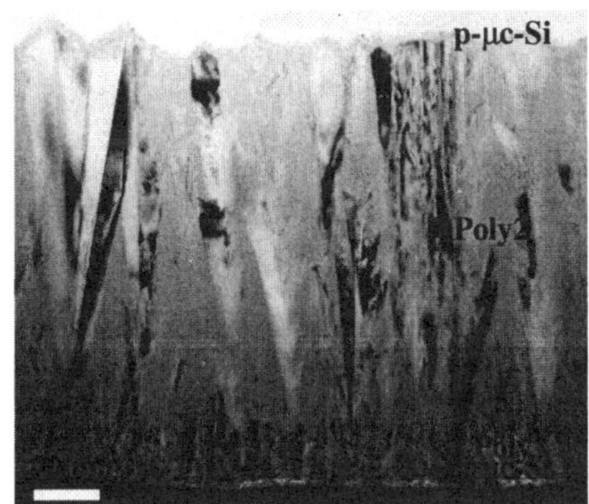

Fig. 4: SS/n-i-p cell. Scale bar 200nm.

575

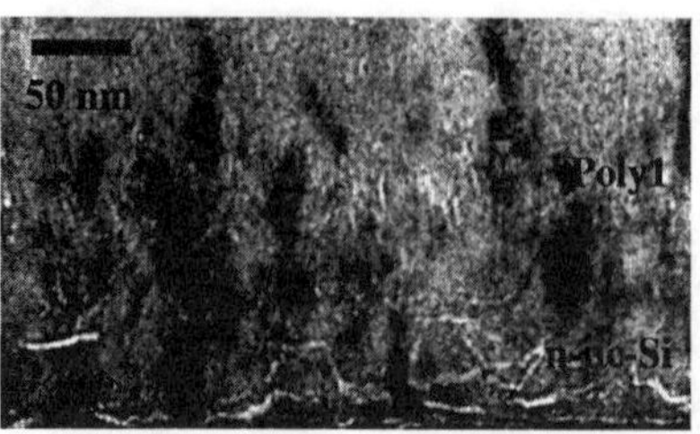

Fig. 5: n-μc-Si layer with horizontal cracks.

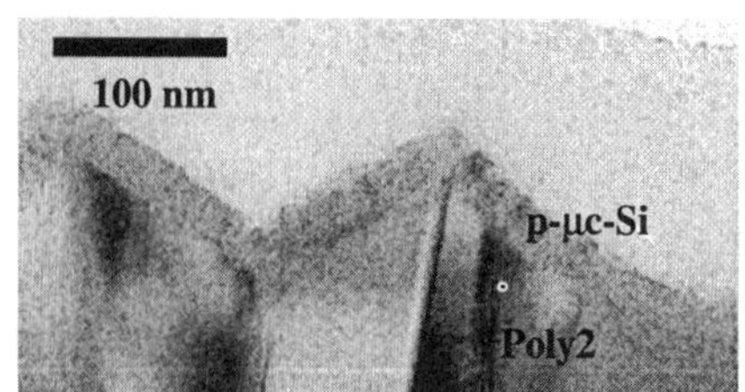

Fig. 6: The p-μc-Si top layer is imaged here.

the XTEM image at under-focussed (BF) condition of the *profiled* layer of Poly2 on Poly1. It is clear that Poly1 is highly porous and that there is a sharp decline in void fraction at the interface with Poly2. Areas of light contrast are areas of low density, i.e., voids. The small dark spots are pollution of the specimen surface. A difference in density between Poly1 and Poly2 is difficult to quantify, since foil thickness is an important factor, and foil surface contamination plays a role. In the Poly1 layer near the substrate, interconnected elongated voids occur. The length of the voids is 10-60 nm. The lateral spacing between voids is 6-20 nm in the cross sectional plane. From the foil thickness of 90 nm the void concentration is estimated to be very high, about 25000 voids/μm^3. In the Poly2 layer two types of voids exist: globular voids all smaller than 15 nm, and elongated voids, often present between large columnar crystals that are 150 to 250 nm wide at the top. But in most cases the crystalline columns are connected. The density is usually about 50/μm^3 for the 5 - 15 nm spherical voids, but larger concentrations up to 1000/μm^3 often occur locally. This is still only ~0.05% volume fraction. In some areas large densities up to 2000/μm^3 of smaller voids (<5nm) occur. The above result clearly shows the compact nature of the Poly2 film. The fact that the void content is directly related to the hydrogen dilution was confirmed from the film which was made by ramping the hydrogen dilution from high to low during deposition. In fact a gradual decrease of the void concentration from the bottom to the top of the film was observed. The oxygen diffusion in Poly1 is due to intrusion of water vapour through the large interconnected voids which is a characteristic for most microcrystalline films. This is responsible for the metastable increase of conductivity when exposed to air. The oxygen incorporation is a post-deposition effect which can be prevented by applying a void free compact capping layer such as Poly2 layer. In order to make device-quality poly-Si:H films without large void content and avoiding any amorphous incubation phase, *profiled* poly-Si:H layers were made by integrating the two growth regimes of high hydrogen dilution and low hydrogen dilution. The thickness of the void rich Poly1 layer (which is used as seed layer) is kept as small as possible. The rest of the layer (top layer) is made at low hydrogen dilution.

Solar Cell

A solar cell structure on stainless steel (SS) substrate was made in the substrate configuration SS/n-μc-Si:H(PECVD)/i-poly-Si:H(HWCVD)/p-μc-

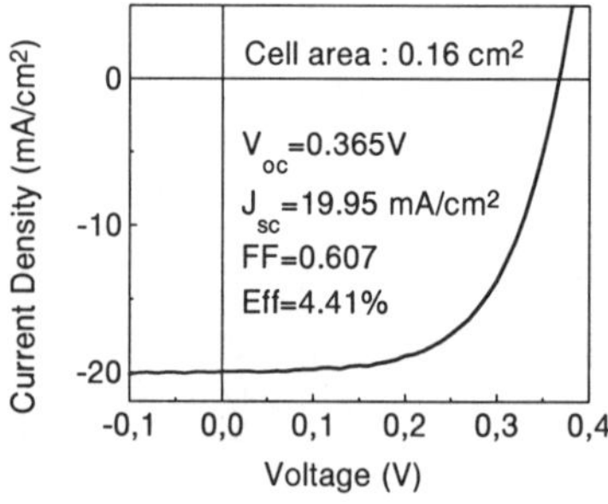

Fig. 7: I-V characteristics of the n-i-p cell with HWCVD poly-Si:H i-layer.

Si:H(PECVD)/ITO. The n-μc-Si:H and p-μc-Si:H films were made by PECVD at a substrate temperature about 200 °C whereas the i-layer was deposited by HWCVD in the same multichamber system. *Profiled* poly-Si:H with 20 nm seed layer was used as i-layer. A thick n-layer was used (to isolate the i-layer from metal diffusion) without sacrificing the low series resistance. The thickness of the ITO layer is 80 nm for minimum reflection. The ITO is a limiting factor of the cell performance due to its high sheet resistance (>100 Ω/cm^2) which resulted in large series resistance of the cell. To overcome this

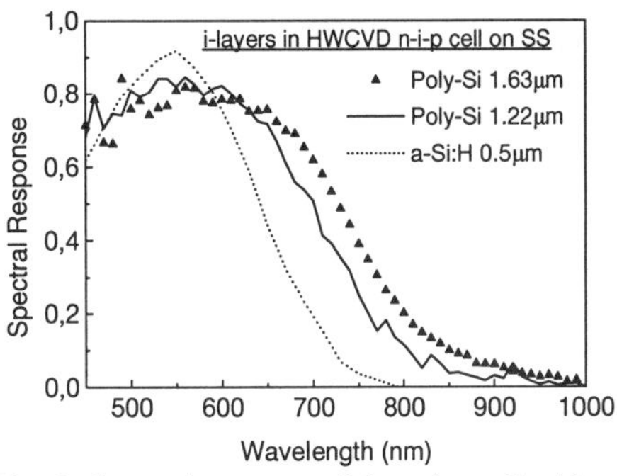

Fig. 8: Spectral response of the n-i-p cell with different i-layers.

problem gold grid lines were deposited on ITO which significantly decreased the series resistance and improved the FF of the cell. Figure 4 shows the XTEM image of the cell on SS substrate. The thick poly-Si layer consists of V-shaped crystals with small inverse V shaped amorphous areas near the bottom of this layer. The amorphous nature of these regions was deduced from tilting experiments in BF and local diffraction. This result is significantly different from the structure at the Poly1/Poly2 interface region of similarly *profiled* layers on glass substrates [3] where there were no noticeable amorphous regions. At present we do not have proper explanation for the differences in the structure of the interface region observed for glass or steel substrates, though it may be speculated that the conducting nature of the substrate has an effect on the crystalline growth kinetics. More experiments are needed to understand the possible effect of charging on the nucleation process. The n-microcrystalline layer and the i/n interface are shown in Fig. 5. The poly-Si seed layer may be recognised by a zone of vertical voids. Some cracks are observed in the n-layer in the horizontal direction. This may be attributed to the stress induced by the poly-Si i-layer deposition at high temperature on a comparatively low temperature deposited (~200 °C) n-layer. This may have a negative effect on the cell performance. A high temperature n-layer may be necessary to solve this problem. However, the p-microcrystalline layer and the p/i interfaces shown in Fig. 6 are very smooth. Figure 7 shows the cell characteristics of the poly-Si cell. A current density (J_{sc}) of 19.95 mA/cm^2 for a 1.22 μm i-layer shows the light harvesting potential. An efficiency of 4.41% and a FF of 0.607 at AM1.5 illumination have been achieved. However, V_{oc} shows room for improvement.

Fig.8 shows the spectral response of the cells. The spectral response at short circuit condition of the cell with 1.22 μm poly-Si i-layer is the same in the dark as under AM1.5 bias light and remains unchanged at a negative voltage bias of −2V. This shows that the photogenerated carriers are completely collected in the cells. It is also observed from the spectral response that most of the current is generated in the high energy region and the current contribution from the low energy region is only moderate. However, it should be noted that SS is a poor reflector and that no special back reflector is used (near the n-layer) in our cell. XTEM was used to estimate the roughness of the steel substrate. The steel substrate has a roughness (maximal height difference) of ~200 nm at the intersection of grain boundary in the steel substrate and the steel/Si interface. Usually the roughness is 30-80 nm with a width of the order of 0.5-2μm. The rather small angle of reflection from the steel substrate can only contribute nominally to scattering at the back surface. A comparison between cells (in the same configuration) with amorphous silicon (HWCVD) i-layer and poly-Si i-layer, however, shows the higher current generating potential of the poly-Si cell in the low energy region. The response in the long wavelength further increases with increasing thickness of the i-layer to 1.63 µm. Future work is aimed at using thicker i-layers without sacrificing FF or Voc. A detailed SIMS study of the impurity levels in the cell is done to determine the extent of possible contaminants (Fe,Cr,Ni) from the SS substrate and their diffusion at the deposition temperature used (500 ^{0}C). The roughness of the substrate makes the SIMS analysis rather difficult, particularly the depth resolution. However an upper limit to the contaminants in the bulk of the poly-layer were calculated and they are as follows: $C_{Fe}=1.6x10^{17}$ cm^{-3}, $C_{Cr}=1.8x10^{15}$ cm^{-3} and $C_{Ni}=3.7x10^{16}$ cm^{-3}. The actual impurity content could be much lower than these values. However, considerable contamination of the n-layer can not be ruled out. The roughness of the substrate makes it difficult to ascertain whether there is in-diffusion into the n-layer or not.

CONCLUSION

Compact poly-Si:H films of high optoelectronic quality have been deposited by the HWCVD method in a carefully defined growth regime. The high oxygen content in the films made at high hydrogen dilution is correlated with the presence of a large number of interconnected voids. The device-quality poly-Si films made at low hydrogen dilution have a much smaller void density. Device-quality poly-Si:H *profiled* layers containing small density of microstructural defects and without amorphous incubation phase were made combining different regimes of growth, i.e., low and high hydrogen dilution. A solar cell made in the configuration SS/n-µc-Si:H(PECVD)/i-poly-Si(HWCVD)/p-µc-Si:H(PECVD)/ITO generated a current density of 19.95 mA/cm^2 with an i-layer of only 1.22 µm thickness. An efficiency of 4.41% has been achieved at a growth rate of about 5Å/s.

ACKNOWLEDGEMENT

This research was partially funded by NOVEM (Netherlands Organisation for Energy and Environment). We thank P. Alkemade of TU Delft for SIMS analysis, Gert Hartman and Karine van der Werf for deposition of all the films.

REFERENCES

1. J.Meier, P.Torres, R.Platz, S.Dubail, U.Kroll, J.A.Anna Selvan, N.Pellaton Vaucher, Ch. Hof, D.Fischer, H.Keppner, A.Shah, K.-D.Ufert, P.Giannaoules, J.Koeler, in "Amorphous Silicon Technology-1996", Edited by M.Hack, E.A.Schiff, S.Wagner, R.Schropp and A.Matsuda, Mat. Res. Soc. Symp. Proc., **420**, 3 (1996).
2. J.K.Rath, H. Meiling, R.E.I.Schropp, Jpn. J. Appl. Phys., **36**, 5436 (1997).
3. J.K.Rath, F.D.Tichelaar, H.Meiling, R.E.I.Schropp, in "Amorphous and Microcrystalline Silicon Technology-1998", Edited by R.Schropp, H.Branz, S.Wagner, M.Hack and I.Shimizu, Mat. Res. Soc. Symp. Proc., **507**, 879 (1998).
4. R.E.I.Schropp, K.F.Feenstra, E.C.Molenbroek, H.Meiling, J.K.Rath, Phil. Mag. B., **76**, 309 (1997).
5. J.R.Patel, Semiconductor Silicon 1981 (Electrochem. Soc. Pennington, 1981) p189.

MICROCRYSTALLINE SILICON TUNNEL JUNCTIONS FOR AMORPHOUS SILICON-BASED MULTIJUNCTION SOLAR CELLS

A. S. FERLAUTO, Joohyun KOH, P. I. ROVIRA, C. R. WRONSKI, and R. W. COLLINS
Center for Thin Film Devices, The Pennsylvania State University, University Park, PA 16802.

ABSTRACT

The formation of tunnel junctions for applications in amorphous silicon (a-Si:H) based multijunction n-i-p solar cells has been studied using real time optics. The junction structure investigated in detail here consists of a thin (~200 Å) layer of n-type microcrystalline silicon (μc-Si:H) on top of an equally thin layer of p-type μc-Si:H, the latter deposited on thick (~2000 Å) intrinsic a-Si:H. Such a structure has been optimized in an attempt to obtain single-phase μc-Si:H with a high crystallite packing density and large grain size for both layers of the tunnel junction. We have explored the conditions under which grain growth is continuous across the p/n junction and conditions under which renucleation of n-layer grains can be ensured at the junction. One important finding of this study is that the optimum conditions for single-phase, high-density μc-Si:H n-layers are different depending on whether the substrate is a μc-Si:H p-layer or is a H_2-plasma treated or untreated a-Si:H i-layer. Thus, the top-most μc-Si:H layer of the tunnel junction must be optimized in the multijunction device configuration, rather than in single cell configurations on a-Si:H i-layers. Our observations are explained using an evolutionary phase diagram for a-Si:H and μc-Si:H film growth versus thickness and H_2-dilution ratio, in which the boundary between the two phases is strongly substrate-dependent.

INTRODUCTION

The optimization of p/n and n/p tunnel junctions in amorphous silicon (a-Si:H) based n-i-p and p-i-n multijunction solar cells poses a considerable challenge when very thin (~100 Å), microcrystalline silicon (μc-Si:H) doped layers are used for high conductivity and reduced optical absorption [1]. Few studies have been published that provide insights into the fabrication of tunnel junctions, and as a result optimization usually proceeds by trial and error. There are two key steps in tunnel junction formation. For the case of the p/n junction of an n-i-p multijunction solar cell, for example, one must first ensure immediate, high-density nucleation of a single-phase μc-Si:H p-layer on the underlying i-layer. One must then overdeposit a μc-Si:H n-layer to form a contact having characteristics such that junction recombination is favored over trapping within the doped layers [2]. Rath et al. have found that for high current-generating component cells in the p-i-n configuration, an optimized tunnel junction consists of ~200 Å μc-Si:H n and p layers with an ultrathin intervening oxide layer at the junction. They have proposed that the low mobility gap of the μc-Si:H and the high defect density at the oxide junction are critical to obtaining high recombination rates. Although recent studies have characterized the nucleation and growth of intrinsic, p, and n type μc-Si:H on a-Si:H i-layers [3-7], few studies have characterized the deposition of μc-Si:H layers on underlying μc-Si:H either with plasma termination or with intervening interface layers. In this study, we have applied real time spectroscopic ellipsometry (RTSE) to characterize the nature of μc-Si:H p/n interface formation.

EXPERIMENTAL DETAILS

In fabricating the μc-Si:H p/n tunnel junctions, the substrate temperature during all processing was 200°C. The underlying a-Si:H i-layer was prepared first using rf plasma-enhanced chemical vapor deposition (PECVD) with pure SiH_4 (i.e., with $R=[H_2]/[SiH_4]=0$), a flow rate of 5 sccm, a plasma power of 70 mW/cm^2, and a pressure of 0.07 Torr. The n-layer growth was performed on two different types of p-layers [7]. For both p-layers, a 150 s H_2-plasma pretreatment of the underlying a-Si:H i-layer was applied, using a H_2 flow rate of 200 sccm, a power of 700 mW/cm^2, and a pressure of 0.9 Torr. In one case, the p-layer was prepared with R=200, a trimethyl boron doping level of $D=[B(CH_3)_3]/[SiH_4]=0.01$, a power of 230 mW/cm^2, and a total pressure of ~0.9 Torr. In the other case, the p-layer was prepared with R=200, a boron trifluoride doping level of $D=[BF_3]/[SiH_4]=0.05$, a power of 700 mW/cm^2, and a pressure of ~0.9 Torr. Both sets of conditions

Mat. Res. Soc. Symp. Proc. Vol. 557 © 1999 Materials Research Society

led to immediate nucleation of single-phase nanocrystalline silicon films which coalesced at thicknesses as low as ~25-40 Å [7]. In addition, these p-layers exhibited well-defined absorption onsets near 2.4 eV, characteristic of the presence of nanocrystals, and a low relative void volume fraction. In Ref. [7], the range of gas phase doping level D for optimum structure was found to be narrow for $B(CH_3)_3$, but very broad for BF_3. We present results for these latter films owing to the relative insensitivity to the deposition process.

For the tunnel junctions, two different n-layer conditions using R=50 (at 0.5 Torr pressure) and R=200 (at 0.9 Torr) were explored in these studies. The fixed conditions included a doping level of $D=[PH_3]/[SiH_4]=0.012$ and a plasma power of 700 mW/cm^2. The latter conditions were obtained in optimization studies using H_2-plasma-treated a-Si:H i-layer substrates as described in the next section. Amorphous n-layers 50 Å thick used at the tunnel junction interface were prepared using R=0, D=0.02, a plasma power of 70 mW/cm^2, and a pressure of 0.1 Torr. Finally, ultrathin oxides 10 Å thick used at the tunnel junction interface were obtained by exposing the completed p-layer to a pure O_2 plasma for 2 min using a flow rate of 5 sccm, a power of 70 mW/cm^2, and a pressure of 0.09 Torr.

RTSE measurements were performed using a rotating compensator multichannel ellipsometer with a spectral range from 1.4 to 4.5 eV [8]. Spectra were collected during p-layer, interface layer (if used), and n-layer growth with acquisition and repetition times of 1 and 5 s, respectively. Realistic multilayer models were employed to analyze the RTSE data that incorporate the effects of substrate modification, substrate surface roughness filling, nucleation, and growth [3].

RESULTS

The initial experiments performed in this study involved optimization of n-layers on a-Si:H i-layers for single-phase, high-density, n-type μc-Si:H. Both H_2-plasma treated and untreated i-layer substrates were used and n-layers were deposited with a range of R values (50-200), different doping levels (D=0.006, 0.012), and different plasma power levels (P=230, 700 mW/cm^2). Figure 1 shows results for the change in surface roughness layer thickness Δd_s versus bulk layer thickness d_b for three n-layers prepared on H_2-plasma treated a-Si:H i-layers at different R with the other parameters held constant (D=0.006, P=700 mW/cm^2). Because the initial roughness layer thickness depends on the deposition history of the underlying substrate, it is subtracted from the data. For R=50 and 100, the relatively stable surface with gradual smoothening in the initial stages of growth is attributed in this case to conformal coverage of the i-layer by a-Si:H:P. The abrupt roughening transitions after 75 Å for R=50 and 40 Å for R=100 are consistent with the development of nanocrystals which then propagate throughout the film with increasing thickness d_b [9]. The behavior for R=200 is the opposite; in the early stages of growth, d_s gradually increases and then decreases with increasing d_b. This behavior is characteristic of immediate nucleation of nanocrystals on the i-layer. The optical properties of the n-layers obtained in the different stages of growth also support such interpretations of Fig. 1. Figure 2 shows the imaginary parts of the dielectric functions ε_2 for the films of Fig. 1 with R=50 and 200, measured at thicknesses of 130 Å. Both reveal the characteristics of nanocrystalline Si; however, for R=200 the E_1 and E_2 crystalline Si features near 3.4 eV and 4.2 eV are more strongly developed and the amplitude is significantly larger. These characteristics indicate a larger grain size and a higher-density for the R=200 film. To summarize Figs. 1 and 2, it is suggested that R=200 is optimum for μc-Si:H n-layer growth on an a-Si:H i-layer. Using a similar analysis strategy, the power level of 700 mW/cm^2 was found to be optimum, whereas the other deposition parameters explored here, including substrate treatment and doping level, exert weaker influences on the growth process.

Figure 3(a) shows the microstructural evolution including d_b and d_s as a function of time for the growth of the optimum R=200 μc-Si:H n-layer on the H_2-plasma-treated i-layer from Fig. 1. Also shown are the results (b-d) for the growth of three n-layers under identical conditions as in (a), but for different tunnel junction interfaces: (b) 50 Å a-Si:H:P n-layer at the junction interface; (c) 10 Å SiO_2 layer at the interface; and (d) no interface layer (i.e., growth directly on the top of the μc-Si:H p-layer). Analyses of the optical properties show that all n-layers in Fig. 3 grow in the crystalline phase from the start. The results for n-layer growth on the a-Si:H:P interface in the tunnel junction [Fig. 3(b)] show similarities to those for n-layer growth on the i-layer [Fig. 3(a)]. This behavior is expected since both substrate films are amorphous Si. For both depositions of Figs. 3(a-b), nanocrystal nucleation and interface roughness filling occur initially over a period of 2-3 min with d_s

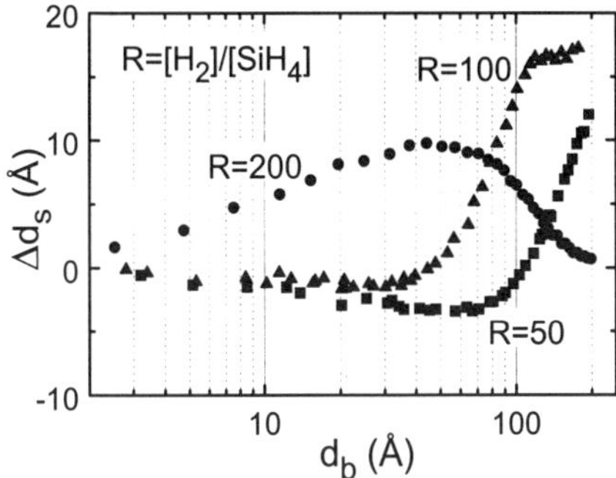

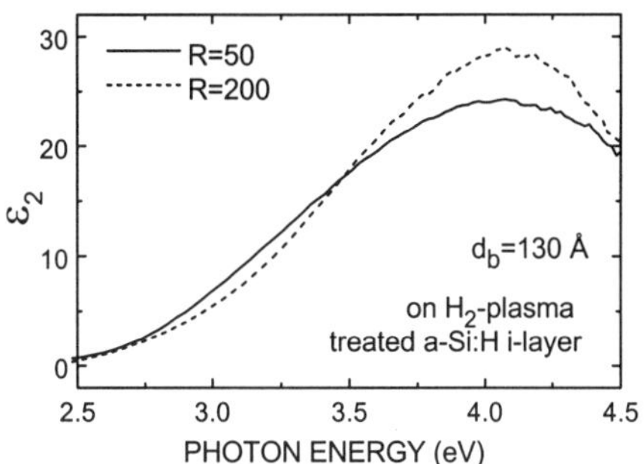

FIG. 1: Change in the surface roughness layer thickness (Δd_s) versus bulk layer thickness (d_b) for n-layers prepared at different R values on H$_2$-plasma treated a-Si:H i-layers. Other n-layer deposition parameters are held constant.

FIG. 2: (above) Imaginary parts of the dielectric functions of the n-layers of Fig. 1 prepared with R=50 (solid line) and R=200 (broken line) on H$_2$-plasma treated a-Si:H i-layers, measured at 130 Å and a temperature of 200°C.

FIG. 3: (right) Microstructural evolution including surface roughness (d_s) and bulk (d_b) layer thicknesses vs. time for the growth of (a) a nanocrystalline n-layer with R=200 on a H$_2$-plasma-treated a-Si:H i-layer. Also shown (b-d) are the results for the growth of three n-layers under identical conditions as (a), but for different tunnel junction structures: (b) 50 Å a-Si:H:P n-layer at the tunnel junction interface; (c) 10 Å SiO$_2$ layer at the interface; and (d) no interface layer (i.e., growth directly on the μc-Si:H p-layer).

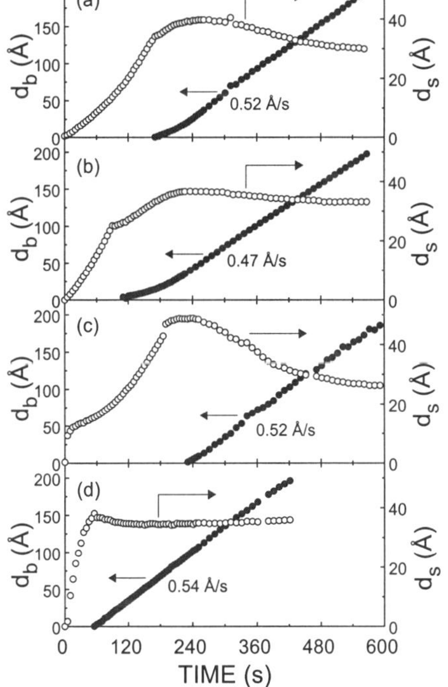

increasing to 40 Å before nuclei coalescence and surface smoothening occur. In both cases, there is an initial slow bulk layer growth period attributed to simultaneous substrate etching. This effect is larger for the untreated a-Si:H:P interface layer in comparison with the H$_2$-plasma treated a-Si:H i-layer. For n-layer growth on the SiO$_2$ interface layer in Fig. 3(c), nanocrystal nucleation and interface roughness filling occur over a longer period (4 min) with d_s increasing to 50 Å before bulk layer growth. In this case, the bulk layer induction period is absent, indicating that the SiO$_2$ layer promotes nanocrystal nucleation while arresting substrate etching. For n-layer growth directly on the p-layer without an interface layer in Fig. 3(d), the structural evolution is characteristic of interface roughness filling *without renucleation*. This is indicated by the constant value of d_s after interface filling (0-1 min), and the very short period before linear bulk film growth occurs (1 min) compared to the two cases (b-c) in which renucleation occurs (3-4 min).

The dielectric functions of these films also exhibit clear trends, as shown by the spectra for ε_2 in Figs. 4 and 5. The n-layers deposited on the H$_2$-plasma treated a-Si:H i-layer [broken line in Fig. 4] and the a-Si:H:P interface layer of the tunnel junction [broken line in Fig. 5] reveal results typical of a high density nanocrystalline structure with sharp absorption onsets at 2.4-2.5 eV [7]. In contrast,

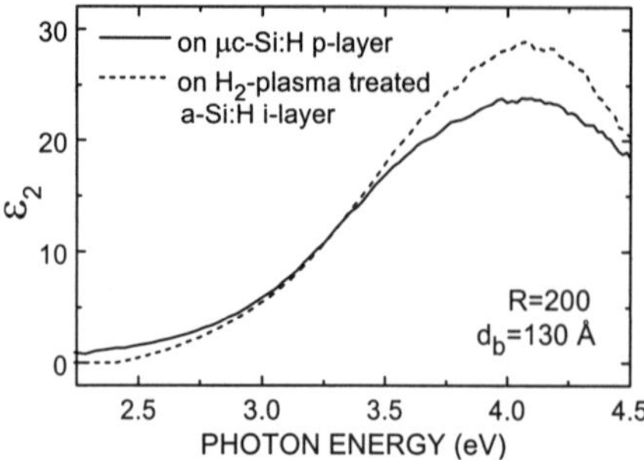

FIG. 4: Imaginary parts of the dielectric functions at 200°C for the n-layers of Figs. 3(a) and (d), including the n-layer deposited on the H_2-plasma treated i-layer (broken line) and the n-layer deposited directly on the μc-Si:H p-layer (solid line). The bulk n-layer thicknesses for these measurements are 130 Å.

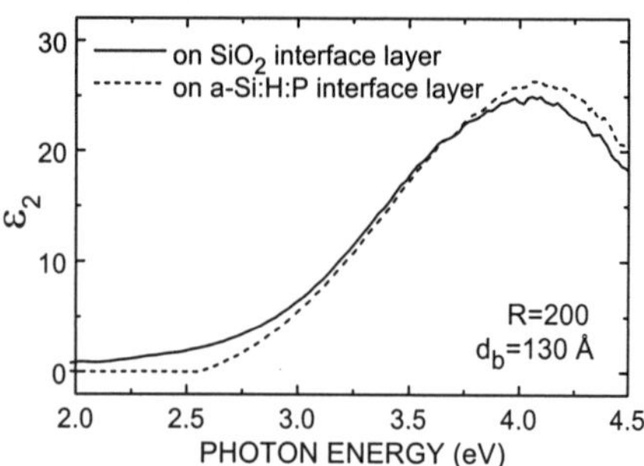

FIG. 5: Imaginary parts of the dielectric functions at 200°C for the n-layers of Figs. 3(b) and (c), including the n-layer deposited on the a-Si:H:P interface layer of the tunnel junction (broken line) and the n-layer deposited on the SiO_2 interface layer (solid line). The bulk n-layer thicknesses for these measurements are 130 Å.

the n-layers deposited directly on the μc-Si:H p-layer [solid line in Fig. 4] and on the SiO_2 interface layer [solid line in Fig. 5] reveal results typical of a lower-density but larger-grained μc-Si:H structure, with an absorption tail increasing gradually with increasing energy throughout the visible as in bulk crystalline Si. We suggest that the ~10 vol.% higher void fraction for the n-layer deposited directly on the p-layer compared with the n-layer deposited on the H_2-plasma treated i-layer (see Fig. 4) is related to the development of the larger grains in the n-layer continuously from grains in the p-layer. The etching that occurs at R=200 leaves significant void structures in the grain boundary regions of this larger-grained film. The voids can be eliminated by reducing R. Although a value of R=50 for the n-layer leads to an initial 75 Å amorphous layer when deposition occurs on the H_2-plasma treated i-layer, this lower dilution level also leads to continuous grain growth across the tunnel junction when deposition occurs directly on the μc-Si:H p-layer (without an interface layer). The structural evolution for R=50 in this latter case is almost identical to that for R=200 in Fig. 3(d), except that the interface filling period is now one-half as long (~0.5 min), and the bulk layer growth rate is more than twice as fast (1.25 Å/s). Figure 6 shows the improvement in ε_2 that occurs from R=200 to R=50 for the two n-layers deposited directly on the μc-Si:H p-layers. These results should be compared with those in Fig. 2, in which the substrate is a H_2-plasma treated a-Si:H i-layer, and the reverse trend occurs from R=200 to R=50. Based on the results in Figs. 2 and 6, we conclude that optimization studies of μc-Si:H layers for tunnel junction components must be made in the tunnel junction configuration. These results also lead to a general suggestion that although very high R is required for optimum nucleation of nanocrystals on an a-Si:H i-layer, once these nanocrystals form, a much lower R value can be used to propagate a higher-density, larger-grained μc-Si:H structure at higher rates.

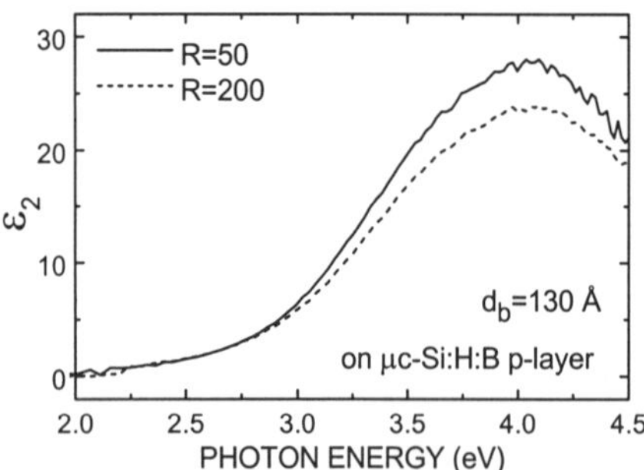

FIG. 6 Imaginary parts of the dielectric functions of n-type μc-Si:H films prepared with R=50 (solid line) and 200 (broken line) directly on μc-Si:H p-layers, measured at a thickness of 130 Å and a sample temperature of 200°C.

DISCUSSION

A key issue raised in this article is the substrate dependence that must be considered in the optimization of thin layers, particularly when high H_2-dilution is used. In order to obtain insights into such effects, the evolutionary phase diagram for the growth of a-Si:H and μc-Si:H versus bulk layer thickness d_b and H_2-dilution ratio R is presented in Fig. 7. This phase diagram has been applied earlier to demonstrate that optimum a-Si:H i-layers for solar cells are obtained under deposition condition as close as possible to the boundary, but on the amorphous side *versus thickness* [9]. In Fig. 7, three different phase boundaries are shown appropriate for untreated a-Si:H (R=0) substrate films (solid line), native-oxide (~22 Å) covered c-Si substrates (dashed line), and optimum 200 Å μc-Si:H p-layer substrate films (dotted line). The Si film deposition conditions in all cases include T=200°C, no doping (D=0), and low plasma power (70 mW/cm^2). Figure 7 shows that for an a-Si:H substrate and for R=50, 75 Å of a-Si:H develops before crystallites nucleate. In fact, R must be set to a value well above 100 to obtain immediate nucleation of crystallites. In contrast, for growth on the native SiO_2 of c-Si (which remains intact in the growth process), immediate nucleation can occur for R=40 (open circle with arrow). For growth on a μc-Si:H p-layer substrate film, μc-Si:H propagates even for R as low as 10 and for thicknesses as high as 3000 Å (open square). For these latter substrate films further efforts must be made to identify the exact position of the phase boundary. The arrows on the vertical dotted line in Fig. 7 indicate that this boundary lies between R=0 and R=10. In this case, the phase boundary may be inverted, i.e., show a positive slope rather than a negative slope. This would occur if the μc-Si:H substrate forced local epitaxy of Si even at very low R (e.g., R~5), but with a transition *to a-Si:H* with increasing thickness [10]. Generally, the phase boundary behavior of Fig. 7 can explain the fact that the μc-Si:H n-layer depositions directly on the μc-Si:H p-layers and on the oxidized μc-Si:H p-layers achieve larger grain sizes than an identical deposition on the a-Si:H:P interface layer as shown in Figs. 4 and 5 (note the differences in the low energy tails of ε_2).

Figure 8 shows the corresponding evolutionary phase diagram for n-type Si films prepared on both H_2-plasma treated (solid points on solid line phase boundary) and untreated a-Si:H i-layers (open points). Conditions for growth include T=200°C, D=0.006-0.012, and high power (700 mW/cm^2). In this case, there is no significant effect of the H_2-plasma treatment of the a-Si:H i-layer substrate in shifting the phase boundary. In addition, the phase boundary for the n-layers with 50≤R<200 is similar to that for undoped Si depositions at low power (compare Figs. 7 and 8). It should be noted that in contrast to n-type doping with PH_3, p-type doping with $B(CH_3)_3$ generates a large shift in the phase boundary to higher R such that a film prepared with R=200 on an untreated a-Si:H i-layer cannot initially nucleate in the form of a crystalline film and grows as an amorphous film to a thickness of 120 Å (open square in Fig. 8). The lower limit of the phase boundary for p-layers doped using $B(CH_3)_3$ is shown in Fig. 8 as the dotted line. By performing a H_2-plasma treatment of the underlying a-Si:H i-layer, however, the phase boundary for these p-layers can be shifted to much lower R (close to the boundary for the n-layer), apparently owing to the role of the H_2-plasma in

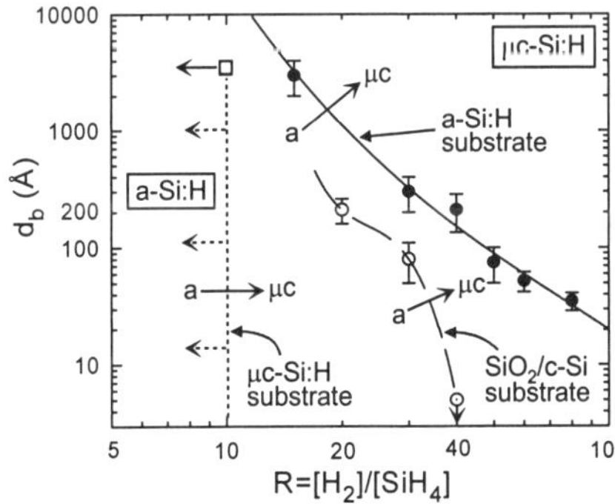

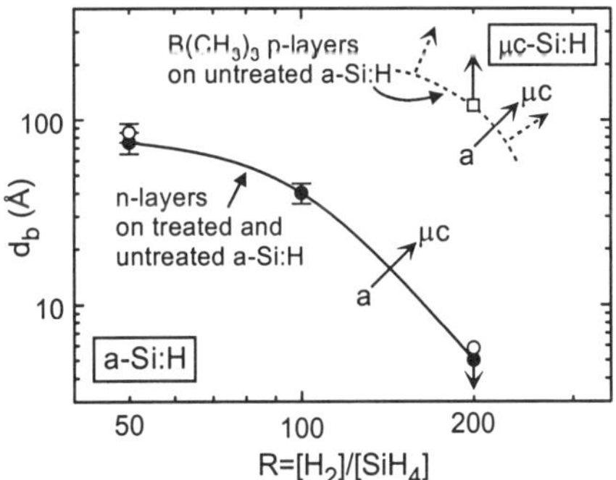

FIG. 7: Evolutionary phase diagram for the growth of a-Si:H and μc-Si:H versus bulk layer thickness d_b and H_2-dilution ratio R. Three different phase boundaries are shown, appropriate for untreated a-Si:H substrates (solid line and filled points), native-oxide covered c-Si substrates (dashed line and open points), and 200 Å μc-Si:H p-layer substrate films (dotted line and square).

FIG. 8 Evolutionary phase diagram for n-type Si films prepared using PH_3 doping gas on both H_2-plasma treated (solid points on solid line phase boundary) and untreated (open points) a-Si:H i-layers. Also shown is the estimated lower limit of the phase boundary for p-layers prepared from $B(CH_3)_3$ on untreated a-Si:H i-layers (open square, dotted line).

generating nanocrystals at the i-layer surface [7]. For the n-layers, direct deposition on the μc-Si:H p-layers also leads to a boundary at very low R for a similar reason; namely, once the crystallites form, the high value of R is not needed to sustain their growth. This is shown most clearly in the phase diagram in Fig. 7 in which μc-Si:H growth at R=10 can be sustained for a μc-Si:H p-layer substrate film.

SUMMARY

The preparation of fully μc-Si:H p/n tunnel junctions for multijunction n-i-p solar cells has been studied using RTSE. In the initial studies, p and n layer recipes have been used based on optimized depositions with $R=[H_2]/[SiH_4]=200$ on a-Si:H i-layer substrates (e.g., those used for single junction n-i-p and p-i-n solar cells). For the μc-Si:H n-layer depositions on μc-Si:H p-layers, the role of different interface layers has been studied. We have found that for a thin a-Si:H:P n-type interface layer at the tunnel junction, n-layer growth follows a very similar pattern to that on an a-Si:H i-layer; namely, crystallite nucleation is immediate and leads to a single-phase, high-density nanocrystalline n-layer after 100 Å. In contrast for an ultrathin SiO_2 interface layer, nucleation occurs at somewhat lower density and leads to a larger-grained μc-Si:H n-layer with a higher volume fraction of voids. For n-layer deposition directly on the μc-Si:H p-layer without an interface layer, renucleation does not occur and grain growth is continuous across the interface. The resulting larger grain size in this latter case is also accompanied by a larger void volume fraction. Surprisingly, the final microstructure of the n-layer is similar for no interface and for the ultrathin SiO_2 interface layer. This observation is a testament to the ability of ultrathin (10 Å) SiO_2 layers to nucleate large grain-size μc-Si:H immediately. For the n-layer deposited on the p-layer to form a tunnel junction with no discontinuity at the interface, improved microstructure can be obtained by deviating from the conditions found to be best for growth on a-Si:H i-layers. It is found that when the H_2-dilution ratio is decreased from R=200 (which is found to be optimum for initial growth on an a-Si:H i-layer) to R=50, μc-Si:H growth from the p to the n-layer remains continuous, but the resulting n-layer shows a large grain size *and* a dense structure. Presumably this behavior is due to a reduction in the etching that occurs at high R. These results for p/n tunnel junctions in turn suggest a new approach for the deposition of improved single μc-Si:H doped layers on i-layers. In this approach the value of R is reduced after 50 Å has been deposited. In this way, the deposition process is maintained closer to the a-Si:H$\rightarrow\mu$c-Si:H phase boundary, but on the μc side versus thickness, in order to avoid excessive etching that occurs at higher R and to maintain higher deposition rates.

ACKNOWLEDGEMENTS

This research was supported by National Science Foundation Grant No. DMR-9820170 and by National Renewable Energy Laboratory Subcontract No. XAF-8-17619-22.

REFERENCES

[1] S. Guha, J. Yang, P. Nath, and M. Hack, Appl. Phys. Lett. **49**, 218 (1986).
[2] J.K. Rath, F. Rubinelli, and R.E.I. Schropp, J. Non-Cryst. Solids **227-230**, 1282 (1998).
[3] J. Koh, H. Fujiwara, C.R. Wronski, and R.W. Collins, Sol. Energy Mater. Sol. Cells **49**, 135 (1997).
[4] P. Pernet, M. Goetz, H. Keppner, and A. Shah, Mater. Res. Soc. Symp. Proc. **452**, 889 (1997).
[5] L.L. Smith, W.W. Read, C.S. Yang, E. Srinivasan, C.H. Courtney, H.H. Lamb, and G.N. Parsons, J. Vac. Sci. Technol. A **16**, 1316 (1998).
[6] P. Roca i Cabarrocas and S. Hamma, Thin Solid Films **337**, 23 (1999).
[7] J. Koh, H. Fujiwara, R. Koval, C.R. Wronski, and R.W. Collins, J. Appl. Phys. (in press, April 1999).
[8] J. Lee, P.I. Rovira, I. An, and R.W. Collins, Rev. Sci. Instrum. **69**, 1800 (1998).
[9] J. Koh, Y. Lee, H. Fujiwara, C.R. Wronski, and R.W. Collins, Appl. Phys. Lett. **73**, 1526 (1998).
[10] C.C. Tsai, in *Amorphous Silicon and Related Materials*, Vol. A, edited by H. Fritzsche, (World Scientific, Singapore, 1989) p. 123.

PLASMA ETCHING CONDITIONING OF TEXTURED CRYSTALLINE SILICON SURFACES FOR a-Si/c-Si HETEROJUNCTIONS

R. DE ROSA, M.L. ADDONIZIO, E. CHIACCHIO, F. ROCA, M.TUCCI.
ENEA - Research Centre, Località Granatello-80055 Portici (Na), Italy

ABSTRACT

The development of a hybrid heterojunction fabricated by growing ultrathin amorphous silicon by Plasma Enhanced Chemical Vapor Deposition using temperatures below 250°C offers the potential of obtaining high efficiency solar cells deposited on glassy substrates. The surface preparation represents one of the most critical steps. The first aim of etching is to remove the native oxide layer from the surface of the crystalline wafer, before amorphous layer deposition. The possibility of obtaining this goal with a dry procedure that reduces the exposure of the sample to the environment is not trivial.

We performed several dry etching processes but the best results were obtained using an etching process involving CF_4/O_2 gases. We have found evidence that plasma etching acts by removing the native oxide and the damaged surface of textured silicon and by leaving an active layer on silicon surface suitable for the emitter deposition. SEM analysis has confirmed that it is possible to find plasma process conditions where no appreciable damage and changes in surface morphology are induced. Detailed investigation was performed to find compatibility and optimization of amorphous layer deposition both on flat and textured cast silicon by changing the plasma process parameters. By using this process we achieved on cast silicon for solar applications photovoltaic conversion efficiencies of 12.9% on 51 cm^2 and 9.2% on 45 cm^2 active areas for amorphous crystalline heterostructure devices realized on monocrystalline and polycrystalline silicon respectively. We also investigated the compatibility of the process with industrial production of large area devices.

INTRODUCTION

The realization of electronic devices based on c-Si heterojunctions using amorphous silicon grown by Plasma Enhanced Chemical Vapor Deposition (PECVD) appears to be a very attractive and low cost technology, especially for large area applications such as solar cells [1,2]. In fact, due to the low temperature processes used during deposition of amorphous silicon, this junction forming technique can be efficiently applied also on substrates that degrade at high temperature, such as glass [3]. Moreover, due to the very low thickness of the amorphous silicon layers, these heterostructures do not show any light induced degradation as in the case of purely a-Si solar cells. Thus they can achieve high and stable conversion efficiencies.

The fabrication process of an a-Si/c-Si heterojunction would seem very easy to perform using the well developed techniques for the realization of crystalline and amorphous silicon solar cells. Nevertheless particular care is needed in some critical steps of the process. The surface of the substrate must be regular enough to allow conformal deposition of the thin a-Si layer in order to avoid shorts between the substrate and other layers. Any high temperature step must be avoided after the deposition of amorphous silicon in order to preserve the junction integrity.

One of the most critical steps in fabrication of the a-Si/c-Si heterostructure is the preparation of the crystalline silicon surface before the amorphous silicon deposition. When the

Mat. Res. Soc. Symp. Proc. Vol. 557 © 1999 Materials Research Society

a-Si layer is deposited on c-Si, without any treatment before the deposition, the correct rectifying behavior of the device is not produced due to the presence of an insulating silicon oxide layer on the surface of the wafers. A very effective method to reduce this SiO_2 layer is in a bath in HF:DW, (1-5):100 for few minutes [4]. Other wet etching methods are available: in particular a very useful treatment for textured cast silicon is a BHF bath with the following recipe: $HF:HNO_3:DW$ at 1.5:1:30. This bath seems effective in removing not only the oxide layer but also the surface damage due to texture formed in previous baths. In any case, the wet treatment does not represent a good choice, especially for transfer to the industrial production. In production dry etching treatment is preferable.

In this work. we experimented with the applicability of a dry etching process based on a CF_4/O_2 gas mixture to prepare the crystalline silicon surface for the plasma deposition of amorphous silicon. We studied this plasma etching on several substrates, including textured cast polysilicon, in order to ensure application to standard solar cell technology. In particular we analyzed the effect of the CF_4/O_2 plasma on the silicon surface, as well as the behavior of the a-Si/c-Si heterojunction after different plasma etching treatments. Good rectifying behavior as well as an encouraging conversion efficiency were found for the solar cell devices made on conditioned crystalline silicon surface.

EXPERIMENT

The CF_4/O_2 plasma etching was performed in a commercial 13.56 MHz capacitive coupled glow discharge system with single UHV chamber. We used the same system for PECVD of amorphous silicon. The etching and junction properties were tested both on flat <100> oriented Czochralski silicon wafers with p type boron doped, 0.7-1.3 Ω·cm resistivity, and on textured cast silicon with the same orientation and doping. We also investigated the effect of this treatment on cast polysilicon.

For plasma etching we used a CF_4/O_2 mixture with 8% oxygen content, using the following process parameters: 200 mTorr working pressure, 30 sccm gas flow, 98 mW/cm^2 glow discharge power density, and temperatures in the range 40 - 320 °C. We first measured the etching rate of CF_4/O_2 on flat crystalline silicon at fixed temperature, and in particular we found an etching rate of 200 nm/min at 240 °C. Then we applied this treatment on textured cast silicon before heterojunction formation with different etching times before the emitter deposition. Reflectance measurements using a Perkin-Elmer Lambda 9 spectrophotometer in the range of 300-1100 nm and SEM analysis of the etched surface were done to investigate the optical and morphological properties of the wafer.

A 30 nm n$^+$ layer of amorphous silicon was deposited using a 200 mTorr working pressure, 320°C substrate temperature, 20 sccm SiH_4 and 10 sccm PH_3 gas flows, and 8.4 W/cm^2 power density. A grid shaped silver contact was used as the front electrode of the solar cell. No ITO collecting layer was applied to avoid possible contamination. The back contact was ensured by a screen printed deposition of an aluminum silver alloy. An annealing procedure for 10 min at 250°C was performed before the measurements [5]. Photovoltaic characteristics, under standard AM1.5 conditions using a Spectrolab solar simulator, were performed to extract information about the status of the heterojunction after the different etching time.

By using the cleaning procedure described here we were able to fabricate large area heterojunction solar cells at low temperature using steps compatible with industrial production. In this case we realized the front contact of the devices with a collecting window consisting of a 90 nm Indium Tin Oxide layer and a screen printed silver grid optimized for low temperature conditioning. Preliminary results and applications of this cleaning technique for heterojunctions between amorphous silicon and cast polycrystalline textured silicon are finally reported.

RESULTS AND DISCUSSION

In the initial stages the effect of etching on the textured crystalline silicon surface was monitored by reflectance measurements. In particular, we performed direct and diffuse reflectance measurements on samples treated from 1 min up to 20 min, and we found that the diffuse component represents the major contribution to the total reflectance because of the pyramidal shape of the surface texture. In Fig. 1 the linear behavior of the average diffuse reflectance in the range of 300-1100 nm as a function of etching time is reported. The increasing values suggest an anisotropic effect of the etching process, since a change in morphology of the surface occurs.

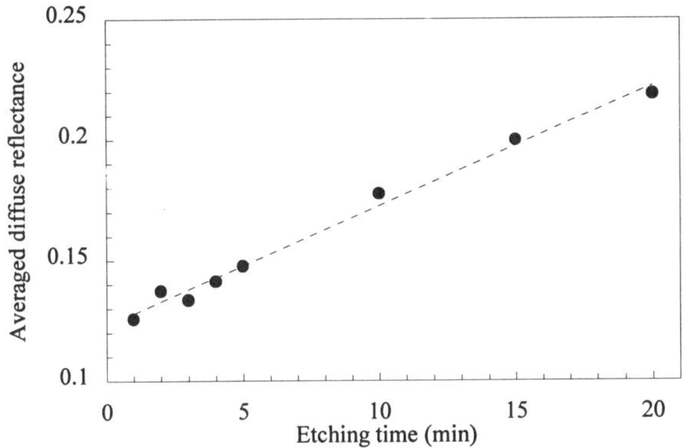

Fig. 1: Averaged diffuse reflectance in the range of 300-1100 nm as a linear function of the etching time in the CF_4/O_2 gas mixture.

SEM analysis performed on c-Si textured samples confirmed the previous results, showing that the incoming flux of CF_4/O_2 plasma species act only on the lateral faces, and not on the peaks and the valleys of the pyramids. We observed a change in the shape of the textured profile. The effect increases with time, and after 20 min a planar area within the valley and sharper tip on the top of the pyramids start to appear. In Figs. 2a,b,c and d the effects of increasing the time of etching on the surface of the pyramids after 1, 5, 10 and 20 minutes are depicted respectively. One can also can note an increase in hole formation on the pyramidal faces due to the presence of impurities that enhance the erosion process. This explains the observed decrease of photovoltaic parameters in the case of high etching time as reported in Figs. 3 a, b, c and d. The flatness in the valley increases the reflectance of the surface, and the sharpening of the peak produces electrical shorts through the amorphous silicon leading to a reduction in both V_{OC} and J_{SC}. Moreover for shorter etching times a decrease in fill factor occurs due to carrier recombination via defects that form at the interface between the amorphous and crystalline materials as reported in Fig 3d. Those defects can be related both to a residual oxide layer at the interface and to a damaged silicon surface. During the etching process, two effect must be taken into account. The first effect is the native oxide removal and the second is the silicon oxidation and removal, the latter leaving a thinner oxide layer on the surface than the native one [6]. Then a trade off is needed in the choice of etching time to obtain the best conditioning of the surface. In fact, we seek to reduce the oxide thickness and remove the damaged silicon wafer surface present after the cutting and texturing steps, but we also need to preserve the low reflectance of the surface. A reduction in defect formation can avoid a pinning of the Fermi level at the crystalline silicon surface in the midgap position corresponding to a maximum of the recombination rate of the photogenerated carriers. Then increasing depletion in the crystalline region after junction formation is expected [7].

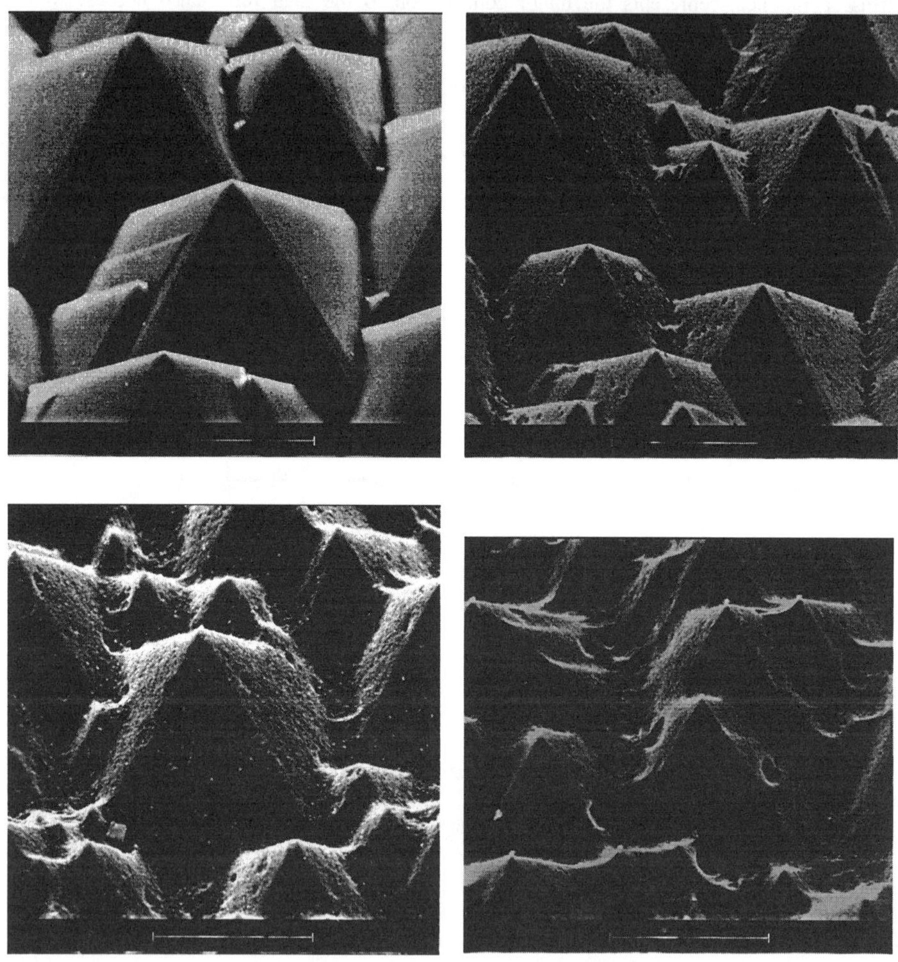

Fig. 2: SEM pictures of the textured silicon surface respectively after (a) 1min, (b) 5min, (c) 10min, (d) 20min etching time in CF_4/O_2 gas mixture.

Fixing the etching time at 3 min for this cleaning procedure, we obtain a good compromise between the short circuit current and open circuit voltage of the solar cells.

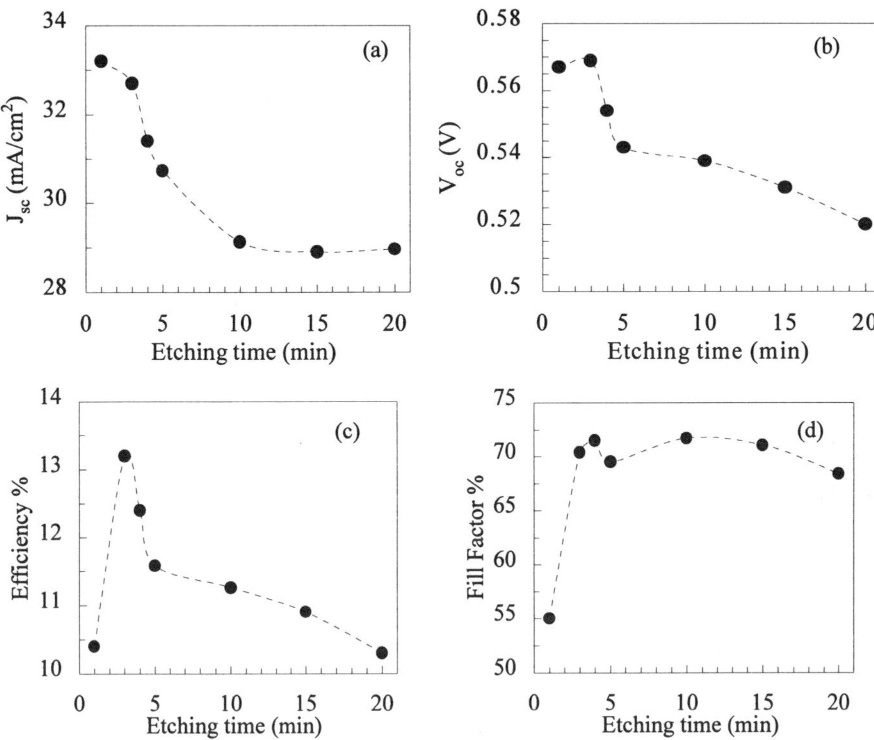

Fig. 3: (a) Short circuit current, (b) Open circuit voltage, (c) Efficiency and (d) Fill Factor vs etching time in a CF_4/O_2 gas mixture.

In Fig. 4 a comparison between different heterostructure solar cell samples fabricated using this cleaning procedure is reported. In particular Fig. 4 shows the internal quantum efficiency of two large area heterojunctions. One is grown on a monocrystalline textured substrate, and the other on cast textured polysilicon. In Table I the photovoltaic parameters, the active area and the crystalline material used in the samples are summarized. Optimization of the grid design is needed to increase the conversion efficiency.

Table I. Silicon substrate, active area and photovoltaic parameters of the samples reported in Fig. 4.

Si Substrate	active area (cm^2)	V_{OC} (mV)	J_{SC} (mA/cm^2)	FF %	Eff %
monocrystalline	51	591	34.79	62.58	12.9
polycrystalline	45	554	31.54	52.67	9.2

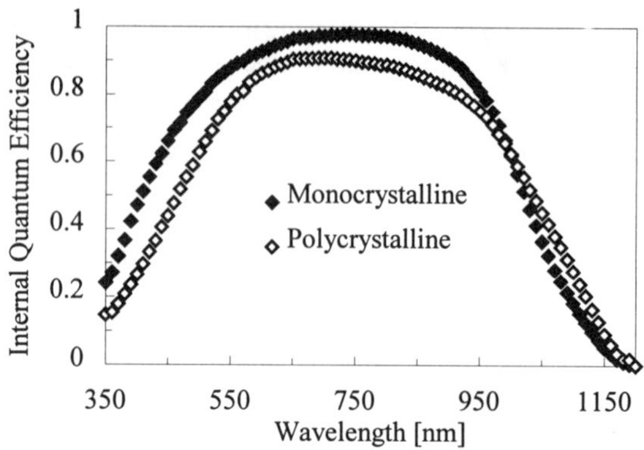

Fig. 4. Comparison of the Internal Quantum Efficiency of heterostructure solar cells grown on monocrystalline and polycrystalline silicon.

CONCLUSIONS

In this work, we examined the possibility of obtaining all low temperature steps for a-Si/c-Si heterojunction solar cell fabrication easily applicable to industrial processes. A dry cleaning procedure is proposed to reduce the effect of the native oxide layer and remove damage from the silicon surface. With this technique, it is possible to achieve high quality heterojunctions between amorphous and cast textured crystalline silicon and polysilicon. We found that care is needed in the etching process in particular with regards to the choice of the etching time in order to optimize the photovoltaic properties of the heterojunctions. SEM analysis and reflectance measurements of the treated surface are reported for evaluation of the behavior of the cleaning procedure. Finally we suggested a 3 minute plasma treatment in a CF_4/O_2 gas mixture. Encouraging values of the photovoltaic parameters were found especially in the case of large area devices.

REFERENCES

1. K. Tanaka, M. Taguchi, T. Takahama, T. Sawada, S. Kuroda, T. Matsuyama, S. Tsuda, A. Takeoka, S. Nakano, H. Hanafusa, Y. Kuwano, Progress in Photovoltaics: Research and Applications, 1, p. 85 (1993).

2. H. Matsuura and H. Okushi, in "Amorphous and Microcrystalline Semiconductor Devices", Jerzy Kanicki Editor, 2, p. 517 (1992).

3. F. Roca, R. De Rosa, M.L. Grilli, G. Sasikala, M. Tucci, *Solid State Phenomena*. Scitec Publications Ltd, Zug, (1998) in press.

4. S. Higashi, Y. J. Chabal, G. W. Trucks and Krishnan Raghavachari, *Appl. Phys. Lett.* 56, p. 656 (1990).

5. R. De Rosa, P. Grillo, G. Sinno, F. Roca, M.Tucci, 2nd *World Conf. Exhib. On Photovol. Solar Eenergy Convers Proc.* 2, p.1583 (1998).

6. T. Shirafuji, W.W. Stoffels, H. Moriguchi, K. Tachibana, *J. Vac Sci Tech.* 15, p. 209 (1996).

7. M.Tucci *Solar Energy Materials and Solar Cells,* 57, p. 249 (1999).

ROLE OF BANDGAP GRADING FOR THE PERFORMANCE OF MICROCRYSTALLINE SILICON GERMANIUM SOLAR CELLS

M. KRAUSE, R. CARIUS, H. STIEBIG, F. FINGER, D. LUNDSZIEN, H. WAGNER
Forschungszentrum Jülich GmbH, ISI-PV, 52425 Jülich, Germany, m.krause@fz-juelich.de

ABSTRACT

The optical absorption of microcrystalline silicon germanium alloys (μc-Si$_{1-x}$Ge$_x$:H) increases in the near infrared region with increasing germanium content. Therefore, these alloys are promising candidates for the application as intrinsic absorber material in thin film solar cells with tandem and triple structure or IR-detectors. The material properties for a germanium content (x) up to 0.6 and the performance of solar cells based on this material were investigated. Graded bandgap structures are used to optimize the cell performance. For cells with x > 0.3 a continuously graded bandgap in the rear 20 nm of the i-layer (at the i/n interface) results in an enhancement of the open circuit voltage by 40-80 mV compared to an abrupt bandgap discontinuity. When the design of the p/i interface region is changed in a similar way no effect on V_{oc} is observed.

INTRODUCTION

For stacked thin film solar cells and infrared sensors low gap materials are required. Due to the low absorption of μc-Si in the energy range of interest a thickness of some μm is needed and with regard to the low deposition rates long deposition times are required. To overcome these restrictions μc-Si$_{1-x}$Ge$_x$:H alloys prepared by very high frequency PECVD were investigated as an alternative material with high absorption in the long wavelength region. While microcrystalline silicon germanium alloys have already been prepared during the development of amorphous SiGe alloys [1], only few papers reported about the employment of such films as absorption layers in devices [2,3].

We present a study on very thin μc-Si$_{1-x}$Ge$_x$:H pin-diodes with different bandgaps and different interface designs. Since bandgap discontinuities at the interfaces are discussed as possible explanation for the deterioration of the performance [3], the influence of the bandgap gradings is examined in detail.

EXPERIMENT

The μc-Si$_{1-x}$Ge$_x$:H films and solar cells were deposited in a multi-chamber PECVD system at 200°C substrate temperature, a plasma excitation frequency of 95.5 MHz, a pressure of 200 mTorr and a power density of 60 mW/cm^2. To obtain microcrystalline growth of the samples we diluted the mixture of disilane Si$_2$H$_6$ and germane GeH$_4$ in H$_2$ with a ratio between 1:80 and 1:320. Deposition rates between 0.4 and 1.2 Å/s were achieved. The optical absorption coefficient α(E) was determined from PDS and simultaneous transmission measurements, which yields interference free absorption values [4]. The structural properties of the samples are characterized by Raman spectroscopy in a near-back scattering configuration at $\lambda_0 = 488$ nm.

The solar cells were prepared on glass coated with textured TCO (AsahiU type) as stacks glass/TCO/μc-Si:H (p)/ μc-Si$_{1-x}$Ge$_x$:H/ a-Si:H (n) /Ag. The total thickness of all i-layers was 150 nm. Bandgap gradings in the interface regions of 20 nm were achieved by variation of the [GeH$_4$]/[Si$_2$H$_6$] ratio keeping the hydrogen dilution fixed. After preparation the cells were

Mat. Res. Soc. Symp. Proc. Vol. 557 © 1999 Materials Research Society

tempered for half an hour at 150°C and 175°C. The open circuit voltage (V_{oc}), short circuit current (I_{sc}), fill factor (FF) and efficiency (η) were measured under AM1.5 light illumination. The quantum efficiency (QE) was determined by DSR-measurements under short circuit conditions.

MATERIAL PROPERTIES

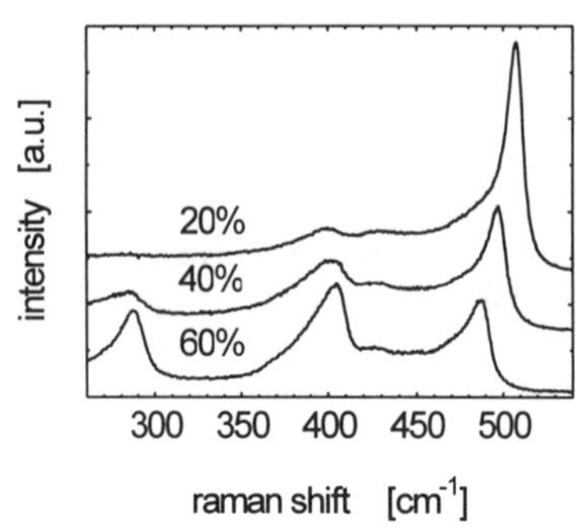

Fig. 1: Raman spectra of μc-SiGe:H with indicated germane content.

In Fig. 1 Raman spectra of μc-Si$_{1-x}$Ge$_x$ alloys deposited with different germane content are plotted. Three strong peaks are observed which are attributed to Si-Si (~500 cm^{-1}), Si-Ge (~400 cm^{-1}), Ge-Ge (~300 cm^{-1}) optical modes according to [5]. The variation of the width of the peaks can either be caused by differences in disorder, differences in strain distribution or by size effects of small crystallites. The Raman spectra show no signal of an amorphous phase. Moreover, since no peak at 520 cm^{-1} is detected no indication of silicon clusters is found in the investigated samples. Because a linear relation between the shift of the position of the Si-Si mode and the germanium content in the solid phase is reported for (poly-)crystalline SiGe [6,7] and μc-SiGe:H [3], we used the Si-Si peak positions for the evaluation of the composition (Tab. 1). The content in the solid phase is smaller than the content of GeH$_4$ and decreases with higher germanium content. No influence of the H$_2$-dilution on the Raman peak position in the region between 1:120 and 1:320 has been found.

Tab. 1: Germanium content in the solid phase determined from Raman spectroscopy for different germane content in the gas phase.

Germane content in the gas phase	20%	40%	60%	70%
Germanium content x in the solid phase	0.21	0.37	0.51	0.56

Optical absorption spectra derived by PDS spectra are shown in Fig. 2. An increase of the absorption is observed over the whole region of energy with rising germanium content. Since a plot of $\sqrt{(\alpha)}$ as a function of E does not always yield straight lines over a significant energy range the gap is defined as $\alpha(E_g) = 10$ cm^{-1} which is close to the indirect bandgap [3]. The evaluation of the optical gap is further complicated by an enhanced absorption below the gap. Similar subgap absorption is found in μc-Si:H, but its origin is not yet clear. Defect states, strain and potential fluctuations are likely causes for this absorption. Neglecting these deviations we can evaluate a bandgap of 0.95 eV for the sample prepared with 40% germane in

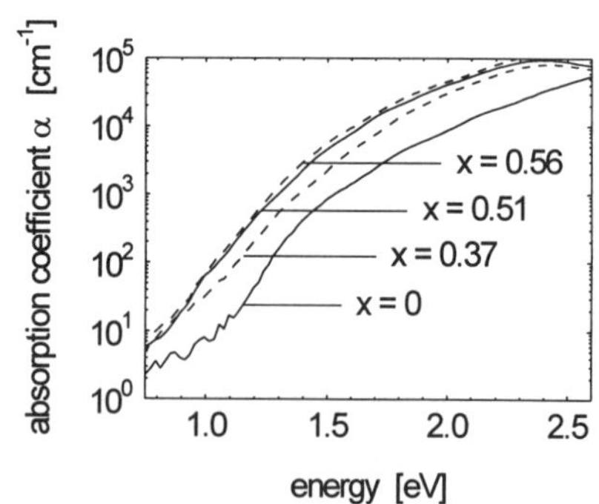

Fig. 2: Optical absorption spectra for different composition.

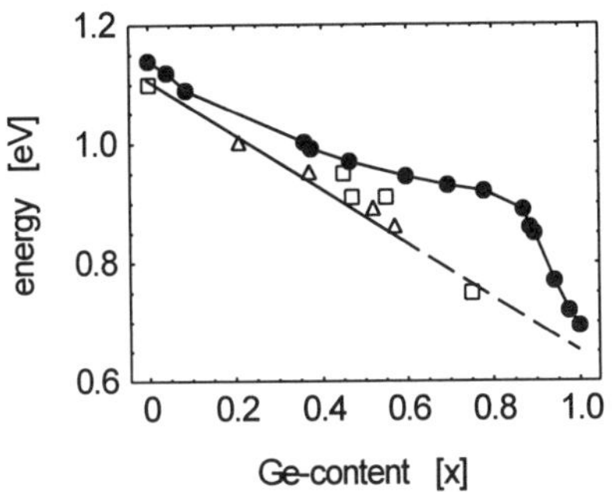

Fig. 3: Optical gap as a function of germanium content.

the gas phase and a bandgap of 0.89 eV for the 60% germane films. When the H_2 dilution is enhanced for samples with a constant germanium content the deviation of the absorption spectra from a square root behavior near the bandgap increases which we attribute to an increase of the internal stress. With decreasing H_2 dilution an increase of the absorption coefficient at higher energies is observed caused by a contribution of an amorphous phase.

Taking into account the evaluated data from PDS and Raman measurements, the bandgap of the prepared samples is determined as a function of the germanium content. A comparison of the bandgap of $\mu c\text{-}Si_{1-x}Ge_x$:H using disilane ($\triangle$) and monosilane ($\square$) [3] with literature data of polycrystalline silicon germanium ($\bullet$) determined by photoluminescence measurements [8] is shown in fig. 3. The relation between the energy gap and the germanium content prepared by a PECVD process does not follow the behavior of (poly-)crystalline SiGe and exhibits an almost straight line. For the investigated samples no indication of a distinct transition from the indirect bandgap of the Si-like [100] symmetry to the indirect bandgap of a Ge-like [111] symmetry known from crystalline materials [8] is found.

$\mu c\text{-}Si_{1-x}Ge_x$:H-SOLAR CELL PARAMETERS AS FUNCTION OF GERMANE CONTENT

In a next step a series of solar cells with different germanium content in the absorption layer was prepared. The energy gap of the i-layer was varied by changing the ratio of the process gases $[GeH_4]/[Si_2H_6]$. Since the transition region from amorphous to microcrystalline growth is shifted towards higher hydrogen dilution with increasing germane content, the dilution of the deposition gases in H_2 is enhanced from 1:160 for a germane content lower than 60% to 1:320 for higher germane content to ensure microcrystalline growth. In Fig. 4 the cell parameters V_{oc}, I_{sc} and FF are plotted as functions of the germane content in the gas phase. For a μc-Si solar cell a V_{oc} of 501 mV, a I_{sc} of 9.07 mA/cm^2 and a FF of 70% are measured. With increasing GeH_4

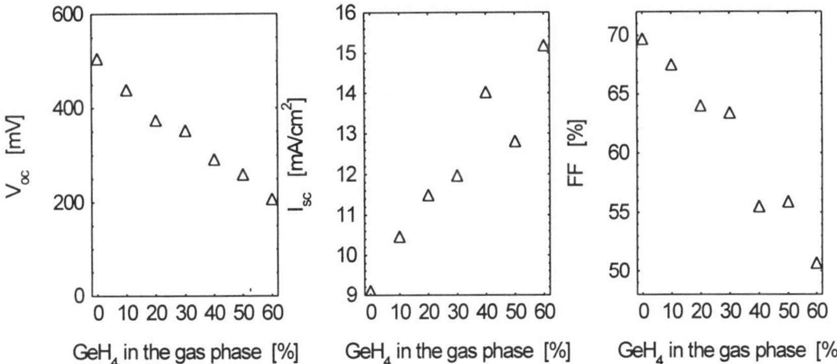

Fig. 4: Cell parameters V_{oc}, I_{sc} and FF as a function of the germane content in the gas phase.

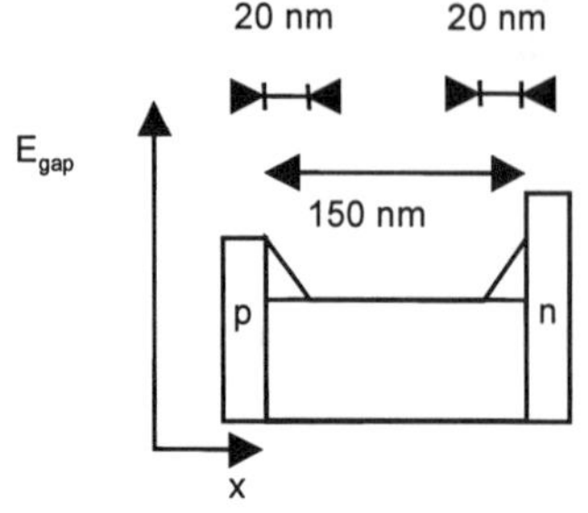

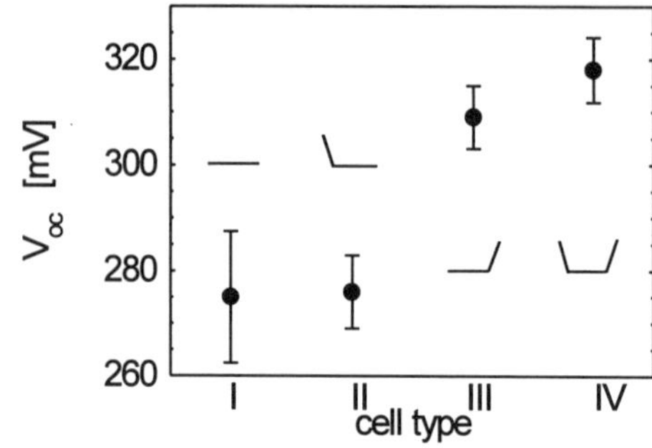

Fig. 5: Scheme of the graded cell structure.

Fig. 6: V_{oc} as a function of the cell types for a germane content of 40%.

content in the gas phase V_{oc} reduces almost linearly to 200 mV for the diode deposited with a germane content of 60%. However, the decrease of the V_{oc} is larger than expected from the variation of the optical gap that declines from 1.1 eV (0% GeH_4) to 0.89 eV (60% GeH_4). The fill factor also decreases with increasing germanium content caused by the poorer electronic quality of the material, a reduced built-in potential and the influence of the bandgap discontinuities on the carrier collection. The short circuit current increases with increasing germane content due to the higher absorption. This behavior will be discussed in the following.

OPTIMIZATION BY BANDGAP ENGINEERING

In order to investigate the effect of bandgap discontinuities at the p/i and the i/n interfaces µSiGe solar cells were prepared with a germane content of 40% and different gradings. The gradings were achieved by changing the $[GeH_4]/[Si_2H_6]$ ratio within a region of 20 nm. Figure 5. shows schematically the examined cell types. Type I denotes the cell without any gradings at the interfaces. Type II has a grading at the p/i interface and type III at the i/n interface. Type IV is characterized by gradings at both junctions. The V_{oc} values of the cell types are plotted in Fig. 6. First, a grading at the i/n interface

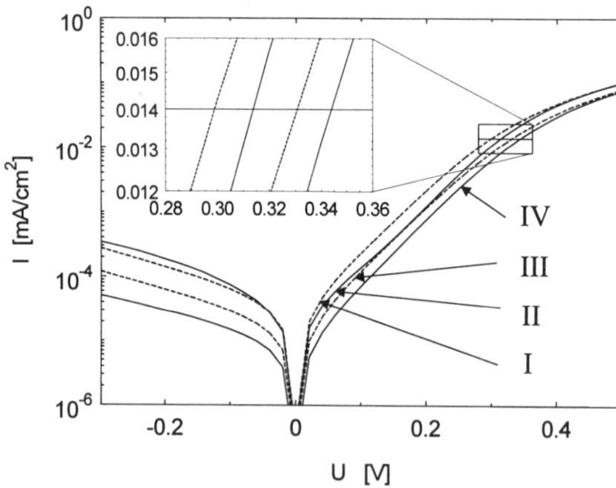

Fig. 7: Dark I/V characterisitics of the samples with different bandgaps. The insert shows the intersection points.

Tab. 2: Intersection points of the dark IV curves and a constant photocurrent at V_{oc}^* versus measured V_{oc}.

	I	II	III	IV
V_{oc}^*	298	313	330	344
V_{oc}	267	274	306	319

(type III and IV) leads to a V_{oc} that is 40 mV higher than for the cells without a grading at the i/n interface (type I and II). Second, no influence of the grading at the p/i interface on V_{oc} is observed. This shows that the different nucleation and growth conditions, due to the different germane concentration do not influence the V_{oc} behavior. The short circuit current of the prepared cells is nearly constant (~14 mA/cm^2) and the FF is nearly independent of the cell type. The different V_{oc} behavior is also reflected in the dark I/V characteristics of the cells (Fig. 7). In the insert the intersection of the dark I/V curves and a constant photocurrent line at 14mA/cm^2 (I_{sc}) is shown (V_{oc}^*), which determines V_{oc} in

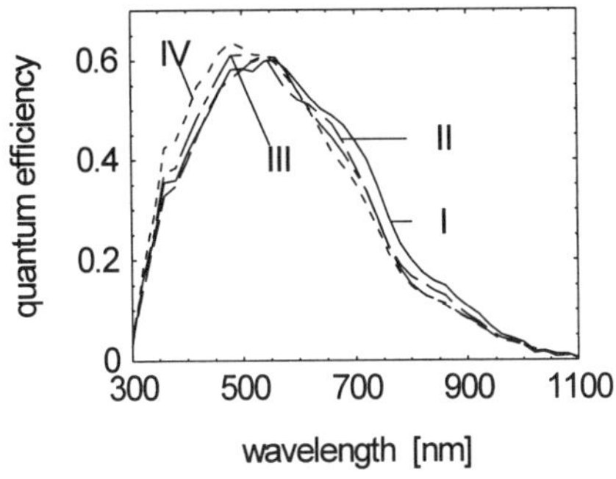

Fig. 8: Quantum efficiency of the graded solar cells (type I-IV).

a crystalline solar cell if recombination is neglected. The order of the intersection points corresponds to the order of the measured V_{oc} as depicted in Tab. 2. This behavior remains valid for the voltage range from 250 - 400 mV.

Besides the dark I/V-behavior also the collection efficiency is affected by the bandgap grading. Figure 8 shows the quantum efficiency (QE) determined from DSR-measurements. In the short wavelength region the QE of the cell types III and IV exceeds the others indicating that the collection of holes is not limited by the bandgap discontinuity at the p/i interface. In the long wavelength region a slight enhancement of the QE is observed for the cell structures I and II. Especially the structure without gradings shows the highest red response due to the lowest average gap. The results also reflect the nearly constant current of the cells by the equality of the convolution of the QE curve and the sun spectra of AM1.5 for the four cell types.

The advantageous effect of the grading at both interfaces is also observed for pin diodes prepared with 60% germane. In Table 3 the solar cell parameters for a structure without any gradings (type I) and with gradings at both interfaces (type IV) are summarized. The V_{oc} value of type IV is 70 mV higher and the I_{sc} value is slightly lower than for type I, the fill factor of the graded structure is 54.1% and exceeds the fill factor of the non-graded structure of 50.6%. So, the behavior of V_{oc} is similar to the 40% grading series and dark I/V measurements exhibits the earlier mentioned relation between V_{oc}^* and V_{oc}. Comparison of the results of the 40% and the 60% grading series shows that the difference in V_{oc} between the graded and the ungraded structure is enhanced with increasing germane content. A comparison of the bandgap shrinking and the reduction of V_{oc} for solar cells with graded interface regions with increasing germane content exhibits similar values. However, higher V_{oc} values should be accessible, taking into account the V_{oc} behavior of a-Si:H based solar cells with varied bandgaps of the i-layer.

Tab. 3: Cell parameters for different cell structures with 60% germane content in the gas phase.

	cell design	I_{sc} [mA/cm^2]	V_{oc} [mV]	FF
I	no gradings	15.15	204	50.6
IV	gradings at p/i and i/n interface	14.85	273	54.1

Figure 9 shows the QE curves of optimized solar cells prepared with 0%, 20%, 40% and 60% germane content in the gas phase and gradings at both interfaces to ensure a high V_{oc}. With increasing GeH$_4$ in the gas phase the QE in the long wavelength region rises and the maximum

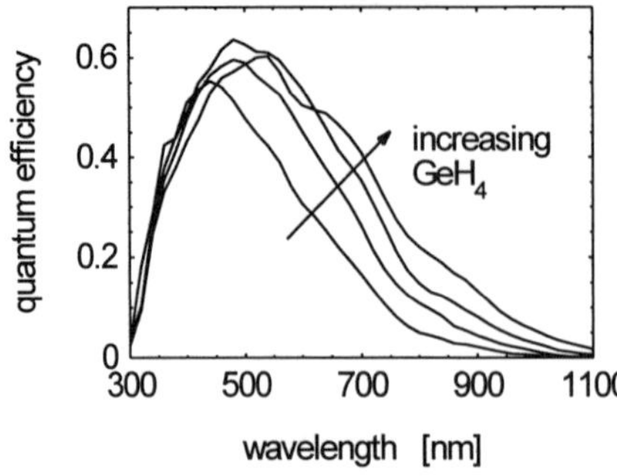

Fig. 9: Quantum efficiency of the graded series prepared with different germane content in the gas phase (0%, 20%, 40%, 60%).

of the QE shifts towards longer wavelengths, due to the enhanced absorption coefficient. The difference of the QE at 500 nm can also be attributed to the difference in the absorption coefficents, since the thickness of 150 nm is insufficient to absorbe all incident photons. The QE at 900 nm is a suitable measure for the applicability of the absorber material as an infrared detector. The QE of 900 nm increases from 2.1% of the μcSi-diode to 13% of the diode deposited with 60% GeH$_4$ in the gas phase. The combination of these structures with an optimized back reflector could result in a significant enhancement of the infrared signal.

CONCLUSIONS

μc-SiGe:H films have been prepared by PECVD. Raman spectroscopy demonstrates the good microcrystallinity and by PDS measurements an enhanced absorption in the near infrared region is observed. p-i-n solar cells with an increasing germanium content in the i-layer exhibit a linearly decreasing V_{oc} and an increasing I_{sc} due to the lower energy gap and the higher absorption. To avoid bandgap discontinuities at the interfaces an optimized cell structure with bandgap gradings at the interfaces was realized. The V_{oc} of this structure is enhanced e.g. by 70 mV for a cell deposited with 60% germane content in the gas phase in comparison to cells without gradings. These results encourage the utilization of μc-SiGe:H as the absorber material for solar cells and infrared sensors.

ACKNOWLEDGMENTS

We would like to thank W. Reetz for the DSR-measurements and J. Klomfaß for PDS-measurements and helpful discussions. This work is supported by the Ministerium für Bildung, Forschung und Technologie (BMBF).

REFERERENCES

1. C.M. Fortmann, D.E. Albright, I.H. Campbell, P.M. Fauchet, Mat. Res. Soc. Symp. Proc. Vol. **164**, 315 (1990).
2. G. Ganguly, T. Ikeda, T. Nishimiya, K. Saitoh, M. Kondo, A. Matsuda, Appl. Phys. Lett. **69**, 4224 (1996); G. Ganguly, T. Ikeda, K. Kajiwara, A. Matsuda, Mat. Res. Symp. Proc. Vol. **467**, 681 (1997).
3. R. Carius, J. Fölsch, D. Lundszien, L. Houben, F. Finger, Mat. Res. Symp. Proc. (1998), in print.
4. M. Tzolov, R. Carius, unpublished.
5. J. Weber, M.I. Alonso, Phys. Rev. B **40**, 5683 (1989).
6. W.J. Brya, Sol. State. Comm. Vol. 12, **253** (1973).
7. M.I. Alonso, K. Winer, Phys. Rev. B **39**, 10056 (1989).
8. R. Braunstein, A.R. Moore, F. Herman, Phys. Rev. Vol. **100**, No. 3, 695 (1958).

LONG-TERM STABILITY OF MICROCRYSTALLINE SILICON P-I-N SOLAR CELLS EXPOSED TO SUN LIGHT

P. SANGUINO[1], S. KOYNOV[1], R. SCHWARZ[1], M. FERNANDES[2], M. VIEIRA[2], R. MANSO[3], A. JOYCE[3], and M. COLLARES-PEREIRA[3]

(1) Instituto Superior Técnico IST, Physics Department, P-1096 Lisbon, Portugal
(2) Instituto Superior de Engenharia de Lisboa ISEL, Electronics Dept., P-1900 Lisbon, Portugal
(3) Instituto National de Energia e Tecnologia Industrial INETI, Renewable Energy Dept., P-1200 Lisbon, Portugal

ABSTRACT

The performance of an entirely microcrystalline p-i-n solar cell was monitored during a long-term outdoor test in Lisbon starting in September 1998. A small decrease of the short circuit current was observed after 5 months of operation. The open-circuit voltage remained stable around 400 mV. From the analysis of the I-V characteristic in dark and under illumination we could identify the weak points of the test structure, like large series resistance, high recombination rate, and intensity-dependent collection efficiency.

1. INTRODUCTION

Research devoted to the application of microcrystalline silicon (μc-Si:H) in thin film solar cells has recently seen a vast revival. The interest in this material comes from its suitable optical and electronic properties, and especially from the hope that the μc-Si:H films of high crystalline fraction could be stable under prolonged illumination [1] in contrast to a-Si:H films.

Several new preparation methods like Very-High-Frequency CVD (VHF-CVD) [2], Hot-Wire CVD (HW-CVD) [3] and Closed-Chamber CVD (CC-CVD) [4] have been developed recently for deposition of highly crystalline μc-Si:H films at a drastically increased growth rate, when compared with the conventional "H-dilution" PE-CVD technique. For entirely μc-Si:H solar cells, prepared by VHF-CVD, a conversion efficiency as high as 8,5 % was demonstrated, while tandem "micromorph" cells obtained by this technique are approaching 12% efficiency [6].

It is important to note that these encouraging results were obtained after applying an additional purification of the source gases [7]. The avoidance of any pollution caused by tracks of O_2 is of crucial importance due to the high doping efficiency of the crystalline material. It was also observed that μc-Si:H material, after prolonged exposure to air (humidity), is post-oxidized along grain boundaries [7]. Thus, tests of the μc-Si:H solar cells under natural outdoor conditions can be useful for proving their stability with respect to both prolonged sunlight exposure and post-oxidization effects.

2. EXPERIMENTAL ASPECTS

The entirely μc-Si:H p-i-n solar cells were deposited by the closed-chamber chemical vapor deposition technique (CC-CVD) as described in previous publications [4,5]. As substrates we used Corning 7059 glass, coated with ZnO acting as transparent front contact. The microcrystalline p-, i-, and n-layers had thicknesses of 0.017-0.035 μm (variable), 1.3 μm, and 0.05 μm, respectively. The back

Mat. Res. Soc. Symp. Proc. Vol. 557 © 1999 Materials Research Society

contact was Ag. Before the outdoor test, the solar cell characterization was done by the conventional I-V dark and light curves (see Fig. 1 and 2).

In order to protect the solar cell from adverse atmospheric conditions, a rudimentary encapsulation of the solar cell was performed. The encapsulation was made with transparent silicon rubber over the backside of the solar cell. In this way, we expect to protect the wire contacts from the rain and to some degree, we also hope to protect the solar cell from oxidation. The wires were connected to the solar cell metal contacts with the help of silver paste. The possibility of chemical interaction between the silicon glue, the silver paste, and the solar cell material was not studied.

For the outdoor test, the encapsulated solar cell was placed in INETI in Lisbon, which has a high average solar irradiation during all year. The solar cell is positioned viewing south and inclined to the latitude of the site. To check if the solar cell is degrading under solar radiation and open air, four parameters are measured: short circuit current (I_{sc}), open circuit voltage (V_{oc}), temperature of the solar cell (T_{cell}), and intensity of sunlight (I_{solar}). For completeness, the ambient temperature was also stored. All parameters are measured simultaneously every 30 seconds and an average is stored after 5 minutes of acquisition. The monitoring started on 22nd of September of 1998 and is still going on.

3. RESULTS

A typical dark I-V characteristic of a cell, measured prior the outdoor test is shown in Fig. 1. The initial I-V characteristics under illumination of two cells with different p-layer thicknesses (SC4 – 0.017 µm and SC6 – 0.035 µm) are compared in figure 2. The cells were not optimized and had a modest conversion efficiency of 1 to 3 %, a short circuit current I_{sc} of 5 to 10 mA/cm^2, and an open-circuit voltage V_{oc} of 370 to 410 mV under 80 mW/cm^2 illumination from a tungsten-halogen lamp.

Typical data acquired during one day is presented in Fig. 3. The short-circuit current I_{sc} follows the variations in I_{solar} directly, while T_{cell} shows some delay especially in the afternoon. Based on this data we will discuss the behavior of the solar cell under different conditions of light and temperature.

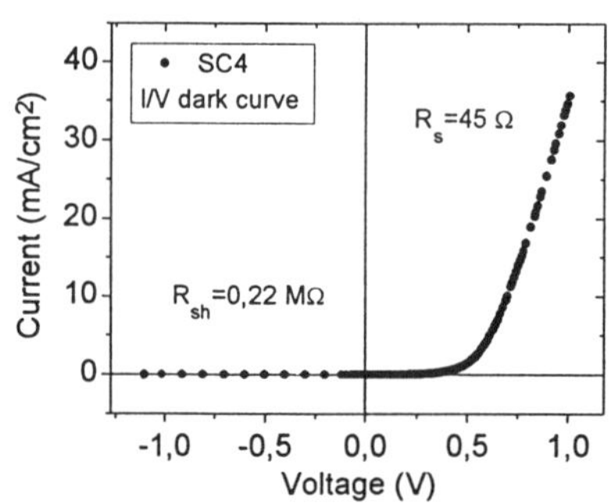

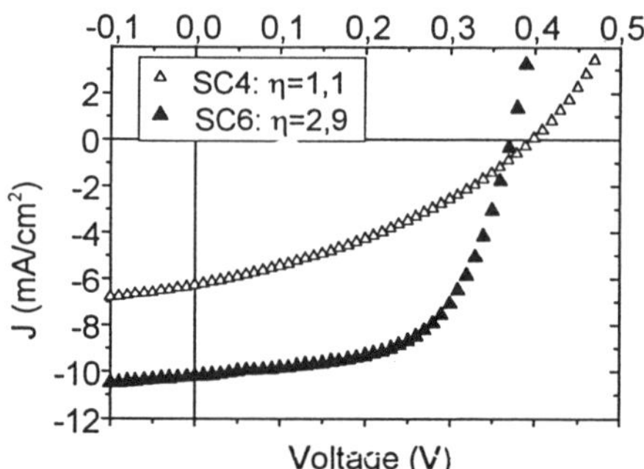

Fig. 1: Solar cell I-V dark curve.

Fig. 2: Solar cells I-V curves under illumation with 80 mW/cm^2.

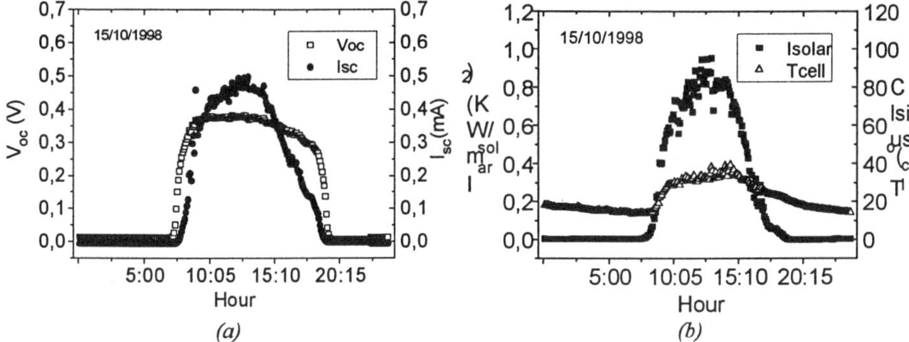

Fig. 3: *A typical day of data acquisition.*

In Figs. 4 and 5 we have plotted the dependence of I_{sc} and V_{oc}, respectively, on the intensity of solar light I_{solar}. Apart from the low-intensity region the intensity dependence of I_{sc} is linear, whereas V_{oc} shows the expected logarithmic dependence. In both cases the effect of increasing temperature is an undesired decrease (if our goal is to produce energy) of these parameters.

The influence of T_{cell} on I_{sc} and V_{oc} is presented in Fig. 6. The value of I_{solar} of 80 mW/cm^2 was chosen. While V_{oc} exhibits a clear linear dependence, the temperature dependence of I_{sc} shows some fluctuations, possibly due to the chosen sequence during data acquisition.

The relationship between I_{sc} and V_{oc} is plotted in Fig. 7, suggesting an exponential relation as in a usual p-n diode (see the discussion).

Finally, we present the long-term performance of the test solar cell. Fig. 8 shows V_{oc} and I_{sc} (for different intensities) as a function of outdoor exposure. Even after 140 days V_{oc} does not seem to change (see Fig. 8(a)), while from Fig. 8(b) we notice a small decrease of I_{sc}.

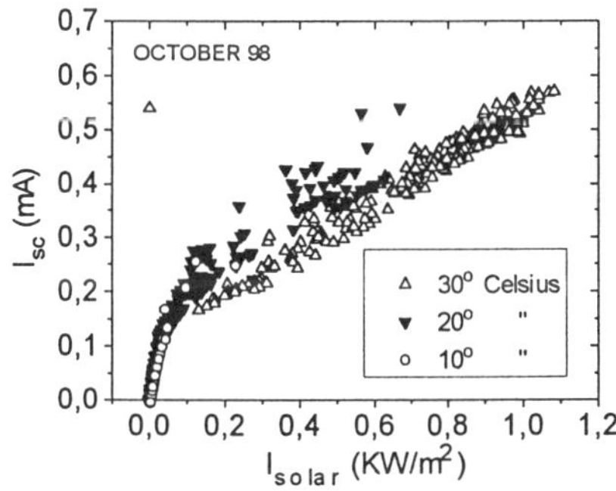

Fig. 4: *The dependence of* I_{sc} *on* I_{solar}
The data was taken during October 1998.

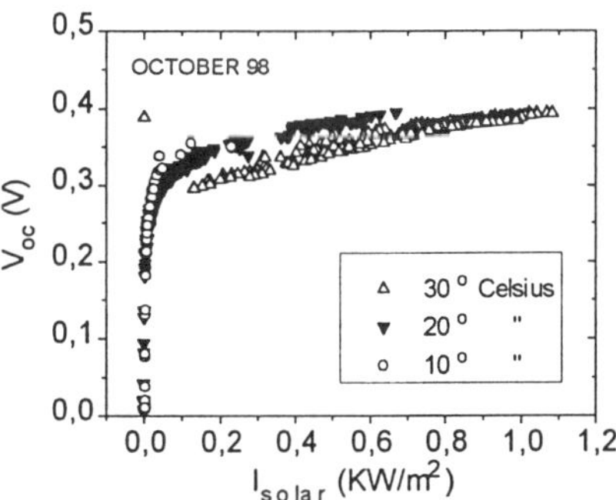

Fig. 5: *The dependence of* V_{oc} *on* I_{solar}
The temperature dependence is also shown.

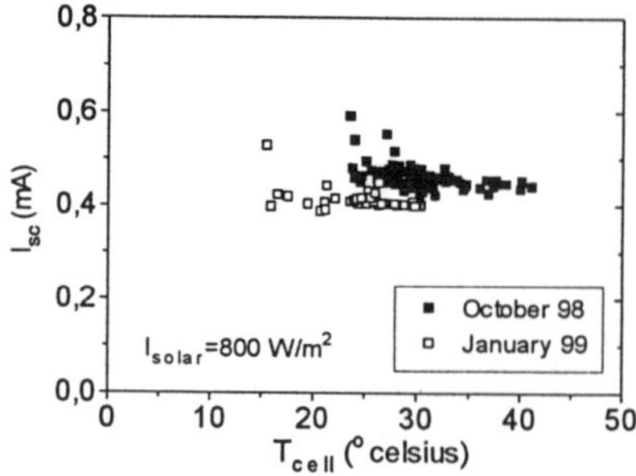
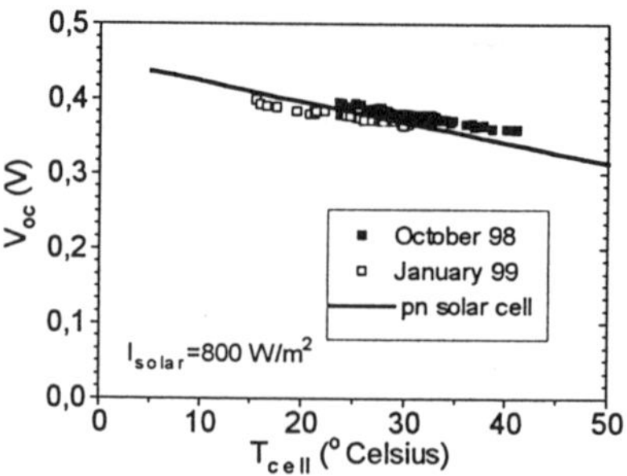

Fig. 6: Temperature dependence of I_{sc} an V_{oc}.

4. DISCUSSION

From Figs. 1 and 2 we see that the solar cells used for this test are well below the best performing solar cells reported in the literature [6]. We can point out at least two general reasons for the modest conversion efficiencies: 1) we did not use any additional purification of the source gases during the CC-CVD process – i.e. unintentional doping of the μc-Si:H films by oxygen is not ruled out, and 2) the solar cells are semitransparent in the visible spectral range, i.e. their thickness is too low to absorb efficiently the solar light. However, performance is not an important characteristic for this study, since we are trying to investigate the possible degradation of a particular solar cell structure, which, in this case, is composed entirely of μc-Si material deposited by the CC-CVD method.

Assuming that the light-induced current is independent of voltage, one can extract the following relation between I_{sc} and V_{oc}:

$$I_{sc} = I_{th}(e^{\frac{qV_{oc}}{nKT}} - 1).\qquad\qquad(1)$$

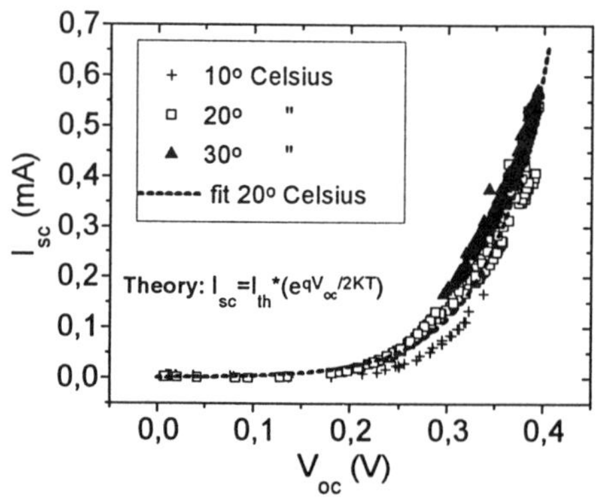

Fig. 7: Relationship between I_{sc} and V_{oc}. The data is best described by eq. (1) with an ideality factor of $n = 2$, indicative of strong recombination.

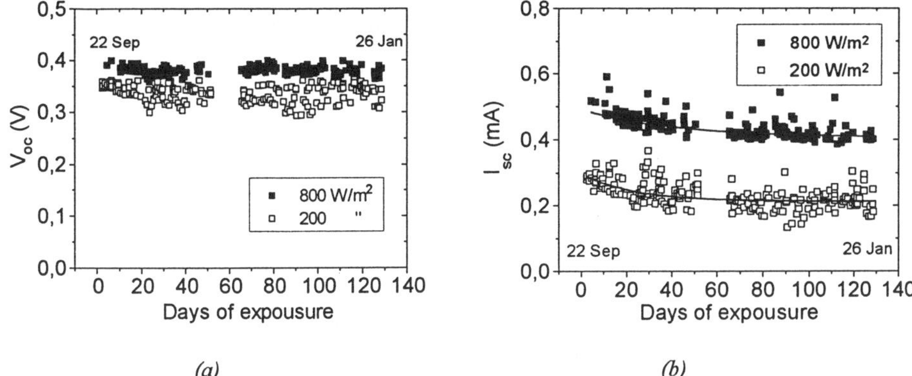

(a)   *(b)*

Fig. 8: *Long term performance of I_{sc} (a) an V_{oc} (b) for different intensities of sunlight.*

The dashed line in Fig. 6 shows that an exponential fit to the data with an ideality factor of $n = 2$ is to be preferred over $n = 1$. This is an indication for carrier collection controlled by recombination. A high recombination rate is possible in microcrystalline layers of high defect density – which would shrink the space-charge region and impede carrier collection - and in disordered transition regions located probably near the p/i-interface.

The temperature dependence of V_{oc} shows a remarkably similar behavior with the one that is expected from the theory of crystalline p-n solar cells [8]. The fit shown in Fig. 8 is based on the following equation:

$$V_{oc} = \frac{2kT}{q}\ln\left(1 + \frac{J_{light}}{J_{thermal}}\right). \tag{2}$$

Intuitively, this equation can be understood as the balance of carrier diffusion in the junction by the increase in thermal generation current with rising temperature. This will produce a decrease of the built-in field of the junction, and therefore a decrease in V_{oc}. The increase of $J_{thermal}$ is mainly due to the exponential increase of the intrinsic carrier concentration with temperature.

Can we conclude that the small decrease of I_{sc} after 140 days of outdoor exposure is caused by a critical degradation of the solar cell? The problem is not simple, because a variety of processes might be responsible for this overall effect. First, the test covered a period with seasonal variations of the solar spectrum. We have to see whether the decrease of I_{sc} will recover to its initial value after a full year of data acquisition. Accumulation of dust on the solar cell cannot be a significant reason for the I_{sc} decrease because it was cleaned by rain or manually.

Some stabilization effect seems to be present. This behavior is similar for all intensity values of the sunlight. The data are well fitted by an exponential decay function (gray curves in fig.8(b)),

$$I_{sc} = I_{stab} + A\exp(-t/\tau). \tag{3}$$

having as parameters: a characteristic time t » 30 days, in which the major changes occurred and a saturation value I_{stab} » 83% of the initial value.

Obviously, of primary concern is the question whether the active mc-Si:H structure degrades due to the creation of metastable electronic defects or because of a post-oxidation. The exponential decay fit in fig. 8(b) indicates that the I_{sc} deterioration may be a result of a diffusion-barrier-limited chemical process – e.g. oxidation of the μc-Si:H film, of the Ag contacts or of the external connections to the cell (silver paste). On the other hand, a power law decay, typical for metastable defect creation, can also fit the data reasonably well (Fig. 8(b)). A better discrimination between these two cases will be possible after completion of the whole year cycle of the outdoor tests.

5. CONCLUSIONS

In summary we have shown that entirely microcrystalline Si thin film solar cells can be prepared by the CC-CVD technique. The non-optimized p-i-n structures show V_{oc} = 400 mV, I_{sc} = 10 mA/cm^2, and efficiency η = 2,9%. These modest results could be improved by purification of the source gases during the preparation process and by increasing the absorption of solar light in the cell (thicker i-layer, light trapping). The cell structure shows a too large series resistance, which deteriorates the cell performance especially at high light intensities. This shortcoming could be avoided by soldering the external connections to the active structure instead of using silver paste. After such improvements the cells would certainly be better adapted for degradation tests.

The degradation of the test cell at outdoor conditions for more then 140 days is seen as a small reduction of the short-circuit current (less than 17%) without change of V_{oc}. The character of the I_{sc} decrease with the time of exposure (exponential decay) indicates a diffusion-limited chemical process, such as oxidation, rather than a creation of metastable electronic defects in the mc-Si:H film. This is in agreement with the observations of the Neuchatel group that no degradation was observed of the electronic properties of a high-efficiency microcrystalline solar cell prepared by VHF-CVD [7].

Acknowledgment: S.K. would like to acknowledge a NATO fellowship. The work at IST is supported by the Fundação da Ciência e Tecnologia under project no. PRAXIS/PCEX/P/FIS/7/98. The groups at IST and ISEL are both member of the TIMOC network (project no. ERB IC15 CT98 0819) funded under the COPERNICUS programme of the EC. In particular, we want to thank F. Koch, J. Völkl, and M. Stanger from the Technical University of Munich in Garching, Germany, for maintaining the infrastructure necessary for the deposition of the samples used in this study.

References

[1] A.H. Mahan, B.P. Nelson, S.J. Salamon, and R.S. Crandall, Mat. Res. Soc. Symp. Proc. **219** (1991) 673.
[2] H. Keppner, U. Kroll, J. Meier, A. Shah, Solid State Phenomena **44-46** (1995) 97.
[3] H. Meiling and R.E.I. Schropp, Appl. Phys. Lett. **69** (1996) 1062.
[4] S. Koynov, R. Schwarz, T. Fischer, S. Grebner, H. Muender, Jpn. J. Appl. Phys. **33** (1994) 4534;
[5] R. Krankenhagen, M. Schmidt, W. Henrion, I. Sieber, S. Koynov, S. Grebner, and R. Schwarz, Solid State Phenomena **47** (1995) 607.
[6] J. Meier, H. Keppner, S. Dubail, U. Kroll, P. Torres, P. Pernet, Y. Ziegler, J.A. Anna Selvan, J. Cuperus, D. Fischer, and A. Shah, Mat. Res. Soc. Symp. Proc. **507** (1998) 139.
[7] J. Meier, P. Torres, R. Platz, S. Dubail, U. Kroll, J.A.A. Selvan, N.P. Vaucher, C. Hof, D. Fischer, H. Keppner, and A. Shah, Mat. Res. Soc. Symp. Proc. **420** (1996) 3.
[8] M.S. Sze, *Physics of Semiconductor Devices*, (Wiley, New York, 1981).

HIGH QUALITY MICROCRYSTALLINE SILICON-CARBIDE FILMS PREPARED BY PHOTO-CVD METHOD USING ETHYLENE GAS AS A CARBON SOURCE

Seung Yeop MYONG, Hyung Kew LEE, Euisik YOON, and Koeng Su LIM
Department of Electrical Engineering, Korea Advanced Institute of Science and Technology
373-1 Kusong-dong, Yusong-gu, Taejon 305-701, Korea
Tel: +82-42-869-8046, 8027/ FAX: +82-42-869-8530/ E-mail: myong@cais.kaist.ac.kr

ABSTRACT

Hydrogenated boron-doped microcrystalline silicon-carbide (p-μc-SiC:H) films were grown by a photo chemical vapor deposition (photo-CVD) method from silane (SiH_4), hydrogen (H_2), diborane (B_2H_6), and ethylene (C_2H_4) gases. Since the photo-CVD is a mild process ($\sim 10 mW/cm^2$), we can avoid the ion damage of the film, which is inevitable during the deposition of μc-SiC:H employing conventional PECVD technique. A dark conductivity as high as 5×10^{-1} S/cm, together with an optical bandgap of 2 eV, was obtained by the C_2H_4 addition, which is the first approach in photo-CVD systems. From the Raman and FTIR spectra, it is clear that our p-μc-SiC:H films are made up of crystalline silicon grains embedded in amorphous silicon-carbide tissue. We investigate the role of the hydrogen dilution and ethylene addition on the electrical, optical, and structural properties of p-μc-SiC:H films.

INTRODUCTION

We have reported on the amorphous silicon (a-Si:H) solar cells and the a-Si:H thin film light emitting diodes (TFLEDs), which have the superstrate configuration of glass/TCO/p-i-n, fabricated by a photo-chemical vapor deposition (photo-CVD) technique [1-2]. We have used the hydrogenated p-type amorphous silicon-carbide (p-a-SiC:H) films having a conductivity of 10^{-6} S/cm as a window layer material. The incorporation of the thin p-a-SiC:H into these devices as a window layer material has been one of the major advances in this technology [3]. Since the increase of the optical bandgap of p-a-SiC:H film through the introduction of carbon increases the series resistance of the film, the overall performance of such films remains severely limited [4]. In general, the most necessary conditions of good window layer material are for the material to have a wide bandgap and high electrical conductivity. Wide optical bandgap window p-layer of a-Si:H solar cell is essential to minimize light absorption in this layer. On the other hand, high electrical conductivity of the p-layer is indispensable to form a high built-in potential and to generate a strong electrical field in the active intrinsic layer of a-Si:H solar cells. In the case of a-Si:H TFLEDs, wide bandgap p-type carrier injector into the luminescent i-layer is the key to improve the brightness. From this point of view, p-type microcrystalline silicon-carbide (p-μc-SiC:H) thin films are expected as better window layer materials due to the higher electrical conductivity, optical transmittivity, carrier mobility, and dopability compared to p-a-SiC:H thin films.

Plasma enhanced chemical vapor deposition (PECVD) [4-6] and electron cyclotron resonance chemical vapor deposition (ECR CVD) techniques [7] have been the conventional growth methods of μc-SiC:H thin films. However, the PECVD technique has the problem of ion-damage on the films and interfaces because of involving high power density ($>100 mW/cm^2$). One the other hand, the ECR CVD technique has the problem of non-uniform deposition over large area and the requirement of very complex system.

Dasgupta *et al.* [8] have reported p-type μc-SiC:H having conductivity as high as 2.0×10^{-2} S/cm using acetylene (C_2H_2) as a carbon source by the photo-CVD technique. In this paper, we have studied the properties of p-type μc-SiC:H as a window material of a-Si:H solar cells using ethylene (C_2H_4) as a carbon source by a photo-CVD technique. Mercury-sensitized photo-CVD by UV light of wavelength 184.9nm and 253.7nm was used as a deposition method. Optical absorption cross section of C_2H_4 at wavelength 193nm is far larger ($10^{-20} \sim 10^{-19} \text{Å}^2$) than C_2H_2 and CH_4. Therefore, C_2H_4 is more suitable to this method.

603

EXPERIMENT

Films were grown in a mercury sensitized photo-CVD system described elsewhere [1-2] using the mixture of SiH_4, H_2, B_2H_6, and C_2H_4 reactant gases. A low-pressure mercury-sensitized lamp with resonance lines of 184.9nm and 253.7nm was used as a UV light source to dissociate the mixture gases. In all deposition, the diborane doping ratio (B_2H_6/SiH_4), chamber pressure, mercury bath temperature and substrate temperature were kept at 3000 ppm, 0.46 torr, 20℃ and 250℃, respectively. To investigate the film properties, we deposited about 900 ~ 1400Å-thick films onto Corning 7059 glass substrates.

We inspected the results of Raman spectroscopy measurements and the deposition ratio of p-μc-SiC:H films to elucidate the effects of C_2H_4 addition and hydrogen dilution on the formation of the silicon-carbide film. Dark conductivity (σ_d) was measured using Al coplanar electrode configuration. Optical absorption coefficients and film thickness were measured by a Spectroscopic Ellipsometer (Jobin-Yvon, UNISEL).

RESULTS AND DISCUSSION

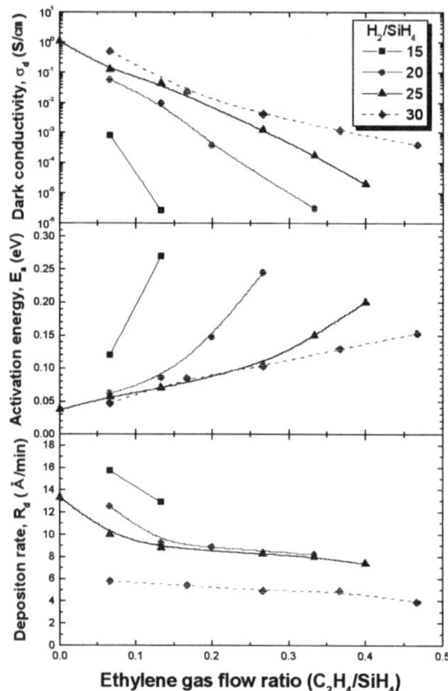

Figure 1. Effect of variation of the C_2H_4/SiH_4 on the dark conductivity, activation energy, and deposition rate of films.

Figure 1 presents the effect of variation of the ethylene gas flow ratio (C_2H_4/SiH_4) against the dark conductivity (σ_d), activation energy (E_a), and deposition rate (R_d) of p-μc-SiC:H films. The σ_d of the films decreases as the C_2H_4/SiH_4 increases. In addition, for the higher hydrogen dilution ratio (H_2/SiH_4), the σ_d decreases more slowly with the increase of C_2H_4/SiH_4. In contrast, the E_a of the films increases with increasing the C_2H_4/SiH_4. In the lower ethylene addition regime, the E_a depicts the fast increase. However, it enhances slightly when the C_2H_4/SiH_4 is more than 0.13. Besides, higher H_2/SiH_4 reveals the slower increase in the E_a. In the case of R_d of films, it shows the opposite tendency of the E_a. It decreases as the C_2H_4/SiH_4 increases, but the extent of decrease is small when the C_2H_4/SiH_4 is more than 0.13.

Figure 2 shows the absorption spectra of p-μc-SiC:H films fitted by the effective medium approximation (EMA) method [9]. From these series of figures, it is found that the optical bandgap of E_{04} improves as the C_2H_4/SiH_4 increases due to the decrease in the optical absorption in the visible region. Also, it is found that the E_{04} enhances slightly with increasing the H_2/SiH_4.

Figure 3 (a) depicts the dependence of Raman spectra of films on the C_2H_4/SiH_4. Each spectrum can be decomposed into two peaks; one is the crystalline Si peak around 520cm^{-1} and the other is the much broader amorphous Si peak around 480cm^{-1}. From these spectra, it is clear that these microcrystalline films consist of c-Si grains embedded in amorphous matrix. The degree of microcrystallinity is indicated by the crystal volume fraction. It is the ratio of the area under c-Si peak to the sum of that and the area under the amorphous Si peak, corrected by the ratio of the Raman cross section for c-Si to a-SiC:H [4]. In this paper, we used the non-corrected crystal volume fraction (X_{nc}). In other words, we assume the ratio of the Raman cross section as 1 because crystalline grains are few nanometer sizes in diameter, as estimated from Raman line shape [10]. Therefore, we can calculate easily the respective volume fraction of the films from the Raman spectra, which is represented in Fig. 3 (b). When the C_2H_4/SiH_4 is 0.07, the X_{nc} enhances drastically with increasing the H_2/SiH_4 due to the transition from amorphous phase to microcrystalline. Contrary this, the X_{nc} decreases severely with increasing the C_2H_4 addition.

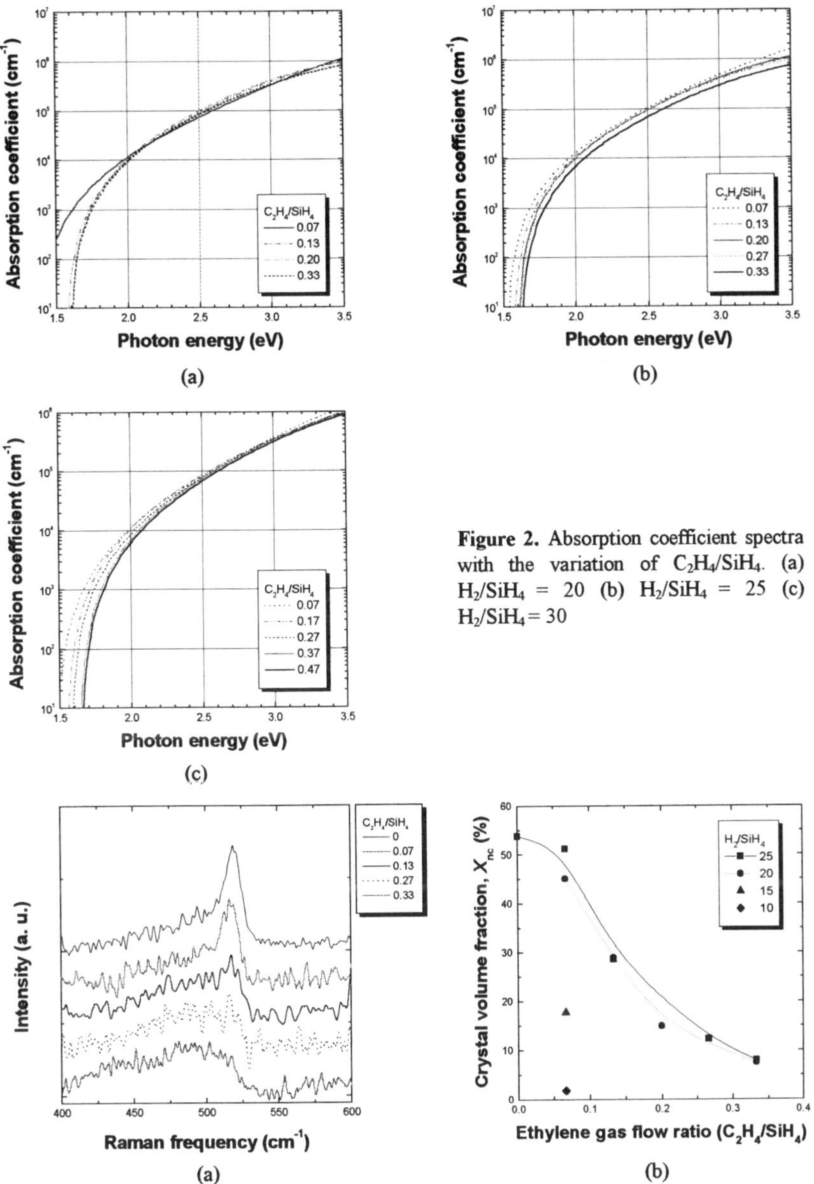

Figure 2. Absorption coefficient spectra with the variation of C_2H_4/SiH_4. (a) $H_2/SiH_4 = 20$ (b) $H_2/SiH_4 = 25$ (c) $H_2/SiH_4 = 30$

Figure 3. Dependence of Raman spectra of p-SiC:H films on the C_2H_4/SiH_4. (a) Raman spectra when the H_2/SiH_4 is 25 (b) crystal volume fraction

Table I. Wavenumbers and assignments of FTIR transmission modes.

Wavenumber (cm^{-1})	Mode of vibration assignments
630	SH_n rocking/wagging
670	Si-C stretching
780	$Si-CH_3$ rocking/wagging and/or Si-C stretching
845-900	$(SiH_2)_n$ scissors bending/wagging
1000	CH_n rocking/wagging
1200-1500	CH_n bending
2000	SiH stretching
2100	SiH_2, C-SiH stretching
2800-3000	CH_n stretching

From these results, we can find the roles of the hydrogen dilution and ethylene addition on the growth mechanism of p-type microcrystalline silicon-carbide films prepared by photo-CVD.

Hydrogen dilution of silane causes the hydrogen coverage on the growing surface of film. Such the coverage by atomic hydrogen enhances the surface diffusion coefficient of the precursors [11-12] decomposed by the UV light, and thereby the precursors adsorbed at the growing surface are able to find energetically suitable sites. This results in the nucleation of microcrystallites by decreasing the structural disorder and retarding the deposition rate. In addition, hydrogen atoms are thought to be responsible for microcrystallization by breaking weak Si-Si bonds. This structural improvement gives rise to the significant enhancement in the σ_d and the slight increase in the optical bandgap. Also, the extent of the C_2H_4/SiH_4 that can bring about the microcrystalline growth is extended with the increase in the H_2/SiH_4. This fact provides us with possibility of more high quality p-μc-SiC:H films being prepared.

On the other hand, it is well known feature that high carbon contents hinder the microcrystallization of SiC:H films. The photodecomposition of the ethylene is more difficult than that of silane because the bonding energy of ethylene is higher than that of silane. Thus, at the higher C_2H_4/SiH_4, considerable portion of energy supplied by UV light is expended on disassociating the ethylene molecules to precursors. This in turn decreases the precursors from silane. Therefore, the R_d decreases with increasing the C_2H_4/SiH_4. However, these phenomena are limited at the higher C_2H_4/SiH_4 without increasing UV light power or the mercury bath temperature due to our mercury-sensitized photo-CVD bounds. From the Figs. 1 and 2, it is found that the carbon content (C_c) in film enhances with increasing the C_2H_4/SiH_4. Since the conductivity of carbon (C) is lower than that of silicon (Si), the σ_d decreases considerably with the increase in the C_c. However, the optical bandgap of E_{04} improves as the C_2H_4/SiH_4 increases. It can be attributed for the lower optical absorption in visible range of C than that of Si.

We performed the FTIR analysis of a 4200 Å -thick film deposited onto a bare Si wafer in order to obtain the

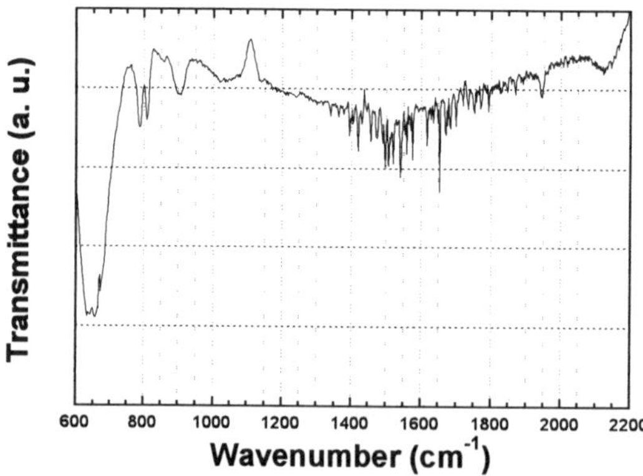

Figure 4. FTIR transmission spectrum of a 4200 Å -thick p-μc-SiC:H film (C_2H_4/SiH_4 = 0.07, H_2/SiH_4 = 20).

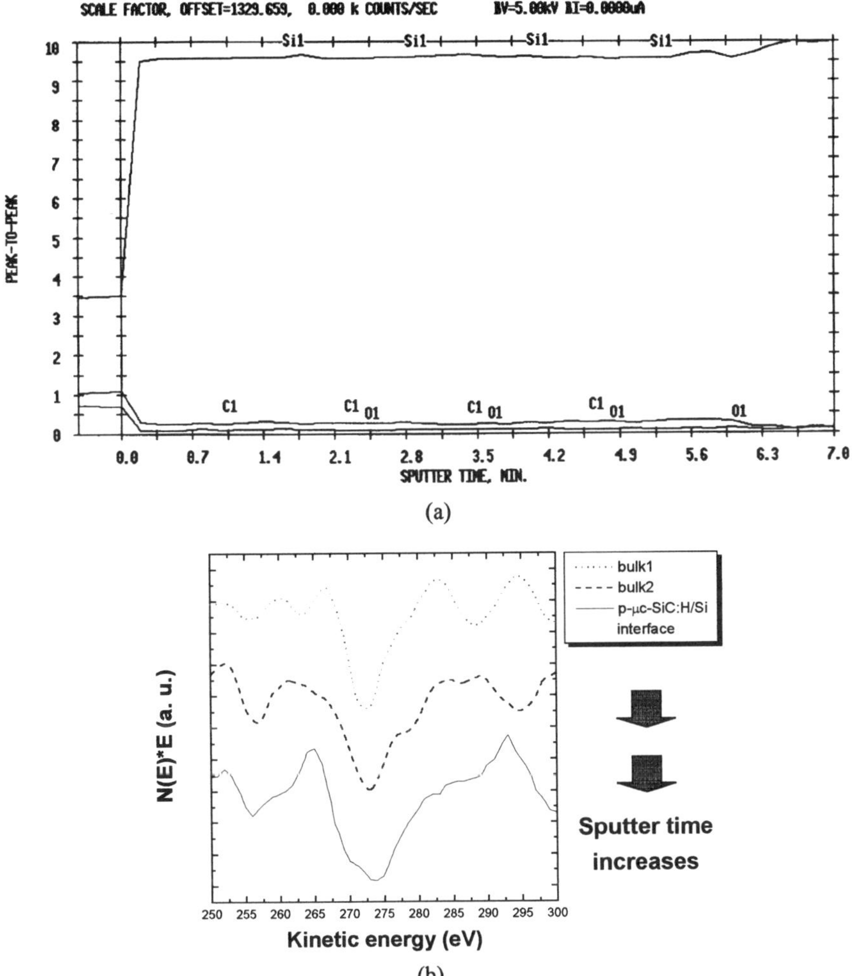

Figure 5. Results of Auger analysis (C_2H_4/SiH_4 = 0.07, H_2/SiH_4 = 20).
(a) depth profile (b) change of carbon peak near the SiC/Si interface

evidence of SiC:H formation. The C_2H_4/SiH_4 and H_2/SiH_4 are 0.7 and 20, respectively. We represent the FTIR transmission spectrum of this film in Fig. 4. Also, we provide the typical peak assignments of SiC:H film in table I [3][8][13]. The shoulders at around 657 cm^{-1} and 785 cm^{-1} confirm the formation of SiC:H.

We also performed the Auger analysis of a 2000 Å-thick film deposited onto a bare Si wafer. The C_2H_4/SiH_4 and H_2/SiH_4 are 0.7 and 20, respectively. The depth profile is shown in fig. 5 (a). It is found that the Si content (C_{Si}), oxygen content (C_O), and C_c are maintained each fixed quantity at the bulk. However, these contents change at the SiC/Si interface. The C_{Si} enhances because the substrate material is Si. In contrast, the C_c decreases. This film possess about 1 at. %C, which roughly determined from the Auger results. Figure 5 (b) presents the change of C peak near the SiC/Si interface. As closing the interface, the sharpness of C peak reduces due to the decrease in the C_c. This is also the evidence of SiC:H formation.

CONCLUSIONS

In this paper, highly conductive boron-doped μc-SiC:H films were successfully prepared for the first time by the photo-CVD using ethylene gas as a carbon source. A dark conductivity as high as 5×10^{-1} S/cm, together with an optical bandgap of E_{04} of 2 eV, was obtained by ethylene addition. We investigated the effects of the hydrogen dilution and ethylene addition on the formation of microcrystalline silicon-carbide films by inspecting the electrical, optical, and structural properties. By virtue of these efforts, we found that the hydrogen and carbon atoms introduced to the growing surface play an important role on the film properties. It is clear from the enhanced dark conductivity, optical bandgap, and crystal volume fraction that high hydrogen coverage at the growing surface improved the microcrystalline growth. Besides, high carbon content suppresses the nucleation of microcrystallites.

ACKNOWLEDGMENTS

This work was supported by the Korea Electric Power Research Institute (KEPRI).

REFERENCES

1. J. H. Jang and K. S. Lim, Jpn. J. Appl. Phys. **36**, L1082 (1997); **36**, 6230 (1997); Appl. Phys. Lett. **71**, 1846 (1997).
2. J. W. Lee and K. S. Lim, Appl. Phys. Lett. **68**, 1031 (1996); **69**, 547 (1996); J. Appl. Phys. **81**(5), 2432 (1997).
3. Y. Tawada, K. Tsuge, M. Kondo, H. Okamoto, and Y. Hamakawa, J. Appl. Phys. **53**, 5273 (1982) .
4. B. Goldstein and C. R. Dickson, Appl. Phys. Lett. **53**, 2672 (1988).
5. F. Demichelis, C. F. Pirri and E. Tresso, Phil. Mag. B **67**, 331 (1993).
6. R. Fluckiger, J. Meier, A. Shah, J. Pohl, M. Tzolov and R. Carius, Mater. Res. Soc. Symp. Proc. **358**, 793 (1995).
7. Y. Hamakawa, Y. Matsumoto, G. Hirata and H. Okamoto, Mater. Res. Soc. Symp. Proc. **164**, 291 (1990).
8. A. Dasgupta, S. Ghosh and S. Ray, J. Mater. Sci. Lett. **14**, 1037 (1995).
9. W. Y. Cho and K. S. Lim, Jpn. J. Appl. Phys. **36**, 1094 (1997).
10. I. H. Campbell and P. M. Fauchet, Solid State Commun. **58**, 739 (1986); CRC Crit. Rev. Solid State Mater. Sci. **14**, S79 (1988).
11. S. Ghosh, A. De, S. Ray and A. K. Barna, J. Appl. Phys. **71**, 5205 (1992)
12. A. Matsuda, J. Non-Cryst. Solids **59&60**, 767 (1983)
13. G. Ambrosone, S. Catalanotti, U. Coscia, S. Mormone, A. Cutolo, and G. Breglio, presented at the 2nd World Conf. and Exhib. on PVSEC, Vienna, Austria, 1998 (unpublished).

HIGH RATE DEPOSITION OF MICROCRYSTALLINE SILICON USING RESONANCE PLASMA SOURCE (HELIX) – PLASMA PROPERTIES AND DEPOSITION RESULTS

H. Grüger, R. Terasa, A. Haiduk and A. Kottwitz
Semiconductor and Microsystems Technology Laboratory
Dresden University of Technology, Dresden, GERMANY

ABSTRACT

Microcrystalline silicon layers have been deposited by PECVD using a resonance plasma source (Helix) operating at frequencies of 46, 68, 113 and 163 MHz. The plasma discharges in hydrogen and various hydrogen/silane mixtures were investigated by optical emission spectroscopy (OES) and mass spectroscopy (MS). Growth rate, crystalline fraction and hydrogen content of the layers were studied for different gas compositions, excitation frequencies and plasma powers.

Plasma monitoring revealed preferably the formation of positive ions. The density of positive hydrogen ions increased steadily 15 times by raising the frequency from 46 MHz to 163 MHz, whereas the ion energy was reduced from 75 eV to 23 eV and the radiation from hydrogen decreased to 50 %. Growth rates up to 1.5 μm/h have been achieved for microcrystalline layers at 230 °C deposition temperature. The content of hydrogen was below 15 at.%. Raman spectroscopy measurements reveal that 20 to 70 % of the silicon was crystalline depending on the silane concentration.

INTRODUCTION

In comparison to amorphous hydrogenated silicon (a-Si:H) microcrystalline silicon (μc-Si) has some interesting properties concerning doping, conductivity or band gap for devices like solar cells, x-ray detectors or flat panel displays. The low deposition rates of μc-Si, approximately one order of magnitude lower than that of a-Si:H, are a economic problem for the establishment in device production. High rate deposition of device quality μc-Si, i.e. microcrystalline silicon with a low defect density and a limited hydrogen incorporation, seems to be the key to the invention of μc-Si technology.

Plasma enhanced chemical vapor deposition (PECVD) starting from silane (SiH$_4$) highly diluted in hydrogen is the mainly used technique to deposit μc-Si thin films. High excitation frequencies have shown great advantages for the high rate deposition [1-3]. The capacitively coupled parallel plate plasmas have the disadvantage that the plasma sheath produces high energy ions that bombard the substrate surface leading to the formation of dangling bonds and the reduction of the crystallite size, finally resulting in amorphous silicon. This problem can be solved using alternative plasma sources with mostly inductively coupled plasma power or remote deposition techniques. One very promising plasma source is the Helix resonance plasma source. Details of this plasma source have been described earlier. [4,5].

EXPERIMENT

The Helix plasma source (Fig. 1) uses very high frequency (VHF) resonance to generate high axial and radial electrical and magnetic fields for the plasma excitation. The coil layout defines the resonance frequency. Besides the λ/4 resonance it is also possible to use

Mat. Res. Soc. Symp. Proc. Vol. 557 © 1999 Materials Research Society

higher harmonics like the 3λ/4 resonance to achieve a higher frequency range with suitable source sizes. Harmonic resonances enhanced inductively coupled plasma power on the top side of the plasma source near the substrate.

Experiments have been carried out using two different coils with five and seven turns operating at λ/4 resonance frequencies of 46 and 68 MHz and their first harmonics (3λ/4) of 113 and 163 MHz. VHF-Power up to 100 W was fed into the plasma discharge.

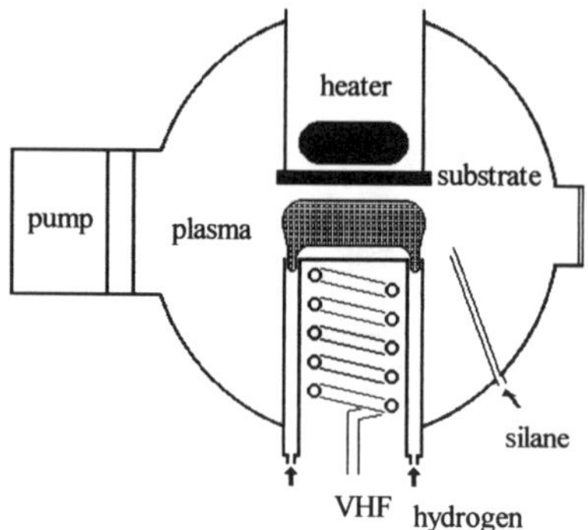

Fig. 1: Helix – setup

The plasma source was adopted to an ultra high vacuum chamber, where the substrates have been placed 20 mm above the helix top. The hydrogen gas flow was led through the helix whereas the silane was led directly to the substrates through an additional inlet. The substrates were mounted to a heater enabling temperatures up to 550 °C.

The plasma has been investigated by OES and MS. Optical spectra in the wavelength range from 200 nm to 750 nm were taken above the helix, side-on at different radii from the plasma z-axis through a window in the reactor chamber and analyzed by the spectrometer setup. Mass spectra were obtained with a Hiden EQP plasma monitor through an orifice with 100 μm width placed at the substrate position. In this way the mass spectrometer setup represented the same geometry and potential to the plasma like the substrate holder to minimize the variations in the plasma. Spectra could be obtained up to mass number 500, quantitative measurements were performed as a function of particle energy for constant mass number.

Layers have been deposited on glass (Corning 7059) and crystalline silicon substrates. In series of previous experiments a set of "standard deposition parameters" has been fixed. "Standard parameters" and parameter ranges are summarized in Table 1.

Table I: "Standard parameters" and parameter range

	Standard	Range
Temperature	230 °C	50 – 550 °C
Pressure	40 Pa	40 – 120 Pa
Plasma power	50 W	22 - 100 W
Gas flow	500 sccm	190 – 500 sccm
Content of silane	3,8 %	2 – 33 %
Deposition time	2 h	20 min – 4 h

RESULTS

Plasma characterization in discharges in pure hydrogen

OES and MS were performed in discharges in pure hydrogen first, scanning the whole parameter range possible from 46 to 163 MHz, 10 to 40 Pa and 22 to 100 W to gain a wide data basis for the Helix discharge and to understand the discharge behavior without plasma chemical aspects due to the silane chemistry. OES signals were averaged over the radii. The intensities of H_α (656,2 nm), H_β (486.1 nm) and a strong molecule transition at 611.6 nm have been evaluated. Quantitative MS scan were performed versus ion energy at mass 3 amu for the H_3^+ ion most common in the plasma. Negative ions occurred only in plasmas with weak excitation. The integrated count rates and the energy value of the most populated energy interval (width 0.2 V) were evaluated.

H_3^+ ion densities and energies retained as expected in dependence of pressure and plasma power. Higher power and lower pressure lead to an increase in ion density and energy. The dependence of the excitation frequency (Fig. 2) showed that the densities increase and the energies decrease significantly with higher frequencies.

Spectral emission of the plasma discharge also revealed the expected behavior. The intensities incrcascd rapidly with increasing power and reduced pressure. The dependence upon frequencies was more interesting (Fig. 3): The molecule emission was reduced by increasing the frequencies, the atom emission first increased due to the increasing hydrogen radical density and then decreased similar to the molecule emission due to the reduced plasma excitation.

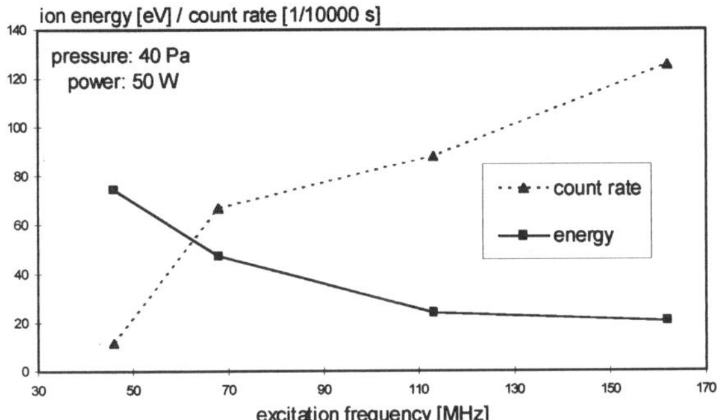

Fig. 2: Hydrogen ion (H_3^+) densities and energies in Helix discharges at various frequencies.

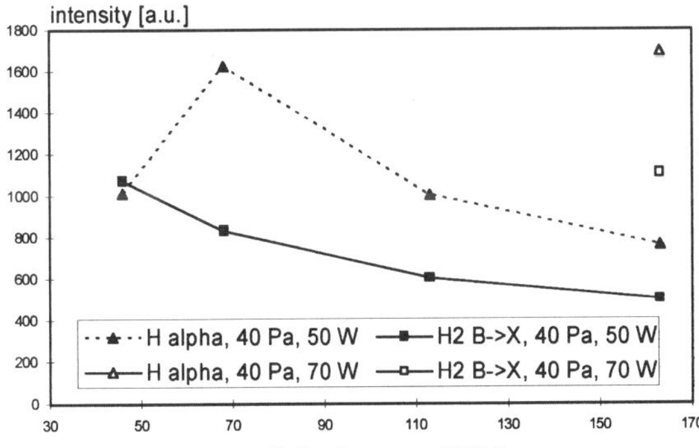

Fig. 3: Intensities of the OES measurements in various Helix discharges.

Most important was to investigate the changes in the hydrogen plasma from silane admixture typically used for the deposition, i. e. 2 – 10 % silane in hydrogen. Within the normal deposition conditions little changes in the mass spectra have been seen using 2 - 10 % silane admixture. Using high plasma power and low pressure especially at low excitation frequencies the formation of polysilane species was detected by MS either in positive (Fig. 4) or negative ion mode.

The silane admixture has strong effects on the optical spectrum of the discharge. Starting from an admixture of 2 % silane the emission of hydrogen decreased rapidly with increasing silane content. A almost constant emission of the atomic silicon line at 288.2 nm was detected. The emission of H_β vanished, no emission of SiH bands was observed.

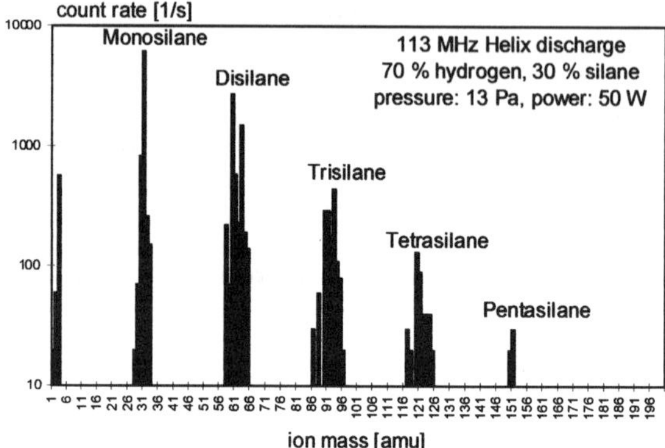

Fig. 4: Positive ion mass spectrum of a discharge in silane/hydrogen gas mixture

Deposition results

Layers have been deposited starting from the "standard conditions" mentioned in table 1. In various series of experiments single parameters have been varied to study their effects on the layer deposition in detail. The deposition time was varied in order to achieve layers with a thickness suitable for the characterization methods, typically 1 to 1.5 μm. Layer thickness was measured optically, crystallinity by raman spectroscopy and hydrogen incorporation by infrared spectroscopy.

Very important was the transition between microcrystalline and amorphous phase. This transition depended most on the silane concentration and less on pressure, plasma power and deposition temperature. The transition is different for each deposition method.

Using the Helix plasma source, the growth rate was rising with the silane concentration (Fig. 5). Above 10 % silane in hydrogen the layers became amorphous but no unsteady change in the growth rate appeared. Up to 100 W plasma power increasing growth rates have been found for a plasma excited with 163 MHz and a silane content of 3.8 %. Transmission electron microscopy of a sample deposited at 112 MHz with 2.5 % silane content revealed crystallite sizes of 40 – 50 nm and crystalline fractions 10 to 20 % higher than the raman results.

The influence of the deposition temperature showed that low temperatures as 50 °C led to low deposition rates and crystalline fractions. Raising the temperatures from 230 °C of the standard conditions up to 550 °C the deposition rate increased 10 %, the crystallinity stayed within the margin of error.

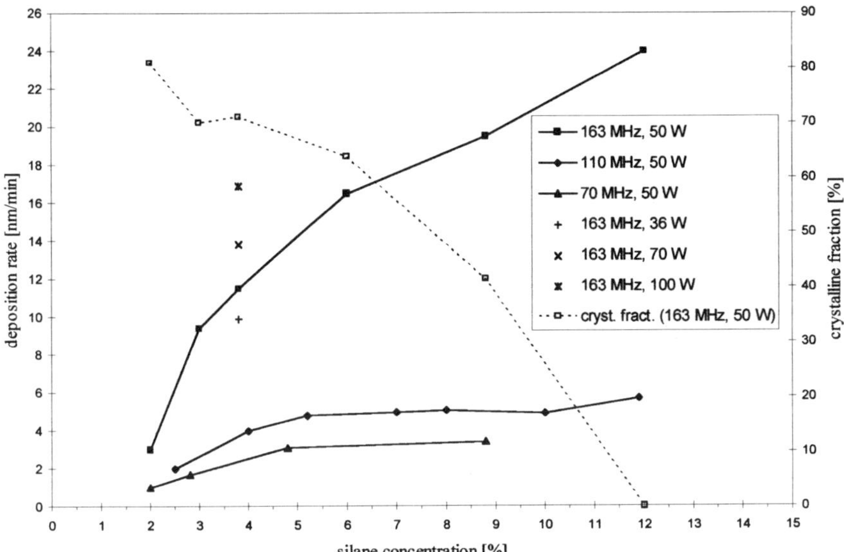

Fig. 5: Growth rate and crystal fraction as function of silane content for different frequencies and plasma powers.

The total gas flow has only minor influence to deposition rate and layer crystallinity. Reduction from 520 to 190 sccm total gas flow led to 10 % lower growth rates and an unessential higher crystallinity.

The hydrogen content of the layers depended most on the silane concentration in the gas phase. Below 5 % silane concentration the hydrogen incorporation was about 10 at.-% mainly bonded in SiH_2 or SiH_n. Higher silane concentrations led to the formation of SiH bonds, the total hydrogen content reached values of 15 at.-%.

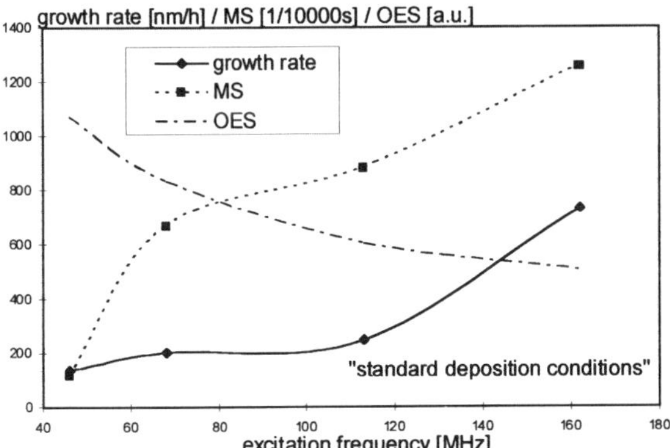

The comparison of growth rate with the results of MS for H_3^+-ions and OES for the H_2 B→X transition (Fig. 6) showed that the growth rates are connected with the ion densities for constant silane concentrations. The optical emission seemed to be a loss of energy for the plasma.

Fig. 6: Comparison of growth rate with MS and OES results

CONCLUSIONS

The plasma diagnostic investigation of the Helix discharge revealed 15 times higher ion densities with particle energies reduced from 75 to 23 eV and 50 % lower optical emission when raising the plasma excitation frequency from 46 to 163 MHz under constant conditions. Higher plasma power led to higher energies and higher particle densities.

Deposition results showed that the deposition rate was increased four times when using 163 MHz instead of 46 MHz excitation frequency at the same power, pressure and gas mixture. The transition from amorphous to microcrystalline silicon was found at about 10 % content of silane in the gas phase. Increasing plasma power from 50 to 100 W led to 64 % higher deposition rates.

Summarizing suitable for the high rate deposition were conditions that produced high densities of activated particles with low energies. This is possible by using high excitation frequencies. Also the use of harmonic resonances of the Helix source led to additional inductively coupled plasma power and thus to higher densities and lower energies of activated particles.

Promising results have been achieved by fine tuning of the Helix resonance, leading to an increase of 50 % in the deposition rate with the same gas mixture, power and pressure. Details will be published soon elsewhere.

ACKNOWLEDGEMENTS

The authors would like to thank Dr. Mensch for the TEM analysis and Mr. Sandmann for the raman measurements.
This work was supported by the BMBF under contract 0329563B.

REFERENCES

[1]   P. Torres, H. Keppner, J. Meier, U. Kroll, N. Beck, A. Shah
       Phys. Stat. Sol. A **163**, R9 (1997)

[2]   P. Hapke, PhD Thesis, RWTH Aachen (1995)

[3]   P. Hapke, F. Finger, J. Non Cryst. Solids **227-230**, 862 (1998)

[4]   Patent DE 43 37 119 C2 (1996)

[5]   U. Stephan, J. Kuske, K. Schade, MRS Symp. Proc. **336**, 79 (1994)

Structured Polysilicon for Photonic Applications

J.G.Fleming, Shawn-Yu Lin
Sandia National Laboratories, Albuquerque NM 87185, fleminjg@sandia.gov

Abstract

Three-dimensional photonic lattices have been fabricated in the infrared using a combination of advanced silicon processes. The structures display bandgaps centered at 12μ and 1.55μ.

Introduction

The drive for miniature photonic devices has been hindered by our inability to tightly control and manipulate light. Moreover, photonics technologies are typically not based on silicon and, up till now, only indirectly benefited from the rapid advances being made in silicon processing technology. In this work, we overcome some of the disadvantages of silicon inherent in its electronic structure through the use of a 3-D silicon photonic lattice. This advance has been made possible through a combination of integrated circuit fabrication technologies and may enable the development of entirely new Si photonic devices.

The ability to confine and control light in three dimensions would have important implications for quantum optics and quantum-optical devices: the modification of black-body radiation, the localization of light to a fraction of a cubic wavelength, and thus the realization of single-mode light-emitting diodes, are but a few examples [1-3]. Photonic crystals- the optical analogues of semiconducting crystals- provide a means of achieving these goals. The photonic band structure results when light encounters a well-controlled repeating arrangement of materials with differing refractive indexes. This behavior is entirely independent of the electronic properties of the material making up the lattice, as long as it does not strongly adsorb. When correctly designed and fabricated, such structures can exhibit the property that photons with the bandgap energy may not penetrate the lattice, regardless of their angle of incidence. The existence of photonic bandgaps was proposed over a decade ago by Yablonovitch (currently of UCLA) and was quickly demonstrated at millimeter wavelengths using macroscopic repeating structures made of alumina rods [1-2]. However, since the critical dimensions of the lattice scale with the wavelength of the light, a reduction in wavelength to the infrared requires structures with minimum feature sizes on the order of a micron. As a result, fabrication difficulties have stymied research in this area.

Design

The first ever silicon 3-D photonic crystal yielding a photonic bandgap at infrared wavelengths, 10-15μ, has been created at Sandia [4]. The design that we have fabricated was proposed by researchers at Iowa State University [5-8] and consists of a series of dielectric rods all aligned parallel to each other with a pitch equal to roughly half the

Mat. Res. Soc. Symp. Proc. Vol. 557 © 1999 Materials Research Society

wavelength of the mid point of the bandgap. The next layer is laid on top of, and orthogonal to, the first. The third level is aligned parallel with the first, but translated by a distance equal to half the pitch. The fourth level is parallel to the second, but is again translated by a distance of half the pitch. The sequence then begins to repeat itself. The resulting structure has a face-centered-tetragonal lattice symmetry of which face-centered-cubic is a special case. In the work described here, the width of the rods was targeted initially at 1.2μ, the height of the rods was targeted at 1.6μ and the pitch of the rods was targeted at 4.2μ. In our more aggressive structure the width was targeted at 0.18μ, the height at 0.22μ and the pitch at 0.65μ. The gap size is determined by the dielectric contrast of the two different materials that constitute the 3-D structure and by the filling fraction of the higher dielectric constant material. The index contrast in our case is between polysilicon and the surrounding air and is ~3.6:1.

FABRICATION

Two different processes were developed in the course of this work. The first process combined a series of mold, filling and CMP (Chemical Mechanical Polishing) steps to create parts with a minimum feature sizes down to 0.5 microns. The second process enabled us to fabricate parts with minimum feature sizes of 0.18 microns. This was performed on a stepper with a 0.5 μ limit of resolution and involved a complicated sequence of, fillet processing, reactive ion etching and CMP. In both cases, processing proceeded in a layer-by-layer fashion. The first process is schematically rendered in Fig.1. There are a number of possible modifications to this basic flow, for example the patterning of the as-deposited poly, or the filling of trenches between poly lines. However, in every case, the elimination of previous level topography by CMP is a critical step.

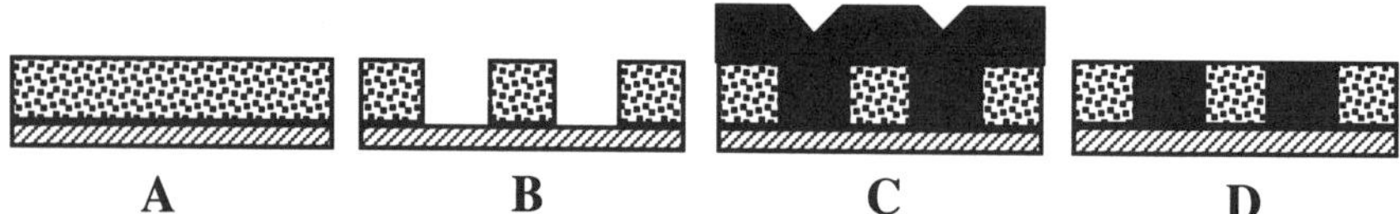

A   B   C   D

Figure 1.

Schematic of the mold fabrication process.

The first step of the mold process was the deposition of a film of silicon dioxide on the surface of the wafer with a thickness equal to that of the desired height of one layer of the lattice [1A]. A mold was then formed by photolithographic patterning and etching through the oxide down to the substrate [1B]. The mold was then filled with a deposition of polysilicon [1C]. The polysilicon was be chemically mechanically polished back to the original oxide surface [1D]. There is excellent, ~100:1, CMP selectivity between the polysilicon and the silicon dioxide. Alternatively, the poly thickness can be made equal to the layer thickness. In this case, the second layer can be formed by patterning this film.

Gaps between the rods can then be filled with silicon dioxide. The silicon dioxide was then planarized by CMP. The entire process could then be repeated to add additional layers. In this manner, the number of levels was limited only by the generation of wafer stress. At the end of the process, the silicon dioxide was selectively removed using a HF based solution. Figure 2 shows a 7 layer structure fabricated using this process.

Figure 2.

Oblique electron micrograph showing a 7 layer photonic lattice fabricated in polysilicon with stop gap between 10 and 15 microns.

The second, more aggressive process is outlined in Fig. 3. In the first step of this process a thin film of polysilicon was deposited having the thickness of the desired final height of the line, 0.22 microns. The poly silicon was then capped with a thin film of silicon nitride which acts as a combination etch and CMP stop [3A]. The sacrificial step material was then deposited [3B]. A plasma enhanced CVD (chemical vapor deposition) silicon dioxide was used in this step. The layer was then photopatterned. Since the minimum dimension was attained using the fillet process, the minimum stepper feature size required was now only slightly smaller that of the pitch of the array, 0.65 microns. After photo patterning, the oxide was anisotropically etched to just above the level of the silicon nitride layer [3C].

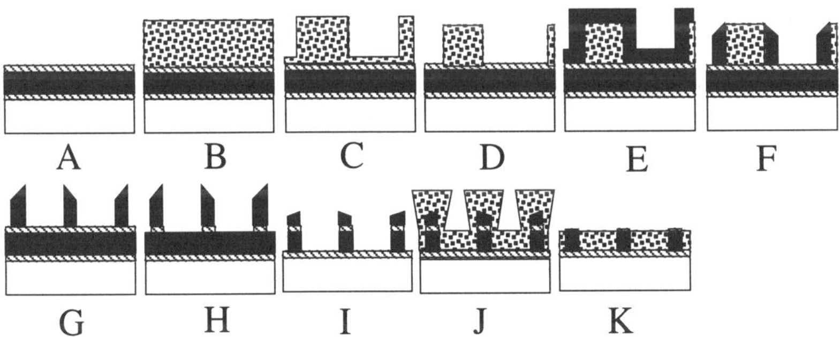

Figure 3.

Following patterning, the oxide was densified using a 1050 C, 20 second rapid thermal anneal in a nitrogen/ oxygen mixture. A 30 second wet etch in a room temperature 6:1 (Ammonium fluoride: Hydrofluoric acid) mixture was used to isotropically remove ~0.1

microns of silicon dioxide [3D]. This was done to ensure that the silicon nitride was exposed and to slightly relax the minimum feature requirement of the photolithography. Polysilicon was used to form the fillet. Since the deposition process has excellent step coverage, the width of the fillet was essentially the same as the thickness of the film deposited [3E]. The silicon was then etched in a high density plasma source system using a HBr/O_2 chemistry [3F]. Following fillet formation the sacrificial oxide is stripped in 6:1 mixture [3G]. The fillet was used as a mask for the etch of both the underlying silicon nitride and silicon layers. The silicon nitride was isotropically etched in a 180C bath of refluxed phosphoric acid [3H]. The underlying ploy silicon was etched in a high-density plasma system using a HBr/O_2 chemistry [3I]. Since the height of the polysilicon fillet was greater than that of the layer being etched, the fillet itself served as the mask. Subsequent processing involved the use of CMP to maintain planarity. The first step was to fill the gaps between the lines of polysilicon with a 0.3 micron deposition of silicon dioxide [3J]. This was done using a plasma enhanced TEOS (TetraEthOxySilane) process commonly used in back end IC technologies. The wafers were then planarized back to the silicon nitride stopping layer using CMP [3K]. Again, CMP processing is critical since it prevents topography generated in the first levels from being replicated in each subsequent level. At this point the entire process was repeated to form the subsequent layers of the structure. After completion of the desired number of layers, the silicon dioxide between the polysilicon lines was removed in a concentrated HF/water solution with excellent selectivity between polysilicon and silicon dioxide. Figure 4 gives an example of a four-layer structure.

To date, we have found the performance of these devices to be relatively insensitive to most processing parameters. Level-to-level alignment, layer width and height can all deviate significantly without seriously compromising performance. However, care must be taken to ensure that neighboring levels make physical contact.

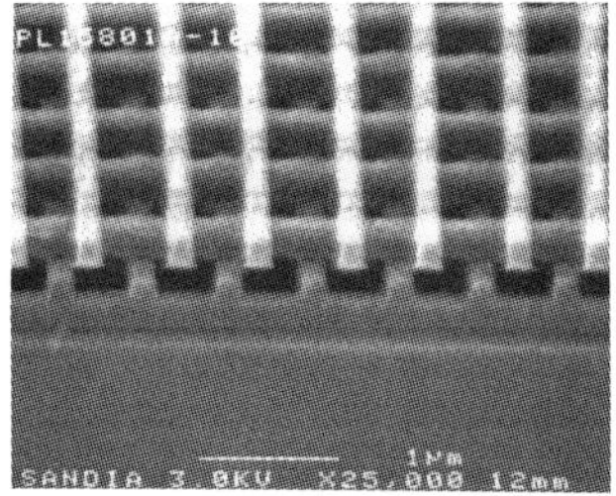

Figure 4

An oblique view of a cross section through a device formed by fillet processing. Processing defects had little effect on device performance.

TESTING

To test the transmission properties of the 12µ devices we used a standard room temperature Fourier-transform infrared measurement system with a broad spectral response from 2.5 to 25 microns. The more aggressive device was tested with a similar system having a spectral range from 1-3µ. Before measurement, the backs of the Si substrates were polished to a smoothness of better than 0.3 microns to avoid significant light scattering. The sampling beam was collimated to within a 10 degree divergence angle and had a spot size of ~3mm, the infrared light was incident along the stacking direction of the 3D photonic crystal and was unpolarized. To find the absolute transmittance, a reference spectrum from a bare Si wafer was first obtained along with the spectrum taken from a wafer with a 3-D photonic crystal built on it. By ratioing the signal to the reference spectrum, the system's detector response was normalized, thereby eliminating the small, unwanted absorption and other losses in the silicon substrate. The absolute transmission spectrum of light propagating along the <001> direction of the 3D photonic crystal, (normal to the substrate) is plotted on a logarithmic scale as a function of wavelength in Figs. 5 and 6. Between 10 and 14 microns, a strong transmittance dip is observed, signifying the existence of a photonic bandgap in the 3-D structure, Fig 5. A similar phenomena is found in Fig. 6 between 1.35 and 1.95μ.

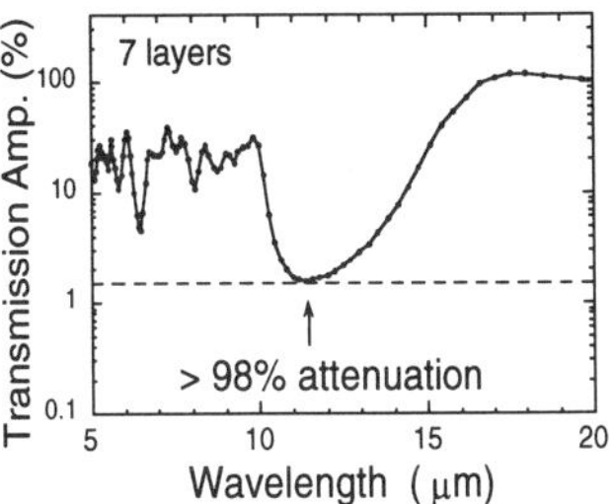

Figure 5

Plot of transmission amplitude as a function of wavelength for the structure shown in Fig. 2. At a wavelength of 11 microns there is ~98% attenuation. The lattice itself is only 10.5 microns thick.

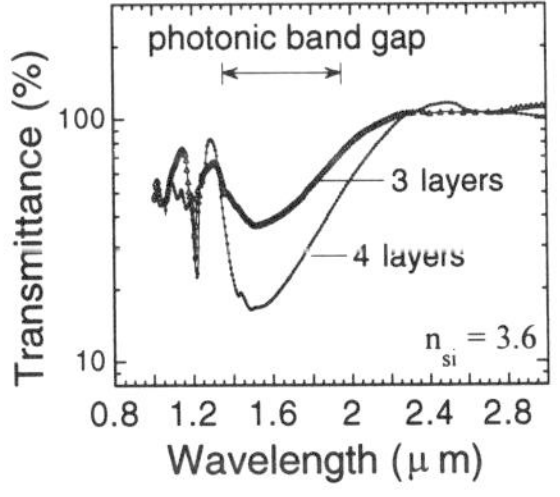

Figure 6

A plot of transmission as a function of wavelength for 3 and 4 layers of crystal with gap centered at 1.6 microns. The development of the gap with the addition of layers can be seen.

At a thickness equal to only one wavelength of light, a 7-layer, 3D crystal rejects 98% of the incident light, Fig.5. Such performance is 10 times more effective than the 1D distributed-feedback-mirrors (DBR), widely used in VCSEL (Vertical Cavity Surface Emitting Laser) technology. Indeed, a 3D crystal is equivalent to a 3D DBR that controls and manipulates light in all three dimensions. The structures also display less adsorption compared with metallic mirrors.

POTENTIAL APPLICATIONS

The successful demonstration of this technology is expected to herald a new revolution in both commercial and military photonics. Military applications in the infrared include, emissivity modification, and bandpass filters for thermal image control, target identification and signature recognition. Commercial applications targeted at the wavelengths used for optical communications at 1-2 micron wavelengths will lead to a totally new generation of Si photonic devices that can be integrated with current silicon microelectronics. Examples include high performance, low cost lasers, and light emitting diodes, mirrors, waveguides and optical filters.

ACKNOWLEDGEMENTS

These parts were fabricated at Sandia National Laboratories Microelectronics Development Laboratory. The authors especially acknowledge the efforts of Tony Farino, Brad Smith, Dale Hetherington and Pat Shea. This work was supported by the United States Department of Energy under contract DE-AC04-94AL85000. Sandia is a multiprogram laboratory operated by Sandia Corporation, a Lockheed Martin Company, for the United States Department of Energy.

REFERENCES

1   E. Yablonovitch, J. Opt. Soc. Am. B **10**, 283-295 (1993).

2   J. Joannoupoulos, R. Meade, and J. Winn, *Photonic Crystals*, (Princeton, New York, 1995).

3   J. Foresi, P.R. Villeneuve, J. Ferrera, E.R. Thoen, G. Steinmeyer, S. Fan, J.D. Joannopoulos, L.C. Kimerling, Henry I. Smith and E.P. Ippen, Nature, **390**, 143-145 (1997).

4   S.-Y. Lin, J/G. Fleming, D.L. Hetherington, B.K. Smith, R. Biswas, K.M. Ho, M.M. Sigalas, W. Zubrzycki, S.R. Kurtz and Jim Bur, Nature, **394**, 251-253 (1998).

5   K.-M. Ho, C.T Chan, C.M. Soukoulis, R. Biswas and M. Sigalas, Solid State Comm. **89**, 413-416 (1994).

6   H. Sozuer, and J. Dowling, Modern Optics, **41**, 231-239 (1994).

7   S. Fan, P.R. Villeneuve, P. Meade and J.D. Joannopoulos, Appl. Phys. Lett. **65**, 1466-1468 (1994).

8   E. Ozbay, A. Abeyta, G. Tuttle, M. Tringides, R. Biwas, C.T. Chan, C.M. Soukoulis and K.M. Ho, Phys. Rev. B **50**, 1945-1948 (1994).

Part VIII

Thin-Film Transistors
and Displays

LASER CRYSTALLIZED POLYSILICON TFT'S USING LPCVD, PECVD AND PVD SILICON CHANNEL MATERIALS—A COMPARATIVE STUDY

R.T. FULKS*, J. B. BOYCE*, J. HO*, G. A. DAVIS**, V. AEBI**
*Xerox Palo Alto Research Center, Electronic Materials Laboratory, Palo Alto, CA 94304
**Intevac Corporation, 3550 Basset Street, Santa Clara, CA 95054

ABSTRACT

In this work polysilicon TFT's were fabricated by excimer laser crystallization of active channel silicon which was deposited by three different methods: 1) LPCVD at 550 °C; 2) PECVD at 225 °C; and 3) PVD at room temperature. CMOS devices were produced with the same low temperature (less than 600 °C) top gate process and the laser anneal condition was optimized for the material type and thickness. For PECVD material a pre-anneal step of 450 °C for 1 hour was required before crystallization to avoid bubbling and ablation due to hydrogen evolution, but no such anneal was required for either LPCVD or PVD material due to their low hydrogen and Ar content. For 50 nm films, laser energy densities were typically in the range of 300-400 mJ/cm^2. Excellent device results were obtained for both LPCVD and PECVD material with n-channel field effect mobilities greater than 100 cm^2/Vs and on/off ratios greater than 10^8 at 5 V drain bias. Good results were also obtained for PVD films that can be further improved by optimizing deposition and anneal conditions. In moving toward very low temperature polysilicon processing (less than 230 °C) both PECVD and PVD channel films appear to be viable candidates.

INTRODUCTION

Low temperature, glass-compatible, polysilicon thin film transistor (TFT) technology using laser crystallization has become increasingly important for flat panel display and imaging applications because it enables integrated drivers and higher resolution [1-3]. In many research laboratories, the channel material for crystallization was deposited by Low Pressure Chemical Deposition (LPCVD) and very high performance transistors have been obtained [4]. Plasma Enhanced Chemical Vapor Deposition (PECVD) films have been used in polysilicon display production lines due to the availability of PECVD equipment for large area amorphous silicon processes [5]. In addition, PVD (Physical Vapor Deposition), i. e. sputtered, silicon is a candidate for crystallization where very low temperature processes (e.g. on plastic substrates) [6]. In evaluating the suitability of these materials it is often difficult to compare them due to differences in either TFT architecture or TFT processing used by various researchers.

In this work we compare TFT characteristics fabricated with the same low temperature (600 °C) process using channel silicon deposited by three different methods and laser crystallized. To eliminate the substrate as a variable, fabrication was done on quartz coated with a 700 nm buffer oxide.

DEPOSITIONS

Table I shows the deposition conditions for the active layer silicon for this work. For comparison purposes the silicon thickness in this study was 50 nm, which is also a common silicon thickness in an amorphous silicon TFT process.

The LPCVD material was deposited in a standard quartz low-pressure system made by
Thermco. The silicon is deposited in an amorphous phase that has also been found suitable for

Table I. Silicon Deposition Parameters

Method	Temp (°C)	Rate (nm/min)	Source	Pressure (mtorr)
LPCVD	550	3.0	SiH_4	350
PECVD	225	6.4	SiH_4	230
PVD	RmT	24.0	Si	3.0

solid phase crystallization processes [1,2] There is very little hydrogen in such films and the
film purity is expected to be very high in the quartz environment.

For the PECVD films an Ulvac in-line system with loadlock was used for deposition. A
low temperature deposition process at 225 °C was found suitable for laser processing, although
a 450 °C, 1 hr., dehydrogenation anneal after deposition was required to prevent laser ablation
due to outgassing of hydrogen present in the as-deposited films. Film contamination in such a
system could be a problem, but we have attempted to minimize this risk by using a low
temperature and a low power plasma process (25 mW/cm^2).

PVD silicon was deposited by dc sputtering in an Intevac system at room temperature. A
lightly doped, 60-ohm-cm p-type silicon target was the deposition source.

LASER CRYSTALLIZATION

An excimer laser made by XMR was used for crystallization of the deposited amorphous
silicon films. This laser operates at a wavelength of 308 nm and has a 30 ns pulse length. A
$24x2.4$ mm^2 beam was scanned in one direction across the substrate with an overlap of 99%.
This means that a given point on the wafer received 100 shots.

For each film type, a matrix of laser energy densities was applied to each sample to
determine the crystallization threshold and the threshold of ablation. As mentioned previously,
the PECVD material required a furnace anneal to remove hydrogen and prevent bubbling or
premature ablation. Such an anneal was not required for the LPCVD or PVD films. Even
though there may be trace amounts of Ar trapped in the PVD films, no particular problems were
observed in the laser annealing of these films.

X-ray diffraction experiments were done to locate the energy density for maximum grain
size [7]. Then a matrix of 3 laser conditions near this point was applied to a four-inch wafer in 3
separate 24 mm wide stripes and TFT's were fabricated. In this way an optimum range of laser
conditions was established for each material type.

TFT FABRICATION

The laser crystallized samples all received the same top gate polysilicon TFT process as
shown in Fig. 1. After island patterning, a 100 nm LPCVD oxide was deposited at 420 °C,
followed by a 350 nm silicon gate deposition at 550 °C. The gate was patterned and either a
phosphorus implant (2×10^{15} cm^{-2}, 65 keV) or boron implant (2×10^{15} cm^{-2}, 30 keV) was
performed to create the source-drain regions of n-MOS or p-MOS devices, respectively. A
furnace anneal (600 °C, 2 h) was used to activate the source-drain implants. The devices were
then soaked in an rf hydrogen plasma for several hours at 350 °C to ensure proper defect
passivation. After deposition of a 700 nm isolation oxide, contacts were etched for the source,
drain and gate. A TiW/AlCu metallization was deposited and patterned to form the contact

interconnect and the device characteristics were measured. Since the same process was used for each channel silicon type, the device characteristics should reflect primarily the quality of the channel film and its deposition and laser treatment conditions.

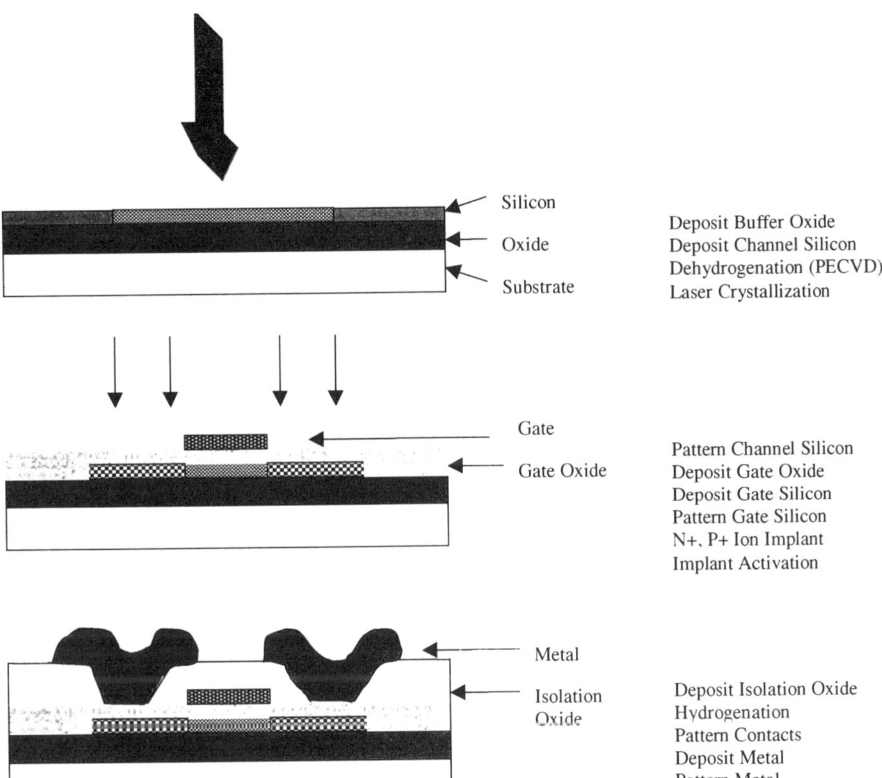

Figure 1. Process flow for CMOS Polysilicon TFT Fabrication

DEVICE RESULTS

Transfer characteristics for n-channel TFT's fabricated in LPCVD silicon and PECVD silicon are shown in Figure 2. These transistors have a linear region field effect mobility (μ) of 150 cm^2/V-sec, a minimum leakage (I_{min} at 5V) current of 0.01 pA/um and a subthreshold slope (S) of 0.25 V/decade. These very high performance TFT's have an on/off ratio of nearly 10^9. Dual gate devices with W/L =10/5,5 made in these materials have an on/off ratio of 10^{10} and leakage current at 5 V drain bias of 25 fA. This makes them suitable not only for display applications, but possibly for X-ray imagers as well where pixel leakage current is critical [8]. P-channel devices have similar leakage and turn-on characteristics. The threshold voltage (V_{th}) is -1.2 V and the mobility is 55 cm^2/Vs which results in a slightly smaller on/off ratio.

Output characteristics of the n-channel devices are shown in Figure 3. The TFT's are well behaved and have good saturation characteristics. The higher drain current in the PECVD device is due to its slightly higher mobility.

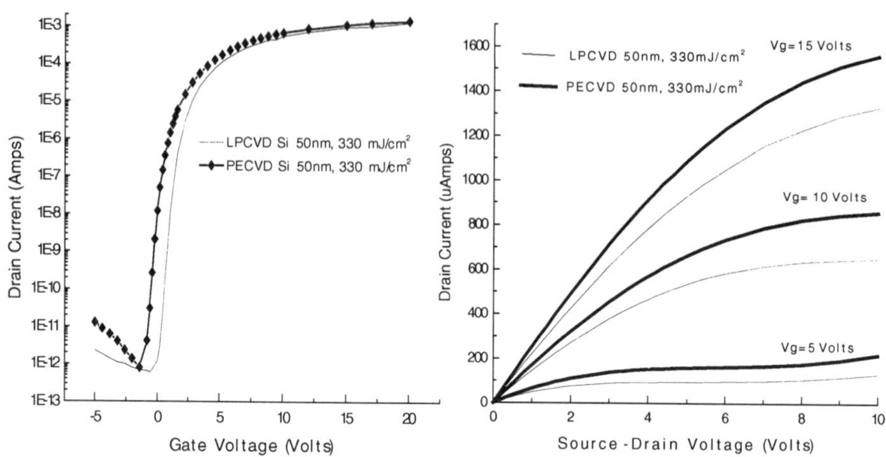

Figure 2. Transfer characteristics of n-channel TFT's fabricated in LPCVD and PECVD channel silicon. Vds = 5 V, W/L = 50/15.

Figure 3. Output characteristics of n-channel TFT's fabricated in LPCVD and PECVD channel silicon for Vg = 5, 10 and 15 V. W/L=50/15.

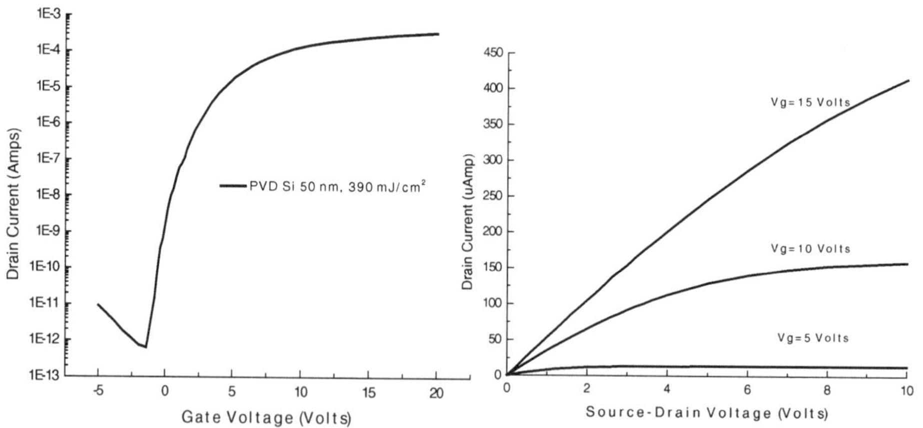

Figure 4. Transfer characteristics of n-channel devices fabricated in 50 nm PVD silicon. V_{ds} = 5 V, W/L = 50/15.

Figure 5. Output characteristics of n-channel devices fabricated in 50 nm PVD silicon for V_g = 5, 10 ,15 V. W/L=50/15.

The transistor n-channel characteristics in Figures 4 and 5 are for 50 nm laser crystallized PVD channel material. These devices have a field effect mobility of about 50 cm^2/Vs and minimum leakage current of 0.04 pA/μm. The subthreshold slope is about 0.40 Volts/decade. Again, the output characteristics are well saturated. P-channel devices are currently under fabrication. Although the performance of these devices is not as good as the LPCVD and PECVD channel devices, they are adequate for many display applications. Sputtering also has the advantage of being a room temperature process that is important for very low temperature polysilicon TFT technology. In addition, the PVD films do not require an outgassing step.

Further improvement of the PVD devices is expected as deposition and laser conditions are optimized for this material. One important consideration is the choice of target materials for sputter deposition. Another is the minimization of contamination from deposition of unwanted materials during the sputter process itself.

A comparison of device parameters for the three starting materials is shown in Table II.

Table II. Comparison of measured n-channel device parameters for LPCVD, PECVD, and PVD channel silicon for a transistor with W/L = 50/15.

Material	μ (cm^2/Vs)	I_{min}, 5V (pA)	V_{th} (V)	S (V/decade)
LPCVD	120-160	0.2-0.8	0.2-0.5	0.24-0.28
PECVD	100-160	0.3-1.0	0-0.3	0.25-0.30
PVD	50-100	0.6-30	0.2-3.0	0.35-0.50

CONCLUSION

Thin film transistors have been fabricated using LPCVD, PECVD and PVD silicon as channel materials in a polysilicon laser crystallization process. High performance devices were produced with both LPCVD and PECVD starting material and somewhat lesser performance was achieved with PVD silicon. The PVD material looks very promising for low temperature polysilicon processing and optimization of the deposition and laser conditions should improve device performance.

ACKNOWLEDGEMENTS

The authors wish to acknowledge Yu Wang, Vicki Geluz-Aguilar and Janice Arroyo for laser and TFT processing in the Xerox PARC Research Line. They also thank Jeng-Ping Lu, Ping Mei and Francesco Lemmi for useful measurements and discussions. The authors also wish to acknowlege support from NIST (Contract No. 70NANB7H3007) for partial funding of this work.

REFERENCES

1. I.-W. Wu, Solid State Phenomena, **37-38**, pp. 553-564 (1994).

2. F. Plais, P. Legagneux, C. Reita, O. Huet, F. Petinot, D. Pribat, B. Godard, M. Stehle, and E. Fogarassy, Microelectron. Eng., **28**, pp. 443-446 (1995).

3. K. Shimizu, O. Sugiura, and M. Matsumura, Jpn. J. Appl. Phys., **29**, pp. L1775-L1777 (1990).

4. G. K. Giust, T. W. Sigmon, J. B. Boyce, and J. Ho, IEEE Electron Dev. Lett., **20**, pp. 77-79 (1999).

5. T. Monta, Y. Yamamoto, M. Itoh, H. Yoneda, Y. Yamane, S. Tsuchimoto, F. Funada, and K. Awane, *Proceedings, 1995 Int. Electron Devices Meeting (IEDM '95)*, pp. 841-843, Washington, D. C., December, 1995.

6. G. K. Giust, T. W. Sigmon, P. G. Carey, B. Weiss, and G. A. Davis, IEEE Electron Dev. Lett., **19**, pp. 343-344 (1998).

7. J. B. Boyce, P. Mei, R. T. Fulks, and J. Ho, Phys. Stat. Sol. (a), **166**, pp. 729-741 (1998).

8. R. A. Street, private communication.

INTEGRATED AMORPHOUS AND POLYCRYSTALLINE SILICON TFTs WITH A SINGLE SILICON LAYER

K. PANGAL, Y. CHEN, J.C. STURM AND S.WAGNER
Dept. of Electrical Engineering, Princeton University, Princeton, NJ 08544

ABSTRACT

Selective exposure of an a-Si:H film to a room temperature hydrogen plasma using a patterned SiN_x capping layer and a subsequent anneal at 600°C, resulted in polycrystalline and amorphous silicon regions in a single silicon layer on the same glass substrate. Top-gate non-self-aligned TFTs were fabricated in both the amorphous and polycrystalline regions with all shared processing steps and no laser processing using a re-hydrogenation step. The TFTs had good characteristics, with field-effect mobilities upto 1.2 cm^2/Vs and 15 cm^2/Vs for the a-Si:H and the poly-Si TFTs, respectively, and ON/OFF ratios >10^5 in either case.

INTRODUCTION

For large-area electronics, there has been considerable interest to integrate both polycrystalline and a-Si:H TFTs. This can be done by selective laser crystallization of a-Si:H [1], or using two layers of silicon, one polycrystalline and second amorphous [2], to integrate the a-Si:H and the poly-Si TFTs on the same substrate. In active-matrix flat panel displays, a-Si:H TFTs provide low leakage in the OFF state and poly-Si TFTs high drive currents. Integration of a-Si:H and poly-Si TFTs is traditionally difficult for three reasons. First, the conventional a-Si:H TFT fabrication process is a low temperature process (<300°C) [3], while the poly-Si TFT fabrication requires a 600°C anneal starting from a-Si:H if laser processing is not used. Second, one would like to deposit only a single Si layer instead of two (a-Si:H and poly-Si) to save cost, and third, the structure and fabrication sequence of a-Si:H TFTs and poly-Si TFTs are traditionally very different (e.g. bottom gate vs. top gate process), so that few process steps can be shared.

In this paper we demonstrate a method for integrating such transistors together starting with a single Si layer without laser processing. The work is based on the selective crystallization of amorphous silicon to polysilicon by locally seeding the crystallization with a masked RF hydrogen plasma at room temperature, followed by subsequent crystallization anneal at 600 °C [4]. All of the fabrication steps between the two transistors are combined to simplify the process. Using a critical re-hydrogenation step after the 600 °C annealing, top-gate a-Si:H TFTs with a field-effect mobility as high as 1.2 cm^2/Vs and an ON/OFF ratio of 10^6, and polysilicon TFTs with a field-effect mobility of 15 cm^2/Vs and an ON/OFF ratio of 10^5 can be achieved.

SELECTIVE CRYSTALLIZATION

Hydrogenated amorphous silicon (a-Si:H) films 150 nm thick, were deposited by PECVD using pure silane, on 1737 glass substrates at a substrate temperature of 150 °C, at RF power of ~0.02 W/cm^2. The usual process [3] for growth of a-Si:H for TFT fabrication is the same but the substrate temperature is 250°C. The substrate temperature was lowered to 150°C in this case because, due to the higher hydrogen content of the a-Si:H of 15 at.%, the lower growth temperature ensures a longer incubation time during crystallization anneal [5,6]. The subsequent

Mat. Res. Soc. Symp. Proc. Vol. 557 © 1999 Materials Research Society

exposure to atomic hydrogen was done in a parallel plate Reactive Ion Etcher (RIE) at room temperature. The RF power was 200 W, RF frequency was 13.56 MHz, the chamber pressure was 50 mtorr, the exposure time was 60 min, and the RIE electrode area was 250 cm^2. The samples were then annealed in a furnace at 600 °C in N_2. UV reflectance measurement [7] was done on all samples to monitor the crystallization process.

Exposing a-Si:H to room temperature plasma before the 600°C anneal greatly reduced the time taken to crystallize the amorphous film from the ~20 h required for an untreated film [4]. A hydrogen plasma has the largest effect, with the crystallization time reduced by a factor of five, and argon plasma had the smallest effect (Fig. 1). Exposure of an a-Si:H film to the hydrogen plasma leads to hydrogen abstraction from the surface of the film resulting in Si dangling bonds [5]. This facilitates the formation of silicon crystal nuclei, which act as seeds during subsequent annealing. This hydrogen-plasma-induced crystallite formation is limited to the top 30-40 nm of the surface [5]. The evidence of crystallite formation is indirect, because Raman spectrum analysis of the films was inconclusive. Most likely the crystallite size and number is too small to be detected by Raman spectroscopy.

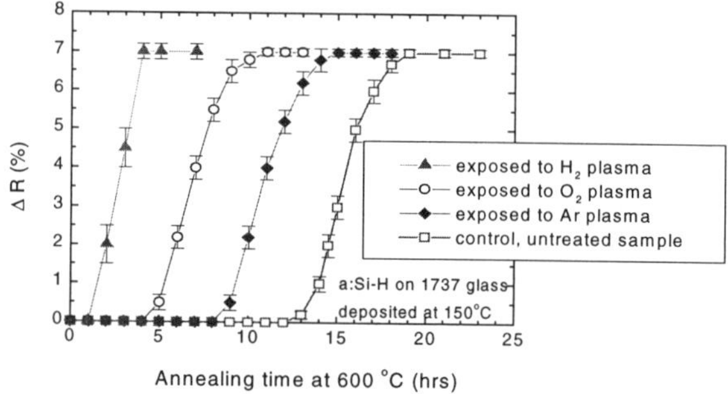

Figure 1. Change in UV reflectance at 276 nm as a function of annealing time for samples exposed to H_2, O_2 or Ar plasmas. Untreated films shown for comparison [4].

The principle of area-selective crystallization is to protect selected areas of the a-Si:H precursor film from the plasma exposure by masking [4,5,7,8]. A 120 nm thick SiN_X mask film was deposited by PECVD using SiH_4, H_2 and NH_3, at substrate temperature of 200 °C and RF power of ~ 20 W [9], on a 150 nm thick a-Si:H precursor layer deposited at 150°C on 1737 glass substrate. The SiN_X was patterned by etching in dilute hydrofluoric acid (HF) and the sample was exposed to the hydrogen plasma. Then the samples were annealed at 600°C in a N_2 ambient, with the SiN_x capping the a-Si regions to minimize the out diffusion of hydrogen during the anneal.

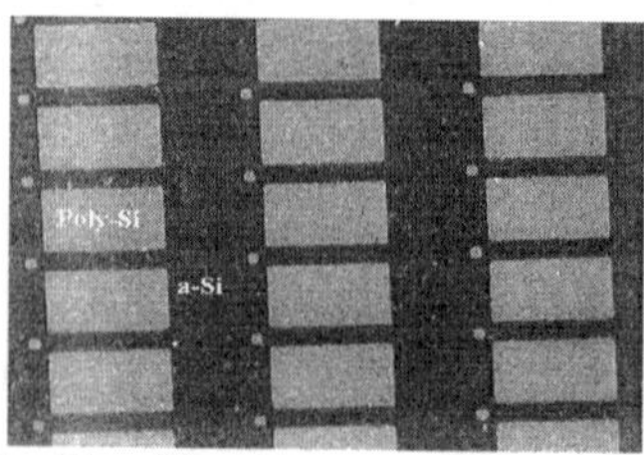

Figure 2. Optical micrograph of a silicon film with patterned areas (100x200 μm^2) of polycrystalline silicon (light) surrounded by amorphous film (dark) [4].

After ~4 hrs of anneal the portion of the a-Si:H film that had been exposed to the hydrogen plasma was completely crystallized (confirmed by UV reflectance measurement) while the unexposed regions were still amorphous. The SiN$_x$ capping layer was then removed by etching in dilute HF. Both amorphous and polycrystalline silicon were obtained in a single silicon layer as can be seen in Fig. 2.

HYDROGEN EFFUSION AND NEED FOR RE-HYDROGENATION

A critical issue for the fabrication of a-Si:H devices in such a process is the loss of hydrogen from the amorphous regions during the 600 °C anneal. The SiN$_x$ layer is a good diffusion barrier, but not sufficient to prevent out-diffusion of hydrogen in the film. To measure hydrogen content, 250 nm a-Si:H deposited on SiO$_2$/Si substrates (Si to allow infra-red measurement and thin SiO$_2$ to prevent crystallization during the anneal) were used. The atomic hydrogen content in the films was deduced from integrated absorption near 630 cm^{-1}, which is due to the Si-H and Si-H$_2$ wagging modes [10], with a conversion factor of 4.2x10^{-4} cm. The atomic hydrogen content in the as-grown a-Si:H was ~15 at.% and after annealing it reduced to about ~0.3 at.%. The defect state density, which results in mid-gap states, increased due to the hydrogen out-diffusion resulting in increase in sub-gap absorption as measured by photo-thermal deflection spectroscopy (Fig. 3). A similar loss of hydrogen is evident in Si-H stretching mode IR absorption at ~2000 cm^{-1} (Fig. 4). The a-Si:H top-gate transistors made with this film had electron mobilities of only 0.01 cm^2/Vs with an ON/OFF ratio of only ~10^4.

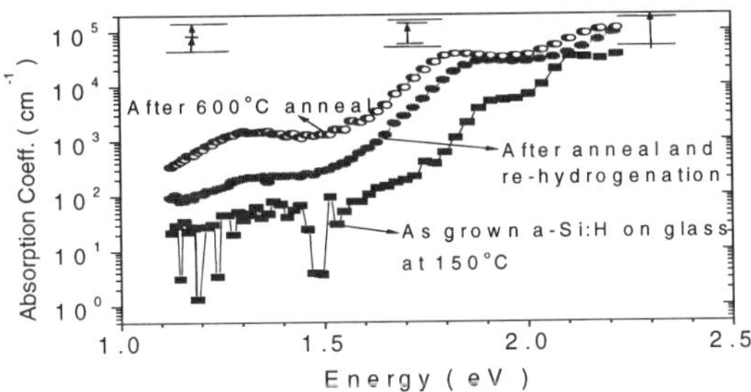

Figure 3. Photothermal deflection spectra of the optical absorption coefficient of a-Si:H films in the as-deposited, annealed with SiN$_x$ cap layer at 600 °C, and re-hydrogenated states. The inset shows the relevant transitions at different energies in a band diagram.

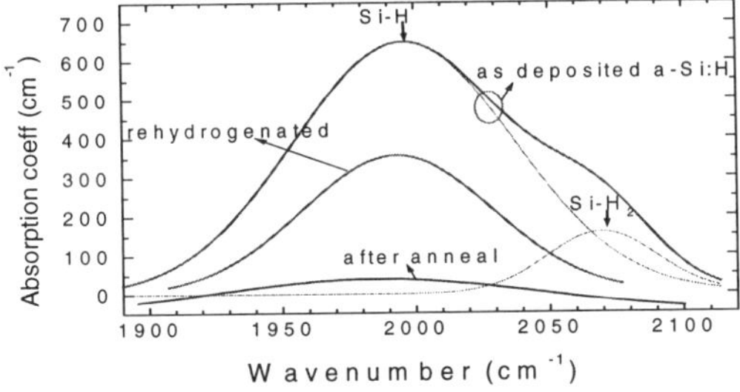

Figure 4. Infrared absorption specturm of the a-Si:H films around 2000 cm^{-1} (Si-H stretch vibration). The spectrum of the as-grown film is decomposed into the Si-H and Si-H$_2$ components. The total H contents of the as-deposited film is ~15 at.%, of the annealed film (with SiN$_x$ cap) ~0.3 at.%, and of the re-hydrogenated film ~4.4 at.%.

Re-hydrogenation is necessary to reduce the defect state density in both the amorphous and polycrystalline regions of the film. This was done after stripping the patterned SiN_x cap layer with dilute HF. The hydrogen plasma conditions were chosen such that hydrogen abstraction and etching are minimal and hydrogen insertion is the dominant mechanism. This is realized by performing the re-hydrogenation in a plasma deposition system with lower sheath bias voltage, at an elevated substrate temperature of 350°C, a low RF power of ~0.2 W/cm^2 and at a high pressure of 1 Torr, so that the hydrogen ion energies are small (<30 eV) when compared to the hydrogen ion energy (~500 eV as estimated by the DC self bias) during the crystallization seeding process. The re-hydrogenation increased the hydrogen content of the film to ~4.4% as measured by the IR absorption at 630 cm^{-1} for an exposure time of 75 min. The increase in hydrogen content resulted in the passivation of the Si dangling bonds and therefore led to a decrease in sub-gap absorption as can be seen in figure 3.

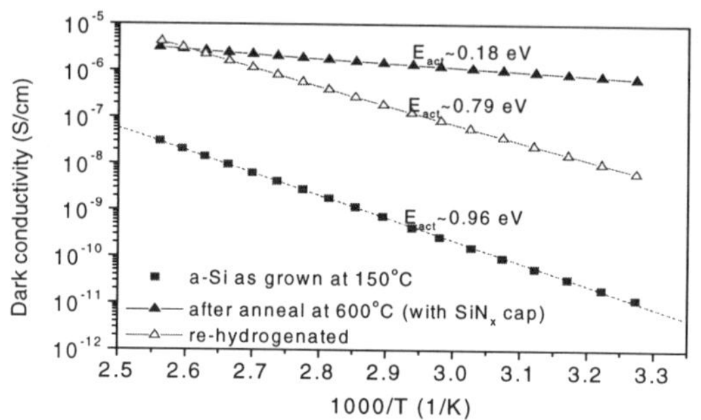

Figure 5. Dark conductivity measurement of the a-Si:H samples. A high activation energy reflects conduction above the mobility edge of low-defect density a-Si:H. Low activation energies reflect activated hopping through a high density of defect states.

Re-hydrogenation, in addition to increasing the hydrogen content of the a-Si:H improves the stability and the electrical characteristics of the film, as the hydrogen predominantly bonds in the form of Si-H in contrast to the as-grown a-Si:H film deposited at 150°C (Fig. 5) which had a considerable Si-H_2 content. The Tauc optical gaps calculated from the optical transmission spectra above the band-gap show that the optical band-gap of the film drops to 1.6 eV from 1.8 eV during annealing and that re-hydrogenation raises the band-gap value to 1.7 eV. Dark conductivity measurement shows that the conductivity of the a-Si:H film increases after the anneal and that the films have low activation energies indicating conduction through defect states. Re-hydrogenation brings the activation energy of the film closer to $E_g/2$ indicating a reduced defect density. But conductivity values are still not as small as that of the as-grown a-Si:H, which may explain the slightly higher leakage currents of the transistors (Fig. 8). Overall the data show that the capped a-Si:H film remains amorphous after the anneal and that the re-hydrogenation results in a device quality a-Si:H film.

TFT FABRICATION AND RESULTS

After the re-hydrogenation step, ~50 nm of n^+ microcrystalline (μc-Si:H) silicon was deposited by PECVD using SiH_4, H_2 and PH_3, at pressure of 900 mTorr, RF power of ~0.3 W/cm^2, and a substrate temperature of 340°C. Device islands were then defined by dry etching in a SF_6/CCl_2F_2 plasma and future channel regions were defined by dry etching just the n^+ μc-Si:H layer in a separate etching step using a CCl_2F_2/O_2 plasma.

The gate dielectric, ~180 nm of SiO_2, was then deposited by PECVD at a substrate temperature of about 250 °C using SiH_4 and N_2O at a RF power density of ~0.1 W/cm^2. Etching in dilute HF opened contact holes to the source and drain regions. Aluminum was then evaporated and patterned to form the gate, and source and drain contacts (Fig. 6). The samples were then annealed at 200°C in forming gas to reduce contact resistance. The same process was simultaneously applied in both polycrystalline and amorphous silicon regions. Note this is not a self-aligned process; the source/drain contacts process are similar to those in a standard bottom gate staggered a-Si:H TFT fabrication.

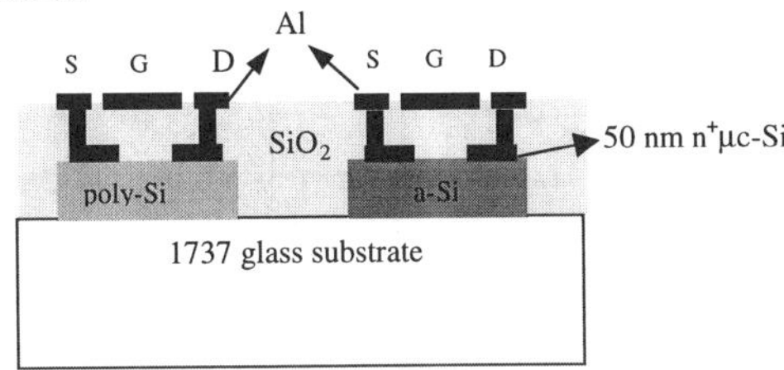

Figure 6. Cross-section of top gate integrated a-Si:H and poly-Si TFTs on glass substrate.

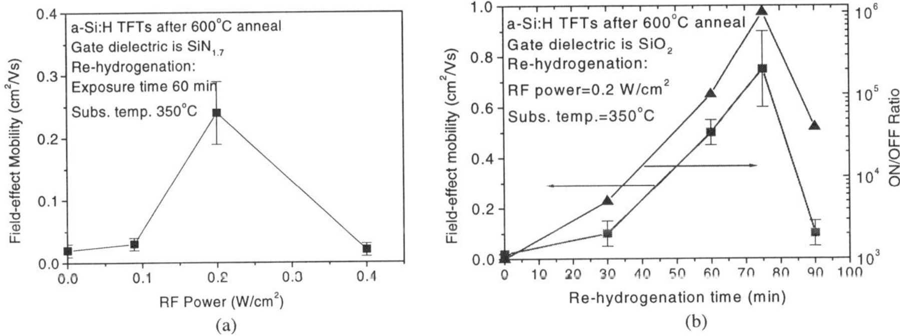

Figure 7. Field-effect mobiltiy of the a-Si:H TFTs for different re-hydrogenation conditions, (a) as a function of RF power, and (b) as a function of exposure time.

Using a slightly different process, polysilicon TFTs with field-effect mobility as high as 75 cm^2/Vs and ON/OFF ratio of ~10^7 with a maximum process temperature of 600 °C were achieved [11]. The properties of the a-Si:H TFTs were critically dependent on the re-hydrogenation conditions (Fig. 7). The best a-Si:H TFT performance was achieved for crystallization anneal of 625 °C and re-hydrogenation time of 75 min at RF power density of 0.2 W/cm^2, resulting in a field-effect mobility of ~1.2 cm^2/Vs and ON/OFF >10^6 (Fig. 8(a)) [6].

For the optimized case with both a-Si:H and polysilicon TFTs integrated together, the field-effect mobilities were ~0.7 and ~15 cm^2/Vs for the a-Si:H and poly-Si TFTs respectively. The leakage current was ~10 fA/μm at V_{DS} = 10 V for the a-Si:H in the OFF state. The ON current of the poly-Si TFT was $\geq$0.5 μA/μm, with ON/OFF ratios of both types of devices $\geq10^5$ (Fig. 8(b)). This is the first demonstration of the integration of a-Si:H and poly-Si TFTs on the same substrate in a single layer without laser processing. These results compare favorably with other work on integrated a-Si and poly-Si TFTs using laser processing which resulted in a-Si TFTs with field-effect mobility of ~0.9 cm^2/Vs and poly-Si TFTs with field-effect mobility of ~20 cm^2/Vs [1]. These are also the best results for a-Si:H top gate TFTs after a 600 °C anneal step .

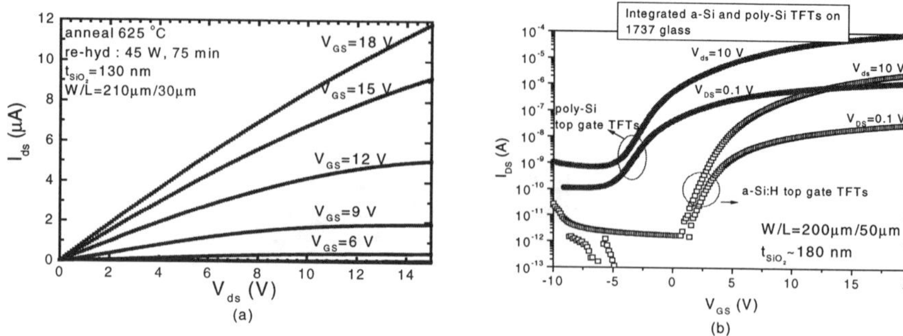

Figure 8. TFT characteristics of (a) the best a-Si:H top-gate TFT fabricated after 600 °C anneal with linear field-effect mobility of ~1.2cm^2/Vs, and (b) optimized integrated a-Si:H and poly-Si TFTs on the same glass substrate.

CONCLUSION

We have successfully integrated high performance a-Si:H and poly-Si TFTs in a single Si layer on the same glass substrate, using a hydrogen-plasma selective crystallization technique. The TFTs share all processing steps and no laser processing is involved. Re-hydrogenation is critical after the crystallization anneal at 600 °C to obtain a device quality a-Si:H film.

ACKNOWLEDGEMENTS

This work was supported by DARPA and the Princeton Program in Plasma Science and Technology (DOE contract no. DE-AC02-76-CHO-3073).

REFERENCES

[1] P. Mei, J.B. Boyce, D.K. Fork, G. Andersen, J. Ho, J. Lu, M. Hack and R. Lujan, Mat. Res. Soc. Symp. Proc. **507**, 3 (1998).
[2] T. Aoyama, K. Ogawa, Y. Mochizuki and N. Konishi, Appl. Phys. Lett. **66**, 3007 (1995).
[3] S. D. Theiss and S. Wagner, Mat. Res. Soc. Symp. Proc. **424**, 65 (1997).
[4] K. Pangal, J.C. Sturm and S. Wagner, Mat. Res. Soc. Symp. Proc. **508**, 577 (1998).
[5] K. Pangal, J.C. Sturm and S. Wagner, J. Appl. Phys. **85**, 1900 (1999).
[6] K. Pangal, J.C. Sturm and S. Wagner, to be published.
[7] A. Yin and S.J. Fonash, Mat. Res. Soc. Symp. Proc. **321**, 683 (1994).
[8] A. Yin and S.J. Fonash, US Patent No. US5624873 (1995).
[9] H. Gleskova, S. Wagner and Z. Suo, Mat. Res. Soc. Symp. Proc. **507**, 577 (1998).
[10] C.C. Tsai and H. Fritzsche, Solar Energy Materials **1** 29 (1979).
[11] K. Pangal, J.C. Sturm and S. Wagner, to be published.

STRUCTURE SENSITIVE HYDROGENATION EFFECTS IN POLYSILICON HIGH VOLTAGE THIN FILM TRANSISTORS

F.J. CLOUGH, Y.Z. XU, E.M. SANKARA NARAYANAN and R. CROSS, Emerging Technologies Research Centre, Department of Electrical and Electronic Engineering, De Montfort University, Leicester UK, fjclough@dmu.ac.uk.

ABSTRACT

Hydrogen passivation of grain boundary and in-grain defects is a key process step in the fabrication of high quality poly-Si TFTs. The sensitivity of the hydrogenation process to device geometry is therefore an important consideration. The effects of rf-plasma hydrogenation on the operating performance of a range of self-aligned and offset drain (L_{off} = 5 to 40 μm) poly-Si TFT configurations is reported. The hydrogenation of offset drain structures results in a predictable increase in the pre-threshold slope and a reduction in the device threshold voltage. However, extended hydrogenation (up to 12 h) can result in a significant reduction in the device drive current (by up to 2 orders). A similar effect is observed in metal field plate HVTFTs in which some portion of the offset region is un-modulated by the additional electrode. The on state conduction in the offset region is examined as a function of hydrogenation time, temperature and planar electric field. The increase in the on resistance is attributed to a reduction in the poly-Si defect density, which moderates carrier transport through the offset region.

INTRODUCTION

Poly-Si TFT technology is a leading candidate for the realisation of large-scale system integration on insulating substrates and is currently used to implement both the pixel switches and on-board peripheral circuitry for active matrix liquid crystal displays [1]. A key process step in the realisation of high quality devices is the passivation of grain boundary and in-grain defects by hydrogenation, typically from a plasma source.

Conventional poly-Si TFT technology is based on the self-aligned configuration shown schematically in Fig. 1 (a). During the fabrication process, the edges of the source and drain regions are perfectly aligned with the edges of the gated channel region. The hydrogenation process in such structures has well documented beneficial effects on both the dc (increased pre-threshold slope and field effect mobility, reduced threshold voltage and off state leakage) [1][2][3] and transient (faster switching) [4] performance of the device.

A number of studies have been carried out to investigate the pathways by which atomic hydrogen is introduced in to the channel of self-aligned poly-Si TFTs. The diffusion of atomic H through poly-Si is slow as the grain boundaries act as efficient hydrogen traps. In contrast, H diffusion through silicon oxide is relatively fast. Mitra et al. suggested that the preferred route for atomic H, to the TFT channel region, is by lateral diffusion through the glass (or thermally oxidised silicon) substrate, to the centre of the device [5]. The H can then diffuse upwards, through the poly-Si layer, to the channel region. This pathway is shown as P1 in Fig. 2 (a).

More recent work has confirmed the poor diffusion of H through poly-Si but suggests that significant upward diffusion of H, from the substrate to the TFT channel, is unlikely [6]. An alternative pathway has been proposed, shown as P2 in Fig. 2 (a), in which the H diffuses downwards through the overlying oxide layer(s) and reaches the channel region after lateral diffusion through the gate oxide [6]. This mechanism is supported by the observation of reduced hydrogenation times for poly-Si TFTs with a thicker gate oxide layer and shorter channel length

Mat. Res. Soc. Symp. Proc. Vol. 557 © 1999 Materials Research Society

[7]. An increased gate oxide thickness offers a larger entrance window for diffusing H atoms and a shorter channel length means that lateral diffusion distances are reduced.

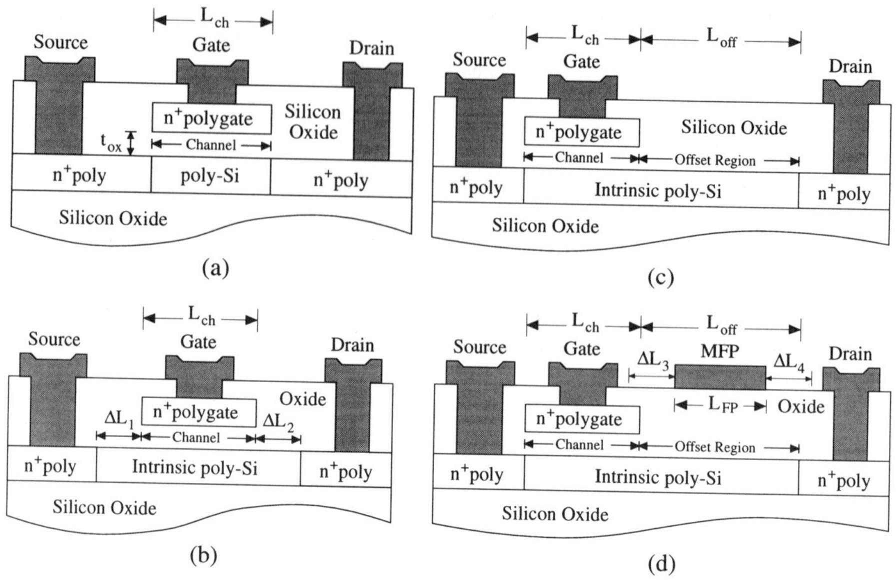

<u>Fig. 1</u>. Self-aligned and offset drain poly-Si TFT structures.

The conventional, self-aligned poly-Si TFT suffers from a number of important operating limitations, such as high off state leakage current and poor saturation of the drain current at high drain bias (the 'kink' effect) [1]. To minimise these effects a range of TFT structures have been proposed which incorporate un-doped [8] and doped [9] offset regions between the source and drain contacts and the edges of the gated channel. A general poly-Si TFT structure, with gate/source and gate/drain offsets, is shown schematically in Fig. 1 (b). Offset structures, with ΔL_1 and ΔL_2 in the range 0.2 to 10 μm, have been shown to reduce the peak electric field at the drain end of the TFT which is known to lower the off state leakage current [10].

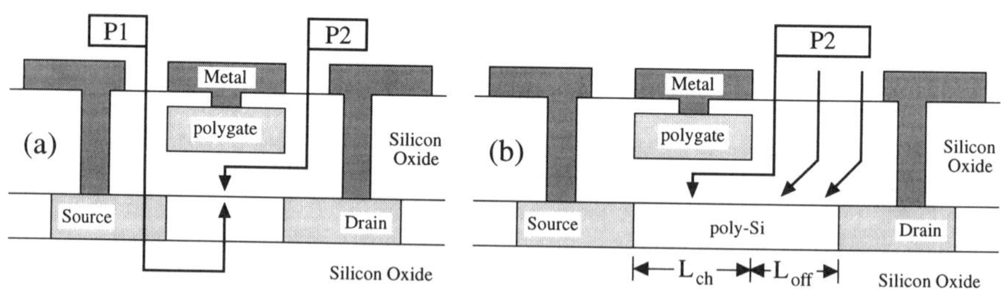

<u>Fig. 2</u> (a). Hydrogenation pathways in conventional self-aligned TFT structures.

<u>Fig. 2</u> (b). Hydrogenation pathway in offset-drain TFT structures.

636

A gate to drain offset region has also been used to realise poly-Si TFTs suitable for high voltage applications (> 200 V). Variations on the standard Offset Drain (OD) HVTFT structure [11] include the integration of a metal field plate (MFP HVTFT) to modulate the conductivity of the offset region [12]. The OD and MFP HVTFT structures are shown schematically in Fig. 1 (c) and Fig. 1 (d) respectively. A general MFP HVTFT structure is shown in which the coverage of the offset region by the metal field plate is adjustable (i.e. the MFP will 'underlap' the offset region when $L_{FP} < L_{off}$ and 'overlap' the offset region when $L_{FP} > L_{off}$). For typical HVTFT structures, the offset length (L_{off}) can vary from 5 to 40 µm.

To date, no study, comparable to those carried out on self-aligned TFT structures, has been reported which describes the relationship between the hydrogenation conditions and the performance of poly-Si TFTs with an integrated offset region. In this report, we investigate the efficiency of the hydrogenation process for a range of offset structures. In particular, the critical role played by the offset region in determining device performance is discussed.

EXPERIMENT

Self-aligned LV TFTs (W = 100 µm, L_{ch} = 5, 10, 15, 20 µm), OD HVTFTs (W/L_{ch} = 100 µm/15 µm, L_{off} = 5, 10, 20, 40 µm) and MFP HVTFTs (W/L_{ch} = 100 µm/15 µm, L_{off} = 10, 20, 30, 40 µm) have been fabricated. All devices presented here are n-channel and were manufactured on the same thermally oxidised (2µm) c-Si wafer, using the low temperature (< 615°C) SPC poly-Si (grain size ≈ 50 nm) process reported previously [13]. Separate phosphorus implantations were used to define the conductivity of the n^+ poly-gate and the n^+ source and drain regions. During the implantation steps the TFT offset region was protected by the either the poly-gate layer or by a thick layer of photoresist. The offset region is therefore intrinsic.

The final stage in the fabrication process was the plasma hydrogenation of samples using a capacitively coupled, rf-PECVD system. Devices were hydrogenated for between 30 min. and 12 h. All hydrogenations were carried out at 350°C, 500 mTorr and 0.1 mW/cm^3 using 100 sccm of pure H_2. Current-voltage measurements were made at a range of temperatures up to 373 K using a pc-controlled HP4140B picoammeter and voltage source.

RESULTS AND DISCUSSION

Figs. 3 (a) and (b) show typical transfer characteristics for self-aligned LV TFTs (L_{ch} = 10 and 20 µm) which have been exposed to 1 h and 12 h hydrogenations respectively. For easy analysis the characteristics have been normalised to a standard W/L_{ch} ratio of 100 µm / 5 µm. Normalised characteristics of the un-hydrogenated devices were identical and one is included in each plot for comparison.

Inspection of Figs. 3 (a) and (b) confirms the beneficial effects of plasma hydrogenation on the transfer characteristics of self-aligned TFT structures. The observed improvements in device pre-threshold slope, threshold voltage and field effect mobility arise as a result of the reduction in the poly-Si deep and tail state defect densities. The previously reported channel length dependence of the hydrogen passivation process can also be observed. The shorter channel length TFT (L_{ch} = 10 µm) shows a more rapid improvement after 1 h exposure to atomic hydrogen. After an extended 12 h hydrogenation the transfer characteristic of the longer channel length device (L_{ch} = 20 µm) reaches parity with the shorter channel device. This is further strong evidence in support of hydrogenation pathway P2, as shown in Fig. 2(a).

The effect of the hydrogenation process on a standard offset drain (L_{off} = 5 µm) TFT structure (see Fig. 1 (c) or Fig. 1 (b) with ΔL_1 = 0) is shown in Fig. 4. As the hydrogenation exposure time is increased from 0.5 to 12 h a steepening of the pre-threshold slope is again evident, consistent with a reduction in the poly-Si deep defect density. However, the most striking observation is the reduction in the TFT on current by almost 2 orders of magnitude.

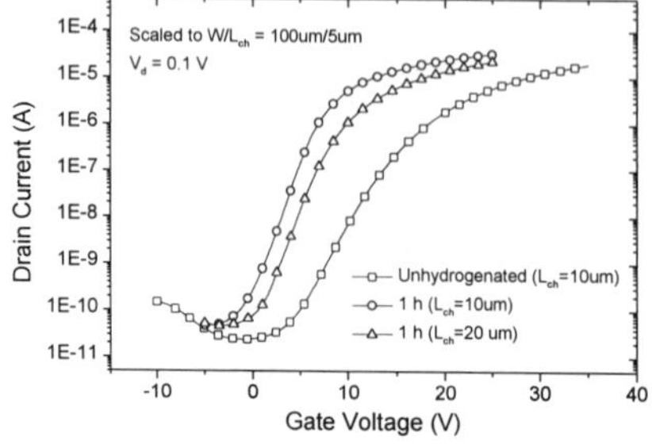

Fig. 3 (a). LVTFTs (1 h hydrogenation).

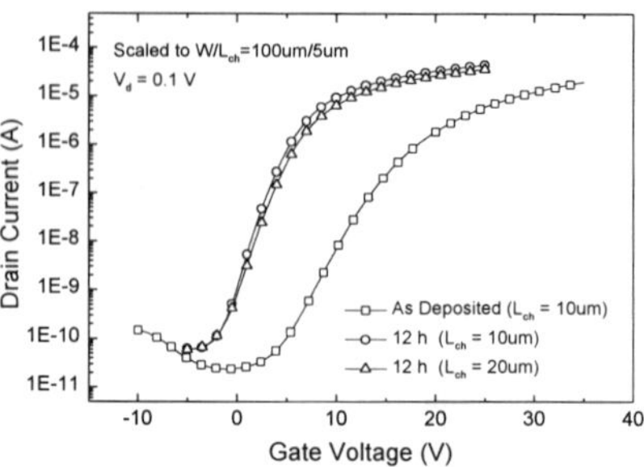

Fig. 3 (b). LV TFTs (12 h hydrogenation).

The reduction in the on current is further illustrated in Fig. 5, which shows typical transfer characteristics for hydrogenated (0.5, 1 & 12 h) OD HVTFTs with offset lengths (L_{off}) of 20 and 40 μm. The hydrogenation induced improvements in the pre-threshold slope are clearly independent of the offset length. For a fixed drain bias ($V_d = 10$ V) and hydrogenation time, a longer offset length results in a lower on current. The 12 h hydrogenation again results in the most significant degradation in the TFT drive current.

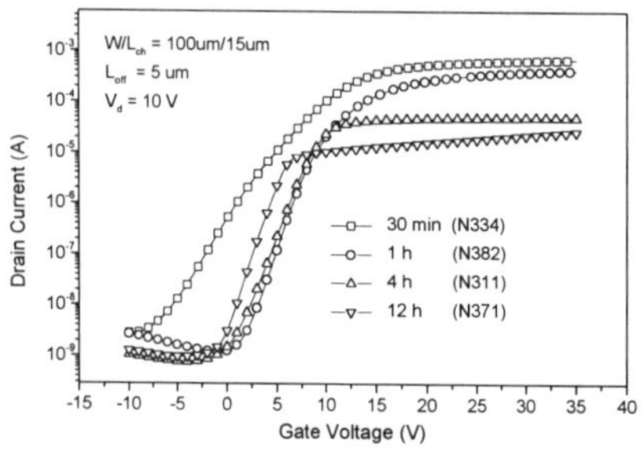

Fig. 4. Transfer characteristics of hydrogenated OD HVTFTs (0.5, 1, 4, 12 h).

Fig. 5. Transfer characteristics of hydrogenated OD HVTFTs with various L_{off}.

The consistent improvement in the pre-threshold slope for OD structures can be simply understood in terms of a reduction in the poly-Si deep defect density in the region under the gate. Hydrogenation of the channel region in OD devices proceeds as for conventional, self-aligned TFTs. The key considerations are therefore the ease of hydrogen access through the overlying oxide layers and the channel length (L_{ch}). The L_{off} independence is therefore unsurprising. Comparison of Figs. 4 and 5 with Figs 3 (a) and (b) suggests that hydrogenation of the channel region in OD devices may actually proceed more rapidly. This can be understood in terms of the improved access to diffusing hydrogen, which is a direct result of offsetting the drain metallisation. This is illustrated in Fig. 2 (b).

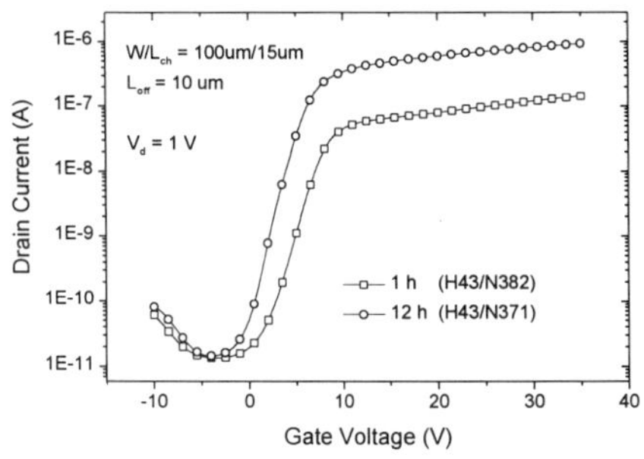
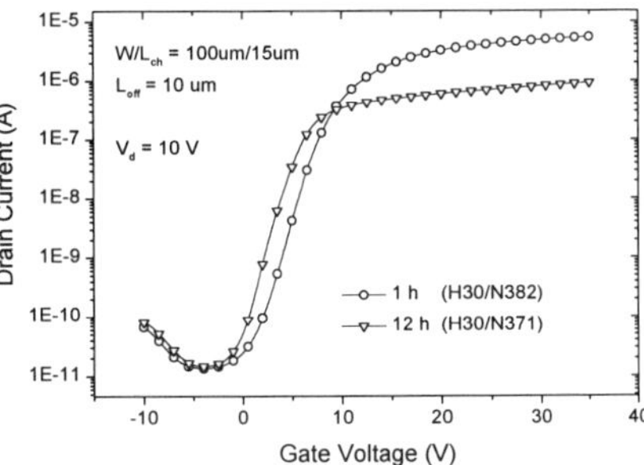

Fig. 6 (a). 'Overlapped' MFP HVTFT.   Fig. 6 (b). 'Underlapped' MFP HVTFT.

A similar reduction in the post-hydrogenation drive current has been observed for MFP HVTFT structures. Figs. 6 (a) and (b) demonstrate the effect of an extended hydrogenation on 'overlapped' ($L_{FP} = 3$ μm $+ L_{off} + 3$ μm) and 'underlapped' ($L_{off} = 3$ μm $+ L_{FP} + 3$ μm) MFP devices respectively ($V_{FP} = 30$ V). The characteristics of 'overlapped' devices show a consistent improvement with increasing hydrogenation time. However, 'underlapped' devices show a degradation in the on current similar to that observed for OD structures with no metal field plate.

The observation of an improved pre-threshold slope is not consistent with device degradation by electrostatic charging damage [14]. The reduction in the post-hydrogenation drive current of OD and 'underlapped' MFP structures is indicative of an increase in the series resistance of the un-modulated offset region (i.e. a decrease in the conductivity of intrinsic poly-Si after hydrogenation). Space charge limited conduction (SCLC) has been proposed as a possible mechanism for charge transfer in the offset region. Hydrogenation will reduce the poly-Si trap density. This would result in an increase in space charge limited conduction [15], contrary to our observations.

Saito et al. have observed a similar increase in the post-hydrogenation resistivity of un-doped poly-Si [16]. This is attributed to a conduction mechanism which is a combination of a conventional thermionic (or drift) current (via excitation to the mobility band edge) and hopping current (through the band tails).

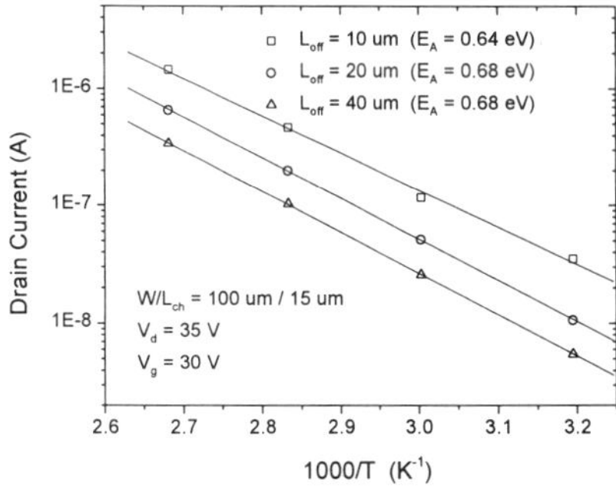

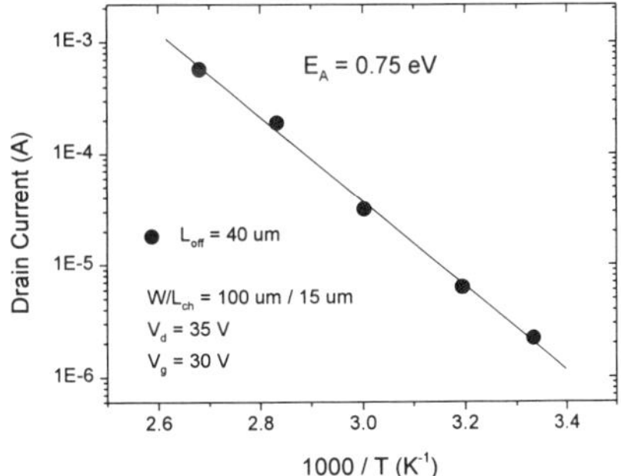

Fig. 7 (a). Activation energy plot for unhydro-genated OD TFTs with various L_{off}.

Fig. 7 (b). Activation energy plot for 12 hydrogenated OD TFT ($L_{off} = 40$ μm).

We have measured the activation energy of the drain current (E_A) of hydrogenated (12 h) and un-hydrogenated OD HVTFTs (25 to 100°C). The gate voltage was set to 30 V to ensure that the drain current is limited by conduction through the offset region. Fig. 7 (a) shows an activation energy plot for un-hydrogenated OD HVTFTs. The measured activation energy of 0.66 ± 0.02 eV is independent of L_{off}. Conduction through the offset region of un-hydrogenated devices can be interpreted in terms of thermal excitation from the high trap density around the Fermi level to the conduction band mobility edge (n-channel device). Within the temperature range investigated there is no evidence of a significant current component due to hopping. A similar E_A value, for thermionic drift at room temperature and above, has been reported previously and was attributed to the p-type nature of intrinsic poly-Si [15]. The increase in the current as L_{off} is reduced is a result of the higher drift field (V_d is constant $= 35$ V).

Fig. 7 (b) shows the activation energy plot of an OD HVTFT structure ($L_{off} = 40$ um) after a 12 h hydrogenation. The activation energy for conduction through the offset region increases to 0.75 eV. This can be understood by a preferential decrease in the density of defect states above the un-hydrogenated Fermi level. This results in a shift in the Fermi level towards the valence band thereby increasing the activation energy for electron conduction. Hence the observed increase in the resistivity of the offset region. This effect is highlighted in OD structures due to the easy access of atomic hydrogen to the offset region shown in Fig. 2 (b).

CONCLUSIONS

The sensitivity of the hydrogenation process to the geometry of self-aligned and offset drain TFT structures has been clearly demonstrated. Degradation in the on state performance of OD TFT structures (with some part of the offset region left un-modulated by metal field plate) can be severe following an extended hydrogenation (1 to 2 orders reduction). The observed increase in the resistivity of the intrinsic poly-Si offset region has been attributed to the efficient passivation of the offset region during the hydrogenation process. This leads to a much reduced defect density and a modification of the conduction mechanism through this the offset region.

REFERENCES

[1]   S.D. Brotherton, Semiconductor Science & Technology, **10**, 721, (1995).
[2]   T.I. Kamins and P.J. Marcoux, IEEE Electron Device Letts., **EDL-1**, 159, (1980).
[3]   I-W. Wu, A.G. Lewis and A. Chiang, IEEE Electron Device Lett., **EDL-10**, 123, (1989).
[4]   Y.Z. Xu, F.J. Clough, Y. Chen, E.M. Sankara Narayanan and W.I. Milne, IEEE Electron Device Letts., **EDL-20**, 80, (1999).
[5]   U. Mitra, B. Rossi and B. Khan, J. Electrochem. Soc., **138**, 3420, (1991).
[6]   W.B. Jackson, N.M. Johnson and D. Smith, Appl. Phys. Letts., **61**, 1670, (1992).
[7]   K-Y. Choi, J-S. Yoo, M-K. Han and Y-S. Kim, Jpn J. Appl. Phys., **35**, 915, (1996).
[8]   T. Unagami and B. Tsujiyama, IEEE Electron Device Letts., **3**, 167, (1982).
[9]   S. Seki, O. Kugure and B. Tsujiyama, IEEE Electron Device Letts., **8**, 434, (1987).
[10] J.G. Fossum and A. Ortiz-Conde, IEEE Trans. Electron Devices, **32**, 1878, (1985).
[11] K. Tanaka, H. Arai and S. Kohda, IEEE Electron Device Letts., **9**, 23, (1988).
[12] T-Y. Huang, I-W. Wu and R.H. Bruce, IEEE Electron Device Letts., **11**, 244, (1990).
[13] F.J. Clough, E.M. Sankara Narayanan, Y. Chen, W. Eccleston and W.I. Milne, Appl. Phys. Letts., **71**, 2002, (1997).
[14] K-Y. Lee, Y-K. Fang and S.G. Wuu, IEEE Electron Device Letts., **EDL-18**, 187, (1997).
[15] D.R. Lamb, 'Electrical conduction mechanisms in thin insulating films', (Methuen & Co. Ltd, 1967).
[16] Y. Saito, M. Aomori and H. Kuwano, J. Appl. Phys. **81**, 754, (1997).

QUASI DRIFT AND QUASI DIFFUSION: A GRAIN SIZE DEPENDENT POLYSILICON TFT MODEL

W. ECCLESTON

Department of Electrical Engineering and Electronics, University of Liverpool, Liverpool L693BX, UK

ABSTRACT

Most of the models for Thin Film Transistors take no account of the way that electrons move across grain boundaries. The purpose of this work is to produce an analytical model, physically based, that takes account of both the change in height of the potential barriers that separate the grains, and the change of carrier density in the grains with gate voltage. The high current and subthreshold regions are treated. The model enables the change of field effect mobility with grain size and temperature to be determined. Two closed form expressions are provided which should be of value to both device and circuit designers, as well as providing insight to the physical processes occurring in such devices. They form a close analogy with the electrical characteristics of MOSFETs on crystalline silicon, diffusion and drift being replaced by quasi-diffusion and primary quasi-drift.

INTRODUCTION

The earliest attempts to represent the electrical characteristics of field effect transistors were developed in the 1970s. It was then that the gradual channel approximation became widely accepted for the high current region, and diffusion was accepted as the primary mechanism controlling current flow in subthreshold. The same basic equations have, since then, been used for MOSFETs, and for Thin Film Transistors (TFTs) with only minor modification. The channel regions of these two types of devices are, however, very different. In the TFT there are grains, usually smaller then the channel length, and grain boundaries containing traps which change their occupancy with gate voltage. Potential barriers exist therefore at the boundaries and are always of such a polarity as to impede electron flux down the channel. This is a two dimensional system suitable for numerical solutions(1)-(3).

Most TFTs are made in intrinsic polycrystalline silicon so that conventional depletion layers do not exist along the channel(4). Increase in gate voltage increases the density of electrons in traps and the grain. The latter is closely related to the Debye Length. In intrinsic silicon the Debye Length is always much longer (5) than the grains. There is, therefore, no dip in the conduction band at the centre of the grain.

With increase of gate voltage there is a rapid reduction in the Debye Length and the electrons sit at the lowest, most positive point of the conduction band, in the absence of a drain voltage. This is the dominant feature of the conduction process in the subthreshold region.

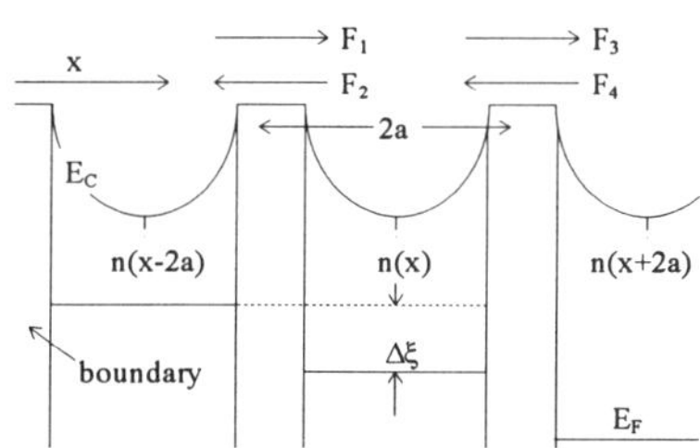

Fig 1 The four fluxes that contribute to the net flux of electrons into a grain.

Mat. Res. Soc. Symp. Proc. Vol. 557 © 1999 Materials Research Society

As the gate voltage is further increased, and with it the droop conduction band edge at the centre of the grain, it approaches the fermi level and further band bending is impeded. The barriers to electron flux gradually disappear, at this point, and have little control on the current flow. This is the gradual channel regime.

SUBTHRESHOLD REGION

Consider first the subthreshold region. Fig.1 shows schematically the variation of the conduction band with distance across a series of grains. One grain is centred at distance x and adjacent grains are centred at x+a and x-a, so that the grain dimension is 2a.

Conventionally the subthreshold region is most easily studied with a sufficient drain voltage to just cause current flow. Because the current is very low the potential drop at any point in the channel is too small to significantly affect the gate field and hence the surface potential. The carrier density is significantly changed by increase in gate voltage so that there must be a change in the fermi level down the channel to sustain the current, and this change must be from grain to grain, whilst remaining almost totally flat within each grain. (For a given current density the slope of the fermi. level is, in any device, inversely proportional to carrier density.) The carrier concentrations n(x-2a)...n(x)...n(x+2a) for any energy level at, or above the conduction band edge, are different and there is a reducing concentration gradient down the channel region from grain to grain. This is analogous to the carrier concentration gradient seen in the subthreshold region of a conventional MOSFET(6).

We have for any semiconductor, or any assembly of particles, a relation ship of the form:

$$\frac{dF}{dx} = -\frac{dn}{dt} = -v^* n$$

which when applied to granular material, with dx equal to the grain size 2a:

$$\Delta F = -2va\Delta n$$

where v^* is a constant having units of t^{-1}. ΔF is the difference in the fluxes across the two grain boundaries on either side of the grain at x:

$$\Delta F = F_1 - F_2 - (F_3 - F_4)$$

We have several fluxes, F_1 to F_4, defined clearly in Fig.1 related by four equations

$$F_1 = -2va[n(x-2a) - n(x)], \quad F_2 = -2va[n(x) - n(x-2a)],$$
$$F_3 = -2va[n(x) - n(x+2a)], \quad F_4 = -2va[n(x+2a) - n(x)],$$

It is readily shown that for a granular material:

$$\Delta F = -va^2 \exp-\left(\frac{q\Delta\phi_B}{kT}\right).\left[\frac{dn(x)}{dx}\right] \quad (2)$$

The exponential term is introduced so that the carrier densities, and flux densities related to the energy level of the carriers passing across barrier. n_0 is the electron concentration at the barrier, immediately adjacent to the grain boundary, as shown.

$$n_0 = n_i \exp\left[\frac{q\phi_0(y)}{kT}\right]$$

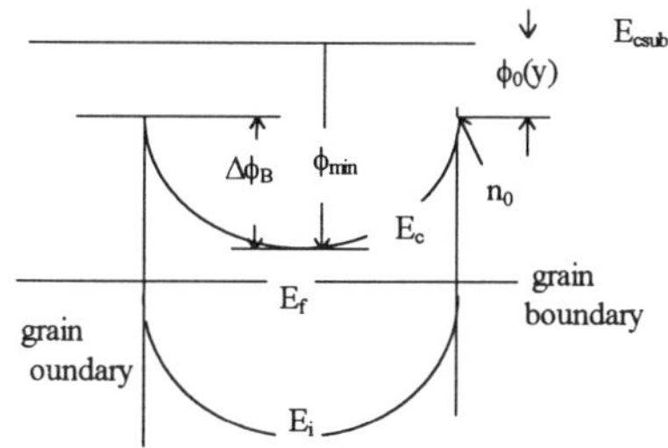

Fig. 2 The definition of various energy levels associated with the grain and the grain boundary.

$\Delta\phi_B$ is the height of the barrier to electron flow also shown in Fig.2. The quantity v has units s^{-1} and is different from v^* because it includes numerical factors which allow for the movement of electrons out of the plane of Fig.1.

$\phi_0(y)$ is the band bending at the surface with respect to the conduction band at the back of the film. E_C and E_{Csub} the conduction band edges at the top and bottom of the the film and E_V is the valence band edge at the front of the film valence band edges respectively and E_f is the fermi level. The quantity n_i is an intrinsic carrier concentration associated with material of the grain. If we were to assume that the grain boundary is made up of amorphous material then this quantity would be the intrinsic carrier density of αSi. Let the field in the y direction by $E(y)$ and the field when y=0 be $E_0(y)$. The current is obtained by integrating, through the surface region, at the edge of the grain immediately adjacent to the grain:

$$I_x = (W/L)qva^2 \int \frac{n_0}{E(y)} d\phi(y) \qquad (3)$$

where W is the width of the channel and L is its length. W/L is the aspect ratio.
For a grain boundary of width comparable to the thickness of the gate dielectric Poisson's equation in one dimension can be applied:

$$\frac{d^2\phi(y)}{dy^2} = \frac{qN_t(y)}{\varepsilon}$$

where y is the direction normal to the surface, $N_T(y)$ is the density of charged traps within the grain boundary, and ε is the total permittivity of the material.. We have made a number of assumptions here. The first is that the grain boundary is sufficiently wide to quote a volume density of charge to which one dimensional electrostatics can be applied. The resulting N_T is then an effective density since this approximation will not always apply. Implicit in this development is the assumption that all the grains are equal in size. In real polysilicon this is not the case. The carriers find a route from source to drain through the valleys where the conduction band dips to its lowest points due to the Debye Length becoming comparable with the grain size. The values of the electrical device width, and aspect ratio, are not then the same as the geometrical values. To take this into account is well beyond the scope of this analysis. We assume that the traps are acceptor like, being negative when filled with electrons but otherwise neutral. This is consistent with the material being intrinsic, since these traps would be neutral with the fermi level near to the centre of the energy gap. In order to simplify the algebra we define a density of negatively charged traps $N_T(y)$ which increases with   surface potential $\phi(y)$ according to a relationship of the form:

$$N_t(y) = N_0 \exp(\frac{q\phi(y)}{kT_C}) \qquad (4)$$

The quantity T_C is simply a way of specifying the rate at which the density changes. We have termed it the characteristic temperature of the distribution. Its definition is different from that used elsewhere. N_0 is the small density of negative charge when there is no band bending.

More realistically it is the extrapolated value, back from the level at which the trapped charge first becomes significant.

The field E(y) is obtained by the integration of Poisson's equation:

$$E\ (y) \cong (\frac{qkT_C N_o}{\varepsilon})^{1/2} \exp(\frac{q\phi(y)}{2kT_C})$$

Applying Gauss' Law across the interface between the gate oxide and the grain boundary:

$$C_0[V_{GT} - \phi_0(y)] \cong (qkT_C N_0\varepsilon)^{1/2} \exp[\frac{q\phi_0(y)}{2kT_C}]$$

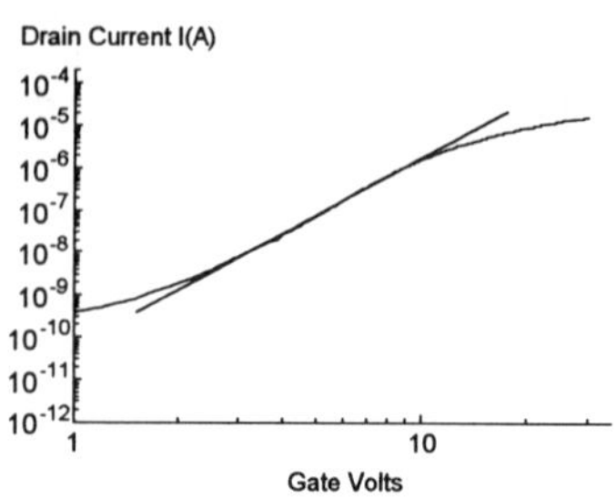

Fig. 3 The variation of drain current with gate voltage for the subthreshold region. The straight line is the best fit to the curve for the power law obtained from the quasi diffusion analysis

where C_0 is the capacitance per unit area of the gate electrode. V_{GT} is gate voltage (less the threshold voltage) associated with work function difference, oxide charge and interface states, but not charge associated with the grain boundary.

$$I_D = (W/L)qn_i v\alpha^2 (\frac{qkT_C N_0}{\varepsilon})^{-1/2}(\frac{2kT_C}{qC})\exp(\frac{q\phi_0(y)C}{2kT_C})\ (5)$$

after integrating between $\phi_0(y)$ and $\phi_0(y)=0$. We have assumed that $q\phi_0(y)C>>2kT_C$ an approximation that will be seen to be self-consistent later. From equation (4) if $V_{GT}>>\phi(0)$

$$[C_0V_{GT}\varepsilon^{-2}(\frac{qkT_C N_0}{\varepsilon})^{-1/2}]^C = \exp[\frac{q\phi(0)}{2kT_C}]$$

$$I_D = \left(\frac{W}{L}\right)qn_i v\alpha^2\left(\frac{qkT_C N_0}{\varepsilon}\right)^{-1/2}\left(\frac{2kT_C}{qC}\right)\left(C_0\varepsilon^{-2}(\frac{qkT_C N_0}{\varepsilon})^{-1/2}\right)^C V_{GT}{}^C \qquad (6)$$

Fig. 3 for a real polysilicon device shows a value of C~4.5 giving T_C=650K if V_T is taken as 0V. Note that T_C can vary over perhaps 100K between devices. It is, in any case, difficult to measure because of the onset of quasi drift at higher current and temperature activated leakage at lower currents.(Fig.4) A high value of T_C, in the quasi-diffusion region, is an indication of trapping at levels which increase exponentially in density in the sub-threshold region. From the practical point of view polysilicon transistors belong to the silicon on insulator class of device, which are prone to internal heating [8] because of the high thermal insulation of the substrate. In analytical terms we would expect from the earlier theory that the characteristic temperature would stay constant with measurement temperature for the sub-threshold regime There is

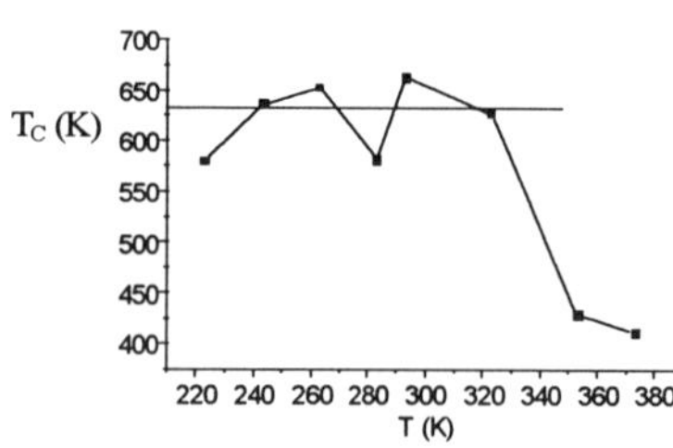

Fig .4 The variation of the characteristic temperature of the trap distribution with the temperature of the measurement, excluding joule heating.

some difficulty in locating this region at higher temperatures since the off-current increases rapidly. The slope of the sub-threshold plot, therefore, becomes skewed towards lower values, and so approaches the measurement temperature. Note that at lower measurement temperatures, however, the value of T_C is relatively constant (~625K)as predicted by the quasi-diffusion analysis where the slope of the sub-threshold region is less distorted by the high off current.

QUASI DRIFT

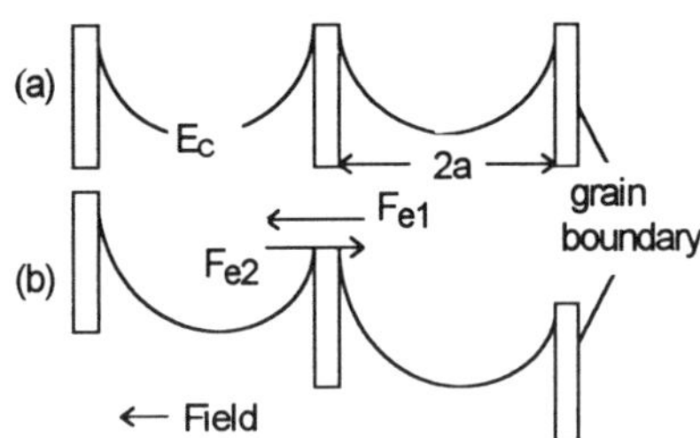

Fig 5 The band diagram for the polysilicon region in the gradual channel regim e. (a) is w ith no ap plied voltage and (b) is w ith applied drain voltage

We have also applied this kind of analysis to the gradual channel region (7). This is the region where the reistive potential drop, associated with current flow, modifies the oxide field and hence the surface potential. The resulting band diagram is shown in Fig. 5 The effect of the field is to reduce the barrier for electrons moving to the positive drain. This has the effect, providing the barrier across each grain is lowered by no more than a few kT of giving ohmic conduction. The fermi level remains horizontal in each drain, and equidistant from the minimum. The voltage drop between the source and drain may still be quite large so this assumption is still useful for real device characteristics. The procedure is then to develop expressions for the current voltage characteristics in the same way employed above for quasi diffusion region. The result is an equation of the form:

$$ I = (W / L)\left(\frac{2A}{C(C+1)D^C}\right)\left[\left(V_{GT} - V_D\right)^{C+1} - V_{GT}^{C+1}\right] \qquad (7) $$

where $A = \dfrac{2\mu_j\, n_i kT_C}{K}$, $\quad C = \dfrac{2T_c}{T} - 1 \qquad D = \left(\dfrac{\varepsilon\, x_t K}{\varepsilon_o}\right)^C$ and $K = (2kT_C N_0 / \varepsilon)^{1/2}$.

where T_C and N_0 are different from the values needed to fit the subthreshold characteristics.

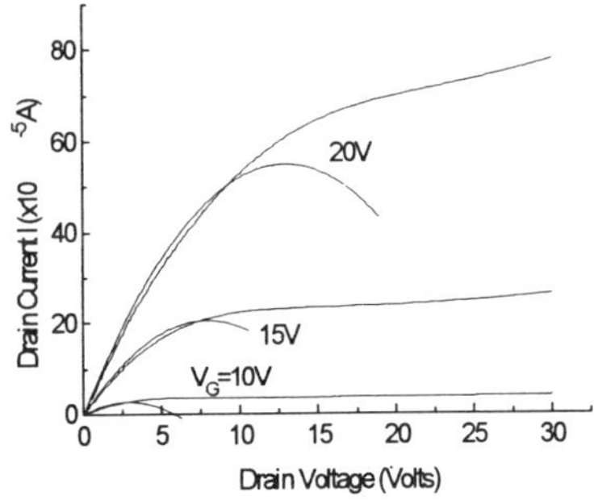

Fig.6 The output characteristic of a polysilicon T F T com pared w ith the theoretical curve

μ_j is the 'jump mobility', to be discussed below. When T_C=T this equation has the form:

$$ I = \mu\, C_0\left(\frac{W}{L}\right)\left[V_{GT}V_D - \frac{V_D^2}{2}\right] \qquad (8) $$

the classical gradual channel relationship developed first for the CdS transistor but later borrowed by silicon, modified, but now returning in the context of Silicon on Insulator. The fit to real device characteristics is shown in Fig.6 The downturn in the current for the theoretical curve occurs for all one dimensional equations and is associated with pinch off condition. For the purpose of circuit design an equation in V_{GT} only can be developed.

There is the possibility of predicting the temperature dependence of the field effect mobility. This is not the same as μ_j. The simplest approach is to obtain μ in equation (8) by equating it with (7). At large grain size, when it becomes larger than the channel length, the mobility will be dominated by the surface properties of the material of the grain. The field effect mobility μ_{FE} is given by:

$$\frac{1}{\mu_{FE}} = \frac{1}{\mu_{FE\infty}} + \frac{1}{\mu_{FEa}}$$

where μ_{FEa} is the value corresponding to a grain size a, and $\mu_{FE\infty}$ is the value for grain larger than the channel area. By fixing one point from a real device made on polysilicon of about 0.1μm, we have computed the predicted dependence of field effect mobility on grain size as shown in Fig7.

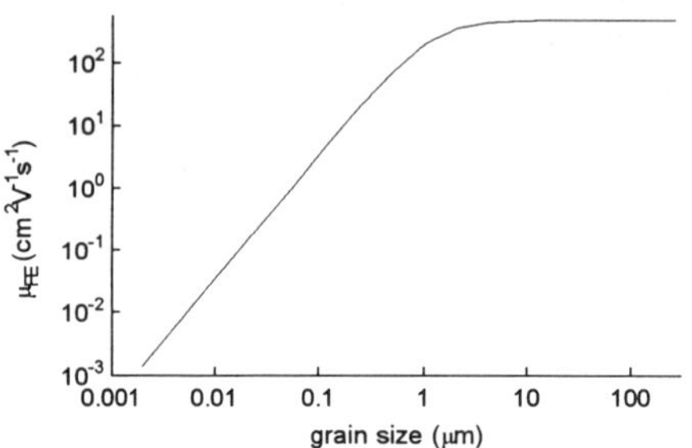

Fig. 7 The variation of field effect mobility with grain size as predicted by the quasi drift equation.

CONCLUSIONS

A model is proposed for the movement of electrons between source and drain in polysilicon, in the sub-threshold region. It is based on the principle of quasi-diffusion. A concentration gradient exists between the electrons in adjacent grains which fall monotonically from the centre of the grain immediately adjacent to the source to that adjacent to the drain. Because the flux of the electrons over the boundaries depends on the concentrations within the grains, the behaviour is very similar, at least in terms of the equations, to that encountered with pure diffusion. Results for real devices are examined in the light of the analysis, and it is deduced that the characteristic temperature of the traps is not the same as the temperature as the measurement. There is strong trapping in the grain boundary across the wide range of surface potential represented by the sub-threshold region. This, in contrast to the low current region which was analysed earlier on a similar basis, gives a characteristic temperature equal to actual temperature of the measurement. The simplest interpretation in the latter case was that there is little trapping but the carrier mobility is much reduced.

REFERENCES

[1] G. Baccarini and B. Ricco, J.Appl.Phys., 49,(11), pp. 5565-5570, November, 1978.

[2] A. N. Khondker, D. M. Kim, S. S. Ahmed and R. R. Shah, IEEE Transactions on   Electron Devices, vol. ED31, No. 4, pp. 493-500, April, 1984.

[3] J. G. Fossum and A. Ortiz-Conde, IEEE Trans. on Electron Dev., vol. ED-30, No. 8, 933 1983.

[4] J. Y. W. Seto, Journal of Applied Physics, 46, No. 12, pp. 5247, 1975

[5] G. Fortunato, D. B. Meakin and P. Migliorato, Japanese Journal of Applied Physics, vol. 27, No. 11, pp. 2124, Nov., 1988.

[6] Stuart S. Eccleston W. Electron. Lett., 8, 9, 1972, 225-7.

[7] Eccleston W. Mat.Res.Soc.Symp.Proc. 471,149-153 1997

[8] McDaid, L, Hall, S., Eccleston, W., Alderman, J. C. Electron. Lett., Vol. 27, No. 2, 2006-2007, 1991.

APPLICATION OF SELF-ALIGNED AMORPHOUS SI THIN-FILM TRANSISTORS

J.P. LU, P. MEI, C. CHUA, J. HO, Y. WANG, J. B. BOYCE, R. LUJAN*
Xerox Palo Alto Research Center, Palo Alto, CA
*Xerox dpiX, Palo Alto, CA.

ABSTRACT

We have successfully used self-aligned Amorphous Si Thin-Film Transistors, fabricated by a laser doping/annealing process, to construct dynamic shift register circuits, which can be used as gate-line drivers or in other peripheral circuits for flat-panel displays and imagers. Taking advantage of easily scaling down the TFT channel length in a self-aligned process, much higher circuit speeds can be achieved compared to that of circuits using conventional TFTs. We have successfully demonstrated a four-phase dynamic shift register, operating at a clock speed higher that 250 kHz (1 μs for each clock phase) built on 3 μm channel length TFTs. This new technology opens up possibilities for integrating peripheral circuits in flat-panel displays and imagers based on a-Si TFTs.

INTRODUCTION

Amorphous silicon thin-film transistors (a-Si TFTs) are widely used as switch elements for both large area active-matrix liquid crystal displays and large-area image sensors, because of their excellent off-state current, high on/off current ratio and compatibility with low-cost, large-area fabrication process. However, there are two major factors limiting the performance of conventional a-Si TFTs. First, there is significant parasitic capacitance between the source/drain electrodes and the gate. And second, it is difficult to manufacture short-channel TFTs. The significant parasitic capacitance between the source/drian electrodes and the gate can cause image flicker and sticking in displays, poor noise performance for flat-panel imagers, and higher propagation delay when used in digital circuits. In our previous study, we have demonstrated that using self-aligned a-Si TFTs to eliminate the parasitic capacitance can improve the propagation delay of a-Si TFT inverters by over 35%. As demonstrated by the main-stream single-crystalline-Si technology, scaling down the transistor channel length provides tremendous advantage for improving circuit speed. The difficulty of manufacturing short channel a-Si TFTs with the conventional process prevents a-Si TFT technology from realizing this advantage to making circuits that can be operated at a useful speed. Our previous study of short channel a-Si TFT ring oscillators demonstrated propagation delays as short as 1 μs per logic stage, which is sufficiently high for most applications.

In this paper, we demonstrate that using self-aligned, short channel a-Si TFTs and properly designed circuits, four-phase dynamic shift registers can be operated at higher than 250 kHz. On the same wafer, TFTs with longer channel length preserve excellent properties such as very low off-state current, sharp turn on, and high on-state current, and are ideal for pixel switches. This demonstration indicates that some of the peripheral circuits can be easily integrated in flat-panel displays and imagers using the self-aligned a-Si TFTs technology.

Mat. Res. Soc. Symp. Proc. Vol. 557 © 1999 Materials Research Society

EXPERIMENTS

<u>Device structure and characteristics</u>

Figure 1a shows the typical structure of a conventional bottom-gate a-Si TFT. In the standard process for this kind of device, source/drain electrodes and gate are defined in different masking steps. Therefore, a reasonably large overlap between the channel and the source/drain electrodes must be allowed in order to avoid stringent lithography requirement. This overlapping area is responsible for both the significant parasitic capacitance and the difficulty of scaling down the channel length.

In order to improve the performance of a-Si TFTs, self-aligned processes have been proposed to eliminate the necessity for this overlapping area [1,2,3,4]. Different processes, including ion implantation followed by laser annealing [1], ion doping and chromium silicide formation [2,3], and laser doping [4] have been developed to eliminate the requirement for the overlapping region between the source/drain electrodes and the channel. Figure 1b shows the device structure of our self-aligned a-Si TFTs fabricated by the laser doping/annealing technology [4]. The top passivation island was formed by the backside exposure process as found in a conventional bottom-gate TFT process, and is self-aligned to the gate. Dopant (phosphor) was then introduced into the material by either ion implantation or deposition. A XeCl excimer laser [5] was used to anneal and to active the source/drain contacts region. In this process, the channel region was protected by the top island and, therefore, self-aligned to the gate. Figure 2 shows the typical transfer characteristic of a-Si TFTs fabricated by this process. Excellent TFTs with channel length as short as 3 μm can be easily achieved by this process. The on-state current and the subthreshold slope of these TFTs with channel lengths ranging from 10 μm to 3 μm are similar to each other and are similar to the characteristics of conventional a-Si TFTs. The off-state current of the long channel device (channel length L=10 μm) is as good as

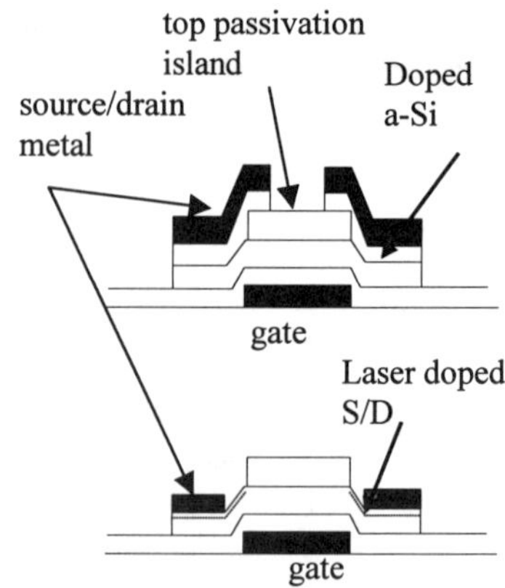

Figure 1. Schematic of (a) conventional a-Si TFT with doped a-Si source/drain contacts, and (b) a self-aligned a-Si TFT with laser doped source/drain contacts.

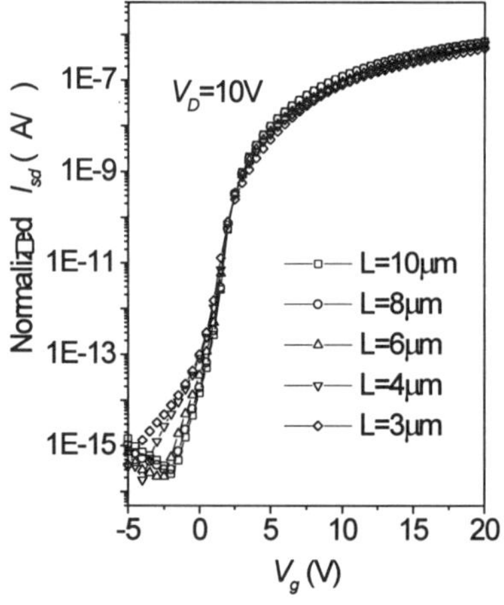

Figure 2. Typical TFT transfer curves for channel length of 3, 4, 6, 8 and 10 μm. The source/drain currents have been normalized to W/L=1 for direct comparison.

state-of-the-art conventional a-Si TFTs (less than 0.03 fA/µm). The leakage current increases slightly as the channel length decreases. However, for most electronic circuits, especially for driver circuits, the requirement for the off-state current is much relaxed.

Shift registers based on self-aligned a-Si TFTs

The application of a-Si TFTs in digital circuits is hindered by three limitations [6]: the low mobility, the significant

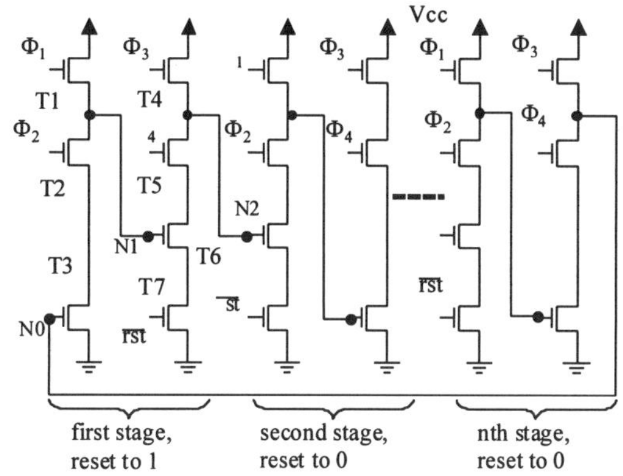

Figure 3. Circuit diagram of a standard shift register based on four-phase dynamic logic.

source/drain to gate parasitic capacitance, and the lack of suitable pull-up device. The problem of low mobility can be alleviated partially by scaling down the device channel length, enabled by the self-aligned TFT process. The significant parasitic capacitance is also efficiently reduced by the self-aligned source/drain contacts. The problem of the lack of a suitable pull-up device can be addressed by choosing the proper type of circuit.

Figure 3 shows the shift register circuits used in this study, which are based on a standard four-phase dynamic logic circuit found in many text books [7]. The clock signal, Φ_1, the simulated output waveform, and the measured result of a five-stage shift register based on self-aligned, short channel a-Si TFTs are shown in Fig. 4. The other three of the four clock signals, Φ_2 to Φ_4 are shifted by 90°, 180°, and 270° respectively from Φ_1. Each stage of this type of shift register consists of seven TFTs, including TFT T7 as the reset transistor. During the normal operating mode, these reset transistors are always kept on by applying Vcc to $\overline{rst}$. The four clock lines, Φ_1 to Φ_4 are sequentially turned from ground to Vcc.

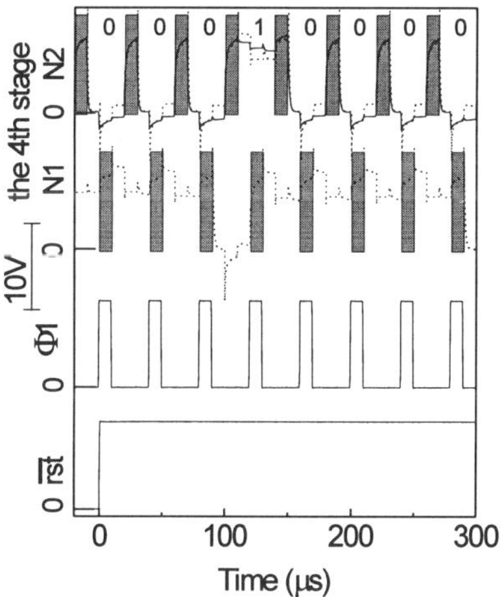

Figure 4. Waveforms of clock signal Φ_1, the reset signal, and the simulated (dotted curve) and the experimental data (solid curve) of a four-phase dynamic shift register. Clock signals Φ_2 to Φ_4 are phase shifted from Φ1 by 90°, 180° and 270° respectively. The shaded regions mark the precharging phase, where the output is not valid.

Therefore, TFTs T1, T2, T4, and T5 are sequentially turned on by the four clock signals. When T1 is on, node N1 is precharged to Vcc. After T1 is turned off and T2 is turned on, the state of node N1 is discharged only if T3 is also on. Therefore, the state of node N1 is the inverse of node N0 and is kept valid until the next Φ_1 cycle. T4, T5, and T7 form the next half stage, similar to T1 to T3. Node N2 is the inverse of N1 at Φ_4 cycle and kept valid until the next Φ_3 cycle. Thus, the net output of the first shift register stage, node N2, is a copy of the input node, N0, delayed by a whole clock cycle. During the reset mode, clocks Φ_1 to Φ_4 are kept running but with $\overline{rst}$ pulled down to ground. As a result, node N2 can not be discharged during Φ_4, regardless the state of node N1. The output of the first stage is said to be "reset" to logic 1. Note that for this type of dynamic logic circuit, the pull-up transistors and the pull-down transistors are never turned on simultaneously. Problems related to the lack of suitable pull-up devices have been effectively avoided by this approach.

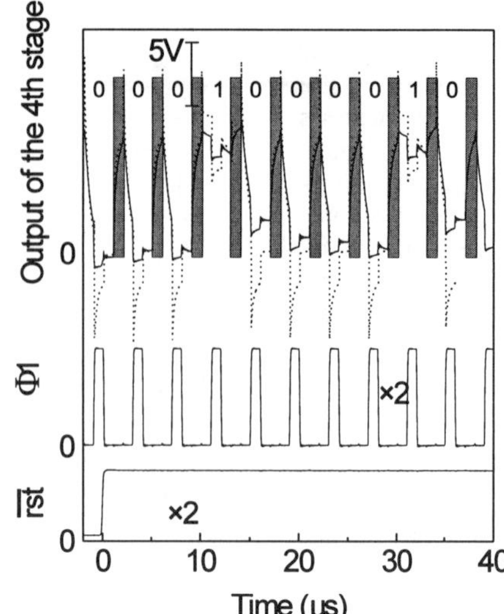

Figure 5. Output waveform of a shift register running at 250 kHz, which is the maximum frequency of our measurement set up. The solid curve is the experimental data and the dotted curve is an AIM-Spice simulation. The shaded regions mark the precharging phase, where the output is not valid.

We have built test circuits of this type based on our self-aligned a-Si TFT technology. TFT channel lengths, L, ranging from 10 μm to 2 μm have been used. The TFT channel width, W, are designed such that the W/L ratios are kept the same for corresponding TFTs with different L to ensure a fair comparison. The sizes of representative TFTs T1, T2, T3, and T7 are W/L= 30, 60, 30, 120 respectively. The output waveform of the circuits are measured by a GGB Pico-probe, (model 18b) and recorded on a digital oscilloscope. The small input capacitance (0.02 pf) and high input impedance (10^{14} Ω) of the picoprobe ensures that the measurement setup did not perturb the circuit operation. The circuits are operated at V_{cc}=10V for all the experiments.

Clock rate over 250 kHz, which correspond to 1μs for each clock phase, can be achieved by combining both the short-channel devices and the dynamic logic technology described above. Figure 5 shows the output waveform of a circuit using L=3 μm TFTs. In order to achieve this speed, the four clock input lines are operating at V_{pp}=15V to further compensate for the inefficiency of using n-channel TFTs as pull-up devices (this arrangement can be readily achieved by external circuits). The simulation results for this circuit are shown with the experimental data in Fig. 5. Excellent matching between the simulation results and the experimental data are observed. A slightly higher overshoot is observed in the simulated waveform, and is currently not fully understood. The simulation results were calculated using AIM-spice [8] with no adjustable parameters. The a-Si TFT model was extracted from a real TFT on the same wafer, with the same channel length L=3 μm. Based on our previous study [9], even with the self-aligned TFT structure, the source/drain contacts to gate parasitic capacitance is not strictly zero. Instead, there seems to be some residue parasitic capacitance equivalent to about 0.9 μm source/drain overlap with the gate for the samples we studied. Slightly over-etched top passivation island and fringing field effects are the two most plausible sources responsible for this residue parasitic capacitance. We included this parasitic capacitance explicitly as two capacitors across source/drain contacts and gate in order to get the correct simulation results.

The maximum achievable clock frequencies for the same type of circuit with different TFT channel lengths are plotted against the channel lengths in

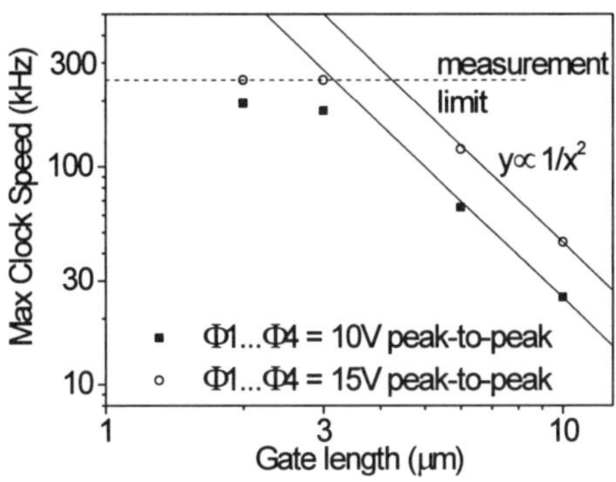

Figure 6. Log-log plot of the observed maximum operating frequency of the four-phase dynamic shift register *vs.* TFT channel length. The open circles represent the maximum frequency observed by using 15V clock signal.

Fig. 6. These circuits are designed with exactly the same W/L ratios for the corresponding TFTs to ensure a fair comparison. The advantage of scaling down the TFT size is obvious. As a simple scaling consideration, the fall time of a logic circuit can be approximated by:

$$t_f \approx 4 \cdot \frac{C_L}{\beta \cdot V_{cc}},$$

where t_f is the fall time, C_L is the load capacitance, β is the TFT gain factor defined as $C_0 \cdot \mu \cdot \left(\frac{W}{L}\right)$, C_0 is the capacitance per unit area for gate, and V_{cc} is the power line voltage [7]. If W/L is kept the same for all the corresponding TFTs, t_f will essentially scale with the square of the TFT channel length since C_L is mainly determined by the input capacitance of the next stage and is proportional to $L \cdot W = L^2 \cdot \left(\frac{W}{L}\right) \propto L^2$. The rise time of a circuit scales with similar functional form. Figure 6 clearly shows this trend. The data points corresponding to L=10 and 6 μm match nicely to the expected value from the scaling. The L=3 μm circuit is slower than the expected value. This is understandable because the residue source/drain capacitance is not scaled with L^2, but instead, scaled with L (0.9 μm×W). The experimental data corresponding to the L=2μm device is only slightly faster than the L=3μm device. This is actually caused by the extra load of the input capacitance of our measurement instrument. For the L=2 μm device, the input capacitance of a W/L=30 TFT is about 0.025 pf, comparable to the input capacitance of the picoprobe. Currently, our measurement setup is limited to 250 kHz for the four-phase clock sequence. The actual maximum working frequency for L=2 μm and L=3 μm circuits can be even higher than 250 kHz.

SUMMARY

We have demonstrated self-aligned a-Si TFT circuits fabricated by laser doping/annealing technology. Shift registers based on four-phase dynamic logic can be operated at clock speed higher than 250 kHz. The high circuit speed is achieved by taking advantage of the smaller parasitic capacitance between source/drain contacts and gate, by scaling down the TFT size without the requirement of stringent lithography technology, and by choosing proper circuit type to avoid the problems related to the lack of a suitable pull-up device. SPICE simulations for the circuit are shown and found to be in excellent agreement with the experimental data. The longer channel length a-Si TFTs on the same wafer preserve all the excellent properties of conventional TFTs, such as low leakage current and sharp turn on. This study demonstrates the possibility of integrating part of the peripheral driver circuits for active matrix flat panel displays and imagers based on the self-aligned a-Si TFT technology.

The authors are grateful for the help of the staff of the Xerox processing line. This work is partially supported by NIST (70NANB7H3007).

REFERENCES

1. A. Saitoh, C. D. Kim, T. Sakoda, and M. Matsumura, Thin Film Transistor Technologies III, Electrochemical Soc. Proc., PV 96-23, p123 (1997).

2. S. Nishida, H. Uchida, and S. Kaneko, MRS Symposium Proceedings., V219, p303, (1991).

3. M. J. Powell, C. Glasse, J. E. Curran, J. R. Hughes, I. D. French and B. F. Martin, MRS Symposium Proceedings, 1998.

4. P. Mei, G. B. Anderson, J. B. Boyce, D. K. Fork, and R. Lujan, Thin Film Transistor Technologies III, Electrochemical Soc. Proc., PV 96-23, p51 (1997).

5. R. Steward, SID symposium digest, 89 (1995).

6. M. S. Shur, H. C. Slade, T. Ytterdal, L. Wang, Z. Xu, M. Hack, K. Aflatooni, Y. Byun, Y. Chen, M. Froggatt, A. Krishnan, P. Mei, H. Meiling, B. H. Min, A. Nathan, S. Sherman, M. Stewart, and S. Theiss, MRS Symposium Proceedings, V467, p831, 1997.

7. N. Weste and K. Eshraghian, Principles of CMOS VLS Design (Addison-Wesley publishing company).

8. XMR Inc, Fremond, California.

9. J.P. Lu, P. Mei, R. Lujan and J. B. Boyce, submitted to Thin Film Transistor Technologies IV, Electrochemical Soc. Proc. (1998).

RUGGED a-Si:H TFTs ON PLASTIC SUBSTRATES

H. GLESKOVA[1], S. WAGNER[1], Z. SUO[2]
Princeton University, [1]Department of Electrical Engineering, and [2]Department of Mechanical and Aerospace Engineering and Princeton Materials Institute, Princeton, NJ 08544

ABSTRACT

Much of the mechanical strain in semiconductor devices can be relieved when they are made on compliant substrates. We demonstrate this strain relief with amorphous silicon thin-film transistors (a-Si:H TFTs) made on 25-μm thick polyimide foil, which can be bent to radii of curvature R down to 0.5 mm without substantial change in electrical characteristics. At $R = 0.5$ mm the channel area of the TFTs is strained by ~ 1%. The reduction in bending radius, from $R \simeq 2$ mm on steel foil of the same thickness, agrees with the theoretical prediction that changing from a stiff to a compliant substrate can reduce the bending strain in the device plane by a factor of up to 5.

INTRODUCTION

The main failure mechanism of laptop computers, cellular phones and similar portable devices is breakage of the glass of the display. To solve this problem much research is focused on the fabrication of thin-film electronics on plastic substrates [1-8]. Plastic substrates have Young's moduli about 100 times smaller than those of glass or stainless steel, and therefore are flexible. Their flexibility makes them ideal for roll-to-roll processing and also for large-area flexible electronics. However, the transition from glass to plastic substrates has important consequences for fabrication and mechanical performance. For example, the glass transition temperature of most plastics lies below 200°C, which restricts the maximum process temperature to $\leq$ 150°C. Due to the lower Young's modulus of plastic substrates, the mechanical behavior of the circuits fabricated on them is different from that on the rigid glass substrates. We have analyzed theoretically this transition from stiff to compliant substrate materials [9]. Here we report measurements that show to what extent amorphous silicon (a-Si:H) thin-film transistors (TFTs) made on polyimide foil indeed are insensitive to rolling and bending.

In a thin-film circuit / substrate couple, the substrate becomes compliant when its stiffness becomes comparable to the stiffness of the circuit films. Because Young's moduli Y_f of the a-Si:H TFT materials are ~ 200 GPa, Young's modulus Y_s of the polyimide Kapton E is ~ 5 GPa, and the film thickness d_f typically is ~ 1 μm, the compliant condition is established for substrate thickness $d_s \lesssim 300$ μm. To test this condition, we have made a-Si:H TFTs on 1-mil (25 μm) thick Kapton E foils. A series of mechanical tests performed on the TFTs in conjunction with their electrical evaluation indeed demonstrate their extraordinary ruggedness.

TRANSISTOR FABRICATION

Because the glass transition temperature of plastics ($\leq$ 200°C) is much lower than that of the conventional substrates, Corning 7059 and 1737 glass ($\geq$ 600°C), the deposition of all a-Si:H TFT layers had to be re-optimized for 150°C to achieve electrical properties comparable to

Mat. Res. Soc. Symp. Proc. Vol. 557 © 1999 Materials Research Society

those obtained at the conventional process temperature of 250°C to 350°C [2]. All silicon layers were deposited using a three-chamber rf-excited plasma enhanced chemical vapor deposition system. The gate SiN_x was deposited from a mixture of SiH_4, NH_3 and H_2, the undoped a-Si:H from a mixture of SiH_4 and H_2, and the (n^+) a-Si:H from a mixture of SiH_4, PH_3 and H_2.

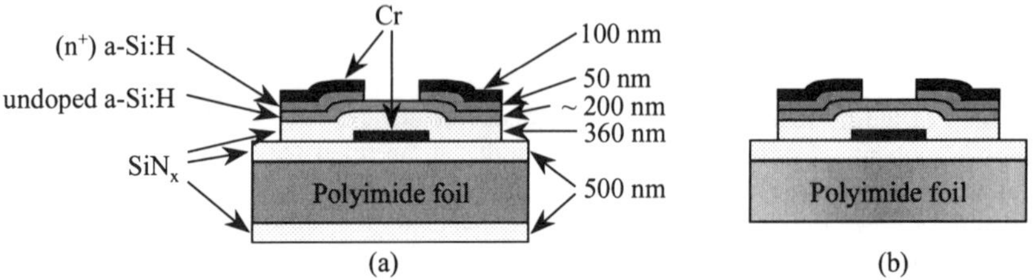

Fig. 1.   Cross-section of a-Si:H TFT on 25 μm thick Kapton foil. (a) As-fabricated sandwich structure; (b) Back SiN$_x$ removed for the bending tests of Fig. 4.

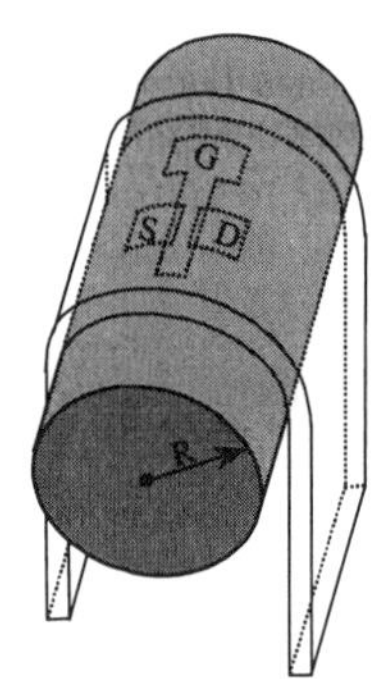

Fig. 2. TFT bent inward.

The TFTs have the bottom gate, back-channel etch structure shown in Fig. 1. We began by passivating a 25-μm-thick Kapton E substrate on both sides with 0.5 μm thick layers of SiN_x. These layers serve as a barrier against the solvents, bases and acids used during photolithography. An ~ 100 nm thick Cr layer was thermally evaporated and wet etched to create the gate electrode. The TFT sequence was 360 nm of SiN_x, ~ 200 nm of undoped a-Si:H, and ~ 50 nm of (n^+) a-Si:H. An ~ 100 nm thick Cr layer was thermally evaporated. We wet etched the Cr source-drain pattern and dry etched the (n^+) a-Si:H in CF_4 gas. Then the undoped a-Si:H was dry etched to define the transistor island. In the last photolithographic step, we dry etched windows into the SiN_x to open access to the gate contact pads. Arrays of TFTs with gate length $L = 15$ μm and width $W = 210$ μm were fabricated on 1.5 x 1.5 sq. in. substrates. Four substrates were processed simultaneously.

The as-fabricated TFT/substrate structure had a built-in radius of curvature R of 18 mm, with the TFTs on the outside, showing that the as-fabricated transistors were under compressive stress. Before bending the TFTs we removed the stiff SiN_x layer from the back of the substrate. This reduced the radius of curvature of the structure to 8 mm, with the TFTs still on the outside. The stiff SiN_x layer on the back of the substrate stiffens the whole structure, thus masking the compliancy of the plastic substrate. Removing the back SiN_x layer renders the plastic truly compliant. (See the Results and Discussion section for further discussion.)

Individual transistors were stressed mechanically by bending inward (the devices facing in), which is shown schematically in Fig. 2, or outward (the devices facing out). Single TFTs were bent to decreasing R, beginning with $R = 4$ mm down to $R = 0.5$ mm. The TFT was stressed for one minute at each bending radius, and then was released and remeasured.

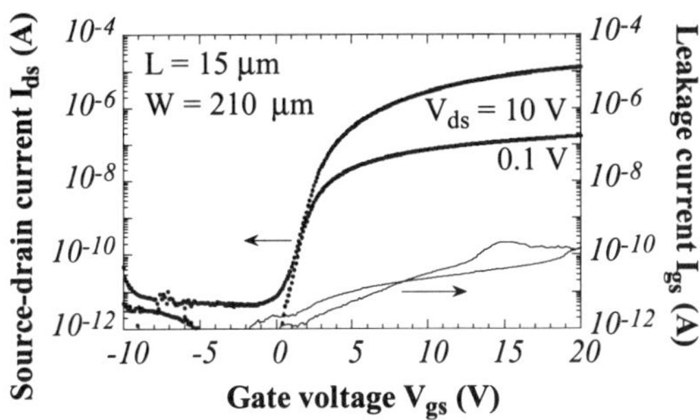

Fig. 3. Transfer characteristics of a-Si:H TFT on 25 μm thick Kapton.

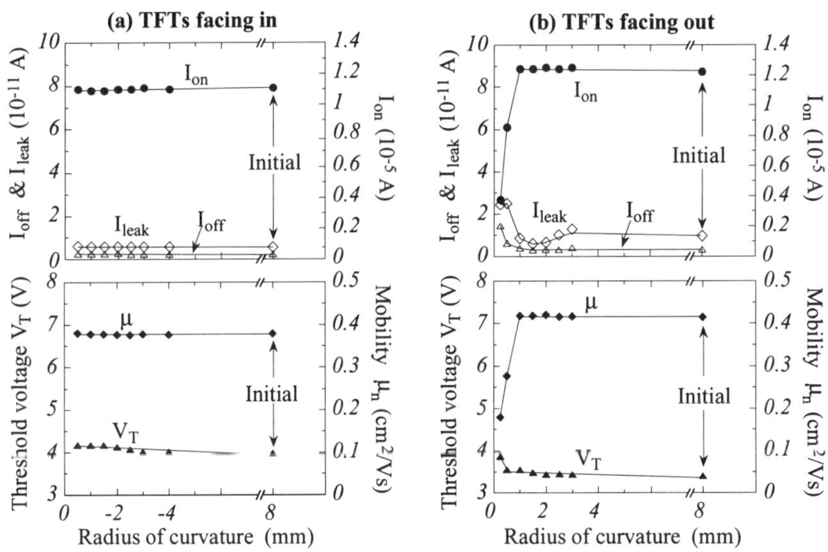

Fig. 4. On-, source-gate leakage, off-currents, electron mobility, and threshold voltage in the saturation regime as functions of bending radius. The initial, built-in radius is 8 mm. Outward bending R is defined positive, inward bending negative. Differences between the "Initial" characteristics reflect spread between as-fabricated TFTs.

RESULTS AND DISCUSSION

a-Si:H TFT Performance Upon Bending

The TFT transfer characteristics I_{ds} vs. V_{gs}, for $V_{ds} = 0.1$ V and 10 V, are shown in Fig. 3. The off-current is $\sim 5 \times 10^{-12}$ A ($\sim 2.4 \times 10^{-14}$ A/μm) and the on-off current ratio is $> 10^6$.

In the saturation regime the source-drain current is given by:

$$I_{ds} = \mu_n C_{SiN} \frac{W}{2L}\left(V_{gs} - V_T\right)^2 \tag{1}$$

where μ_n is the effective electron mobility, C_{SiN} the capacitance of the gate insulator, W the channel width, L the channel length, V_{gs} the gate voltage, and V_T the threshold voltage. The dielectric constant of our 150°C nitride measured at 1 MHz is 7.46 and we calculate $C_{SiN} = 1.83 \times 10^{-8}$ F cm^{-2}. At $V_{ds} = 10$ V we obtain $V_T \sim 2.7$ V and a saturated mobility of ~ 0.4 cm^{2}V^{-1}s^{-1} for the TFT of Fig. 3.

Fig. 4 summarizes the results of the inward (a) and outward (b) bending tests performed on TFTs fabricated on 25 µm thick Kapton foil. The top graphs show the on-current I_{on}, the off-current I_{off}, and the gate-leakage current I_{leak} as functions of the radius of curvature. The definition of these currents is as follows: the off-current is the smallest source-drain current at $V_{ds} = 10$V, the on-current is the source-drain current for $V_{ds} = 10$ V and $V_{gs} = 20$ V, and the leakage current is the source-gate current for $V_{ds} = 10$ V and $V_{gs} = 20$ V. No change in these parameters for bending down to $R = 0.5$ mm was observed for inward bending and only a minor change for $R \leq 0.5$ mm for outward bending. For outward bending we also show some data for $R = 0.25$ mm but these should be taken with some reservation because the radius of 0.5 mm was the smallest controlled value we could set reliably. The bottom graphs show the threshold voltage V_T and the saturated electron mobility μ_n, calculated using Eq. (1), as a function of the bending radius. The slight monotonous rise in the threshold voltage, observed in both cases, does not result from bending because a similar shift was observed also in repeated measurements of unstressed TFTs. The mobility remains constant for inward bending, but for outward bending a decrease is observed for $R \leq 0.5$ mm. No catastrophic failure occurs.

Strain in the Channel of the TFT

Both the fabrication process and the externally applied bending moment cause strain in the TFT structure. The difference in the coefficient of thermal expansion between Kapton and the TFT layers, and strain built in during deposition cause the as-fabricated TFTs to be under compressive strain (the radius of curvature of the substrate, coated on both sides with SiN$_x$, is 18 mm and the TFTs are on the outside). Let us analyze the TFTs when we bend them inward. In that case the TFTs are put under additional compressive strain. We calculate the strain that is produced during our bending experiment in the channel of the TFT. For the calculation of this additional strain we assume that the as-fabricated structure is free of strain. Therefore, the structure is assumed to be flat before the external bending moment is applied. The substrate has thickness d_s and Young's modulus Y_s. It is covered on both sides with stiff films with Young's modulus Y_f and thicknesses d_{f1} and d_{f2}, as shown in Fig. 5.

When an external bending moment is applied to this structure, the film on the outside is under tension while the film on the inside is under compression. Geometry dictates that the strain ε_x in the bending direction x is linear in z, namely,

$$\varepsilon_x = z / R \tag{2}$$

because the neutral plane is free of strain [9]. Here R is the radius of curvature and z is measured from the neutral plane. Each layer of the material is taken to be an isotropic elastic solid with Young's modulus Y. The film and the substrate are dissimilar materials, and Y is a

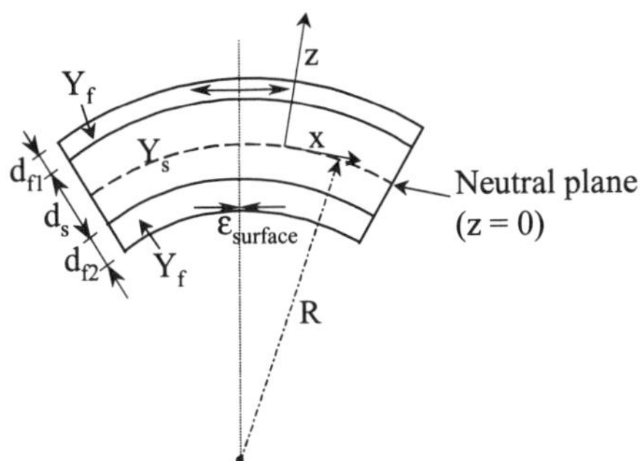

Fig. 5. Films-on-foil structure bent to a cylindrical roll.

known function of z. No external force is applied in the plane of the foil. Force balance requires that

$$\int \sigma_x dz = 0 \qquad (3)$$

Hooke's law relates the stress σ to the strain ε via

$$\sigma_x = Y\varepsilon_x \qquad (4)$$

By inserting Eqs. (2) and (4) into Eq. (3) and integrating we obtain the position of the neutral plane with respect to either of the top surfaces. This allows us to calculate the strain $\varepsilon_{surface}$ on the surface of the TFT film (the inside film) using Eq. (2), namely,

$$\varepsilon_{surface} = \frac{d_s + d_{f1} + d_{f2}}{2R} \cdot \frac{\chi(\eta_1^2 + \eta_2^2) + 2(\chi\eta_1 + \chi\eta_1\eta_2 + \eta_2) + 1}{\chi(\eta_1 + \eta_2)^2 + (\eta_1 + \eta_2)(1 + \chi) + 1} \qquad (5)$$

where $\chi = Y_f/Y_s$, $\eta_1 = d_{f1}/d_s$ and $\eta_2 = d_{f2}/d_s$.

We used Eq. (5) to calculate the value of the compressive strain in the channel of our TFTs for each bending radius R. We calculate $\varepsilon_{surface}$ for two structures: one with the back SiN_x layer, i.e., the as-fabricated sandwich structure of Fig. 1(a), and the other without this SiN_x layer, as employed in the bending experiments, Fig. 1(b). Table 1 summarizes the results. Removing the back SiN_x layer makes the substrate more compliant and reduces the strain in the TFT channel area by a factor of ~ 2. From experiment and the calculated strain we conclude that TFTs on 25 µm thick Kapton continue to function, without any substantial change in their electrical performance, down to $R = 0.5$ mm, which corresponds to a bending strain of ~ 1%. Previous experiments with bending TFTs on 25 µm thick steel foil (a stiff substrate) have shown that a-Si:H TFTs can be strained by ~ 0.5 % before failing mechanically at a radius of curvature R of ~ 2 mm [10]. This lower attainable strain in TFTs on steel may be caused by film delamination due to inadequate adhesion between the substrate and the TFT structure.

Table 1. Bending strain $\varepsilon_{surface}$ in the channel area of the TFT calculated from Eq. (5). ($d_s = 25$ µm, $Y_f = 200$ GPa, $Y_s = 5$ GPa)

R (mm)	Strain (%) (sandwich structure)	Strain (%) (back SiN_x layer removed)
0.25	4.15	2.20
0.5	2.08	1.10
1	1.04	0.55
1.5	0.69	0.37
2	0.52	0.27
2.5	0.42	0.22
3	0.35	0.18
4	0.26	0.14

SUMMARY

Earlier experiments with bending TFTs on 25-μm thick steel foil (a stiff substrate) have shown that a-Si:H TFTs can be strained by at least 0.5 % before failing mechanically at a radius of curvature R of ~ 2 mm [10]. Our present observation that TFTs made on the same thickness of Kapton foil (a compliant substrate) function down to R = 0.5 mm provides direct evidence for the strain-relieving effect of a compliant substrate. This observation agrees with the theoretical prediction that the strain in the plane of the circuit can be reduced by a factor of up to 5 when going from stiff to a compliant substrate [9]. Thus both experiment and theory suggest that nearly-foldable thin-film circuits are a realistic prospect.

ACKNOWLEDGMENTS

We gratefully acknowledge support from the DARPA HDS program and thank DuPont for donating Kapton E foils.

REFERENCES

1. G.N. Parsons, C.S. Yang, C.B. Arthur, T.M. Klein, and L. Smith, in *Flat Panel Display Materials - 1998*, edited by G.N. Parsons, C.C. Tsai, T.S. Fahlen, and C.H. Seager, (Mat. Res. Soc. Symp. Proc. **508**, Pittsburgh, PA, 1998), pp. 19-24.

2. H. Gleskova, S. Wagner and Z. Suo, in *Flat Panel Display Materials - 1998*, edited by G.N. Parsons, C.C. Tsai, T.S. Fahlen, and C.H. Seager, (Mat. Res. Soc. Symp. Proc. **508**, Pittsburgh, PA, 1998), pp. 73-78.

3. J.N. Sandoe, Digest of Technical Papers of the Society for Information Display, Vol. **29**, 1998, pp. 293-296.

4. E. Lueder, M. Muecke, S. Polach, Proc. of the Int. Display Research Conf., Asia Display '98, 1998, pp. 173-177.

5. A. Constant, S.G. Burns, H. Shanks, C. Gruber, A. Landin, D. Schmidt, C. Thielen, F. Olympie, T. Schumacher, and J. Cobbs, The Electrochemical Society Proceedings, Vol. **94-35**, pp. 392-400.

6. S.G. Burns, H. Shanks, A. Constant, C. Gruber, D. Schmidt, A. Landin, and F. Olympie, The Electrochemical Society Proceedings, Vol. **96-23**, pp. 382-390.

7. S. D. Theiss, P. G. Carey, P.M. Smith, P. Wickboldt, T.W. Sigmon, Y.J. Tung, and T.-J. King, IEEE Int. Electron Devices Meeting 1998, Technical Digest, pp. 257-260.

8. D.B. Thomasson, M. Bonse, J.R. Huang, C.R. Wronski, and T.N. Jackson, IEEE Int. Electron Devices Meeting 1998, Technical Digest, pp. 253-256.

9. Z. Suo, E.Y. Ma, H. Gleskova, and S. Wagner, Appl. Phys. Lett. **74**, 1999, pp. 1177-1179.

10. E.Y. Ma and S. Wagner, in *Flat Panel Display Materials - 1998*, edited by G.N. Parsons, C.C. Tsai, T.S. Fahlen, and C.H. Seager, (Mat. Res. Soc. Symp. Proc. **508**, Pittsburgh, PA, 1998), pp. 13-18.

THIN-FILM TRANSISTORS BASED ON
HOT-WIRE AMORPHOUS SILICON ON SILICON NITRIDE

B. STANNOWSKI, H. MEILING, A. M. BROCKHOFF, AND R. E. I. SCHROPP
Debye Institute, Interface Physics, Utrecht University
P.O. Box 80000, 3508 TA Utrecht, The Netherlands

ABSTRACT

We present state-of-the-art thin-film transistors (TFTs) incorporating amorphous silicon i-layers deposited by hot-wire chemical vapor deposition. The TFTs are deposited on glow-discharge silicon nitride as well as on thermally-grown silicon dioxide. The devices on silicon nitride have a field-effect mobility above 0.7 cm^2/Vs, a threshold voltage around 2 V and a sub-threshold slope as low as 0.5 V/dec. As commonly observed, the TFTs on silicon-dioxide have higher values for the threshold voltage and the sub-threshold slope. In the annealed state the hot-wire TFTs show almost the same properties as TFTs deposited by conventional plasma-enhanced chemical vapor deposition. Nevertheless, the stress-time dependent behavior under prolonged gate-voltage stress at elevated temperature is different from that of the glow-discharge devices. The hot-wire TFTs are clearly more stable than their glow-discharge counterparts. Furthermore, we found differences in the stress behavior of the hot-wire TFTs deposited on silicon nitride and silicon dioxide.

INTRODUCTION

Thin-Film Transistors (TFT) based on hydrogenated amorphous silicon (a-Si:H) are used for large-area applications such as active-matrix flat-panel displays or sensor arrays. Amorphous silicon is commonly deposited by plasma-enhanced chemical vapor deposition (PECVD) at a plasma-excitation frequency of 13.56 MHz. Although these TFTs have suitable properties, like high switching ratios with a low off-current, the limited device stability remains an issue. Namely, the TFTs exhibit a threshold-voltage shift after prolonged application of gate voltage. For moderate voltages this is ascribed to the creation of dangling-bond defects in the a-Si:H within the channel region. For higher voltages also charge trapping in the gate dielectric is expected [1-3]. Both effects imply a variation of the space-charge layer in the channel, thus resulting in a shift of the TFT on-set voltage. The effect can be reversed by annealing at high temperatures. In order to investigate this metastability of a-Si:H several groups have extensively studied a-Si:H TFTs over the last two decades ([4-6] and references therein). By applying gate-voltage stress, the Fermi level in a-Si:H can be controlled and shifted without the influence of any doping atoms. The equilibrium defect distribution in a TFT changes due to this Fermi level shift, as described by the defect-pool model [7].

So far, only a few groups reported on hot-wire CVD (HWCVD) material included as active layer in a-Si:H TFTs [8-10]. We found for such HW-TFTs incorporating a-Si:H or heterogeneous-Si:H i-layers an essentially higher stability compared to their PECVD counterparts [11,12]. Also, deposition rates higher than 1.5 nm/s were readily obtained, which makes this deposition technique promising for industrial application. In this paper we report on TFTs including HW a-Si:H i-layers deposited on PECVD silicon-nitride and thermally grown silicon dioxide (SiO$_2$). By comparing the two types of TFTs with different insulators, the influence of the dielectric material on the device performance is studied.

Mat. Res. Soc. Symp. Proc. Vol. 557 © 1999 Materials Research Society

EXPERIMENTAL

TFT preparation

We studied three kinds of TFTs of the inverted-staggered type including different types of a-Si:H layers. The device structure is sketched in Fig. 1. Two samples (A, B) have an intrinsic a-Si:H layer deposited by HWCVD. For sample A the a-Si:H layer is deposited on 300 nm silicon nitride (a-SiN$_x$:H), for sample B it is deposited on 160 nm SiO$_2$. As reference we used a TFT with a layer of device quality a-Si:H deposited by PECVD at 50 MHz on silicon nitride (C). The deposition of the devices was performed in two multi-chamber ultrahigh-vacuum CVD systems. We used our

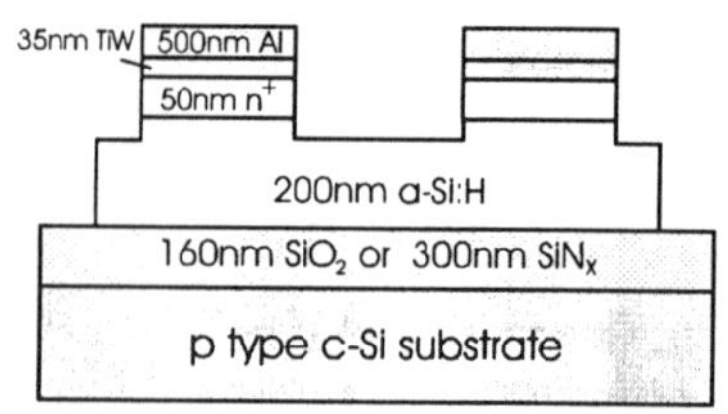

Fig.1 Cross section of a TFT. The highly doped substrate is the gate (not to scale).

system ASTER [13] for all silicon nitride depositions and a-Si:H deposition for sample C. The HW i-layers and glow-discharge n$^+$-contacts for samples A and B were deposited in our system PASTA [14]. In contrast to samples A and B, sample C was completely deposited within one system without exposure of the dielectric to air.

As substrate we used highly doped p-type c-Si wafers (0.010 - 0.018 Ωcm), which form also the gate contacts. In order to remove the native oxide before a silicon-nitride deposition, the wafer was dipped in 0.5 % hydrofluoric acid before loading into the deposition system. The a-SiN$_x$:H deposition was performed at a substrate temperature of 400°C, an excitation frequency of 50 MHz and a power density of 225 mW/cm^2. As source gases ammonia (NH$_3$) and silane (SiH$_4$) with a gas-flow ratio NH$_3$/SiH$_4$ = 30 and a pressure of 30 Pa were used. Thereby, we achieved deposition rates above 0.3 nm/s. After silicon-nitride deposition for sample A it was taken out of the system, again dipped in hydrofluoric acid for one minute, and loaded into the PASTA system. For HWCVD the source gases are catalytically dissociated at two tantalum wires, which are positioned parallel to the substrate in the deposition chambers. The wire temperature ranges between 1800°C and 1900°C. In order to achieve a passivated insulator-silicon interface, a treatment with atomic hydrogen was performed before starting the a-Si:H deposition by exposing the substrate to a hydrogen flow with the wire current switched on for 15 minutes. The intrinsic HW a-Si:H layer was grown to a thickness of 200 nm at a substrate temperature of about 400°C. Deposition rates as high as 1.7 nm/s are reached. Finally, a 50 nm thick n$^+$-type a-Si:H layer was deposited on top, in order to achieve hole-blocking contacts. This was performed with PECVD at 13.56 MHz at a substrate temperature of 200°C within a separate chamber of the system.

In order to achieve thermally stable source and drain contacts, we used sputtered TiW and Al for samples A and B. In contrast, sample C has pure Al contacts. Finally, the TFT structure is defined by optical lithography and structured with conventional back-channel etching in a two-step process using wet etching for the metal contacts and plasma etching for an accurate definition of the channel dimensions.

Material properties

We extensively characterized the silicon nitride. The refractive index n and the optical band gap of the silicon-nitride layers, both measured with ellipsometry, amount to 1.89 and 4.7 eV, respectively. Rutherford Back Scattering (RBS) experiments gave a [Si]/[N] ratio of 0.68. This value is consistent with the correlation $n = 1.4 + 0.7 \cdot$[Si]/[N] found by Claassen *et al.* [15]. The hydrogen content, measured with Elastic Recoil Detection (ERD) and cross-checked by Fourier-Transform Infrared Spectroscopy (FTIR), amounts to 21 %. In the FTIR spectra we detected no

Si-H stretching-mode absorption peak near 2200 cm^{-1}, while a large N-H absorption peak at 3340 cm^{-1} was found. Electrical characterization of a 200nm-thick a-SiN$_x$:H layer deposited on p-type c-Si gave a breakdown-field strength above 5 MVcm^{-1}. The electrical conductivity, measured at 1 MVcm^{-1}, is below 10^{-12} $\square^{-1}$cm^{-1}. Capacitance-voltage measurements at 1 MHz resulted in a density of fixed charge of 6.9·10^{11} cm^{-2}. The density of trapped charge, extracted from the shift in the C-V curve upon changing the voltage-sweep direction, amounts to only 7·10^{10} cm^{-2}. The relative dielectric constant $\square$ is found to be 5.

The hot-wire silicon layers used for the TFTs are purely amorphous. From the intensity of the Si-H rocking mode (640 cm^{-1}) in the infrared-absorption spectrum, a hydrogen content of 8 at.% is derived. The microstructure factor R* = I$_{2100}$/(I$_{2100}$ + I$_{2000}$) is below 0.1, which indicates a compact material with a low void fraction. The intrinsic material has a photosensitivity σ_{ph}/σ_d above 10^5, while the optical band gap is E_{tauc} = 1.74 eV. From ESR measurements on HW a-Si:H layers comparable to those used here a defect density of 3·10^{16} cm^{-3} was found.

<u>TFT measurements</u>

The TFTs characterization and stressing experiments were performed in a computer-controlled measurement set-up including a Keithley 283 High-Current Source-Measure Unit. This facilitated I-V measurements with a low noise level and an automated stressing procedure with well-defined stress duration and a stable sample temperature. All TFT characteristics were measured at room temperature. In order to remain in the linear regime we applied a source-drain voltage of V_{sd} = 0.2 V for all transfer characteristics. The field-effect mobility μ_{FE} and the threshold voltage V_{th} are calculated from the equation $I_{sd} = W/L \cdot \mu_{FE} \cdot C_i \cdot (V_{gate} - V_{th})V_{sd}$, where C_i is the specific capacitance of the insulator. All devices have a channel width of W = 300 μm and length of L = 60 μm. We used a standard fitting procedure for the different TFT characteristics in order to exclude errors related to the choice of the fitting range.

Before measuring, all devices were annealed at 180°C for at least 30 minutes, which was performed in the same set-up. During application of gate-voltage stress the source and drain contacts were both grounded.

RESULTS AND DISCUSSION

<u>TFT Characteristics</u>

Figure 2 shows the output characteristics of sample A. It exhibits good current saturation with no detectable charge trapping in the insulator for all applied gate voltages. A low contact resistance is apparent from the absence of current crowding during current rise. Samples B and C have comparable output characteristics (not shown here).

Figure 3 shows the linear transfer characteristics of samples A, B and C. All TFTs exhibit off-currents between 1 and 10 pA and switching ratios above 10^5. The TFT parameters calculated from this transfer characteristics are summarized in Table 1. Values for threshold voltage and field-effect mobility are comparable to those calculated from the saturation-transfer characteristics, measured with $V_{gate} = V_{sd}$. The devices show typical properties of state-of-the-art a-Si:H TFTs. Only minor differences are seen in the characteristics of sample A and C, both with a silicon-nitride gate dielectric and a-Si:H layers grown by different deposition techniques. The higher on-current of sample C can be attributed to the different source- and drain-contact material. The sub-threshold slope, which is mainly determined by the sub band gap defect density in the channel region, is slightly lower (0.48 V/dec), while the field-effect mobility is higher (0.89 cm^2/Vs) than for sample A. This indicates a better PECVD a-Si:H material in the channel, with a lower defect density and a steeper valence-band tail in sample C. However, the

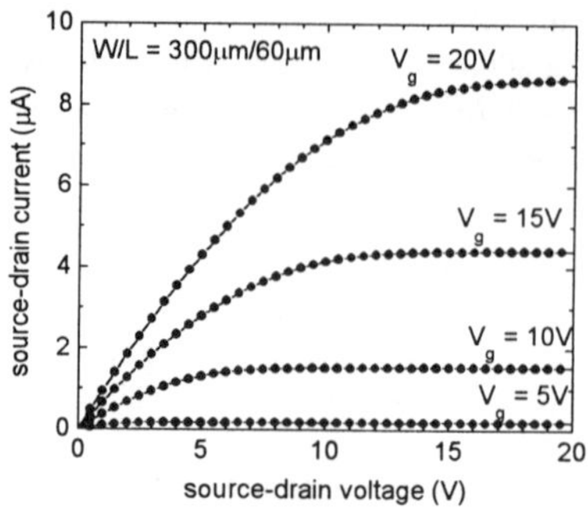

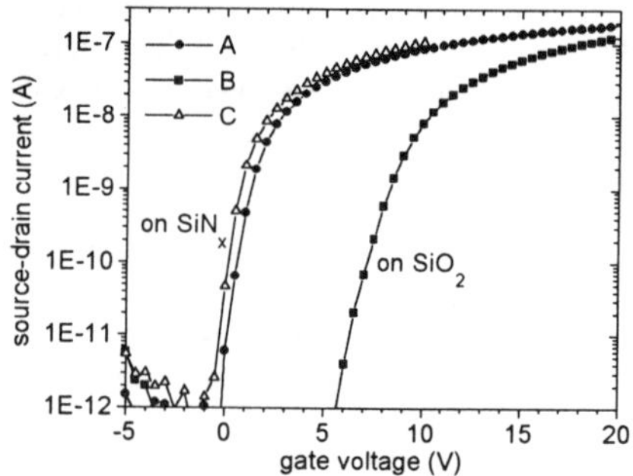

Fig. 2 Output characteristics of the TFT with HW a-Si:H on a-SiN$_x$:H (A).

Fig. 3 Linear transfer characteristics ($V_{sd} = 0.2$ V) of sample A, B and C (compare Table 1).

different surface pre-treatments of the silicon nitride could also account for these differences (during preparation sample A was exposed to air).

In contrast, sample B shows a different behavior. It has a much higher threshold voltage of 10.7 V. This is in accordance with our earlier results [11]. It can be explained by a higher equilibrium defect distribution in the upper half of the band gap of the a-Si:H layer near the interface, because of a lower position of the equilibrium Fermi level at the interface due to a more negative trapped-charge density in the SiO$_2$ compared to a-SiN$_x$:H [7]. The higher defect density also results in a higher sub-threshold slope, found to be 0.94 V/dec in our measurements. However, the field-effect mobility (0.66 cm^2/Vs), mainly determined by the width of the valence-band tail rather than by the defect density, is lower than that of sample A. This indicates an inferior interface quality. Apparently the type of substrate (SiO$_2$ or a-SiN$_x$:H) influences the quality of the first a-Si:H layers grown on top of it. Its quality appears to be better on silicon nitride.

<u>TFT stability</u>

We investigated the TFT stability by applying gate-voltage stress of 25 V for different stress times up to $3\cdot10^5$s. This is carried out at a temperature of 80□C. In contrast to defect creation under light exposure where the energy for defect creation originates from recombination of photo-generated carriers, the defect creation in a-Si:H TFTs under gate-voltage stress is a thermally activated

	sample	V_{th} (V)	μ_{FE} (cm^2/Vs)	*slope* (V/dec)
A	HW a-Si:H / a-SiN$_x$:H	2.2	0.74	0.53
B	HW a-Si:H / SiO$_2$	10.7	0.66	0.94
C	GD a-Si:H / a-SiN$_x$:H	1.9	0.89	0.48

Tab. 1 Threshold voltage, field-effect mobility and sub-threshold slope, calculated from the linear transfer characteristic after annealing at 180°C (Fig. 3).

process. Recently, Deane *et al.* [16] described this in a theory on the unification of stress time *t* and temperature *T*. Therein, the effect of both is combined to one physical parameter called thermalization energy $E_{th} = kT\cdot\ln(\nu t)$. By curve fitting they found an attempt-to-escape frequency for defect creation of $\nu = 10^{10}$ s^{-1}. Therefore, our stressing at 80□C can be compared to stressing with tremendous long stress times at room temperature.

Figures 4 and 5 demonstrate the linear transfer characteristics of the three TFTs after gate-bias stress for 100 s and 10^4 s, respectively. After 100 s of stress only a minor threshold-voltage shift (< 1 V) is found for the HW TFT on SiO_2 (B). In contrast, samples A and C, both deposited on silicon nitride, already show a clear shift of about 3.4 V. For all TFTs hardly any degradation of the field-effect mobility is found and only a minor increase of the sub-threshold slope. After 10^4 s of stress it becomes obvious that the threshold voltage of the glow-discharge TFT (C) degrades much more than V_{th} of the HW TFTs (A, B).

In order to account for the different initial threshold voltages V_{th}^{ini} due to the different gate-dielectric materials, we compare the relative threshold voltage shifts $\Delta V_{th}^{rel} = (V_{th} - V_{th}^{ini})/(V_{stress} - V_{th}^{ini})$. For infinitely long stress times ΔV_{th}^{rel} approaches the saturation value of 1, which means that $V_{th} = V_{stress}$. In Fig. 6 the GD TFT (C) shows the same stress-time dependent behavior which was found for other PECVD a-Si:H TFTs [16]. The curve of the HW TFT on SiO_2 (B) shows a very similar behavior but it is shifted towards longer stress times by about one order of magnitude. This clearly indicates a higher stability. The HW TFT on a-SiN$_x$ (A) shows a more stretched behavior. While for short stress times both silicon-nitride TFTs (A, C) have higher ΔV_{th}^{rel} than B, for longer times A and C behave clearly different. Sample A then shows the more stable characteristic of sample B. Therefore, for short stress times the behavior can be attributed to the gate-dielectric, while it seems to be determined by the type of a-Si:H i-layer for longer stress times.

The reason for this effect is not yet clear. Different defect-creation kinetics could be an explanation. However, it cannot be excluded that charge trapping within the silicon nitride or in the a-SiN$_x$/a-Si:H interface contributes to the threshold-voltage shift. This means that charge trapping dominates for short stress times, resulting in the higher ΔV_{th}^{rel} for A and C, while defect creation becomes dominant later on. Also defect creation in the silicon nitride should be taken into account. In order to resolve this, further investigations, i. e. varying the stress voltage and the stress temperature, need to be carried out.

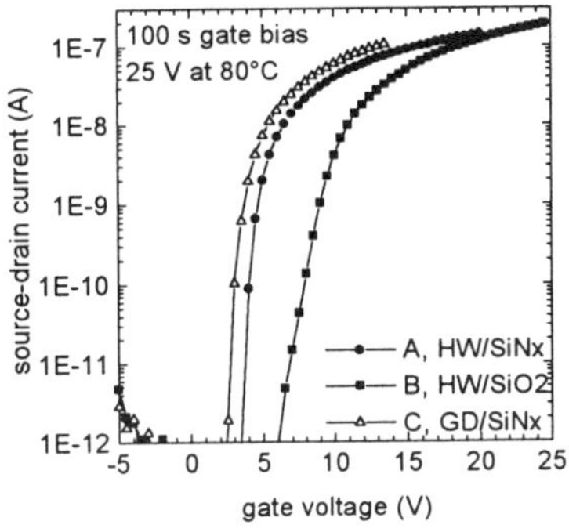

Fig. 4 Linear transfer characteristics ($V_{sd} = 0.2$ V) after 100s gate bias stress.

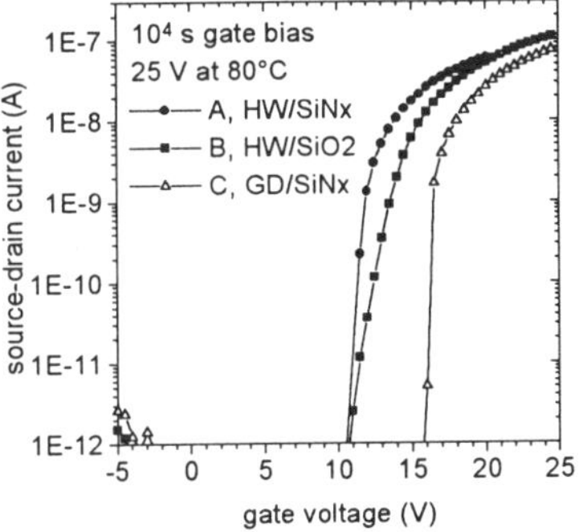

Fig. 5 Linear transfer characteristics ($V_{sd} = 0.2$ V) after 10^4 s gate-bias stress.

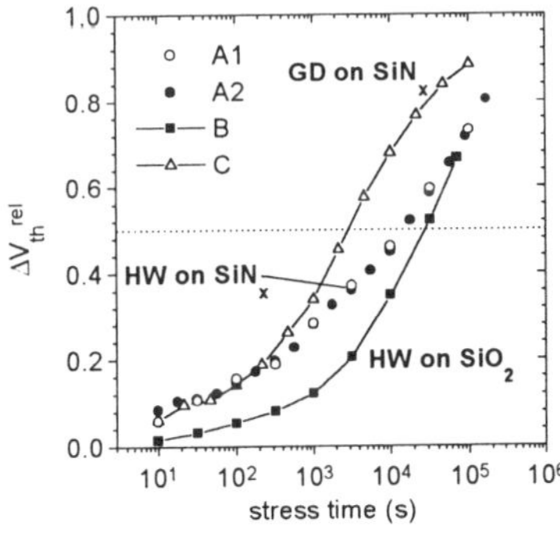

Fig. 6 Relative threshold-voltage shift after 25 V gate-bias stress at 80□C for samples A, B and C. A1, A2 are results from different TFTs on the same wafer.

CONCLUSIONS

We presented state-of-the-art thin film transistors incorporating an a-Si:H i-layer deposited by hot-wire CVD on glow-discharge silicon nitride. The characteristics and the stability of the HW TFTs on silicon nitride were compared to those deposited on SiO_2. Both showed initial characteristics comparable with glow-discharge TFTs, but have a superior stability upon prolonged gate-voltage stress. A different behavior under stress is found for the HW TFT on silicon nitride compared to that of the HW TFT on SiO_2, apparent in a more stretched increase of the threshold-voltage shift with increasing stress time. This suggests that more than one mechanism is responsible for the threshold-voltage shift: an initial mechanism related to the silicon nitride and a long term mechanism related to the HW a-Si:H.

ACKNOWLEDGMENTS

For TFT preparation R. Dekker (MESA Inst., Univ. of Twente) and for layer deposition J. Groenewoud and C. H. M. van der Werf are gratefully acknowledged; A. Nascetti ('La Sapienza' Univ. of Rome) for stimulating discussions and assistance in TFT measurements. E. van der Wal and W. M. Arnoldbik assisted in RBS and ERD, L. C. Jacobs performed ellipsometry measurements. B. Stannowski acknowledges financial support from NWO (Netherlands Organization for Scientific Research) and STW Technology Foundation.

REFERENCES

[1]   A. R. Hepburn, J. M. Marshall, C. Main, M. J. Powell, and C. van Berkel, Phys. Rev. Lett. **56** (20) p. 2215 (1986).

[2]   C. van Berkel and M. J. Powell, Appl. Phys. Lett. **51** (14) p. 1094 (1987).

[3]   M. J. Powell, C. van Berkel, and J. R. Hughes, Appl. Phys. Lett. **54** (14) p. 1323 (1989).

[4]   M. J. Powell, S. C. Deane, I. D. French, J. R. Hughes, W. I. Milne, Philos. Mag. B **63** (1) p. 325 (1991).

[5]   R. E. I. Schropp and J. F. Verwey, Appl. Phys. Lett. **50** (4) p. 185 (1987).

[6]   N. Nickel, W. Fuhs, H. Mell, Philos. Mag. B **61** (2) p. 251 (1990).

[7]   M. J. Powell, C. van Berkel, A. R. Franklin, S. C. Deane, and W. I. Milne, Phys. Rev. B **45** (8) p. 4160 (1992).

[8]   H. Meiling and R. E. I. Schropp, Appl. Phys. Lett. **69** (8) p. 1062 (1996).

[9]   H. Matsumura, Proc. Electrochem. Soc. *Thin Film Transistor Technologies IV*, Boston (1998).

[10] V. Chu, J. Jarego, H. Silva, T. Silva, M. Reissner, P. Brogueira and J. P. Conde, Appl. Phys. Lett. **70** (20) p. 2714 (1997).

[11] H. Meiling and R. E. I. Schropp, Appl. Phys. Lett. **70** (20) p. 2681 (1997).

[12] A. M. Brockhoff, H. Meiling, F. H. P. M. Habraken, and R. E. I. Schropp, Proc. Electrochem. Soc. *Thin Film Transistor Technologies IV*, Boston (1998).

[13] H. Meiling, W. G. J. H. M. van Sark, J. Bezemer, and W. F. van der Weg, J. Appl. Phys. **80** p. 3546 (1996).

[14] R. E. I. Schropp, K. F. Feenstra, E. C. Molenbroek, H. Meiling, J. K. Rath, Philos. Mag. B **76** p. 309 (1997).

[15] W. A. P. Claassen, W. G. Valkenburg, F. H. Habraken , and Y. Tamminga, Proc. Electrochem. Soc. **83** (8) p. 430 (1983).

[16] S. C. Deane, R. B. Wehrspohn, and M. J. Powell, Phys. Rev. B **58** (19) p. 12625 (1998).

THIN FILM TRANSISTORS OF MICROCRYSTALLINE SILICON DEPOSITED BY PLASMA ENHANCED-CVD

Y. Chen and S. Wagner
Department of Electrical Engineering and
Center for Photonics and Optoelectronic Materials
Princeton University, Princeton, NJ 08544

Abstract

We fabricated top gate TFTs of microcrystalline silicon (μc-Si) deposited at 360 $^\circ$C. The TFTs have field-effect electron mobilities of up to 7.9 cm^2/Vs in the saturation regime and 5.8 cm^2/Vs in the linear regime. The highest I_{ON}/I_{OFF} ratio is 10^5. Typical values for V_{th} is 6.5 V and for the subthreshold slope is 1.7 V/decade. The μc-Si is grown by PE-CVD from a source gas mixture of SiH_4, SiF_4 and H_2, with a typical flow ratio of 1:20:200, at a pressure of 120 Pa and a power density of 160 mW/cm^2. The TFT structure is built on un-passivated Corning 7059 glass, with 300 nm μc-Si, 60 nm n$^+$ μc-Si source and drain contact layers, 200 nm SiO_2 or 300 nm SiN_x gate insulator, and 100 nm Al gate, source and drain electrodes.

Introduction

Silicon thin film transistors (TFTs) are widely used in large-area electronics. a-Si:H TFTs are employed as the switching devices for pixels in active matrix liquid crystal displays (AMLCDs) and polycrystalline Si (poly-Si) TFTs are used for switches and driver electronics of AMLCDs. However, the high crystallization temperatures of 600°C employed for poly-Si TFTs motivate the search for an alternative CMOS-capable Si TFT technology. High carrier mobility and the capability for bipolar/CMOS circuits are the two principal motives for developing silicon thin film transistors with microcrystalline silicon (μc-Si) channels. Because direct deposition typically is conducted at the relatively low temperature of plasma-enhanced CVD for a-Si:H, it is an attractive alternative for preparing the microcrystalline silicon film [1][2]. The low temperature of typically 300°C makes μc-Si accessible to a wider variety of substrates than furnace and rapid thermal annealing. μc-Si also retains the advantages of a-Si:H such as good film uniformity over large deposition areas. Therefore, μc-Si TFTs are promising for making both driving and switching circuits for large area displays. Transistors made from directly deposited μc-Si have been reported over the years [3][4][5]. But none of them have yet suggested a practical technology, so that further research needed to be carried out to improve the TFT performance.

Using the conventional PE-CVD technique, we fabricated μc-Si TFTs with a top gate structure. We used both SiN_x and SiO_2 as the gate insulators. The TFTs have saturated mobilities up to 7.9 cm^2/Vs and linear mobilities up to 5.8 cm^2/Vs. The highest I_{ON}/I_{OFF} ratio is 10^5. Typical values for V_{th} is 6.5 V and for the subthreshold slope is 1.7 V/decade.

Experimental procedures

The μc-Si films were grown on unpassivated 1" x 3" Corning 7059 glass slides with a thickness of 1.1mm. The glass slides were cleaned with the MICRO series 8790 glass-cleaning fluid (International Products Corporation) before deposition. The film growth conditions are

Mat. Res. Soc. Symp. Proc. Vol. 557 © 1999 Materials Research Society

listed in Table I. The µc-Si layers were deposited using DC-excited PE-CVD from a gas mixture of SiH_4, SiF_4, and H_2. Adding SiF_4 to the source gas for µc-Si deposition provides a larger deposition space than deposition from H_2-diluted SiH_4 alone [6][7]. The typical growth rate of the µc-Si layer is 0.6 Å/s. We fixed the thickness of the µc-Si layer for the TFT at 300 nm for better comparison among the TFTs. Earlier we had found that the conductivity of the PE-CVD µc-Si is a strong function of the film thickness [8].

The TFT was made in the top-gate configuration to utilize the top surface of the µc-Si layer as the conducting channel. Figure 1 shows the schematic cross section of our TFT structure together with the layer thicknesses. The size of the crystal grains of µc-Si grows from the substrate to the film surface, so that the top of the film has the highest crystallinity [9] and electron mobility. A four-level mask process with specially-designed masks was used to define the TFTs. The semiconductor layers and SiN_x layer were all deposited using a four-chamber Solarex S900 PE-CVD system. The SiO_2 layer was deposited using a Plasma Therm PECVD system. The Al layer was evaporated in an Edwards 306A thermal evaporator. The semiconductor and SiN_x layers were patterned by reactive ion etching with a 10% O_2 and 90% CF_4 mixture. The SiO_2 layer was patterned by 10:1 buffered oxide etch. Al was etched with a wet-etchant formed by 1 part of HNO_3, 1 part of CH_3COOH, 2 parts of H_2O and 16 parts of H_3PO_4.

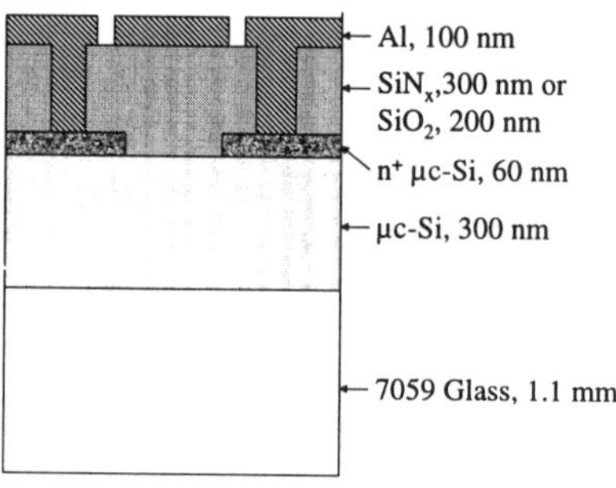

Figure 1. Schematic cross section of TFT, with layer materials and thicknesses

Table I. Deposition parameters for µc-Si, n$^+$ µc-Si and gate insulator layers

Layer	SiH_4 (sccm)	H_2 (sccm)	SiF_4 (sccm)	PH_3 (sccm)	N_2O (sccm)	NH_3 (sccm)	Power density (mW/cm^2)	Pressure (mTorr)	Temp (°C)
µc-Si	1	200	20	0	0	0	160	900	360
n$^+$ µc-Si	2	100	0	12	0	0	324	900	280
SiO$_2$	35	0	0	0	160	0	85	400	250
SiN$_x$	13	0	0	0	0	130	22	500	280

We begin by growing the μc-Si and n⁺ μc-Si layers without breaking vacuum, then pattern the intrinsic layer island and follow by patterning the n⁺ μc-Si source and drain. The next step is the deposition of the gate insulator. After patterning the gate insulator we evaporate the Al and then pattern the gate, source and drain electrodes. The channel length L is 45 μm and the width W is 180 μm. The process sequence is shown in Figure 2.

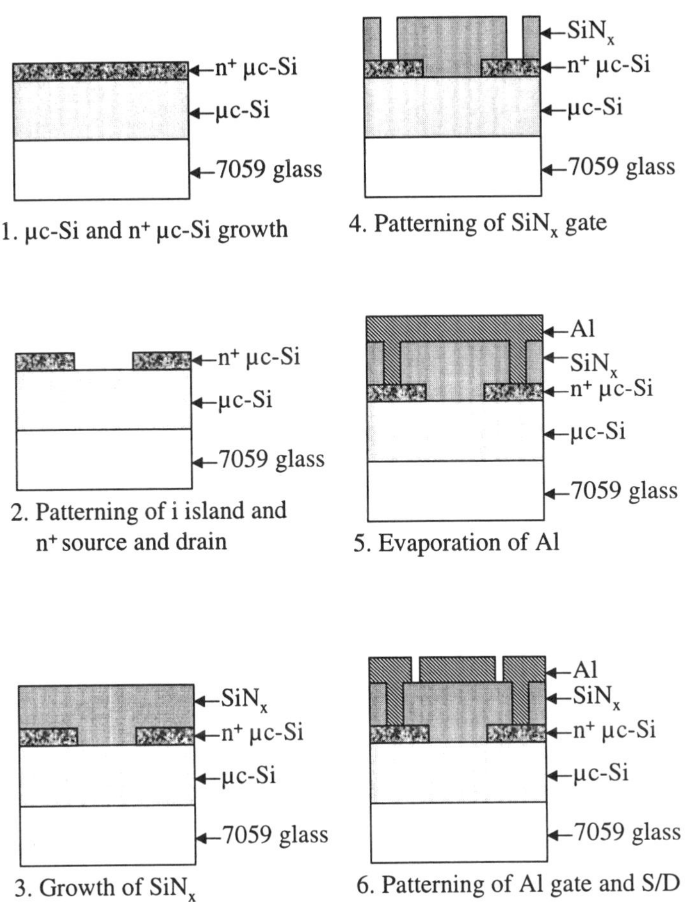

Figure 2. TFT process sequence with SiN_x gate insulator. Sequence for SiO₂ insulator is identical.

Results and Discussion

The fabricated TFTs were annealed at 170 °C in vacuum for 1 hour and then measured using a HP 4155 parameter analyzer. Figure 3 shows the transfer characteristics of the μc-Si TFT with 300 nm SiN_x as the gate insulator and with 200 nm SiO_2 as the gate insulator. Figure 4 shows

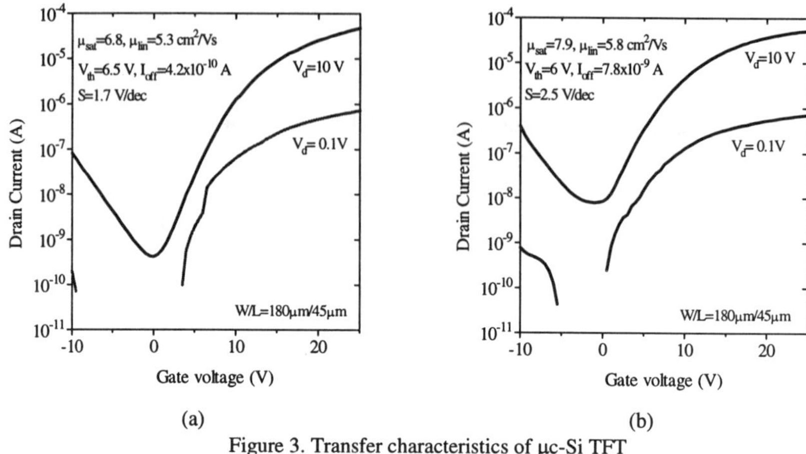

Figure 3. Transfer characteristics of μc-Si TFT
(a) using a 300-nm thick SiN_x gate insulator
(b) using a 200-nm thick SiO_2 gate insulator

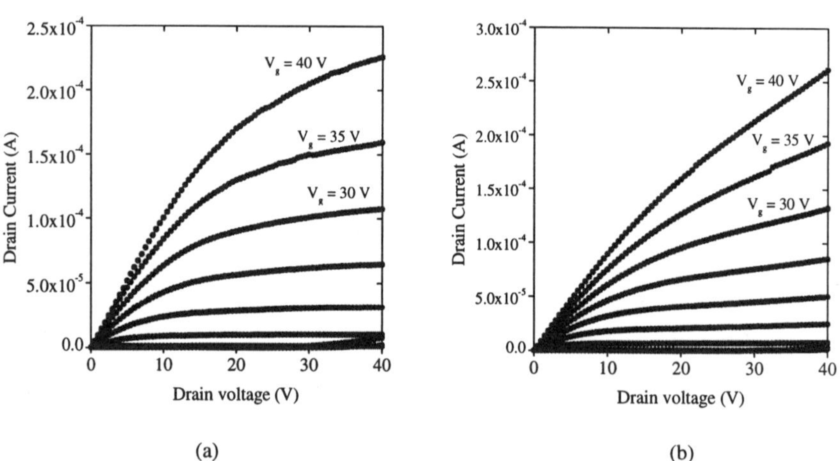

Figure 4. Output characteristics of μc-Si TFT
(a) using a 300-nm thick SiN_x gate insulator
(b) using a 200-nm thick SiO_2 gate insulator

the output characteristics of the μc-Si TFT with 300 nm SiN$_x$ as the gate insulator and with 200 nm SiO$_2$ as the gate insulator.

The electron mobility in the linear region μ_{lin} was extracted from the slope of the I_d vs V_g plot measured at $V_{ds} = 0.1$ V using the following equation [10],

$$I_d = \mu_{lin}C_{ins}(W/L)(V_g-V_{th})V_{ds} \tag{1}$$

where I_d, μ_n, C_{ins}, W, L, V_g, V_{th} and V_{ds} are drain current, electron linear field effect mobility, gate capacitance per unit area, gate width, gate length, gate voltage with respect to source, threshold voltage and voltage between drain and source, respectively. The electron mobility μ_{sat} in the saturated region and the threshold voltage V_{th} were extracted from the plot of $I_d^{1/2}$ vs. V_g measured at $V_g = V_d$ using the following equation [10],

$$I_d^{1/2} = (\mu_{sat}C_{ins}W/2L)^{1/2} (V_g-V_{th}) \tag{2}$$

and the subthreshold slope $S = \partial V/\partial(\log_{10}I_d)$ was obtained from the reciprocal slope of the linear region in the semi-log I_d vs V_g plot measured at $V_{ds} = 10$ V. The OFF current I_{OFF} is defined as the lowest current in the I_d vs V_g plot measured at $V_{ds} = 10$ V.

The TFT with a 300 nm SiN$_x$ gate insulator has a saturated electron mobility of 6.8 cm^2/Vs, a linear electron mobility of 5.3 cm^2/Vs, an OFF current of 4.2 x 10^{-10}A, an I_{ON}/I_{OFF} ratio of 10^5, a threshold voltage of 6.5 V and a subthreshold slope of 1.7 V/dec. The TFT using a 200 nm SiO$_2$ gate insulator has a saturated electron mobility of 7.9 cm^2/Vs, a linear electron mobility of 5.8 cm^2/Vs, an OFF current of 7.8 x 10^{-9}A, an I_{ON}/I_{OFF} ratio of 10^4, a threshold voltage of 6 V and a subthreshold slope of 2.5 V/dec. The TFT with the SiO$_2$ gate dielectric has a higher electron mobility than that of SiN$_x$ gate dielectric, which suggests that SiO$_2$ may have a better interface with μc-Si in the top gate configuration than SiN$_x$. The OFF current of the TFT with the SiN$_x$ gate dielectric is lower than with the SiO$_2$ gate dielectric. The lower threshold voltage for the TFT with SiO$_2$ gate dielectric suggests a larger density of positive charges in the SiO$_2$ than SiN$_x$.

Summary

We fabricated top gate TFTs made with microcrystalline silicon (μc-Si) directly deposited at 360 °C. The μc-Si is grown by PE-CVD from a source gas mixture of SiH$_4$, SiF$_4$ and H$_2$. We used 200 nm SiO$_2$ or 300 nm SiN$_x$ as the gate insulators. The TFT using the SiO$_2$ gate insulator has a higher saturated electron mobility of 7.9 cm^2/Vs but a higher OFF current as well. The TFT using the SiN$_x$ gate insulator has a saturated electron mobility of 6.8 cm^2/Vs, but the OFF current is one order of magnitude lower than in the SiO$_2$/μc-Si TFT.

Acknowledgements

We would like to thank the New Jersey Commission on Science and Technology for supporting this research.

References

1. Y. Okada, J. Chen, I. H. Campbell, P. M. Fauchet and S. Wagner, Mat. Res. Soc. Symp. Proc. **149**, 93 (1989)
2. M. Taguchi and S. Wagner, Mat. Res. Soc. Symp. Proc. **358**, 739 (1995)
3. T. Nagahara, K. Fujimoto, N. Kohno, Y. Kashiwagi and H. Kakinoki, Jpn. J. Appl. Phys. **31**, 4555 (1992)
4. J. Woo, H. Lim and J. Jang, Appl. Phys. Lett. **65**, 1644 (1994)
5. H. Meiling, A. M. Brockhoff, J. K. Rath and R.E.I. Schropp, Mat. Res. Soc. Symp. Proc. (1998) to be published
6. H. Kakinuma, M. Mohri and T. Tsuruoka, Jpn. J. Appl. Phys. **77**, 646 (1995)
7. M. Nakata, Ph.D. Thesis, Tokyo Institute of Technology (1992)
8. Y. Chen, M. Taguchi and S. Wagner, Mat. Res. Soc. Symp. Proc. **424**, 103 (1997)
9. Y. Chen and S. Wagner, Presentation Y8, Session Y, Electronic materials conference, Fort Collins, Colorado, June 25, 1997
10. R. Muller and T. Kamins, Device electronics for integrated circuits, p. 428, John Wiley and Sons, New York (1986)

TEMPERATURE DEPENDENT TRANSIENT LEAKAGE CURRENTS IN AMORPHOUS SILICON THIN FILM TRANSISTORS

F. LEMMI, R. A. STREET
Xerox Palo Alto Research Center, 3333 Coyote Hill Road, Palo Alto, CA, 94304

ABSTRACT

The transient response of a-Si:H thin film transistors (TFT) provides information about gap states in the channel and about the TFT conduction mechanisms. We present results of a detailed parametric study of transient leakage currents in TFTs, extending from 10^{-3} sec to >100 sec and at temperatures up to 400K. The form of the transient decay depends most notably on the off-state gate voltage, which determines the leakage current mechanism and magnitude. The measurements use chains of many TFTs, such that we can measure dc on/off ratios of >10^{11}.

In the sub-threshold region (e.g. Vg~-2V) some samples exhibit non-monotonic decay of the drain current, with a minimum followed by a slow increase to a steady state. Typically this effect is barely visible at room temperature due to the extremely low values of the dc current. Higher temperatures enhance the phenomenon, which has activation energy of about 1 eV for the current magnitude and time constant. We attribute the effect to back-channel electron conduction. This explanation is consistent with the form of the TFT transfer characteristics, and the weak drain-source voltage dependence excludes a contact injection mechanism.

At more negative gate off-state voltages, some samples exhibit increasing leakage currents, indicative of hole conduction. These samples feature a monotonically decreasing transient decay with a pronounced drain-source voltage dependence, consistent with contact injection being enhanced by a hole-enriched channel in the high field drain region.

Effects of on-state gate pulse width and amplitude are also discussed, as well as the effect of high temperature stress on the TFT transients.

INTRODUCTION

Hydrogenated amorphous silicon thin film transistors (TFTs) are used as pixel switches in flat panel liquid crystal displays and large area sensor arrays. The latter application requires the leakage current of the TFTs to be extremely low in order to minimize the dark signal and achieve high sensitivity, especially important for X-ray medical imaging applications [1] where low radiation doses are obviously desirable.

Large area sensor arrays typically integrate charge on the sensor capacitance and rely on the isolation provided by the corresponding TFT once it has been switched off to preserve it. As a result of the existence of deep states in the amorphous silicon channel of the TFT, the switching off of the device typically takes long times (several seconds and more) to actually reach an optimum isolation [2]. An understanding of the leakage current origin and time dependence is desirable, since large area high-sensitivity imagers can integrate charge for up to several seconds.

Currents are typically very small and a readable signal requires several TFTs to be connected in parallel. Several mechanisms can contribute to the transistor leakage, not independently from each other.

Mat. Res. Soc. Symp. Proc. Vol. 557 © 1999 Materials Research Society

In this paper we try to isolate the various sources of leakage by monitoring temperature and bias dependence of transient and steady state currents.

EXPERIMENTAL RESULTS AND DISCUSSION

Inverted staggered a-Si:H TFTs are grown in a standard PECVD system. Devices include sets of 50 to 100 TFTs connected in parallel and measurements are performed using a Keithley 427 Current amplifier capable of gains up to 10^{11} V/A with a digitizing oscilloscope for short (ms to seconds) time response. A Keithley 617 Electrometer for extremely high sensitivity measurements (down to the fA range) is used at longer times. Temperature control is obtained with an MMR three-probe stage that holds samples in a vacuum chamber and controls their temperature by resistive heating and Joule-Thompson cooling. Temperature control is accurate to better than 0.1 °C and the whole setup is computer-assisted.

Typically fast and slow time constants are associated with the transient of TFTs and the term "steady-state" must be clarified. In typical transfer characteristic (I-V) measurement, one minute per point is found to be acceptable, normally cutting off only longer time variations related to sample stressing. Stressing is defined as a long time variation of the device performance that can be fully recovered in reasonable times using higher temperature annealing, as opposed to electronic transient, promptly recovered at room temperature.

Steady state characteristics

Fig. 1 shows I-V curves taken from a set of 100 TFTs with W/L=65/15 μm as a function of the sample temperature and scaled to a single device. The drain-to-source voltage V_{DS} was kept at 10V and the gate voltage V_{GS} was varied from -5V to +15V. The inset shows a cross-section of the device structure together with a distributed equivalent circuit describing the channel capacitance and resistance [3] and a back channel leakage path to be described in the following. The temperature was kept to a maximum of 390°K to avoid metastable effects during the measurements. The data at 300°K is highlighted with markers. At temperatures lower than 300°K, when the gate off-voltage is negative enough, the sample does not reach a full equilibration during the measurement and apparent negative currents are measured (plots show the modulus). The re-equilibration of the channel potential after the gate voltage step is discussed later in greater detail.

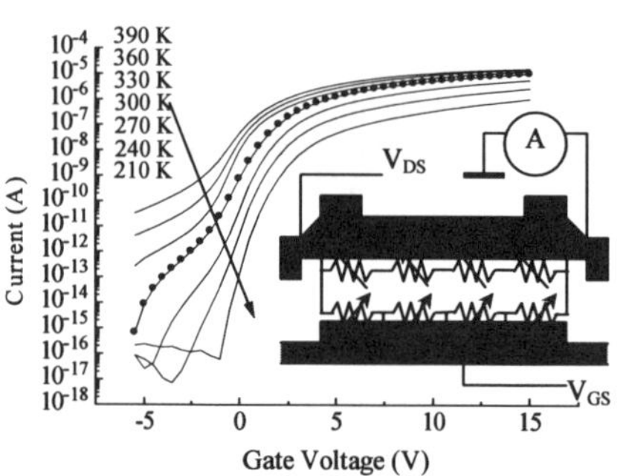

Fig. 1. Temperature dependence of TFT transfer characteristics. The inset shows a cross section of a device and an equivalent circuit description. Currents have been scaled down to one device.

The data show a monotonic decrease of the steady-state current in the negative V_{GS} range and the current increases with temperature by approximately one decade per 30°K. An activation energy of about 1 eV at V_{GS}=-5V and one of 0.13 eV at V_{GS} = +10V are extracted. The former is consistent with thermal emission from deep states and the latter is in agreement with what is commonly observed with the TFT in its conductive state.

Temperature dependence of transient decay

The temperature dependence of the transient current was studied after switching off the TFT, from a pulse voltage of +15 V with duration 1 ms, to –5V. Measurements shown in Fig. 2 are made at and above room temperature, since lower temperatures gives an extremely small signal. No dependence of the decay was found as the on-pulse width was varied from less 100 μs to 1 second. The on-pulse amplitude was also found to be not critical as long as the transistor is driven well into its conductive state

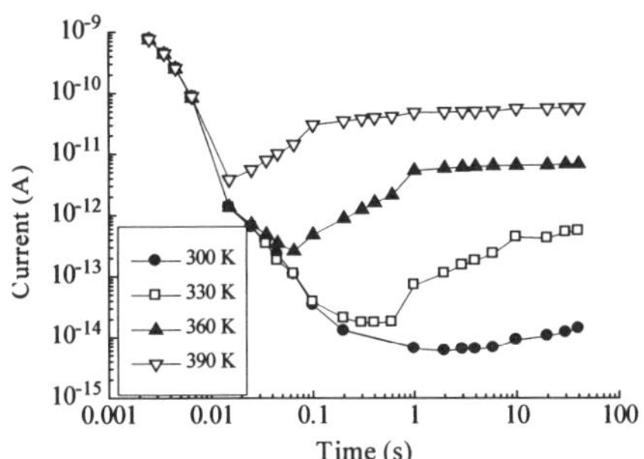

Fig. 2. Transient leakage current of TFT being switched off from +15V to –5V gate voltage as a function of sample temperature.

The data exhibit a non-monotonic current transient with an initial decay followed by an increase to a final steady state. The activation energy of about 1 eV for the steady state current has already been noted. It appears that the onset of the current increase has approximately the same time constant. Two mechanisms appear to compete. The initial decay is due to trapped charge in the channel being collected at the readout electrode as discussed in Ref. 2 and is expected to be almost temperature independent. There is a concurrent phenomenon that features a current increase that is instead temperature dependent. We attribute this current increase to conduction through the top part of the channel, next to the interface with the top nitride insulator, in the following referred to as "back-channel". As the main channel is being depleted, the positive charge left reduces the field effect on the back-channel, screening the electric filed. An increase of the leakage current in time through this path is therefore to be expected. It is worth also pointing out that the existence of states at that interface has been reported [4]. The activation energy of 1 eV confirms that such conduction can be arising from deep states.

Repeated cycling of the gate pulse at room temperature confirms the electronic origin of the current increase, separating it from bias stress effects that take place in considerably longer times (days). This point will be discussed later in further detail.

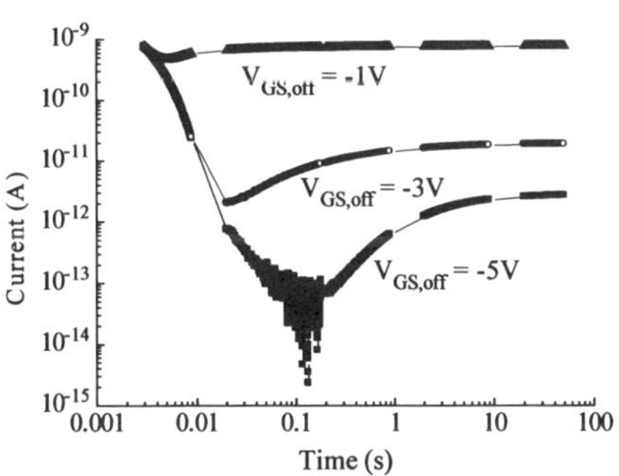

Fig. 3. Transient leakage currents of TFTs using different gate off voltages, at 350 °K temperature.

The gate off-state voltage $V_{GS,off}$ does not affect significantly the short time decay at room temperature. To better understand the actual effect we present results obtained using different $V_{GS,off}$ with the sample kept at 350 °K. The temperature was chosen to increase the visibility of the leakage signal.

Steady-state currents are clearly smaller when $V_{GS,off}$ is more negative, as consistent with a deeper depletion of the channel when the electric field is stronger.

Hole conduction

The dependence of the TFT leakage on $V_{GS,off}$ is found to be different in some samples, particularly at high negative gate voltages and when hole conduction dominates. This is allowed by injection at the high field drain contact and moves the Fermi level below mid-gap.

Fig. 4a shows transfer characteristics from one of such samples (50 TFTs scaled to one in the plots) for drain-to-source voltages V_{DS} of +10V and +1V. At +10V there is an increasing

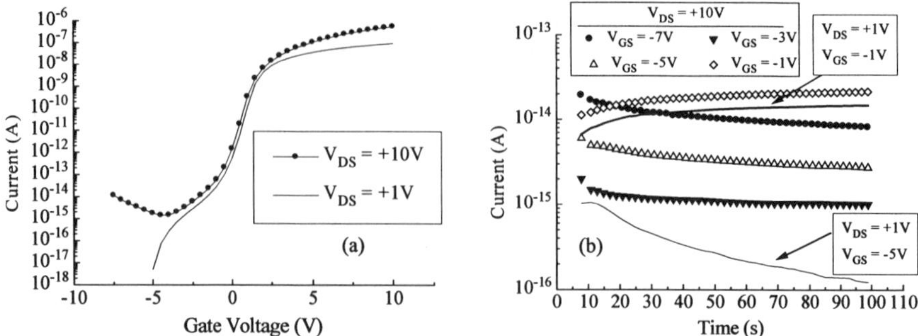

Fig. 4. (a) Transfer characteristics of TFT with hole conduction leakage with contact injection effect. (b) Long time transients for different gate voltages with different leakage mechanisms.

leakage current as the gate voltage becomes more negative (not an effect of a non-equilibrium measurement).

Such gate voltage dependence is evidence of hole conduction and contact injection. This is also confirmed by the comparison with the curve at $V_{DS} = +1V$. A smaller field does not allow hole injection and clearly suppresses the phenomenon. The region where current is given by collection of thermally emitted carriers from states in the main channel and in the back-channel (small negative V_{GS}) is, on the contrary, almost unaffected.

A closer look at the leakage currents at long times was made using the Keithley electrometer, giving useful results even at room temperature (normal operation) in the tens of seconds range, as shown in Fig. 4b. Transient currents show a different behavior in the thermal emission back-channel region and in the hole conduction region. While the former case confirms the increasing current feature discussed before, the latter gives a decreasing transient. This difference is found in all the samples with hole conduction and can be roughly associated with the position of $V_{GS,off}$ with respect to the minimum of the I-V plot. When $V_{GS,off}$ is above the minimum, the transient current increases towards the steady state and below the minimum it decays monotonically (hole conduction region). The effect is illustrated in Fig 4b for $V_{DS} = +10V$ in the curves with markers.

In the hole conduction region, the transient component of the current is still attributed to the depletion of occupied states as the Fermi level sinks deeper into the gap; it reaches a position below mid-gap, as allowed by holes being injected at the drain contact. The steady state value reflects the actual enrichment of holes, giving a larger current for a Fermi level final position closer to the valence band edge when the gate field is stronger.

Channel capacitance effects

Transients using $V_{DS} = +1V$ are shown in Fig. 4b for two values of gate voltage. The data for $V_{GS} = -5V$ exhibits a negative measured current (absolute value plotted) for very long times, eventually turning positive later than 100 s. The phenomenon is explained by the stored charge in the channel according to the energy band diagram sketched in Fig. 5 at high and low V_{DS}. The conduction band edge along the channel is shown at different times after the gate voltage switches under a simplifying assumption of linear potential variation. The flow of emitted electrons is shown, as driven by the existing electric fields.

The rate of change of channel potential depends on the product of the distributed capacitance and the distributed resistance of the channel, (see the equivalent circuit in the inset of Fig.1). The channel potential experiences a 20 V drop when the gate voltage is switched. The displacement

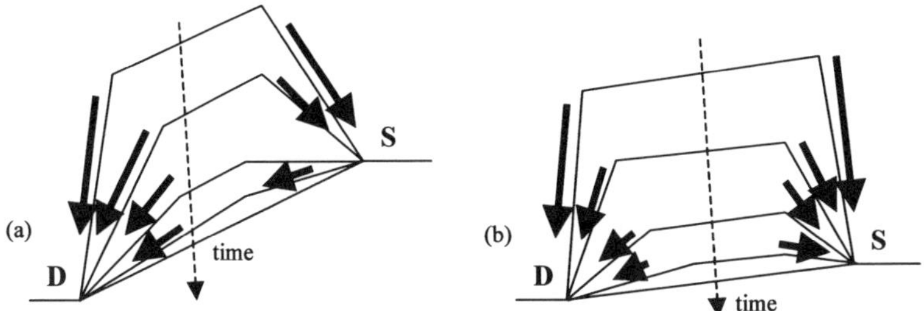

Fig. 5. (a) Energy band diagram describing the conduction band edge as a function of position along the channel. Qualitative time dependence is shown for high (a) and low (b) V_{DS}.

current starts charging the channel capacitance, changing the potential along the channel, and the current at the source contact S is negative.

While the large abrupt potential drop is associated with large negative currents at very short times, the final adjustment takes place in conditions of a very high resistance channel and at considerably longer times. In case (a) for high V_{DS}, the field at the source contact is promptly reversed and regular positive currents are measured during depletion up to the steady state. In case (b) for low V_{DS}, the inversion of the field at the source contact requires more complete channel relaxation. This condition occurs close to the steady state with an extremely resistive channel and long (several seconds) RC time constants. The RC time is 400 seconds in a TFT with a 40 fF channel capacitance and a distributed resistance of order 10^{16} Ohm.

Confirmation of this behavior was obtained by separating the two current components that flow into the source contact from the gate and the drain. The gate current is associated with the change in channel potential and ultimately to the long time re-equilibration of trapped charge. The components were extracted by measuring the transients at $V_{DS} = +1V$ and $V_{DS} = -1V$ and separating the even and odd terms as illustrated in Fig. 6.

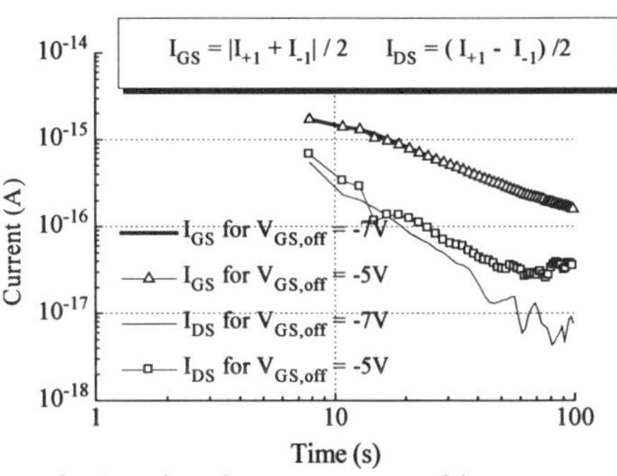

Fig. 6. Drain and gate components of the source current for long time transients at low V_{DS}

Results plotted in the figure show that the gate current (negative as measured at the source) still prevails after 100 s. The direct drain-to-source current decays and the conductivity of the channel decreases and stabilizes after about 50 sec. The gate current decays with a 1/t law, typical of thermal emission currents [2,5], confirming the validity of the proposed physical mechanism. This finally explains the data presented in Fig, 1 in the large negative V_{GS} region.

Bias stress effects

Negative bias stress at high temperature (440 °K) was applied to some samples (50 TFTs in parallel) to investigate the effects on the transient currents of a regular operation of TFTs in large area imagers. Same effects are visible at room temperature with much longer stress times.

Fig. 7 shows an example of negative bias stress ($V_{DS} = +10V$, $V_{GS} = -5V$) for 10 minutes at 440 K. The change in the steady state curve (scaled to one device) in Fig. 7a is accompanied by

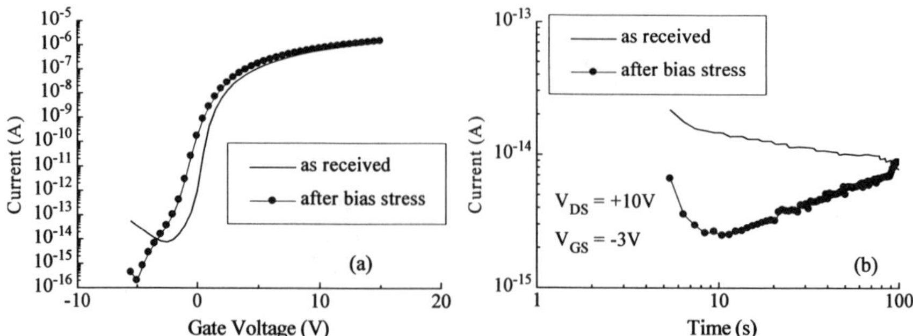

Fig. 7. (a) Transfer characteristics of TFT before and after bias stress. (b) Long time transients for –3V gate voltage. About the same final value is associated with different transient slopes.

a change in the transient current shown Fig. 7b when $V_{GS,off}$ is –3V. The results support the above interpretations discussed so far. The decreasing or increasing transients (after stress) are associated to hole and back-channel conduction respectively, as evident from the steady state transfer curve shape. The change in mechanism may either be attributed to a positive charge trapping in the nitride or to defects creation in the channel. The former case would oppose the onset of contact injection by mean of electrostatic shielding provided by the positive trapped charge and is confirmed by comparison with preliminary light soaking/annealing cycles of the samples. More investigation on this subject is still required.

CONCLUSIONS

We have presented studies of the transient and steady state leakage currents after the TFT has been switched off together with the temperature dependence. The transient decay shows very clearly the role of back-channel conduction in the sub-threshold region. The gate voltage dependence reveals a clear distinction between electron and hole conduction. The former is related to thermal emission of trapped carriers, while the latter is associated to high field effects in the drain region, as confirmed by the different drain voltage dependence. In some cases negative currents are measured at the source contact up to long times and we show evidence of their origin in the channel capacitance. The effect of on-pulse width was discussed as well as effects of high temperature bias stress.

ACKNOWLEDGMENTS

The authors wish to thank J.P. Lu and K. Van Schuylenbergh for helpful discussion. This work is partially sponsored by NIST (70NANB7H3007).

REFERENCES

1. L.E. Antonuk, Y. El-Mohri, A. Hall, K.-W. Jee, M. Maolinbay, S.C. Nassif, X. Rong, J.H. Siewerdsen, Q. Zhao, R.L. Weisfield, Proceedings of SPIE 3336, p. 2 (1998).
2. F. Lemmi, R.A. Street, MRS Symp. Proc. **507**, p.79 (1998).
3. M.S. Shur, M.D. Jacunski, H.C. Slade, A.A. Owusu, T. Ytterdal, M. Hack, Proceedings of the Third Symposium on Thin Film Transistor Technologies, **96-23**, p. 242 (1996).
4. R.A. Street, C.C. Tsai, Appl. Phys. Lett. **48** (24), p. 1672 (1986).
5. R.A. Street, Philos. Mag. B, **63** (6), p. 1343 (1991).

OPTICAL FILTER FOR FABRICATING SELF-ALIGNED AMORPHOUS Si TFTS

P. MEI, J. P LU, C. CHUA, J. HO, Y. WANG, AND J. B. BOYCE
Xerox Palo Alto Research Center, Palo Alto, CA

ABSTRACT

Self-aligned structures for bottom-gate amorphous Si TFTs provide a number of advantages, including reduced parasitic capacitance, smaller device dimensions, and improved uniformity in device performance for large-area electronics. A difficult challenge in making self-aligned TFT structures is the necessity of making source/drain contacts that exhibit low contact resistances and that are precisely aligned relative to the gate electrode. In this article, we describe a novel process for fabricating self-aligned amorphous Si TFTs. This process utilizes a pulsed excimer laser (308 nm) to dope or to activate dopants in a-Si to form the source/drain contacts. An important feature of the device design is an optical filter to protect the a-Si channel region from radiation damage during the 308 nm laser process. However, the optical filter allows the transmission of the uv light for lithography exposure from the backside of the substrate to align the channel region with the gate electrode. This new process enables the fabrication of high performance self-aligned a-Si TFTs with poly-Si source and drain contacts.

INTRODUCTION

Bottom-gate a-Si:H TFTs are widely used as pixel switches in active matrixes for large-area display and imaging applications due to their low off current, high on/off current ratio, and low cost of fabrication. In the conventional bottom-gate a-Si:H TFT fabrication process, the channel region is defined by a self-aligned passivation island, formed by backside lithography. The source/drain contacts are formed by depositing doped a-Si:H with subsequent patterning. Since it is difficult to selectively etch doped a-Si over the intrinsic a-Si, the top passivation island is utilized as the etch-stop to form the source/drain electrode (Fig. 1). As a result, there are overlaps of the source/drain electrodes with the channel region, causing an extra parasitic capacitance between the gate and the source/drain electrodes. The high parasitic capacitance results in a feed-through voltage on the pixel electrode, producing image flicker and sticking [1].

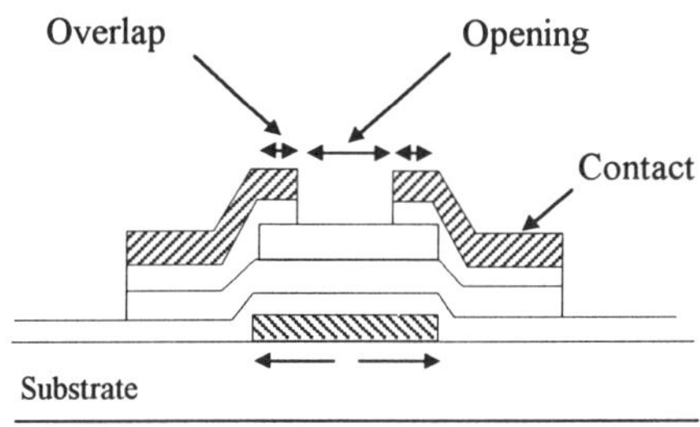

Figure 1. Device structure of a conventional bottom-gate a-Si TFT.

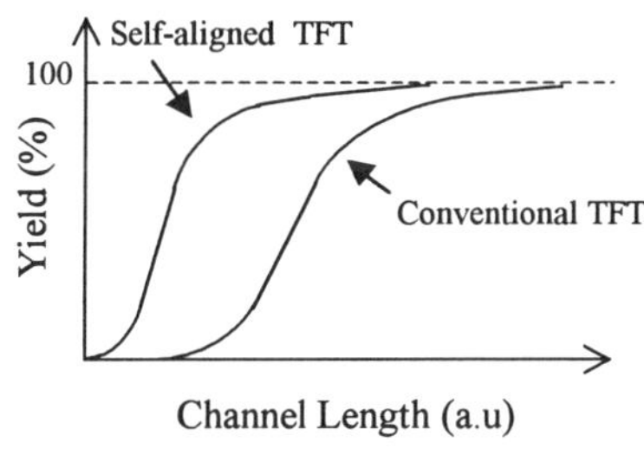

Figure 2. Sketch of the manufacturing yield as a function of the channel length for conventional and self-aligned a-Si TFTs.

Mat. Res. Soc. Symp. Proc. Vol. 557 © 1999 Materials Research Society

Also, the overlap introduces a variation of the parasitic capacitance among the pixels, resulting in non-uniform gray-level performance.

In addition to the problem of the parasitic capacitance, the non-self-aligned structure limits scaling down the channel length. The following analysis describes how the yield of large-area manufacturing process is affected by scaling the channel length.

In the conventional manufacture process for the bottom-gate a-Si TFTs, the yield for the channel length definition is governed by:

$$Y = y_{op} * y_{al} * y_{etch} * y_{self} \tag{1}$$

where y_{op} is the yield factor related to the opening between source/drain contacts, y_{al} is the factor of the alignment accuracy of the source/drain contact versus the passivation island, y_{etch} is the factor related to the etching accuracy of the source/drain contact, and y_{self} is the factor related to the formation of the self-aligned passivation island. All of these factors, except y_{self}, depend on the channel length. As depicted in Fig. 2, that the manufacturing yield decreases with the reduction of in channel length.

This problem can be alleviated by forming self-aligned source/drain contacts with a pulsed laser doping technique [2]. A difficult challenge in making self-aligned TFT structures is the necessity of making source/drain contacts that exhibit low contact resistances and that are precisely aligned relative to the gate electrode. In this paper, we describe a novel process which utilizes a pulsed excimer laser (308 nm) to dope or to activate dopants in a-Si to form the source/drain contacts. An important feature of the device design is an optical filter to protect the a-Si channel region from radiation damage during the 308 nm laser process. However, the optical filter allows the transmission of the uv light for lithography exposure from the backside of the substrate to align the channel region with the gate electrode. This process not only reduces the TFT parasitic capacitance, but also improves the yield factor by avoiding yield factors of y_{op}, y_{al}, and y_{etch} in formula (1) (Fig. 2).

PROXIMITY LASER DOPING TECHNIQUE

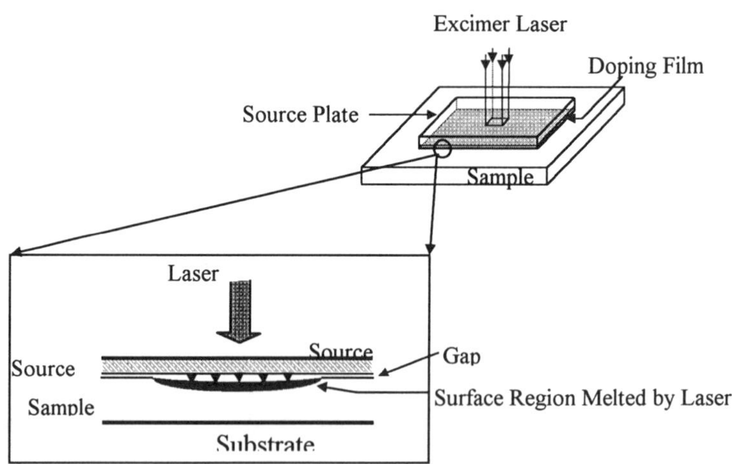

Figure 3. Schematic of the proximity laser doping technique [3].

We developed a large-area electronics manufacturing process for a highly efficient and high throughput process of the proximity laser doping technique which is illustrated in Fig. 3. In this process, the doping source is a thin film deposited on a transparent substrate (referred to as the doping plate). The doping source for a-Si TFT source/drain contacts is a silicon and phosphorus alloy deposited by a PECVD process. The doping plate and the device plate are contacted face to face in a close distance. A pulsed excimer laser is used to generate doping species and to drive the dopant into the device film. Since the doping film is thin (less than 10 nm) and it absorbs photons at 308 nm wavelength, the film is readily ablated by a laser pulse at the energy of a-Si surface melting threshold. Meanwhile, the transmission of the same laser pulse melts the surface of the device plate. The surface melt duration depends on the absorbed energy and the substrate thermal transport properties. Typically, it may last 50 - 100 ns. If the ablated dopant species arrive at the molten surface of the device plate before it solidified, the dopant are incorporated in the device film, since the dopant diffusion in molten silicon is very fast [4]. The dopant are activated during the solidification process. The doping depth is controlled by the laser energy, and the dose is dependent on the laser shot density and the doping film thickness.

Material properties of the laser doped Si films were studied by SIMS measurements, 4-point probe measurements, and transmission electron microscopy (TEM) [5]. SIMS measurements of phosphorus depth profile shows the correlation between the doping depth and the Si film melting depth during the laser radiation. The doping level depends on both the laser energy density and the shot density. The doping efficiency increases linearly with the energy when the laser energy exceeds the Si surface melting threshold of about 200 mJ/cm^2.

For laser doped Si films, the electrical properties relate directly to the structural properties. The structure of the silicon film, starting as amorphous, is converted to polycrystalline by the laser irradiation during the doping process. The spread sheet resistance measurements on these doped Si films indicate that most of the doped phosphorus are electrically activated after the film is solidified [5]. A sheet resistance below 1 kΩ/sqr is readily achievable with a laser energy density slightly above the Si melt threshold.

DEVICE FABRICATION AND OPTICAL FILTER

(a)   **(b)**

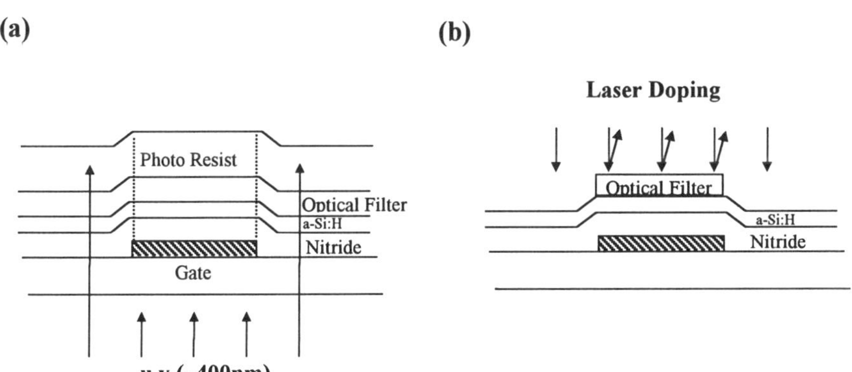

Figure 4. Structure of the optical filter for the self-aligned laser doping mask.

A self-aligned and short-channel a-Si:H TFT structure has been realized with the laser doping technique as described above. The first few fabrication steps for the self-aligned TFT are the same as that of the conventional process. The metal gate is patterned on a substrate. On top of the gate electrode, a layer of gate dielectric of silicon nitride, an intrinsic a-Si:H, and an optical filter are deposited sequentially by PECVD.

The optical filter is used as the self-aligned doping mask. This mask is transparent at the wavelength (~ 400 nm) for the lithography process to allow the backside uv exposure in forming the self-aligned island (Fig. 4a). On the other hand, the mask is reflective at the wavelength (308 nm for XeCl excimer laser) of the laser doping to protect the a-Si channel region (Fig. 4b). This optical filter is a stack of silicon nitride and silicon dioxide layers with the thickness of 1/4 wavelength of the excimer laser (308 nm). Fig. 5a shows a simulation result of the reflectance at the wavelength of 308 nm as a function of the number of the oxide/nitride pairs used for the optical filter [6]. With 4 pairs, the reflectance of the filter approach 95%. Figure 5b shows the reflectance spectrum for 3 pair optical filter. The reflectance at the wavelength of 308, and 400 nm are about 93% and 5% respectively. This difference in the reflectance at the two wavelengths makes it possible for the mask to protect the a-Si channel region from excimer irradiation and to allow backside exposure of the lithographic uv photo to make the self-aligned channel definition.

The laser doping is then performed with the self-aligned optical filter as the doping mask. Since the doping only requires a few laser pulses, the laser doping throughput is high. Also, this process has a wide process window in terms of laser energy density, laser beam profile and laser pulse stability, and doping plate uniformity and reproducibility.

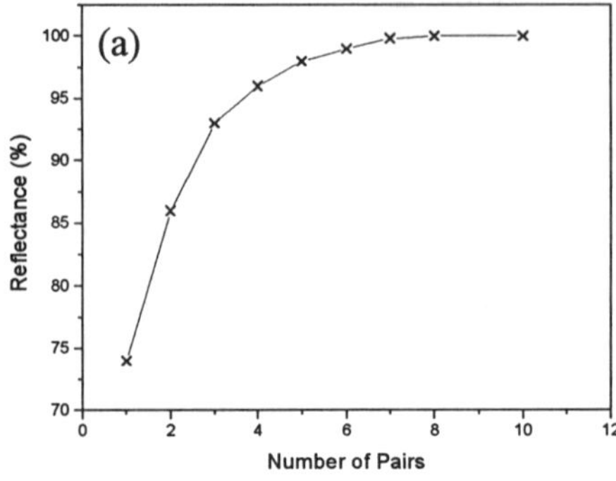

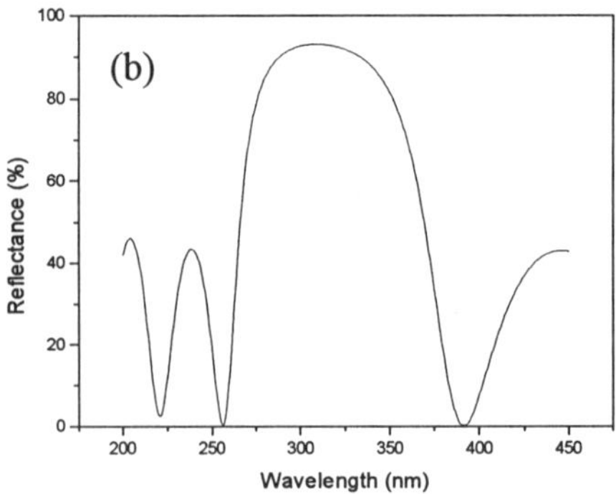

Figure 5. Simulation results for the SiO2/SiN optical filter. (a) Reflectance as a function of the number of pairs at the wavelength of 308 nm; (b) Reflectance spectrum of a optical filter with 3 oxide and nitride pairs.

After the laser doping, a via to contact the gate metal is patterned and etched and the source/drain metal is deposited and patterned . The distance between the edge of the metal and the edge of the self-aligned island can be as far as 5-10 μm apart without having a contact problem.

DEVICE PERFORMANCE

The efficiency of the optical filter as the protection mask during laser radiation was investigated by making self-aligned a-Si TFTs with optical filters of 2 and 3 pairs. In this study, the TFT source/drain contacts were made by phosphorus ion implantation at 10 keV with a dose of 3×10^{15} ions/cm^2 followed by a laser anneal at 240 mJ/cm^2. A sheet resistance of 350 Ω/sqr was obtained after the laser anneal. Figure 6 shows the TFT transfer characteristics. The TFT with 2 pairs of the optical filter exhibits a reduced on-current, a high leakage current and poor subthreshold slope. The poor performance is primarily due to laser damage to the a-Si channel during the laser annealing process. As it can be seen that, 3 pairs of the optical filter protects the channel region well. The TFT has a normal on-current and a low leakage current.

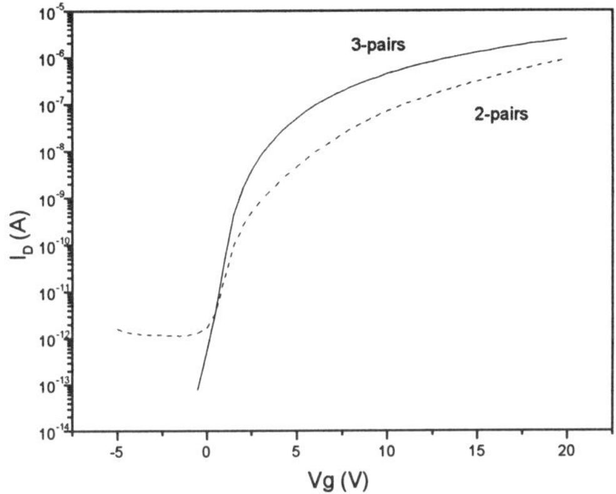

Figure 6. Transfer characteristics of TFTs with 2 and 3 pairs of optical filters. The TFTs are 5 μm long and 10 μm wide. The characteristics were measured at source/drain voltage of 10 V.

Performance of the self-aligned a-Si:H TFTs made by the proximity laser doping technique has been evaluated under DC and AC operation conditions. Under DC operation conditions, the device shows a low source/drain contact resistance. The source/drain contact resistance in the linear region was determined from the inverse of the output conductance as a function of the TFT channel length [7]. It has been found that the contact resistance for the laser doped source/drain contact is more than 20 times lower than that of the conventional a-Si TFT with doped a-Si contacts [8]. This low contact resistance has a significant impact on the performance of short-channel TFTs. For short-channel TFTs, the source/drain contact resistance needs to be scaled down with channel length reduction in order to maintain a low ratio of the contact resistance to channel resistance. It is commonly seen that when the contact resistance is poor, the TFT field effect mobility decrease with channel length reduction [9]. This problem is solved with our proximity laser doping technique in that the TFT mobility maintains its value when the channel length is reduced to 3 μm [5].

Performance of the self-aligned TFT under AC operation conditions is described in reference [10, 11]. The measurements of the delay time of ring oscillators proved that the reduction of the TFT parasitic capacitance for self-aligned TFTs is determined by the geometric factor of the overlap of the source/drain contact versus the channel region. Therefore, self-aligned TFTs provide a minimum value of the parasitic capacitance. The self-aligned TFTs are excellent pixel switches in an active matrix which will provide low feed-through voltages and the minimum cross talk noise.

SUMMARY

We have developed an optical filter for the proximity laser doing process to fabricated self-aligned a-Si TFTs. The proximity laser doping technique enables a highly efficient, large-area manufacturing doping process. The optical filter provides a self-aligned protection mask for the laser doping process. Device performance of self-aligned and short-channel TFTs has been analyzed. The TFT shows excellent performance under DC operation conditions. Compared with the conventional source/drain contact, there is more than 20 times reduction in the contact resistance for the self-aligned TFT. The low contact resistance leads to a constant field effect mobility when the TFT channel length is scaled down. Analysis of AC performance exhibits reduced parasitic capacitance in the self-aligned TFTs. This new self-aligned TFT process has been implemented in 2-dimentional imaging arrays and a successful fabrication of these arrays has been demonstrated [12].

ACKNOWLEDGEMENTS

The authors are grateful for the help of the staff of the Xerox processing line. This work is partially supported by NIST (70NANB7H3007).

REFERENCES

1. S. Nishida, H. Uchida, and S. Kaneko, Mat. Res. Soc. Symp. Proc. 219, 303 (1991).
2. P. Mei, G. Anderson, J. B. Boyce, D. K. Fork, and R. Lujan, Thin Film Transistor Technologies III, Electrochemical Soc Proc., PV 96-23, p51 (1997).
3. P. Mei, R. A. Lujan, and J. B Boyce, US Patent 5,871,826 (1999).
4. D. L. Kkendall and D. B. DeVries, Diffusion in Silicon., R. R. Haberecht and E. L. Kern, Eds., Semiconductor Silicon, Electrochemical Society, New York,. 358 (1969).
5. P. Mei, J. B. Boyce, D. K. Fork, G. Anderson, J. Ho, J. Lu, M. Hack, and R. Lujan, Mat. Res. Soc. Symp. Proc. 508, 3 (1998).
6. The reflectance simulation was performed using Vertical software package developed by Optical Concepts, Inc.
7. S. Luan and G. Neudeck, J. Appl. Phys. 72(2), 766 (1992).
8. P. Mei, J. B. Boyce, R. Lujan, D. Fork, and G. Anderson, Digest of Technical Papers, AM-LCD, 75 (1997).
9. M. J. Powell, C. Glasse, J. E. Curran, J. R. Hughes, I. D. Frech and B. F. Martin, Mat. Res. Soc. Symp. Proc. 507, 91 (1998).
10. J. P. Lu, P. Mei, R. Lujan, and J. B. Boyce, to be published in Thin Film Transistor Technologies III, Electrochemical Soc Proc. (1998).
11. J. P. Lu, P. Mei, C. Chua, J. Ho, Y. Wang, J. B. Boyce, and R. Lujan, submitted to Mat. Res. Soc. Symp. Proc. (1999).
12. J. T. Rahn, F. Lemmi, P. Mei, J. P. Lu, J. B. Boyce, R. A. Street, R. B. Apte, S. E. Ready, K. F. Van Schuylenbergh, J. Ho, R. Fulks, R. L. Weisfield, submitted to Mat. Res. Soc. Symp. Proc. (1999).

THIN FILM TRANSISTORS BASED ON MICROCRYSTALLINE SILICON ON POLYIMIDE SUBSTRATES

A.P. CONSTANT, T. WITT, K.A. BRATLAND, H.R. SHANKS, A.R. LANDIN
Microelectronics Research Center, Iowa State University, Ames Iowa

ABSTRACT

This study reports on the fabrication of inverted gate TFT devices and circuits using a low temperature microcrystalline silicon deposition process compatible with polyimide substrates. Results from the electrical and material characterization of μ-Si:H based TFT's are presented. Device performance is compared with that of α-Si:H based TFT's constructed on polyimide. Results indicate that the anticipated improvement in device performance due to an increase in the μ-Si:H Hall mobility (~10 cm^2/V-sec) over that of α-Si:H (~1 cm^2/V-sec) is not realized. Properties of μ-Si:H films are related to deposition parameters.

INTRODUCTION

Microelectronic devices based on thin film transistors (TFT's) are currently being manufactured on glass substrates for various LCD display technologies. Polycrystalline silicon is typically used as the active layer to achieve switching speeds and drain currents approaching those of crystalline silicon. However, efforts to build similar devices on flexible polymeric substrates have been hampered by the limits placed on processing temperatures by the polymers (< 350°C). Devices fabricated on polyimides therefore currently employ amorphous silicon (α-Si:H) which can be deposited at suitably low temperatures [1-3]. These α-Si:H based devices exhibit lower performance than their polycrystalline counterparts, due in part to the significantly lower mobility of α-Si:H (~1 cm^2/V-sec) compared with that of poly-Si (~ 100 cm^2/V-sec). Switching speeds for these devices are on the order of 1 msec with typical I_{DS}=100 μA (at V_{DS}=10V, V_{GS}=10V) for devices with L=2 μm and W/L =10. This study reports on the fabrication of inverted gate TFT devices using a low temperature microcrystalline silicon (μ-Si:H, ~10 cm^2/V-sec) deposition process compatible with polyimide substrates.

EXPERIMENT

The devices and circuits fabricated in this study employ an inverted gate TFT structure. A cross-sectional diagram is given in Figure 1. The entire process requires 7 photolithographic masking steps. For devices constructed on polyimide, a conformal coating of 6 μm thick polyimide is applied onto the 4-inch wafer using the spin–on technique. The wafer provides rigid backing for lithography alignment. The film is then cured in several stages up to 350°C. The polyimide is next passivated by depositing 1000 Å of PECVD oxide (Si_xO_y:H) using SiH_4 and N_2O at 275°C.

The first step in TFT fabrication is the deposition, patterning and etching of a 500-1000 Å thick chromium layer to create the inverted gate electrode. The gate dielectric is deposited in a two step process. A 1000 Å layer of PECVD oxide is deposited first followed by a 300-500 Å layer of PECVD oxy-nitride ($Si_xO_yN_{1-y}$:H), both at 275°C. The oxy-nitride layer is added because it has been shown to yield a lower surface defect density at the α-Si:H/gate dielectric

Mat. Res. Soc. Symp. Proc. Vol. 557 © 1999 Materials Research Society

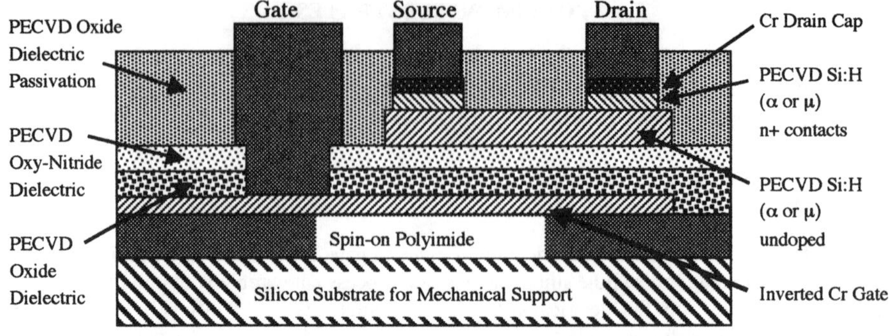

Figure 1. Inverted gate TFT structure employed in this study.

interface and hence higher channel mobility [4,5]. Immediately after forming the gate dielectric the wafer is transferred into a PECVD silicon deposition system.

The microcrystalline and amorphous films are deposited in a parallel plate R-F plasma enhanced reactor. The substrates are mounted on a heated platen which acts as the ground electrode. The powered electrode area is approximately 250 cm^2 with an inter-electrode spacing of 2.5 cm. Intrinsic Si:H films are deposited using a feed gas of 10% silane pre-mixed in H_2 with additional hydrogen added for microcrystalline films. N-type material is produced by the introduction of phosphine/H_2 mixtures, while p-type films are made with diborane/ H_2 additions. All films are deposited at 275°C, a temperature shown to be compatible with polyimide processing.

Many films reported here were deposited on glass or oxidized silicon substrates for materials testing and systems calibration, but all were deposited under conditions suitable for polyimide. The H_2/SiH_4 ratio and R-F power density were varied to manipulate the properties of resulting films. Films intended for the fabrication of TFTs were generally deposited at 300 mTorr with 50 mW/cm^2 for amorphous films and 500 mTorr at 136 mW/cm^2 for microcrystalline.

To realize the active layer, 1000-1500 Å of undoped Si:H is deposited followed immediately by a 500 Å contact layer of n+ Si:H. The remaining processing steps complete the structure shown in Figure 1 and have been discussed elsewhere [1]. Figure 2 illustrates a portion of a shift register containing several TFT's fabricated by this process.

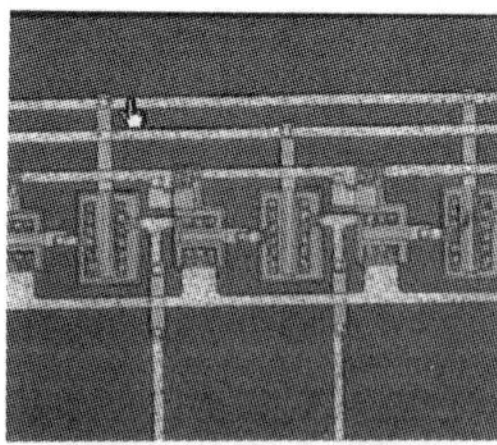

Figure 2. Shift register showing individual TFT's on polyimide. Line width ~ 5μm.

TFT transistors and circuits were tested to measure device characteristics and performance. Samples of the amorphous and microcrystalline phases of Si deposited on glass substrates where used for 4-point probe, Raman and Hall mobility measurements. In some cases, resistors where fabricated on thermally oxidized wafers to measure the resistivity of these materials.

RESULTS

The properties of the deposited Si:H films were found to vary most strongly as functions of H_2/SiH_4 ratio and R-F power density. Sensitivities to temperature and total pressure over the rather narrow ranges examined were not significant. A series of n-type (P doped) films was made to evaluate variation with these parameters. For this series of films, the temperature, total pressure, and phosphine flow rate were held constant. Power density was varied from 50 to 136 W/cm^2 and the gas ratio from 10 to 92. A gas ratio of 10:1 was the lowest possible using the pre-mixed gases available. The electrical resistivity of these materials is shown as a contour plot in Figure 3. The contour lines on the plot are at every half log unit resistivity. The most notable feature of this plot is the division of the field into essentially two plateaus with a sharp transition range (i.e. high contour density). There are 3-4 orders of

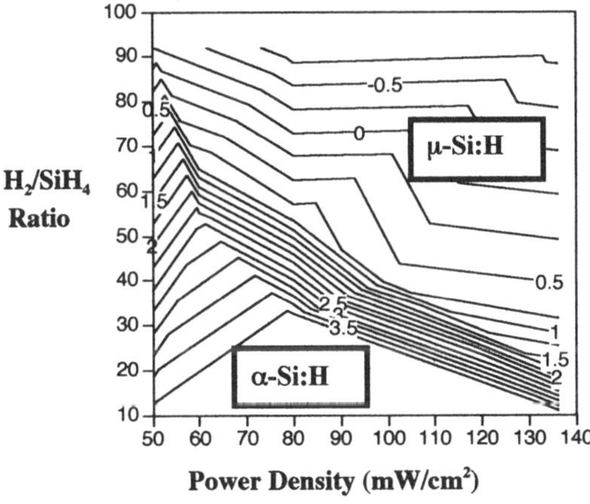

Figure 3. Contour plot showing resistivity of P-doped PECVD silicon as a function of H_2/SiH_4 ratio and power density. Contours are log ρ. The μ-Si:H and α-Si:H regimes are shown.

magnitude change in resistivity over this narrow transition range. Low power and low gas ratio favor a highly resistive material while high power and gas ratio yields much more conductive material. These two regions have been identified as amorphous and microcrystalline respectively. An additional sample, not plotted here, at a gas ratio of 190 and power density 136 W/cm^2 shows the microcrystalline plateau to be quite flat at much higher gas ratio. While this map is valid for a specific pressure, temperature, and dopant level; a very similar contour might be expected under different combinations of these parameters.

The transition from amorphous to microcrystalline properties with varying growth parameters was observed to be quite sensitive. Reproducibility within this region was poor and samples grown simultaneously on separate substrates mounted in different regions of the reactor showed differences in electrical resistivity of as much as 2-3 orders of magnitude while differences in similar samples grown within the "plateau" regions were typically within a factor of 2. The mechanism(s) involved in this high sensitivity to growth parameters are uncertain at this time and await more detailed analysis of materials grown in the transition region.

The attribution of the high resistivities to an amorphous condition and low resistivity to microcrystallinity have been reinforced by Raman spectrographic data. Figure 4 shows the Raman spectra for films grown at various power densities at two gas ratios. A silicon crystalline peak occurs at a wave number of 522 cm^{-1}. This peak is seen in Figure 4 to develop at high power density, possibly maximizing at an intermediate power. At lower power densities there is no sign of a peak at this wave number. Some changes were made in the reactor configuration between the collection of the data in Figure 4 and those in Figure 3, so a direct comparison at equal power levels and gas ratios is probably not valid. The trends in the

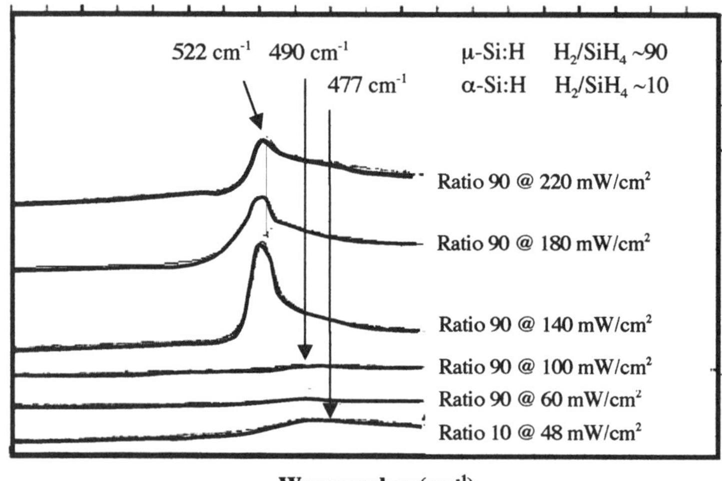

Figure 4. Raman spectra of α to μ transition in PECVD silicon as a function of deposition conditions. The peak at 522 cm^{-1} indicates the presence of crystalline bonding.

data in both figures, however, confirm the conclusion that the low resistivity material deposited does indeed have a crystalline component. X-ray diffraction data have also confirmed presence of crystallinity for samples grown on the μ-Si:H plateau [6].

Results for both α-Si:H and μ-Si:H based TFT's are given in Figures 5 and 6. Figure 5 plots the drain-source current (I_{DS}) vs the drain-source voltage(V_{DS}) at various gate-source voltages (V_{GS}) for a transistor with W/L ratio of 4780/4 for both types of devices. The on-off current ratios for these same devices are plotted in Figure 6. Threshold voltages, V_T, were obtained by extrapolation of the linear portion of $I_{DS}^{1/2}$ vs V_{GS} plots for both types of TFT's and were found to equal 3.5 V (μ-Si:H) and 2.2 V (α-Si:H). Comparison of the curves given in

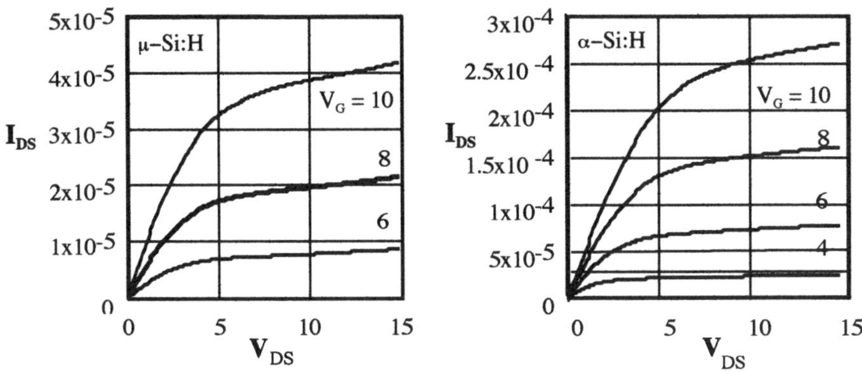

Figure 5. Transistor behavior of α-Si:H and μ-Si:H based TFT's [6].

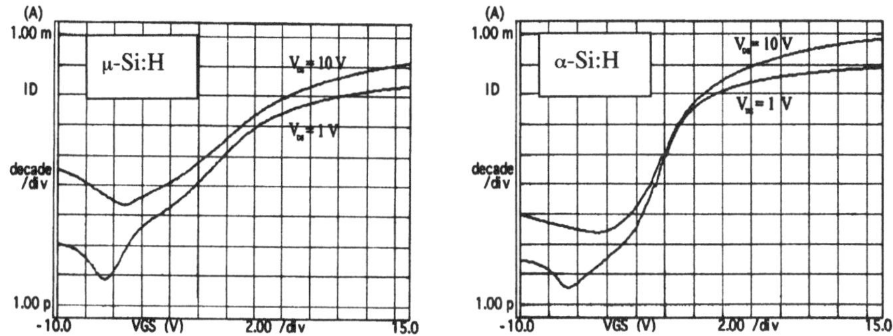

Figure 6: On-off current ratio of α-Si:H and μ-Si:H based TFT's [6].

Figure 5 indicates that α-Si:H based devices exhibit nearly one order of magnitude higher drain currents than similar μ-Si:H based devices. Channel mobilities were calculated according to:

$$\mu = (2L/C_{ox}W)(I_D/V_{GS}-V_T)^2 \qquad (1)$$

This yields channel mobilities (μ) of 0.042 cm^2/V-sec and 0.28 cm^2/V-sec for μ-Si:H and α-Si:H, respectively. The fact that the α-Si:H mobility is somewhat lower than those routinely reported in the literature for similar device structures is not unexpected given the lower deposition temperatures used in this work. The apparent reduction in channel mobility and current carrying capability in the case of μ-Si:H based devices, as compared to the α-Si:H, is unexpected though. Clearly, as discussed in the previous section, the silicon resistivity decreases drastically when the microcrystalline phase is formed. We should therefore expect to see an increase in drain current for the μ-Si:H based devices, corresponding to an increased carrier mobility in the silicon channel. From Figure 6 the on-off current ratios for α-Si:H and μ-Si:H based devices are estimated to be 2.9x10^6 and 6.3x10^4, respectively. This might imply that the μ-Si:H based devices have a much larger defect density than comparable α-Si:H based devices. These defects act as additional leakage paths in the off state of the μ-Si:H devices. In fact, both of these results taken together suggest that the μ-Si:H based TFT's are being limited by defects that are associated with the device structure, and not bulk properties of the material, which could include interfacial defects and/or poor drain and source contacts.

CONCLUSIONS

The anticipated improvements in device performance by employing μ-Si:H as the active layer in TFT's fabricated on polyimide substrates at low temperature are not observed. Results indicate that defects inherent in the device structure, and not related to the bulk properties of μ-Si:H, may be responsible for the degradation in performance observed.

REFERENCES

1. A. Constant, S. G. Burns, H. R. Shanks, C. Gruber, A. Landin, D. Schmidt, C. Thielen, F. Olympie, T. Schumacher, and J. Cobbs in *Development of Thin Film Transistor Based Circuits on Polyimide Substrates*, (Proceedings of the II Symposium on Thin Film Transistor Technology, Miami, FA, 1994) pp. 392-400.

2. S. G. Burns, H. R. Shanks, A. P. Constant, C. Gruber, D. Schmidt, A. Landin, and F. Olympie in *Design And Fabrication Of High Current Switching TFTs On Flexible Polyimide Substrates*, edited by Y. (Proceedings of the III Symposium on Thin Film Transistor Technology, San Antonio, TX, 1996) pp. 382-390.

3. S. G. Burns, H. R. Shanks, A. Constant, C. Gruber, D. Schmidt, A. Landin, C. Thielen, F. Olympie, T. Schumacher, and J. Cobbs in *Design and Fabrication of a-Si:H Based EEPROM Cells*, (Proceedings of the II Symposium on Thin Film Transistor Technology, Miami, FA, 1994) pp. 370-380.

4. W.B. Jackson, J.M. Marshall and M.D. Moyer, Phys. Rev. B, <u>39</u>, 1164 (1989).

5. A. Sanjoh, N. Ikeda, K. Komaki, J. Electrochem. Soc., <u>138</u>, 1474 (1991).

6. K. A. Bratland, Masters Thesis, Iowa State University, 1998.

THE RECRYSTALLIZATION DEPTH CONTROL OF THE EXCIMER-LASER-RECRYSTALLIZED POLY-CRYSTALLINE SILICON FILM

KEE-CHAN PARK, KWON-YOUNG CHOI*, JAE-HONG JEON, MIN-CHEOL LEE AND MIN-KOO HAN
School of Electrical Engineering, Seoul Nat'l Univ., Seoul, KOREA.
*AMLCD Division, Samsung Electronics co., KOREA

ABSTRACT

A novel method to control the recrystallization depth of amorphous silicon (a-Si) film during the excimer laser annealing (ELA) is proposed in order to preserve a-Si that is useful for fabrication of poly-Si TFT with a-Si offset in the channel. A XeCl excimer laser beam is irradiated on a triple film structure of a-Si/ thin native silicon oxide (~20Å)/ thick a-Si layer. Only the upper a-Si film is recrystallized by the laser beam irradiation, whereas the lower thick a-Si film remains amorphous because the thin native silicon oxide layer stops the grain growth of the poly-crystalline silicon (poly-Si). So that the thin oxide film sharply divides the upper poly-Si from the lower a-Si.

INTRODUCTION

Polycrystalline silicon thin film transistor (poly-Si TFT) is a promising device for active matrix liquid crystal display (AMLCD) due to superior current driving capability. However, poly-Si TFT has several critical problems such as large leakage current and long-term instability. In order to reduce the leakage current, various methods have been investigated such as offset gate [1] and lightly doped drain (LDD) [2]. Poly-Si TFT employing partial a-Si channel region near source and drain was reported to be quite effective in reducing the leakage current [3, 4]. The insertion of a-Si in the channel of poly-Si TFT requires additional deposition and photolithography steps.

The purpose of this new method is to preserve a-Si intentionally by precise recrystallization depth control during the excimer laser annealing of a-Si. The proposed method makes it possible to decide the location of the boundary between a-Si and poly-Si and does not require additional photolithography process. The new method allows large process margin in the fabrication of poly-Si TFT with uniform device characteristics in large area applications.

EXPERIMENT

The process sequence of the proposed method is shown in Fig. 1. a-Si films of various thickness (800Å, 2500Å, 4000Å) were deposited on the wet oxidized silicon wafer substrate by low pressure chemical vapor deposition (LPCVD) at 550°C or by plasma enhanced chemical vapor deposition (PECVD) at 300°C. Then the upper surface of a-Si was oxidized for 10 minutes in the cleaning solution of sulfuric acid (97% purity) and hydrogen peroxide (35% purity) with the volume ratio of 2:1 at 120°C. The thickness of the native silicon oxide film grown during the cleaning process was about 20 Å, which was measured by transmission electron microscopy (TEM). On the oxidized a-Si, 400Å thick a-Si film was deposited again by LPCVD or PECVD respectively. PECVD a-Si film was dehydrogenated at 450Å in N_2 ambient

Mat. Res. Soc. Symp. Proc. Vol. 557 © 1999 Materials Research Society

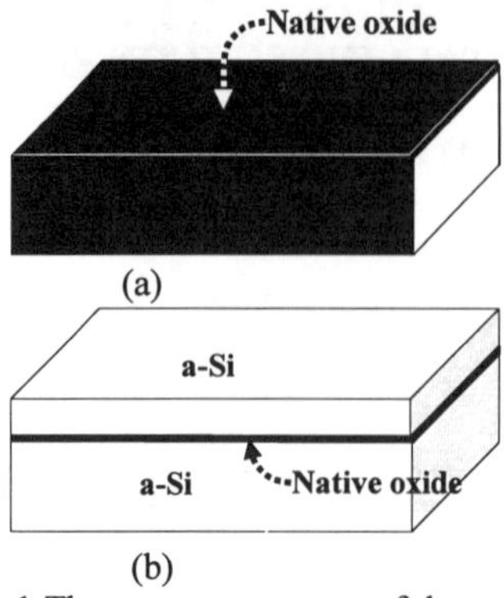

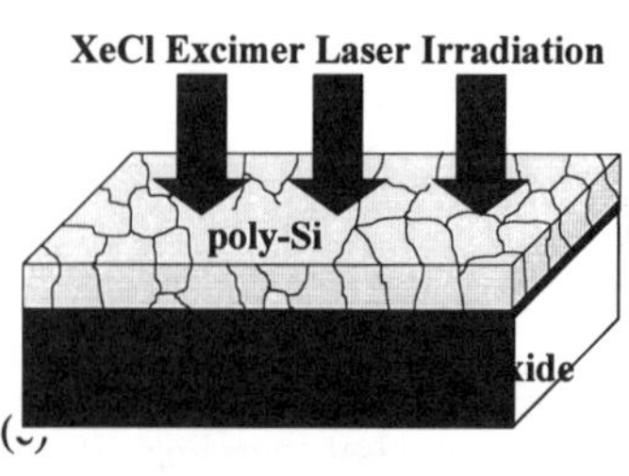

Fig. 1 The process sequence of the proposed method. (a) Native oxide is grown by dipping in the solution of H_2O_2 and H_2SO_4 at 120 f o r 10 minutes. (b) a-Si (400 Å) / thin native silicon oxide (~20Å)/thick a-Si (800 Å, 2500 Å, 4000 Å) structure. (c) Recrystallization of upper a-Si film by XeCl excimer laser irradiation (λ=308nm, E_{laser}=230 mJ/cm^2 for LPCVD a-Si and E_{laser}=280 mJ/cm^2 for PECVD a-Si).

for 3 hours. Then XeCl (λ=308nm) excimer laser irradiation was carried out at the energy density of 230 mJ/cm^2 for LPCVD a-Si and 280 mJ/cm^2 for PECVD a-Si. Cross-sectional sample structure was investigated by TEM observation.

RESULTS

The cross-sectional TEM image of the poly-Si / a-Si double layer fabricated by the proposed method is shown in Fig. 2. As shown in Fig. 2 (a), the grain growth is blocked at the thin native oxide film in all cases and the oxide film sharply divides the upper poly-Si film from the lower poly-Si film in all cases. From this figure it is evident that the thin native oxide blocks the successive grain growth of poly-Si and the poly-Si grains are discontinuous at the native oxide layer. This is due to the crystallographic differences between silicon and SiO_2 such as lattice constant mismatch and crystal structure difference.

When the laser beam is irradiated on the a-Si film, most of the laser energy is absorbed at shallow depth (~200Å) of a-Si film [5]. The absorbed energy transforms the upper part of the a-Si film into liquid phase. Then nucleation takes place and poly-Si grains begin to grow from these nuclei. The resolidified poly-Si releases latent heat and this heat melts underlying a-Si. The consecutive melting and resolidificaton of underlying a-Si repeats itself. This phenomenon is called explosive crystallization by self-propagating liquid silicon layer [6, 7].

Contrary to the previous results that reported that fine-grained poly-Si is produced by the self-propagating liquid silicon layer in case of crystalline silicon substrate [6, 7], Fig 2. (b) shows that solid phase grain growth also takes place when the substrate is SiO_2. Poly-Si grains in the lower a-Si layer in Fig 2. (b) seem to have grown from the bottom, which is impossible if the self-propagation of the liquid layer is the unique recrystallization mechanism of fine-grained poly-Si. The solid phase grain growth occurs due to the thermally insulating SiO_2 and the latent heat released from the resolidified poly-Si.

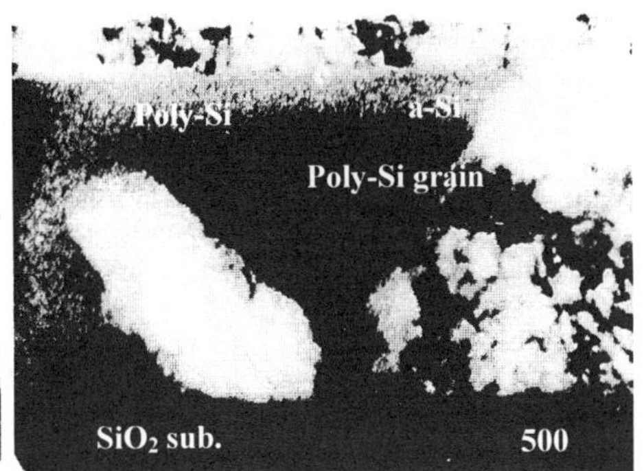

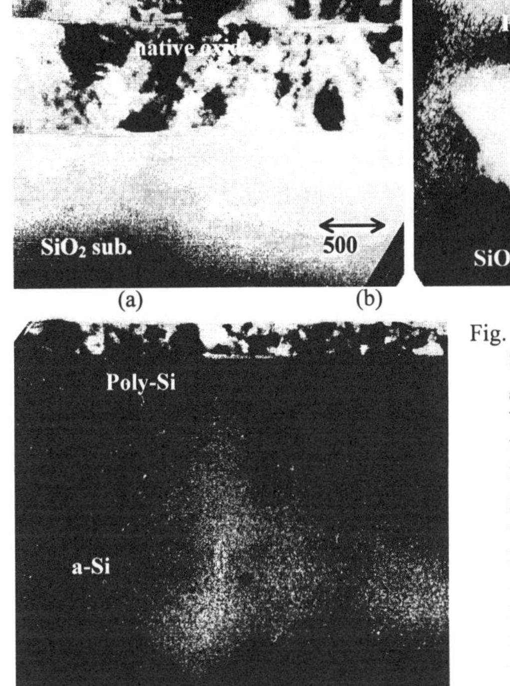

Fig. 2 TEM images of fabricated poly-Si / a-Si double layer. E_{laser}=230mJ/cm^2, $T_{upper\ a-Si}$=400Å for all samples. All a-Si films were deposited by LPCVD. (a) The upper a-Si and 1000Å thick lower a-Si is fully recrystallized. The successive grain growth is blocked by thin native oxide. (b) 2500Å thick lower a-Si is partially recrystallized and poly-Si grains grow in solid phase from the bottom a-Si / SiO$_2$ interface. (c) In 4000Å thick lower a-Si no poly-Si grains are seen and the thin native oxide clearly divide the lower a-Si from the upper poly-Si.

Figure 2 also shows that the lower a-Si is preserved if this layer is sufficiently thick, i.e. when it has sufficient thermal capacity so that the latent heat from upper poly-Si can be absorbed in lower a-Si without recrystallizing it. For the same laser energy density of 230 mJ/cm^2, the 1000Å thick lower a-Si is completely recrystallized (Fig. 2 (a)) and the 2500Å thick lower a-Si is partially recrystallized (Fig. 2 (b)), while the 4000Å thick lower a-Si is perfectly preserved (Fig. 2 (c)).

Figure 3 shows the TEM image of the poly-Si / a-Si double layer in case of PECVD deposited a-Si film. For PECVD a-Si film, the lower a-Si is preserved even though the thickness of the lower a-Si film is 2500Å, which is much thinner than the lower LPCVD a-Si film thickness for perfect a-Si preservation (4000Å). This is due to the process temperature difference between PECVD (300°C) and LPCVD (550°). The LPCVD a-Si film structure is denser than the PECVD a-Si film since the deposition temperature is higher. Therefore the activation energy barrier for PECVD a-Si nucleation is higher than that of LPCVD a-Si.

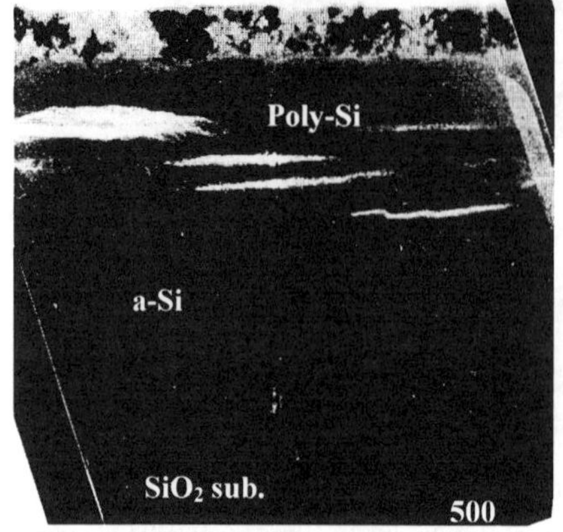

Fig. 3 TEM image of fabricated poly-Si / a-Si double layer. E_{laser}=280mJ/cm^2, $T_{upper\ a\text{-}Si}$=400Å, $T_{upper\ a\text{-}Si}$=2500Å. All a-Si films were deposited by PECVD. No poly-Si grains are seen in the lower a-Si film even though the thickness is 2500Å in case of PECVD a-Si film. White cracks are due to poor film quality of PECVD a-Si.

Since the activation energy of grain growth is smaller than that of the nucleation, the poly-Si grains keep growing even at relatively lower temperature than the nucleation temperature. When the grain growth reaches the native oxide, the solid phase grain growth is blocked by thin native oxide. This is due to the facts that the lattice constant and the lattice structure of SiO$_2$ (hexagonal structure composed of SiO$_4$ tetrahedra) are different from those of silicon (diamond structure) and that the activation energy for SiO$_2$ recrystallization is much larger than that of silicon. Accordingly thin native oxide is not recrystallized and the amorphous silicon oxide film does not lower the activation energy barrier height for recrystallization of underlying a-Si even though growing poly-Si grain does lower the barrier height for recrystallization of neighboring a-Si. In sum, the successive grain growth of poly-Si is blocked at the native oxide barrier because there is nothing to serve as a silicon grain seed beyond it as shown in Fig. 4 and the lower thick a-Si can be preserved even though it also absorbs the transferred heat from the resolidified upper poly-Si.

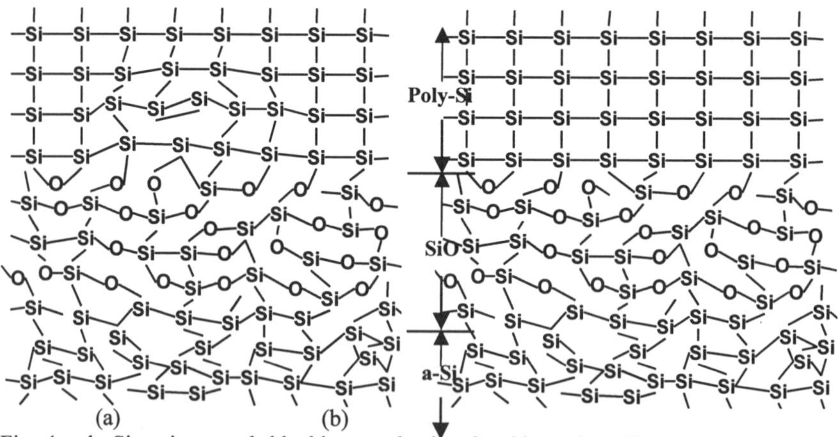

Fig. 4 poly-Si grain growth blocking mechanism by thin native silicon oxide. (a) a-Si adjacent to poly-Si grain is easily recrystallized. (b) a-Si below the thin oxide barrier is not recrystallized because there is no silicon grain seed.

CONCLUSION

We have successfully fabricated the poly-Si / a-Si double layer at low temperature ($\leq$550°C for LPCVD a-Si and $\leq$300°C for PECVD a-Si) employing the native oxide barrier and the excimer laser recrystallization. The TEM images show that thin native oxide can block the successive grain growth of poly-Si and that the oxide is thin enough ($\sim$20Å) for electrons to tunnel through. In order to preserve a-Si, it is also important that the lower a-Si should have sufficient thermal capacity.

The control of grain growth depth in laser crystallization of a-Si is very useful in the fabrication of high-performance poly-Si TFT. For example, poly-Si in poly-Si / a-Si double layer can be used as poly-Si channel of TFT and a-Si can be utilized as a-Si offset between poly-Si channel and source / drain in order to suppress the electron-hole pair generation via poly-Si grain boundary traps in drain depletion region.

REFERENCES

1. S. Seki, O. Kogure, and B. Tsujiyama, IEEE Electron Device Lett. **8,** 434 (1987)

2. K. Nakazawa, K. Tanaka, S. Suyama, K. Kato and S. Kohda, SID '90 Technical Digest, 311 (1990)

3. J. H. Jeon, C. M. Park , J. S. Yoo, C. H. Kim and M. K. Han, Jpn. J. Appl. Phys. **37,** 1064 (1998)

4. K. H. Lee, W. Y. Lim, J. K. Park, J. Jang, SID '97 Technical Digest, 481 (1997)

5. Cody, G. D., B. Abeles, C. Wronski, C. R. Stephens, and B. Brooks, *Optical characterization of amorphous silicon-hybrid films*, Solar Cells, vol. 2, p. 227, 1980

6. Michael O. Thompson, G. J. Galvin, and J. W. Mayer, Phys. Rev. Lett. **52,** 2360 (1984)

7. J. Narayan, C. W. White, O. W. Holland, and M. J. Aziz, J. Appl. Phys. **56** 1821 (1984)

8. M. K. Moravvej-Farshi and M. A. Green, Solid State Electronics, **30** 1053 (1987)

HIGH-PERFORMANCE DAMASCENE-GATE THIN FILM TRANSISTORS

EUGENE MA[*], SIGURD WAGNER
Department of Electrical Engineering, Princeton University, Princeton, NJ 08544
[*]New Address: Printed Transistor Inc., 66 Witherspoon St., No.131, Princeton, NJ 08542,
eugenema@printedtransistor.com

ABSTRACT

We report a novel TFT structure where the gate metal is embedded into a SiN_x passivation layer. This allows the subsequent gate dielectric layer to be much thinner than in conventional bottom-gate structures. thereby reducing the threshold voltage and the sub-threshold slope. TFTs employing these damascene-gate structures were fabricated with SiN_x gate dielectrics as thin as 50 nm. Such devices exhibit threshold voltages of 0.9 V, sub-threshold slopes of 0.1 V/dec, I_{ON}/I_{OFF} current ratios of 10^6 and linear region field-effect mobilities of 0.6 cm^2/V·s.

INTRODUCTION

In the push towards larger flat panel display sizes and higher resolutions, it becomes necessary to develop ever faster active-matrix thin film transistor (TFT) technologies. One way this can be accomplished is by lowering the gate line resistance in the transistor array. Another is by reducing the sub-threshold voltage swing required to turn on a display transistor.

One can reduce the sub-threshold voltage swing by minimizing the thickness of the gate dielectric in the transistor. However, since amorphous silicon transistors currently used in the design of active-matrix liquid crystal displays are bottom gate structures, there is the problem of step coverage. It is most critical that the gate dielectric layer adequately cover the gate metal to prevent leakage current between the gate and the source/drain. Such a requirement puts a lower limit on the thickness of the gate dielectric.

Gate line resistance can be reduced by using thicker gate lines or by using low resistivity metals such as copper. Using thicker gate lines, however, only increases the step coverage requirement described above, requiring the deposition of a thicker gate dielectric. To circumvent this problem, techniques have been developed which "taper" the edges of the gate[1], thereby eliminating any "step" in the topology of the gate. However, this can only be done with certain metal alloys, namely Mo/Ta, which restricts our choice of gate metals. Mo/Ta has sufficiently low resistivity for current display sizes, but as display sizes and resolutions increase, lower resistivity metals such as copper will be necessary.

The approach presented in this paper is to embed the gate metal into a trench made into a passivating layer above the substrate so that the top of the gate metal is level with the surface of the passivating layer. This structure has several advantages. It:
(1) reduces gate line resistance by allowing for thick gate lines,
(2) does not increase gate line capacitance since there is no increase in the effective area,
(3) allows the use of low-resistivity metals such as copper which would be encapsulated by the
    nitride sidewalls and a metal cap-layer, and
(4) permits the use of thin SiN_x gate dielectrics (< 200 nm).
A cross-sectional schematic of this damascene-gate TFT is illustrated in Fig. 1.

Mat. Res. Soc. Symp. Proc. Vol. 557 ©1999 Materials Research Society

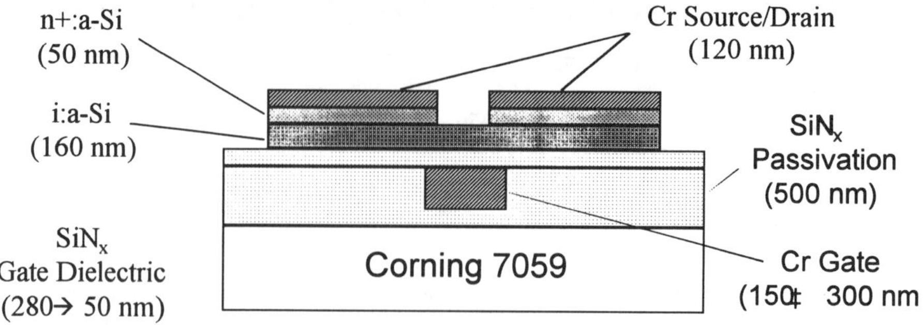

Fig. 1 Cross-sectional schematic of a damascene-gate TFT (dTFT).

The minimum gate dielectric thickness in this structure is primarily determined, not by the gate step coverage requirement, but by the breakdown field strength of the insulator. Bottom-gate a-Si:H TFTs currently used in flat panel displays typically have 100 nm thick gate lines and 200-300 nm gate dielectric layers. Thinner gate insulators have been demonstrated in top-gate structures but in almost all cases the substrate is single-crystal silicon and device fabrication is costly, involving ion implantation[2] and in some cases chemical-mechanical polishing (CMP)[3].

EXPERIMENT

Damascene-gate thin film transistors (dTFTs) were fabricated on Corning 7059 glass substrates coated with a 500 nm SiN$_x$ passivation layer. 150 nm trenches were etched into this passivation layer through a photoresist etch mask using RIE. The photoresist etch mask remains, and 150 nm of Cr were e-beam evaporated to fill the trenches. Cr not in the trench is then lifted-off.

There is no polishing (CMP) step in our fabrication procedure. Also, no extra mask steps are required over those used in the fabrication of conventional structures. The mask pattern used to define the trench is left on for the metal lift-off step used to define the gate. When one considers the case of metal[4] or even plastic[5] substrates where insulation/passivation of the substrate may be required, then the only additional fabrication step required is the etching of the trench which is a relatively fast process.

SiN$_x$ gate insulators of thicknesses 280 nm, 180 nm, 140 nm, and 90 nm were deposited on different glass substrates. The active and doped layers are intrinsic and n$^+$ amorphous silicon of thicknesses 160 nm and 50 nm, respectively. The source/drain metal is 120 nm of e-beam evaporated Cr. All films were grown using a multi-chamber RF-PECVD system.

RESULTS

It must first be established whether such a gate structure can even be fabricated in the manner described. As shown in Fig. 2, Dektak profilometry performed after formation of the damascene gate reveals that the surface topology is quite smooth with mismatch between the top of the gate metal and the top of the surrounding passivation layer typically around ±10 nm. Having shown that such damascene-gate structures can indeed be made, working transistors must next be demonstrated The transfer characteristics of a dTFT with a 280 nm thick gate dielectric are shown in Fig. 3. TFT performance metrics such as the I_{ON}/I_{OFF} ratio, threshold voltage, sub-threshold slope, and linear field-effect mobility are similar to those of conventional bottom-gate TFTs with comparable gate dielectric thickness.

A comparison of dTFT transfer characteristics in the transistor switching regime for the five gate dielectric thicknesses used is shown in Figs. 4 (a) and (b). In these plots, a reduction in gate dielectric thickness is accompanied by a lowering of both the threshold voltage and the sub-threshold slope. I_{ON}/I_{OFF} ratios between 10^6 and 10^7 (with I_{ON} measured at $V_{GS} = 20V$) are obtained for all devices regardless of gate dielectric thickness. The variation in I_{OFF} observed in Fig. 4(a) does not correspond directly to the use of thinner gate dielectrics as one might expect. Careful inspection of the data reveals, in fact, that the dTFT devices in order of *increasing I_{OFF}* have gate dielectrics of thickness: 180 nm, 280 nm, 50 nm, 90 nm, 140 nm.

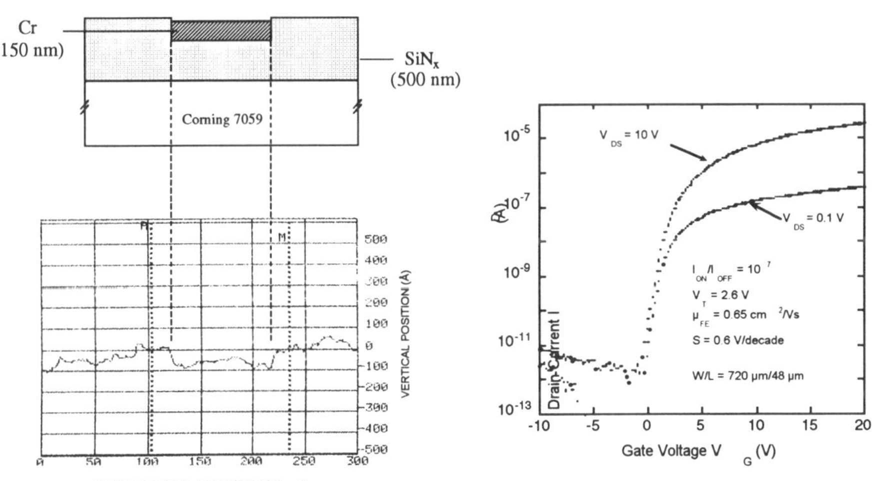

Fig. 2 Cross-sectional schematic of the damascene-gate structure alone with Dektak profilometer measurement.

Fig. 3 Transfer characteristics for dTFT with 280 nm thick gate dielectric and 150 nm thick Cr gate.

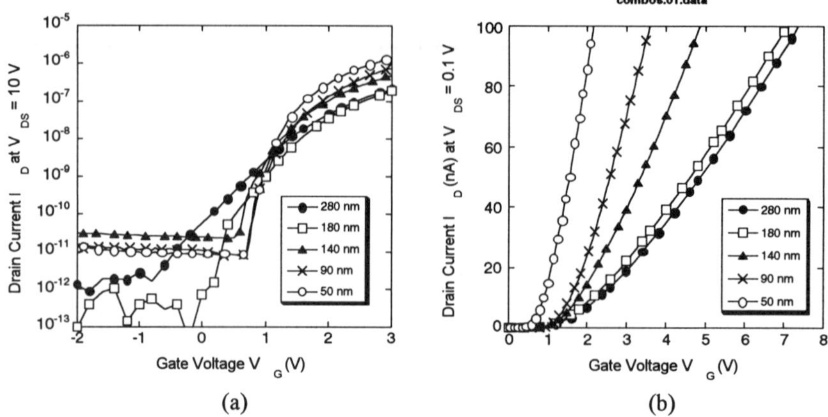

Fig. 4 dTFT performance in the transistor switching regime for five gate dielectric thicknesses: (a) semi-log plot used to determine sub-threshold slope, and (b) linear plot used to obtain threshold voltage and field-effect mobility.

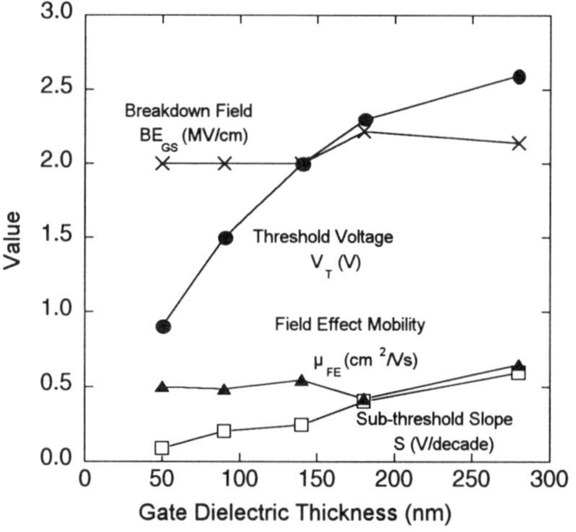

Fig. 5 Performance metrics for dTFTs as a function of gate dielectric thickness.

Performance metrics for dTFTs are extracted from Fig. 4 and summarized in Fig. 5. Since the channel itself is the same for all devices, regardless of gate dielectric thickness, we expect that mobility remains relatively unchanged. Also, nitride growth conditions are the same for all samples, and therefore we do not expect any change in breakdown field strength. This is confirmed by the data which show that these two metrics remain relatively unaffected. On the other hand, it is clear that both the threshold voltage and the sub-threshold slope decrease with a reduction in gate dielectric thickness.

However, while the sub-threshold slope decreases linearly with gate dielectric thickness as expected, the threshold voltage decreases non-linearly. One possible explanation is the existence of fixed charges in the SiN$_x$ gate dielectric and at the interface. Assuming for simplicity that the bulk charge distribution ρ is constant through the thickness of the dielectric, the flat band voltage (and thus the threshold voltage) is shifted[6] by an amount

$$\Delta V_{FB} = -\frac{\rho}{2\varepsilon_{SiN}}\, d^2_{SiN} \tag{1}$$

where ε_{SiN} and d_{SiN} are the dielectric constant and thickness of the SiN$_x$ gate dielectric. A fixed charge density, Q_i, located at the nitride-semiconductor interface introduces a linear dependence on gate dielectric thickness:

$$\Delta V'_{FB} = -\frac{Q_i}{C_{SiN}} = -\frac{Q_i}{\varepsilon_{SiN}}\, d_{SiN} \tag{2}$$

In the "ideal" case, when there is no fixed charge, the threshold voltage is equal to the work function difference, Φ_{ms}, between the gate metal and the semiconductor. For these data[7,8] $V_T^{ideal} = \Phi_{ms} \square\ 0\ \mathrm{V}$. Therefore, according to this model,

$$V_T^{actual} = -A \cdot d_{SiN} - B \cdot d^2_{SiN} \tag{3}$$

where $A = \dfrac{Q_i}{\varepsilon_{SiN}}$ [V/cm] and $B = \dfrac{\rho}{2\varepsilon_{SiN}}$ [V/cm^2]. As shown in Fig. 6, plotting Equation (3) as a function of d_{SiN}, using appropriate fitting parameters A and B, produces a close fit between the modeled and measured values of V_T.

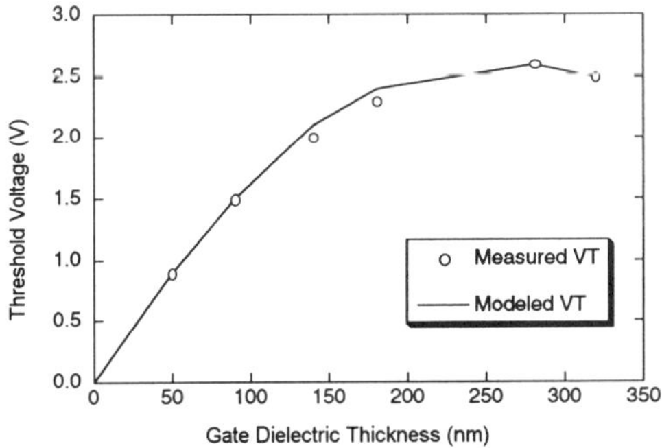

Fig. 6 Results of a model used to show how fixed charges in the SiN$_x$ gate dielectric and at the SiN$_x$/a-Si interface affect the threshold voltage for TFTs. The 320 nm data point is obtained from a conventional (non-damascene) inverted-staggered TFT.

From the fitted values of $A = -2 \times 10^5$ [V/cm] and $B = 3.8 \times 10^9$ [V/cm^2], the interface state density and bulk charge number density are then extracted and found to be $n_i = \dfrac{Q_i}{-q} = 7.5 \times 10^{11}$ [acceptors states/cm^2] and $n = \dfrac{\rho}{q} = 2.86 \times 10^{16}$ [cm^{-3}], respectively.

CONCLUSIONS

We have fabricated high-performance damascene-gate amorphous silicon TFTs which have several advantages over conventional bottom-gate TFTs. The use of an inlaid gate effectively eliminates any step coverage requirement and makes possible the use of thin gate dielectrics. This, in turn, reduces the threshold voltage and sub-threshold slope, without affecting the field-effect mobility and breakdown field strength. This is all accomplished without having to modify existing mask sets or film recipes. Additionally, by modeling the observed non-linear dependence of threshold voltage on gate dielectric thickness, one can extract values for the nitride bulk and interface densities. Our results suggest that such structures can be used in giant area TFT displays where low gate line resistance is necessary and in portable electronic displays where low threshold voltage and sub-threshold swing are advantageous from a power consideration.

ACKNOWLEDGMENTS

The authors wish to thank the DARPA HDS program and Princeton Plasma Physics Laboratory for their support of this research.

REFERENCES

1. M. Dohjo, T. Aoki, K. Suzuki, M. Ikeda, T. Higuchi, Y. Oana, "Low-Resistance Mo-Ta Gate Line Material for Large Area a-Si TFT-LCDs," *SID International Symposium Digest of Technical Papers*, 1988, pp. 330-333.
2. M. Hatano, H. Akimoto, and T. Sakai, "A Novel Self-aligned Gate-overlapped LDD Poly-Si TFT with High Reliability and Performance," *IEDM Tech. Dig.*, 1997, pp. 523-526.
3. A. Kumar K.P. and J.K.O. Sin, "A Simple Polysilicon TFT Technology for Display Systems on Glass," *IEDM Tech Dig.*, 1997, pp. 515-518.
4. S.D. Theiss and S. Wagner, "Amorphous Silicon Thin-Film Transistors on Steel Foil Substrates," IEEE Electron Dev. Lett. **17**, 578-580 (1996).
5. H. Gleskova, S. Wagner and Z. Suo in *Flat-Panel Display Materials - 1998*, edited by G.N. Parsons, C.C. Tsai, T.S. Fahlen, and C.H. Seager (Mater. Res. Soc. Proc. **508**, Pittsburgh, PA, 1998), pp. 73-38.
6. R. Muller and T. Kamins, Device Electronics for Integrated Circuits, 2nd Edition, p. 401, Wiley, New York (1986).
7. CRC Handbook of Chemistry and Physics, D.R. Lide, Editor, p. **12**-113, CRC Press, Boca Raton, FL (1994).
8. R. Street, *Hydrogenated Amorphous Silicon*, p. 331, Cambridge University Press, Cambridge, England (1991).

FLEXIBLE A-SI BASED SOLAR CELLS WITH PLASTIC FILM SUBSTRATE

**Yukimi Ichikawa, Shinji Fujikake, Katsuya Tabuchi, Toshiaki Sasaki, Toshio Hama,
Takashi Yoshida, Hiroshi Sakai and Misao Saga**
New Energy Laboratory, Fuji Electric Corporate Research and Development, Ltd.
2-2-1, Nagasaka, Yokosuka, Kanagawa 240-01, Japan

ABSTRACT

A flexible amorphous silicon (a-Si) based photovoltaic (PV) module has been developed. The a-Si solar cell fabricated on a heat-resistance plastic film with a thickness of 50μm has a new monolithic series-connected structure named *SCAF* to obtain a high output voltage required for practical use. Applying a-Si/a-SiGe double-junction structure to 40cm x 80cm *SCAF* solar cells, we obtained a stabilized efficiency of 9% with an excellent reproducibility. Flexible PV modules fabricated from the *SCAF* solar cells have been tested in several demonstration sites to study their durability under outdoor conditions.

INTRODUCTION

We have developed amorphous silicon (a-Si) based solar cells formed on thin flexible plastic film. There are two principal reasons for adopting the plastic film as a substrate; one is to reduce the production cost, and the other is to realize a flexible PV module. With regard to the former, employing a roll of plastic film as a substrate, we can apply a roll-to-roll system to its production processes. This is very advantageous to vacuum processes such as plasma CVD and sputtering because the apparatuses are maintained under vacuum until completion of the process for a roll. This results in improvement of throughput and reproducibility compared with the conventional batch processes. The latter gives an advantage to the a-Si solar cell from the viewpoint of practical application. At present, no other solar cells than the a-Si base have succeeded in practical flexible devices. They are very light and easy to use and are expected to open a door for new applications in the residential and building fields.

From the above reasons, we concluded that flexible a-Si solar cells formed on plastic film have the highest potentiality as the next generation a-Si solar cell. On the basis of previously developed technology, [1] we started developing plastic film-substrate solar cells in 1992. The key technologies to be developed for low-cost, high-throughput production of the film substrate solar cell are the following two items:

a) Making a monolithic series-connected structure on large-area devices; this provides a high output voltage from single solar cell, and results in a PV module with high usability for system construction.

b) Applying a roll-to-roll system to all the fabrication processes.

We proposed a new monolithic series-connected structure named *SCAF* (<u>S</u>eries-<u>C</u>onnection through <u>A</u>pertures formed on <u>F</u>ilm) for the film substrate solar cells as well as its fabrication processes based on roll-to-roll systems.[2] Then, in 1996, we completed a prototype production line for the *SCAF* structure solar cells. The line had a capacity of several hundred kWp/year for a 40cm x 80cm *SCAF* cell with a stabilized aperture-area efficiecy of about 8%. In 1997, we started a new challenge to raise the conversion efficiency to 10% in 4 years under a national project, The New Sunshine Program. For this purpose, we decided to adopt a-SiGe to the bottom cell of the double junction structure.

Mat. Res. Soc. Symp. Proc. Vol. 557 © 1999 Materials Research Society

In this paper, first, we present the basic concept of our flexible a-Si PV modules and fabrication processes for that, and then describe the present status of the *SCAF* cells with the a-Si/a-SiGe double junction structure. Finally we show some demonstration PV systems using our flexible modules.

STRUCTURE AND FABRICATION PROCESS OF FLEXIBLE CELL

SCAF Structure

Processing plastic film such as cutting and making holes is very easy compared with glass. We used this advantage to develop a new breakthrough technology for low-cost mass production with plastic film substrates. An idea for the monolithic structure was incubated on this advantage. We are able to make holes on the film substrate easily by punching out, and used them for through-hole-contact. The concept of the new monolithic structure we proposed is shown in Fig. 1. The structure was named *SCAF* (Series-Connection through Apertures formed on Film). In this structure, a substrate type a-Si:H based solar cell is fabricated on one surface of the plastic film; the cell is composed of metal electrode, a-Si:H based layers, and transparent electrode stacked in the order on the substrate. The plastic films used were heat-resisting plastics such as poly-imide with a thickness of several tens of micrometers.

A remarkable feature of the *SCAF* structure is a number of holes made on the substrate. These holes are divided into two groups and have different functions. One function is through-hole contact between the transparent electrode and the backside electrode formed on the other surface of the substrate; these holes are named *current-collection holes* and are distributed in the active-area (the transparent electrode) region. The other is for the connection between the backside

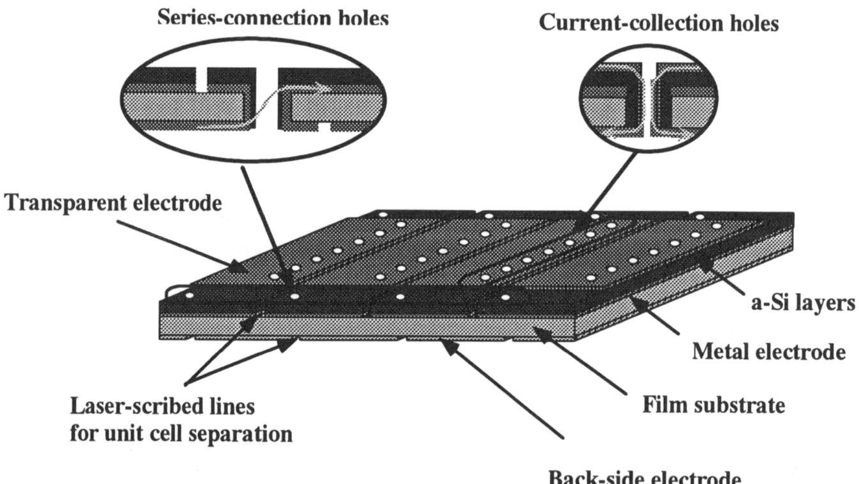

Fig. 1 Schematic diagram of *SCAF* structure solar cell.

electrode and the metal electrode; these holes are named *series-connection* holes and are on both sides out of the transparent electrode region.

As shown in Fig. 1, the *SCAF* solar cell consists of a number of divided unit cells. The current generated from the unit cells has to flow in the transparent electrode with relatively large sheet resistance. To avoid power loss there, the current-collection holes collect the current generated near them and transport it to the backside electrode with low sheet resistance. The current collected into the backside electrode flows to both sides with little joule loss, and then moves into the metal electrode of the next unit cell through the series-connection holes. In the *SCAF* structure, a number of unit cells are connected in series successively by this manner and required output voltages are obtained by adjusting the number of unit cells.

Multi-column *SCAF* Cell Structure

In the grid-connected PV systems, the output power from the PV modules is fed to an inverter, and typical input voltage for the inverter is about 200V. A conventional method to obtain such a high voltage is to connect several modules in series. This makes wiring process complicated in PV systems. In principle, we can obtain even 200V from single solar cell by the *SCAF* structure. It is, however, of no practical method because the unit cells have to be narrowed down extremely. This results in reduction of the active-area of the cells.

To avoid this difficulty, we proposed a multi-column *SCAF* cell structure as shown in Fig. 2. The solar cell is divided into several rectangular regions (columns); in this example, it consists of four regions. Each column has the *SCAF* structure as shown in Fig. 1, but the direction of series-connection of two columns placed side by side is opposite each other. No additional process is needed for making this structure except for modification of the pattern such as the position of holes and scribed lines. As is described later, this structure works successfully and we obtained even 200V from single SCAF solar cell of 40cm x 80cm.

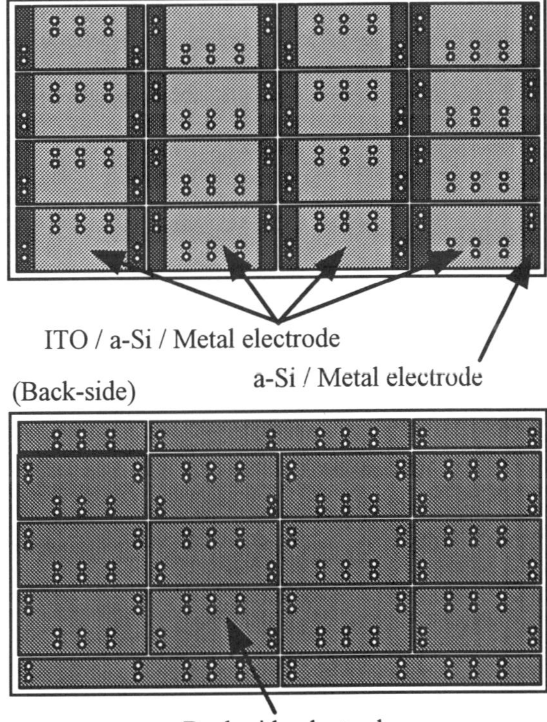

Fig. 2 Schematic diagram of multi-column SCAF cell.

<u>**Stepping-roll Process for Thin Film Deposition**</u>

Plasma CVD process for a-Si based film deposition is the most important key technology for determining not only cell performance but also production throughput. With regard to throughput, roll-to-roll processing is a very attractive method for flexible a-Si solar cell production. However, the conventional roll-to-roll system has several problems. For instance, isolation of raw material gases among reactors is not perfect in an in-line multi-reactor system. As a result, interdiffusion of material gases is not avoidable, and also the gas pressures and the deposition rates are not chosen for each reactor independently. In addition, making band profiling i layers or graded gap buffer layers is very difficult. These difficulties form barriers to improvement of solar cell efficiency.

What we proposed for the plasma CVD to solve these difficulties is a combination of the conventional roll-to-roll system and a batch system; we named it *Stepping-Roll (SR) System.* [3] Figure 3 shows the schematic diagram of the SR system. The basic configuration of the apparatus consists of a common chamber having several film deposition reactors in it. Each reactor has a movable grounded electrode (GE). When the GE is at down position and presses the frame of reactor, the film substrate is held between them. Since their surfaces contacting with the film have O-rings, the reactors are completely isolated from the common chamber. On the other hand, when GE is at upper position, the film-substrate is released and can be wound up as the conventional roll-to-roll system. Using this system, we can deposit various film layers simultaneously according to the following sequence:
1) Pressing down the GE's to isolate the reactors from the common chamber.
2) Film deposition
3) Moving up the GE's
4) Winding up film by one flame (reactor length of film)

As one can understand easily, the SR system has the following advantages compared with the conventional roll-to-roll system:
1) Interdiffusion of raw material gases among reactors is completely prevented.
2) The configuration of the reactors is compact and simple because no intricate mechanisms for transport of the substrate are required.
3) Precise control of the composition in the depth direction can be made.
4) Deposition conditions such as gas pressure and deposition rate can be chosen independently for each reactor.

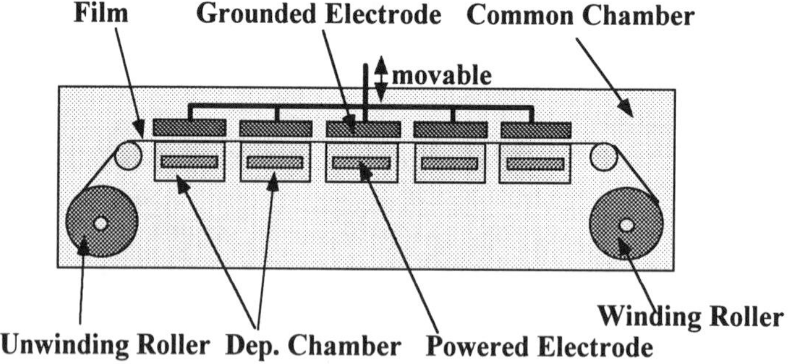

Fig . 3 Basic concept of stepping-roll film deposition apparatus

In addition to these, the SR system can have plasma CVD reactors and sputtering reactors in a common chamber. This feature is very convenient for the *SCAF* structure solar cell as described in the following section.

Process Flow for *SCAF* Solar Cells

The entire process for *SCAF* cell production is schematically shown in Fig. 4. To complete the *SCAF* solar cell, there are four processes, and all the apparatuses used are based on roll-to-roll processes. First, holes are made on the plastic film by a punching unit and then the substrate is cleaned by a cleaning unit in a pretreatment apparatus. After that, the roll is transferred to a roll-to-roll sputtering apparatus to deposit metal electrodes. Then the pretreated film-substrate roll is set in the SR apparatus composed of plasma CVD reactors and sputtering reactors to deposit the a-Si layers, the transparent electrode and back-side electrode. Finally, the layers formed on both surfaces of the film are laser-scribed to make unit cells by a roll-to-roll laser-scriber.

The fabrication process for the *SCAF* solar cell presented here is further simplified compared with that for the glass substrate solar cells. Moreover since all the process units are based on roll-to-roll systems, it is easy to construct an automated production line with these processes.

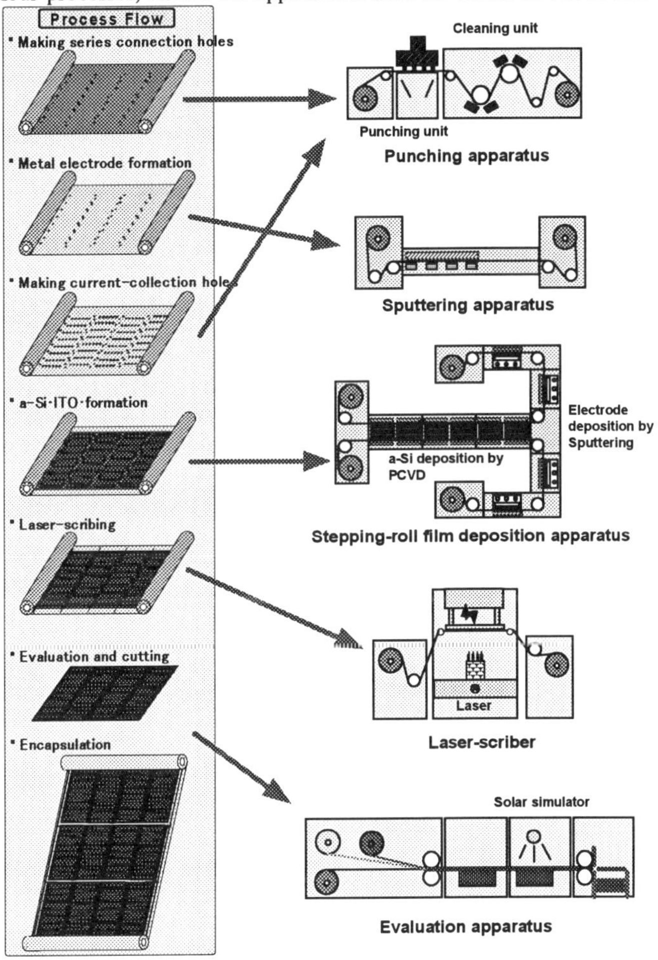

Fig. 4 Fabrication process flow for SCAF solar cells based on roll-to-roll apparatuses.

FABRICATION OF *SCAF* SOLAR CELLS

A-Si/a-Si Double Junction Structure

To brush up the developed process and improve the performance of *SCAF* cells, we have carried out trial production of 40cm×80cm *SCAF* cells. The present *SCAF* cell was designed to generate an optimum output voltage of about 200V for the a-Si/a-Si tandem structure; to generate this voltage, 152 unit cells were connected in series with four columns having 38 unit cells. The advantage to generate high voltages is to raise usability of the *SCAF* solar cell as well as to avoid the joule loss due to high current.

The present production line fits for 0.5m-wide film, and 40cm x 80cm *SCAF* solar cells are formed on it one after another like picture frames on a photographic film. Firstly, we studied a-Si/a-Si double junction *SCAF* cells, and obtained a stabilized aperture-area efficiency of 8.1% in a 40cm×80cm *SCAF* solar cell. To achieve this, we applied various techniques to the fabrication processes, including optimum design of ITO thickness, low resistive ITO layer, high quality a-SiO:H p-type layer, a-SiO:H p/i interface layer, wide gap a-Si:H top intrinsic layer, narrow gap a-Si:H bottom intrinsic layer, metal electrode with high reflectivity and textured morphology, and precise process temperature control.

A-Si/a-SiGe Double Junction Structure

For farther improvement of efficiency, we have applied a-SiGe alloy to the bottom cell. To raise the efficiency, first we studied with single junction a-SiGe solar cell. An optical gap of 1.5 eV was chosen for our a-SiGe cell. We focused on hydrogen dilution ratio of the raw material gas for the i-layer deposition and the p-layer material. Figure 5 shows the photovoltaic characteristics of the a-SiGe cell (1cm^2) as a function of hydrogen dilution ratio of raw material gas, $H_2/(SiH_4+GeH_4)$, for a-SiGe deposition. The cells fabricated have a structure of Ag/a-Si(n)-a-Si(i/n)-a-SiGe(i)-a-Si(p/i)-a-SiO(p)/ITO as shown in Fig. 6. The initial and stabilized characteristics

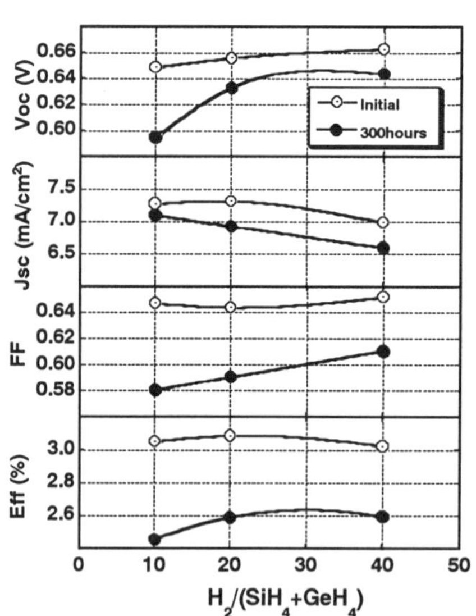

Fig. 5 Characteristics of a-SiGe cell as a function of H$_2$ dilution ratio under red light condition.

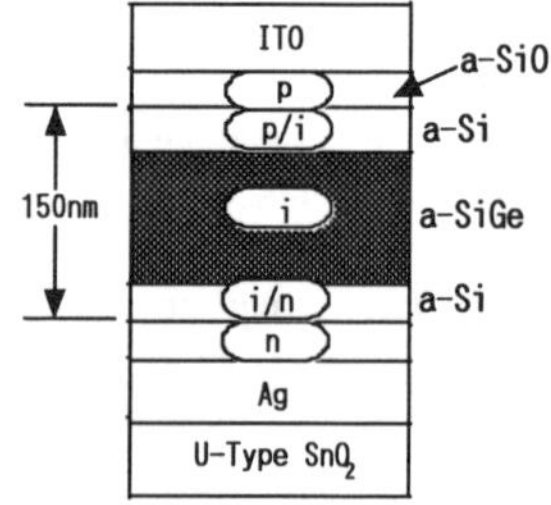

Fig.6 Structure of a-SiGe cell.

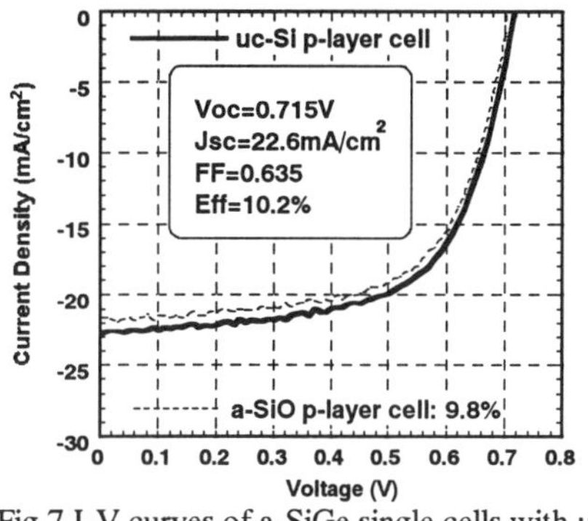

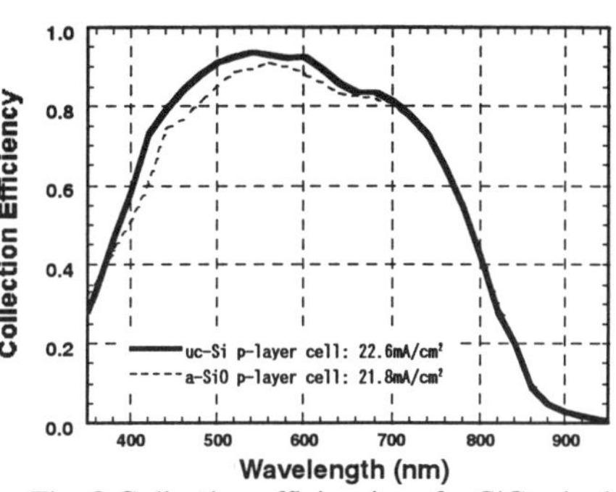

Fig.7 I-V curves of a-SiGe single cells with μc-Si and a-SiO p-layers.

Fig. 8 Collection efficiencies of a-SiGe single cells with μc-Si and a-SiO p-layers.

under red light condition with a cut-off wavelength of 650nm are shown. Better stabilized efficiencies are obtained under higher dilution ratios.

The p-layer is another key for achieving higher efficiencies. We tried to apply microcrystalline silicon (μc-Si) to the p-layer, though our standard p-layer consists of a-SiO as shown in Fig.6. We optimized its deposition condition. The result shows the substrate temperature during μc-Si deposition affects the performance very strongly. Figure 7 and 8 shows the initial I-V curves and collection efficiencies of the optimized μc-Si p-layer cell and a conventional a-SiO p-layer cell. From this result, we see that the collection efficiency for the short wavelength range is enhanced in the μc-Si p-layer cell.

We applied these technologies to the a-Si/a-SiGe *SCAF* solar cell of 40cm x 80cm, and the best result obtained so far is a stabilized efficiency of 9.0% with a initial of efficiency of 10.1% in aperture area as shown in Figure 9.

The reproducibility of the production process is one of the most important issues. To confirm

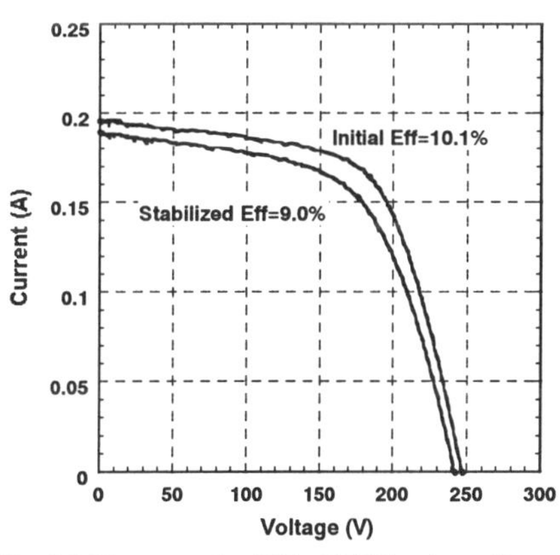

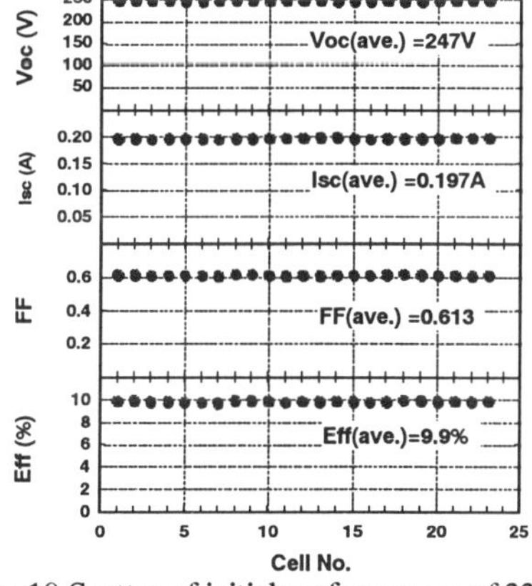

Fig. 9 I-V curve of a-SiGe SCAF solar cell.

Fig. 10 Scatter of initial performances of 23 a-SiGe SCAF solar cells.

that for the a-Si/a-SiGe *SCAF* cell production, we carried out several trial productions. Figure 10 shows an example of those results; the initial performance of 23 a-Si/a-SiGe *SCAF* cells formed successively on a film roll under the same deposition condition. From this result, it is demonstrated that our process has a very high potentiality for mass production.

FLEXIBLE PV MODULE

Module Structure

Encapsulation is the final process for finishing PV modules. This process is greatly behind in automation compared with other fabrication processes and a kind of handicraft manufacturing is applied at present. Thus in the usual PV module production, this process is a bottleneck for mass production. In the encapsulation of the *SCAF* cells, however, an automated encapsulation process can be constructed easily. As described in the previous section, the *SCAF* cell can be designed to generate the most suitable voltage required for each PV system. This means no series-connection among cells is required. Another concern is flexibility of the cell. Combining these advantages, we have been developing "flexible PV module belt" and encapsulation process for it.

The PV module belt under development is about 1m-wide and 200m-long with a weight of about 1kg/m^2 and rolled up around a cylindrical core. In the belt, a number of 40cm x 80cm *SCAF* cells are arranged side by side and connected in parallel by a pair of metal tapes wired in both sides of the module (see Fig. 11). Using this module, we can cut it and obtain a flexible module with required length.

To fabricate the PV belt, we have been developing an encapsulation process. Flexibility and simplified wiring of the module enable us to apply a roll-to-roll process to it. Figure 12 shows the schematic diagrams of the cross-sectional structure of PV belt (a) and a prototype laminator developed (b). The apparatus is composed of two unwinder for top and

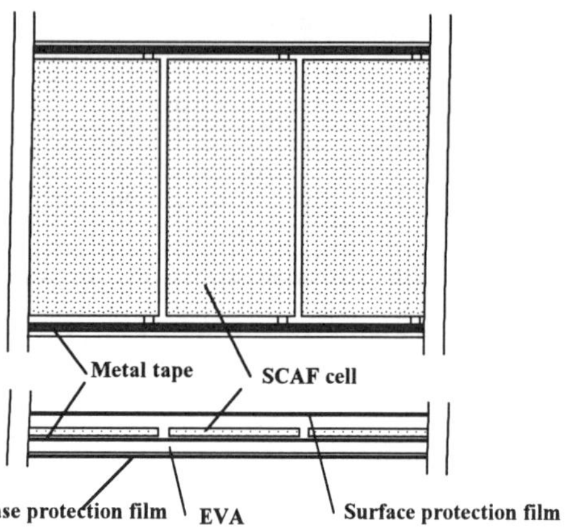

Fig. 11 Structure of flexible PV module.

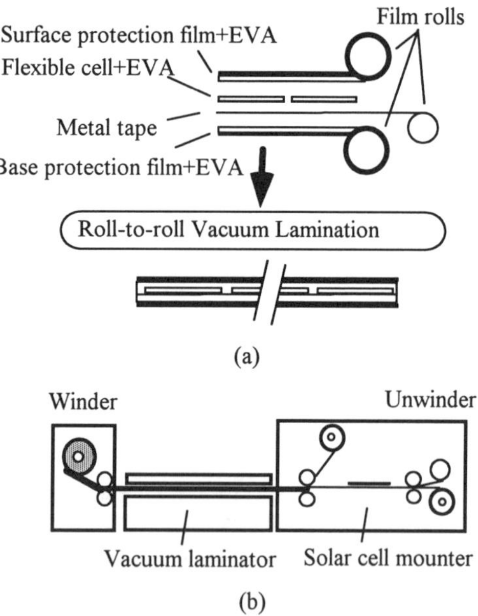

Fig. 12 Schematic diagram of the lamination process (a) and laminator (b).

bottom resin sheet, two reels for metal tape, a solar cell mounter, a vacuum laminator and a winder for PV belt. The resin sheet has a thickness of about 0.5mm and composed of a surface protection film adhering to EVA sheet. The *SCAF* cells of 40cm x 80cm are placed on the bottom resin sheet successively by the solar cell mounter, and after sandwiched by the top resin sheet, it is heated and pressed and cured in the vacuum laminator. The PV belt fabricated by this process has sufficient flexibility, and curvature radius allowed for bending is about 0.5cm.

Applications

The flexible PV modules described above are expected to be an innovative product for expanding PV market. We have now studied their applications. The pictures of some demonstration systems for the flexible modules are presented in Figs. 13-16.

Residential application is an attractive market for its potentialities. For this application, we have developed roofing material integrated PV modules named Solar Roofing in cooperation with Misawa Homes Co., Ltd under the "New Sunshine Program." Figure 13 shows the roof structure with Solar Roofing. [4] To finish a roof, Solar Roofing is spread between glass supporting rails on the roof board and tempered glass is placed above it with an appropriate gap. Figure 14 shows the picture of the demonstration house having a roof with Solar Roofing of 1.4kW.

Since the flexible modules are thin and light, we can paste them on a building wall. Figure 15 shows one of such applications. In this demonstration systems, 29 modules consisting of five 40cm x 80cm *SCAF* cells were pasted on a vertical wall made of concrete. The merit of this installation method is that no special facility for mounting the PV modules is needed, and this results in cost reduction of the system and expansion of installation site.

Flexibility of the modules is useful for installing them on curved construction. Figure 16 shows the picture of the roof of a parking lot with flexible modules on it. In this demonstration system, The power

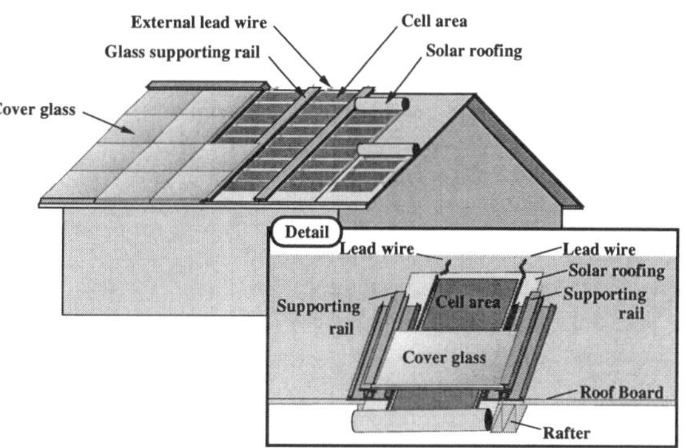

Fig. 13 Structure of roof with Solar Roofing.

Fig. 14 Picture of demonstration house for Solar Roofing.

generated from the modules is stored in batteries and is used for lighting at night.

CONCLUSION

We have developed a flexible a-Si PV module with a new monolithic series-connected structure named *SCAF*. This module is fabricated by automated apparatus based on roll-to-roll processes. Development of the apparatus and the fundamental fabrication process was almost completed, and now studies for improving performances have been carried out. Applying a-Si/a-SiGe double junction structure to the *SCAF* cell of 40cm x 80cm, we obtained a stabilized efficiency of 9.0%. Flexible PV modules fabricated with the *SCAF* solar cells are expected to create new applications such as roofing material integrated PV modules in the near future.

ACKNOWLEDGMENTS

This work was supported by the New Energy and Industrial Technology Development Organization under the New Sunshine Program of the Ministry of International Trade and Industry.

Fig. 15 Demonstration system of PV modules pasted on concrete wall.

Fig. 16 PV modules installed on curved roof.

REFERENCES

1. Ichikawa, Y et al. (1993) "Large-area Amorphous Silicon Solar Cells with High Stabilized Efficiency and Their Fablication Technology", 23rd IEEE Photovoltaic Specialists Conference, 10-14 May 1993, Lousville, USA.
2. Ichikawa, Y et al. (1994) "A New Structure a-Si Solar Cell with Plastic Film Substrate", 1st World Conference on Photovoltaic Energy and Conversion, 5-9 December 1994, Waikoloa, Hawaii.
3. Ichikawa, Y et al. (1996) "Film-substrate a-Si Solar Cells with a new monolithic Series-connected Structure", Journal of Non-crystalline Solids, *198-200* (1996), 1081-1086.
4. Horiguchi, M et al. (1996) "A Flexible PV Module for Residential Applications", International PVSEC-9, 11-15 Nov. 1996, Miyazaki, Japan.

NOVEL AMORPHOUS SILICON SOLAR CELL USING A MANUFACTURING PROCEDURE WITH A TEMPORARY SUPERSTRATE

R.E.I. SCHROPP [1], C.H.M. VAN DER WERF [1], M. ZEMAN[2], M.C.M. VAN DE SANDEN[3], C.I.M.A. SPEE[4], E. MIDDELMAN[5], L.V. DE JONGE-MESCHANINOVA[5], P.M. G.M. PETERS[5], A.A.M. VAN DER ZIJDEN[5], M.M. BESSELINK[5], R.J. SEVERENS[5], J. WINKELER[5], and G.J. JONGERDEN[5],

[1] Debye Institute, Utrecht University, P.O. Box 80 000, 3508 TA Utrecht, The Netherlands;
[2] Delft University of Technology, DIMES, P.O. Box 5053, 2600 GB Delft, The Netherlands;
[3] Eindhoven University of Technology, Den Dolech 2, 5600 AB Eindhoven, The Netherlands;
[4] Netherlands Organisation for Applied Scientific Research, P.O. Box 595, 5600 AB Eindhoven, The Netherlands;
[5] Akzo Nobel Chemicals and Coatings Research Center, P.O. Box 9300, 6800 SB Arnhem, The Netherlands.

ABSTRACT

We developed a novel temporary-superstrate concept for SnO_2/p-i-n amorphous silicon solar cells. This concept combines roll-to-roll manufacturability, the utilization of light scattering, the conductive properties of SnO_2:F, and the possibility of monolithic series connection. To this purpose, the deposition of high quality textured SnO_2:F layers on aluminum foil has been developed using APCVD. Layers have been obtained with a transmission of 80 % and a sheet resistance of 12 $\Omega/\square$. After definition of the entire p-i-n cell structure including the back contact, the cell is laminated with a permanent carrier foil. By removing the aluminum foil, light can enter the cell through the textured TCO layer. A wide variety of transparent encapsulants can be used to finalize the (flexible or preshaped) product. Preliminary cells made according to this scheme have an efficiency of 6 %.

INTRODUCTION

The Helianthos project was started in 1997 with the purpose to develop a roll-to-roll production method for amorphous silicon thin film solar cells at a manufacturing cost below 0.50 \$/Wp [1,2]. The Helianthos consortium is sponsored by the Economy, Ecology and Technology program of the Dutch Government and consists of Utrecht University, the Delft and Eindhoven Universities of Technology, the Netherlands Organisation for Applied Scientific Research, and the multinational corporation Akzo Nobel. In our opinion, a break through for photovoltaics will take place when price competitiveness with grid electricity in western countries is within reach, i.e. photovoltaic systems should be capable of generating electricity at $\approx$ \$ 0.10/kWh (in Europe and Japan grid electricity is more expensive than in the U.S.A.). The development is based on a proprietary manufacturing procedure [3], in which only abundant and relatively cheap source materials are used. Cost calculations showed that the envisaged cost level comes within reach using, e.g., a-Si:H/a-SiGe:H tandem cells, if a stabilized efficiency of 7 % is achieved in roll-to-roll production technology over a foil width of 130 cm.

713

Mat. Res. Soc. Symp. Proc. Vol. 557 © 1999 Materials Research Society

NOVEL SOLAR CELL MANUFACTURING PROCEDURE

<u>General Background</u>

In amorphous and microcrystalline thin film silicon solar cells, light trapping is of crucial importance [4]. In amorphous silicon alloy cells, efficient light management allows the active photovoltaic junction to be thin which enhances stable operation. In thin film micro- or polycrystalline components, that are more stable by nature, enhanced effective optical absorption is necessary to minimize the film thickness. This reduces the recombination losses and the time required for film deposition.

The fabrication of amorphous silicon cells on glass in a *superstrate* configuration, where light scattering is achieved with a textured SnO_2:F layer, is a well established procedure (e.g. Solarex, Sanyo). The textured SnO_2:F is usually made by Atmospheric Pressure Chemical Vapor Deposition (APCVD). This technology results nowadays in amorphous silicon panels with $\approx$ 6 - 8 % stabilized efficiencies, which have high durability in the field, because glass provides a good shield against environmental effects. However, as the glass superstrates are coated directly with the silicon thin films by Plasma Enhanced CVD (PECVD), the heating and cooling times, as well as the handling of the heavy and rigid glass panes in a vacuum system may add too much to the total cost per Wp.

In contrast, roll-to-roll technologies offer the feasibility of a compact deposition system with high throughput even at relatively low deposition rates, and reduced cost due to less elaborate handling and processing schemes, better uniformity of the coatings, and better overall yield. Roll-to-roll methods further offer opportunities for advanced automization. The finished product may still contain glass as an encapsulation material. Roll-to-roll manufacturing of thin film solar cells can be found in (pilot-)production (Unisolar, Canon, ITFT, Sanyo, Fuji Electric).

The design of an efficient light trapping scheme in a p^+-i-n^+ type solar cell combined with roll-to-roll production would require either a highly transparent flexible foil that is temperature-resistant (for APCVD of SnO_2:F), or the preparation of the whole cell, including the front electrode, at low temperature. To our knowledge, no suitable temperature-resistant transparent foils are available at sufficiently low cost. The preparation of a good quality complete cell at low temperature (< 150 °C) is only possible at relatively low deposition rates for both the textured front electrode (e.g. by sputtering) and the silicon semiconductor structure. If the solar cell structure is inverted, i.e., if an n^+-i-p^+ type solar cell is produced in a *substrate* configuration, a wider choice of flexible (temperature resistant, though non-transparent) foils is available (e.g., polyimid, or metal). Here, optical light trapping can be achieved by utilizing a textured back reflector, but a transparent low-temperature electrode is still required as the window electrode. Various industries are actively developing such structures. World-record efficiencies have been achieved by Unisolar in multijunction cells of n^+-i-p^+ structures on metal/textured back reflector substrates [5]. However, *substrate* modules on conducting carriers cannot be monolithically interconnected.

For the deposition of a textured TCO (Transparent Conductive Oxide) electrode the APCVD technique is very attractive, as it is a high deposition rate technique (up to 1000 Å/s), with very efficient source material utilization, and, at 500 - 600 °C, it produces textured coatings in one step, that have excellent chemical stability, light scattering capability, temperature durability, sheet conductivity, and very low absorption of visible light.

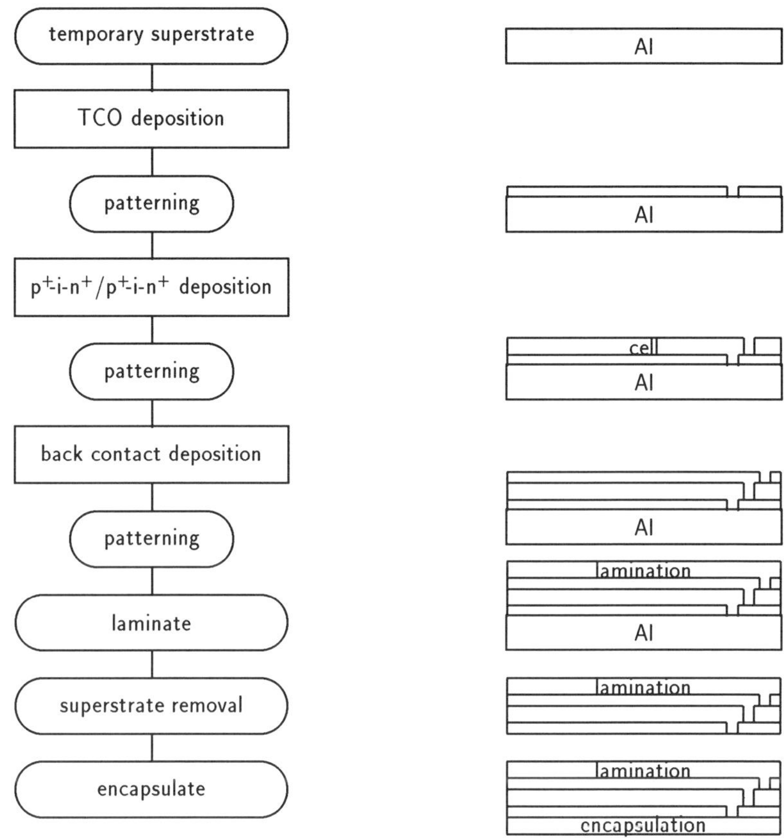

Figure 1: Processing sequence for temporary superstrate solar cell.

New processing sequence based on a temporary superstrate

On the basis of the above considerations, we came to the following conceptual solar cell and manufacturing procedure, depicted in Figure 1. Series connection of solar cells is achieved monolithically by laser scribing and other patterning steps. The temporary superstrate can be a metal foil such as aluminium, steel, or copper with a thickness of 50-100 μm and a few km long. Onto this metal foil, a textured transparent conductive oxide is deposited in a roll-to-roll fashion with a *modified* APCVD method. After laser scribing, the multijunction p$^{\pm}$-i-n^{+}-p$^{\pm}$-i-n^{+} structure is deposited in another roll-to-roll system on the continuously moving TCO-coated foil, without breaking vacuum between any of the

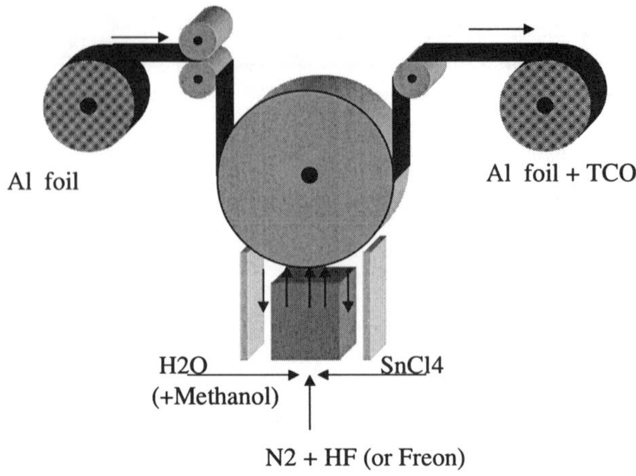

Figure 2: Schematic drawing of an APCVD set up for roll-to-roll deposition of SnO_2:F.

individual layers. After laser patterning the semiconductor structure, the back contact is deposited. The back contact can be a single metal layer or an enhanced back reflector double layer structure, such as ZnO/Ag. The final patterning step completes the series connection of individual strip subcells. Then, the foil with the interconnected cells is laminated with a permanent carrier foil. Next, the temporary superstrate is removed, e.g., by wet etching, and replaced by a transparent encapsulant. This encapsulant can be either a foil (suitable for applications where flexibility, light weight, and ease of handling is advantageous) or a rigid superstrate (for flat panels or applications with a fixed curvature, demanding rigidity). In this final stage of the process, any desired material can be chosen as the superstrate, without any constraints with respect to vacuum or temperature compatibility.

RESULTS

The fact that the temporary carrier foil has to be removed at a later stage of the sequence imposes the requirement that the front window electrode is chemically as well as mechanically stable. Polycrystalline SnO_2:F is capable of fulfilling these requirements. In order to optimize the SnO_2:F deposition process for roll-to-roll application on a moving metal foil, a dedicated set up has been built at Akzo Nobel for depositions over a width of 25 cm (Figure 2). In the second-generation APCVD roll coater, we have achieved wrinkle-free deposition of SnO_2:F with appropriate properties from HF, $SnCl_4$, H_2O and CH_3OH at a superstrate temperature between 500 and 550 °C. The presently deposited layers have a transmission of 80 % for visible light (400 - 800 nm) and a sheet resistance of 10 - 20 $\Omega/\square$. The layer is continuous and has pyramidal texture with sharp crystalline facets (Figure 3). The typical peak-to-valley roughness is 100 - 120 nm, which produces suitable scattering

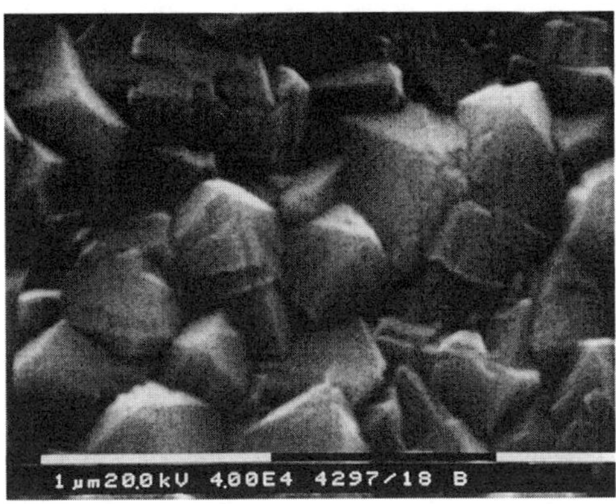

Figure 3: A Scanning Electron Micrograph of Akzo Nobel SnO$_2$:F deposited on Al foil.

of the incident light.

Utrecht University and Delft University of Technology have fabricated "standard" p-i-n/Ag structures on these superstrates. The Al superstrate could subsequently removed by etching in NaOH or in H$_3$PO$_4$-based etchants. For the preparation of test cells, the superstrate is removed in approximately 3 minutes at an etch rate of 0.4 μm/s.

Cells have been completed according to the temporary-superstrate concept as well as reference cells. The reference cells were built on thin glass sheets ("permanent-superstrate" cells) that were moved through the same roll coater using the same processing parameters. Table 1 shows the performance of these two devices. It is seen that the temporary-superstrate sequence causes limitations to the short circuit current J_{sc}. Part of the difference is due to the fact that the cell is not encapsulated, i.e., the exposed surface is bare SnO$_2$:F, which has a larger refractive index than glass (glass acts as an antireflective coating in permanent-superstrate cells). J_{sc} shows a variation between 10 to 14 mA/cm^2 and we suspect that this variation is primarily due to the difficulty of active area definition.

cell type	V_{oc} (V)	J_{sc} (mA/cm^2)	FF	η (%)
temporary superstrate	0.827	10.71	0.678	6.01
permanent superstrate	0.80	15.4	0.72	8.7

Table 1: Performance of devices made according to the temporary-superstrate concept and a reference cell (conventional "permanent-superstrate" method). In both cases the TCO was made by Akzo Nobel in a roll coater suitable for deposition on Al foil. The area of both devices is 0.25 cm^2 and the thickness of the cell is approximately 450 nm.

CONCLUSIONS

The feasibility of the new temporary-superstrate concept for roll-to-roll processing of amorphous silicon based solar cells has been demonstrated. The most critical process, the deposition of textured SnO_2:F on Al foil, has been proven to be possible. After deposition of the complete solar cell, and lamination with a suitable carrier (e.g. PET), the Al foil can be removed by wet etching. Due to its resistance against most etchants, the SnO_2:F acts as an effective etch barrier. The new processing scheme using a sacrificial carrier foil can be applied also to other thin film photovoltaic technologies, e.g., CdTe, CI(G)S, and dye-sensitized TiO_2.

The intended roll-to-roll processing sequence has undergone a detailed cost analysis. Due to the utilization of cheap, abundant materials only, and the continuous processing method a level of $ 0.50/Wp for the manufacturing costs can be obtained. The temporary-superstrate method offers versatility in system integration of PV. The PV product can be made flexible or rigid. The rigid products can be made flat or pre-shaped.

A roll-to-roll pilot plant will be developed within the next 3 - 4 years.

ACKNOWLEDGEMENT

The Helianthos project has been partially supported by the Netherlands Agency for Energy and Environment and the Economy, Ecology and Technology program of the Dutch Government.

REFERENCES

1. E. Middelman, E. van Andel, P.M.G.M. Peters, L.V. de Jonge-Meschaninova, R.J. Severens, G.J. Jongerden, R.E.I. Schropp, H. Meiling, M. Zeman, M.C.M. van de Sanden, A. Kuipers, and C.I.M.A. Spee, 2nd World Conference and Exhibition on Photovoltaic Solar Energy Conversion, July 6-10, 1998, Vienna, Austria, pp. 816-820.
2. R.E.I. Schropp, H. Meiling, C.H.M. van der Werf, E. Middelman, E. van Andel, P.M.G.M. Peters, L.V. de Jonge-Meschaninova, J. Winkeler, R.J. Severens, G.J. Jongerden, M. Zeman, M.C.M. van de Sanden, A. Kuipers, and C.I.M.A. Spee, 2nd World Conference and Exhibition on Photovoltaic Solar Energy Conversion, July 6-10, 1998, Vienna, Austria, pp. 820-822.
3. Patent application number WO 98/13882.
4. R.E.I. Schropp and M. Zeman, *Amorphous and Microcrystalline Solar Cells: Modeling, Materials and Device Technology*, Kluwer Academic Publishers, ISBN 0-7923-8317-6 (Boston/Dordrecht/London, 1998).
5. J. Yang, A. Banerjee, and S. Guha, Appl. Phys. Lett. **54** (1997) 2975.

REAL TIME OPTICS OF AMORPHOUS SILICON SOLAR CELL FABRICATION ON TEXTURED TIN-OXIDE-COATED GLASS

P. I. ROVIRA*, A. S. FERLAUTO*, Ilsin AN**, H. FUJIWARA***,
Joohyun KOH*, R. J. KOVAL*, C. R. WRONSKI*, and R. W. COLLINS*

* Center for Thin Film Devices, The Pennsylvania State University, University Park, PA 16802.
** Department of Physics, Hanyang University, Ansan, KOREA.
*** Thin Film Silicon Solar Cells Superlaboratory, Electrotechnical Laboratory, Tsukuba, JAPAN.

ABSTRACT

A rotating-compensator multichannel ellipsometer has been used to measure the four spectra (1.4-4.0 eV) that describe the Stokes vector of the light beam reflected from the surface of an amorphous silicon (a-Si:H) p-i-n solar cell during fabrication on textured tin-oxide (SnO_2) coated glass. The Stokes vector elements include the irradiance in the reflected beam (or the reflectance) and the parameters $\{(Q, \chi), p\}$ of the reflected beam, where Q and χ are the tilt and ellipticity angles of the polarization ellipse and p is the degree of polarization. The value of p deviates from unity in part due to the non-uniform nature of the textured SnO_2 substrate film. An analysis of Q and χ that neglects the effects of the texture can provide the time evolution of the thicknesses, microscopic structure, and the optical properties of the component layers of the a-Si:H solar cell. Deviations of the measured reflectance spectra from those predicted on the basis of the (Q, χ) analysis provide the thickness dependence of the scattering and the evolution of the macroscopic structure of the solar cell. The measurement and analysis approach is important because of its potential application for real time monitoring of solar cell production. The analysis results also provide realistic inputs for optical modeling of the effects of texture in light trapping for solar cell efficiency enhancement.

INTRODUCTION

An understanding of the optics of textured a-Si:H thin film structures is important for two reasons. With such an understanding, optical probes can be applied for the characterization of solar cell structures in the configuration used in production. Furthermore, improved models can be developed to simulate the scattering processes that lead to light trapping and efficiency enhancement in a-Si:H solar cells. The surface roughness associated with the texture exists on a wide range of scales from microscopic to macroscopic. For microscopic roughness (i.e., roughness with correlation lengths L much smaller than the wavelength λ), an effective medium theory (EMT) can be used to model the optical response of the roughness on surfaces; however, the EMT fails as L approaches λ [1]. For macroscopic roughness with L~λ, numerous theories have been developed of varying complexity from scalar diffraction [2] to first-principles analyses [3]. It is clear that a synthesis of theories, covering the full range of scales, will be needed to fully understand the effect of texture.

In this paper, we have addressed the problem of real time optical characterization of textured a-Si:H solar cell structures. We have applied a novel multichannel ellipsometer in the rotating-compensator configuration [4] to measure spectra from 1.4 to 4.0 eV in the unnormalized Stokes vector of the light beam reflected from a textured a-Si:H p-i-n solar cell with 0.8 s time resolution during deposition. This provides sufficient information to extract a number of solar cell characteristics including layer thicknesses, layer optical properties, microscopic roughness, macroscopic roughness, and spectrally-resolved scattering integrated over all solid angles. This work opens up the possibility of monitoring and controlling the fabrication of solar cells in production.

EXPERIMENTAL PROCEDURES

The a-Si:H p-i-n solar cell studied here was prepared by rf plasma enhanced chemical vapor deposition (PECVD) onto Asahi U-type textured SnO_2 held at 200°C. The a-$Si_{1-x}C_x$:H:B (x~ 0.05) p-layer was prepared to an estimated thickness of 200 Å using 6:4:0.02:1 flow ratios of SiH_4:CH_4:$B(CH_3)_3$:He. The a-Si:H i-layer was prepared to 5000 Å using 20 sccm of SiH_4, and the a-Si:H:P n-layer was prepared to 300 Å using a 10:0.12 flow ratio of SiH_4:PH_3. The plasma power for the three layers was 70 W/cm^2, and the total gas pressures were in the range 0.1-0.2 Torr. Only

Mat. Res. Soc. Symp. Proc. Vol. 557 © 1999 Materials Research Society

the analysis results for the p and i layer are presented in this paper. During deposition, real time optical measurements were performed using a rotating-compensator multichannel ellipsometer [4]. This instrument consists of a Xe source, fixed polarizer, sample in the PECVD reactor, rotating-compensator, fixed analyzer, spectrograph, and photodiode array detector. With this instrument, four spectra from 1.4 to 4.0 eV in $\{(Q, \chi), p, R\}$ are collected simultaneously with a total acquisition time of 0.8 s. The repetition time ranges from 3.5 s for the p-layer and early stages of the i-layer, to 50 s in the later stages of i-layer growth. (Q, χ) are the tilt and ellipticity angles of the ellipse of polarization, p is the degree of polarization, and R is the reflectance of the sample for incident light polarized at an angle of $P'=45°$ measured with respect to the plane of incidence.

Analysis of the uncoated SnO_2 film and a-Si:H solar cell fabrication involves breaking the optical problem into two parts (see Fig. 1 for the case of the uncoated SnO_2) [5]. First, (Q, χ) are converted to the ellipsometry angles (ψ, Δ) assuming an ideal sample. Then (ψ, Δ) are interpreted to deduce the layer thicknesses, microscopic roughness $(L<<\lambda)$, and layer optical properties in the same way as for a specular sample [see Fig. 1(b)]. Second, with the microscopic optical problem solved, the reflectance R that would be predicted for a specular structure is calculated. The differences between the predicted and measured spectra in R are attributed to integrated scattering losses generated by macroscopic roughness with $L\sim\lambda$. The reflectance loss can be modeled using a scalar diffraction theory [2,5] that predicts a dampening of the Fresnel amplitude reflection and transmission coefficients in accordance with $c=c_0\exp(-\sigma_s\Delta k_c/2)$ (c=r,t) for Lorentzian-distributed macroscopic roughness, i.e., for a surface distribution w(z) given by $w(z)=2\sigma_s[\pi(4z^2+\sigma_s^2)]^{-1}$, where z is the height from the average surface plane and σ_s is the full-width at half-maximum [see Fig. 1(c)]. Δk_c (c=r,t) is the change in propagation vector normal to the surface.

For this analysis approach to be valid, three conditions must apply [6]: (i) light scattering from the textured film must be independent of the incident polarization and must be rejected by the high f/# detection system; (ii) the detected beam must be describable as an incoherent superposition of components from different areas of the SnO_2 surface, i.e., from areas that are separated by distances larger than the lateral coherence of the beam, and (iii) the resulting incoherent superposition from areas with different SnO_2 and/or a-Si:H thicknesses must provide polarization information that accurately represents the average thickness and the optical properties. Several studies [4,5] along with corroborative measurements of the deduced microstructure suggest that these conditions do apply for the textured SnO_2 and for overlying a-Si:H solar cell fabrication. In the future, it may also be possible to extend the analysis further by addressing a third aspect of the reflected wave, namely, the degree of polarization p which deviates from unity in part owing to the non-uniformity of the SnO_2 [4]. In principle, analysis of p can provide information on the thickness distribution having correlation lengths larger than the lateral coherence of the beam $(L>>\lambda)$.

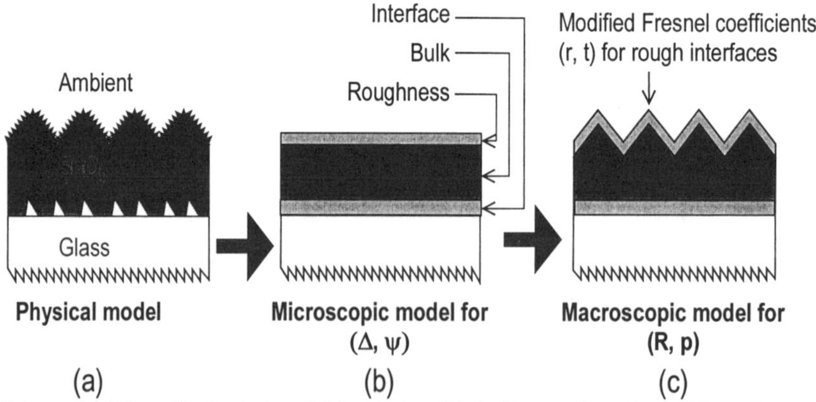

FIG. 1   (a) Schematic physical model for a textured SnO_2 film on glass, along with the decomposition into (b) the microscopic model used in the analyis of (ψ, Δ) and (c) the macroscopic model used in the analysis of the polarized reflectance R.

RESULTS AND DISCUSSION

Figure 2 shows the microstructural model ($L<<\lambda$) for p-layer deposition on the textured SnO_2. An initial study of the textured SnO_2 film similar to that described elsewhere [5] yields a three layer microscopic model consisting of (i) a low density layer of thickness $d_i(SnO_2)$=340 Å and material volume fraction $f_i(SnO_2)$=0.78 at the interface to the glass, (ii) a bulk layer of thickness $d_b(SnO_2)$=6180 Å and full density [$f_b(SnO_2)\equiv 1$], and (iii) a microscopic surface roughness layer of thickness $d_s(SnO_2)$=450 Å with material fraction $f_s(SnO_2)$=0.57. The dielectric function ε of the SnO_2 includes an ε_2 spectrum expressed as the sum of a Tauc-Lorentz oscillator for the interband transitions and a Drude term for the intraband transitions. ε_1 is obtained from ε_2 using the Kramers-Kronig transformation [7]. Analysis of the SnO_2 reflectance provides a macroscopic roughness distribution characterized by σ_s=770 Å. To analyze film growth on the SnO_2 surface, three photon energy independent parameters are used: (i) $f_i(a\text{-}Si_{1-x}C_x\text{:}H)$ which describes the filling of the microscopic modulations in the SnO_2 surface by depositing p-layer material; (ii) $d_s(a\text{-}Si_{1-x}C_x\text{:}H)$ which describes the thickness of the p-layer extending above the SnO_2 modulations; and (iii) $f_s(a\text{-}Si_{1-x}C_x\text{:}H)$ which describes the volume fraction of p-layer material within the layer of thickness $d_s(a\text{-}Si_{1-x}C_x\text{:}H)$. The dielectric function of the p-layer is obtained simultaneously in the analysis along with these three parameters.

Figure 3 shows the best fit results for the microstructural evolution of the p-layer, including the parameters $d_s(a\text{-}Si_{1-x}C_x\text{:}H)$ and $f_i(a\text{-}Si_{1-x}C_x\text{:}H)$. The parameter $f_s(a\text{-}Si_{1-x}C_x\text{:}H)$ is fixed at 0.35 since this best fit value varies little over time. The parameter $d_{eff}(a\text{-}Si_{1-x}C_x\text{:}H)$ in Fig. 3 is the effective thickness of the p-layer; i.e., the p-layer thickness that would be obtained for uniform, microscopically-smooth film growth on an ideal substrate such as c-Si. In general for material x, d_{eff} is given by $d_{eff}(x)=f_i(x)d_i(x)+f_s(x)d_s(x)$. The deduced effective thickness for the p-layer of 201 Å in Fig. 3 is remarkably close to the intended thickness of 200 Å. Obtaining the intended d_{eff} requires careful deposition timing using the the p-layer rate measured from a c-Si wafer substrate. It is surprising that the effective thickness of the 200 Å p-layer can be determined to better than 1% when growth occurs on a SnO_2 surface with 450 Å of microscopic roughness (which is filled in by the p-layer) and an even thicker layer associated with macroscopic roughness (which is conformally covered by the p-layer).

Figure 4 shows the corresponding results for i-layer growth. Here, four parameters are plotted; in addition to $d_s(a\text{-}Si\text{:}H)$ and $f_i(a\text{-}Si\text{:}H)$ as in Fig. 3, the surface layer material fraction $f_s(a\text{-}Si\text{:}H)$ and the bulk layer thickness $d_b(a\text{-}Si\text{:}H)$ are shown. The bulk i-layer thickness increases above zero in this case once the roughness at the p/i interface is completely filled by i-layer material [$f_i(a\text{-}Si\text{:}H)\rightarrow 0.65$]; thereafter, $f_i(a\text{-}Si\text{:}H)$ is fixed such that $f_s(a\text{-}Si_{1-x}C_x\text{:}H)+f_i(a\text{-}Si\text{:}H)=1$ (horizontal broken line in Fig. 4, upper panel). The effective layer thickness in this case is given by $d_{eff}=d_b+f_id_i+f_sd_s$, where all quantities pertain to a-Si:H. We note that values of $d_b(a\text{-}Si\text{:}H)$ greater than 4500 Å cannot be determined due to the nearly complete loss of interference fringes over the spectral range of analysis. The extrapolated final effective layer thickness is 4892 Å [see solid line for $d_b(a\text{-}Si\text{:}H)$ in Fig. 4, lower panel], in very good agreement with the intended value of 5000 Å. One observation in Fig. 4 that differs from corresponding analysis results for a-Si:H p-i-n solar cell deposition [8] on much

<table>
<tr><td>

FIG. 2   Schematic microstructural model (left) and microscopic optical model (right) used in the analysis of (ψ, Δ) spectra for the growth of the a-$Si_{1-x}C_x$:H p-layer of an a-Si:H solar cell on textured SnO_2-coated glass. The parameters d_s and f_s are the thickness and material volume fraction for the surface roughness layer; d_i and f_i are the thickness and overlying material volume fraction for the interface layers; and d_b is the bulk layer thickness. All surface and interface roughness layers are assumed to be microscopic, meaning that the correlation lengths of the structure are less than the wavelength of the light.

</td><td>

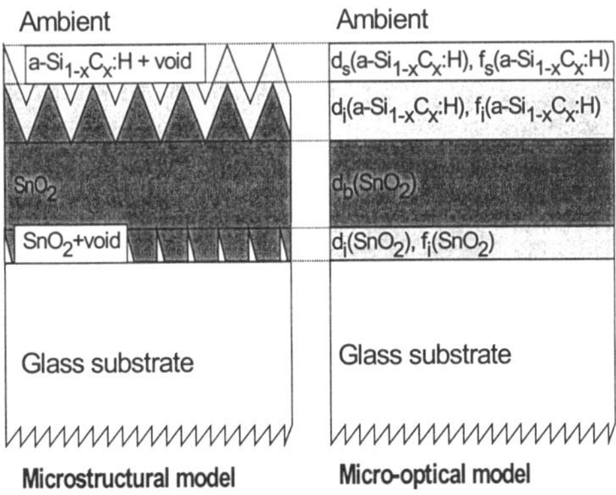

</td></tr>
</table>

smoother specular SnO_2 [with $d_s(SnO_2)$=100 Å] is the extensive microscopic roughness on the a-Si:H even at the end of the i-layer deposition. Such microscopic roughness, modeled with the Bruggeman EMT, must be included in any realistic optical model of the solar cell. Figure 5 shows a comparison between the evolution of the macroscopic roughness σ_s(a-Si:H) as deduced from an analysis of R and the microscopic roughness d_s(a-Si:H) deduced from the analysis of (ψ, Δ). Both are plotted as a function of time during i-layer deposition after complete interface filling has occurred. It is interesting to note that although the microscopic roughness decays by more than a factor of four relative to the initial starting value for the SnO_2 (horizontal lines in Fig. 5), the macroscopic roughness decays by only ~20%.

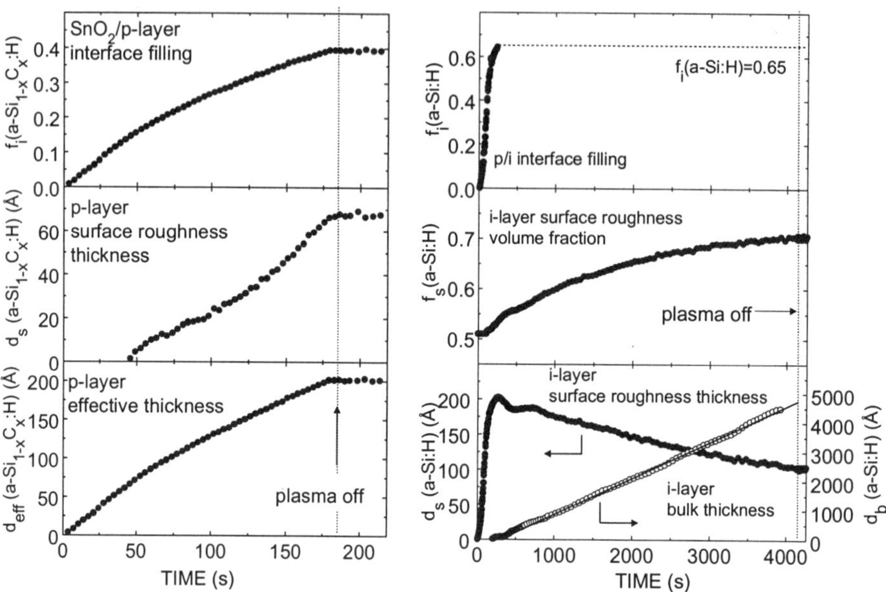

FIG. 3 Evolution of the microstructural parameters for the growth of an a-Si$_{1-x}$C$_x$:H p-layer on textured SnO$_2$-coated glass. The parameter d_s(a-Si$_{1-x}$C$_x$:H) is the thickness of the p-type a-Si$_{1-x}$C$_x$:H surface roughness layer that extends above the SnO$_2$ substrate layer. In this analysis the p-type a-Si$_{1-x}$C$_x$:H volume fraction within this surface layer is fixed as f_s(a-Si$_{1-x}$C$_x$:H)=0.35. Here f_i(a-Si$_{1-x}$C$_x$:H) is the p-type a-Si$_{1-x}$C$_x$:H volume fraction filling the interface roughness layer between the SnO$_2$ and the p-layer. The thickness of this layer is given by d_i(a-Si$_{1-x}$C$_x$:H) = d_s(SnO$_2$) = 450 Å as determined in an analysis of the starting SnO$_2$ substrate film. The parameter d_{eff}(a-Si$_{1-x}$C$_x$:H) is the effective thickness of the p-layer.

FIG. 4 Evolution of the microstructural parameters for the growth of an a-Si:H i-layer on top of a glass/SnO$_2$/p-layer structure. The parameters d_s(a-Si:H) and f_s(a-Si:H) are the thickness and material volume fraction in the surface roughness on the a-Si:H i-layer. The parameter f_i(a-Si:H) is the a-Si:H i-layer volume fraction filling the interface roughness layer between the p and i layers, and d_b(a-Si:H) is the a-Si:H bulk i-layer thickness. Once the p/i interface roughness layer is established as a 0.35/0.65 volume fraction mixture of p-layer/i-layer materials, the bulk layer begins to form. In the bulk layer growth regime f_i(a-Si:H) is fixed at 0.65, representing a completely filled p/i interface roughness layer.

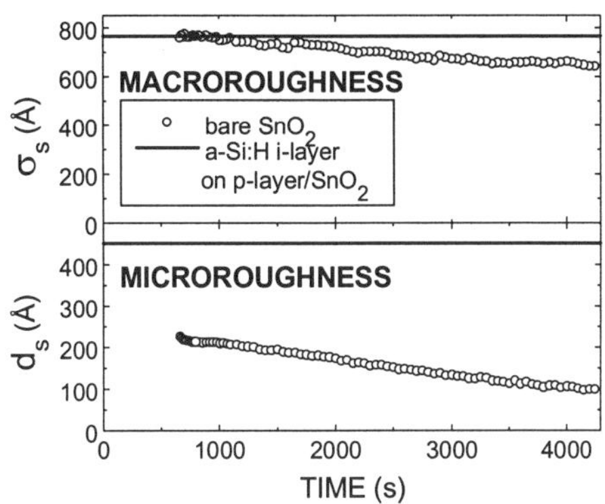

FIG. 5 Evolution of the microscopic and macroscopic roughness layer thicknesses deduced in analyses of (ψ, Δ) and R, respectively, for the later stages of growth of the i-layer on the textured SnO_2/p-layer structure. The macroscopic roughness σ_s(a-Si:H) is defined as the width of the Lorentzian distribution describing the a-Si:H surface profile in the random diffraction model. The microscopic roughness d_s(a-Si:H) describes the thickness of the a-Si:H/void mixture in the EMT. The horizontal lines designate the corresponding values for the uncoated SnO_2 surface.

In order to assess the validity of the roughness layer thicknesses on the uncoated and over-deposited SnO_2, these samples were measured ex situ by atomic force microscopy (AFM). Figure 6 shows the distribution of the maximum peak-to-valley vertical distances d_{p-v} for the uncoated textured SnO_2 and the final solar cell on the SnO_2 [5]. Each count plotted on the ordinate in Fig. 6 was obtained by considering the surface topography within one of a large number of 0.1×0.1 μm^2 boxes excised from 5×5 μm^2 images in order to restrict the analysis to the microscopic roughness. (Note that a correlation with root-mean-square values from AFM would be ambiguous.) Excellent agreement is obtained between the optical measurement of microscopic roughness (d_s) and the AFM measurement ($<d_{p-v}>$, the average value of d_{p-v}): 450 versus 430 Å for the uncoated SnO_2 and 100 Å versus 120 Å for the completed solar cell. Figure 7 shows 5×5 μm^2 images of (a) the uncoated SnO_2 surface and (b) the solar cell surface along with (c) one dimensional Fourier analyses along the diagonal lines indicated on the images. It is clear that upon solar cell deposition on the SnO_2, the microscopic roughness is suppressed to a greater degree than the macroscopic roughness, as also indicated in Fig. 5. Further quantitative corroboration of the macroscopic roughness is needed to support the simple diffraction-based optical model used here.

Although the dielectric function data have not been presented here, the results for the p and i layers deposited on the textured SnO_2 are in good agreement with the corresponding results for deposition on specular SnO_2 described previously [8]. First, these observations suggest that the assumptions built into the optical analysis are correct. Specifically, we conclude that it is possible to break the optical analysis into two independent parts: a polarization analysis of the reflected beam that is unperturbed by the texture, and a reflectance loss analysis which provides information on the texture using as input the microstructural model deduced in the polarization analysis. Second, these

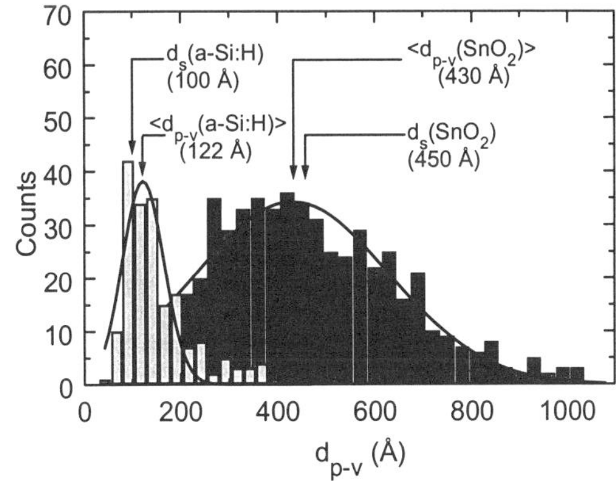

FIG. 6 Distributions describing the maximum peak-to-valley excursions d_{p-v} within 0.1×0.1 μm^2 regions of 5×5 μm^2 AFM images for the uncoated textured SnO_2 surface (dark shading) and for the completed textured a-Si:H solar cell surface (light shading). The solid lines depict fits of these distributions to Gaussian functions. The arrows indicate the average values $<d_{p-v}>$ placed at the center of the Gaussian functions, in comparison with the d_s values deduced in the optical analysis of the microstructure.

FIG. 7   Atomic force microscopy images over 5x5 μm^2 areas for (a) the uncoated textured SnO_2 and (b) the completed textured a-Si:H solar cell.   In part (c), one-dimensional Fourier analyses along the diagonals of the images are shown.   In these analyses, the root-mean-square amplitude is plotted as a function of the spatial frequency (or surface profile wavelength) for the two samples.

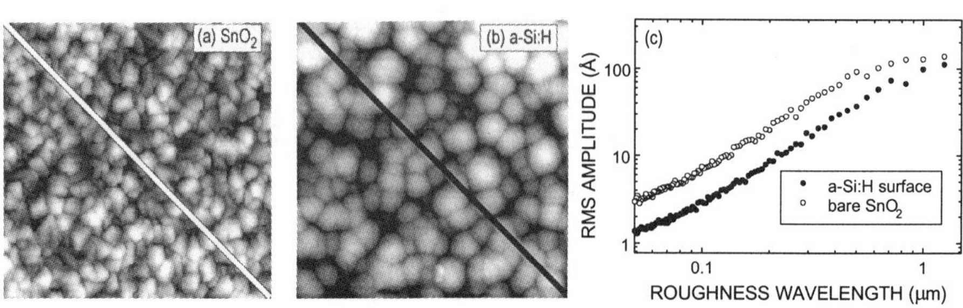

observations demonstrate that the coverage of the texture by the p and i layers is truly conformal, and additional voids are not introduced into these layers due to the texture.

SUMMARY

We have successfully deduced the time evolution of the thicknesses, microscopic and macroscopic structure, and optical properties of the component layers of a-Si:H solar cells from Stokes vector spectroscopy performed in real time during fabrication on textured SnO_2-coated glass. The analysis of the microstructure, i.e., roughness with correlation lengths L much less than the wavelength λ, is performed using the Bruggeman effective medium theory, whereas the analysis of the macrostructure, i.e., L~λ, is performed using a random diffraction theory. The validity of the analysis developed here is supported by a number of observations: (i) excellent agreement between measured and intended layer thicknesses; (ii) excellent agreement in the microscopic roughness between optical and AFM results; (iii) qualitative agreement between the optical and AFM results in the comparative evolution of the microscopic and macroscopic roughness layers; and (iv) component layer optical properties that are in good agreement with corresponding studies of solar cell preparation on specular surfaces.

ACKNOWLEDGMENTS

This research was supported by National Science Foundation Grant No. DMR-9820170 and by National Renewable Energy Laboratory Subcontract No. XAF-8-17619-22.

REFERENCES

[1]   G. A. Niklasson, C. G. Granqvist, and O. Hunderi, Appl. Opt. **20**, 26 (1981).
[2]   P. Beckmann and A. Spizzichino, *The Scattering of Electromagnetic Waves from Rough Surfaces*, (Pergamon, Oxford, 1963).
[3]   W. L. Mochan and R. G. Barrera, Phys. Rev. B **32**, 4984 (1985); **32**, 4989 (1985).
[4]   J. Lee, P. I. Rovira, I. An, and R. W. Collins, Rev. Sci. Instrum. **69**, 1800 (1998).
[5]   P. I. Rovira and R. W. Collins, J. Appl. Phys. **85**, 2015 (1999).
[6]   M. D. Williams and D. E. Aspnes, Phys. Rev. Lett. **41**, 1667 (1978).
[7]   G. E. Jellison, Jr., and F. A. Modine, Appl. Phys. Lett. 69, 371 (1996); **69**, 2137 (1996).
[8]   J. Koh, Y. Lu, S. Kim, J. S. Burnham, C. R. Wronski, and R. W. Collins, Appl. Phys. Lett. **67**, 2669 (1995).

EFFECT OF INTERFACE ROUGHNESS ON LIGHT SCATTERING AND OPTICAL PROPERTIES OF a-Si:H SOLAR CELLS

M. ZEMAN*, R.A.C.M.M. VAN SWAAIJ*, M. ZUIDDAM*, J.W. METSELAAR*, and R.E.I. SCHROPP**
*Delft University of Technology - DIMES, P.O. Box 5053, 2600 GB Delft, The Netherlands, email: m.zeman@its.tudelft.nl
**Utrecht University, Debye Institute, P.O. Box 80 000, 3508 TA Utrecht, The Netherlands.

ABSTRACT

The effect of interface roughness on the optical properties of amorphous silicon (a-Si:H) solar cells was investigated using *rms* roughness measurements and computer modeling. We deposited four single junction a-Si:H solar cells on Asahi U type substrate each with a different intrinsic layer thickness. The roughness of the substrate and of the subsequent interfaces of the cells was measured by Atomic Force Microscopy. The relations between the computer input parameters, which describe the diffuse part of reflected and transmitted light at a rough interface, and the *rms* roughness of the interfaces in a-Si:H solar cell are presented. After obtaining a good matching between the simulated and measured external quantum efficiencies (*QE*) of the four cells, we investigated the effect of interface roughness on the absorption in all individual layers of the a-Si:H solar cell using the determined scattering parameters.

The *rms* roughness of the Asahi U-type substrate surface is determined to be 40 nm. Deposition of a 9 nm thick p-type a-SiC:H layer on the substrate has not changed the surface roughness. Due to a high refractive index of n-type a-Si:H the back contact interface acts as a nearly perfect diffuser for the reflected light when the *rms* roughness of the interface is higher than 25 nm. The *rms* roughness of the back interface of the four cells is found to be dependent on the intrinsic layer thickness and is larger than 25 nm.

INTRODUCTION

In order to enhance the light absorption of solar cells, textured substrates arc uscd in the fabrication of a-Si:H solar cells. As a result of this texture all subsequent interfaces in the cell are rough. For optimal optical design of a-Si:H solar cells, in which computer simulations play an important role, it is necessary to establish a relation between the physical interface roughness and corresponding descriptive scattering parameters of computer models. Several authors have used optical computer modeling to study an absorption enhancement in a-Si:H solar cells taking into account scattering at the rough front and back contact interfaces [1-4]. In their work, the diffuse part of reflected and transmitted light has not been related to a morphology of the rough interfaces. Recently, the roughness of the TCO substrate was taken into account to determine the wavelength dependent diffuse reflectance [5,6]. In this work we present the relations between the computer input parameters, which describe the diffuse part of reflected and transmitted light, and the interface roughness measured by Atomic Force Microscopy (AFM). After obtaining a good matching between the experimental and simulated characteristics of the fabricated single junction a-Si:H solar cells, we used the determined scattering parameters to investigate the effect of interface roughness of the front and back contact on the absorption in all individual layers of the a-Si:H solar cell.

Mat. Res. Soc. Symp. Proc. Vol. 557 ©1999 Materials Research Society

EXPERIMENT

1. Deposition and characterization of a-Si:H solar cells

Four single junction a-Si:H solar cells each with different thickness of the intrinsic layer were deposited on Asahi U type substrate. The solar cells had the following structure: Asahi U-type/ p-type a-SiC:H (9 nm)/ intrinsic a-Si:H (160 to 610 nm)/ n-type a-Si:H (20 nm)/ Ag (300 nm). The solar cells were characterized by dark and illuminated I-V measurements and by spectral response measurements.

2. Characterization of interface roughness

The following parameters are used to describe a rough surface: the root mean square (*rms*) roughness, σ_r, the *rms* slope, and the scattering data, which include the haze parameter that is defined as the ratio between the diffuse reflectance (transmittance) to the total reflectance (transmittance) and the angle distribution of diffuse light. We used scanning electron microscopy, transmission electron microscopy (TEM), and AFM to characterize the roughness of the interfaces in a-Si:H solar cell.

The value of the *rms* roughness of the Asahi U-type TCO substrate is found to be dependent on the scanned area measured with AFM and is higher for larger scanned area. This can indicate that a thickness variation is present in the substrate over a distance of several micrometers. The TEM measurement on this substrate revealed a thickness variation from 620 to 740 nm. Under the assumption that for light in the wavelength range of interest (300 to 900 nm) only a local roughness over an area less than 1 μm^2 is important we have chosen $\sigma_r = 40$ nm to be used in our simulations for the TCO/p interface. This value is an average from 20 measurements carried out on 3×3 μm^2 scanned area.

Deposition of a 9 nm p-type a-SiC:H has not changed the surface roughness. Deposition of 150 nm thick intrinsic layer resulted in a surface characterized by $\sigma_r = 36$ nm and 300 nm thick layer resulted in $\sigma_r = 33$ nm. We expect that the deposition of 20 nm n-layer does not change the surface roughness significantly.

SIMULATIONS

We used the *GENPRO2* optical program [7] in combination with the SR measurements for extracting the scattering parameters of the rough interfaces in the a-Si:H solar cell. The optical properties of the individual a-Si:H layers were determined from Reflection and Transmission measurement, Dual beam photoconductivity, and Photothermal Deflection Spectroscopy and are reported in reference [8, p. 151]. It has been reported that the evaporated Ag back metal contact intermixes with silicon which lowers the expected high reflectivity [6]. Therefore for simulation purposes a thin 1.5 nm Al layer is introduced between the n-type a-Si:H and Ag [6] to account for the lower reflectivity of the real Ag back contact.

With the *GENPRO2* program we calculated the absorption profile in a-Si:H solar cells with textured interfaces. The absorption profile in the individual layers of the cell was integrated over thickness of the layer in order to obtain the total absorption. The total absorption in the intrinsic a-Si:H layer calculated per incident photon is related to the external quantum efficiency, *QE*, [8, p. 188].

1. Determination of diffuse reflectance

GENPRO2 program requires input data, which characterize the scattering of light at a rough interface. The haze parameter of the reflected and transmitted light and the angle distribution of the diffuse light are of major importance. Bennett and Porteus [9] demonstrated that in case of normal incidence the specular part of the total reflectance, R_s, is related to σ_r by:

$$R_s = R_0 \exp\left[-\left(4\pi\sigma_r n_0/\lambda\right)^2\right] \tag{1}$$

where R_0 is the reflectance of a perfectly smooth surface and n_0 is the refractive index of the medium of incidence. Under the assumption that the sum of the specular and diffuse reflectances of a rough interface is equal to the reflectance of a perfectly smooth interface between the same materials ($R_s + R_d = R_0$) the diffuse reflectance can be determined from Eq. 1. Fig. 1. presents the ratio of R_d to R_0 as calculated from Eq. 1 for the Asahi U-type TCO material and the n-type a-Si:H for several values of the *rms* roughness. An important result, which can be noticed from Fig. 1b, is that the back contact interface acts as a nearly perfect diffuser for the reflected light in the wavelength range of interest when the *rms* roughness of the interface is above 25 nm.

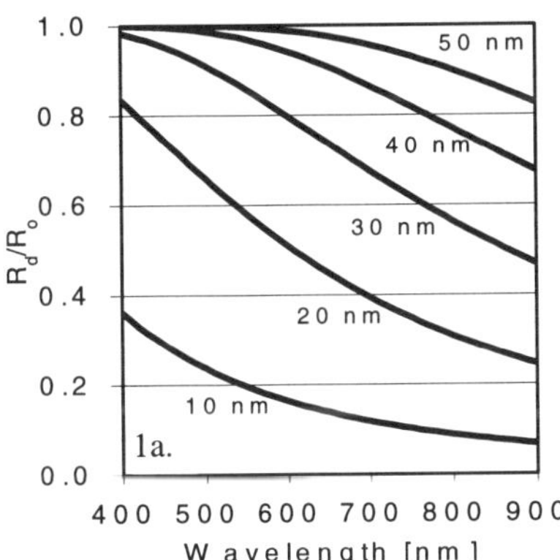

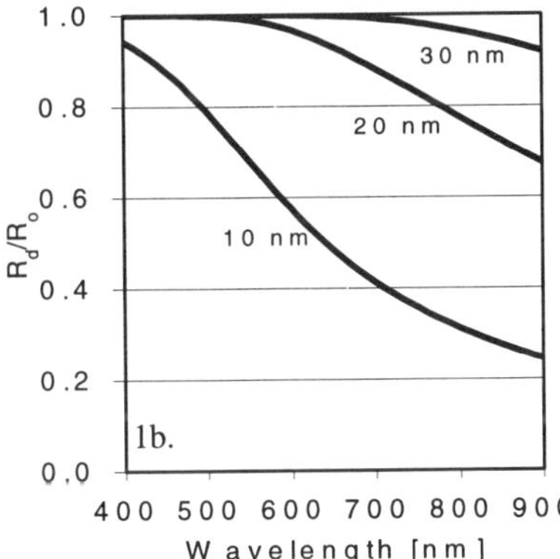

Fig. 1: The ratio of R_d to R_0 as function of wavelength for several values of the *rms* roughness of the rough interface for: a) Asahi U-type TCO material and b) n-type a-Si:H.

2. Determination of diffuse transmittance

From matching the simulations with measurements of the total integrated transmission of a simple optical system consisting of Asahi U-type substrate / medium / glass, where the medium is air and water, we extracted the diffuse transmittance of the rough Asahi U-type interface as a function of the wavelength. Eq. 1 was used to calculate the diffuse reflectances at the rough TCO/medium interface. We could fit the wavelength dependence of the extracted diffuse transmittance using the following formula:

$$T_d = T_0\left(1 - \exp\left[-\left(4\pi\sigma_r C|n_0 - n_1|/\lambda\right)^3\right]\right) \tag{2}$$

where T_d is the diffuse transmittance, T_0 is the transmittance of a perfectly smooth interface between two media, which are characterized by refractive indices n_0 and n_1, and C is a factor that depends on the two media. We used Eq. 2 to calculate the diffuse transmitance for the rough interfaces in a-Si:H solar cells. Fig. 2 shows the ratio of T_d to T_0 for the TCO/p and p/i interfaces for $C = 1$ and several values of σ_r.

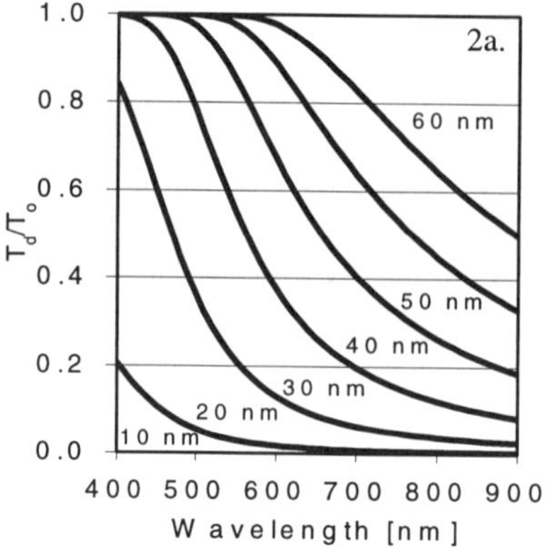

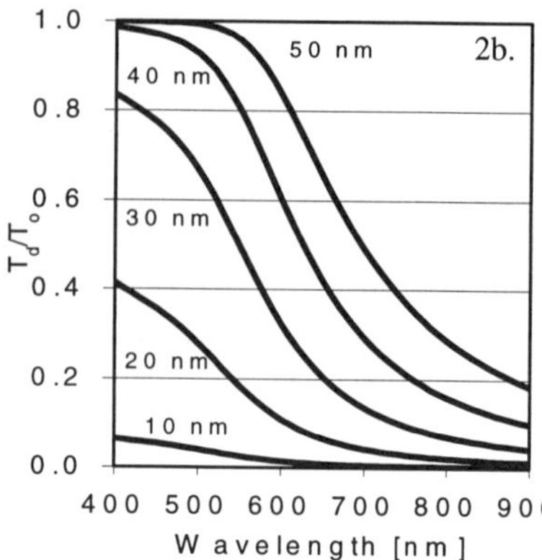

Fig. 2: The ratio of T_d to T_0 as function of wavelength for several values of σ_r for: a) the TCO/p interface and b) the p/i interface.

3. Simulations of External Quantum Efficiency

Using the experimentally determined values of σ_r of the rough interfaces in the cells and the above described approach to model the diffuse reflectance and transmittance we calculated QE for each of the four deposited a-Si:H solar cells. We assumed that light is scattered at all rough interfaces in the cell including the p/i and i/n interfaces except at the glass/TCO interface. We obtained a good agreement between the simulated QE and the QE measured at -1 V for all four cells as shown in Fig. 3 (curve 4). There is no interference pattern in the simulated QE curves since the *GENPRO2* program is based on incoherent propagation of light. The advantage of modeling is that it enables us to investigate solar cell structures that are difficult to fabricate in practice and evaluate the effect of particular model parameters separately. To illustrate the effect of the front and back contact texture on the QE we simulated the QE of the solar cell assuming that 1) all interfaces are flat, 2) only the front interfaces of the cell (TCO/p and p/i) are rough, and 3) only the back interfaces (n/metal and i/n) are rough. The results are presented in Fig. 3. From the QE curves we calculated the photogenerated current corresponding to AM1.5 standard illumination spectrum. The values of the photogenerated current, J_{ph}, are included in Fig. 3.

RESULTS AND DISCUSSION

Fig. 3 clearly demonstrates that implementation of rough interfaces in a-Si:H solar cells leads to an enhanced absorption in the intrinsic layer. In case of our solar cells a gain in photogenerated current due to use of textured substrate is around 2.4 mA cm^{-2}. This value is only slightly dependent on the thickness of the intrinsic layer.

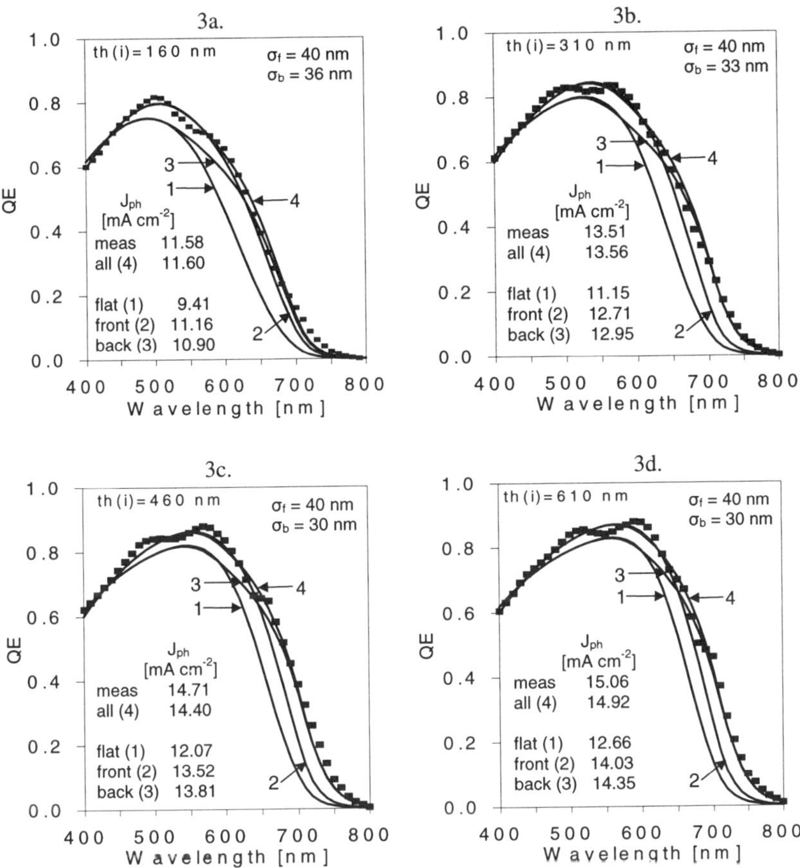

Fig. 3: The measured (points) and simulated (lines) *QE* of the four single junction a-Si:H solar cells each with a different thickness of the intrinsic layer: a) 160 nm, b) 310 nm, c) 460 nm, and d) 610 nm. The simulated *QE* curves: 1) all interfaces are flat, 2) only the front interfaces (TCO/p and p/i) are rough, 3) only the back interfaces (i/n and n/metal) are rough, and 4) all interfaces are rough. σ_f (σ_b) is the *rms* roughness of the front (back) interfaces of the cell.

The simulation results reveal the trends in optical behavior of the solar cell when scattering is applied at various interfaces in the solar cell. It is convenient to discuss the influence of scattering for two wavelength regions: the short wavelength region from 350 nm to 520 nm and the long wavelength region from 520 nm to 900 nm. In the short wavelength region the back contact roughness has no effect on the *QE*. Scattering at the front rough interfaces causes a decrease in the *QE* below 450 nm due to a strong absorption in the TCO and p-layer. Above 450 nm scattering at the front rough interfaces contributes to a large absorption enhancement. The major contribution of scattering light at the front interfaces is in the wavelength range between 450 to 650 nm. Above 650 nm scattering at the back contact begins to play a dominant role.

Scattering at the back contact of the cell increases the QE in the long wavelength region but causes also an increased absorption in the n-layer and metal. The major loss in the long wavelength region due to scattering is the absorption in the metal contact layer. Therefore, a further enhancement of light absorption in the intrinsic layer can be expected from optimizing the n-layer thickness and increasing the reflectivity of the back metal contact.

CONCLUSIONS

We investigated the effect of interface roughness on the optical properties of amorphous silicon (a-Si:H) solar cells by combining σ_r measurements and computer modeling. The relations between the computer input parameters, which describe the diffuse part of reflected and transmitted light at a rough interface, and σ_r of the interfaces in single junction a-Si:H solar cells are presented.

The σ_r of the Asahi U-type substrate surface is determined to be 40 nm. Deposition of a 9 nm thick p-type a-SiC:H layer on the substrate has not changed the surface roughness. Due to a high refractive index of n-type a-Si:H the back contact interface acts as a nearly perfect diffuser for the reflected light when σ_r is above 25 nm. The σ_r of the back interface is found to be dependent on the intrinsic layer thickness, but in all cases larger than 25 nm.

Increased scattering at the front interfaces causes a decrease in the QE below 450 nm due to a strong absorption in the TCO and p-layer. Above 450 nm scattering at the front interfaces contributes to a large absorption enhancement. Scattering at the back contact of the cell increases the QE in the long wavelength region (above 520 nm) and plays a dominant role in the improvement of the QE above 650 nm.

ACKNOWLEDGMENTS

Financial support from the Netherlands Agency for Energy and the Environment (NOVEM), the Economy, Ecology and Technology (EET) Program of the Dutch Government, the Dutch Council for Scientific Research (NWO), and the multinational company Akzo Nobel is gratefully acknowledged. The authors also like to thank Akzo Nobel Chemicals and Coatings Research Center Arnhem for their assistance in carrying out the optical characterizations.

REFERENCES

1. H. Schade and Z.E. Smith, J. Appl. Phys. **57** (2), p. 568 (1985).
2. J. Morris, R.R. Arya, J.G. O'Dowd, and S. Wiedeman, J. Appl. Phys. **67** (2), p. 1079 (1990).
3. F. Leblanc, J. Perrin, and J. Schmitt, J. Appl. Phys. **75** (2), p. 1074 (1994).
4. M. Zeman, J.A. Willemen, L.L.A. Vosteen, G. Tao, and J.W. Metselaar. Solar Energy Materials and Solar Cells **46**, p. 81 (1997).
5. G. Tao, M. Zeman, and J.W. Metselaar. in Amorphous Silicon Technology-1994, ed. by M. Hack et al., (Mater. Res. Soc. Proc. **336**, San Francisco, CA 1994), p. 705-710.
6. H. Stiebig, A. Kreisel, K. Winz, N. Schultz, C. Beneking, Th. Eickhoff and H. Wagner, Proceedings of WCPC-1, Hawaii, (1994), p. 603-606.
7. G. Tao, M. Zeman, and J.W. Metselaar. Solar Energy Materials and Solar Cells **34**, p. 359 (1994).
8. Ruud Schropp and Miro Zeman, *Amorphous and Microcrystalline Solar Cells: Modeling, Materials, and Device Technology*, 1st ed. (Kluwer Academic Publishers, 1998).
9. H.E. Bennett and J.O. Porteus. J. Opt. Soc. Am. **51** (2), p. 1234 (1961).

INVESTIGATION OF TEXTURED BACK REFLECTORS FOR MICROCRYSTALLINE SILICON BASED SOLAR CELLS

O. KLUTH, O. VETTERL, R. CARIUS, F. FINGER, S. WIEDER, B. RECH, H. WAGNER
Institut für Schicht und Ionentechnik – Photovoltaik, Forschungszentrum Jülich,
D-52425 Jülich, Germany; o.kluth@fz-juelich.de

ABSTRACT

Microcrystalline silicon (μc-Si:H) solar cells require an effective light trapping in the near infrared (NIR) to enhance the long wavelength spectral response. For this purpose we investigated back reflectors based on texture-etched ZnO/Ag stacks prepared on glass substrates by magnetron sputtering. With decreasing sputter pressure the resulting surface texture of the glass/Ag/ZnO substrates after etching exhibits a larger feature size and root mean square roughness. The increase in feature size corresponds to an increase of diffuse reflectivity. Applied in microcrystalline solar cells prepared by VHF plasma enhanced chemical vapour deposition (PECVD), the reflectors showing the largest feature size (prepared at the lowest possible sputter pressure) yielded the highest long wavelength spectral response. The μc-Si n-i-p cells prepared on the latter back reflector exhibited efficiencies of 6.9 % (short circuit current density j_{sc}= 18.8 mA/cm^2) and 7.5 % (j_{sc}=25 mA/cm^2) for an i-layer thickness of 1 μm and 3.5 μm, respectively.

INTRODUCTION

High efficiency amorphous and microcrystalline silicon based solar cells rely on an improvement of the light absorption due to an optical confinement of the long wavelength light (light trapping). In the ideal case, the solar radiation is scattered, repeatedly reflected within the multi-layered thin film solar cell and almost completely absorbed after multiple passes through the intrinsic layer which generates the photo-current. An effective light trapping yields high currents and consequently high efficiencies even if thin i-layers are used. In state-of-the-art n-i-p (substrate) technology, the front TCO film provides low absorption and reflectance losses, while the light scattering is provided by the natural surface texture of the intrinsic silicon film [1] or/and a textured back reflector [2,3] for which a combination of ZnO and Ag is often used. In the latter case, the texture of the ZnO/Ag back reflector strongly influences the device performance. The required texture for optimized scattering can be obtained by depositing Ag and ZnO at high temperatures [2,3]. Textured ZnO/Ag back reflectors can also be realized by post deposition chemical etching of sputtered ZnO films [4,5]. Chemical etching of sputtered ZnO:Al films e.g. in diluted hydrochloric acid (HCl) allows to adjust the surface morphology depending on the structural properties of the initial film and the etching time [6-8].

This paper deals with the preparation and characterization of texture-etched ZnO/Ag back reflectors for μc-Si based n-i-p solar cells. The following questions were addressed: Firstly, how does the surface morphology (feature size) of texture-etched ZnO/Ag back reflectors depend on the deposition conditions of the ZnO sputter process? Here, the deposition pressure turned out to be the key parameter. Secondly, how do the resulting different surface textures obtained after etching influence the diffuse reflectivity of the ZnO/Ag back reflectors? The final goal is to find correlations between the structural and optical properties of differently textured ZnO/Ag back reflectors on the one hand, and I-V parameters and spectral response data of microcrystalline n-i-p solar cells prepared on these reflectors on the other hand.

EXPERIMENTAL

All ZnO and Ag films for the back reflectors and the 80 nm thick ZnO:Al antireflectance (AR) solar cell top contacts were deposited in a Lesker high vacuum sputtering system (base

Mat. Res. Soc. Symp. Proc. Vol. 557 © 1999 Materials Research Society

pressure $1 \cdot 10^{-7}$ Torr) with 6" circular cathodes. The 700 nm thick ZnO:Al layers were reactively rf-magnetron sputtered in O_2/Ar-gas mixtures from a ZnO:Al_2O_3 (98/2wt%)-target on glass substrates. The substrate temperature during ZnO sputtering was 270°C. The Ag films were dc-sputtered without intentional heating and showed no light scattering effect. To obtain the surface texture all presented ZnO films were etched in diluted HCl for 15s (0.5% HCl in H_2O). Optical reflectance, both specular and total, was recorded with a double beam spectrometer. The diffuse reflectance was calculated as the difference of total and specular reflectance. Photothermal deflection spectroscopy (PDS) was performed to identify absorption losses. The surface morphology was investigated by an atomic force microscope (AFM) and a scanning electron microscope (SEM). The microcrystalline solar cells were prepared in a 6 chamber PECVD system using a VHF plasma excitation frequency of 95MHz (solar cell recipes can be found in [9]). I-V curves were recorded under standard conditions (25°C, 100mW/cm^2, AM1.5). Quantum efficiency measurements (QE) were performed.

RESULTS

Back Reflector Characterization

Previous investigations on rf [6] and dc [8] magnetron sputtered ZnO:Al have shown that the sputter pressure is a crucial parameter for the structural properties of the ZnO films grown on glass and for the resulting surface morphology after the etching step. To investigate the influence of the substrate on the ZnO:Al film growth, films were deposited on glass/Ag and glass substrates. Fig. 1 (right column) shows SEM micrographs of selected ZnO:Al films co-deposited with 4% O_2 in Ar at different deposition pressures on glass/Ag substrate (left side) and glass substrate (right side) after 15 s etching in diluted HCl. AFM profiles of the texture-etched glass/Ag/ZnO back reflectors were recorded and are also plotted in Fig. 1 (left column). The SEM micrographs clearly show that the growth of these ZnO:Al films strongly depends on the substrate. Especially at lower sputter pressures the ZnO films prepared on Ag substrates exhibit a remarkably rougher surface as compared to the ones deposited on glass. In the following we focus on the texture-etched glass/Ag/ZnO stacks. The structural properties determined from SEM and AFM measurements are summarized in Table I. In the pressure range between 0.4 and 10mTorr a surface structure consisting of craters with different sizes and characteristic opening angles of 120° develops upon etching. Increasing the sputter pressure in this range leads to a decrease in the root mean square roughness δ_{rms} and the feature size. The surface appearance changes in the pressure region between 10 mTorr and 15 mTorr. At 15 mTorr a regular structure characterized by hills with a small feature size and a remarkably reduced δ_{rms} is obtained.

The structural differences of the texture-etched ZnO/Ag stacks shown in the previous section are also reflected in different light scattering properties. In Fig. 2 the diffuse spectral reflectance is plotted for the three ZnO/Ag films shown in Fig. 1. The diffuse reflectance increases with increasing feature size and surface roughness (decreasing sputter pressure

Table I: Surface appearance and root mean square roughness δ_{rms} (determined by AFM) for ZnO:Al films sputtered with 4% O_2 in Ar on Ag coated glass at different sputter pressures.

ZnO type	Appearance	δ_{rms} (nm)
0.4 mTorr	Crater	135
5mTorr	Crater	137
10mTorr	Crater/Hills	102
15mTorr	Hills	60
30mTorr	Hills	64

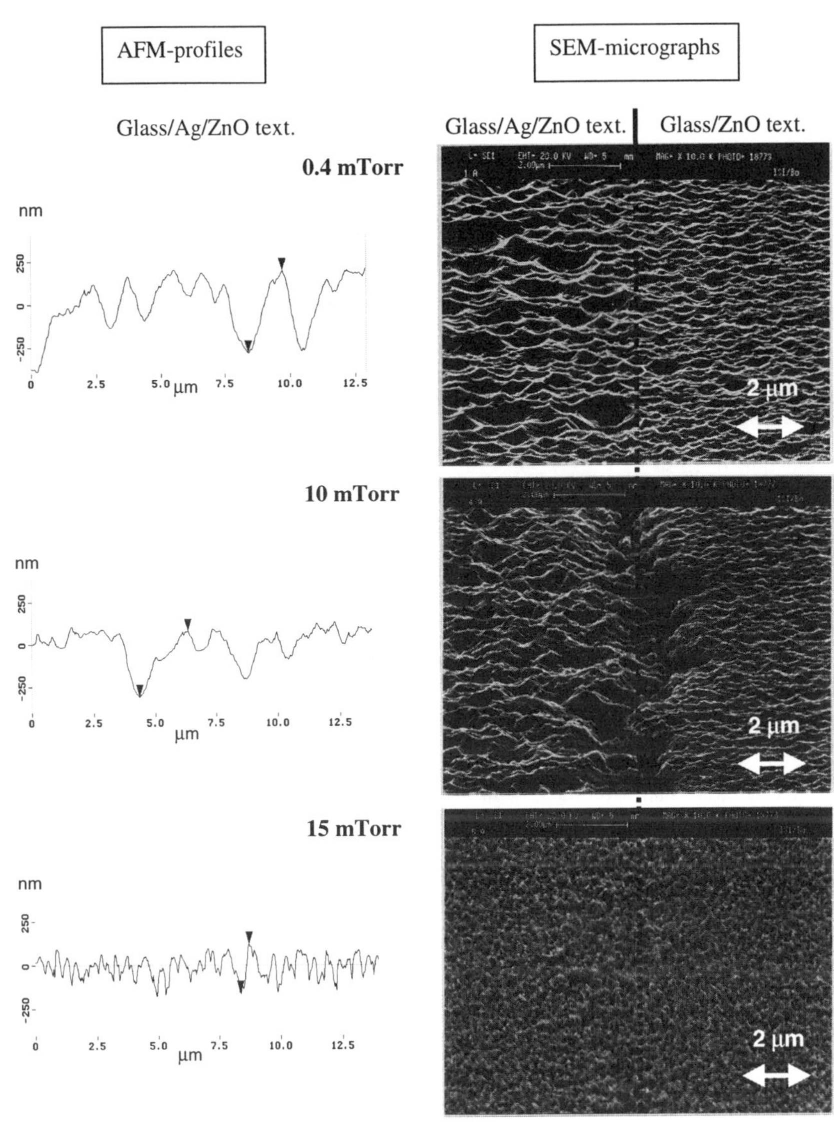

Fig. 1: SEM micrographs of ZnO:Al films co-deposited with 4% O_2 in Ar at different deposition pressures on a glass/Ag substrate (left side) and a glass substrate (right side), after 15 s etching in diluted HCl. AFM profile scans of the ZnO/Ag back reflectors are shown in the left column.

during the ZnO:Al deposition). The texture-etched glass/Ag/ZnO reflector (ZnO deposited at 15 mTorr, regular structure and small feature size) shows light scattering only for short wavelengths. In contrast, the texture-etched ZnO film with the largest feature size and roughness

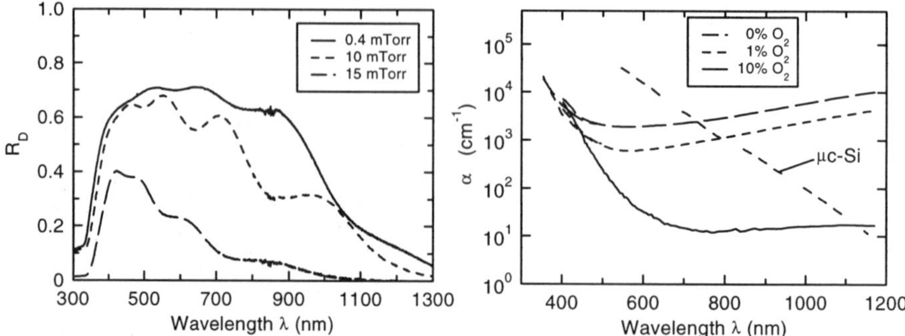

Fig. 2: Diffuse reflectance R_D of three representative ZnO/Ag substrates (4% O_2 in Ar) from the sputter pressure series after an etching time of 15 s.

Fig. 3: Absorption coefficient α as function of the wavelength for µc-Si and three ZnO:Al films reactively sputtered on glass with different dilutions of O_2 in Ar.

(deposited at the lowest sputter pressure) exhibits an uniformly high diffuse reflectance of more than 60 % for 400 nm $\leq \lambda \leq$ 900 nm.

In addition to the optimization of the light scattering properties absorption in the back reflector has to be reduced to values much lower than the intrinsic absorber layer in solar cells. To study the absorption in the TCO layer itself 700nm thick ZnO:Al layers were deposited on glass substrates with different O_2 contents in Ar. The absorption losses were detected by PDS since this technique is sensitive for determining even low absorption coefficients. Fig. 3 shows the spectral absorption coefficient of these ZnO:Al films. Since the absorption coefficient $\alpha_{\mu c-Si}$ of microcrystalline silicon becomes relatively low for wavelengths above 600 nm (see Fig. 3), the absorption coefficient of the ZnO:Al films prepared without or with low O_2 content is similar or higher than $\alpha_{\mu c-Si}$ for $\lambda \geq$ 800 nm. By increasing the O_2 to Ar gas flow ratio during the sputter process the absorption coefficient of the ZnO films has been reduced by more than two orders of magnitude in the important wavelength region between 600 and 1100 nm. The structural properties of the ZnO:Al films were almost not affected by the oxygen.

Solar Cells

Microcrystalline silicon n-i-p solar cells with 1 µm i-layer thickness were prepared on all texture-etched ZnO/Ag back reflectors (pressure series). The corresponding light I-V parameters and i-layer thicknesses are listed in Table II. The Cr coated SnO_2/glass substrate serves as reference back reflector with very poor reflectivity. In addition, the results of a 3.5 µm thick n-i-p cell on a texture-etched ZnO/Ag coated glass substrate (0.4 mTorr ZnO sputter pressure) are included. Within the pressure series, a systematic increase in short-circuit current densities from 12.3 to 18.8 mA/cm^2 could be observed upon decreasing the deposition pressure during sputtering from 30 mTorr to 0.4 mTorr. The corresponding variations in open-circuit voltage and fill factor are not systematic and are probably related to some irreproducibilities of the sputter and/or PECVD process. The highest efficiency of 7.5 % (j_{sc} = 25 mA) for a 3.5 µm thick n-i-p solar cell was obtained on a 0.4 mTorr reflector (1% O_2 content in O_2/Ar sputter gas). Spectral response measurements (1 µm thick cells) reveal that the increase in j_{sc} with decreasing deposition pressure is fully related to an increase of the long wavelength quantum efficiency (λ > 500 nm). Fig. 4 shows the quantum efficiency obtained on three representative texture-etched ZnO/Ag reflectors and on the Cr/SnO$_2$ reference reflector.

Table II: Light I-V parameters and i-layer thicknesses d of μc-Si solar cells prepared on different glass/Ag/ZnO substrates (etching time 15 s, 4% O_2 in Ar) and on an SnO_2/Cr substrate (not etched).

ZnO type	d (μm)	j_{sc} (mA/cm^2)	FF (%)	V_{oc} (V)	η (%)
0.4 mTorr	1	18.8	70.8	518	6.9
5 mTorr	1	17.1	61.4	403	4.4
10 mTorr	1	17	65.4	436	4.9
15 mTorr	1	15.3	67.3	452	4.8
30 mTorr	1	12.3	70.9	471	4.8
SnO_2/Cr	1	11.1	66.3	446	3.3
0.4mTorr	3.5	25	66	454	7.5

The improved long wavelength QE achieved for the almost negligible diffuse reflecting 15 mTorr reflector as compared to the Cr reflector can essentially be explained by the high reflectivity of ZnO/Ag enhancing the optical pathlength by a factor of two as compared to the highly absorbing Cr. The further increase in the long wavelength quantum efficiency with decreasing deposition pressure can be attributed to a higher amount of light trapped within the solar cell. This is also confirmed by measurements of the total reflectivity performed on the latter cells (see Fig. 5). For the ZnO/Ag reflectors, the increase in quantum efficiency corresponds to a decrease of solar cell reflectance. The best cell (0.4 mTorr) shows a quantum efficiency of 0.2 and a reflectance of 0.4 at 900 nm. Although 40 % of the photons at 900 nm are lost due to reflectance still another 40 % are absorbed within the device and do not contribute to the photo-current. This experimental result is rather difficult to explain since absorption losses in the top ZnO films are largely minimized (top contact thickness 80 nm). Absorption losses in the thin doped μc-Si n- and p-layers can only partly explain this finding. Possible other reasons are the remaining absorption losses in the ZnO layer and especially at the ZnO/Ag interface, which is indicated by additional PDS-measurements (not shown here). Further experiments are necessary to identify the reasons for this gap between quantum efficiency and reflectance measurements.

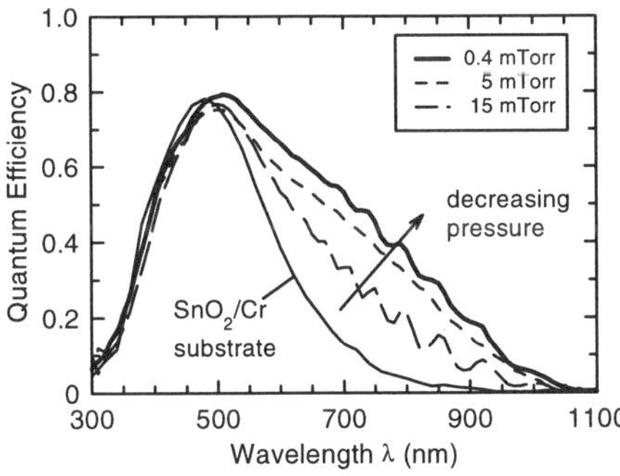

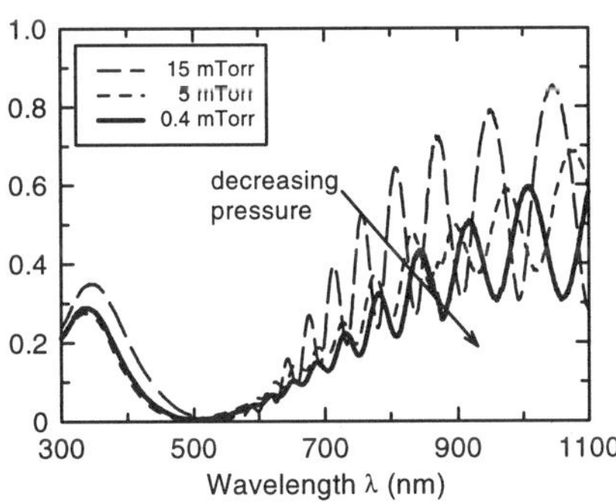

Fig. 4: Quantum efficiency of 1 μm thick nip μc-Si cells deposited on three characteristic ZnO/Ag substrates prepared at different sputter pressures and on a SnO_2/Cr reference substrate without light trapping properties.

Fig. 5: Total reflectance R_T of the solar cells of Fig. 4. The 80 nm thick ZnO AR top contact leads to a minimum total reflectance at λ ≈ 500 nm.

CONCLUSIONS

The method of texture-etching sputtered ZnO films was applied to develop optimized ZnO/Ag back reflectors for µc-Si based n-i-p solar cells. The surface morphology of the ZnO films obtained after etching strongly depends on the deposition pressure during ZnO sputtering. Moreover, different ZnO surface morphologies were obtained after etching of ZnO films prepared on glass or on Ag coated glass indicating a strong substrate dependence of the ZnO film growth. In general, with decreasing sputter pressure the resulting surface texture of the glass/Ag/ZnO substrates after etching exhibits a larger feature size and root mean square roughness. The increase in feature size corresponds to an increase of diffuse reflectivity. By introducing significant amounts of oxygen during the ZnO sputter process absorption losses in the ZnO films applied for the back reflector could be largely minimized as confirmed by PDS-measurements. Applied in microcrystalline solar cells, the reflectors exhibiting the largest feature size (prepared at the lowest possible sputter pressure) yielded the highest long wavelength spectral response. The µc-Si n-i-p cells prepared on the latter back reflector exhibited efficiencies of 6.9 % (j_{sc}= 18.8 mA/cm^2) and 7.5 % (j_{sc}=25 mA/cm^2) for an i-layer thickness of 1 µm and 3.5 µm, respectively.

ACKNOWLEDGEMENTS

This work was partly funded by the E.U. under contract JOR3-CT97-0145 and by the BMBF (German Federal Ministry of Education and Research). The authors thank J. Klomfaß, A. Lambertz, J. Wolff, W. Reetz, and H. Siekmann for experimental assistance and H.P. Bochem for performing SEM measurements.

REFERENCES

1. K. Yamamoto, A. Nakajima, T. Suzuki, M. Yoshimi, Jpn. J. Appl. Phys. **36** (1997) L569
2. A. Banerjee and S. Guha, J. Appl. Phys. **69** 1030 (1991)
3. X. Deng and K. L. Narasimhann, *1*st *IEEE World Conference on Photovoltaic Solar Energy Conversion*, Waikoloa, Hawaii, 555 (1994)
4. B. Rech, S. Wieder, C. Beneking, A. Löffl, O. Kluth, W. Reetz, H. Wagner: *Conference Record of the 26th IEEE Photovoltaic Specialists Conference*, Anaheim, USA, 619 (1997)
5. N. Wyrsch, P. Torres, M. Goetz, S. Dubail, L. Feitknecht, J. Cuperus, A. Shah, B. Rech, O. Kluth, S. Wieder, O. Vetterl, H. Stiebig, C. Beneking, H. Wagner, *2nd World Conference on Photovoltaic Solar Energy Conversion*, Vienna, Austria, 467 (1998)
6. O. Kluth, A. Löffl, S. Wieder, C. Beneking, W. Appenzeller, L. Houben, B. Rech, H. Wagner, S. Hoffmann, R. Waser, J.A. Anna Selvan, H. Keppner, *Conference Record of the 26th IEEE Photovoltaic Specialists Conference*, Anaheim, USA, 715 (1997)
7. A. Löffl, S. Wieder, B. Rech, O. Kluth, C. Beneking, H. Wagner, *14th European Photovoltaic Solar Energy Conference*, Barcelona, Spain, 2089 (1997)
8. O. Kluth, B. Rech, L. Houben, S. Wieder, G. Schöpe, C. Beneking, H. Wagner, A. Löffl, H. W. Schock, Thin Solid Films, in print
9. O. Vetterl, P. Hapke, O. Kluth, A. Lambertz, S. Wieder, B. Rech, F. Finger, H. Wagner in *Polycrystalline Semiconductors V – Bulk Materials, Thin Films and Devices*, edited by J. H. Werner, H. P. Strunk, H. W. Schock (Solid State Phenomena, Scitech Publ., Uettikon am See, Switzerland, 1999), in print

A NEW METHOD TO CHARACTERIZE TCO/P CONTACT RESISTANCE IN a-SI SOLAR CELLS

Steven S. Hegedus and Michael Gibson, Institute of Energy Conversion, University of Delaware, Newark, DE 19716 USA   ssh@udel.edu
Gautam Ganguly and Rajeewa Arya, Solarex,Toano, VA 23168 USA

ABSTRACT

A method is presented to characterize the TCO/p contact and the TCO sheet resistance in a-Si based p-i-n superstrate devices. It requires having scribed TCO strips, which are electrically isolated before a-Si deposition, then fabricating rows of individual devices on each strip. Analysis of 4-terminal measurements in different V-sensing configurations yields the TCO/p contact and TCO sheet resistance in a straightforward manner.   The method is applied to devices fabricated on 3 brands of commercial SnO_2 substrates. The TCO/p contact resistance is found to be ~2 Ω-cm^2 and the sheet resistance decreases by 2-4 Ω/sq after a-Si deposition on all 3 brands of SnO_2 substrates.

INTRODUCTION

Transparent conductive oxides (TCO) are critical to the optical and electrical performance of a-Si based solar cells. This is especially true for superstrate a-Si p-i-n solar cells which are deposited on a glass/TCO substrate. Optically, the TCO must have high transparency and provide scattering to enhance light trapping. Electrically, the TCO must have low lateral sheet resistance since there are no grids and low contact resistance with the p-layer. The TCO is subjected to all processing steps including H-rich plasmas during the p-layer deposition which can degrade its optical transparency (1,2). The TCO must be robust, inert to subsequent chemical and thermal device processing (3) and also must be able to be laser scribed with high yield (4).   Presently, textured fluorine doped SnO_2 is the standard TCO for commercial applications (5).

Minimizing the resistance between the p-layer and TCO of superstrate p-i-n a-Si devices and modules is a critical issue for utilizing new TCO materials like ZnO and new p-layers like μc-SiC or μc-SiO. Compared to SnO_2, ZnO typically gives higher short circuit density (J_{sc}) due to lower absorption losses but poorer electrical performance (4,6) commonly attributed to the ZnO/p interface, forming a non-ohmic contact. Various schemes to improve the ZnO/p electrical contact have been discussed (7). However, characterization of the TCO/p interface is difficult since it is in series with the dominant p-i-n junction.

We have developed a method to characterize the TCO/p contact based on the work of Shafarman and Phillips (8). This method requires having a two adjacent TCO regions. A device is biased to have standard current flow through its TCO/p contact and TCO region while the voltage is measured on the adjacent TCO pad which is electrically "floating." The second pad is thus a voltage sensing contact giving internal access to the potential in the biased device. An additional benefit to this technique is that the SnO_2 sheet resistance in a completed device structure is also obtained.

SAMPLE FABRICATION

The substrates studied here were textured SnO_2-coated glass made by Asahi Glass (Type U), AFG, or Solarex. The devices were fabricated at Solarex. The SnO_2 on the 3x3 inch square pieces was laser scribed to create long parallel strips approximately 8-10 mm wide and 76 mm (3") long. Then, single junction a-Si p-i-n layers were deposited by PECVD. The p-layers were a-SiC:H. A row of six individual devices were fabricated on each SnO_2 strip by depositing a ZnO/Al back contact (5x5 mm^2) mm through a mask on the n-layer. A low resistance contact to the SnO_2 strip was made with Ag paste which was baked at 200°C.

Mat. Res. Soc. Symp. Proc. Vol. 557 © **1999 Materials Research Society**

ANALYSIS

A top view and side view of the resulting structure is shown in Figure 1. The two SnO_2 strips of width W, with Ag-paste contacts labeled A and B, each have a row of square devices. The current in cells 1, 2, 3, etc. travels in the SnO_2 a distance L from each device to the SnO_2 contact A. The series resistance of the SnO_2 between the device and the contact is

$$R_{TCO} = R_{SH} \times (L/W) \tag{1}$$

where R_{SH} is the sheet resistance of the SnO_2. As L increases, the series resistance of the TCO between the device and its SnO_2 contact increases. Assuming each device is identical, and its TCO/p contact is identical, this increase in R_{TCO} will be the only difference between devices in the same row.

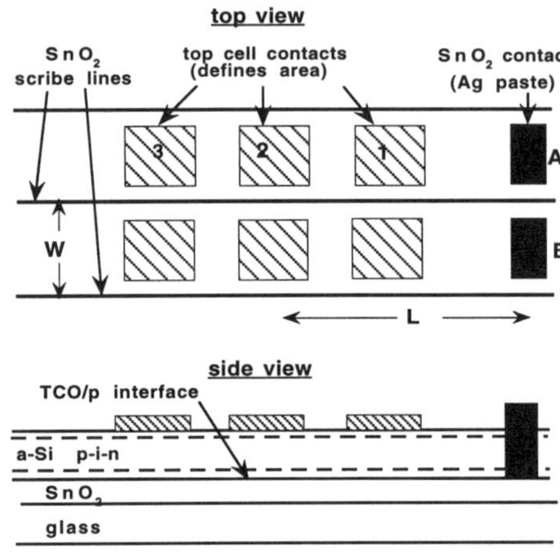

Figure 1. Top view and side view of row of a-Si devices (1-2-3) on scribed SnO_2 strips with contacts labeled A and B.

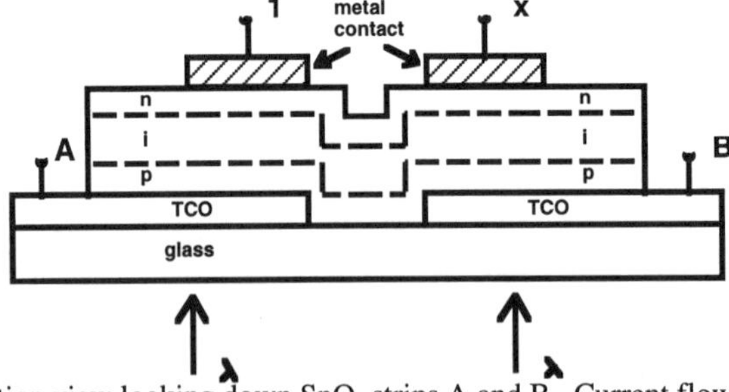

Figure 2. Cross-section view looking down SnO_2 strips A and B. Current flow is established in p-i-n cell 1 between 1 and A while voltage is measured between 1 and A, 1 and B, or A and B. SnO_2 strip B is floating at same potential as the p-layer over strip A.

The three different measurement configurations will be discussed in reference to Figure 2 which shows a cross-sectional view looking down the two SnO_2 strips having contacts A and B. For example, assume that cell 1 in the row of devices labeled 1-2-3 in Figure 1 is to be characterized. For all three measurements, it is biased such that the *current flow* is always between the cell's back contact (1) and the strip's TCO contact (A) as in a standard device JV test. The difference between the 3 measurement configurations is where the *voltage* is measured. In a standard JV measurement, the voltage V_{1A} is measured between the same two contacts where the current is flowing. This voltage V_{1A} includes potential across the p-i-n junction, the TCO/p contact and voltage dropped across the SnO_2 series resistance. When the voltage is instead measured across 1 and B or across A and B, there is no current flow through B so it is floating at the same potential as the p-layer contacting it, i.e. the p-layer over strip A. The SnO_2 strip B is used as a voltage probe giving access to the internal voltage of devices on strip A. Thus, the voltage measured across device 1 and B (V_{1B}) is only the potential *across* the p-i-n junction of a-Si device 1. It excludes the voltage dropped across TCO/p contact and TCO series resistance. Measuring the voltage across adjacent TCO strips (V_{AB}) gives *only* the voltage dropped across TCO/p contact and TCO series resistance since strip B is floating at the same potential as the p-layer above strip A.

We chose to analyze the different JV data in terms of resistances. The three components of resistance and their relation to the three different voltage measurements are given in Equations 2-4. The three dominant resistances are the junction dynamic resistance R_J, the TCO/p contact resistance R_{TCO}/p, and the series resistance through the TCO R_{TCO}, as defined in equation 1. The subscripts 1A, 1B, and AB represent voltage measured between nodes 1 and A, 1 and B or A and B as defined in Figure 2. We assume that the other contact resistances, such as between the cell back contact and the n-layer and between the SnO_2 and Ag contact, are negligible.

$$R_{1A} = dV_{1A}/dJ = R_J + R_{TCO/p} + R_{TCO} \tag{2}$$

$$R_{1B} = dV_{1B}/dJ = R_J \tag{3}$$

$$R_{AB} = dV_{AB}/dJ = R_{TCO/p} + R_{TCO} \tag{4}$$

$$R_J = dV/dJ = (AkT/q)/(J + J_{sc}) \tag{5}$$

A critical assumption is that TCO strip B is floating at the same potential as the p-layer over strip A. If this is correct, equations 2-4 predict that the difference in resistance between cases 1A and 1B should be the same as that measured in case AB. This resistance, R_{AB}, should be the sum of the TCO/p contact and the TCO sheet resistance. From equation 1, R_{AB} should be linear against L/W with a slope of R_{TCO} and an intercept of $R_{TCO/p}$. Thus, measuring the three resistances on a series of devices on the same strip with increasing L/W should allow determination of $R_{TCO/p}$ and R_{TCO}. Hence, equations 2-4 show two *independent* ways to obtain R_{AB}. This allows verification of the assumptions and the measurements.

RESULTS

Figure 3 shows JV curves for all three cases of voltage measurement on one device (#8090-4C-1) which was on Asahi SnO_2. Normal solar cell JV curves result when the V_{1A} or V_{1B} is measured while V_{AB} gives a linear JV curve through the origin. This linear JV relation indicates the TCO/p contact is ohmic according to equation 4. For purposes of analysis, we determine values of the resistances defined in equations 2-4 at V_{OC}. Thus, for the case where V is measured across 1 and B, R_{OC} is dV_{1B}/dJ at V_{OC}. Values of V_{oc}, FF and R_{OC} are listed in Table I. Note that the V_{OC} is the same between configurations 1A and 1B, as expected since the p-i-n junction determines V_{OC}, but the FF increases in configuration 1B since it excludes the TCO series resistance. V_{OC} is zero in configuration V_{AB} as expected from equation 4. (Some devices we have measured on other TCO materials have exhibited small values of V_{oc}~20-40 mV in configuration V_{AB}, indicating a photovoltage was developed at the TCO/p contact. These devices had inferior V_{oc} and FF when measured in the standard configuration.)

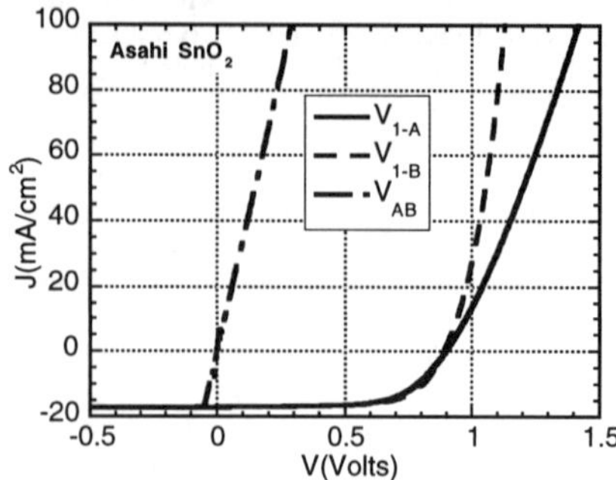

Figure 3. JV curves under illumination for a device (cell 1) with three voltage measurement configurations. Parameters given in Table I.

Table I. FF, resistance at V_{oc}, and V_{oc} for the three curves in Figure 3 obtained from three-voltage measurement configurations on the same device.

test voltage configuration	FF (%)	R_{OC} (Ω-cm^2)	V_{OC} (V)
cell 1 and A	64	8.8	0.89
cell 1 and B	68	6.1	0.89
A and B	0	2.9	0

Figure 4 shows linear JV behavior for voltages measured across AB for three devices on the same strip of TCO having different L/W. Table II shows values of resistance calculated from both methods. The two resistances increase with L/W as expected, and the FF measured in standard configuration (V_{CELL-A}) decreases as L/W increases as expected. Note the close agreement in Table II between resistances determined independently.

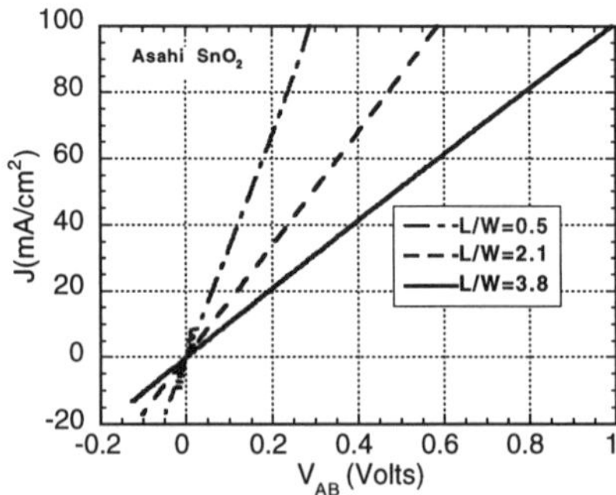

Figure 4. JV measurements for three devices with varying L/W on the same TCO strip A. Voltage was measured across the adjacent strips A and B.

Table II. Resistance and FF for 3 devices with increasing L/W. $R(V_{AB})$ determined from slopes of lines in Figure 4. ΔR_{OC} determined from difference of $R(V_{CELL-A})$ and $R(V_{CELL-B})$ at V_{oc}.

L/W	$R(V_{AB})$ (Ω-cm^2)	ΔR_{OC} (Ω-cm^2)	$FF(V_{CELL-A})$ (%)
0.5	2.9	2.7	64
2.1	5.8	6.0	60
3.5	9.5	10	56

Figure 5 shows the two resistances plotted against L/W for devices on the same strip of Asahi SnO$_2$. ΔR_{OC} is the difference in R_{OC} for JV measured as V_{CELL-A} and V_{CELL-B} while $R(V_{AB})$ is the resistance obtained directly from the slope of V_{AB} as in Figure 4. Clearly, they are the same, verifying equations 2-4 that the difference in resistance between $R(V_{CELL-A})$ and $R(V_{CELL-B})$ is the same as $R(V_{AB})$, i.e., the TCO sheet resistance and TCO/p contact resistances. From the slope, R_{SH} is ~8 Ω/sq and from the intercept, $R_{TCO/p}$ is 2 Ω-cm^2.

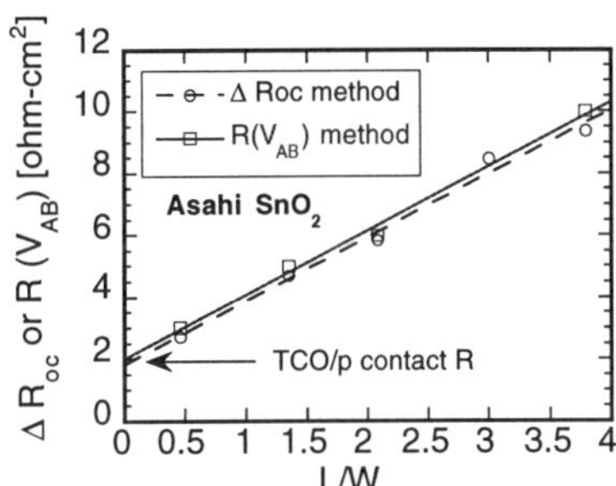

Figure 5. Resistance calculated from two methods vs L/W for row of cells on Asahi SnO$_2$. The slope is the SnO$_2$ sheet resistance and the intercept is the TCO/p contact resistance.

Table III. TCO sheet resistance before and after a-Si deposition and TCO/p contact resistance for three types of SnO$_2$. R_{SH} before deposition obtained from 4 point probe measurements.

TCO	R_{SH} (Ω/sq) before a-Si	R_{SH} (Ω/sq) after a-Si	$R_{TCO/p}$ (Ω-cm^2)
Solarex	~12	9	2.5
Asahi	~12	8	2.0
AFG	~13	9	2.5

Table III shows results from the 3 brands of SnO$_2$ studied. It indicates the SnO$_2$ sheet resistance is lower after a-Si deposition, with the largest change being for the Asahi SnO$_2$. This decrease in SnO$_2$ sheet resistance is consistent with results from devices deposited at IEC on unscribed 1 inch squares of Asahi TCO. A tab-to-tab SnO$_2$ sheet resistance of 8-9 Ω/sq is typically found after fabrication of a device while the sheet resistance is 13-14 Ω/sq before deposition. The Asahi group has reported a decrease in SnO$_2$ resistivity of the same amount (from 12 to 8 Ω/sq) with mild H_2 plasma treatments at 200°C due to increased mobility without any loss in optical transmission (9). Our results from Table III are consistent with these values.

Table III indicates a TCO/p contact resistance of approximately 2 Ω-cm^2 for all three SnO_2 materials. This value also includes any fixed contact or bulk parasitic resistances not accounted for in equations 2-4. However, values of ~1-2 Ω-cm^2 are reported by others (7,10) using special test structures (not p-i-n devices) which confirms that the values in Table III are representative of the TCO/p contact.

CONCLUSIONS

A new procedure has been presented to determine the TCO sheet resistance and TCO/p contact resistance in operating TCO/p-i-n superstrate devices. Devices must be fabricated on scribed TCO regions. Two independent methods give very close agreement, verifying the assumptions and analysis. Three brands of SnO_2 were studied here. Very similar TCO/p contact resistances were found. The sheet resistance of the Asahi brand SnO_2 decreased ~50% with a-Si processing. This is consistent with reports of others. We conclude that this technique can be helpful in evaluating factors which affect the TCO/p contact resistance, such use of new TCO materials or p-layer processing. In particular, it should be useful in solving the ZnO/p contact problem and evaluating the effect of microcrystalline p-layers.

ACKNOWLEDGMENTS

This work was supported under NREL subcontracts #ZAK-8-17619-33 and #ZAK-8-17619-02. We appreciate technical assistance from Ron Dozier in setting up the JV measurements.

REFERENCES

1. H. Shade, Z Smith, J. Thomas, A. Catalano, Thin Solid Films **117**, 1984., p. 149.
2. S. Hegedus, H. Liang, R. Gordon, AIP Proc. **353**, 1995, p. 465.
3. R. Gordon, Mat. Res. Soc. Symp. Proc. **426**, 1996, p. 419.
4. L. Yang, M. Bennett, L. Chen, K. Jansen, J. Kessler, Y. Li, J. Newton, K. Rajan, F. Willing, R. Arya, D. Carlson, Mat. Res. Soc. Symp. Proc. **426**, 1996, p. 3.
5. P. Gerhardinger, R. McCurdy, Mat. Res. Soc. Symp. Proc. **426**, 1996, p. 399
6. S. Hegedus, W. Buchanan, X. Liu, R. Gordon, Proc. 25th IEEE PVSC, 1996, p. 1129.
7. M. Kubon, N. Shultz, M. Kolter, C. Beneking, H. Wagner, Proc. 12th Euro. PVSEC, 1994, p. 1268.
8. W. Shafarman and J. Phillips, Proc. 25th IEEE PVSC, 1996, p. 917.
9. K. Sato, Y. Gotoh, Y. Hayashi, H. Nishimura, Proc. 21st IEEE PVSC, 1990, p. 1584.
10. S. Guse, D. Peros, M. Wagner, M. Bohm, Mat. Res. Soc. Symp. Proc. **426**, 1996, p. 437.

OPTIMIZATION OF HIGH EFFICIENCY AMORPHOUS SILICON ALLOY BASED TRIPLE-JUNCTION MODULES

A. BANERJEE, J. YANG, and S. GUHA
United Solar Systems Corp., 1100 W. Maple Road, Troy, Michigan 48084, U.S.A.

ABSTRACT

A systematic approach has been used to scale up high efficiency $0.25cm^2$ active-area amorphous Si alloy based triple-junction devices to high-efficiency encapsulated modules of aperture area $\sim920cm^2$. In order to analyze the losses involved in the scale-up, intermediate aperture area, $40cm^2$ and $450cm^2$, modules have also been fabricated. The best stable active-area efficiency obtained on the small-area cells is 12.9%. The best initial efficiency of a $\sim920cm^2$ aperture area encapsulated module is 12.1%. National Renewable Energy Laboratory (NREL) has independently light soaked three of the $\sim920cm^2$ modules. They have measured a stable efficiency of 10.5% which represents a new world record. This paper presents various aspects of the large-area module work.

INTRODUCTION

We have previously reported [1,2] on the achievement of a 10.2% stable module efficiency as confirmed by NREL using a-Si/a-SiGe/a-SiGe alloy based triple-bandgap triple-junction structure. Since then we have improved the small-area device to obtain a world record stable efficiency of 13.0% [3,4]. We have developed a new optimization scheme which scales up the $0.268cm^2$ cell results into large-area $\sim920cm^2$ aperture-area modules using identical materials and processes. These materials and processes are the same as those used in the manufacture of our commercial product. Intermediate aperture-area, $\sim40cm^2$ and $\sim450cm^2$, modules have also been fabricated for diagnostic purpose. This approach enables the analysis of the devices at both the cell and module levels. Evaluation of the small-area cells using current-voltage (I-V) and quantum efficiency (Q) measurements provide information about the device without complications of electrical and optical losses associated with modules. The analysis of the performance and uniformity of the intermediate-area modules provides a preview of the module efficiency that can be expected from a $\sim920cm^2$ module taking into consideration all the losses related to modules.

The optimization scheme has culminated in the achievement of an aperture-area efficiency of 12.1% on a $\sim920cm^2$ module. NREL has light-soaked three of these modules and measured a stable efficiency of 10.5% which corresponds to a new world record. Since these modules have been fabricated using production technology, it provides the most realistic practical goal for a-Si alloy commercial product.

EXPERIMENT

A large-area stainless steel substrate is first sputter-coated with a textured Ag/ZnO back reflector layer. The deposition is over an area of ~1 sq. ft. This is followed by the deposition over the same area of a triple-junction triple-bandgap a-Si/a-SiGe/a-SiGe alloy cell using conventional glow discharge technique. The top transparent conducting oxide (TCO) is deposited in two different configurations for: (1) small-area devices and (2) modules.

In the first case, the completed large-area device is cut into 2"x2" substrates. The TCO is deposited through an evaporation mask to yield devices of total area $0.268cm^2$. These devices are used for I-V and Q measurements. Stable cell results are obtained by exposing several of the representative 2"x2" samples from the original $\sim920cm^2$ substrate to one-sun illumination at 50^0C. In the second case, the TCO film is deposited over the entire area. The TCO film is then etched in two different ways: (1) to fabricate a module of aperture area $\sim450cm^2$ and (2) to demarcate an array of nine cells of aperture area $\sim40cm^2$ spread over an area of 9"x9". One of the sections also has a TCO-etched 5x5 array of $0.268cm^2$ devices. The $0.268cm^2$ devices are used for the small-area I-V and Q measurements.

Front surface grids are applied using a wire bonding process and bus bars are connected. The modules are then encapsulated using a stack of Tefzel and EVA. The fabrication steps for the small-area cells and all the different size modules are identical. Therefore, the error involved in extrapolating the results of the small-area devices to the modules of aperture area $\sim920cm^2$ is minimized. I-V measurements of the modules are made using a Spire pulsed solar simulator Model 240A equipped with a peak detector circuit board.

RESULTS

Small-area Cells of Total Area $0.268cm^2$

Small-area cell performance was measured on four 2"x2" substrates, representing the large-area substrate. The initial and the stabilized results after 1067 hours of light soak are shown in Table I. Each value in the table represents the average of several $0.268cm^2$ devices on that substrate. The overall average values of initial efficiency of all the substrates are active-area efficiency = 14.3% and total-area efficiency = 13.3%. The values for the 1067 hours light stabilized case are 12.5% and 11.6%, respectively, and is representative of the large-area 12"x12" substrate. The stable J-V characteristics of the best device and the Q measurements after the light soak are shown in Figures 1 and 2, respectively. The stabilized results are V_{oc} = 2.213V, J_{sc} (from Q) = $8.79mA/cm^2$, FF = 0.661, and active area efficiency $\sim12.9\%$. The current of the top, middle, bottom cells, and the complete device are 8.87, 8.79, 8.82, and $26.48mA/cm^2$. The stable efficiency of 12.9% is close to the world-record value of 13.0% reported earlier [3,4].

Table I. Average results of $0.268cm^2$ small-area triple-junction cells on 2"x2" substrate.

Sample 2B5616	Active area (cm^2)	Total area (cm^2)	Light Soak (Hours)	V_{oc} (V)	J_{sc} (Q) (mA/cm^2)	FF	Efficiency (Q) %	
							Active area	Total area
LE	0.25	0.268	0	2.281	8.84	0.708	14.3	13.3
			1067	2.209	8.64	0.66	12.6	11.7
LF4	0.25	0.268	0	2.279	8.77	0.717	14.3	13.4
			1067	2.208	8.49	0.664	12.5	11.6
LD4	0.25	0.268	0	2.28	8.81	0.713	14.3	13.4
			1067	2.207	8.43	0.666	12.4	11.6
LC1	0.25	0.268	0	2.25	8.77	0.718	14.2	13.2
			1067	2.178	8.49	0.674	12.5	11.6
Average	0.25	0.268	0	2.273	8.8	0.714	14.3	13.3
			1067	2.201	8.51	0.666	12.5	11.6

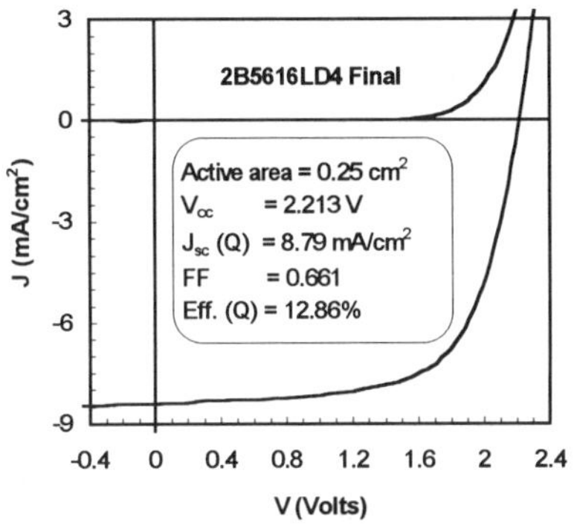

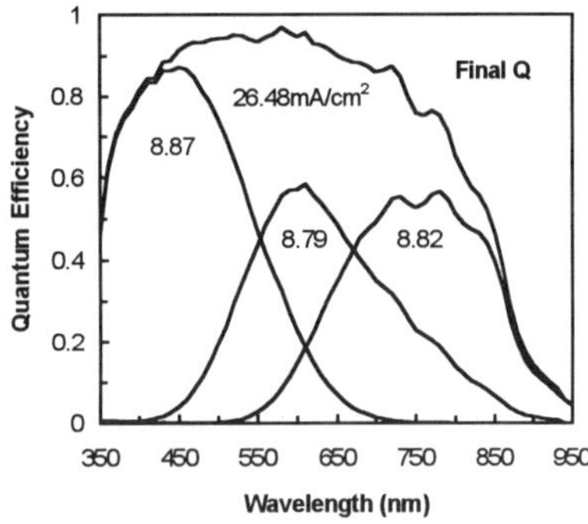

Figure 1. Stable J-V characteristics of triple-junction cell of efficiency 12.9%.

Figure 2. Stable Q characteristics of cell shown in Fig. 1.

Modules of Aperture Area ~40cm^2

An initial efficiency of 13.3% has been obtained on an unencapsulated module of aperture area 44.39cm^2. The V_{oc} = 2.30V, I_{sc} = 0.364A, and the FF = 0.705. The efficiency is the same as the average total area initial efficiency obtained on the small-area cells shown in Table I. The uniformity in efficiency over a 9"x9" area is usually $\pm$2.5% of the average value. The initial and stable values of encapsulated modules after 1056 hours of light soak are summarized in Table II. The initial efficiency is in the range of 12.1-12.6%. The stable aperture area efficiency is 10.9-11.2%. These values project the efficiency of larger area modules since the design and fabrication procedure are identical. Module 5549#E2L exhibits the highest stable efficiency of 11.2%. The aperture area is 41.2cm^2, V_{oc} = 2.21V, I_{sc} = 317mA, FF = 0.655, and P_{max} = 460mW. This is the highest stable aperture-area efficiency obtained on a fully encapsulated module. The value of the stable efficiency compares well with the average total-area stable efficiency of 11.6% obtained on the small-area cells (see Table I) after derating for the encapsulation losses.

Table II. Initial and stable Spire results of encapsulated modules of aperture area ~40cm^2.

2B #	Sample #	Aperture Area (cm^2)	Light exposure (Hours)	V_{oc} (V)	I_{sc} (mA)	FF	P_{max} (mW)	Efficiency (%)
5546	F2L	41.30	0	2.26	319	0.693	501	12.1
			1056	2.20	313	0.656	453	11.0
5549	E2L	41.19	0	2.28	321	0.710	519	12.6
			1056	2.21	317	0.655	460	11.2
5549	E3L	41.54	0	2.28	325	0.686	509	12.3
			1056	2.22	319	0.642	455	10.9
5552	E1L	41.41	0	2.24	320	0.700	502	12.2
			1056	2.19	317	0.656	454	11.0

<u>**Modules of Aperture Area ~ 450cm^2**</u>

Modules of aperture area ~450cm^2 have been fabricated. Initial and stable results after 1000 hours of light soak of three encapsulated modules are summarized in Table III. The initial efficiency of the modules is 12.1-12.3%. Module 2B5632L of aperture area 452.6cm^2 exhibits the highest initial efficiency of ~12.3%; the V_{oc} = 2.30V, I_{sc} = 3.51A, FF = 0.689, and P_{max} = 5.56W. The stable efficiency of the modules is 11.0-11.1%. The best module 2B5632L has stable values of efficiency = 11.1%, V_{oc} = 2.23V, I_{sc} = 3.46A, FF = 0.652, and P_{max} = 5.02W. This is the highest stable efficiency obtained on an encapsulated module of this size. The value of the stable efficiency compares well with the best stable efficiency of 11.2% obtained on the ~40cm^2 modules as shown in Table II.

Table III. Initial and stable results of encapsulated modules of aperture area ~450cm^2.

Module	Aperture area (cm^2)	Light Soak (Hours)	V_{oc} (V)	I_{sc} (A)	FF	P_{max} (W)	Efficiency (%)
2B5632L	452.62	0	2.30	3.51	0.689	5.56	12.3
		1000	2.23	3.46	0.652	5.02	11.1
2B5645L	450.91	0	2.29	3.46	0.693	5.50	12.2
		1000	2.23	3.4	0.657	4.98	11.0
2B5654L	452.62	0	2.30	3.51	0.679	5.48	12.1
		1000	2.22	3.47	0.644	4.96	11.0

<u>**Modules of Aperture Area ~920cm^2**</u>

Several encapsulated modules of aperture area ~920cm^2 have been fabricated by connecting two of the ~450cm^2 modules in series. The results of seven modules are summarized in Table IV. The initial efficiency is in the range of 11.7-12.1%. Module 567577L of aperture area 918.0cm^2 exhibits the highest efficiency of 12.1%; the V_{oc} = 4.60V, I_{sc} = 3.57A, FF = 0.678, and P_{max} = 11.13W. The highest efficiency of 12.1% compares well with the best initial efficiency of 12.3% obtained on the ~450cm^2 modules (see Table III) after derating for the losses associated with the series connection. One of the modules 558485 of initial efficiency 11.5% has been light soaked for 1056 hours. The stable values of the module are efficiency = 10.5%, V_{oc} = 4.41V, I_{sc} =

Table IV. Initial and stable results of encapsulated modules of aperture area ~920cm^2.

Module	Aperture Area (cm^2)	Light Soak (Hours)	V_{oc} (V)	I_{sc} (A)	FF	P_{max} (W)	Efficiency (%)
562194L	918.2	0	4.56	3.51	0.680	10.88	11.9
563946L	919.3	0	4.58	3.51	0.683	10.98	11.9
565398L	919.9	0	4.57	3.55	0.679	11.02	12.0
566179L	922.1	0	4.56	3.59	0.657	10.75	11.7
566566L	907.1	0	4.57	3.51	0.675	10.83	11.9
567577L	918.0	0	4.60	3.57	0.678	11.13	12.1
558485L	915.4	0	4.55	3.41	0.680	10.55	11.5
		1056	4.41	3.42	0.637	9.60	10.5

3.42A, FF = 0.637, and P_{max} = 9.60W. This is the highest value of stable efficiency that has been obtained on a module of this area. Note that the initial efficiency of the other modules is higher.

NREL Measurement of Modules of Aperture Area ~920cm^2

Three of the ~920cm^2 modules shown in Table IV, #s 566179L, 566566L, and 567577L, were sent to NREL for confirmation of the I-V measurements. NREL made three types of measurements: (1) indoor using Spire solar simulator model 240A equipped with a peak detector circuit board, (2) indoor using Large Area Continuous Solar Simulator (LACSS), and (3) outdoor using Standard Outdoor Measurement System (SOMS). A summary of the NREL measurements and the corresponding United Solar Spire measurements is provided in Table V.

All the United Solar Spire measurements were made using the same model solar simulator as the NREL Spire. The table shows that the initial efficiency measured by NREL is lower all across the board. The United Solar efficiency values are 11.7-12.1%; the corresponding NREL numbers are 10.90-11.40%. The NREL efficiency measurements are 4.2-9.9% lower than the United Solar ones. Even the NREL Spire measurements are 5.8-8.3% lower than the United Solar ones. Almost the entire discrepancy can be attributed to the discrepancy in the I_{sc} measurements: 5.7-9.4% between the two Spire measurements and 3.5-7.0% between the United Solar Spire and NREL outdoor values.

Table V. Comparison of module performance at United Solar, NREL, and Sandia Labs.

Module 2B	State	Area (cm^2)	Laboratory	Temp. (^{0}C)	V_{oc} (V)	I_{sc} (A)	FF	Efficiency (%)
566179L	Initial	922.1	United Solar Spire	25	4.56	3.59	0.657	11.7
		915.9	NREL Spire	25.1	4.519	3.2907	0.6753	10.96
			NREL LACSS	29.0	4.422	3.4336	0.6675	11.10
			NREL SOMS @ 964W/m^2	31.7	4.387	3.2179	0.6800	10.90
	Stable		NREL Spire	25.4	4.381	3.2502	0.6491	10.09
			NREL LACSS	26.2	4.331	3.3779	0.6378	10.20
			NREL SOMS @ 1054W/m^2	27.7	4.323	3.5132	0.6438	10.10
566566L	Initial	907.1	United Solar Spire	25	4.57	3.51	0.675	11.9
		905.1	NREL Spire	25.4	4.528	3.3100	0.6769	11.21
			NREL LACSS	27.2	4.488	3.3491	0.6873	11.40
			NREL SOMS @ 988W/m^2	33.9	4.366	3.346	0.6836	11.20
	Stable		NREL Spire	25.1	4.418	3.2996	0.6537	10.53
			NREL LACSS	26.5	4.353	3.2854	0.6598	10.40
			NREL SOMS @ 964W/m^2	28.4	4.327	3.4778	0.6520	10.50
567577L	Initial	918.0	United Solar Spire	25	4.60	3.57	0.678	12.1
		918.0	NREL Spire	25.4	4.535	3.2864	0.6837	11.10
			NREL LACSS	27.0	4.490	3.4215	0.6772	11.30
			NREL SOMS @ 970W/m^2	32.1	4.393	3.2424	0.6840	10.90
	Stable		NREL Spire	25.4	4.408	3.2551	0.6462	10.10
			NREL LACSS	27.9	4.302	3.3228	0.6513	10.10
			NREL SOMS @ 964W/m^2	28.6	4.322	3.5253	0.6413	10.10
563946L	Initial	919.3	United Solar Spire	25	4.58	3.54	0.700	12.3
		919.0	Sandia National	25	4.49	3.5	0.684	11.7

The fourth module in Table V, 563946L, has been measured at United Solar under the Spire solar simulator and Sandia National Laboratories under outdoor conditions. The efficiency measured at United Solar and Sandia are 12.3% and 11.7%, respectively. This corresponds to a discrepancy of 4.9%. Note that the discrepancy in the I_{sc} is only 1.1%. The Sandia value of V_{oc}, 4.49V, is lower than the corresponding United Solar value of 4.58V which suggests that the measurement temperature at Sandia might be higher than 25^0C. Since the modules exhibit negative temperature coefficient of efficiency, it is conceivable that the Sandia number would be higher than the measured value of 11.7%. Still, the Sandia number is much higher than the highest NREL measured value of 11.21%. Efforts are under way to resolve these discrepancies in measurement at the various laboratories.

The first three modules in Table V were light soaked at NREL under one-sun, 50^0C, and maximum power point conditions. The results show that the stabilized values of efficiency measured by all three techniques are in the range of 10.1-10.5%. Module 566566L exhibits the highest efficiency of 10.4-10.5%. This corresponds to a new world record for an a-Si alloy module of this size.

CONCLUSIONS

A systematic study has been carried out to scale up 0.268cm^2 total-area cells to encapsulated modules of aperture area $\sim920\text{cm}^2$. It has been shown that the aperture area efficiency of the $\sim920\text{cm}^2$ encapsulated modules correlate well with the small-area cell results after accounting for the additional module losses. Thus, it is necessary to improve the small-area cell efficiency before any further improvement can be seen in the module results. NREL has confirmed a stable module efficiency of 10.5% which is a new world record for a-Si alloy based modules. Since these modules have been fabricated using materials and processes used in our production line, this efficiency represents a practical target for commercial a-Si alloy based products.

ACKNOWLEDGMENT

We thank K. Lord and B. Yan for helpful discussions, D. Wolf, N. Jackett, J. Edens, J. Noch, A. Mohsin, E. Chen, and G. Hammond for the fabrication of the modules and various measurements. We thank V. Trudeau for her help in the preparation of the manuscript. This work was supported in part by NREL under Subcontract No. ZAK-8-17619-09 and we appreciate their help in the measurement and light soak of some of the modules.

REFERENCES

1. S. Guha, J. Yang, A. Banerjee, T. Glatfelter, K. Hoffman, S.R. Ovshinsky, M. Izu, H.C. Ovshinsky, and X. Deng, in *Amorphous Silicon Technology*, edited by E.A. Schiff, M. Hack, A. Madan, M. Powell, and A. Matsuda (Mater. Res. Soc. Proc. 336, San Francisco, CA 1994) pp. 645-655.
2. J. Yang, A. Banerjee, T. Glatfelter, K. Hoffman, X. Xu, and S. Guha, 1st World Conf. on Photovoltaic Energy Conversion, Waikoloa, Hawaii (1994) pp. 380-385.
3. J. Yang, A. Banerjee, and S. Guha, Appl. Phys. Lett. 70, 2975 (1997).
4. J. Yang, A. Banerjee, and S. Guha, in *Amorphous Silicon Technology*, edited by S. Wagner, M. Hack, E.A. Schiff, R. Schropp, and I. Shimizu (Mater. Res. Soc. Proc. 467, San Francisco, CA 1997) pp. 693-698.

LOW TEMPERATURE DEPOSITION OF AMORPHOUS SILICON BASED SOLAR CELLS

CHRISTIAN KOCH, MANABU ITO, MARKUS SCHUBERT,
AND JÜRGEN H. WERNER
Institut für Physikalische Elektronik, Universität Stuttgart
Pfaffenwaldring 47, D-70569 Stuttgart, Germany
Email: chkoch@ipe.uni-stuttgart.de

ABSTRACT

Our investigations study the influence of various deposition parameters on the structural and optoelectronic properties of a-Si:H deposited at temperatures of 100 °C and below. Despite a significant material quality deterioration at low substrate temperatures, we observe remarkable improvements of charge carrier transport properties due to an increased H_2-dilution of the process gases. In case of doped layers, we restore the electrical conductivity of low temperature n-type a-Si:H to standard values, whereas the p-layer quality is still inferior. We incorporate our optimized low temperature layers into a variety of solar cell types, including single and tandem cell structures. In addition to our conventional *pin*-structures we also successfully develop *nip*-cells for growth on opaque polymer substrates. From *pinpin* tandem cells, we record initial efficiencies of 6.0 % at a deposition temperature of 100 °C and 3.8 % at 75 °C. Interestingly, our *nipnip* tandem structures attain similar values which offers the possibility for deposition on low-cost plastic substrates. The mechanical flexibility of such substrates offers a wide variety of novel applications.

INTRODUCTION

Inexpensive flexible substrates for thin film solar cells would expand the field of photovoltaic applications. Currently used polyimide foils offer thermal stability up to temperatures occuring in standard plasma enhanced chemical vapour deposition (PECVD) of a-Si:H. Unfortunately, these polyimide films are too expensive. Using synthetic low-cost substrates like polyethylene (PE) or polyethyleneterephtalate (PET) would be a real chance for further cost reduction of a-Si:H based solar cells. The low thermal stability of these inexpensive plastic materials, however, limits the deposition temperature to values $T_S < 100$ °C . At such low temperatures, films from standard PECVD suffer from inferior structural quality, which causes a remarkable deterioration of their optoelectronic properties. Low diffusion lengths for both electrons and holes and insufficient electrical conductivities of doped n-type and p-type a-Si:H prevent the use of such films from application in photovoltaics.

This study presents reasonable quality a-Si:H deposited at $T_S = 100$ °C and below that is well suited for solar cell applications. In addition to materials quality optimization, we design, produce, and investigate different cell structures containing single- and multijunctions. The desired low cost synthetics do not offer sufficient transparency for being used as a substrate in *pin*-structures illuminated through the substrate itself. Therefore we develop not only *pin* but also *nip*-structures.

Mat. Res. Soc. Symp. Proc. Vol. 557 © 1999 Materials Research Society

EXPERIMENTAL DETAILS

We deposit the single layers and cell structures examined in this study by standard PECVD (radio frequency $rf = 13.6$ MHz) using silane (SiH_4) diluted by hydrogen (H_2), methane (CH_4) for the p-layer, and the doping gases phosphine (PH_3) and diborane (B_2H_6). The hydrogen dilution ratio is defined as $r_H = [SiH_4]/([SiH_4]+[H_2])$. The radio frequency power density P (plasma pressure p) values range from 50 mW/cm^2 (150 mbar) in case of undiluted growth, to 200 mW/cm^2 (750 mbar) for highly diluted films to keep the growth rate r_d at reasonable values. Light soaking under 2.5 suns conditions is carried out on single layers for 50 hours. We measure the minority carrier diffusion length L_D using Steady State Photocarrier Gratings (SSPG) technique and carry out Constant Photocurrent Method (CPM) on doped and undoped single layers to evaluate the Urbach energies (E_U), and subgap absorption coefficients $\alpha_{1.2eV}$ for defect density (N_D) estimations. Optical Transmission Spectroscopy is used to calibrate CPM-data $\alpha(h\nu)$ and to derive the optical gap τ_c from the Tauc relation.

The cell parameters - efficiency η, open circuit voltage V_{OC}, short circuit current density j_{SC} , and fill factor FF- are determined under AM 1.5-like illumination at room temperature. We deposit pin-structures on SnO_2 Asahi-glass with Ag back contacts defining active area sizes of 0.5 to 1.0 cm^2 and nip-cells with sputtered Indium Tin Oxide (ITO) front contacts on Ag-evaporated Corning 7059 glass. For the p-layers we use a-SiC:H with $\tau_c= 2.1$ eV. The series of nip-cells for our investigations on front contact design have nominal identical structure but different thicknesses of an inserted buffer layer consisting of p-type nc-Si between the absorption layer and p-contact layer deposited at $r_H = 1:40$ and $P = 500$ mW/cm^2. Due to the uncertainty in the size of the ITO contact areas, we can indicate the current density and therefore the efficiency of our nip-structures only with an error of about 20%.

RESULTS AND DISCUSSION

Single Layers

In order to examine the influence of the deposition temperature on the structural characteristics of a-Si:H, we deposit a temperature series under standard conditions. Keeping the pressure at $p = 150$ mbar and the plasma power at $P = 50$ mW/cm^2, we vary the deposition temperature from $T_S = 200\,°C$ down to $T_S = 50\,°C$.

Figure 1 shows r_d and τ_c of undiluted and undoped films as a function of the deposition temperature T_S . We observe a strong increase of the growth rate from $r_d = 1$ to $r_d = 4$ Å/s towards low temperatures. The tauc gap shows a similar behavior and surpasses $\tau_c = 1.9$ eV at $T_S = 50\,°C$. In the following, we explain this behavior by the nature of H-bonding in the material. The empirical relationship between τ_c and the hydrogen content C_H reported by Cody et al.[3]

$$\tau_c(eV) = 1.53 + 0.015\,C_H \text{ (at.\%)} \tag{1}$$

gives values for C_H surpassing 25 % for films deposited at $T_S = 50\,°C$. The relation between SiH_2-bonds and the optical band gap reported by Futako et al.[2] gives a SiH_2-

bond density of about 10% which means that nearly all hydrogen is bonded in SiH_2 at very low T_S . At higher T_S , SiH-bonds

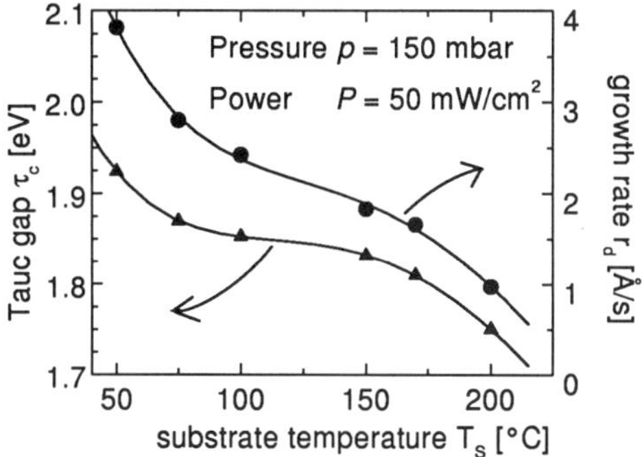

Figure 1: Impact of low substrate temperatures on growth and structure of a-Si:H

seem to be the main origin for the total hydrogen concentration. At $T_S =$ 150 °C the value of C_H is 20%, whereas the SiH_2-bond density is below 5% and, therefore, more than half of the hydrogen is bonded in the form of SiH. In summary, lowering T_S results in a slight increase of C_H, but in a significant increase of the amount of SiH_2 at the expense of SiH which is also reported by Tsu and Lucovsky [5]. They found the transition between SiH- and SiH_2-dominant H-bonding to be at $T_S \approx$ 100 °C. Figure 2 clearly outlines the influence of this behavior on structural and optoelectronic properties of a-Si:H. Without dilution, the sub-gap absorption coefficients increase by two orders of magnitude to values of $\alpha_{1.2eV} > 100$ cm^{-1}. The Urbach energies rise up to 150 meV when we reduce T_S from 150 °C down to 50 °C (cf. Figure 2a). An impressive deterioration of the electronic properties is related to this decrease of structural quality. The $\mu\tau$-product drops by more than four orders of magnitude down to 10^{-11}cm^2/V (cf. Figure 2b). At 100 °C and even more at 75 °C this material is not suited to form an absorber layer in solar cells.

H_2-dilution compensates materials deterioration by lowering the deposition temperature from $T_S = 150$ °C to $T_S = 75$ °C. In order to investigate the impact of H_2-dilution on the a-Si:H - layers, we vary r_H at different T_S for both undoped and n-doped a-Si:H. For undoped a-Si:H we put our interest primarily on hole diffusion lengths and on $\mu\tau$-products which are considered most significant for the absorption layer quality in an amorphous solar cell. For n-type a-Si:H we try to restore the elecrical dark conductivity σ_d . As depicted in Figure 3a, we vary the dilution ratio from r_H

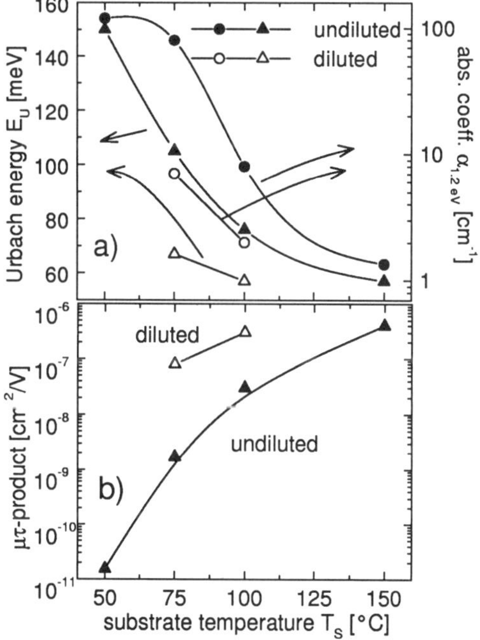

Figure 2: Structural and optoelectronic properties of a-Si:H at different T_S both undiluted and diluted. One sees significant material quality deterioration at low T_S and remarkable improvements by hydrogen dilution.

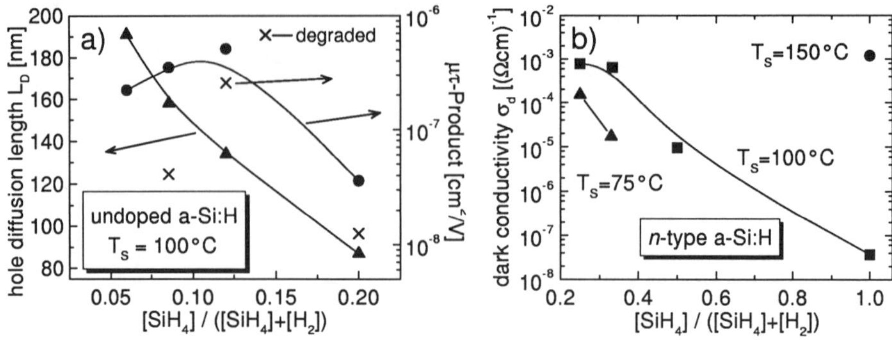

Figure 3: a) and b) depict the influence of various dilution ratios r_H on L_D and the $\mu\tau$-product of undoped films as well as the dark conductivity σ_d of n-type layers. The crosses in a) denote the $\mu\tau$-values of degraded layers.

= 1:5 to 1:20, and observe remarkable gain in the initial (as grown) hole diffusion length from L_D= 85 nm to L_D= 192 nm. This behavior seems to continue to higher dilution ratios, which we, however, did not further investigate because the growth rate drops off. The $\mu\tau$-product exhibits a maximum of 5×10^7cm^2/V at r_H = 1:8. For absorber layers incorporated into solar cells, we achieve the best results by using dilution ratios of r_H = 1:10 at T_S = 100 °C and r_H = 1:15 at T_S = 75 °C with growth rates of 3 Å/s and 2 Å/s, respectively. Figure 3a also shows the $\mu\tau$-product after degradation. High initial values seem to imply low degradation confirming an optimum $r_H \approx$ 1:10.

In addition to high-quality absorber material we need high conductive contact layers for good solar cells. Figure 3b outlines impressive improvements on the electrical conductivity of n-type a-Si:H by hydrogen dilution. A decrease of more than four orders of magnitude in σ_d occurs when lowering the deposition temperature T_S from T_S = 150 °C to T_S = 100 °C in case of undiluted grown films. By a hydrogen dilution of r_H = 1:4 we are able to successfully restore σ_d to the original 150 °C values. We observe this behavior even more markedly at 75 °C.

For p-type a-Si:H we do not attain significant improvements of σ_d by increasing r_H or varying the B$_2$H$_6$- flow. We believe that the amount of hydogen in the low-temperature material impairs electrical conductivity. Pietruszko et al.[4] report that dopant passivation by hydrogen is more pronounced on boron acceptors than on phosphorus donors. While we are able to grow high quality n-type material, for p-type a-Si:H we achieve electrical conductivities $\sigma_d > 10^{-7}$ Ωcm^{-1} at T_S = 100 °C and below.

Optimization of Solar Cells

The series of nip-cells for our investigations on front contact design contain nanocrystalline p-type buffer layers inserted between the ITO contact and the p-type a-Si:H contact. We vary the thicknesses between d_b= 0 nm (i.e. no buffer layer) and d_b= 30 nm. Unexpectedly, any of these additional layers prove to be destructive for the solar cell perfor-

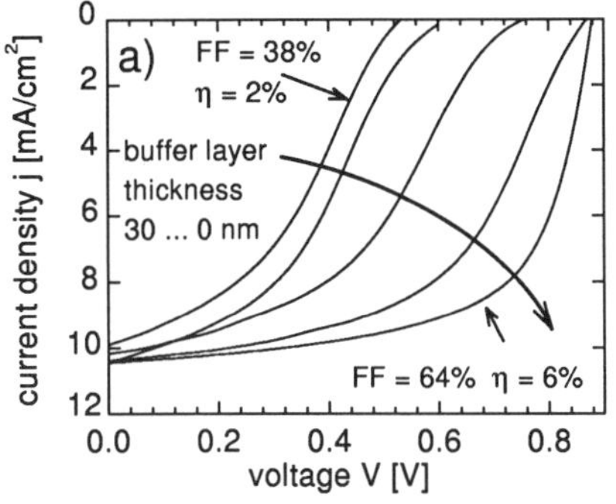

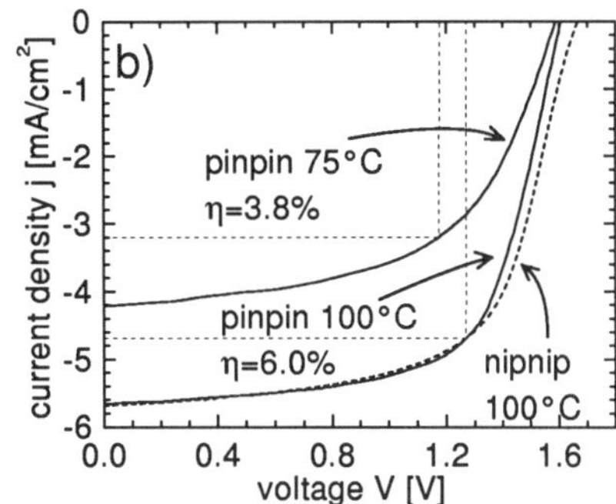

Figure 4: a) shows a series of *nip*-cells with nc-Si buffer layers of different thicknesses whereas b) shows our best results from tandem structures at 75 °C and 100 °C .

mance. Figure 4a shows the $j(V)$-characteristics of this cell series under AM1.5 illumination. Whereas the *nip*-structure with the thickest buffer layer exhibits a very poor fill factor of $FF = 38$ % and an efficiency of $\eta = 2$ %, the *nip*-cell with no buffer layer has a fill factor of $FF = 64$ % and an efficiency of $\eta = 6$ %. The corresponding loss in the open circuit voltage from $V_{OC} = 0.88$ V to $V_{OC} = 0.53$ V points towards the formation of an additional barrier for the photocurrent at the p/ITO-interface upon insertion of a buffer layer.

Table 1: Parameters of our best low-temperature tandem cells deposited at 75 °C and 100 °C. (The uncertainty of j_{SC} and η for the *nipnip*-structure is explained in the text.)

T_S [°C]	structure	V_{OC} [V]	j_{SC} [mA/cm²]	FF [%]	η [%]
100	*pinpin*	1.61	5.67	65.7	6.0
100	*nipnip*	1.67	5-6	62.2	≈ 6
75	*pinpin*	1.58	4.21	55.3	3.8

Figure 4b and table I present our best results on a-Si:H/a-Si:H tandem cells in *nipnip* and *pinpin*-configuration. At $T_S = 100$ °C we attain a fill factor of $FF = 65.7$ % and an efficiency of $\eta = 6.0$ %. The V_{OC} is 1.61 V and for j_{SC} we attain 5.67 mA/cm². While V_{OC} is comparable to state-of-the-art a-Si:H-cells, the value for the fill factor is about 15 % below FF of good high-temperature cells. However, the most striking feature is the comparably small current density of $j_{SC} = 5$-6 mA/cm² for tandem cells. The way the cell properties change, continues for lower deposition temperatures in a similar manner. Lowering the deposition temperature down to $T_S = 75$ °C causes a small decrease in V_{OC} to 1.58 V and a moderate decrease in FF to 55.3 %. The main effect which reduces the efficiency to $\eta = 3.8$ % is the low photocurrent density leading to a $j_{SC} = 4.21$ mA/cm². We primarily attribute this decrease in j_{SC} to a reduced $\mu\tau$-product of the undoped absorber material grown at low temperatures. The deterioration of *FF* originates from the increased series resistance which we determine from dj/dV at $V = V_{OC}$ of the $j(V)$-characteristics to be $R_S = 25 \ \Omega \text{cm}^2$ for the $T_S = 100\,°\text{C}$ *pinpin*-cell and $R_S = 40 \ \Omega \text{cm}^2$ for the $T_S = 75\,°\text{C}$ *pinpin*-cell.

CONCLUSION

We have shown the deterioration of a-Si:H materials quality when lowering the deposition temperature from $T_S = 150\,°\text{C}$ down to $50\,°\text{C}$. Parameters important for solar cell application for both doped and undoped a-Si:H become worse at lower T_S but are significantly improved by hydrogen dilution of the feedstock gas. High dilution ratios reduce the SiH_2-concentration in the deposited layers and hence improve its structural and optoelectronic properties. We also manage to restore the conductivity of low-temperature n-type layers to standard temperature values, but not that of p-type layers. The p-layer conductivity is, however, still sufficient for solar cell applications. For our optimized films we achieve growth rates of $r_d > 2$ Å/s by increasing plasma power P and pressure p which implies net deposition times for typical solar cell structures of less than 40 min even for highly diluted growth at $T_S = 75\,°\text{C}$. We deposit different cell structures in either deposition sequence, namely *pin*- and *nip*-structures in both single- and tandem configuration, and accomplish initial efficiencies of $\eta = 6.0$ % and 3.8 % at $T_S = 100\,°\text{C}$ and $T_S = 75\,°\text{C}$, respectively. For opaque substrate materials our *nip*-deposition sequence offers equivalent performance as compared to our *pin*-cells. ITO sputtered on the p-layer without any buffer layer works well as a front contact to this structure.

ACKNOWLEDGEMENTS

We would like to thank C. Köhler and S. Trottenberg for technical support and F. Pfisterer for carefully reading the manuscript.

References

[1] M. Shima, A. Terakawa, M. Isomura, M. Tanaka, S. Kiyama, and S. Tsuda, Appl. Phys. Lett. **71**, 84 (1997).

[2] W. Futako, K. Yoshino, C. M. Fortmann, and I. Shimizu, J. Appl. Phys. **85**, 812 (1999).

[3] G. D. Cody, B. Abeles, C. R. Wronski, R. B. Stephens, and Br. Brooks, Sol. Cells **2**, 227 (1980).

[4] S. M. Pietruszko, M. Pachocki, and J. Jang, J. Non-Cryst. Sol., **198-200**, 73 (1996).

[5] D.V. Tsu and G. Lucovsky J. Non-Cryst. Sol., **97 & 98**, 839 (1987).

OPTICAL MODELING OF a-Si SOLAR CELLS

BHUSHAN SOPORI*, JAMAL MADJDPOUR*, YI ZHANG*, WEI CHEN*,
SUBHENDU GUHA**, JEFF YANG**, ARINDAM BANERJEE**, STEVEN HEGEDUS***
*National Renewable Energy Laboratory, Golden, CO
**United Solar Systems Corp., Troy, MI
***Institute of Energy conversion, Newark, DE

ABSTRACT

We describe applications of *PV Optics* to analyze the behavior of a metallic back-reflector on an
a-Si solar cell. The calculated results from *PV Optics* agree well with the measured data on solar
cells. Several unexpected results obtained from these calculations are qualitatively explained.

INTRODUCTION

Modern high-efficiency amorphous-Si solar cells use fairly complex optical designs. Typically,
these cells may have three junctions and as many as 12 layers of metal, dielectric, and
semiconductor materials. Furthermore, they use many design features that are very difficult or
even impossible to handle by a simple optical analysis. These features include (i) nonplanar
interfaces, (ii) a combination of thick and thin layers, (iii) multiple semiconductors of different
optical properties, and (iv) dielectric and metal coatings. More mature numerical analysis tools
are needed to deal with the complicated optical design and analyses of these cells and modules.
We have recently developed an optical software package, *PV Optics* that can handle many
aspects of cell and module design. This paper describes some results of applying *PV Optics* to
the design of a-Si solar cells. The intent of this short paper is not to deal with a comprehensive
cell design, but to identify features of the software that can address unique cell-design issues.

BRIEF DESCRIPTION OF PV OPTICS

PV Optics is a software package developed for the design of thick as well as thin multilayer solar
cells and modules. It can handle multilayer structures of highly absorbing semiconductor and
metallic layers with planar and nonplanar interfaces. It uses combined features of ray and wave
optics selected by coherence criteria. This package outputs a variety of data in a form that can be
easily used by a design engineer. These data include plots of reflectance, transmittance,
absorbed photon flux and its distribution within each layer, and the Maximum Achievable
Current Density (MACD) from each active layer. MACD is the value of the photocurrent if each
absorbed photon produces an electron-hole pair and all of the generated carriers are collected.
Further description of the package is given elsewhere [1,2]. Here we will primarily use MACD
as a measure of the cell performance.

RESULTS AND DISCUSSION

We first describe some capabilities of *PV Optics* that make it well suited for a-Si solar cell
analysis and design. We show that if the optical parameters of a cell are known, the theory can
yield results that are in good agreement with the experimental data. Next, we discuss examples
related to back-reflector design, and arrive at some physical insight into the back-reflector optics.

Mat. Res. Soc. Symp. Proc. Vol. 557 © 1999 Materials Research Society

The examples described here are not meant to be exhaustive, but are selected to illustrate mechanisms that must be considered in analysis of back-reflector design of thin devices.

1. Comparison of theory and experiment

Previous publications have shown that for thick, wafer-based Si solar cells, the calculated results from *PV Optics* agree very well with the experimental results [2,3]. Here we will first illustrate a similar correlation between the theory and the experiment for a thin, single-junction cell consisting of a-Si layers on a metal substrate. To accomplish this, single-junction, p/i/n devices were fabricated at United Solar with controlled parameters and thicknesses. One of the objectives of our study is to compare light trapping properties of a planar versus a textured metal interface; these interface configurations were included in the fabricated cells.

Figure 1 illustrates the structure of the cells, the nature of interfaces, and the thickness of each layer. In Figure 1a, all layers have a conformal texture on the metal substrate, while in Figure 1b the cell has a planar metal-buffer interface. All physical and optical parameters, such as thickness of each layer, texture shape and height (h), and refractive indices and extinction coefficients, were measured and input into the *PV Optics*. The calculated MACD and the measured values of J_{ph} are given in Table 1. A good agreement between the theory and the experiment is seen. It may be noted that, because the MACD does not include electronic losses due to carrier recombination, it is expected, in general, to be somewhat higher than the actual value of J_{ph}. From Table 1 we also see that, in this cell structure, the planar buffer interface results in a lower current than when the interface is textured.

Table 1: A Comparison of Theoretical and Experimental Values of Photocurrent from Cells Whose Structures are Shown in Figure 1.

Substrate (+ metal)	Buffer (ZnO)/metal interface	Cell config. (Planar or textured)	Calculated MACD (mA/cm^2)	Jph, United Solar-Cell (mA/cm^2)
stainless steel	N/A	Planar (Fig. 1b)	12.56	12.7
Ag/ stainless steel	Planar	$h_{tex.}$=0.1 µm (Fig. 1a)	17.05	15.7
Ag/ stainless steel	Textured	h_{tex} =0.2 µm (Fig. 1a)	18.0	16.8

(a) 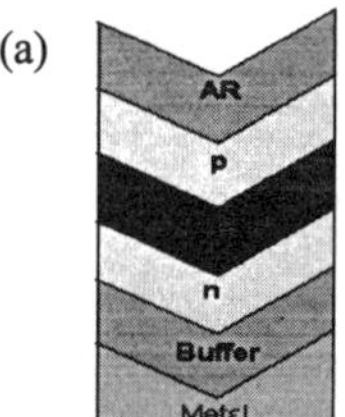

(b) 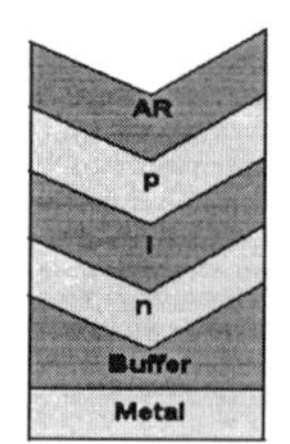

Figure 1. The structures of the cells fabricated for comparison with theory.
Layer thicknesses
AR: ITO/0.069 µm; p-layer: 0.0243 µm; i-layer: 0.2795 µm; n-layer: 0.0333 µm; buffer: 0.5 µm

2. The dependence of J_{ph} on buffer thickness

Although some excellent work has been done toward optimizing the back-reflector leading to fabrication of record efficiency a-Si solar cells, much of this is based on empirical analysis using arguments from conventional optics [4,5,6]. However, because a back reflector involves nonplanar interfaces and highly absorbing layers, the conventional concepts must be used with care. As an example, we consider the concepts involved in qualitatively predicting the effect of the buffer layer thickness on the MACD of the cell illustrated in Figure 1a (with textured buffer-metal interface). From the conventional optical considerations, the effect of a non-absorbing buffer layer may be viewed in two ways. Firstly, from mirror technology it is known that the reflectance from an air-metal interface can be enhanced by depositing a thin layer of a low-index dielectric. By a similar argument, the insertion of the buffer layer between a-Si and the back metal is expected to increase the cell current by lowering the metal loss. Secondly, because the buffer layer has significantly lower refractive index compared to a-Si, one normally expects that light emerging from a high index medium will experience total reflection when the incidence angle at the interface exceeds the critical angle. Thus, from both considerations, it is expected that for a loss-less buffer material (a) as the buffer thickness increases, the cell current should increase, and (b) cell current will also increase if the refractive index of the buffer is reduced.

Now we consider the calculated results from *PV Optics*. Figure 2a shows the calculated MACD as a function of buffer thickness for the cell illustrated in the Figure 1a, with ZnO as the buffer. This figure shows an unexpected dependence of the MACD on the buffer thickness; the MACD value first decreases to a minimum value of 17.2 mA/cm^2, and then increases again toward the value of 18 mA/cm^2. To investigate this further, let us also examine the effect of replacing the ZnO buffer layer with MgF$_2$ whose refractive index is lower than that of ZnO. Figure 2b shows the dependence of MACD on the thickness of MgF$_2$ as a buffer. We see that the MACD is lower with MgF$_2$ compared to ZnO—again, this is not expected from the conventional wisdom.

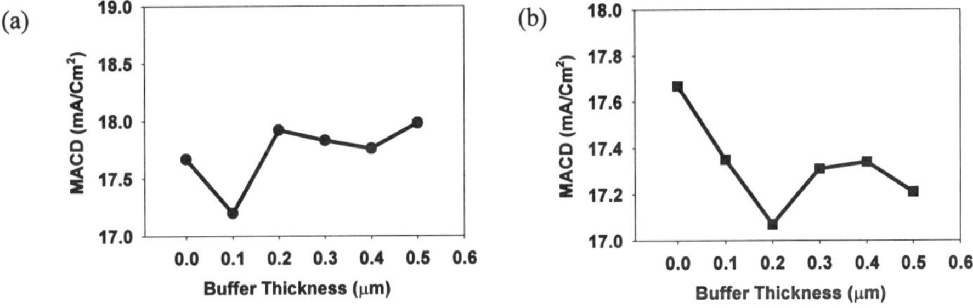

Figure 2. Calculated MACD for the cell shown in Figure 1, as a function of buffer thickness (a) buffer =ZnO, and (b) buffer = MgF$_2$.

To understand the behavior of MACD as a function of the buffer layer thickness, it is fruitful to consider a simpler case of a single a-Si layer on a metal substrate with a ZnO-buffer layer, as

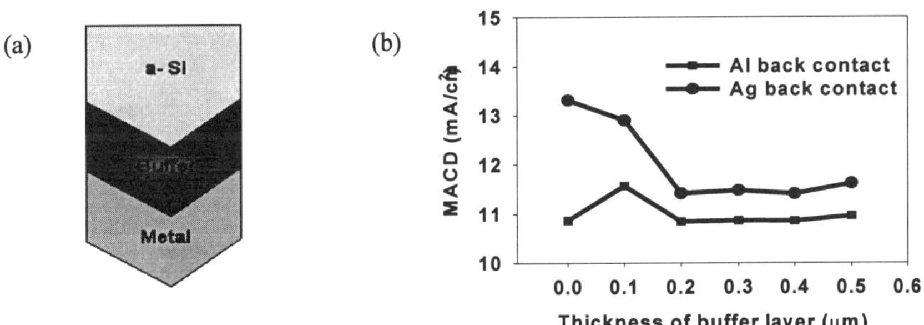

Figure 3. (a) The structure of the cell considered in Figure 3b. (b) Plot of MACD as a function of buffer (ZnO) thickness.

757

shown in Figure 3a. The cell has a planar front and a textured back interface as shown in figure. The a-Si layer has a thickness of 0.4 µm and a texture height of 0.3 µm, respectively. Figure 3b shows the plot of calculated MACD vs. buffer thickness, and for two cases of metal—Ag and Al. We see that, for the Ag back reflector, the MACD value decreases with the buffer thickness, whereas for an Al back reflector, the MACD is lower for all thicknesses of the buffer layer and shows a peak at 0.1 µm buffer thickness. A detailed understanding of this behavior is beyond the scope of this paper, but one can qualitatively visualize it by considering behavior of light at the a-Si-buffer and buffer-Ag interfaces. Figure 4 is drawn to help recognizing the following:

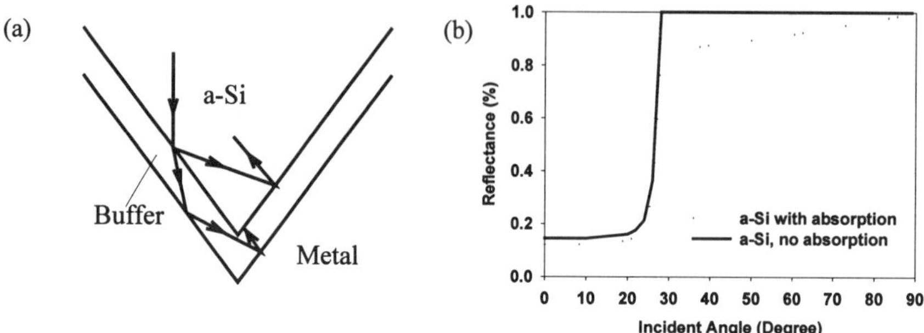

Figure 4. (a) Illustration of optical path at a textured back interface. (b) Calculated reflectance at the a-Si/ZnO interface.

1. Light reaching the backside of the cell will be incident on the a-Si-buffer interface at a large angle, and will experience reflection and transmission. If the buffer thickness is zero, the reflection coefficient at the a-Si-Ag interface is ~ 0.97, and the beam will experience two reflections from the texture before propagating back to the a-Si-air interface (see Figure 4a). Thus, a beam after two reflections at this interface will have a net reflectance of about 0.94.
2. When the buffer layer is present and the angle of incidence exceeds critical angle, it is important to recognize that the reflectance at the a-Si-buffer interface is not unity. Figure 4b shows the reflectance at a-Si-ZnO interface as a function of the incident angle. Also shown is the reflection coefficient for a nonabsorbing a-Si (i.e., where k=0). It is seen that the absorption of a-Si alters its total-internal-reflection behavior; and the reflectance is significantly less than unity, even when the angle of incidence exceeds the critical angle. Thus, a part of the light is transmitted into the buffer, and the absorption in a-Si is also reduced. This is true only for one propagation path of the beam.
3. The light transmitted to the metal interface also has a high reflection coefficient, but the effective reflectance at the a-Si-buffer is lower.
4. One can track the effective reflectance as a function of buffer thickness to qualitatively explain the results of Figure 3. These results will be discussed in detail elsewhere. It should be pointed out for a more lossy metal, such as Al, the results could be quite different from that of Ag.

This behavior of the cell structure in Figure 3a can be contrasted with a similar cell when all the interfaces are made planar. Figure 5 shows MACD as a function of buffer thickness for a planar, 0.4-µm-thick, a-Si cell. The calculations were performed for ZnO and MgF_2 as the buffer, and Ag and Al as the back-reflector layers. It is seen that inserting a buffer layer of a low index material can reduce the metallic loss and hence increase the MACD. Furthermore, the buffer with the lower index offers more effective "isolation" from the metallic layer. The figure also shows the interference effect associated with the buffer layer, manifested as a modulation of MACD with thickness. These results show that the analysis and the design of a textured or rough cell is quite involved and requires more detailed considerations of the optical phenomena.

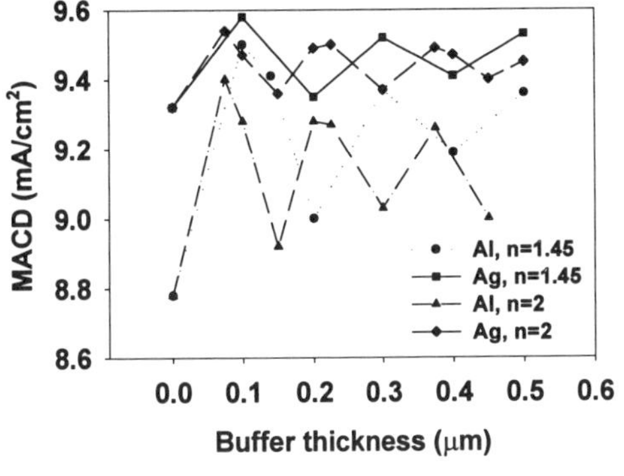

Figure 5. Calculated MACD's of a planar, single-layer cell as functions of buffer thickness on different buffers and back contacts

3. Dependence on the texture parameters

Table 2 lists results of analyses showing the effect of texture height on the MACD. These calculations are performed for the structures shown in Figure 1. We have used two different values of texture height—h= 0.2 μm, and 0.1 μm. The ITO layer is an antireflection coating that results in an increase in the J_{ph}. Interestingly, it is seen that an increase in the texture height has a slightly negative effect on MACD—which is again unexpected. Other calculations (not described here) indicate that the texture angle has a stronger effect on the MACD as compared to the texture height.

Table 2. Effect of light-trapping parameters on the photo-currents in single-junction devices

Substrate (+ metal)	Buffer (ZnO)/metal interface: planar or textured	Texture height	Calculated J_{ph} (mA/cm^2)	Jph, United Solar-Cell (mA/cm^2)
Al/SS	Planar	h = 0.1 μm	14.85 (no ITO) 16.26 (with ITO)	
Al/SS	Planar	h = 0.2 μm	14.77 (no ITO) 16.1 (with ITO)	
Ag/SS	Planar	h = 0.1 μm	16.85 (with ITO)	15.7
Ag/SS	Planar	h = 0.2 μm	16.66 (with ITO)	
Al/SS	Textured	h =0.1 μm	14.86 (no ITO) 16.44 (with ITO)	
Al/SS	Textured	h =0.2 μm	14.8 (no ITO) 16.41 (with ITO)	
Ag/SS	Textured	h =0.1 μm	17.51 (with ITO)	
Ag/SS	Textured	h =0.2 μm	17.45 (with ITO)	16.8

4. The influence of the bottom layer thickness on a three-junction cell

In this example, we consider how the thickness of the bottom layer in a three-junction cell will influence the MACD in all three Si layers. The cell structure used in our calculation is shown in

Figure 6b. The thicknesses of the different a-Si layers are: a-Si-T=0.3 μm, a-Si-M=0.2 μm, and a-Si-B=variable, corresponding to top, middle, and bottom cells, with bandgaps of 1.8, 1.6, and 1.4 eV, respectively. The results are plotted in Figure 6a. As expected, the MACD of the bottom layer increases with the increase of its thickness. But the MACD of the top layer actually decreases when the bottom layer becomes thicker. Also plotted is the total current from the cell, which shows an increase with the bottom- layer thickness. The metal used for this example is Al.

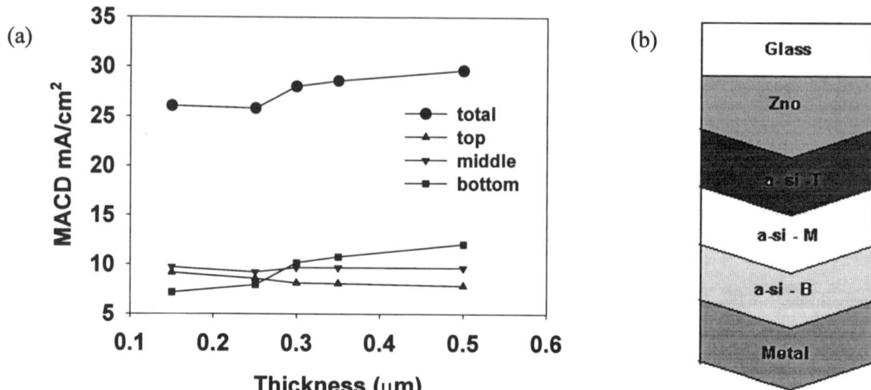

Figure 6. (a) Calculated MACD of a three-junction cell as a function of the thickness of the bottom layer. (b) The cell structure considered in the calculation.

CONCLUSIONS

We have pointed out some capabilities of *PV Optics* in the a-Si cell design to show that it yields results in good agreement with those measured from the a-Si cells. We have also shown that the somewhat-unexpected results from the back-reflector analysis can be understood by considering details of the optical processes. Because these processes are included in the *PV Optics*, it can serve as a tool not only for cell design, but also to develop an insight into the loss mechanisms.

ACKNOWLEDGEMENT

This work was supported by US Department of Energy under Contract #DE-AC36-98-G010337. The work at United Solar was supported in part by NREL under subcontract # ZAK-8-17619-09.

REFERENCES

[1] B. L. Sopori, Laser Focus, **34**, 159 (Feb. 1998).
[2] B. L. Sopori and T. Marshall, Procd. 23rd IEEE Photovoltaic Specialists Conference, 1993 (New York: IEEE).
[3] B. L. Sopori, J. Madjdpour, B. von Roedern, W. Chen, and S. S. Hegedus, MRS Symp. Proc. **467**, 777 (1997).
[4] A. Banerjee and S. Guha, J. Appl. Phys., **69** (2), 1991.
[5] S. S. Hegedus and X. Deng, Proc. 25th IEEE PVSC, Washington DC, 1061(1996).
[6] A. Banerjee, J. Yang, K. Hoffman, and S. Guha, Appl. Phys. Lett., **65**, 472(1994).

MATERIAL REQUIREMENTS FOR BUFFER LAYERS USED TO OBTAIN SOLAR CELLS WITH HIGH OPEN CIRCUIT VOLTAGES

BOLKO VON ROEDERN[*] and GOTTFRIED H. BAUER[**]
[*]National Renewable Energy Laboratory, 1617 Cole Blvd, Golden, CO 80401-3993
[**]Carl von Ossietzky Universität, P.O. Box 2503, D-26111 Oldenburg, Germany

ABSTRACT

This paper discusses material requirements for junction layers needed to obtain solar cells with highest possible open-circuit voltages (V_{OC}). In a typical a-Si:H-based "p/i/n" solar cell, this includes the transparent conductive oxide (TCO) contact layer, the p-layer, a "buffer layer" inserted at the p/i interface, and the surface portion of the intrinsic layer. In HIT-cells, the i-layer between (n-type) c-Si and (p-type) a-Si:H may be regarded as the buffer. Our suggestion to obtain high values of V_{OC} relies on using materials with high lifetimes and low carrier mobilities that are capable of reducing surface or junction recombination by reducing the flow of carriers into this loss-pathway. We provide a general calculation that supports these approaches and can explain why these schemes are beneficial for all solar cells.

INTRODUCTION

The open-circuit voltages of amorphous silicon solar cells are limited to values of about 1.0V. There is an ongoing discussion in the literature whether this limitation arises from bulk properties of the intrinsic layer (Urbach edges), junction recombination, or from the properties of the p- and n-layers used (built-in potential) [1]. Researchers have found that the optimization of junction layers in solar cells is extremely critical and interactive; i.e., in order to reach highest cell performance, layers have to be constantly reoptimized to achieve the best performance, or in production, yield. A unique combination of optimum materials for maximum device performance does not exist. Rather, the interactive nature of the optimization process suggests that, in many instances, materials with less than ideal properties have to be used to achieve maximum device performance.

In this paper, we argue that profiles of quasi-Fermi levels in each device have to be adjusted for optimum performance. In devices that have very high gradients of the quasi Fermilevel, which to some extent emerges in the profile of the electric field, V_{OC} is reduced in comparison with devices that have an optimized field or quasi-Fermi-level profile. The profiling of these gradients is best accomplished by using layers with low recombination (= high lifetimes) by adjusting the transport properties or carrier mobilities of these layers. Similar arguments apply to the Fermi-level or electrical-field gradients near the i/n interface.

The layers suggested to be used for V_{OC} enhancement can be rather resistive because of reduced carrier mobilities. The use of resistive layers is often thought to be undesirable because they introduce series resistance losses into the solar cell. However, we argue that a carefully optimized limited additional series resistance can be tolerated and allows maximizing cell performance. We will review some of the experimental procedures used to optimize V_{OC}, and argue that these observations support our claim and calculate in a generic (nonmaterial specific) way the benefits of using these layers.

Mat. Res. Soc. Symp. Proc. Vol. 557 ©1999 Materials Research Society

REVIEW OF EXPERIMENTAL RESULTS

Doping of the Emitter Layer

We briefly review three experimental observations how of these buffer layers are used in practice. The first observation relates to doping of the emitters of a-Si:H p/i/n cells. Dawson et al. [2] reported that cells made with the least doped (lower conductivity activation energy) a-Si:H p-layers exhibit highest voltages. It is noteworthy that the introduction of carbon alloying for the p layer, which usually makes it easier to obtain high V_{OC}, has the tendency to reduce the electron mobility and introduces a barrier for electron back diffusion. Carbon-alloyed p-layers are typically one to four orders of magnitude more resistive and are known to possess lower electron mobilities compared to boron-doped a-Si:H. A similar observation is true for crystalline Si. Here, larger V_{OC} values could be obtained when a lower phosphorous-doped emitter has been used. The practical requirements are to "overdope" the emitter to allow ohmic contacts between the emitter and the screen-printed collector grids on the cells. Conventional wisdom suggests that overdoping causes Auger recombination losses. We suggest that our concept may be able provide an alternative explanation for voltage loss in all solar cells whenever the emitter is doped "too much."

Use of TCO Bilayer with High Surface Resistivity

In a-Si:H and other thin-film solar cells, buffer layers are empirically used to enhance V_{OC}. It is often possible to enhance V_{OC} by using bilayer TCO films for the contact. This often allows the traditional emitter layer to be thinned without loss of V_{OC}. An example of this being achieved for a-Si solar cells was reported by Yoshida et al. [3]. For $CuInSe_2$- and CdTe-based solar cells, buffer layers have been used in production of PV modules [4]. Here, the TCO bilayer with the resistive surface allows thinning and even elimination of the CdS heterojunction emitter layer without a loss of V_{OC} [5,6]. We note that the concept of using a TCO bilayer can often be beneficial for solar cells made from materials with completely different chemistries. We therefore postulate a general mechanism (in addition to the mechanisms suggested traditionally such as "cleaning up" the interface, avoiding unwanted reactions near the interface, and suppressing diffusion of chemical species) that makes the use of these bilayer TCOs advantageous. It is of interest to note that for both $CuInSe_2$- and CdTe-based solar cells the best results for cells with very thin or no CdS layers are achieved when the resistivity of the resistive TCO layers is about 10^4 Ωcm [4,5]; typically these layers are a few tens of nanometers thick.

Use of "Resistive" Buffer Layers between the Emitter and Absorber Layer

There are many examples in which actual cell structures contain such a "buffer" layer. For a-Si cells, thin undoped a-SiC:H buffer layers, sometimes graded in carbon content, are inserted between the p- and the i-layer of p/i/n solar cells [7]. Another example is the HIT solar cell manufactured by Sanyo [8]. This cell is a crystalline-Si solar cell where the emitter and collector layers are deposited using a-Si techniques. To optimize V_{OC} in this type of cell, thin (<10-nm thick) intrinsic amorphous silicon layers are inserted between the wafer and the p-type emitter or n-type collector. Sanyo proposed that these layers are beneficial because they passivate or improve the quality of the junction. We interpret this approach in terms of using the buffer layers to control the recombination of minority carriers in the doped a-Si:H or μc-Si:H emitter. It is also noteworthy that a buffer layer is unknowingly used in c-Si solar cells prepared by phosphorus diffusion. It is well known that V_{OC} of such cells is sensitive to the doping level of the emitter and the base. Best results are typically obtained with wafers that are boron-doped in the range of 0.1 to 2 Ωcm [9]. We state that most conventional cell analyses may have misdiagnosed this doping requirement of the wafer (base). We suggest that the requirement is dictated by the need to create an appropriate buffer layer between emitter and base layer that limits the recombination of minority carriers in the emitter by providing a barrier for their movement. In c-Si solar cells, this buffer layer is provided by a compensated Si layer that is automatically formed during the phosphorus-diffusion step.

CALCULATION

Because a "resistive," gradual buffer layer appears to enhance V_{OC} in many different types of solar cells with different chemistry, interface, and interface diffusive properties, we offer below some generic (not material specific) calculations that may explain the need for the presence of such layers. Numerically, we have considered several types of buffer layers suitable to prevent photoexcited carriers to move to contacts (low-lifetime regions) and recombine there.

<u>Buffer Layer Scenarios:</u>

Type (1) -   Buffer layer with band alignment (ΔE_C=0, ΔE_V=0) to the absorber and lower minority and ambipolar carrier diffusion length L_m=$(D_m\tau_m)^{1/2}$ than in the absorber; the reduction of L_m is due exclusively to lower diffusion constant D_m, as suggested by one of the authors previously [10]; this means lower mobility μ_m too, and we assume that the lifetime τ_m is unaffected.

Type (2) -   Buffer layer like in (1), however the reduction in L_m results exclusively from a drop in life time τ_m, while D_m and μ_m are unaffected (i.e., a defective layer).

Type (3) -   Buffer layer without band alignment for minorities and with aligning band for the majorities of which the main function consists in blocking minority motion to the contact by a barrier (i.e., back diffusion of electrons in a-Si:H p/i/ns prevented by the carbonized window) and allowing for reasonable majority flow to the contact (most probable case in real approaches).

<u>Method</u>

For the investigation of the potential benefit of buffer layers seen as separating the high recombination regions of contacts from the base or absorber layer, we try an analytically solvable method for the determination of "best cases," and thus show the maximum achievable beneficial effects. The approach is based upon:

-    Determination/estimation of best case V_{OC} in ideal diffusion diodes as an upper limit of V_{OC} behavior in non-ideal (real) diodes.

-    Calculation of local minority carrier concentration $m(x)$ (in $0 \le x \le d$) in a homogeneous (Fig. 1a) as well as in an inhomogeneous absorber (Fig. 1b) via 1-dimensional steady-state continuity equation (exclusively diffusion currents) under $\exp(-\alpha x)$ generation with boundary conditions at x=0 and x=d resulting from surface recombination with velocities S_o=$S(x$=0), and S_d=$S(x$=d) as a function of layer parameters.

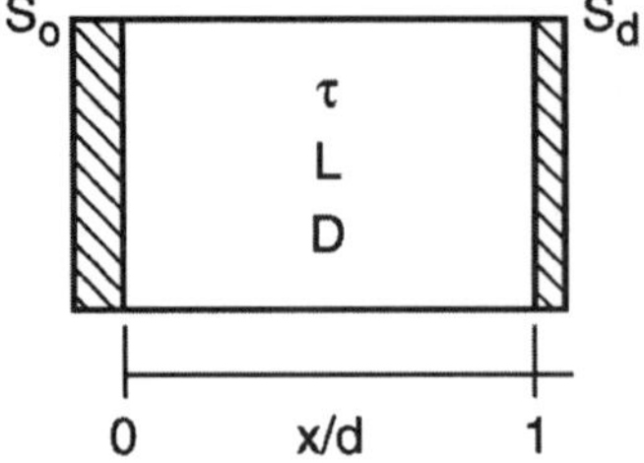

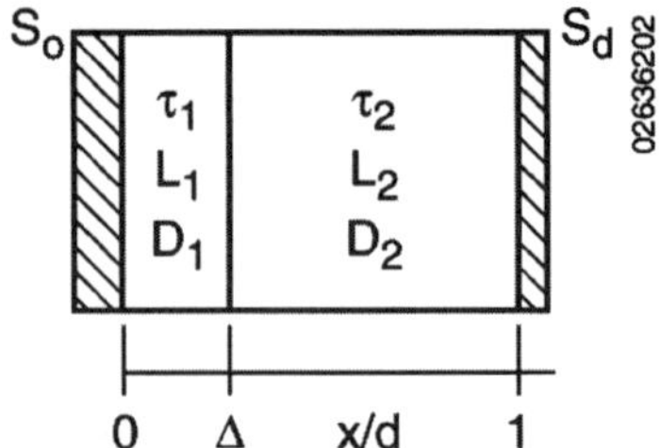

Fig 1a: Schematic of "solar cell" used for calculation.

Fig 1b: Schematic of "solar cell" with a buffer layer of thickness Δ

- The resulting local minority-carrier distribution reads:

$$m_i(x) = A_i \exp[x/L_{m,i}] + B \exp[-x/L_{m,i}] + [(g_o\tau_i)/(1-(\alpha L_{m,i})^2)] \exp(-\alpha x) \qquad (1)$$

(with A_i, B_i being dependent in a complex manner on $L_{m,i}$, τ_i, α, S_o, S_d, Δ, and d; the solution $\alpha L=1$ is excluded for reasons of numerical instability; i=1,2, with i=1 for $0 \leq x \leq \Delta$ and i=2 for $\Delta < x \leq d$).

- The translation of local minority excess carrier densities $m_{phot}(x)$ into the chemical potential and thus the maximum open-circuit voltage V_{OC} is performed via Boltzmann-approximation and the assumption that photogenerated majorities M_{phot} are small compared with their thermal equilibrium concentration M_o:

$$\mu(x) = kT \ln[(m_o+m_{phot})(M_o+M_{phot})/m_oM_o] \approx kT \ln[(m_o+m_{phot})/m_o] \qquad (2)$$

where $\mu(x)$ is an ambiguous and monotonous function of $m(x)$.

- The pn-junction, in which like in the ideal-diode approach, we neglect recombination due to geometrical relations ($w_{jct} << d$).

Results of the Calculation

The introduction of type (1)-buffer layers (decreased μ, unaffected τ) increases the minority-carrier concentration $m(x)$ continuously with decreasing mobility, or increasing buffer thickness Δ, which means displacing the surface region (at x=0) with its high recombination rate (in terms of diffusion lengths) as far as possible from the junction $x(m_{max})$. Formally, this can be achieved by $L_m \to 0$ (except with $x(m_{max}) \to \infty$). Fig. 2a shows the results from introducing effects of a type (1) buffer layer (decreased L, unaffected τ) into a homogenous absorber. The introduction of buffer layers of type (2) (unaffected L, decreased τ), shown in Fig. 2b., results in an optimum position for maximum excess-carrier density and maximum chemical potential as well. Each of these magnitudes strongly depend on the parameters in the individual layers such as diffusion lengths and coefficients, life times, and on thicknesses Δ, d, surface recombination velocities S_o, S_d, and on absorption coefficient α as well. By appropriate adjustment of $L_{m,1}/L_{m,2}$, Δ, and of the position of the junction, an increase in V_{OC} is achieved. Qualitatively, for the composed structures

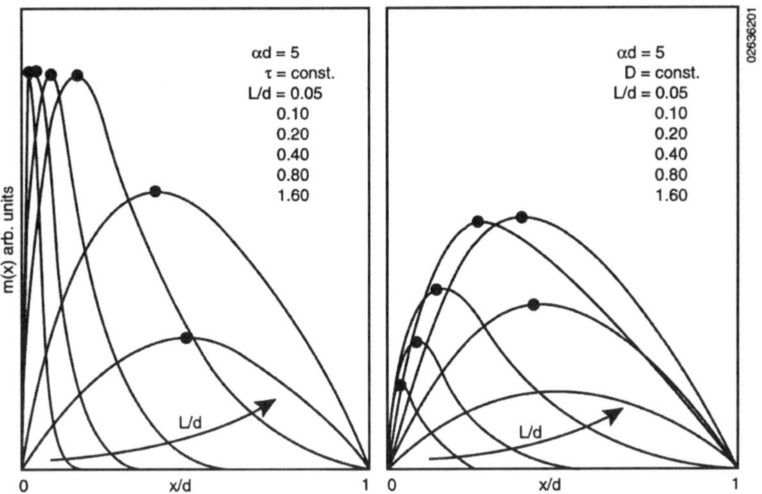

Fig. 2a: Minority-carrier concentration in homogenous structure for various L/d ratios (type (1), τ = const. condition).

Fig. 2b: Minority-carrier concentration in homogeneous structure for various L/d ratios (type (2), $D \sim \mu$ = const. condition).

(buffer/absorber) the profiles of minority carriers m(x) look similar to those in Figs. 2a, b; the introduction of buffer layers with lower L_m shifts m(x) analogous to a decrease of L/d.

While the effects of type (3) buffer layers could not be calculated analytically, this type of buffer, with majority-band alignment and barriers for minorities, will provide an optimum strategy for the prevention of the motion of photoexcited minorities into the "wrong direction" and their recombination at the interface, provided that type (3) buffer layers exhibit values of minority-carrier lifetimes at least as high as those of the absorber. Interestingly, this buffer layers also serve as an optical window with higher band gap, and consequently, lower refractive index for increased coupling of light into the absorber.

Consequences for the Operation of Buffer Layer Cells at Mpp

The operation of solar cells at maximum power point (mpp) conditions requires the extraction of nearly the entire I_{SC} at nearly V_{OC}, which means that the amount of "internally" created (photo-induced) chemical potential $\Delta\mu_{transp.}$ necessary for the transport of minorities to the contacts has to be minimized; because of the introduction of low-mobility buffer layers at current densities according to mpp, some of the internal chemical potential has to be consumed for transport. Our calculations – which due to the large number of parameters have been numerically run only for a limited number of different variables – show that for buffer layers of type (1), the balance of the benefit in V_{OC} equals or is smaller than the losses at mpp.

From an experimental device optimization point, this type of buffer layer leads to a regime in which J_{SC} and fill factor are "traded" for V_{OC}, that is a common observation in many types of cells, and, as an interesting side observation, not sensitive to the V_{OC}, J_{SC}, or efficiency level of the cell.

As the buffers of type (2) show even less V_{OC} increase versus the selection of optimum geometrical and electronic parameters the benefit with respect to output power generally seems even to be negative.

The analysis of type (3) buffer layer effects cannot be accomplished in the framework of this approach which we restrict to analytical solutions. Nevertheless, the general effect of an ideal barrier for minority carriers and the simultaneously assumed unaffected majority transport to be beneficial for V_{OC} is convincingly obvious.

CONCLUSIONS

We have argued that buffer layers play an important part in optimizing V_{OC} in all types of solar cells. The benefit of the buffer layers arises because they essentially remove the contact, which has high minority-carrier recombination losses from the absorber or base layer. Our approach appears to indicate that the best scheme is to employ a type (3) buffer layer (wider bandgap material without band alignment for minorities and with aligning band for the majorities). A secondary benefit of this type of layers, when used near the top of a cell, is optical enhancement because of reduced absorption and reflection losses. Type 1 buffer layers may offer some opportunity for V_{OC} optimization, but generally lead to losses in J_{SC} and FF, which experimentally have to be traded off carefully. Type (2) buffer layers offer no practical possibilities to increase cell efficiency.

ACKNOWLEDGEMENTS

This work was supported by the U.S. Department of Energy under Contract DE-AC36-98-GO10337.

REFERENCES

1. R.S. Crandall and E.A. Schiff, *Proc. of the 13th NREL Photovoltaics Program Review Meeting* (Am. Inst. of Physics Conf. Proc. **353**), 101 (1996).
2. R.M.A. Dawson, S.S. Nag, C.R. Wronski, and N. Maley, *Proc. 23rd IEEE Photovoltaic Specialists Conf.*, 960, (1993).
3. T. Yoshida, S. Fujikake, S. Saito, Y. Ichikawa, and H. Sakai, *Proc 10th European Photovoltaic Solar energy Conference*, 1193, (1991).
4. S. Albright, private communication.
5. L.C. Olsen, H. Aguilar, F.W. Addis, W. Lei and J. Li, *Proc. 25th IEEE Photovoltaic Specialists Conf.*, 997, (1996).
6. X. Li, R. Ribelin, Y. Mahathongdy, D. Albin, R. Dhere, D. Rose, S. Asher, H. Moutinho, and P. Sheldon, *Proc. of the 15th NCPV Photovoltaics Program Review Meeting* (Am. Inst. of Physics Conf. Proc. **462**), 230 (1998).
7. For example, see H. Sakai, T. Yoshida, S. Fujikake, and Y. Ichikawa, *Mat. Res. Soc. Symp. Proc.* **149**, 477, (1989)
8. T. Sawada, N. Terada, S. Tsuge, T. Baba, T. Takahama, K. Wakisaka, S. Tsuda, and S. Nakano, *1st World Conf. on Photovoltaic Energy Conversation (also 24th IEEE PV Spec. Conf.)*, 1219, (1994).
9. M.Y. Ghannam, S. Sivoththaman, H.E. Elgamel, J. Nijs, M. Rodot, and D. Sarti, *Proc. 23rd IEEE Photovoltaic Specialists Conf.*, 106, (1993).
10. B. von Roedern, Proc. *12th European Photovoltaic Solar Energy Conference,* 1354, (1994).

AMORPHOUS SILICON SOLAR CELL TECHNIQUES FOR HIGH TEMPERATURE AND/OR REACTIVE DEPOSITION CONDITIONS

M. Kanbe, T. Komaru, K. Fukutani, T. Kamiya, C.M. Fortmann, and I. Shimizu, Department of Innovative and Engineered Materials, Tokyo Institute of Technology, 4259 Nagatsuta, Midori-ku, Yokohama 226, Japan

ABSTRACT

Several promising new methods for amorphous silicon solar cell preparation involve high substrate temperatures and/or very reactive atmospheres. When incorporated into solar cells, the performance of these layers has often been less than expected due to enhanced diffusion and/or chemical reactions. This poor performance results from the harsh deposition environments. Deleterious effects include darken of TCO coated glass substrates due to hydrogen diffusion to and hydrogen reduction at the TCO interface when solar cells are prepared in the p-i-n deposition sequence. Alternatively, the deposition of TCO layers onto amorphous layers also involves rather harsh oxidizing conditions that have a deleterious effect on the top most amorphous silicon-based p-layers. Strategic use of blocking layers results in remarkably improved solar cell performance. A thin Cr layer (probably becoming Cr_2O_3) shows ability to improve the performance of both n-i-p and p-i-n solar cells by inhibiting both O and H diffusion.

INTRODUCTION

The a-Si:H photovoltaic industry has made remarkable progress. Guha et al. [1] reported that the stabilized efficiency of a triple junction a-Si:H, a-SiGe:H solar cell had achieved 13%. Nonetheless, the present low oil prices continue to drive the search for lower cost and increased performance amorphous silicon-based solar cells. One approach for lower cost is to remove or reduce the amount of expensive germanium used in red light scavenging component cell of the multi junction solar cell. One promising technique by which to accomplish this was reported by Fukutani et al. [2] who found that a high temperature, Ar* treatment technique could deposit relatively stable, narrow band gap (~ 1.52 eV) un-alloyed amorphous silicon materials. Elsewhere, Yamamoto et al. [3] reported that a-Si:H materials prepared from relatively inexpensive $SiCl_2H_2$ offered improved stability relative to standard device quality materials. The entire micro crystalline solar cell (possibly combined with an amorphous silicon solar to make a tandem solar cell) is yet other technique by which to circumvent the use of an amorphous silicon germanium solar cell. Also, the use of micro crystalline silicon doped layers in an amorphous silicon solar cell may increase the *built-in* potential [4] as well as to inhibit some types of species diffusion that occur during processing.

While, the a-Si:H boron diffusion coefficient is much greater than that of crystalline silicon, the a-Si:H hydrogen diffusion coefficient is much less than that of crystalline silicon [5]. Therefore one of the most severe cases for TCO reduction occurs when a microcrystalline silicon p-layer (p-μX-Si:B) is directly deposited onto a TCO coated substrate. Because, a hydrogen rich plasma is used for the deposition of these films, and even as the deposited film becomes thicker it easily transports hydrogen to the TCO interface. However, a substrate that withstands the deposition of p-μX-Si:B should be robust against any further degradation by subsequent hydrogen rich i-layer depositions. In this work both p-μX-Si:B layers and p-μX-Si:B/i-a-Si:H/n-a-Si:H:P optimized solar cells were prepared and analyzed as a function of various blocking layers applied to commercially available Asahi U type TCO. We also studied the use of

Mat. Res. Soc. Symp. Proc. Vol. 557 © 1999 Materials Research Society

amorphous silicon protection layers for solar cells deposited in the n-i-p deposition sequence. Where, in this n-i-p case the goal was to protect an amorphous p-layer from oxidation during the sputter deposition of the ZnO top contact.

EXPERIMENT

P-i-n and n-i-p solar cells were prepared on various substrates using a previously described [6] deposition system which had the ability to deposit amorphous and microcrystalline silicon layers using either RF, VHF or MW remote plasma enhanced chemical vapor deposition. In order to examine and improve the resistance of TCO coated glass substrates p layers, and complete p-i-n and n-i-p solar cells were prepared. The p-i-n solar cells were prepared on TCO coated glass substrates and employed p-μX-Si:B layers. Prior to incorporation into the solar cell the p-μX-Si:B was optimized by analysis of films deposited onto glass substrates using VHF-(144MHz) plasma assisted CVD as a function of temperature. The i- and n-layers of all solar cells studied in this work were prepared by conventional RF-CVD at 200°C. Raman analysis indicated that films deposited with SiH_4 to H_2 ratios of 1% had significant crystalline silicon contents(as well as an a-Si:H matrix). Each substrate was arranged to have a surface area comprised of un-coated Asahi U type SnO_x, Ga-doped ZnO and Cr over coated SnO_x; as well as, both Cr and Ga-doped ZnO over coated Asahi™ U-type SnO_x regions as shown in Fig.1a. In the case of n-i-p solar cells prior to the sputter deposition of Ga-doped ZnO, Cr was e-beam deposited onto a region of the intended solar cell surface as shown in Fig. 1b.

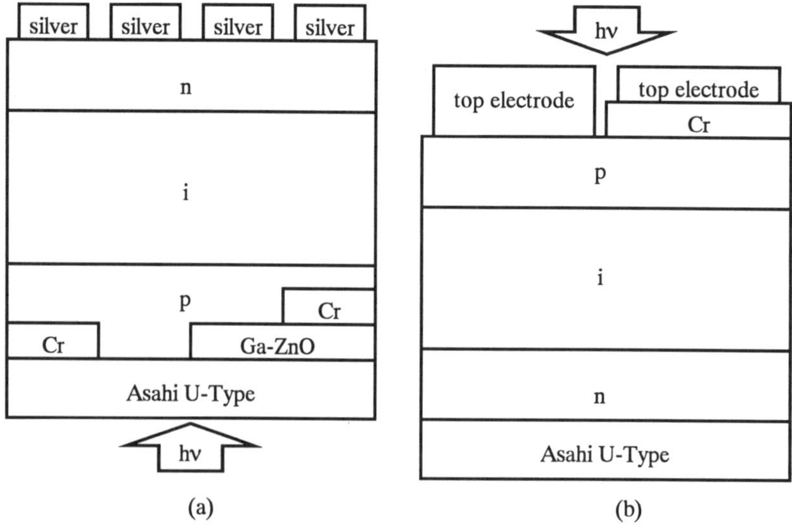

Figure 1 Configuration of SnO_x coated substrates and the various protective layers used in this study for the p-i-n case (a), and the application of Cr protective layers to selected surface areas of p-layers of n-i-p deposition sequence solar cells (b).

RESULTS AND DISCUSSIONS

To identify the hydrogen dilution and other conditions necessary for the deposition of p-μX-Si:B layers were systematically deposited onto glass substrates as a function of dilution ratio and substrate temperature. The Raman spectra of this series of films is shown in Fig. 2. It is noted that all samples contained some amorphous fraction. However, large a-Si:H content is expected in such thin films. For the evaluation of substrate hydrogen plasma resistance and solar cell optimization the 1% SiH_4 in H_2 dilution at 220 and 300 $^{\circ}$C were identified for further study.

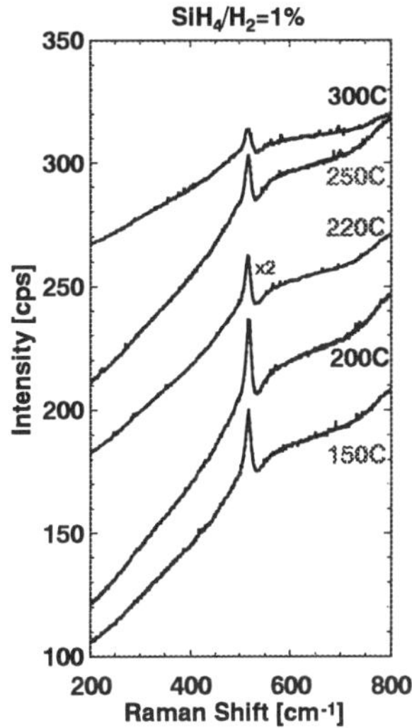

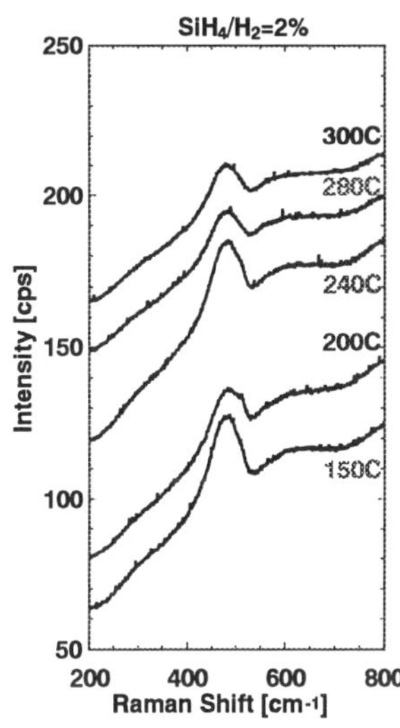

Figure 2 The Raman spectra of p-μX-Si:B layers deposited at 1 and 2% SiH_4 in H2 as a function of substrate temperature (as indicated).

Figure 3 shows the solar cell parameters of p-i-n solar cells having an Asahi™ U-type substrate as a function of the micro-crystalline silicon p-layer deposition time (proportional to the p-layer thickness). It is interesting that in the thinner p-layer cases the V_{oc} was greatest for the Cr over-coated substrates and that the V_{oc} increased with the increasing TCO metal component work function (i.e., the work function Cr is greater than Zn which in turn is greater than Sn). It is also interesting that the Cr coated substrates did not show a decreased J_{sc} as compared to the substrates having only transparent TCO coatings (SnO_x and SnO_x / ZnO). The high J_{sc} in the Cr coated cases suggest that Cr has oxidized to a more transparent Cr_2O_3 during its exposure to air prior to the solar cell preparation, and/or that the presence of a Cr absorbs less light than the metal films that unintentionally form when either SnO_x or ZnO are reduced during hydrogen plasma processing associated with p-layer deposition.

Figure 4 shows the current versus voltage characteristics of solar cells having p-μX-Si:B layers deposited at 220 and 300°C as a function of the protection layers applied to the Asahi™ U-type substrates. In this case the i- and the n-layer for these cells was deposited at 220°C. This result demonstrates that it is possible to prepare reasonably good solar sells using high deposition temperatures and reactive conditions for p-layer preparation.

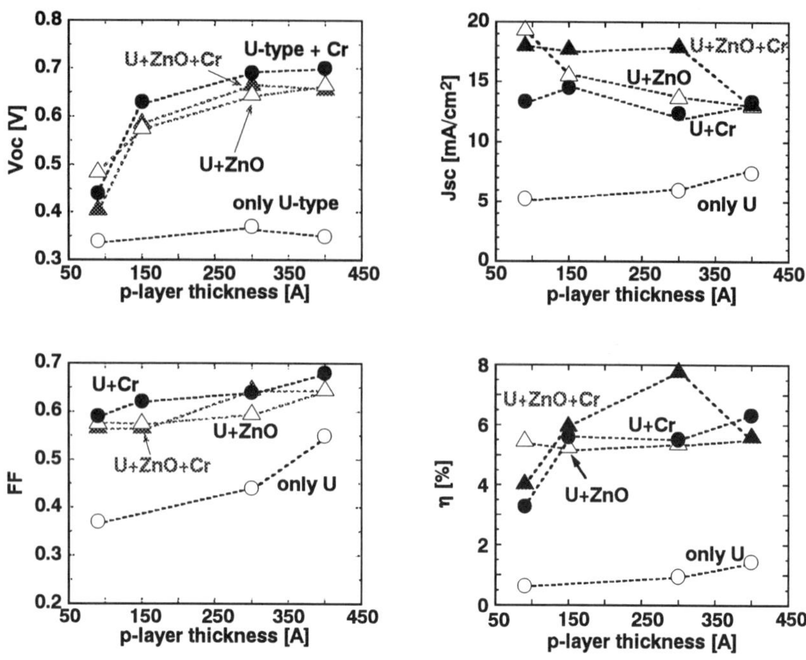

Figure 3　The performance characteristics of p-i-n solar cells as a function of the substrate protection layers and as a function of the thickness of the microcrystalline silicon p-layer deposition time.

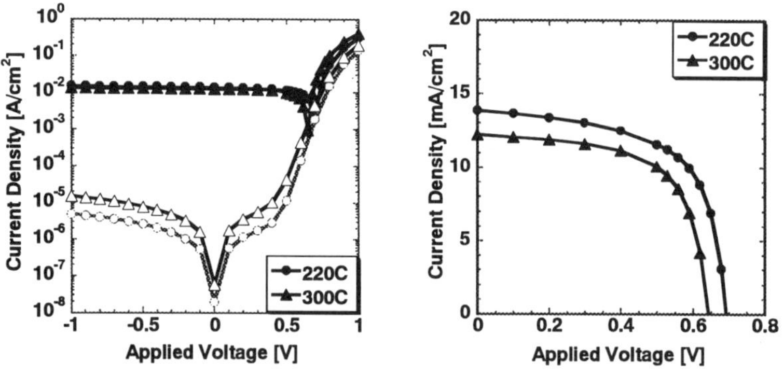

Figure 4 The current versus voltage characteristics of p-i-n solar cells having *Cr* over coated Asahi™ U-type substrate and p-μX-Si:B layers deposited at 220 and 300°C as indicated.

Table 1 shows the performance of solar cells deposited in the n-i-p deposition sequence as a function of the p-layer structure (amorphous or microcrystalline) as well as the surface treatments and contacting schemes used to provide a top contact. Note, that all solar cells in this series suffer from some series resistance losses (low fill factors) as no collection grids were applied. Nonetheless, it is clear that the if a *ZnO* layer is directly deposited onto an a-SiC:H:B p-layer the solar cell performance is significantly lower than the case of when a p-μX-Si:B layer was used. To test whether the relatively poor performance of the amorphous p-layer was due to an oxide layer and/or other deleterious effects arising from the sputter deposition of the *ZnO* layer onto amorphous silicon, some solar cells were prepared having a *Cr* protection barrier e-beam deposited onto the a-SiC:H p-layer prior to the sputter deposition of the *ZnO* layer. As can be seen in Table 1 these solar cells had performance equal to or better than the p-μX-Si:B cases.

TABLE 1

n-i-p Solar Cell Parameters as a Function of p-Layer and Top Contact Method

top electrode	p layer	Cr*	Voc [V]	Jsc [mA/cm^2]	FF	Eff [%]
Pt	a-SiC:H	−	0.62	4.19	0.51	1.35
	μc-Si:H	−	0.68	3.38	0.52	1.20
GZO (120℃)	a-SiC:H	−	0.71	9.01	0.36	2.32
	a-SiC:H	+	0.72	9.01	0.53	3.41
	μc-Si:H	−	0.59	7.80	0.48	2.21
ITO (R.T.)	a-SiC:H	−	0.67	4.29	0.51	1.42
	a-SiC:H	+	0.69	9.90	0.47	3.23
	μc-Si:H	−	0.69	5.79	0.50	2.01

* −:without Cr, +:with Cr

It is constructive to consider the relative stability of the various oxides that may form during the preparation of amorphous and micro-crystalline silicon solar cells. Using thermodynamic values given by Dickerson [7] the reduction reactions for some of the oxides relevant to this work are:

$$SnO_2 + 2H_2 \Leftrightarrow 2H_2O + Sn \qquad \Delta G_{298} = +14.3 \, kcal/mole$$

$$ZnO + H_2 \Leftrightarrow H_2O + Zn \qquad \Delta G_{298} = +21.4 \, kcal/mole$$

$$Cr_2O_3 + 3H_2 \Leftrightarrow 3H_2O + 2Cr \qquad \Delta G_{298} = +86.3 \, kcal/mole$$

$$SiO_2 + 2H_2 \Leftrightarrow 2H_2O + Si \qquad \Delta G_{298} = +83.1 \, kcal/mole$$

It is evident that all of these oxides can be expected to be stable against H_2 reduction. However, the thin, small grained, films of these metal oxides exposed to an atomic, and/or the radical and ionic hydrogen flux generated by a plasma are much more likely to be reduced. The metal oxide with the greatest *built in* thermodynamic barrier to reduction is Cr_2O_3. Interestingly this also means that Cr_2O_3 is the most stable of these metal oxides against the formation of SiO_2 at its silicon interface. The relative resistance of the various metal oxides to hydrogen plasma reduction appears to follow an order predicted by classical thermodynamic in which Sn is the least resistant and SiO_2 is the most resistant. The exposure of an a-SiC:H surface to an oxygen rich plasma during the deposition of a transparent contact for a n-i-p solar cell is expected to produce an interfacial SiO_2 layer. However, by first depositing a Cr layer onto the a-SiC:H p-layer prior to the TCO deposition, the underlying amorphous silicon is protected by a Cr (and/or Cr_2O_3 formed during oxygen plasma exposure) diffusion barrier.

CONCLUSIONS

The strategic placement of semi-transparent Cr layers are an easy and effective way to extend the useful temperatures range and/or to protect against plasma species induced chemical reactions when TCO coated glass substrates are used for p-I-n solar cell fabrication. These thin Cr layers when applied to standard, commercially available SnO_x substrates are as effective as many presently available ZnO over coating technologies. Even greater resistance to chemical breakdown is possible by using the Cr layers in addition to an optimized ZnO layer. Use of a Cr protection layers on the top surface of an a-SiC:H p-layer of solar cells prepared in the n-i-p deposition sequence allows the use of an a-SiC:H p-layer.

The Cr layers in almost all cases, even those using standard deposition condition, caused no observable deleterious effects. In particular standard solar cells with Cr layer interfacial layers showed improved fill factor, open circuit voltage, and surprising even improved the short circuit currents. This result suggests that even under optimized standard deposition practice some reduction of the TCO layers may take place. Special reduction resistant ZnO layers also aid the preparation of improved performance solar cells. These more heat and chemically resistant substrate coatings facilitate the preparation of both a-Si:H and μx-Si layers and solar cells by reducing impurity and dopant diffusion.

ACKNOWLEDGMENTS

We gratefully acknowledge the many useful discussions and substrates provided by the Asahi Glass Company.

REFERENCES

1. J. Yang, A. Banerjee, T. Glatfelter, S. Sugiyama, S. Guha, Conference Record of the 26th IEEE Photovoltaic Specialists Conference 1997, IEEE, Piscataway, NJ p.563
2. K. Fukutani, T. Sugawara, W. Futako, T. Kamiya, C.M. Fortmann, I. Shimizu, Mat. Res. Soc. Symp. Proc. Vol. 507, (1998), p. 211
3. Y. Yamamoto, W. Futako, K. Fukutani, M. Hagino, T. Sugawara, T. Kamiya, C.M. Fortmann, I Shimizu, Mat. Res. Soc. Symp. Proc. Vol. 507, (1998) p. 199
4. J. Gibbons, Stanford University, private communication 1997
5. Richard J. Borg & G.J. Dienes, An Introduction to Solid State Diffusion, 1988, Academic Press, NY 206-207
6. Mika Kanbe, Master Thesis, 1999, Tokyo Institute of Technology
7. R. E. Dickerson, Molecular Thermodynamics, 1969 W.A. Benjamin, Inc., Menlo Park CA

Simulation of Hydrogenated Amorphous Silicon Germanium Alloys for Bandgap Grading

E. Schroten, M. Zeman, R. A. C. M. M. van Swaaij, L. L. A. Vosteen, and J. W. Metselaar
Delft University of Technology, Laboratory of Electronic Components, Technology and Materials – DIMES, P. O. Box 5053, NL-2600 GB Delft, the Netherlands

ABSTRACT

Computer simulations are reported of hydrogenated amorphous silicon germanium (a-SiGe:H) layers that make up the graded part of the intrinsic layer near the interfaces of a-SiGe:H solar cells. Therefore the graded part is approached with a 'staircase' bandgap profile, consisting of three layers within which the material properties are constant. Calibrated model parameters are obtained by matching simulation results of material properties of intrinsic a-SiGe:H single layers to measurements. Using the obtained model parameter sets subsequent simulations of p-i-n devices with intrinsic material similar to the single layers are matched to measured current-voltage characteristics. The changes in parameter values are evaluated as a function of optical gap.

INTRODUCTION

Amorphous silicon germanium (a-SiGe:H) has proven to be a suitable low band gap material for the intrinsic layer of the bottom solar cell in a tandem structure [1]. Incorporation of Ge improves the red response, but deteriorates electronic properties. Excellent overviews of experimental data on a-SiGe:H are given by Stutzman *et al.* [2] and Bauer [3]. Subsequent studies showed that material properties are improved when using strong hydrogen dilution during the deposition [4]. For application in solar cells, band gap grading near the p-i and i-n interfaces - by profiling the Ge content in the intrinsic layer - to accommodate the band offsets, has considerably enhanced the cell characteristics [5]. So far the effect of these graded parts on the electronic behavior of the cell is not fully understood. Computer simulations can be used to give insight in this matter.

Guha *et al.* [5] simulated solar cells with bandgap grading using different profiles and was able to verify trends in experimental observations. Several grading profiles were also studied using computer simulations by Vasanth *et al.* [6]. In that work the whole i-layer was split into five sub-layers to represent the profiles. Recently, Zimmer *et al.* [7] simulated bandgap profiling by variation of the band edges, defect density and Urbach energy. However, the dependence of other model parameters on the Ge content has not been taken into account. In our approach we divide the graded part of the device into several layers in which the material properties do not vary with depth, thereby obtaining a 'staircase' bandgap profile. We assume that the properties in the graded part of the solar cell change in the same way as the properties of a series of single layers in which the Ge content is varied. Wronski *et al.* [8] pointed out that such self-consistent simulation parameters are necessary for a realistic simulation of solar cells, which enables a reliable analysis for optimization. In this study we present how the input parameters of these layers are determined following the calibration strategy proposed by Zeman *et al.* [9]. To assign proper values to the model input parameters simulation results are matched to experimental data from literature as well as from measurements carried out on single layers that are deposited with different germane (GeH$_4$) fraction. In addition, current-voltage characteristics of p-i-n devices with intrinsic material similar to the single layers are simulated and matched with measurement

Mat. Res. Soc. Symp. Proc. Vol. 557 © 1999 Materials Research Society

results. The obtained parameter sets are evaluated as a function of the bandgap.

EXPERIMENTAL

Sample preparation

Three single films with increasing germanium content were deposited using a conventional multichamber plasma enhanced CVD set-up. Silane (SiH_4) and GeH_4 flows were chosen according to the start, an intermediate, and the end condition of a state-of-the-art grading procedure used for the fabrication of an *a*-SiGe:H cell in our laboratory. In all cases strong hydrogen dilution was used in order to obtain high-quality material. The deposition temperature for all films was 190°C. 300 nm thick films were deposited on Corning 1737 glass for optical and electrical measurements. For the latter, contacts were deposited by e-beam evaporation of aluminum, followed by two hours annealing in air at 180°C.

In addition, three p-i-n structures were deposited on Asahi U-type transparent conductive oxide (TCO). The i-layers were formed by 150 nm of *a*-SiGe:H deposited at conditions identical to those of the single layers. In these cells *no* grading near the interfaces was applied. Contacts were deposited by evaporation of 100 nm Ag followed by 200 nm Al by e-beam evaporation. To obtain proper ohmic contacts, thermal annealing was carried out at 150°C in air. The current spreading through the n-layer was suppressed by removing this around the contacts by reactive ion etching.

Measurement procedures

For each film the reflection and transmission (RT) spectrum was recorded simultaneously. These spectra were analyzed using an iterative simulation procedure to obtain the thickness and the absorption spectrum without interference. The dual beam photoconductivity (DBP) method was used to extend the absorption spectrum into the infra red region.

Temperature dependent conductivity in the dark was measured in the temperature range of 290-400 K using an electrometer (Keithley 6517A). During these experiments the sample was immersed in a constant flow of dried air. The photoconductivity, $\sigma_{AM1.5}$, was measured by illuminating the samples through the glass substrate using a solar simulator (Oriel) calibrated for Air Mass 1.5 (AM1.5). Dark and AM1.5 current-voltage curves of the cells were recorded by a parameter analyzer (HP4045).

RESULTS

Measurements

The absorption spectra of the three single films, deduced from both RT and DBP, are shown in Fig. 1. From this spectrum the optical gap, E_{Tauc}, is determined using the Tauc approximation and the Urbach edge, E_{Urb}, from the slope of the exponential region just below the optical gap.

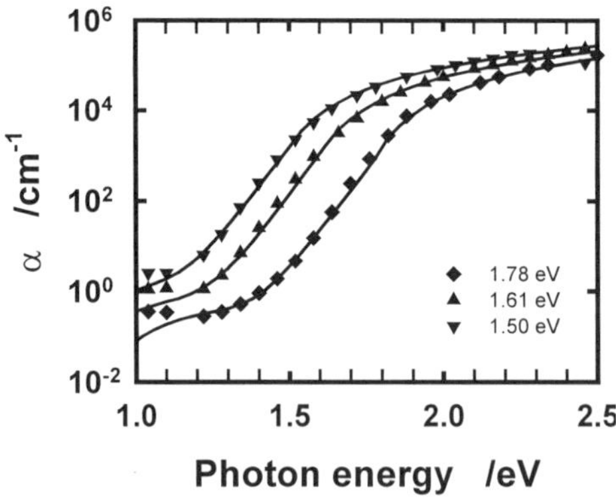

Figure 1. *Measured (data points) and simulated (solid lines) absorption spectra of the standard a-SiGe:H films.*

Table I: Measured and simulated properties of single films.

type	E_{Tauc}	E_{mob}	E_{Urb}	N_{db}	E_{act}	σ_{dark}	$\sigma_{AM1.5}$
	[eV]	[eV]	[eV]	[cm^{-3}]	[eV]	[Ω^{-1}cm^{-1}]	[Ω^{-1}cm^{-1}]
i-type a-Si:H	1.78		0.048	1.2×10^{16}	0.926	2.2×10^{-12}	2.0×10^{-4}
simulated	1.78	1.82		1.2×10^{16}	0.925	2.0×10^{-12}	5.3×10^{-4}
i-type a-SiGe:H	1.61		0.050	2×10^{16}	0.788	1.2×10^{-10}	2.2×10^{-5}
simulated	1.61	1.66		2×10^{16}	0.787	2.7×10^{-10}	0.9×10^{-5}
i-type a-SiGe:H	1.50		0.049	4×10^{16}	0.713	1.9×10^{-9}	2.0×10^{-5}
simulated	1.50	1.55		4×10^{16}	0.714	2.0×10^{-9}	0.1×10^{-5}
p-type a-SiC:H	1.95		0.205		0.45	1×10^{-6}	2×10^{-6}
simulated		1.95		1.5×10^{19}	0.43	2×10^{-6}	3×10^{-6}
n-type a-Si:H	1.76		0.105		0.20	2×10^{-2}	5×10^{-2}
simulated		1.80		2×10^{19}	0.20	4×10^{-2}	4×10^{-2}

The obtained measured and simulated values of film properties are given in Table I. The optical bandgap decreases with increasing GeH$_4$ fraction, whereas E_{Urb} is hardly influenced. The dangling bond density, N_{db}, is deduced from the low energy part of the spectrum following the method described in [10] and is found to increase with the GeH$_4$ fraction. Results of electrical measurements are also given in Table I. The dark conductivity, σ_{dark}, strongly increases with increasing GeH$_4$ fraction, which is consistent with the decreasing activation energy. The photoconductivity appears to become slightly smaller due to the increase of the dangling bond density.

In Fig. 2 the dark and illuminated current-voltage characteristics of the three p-i-n structures are presented. Clearly, as the bandgap of the intrinsic layer is reduced, the open circuit voltage of the cell decreases from 0.80 to 0.56 V. This is accompanied by an enhancement of the short circuit current from 12.1 to 16.4 mA/cm^2. In addition, the fill factor of the cell decreases from 0.69 to 0.56 when lowering the bandgap, which reflects a higher recombination in the device probably as a result of the increasing the dangling bond density. The latter can also be concluded from the shallower slope of the dark current-voltage measurements with the bandgap.

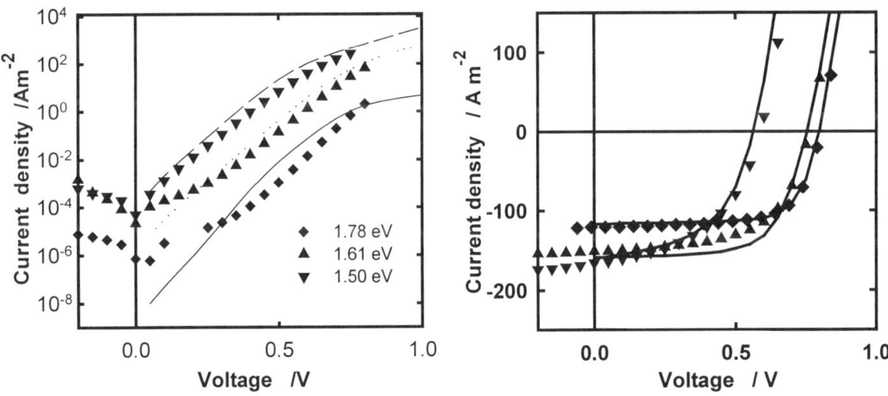

Figure 2. Measured (data points) and simulated (solid lines) current-voltage characterisctics of the p-i-n test structures, in the dark (left) and under AM 1.5 illumination (right).

Simulations

The simulations have been carried out using the simulation package ASA [11]. A standard model for the density of states is used, consisting of parabolic band states and exponential tail states. The gap states are represented by one acceptor-like and one donor-like Gaussian distribution, separated by a correlation energy of 0.4 eV (0.2 eV for the doped layers). The transition points between the extended and tail states are defined as the mobility edges, at which the total density of states N_c^{mob} and N_v^{mob} and its derivatives are continuous. Carriers inside this mobility gap are assumed to be immobile.

The procedure for obtaining a parameter set for a single layer is as follows. First the mobility gap, E_{mob}, of the material is calculated from the Tauc gap and the characteristic energies of the valence and conduction band tail. Therefore it is assumed that the Urbach edge is dominated by the characteristic energy of the valence band tail, $E_{VB,char}$, whereas the characteristic energy of the conduction band tail, $E_{CB,char}$, is derived from combined Time Of Flight and Post-Transit Spectroscopy reported in literature [12]. Subsequently the dangling bond density and position as well as the density of states at the mobility edges is varied to match the calculated absorption spectrum and Fermi-level position to the measured data. Assuming a constant optical matrix element, M, the spectrum is obtained by a convolution of the occupied and unoccupied states. The value for the optical matrix element is calculated by scaling the simulated spectrum to the measured data at a photon energy of 2.0 eV. It appeared that the density of states at the mobility edges can be kept constant for all intrinsic layers.

The obtained model parameter sets were then used for the intrinsic layers of the p-i-n devices. For the doped layers we used simulation parameters that are reported in [9]. To match the simulations to the current-voltage characteristics of the p-i-n devices, the capture rate constants, C, of the dangling bond states needed to be increased with increasing GeH_4 fraction. However, this resulted in a slight mismatch of the photoconductivity of the single layers. The values of the model parameters matching the simulated and measured properties are tabulated in Table II and the corresponding absorption spectra are shown in Fig. 1.

Table II: *The calibrated set of model parameters used in the simulations. Note that the dangling bond energy level is taken relative to midgap. $N_{c,v}^{eff}$ are the effective density of states of the conduction and valence band, respectively.*

		i-*a*-Si:H	i-*a*-SiGe:H	i-*a*-SiGe:H	p-*a*-SiC:H	n-*a*-Si:H
E_{mob}	[eV]	1.82	1.66	1.55	1.95	1.80
N_c^{eff}	[cm^{-3}]	3.5×10^{20}	3.0×10^{20}	2.9×10^{20}	1×10^{20}	1×10^{20}
N_v^{eff}	[cm^{-3}]	1.4×10^{20}	1.4×10^{20}	1.4×10^{20}	1×10^{20}	1×10^{20}
N_c^{mob}	[cm^{-3}eV^{-1}]	8×10^{21}	8×10^{21}	8×10^{21}	2×10^{21}	1×10^{21}
N_v^{mob}	[cm^{-3}eV^{-1}]	4×10^{21}	4×10^{21}	4×10^{21}	1×10^{21}	2×10^{21}
$E_{CB,char}$	[eV]	0.025	0.040	0.045	0.180	0.070
$E_{VB,char}$	[eV]	0.049	0.050	0.050	0.090	0.160
C_{tail}	[m^{3}s^{-1}]	0.5×10^{-15}	0.7×10^{-15}	0.7×10^{-15}	0.7×10^{-15}	0.7×10^{-15}
N_{db}	[cm^{-3}]	1.2×10^{16}	2×10^{16}	4×10^{16}	1.5×10^{19}	2×10^{19}
$E_{db}^{+/0}$	[eV]	0.23	0.28	0.28	0.38	0.5
$C_{db,n}$	[m^{3}s^{-1}]	2×10^{-14}	12×10^{-14}	50×10^{-14}	5×10^{-15}	5×10^{-15}
$C_{db,c}$	[m^{3}s^{-1}]	20×10^{-14}	25×10^{-14}	100×10^{-14}	50×10^{-15}	50×10^{-15}
μ_e	[m^{2}V^{-1}s^{-1}]	2.0×10^{-3}	2.0×10^{-3}	2.0×10^{-3}	2.0×10^{-3}	2.0×10^{-3}
μ_h	[m^{2}V^{-1}s^{-1}]	5.0×10^{-4}	5.0×10^{-4}	5.0×10^{-4}	5.0×10^{-4}	5.0×10^{-4}
M	[m^{5}eV^2]	0.96×10^{-49}	1.2×10^{-49}	1.3×10^{-49}	7×10^{-49}	10×10^{-49}

DISCUSSION

Using the model procedure described above, combined with the stringent restriction of continuous density of states, a reasonable match between the experimental data and the simulation results is achieved. Only a few parameters have to be adjusted to comply with the changes in material properties when increasing the Ge content. It appeared that the simulated conductivities are sensitive for the position of the dangling bonds, $E_{db}^{+/0}$ (Table II, relative to midgap). Changing the position of the dangling bonds leads to a shift of the Fermi-level position, which in turn is reflected in the dark conductivity.

For the simulation of the p-i-n structures we used an equal band offset at the p-i and i-n interfaces. To obtain reasonable fits to the current-voltage characteristics under AM1.5 illumination, an increase is needed of the capture rate constants, $C_{db,n}$ and $C_{db,c}$, of the neutral and charged dangling bond states, respectively, with decreasing bandgap (Figure 3). This is consistent with experimental evidence showing a strong decrease in the $\mu\tau$ product for a-SiGe:H with increasing Ge content [3].

CONCLUSIONS

We have identified seven independent model parameters that, when adjusted, are able to simulate p-i-n devices with bandgap grading: the bandgap, the characteristic energies of the band tails, the dangling bond density, the capture rate constants, and the position of dangling bonds. The first four parameters can be derived from experimental data. For the conduction band characteristic energy, the dangling bond density, and the capture rate constants we found a clear relation with the bandgap consistent with other reports [3].

So far we did not find a clear relation between the dangling bond position and the bandgap. The number of layers used for these simulations should be increased to establish a possible relation. Alternatively, bandgap grading can be modeled by a limited number of layers in which the parameters describing the dangling bonds are kept constant.

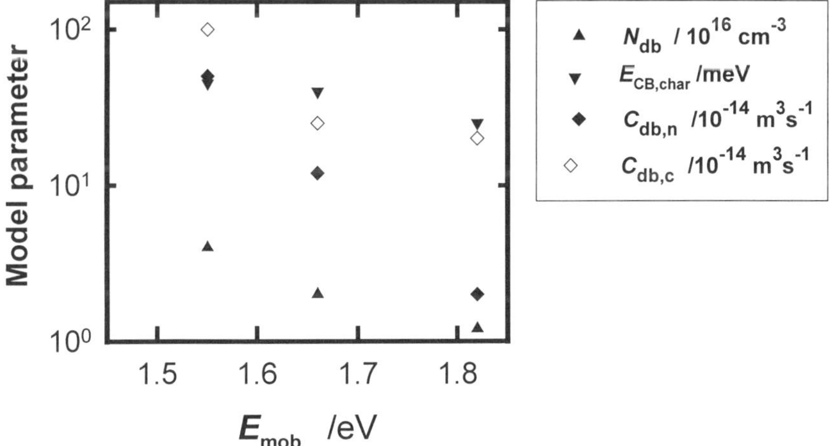

Figure 3. *Bandgap dependence of model parameters.*

ACKNOWLEDGEMENTS

This work is financially supported by the Netherlands Agency for Energy and the Environment (NOVEM) and the Dutch Council for Scientific Research (NWO).

REFERENCES

1. S. Guha, J. S. Payson, S. C. Agarwal, and S. R. Ovshinsky, J. Non-Cryst. Solids **97 & 98**, p. 1455 (1987).
2. M. Stutzman, R. A. Street, C. C. Tsai, J. B. Boyce, and S. E. Ready, J. Appl. Phys. **66**, p. 569 (1989).
3. G. H. Bauer, Sol. State Phen. **44**, p. 365 (1995).
4. M. S. Bennet, A. Catalano, K. Rajan, and R.R. Arya, in *Proc. 21th PVSC* (IEEE, New York, 1990), p. 1653.
5. S. Guha, J. Yang, A. Pawlikiewicz, T. Glatfelter, R. Ross, and S. R. Ovshinsky, Appl. Phys. Lett. **54**, p. 2330 (1989).
6. K. Vasanth, A. Payne, B. Crone, S. Sherman, M. Jakubowski, and S. Wagner, in *Amorphous Silicon Technology*, edited by M. Hack, E. A. Schiff, A. Madan, M. Powell, and A. Matsuda (Mater. Res. Soc. Proc. **377**, Pittsburgh, PA 1995), p. 663.
7. J. Zimmer, H. Stiebig, J. Fölsch, F. Finger, and H. Wagner, *in Proc. 14th EPVSEC*, Barcelona, edited by H. A. Ossenbrink, P. Helm, and H. Ehmann (Bedford, UK, 1997), p. 593.
8. C.R. Wronski, Z. Lu, L. Jiao, and Y. Lee, in *Proc. 26th PVSC* (IEEE, New York, 1997), p. 587.
9. M. Zeman, R. A. C. M. M. van Swaaij, E. Schroten, L. L. A. Vosteen, and J. W. Metselaar, in *Amorphous Silicon Technology*, edited by H. Branz, M. Hack, R. Schropp, I. Shimizu, and S. Wagner (Mater. Res. Soc. Proc. **507**, Pittsburgh, PA 1998), p. 409.
10. C. F. O. Graeff, M. Stutzmann, and K. Eberhardt, Phil. Mag. B **69**, p. 387 (1994). We used the absorption coefficient at a photon energy of 0.5 eV below the $E_{3.5}$ gap and multiplied this by 2×10^{16} cm^{-2} for a-Si:H, and towards 1.3×10^{16} cm^{-2}, which is valid for a-Ge:H.
11. M. Zeman, J.A. Willemen, L.L.A. Vosteen, G. Tao, and J.W. Metselaar, Solar Energy Mat. and Solar Cells **46**, p. 81 (1997).
12. C. E. Nebel, Ph.D. Thesis University of Stuttgart, 1990.

BUFFER LAYERS FOR NARROW BANDGAP A-SIGE SOLAR CELLS

X. B. LIAO, J. WALKER and X. DENG
Department of Physics and Astronomy, University of Toledo, Toledo, OH 43606

ABSTRACT

In high efficiency narrow bandgap (NBG) a-SiGe solar cells, thin buffer layers of unalloyed hydrogenated amorphous silicon (a-Si) are usually used at the interfaces between the a-SiGe intrinsic layer and the doped layers. We investigated the effect of inserting additional a-SiGe interface layers between these a-Si buffer layers and the a-SiGe absorber layer. We found that such additional interface layers increase solar cell V_{oc} and FF sizably, most likely due to the reduction or elimination of the abrupt bandgap discontinuity between the a-SiGe absorber layer and the a-Si buffer layers. With these improved narrow bandgap solar cells incorporated into the fabrication of triple-junction a-Si based solar cells, we obtained triple cells with initial efficiency of 10.6%.

INTRODUCTION

Narrow bandgap a-SiGe materials and solar cell devices have been studied extensively for their use in the spectrum splitting, multiple-junction a-Si based solar cells [1-3]. To achieve high conversion efficiency solar cells, an a-SiGe absorber layer is usually sandwiched between two thin a-Si buffer layers which are in direct contact with the p- and n- doped layers [3,4]. These a-Si buffer layers were found to enhance the performance of a-SiGe solar cells. However, even with these a-Si buffer layers, there are still abrupt discontinuity in the bandgap at the interfaces between these buffer layers and the a-SiGe absorber layer. In this paper, we report our study on the insertion of additional a-SiGe interface layers (with less Ge compared with the absorber) which reduces the bandgap offset.

EXPERIMENTAL DETAILS

The a-SiGe and a-Si materials used in this study were deposited using a ultra high vacuum plasma enhanced chemical vapor deposition (PECVD) system at the University of Toledo (UT). The intrinsic and doped layers were deposited in separated chambers of this multi-chamber, load-locked deposition system, having a base vacuum of 2×10^{-8} Torr. A gas mixture of GeH_4, Si_2H_6 and H_2 were used for the a-SiGe deposition and Si_2H_6 and H_2 for a-Si deposition. The deposition conditions include a substrate temperature range of 300-400 $^{\circ}$C, a chamber pressure of 0.5-0.6 Torr and an rf power of 2.5-3.0 W for the a-SiGe and a-Si layer depositions. Different GeH_4 to Si_2H_6 ratios were used to achieve different bandgaps for a-SiGe material. Graded bandgaps for the a-SiGe absorber layer and the a-SiGe buffer layers were achieved by adjusting GeH_4 flows during growth. The GeH_4/Si_2H_6 gas flow ratio for the a-SiGe absorber layers studied here was 0.88. All of the intrinsic absorber and buffer layers were deposited in a deposition chamber designated for intrinsic layer growth.

Boron doped microcrystalline silicon (μc-Si) p-layer and phosphorus doped a-Si n-layer were deposited in another chamber designated for the growth of doped layers. The device structure used in this study was: SS/Ag/ZnO/n(a-Si)/b(a-Si)/i(NBG a-SiGe)/b(a-Si)/p(μc-Si)/ITO for the standard device and SS/Ag/ZnO/n(a-Si)/b(a-Si)/b(a-SiGe)/i(NBG a-SiGe)/b(a-SiGe)/b(a-

Mat. Res. Soc. Symp. Proc. Vol. 557 © 1999 Materials Research Society

Si)/p(μc-Si)/ITO for the new devices under study, where SS is stainless steel, Ag/ZnO is back-reflector, b(a-Si) and b(a-SiGe) are thin a-Si and a-SiGe buffer and interface layers, respectively. The ITO layer, serving as the top electrode and anti-reflection coating, was deposited using a rf sputtering process. Many of the I-V measurements were done using 0.05-cm^2 device without metal grids. Quantum efficiency (QE) measurements were used to measure spectral response and to obtain short circuit current (J_{QE}) by integrating the QE value over AM1.5 Global spectrum.

RESULTS AND DISCUSSIONS

Figure 1 shows the band structure of the various devices used in this study. Fig. 1a is a schematic of the structure for standard a-SiGe devices, showing two a-Si buffer layers on both sides of a-SiGe absorber layer. Fig. 1b and 1c show the schematics of the device structures with the additionally inserted a-SiGe interface layers having a fixed Ge content (Fig. 1b) and with interface layers having a graded bandgap (Fig. 1c).

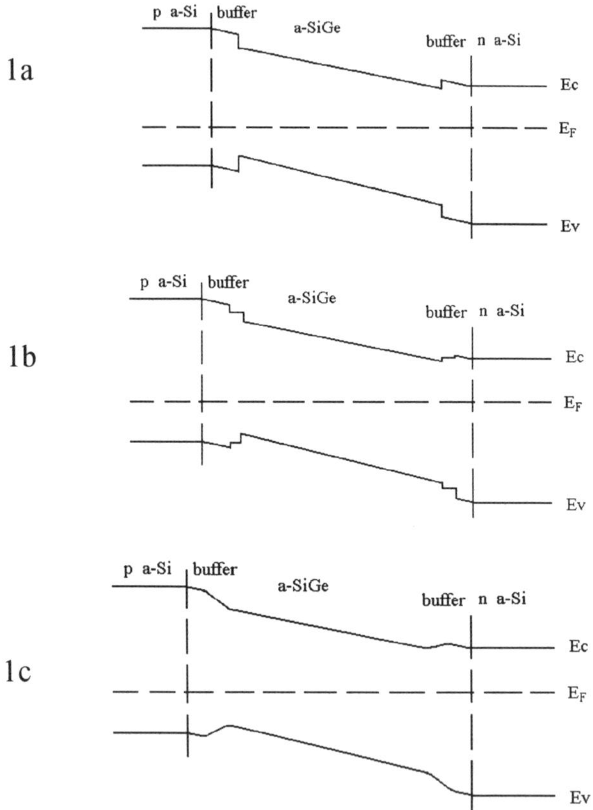

Fig. 1a, 1b and 1c Schematic energy band diagram of a-SiGe solar cells with different interfacial buffer layers under the thermal equilibrium condition.

Table 1 lists the photovoltaic parameters of the open circuit voltage V_{oc}, short circuit current density J_{sc}, fill factor FF and the maximum power output P_{max} of a-SiGe solar cells with different interfacial buffers. GD180 is a standard a-SiGe device (Fig.1a) and GD202 contains a-SiGe buffer layers with a fixed Ge content inserted between the a-Si buffers and a-SiGe i-layer at both the n- and p-interfaces (See Fig.1b), while GD203 and GD209 have Ge-content graded buffer layers at both interfaces (See Fig.1c). It is clearly seen that the V_{oc} and FF have been increased due to the introduction of the additional a-SiGe interfacial layers and the improvements are approximately in the range of 2-5% for V_{oc} and 5-10% for FF. The V_{oc} and FF enhancement was higher for graded buffer layers. For this reason, the J-V characteristics of GD180 and GD209 are compared in Fig.2, and the results of these two devices are discussed further in the following.

Figure 3 shows the QE curves for a-SiGe cells without (GD180) and with (GD203 and GD209) bandgap graded interfacial layers. The spectral responses of these devices are approximately the same in the red region. The small variation in the red could also be due to the variation in the back-reflector. The difference in the blue region of the QE curve is mostly resulted from the variation in the transmission of ITO films.

Table 1 Photovoltaic parameters of solar cells with different buffer layers

Samples	V_{oc} (V)	J_{sc} (mA/cm^2)	FF	P_{max} (mW/cm^2)
GD180 Standard	0.647	23.5	0.506	7.7
GD202 w/ a-SiGe buffer	0.662	21.6	0.534	7.6
GD203 w/ graded a-SiGe buffer	0.680	23.4	0.536	8.5
GD209 w/ graded a-SiGe buffer	0.675	24.3	0.559	9.2

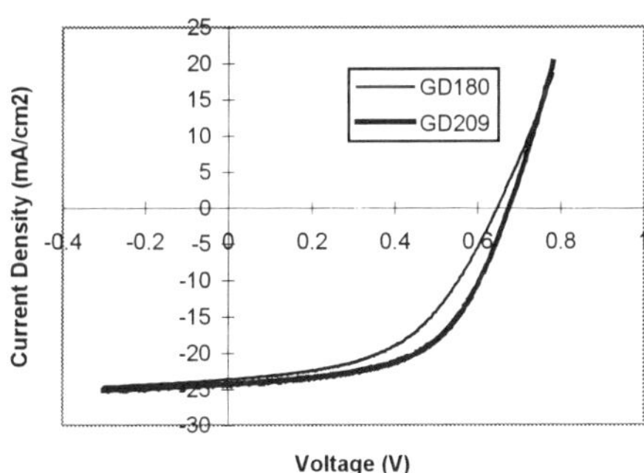

Fig.2 Illuminated J-V characteristics of GD180 and GD209 samples

Since small ITO contacts (0.05 cm^2) were used as the top electrodes, there might be some uncertainty in the active areas of the devices. Therefore, the J_{sc} values listed in Table 1 might not be highly precise. We used the current density values, J_{QE}, obtained by integrating the quantum efficiency curves over AM1.5 Global spectrum, as an independent check of the short circuit current values. Although the accuracy of absolute J_{QE} values from the QE measurements depend on the accuracy of the calibration of QE measurements, J_{QE} values obtained in this way are much more reproducible and precise than the J_{sc} values from the I-V measurements. The obtained J_{QE} and the relative P_{max} calculated from these J_{QE} values are summarized in Table 2. From this table, we find that the short circuit currents of standard a-SiGe sample (GD180) and the a-SiGe samples with graded buffer layers (GD203 and GD209) are approximately the same within the experimental variation. The enhancement in V_{oc} and FF result in a net increase in device P_{max}.

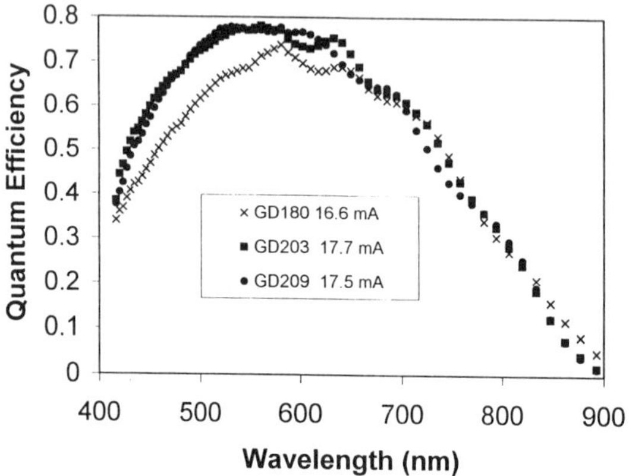

Fig.3 Quantum efficiency curves of a-SiGe solar cells. GD180 is standard solar cell, while GD203 and GD209 are cells with bandgap graded a-SiGe buffers.

Table 2 a-SiGe solar cell device performance

Sample	J_{QE} (mA/cm^2)	QE @800nm	V_{oc} (V)	FF	Relative P_{max} (mW/cm^2)
GD180	16.6	0.288	0.647	0.51	5.5
GD203	17.7	0.305	0.680	0.54	6.5
GD209	17.5	0.315	0.675	0.56	6.6

The influences on the performances of the solar cells from the recombination processes taking place at both interfaces between the intrinsic a-SiGe layer and doped layers could also be observed from the ideality factor (n) of the p-i-n devices. The improved J-V performance in the cells with bandgap graded a-SiGe interfacial layers are reflected also in the improvement of the ideality factor and the reduction in the reverse saturation current density (J_o) of these devices.

It is well known that the J-V characteristics of p-i-n a-Si solar cells under illumination can be approximately described by using transport equation of an ideal pn junction [5]:

$$J_{ph} = J_o [\exp(q(V - J_{ph}R_s) / nkT) - 1] - J_{sc} \qquad (1)$$

where n is the ideality factor of the device under illumination, J_{ph} represents the photocurrent density, $(V - J_{ph}R_s)$, in which J_{ph} could have negative value, represents the voltage at the junction of an ideal diode. The effect of the equivalent series resistance R_s is included in the equation, while the equivalent shunt resistance R_{sh} is not taken into account. In order to obtain n factor directly from experimental measurement, one could use the dependence of V_{oc} on J_{sc} under varying light intensities,

$$V_{oc} = (nkT/q)(\ln(J_{sc}/J_o + 1) \qquad (2)$$

where the effect of R_s is avoided, hence n factor can be determined from the slope of the V_{oc} versus $\ln(J_{sc})$. Figure 4 shows the experimental data of J_{sc} (V_{oc}) measured under various light intensities for samples GD180, with the standard structure of a-SiGe bottom cell, and GD209, with a graded a-SiGe buffer layer at both interfaces. It is seen that there is a nearly linear relationship between $\ln(J_{sc})$ and V_{oc} with R-squared values of 0.9975 and 0.9998 for the two samples studied, consistent with the expectation of Eq. (2). From the slopes of these two lines we deduced the quality factor n to be 2.02 and 1.77, for the samples without (GD180) and with (GD209) graded a-SiGe interfacial layers at the p-i and i-n junction between the a-Si buffer layer and a-SiGe intrinsic layer, respectively. The reverse saturation current density, obtained from these fittings, were 8.3×10^{-8} A/cm^2 and 7.2×10^{-9} A/cm^2 for GD180 and GD209, respectively. The reductions in the n and J_o values indicate an improvement in the pn junction quality and a decrease in the recombination rate of extra minority carriers, as a result of these additional interfacial layers.

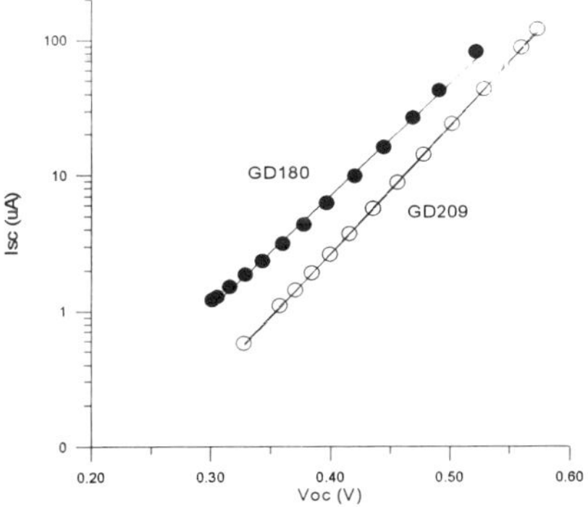

Fig.4 J_{sc} versus V_{oc} under various light intensity for samples GD180 (solid circle) and GD209 (open circle)

We incorporated the improved bottom cell into triple-junction a-Si based solar cells as the bottom cell. Figure 5 is a J-V curve for such a triple-junction solar cell measured at Energy

Conversion Devices, Inc. (ECD) under a solar simulator light, showing a 10.6% initial efficiency. Figure 6 is the QE curve for the triple-cell measured at ECD, showing a high integrated-total current of 23.6 mA/cm^2.

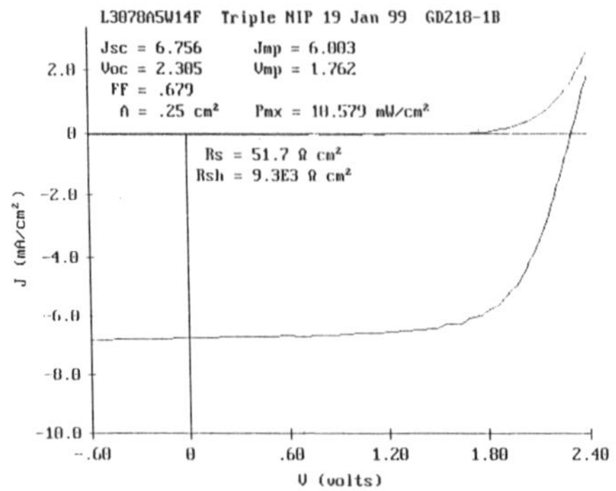

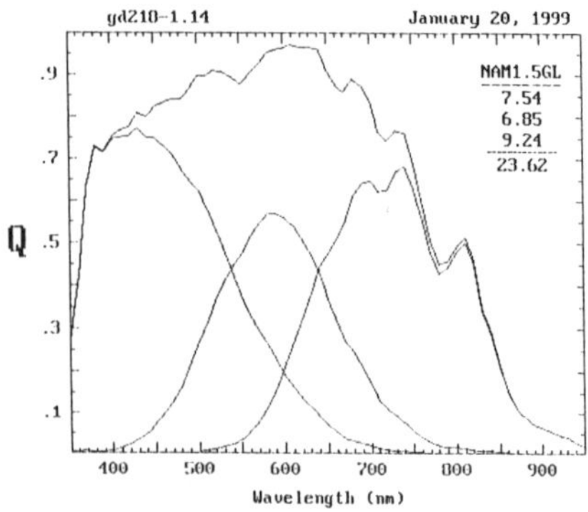

Fig. 5 J-V curve of a triple-junction solar cell incorporating the improved a-SiGe devices, showing an initial efficiency of 10.6%.

Fig. 6 QE curves of a triple cell incorporating the improved a-SiGe devices, showing an integrated J_{QE} of 23.6 mA/cm^2.

SUMMARY

We explored the effect of additional a-SiGe interface layers between narrow bandgap a-SiGe absorber layer and the a-Si buffer layers. The reduction and elimination in the bandgap offset at these interfaces result in sizable increases in device V_{oc} and FF, although the J_{sc} was somewhat unchanged. Incorporating NBG solar cells having such bandgap graded interface layers as the bottom cells, we fabricated triple-junction solar cells with 10.6% initial efficiency.

ACKNOWLEDGMENTS

We would like to thank Drs Al Compaan, Sijin Han, Pratima Agarwal, and Mr. Henry Povolny at UT, and Dr. Scott Jones and Mr. Tongyu Liu at ECD for discussions and help. This work was funded by NREL under the Thin Film Partnership Program (Subcontract No. ZAF-8-17619-14).

REFERENCES

1. See. Review by Y.S. Tsuo and W. Luft, Appl. Phys. Comm, **10**, 71 (1990) and references there in.
2. S. Guha, J.S. Payson, S.C. Agarwal and S.R. Ovshinsky, J. Non-Cryst. Solids 97-98, 1455 (1988).
3. J. Yang, R. Ross, T. Glatfelter, R. Mohr and S. Guha, MRS Poc. 149, 435 (1989).
4. X. Deng, "Development of High, Stable-efficiency Triple-junction a-Si Alloy Solar Cells", *Phase I Annual Subcontract Report*, July 18, 1994-July 17, 1995, prepared under Subcontract No. ZAN-4-13318-11. (February 1996). NREL/TP-411-20687. Available NTIS: Order No. DE9696000530.
5. S.M. Sze, Physics of Semiconductor Devices, 2nd Edition, John Wiley & Sons, New York, 1981.

CHARACTERISTICS OF DIFFERENT THICKNESS a-Si:H/METAL SCHOTTKY BARRIER CELL STRUCTURES—RESULTS AND ANALYSIS

Zhou Lu, Lihong Jiao, Randy Koval, Robert W. Collins and Christopher R. Wronski
Center for Thin Film Devices, The Pennsylvania State University, University Park, PA 16802.

ABSTRACT

Current-voltage, light I-V and internal quantum efficiency characteristics have been investigated in specular TCO/n$^+$(a-Si:H)/i(a-Si:H)/Nickel Schottky barrier cell structures with protocrystalline intrinsic layers. The studies were carried out on structures with different thickness i layers after a degraded steady state had been reached with AM1.5 illumination at 27°C. These characteristics were modeled using the Analysis of Microelectronic and Photonic Structures(AMPS) and a gap state distribution, which includes charged defects, used in the analysis of results of detailed studies on thin films[1]. Fits are obtained to these characteristics for the different thickness cell structures using the same parameters as those used to fit the results on the corresponding intrinsic thin films. The results obtained from this study offer an approach to more reliable modeling of solar cells at their "end of life".

INTRODUCTION

There is considerable interest in the reliable modeling of "end of life" in a-Si:H based solar cells. The distributions for the gap states used in the vast majority of the reported simulations on solar cell characteristics are based on gap state distributions associated with neutral dangling bonds whose densities are obtained from subgap absorption[1] or ESR measurements[2]. Unfortunately there is still a lack of direct correlation between such results on thin films and performance of their solar cells before[3] as well as after light induced degradaton[4]. Since the a-Si:H materials have significant differences in microstructures and hydrogen incorporation[5,6] it can be expected that they have correspondingly significant differences in gap states, not only just in the densities of their neutral dangling bond defects. This however appears not to be appreciated since most often the materials are still characterized by very limited measurements of photoconductivity and subgap absorption, generally by CPM[7]. The very limited information obtained from just those two measurements, and their simple interpretation limits the insights into the nature of the gap states affecting solar cell stability and performance. However, in the last few years the importance of charged defect states on the properties of a-Si:H materials has been recognized[8,9,10] and there have been attempts to characterize them and identify their contributions to the optoelectronic properties of a-Si:H and how they may affect solar cells[11].

The distribution of gap states used in the modeling here is based on that developed from earlier studies on thin films and Schottky barrier cell structures[8,11,12] which found that results, particularly in the degraded states, could not be fitted with just the two gaussian distributions extensively used[13,14]. A three gaussian distribution was proposed and used where in addition to the neutral dangling bond states near midgap, there were positively charged and negatively charged defects, above and below midgap respectively. This greatly improved the ability to model results on films as well as Schottky barrier cell structures[11,12], but there were still problems with fitting all the results selfconsistently. Selfconsistent analysis of the many results from detailed studies on thin films can however be obtained with a distribution which includes two gaussians for each set of charged defects. The large number of gap states parameters that are now present as inputs into modeling introduces both flexibility and the uncertainties which have

Mat. Res. Soc. Symp. Proc. Vol. 557 © 1999 Materials Research Society

to be minimized by the selfconsistent fitting of as many results as possible. Results are reported here on the characteristics of Schottky barrier solar cell structures of different thickness having protocrystalline i layers whose characteristics are fitted with the same gap state parameters as those used in fitting results on corresponding thin film materials[1]. These results are on both the films and the Schottky barrier structures for the degraded steady states reached with AM1.5 illumination similar to that found in corresponding p-i-n cells[15].

EXPERIMENTAL DETAILS

The a-Si:H layers of the glass(7059)/SnO$_2$(specular)/a-Si:H(n$^+$)/a-Si:H(i)/Ni Schottky barrier solar cell structures were prepared in a Tek-Vak MPS 4000-LS multi-chamber rf PECVD system under conditions previously described[15]. The a-Si:H n$^+$ layer were 350Å thick and the intrinsic layers were deposited with hydrogen dilution, R=H$_2$/SiH$_4$=10. The Schottky barriers were formed by thermal evaporation of 150Å thick, semitransparent Ni onto the i layers after etching with buffered HF. Forward bias I-V characteristics were measured from 0 to 2V and the light I-Vs were obtained by illuminating through the Ni contact with intensities of 0.1 to 3AM1.5 from an Oriel 6258 solar simulator. The internal quantum efficiencies(QE) were measured with no bias light under short circuit current conditions and their values obtained using the effective transmission(T$_r$) of the Ni obtained from saturated collection efficiencies under reverse bias[11]. The light induced degradation was carried out by illuminating at 27°C through TCO, primarily with ELH lamps calibrated for AM1.5. The mobility gap and the Ni electron barrier height were measured using internal photoemission[16] and accurate optical absorption coefficients between 0.4 to 0.8µm were obtained from transmission & reflection as well as spectroscopic ellipsometry measurements[17,18].

MODELING

The characteristics of Schottky barrier cells in the degraded steady state were analyzed using AMPS with the distribution of gap states shown in Fig. 1, where in addition to the neutral dangling bond D^0 includes positively charged (D$^+$, D^{+*}) and negatively charged (D$^-$, D^{-*}) defects states which are above and below midgap respectively. Each type of charged defect states is represented by two separate gaussian distributions to take into account not only their amphoteric nature, but also the very large differences in carrier capture cross-sections expected for their respective electron transitions. The positively charged defect D^{-*} states, which represent 0/- transitions and have small electron capture cross-section, are closer to the conduction band than the D$^-$ states, which represent +/0

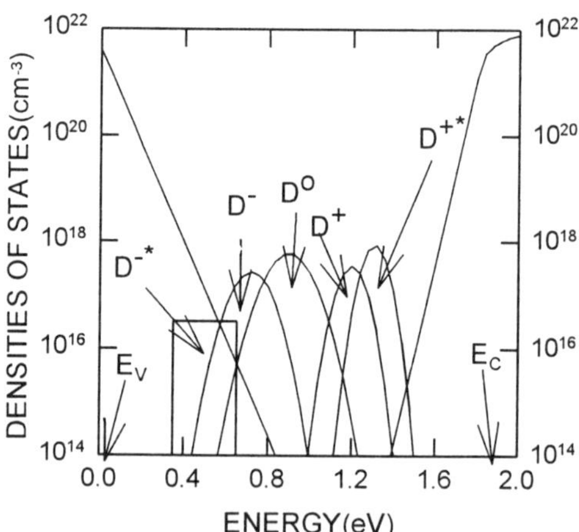

Fig.1 Distribution of gap states used in the analysis of the Schottky Barrier cell results in the AM1.5 degraded steady state. D^0 are the neutral dangling bond states, D$^+$ and D^{+*} are the positively charged and D$^-$ and D^{-*} are the negatively charged states.

transitions and have small hole capture cross-section. The negatively charged defect D^{-*} state, which represent +/0 transitions and have small hole capture cross-section, are closer to the valence band than the D^- states, which represent 0/- transitions and have small electron capture cross-section. The neutral dangling bond defects with their 0/- and +/0 transitions are represented here by a single D^0 gaussian since the differences in electron-hole capture cross-sections and the correlation energy are relatively small. Due to the present limitation in AMPS, the D^{-*} gaussian distribution used in the analysis of the thin film results is replaced by the rectangular one having the same density and twice the width of the corresponding gaussian. This difference is expected to have little effect since these D^{-*} states are shielded by the large valence band tail states with similar capture cross-sections. Because there is a large number of adjustable parameters such as the energy location, halfwidths, densities and carrier capture cross-sections associated with the gap states, a reiterative procedure was adopted first between the parameters obtained from thin film analysis and the forward I-V's. It was then extended to other Schottky barrier characteristics being guided by the relative sensitivity of different results to specific gap state parameters. The particular parameters which have the biggest effects on the analysis of the different results are listed in Table I.

Table I. Most important parameters in analysis of different results.

$\mu\tau$	subgap	forward I-V	light I-V	QE
$D^{-*}, D^0,$ $D^+(E)$	D^0, D^+, D^-	$D^+(E), D^{-*}(N)$	$D^0, D^{-*},$ VB(tail)	α, D^0, D^{-*}, VB(tail)

E: energy location only   N: density only

EXPERIMENTAL RESULTS

Dark I-V characteristics

Forward bias dark current characteristics from 0 to 2V obtained at room temperature are shown in Fig. 2, where the symbols are experimental results for cells with 0.2, 0.5 and 0.7μm thick i-layer in the degraded steady state(DSS). Also shown are results for the 0.7μm cell in the annealed state(AS). At low forward bias, the currents are controlled by the Schottky barrier height for electron, ϕ_{BN}=0.97eV, and have an exponential region before they become limited by the bulk and hence the thickness of the i layer. This can be clearly seen in the case of the 0.7μm cell in the annealed state. Fig. 2 also show that these far forward bias currents decrease by many orders of magnitude after illumination due to the creation of light induced defects in the bulk. This is

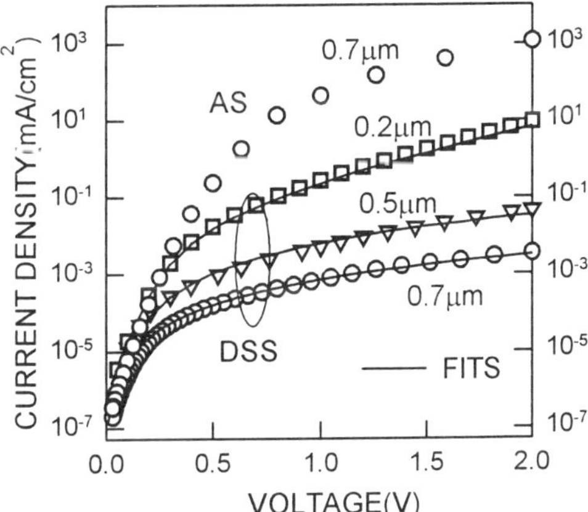

Fig. 2 Forward I-V characteristics of 0.2, 0.5 and 0.7μm n-i Schottky barrier cells in the degraded steady state(DSS). Symbols are experimental results and solid lines are fits. Also shown are results for the 0.7μm in the annealed state(AS).

indicated not only by the thickness dependence of the dark I-V's but also by their large, rapid changes after about 30 seconds of AM1.5 illumination such as observed in the photoconductivity of the thin films[19]. Corresponding illumination with just blue light through the TCO/n⁻ however does not cause any such change in the currents which indicates that the n/i interface regions do not affect the I-V characteristics.

Light I-V characteristics

The light I-V characteristics for the 0.5 and 0.7μm thick Schottky barrier structures of Fig. 2, are shown in Fig. 3. The open circuit voltages, V_{oc}, which are primarily determined by ϕ_{BN}, are independent of the cell thickness and are the same as in the annealed states. The short circuit currents, J_{sc}, which depend on thickness are similar in the DSS unlike the annealed state. The convergence of these values for all the cells in Fig.3 is due to the larger changes in the collection efficiencies which occur as the thickness of the i layer is increased and is discussed later. As expected, the fill factors(FF) which are determined by the bulk depend on the thickness of the cell structures. The degraded steady state FF values of the cells in Fig.3 change from 0.58 to 0.46 as the thickness increases from 0.2 to 0.7μm. Since no contribution of the n/i region is observed in the I-V characteristics, the characteristics of even the 0.2μm thick structure should be dominated by the bulk i layer as found for the corresponding p-i-n cells[20].

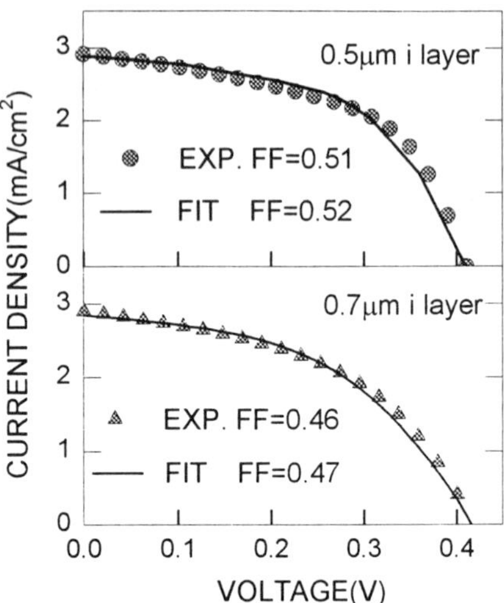

Fig. 3 The light I-V characteristics for the 0.5 and 0.7μm Schottky barrier cells in the degraded steady state. Symbols are experimental results and solid lines are the fits.

Internal quantum efficiency

The internal QE, in which the effective absorption of the Ni contact is taken into account, are shown in Fig. 4 for the two cells of Fig. 3. The distinct differences in both the shapes and magnitudes of the collection efficiencies with thickness of the i layer reflect not only the effects of optical absorption but also the changes occurring in electric fields and recombination directly related to the light induced gap states. Consequently the changes with thickness from AS to DSS systematically increase and result in J_{sc}'s of Fig. 3. These changes are illustrated in Fig.4 where the QE's are also shown in the annealed state as the dashed lines.

ANALYSIS

Refinements were carried out on the original 'operational' parameters derived from the selfconsistent fitting of the thin film results as well as the forward bias and light I-V characteristics of the different thickness structures. When the original gap state parameters were applied to the light I-V characteristics, the simulated FFs were significantly higher than the experimental values. Based on sensitivities, such as listed in Table I, systematic changes were made to the capture cross-sections of the` valence band tail and the D^{+*} states. By adjusting the values of these parameters, while at the same time maintaining good fits to both the film results and the forward I-V characteristics, it was possible to obtain good fits to the light I-V's. Such fits to the forward I-Vs and light I-Vs of cell structures with different thickness are shown as the solid lines in Fig. 2 and Fig. 3 respectively. The good agreement seen in Fig. 3 for the fill factors, within 0.01, and short circuit currents is also obtained for the 0.2μm thick cell structure. The light I-V's in Fig. 3 are for AM1.5 illumination through the Ni contacts and the simulations take into account its effective transmission. Equally good fits are obtained to experimental results when the incident intensity of illumination is changed between 0.1 and 3AM1.5. Even though the current densities in Fig.2 differ by over three orders of magnitude excellent fits are obtained for all these cells. It is important to note here

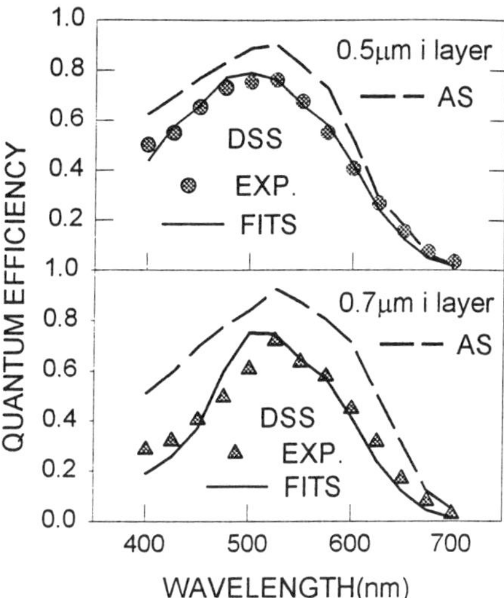

Fig. 4 The internal quantum efficiencies of 0.5 and 0.7μm Schottky barrier cells in the degraded steady state with symbols as experimental results and solid lines as fits. Also shown are the results for the annealed state(AS) as dashed lines.

that in all these simulations no n/i interface region is introduced and that they were all carried out with the **same** gap state parameters as that used to fit the thin film results.

The results for the QE obtained with these gap state parameters are shown as the solid lines in Fig. 4. It is important to note here that the different wavelengths in quantum efficiency measurements probe different regions of the i layers and correspond to a discrete measurement to obtain good fit over the entire wavelength region requires highly accurate values of optical absorption over the entire range as well as reliable gap state parameters. Excellent fits are obtained for both the 0.2 and 0.5μm thick cells over the entire wavelength region. The difference between the simulated and the experimental results for the 0.7μm cell seen in Fig. 4, for wavelengths less than 550nm, could be due to the larger difficulty of fitting thicker cell structures and/or the onset of a phase transition in this i layer(R=10)[21]. Effects of possible nonuniform generation of light induced defects could be discounted as there was no further changes in the light I-V after additional illumination with red light.

CONCLUSIONS

Schottky barrier solar cell structures with different thickness of protocrystalline i layers have been studied in their AM1.5 degraded steady state. Current-voltage, light I-V and internal quantum efficiencies were characterized for i layer thickness from 0.2 to 0.7μm. Analysis of their characteristics was carried out using AMPS and a distribution of gap states previously used in the analysis of results from detailed studies on thin films[1]. Good fits are obtained with the same parameters for all the gap states as those used to fit corresponding thin film results. The 'operational' parameters derived here offer a more reliable approach for modeling corresponding p-i-n and n-i-p solar cells characteristics in their "end of life" degraded state.

ACKNOWLEDGEMENTS

The authors wish to acknowledge the technical help and useful discussion with J. Koh, A.H. Mahan, R.S. Crandal and E.A. Schiff. This work is sponsored by National Renewable Energy Laboratory(NREL) under subcontract number XAF-8-17619-22, Electric Power Research Institute(EPRI) Thin Film Solar Cell Program, and New Energy Development Organization(NEDO) International Joint Research Grant.

REFERENCE

1. L. Jiao, S. Semoushikiana, Y. Lee, C. R. Wronski, Mat. Res. Soc. Proc, 97(1997)
2. H. Dersch, J. Stuke, and J. Beichler, Appl. Phys. Letts. 38, 456
3. J. Yang, X. Xu, and S. Guha, Mat. Res. Soc. Symp. Proc. Vol. 336, 687(1994)
4. L. Yang and L. Chen, Mat. Res. Soc. Symp. Proc., 336, 669(1994)
5. J. Matsuda et. al., Jpn. J. Appl. Phys., 25, L54(1986)
6. R. Collins, et. al., Conf. Record of 25th IEEE PVSC(1996)
7. M. Vanacek, J. Stuchlik, J. Kocka and A. Triska. J. Non-Crys. Solids 78/79, 299(1985)
8. M. Gunes and C.R. Wronski, J. Appl. Phys. 81, 3526(1994)
9. M.J. Powell and S.C. Dean, Phys. Rev. B, 48, 10815(1993)
10. H.M. Branz and M. Silver, Phys. Rev. B 42, 7420(1990)
11. H. Liu, L. Jiao, S. Semoushkina, C.R. Wronski, J. Non-Crys. Solids 198-200(1996)
12. L. Jiao, H. Liu, S. Semoushikina, Y. Lee and C.R. Wronski, Conf. Record of 25th IEEE PVSC(IEEE, NY, 1996), 1073(1996)
13. K. Vasanth and S. Wagner, First WCPEC, 488,(IEEE, 1994)
14. H. Zhu and S. Fonash, Conf. Record of 25th IEEE PVSC(1996)
15. Y. Lee, L. Jiao, H. Liu, Z. Lu, R. Collins and C. Wronski, Conf. Record of 25th IEEE PVSC(1996)
16. C.R. Wronski, S. Lee, M. Hicks and S. Kumar, Phys. Rev. Lett. 63, 1420(1989)
17. N. Maley, Jpn. J. Apl. Phys. 31, 768(1992)
18. R. Dawson, Y. Li, M. Gunes, S. Nag, R.W. Collins, C.R. Wronski, M. Bennett, and Y.M. Li, Conf. Record of 11th E.C. Photovoltaic Solar Energy Conf.(1992)
19. L. Jiao, H. Liu, S. Semoushikina, Y. Lee and C.R. Wronski, Appl. Phys. Lett. 69, 3713(1996)
20. R.J. Koval, J. Koh, Z. Lu, Y. Lee, L. Jiao, C.R. Wronksi, R.W. Collins(These proceeding)
21. J. Koh, Y. Lee, H. Fujiwara, C.R. Wronski, R.W. Collins, Appl. Phys. Lett. 73, 1526(1998)

AMORPHOUS SILICON SOLAR CELLS TECHNIQUES
FOR REACTIVE CONDITIONS

Satoshi Shimizu, Kojiro Okawa, Toshio Kamiya, C.M.Fortmann and Isamu Shimizu
The Graduate School, Tokyo Institute of Technology, 4259 Nagatsuta, Midori-ku,
Yokohama, JAPAN

ABSTRACT

The preparation of amorphous silicon films and solar cells using SiH_2Cl_2 source gas and electron cyclotron resonance assisted chemical vapor deposition (ECR-CVD) was investigated. By using buffer layers to protect previously deposited layers improved a-Si:H(Cl) solar cells were prepared and studied. The high quality a-Si:H(Cl) films used in this study exhibited low defect densities ($\sim10^{15}cm^{-3}$) and high stability under illumination even when the deposition rate was increased to $\sim15A/s$. The solar cells were deposited in the n-i-p sequence. These solar cells achieved V_{oc} values of $\sim 0.89V$ and $\sim 3.9\%$ efficiency on Ga doped ZnO (GZO) coated specular substrate. The a-Si:H(Cl) electron and hole $\mu\tau$ products were $\sim10^{-8}cm^2/V$.

INTRODUCTION

Amorphous silicon solar cells continues to be one of the most promising thin film photovoltaic technologies due to its high performance and its low cost. Guha [1] et al. have reported amorphous silicon based solar cells with stabilized efficiencies of greater than 13%. Even so, further cost reductions and performance increases must be implemented in order to achieve more widespread commercialization. For example, cost reductions can be obtained through increased film deposition rates which increases the cost effective use of production equipment and by using less expensive source materials. The light induced degradation of amorphous silicon materials is still one of the greatest remaining performance issue. Previously Azuma *et al.* [2,3] reported that a-Si:H(Cl) films prepared using less expensive SiH_2Cl_2 source gas could be deposited at high growth rates. These a-Si:H(Cl) films also showed improved stability under illumination compared to that of standard a-Si:H prepared using SiH_4[4]. Also, the defect density of these a-Si:H(Cl) films was $5\times10^{15}cm^{-3}$ in the annealed state, and $2\times10^{16}cm^{-3}$ in the degraded state. The time to reach the saturation degraded state was also short. These results are an indication that a particularly well relaxed, stable network was formed by the ECR-hydrogen plasma technique used in these reports.

This paper focuses on the study and optimization of SiH_2Cl_2 based n-i-p solar cells. The effect of band gap tuning and of the a-Si:H(Cl) i-layers properties on the solar cell performance was systematically examined. Capacitance measurement and the dependence of illumination direction on collection efficiency were used to probe electric transport. Also, the i-layer space charge regions were examined by the analysis of the light intensity dependence of photovoltaic parameters. Solar cell stability was also measured.

Mat. Res. Soc. Symp. Proc. Vol. 557 © 1999 Materials Research Society

EXPERIMENT

Amorphous silicon thin films and solar cells were fabricated in a load locked three chamber system. The a-Si:H(Cl) intrinsic layers were deposited in a ECR-CVD chamber using SiH_2Cl_2 and H_2 gas mixtures. The band gap of the a-Si:H(Cl) i-layers were relatively wide at ~1.85eV. The doped layers and buffer layers used in this work were prepared using conventional parallel electrode, capacitively coupled RF-CVD using SiH_4. Boron doped a-SiC:H or a-Si:H p-layers were used. All n-layer used in this study were P doped a-Si:H. A 40 nm n/i interfacial buffer layer was applied to all solar cells of this study to prevent the reactive species from the high density ECR-hydrogen plasma from degrading the n-layer. N-i-p solar cells were deposited on either textured Ag/ZnO coated stainless steel substrates or specular transparent Ga doped ZnO (GZO) coated Corning 7059 glass substrates. In both cases semi transparent Pt top contacts were used. In the optimized solar cell cases of this work, a more reflective Ag contact was used. Solar cell characterization included dark and photo I-V; as well as, stability measurements which were carried out under halogen lamp ($100mW/cm^2$) and Xe lamp ($300mW/cm^2$) illumination respectively.

RESULTS AND DISCUSSION

For electronic transport analysis a n-i-p solar cell with an i-layer thickness of 300 nm and 40nm n/i interface buffer layer was prepared on a Ag/ZnO coated stainless steel substrate. The effect of the band gap of the n/i interfacial layer was examined. Figure 1 shows the current vs. voltage characteristics of two solar cells having either a 1.75eV or a 1.80eV a-Si:H n/i buffer layer. Under $100mW/cm^2$ white light p-side illumination, the 1.80eV band gap buffer layer case showed a much higher J_{sc}. The spectral response revealed that the J_{sc} increase was due to improved short and medium wavelength response. Both cells showed a V_{oc} value of ~0.80V. This result indicates that reduced energy barriers at the n/i interface for electron transport, and favorable band bending reducing the hole recombination rate at the n/i interface. However, the low fill factor required further optimization, therefore, the i-layer dependence of the fill factor was investigated.

Using a n: a-Si:H(P) /

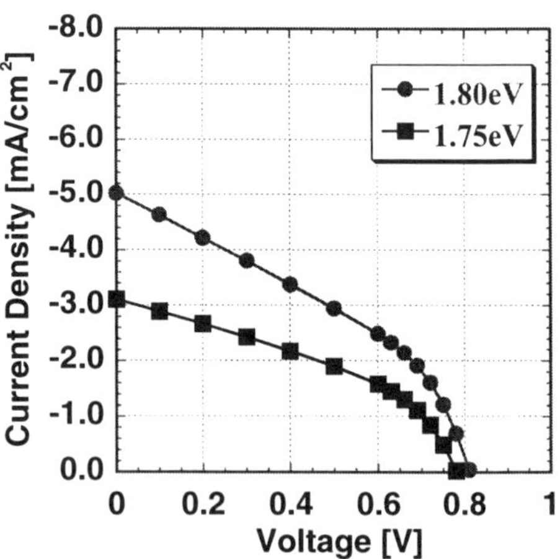

Fig.1 Photovoltaic performance of solar cells with different n/i buffer layers.

buffer: a-Si:H / i: a-Si:H(Cl) (2μm) / p: Si:H(B) solar cell the μτ products were estimated by fitting the measured 520 nm monochromatic current collection versus voltage data using the Hecht relation. Figure 2 shows the electron and hole μτ products versus reciprocal temperature for this n-i-p solar cell. Where, the electron and hole μτ activation energies were 29meV and 169meV respectively, and room temperature electron and hole μτ products of 4.4×10^{-8}cm^2/V and 3.5×10^{-8}cm^2/V were determined respectively. The hole μτ products is comparable to high quality un-alloyed amorphous silicon [5]. However, this electron μτ products is slightly smaller than that reported

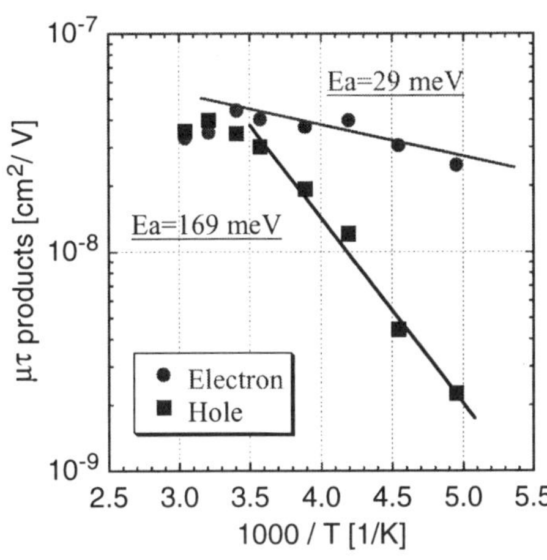

Fig.2 Electron and hole μτ products of a-Si:H(Cl).

for device quality un-alloyed amorphous silicon (2.0×10^{-7}cm^2/Vs)[6]. However, it is noted that in the case of a-SiGe:H solar cells prepared from materials with $\mu\tau_e$ values less than 10^{-7}cm^2/Vs, that it is still possible to prepare solar cells with fill factor greater than 60%[7].

Previous studies [2] indicated that a-Si:H(Cl) films showed that the Cl content, the defect density and the electronic transport were all strong functions of deposition conditions such as the substrate temperature and the SiH$_2$Cl$_2$ and H$_2$ flow rates. For example, high defect densities, large Cl contents and poor electronic transport were found at low T$_s$ values. The substrate temperature and hydrogen dilution ratio were systematically varied; however, the fill factor was found to be unrelated to these parameters. This result suggested that the low fill factor was not related to the i-layer properties. Therefore, other components or the device structures were examined in detail. Using the transparent GZO coated glass substrates and semi transparent Pt electrodes, n and p side illuminated photovoltaic properties were probed [8].

Figure 3 shows the p and the n-side light I-V characteristics of a GZO / n: a-Si:H(P) / buffer: a-Si:H / i: a-Si:H(Cl) (300nm)/ p: Si:H(B) solar cell. A 48% fill factor was mesured for the n-side illumination case;

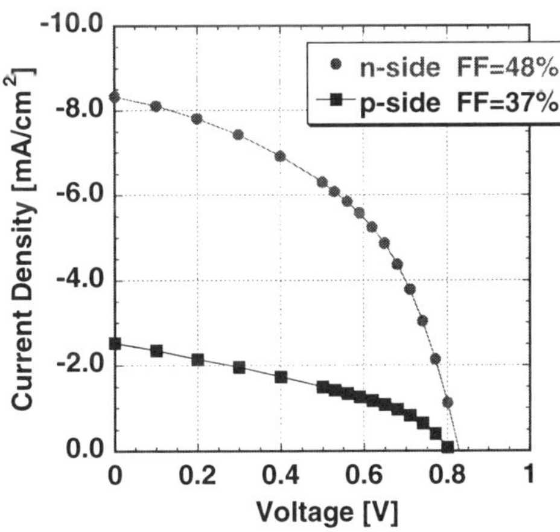

Fig.3 Photovoltaic performances under light illumination incident from n- or p-layer.

while, 37% was measured for the p-side illumination case despite the lower light intensity (and J_{sc}) resulting from the attenuation (~30% transmission) of the incident light by the semi-transparent Pt contact. Figure 4 shows the spectral responses of a GZO / n: a-Si:H(P) / buffer: a-Si:H / i: a-Si:H(Cl) (1μm) / p: Si:H(B) solar cell. The short wavelength response is much reduced in the p-side illumination case even when the response is corrected for the light attenuation in the Pt top contact. The p-side illumination case shows voltage dependent response particularly in the short wavelength region. This voltage dependent short wave-

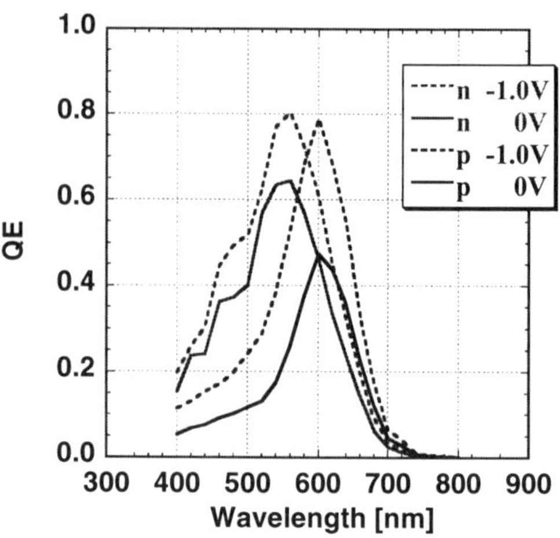

Fig.4 Spectral response under reverse bias as a function of light illumination direction.

length spectral response suggests the electric field distribution is not completely uniform. The a-Si:H(Cl) i-layer appears intrinsic since 40kHz frequency capacitance measurement assuming ε=12, indicated full i-layer depletion. Even so, the electric distribution could be influenced by the weakly p-type character of the i-layer [2,3] or by a defected layer near the p/i interface.

The p/i interfacial region was further optimized. Wide gap B_2H_6 doped a-SiC (B) p-layers were used instead of the more simple p-type a-Si:H(B). As the a-SiC bandgap increased from 1.84eV to 1.96eV, the Voc increased from 0.88V to 0.90V. On the other hand, the fill factor was not correlated to the p-layer band gap over this range and was ~ 46% in all cases. Figure 5 shows the properties of GZO / n: a-Si:H(P) / buffer: a-Si:H / i: a-Si:H(Cl) (250nm) / p: SiC:H(B) solar cell which had a reflective Ag back electrode under n-side illumination. The solar cell characteristics were: Voc=0.89V, Jsc=9.5mA/cm², FF=46% and η=3.9%. This is the highest value yet reported for a solar cell employing an a-Si:H(Cl) i-layer. It was also found that the fill factor was independent of the light intensity which suggests that the depletion region width is not affected by photo

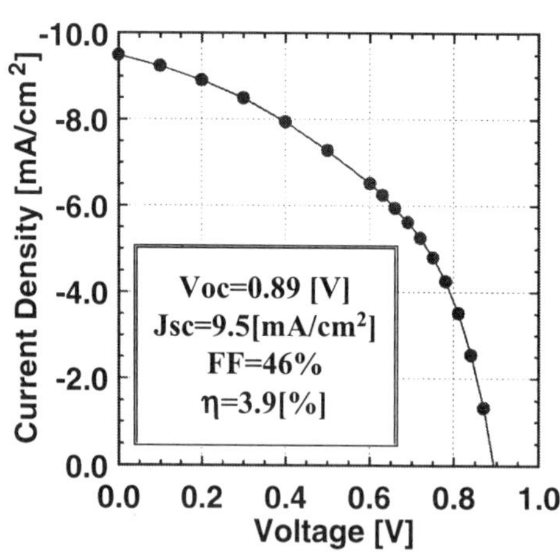

Fig.5 Photovoltaic performance of the n-i-p solar cell.

generated defect charging.

Light induced effects in a-Si:H(Cl) solar cells. were examined. Figure 6 compares the stability of a standard RF-PECVD (300nm i-layer) solar cell to a GZO / n: a-Si:H(P) / buffer: a-Si:H / i: a-Si:H(Cl) (300nm) / p: SiC:H(B) solar cell. Voc and FF are plotted as a function of light soaking time. We compared the optimized SiH_2Cl_2 based solar cell and SiH_4 based conventional one. Here, it should be noted that this SiH_4 based standard solar cell did not have any buffer layers and was not fully optimized. For both standard and a-Si:H(Cl) solar cells, the Voc values were independent on the light soaking, while obvious differences were

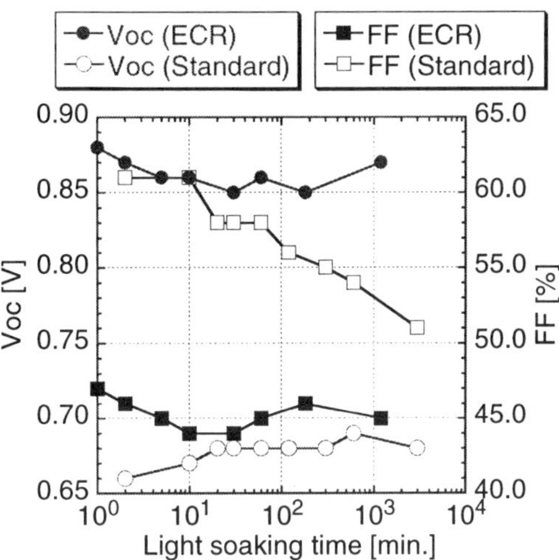

Fig.6 Photovoltaic performance (Voc, FF) stability against light soaking.

seen in the FF values. While the a-Si:H(Cl) solar cell started with a rather low fill factor, it was quite stable. On the other hand, strong degradation was seen in the standard RF a-Si:H solar cell. To confirm the high stability of amorphous silicon layer made from SiH_2Cl_2, we also measured electron and hole $\mu\tau$ products as a function of light soaking. The 2μm thick i-layer n-i-p solar cell was used for this measurement. The electron and hole $\mu\tau$ products were constant at $2\times10^{-8}cm^2/V$ and $6\times10^{-9}cm^2/V$ respectively as a function of light soaking time. which is consistent with the previous film studies of reference [2,3].

CONCLUSION

N-i-p solar cells have been prepared having a-Si:H(Cl) i-layers deposited from SiH_2Cl_2 using an ECR-CVD deposition technique. By implementing improved solar cell processing steps [9], improved solar cell designs with optimized i-, p-, and buffer layer band gaps improved solar cells could be prepared. These solar cells were studied under n and p-side illumination. Using these solar cell structures the room temperature electron and hole $\mu\tau$ products were measured to be $4.4\times10^{-8}cm^2/V$ and $3.5\times10^{-8}cm^2/V$ respectively which is sufficient for the solar cell application. The J_{sc} markedly increased when a wide gap n/i buffer layer was used. The n-i-p solar cells showed V_{oc} values of 0.90V when larger gap a-SiC:H p-layers were used. A photovoltaic performance characterized by Voc=0.89V, Jsc=9.5mA/cm^2, FF=46% and η=3.9% was obtained on a specular substrate. The stability of this solar cell was excellent. Light intensity dependence measurement indicate that i-layer space charge was not the origin of the FF. Further improvements are expected through further optimization of the components.

REFERENCES

1. J.Yang, A.Banerjee, and S.Guha, Appl.Phys.Lett. **70**, 2975 (1997).
2. M.Azuma, T.Yokoi and I.Shimizu, Mater.Res.Soc.Symp.Proc. **336**, 517 (1994).
3. M.Azuma, T.Yokoi and I.Shimizu, J.Non-Cryst.Solids **198-200**, 419 (1996).
4. M.Stutzmann, W.B.Jackson, C.C.Tsai, Phys, Rev. **B32**, 23 (1985).
5. Topics in Applied Physics **55**, The Physics of Hydrogenated Amorphous Silicon 1, edited by J.D.Joannopoulos and G.Lucovsky (Springer-Verlag, Berlin, Heidelberg, New york, Tokyo, 1984), p.99
6. C.M.Fortmann in Plasma Deposition of Amorphous Silicon-Based Materials, edited by G.Bruno, P.Capezzuto and A.Madan (Academic Press, New York, 1995) p.131
7. C.M.Fortmann, in AIP Conf. Proc. **157**, Stability of Amorphous Silicon Alloy Material and Devices, edited by B.L.Stafford and E.Sabisky (American Institute of Physics, New York, 1987) p.103
8. D. Fischer, N. Wyrsh, C.M. Fortmann, A.V. Shah, IEEE PVSEC (1993)
9. M. Kanbe *et al.* this conference

Part X

Detectors, Imagers and Other Devices

APPLICATION OF THIN-FILM MICROMACHINING FOR LARGE-AREA SUBSTRATES

M. BOUCINHA[1], V. CHU[1], V. SOARES[1] AND J. P. CONDE[2]
[1] Instituto de Engenharia de Sistemas e Computadores (INESC), R. Alves Redol, 9, 1100 Lisboa, Portugal, mjgb@eniac.inesc.pt
[2] Department of Materials Engineering, Instituto Superior Técnico (IST), Av. Rovisco Pais, 1049-001 Lisboa, Portugal

ABSTRACT

Surface micromachining is used with amorphous silicon, microcrystalline silicon, silicon nitride and aluminum films as structural materials to form bridge and cantilever structures. Low temperature processing (between 110 and 250 °C) allowed fabrication of structures and devices on glass substrates. Two processes involving different materials as the sacrificial layer are presented: silicon nitride and photoresist. The mechanical integrity of the fabricated structures is discussed. As examples of possible device applications of this technology, air-gap thin film transistors and the electrostatic actuation of bridges and cantilevers are presented.

INTRODUCTION

Microelectromechanical systems (MEMS), which extend the silicon microelectronics technology to the fabrication of 3-dimensional micromachined structures, is a fast growing field and is already competitive against the traditional sensor and actuator field [1-4]. An important advantage of MEMS lies in the possibility of producing a sensor, an actuator and their control electronics together in the same substrate (and, potentially, also the functions of communication) [4,5]. Various mechanisms of sensing and actuation can be used, such as electrostatic, magnetic and electromagnetic, piezoelectric, shape-memory alloys or thermoelectromechanical. MEMS applications to a variety of fields are being developed using these mechanisms: micromotors, biosensors, optical systems, pressure sensors, fluid systems and accelerometers among others [5-11]. Some are already being marketed [10].

The fabrication processes used in MEMS have been heavily based on those of silicon microelectronics with some extra demands [12]. The first similarity is the extensive use of c-Si as the base substrate. From this starting point, a variety of processing techniques are possible, most relevant of which are bulk micromachining and surface micromachining. In bulk micromachining, the mechanical structures are created within the silicon wafer by selectively removing wafer material. For example, the c-Si wafer is covered with a patterned mask film (silicon carbide or silicon nitride) and then it is etched either isotropically (typically, using wet etch with a HF/Nitric acid mixture or dry etch using XeF_2) or anisotropically (typically, using wet etch with KOH, TMAH or EDP) to produce the electromechanical structure [1,13]. In surface micromachining, the c-Si substrate is first coated with a thin-film isolation layer (usually silicon nitride). Then alternate layers of controlled-stress polysilicon (typically deposited at 580 °C by LPCVD and annealed at 1050 °C) and sacrificial silicon dioxide are deposited and photolithographically patterned. The sacrificial layer is removed by HF. In principle, it is possible to fabricate MEMS using surface micromachining with different combinations of structural, sacrificial and etchant materials (for example, in Al structures with a sacrificial polymer layer, removed by plasma etching). The requirement is that the sacrificial layer must be isotropically etched with very high selectivity compared to the structural layer[1, 14,15]. The technology already developed for c-Si

Mat. Res. Soc. Symp. Proc. Vol. 557 © 1999 Materials Research Society

MEMS can, in principle, be transferred to substrates other than c-Si with possible advantage. Namely, surface micromachining could easily be transferred to other substrates if it is possible to find a combination of a structural layer, a sacrificial layer, and an etchant of the sacrificial layer appropriate to the desired substrate material. a-Si:H is today a standard thin-film semiconductor material with wide application in solar cells, flat-panel displays, and linear and 2D imaging devices [16,17]. These applications result from the possibility of depositing device-quality a-Si:H and related materials such as, e.g., microcrystalline silicon and silicon nitride, in large area, inexpensive or exotic substrates at low substrate temperatures. In this paper, a study of surface micromachining of bridge and cantilever structures on Corning 7059 glass is presented. As an example of applications of this technology to a-Si:H air-gap TFTs and the electrostatic actuation of bridges and cantilevers are discussed.

MICROMACHINED BRIDGE AND CANTILEVERS ON GLASS

Fabrication process

Three-dimensional structures made from a-Si:H, amorphous silicon nitride (SiN$_x$), microcrystalline silicon (μc-Si:H) and metal (Al) were fabricated on glass substrates at low substrate temperatures (T$_{sub}$). The basic structures fabricated were bridges and cantilevers. The general steps to make these simple structures are schematically presented in Figure 1. First, the sacrificial layer is deposited and patterned (fig. 1 (a)). Next, the structural layer is deposited and patterned (fig. 1 (b)). The last step is an immersion in an etchant solution to remove the sacrificial layer and to release the structural layer (fig. 1(c)).

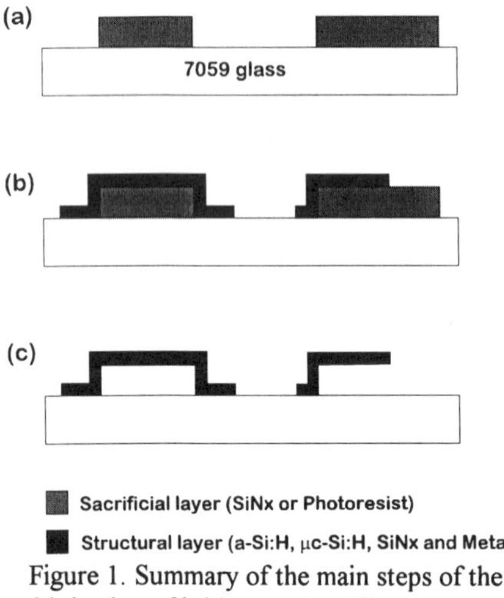

Figure 1. Summary of the main steps of the fabrication of bridges and cantilevers.

Bridge and cantilever structures were obtained using three different fabrication processes (Table I). In the silicon nitride process, SiN$_x$ is used as the sacrificial layer. The maximum processing temperature in this process can be as low as 250 °C. In the photoresist process, photoresist is used as the sacrificial layer. The maximum processing temperature is 110 °C. The modified photoresist process also uses photoresist as the sacrificial layer but the maximum processing temperature reaches 180 °C.

The silicon nitride process was developed for the fabrication of bridge structures using a-Si:H and μc-Si:H as structural layers. a-Si:H is deposited using RF glow discharge of silane, at a substrate temperature of 250 °C, at a process pressure of 100 mTorr, and an RF power of 50 mW/cm^2. μc-Si:H is deposited using 97.5% hydrogen dilution, an RF power of 300 mW/cm^2, a pressure of 100 mTorr and a substrate temperature of 250 °C. The sacrificial layer (silicon nitride) thickness is ~0.5 μm and it is deposited at 250 °C by glow discharge of a mixture of silane and ammonia with a 1:10 gas flow ratio at a process pressure of 100 mTorr and using an RF power of 300 mW/cm^2. These conditions result in a SiN$_x$ with an etch rate of 100 Ås^{-1} in buffered

Table I

Fabrication processes of microstructures on glass

Process ID	Structural Layer	Sacrificial Layer	T_{MAX} (°C)	Etchant of Sacrificial layer
Photoresist modified	Al	Photoresist	180	Photoresist stripper
Photoresist	a-Si:H or a-SiN$_x$	Photoresist	110	Photoresist stripper
Silicon Nitride	a-Si:H or μc-Si:H	low-density SiN$_x$	250	Buffered HF

HF (1:10 BHF). The thickness of the structural layer is typically between 0.2 μm and 0.5 μm.

The photoresist process was developed for fabrication of low-temperature air-gap structures using low-T_{sub} a-Si:H and a-SiN:H as structural materials. The photoresist is spun onto the glass substrate and is subsequently exposed and developed with a pattern which forms the sacrificial island. The thickness of the sacrificial layer is 1.2 μm. The a-Si:H structural layer is deposited either by RF glow discharge at T_{sub}=100 °C using 95% hydrogen dilution, a pressure of 100 mTorr and a power of 50 mW/cm^2, or by hot-wire chemical vapor deposition (HW-CVD) of silane in 60% hydrogen dilution, at T_{sub}=100 °C, a pressure of 20 mTorr, and a filament temperature of 2500 °C. The thickness of the structural layer was varied between 2000 to 6000 Å. The photoresist is removed using a commercial resist stripper.

The modified photoresist process was developed for the fabrication of bridge structures with Al as the structural layer. Al bridges could be useful to avoid intersections in metal interconnections and for air-gap top electrodes. After photoresist (sacrificial layer) patterning, the sample is baked at 180 °C for hardening of the photoresist. A second layer of photoresist is spun and patterned and Al is sputtered onto the substrate. A subsequent immersion in resist stripper serve the dual purpose of defining the metal structure by liftoff of the second photoresist pattern and the release of the structure by the removal of the first photoresist pattern.

Mechanical integrity of bridge and cantilever structures

The characterization of the free-standing bridges and cantilevers was made using optical (OM) and scanning electron microscope (SEM) observations. Bridges with lengths ranging from 1 μm to 50 μm and widths ranging from 1 μm to 50 μm and cantilevers with lengths ranging from 2 μm to 50 μm and widths ranging from 2 μm to 20 μm were fabricated using the different structural layer/sacrificial layer pairs. These structures were observed to find out which geometries led to free standing structures. The maximum free-standing length (L_{max}) is defined as the maximum length for which a visible deformation of the structure is not observed. The OM

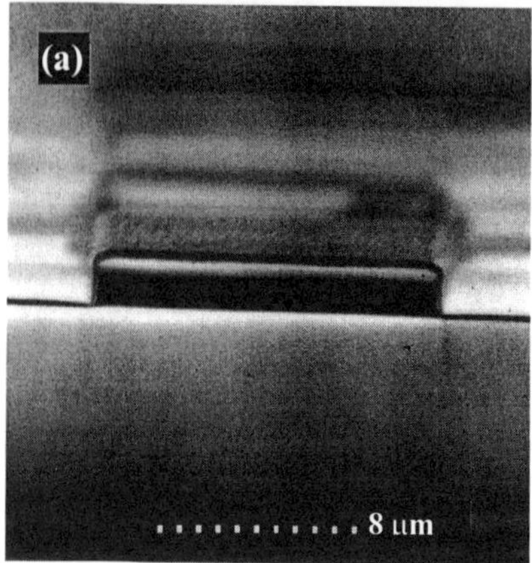

Figure 2. SEM micrographs of (a) a bridge structure fabricated using the photoresist process and using a-Si:H deposited by RF at 100 °C. The length is 15 µm, the width is 5 µm and the thickness is 0.25 µm; and (b) cantilever structures fabricated using the photoresist process and using a-Si:H deposited by HW at 100 °C. The lengths range from 2 µm to 50 µm (up to 5 µm standing) , the width is 5 µm and the thickness is 0.6 µm.

and SEM observations show that there is a strong dependence of this L_{max} on the structural material, on the fabrication process, and on the geometry of the microstructure (type (bridge or cantilever), width, length and thickness). Table II summarizes the results of these observations. Some general comments are possible for the fabrication process developed: (i) $L_{max}(SiN_x)>L_{max}$(a-Si:H)$>L_{max}$(µc-Si:H); (ii) L_{max}(bridges)$> L_{max}$(cantilevers); (iii) L_{max}(thicker structures)$> L_{max}$(thinner structures); and (iv) L_{max}(photoresist process)$> L_{max}$(silicon nitride process). Figure 2 illustrates the photoresist process for bridges using a-Si:H by RF as structural material (fig. 2(a)) and cantilevers using a-Si:H by HW as structural material (fig. 2 (b)). Figure 3(a) shows an Al bridge fabricated using the modified photoresist process. The soft edges result from the rounding of the photoresist pattern during the hardening process. Fig. 3(b) shows a micromachined structure fabricated using the silicon nitride process. In this process, the buffered HF attacks the glass substrate and causes an underetch of the structures, resulting in a removal of the smaller structures and severe yield problems. Preliminary studies show that the use of a buffered HF-resistant coating, such as a-C:H, can eliminate this problem, once problems of adhesion of metal to the coating are solved.

APPLICATIONS

Air-gap thin-film transistors

The fabrication of free-standing structures on glass creates new possibilities for device design. The first demonstration of this technology in an electronic application was the fabrication of an air-gap a-Si:H TFTs [18]. In these devices the dielectric has been replaced by an air gap. Top-gate air-gap TFTs have been fabricated using the modified photoresist process with an Al gate and a RF a-Si:H active layer (fig. 3(a)). Both top-gate and bottom gate TFTs have been fabricated using the silicon nitride process. The top-gate was n$^+$-µc-Si:H, and the bottom gate was Al. Fig. 3(b) shows the bottom-gate TFT, with an RF a-Si:H active layer forming a bridge

Table II

Mechanical integrity of bridge and cantilever structures

Structural Layer	Cantilever L_{MAX} (μm)	Bridge L_{MAX} (μm)
Al (5000 Å)	10	20
RF a-Si:H (2500 Å) at 100 °C	2	W< 10 $\Rightarrow$ 10 10<W<20 $\Rightarrow$ 20 W>20 $\Rightarrow$ 5
HW a-Si:H (6000 Å) at 100 °C	7	W=2 $\Rightarrow$ 10 2<W<40 $\Rightarrow$ 10-50 W>40 $\Rightarrow$ 50
RF SiN$_x$ (3000 Å) at 100 °C	10	$\geq$ 10
RF a-Si:H (2500 Å) at 250 °C	2	W<10 $\Rightarrow$ 0 W>10 $\Rightarrow$ 10
RF μc-Si:H (2500 Å) at 250 °C	0	5
Bilayer Al (2500 Å) / RF a-Si:H (3000 Å) at 100 °C	10	40-50
Bilayer Al (2500 Å) / RF SiN$_x$ (3000 Å) at 100 °C	30	40-50
Bilayer Al (2500 Å) / HW a-Si:H (3000 Å) at 100 °C	10	40 -50

structure. So far the performance of these TFTs has not matched that of conventional a-Si:H transistors: the etching of the substrate when the silicon nitride process is used needs addressing, as well as the step coverage of the metal contacts in the top gate structures.

Electrostatic actuation

The mechanical properties of the structural material played a role in air-gap TFTs - keeping the mechanical integrity of the bridge, but this role was secondary to its electronic properties. Although the electron mobility of a-Si:H is about 1000 times smaller than that of c-Si [16], the elastic modulus is relatively similar (150 GPa for a-Si:H and 180.5-229 GPa for c-Si, depending on the crystalline orientation [19]). This suggests that microstructures with mechanical functions made of a-Si:H could have very similar properties to those fabricated with c-Si or polysilicon, while at the same time allowing vastly improved flexibility in the choice of substrate, processing temperatures, and fabrication facilities.

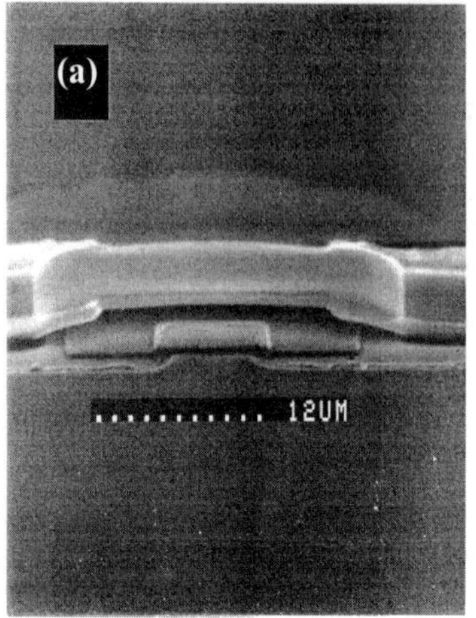

Figure 3. SEM micrographs of (a) a top-gate air-gap TFT with an Al bridge fabricated using the photoresist modified process. The active layer was deposited by RF at 100 °C; and (b) a bottom gate air-gap TFT fabricated using the silicon nitride process. The active layer was a-Si:H deposited by RF at 250 °C and forms a bridge structure. Severe undercutting of the glass by the buffered HF used to remove the sacrificial layer can be observed.

To demonstrate that a-Si:H and related materials can have mechanical functionality, a variation of the simple bridge and cantilever structures has been made in which a metallic gate line is deposited below the bridge (or cantilever). The structure chosen was a bilayer of a-Si:H or SiN_x and Al. If a potential difference is established between the gate and the structure, an electric field will tend to attract the top electrode and the gate. Depending on the dimensions of the micromachined structure and on the applied field, movement may be expected to occur. In these bilayer structures, the a-Si:H or the SiN_x help to keep the mechanical integrity of the bridges or cantilevers (see Table II) while the Al top layer provides the counterelectrode to the gate line. Figure 4 shows a sequence of optical micrographs of a SiN_x/Al cantilever being subjected to a bending voltage. The microscope was focused on the undeformed cantilever (top). It is possible to follow the movement of the cantilever as it leaves the focal plane: at 40 V, the tip bends to touch the substrate, and at 100 V essentially all the cantilever is flat against the bottom electrode. Optical microscope observations are reproducible and electrostatic actuation was confirmed also for a-Si:H/Al bilayer bridges and cantilevers. These observations can be correlated to current-voltage measurements in the same structures. Figure 5 shows the I-V of a 30 µm cantilever and a 50 µm bridge composed of SiN_x/Al. For voltages lower than 70 V (60 V) the current measured is the leakage between the gate and the Al electrode of the bridge (cantilever). The current jump indicates the moment the structure has hit the bottom electrode (at the critical voltage, V_{crit}). After that voltage, the current flows across the SiN_x between the bottom gate electrode and the top electrode of the microstructure. That is confirmed with visual inspection in the optical microscope. After the field is turned off, the structure relaxes back to its initial position. A similar behavior was also observed in a-Si:H/Al structures but, because the critical voltage is above the breakdown voltage of the a-Si:H film, when the structure touches the gate it is destroyed.

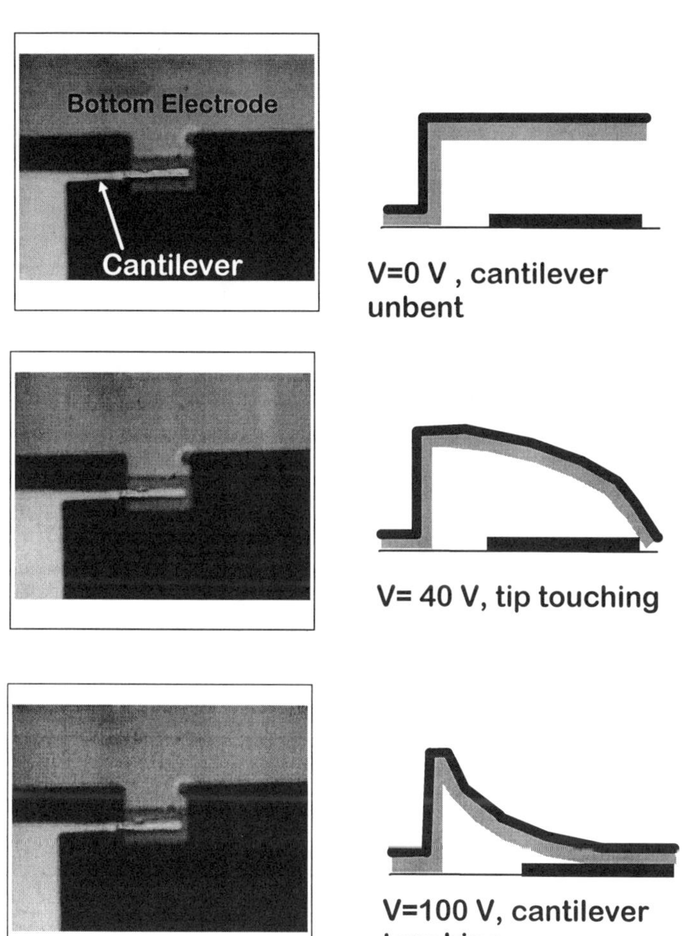

Figure 4. Optical micrograph of a 30 μm-long, 5 μm-wide, SiN$_x$(3000 Å)/Al(2500 Å) cantilever being submited to an electric field from the bottom electrode. The microscope was focused on the undeformed cantilever (top). At 40 V, the tip of the cantilever is defocused, showing that it touched the bottom electrode (middle). At 100 V, most of the length of the cantilever is defocused showing that it is touching the bottom electrode (bottom). This movement can be confirmed by focusing on the bottom electrode instead.

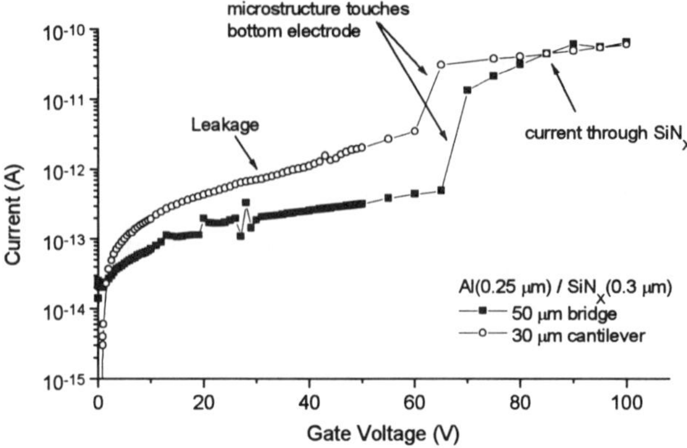

Figure 5. I-V curves for a 30 µm-long, 5 µm-wide, SiN$_x$(3000 Å)/Al(2500 Å) cantilever and 50 µm-long, 5 µm-wide, SiN$_x$(3000 Å)/Al(2500 Å) bridge structure being submitted to a gate voltage from the bottom electrode.

In general, bridges with lengths less than 20 µm did not show any observable deflection within the voltage range used (0 to 200 V) and the same conclusion was obtained for cantilevers with lengths less than 10 µm.

A simple 1-dimensional elastic model can be used to estimate the maximum deflection of a bridge (in the center) or a cantilever (at the tip) due to a uniform density of applied force. For a bridge, maximum deflection is given by $(5.L^4.\omega)/(384.E.I_z)$, where L is the length, E the elastic modulus, I_z the second moment of the area cross section and ω is the linear density of the force [20]. For a cantilever, the numerical factor is 1/8. For a rectangular cross section I_z is given by $(b.h^3/12)$ where b is the width and h is the thickness showing that the deflection is inversely proportional to the third power of the thickness. The values of V_{crit} estimated from the model are in qualitative agreement with the experimental results. For example, the 30 µm cantilever in fig. 5 has a calculated V_{crit} of ~55 V. This model is also useful to suggest how to lower V_{crit}: increasing the length of the structure (L) or decreasing its thickness (h) are the most effective ways. Since this may involve experimental difficulties, decreasing the width, b, and increasing ω by making shallower microstructures may be more useful choices with the current technology.

CONCLUSIONS

Low-temperature surface micromachining of bridge and cantilever structures of amorphous silicon, silicon nitride, microcrystalline silicon and aluminum on glass substrates was developed. Air-gap TFTs demonstrate the use of these 3-dimensional micromachined structures in an electronic application. Steady-state electrostatic actuation of a-Si:H/Al and SiN$_x$/Al bilayers in bridges and cantilevers on glass was achieved, pointing the way to applications in flat-panel displays, functional surfaces, microfluidics, biology, etc.. It is worth mentioning that the small mass of these structures makes them a possible choice for resonators at high frequencies and the confirmation of this expectation will be addressed in the near future. Also in the near future, the development of magnetic and thermoelectromechanical actuation of surface micromachined structures on glass and other large-area substrates will considerably widen the range of movement, forces, and device configurations available.

ACKNOWLEDGEMENTS

The authors gratefully acknowledge F. Silva and J. Bernardo for help in many of the processing steps, F. Vasconcelos and N. Cruz for help in some experimental measurements and to P. Alpuim and P. Brogueira for helpful discussion. M. Boucinha acknowledges a Ph.D. grant and a travel grant from the Fundação para a Ciência e Tecnologia (F.C.T.). This work was partially supported by the F.C.T. through Pluriannual Contracts with UCES/ICEMS (IST) and INESC and by projects PRAXIS/3/3.1/MMA/1775/95 and projects PRAXIS/3/3.1/MMA/1901/95.

REFERENCES

1. W. Lang, Materials Science and Engineering, **R17**, p. 1 (1996).
2. E. M. Kussul, D. A. Rachkovskij, T. N. Baidyk and S. A. Talayev, J. Micromech. Microeng. **6**, p. 410 (1996).
3. J. J. Sniegowsky, Solid State Technology, Feb. 1996, p.83.
4. J. Baliga, Semiconductor International, March 1999, p. 34.
5. R.T. Howe, B.E. Boser, A.P. Pisano, Sens. Actuators A **56**, p.167 (1996).
6. A. N. Cleland and M. L. Roukas, Appl. Phys. Lett. **69**, p. 2653 (1996).
7. J. Hoffman, D. Wagner and A. Dribnak, Solid State Technology, October 1998, p. 123.
8. N. Vandelli, D. Wroblewski, M. Velonis, and T. Bifano, Journal of Microelectromechanical Systems **7**, p. 395 (1998).
9. S. M. Sze, *Semiconductor Sensors*, John Wiley & Sons, New York, 1994.
10. H. L. Kwok, *Electronic Materials*, PWS Publishing, Boston, 1997.
11. R. T. Howe, J. Vac. Sci. Technol. B **6**, p.1809 (1988).
12. D. W. Burns and H. Guckel, J. Vac. Sci. Technol. A **8**, p. 3606 (1990).
13. K. R. Williams and R. S. Muller, Journal of Microelectromechanical Systems **5**, p. 256 (1996).
14. R.T. Howe, R.S. Muller, J. Electrochem. Soc.: Solid State Science and Technology **103**, 1420 (1983).
15. R. Maboudian, R.T. Howe, J. Vac. Sci. Technol. B **15**, p. 1 (1997).
16. R. A. Street, *Hydrogenated Amorphous Silicon*, Cambridge University Press, Cambridge, 1991.
17. *Amorphous and Microcrystalline Semiconductor Devices*, edited by J. Karicki, Artech House, Norwood, MA, 1991.
18. M. Boucinha, V. Chu, and J. P. Conde, Appl. Phys. Lett. **73**, p. 502 (1998).
19. F.C. Marques, P. Wickboldt, D. Pang, J.H. Chen, and W. Paul, J. Appl. Phys. **84**, 3118 (1998).
20. F. P. Beer and E. R. Johnston Jr., *Mechanics of materials*, Mc Graw-Hill, London, 1992.

HIGH RESOLUTION, HIGH FILL FACTOR A-SI:H SENSOR ARRAYS FOR OPTICAL IMAGING

J.T. RAHN*, F. LEMMI*, P. MEI*, J.P. LU*, J.B. BOYCE*, R.A. STREET*, R.B. APTE*, S.E. READY*, K.F. VAN SCHUYLENBERGH*, P. NYLEN**, J. HO*, R.T. FULKS*, R. LAU, R.L. WEISFIELD***
*Xerox PARC, 3333 Coyote Hill Rd., Palo Alto, CA 94304
**Dept. of Physics, University of Stockholm, Sweeden
***dpiX, 3406 Hillview Ave., Palo Alto, CA 94304

ABSTRACT

Amorphous silicon large area sensor arrays are in production for x-ray medical imaging. The most common pixel design works very well for many applications but is limited in spatial resolution because the available sensor area (the fill factor) vanishes in small pixels. One solution is a 3-dimensional structure in which the sensor is placed above the active matrix addressing. However, such high fill factor designs have previously introduce cross talk between pixels.

We present data for a design in which the a-Si:H p-i-n photodiode sensor layer has a continuous i-layer and top p^+-layer, and a patterned n^+-layer contact to the pixel. Arrays of 64 μm and 75μm pitch have been fabricated and are the highest resolution a-Si:H arrays reported to date. The resolution matches the pixel size, and sensitivity has been improved by the high fill factor. Comparison is made between arrays with standard TFTs and TFTs with self-aligned source and drain contacts. Data line capacitance is improved by use of the self-aligned contacts. Measurements are included on the contact to bias capacitance.

The high fill factor design greatly suppresses lateral leakage currents, while retaining ease of processing. Provided illumination levels remain below saturation, the resolution matches expectation for the pixel size.

INTRODUCTION

The potential advantages of amorphous silicon large area sensors, and their operation for x-ray imaging, has been discussed extensively in the literature. Large area arrays presently in production operate using pixels which each have an a-Si:H light sensor and readout thin-film transistor (TFT). X-ray sensitivity is obtained by coupling to a phosphor screen such as Gd_2O_2S:Tb.

This research seeks to address one drawback of the pixel layout used in most flat panel imagers. The layout places isolated sensors adjacent to the readout transistors and data and gate bus lines. The top row of figure 1 shows the standard pixel layout, with the horizontal and vertical bus lines running between the mesa-isolated sensor areas. In this approach, as the pitch is reduced the fill factor (defined as the sensor size divided by the pixel area) quickly vanishes. The fill factor f can be

Standard Pixel Design

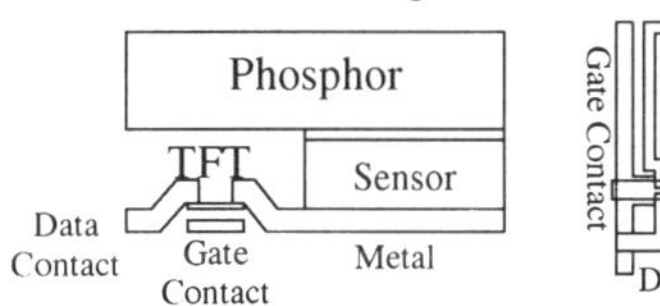

High Fill Factor Pixel Design

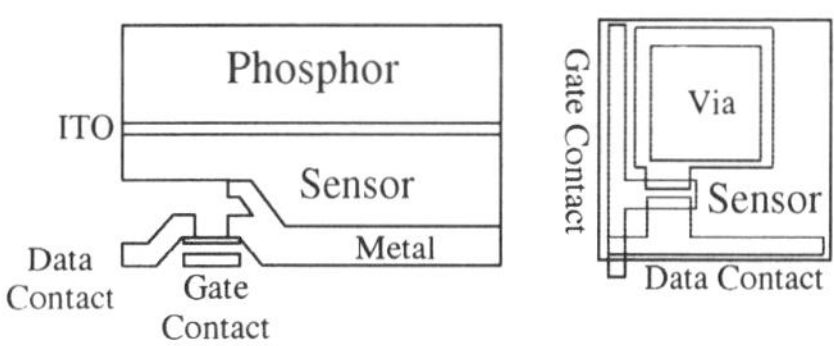

Figure 1. Comparison of standard pixel layout [2] (top row) with high fill factor pixel design. Cross sections and top views of the pixel layout are shown.

estimated from a given set of design rules using the formula:

$$f = \frac{(p - g_1)(p - g_2) - A_{TFT}}{p^2} \tag{1}$$

based on the pitch p, design rules for lost space between adjacent sensors, g_1 and g_2 (horizontal and vertical), and TFT area A_{TFT}. An illustration of the vanishing fill factor is shown in figure 2, which assumes a 30-40 μm gap between pixels and 200 μm^2 area for the TFT. Recently a 127 μm pitch imager achieved a 57% fill factor [3] and a 97 μm pitch imager achieved 45% fill factor [4], as expected from these design rules.

Our approach for improving the fill factor is to extend the sensor into a continuous layer on the top of the array. Such a 3-dimensional structure is shown in the second row of figure 1. This pixel is expected to have light sensitivity across the entire surface. Previous work towards achieving this structure includes using a continuous n$^+$, i, and p$^+$ contact [5]. This structure is observed to have substantial blooming, with charge spreading to adjacent pixels in a fraction of a second. The spreading is modeled as a resistive coupling through the continuous bottom p$^+$ contact.

In another approach [6], a 3-dimensional sensor structure is fabricated with the sensor on top of the addressing circuitry. The sensor is then patterned with grooves parallel to data and gate lines, forming mesa-isolated structures. While this approach eliminates the conduction between pixels, it complicates the processing by requiring additional passivation and metal layers, and a bias connection must cross the sensor. Fill factor is again limited by design rules for the etch process and bias connection width, but improves on the standard mesa-isolated design.

With this paper, we have adopted a patterned bottom metal contact and n$^+$ layer. The sensor then uses continuous i and p$^+$ layers to form the sensor, followed by an ITO (indium tin oxide, a transparent conductor) film for applying the bias voltage. The equivalent circuit for the pixel and the readout electronics are unchanged. The fabrication process is modified to include a passivation layer to isolate the data lines from the sensor contact. The process is designed for compatibility with large-area processing equipment.

In this approach, the pixel fill factor will be at least the fill factor of the n$^+$ contact. Depending on biasing and a-Si:H quality, considerable charge should also be collected from the gaps between n$^+$ contacts. The effective fill factor is expected to be somewhere between the dotted and dashed-dotted lines of figure 2, for typical design rules expected in the large-area process.

The behavior of the sensor in the area between pixels is of particular interest, since the holes generated by light exposure have a considerable distance to travel before being collected by the nearest n$^+$ contact. The continuous sensor can provide a route for charge to flow between n+ contacts, and this effect will be investigated as well.

ARRAY PERFORMANCE

General Properties

The mask uses 512×512 pixels on a 75 μm pitch, with the pitch chosen based on expected requirements of mammography. Ether the standard TFT layout or self-aligned source and

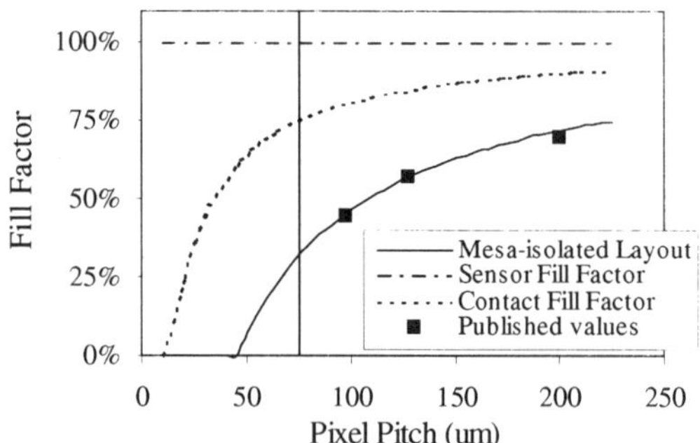

Figure 2. Dependence of fill factor on pixel pitch. Solid line indicates sensitivity for mesa-isolated sensors. A continuous sensor has a contact fill factor defined by the dotted line, although the effective sensor fill factor is assumed to be larger, approaching the dashed-dotted line.

Figure 3. Pictures taken by projecting onto the imager. The imager on the left used standard TFT geometry while the right imager was fabricated using the self-aligned TFT process.

drain contacts could be fabricated. Additional results have been obtained from an older mask set "FF400," with 512×640 pixels and 64 µm pitch [9]. Overall, the array performs as expected, as demonstrated by the images shown in figure 3. These images have been corrected for linear variations in pixel response using a gain-and-offset correction, and bad pixels have been corrected by substituting the average values of adjacent pixels. Images taken with both the standard and self-aligned TFT processing are shown.

There was a concern that placing the a-Si:H sensor above the readout layer might lead to a higher leakage through either the sensor or the TFT [10]. Either leakage can limit frame time. Leakage observed is comparable to values observed in mesa-isolated arrays, allowing frame times approaching 100 s without saturation. This corresponds to < 30 fA per pixel at 3V bias.

<u>Resolution</u>

Tests using visible light projected onto the array demonstrate an improvement in resolution compared to the 127 µm pitch arrays. The line spread function (LSF) is measured by projecting a narrow slit (25 µm) of light onto the array at a small angle with respect to the gate or data lines. The intensity as a function of distance from the center of the slit is the LSF. The resolution matches the expected response for the pitch of the array, as shown by the lowest curve in figure 4. The resolution under x-ray exposure has been studied, with resolution matching expectations, and remains limited by the performance of the x-ray phosphor [11].

While the arrays demonstrate high resolution under a wide range of illumination, bias and frame time, crosstalk is seen if the sensor is allowed to go into saturation. In order to demonstrate this, the LSF was measured as light was increased beyond saturation. LSFs are shown for three light intensities in figure 4, at a level just below that required to saturate the pixel, just above saturation, and about 4 times the saturation level. At the moderate light level below saturation, the LSF shows the expected step response indicating no crosstalk. The vertical lines indicate 75 µm

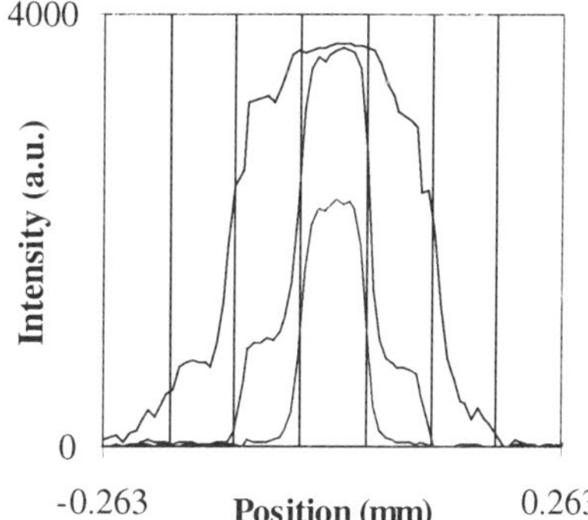

Figure 4. LSF at 3 different light intensities: just below saturation, just above saturation, and about 4 times pixel saturation. Grid indicates pixel edges.

spacing in order to emphasize how the LSF follows the pixel shape. As the light level increases to near the saturation level of the pixel, charge leaks into the neighboring pixels, forming the shoulders on the curves for higher intensities, and even leaking into the next neighbors at the highest light intensity measured. The shape is important, since it demonstrates the crosstalk mechanism operates independently of charge collection from illumination. This crosstalk is analogous to CCD blooming.

Further data was collected for a number of intensity and bias voltage conditions. For each illumination, the average light intensity in the illuminated pixels and their nearest neighbors was observed, shown in figure 5. Several conclusions can be reached from this data. The signal in the illuminated pixel is largely independent of illumination. Note that the readout electronics saturate at a slightly lower charge level than the 3.5 V pixel capacity. A linear extrapolation from low illumination levels would point to pixel saturation at 110 (3.5V), 160 (5.5V), or 210 (7.5V) units of light intensity. Charge collection is not strongly dependent on bias voltage for the relatively high voltages applied here. Second, the nearest neighbors receive much less charge at equivalent illumination when the bias voltage is increased, since the illuminated pixel has a larger well which must be filled before charge will overflow into the neighbor.

A similar crosstalk problem occurs around sensor defects, pixel defects, or broken data or gate lines.

Sensitivity

The optical quantum efficiency has been measured using a monochromatic light source, with the signal observed in the array compared to the incident light power. The quantum efficiency peaks at over 73 ± 8% at 570 nm, as shown in figure 6. Reflectivity at 570 nm was measured to be 7%. This sensitivity is remarkable for a matrix-

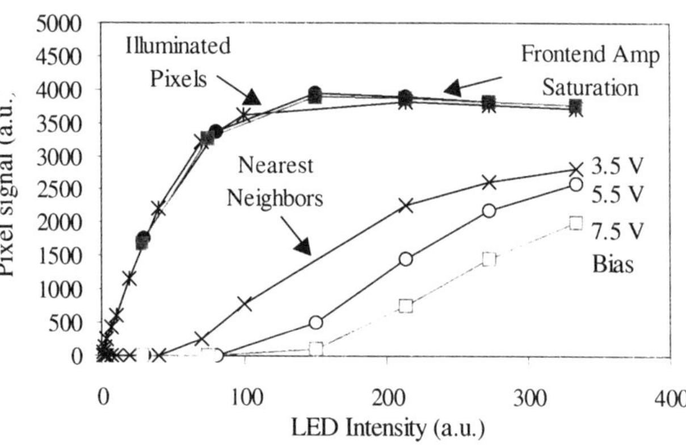

Figure 5. The response of illuminated pixels and neighbors of illuminated pixels, for different bias conditions.

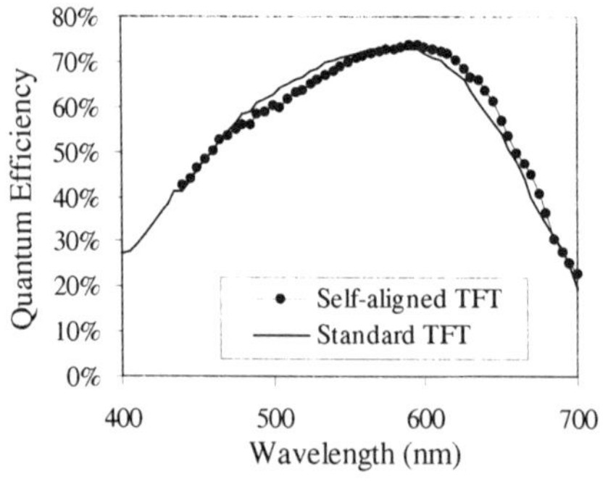

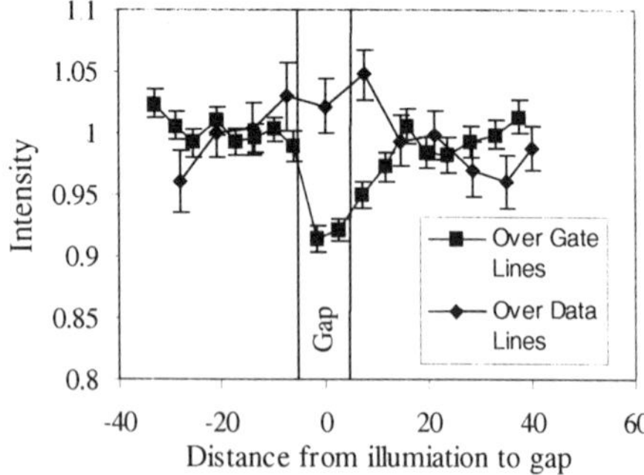

Figure 6. Quantum efficiency array as a function of wavelength, for illumination over a large portion of the array.

Figure 7. Sensitivity of array as a function of distance between illumination and gap between pixels, for gaps over data and gate lines.

addressed array, and is comparable for both transistor architectures.

Fill Factor

The high sensitivity indicates that the array must collect considerable charge from within the gap between pixels. In order to directly measure the sensitivity within the gap, the array was illuminated with a narrow slit of light across the array at a small angle to the readout lines. The total charge collected, by a column of 3 adjacent pixels orthogonal to the illumination, was determined as a function of the distance from the center of the slit to the center of the pixel. If the sensitivity drops considerably between pixels, the charge observed will decrease as the illumination passes between pixels. Only a slight variation was detected for the gaps positioned over gate lines; over data lines, no variation was observed, as shown in figure 7. The fill factor indicated by these measurements is above 97%. Further tests using sensor test structures and laser illumination indicate good charge collection from within the gap region, but require further study.

Image Lag

Since the gaps between pixels could harbor excess trapped charge due to the relatively low fields present there, image lag could result when the trapped charge is released into later frames. Overall, about 2.3% of the illumination level in one frame remained in the next frame when the imager was read out at 0.6 s frame time. The image of a narrow slit, viewed in following frames, should indicate whether the gaps contribute unusually large amounts of charge to the following frames. Figure 8 indicates no differences between line spread function shapes in sequential frames after illumination ended.

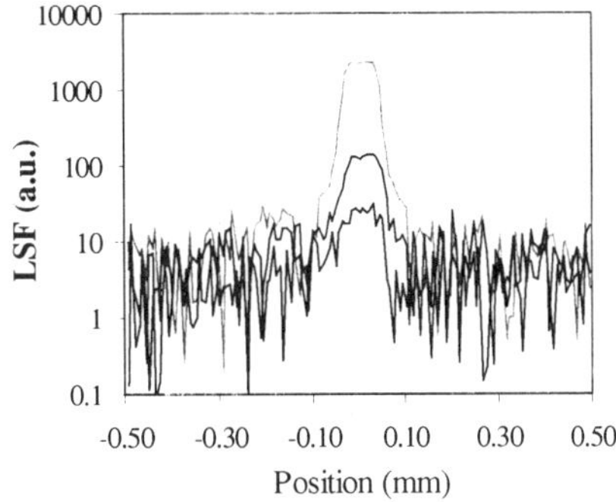

Figure 8. The line spread functions measured for the exposed frame, the next frame after exposure, and the third frame after exposure.

Self-aligned Source and Drain Contacts

This mask set was designed to allow for either the standard TFT design, or a new TFT design which uses self-aligned source and drain contacts [7]. The standard TFT layout places large overlaps across the gate, which contributes to data line capacitance. Data line capacitance directly determines the noise observed by the charge sensitive amplifier [8], either by increasing the capacitive load or by allowing noise to couple to the front-end amplifier. In a typical array, there are three significant sources of data line capacitance: the data to gate crossover, the drain contact of the TFT, and the data to bias capacitance.

General characteristics of array operation were unchanged by the use of self-aligned contacts instead of standard contacts in the TFT. Both versions of the array had uniform response, low sensor and TFT leakage, and good TFT yield. No fixed pattern could be detected which might indicate the laser scanning path.

The data line capacitance was measured for both the standard and self-aligned versions of the array. A 20% decrease in the coupling between gate and data lines was obtained with the self-aligned transistors. Redesigning the crossovers could lead to further improvement.

While the two crossover capacitances and the overlap capacitance would each be about the same in mesa-isolated pixel design, the full fill factor design has considerably higher data line to bias coupling since the sensor and bias completely cover the array. The data line to bias line capacitance was measured to be about 70 fF/pixel, in line with geometrical calculations. There is a clear need for a thicker, lower dielectric constant material to act as the passivation between the sensor and readout layers in a full fill factor array, and this research direction is being persued.

CONCLUSIONS

Advances in amorphous-silicon based flat panel imagers have confirmed their role in moving general radiography from film to digital format. However, the traditional approaches taken thus far will not achieve higher resolution, as would be required for digital mammography. In this paper, we demonstrate a 3-dimensional structure which circumvents the bottlenecks inherent in the present design. By extending the sensor into a continuous film on the top of the pixel, resolution can be substantially increased without reducing the pixel fill factor. Two high fill factor test array designs verify the performance of this approach, using a large-area compatible process. Arrays of 64 μm and 75 μm pitch provide expected resolutions of 7-8 lp/mm, the highest resolution so far reported for an amorphous silicon imager. This resolution is preserved even for long frame times, but is degraded by overexposure. Quantum efficiency for light falling on the array approaches the values expected for a single sensor and was measured to be over 70% at its peak. The high sensitivity is possible because considerable charge is collected from the intrinsic a-Si:H layer several microns from the closest n^+ contact. Measured capacitance values correspond well with expectations, with a reduction obtained by using TFTs with self-aligned source and data contacts. Considerable coupling to the continuous sensor layer is observed, indicating the need for a better passivation layer.

ACKNOWLEDGEMENTS

The authors are grateful for the help of the staff of the Xerox processing line. This work is partially supported by NIST (70NANB7H3007) and NIH (R01-CA56135).

REFERENCES

1. J.A. Rowlands and S.O. Kasap, Physics Today **50**, 24-30 (1995).
2. R.A. Street, X.D. Wu, R. Weisfield, S. Ready, R. Apte, M. Nguyen, P. Nylen, Nucl. Instr. Meth. **A380**, 450 (1996).
3. R.L. Weisfield, M.A. Hartney, R.A. Street, R.B. Apte, Proc. SPIE Conf. On Physics of Medical Imaging, **3336**, 444 (1998).
4. L.E. Antonuk, Y. El-Mohri, A. Hall, K.-W. Jee, M. Maolinbay, S. C. Nassif, X. Rong, J. H. Siewerdsen, Q. Zhao, and R. L. Weisfield, Proceedings of SPIE **3336**, 2 (1998).
5. R.A. Street, X.D. Wu, R. Weisfield, S. Ready, R. Apte, M. Nguyen, and P. Nylen, Mat. Res. Soc. Symp. Proc. **377**, 757-765 (1995).
6. M.J. Powell, C. Glasse, I D. French, A.R. Franklin, J.R. Hughes. and J.E. Curran, Mat. Res. Soc. Symp. Proc. **467**, 863 (1997).
7. P. Mei, J.B. Boyce, D.K. Fork, G. Anderson, J. Ho, J. Lu, M. Hack, R. Lujan, Mat. Res. Soc. Symp. Proc. **507**, 3-12 (1999).
8. R.B. Apte, R.A. Street, S.E. Ready, D.A. Jared, A.M. Moore, R.L. Weisfield, T.A. Rodericks, T.A. Granberg, Proceedings of SPIE **3301**, 2 (1998).
9. J.T. Rahn, F. Lemmi, J.P. Lu, P. Mei, R.B. Apte, R.A. Street, R. Lujan, R.L. Weisfield, J. Heanue, submitted to IEEE Transactions on Nuclear Science.
10. B. Park. R.V.R. Murthy, A. Sazonov, A. Nathan, S.G Chamberlain, Mat. Res. Soc. Symp. Proc. **507**, 237-242.
11. Jeffrey T. Rahn, Francesco Lemmi, Richard L. Weisfield, Rene Lujan, Ping Mei, JengPing Lu, Jackson Ho, Steve E. Ready, Raj B. Apte, Per Nylen, James Boyce, Robert A. Street, Proc. SPIE Conf. On Physics of Medical Imaging **3336** (1999).

UV IMAGER IN TFA TECHNOLOGY

F. MÜTZE[1], K. SEIBEL[1], B. SCHNEIDER[2], M. HILLEBRAND[2], F. BLECHER[2], T. LULÉ[1], H. KELLER[1], P. RIEVE[1], M. WAGNER[1], M. BÖHM[1,2]

[1] Silicon Vision GmbH, D-57078 Siegen, www.siliconvision.de

[2] Institut für Halbleiterelektronik, Universität-GH Siegen, D-57068 Siegen, www.uni-siegen.de/~ihe/

ABSTRACT

An image sensor with enhanced sensitivity for near ultraviolet radiation (UVA) has been fabricated in TFA (Thin Film on ASIC) technology. The device employs an amorphous silicon pin detector optimized for UV detection by carbonization and layer thickness variation. The front electrode consists of an Al grid or TCO. Measurements show a peak responsivity of $90\,mAW^{-1}$ at 380 nm. The UV Imager prototype consists of 128×128 pixels with a size of $25\,\mu m \times 25\,\mu m$ each, fabricated in a 0.7 μm CMOS process. Global sensitivity control serves to achieve a dynamic range in excess of 80 dB. The sensor can be used in fields such as chemical, medical and astronomical applications. Furthermore, a UV monitor has been developed, suited to warn of excessive sunlight exposure, considering skin type and sun protection factor.

INTRODUCTION

Amorphous silicon based thin films with fixed or variable spectral sensitivity within the visible light range have been widely employed for image sensors in TFA technology [1]. Due to their vertical integration of detector and circuitry TFA sensors have an inherently higher fill factor than conventional CCD and CMOS imagers. Furthermore, independent optimization of the two components is possible. Generally speaking, the detector system can be further developed after the completion of the ASIC, or vice versa an optimized standard detector can be employed for different sensor types with similar operating conditions.

Previous investigations demonstrate that a-Si:H is suited for the fabrication of ultraviolet sensitive detectors likewise [2]. Starting with the fundamental pin structure for a visible light detector, bandgap engineering and thickness variation allow to shift the spectral sensitivity to the near UV range. UV optimized detectors employing aluminium or TCO front electrodes have been fabricated and evaluated. While "solar blindness" as postulated in [3] would be useful in case a broadband light source is employed, filters are necessary to suppress visible light in practical applications. The availability of a-Si:H detectors optimized for the UV range and qualified filters allows the fabrication of TFA sensors for a variety of medical, biotechnical and astronomical imaging applications.

The first section of this paper describes the device structures and properties of UVA optimized a-Si:H detectors with aluminium and TCO front electrodes. In the second section a UV imager in TFA technology is presented, based on the design of a previously developed visible light sensor. A specialized UV monitor for the control of sunlight exposure is outlined in the subsequent section.

UV SENSITIVE THIN FILM DETECTORS

The UHV cluster system for PECVD depicted in Fig. 1 allows to fabricate a variety of high-quality amorphous silicon based thin film detectors. A pin layer system with an i-layer thickness of 0.5 - 1 μm exhibits a maximum sensitivity at 550 - 600 nm and is therefore well-suited for visible light detection. Bandgap engineering and layer thickness variation, along with the selection of suitable electrode materials and thicknesses, allow to optimize such a basic device structure for UV detection.

Mat. Res. Soc. Symp. Proc. Vol. 557 © 1999 Materials Research Society

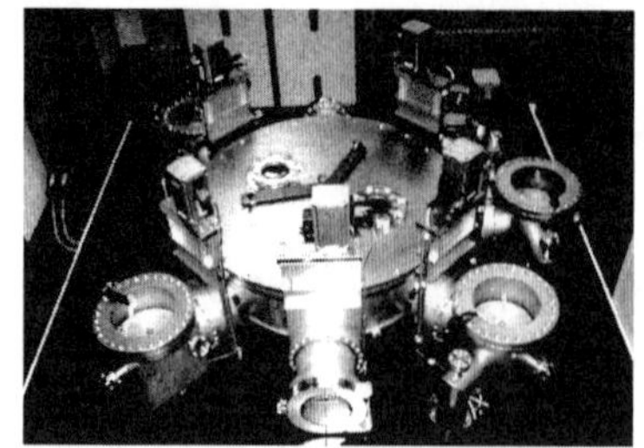

Fig. 1   UHV cluster tool for PECVD

Layer		Material	Thickness [nm]
Rear electrode		Cr	200
n		a-Si:H	50
i		a-SiC:H	40
p		a-SiC:H	6
Front electrode	a)	TCO	300
	b)	Al	9.5

Tab. 1   UV diode layer properties

A wide bandgap boron doped a-SiC:H front layer is used to increase the spectral response for small wavelengths. For the sake of high UV transparency the thickness of this layer should be minimized, however, for layers of less than 4 - 5 nm the dark current rises and the breakthrough voltage sinks significantly. Experiments revealed a thickness of 5 - 6 nm for the a-SiC:H p-layer and 40 - 50 nm for the a-SiC:H i-layer as a compromise between high UV responsivity and steady device operation. Tab. 1 lists the layer properties of the optimized UV diodes.

A crucial issue is the choice of appropriate front and rear electrodes. The rear electrode connects the detector to the pixel circuitry, thereby defining the effective pixel area. Chromium was chosen for this purpose, whereas aluminium was found to be less qualified due to its tendency to promote recrystallization and hillock formation during the thin film deposition, leading to diode short circuits. The basic requirements of high UV transparency and conductivity for the continuous front electrode are best fulfilled by transparent conductive oxides (TCO). The transmittance, however, of the available Al-doped ZnO strongly decreases for wavelengths below 350 nm. As an alternative solution, a thin metal layer can be employed, for instance 10 nm aluminium with a transmission rate around 50 % in the 200 - 400 nm range.

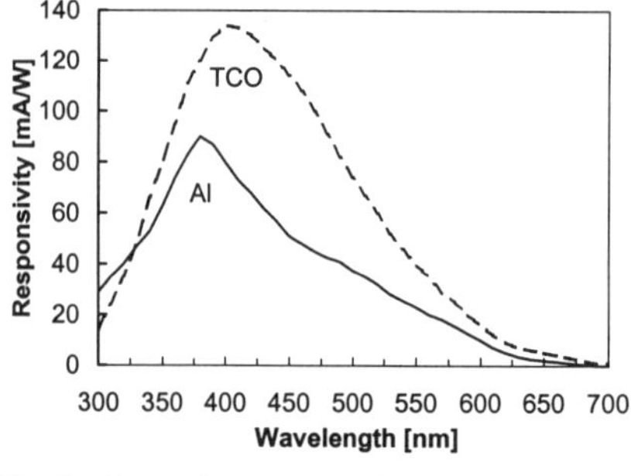

Fig. 2   Spectral responses of UV detectors with TCO and Al front electrodes

Fig. 3   Quantum efficiencies of UV detectors with TCO and Al front electrodes

Measurements of ultrathin UV diodes have to be performed very carefully, since the measuring probe or bond wire may easily cause a short circuit. The photoelectric properties of UV detectors with Al and with TCO electrodes are depicted in Fig. 2 throughout Fig. 5. Either sample has a size of 1 mm^2. The maximum spectral response is 134 mAW^{-1} at 400 nm for the TCO sample, the Al electrode yields 90 mAW^{-1} at 380 nm (Fig. 2). The maximum quantum efficiencies are 41 % for the TCO and 29 % for the Al electrode, respectively (Fig. 3). The TCO diode covers the UVA range (320 - 400 nm) well, while the Al diode, despite its lower overall response, is more sensitive in the UVB region (280 nm - 320 nm). The dark current density of either device was found to be less than 10^{-9} Acm^{-2} at reverse voltages up to 1.5 V (Figs. 4,5), yielding a dynamic range of more than 60 dB for 1000 lx halogen lamp illumination. A setup for UV intensity measurements is currently being developed.

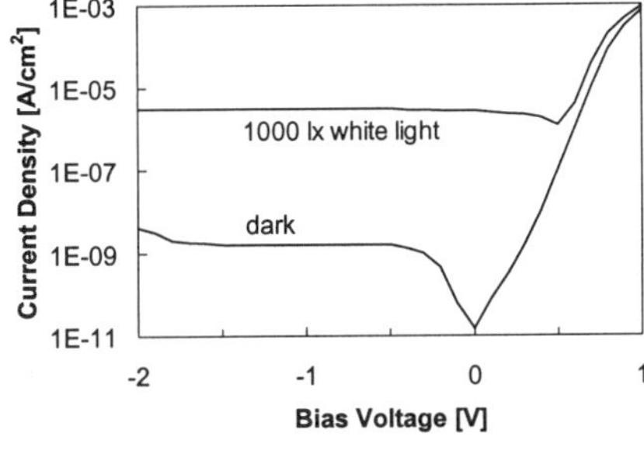

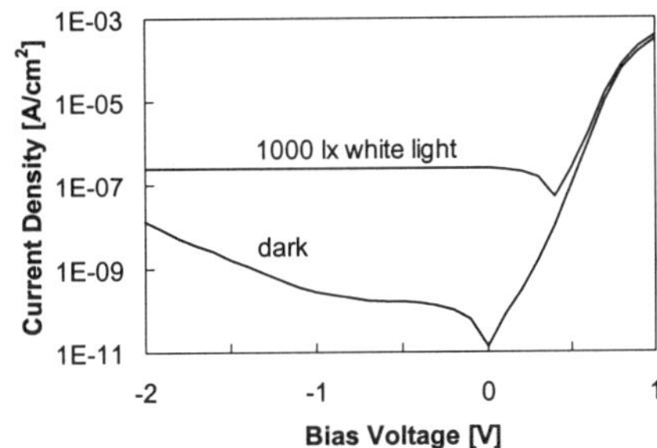

Fig. 4   Photo- and dark current of a UV detector with TCO front electrode

Fig. 5   Photo- and dark current of a UV detector with Al front electrode

Either type still shows a response for visible light. Hence, in case of broadband light sources such as Xe or Hg arc lamps or the solar spectrum, filtering is inevitable to eliminate the non-UV portion of the radiation. A UG11 glass absorption filter can be used for this purpose, resulting in an out-of-band signal of less than 3 % for the AM1.5 solar spectrum.

UV IMAGER

Flexibility is one of the major benefits of TFA technology, implying that different thin film systems can be deposited onto the same ASIC. The UV imager discussed in the following combines the novel UV sensitive detector with TCO electrode and an existing ASIC design previously presented as AIDA (**A**nalog **I**mage **D**etector **A**rray) [4]. This procedure enabled significant development cost and time savings for the UV imager prototype.

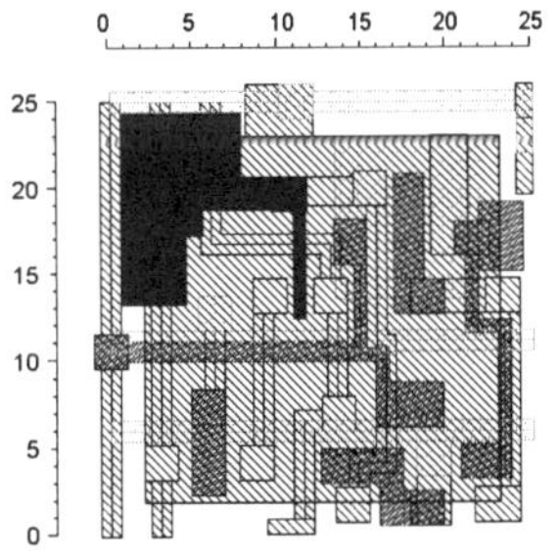

Fig. 6 Pixel layout of UV Imager

Fig. 7 Die photograph of UV imager

The AIDA pixel circuit keeps the detector rear electrode at constant potential so as to suppress lateral balance currents which may flow within the highly doped n-layer in a visible light pin detector and a UV optimized pin detector likewise. In this way the local contrast of the sensor array can be enhanced. Furthermore the integration time can be controlled globally, yielding an overall illumination range of more than 80 dB. The prototype consists of a 128×128 pixel array with a pixel size of 25 μm $\times$ 25 μm. Fig. 6 shows the pixel layout. A die photograph is depicted in Fig. 7.

For a demonstration of the enhanced UVA sensitivity compared to conventional visible light sensors, the UV imager was illuminated through an overhead transparency on which a sun was sketched with clear nail paint. The object size was 16 cm^2, the distance to the camera lens was 25 cm. A Philips HP A 400 table top solarium with built-in UVB blocking filter served as the UVA source. An additional diffusor in the form of a 2 mm sand blasted glass plate provided homogeneous illumination of the object.

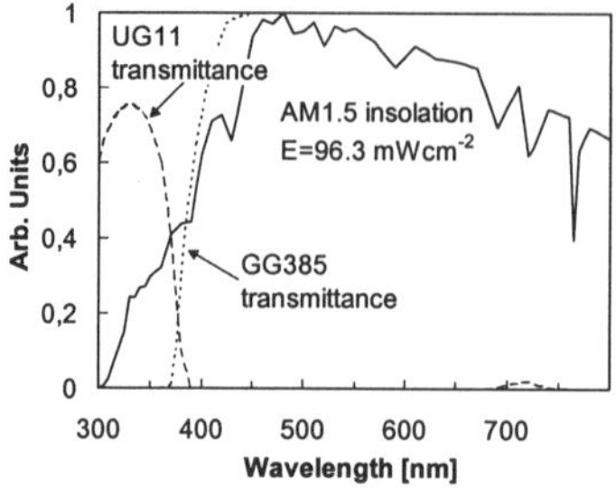

Fig. 8   Solar spectrum and filter transmission spectra

Fig. 9   UV + red image (UG11 filter only)

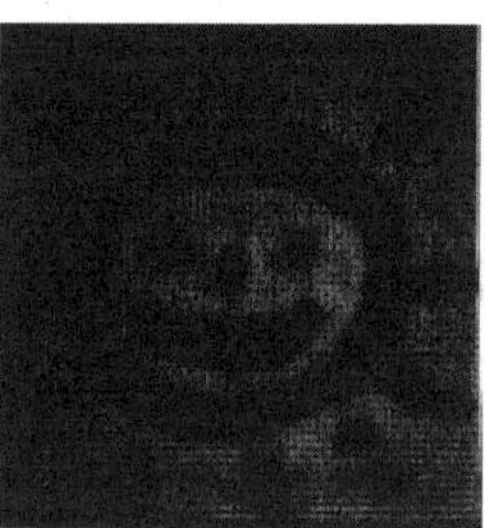

Fig. 10   Red image (UG11 and GG385 filters)

Color glass filters are the best choice to cover the UVA range at a high transmittance. However, the employed UG11 filter exhibits a second responsivity peak for red light (Fig. 8). Fig. 9 shows the image taken with the UG11 only. The unwanted red portion of this image can be separated employing a UV absorption filter GG385 and UG11, and is depicted in Fig. 10. This red image has to be subtracted from the UG11 image in order to eliminate the red sensitivity of UG11 and obtain the UV portion only. Fig. 11 shows the result of the subtraction. The UV image in Fig. 11 demonstrates that the nail paint absorbs UV radiation, whereas the rest of the object is transparent to UV light. Reference images taken with visible light sensors (Fig. 12: flatbed scanner, Fig. 13: visible light AIDA) reveal that the nail paint image is transparent in this spectral range, and its edges can hardly be distinguished from the background.

Fig. 11   Reconstructed UV image

Fig. 12   Scanned reference image

Fig. 13   Reference image taken with visible light AIDA

UV MONITOR

UV radiation undisputedly has a harmful influence on biological systems, in particular the human skin, reflected in symptoms ranging from erythema (skin reddening) to malignant melanoma (skin cancer) in the worst case. A local UV radiation forecast based on time of the day, season, geographical latitude, altitude, absorption by ozone, clouds and surface reflection may provide the necessary information to avoid excessive sunlight exposure. Still, a UV monitor measuring the actual terrestrial UV intensity and also taking into account the individual factors of skin type and sun protection is much more efficient and convenient [5]. Likewise, such a device can be used to monitor the radiation dose in a solarium or in phototherapy.

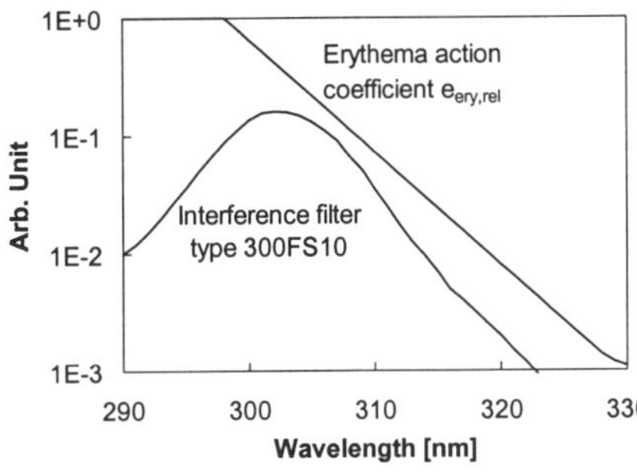

Fig. 14   Erythema action coefficient and transmission of 300 nm filter

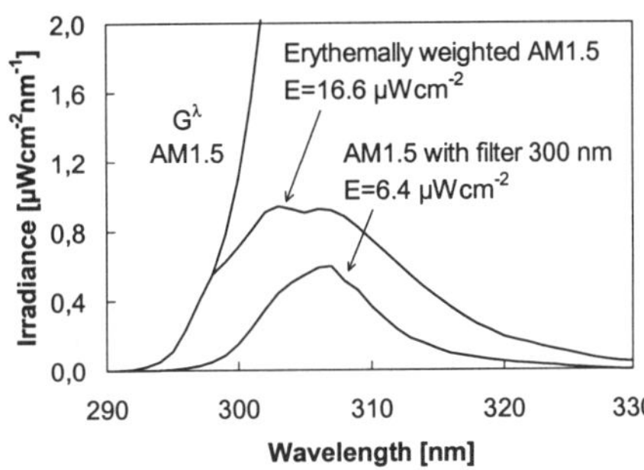

Fig. 15   Erythemally weighted and filtered AM1.5 spectra

In order to warn of skin damage under different radiation spectra, for a UV monitor it is essential to closely match the action spectrum for UV-induced erythema that describes the absorption of radiation inside the skin. Fig. 14 shows that the relative erythema action coefficient decreases exponentially for wavelengths above 300 nm, unlike the characteristic of usual UV detectors. However, it is possible to employ a combination of the UVB sensitive diode with Al front electrode and an interference filter 300FS10 that exhibits an approximately exponential decrease of its transmission above 300 nm. A comparison of erythemally weighted and filtered spectra in case of AM1.5 illumination is depicted in Fig. 15. The photocurrent of the UV monitor being approximately proportional to the erythemal illumination power density, it may be used to calculate the UV index. The dimensionless UV index (UVI) was introduced to provide a comprehensible measure for harmful UV radiation intensity:

$$UVI = \frac{\int G^\lambda \cdot e_{ery,rel}\, d\lambda}{2.5\ \mu Wcm^{-2}}$$

G^λ is the spectral power density, $e_{ery,rel}$ the erythema action coefficient. The erythemally weighted AM1.5 spectrum in Fig. 15 yields 16.6 µWcm⁻² and a UV index of 6.64. The UVI is commonly rounded to an integer value and ranges from 0 in the dark to 15 in tropical regions.

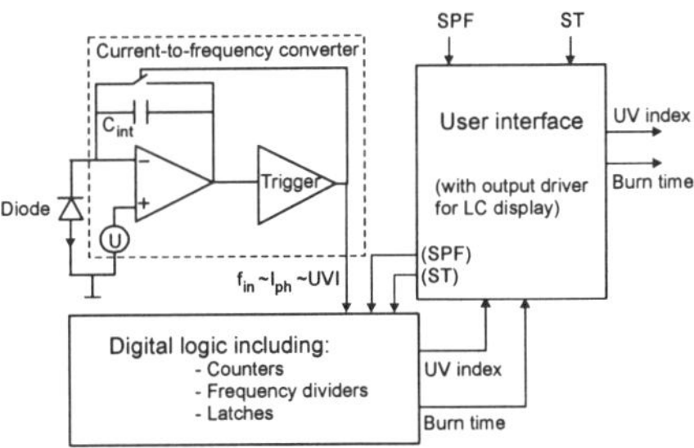

Fig. 16   Block diagram of UV monitor

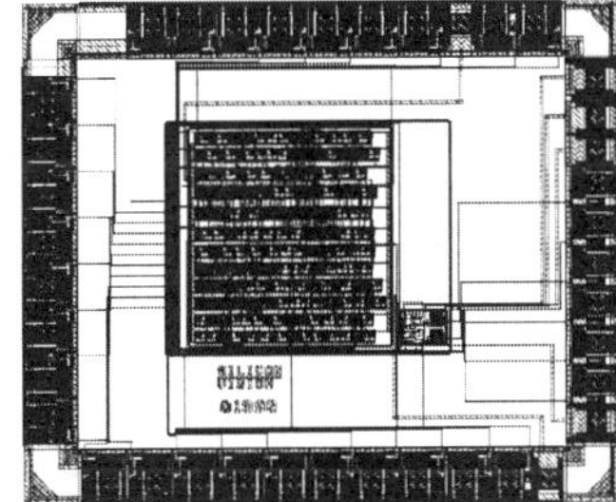

Fig. 17   Chip layout of UV monitor

The block diagram in Fig. 16 illustrates the function principle of the UV monitor. The integration of the photocurrent is stopped as soon as the threshold of the subsequent trigger is exceeded, yielding a periodic voltage ramp with a frequency proportional to the photocurrent. The digital logic calculates the UV index and the recommended remaining burn time. An alarm signal is initiated as soon as the burn time is zero. A user interface allows to set the user's individual sun protection factor (SPF) and skin type (ST). An output driver to display the UV index or burn time on an LCD is also included. The logic can be reset in case the sun exposure is suspended or the UV monitor is handed to another person.

The UV monitor is designed to be integrated into a wrist watch. A chip layout is depicted in Fig. 17. The chip size being 2.9×2.4 mm^2 in 0.8 μm CMOS technology for the prototype can be significantly reduced employing a smaller feature size and custom pads. Low power consumption is achieved using a maximum clock frequency of only 1.28 kHz at UVI = 10. The sole supply voltage of 3 V can be provided by a lithium battery.

CONCLUSION

Amorphous silicon thin film detectors can be optimized for a maximum response in the UVA region and still a considerable responsivity for UVB. Two different versions have been fabricated and tested, one broadband UVA detector with a TCO front contact and one UVA/UVB detector with an aluminium front contact. The first detector type has been employed for a TFA UV imager offering enhanced performance for medical and chemical analysis along with the inherent benefits of TFA technology such as wide detector optimization range, high fill factor and use of low-cost standard technologies. Subsequent prototypes will profit from the general TFA development in that similar circuit concepts as for visible light sensors can be employed. A new UV imager prototype using a recent high-resolution ASIC is currently being prepared. The second detector type is especially suited for a UV monitor. An additional line filter serves to replicate the dermatological erythema action coefficient and to provide a reliable measure for excess sunlight exposure and potential subsequent skin damage.

ACKNOWLEDGEMENT

The authors wish to thank S. Coors, A. Eckhardt and J. Sterzel of the Institut für Halbleiterelektronik (IHE), Universität-GH Siegen, for useful discussions and for proofreading.

REFERENCES

[1] M. Böhm, F. Blecher, A. Eckhardt, K. Seibel, B. Schneider, J. Sterzel, S. Benthien, H. Keller, T. Lulé, P. Rieve, M. Sommer, B. van Uffel, F. Librecht, R. C. Lind, L. Humm, U. Efron, E. Roth, MRS Spring Meeting, San Francisco, 1998.

[2] G. Naletto, P. Nicolosi, E. Pace, G. de Cesare, F. Irrera, F. Palma, Proc. SPIE Vol. 2808, pp. 605-612, 1996.

[3] D. Caputo, G. de Cesare, F. Irrera, F. Palma, IEEE Trans. Electron Devices 43 (9), pp. 1351-1356, 1996.

[4] B. Schneider, H. Fischer, S. Benthien, H. Keller, T. Lulé, P. Rieve, M. Sommer, J. Schulte, M. Böhm, Tech. Digest IEDM, pp. 209-212, 1997.

[5] P. Knuschke, J. Barth, Journal of Photochemistry and Photobiology B: Biology 36, pp. 77-83, 1996.

RESISTLESS PATTERNING OF HYDROGENATED AMORPHOUS SILICON FILMS

Russell E. Hollingsworth,* Mary K. Herndon,** Reuben T. Collins,** J.D. Benson,*** J.H. Dinan,*** and J.N. Johnson****
*Materials Research Group, Inc., 12441 W. 49th Ave., Suite #2, Wheat Ridge, CO 80033
**Colorado School of Mines, Physics Department, Golden, CO
***Night Vision and Electronic Sensors Directorate, Ft. Belvoir, VA
****E-OIR Measurements, Spotsylvania, VA

ABSTRACT

Practical methods for directly patterning hydrogenated amorphous silicon (a-Si:H) films have been developed. Direct patterning involves selectively oxidizing the hydrogen passivated a-Si:H surface or laser crystallization of the bulk. The oxide or polycrystalline layer formed in this way then becomes a mask for subsequent hydrogen plasma etching. Methods for selective oxidation of the a-Si:H surface have been extensively studied. Examination of the pattern generation threshold dose for excitation wavelengths from 248 to 633nm provides indirect evidence for electron-hole recombination breaking of the silicon-hydrogen bond. An additional hydrogen removal mechanism was observed whereby simple proximity of a tapered fiber optic probe less than 30nm from the sample surface resulted in pattern generation. Patterns were generated in both intrinsic and doped a-Si:H films by several means, including contact printing with a mask aligner, *in situ* projection lithography with an excimer laser, and direct writing with a near-field scanning optical microscope (NSOM). Direct patterning of a-Si:H films has a wide range of potential applications. We have demonstrated a-Si:H as an *in situ* photoresist material for patterning HgCdTe infrared detector arrays with all process steps done in vacuum. We have also demonstrated 100nm line widths using NSOM writing with a photolithography goal. Direct patterning of a-Si:H could simplify the manufacturing of thin film transistors, or other devices that require patterned silicon films.

INTRODUCTION

In one vision of the semiconductor factory of the future, wafers are processed entirely in vacuum chambers and are transported between chambers via vacuum interlocks.[1] The potential advantages of such an approach include inherent flexibility with respect to product mix, higher manufacturing yield, and the enabling of advanced device structures not possible on conventional lines. What is lacking for a full wafer-in, array-out demonstration is a photoresist process in which deposition, delineation, and removal of the resist material are accomplished by vapor phase processes. Hydrogenated amorphous silicon can meet all of these requirements. The plasma enhanced chemical vapor deposition (PECVD) technique is already widely used in semiconductor manufacturing. Two methods of modifying the a-Si:H film to generate patterns (optically enhanced oxidation and laser crystallization) have the potential to meet the throughput requirements of modern factories. In the first method, selective removal of the surface hydrogen passivation by optical excitation allows a thin (~1nm) oxide to grow. Hydrogen plasmas have been shown to have extremely high etch selectivity (>500:1) between a-Si:H and SiO_x, allowing pattern development even with a very thin surface oxide.[2] The patterned a-Si:H film can then be used as a photoresist, or left in place as an active part of devices such as thin film transistors.

The ability to sustain submicron feature sizes is a critical aspect of any photoresist material for microelectronics applications. Near field scanning optical microscopy (NSOM) provides an ideal technique for prototyping submicron optical resists and devices.[3] NSOM is a

Mat. Res. Soc. Symp. Proc. Vol. 557 © 1999 Materials Research Society

scanning probe technique that allows optical excitation of materials at spatial resolutions well below the diffraction limit. An optical fiber is tapered to give an aperture typically 100nm or less in diameter. The fiber tip is scanned across the sample surface while maintaining a fixed tip-sample separation using shear force feedback.[4] The separation and aperture are much less than the wavelength of light and, hence, in the near-field with lateral resolution comparable to the aperture diameter. This allows NSOM to be used as a direct write tool for photolithography on the submicron length scale. Amorphous silicon has been shown to be a potentially useful resist for nanoscale lithography based on scanning probe techniques.[3,5,6,7] Madsen *et al.* have shown that NSOM can be used to pattern a-Si:H optically with dimensions less than 100nm.[8] They have also observed that simple proximity of the fiber tip to the a-Si:H surface can lead to patterning even in the absence of optical illumination.

EXPERIMENT

In this work, a-Si:H films were grown on crystalline silicon wafers, Corning 7059 glass, or HgCdTe to a thickness of 100-800 nm. A Materials Research Group multichamber, parallel plate, capacitively coupled plasma enhanced chemical vapor deposition (PECVD) system operated at 13.56 MHz was used for both deposition with pure silane and etching with pure hydrogen. Film thickness was monitored *in situ* by reflecting a HeNe laser off the surface and counting interference fringes. The films were allowed to cool from the deposition temperature to room temperature in vacuum in order to minimize native oxide growth. Laser sources used for optical irradiation included a KrF excimer laser as the source of 248 nm light with a nominal pulse length of 20 ns and Ar ion and HeNe lasers for continuous wave (cw) irradiation from 350 to 633nm. A Cobilt mask aligner with peak emission at 360nm has also been used to generate patterns. Optical irradiation occurred with the sample in various atmospheres, including air, 5% O_2 in Ar, and nitrous oxide (N_2O).

NSOM tips were prepared both by chemical etching and by pulling with a Sutter Instruments P2000 micropipette puller. The sides of the probes were coated with approximately 100nm of aluminum. For proximity effect studies uncoated tips were also used. The fiber optic probe was attached to a piezoelectric tube which was oscillated (dithered) parallel to the sample surface. As the tip approached the sample surface, shear force damping of the amplitude provided the necessary signal for maintaining a constant probe-sample separation of approximately 5-20nm. The dither amplitude was monitored using conventional optical detection of laser light scattered off the tip.[4] Since the relationship between the absolute dither amplitude and this signal depends on factors such as the tip taper and the location of the light on the tip, only relative dither amplitudes were determined. For a given tip, however, this method allows relative changes in dither amplitude to be easily determined.

RESULTS AND DISCUSSION
Laser Crystallization

Selective crystallization of the bulk a-Si:H film provides one means of generating a pattern with etch resistance different than the as-deposited amorphous film. Any laser with high enough absorbed power density can be used. A KrF excimer laser operating at 248nm with nominal 20ns pulse length was used in this work. Both p-type and intrinsic a-Si:H films grown on Corning 7059 glass were used. P-type films were grown with two doping concentrations and three substrate temperatures. The dopant gas used was trimethylboron (TMB) in helium with the gas flows adjusted to give a TMB to SiH_4 ratio of 0.1% or 1.0% in the gas phase. For conductivity measurements, excimer irradiation was done in air with the raw 1cm x 2.5cm beam. In calculating the conductivity, the full film thickness was assumed to be crystallized. In reality, the crystallized thickness will vary with power and probably will always be less than 200nm based on literature results, while the films were 500nm thick. True conductivities of the crystallized films will therefore be higher than the values reported here as the conductivity calculation involves dividing by thickness.

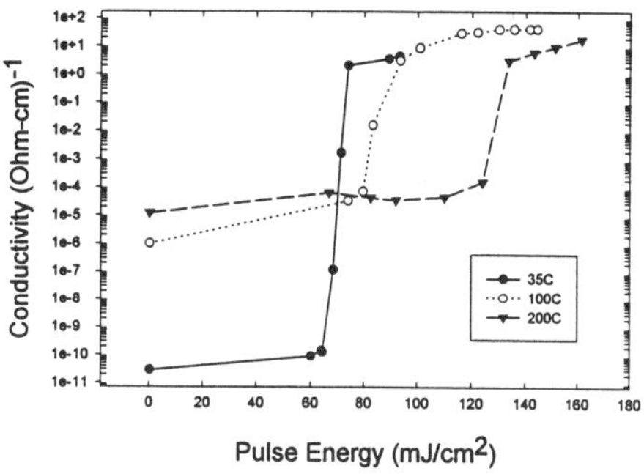

Figure 1 shows the conductivity of p-type films grown with 1% TMB as a function of laser energy. Several features are common for all of the deposition conditions: 1) the conductivity is roughly constant for low laser power, 2) a sharp transition of 5-10 orders of magnitude occurs, 3) a slow increase in conductivity occurs above the threshold, and 4) the threshold energy increases as the film deposition temperature increases. The sharp transition indicates the onset of surface melting, while the slow increase above the threshold is probably due to increased melt depth. The threshold energy is determined primarily by the film deposition temperature, with little or no dependence on doping level. Patterns were generated by imaging a mask onto the film surface.

Figure 1. P-type film conductivity is shown as a function of excimer laser energy for three different film deposition temperatures. The lines are to guide the eye.

Selective Oxidation

Selective oxidation of the a-Si:H surface provides an alternative patterning approach that allows incident illumination power densities to be orders of magnitude smaller than the laser crystallization approach. The optical dose required to induce oxidation of the intrinsic a-Si:H surface in air was measured as a function of illumination wavelength as shown in figure 2. For this study, a 100nm thick a-Si:H film was exposed to a light source through an aluminum on quartz shadow mask. The power densities used at each wavelength were generally kept low (<3W/cm^2) to avoid thermal heating of the sample. The threshold dose was defined as the dose which left the exposed region untouched when the unexposed region was completely removed by the etch. This threshold definition leads to an exposure that would give rise to well defined features in lithographic applications. We note, however, that partially developed patterns could be observed for doses an order of magnitude less than threshold. While we cannot rule out the possibility that high peak intensities due to pulsed operation may contribute to reduced threshold at 248nm, figure 2 makes it quite clear that shorter wavelengths strongly decrease the required dose.

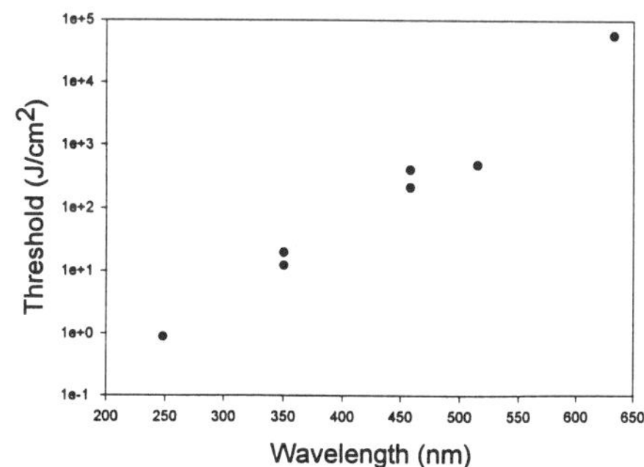

Figure 2. Threshold dose for exposing a-Si:H in air as a function of irradiation wavelength.

823

The dependence on gas ambient during optical illumination has also been briefly examined. Samples were irradiated with 350 nm cw light under a high vacuum of 1 x 10^{-7} torr, and then exposed to air prior to hydrogen plasma etching. No patterns were generated even for a dose an order of magnitude greater than the air threshold. Other samples were irradiated with pulsed 248nm light under various low pressure oxygen containing ambients. For an ambient of 5% O_2 in Ar at a total pressure of 1 torr, the threshold dose increased by roughly a factor of 10 over the value in air.[9] Pure nitrous oxide (N_2O) was also used as an ambient over a pressure range of 1-100 torr. The threshold dose in N_2O was found to be independent of pressure and comparable to the value in O_2/Ar. At this point, it is not clear whether the increased thresholds in low pressure O_2 or N_2O ambients are due to a pressure effect or the associated reduction in water concentration. However, it is clear that some oxygen containing gas must be present simultaneously with the optical irradiation in order to grow an oxide.

Bond breaking through free carrier recombination appears to be the mechanism for optically enhanced oxidation. Direct photo induced hydrogen desorption can be ruled out because patterns are generated with photon energies far below the Si-H bond strength. The strong threshold dependence on photon energy can qualitatively be accounted for by the absorption coefficient variation of the a-Si:H film. As the absorption coefficient increases (shorter wavelength), a larger fraction of the generated carriers will recombine at the surface where an oxide can form.

<u>NSOM Patterning</u>

We have examined both proximity effect and optical patterning of a-Si:H using NSOM.[3] In agreement with Madsen *et al.* we observed that proximity of the NSOM tip to the a-Si:H surface resulted in the oxidation of the surface and pattern generation even in the absence of light.[8] Figure 3, for example shows AFM images of serpentine patterns written in an a-Si:H layer using an NSOM tip with no light. To explore this effect further, patterns were generated while varying NSOM scan conditions. Line widths and heights were not significantly affected by varying scan speed (1-100µm/s) or tip-sample separation (~4-20nm). Dither amplitude, however, strongly influenced line width and height. The trace in figure 3(a) was written with no light at a scan speed of 1µm/s using a small dither amplitude. The lines are ~140nm wide and 15nm tall. The trace in figure 3(b) was drawn after increasing the dither amplitude to a medium value. Its lines are ~320nm wide and 45nm tall. The resulting line widths for a complete series of patterns are shown in figure 4 as a function of dither amplitude. For the largest dither amplitudes, the lines are 75nm tall which is equal to the as-deposited a-Si:H film thickness. It seems likely that the line widths are to first order determined by the probe movement and the diameter of the probe.

Given the dependence of line width and height on dither amplitude during proximity effect patterning, it is natural to speculate that the shear force interaction removes the hydrogen surface termination in the absence of light. The origin of the shear force interaction is, at present, poorly understood. Proposed explanations include long range van der Waals forces,

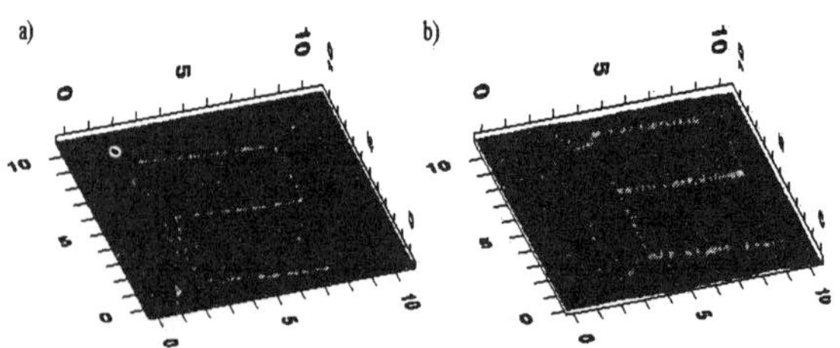

Figure 3. a-Si:H lines on crystalline silicon written with different NSOM dither amplitudes are shown. The dimensions are in microns.

contamination layers, image-charge current dissipation, and physical contact between the probe and sample.[10] Shear force has been shown to depend on ambient conditions and on humidity in particular.[4,11] If the probe were, for example, interacting with a surface water layer, this layer could be responsible for removing hydrogen.

Using a medium dither amplitude and 458nm cw illumination, true optical exposure of the resist was explored by writing a series of patterns while all parameters were held constant except for light intensity. The optical dose was varied from 0 to ~170J/cm^2, roughly the far-field threshold dose at 458nm as shown in figure 2. We note that dose estimates were based on the far-field intensity emitted by the

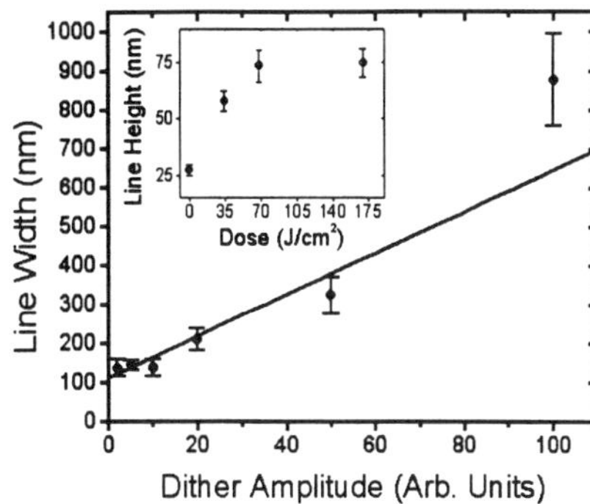

Figure 4. Lithographic line width plotted as a function of dither amplitude for a series of patterns. The inset shows line height as a function of optical dose.

probe, and the actual dose may have been higher. As dose increased from zero to the maximum value, line width increased by only a few percent. Line height, however, changed dramatically as shown in the inset to figure 4. The line height of 75nm for the maximum dose is comparable to the original film thickness.

<u>HgCdTe Device Patterning</u>

Separate calibration runs showed the selectivity between HgCdTe and a-Si:H to be 18:1 using ECR etch conditions optimized for removal of HgCdTe. An a-Si:H film was grown to a thickness of 150nm on a HgCdTe double layer structure and an 8 x 8 pattern of 50μm squares in the a-Si:H was generated and developed. An ambient of 1 torr N$_2$O was used during irradiation with 5000 pulses of 248 nm light at 30 mJ/cm^2. The a-Si:H thickness was chosen so that the mask would be completely removed once the target depth into the HgCdTe was reached. An ECR plasma etch then formed the mesas in the HgCdTe and simultaneously stripped the a-Si:H. Metallization of the devices was done with conventional photolithography. P-type contacts were made using gold deposited at 70°C with a thickness of 300nm. n-type contacts were made using indium deposited at 70°C with a thickness of 800nm. Current voltage characteristics were measured at 77K with manual probing used to contact the devices.

A current-voltage curve for a mid-wavelength infrared (MWIR) HgCdTe diode created using an 8 x 8 a-Si:H mask and ECR plasma etching is shown in figure 5. The R$_o$A value is 3000. This value is equivalent to those obtained for diodes fabricated at NVESD using conventional photolithography and wet-chemical processing, but is below the state of

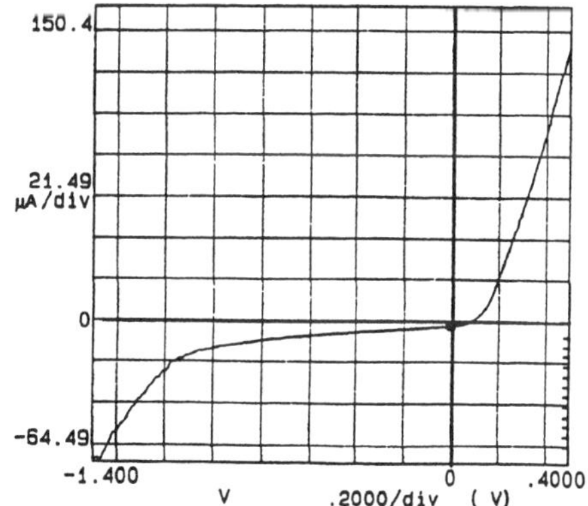

Figure 5. Current vs voltage at 77K for a photovoltaic mesa diode created by etching a HgCdTe double layer through an a-Si:H mask using an ECR plasma.

the art obtained at other laboratories for similar HgCdTe material. We attribute this difference to a lack of passivation of the mesa side walls in these devices. Without state of the art devices, it is not possible to say conclusively that the a-Si:H photoresist process causes no damage to the HgCdTe, but the equivalence with conventional processing is an encouraging result. With the recent attachment of an a-Si:H *in situ* photolithography system to the NVESD microfactory[12] and improvements to the device fabrication process, a definitive answer to the usefulness of a-Si:H as a photoresist for HgCdTe patterning should be obtained in the near future.

CONCLUSIONS

Practical methods of directly patterning hydrogenated amorphous silicon films with feature sizes down to 100nm have been demonstrated. Laser crystallization and optically enhanced oxidation can be used to generate patterns in p-type, n-type, and intrinsic a-Si:H films. Hydrogen plasma etching can remove the untreated a-Si:H with extremely high selectivity. The patterned a-Si:H film can be used as a photoresist for subsequent processing, or can be used directly as part of an active device.

ACKNOWLEDGEMENTS

We would like to thank Dr. Jose Arias of the Rockwell International Science Center for providing HgCdTe double layer samples and Dr. Matt Johnson of the University of Oklahoma for AFM images. This work was supported in part by the U.S. Army through SBIR contract No. DAAB07-96-C-J013, the Defense Advanced Research Projects Agency through STTR contract No. DAAH01-97-C-R225, and the National Science Foundation through grant No. DMR-9704780.

REFERENCES

[1] S.Harrel, J. Vac.Sci. Technol. **B13**, 1879 (1995).

[2] R.E. Hollingsworth, C.DeHart, Li Wang, J.H. Dinan, and J.N. Johnson in *Amorphous and Microcrystalline Silicon Technology-1997*, edited by Sigurd Wagner, Michael Hack, Eric A. Schiff, Ruud Schropp, Isamu Shimizu (Mater. Res. Soc. Proc. **467**, Pittsburg, PA, 1997) pp. 961-966.

[3] M.K. Herndon, R.T. Collins, R.E. Hollingsworth, P.R. Larson, and M.B. Johnson, Appl. Phys. Lett. **74**, 141 (1999).

[4] E. Betzig, P.L. Finn, and J. S. Weiner, Appl. Phys. Lett. **60**, 2484 (1992).

[5] N. Kramer, H. Birk, J. Jorritsma, and C. Schonenberger, Appl. Phys. Lett. **66**, 1325 (1995).

[6] S.C. Minne, Ph. Flueckiger, H.T. Soh, and C.F. Quate, J. Vac. Sci. Technol. **B13**, 1380 (1995).

[7] N. Kramer, J. Jorritsma, H. Birk, and C. Schonenberger, J. Vac. Sci. Technol. **B13**, 805 (1995).

[8] S. Madsen, M. Mullenborn, K. Birkellund, and F. Grey, Appl. Phys. Lett. **69**, 544 (1996); S. Madsen, S.I. Bozhevolnyi, K. Birkelund, M. Mullenborn, J.M. Hvam, and F. Grey, J. Appl. Phys. **82**, 49 (1997).

[9] R.E. Hollingsworth, C. DeHart, Li Wang, J.N. Johnson, J.D. Benson and J.H. Dinan, J. Electron. Mat. **27**, 689 (1998).

[10] C. Durkan and I.V. Shvets, J. Appl. Phys. **80**, 5659 (1996).

[11] T. Okajima and S. Hirotsu, Appl. Phys. Lett. **71**, 545 (1997).

[12] J.H. Dinan, J.D. Benson, A.B. Cornfield, M. Martinka, J.N. Johnson, J. Bratton, and P. Taylor, *Proc. 1996 IEEE/CPMT Intl. Electronics Manufacturing Technology Symp.*, Austin, TX, October, 1996, p. 205.

DEPENDENCE OF IMAGE TRANSIENT BEHAVIOR ON OPERATING PARAMETERS OF AMORPHOUS SILICON IMAGE SENSORS

Koorosh Aflatooni, Richard Weisfield, Fan Zhong, Jon Bornstein, Mike Hack, and Carlos Gomez
dpiX, A Xerox New Enterprise Co., Palo Alto, CA

Abstract

The growing interests for real-time applications of our amorphous silicon X-ray image sensors requires a better understanding of transient behavior in imaging characteristics. In our pixel structure, which consists of a p-i-n photodiode and an addressing thin film transistor (TFT), the image lag can stem from a number of mechanisms, including charge retention on the photodiode, TFT charge transfer efficiency, and data-line charge sharing. The exact prevailing mechanism depends on the array design and the operating conditions of the sensors. This paper describes the results of our systematic studies on image lag, and its dependence on various operating conditions of the image sensors, such as applied bias, light intensity, frame time, and applied gate voltages. We find that the charge retention in the intrinsic region of the photodiode is an important source of the image lag, and it is correlated to the forward transient characteristics of the diode. This correlation can be explained by the common underlying mechanisms leading to depletion region formation and charge transport through the diode. Furthermore, we notice that the time dependent decay of the image lag, originated mainly from charge retention in the diode, presents power-law time dependence, which is consistent with the release of trapped charges from a continuous distribution of defect states. We also find only a small variation in the image lag, <%3, for a wide range of gate on-periods, 10μs- 80μs, and gate on-voltages, 15V- 30V, which indicates an efficient charge transfer by the switching TFT. These results demonstrate that in our devices, the image lag is controlled by the intrinsic defect levels in the a-Si:H photodiode.

Introduction

Digital X-ray imaging is a relatively new area that enjoys a dramatic boost by introduction of hydrogenated amorphous silicon (a-Si:H) X-ray image sensors. Presently, our X-ray image sensors have found their ways into a variety of medical and industrial applications, where they are mainly used for capturing snap shot images. Subsequently, the captured digital-information can be retrieved, analyzed, and stored on a computer. However, the growing demand for real time applications requires a new generation of high-performance a-Si:H image sensors. Such high performance image sensors are required to have high sensitivity, high-speed signal read-out, and more importantly, small image lag [1,2]. Here, the image lag is referred to as the persistence of a previously taken image to appear in subsequent images, and it is defined by the ratio between the left over charge to the initial signal value. High-speed applications require an image sensor with a minimum memory of the previously taken images; therefore, a complete discharge of the pixel after each read-out. There are a variety of mechanisms that can contribute to incomplete discharge of the pixel and image lag. Although charge trapping is an intrinsic property of a-Si:H devices, some of these mechanisms are related to the design of the array and its operating parameters. Thus, an optimized design can greatly enhance the image lag. Some of the important mechanisms that can contribute to the image lag include thin film transistor (TFT) charge transfer efficiency, charge sharing, and charge trapping in the photodiode. TFT charge transfer efficiency is an important mechanism that defines the capability of the switching TFT to discharge the collected charge. Hence, the TFT efficiency is related to parameters such as the size

827

Mat. Res. Soc. Symp. Proc. Vol. 557 © 1999 Materials Research Society

of the TFT, applied gate voltage, gate-on period, mobility, and threshold voltage. Charge sharing describes the residual charge on the photodiode as a result of charge redistribution between the data-line and photodiode capacitances. Here, the image lag is related to the ratio between the data-line and pixel capacitance. The photodiode charge retention is another important mechanism that is related to charge trapping in a-Si:H, and depends on a variety of parameters including applied bias, light intensity, quality of the material, and frame time. In this paper, we present the results of our studies of short time lag and its dependence on different operating parameters of the image sensors, including gate on-voltage, gate on-time, applied bias, integration time, read-out delay, and light intensity. In the next section, the measurement setups are described. The effects of different operating parameters of the TFT on lag are discussed in the next section. Following this, the effect of charge sharing, and charge retention will be discussed.

Experimental Methods

The pixel structures are formed by integration of an a-Si:H p-i-n diode and a switching TFT [2], similar to our commercially available image sensors. The measurements are performed on single pixel structures using light pulses from an LED. The LED emission was centered at a wavelength of 550nm with an intensity of $60\mu W/cm^2$. The light pulses were typically extended for a period of 1ms. The signal read-out started about 1ms after the light pulses by turning on the gate for about $20\mu s$. The period between light pulses, or frame time, was typically kept at 30ms consistent with 30 frame per second. Typically the charge trapping was saturated by eight consecutive light exposures, and then the decay of the charge was monitored with time. The photodiode was operated under a reverse bias of 5V. For read-out of charge, a gate voltage of 15V was applied to the switching TFT.

Similar measurements were also performed on large array of pixels, using X-ray exposures and a scintilator layer. The X-ray exposures continued for a period of 60ms and then the signal read-out was started about 1ms after X-ray exposure. For arrays, the frame time was limited by the read-out electronics to about 3s. Similar biasing conditions to single pixel experiments were also used for array operation.

TFT Charge Transfer Efficiency

The thin film transistors serve as switching elements in our image sensors, where they provide isolation of the photodiode during charge integration period and discharge of the pixel during readout. Therefore, the efficiency of TFT in transferring the charge is an important parameter in determining the residual charge on the pixel, i.e., image lag. Here, the key parameter is the ratio between the pixel time constant and the gate on-time, and the residual charge (Q_{res}) is defined by,

$$Q_{res} = Q_{col} \exp\left(-\frac{t_{on}}{\tau_{pixel}}\right),\qquad(1)$$

where Q_{col} is the initial charge collected by the photodiode, t_{on} is the gate on-time, and τ_{pixel} is the pixel time constant. The time constant of the pixel is defined by

$$\tau_{pixel} \approx R_{on}C_{pixel},\qquad(2)$$

where R_{on} is the on resistance of the TFT and C_{pixel} is the photodiode capacitance. The on-resistance of the TFT is related to parameters such as gate on-voltage (V_g), ratio of channel width to length (β), threshold voltage (V_{th}), mobility (μ), and can be approximated by

$$R_{on} \approx \frac{1}{\mu C_g \beta (V_g - V_{th})},$$ (3)

where C_g is the gate capacitance. From (1) and (3), the dependence of lag on gate voltage can be calculated, which is shown in Fig. 1 and compared with measured results. Figure 2 depicts the dependence of lag on gate on-time. The discrepancy in Fig. 2 between calculated and measured lag results for $t_{on} >> \tau_{pixel}$ arises, where other mechanisms contributing to lag become dominant. Thus, further increases in gate on-time or gate on-voltage do not lead to a reduction of lag.

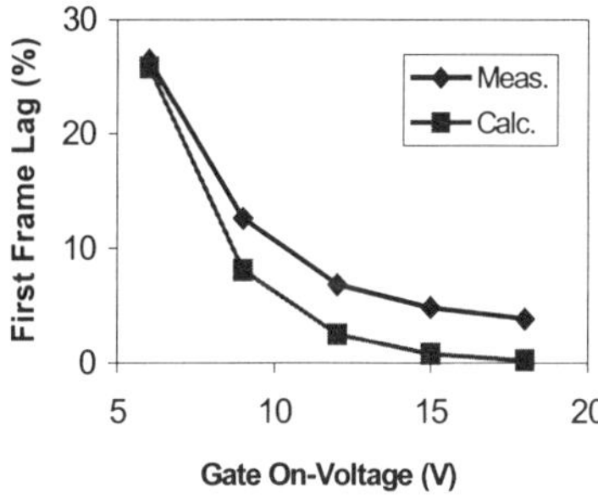

Fig. 1: Dependence of lag on gate on-voltage.

Fig. 2: Dependence of lag with gate on-time.

Inspection of (1), (2), and (3) reveals the trade off between the gate on-time and the gate on-voltage. For faster read-out, comparable charge transfer efficiencies can be achieved using a larger gate on-voltage.

Charge Sharing

After connecting the pixel to the data-line, the collected charge redistributes between the photodiode and the data-line capacitances. The charge remaining on the photodiode is directly related to the ratio of the capacitances (η) and can be shown as [3]

$$Q_{cs} \approx Q_{col} \frac{C_{pix}}{C_{pix} + C_{data}},$$ (4)

where Q_{cs} is the charge left on the photodiode due to charge sharing, Q_{col} is the collected charge, C_{pix} is the pixel capacitance, and C_{data} is the data-line parasitic capacitance. Figure 3 depicts the dependence of lag with the ratio of photodiode and dataline capacitances (η) that is defined in (4) for several different array designs.

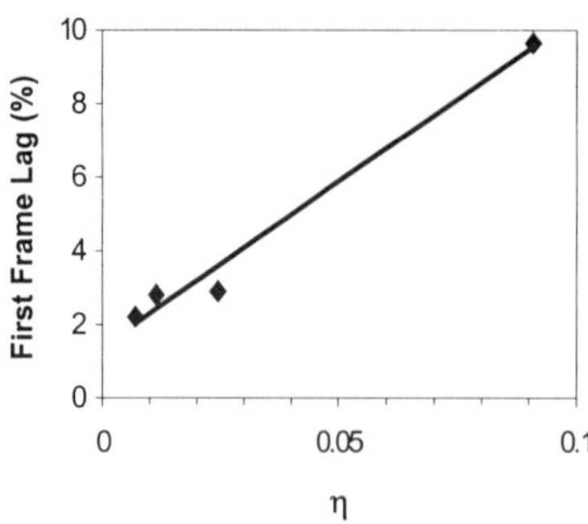
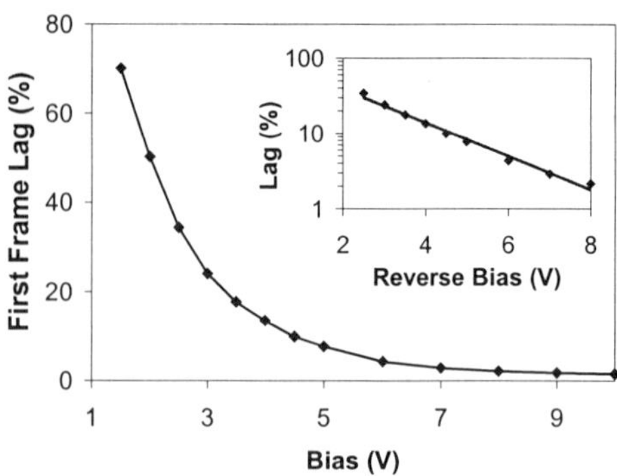

Fig. 3: Dependence of lag on the ratio of pixel to data-line capacitances for several different array design.

Fig. 4: Variation of lag with applied bias for a constant light exposure. The inset shows the exponential dependence between lag and reverse bias for sufficiently large electric fields.

Charge Retention on Photodiode

Charge retention on the photodiode is a dominant component of image lag for imagers with small photodiode capacitances. Here, charge retention mainly originates from charge trapping in mid-gap states intrinsic to a-Si:H. Various parameters can control the charge collection, which in turn effects the charge retention on the photodiode. Some of these parameters are applied bias, density of trap states, initial light level, and storage time. For example the applied bias determines the charge separation and collection, and a strong electric field along the diode can effectively collect the photo generated charge and prevent charge trapping. In competing with electric field, the mid-gap states enhance the charge trapping; therefore, increasing the charge retention on the diode. The dependence between retained charge, the electric field, and the density of mid-gap states can be expressed by [4]

$$Q_{col} = Q_o\left[1 - \exp\left(-\frac{\mu\tau}{L}F\right)\right],$$ (5)

where Q_{col} is the charge retrieved from the diode, Q_o is the total charge generated by the diode, μ is the mobility, τ is the carrier life time, L is the thickness of the photodiode, and F is the electric field. The $\mu\tau$ product defines the quality of the intrinsic layer, and is inversely related to the density of mid-gap states. Typically, a $\mu\tau$ value in the range of 10^{-8}-10^{-7}cm^2/V represent a high quality a-Si:H [4]. Equation (5) is based on the assumption of a uniform electric field along the diode. For the case that the intrinsic layer is completely depleted, the electric field can be approximated as

$$F_f = \frac{V_a + V_b - Q_{col}/C_{pix}}{L},$$ (6)

where Q_{col} is the collected charge, and C_{pix} is the pixel capacitance. Equation (6) relates the decay of the electric field and charge trapping to the photo-generated charge. The dependence of the retained charge with applied reverse bias for a constant light level is shown in Fig 4. Similarly, the dependence between the retained charge and the X-ray exposure level is shown in Fig. 5. For the case that the intrinsic region is partially depleted, the electric field becomes non-uniform along the diode. In this situation the collection efficiency degrades significantly, and

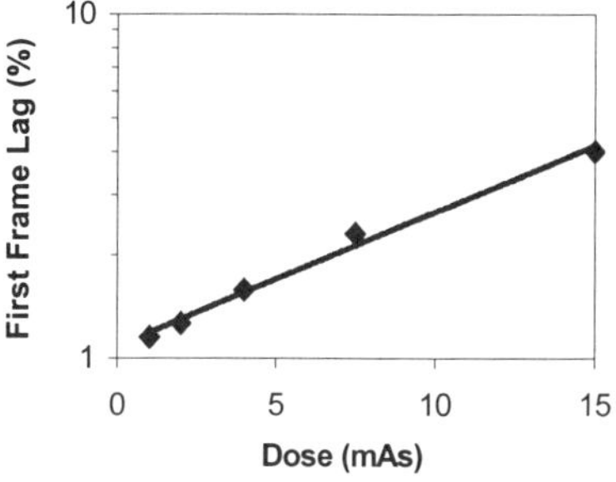

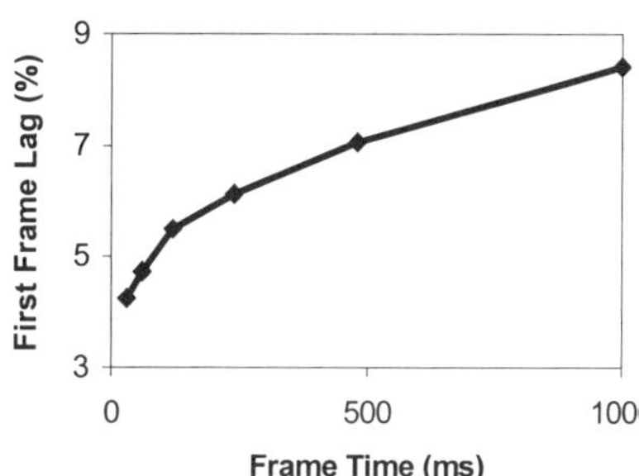

Fig. 5: Dependence of lag with exposure level.

Fig. 6: Dependence of lag with frame time.

charge trapping in the mid-gap states yields severe charge retention in the photodiode. In this case, trapped charges contribute to the signal as they escape, and form the image lag. The charge release is a relatively slow process that presents power law dependence with time as [5],

$$Q \propto t^{-\alpha}, \tag{7}$$

where t is the time after exposure, and α defines the ratio between the characteristic slope of the mid-gap states and the temperature. The released charge contributes to the signal over many readout cycles. Fig. 7 shows the decay of the released charge with time and the fitting results

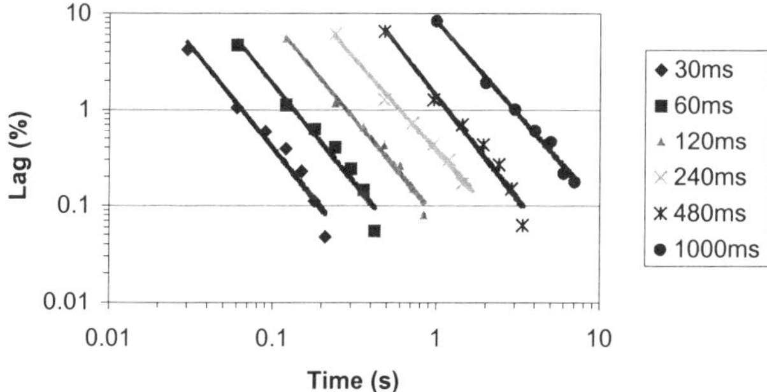

Fig. 7: Release of trapped charges in photodiode presents power law dependence.

obtained by using relation (7). The parameter α reflects the average depth in energy that the charge is trapped and it depends on the light intensity and the frame time.

The initial state of the photodiode will only be attained if all the trapped charges escape or recombine and the depletion layer restored to its initial conditions. In many ways, this is similar to forward transient behavior of the pin diodes. Here, the diode only starts to conduct after the electrons reach the p-i interface, and cause a collapse of the depletion region [6,7,8]. The diodes with slow forward transient suffer from slow transport of electrons that is caused by large density of defects and small $\mu\tau$ values. In Fig. 8 transient behaviors are compared between two different groups of diodes with different processing conditions that yield different levels of charge trapping.

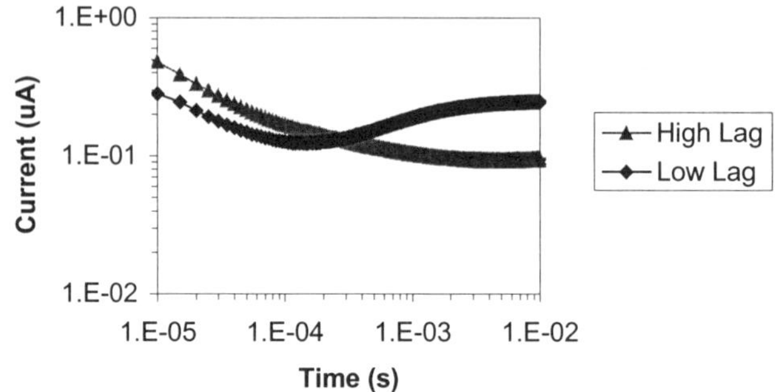

Fig. 8: The correlation between fast forward transient response and low lag.

Conclusion

A variety of mechanisms can contribute to image lag, e.g., TFT charge transfer efficiency, charge sharing with data-line, or charge retention on photodiode. The exact prevailing mechanism depends on the array design and its operating conditions. In an optimally designed pixel, the density of mid-gap states in intrinsic layer is the limiting factor. Our recent fabrication improvements demonstrated that a reduction in the density of mid-gap states could significantly improve the image lag. Further lag improvements may be achieved by using; for example, p+ μc-Si that yields a larger built in voltage and pushes the collapse of depletion region to higher exposure levels. It is also demonstrated that the decay of image lag follows a power law decay, which can be used for lag correction algorithms.

References

1) R. L. Weisfield, M. A. Hartney, R. Schneider, K. Aflatooni, and R. Lujan, SPIE Physics of Medical Imaging, in press (1999).
2) R. L. Weisfield, M. A. Hartney, R. A. Street, and R. B. Apte, SPIE Physics of Medical Imaging **3336**, p. 444 (1998).
3) I. Fujieda, S. Nelson, R. A. Street, and R. L. Weisfield, IEEE Trans. Nucl. Sci. **39**, p. 1056 (1992).
4) R. A. Street, *Hydrogenated Amorphous Silicon*, Cambridge University Press, London, 1991.
5) T. Tiedje, and A. Rose, Solid State Communication **37**, p. 49 (1980).
6) R. A. Street, and M. Hack, Mat. Res. Soc. Symp. Proc. **258**, p. 315 (1992).
7) M. Hack, and R. A. Street, J. Appl. Phys. **72**, p. 2331 (1992).
8) D. Han, K. Wang, and M. Silver, J. Non-Cryst. Solids **164-166**, p. 339 (1993).

MORE INSIGHT INTO THE TRANSIENT PHOTOCURRENT RESPONSE OF THREE-COLOR DETECTORS

H. STIEBIG[1], B. STANNOWSKI[2], D. KNIPP[1], AND H. WAGNER[1]
[1] Forschungszentrum Juelich, D-52425 Juelich, Germany, h.stiebig@fz-juelich.de
[2] Debye Inst., Interface Physics, Univ. Utrecht, POBox 80000, 3508TA Utrecht, Netherlands

ABSTRACT

Nipiin structures with controlled bandgap and mobility-life-time product exhibit excellent stationary properties. However, the photocurrent transients of these devices show reasonable delay times in the range of tens of milliseconds before reaching steady state. Hence, the transient behavior after light switching of three color detectors based on a nipiin structure was studied by a detailed analysis of experimentally determined and simulated data. The numerical simulation reproduces well the characteristic features. The delayed current onset and the sign reversal of the transient photocurrent response are explained by trapping of holes in the vicinity of the central p-layer.

INTRODUCTION

Amorphous silicon (a-Si:H) based sensors are promising candidates for detection of the fundamental components of visible light. The strong wavelength dependent penetration depth of the incident light in a-Si:H and its alloys in the visible range, can be used as a tool for color separation. The wavelength dependent generation profile in combination with a voltage controlled spatial resolution of carrier collection results in color detection in the depth of the structure. Thus, several two terminal detector concepts to separate the color information have been proposed: pin-diodes with modified absorption layers [1,2], pinip or nipin-structures [3,4] or more complex layer sequences [5]. The two terminal devices, which must be read out sequentially to generate a RGB-(red-green-blue)-signal, show pronounced stationary properties like good color separation, high linearity between photocurrent and the intensity of incident light and a high dynamic range exceeding 95 dB (ratio between the current under 1000 lx white light illumination and the dark current). It can be prepared on a conventional read out electronic. However, the transient behavior of these detectors after light and voltage switching limits their application in the field of electronic imaging. The experimentally found transient photocurrent response of a-Si:H based devices after switching on monochromatic light at different bias shows resonable delay times between a few hundred microseconds and some milliseconds before reaching steady state. Therefore, in the following a detailed analysis of the transient photocurrent response of a nipiin structure after light switching will be discussed. Under certain illumination conditions a delayed photocurrent onset or a sign reversal during the photocurrent onset is observed depending on the applied bias voltage and the wavelength [6]. By comparing numerically simulated and experimentally determined data a study of time dependent processes within the device like charge transport and storage in localized states is presented.

EXPERIMENTAL RESULTS

Nipiin structures in principle consist of two antiserielly connected diodes. The spatial distribution of the bandgap of the detector is given schematically in Fig. 1. The optical bandgap of the different intrinsic absorption layers decreases from 2 eV in the top diode to 1.9 eV and 1.55 eV in the bottom diode. The nipiin layer sequence consists of six layers with a total thickness of around 500 nm. Due to the voltage controlled collection efficiency the spectral sensitivity can be shifted from blue to green and red (Fig.2). At positive bias the top diode is reverse biased and blue light is detected. In this case the top diode determines the current of the detector, whereas the bottom diode is forward biased. At low negative bias the top diode is forward biased and the photo-generated carriers in the front part of the bottom diode are extracted. Charge carriers generated in the rear part of the device recombine due to the low electric field and the low mobility-life-time product. In the case of higher negative bias the overall generated

Mat. Res. Soc. Symp. Proc. Vol. 557 © 1999 Materials Research Society

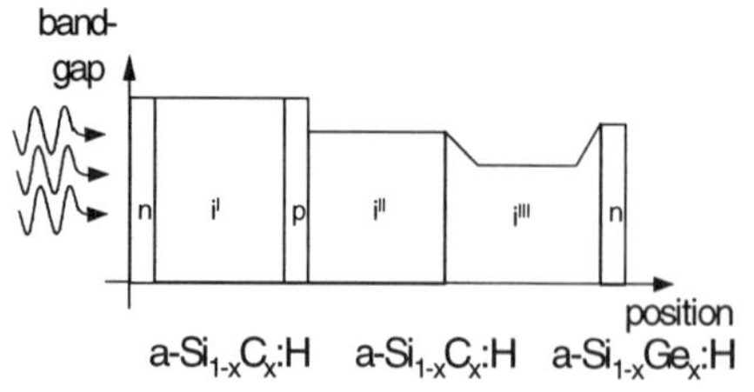

Fig. 1: Bandgap profil of the three color nipiin detector (shown schematically).

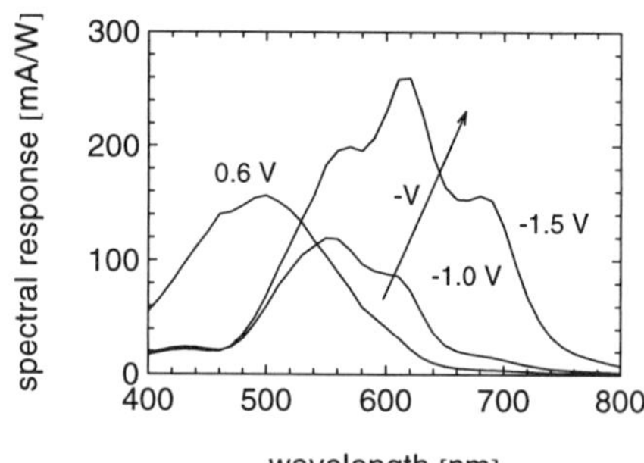

Fig. 2: Measured spectral response of a nipiin-structure for different bias voltages.

carriers in the bottom diode are collected and the maximum of the spectral response shifts to longer wavelengths. The developed device design of the detector results in maxima of the spectral response at 480 nm (blue), 550 nm (green), and 620 nm (red) at 0.6 V, -1.0 V and -1.5 V, respectively.

Beside the stationary behavior the transient response is a further important criterion to characterize the detector performance. In Fig. 3 the transient photocurrent response of a nipiin detector after switching blue (λ = 450 nm, Fig. 3a) and red (λ = 650 nm, Fig. 3b) illumination (ϕ = 10^{14} cm^{-2}s^{-1}) is shown, respectively, with a preceeding dark time of 10 s. The transient response measurements were performed using red or blue LED-switching modules with a rise time less than 10 μs. For the typical color voltages of 0.6 V, -1.0 V and -1.5 V the curves show delay times up to several milliseconds before steady state is reached. The delay times are comparable with those found after voltage switching under constant illumination with similar photon flux [6]. Under blue illumination (Fig. 3a) and positive applied bias a delayed current response is observed, whereas at negative bias a sign reversal of the photocurrent is detected. The opposite behavior is found for red light (Fig. 3b). The delayed monotonously current rise after around 10^{-4} s of light exposure appears at high negative voltages. In the following the different time regimes are considered in detail. Therefore, dot 1 in Fig. 3b denotes the I/V values of the top and bottom diode in the dark. The current is low due to the low rate of thermally excited carriers in the reverse biased diode. The thermally generated electrons are collected at the n-layer of the reverse biased diode whereas the holes recombine in the vicinity of the p-layer with injected electrons from the forward biased diode. After about 10^{-4} s (Dot 2) the photocurrent shows a quasi-stationary value, which is nearly independent of the applied voltage. This distinct plateau is

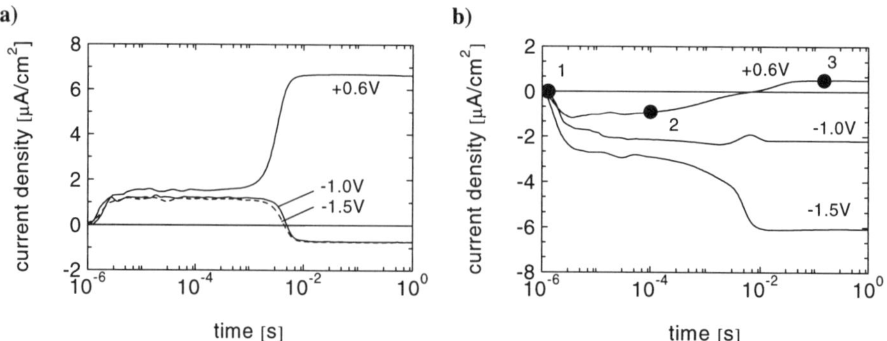

Fig. 3: Measured transient photocurrent response of a nipiin three color detector after switching-on of blue (λ = 450 nm, Fig. 3a) and red (λ = 650 nm, Fig. 3b) illumination (ϕ = 10^{14}cm^{-2}s^{-1}).

more pronounced for blue than for red light. The steady state (Dot 3) is reached after some milliseconds and a sign reversal or a step wise rise of the photocurrent depending on the applied bias and illumination is observed.

DISCUSSION

To explain the transient properties of nipiin detectors numerically calculated transients are compared with experimental data. The simulation program is based on a spatial defect distribution resembling the defect pool model and the complete set of time dependent semiconductor equations for electrons and holes together with the rate equation, which describe the interchange of charge carriers between extended and localized states, are solved numerically. In the equilibrium the defect density is dominated by negatively charged defect states in the n-layers (D^n) and by positively charged states (D^p) in the central p-layer (Fig. 4). Since a-SiGe:H with a bandgap of 1.55 eV shows a higher defect density in comparison to high quality wide bandgap material, an enhanced negatively charged defect density is calculated in the rear part of the bottom cell. More detailed information concerning the numerical model is given elsewhere [7]. Previous investigations have shown that the enhanced defect density in the p-layer has a dominat influence on the I/V-characteristics [8]. Via the enhanced defect density in the vicinity of the central p-layer the collected holes of the reverse biased diode and injected through the p-layer recombine with electrons injected at the contacts.

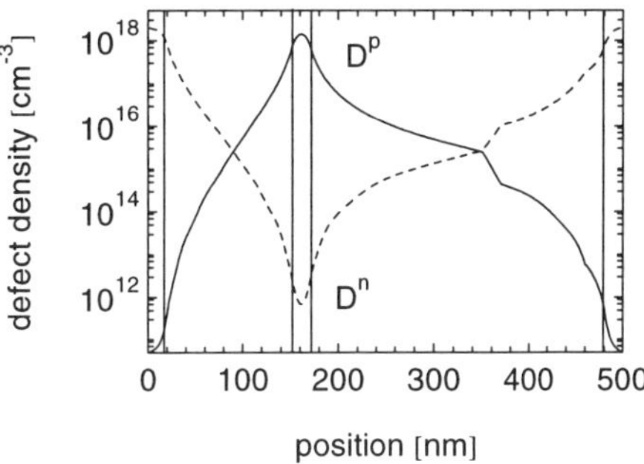

Fig. 4: Self-consistent calculated spatial distribution of the charge defect density of the color detector in the equilibrium.

The measured transient responses under blue and red illumination are well reproduced by numerical simulation (Fig. 5). All the characteristic features of the photocurrent and the time, voltage and illumination dependence of the current can be described by numerical modeling. The good agreement of the data allows deeper insight into the transport and recombination behavior of carriers in nipiin detectors. Thus, in the following the sign reversal as well as the step wise rise of the photocurrent will be discussed.

First, the sign reversal of the photocurrent under a bias voltage of 0.6 V and switching-on red illumination is considered (Fig. 3b). In order to explain the I/V behavior of the device, the color

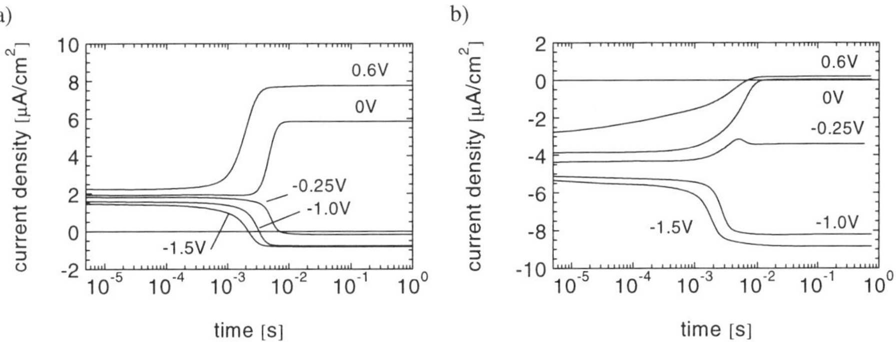

Fig. 5: Simulated transient photocurrent response of a nipiin three color detector after switching-on of blue ($\lambda = 450$ nm, Fig. 3a) and red ($\lambda = 650$ nm, Fig. 3b) illumination ($\phi = 10^{14}\,cm^{-2}s^{-1}$).

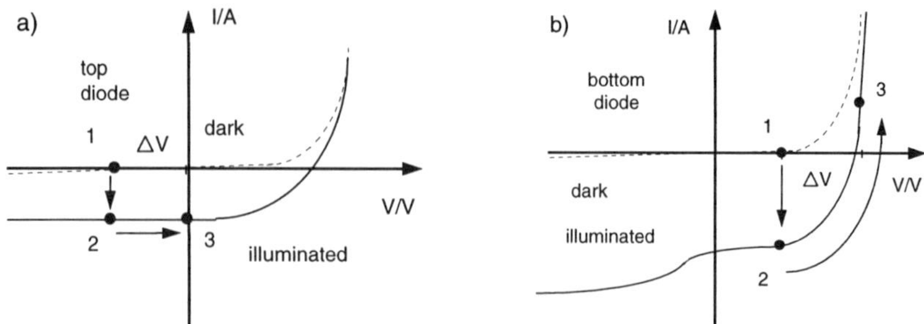

Fig. 6: Schematic I/V behavior of the top (a) and bottom diode (b) of a nipin three color sensor after switching on red illumination: in the dark (Dot 1), after 10^{-4} s (Dot 2), in the steady state (Dot 3). (V = 0.6 V).

sensor is regarded as two antiseriell connected diodes. Fig. 6 schematically exhibits the I/V behavior of the individual diodes. The three marking points of Fig. 6 correspond to the dots in Fig. 3b. Before light switching (Dot 1) the top diode of the nipiin structure is reverse biased (Fig. 6a) and the bottom diode is forward biased (Fig. 6b). The total current is positive and is determined by the dark current (thermal generation of carriers) of the top diode. After light exposure the operation point of the individual diodes switchs from „Dot 1" to „Dot 2" . The top diode is still reverse biased while the operation point of the forward biased bottom diode shifts into the active quadrant. Thus, the bottom diode also generates a photocurrent. At this point of time, the photocurrent of the bottom diode overcompensates the photocurrent of the top diode due to the high absorption of red light in the rear part of the bottom diode. This fact explains the negative values of the current. In the transition regime between Dot 2 and Dot 3 a potential shift in the central p layer by ΔV of around 0.15 V is observed by numerical modeling. Due to the potential shift the bottom diode switches from blocking to permeable. Consequently the total current changes from negative to positive values and in the steady state the current is again determined by generated carriers in the top diode. While the potential difference between the front and rear contact is given by the externally applied voltage and thus is constant in time, the simulation reveals that the individual voltages at the top and bottom diode are time dependent. Fig. 6 also explains the more complex transient behavior under red illumination in comparison to blue light exposue. Since a high defect density is introduced in the rear part of the bottom diode, a more „s"-shape I/V-behavior is expected. Thus, in the transition region from a reverse biased to a forward biased diode, the operation point of the bottom diode must follow the „s"-shape of the I/V characteristic.

More insight into this transient behavior is gained from the spatial distribution of the electron and hole current density regarded at 10^{-4} s, 10^{-3} s, $2*10^{-3}$ s and 10^{-2} s after switching on red illumination. In case of a positive hole current, holes move in x-direction, whereas a positive electron current means that electrons move in reverse direction. Since red light is absorbed dominantly in the rear part of the bottom diode, mainly carrier collection takes place in this region. Fig . 7a shows that in the i-layer of the top diode the current is dominated by the hole current and that neither the electron current nor the hole current varies significantly in time in this diode. For early times, in the bottom diode electrons move towards the n-layer and holes towards the central p-layer. Thus, carrier collection by the electric field occurs in both diodes and the current is drift controlled in both diodes. This behavior is indicated by the high values of the current densities which monotonously increase for electrons and decrease for holes in the rear part of the structure. Furthermore, a nearly constant value for the hole current density in the first absorption region of the bottom diode is observed at early times. After light switching the negative current of the bottom diode overcompensate the positive current of the top diode. For later times ($t > 10^{-4}$ s) the carrier transport in the bottom diode changes. Due to the potential shift the electric field strength within the device reduced and the photo-generated carriers in the bot-

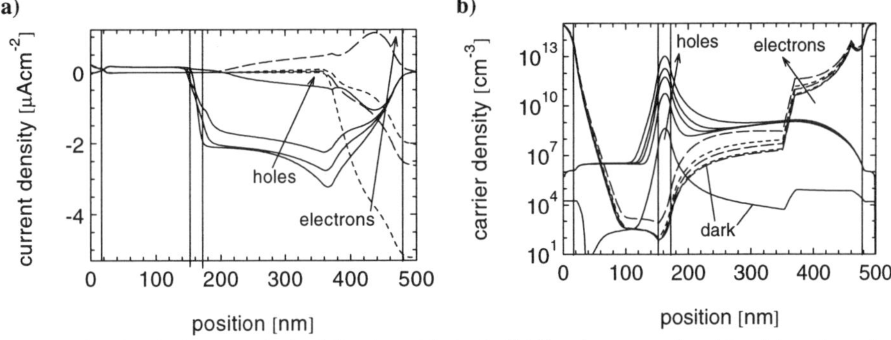

Fig. 7: Simulated electron (dashed lines) and hole (solid lines) current densities (a) and carrier densities (b) after 10^{-4} s, 10^{-3} s, $2*10^{-3}$ s, and 10^{-2} s after switching on red illumination ($\lambda = 650$ nm). In Fig. 7b also the carrier density of electrons and holes in the dark is plotted. (V = 0.6 V).

tom diode recombine. In the steady state the electron current is dominated by a diffusion current. Thus, the sign reversal of the photocurrent can be attributed to a delayed onset of recombination in the rear part of the bottom diode and a change from a drift to a diffusion controlled current.

Further information concerning the potential shift can be achieved from the analysis of the time dependence of electron and hole concentration. In the dark as well as under light exposure the electron density is high in the n-layers, where electrons are majority carriers, and low in the central p-layer, where they are minority carriers. Moreover, one finds only a slight variation of electron concentration in time. In contrast, the hole density in the central p-layer increases over more than 4 orders of magnitude after light exposure. In the dark the strong electric field strength in the top diode results in an extraction of holes in the whole structure and a depletion of the p-layer. Therefore, the hole concentration is reduced and positively charged defect states are neutralized compared to the equilibrium state (Fig. 4). The positive charge which partly compensates the space charge of the negatively charged acceptor states originating from doping, is reduced resulting in a temporary enhanced built-in potential in the top and bottom diode. This enhanced built-in potential explains the high electric field strength in the rear part of the bottom diode, which leads to the extraction of photo-generated carriers. Under illumination the p-layer is refilled by photo generated holes drifting from the bottom diode towards the p-layer. The increase of the positive space charge is responsible for the shift of the electric potential in the p-layer by ΔV. In the steady state the recharging process is completed. Thus, trapping of holes in the defect states in and near the central p-layer is accompanied with the potential shift in the central p-layer due to the increase of positive space charge. The potential shift is also responsible for the second aspect of the photocurrent onset, namely the delayed current onset, which is observed at high negative bias or under different illumination conditions (blue light). In the following the electron and hole current after blue light exposure is investigated. In agreement with the simulation shown above, a bias voltage of 0.6 V is applied.

Blue light is dominantly absorbed in the top diode, leading to high current density values in this diode, which are nearly constant in time. In contrast to the current transport in the top diode at 10^{-4} s the electron and hole current in the bottom diode is very low. With increasing time

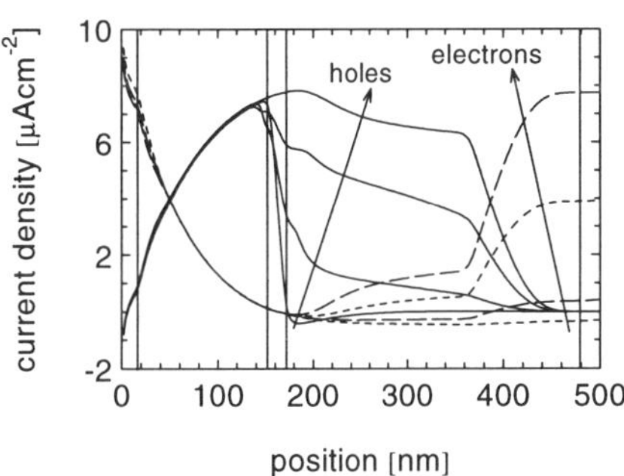

Fig. 8: Electron and hole current densities after 10^{-4} s, 10^{-3} s, $2*10^{-3}$ s and 10^{-2} s after switching on blue illumination. (V = 0.6 V).

the current densities in this diode rise. The enhanced current in the bottom diode is correlated with the increase of the total current (Fig. 3a). In this case, the lack of holes in the p-layer is dominantly refilled by holes collected in the top diode. Consequently, in the time interval from 10^{-4} s to 10^{-2} s, the bottom diode switched from a blocking to a permeable state. Increasingly more holes injected from the top diode through the p-layer into the bottom diode and more and more electrons from the back contact are accumulated in the bottom diode. In the rear part of the structure (Fig. 4) the injected holes recombine with injected electrons, due to the high defect density. Therefore, the increase of the total current is correlated with an increase of the recombination in the bottom diode. The total current in the bottom diode changes from a drift to a diffusion current. The simulation reveals that the transient behavior including the sign reversal or the delayed current onset is a device structure inherent phenomena originating from the depletion of the p-layer in the dark upon an applied bias voltage.

Moreover, the refilling process explains also the time dependence of the rise time on the light exposure. Previous investigations have shown, that the delay time after light switching decreases reciprocally with increasing photon flux over more than 5 orders of magnitude [1]. Under higher illumination, a larger amount of holes drifts to the p-layer in a certain time interval and, therefore, the recharging rate is enhanced.

CONCLUSIONS

A detailed analysis of the transient photocurrent response after switching on monochromatic light by a comparison of simulated and experimentally determined data is presented. The sign reversal of the photocurrent under certain illumination conditions can be attributed to a overcompensation of the photocurrent of the reverse biased diode by the presumably forward biased diode and the delayed onset of the photocurrent is caused by the retarded injection of holes though the p-layer. These device inherent characteristics are caused by the depletion of the p layer in the dark upon an applied bias voltage and results from a remarkable potential shift in the central p-layer upon illumination. The strong illumination dependence of the transient response depending on the recharging rate complicates the application of nipiin structures in the field of electronic imaging.

ACKNOWLEDGMENTS

This work was supported by the Bundesministerium für Bildung, Forschung und Technologie (BMBF).

REFERENCES

1. D. Knipp, H. Stiebig, J. Fölsch, R. Carius, H. Wagner, Mater. Res. Soc. Proc. **467**, Pittsburgh, PA (1997) p. 735.
2. T. Neidlinger, M.B. Schubert, G. Schmid H. Brummack, Mater. Res. Soc. Proc. **420,** Pittsburgh, PA (1996) p. 147.
3. H.-K. Tsai, S.-C. Lee, Appl. Phys. Lett. **52,** 275-277 (1988)
4. H. Stiebig, J. Giehl, D. Knipp, P. Rieve, M. Böhm, Mater. Res. Soc. Proc., **377**, Pittsburgh, PA (1995) 815.
5. D. Caputo, F. Irrera, F. Palma, S.R. Rachele, M. Tucci, J. Non-Cryst. Solids **198-200**, 1172-1175 (1996).
6. H. Stiebig, C. Ulrichs, T. Kulessa, J. Fölsch, F. Finger, H. Wagner, J. Non-Cryst. Solids **198 - 200**, 1185 (1996).
7. B. Stannowski, H. Stiebig, D. Knipp, H. Wagner, J. Appl. Phys. **85** (7) 3904-3911 (1999).
8. H. Stiebig, M. Böhm, J. Non-Cryst. Solids **164-166**, 785 (1993).

NEAR INFRARED RESPONSE OF AMORPHOUS SILICON DETECTOR GROWN WITH MICROCOMPENSATED ABSORBER LAYER

D. CAPUTO*, G. DE CESARE*, A. NASCETTI*, F. PALMA*, M. TUCCI**
* Department of Electronic Engineering, University of Rome "La Sapienza", via Eudossiana 18, 00184 Rome (Italy)
** ENEA, Research Centre, Località Granatello, 80055, Portici, Napoli, (Italy)

ABSTRACT

In this work we demonstrate that radiation up to 2 µm induces photocurrent in a single junction amorphous silicon structure at room temperature. The absorber layer is a microcompensated film deposited using very low concentrations of dopant species. Device operation is based on optical excitation of thermal generated carriers from trap states toward valence and conduction band in the high electric field region of the structure. Transient and frequency response under different bias voltages and illuminations conditions are presented. The possibility to use the infrared sensor in low bit rate communication systems has been demostrated by including our detector in a front-end system and measuring its frequency responce.

Quantum efficiency measurement have been reproduced with a numerical model, able to take into account sub-band gap absorption into single films. Model results indicate the presence of a large valence band tail and a high number of dangling bonds and shallow defects ascribed to the presence of dopant atoms.

INTRODUCTION

Single junction p-i-n amorphous silicon (a-Si:H) structures have been extensively used as solar cells, with quantum efficiency response up to 750 nm or 900 nm when the absorber region is an intrinsic amorphous silicon layer or an intrinsic amorphous silicon-germanium layer, respectively [1, 2, 3]. Constant photocurrent method and photothermal deflection spectroscopy showed that the presence of defects in the forbidden gap allowed to obtain photocurrent response up to 1500 nm in intrinsic a-Si:H films. The first detection of radiation up to 2400 nm by a-Si:H a p-i-n diode, operating under forward bias conditions at low temperature (198 K), has been reported by Wind and Müller [4].

More recently, it has been demonstrated the possibility to achieve near and medium infrared detection at room temperature by using single junction a-Si:H structure with microcompensated absorber zone, whose operation is based on the transitions between the high number of defect density present in this material and the extended states [5, 6]. Here, we present a detailed investigation of the near infrared responce of this device focusing the attention on the detection mechanism and on its transient and frequency responce. Altough near infrared a-Si:H detectors are orders of magnitude slower than detectors based on the III-V technology, they could represent a complementary approach, because a-Si:H technology, which requires low deposition temperature, is suitable for almost any kind of substrate and, in particular, it is compatible with silicon IC and integrated optics technology [7].

EXPERIMENTAL RESULTS

On a transparent conductive oxide coated glass the deposition of the amorphous layers is performed as the following steps:
1) 300 Å thick p-doped amorphous silicon carbide layer;
2) 7000 Å thick microcompensated amorphous silicon layer;
3) 500 Å thick n-doped amorphous silicon layer.
The microcompensated zone has been deposited with 0.25 ppm of PH_3 and 18 ppm of B_2H_6 in the gas mixture. This low dopant concentrations justifies the name of this layer. The front electrode was aluminum deposited by evaporation and then shaped as a thin finger grating, with spacing of 50 µm. The area of the fabricated devices ranged from 25 mm^2 to 1 cm^2.

Mat. Res. Soc. Symp. Proc. Vol. 557 © 1999 Materials Research Society

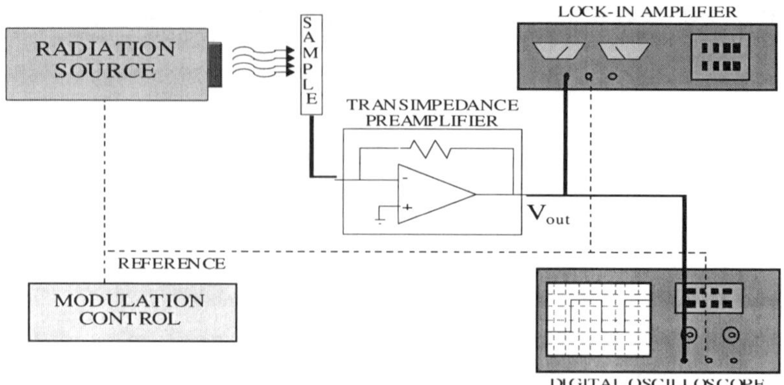

Figure 1. Measurement set-up used in the transient, frequency and quantum efficiency response. The utilized radiation sources were a LED emitting at 950 nm, a Nd:Yag laser emitting at 1.32 μm and a Nernst ceramic element filtered with a monochromator in the range 800 nm - 4.5 μm.

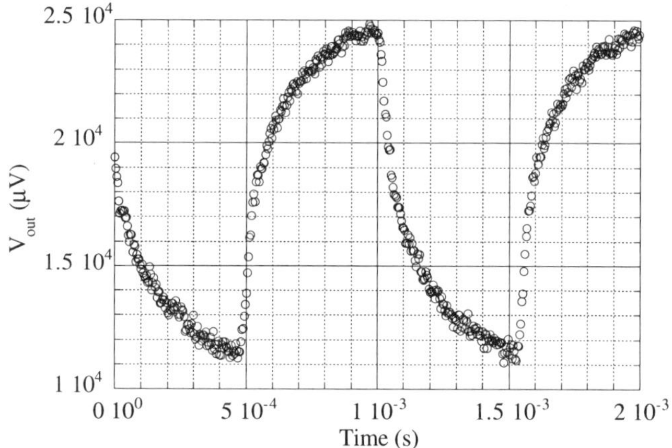

Figure 2. Detector transient response when the power inpinging on the sample is equal to 10mW. The radiation comes from a a Nd:Yag laser emitting at 1.32 μm. The shape of the waveformindicates that a photocurrent signal is detected.

Data reported in this paper refer to 25 mm^2 devices. The radiation is impinging from the aluminum grid, to avoid absorption of the IR radiation by the glass substrate.

Transient response characterization

The measurement set-up for the transient response is reported in figure 1. The radiation source is a Nd:Yag laser emitting at 1.32 μm. The transimpedance amplifier has been realized with a three stages preamplifier using low noise electronic components and having a gain of 10^6 V/A. The output signal coming from the transimpedance amplifier is read with a EG&G 5210 lock-in amplifier and is simultaneously displayed on a 9100 LeCroy digital oscilloscope.

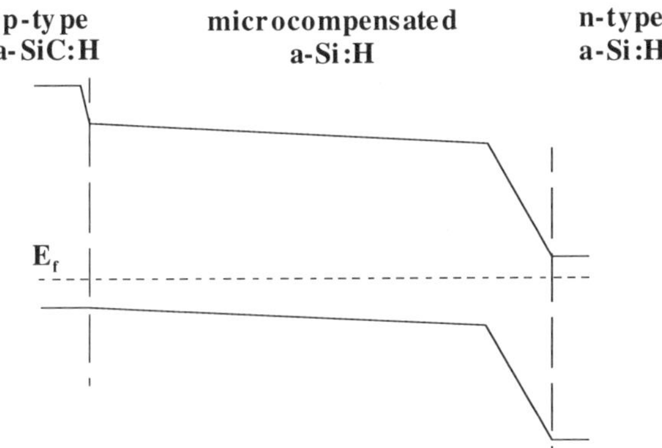

Figure 3. Band diagram inferred from quantum efficiency measurements in the visible range.

In figure 2 the transient response, with incident power equal to 10mW, is reported. The period of the radiation was 1ms with 50% duty cycle. The shape of the waveform indicates that a photocurrent signal is detected. The slow speed response, observable in fidure 2, indicates that inour device photoconduction is strongly affected by trapping mechanisms. Since the energy of the incident photons (0.94 eV) is lower than the band gap of a-Si:H (1.72 eV), the observed signal has to involve sub-gap transitions. On the other side, photocurrent in a junction implies transport of both type of carriers, and then the incident radiation induces transitions between localized and extended states to generate free holes and free electrons. In particular, photocurrent in the device is due to the combined effect of photo-induced and thermal induced transitions from localized states toward valence and conduction band. From quantum efficiency measurements in the visible range with light impinging on the p-side or on the n-side [6], it has beeen inferred that the band diagram of the device is the one reported in figure 3. It can be seen that the microcompensated layer behaves as p-type region near the p-doped a-SiC:H layer and that the high electric field region exists only close to the n-doped zone. Since minority carriers have a very low diffusion length in highly defective amorphous materials as are doped films, photogeneration is effective only in the region near the microcompensated-n interface.

Frequency characterization

In figure 4 detector response as a function of a 950nm LED modulation frequency is shown. The different curves refer to different average incident power achieved by applying different bias voltage to the LED. For each curve the amplitude of the modulated signal is kept constant. It can be observed that frequency response depends on the light intensity. In particular, the higher is the light intensity, the higher the cut-off frequency. This behavior can be explained taking into account that an increase of the bias light intensity causes occupation of defects in the forbidden gap and then a decrease of the number of states which are effective in the trapping process.

Results of figure 4 have been obtained by keeping the detector in short circuit conditions. A further frequency characterization has been performed applying a forward bias voltage to the sensor. Results are reported in figure 5. An increase of the photocurrent is observed as the forward bias level is increased. A similar behavior has been found by Wind and Müller on a-Si:H p-i-n diodes at 198 K, who ascribed the effect to IR induced re-excitation of excess carriers trapped in the band tails, during forward bias injection of charge carriers, toward extended states.

Applying a reverse bias voltage to the device, only a sligthly increase of the frequency responce is achieved, due to the presence of the high defect density in the absorber layer, which screens off the electric field due to the applied voltage.

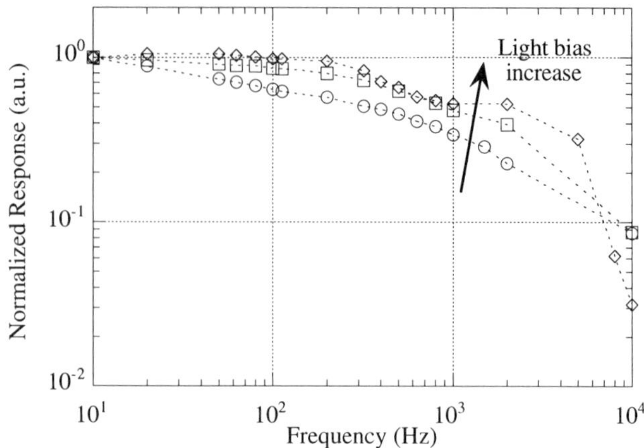

Figure 4. Normalized photocurrent versus frequency of a radiation incident from a 950nm LED. For each curve the amplitude of the modulated signal is kept constant.

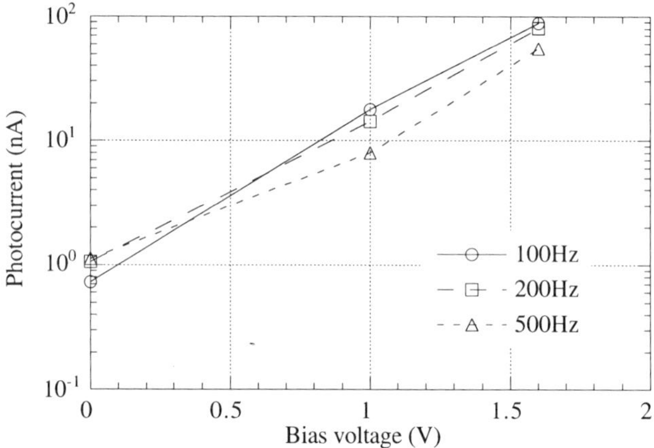

Figure 5. Photocurrent signal as function of the forward bias voltage to the detector. Different curves refer to different modulation frequencies of the incident radiation.

Results of figure 4 and 5 show that the frequency responce of our detector is quite poor (not higher than 500 Hz). In order to overcome this limitation we included our detector in a front-end system realized with discrete electronic component. The operation of the whole system is based on pole-zero compensation and allows to obtain a bandwidth of 100 kHz. The measured frequency response allows the use of the detector in communication system at low bit rate (up to 125 kbit/sec) [8], as remote control systems and IR interface for hand-pocket calculators.

QUANTUM EFFICIENCY CHARACTERIZATION AND MODELING

A better insight in the detection mechanism can be deduced by measuring the sensor response at different wavelengths as reported in figure 6. In this case the radiation source was a Nernst ceramic element filtered by a monochromator in the range 800 nm – 2 μm. Photocurrent values are read by the lock-in amplifier. Altough the photocurrent signal decays almost exponentially, the detector response is detectable up to 2 μm. Performing the same measurement on a p-i-n device, where the absorber layer is an intrinsic film, the photocurrent is always beyond the noise introduced by the measurement set-up.

In order to have quantitative informations on the defect distribution involved in the detection mechanism the quantum efficiency response has been reproduced by a numerical model of sub-band gap absorption in single films. The model includes both thermal and optical induced transitions. The non-equilibrium condition is simulated by assuming excess values of electron and hole densities , which determine lectron and hole quasi Fermi levels. Exponential band tails and Guassian distributions of dangling bonds (DB) and charged defects (CD) have been used. Occupancy function for the tail states were derived using the Shockley-Read-Hall recombination theory, while the occupancy function for DB and CD were derived taking into account their three possible states of occupancy. Solid line in figure 6 is the model reproduction of the experimental quantum efficiency data. An good agreement is achieved. Modeled results have been obtained by using a slope of 950 K for the valence band tail, DB distributions positioned at 0.80 eV and 0.95 eV above the valence band with densities equal to 10^{17} cm^{-3} and $5 \cdot 10^{17}$ cm^{-3}, and CD distribution lying at 0.45 eV above the valence band with density equal to $5 \cdot 10^{18}$ cm^{-3}. The ratio of capture cross section for charged and neutral states are 1 and 100 for DB and CD, respectively. Transitions involved in the investigated wavelength range do not allow to achieve informations on states lying less than 0.6 eV below the conduction band. The high values of modeled density of states can be ascribed to the effect of dopant impurities which dramatically increase the number of band tail states and both deep and shallow defect states [9, 10, 11].

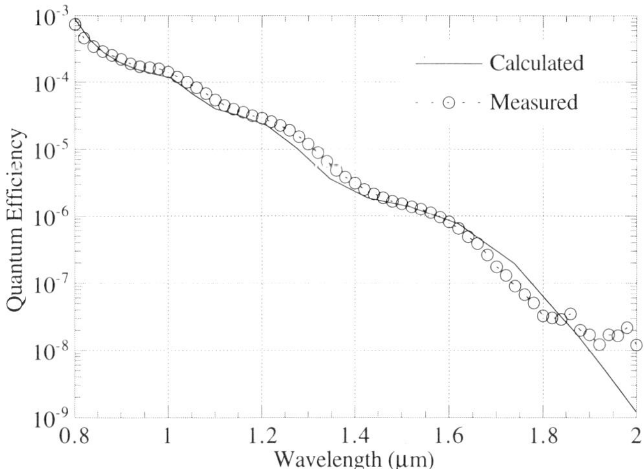

Figure 6. Quantum efficiency response obtained by using as radiation source a Nernst ceramic element filtered by a monochromator

CONCLUSIONS

Room temperature photocurrent caused by near infrared radiation has been observed in a single junction a-Si:H based on microcompensated amorphous silicon layer. Sensor response has been characterized in detail by transient, frequency and quantum efficiency measurents. Modeling of quantum efficiency response by a numerical model of amorphous silicon films suggests the presence of a high number of defects both around mid-gap and in shallow position. An appropriate design of a front-end circuit allows to use the detector in communication systems up to 125 kbit/sec.

REFERENCES

1. S. Guha, J. Yang, A. Banerjee, T. Glatfelter, K. Hoffman, S. R. Ovshinsky, M.Izu, H. C. Ovshinsky, X. Deng in *Amorphous Silicon Technology*, edited by E.A. Schiff, M. Hack, A. Madan, M. Powell and A. Matsuda (Mat. Res. Soc. Symp. Proc. **336**, Pittsburgh 1994), p. 645-655.

2. P.P. Deimel, B.Heimhofer, G. Krötz, H.J. Lilenhof, J. Wind, G. Müller, E. Voegs, IEEE Photon. Technol. Lett., **2**, 499, (1990).

3. Y.K.Fang, S.B. Hwang, K.H. Chen, C.R. Liu, L.C. Kuo, IEEE Trans. Electron Devices, **39**, 1350, (1992).

4.J. Wind, G. Müller, Appl. Phys. Lett. **59**, 956, (1991).

5. D. Caputo, G. de Cesare, A. Nascetti, F. Palma, M. Petri, Appl. Phys. Lett., **72**, 1229, (1998).

6. D.Caputo, A.Nascetti, F.Palma, IEEE Photon. Tech. Lett., **10** , 1147, (1998).

7. H. Fisher, J. Schulte, P. Rieve, M. Bohm in *Amorphous Silicon Technology*, edited by E.A. Schiff, M. Hack, A. Madan, M. Powell and A. Matsuda (Mat. Res. Soc. Symp. Proc. 336, Pittsburgh 1994), p. 867-872.

8. A. Buchwald, K. Martin, Integrated Fiber-Optics Receivers, (Kluwer Academic Publishers, Boston, 1995), p. 168.

9 . R.A. Street, D.K. Biegelsen, J.C. Knights, Phys. Rev. B, **24**, 969, (1981).

10 . W.B. Jackson, N.M. Amer, Phys. Rev. B, **25**, 5559, (1982).

11. D. Caputo, G. de Cesare, F. Palma, M. Tucci, C. Minarini, E. Terzini, J. Non-Cryst. Solids, **227-230**, 380, (1998).

EFFECTS OF MATERIAL PROPERTIES
IN AMORPHOUS SILICON COLOR DETECTORS

D. Caputo°, L. Colalongo*, F. Irrera°, F. Lemmi^, F. Palma°, M. Tucci⁺

°Dept. of Electronic Engineering, University of Rome "La Sapienza", via Eudossiana 18, 00184 Roma (Italy)
*Department of Electronic I.S., University of Bologna, Piazza Risorgimento, Bologna (Italy)
^Xerox Parc, Palo Alto (CA)
+ ENEA, Research Centre, Località Granatello, 80055, Portici, Napoli, (Italy)

ABSTRACT

In this work we study the effects of material properties on the reading process of color detectors by using a two dimensional simulator for the transient regime. In particular, starting from the potential and charge distribution in the device, we describe effects of the density of defects on the self-bias process. Results show the possibility to engineer materials in order to optimize response speed of the device.

INTRODUCTION

Red, green and blue colors are fundamental components of the visible spectrum which can be easily detected by using commercial detectors and/or bandpass filters. While the process of light filtering is rather complex in an array integration, employing distinct wavelength-optimized structures is rather expensive. Therefore, the possibility to obtain a RGB-signal directly from a single detector is very attractive. By using a-Si:H stacked devices, optical filters are dispensable. In fact, in a stacked structure, information on the spectrum corresponds to information on where the radiation is absorbed. The detector structure itself behaves like a filter. In two-terminal structure, photocurrent cannot contain color information, unless a sampling of the absorption region is obtained by optimization of the detector structure, extracting at different times integrated information on radiation absorbed in different portions of the device. Complexity of electrical connection is reduced and the voltage waveform carries the color information.

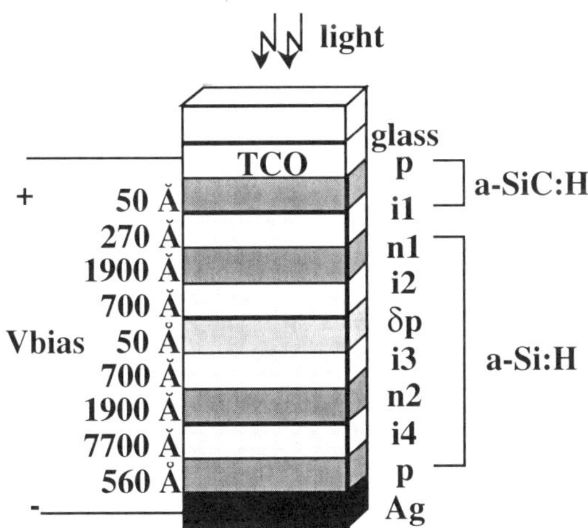

Fig. 1 Structure of the ATCD detector.

Mat. Res. Soc. Symp. Proc. Vol. 557 © 1999 Materials Research Society

The Adjustable Threshold three Color Detector (ATCD) is one of the most promising devices for large area color detection. It consists of three different junctions for sensing light absorbed in different portions (the three absorber layers). Its structure has been extensively discussed in previous papers [1] and will be quickly recalled in the following. With reference to fig.1, ATCD is formed by three stacked junctions: a p-i-n, a n-i-p-i-n and a n-i-p cell. The thin, lightly doped p-layer in the inner n-i-p-i-n cell is designed as the base of a bulk buried phototransistor [2], driven by the photocurrent generated by green light absorption. The n-i-p-i-n cell is a phototransistor and, in dark conditions, lies in a high resistance state whose current grows exponentially with bias voltage. Under strong absorption, it behaves as a current generator in the range 0-1V, beyond which the device experiences a soft breakdown, due to an exponential increase of gain. In fig. 1 thicknesses and materials of the layers are indicated. In particular, thin a-SiC:H layers were used in the front diode (D_1) in order to enhance light transmission.

Referring to fig.1, with negative values of Vbias in the range 0-1 V the ATCD is sensitive to green light, negative values of Vbias above 1V make the ATCD sensitive to blue light and under positive values of Vbias the ATCD is sensitive to the red light.

THE CHARGE INTEGRATION REGIME

Practical use of color sensors in large area arrays requires periodic readout of the photo-charge stored in the parasitic capacitance of the device by a transient technique of sensing [3]. The duration of charge integration and data extraction from columns determines the frame rate. This reading technique is referred to as "charge integration regime". Both integration and readout times are strictly dependent on the electronic characteristics of photodetectors.

In a most general sensing circuit the ATCD is connected to the readout circuit through a switch controlled by the signal Φ_1(refer to [4] for more detailed explanation). When the switch is in the OFF-state (integration time-t_i-), the junctions forming the ATCD lose charge due to photocurrent. When the switch is in the ON-state (readout time-t_r-), charge is fed to the structure. In our devices, t_i is a few milliseconds and t_r a few microseconds. The parasitic capacitance of the matrix column (C_H) is in electrical series with the junctions and thus during t_r it receives the same charge as generated by the detector. C_H holds this charge until it is connected to the sensing amplifier. For an RGB signal, the sensing cycle must be repeated for the three values of the bias voltage.

At a fixed value of the external bias V_{bias}, when the sensing cycle is repeated the charge stored in the three junctions changes, either in the dark or under illumination. During the integration time the three junctions independently lose charge since the device is under open circuit conditions: in fact, the reverse-biased diode discharges its parasitic capacitance through the reverse current, whereas the forward biased diode discharges through recombination of the excess carriers. During a reading pulse, the three capacitances receive the same amount of charge which, before equilibrium, is different from the charge lost during the subsequent integration time. The net balance between the lost and re-fed charge is different from zero. The balance between lost and re-fed charge determines the number of integration/readout cycles necessary to reach a stationary state. In order to avoid unacceptable waiting time, a reduction of the number of cycles necessary to reach equilibrium is extremely important.

To this aim a deeper insight into the process of charging and discharging is following. Denoting with $t_i=0+$ the beginning of the integration time, and with Q_1, Q_{nipin}, and Q_2 the charge stored in $t_i=0+$ by the parasitic capacitances of the front, central and back junction (say, C_1, C_{nipin}, and C_2) it can be stated that:

$$V_{bias}= V_1 + V_{nipin}+ V_2 = \frac{Q_1}{C_1} + \frac{Q_{nipin}}{C_{nipin}}+ \frac{Q_2}{C_2} \tag{1}$$

During the integration time each junction loses its charge indepedently from the others, since the device is under open circuit. The reverse biased junction shows an I-V characteristic which is independent on the voltage drop, since it is the photocurrent. If we assume that junction D_2 is the reverse biased one, then the photo-current is due to red light, and:

$$\Delta Q_2 = I_{red}\, t_i \tag{2}$$

where I_{red} is proportional to the number of absorbed red photons in the junction. In this case, nothing can be said about the other two exchanged charges ΔQ_1 and ΔQ_{nipin}.

During the reading time, the whole structure is recharged at V_{bias}. Each capacitance receives the same amount of charge, say ΔQ_0, so that each junction after the reading time has a total charge slighlty different from the initial one.

$$Q_1' = Q_1 - \Delta Q_1 + \Delta Q_0$$
$$Q_{nipin}' = Q_{nipin} - \Delta Q_{nipin} + \Delta Q_0 \tag{3}$$
$$Q_2' = Q_2 - \Delta Q_2 + \Delta Q_0$$

Of course, the condition that an external voltage V_{bias} is applied to the structure during the reading pulse must be satisfied:

$$V_{bias} = V_1' + V_{nipin}' + V_2' = \frac{Q_1'}{C_1} + \frac{Q_{nipin}'}{C_{nipin}} + \frac{Q_2'}{C_2} \tag{4}$$

From eqs.1-4 we obtain the value of the charge ΔQ_0 exchanged during the reading time:

$$\Delta Q_0 = \frac{\Delta Q_1}{1 + \dfrac{C_1}{C_2} + \dfrac{C_1}{C_{nipin}}} + \frac{\Delta Q_{nipin}}{1 + \dfrac{C_{nipin}}{C_1} + \dfrac{C_{nipin}}{C_2}} + \frac{I_{red}\, t_i}{1 + \dfrac{C_2}{C_1} + \dfrac{C_2}{C_{nipin}}} \tag{5}$$

As a result, in the same sensing cycle the three stacked cells undergo different variations of their voltage drop. In particular, the charge lost by a forward biased junction decreases during subsequent cycles because the forward bias is dropping, whereas the reverse biased junction supplies the same photocurrent.

The equilibrium is reached after a few cycles, when charge lost in a cycle is the same in the three cells, and equal to the re-fed one. This charge is fixed by the reverse biased junction, which exchanges the same charge, even if the operating point slightly changes. This mechanism is referred to as a *self-bias process* [5].

The fact that only one junction is reverse biased makes the equilibrium value of the photocurrent linearly dependent on the incoming radiation flux (F) (see the linearity measurements of the ATCD [6] obtained using the actual three-fold waveform). Eq. 5 also indicates that the amount of charge variation (ΔQ_0) depends on variations of the charge stored in all the three junctions. Therefore, this quantity determines the delay time introduced by the self-bias process.

SIMULATION RESULTS

We used a well-known 2D simulator (HFIELDS [7]) in order to investigate the effect of material properties on the self-bias process. The goal is to get indications about the possibility to reduce the number of readout/integration cycles for the equilibrium starting from the technology, once the illumination conditions are fixed we simulated the self-bias process using reading pulses of 8 µs, and 2 ms integration times.

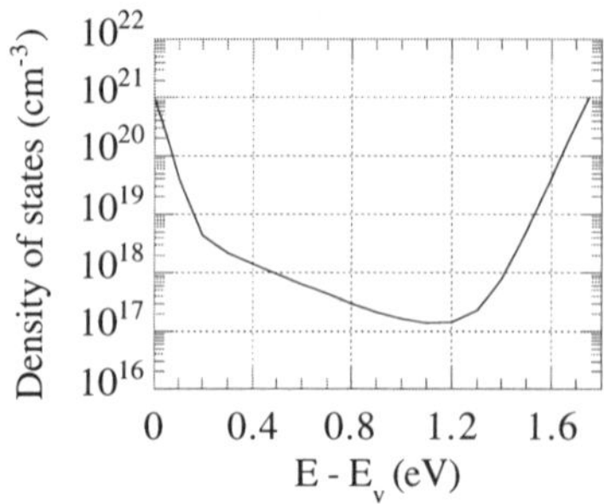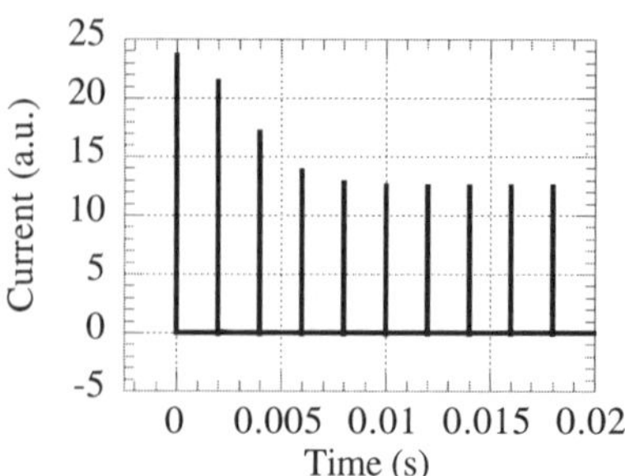

Fig.2 a) Defect distribution used in the simulation; b) simulation of the self-bias process referring to a).

First results are reported below, where a p-i-n-i-p structure (two-color detector) and red light illumination (@ 0.1mW/cm2) have been considered for simplicity. The external voltage in simulations is 4V, so that D2 is reverse biased and D1 forward biased. Referring to the defects distribution shown in fig.2a, As one can see (fig. 2b), six cycles are necessary to reach equilibrium in this case, which means a delay time longer than 10 ms. We verified that variations to the acceptor/donor tail peaks or to the acceptor/donor tail slopes do not affect appreciably the self-bias process. On the contrary, decreasing the slope (and thus the density) of the acceptor/donor deep profile by a factor of three we obtained the results displayed in fig.3. Equilibrium is now reached after only three cycles. In the same figure, data obtained in the dark have been also reported in order to outline the fact that there is a consistent difference between the signal amplitude in the dark and under illumination. On the contrary, increasing the peak of the acceptor deep profile by a factor of five respect to fig.2a, we found that the dark current overcomes the self-bias process, and the photocurrent of the reverse biased junction does not play a role. Results are shown in fig.4. In this case the device operates out of linearity, and color reconstruction is impossible unless the radiation intensity is increased.

As indicated by eq. 5 the *self bias process* process depends in a complex way on either the changes on charge accumulated at the junctions and on their differential capacitances. While for the reverse biased junction these quantities can be easily calculated, in the forward biased one the values depends on the injection level and on the density of defects. In fig. 5 distribution of the total carriers (given by the trapped and free ones) is displayed (dashed line) together with its variation at the end of one integration time. In the forward biased junction excess carriers decrease in the intrinsic zone while trapped charge increases in the depleted zones. In the reverse biased junction the main effect relies on a reduction of the charge in the

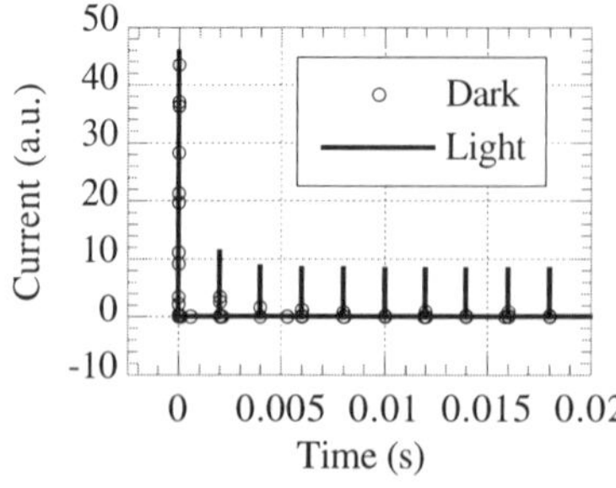

Fig.3. Self-bias process obtained decreasing the slope of the deep defect profile.

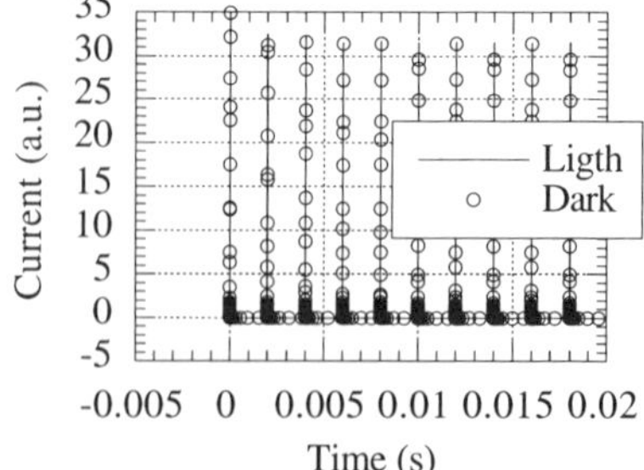

Fig.4 Self-bias process obtained increasing peak values of the deep defect profile.

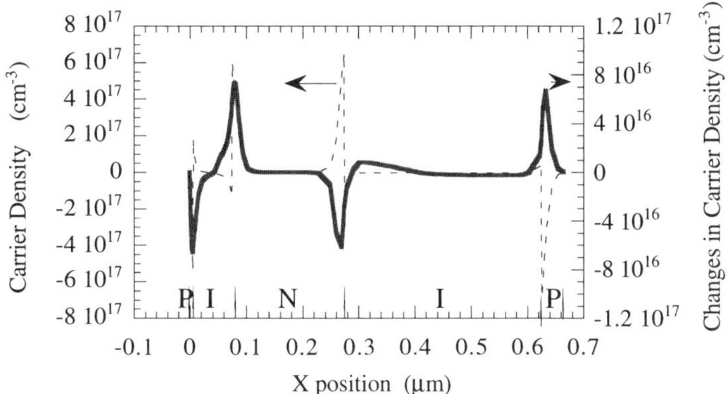

Fig. 5 Total density of carrier (dashed line) and variation of total density of carrier during the integration time (continuos line).

depleted regions. One can notice that charge variation in the forward biased junction is comparable or greater than the charge variation in the reverse biased junction.

Eq. (5) can be simplified if we notice that $C_2 << C_1$, since diode D_1 is the reverse biased one. In the case of two color p-i-n-i-p detector under study, eq.(5) becomes:

$$\frac{\Delta Q_0}{C_2} = \frac{\Delta Q1}{C_1} + I_{red}\, t_i$$

(6)

$$\frac{\Delta Q_0}{C_2} = \Delta V_1 + \Delta V_2$$

(7)

Eq. 7 indicates that the self-bias process is strictly related to the potential variations in the device. In fig. 6 the differential potential distribution along the device is drawn at different times: in $t=0^+$ and in $t-t_i$. Moving along the X coordinate from X=0 we first have D_1 and then D_2. As one can see, after the integration time diode D_1 undergoes an increase of the differential potential drop while diode D_2 undergoes a decrease of the absolute value of the differential potential drop.

This is due to that the fact that the reverse biased junction (D_2) is under charge integration while the forward biased junction (D_1) tends to the built-in potential through excess carrier recombination.

Results from simulations indicate that there is a margin for reducing the delay time introduced by the self bias process, if most care is devoted to the material properties rather than to optical optimization. This means, for example, that the intrinsic layers of the the junctions should be made of a-Si:H and not of a-SiC:H which introducs defects, even if a-SiC:H enhances light transmission to the rear junctions. This means that the circuitry timing can be improved. In order to approach the appealing goal of video-rate applications of an ATCD matrix, we can consider the possibility to further enhance the overall frame rate by an interlaced driving process. In this architecture a reading pulse is applied to all the rows, except row i where the (i,j) pixel is in its charge integration time.

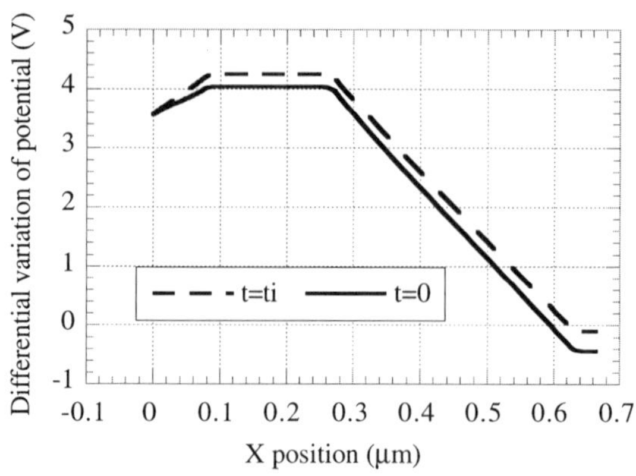

Fig.6 Potential distribution along the device at different times:a) tint=0$^+$, b) t_i.

CONCLUSIONS

Characterization of an ATCD three color detector in the charge integration regime has been presented. In particular, we focused on the self-bias process which the photodetector undergoes in order to reach equilibrium between the charge lost and re-fed in a readout/integration cycle by the three parasitic capacitances. This charge is fixed by the reverse biased junction, and is due to the photocurrent, while the forward biased junctions exchange charge through the recombination current. Starting from the defect distribution in the intrinsic layer we demonstrated that the number of readout/integration cycles necessary to reach equilibrium are greatly affected by the material properties of the parasitic capacitances. In this sense, carbon should be avoided in the front diode intrinsic layer, in order to reduce the delay in detection of the blue color, altough the use of a-SiC:H implies an higher transmission of light, which is a key point in the device operation. We also simulated particular cases in which the device works out of linearity, with the dark current overcoming the photocurrent, depending both on the material quality and on the radiation intensity. As a result, these simulations prompt for the design of ATCDs, with the focus on the material properties rather than on the optical optimization.

ACKNOWLEDGMENTS

This work is supported by the Consiglio Nazionale delle Ricerche under the Progetto Finalizzato MADESS II.

REFERENCES

1. G. de Cesare, F. Irrera, F. Lemmi, F. Palma, Appl. Phys. Lett., 66, 1178, (1996).
2. G. Masini, G. de Cesare, F. Palma, J. Appl. Phys. 77, 1133, (1995).
3. R. L. Weisfield in *Amorphous Silicon Technology*, edited by M.J. Thompson, Y. Hamakawa, P.G. LeComber, A. Madan and E.A. Schiff (Mat. Res. Soc. Symp. Proc. **258**, Pittsburgh 1992) p. 1105-1114.
4. F. Irrera, F. Lemmi, F. Palma, IEEE-Trans. El. Devices, **44**, 1410, (1997).
5. F. Irrera, F. Lemmi, F. Palma, J. Non-cryst. Solids, **227-230**, 1340 (1998).
6. F. Irrera, F. Lemmi, F. Palma, M. Diotallevi in *Amorphous and Microcrystalline Silicon Technology* edited by R. Schropp, H.M. Branz, M. Hack, I. Shimizu, S. Wagner Schiff (Mat. Res. Soc. Symp. Proc. **507**, Pittsburgh 1998) p. 309-314.
7. G. Baccarani, R. Guerrieri, P. Ciampolini, M. Rudan, Proc.Conf. Nasecode IV, editor J.J. H. Miller, 3, (1985).

SIGNAL AMPLIFICATION AND LEAKAGE CURRENT SUPPRESSION IN AMORPHOUS SILICON P-I-N DIODES BY FIELD PROFILE TAILORING

W.S. HONG*, F. ZHONG, A. MIRESHGHI** and V.PEREZ-MENDEZ
Physics Division and *EngineeringDivision, Lawrence Berkeley National Laboratory, Berkeley, CA 94720, U.S.A., **Sharif University, Tehran, Iran

ABSTRACT

The performance of amorphous silicon p-i-n diodes as radiation detectors in terms of signal amplitude can be greatly improved when there is a built-in signal gain mechanism. We describe an avalanche gain mechanism which is achieved by introducing stacked intrinsic, p-type, and n-type layers into the diode structure. We replaced the intrinsic layer of the conventional p-i-n diode with i_1-p-i_2-n-i_3 multilayers. The i_2 layer (typically 1~3 μm) achieves an electric field $> 10^6$ V/cm, while maintaining the p-i interfaces to the metallic contact at electric fields $< 7 \times 10^4$ V/cm, when the diode is fully depleted. For use in photo-diode applications the whole structure is less than 10 μm thick. Avalanche gains of 10~50 can be obtained when the diode is biased to ~500 V. Also, dividing the electrodes to strips of 2 μm width and 20 μm pitch reduced the leakage current up to an order of magnitude, and increased light transmission without creating inactive regions.

INTRODUCTION

Amorphous silicon has been drawing attentions for applications in radiation detection due to its radiation hardness, large area capability and low fabrication cost. In detecting minimum ionizing particles, an amorphous silicon layer, at least 50 μm thick, is required to obtain a large enough signal above noise. However, deposition of such a thick layer is time consuming and difficult due to the high residual stress (~400 MPa). Also, an extremely high reverse bias voltage (>1500 V) is needed to fully deplete the 50 μm thick p-i-n diodes. Therefore, it is desirable to have an intrinsic signal amplification mechanism in the form of avalanche multiplication.

Single gamma-rays (Eγ > 70 KeV) can also be detected by scintillator-coupled avalanche diodes with light transparent contact layers. Also, the increasing use of optical fibers in communication and data processing systems requires the availability of low cost semiconductor avalanche photodiodes. A moderate gain of 10-50 is often sufficient for these applications. However, the present relatively high cost of avalanche photodiodes, which are fabricated at high temperatures on crystalline silicon, limits their general use. Although the amorphous silicon avalanche diode cannot compete with its crystalline counterpart in terms of performance, it can still be adopted when a moderate gain and a low cost are required.

Our previous studies and those of others on the high-field behavior of a-Si:H p-i-n devices indicate that avalanche multiplication does not occur even when the p-i-n diodes were operated at maximum reverse bias ($6 \times 10^5 < V_R < 10^6$ V/cm)[1]. Thus, a lower limit to the multiplication threshold appears to be more than 10^6 V/cm. However, in conventional a-Si:H p-i-n diodes the dark current increases rapidly over about 6×10^5 V/cm and leads to breakdown. This breakdown field is too low to observe the avalanche photocurrent multiplication. Therefore, the p-i-n diode design needs to be modified if avalanche multiplication is to be realized. It has been reported that breakdown is likely to occur near the metal / p-layer interface and can be effectively suppressed by moving the peak field region away from this interface[2].

Mat. Res. Soc. Symp. Proc. Vol. 557 © 1999 Materials Research Society

Avalanche multiplication has been observed in a-Si:H/a-Si:C:H superlattice structures having step-like multilayer barriers[3-5]. In this structure, the impact ionization threshold energy is reduced by an amount equal to the energy band barrier formed by the a-Si:C:H. An optical gain as high as ~200 was reported. However, this multilayer structure was difficult to fabricate and thus the result was not highly reproducible. Therefore, we attempt to develop a new structure which is easier to fabricate than the superlattices. Our approach is to form a high electric field region inside the intrinsic layer of p-i-n diodes, where the carriers gain enough kinetic energy for impact ionization.

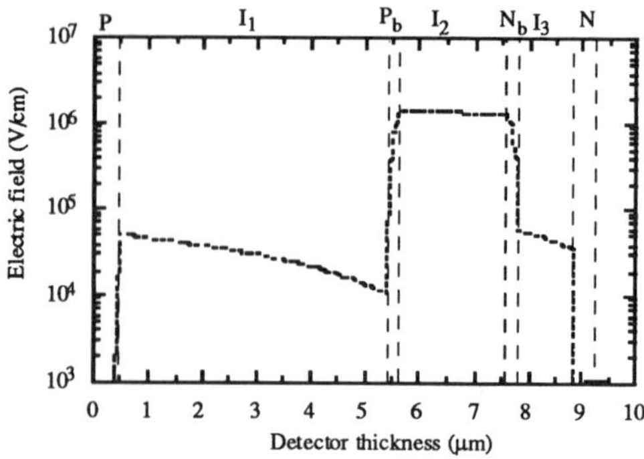

Fig. 1   Electric field profile inside a fully depleted a-Si:H multilayer avalanche photodiode

DEVICE DESIGN

In order to achieve a high field ($>10^6$ V/cm) inside the diode and high breakdown voltage, we have designed a multilayer a-Si:H p-i-n photodiode. The electric field profile inside this device is shown in Fig.1. The intrinsic layer, normally present in a p-i-n device, is divided into three regions by a buried p-type layer, p_b, and an n-type layer n_b. The first intrinsic region, i_1 acts as a charge collection and drift region with low field. The second intrinsic region, i_2, acts as the avalanche region with high field, while the third region, i_3, is used to protect the avalanche region against breakdown, as discussed earlier. The width of i_1 is 2 μm for detecting the photons with wavelength of 500 nm to 650 nm, and 5~10 μm is desired for charged particles. The width of i_2 can be varied from 0.5 μm to 3.0 μm. If this region is smaller than 0.5 μm, there is no avalanche multiplication. If this region is over 3 μm thick, a larger reverse bias has to be applied than can be conveniently handled by realizable doping levels in the p region. The width of i_3 should be between 0.5 μm to 1.0 μm. The doping level and width of buried p_b and n_b layers are important parameters to determine the maximum field in the avalanche region.

The doping levels and the thicknesses of the buried layers, p_b and n_b, are selected in such a way that the maximum field in avalanche region could reach $> 1\times10^6$ V/cm and that in i_3 be brought back to $< 1\times10^5$ V/cm when the diode is fully depleted. According to our model, if the gas phase doping level in the p_b layer is 700 ppm and the thicknesses of the i_1, p_b, i_2, and i_3 layers are 5 μm, 100 nm, 2 μm, and 0.5 μm, respectively, the diode is fully depleted at about 200 V. In this case the maximum field at p-i_1 interface is about 7×10^4 V/cm, well below the breakdown voltage of a normal p-i-n diode.

EXPERIMENTAL PROCEDURE

Amorphous silicon multilayer p-i-n diodes with varying thicknesses of different layers, as discussed in the previous section, were deposited on Corning 7059 glass substrates coated with a thin (100Å) Cr layer, using our plasma enhanced chemical vapor deposition (PECVD) equipment. A 100Å-thick circular Cr dot of 3.3 mm in diameter was sputter deposited as a top electrode. The PECVD deposition parameters were: RF frequency = 85 Mhz, RF power density

Table I Thicknesses of the various layers in multilayer avalanche diodes

Sample ID	p-layer (μm)	i_1 (μm)	p_b (μm)	i_2 (μm)	n_b (μm)	i_3 (μm)	n (μm)	Calculted maximum field at full depletion (10^6 V/cm)	Comment
AV498	0.25	5	0.2	1	0.2	1	0.25	2.2	
AV520	0.25	5	0.2	1	0.2	1	0.25	2.2	840 ppm in p_b
AV524	0.35	5	0.2	1	0.2	1	0.25	2.1	
AV529	0.5	5	0.2	1.5	0.2	1	0.25	2.1	
AV532	0.5	5	0.2	1	0.2	1	0.25	2.6	1200ppm in p_b
AV535	0.5	5	0.2	1.5	0.2	0.75	0.25	2.3	900 ppm in p_b
AV538	0.67	5	0.2	1.5	0.2	0.5	0.25	1.0	670 ppm in p_b
AV547	0.5	5	0.1	2	0.1	0.5	0.25	1.1	
AV549	0.5	5	0.1	3	0.1	0.5	0.25	1.1	
AV557	0.75	5	0.1	3	0.1	0.5	0.25	1.0	He-diluted
AV560	0.6	5	0.1	3	0.1	0.5	0.25	1.0	He-diluted
AV561	0.75	5	0.1	3	0.1	1	0.25	1.0	H_2-diluted
AV562	0.75	5	0.1	3	0.1	2	0.25	1.0	H_2-diluted
D542	0.25			7			0.25		plain p-i-n

= 40 mW/cm^2, substrate temperature = 190°C, chamber pressure = 300 mTorr, and SiH$_4$ flow rate = 40 sccm. Sample dimensions are summarized in Table I.

Light pulses of 20 ns ~ 1 μs duration with wavelength of 510 nm ~ 760 nm from an N$_2$ - Dye laser were used to illuminate the p-side of this sample. The diodes were reverse biased via a 10 MΩ resistor and the electron signal was fed to Tennelec TC 170 preamplifier and TranLamp shaping amplifier, and monitored by an oscilloscope. We also illuminated the same light pulse on the n-side of the sample to examine the hole collection. Signal amplitudes and leakage current were recorded with increasing the bias voltage until breakdown.

RESULTS

The results on the diode charge collection measurement are shown in Fig.2. Both electron and

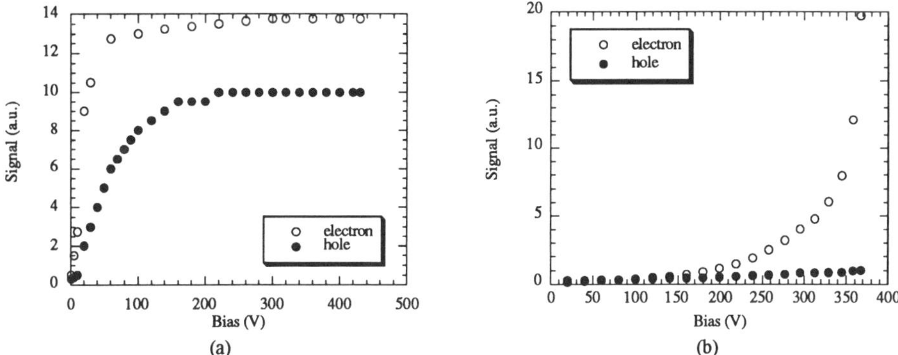

Figure 2. Light signal amplitude as a function of the reverse bias for the 10 μs light pulse of 510 nm wavelength: (a) conventional p-i-n diode (b) avalanche diode (AV547)

hole signals of the avalanche diode (AV547) were compared with those of the conventional p-i-n diode (D542). Signal rise time is much shorter than the light pulses and the charge collection is done rapidly. In the plain p-i-n diode, as shown in Fig.2(a), both the electron and the hole signals initially increased with bias and then started to saturate after the diode is fully depleted. The signal amplitudes did not change until breakdown. The maximum electric field inside the intrinsic layer at the breakdown voltage, $V_b = 400$ V, was calculated to be 6×10^5 V/cm.

In the avalanche diode, we observed a significant increase of the electron charge collection when the applied bias is larger than 200 V as shown in Fig.2(b). Under the maximum reverse bias (about 400 V), the electron charge collection is enhanced by a factor of fifteen over the lower bias voltage charge collection. Complete saturation of the hole signal occurs at around 250 V, whereas the electron signal increases with increasing bias. Thus we have demonstrated an avalanche gain of 15 on these prototype diodes with a bias of 400 V.

Although not very well established, the ratio of the signal amplitudes at the maximum applicable voltage and at the avalanche onset voltage can be a good approximation to estimate the gain in the avalanche diodes. Since the avalanche onset voltage, hole signal saturation voltage, and the calculated full depletion voltage are in reasonably good agreement, we believe that our model described in the previous section is valid and the observe rapid increase in the electron signal is due to the avalanche multiplication. The experimental results of the various diodes are summarized in Table II.

Table II. Performance of the various avalanche diodes

Sample ID	Maximum gain	Breakdown voltage (V)	Avalanche onset voltage (V)	Leakage current at breakdown (A/cm^2)
AV498	2.5	270	200	1×10^{-4}
AV520	2.5	270	220	2×10^{-3}
AV524	3	350	280	1×10^{-3}
AV529	3.5	360	300	4×10^{-4}
AV532	3	270	120	1×10^{-3}
AV535	2	270	230	1×10^{-3}
AV538	2.5	360	270	6×10^{-4}
AV547	15	400	200	3×10^{-5}
AV549	12	500	320	7×10^{-5}

Samples AV498 through AV538 have avalanche region shorter than 1.5 μm and exhibited low gains (<4). All these samples except AV538 were designed such that an electric field higher than 2×10^6 V/cm could be obtained at full depletion. If the gain is solely related to the electric field strength, these samples must have produced high gains and low avalanche onset voltage. However, since their breakdown voltages appeared to be rather low and the leakage current densities at break down are high, it is inferred that a certain distance is required for electrons to travel in order to develop avalanche multiplication. In other words, formation of high electric field region without ensuring sufficient drift distance for electrons may cause breakdown before the avalanche process matures. Samples AV547 and AV549, having the avalanche regions longer than 2 μm and the field strength of 1×10^6 V/cm at full depletion, exhibited moderately good gains with relatively low leakage current.

STRATEGES TO IMPROVE AVALANCHE GAIN

Dilution of the Source Gas

We have previously shown that dilution of SiH_4 with He or H_2 improves the electrical and mechanical properties of the thick (~50 μm) a-Si:H p-i-n diodes for charged particle detection[6]. Both He and H_2 are known to decrease the density of ionizable dangling bonds in the a-Si:H film, which act as traps. Hydrogen also enhances the carrier mobility by forming a microcrystalline

Table III. Performance of the avalanche diodes fabricated from He or H_2 dilution

Sample ID	Maximum gain	Breakdown voltage(V)	Avalanche onset voltage (V)	Leakage current at breakdown (A/cm^2)
AV557	48	620	330	1.5×10^{-4}
AV560	38	570	280	1.0×10^{-4}
AV561	4.5	350	150	2×10^{-5}
AV562	12	450	180	1×10^{-4}

temperature of 160°C were used. For samples AV561 and AV562, only the i_2 layers were deposited from a 6% SiH$_4$ - 94% H$_2$ mixture because the large residual stress made successful growth of layers thicker than 5 μm very difficult. RF power density of 60 mW/cm^2, chamber pressure of 1,000 mTorrr and substrate temperature of 160°C were used. All other parameters were the same as described in the previous section. The results are summarized in Table III.

The He-diluted samples exhibited high gains and high breakdown voltages. The avalanche gain behavior of the He-diluted sample (AV557) is compared with that of the same thickness,

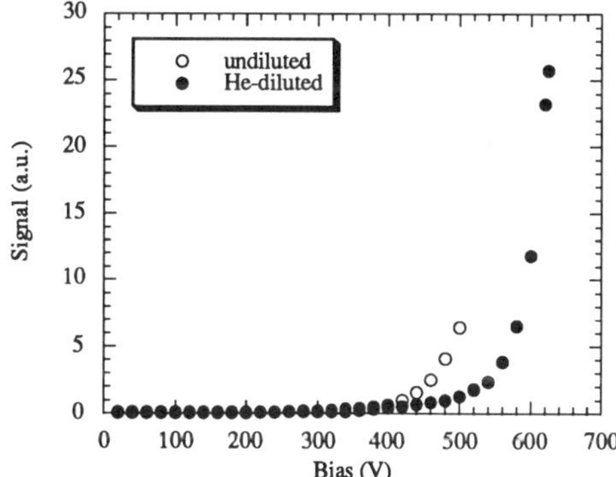

Figure 3. Charge collection of the same thickness He-diluted and undiluted samples vs. bias voltage.

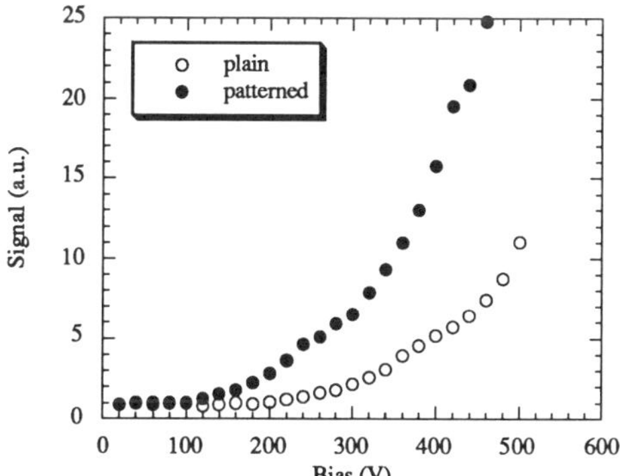

Figure 4. Charge collection of the samples having plain electrode and patterned electrode on the n-side.

phase in the amorphous network and reducing the width of the tail states[7]. Therefore, He or H_2 dilution may increase the avalanche gain by enhancing the mean free path and mobility of the carriers.

Samples AV557 and AV560 in Table I were deposited from a 40% SiH$_4$-60% He mixture. RF power density of 90 mW/cm^2, chamber pressure of 500 mTorr and substrate

undiluted sample (AV547) in Fig. 3. Although the electron signal of the undiluted sample increases faster than that of the He-diluted sample, the maximum gain in that sample is smaller because of the breakdown occurs earlier. This result suggests that He dilution contributes to the gain by allowing a high voltage capability, probably stabilizing the microstructure of the a-Si:H.

Performance of the H$_2$-diluted sample is worse than that of the He-diluted or the same thickness undiluted sample. In most cases, the H$_2$-diluted films are under high residual stress (~700 Mpa), and tend to form void-rich or columnar structures. These structural defects may lead to the poor performance of the H$_2$-diluted diodes.

Electrode Patterning

A high electric field in the avalanche region can be induced by replacing the n-side electrode with strips. As in multiwire proportional counters or in microstrip gas chambers, electric field concentration due to the electrode geometry enables us to obtain a high field without an extra bias voltage[8]. We patterned the n-side electrodes of the samples AV549 (undiluted), AV560 (He-diluted), and AV562 (H$_2$-diluted), to 4 μm wide, 0.5 cm long strips spaced 16 μm apart, and etched the n-layer using the electrode pattern as a mask.

Figure 4 shows the charge collection of AV562 (H_2-diluted sample) with a plain electrode and with a patterned electrode on the n-side. The patterned electrode increased the gain by almost a factor of two. Undiluted and He-diluted samples also showed a gain increase by factors of 1.5~2.5. However, the results of the patterned samples were not highly reproducible.

LEAKAGE CURRENT SUPPRESSION

We applied the concept similar to that described in the previous section to reduce the leakage current. Patterning the p-side electrode can decrease the magnitude of the leakage current by reducing the actual contact area. We patterned the p-side electrode of a plain p-i-n diode and etched the p-layer and an additional 1 μm into the i-layer to minimize the leakage along the surface. The total patterned area was 0.5 cm x 0.1 cm. The leakage current was decreased almost by an order of magnitude.

Dividing the electrode into strips also increased the signal amplitude by providing "windows" for the light to pass without absorption. This result suggests that we can achieve both the high signal and low leakage current by careful design of the electrodes of the avalanche diode. Also, strip electrodes on both p- and n-side, perpendicular to each other, may enable us to realize 2-dimensional readout.

SUMMARY

We designed a new multilayer a-Si:H p-i-n diodes, and demonstrated avalanche multiplication. A maximum gain of ~50 was achieved by replacing the intrinsic layer of the conventional p-i-n diode with i-p-i-n-i multilayers. An electric field $> 10^6$ V/cm in the active avalanche region was obtained, while maintaining the p-i interfaces to the metallic contact at electric fields $< 7 \times 10^4$ V/cm, when the diode was fully depleted. He-diluted samples were more stable than undiluted samples and exhibited high gains by allowing high breakdown voltages. Possibilities for obtaining an improved gain and reduced leakage current were demonstrated by appropriate electrode shaping.

ACKNOWLEDGEMENT

This work was supported by the Director, Office of the Energy Research of the U.S. Department of Energy under Contract No. DE-AC03-76SF00098.

REFERENCES

[1] J.B.Chevrier and B.Equer, J. Appl. Phys. **76** p.7415 (1994)

[2] I.Fujieda, G.Cho, M.Conti, J.Drewery, S.Kaplan, V.Perez-Mendez, S.Qureshi and R. Street, IEEE Trans. Nucl. Sci., **37** p.124 (1990)

[3] S.Jwo, M.Wu, Y.Fang, Y.Chen, J.Hong, C.Chang, IEEE Trans. Elec. Dev., **35** p.1279 (1988)

[4] J.Hong, W.Laih, Y.Chen, Y.Fang, C.Chang, J.Gong, IEEE Trans. Elec. Dev., **37** p.1804 (1990)

[5] T.Jing, J.S.Drewery, W.S.Hong, H.Lee, S.N.Kaplan, A.Mireshghi, V.Perez-Mendez, SPIE Proc., Vol. 2366 p.345 (1994)

[6] W.S.Hong, J.S.Drewery, T.Jing, H.-K.Lee, S.N.Kaplan, A.Mireshghi and V.Perez-Mendez, Nucl. Instr. Meth., **A356** p.239 (1995)

[7] A.Mireshghi, W.S.Hong, J.Drewery, T.Jing, S.N.Kaplan, H.K.Lee and V.Perez-Mendez, MRS Symp. Proc., Vol. 336 p.377 (1994)

[8] A.Oed, Nucl. Instr. Meth., **A263** p.351 (1988)

DIRECTIONAL BREAKDOWN OF METAL/a-Si:H/c-Si HETEROSTRUCTURES AND ITS APPLICATION TO PROMs

Hong Zhu, Ali Kaan Kalkan, Joseph Cuiffi, and Stephen J Fonash
Electronic Materials and Processing Research Laboratory, The Pennsylvania State University, 227 Hammond Bldg, University Park, PA16802

ABSTRACT

This paper gives the first report of a directional breakdown phenomenon present in specially constructed amorphous silicon / crystalline silicon heterojunction structures. After breakdown, the forward current can be greatly increased, but the reverse leakage current remains unchanged or even lowered. We use the Analysis of Microelectronic and Photonic Structures (AMPS) [1] device simulation tool to study this phenomenon and show that the structures can provide a new approach to producing Programmable Read Only Memory (PROM).

INTRODUCTION

Anti-fuse technology has been used in recent years [2,3] for PROM devices for its simplicity of fabrication and small geometry. It holds the promise of replacing nichrome or polysilicon fuse-based logic blocks. Anti-fuse technology requires, however, the use of a diode or transistor at each node in the logic block. The new approach we introduce here, using a phenomenon we term "directional breakdown", combines the functionality of the anti-fuse and diode at each node into one device and combines the fabrication into one simple process. This technique also allows the user to break the selected node in a very simple way.

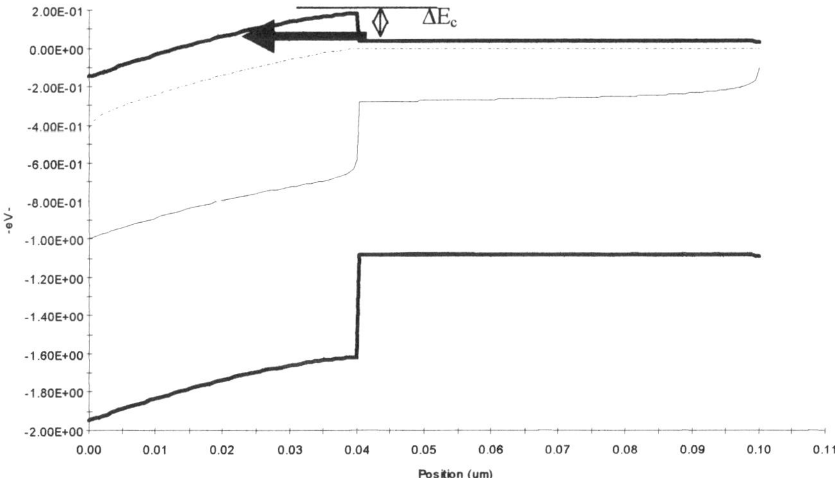

Fig1 Band diagram of metal/intrinsic a-Si:H (50nm)/n^+ c-Si, at 1 volt forward bias. Midgap defects in intrinsic a-Si:H is 10^{16}/cm^3. Dark arrow shows the possible shunting path.

Mat. Res. Soc. Symp. Proc. Vol. 557 ©1999 Materials Research Society

In a recent study we showed that there is a conduction band offset ΔE_c between intrinsic a-Si:H and single crystal silicon (see Fig 1) [4]. In addition to the barrier presented by ΔE_c we found that an even higher barrier could arise at forward bias due to a virtual cathode created by the electrons trapped in the midgap defect states. We call this barrier the space charge limited current (SCLC) barrier. In the metal/a-Si:H/c-Si heterostructures of this report, we make use of this off-set ΔE_c, the SCLC barrier, and the metal/a-Si:H barrier height to create our PROM devices. We note that the offset ΔE_c alone can control the on-current of electrons from n-type c-Si (see Fig 1) if the defect density is low and the intrinsic a-Si is thin. However, the reverse leakage current is totally controlled by the Ni/a-Si barrier height, which we measured as about 0.85eV. Since our devices are n-i-Schottky structures, the hole current is negligible.

With proper stressing, in a carefully designed M/i a-Si/n+ crystal Si structure, we have found that forward current can be increased significantly and highly rectifying diode behavior can be obtained. This anti-fuse/diode behavior is accomplished by using the stress to create a path shunting ΔE_c as seen in Fig 1. The final value of the forward current after stress I_{stress} has a value that depends on the ΔE_c and can be expressed as:

$$\frac{I_{stress}}{I_{initial}} = \exp\left(\frac{\Delta E_c}{kT}\right) \tag{1}$$

Here $I_{initial}$ is the initial current. We found this phenomenon is material dependent and that the thickness of the intrinsic a-Si:H is critical. With the proper stress, the metal barrier height is not decreased and the off-current is preserved (or decreased). With the proper stress anti-fuse behavior in series with diode behavior is achieved

EXPERIMENTS

The intrinsic a-Si:H in our device was deposited by Plasma Enhanced Chemical Vapor Deposition (PECVD) at 250°C on a n-type silicon wafer. The films we tested were made at 130Å, 260Å and 400 Å. We found that devices with 400Å intrinsic a-Si:H work better than the other two thicknesses when judged in terms of stability as well as in terms of the after-stress current ratio. After a-Si deposition, nickel was evaporated on top of the a-Si:H film and patterned for the Schottky contact, and aluminum was evaporated at the back of the wafer as the ohmic contact. The device stress used to create the anti-fuse/diode behavior started from 1volt and rose by 1 volt for each new stress until we observed breakdown; i.e., until we observed the increased forward current with diode behavior. The current-voltage (I-V) characteristic was recorded for each stressing during the stress applications required for the I-V evolution into diodes. The voltage required to break these diodes was found to range from 7-20volts. Some unstressed and stressed diodes were then annealed at 120°C for half an hour to determine if there is any recovery or if there is any increase in the initial current. The results are seen in Fig 2a,b.

We also found that the ratio between $I_{stress}/I_{initial}$ and the directional breakdown behavior are highly dependent on the a-Si:H deposition conditions and on the rate at which the Ni was evaporated. The ratio, for example, was found to vary from 10^2 to nearly 10^5. For the devices with the better $I_{stress}/I_{initial}$ ratios, a-Si:H was processed at 50mTorr with H_2/SiH_4=40sccm/4sccm and the substrate temperature held at 250°C. The power we used for this process was 40W.

We further tested the reliability of the unbroken devices (i.e., non-conducting anti-fuse nodes) at 5volts. We selected this voltage because we believe it would be the highest voltage seen in most applications. We tested those devices for 1 hour at room temperature, 50°C, or 100°C.

RESULTS AND DISCUSSION

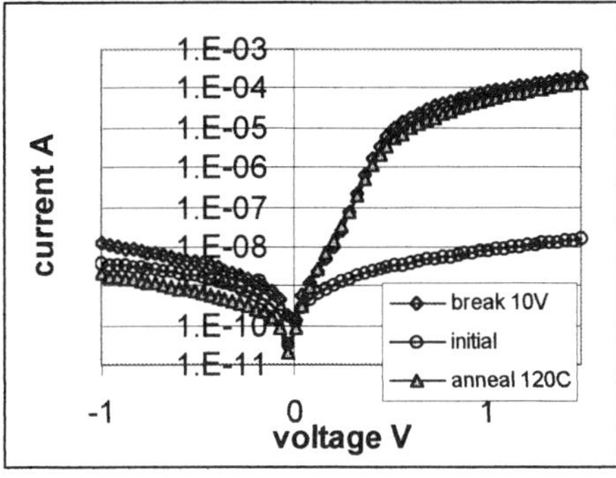

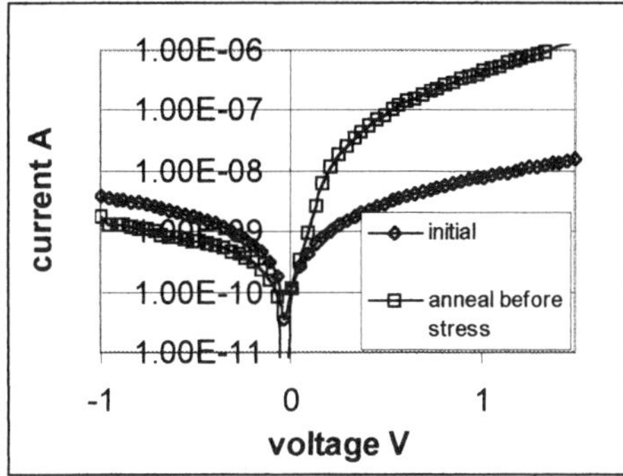

Fig 2a I-V behavior of this metal/a-Si/n c-Si at initial state, after 10V directional breakdown, and 120°C anneal

Fig 2b annealing effect on the unbroken devices

The time and voltage required for a structure to be stressed into a diode with stable characteristics varied slightly from device to device. Most structures reached the stabilized diode behavior condition in a few seconds; however, a few devices took a couple of minutes under high voltage stress. Fig 2a compares the device I-V at initial, after stress and after 120°C annealing. The result shows that, once broken into a diode, this device can not be recovered; i.e., the shunting path is permanently registered in the device. There is no significant increase in reverse current with this stress. Furthermore, Fig 2a shows that the reverse leakage current which may show increase with the stress, can be reduced by annealing. Fig 2b makes the important point that that annealing alone can affect the forward current of an unstressed structure. As seen, the $I_{stress}/I_{initial}$ ratio of the structure selected increases by some 2 orders of magnitude with simply an anneal.

In our AMPS 1D computer modeling of these nickel/intrinsic a-Si:H/n c-Si structures the Schottky barrier height was taken as 0.85eV from our measurements [4] and the ΔE_c was taken as 0.15eV at the a-Si:H/c-Si interface, from our measurements and modeling [4]. The bandgap of a-Si:H was taken to be 1.8eV [5]. Even without modeling it is clear from the band diagram that the current density in this structure is controlled by the doping density of the c-Si and by ΔE_c. However, with AMPS modeling we found an additional point: a space charge limited current (SCLC) barrier can exist at the a-Si:H/c-si region and, in some devices, this barrier can control the on-current. Our modeling showed that SCLC barrier width and height is greatly dependent on the a-Si:H thickness and defect density. We believe that hydrogenated a-Si:H midgap defect density can vary from 10^{15}-10^{19}/cm^3 depending on material history and deposition conditions. A demonstration of SCLC barrier variation with the a-Si:H film defect density is shown in Fig 3. Depending on the a-Si material, the interface, and the stressing parameters, an on-current stress can create a great number of defects at the interface of a-Si:H and c-Si. Those defects created by stress can establish a multi-step tunneling path that shunts the barrier presented by ΔE_c and SCLC. The possibility that the on-current changes after stress are due to disappearance of the barrier due to hydrogen out-diffusion and resultant bandgap lowering is not considered probable. This is because the breakdown happens in very short time (seconds) and the temperature (20°C) is low compared with that used for annealing a-Si:H after Staebler-Wronski effect changes. Therefore, the breakdown phenomenon is proposed to be due to shunting of the $\Delta E_{effective}$ presented by the virtual cathode. As a result, equation 1 should be modified into:

$$\frac{I_{stress}}{I_{initial}} = \exp(\frac{\Delta E_{effective}}{kT})$$

(2)

Here $\Delta E_{effective}$ represents the sum of ΔE_c and SCLC barrier.

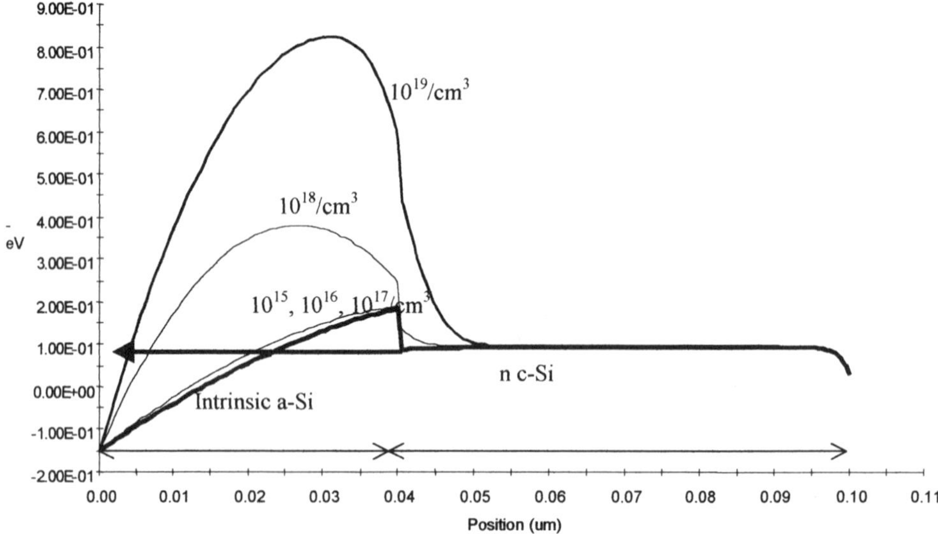

Fig3 effective barrier height increases with different midgap defect density increases from 10^{15}-10^{19}/cm^3. Diagram created by AMPS.

Our modeling further shows that thermionic emission over the Schottky controls the reverse leakage current and it could be defect density dependent as well. Generation current from the a-Si:H layer is found to be negligible when the defect density is "low" and also because the bandgap is 1.8eV. In fact using AMPS we varied the defect density from 10^{15} to 10^{18}/cm^3 for Schokley-Read-Hall generation and found virtually no change in the reverse current at the barrier height (0.85eV) used (see Fig 4). We noticed experimentally that, for most diodes, the reverse leakage current did not change much (within one order) and we believe that result underscores the role of the Schottky barrier height with perhaps some shunting due to defect-assisted tunneling. For example, a high density of defects at the conduction band edge and spatially close to the Schottky contact may help electrons to tunnel through the barrier, so that the effective barrier height is lowered. Experimentally, we found that that small amount increase in reverse leakage current can be annealed out at 120°C, as seen it Fig 2a, b.

Some simple reliability tests were conducted at 5 volts. This voltage was chosen since we believe this is a reasonable voltage limit for CMOS circuits. Fig 5 shows the IV characteristics of structures before they have become a diode; i.e., before being stressed to breakdown. These structures were subsequently stressed at 5 volts for 1 hour at room temperature, 50°C or 100°C. These structures remained very stable during the stress. To fully understand the lifetime of these devices, we need to conduct stress for a much longer time or find a proper process to accelerate the test. However, we have not conducted such test, nor have we found a proper definition for the acceleration test.

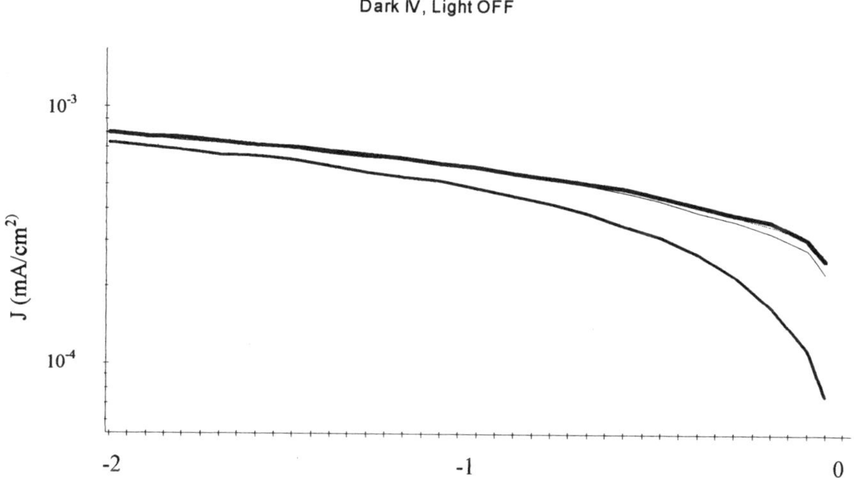

Fig 4 AMPS reverse current simulation of the Schottky structure with defect density from 10^{15}-10^{18}/cm^3

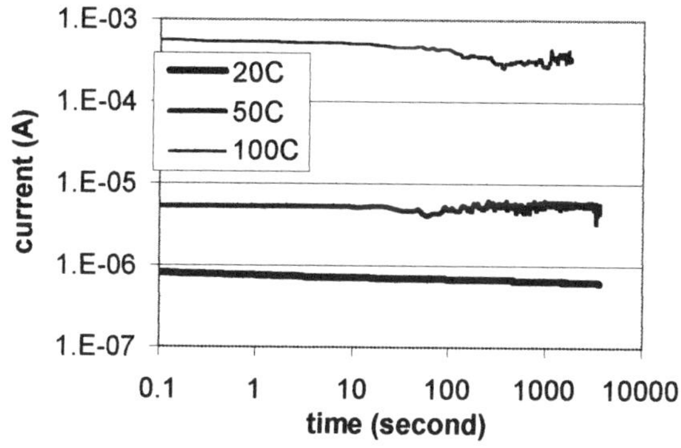

Fig 5 evolution of device characteristic under 5V-1 hour stress

We propose this phenomenon of stress-induced conversion from non-conducting to diode behavior can be very advantageously applied to PROM technology. To do so would require that our metal/a-Si/c-Si hetero-structures be located at the each array node. Only these structures would be used; ie, there is no need for any fuse or anti-fuse structure. To program an array, selected nodes would only need to be broken down. The operation of this PROM is the same as any other ROM after this "writing" step.

CONCLUSION

We have found a new approach to PROMs which combines the anti-fuse and diode functions into one simple structure. This new approach requires the development of a shunting (e.g, tunneling) path to shunt an $\Delta E_{effective}$ barrier set-up in our specifically designed structures. Therefore, any material that may give a higher offset or rise significant SCLC barrier at the carrier supply interface might be a good candidate for this application. Any material that has a larger ΔE_c will enhance the pre- (Initial) and post- (After Stressing) ratio. This may be important for some applications. An obvious candidate is a-SiC:H which has a bandgap of 2eV.

ACKNOWLEDGMENT

The authors thank Dr Sanghoon Bae, Anand Krishnan and Terrence Kuzma for making the samples and helpful discussions.

REFERENCES

1. P. J. McElheny, J. K. Arch, H. S. Lin, and S. J. Fonash, Journal of Applied Physics, 64, 1254 (1988)
2. R. Wong and K Gordon, Evaluating the Reliability of the Quicklogic Antifuse Electronic Engineering 56, P49 June (1992)
3. S. Chiang Antifuse Strucutre Comparison for Field Programmable Gate Arrays IEDM Meeting P611 (1992) San Francisco
4. Hong Zhu and Stephan J. Fonash Band-offset determination for intrinsic a-Si/p$^+$ uc-Si or intrinsic a-Si/n$^+$ uc-Si heterostructure MRS spring conference A5.2 (1998)
5. X.Xu, J.Yang, A.Banejee, and S. Guha, K.Vasanth and S Wagner Band edge discontinuities between microcrystalline and amorphous hydrogenated silicon alloys and their effect on solar cell performance Appl Phys Lett 67(16) Oct P2323 (1995)

INNOVATIVE DIODES BASED ON AMORPHOUS-POROUS SILICON HETEROJUNCTION

R. DE ROSA[*], V. LA FERRARA[**], G. DI FRANCIA[*], L. QUERCIA[*], F. ROCA[*], M. TUCCI[*]
[*]ENEA - Research Centre Portici, Località Granatello-80055 Portici (Na), Italy
[**]INFM - Physics Science Department, Mostra D'Oltremare, Pad. 19 Napoli, Italy

ABSTRACT

In this paper we present an innovative diode based on the heterojunction between amorphous silicon and porous silicon grown on crystalline silicon. The device architecture gives several advantages. Deposition of amorphous silicon on porous material realises high performance junction at temperature less than 250°C and it passives the porous layer against the natural oxidation due to ageing in the environment. Porous technology allows to obtain a controlled textured silicon surface independently from crystalline silicon orientation just to give the opportunity to reduce surface reflectivity and the blue shift of the absorption spectra in solar cell application.Solar cells were characterised by I-V dark/light and quantum yield measurements. Under standard AM 1.5 light we obtained photovoltaic conversion efficiency greater than 10%. Change in photoluminescence in different gas environments showed for gas sensor applications give rise encouraging results. In dark condition we found the typical diode behaviour.

INTRODUCTION

Thin film materials and low temperature processes are currently widely investigated to produce a new generation of electronic devices. Since many problems still need a solution and further studies, heterojunctions represent a special way to introduce new family of devices involving different properties of various materials. In particular heterostructure with thin amorphous silicon film seems to be very appealing since large area and low temperature are involved to obtain a junction on base material. However amorphous silicon chemical and physical compatibility with crystalline wafer and thin film silicon does this material a good candidate to solve problems introduced when porous layer is produced on substrate; such as adhesion and electrical problems of metal contact on porous material.

Porous silicon is usually obtained by an electrochemical etching of a silicon substrate. This material consists of a silicon nano-crystallites network (similar to a sponge) which shows interesting properties for both optical and electrical fields. Its strong room temperature visible photoluminescence has stimulated considerable interest in view of its possible use in silicon based optoelectronics.

In this work we studied the compatibility of amorphous thin film technology with the porous silicon in case of large area applications like solar cells and sensors.

For solar cells the problem of light confinement and collection is still open, since the proposed technology represents the way to enhance the photovoltaic conversion efficiency. Many solutions are showed for this aim: some implying steps derived from microelectronics technology [1], other involving easier chemical bath. These solutions need improvements, the first for high cost process, the latter for low effective light confinement.

Recently a new approach to produce textured silicon surfaces has been shown. This method consists of p type silicon electrochemical etching in HF based solution obtaining the formation of a macroporous layer (with micrometer size pore) at the surface of crystalline Si. Because of its

Mat. Res. Soc. Symp. Proc. Vol. 557 © 1999 Materials Research Society

low reflectivity the proposed textured silicon can be used in solar cells for photovoltaic application [2,3]. Moreover porous silicon was used as "active" layer in sensor field because of the high surface to volume ratio (even 500 m^2/cm^3) and its high reactivity, then porous technology on silicon materials is currently investigate to produce gas sensor for mass production [4]. Heterojunctions with amorphous silicon could represent a promising solution for the driving problems of porous films.

EXPERIMENT

<u>Porous layer preparation</u>

Several porous silicon series have been prepared by electrochemical etching of p-type CZ silicon wafers, with <100> orientation and 10 Ωcm resistivity by using a conventional apparatus [5]. Aluminium deposition was performed on back side of the wafer to ensure ohmic contact. A current density of 10 mA/cm^2 for 10 minutes was applied between this contact (anode) and a platinum electrode (cathode) in 2M HF/acetonitrile, obtaining about 1 micron thickness. Usually porous silicon can be obtained in a large variety of morphologies depending on substrate type, doping and etching conditions. These morphologies can be classify into 3 groups according to the pore average dimension, Wp: macroporous Wp>500Å, mesoporous: 20Å<Wp<500Å, nanoporous: Wp<20Å. Only the last two classes show a room temperature photoluminescence. Our porous silicon samples exhibit a macroporous phase with pore average dimension of about 1 micron with a superimposed nanoporous phase, clearly observable under UV lamp. These different morphologies have been used to fabricate two kind of devices: macroporous in solar cells field, nanoporous in gas sensors.

<u>Solar cells</u>

In case of solar cell applications we removed the nanoporous phase by a 1M NaOH solution in order to investigate only the behaviour of macroporous layer as texturisation profile. The heterojunction between amorphous and crystalline was realised on porous silicon with plasma enhanced chemical vapour deposition system using a 13.56 MHz capacitive coupled commercial UHV single chamber system.

Before amorphous layer deposition a CF_4/O_2 plasma etching conditioning of the sample was performed on the porous silicon surface in order to remove the thin native silicon oxide layer and prepare the surface to the amorphous silicon deposition [6,7]. The process parameters are summarised below: 200 mTorr working pressure, 30 sccm gas flow, 98 mW/cm^2 glow discharge power density and 240 °C substrate temperature.

An n^+ type amorphous silicon layer of 30 nm thick was deposited with the following deposition conditions: 240°C substrate temperature, 200 mTorr chamber pressure, 14 mW/cm^2 RF power density, 20 sccm SiH_4 and 10 sccm PH_3 flow rates [8]. Grid shaped ohmic contacts were ensured on amorphous layer by an e-gun evaporation of silver.

Several characterisations were used to extract informations about the heterostructure samples. For solar cells we measured the photovoltaic parameters under AM 1.5 standard condition, voltage behaviour and quantum yield. The active area of devices were 1.26 cm^2.

<u>Gas sensors</u>

In case of gas sensors we built the heterostructure using the same procedure, reported above, for cleaning step and amorphous layer deposition, without removing the nanoporous phase from crystalline silicon surface. The top contact was ensured by e-gun evaporated chromium grid.

Further step was applied on this heterostructure in order to obtain a porous surface directly exposed to the environment. By using a reactive ion etching treatment we removed the amorphous silicon layer in the area not covered by the chromium grid.

Dark current voltage and photoluminescence behaviour were evaluated on this samples in order to investigate the quality of porous film after plasma treatments as well as the responsivity to different gas exposure. In particular the photoluminescence measurement has been carried out using an excitation source at 442 nm line of a HeCd laser. Incident power on the sample was always 1 mW/cm^2. A monochromator blazed at 500 nm and a CCD detector have been used to collect the spectra. Photoluminescence response in different environment was obtained mounting samples in a testing chamber equipped with a quartz window and vacuum connectors, where gases can be introduced monitoring pressure and flow rate.

RESULTS AND DISCUSSION

Solar cells

In Figs. 1a) and b) we reported a comparison between external quantum yield and current voltage under AM1.5 of heterojunction made on macroporous silicon and on flat silicon surface without any porous treatment.

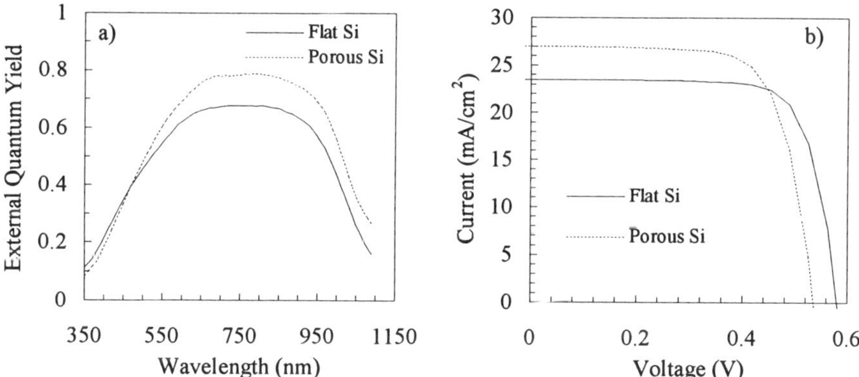

Fig. 1a), b) Comparison between external quantum yield and current vs voltage under AM1.5 of heterojunction made on porous and flat silicon surface.

A remarkable increase in quantum yield in whole range of wavelengths was obtained on macroporous silicon devices. Current voltage measurements in standard AM 1.5 conditions also confirmed this increment, in particular for the short circuit current value we observed an increase of 15%. This improvement can be ascribed to the reduced reflectance of the silicon surface due to texturisation imposed by macroporous layer. On the other hand looking quantum yield in the short wavelength we notice an increase of recombination effect at interface that affects the fill factor and V_{OC} values. This effect is probably due to the morphology of the macroporous surface, since amorphous layer deposition seems not effective in conformal covering the pores.

Then a trade off is needed in optimising porous layer thickness. In Table I photovoltaic parameters of the different substrate heterojunctions are summarised. We remark that no collecting and antireflective layer (i.e. ITO) was applied on the top of these cells.

Table I. Photovoltaic parameters of the different substrate heterojunctions.

Substrate	V_{OC} (mV)	J_{SC} (mA/cm^2)	FF (%)	Eff (%)
Flat Si	583	23.5	75.5	10.3
Macroporous Si	538	27	71.7	10.4

<u>Gas sensors</u>

These heterostructures were characterised by photoluminescence measurements in order to investigate the behaviour of the emission spectra before and after the plasma treatments of porous surface and amorphous layer depositions. First we measured the photoluminescence of the as grown porous silicon. A maximum value at about 760 nm was found. By using a fitting procedure on porous silicon photoluminescence line shape [9] we evaluated the average width L_0 of the emitting nanostructure classes and the porosity of the surface ε. In particular we found:

$$L_0 = 3.3 \text{ nm and } \varepsilon = 44\%.$$

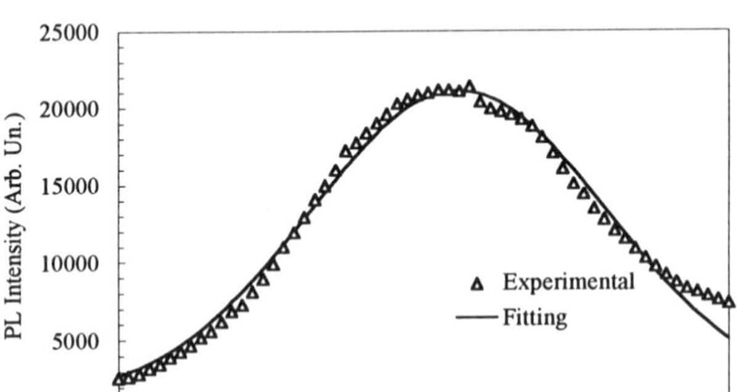

Fig. 2. Experimental PL spectrum for the ε = 44% sample (up triangles) and the calculated line shape (full line).

In Fig. 2 the experimental result and the fitting shape are shown. After the amorphous n$^+$ layer deposition, change in shape and intensity of photoluminescence signal was observed. This is due to a double effect: the plasma action that reduces the nanoporous phase leaving a larger width of the pores and the amorphous film coverage that decreases the quantum confinement inside columnar wire and then the efficiency of the emission process.

Further reduction in intensity of the photoluminescence is observed after the chromium deposition and grid lithography. This phenomenon is easily explained by the shadowing of the metal grid. After the amorphous silicon etching in the area not covered by chromium a slight increase in photoluminescence intensity was found. In Fig. 3 the previous discussed results are summarised.

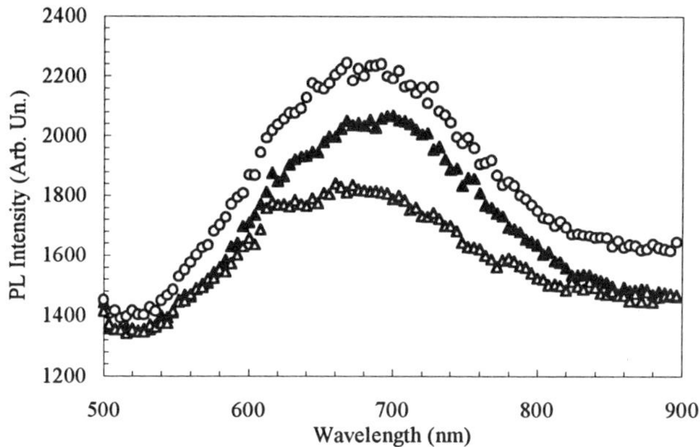

Fig. 3. Measured photoluminescence of the amorphous nanoporous silicon heterostructure (O), the same structure after the chromium grid evaporation (ρ) and after the amorphous layer etching (π).

Finally these heterostructure gas sensors were exposed to different gases: nitrogen and dry air. In Fig. 4 the comparison of the two photoluminescence signals is reported. We also remark that after dry air exposure no recovery of the photoluminescence was observed. This phenomenon is ascribed to the prolonged illumination time to laser source in dry air. In this condition the adsorbed O_2 at silicon dangling bonds dissociates and inserts into Si-Si bond resulting in a not reversible chemisorption process [10]. No appreciable difference in emission spectra was observed on samples in which the amorphous layer completely covered the porous phase.

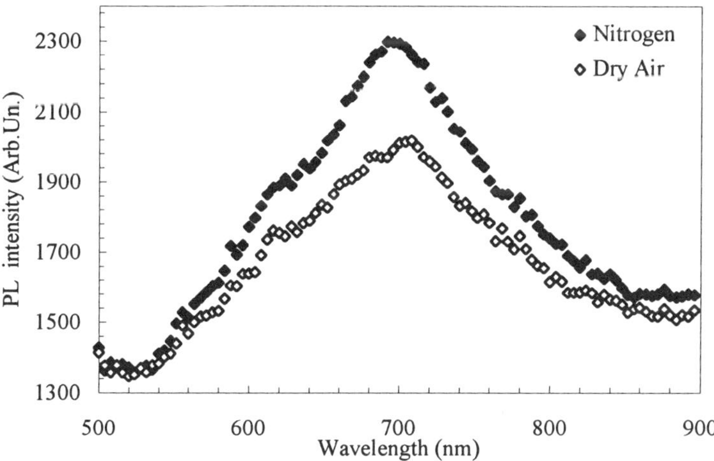

Fig. 4. Comparison between the measured photoluminescence of the heterostructure in two different environment: nitrogen and dry air.

Current voltage measurements of the heterostructure in dark condition ensures the rectifying behaviour. In Fig. 5 the I-V characteristic is reported. The total area of the device is 0.5 cm^2. The grid shadowing is about 50%.

High resistivity of the porous layer and recombination mechanisms at the heterojunction interface strongly affect the current values measured in forward bias conditions. Anyway a rectifying ratio of about two order of magnitude is observed at 0.5 V thanks to a low value of reverse current.

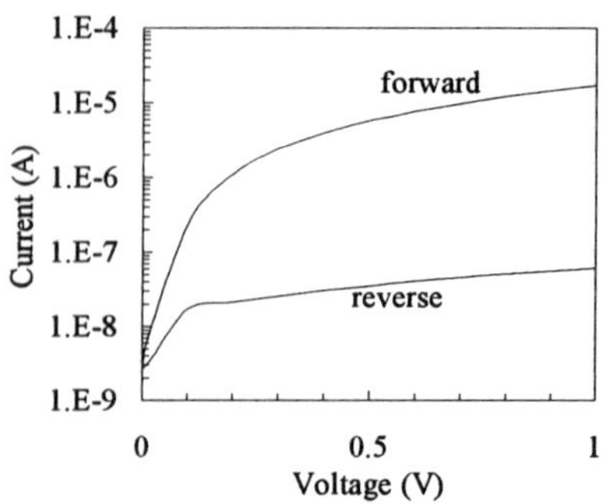

Fig. 5. Dark current versus voltage of heterojunction on nanoporous silicon.

CONCLUSIONS

In this work we investigate the possibility to obtain heterojunctions between amorphous silicon and macroporous or nanoporous silicon. In the first case we obtain alternative texturisation procedure of the silicon wafer surface giving rise to an increase of short circuit current. This process is not dependent by the orientation of silicon substrate, then it is easily transferable on multicrystalline silicon. Further investigations are required to reduce recombination effects in order to increase photovoltaic conversion efficiency.

Amorphous silicon on nanoporous phase structure still maintains photoluminescence properties of the porous material as well as the gas sensitivity if the amorphous layer is removed from the exposed area of the device. Moreover rectifying behaviour is retained on this heterostructure device that can be useful for future applications.

ACKNOWLEDGEMENTS

The authors are very grateful to A. Citarella for helpful collaboration.

REFERENCES

1. M. Green, J. Zhao, A. Wang *2nd World Conf. Exhib. On Photovol. Solar Energy Convers. Proc. Vol.* **2**, p.1187 (1998).
2. Ponomarev, S. Bastide, Q.N. Le, D. Sarti, C.Levy-Clement, *14TH European Photovoltaic Solar Energy Conference, Proc.* **1**, p. 804 (1997).
3. Propst and P.A. Kohl, *J. Electrochem.* Soc. **141**, p. 1006 (1994).
4. M. Sailor in *Properties of Porous Silicon*, L. Canham Ed. INSPEC, p. 364 (1997).
5. Jung, S.Shih, D.L. Kwong, *J. Electrochem. Soc.*, **140**, p. 3046 (1993).
6. De Rosa, E. Chiacchio, T. Fasolino, P. Grillo, M.L. Grilli, M. Tucci, F. Roca. *2nd World Conf. Exhib. On Photovol. Solar Energy Convers. Proc*, **2**, p. 2440 (1998).
7. De Rosa, M.L. Addonizio, F. Roca, E. Chiacchio, M.Tucci, *this conference*, A9.4
8. F. Roca, G. Sinno, G. Di Francia, P. Prosini, G. Fameli, P. Grillo, A. Citarella, F. Pascarella, D. della Sala, *Solar Energy Material and Solar Cells*, **48**, p.15 (1997).
9. Di Francia, G. Iadonisi, P. Maddalena, M. Migliaccio, D. Ninno, E. Santamato, *Optics Communications* **127**, p. 44 (1996).
10.J. Harper and M. Sailor, *Langmuir*, **13**, p. 4652 (1997).

NOISE OF a-Si:H PIN DIODE PIXELS IN IMAGERS
AT DIFFERENT OPERATING CONDITIONS

F. Blecher, B. Schneider, J. Sterzel, M. Böhm,
Universität-GH Siegen, Institut für Halbleiterelektronik (IHE), D-57068 Siegen, Germany.

ABSTRACT

The noise current spectral density of an a-Si:H pin diode can be calculated with experimentally determined flicker noise coefficients by superposition of the shot and flicker noise spectra of photocurrent and dark current. The dependence of the flicker noise current spectral density in pin diodes on the pixel area is calculated with our expansion of Hooge's law for flicker noise. We propose a new method for the calculation of dynamic range (DR) and signal-to-noise ratio (SNR) in pin diode pixels as a function of pixel area, dark current, photocurrent and the integration time of the imager. DR and SNR of the pin diodes are calculated for Thin Film on ASIC (TFA) image sensors.

INTRODUCTION

Amorphous silicon pin diodes are widely used for large area sensor arrays as well as for Thin Film on ASIC (TFA) image sensors [1] to convert the incident radiation to a current. The dc signal current of the a-Si:H detector is superimposed by noise due to number fluctuations of free carriers and mobility fluctuations. TFA imagers usually operate in the charge storage mode so that the dc current and the noise current are integrated in the capacitance C_{int} and read out after the integration time t_{int}. The signal quality and the detection limit of TFA imagers are defined as SNR and DR, respectively, and are determined by the noise of the a-Si:H detector and the noise of the ASIC. This paper presents a method for the calculation of the noise in the a-Si:H pin diode which limits the precision to which the TFA sensor can detect an image.

The quantitative determination of the noise is necessary for the device characterization and the optimization of measures to suppress noise. Noise is a critical device parameter, especially for low irradiance or short integration time. The quantitative description of the flicker noise is especially important for image sensors operated in charge storage mode, since noise at low frequencies may contribute significantly to the total noise of these sensors.

SUPERPOSITION OF NOISE

The noise current spectral density of the diode current is calculated by superposition of the shot and flicker noise spectra of photocurrent and dark current and the thermal noise spectrum of the parallel leakage resistance [2]. These five noise current sources are connected in parallel with the diode capacitance and have a common internal resistance which is given by the parallel connected leakage resistance and differential resistance of the diode. The noise spectra of the diode current under illumination and in the dark are given by equations (1) and (2), respectively.

$$\overline{i_d'^2}(f) = \overline{i_{shot,ph}'^2} + \overline{i_{shot,dark}'^2} + \overline{i_{1/f,ph}'^2}(f) + \overline{i_{1/f,dark}'^2}(f) + \overline{i_{Rp}'^2} \tag{1}$$

$$\overline{i_{dark}'^2}(f) = \overline{i_{shot,dark}'^2} + \overline{i_{1/f,dark}'^2}(f) + \overline{i_{Rp}'^2} \tag{2}$$

The thermal noise source of the pin diode series resistor has to be treated separately, since its internal resistance is relatively low and connected in series with the diode capacitance. The noise of the series resistance contributes to the thermal noise of the pixel input stage in the ASIC and is not considered in this paper.

Mat. Res. Soc. Symp. Proc. Vol. 557 © 1999 Materials Research Society

1/f NOISE

Hooge's empirical law (3) describes 1/f noise in homogenous materials [3]. The constant α is a measure for the relative flicker noise. N is the number of electrons in the sample and therefore a measure of the sample size.

$$\overline{i_{1/f}^2} = \frac{\alpha}{N} \cdot I^2 \cdot \frac{1}{f} \tag{3}$$

We measured the flicker noise current spectral densities of pin diodes (Fig. 1) which are described by relation (4) [2]. The frequency coefficient γ of the flicker noise is not always close to unity, however, we still refer to this noise as 1/f noise.

$$\overline{i_{1/f}^2} \propto |I|^{2\beta} \cdot \frac{1}{f^\gamma} \qquad \text{with} \qquad \beta, \gamma \approx 1 \tag{4}$$

In order to describe the dependence of the 1/f noise on the pixel area A for current exponent coefficients smaller than unity ($\beta < 1$) we developed an equation similar to Hooge's Law, including the relation (4) and magnitude coefficients c_i [4]:

$$\overline{i_{1/f,\,i}^2} = \frac{c_i}{A^{2\beta_i - 1}} \cdot |I_i|^{2\beta_i} \cdot \frac{1}{f^{\gamma_i}} \tag{5}$$

The operating point dependent flicker noise coefficients of Tab. I according to equation (5) are calculated with the data of Fig. 1 in consideration of the dependence of the flicker noise on the area. The relation between the flicker noise and the pixel area of equation (5) is based on the suppositions that the noise of a pixel is not correlated to the noise of neighboring pixels and that the steady state current density does not depend on the position. Consequently the flicker noise power of a pixel is proportional to the pixel area, when the current density remains constant and proportional to $A^{-\delta}$ with $\delta = 2\beta_i - 1 \leq 1$, when the current is kept constant.

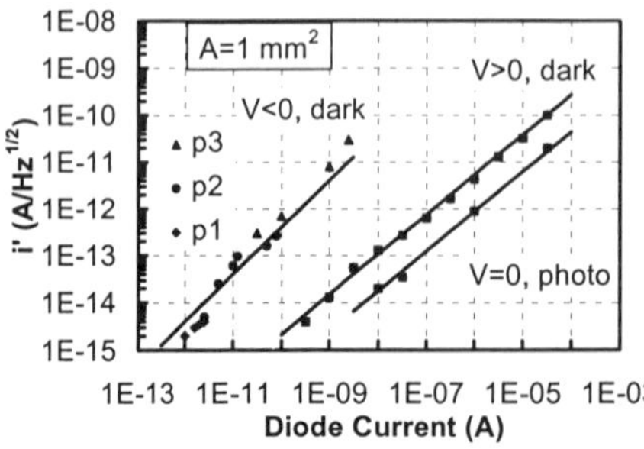

Fig. 1: Noise current density at 1 Hz vs. current.

Tab. I: Magnitude coefficients c_i (cm$^{4\cdot\beta_i-2}$ A$^{2-2\cdot\beta_i}$ Hz$^{\gamma_i-1}$), frequency coefficients γ_i and current exponent coefficients β_i of the flicker noise at different operating points.

I_i	c_i	γ_i	β_i
short circuit photocurrent	$c_{ph}=5\cdot10^{-16}$	$\gamma_{ph}=1$	$\beta_{ph}=0.85$
reverse dark current	$c_{dr}=2\cdot10^{-7}$	$\gamma_{dr}=1$	$\beta_{dr}\approx1$
forward dark current	$c_{df}=2\cdot10^{-14}$	$\gamma_{df}\leq1$	$\beta_{df}=0.85$

THERMAL NOISE AND SHOT NOISE

The thermal noise of the parallel resistance is relatively small since the specific parallel resistance R'_P is in the range of $10^{10}\ \Omega\ \text{cm}^2$.

$$\overline{i_{Rp}^2} = \frac{4 \cdot kT}{R_P} \qquad \text{with} \quad R_P = \frac{R'_P}{A_{pin}} \tag{6}$$

Statistically independent and identical current pulses with respect to charge and transit time cause shot noise which is white noise for frequencies smaller than the reciprocal transit time. In a-Si:H pin diodes we find a decrease of the shot noise with rising frequency and attribute this to trapping and recombination in the intrinsic layer. In this paper we consider the maximum shot noise level in the relevant frequency range up to 10^5 Hz:

$$\overline{i'^2_{shot}} = \overline{i'^2_{shot,ph}} + \overline{i'^2_{shot,dark}} = 2e \cdot \left(\left| I_{ph} \right| + \left| I_{dark} \right| \right) \tag{7}$$

SIGNAL-TO-NOISE RATIO AND DYNAMIC RANGE

The noise of the diode current can be calculated with equations (1), (5), (6), (7) and the data of Tab. I. When the sensor operates in charge storage mode, the dc current and the noise current are integrated in the paralleled diode capacitance and external capacitance. Noise components with a frequency of the reciprocal integration time are extinguished and noise components of higher frequency are reduced due to the integration. The calculation of the noise voltage u_c on the capacitance after the integration takes into account the frequency dependent reduction of the noise spectrum by the integration:

$$u_{c,\,rms} = \sqrt{\int_{f_{min}}^{\infty} \overline{u'^2_c}\, df} = \sqrt{\int_{f_{min}}^{\infty} \overline{\left(\frac{1}{C_{int}} \cdot \int_{t}^{t+t_{int}} i'(t,f)\, dt \right)^2}\, df} = \sqrt{\int_{f_{min}}^{\infty} \frac{1}{C^2_{int}} \cdot f \cdot \overline{\int_{t=0}^{t=1/f} \left[\int_{t}^{t+t_{int}} i'(t,f) dt \right]^2 dt}\, df} \tag{8}$$

The noise current spectral density is described as a sine function with random phase φ_r and frequency dependent magnitude:

$$i'(t,f) = \sqrt{2 \cdot \overline{i'^2(f)}} \cdot \sin(2\pi f \cdot t + \varphi_r) \tag{9}$$

$$u_{c,\,rms} = \frac{1}{C_{int}} \cdot \sqrt{\int_{f_{min}}^{\infty} \overline{i'^2(f)} \cdot \frac{\sin^2(\pi t_{int} \cdot f)}{(\pi \cdot f)^2}\, df} \tag{10}$$

Since the minimum frequency of the flicker noise f_{min} has not been found yet we suggest a minimum frequency of $1\,\mu Hz$. With equation (5), (7) and (10) we obtain the noise voltage components for white noise and flicker noise with Euler's constant C:

$$u_{c,i,\,rms} = \begin{cases} \sqrt{\dfrac{t_{int} \cdot 2e \cdot \left| I_i \right|}{2 \cdot C^2_{int}}} & \text{for shot noise} \\[4mm] \dfrac{t_{int}}{C_{int}} \cdot \sqrt{\dfrac{c_i \cdot \left| I_i \right|^{2\beta_i}}{A^{2\beta_i - 1}}} \cdot \sqrt{1 - C - \ln(2\pi t_{int} f_{min})} & \text{for flicker noise} \end{cases} \quad \text{with C=0.5772} \tag{11}$$

SNR and DR of the integrated voltage are calculated with the dc signal U_C and the noise voltage components of equation (11) in analogy to equations (1) and (2):

$$SNR = 10 \log \frac{U^2_C}{u^2_{c,1/f,dark} + u^2_{c,1/f,ph} + u^2_{c,shot,dark} + u^2_{c,shot,ph} + u^2_{c,Rp}} \tag{12}$$

$$DR = 10 \log \frac{U^2_C}{u^2_{c,1/f,dark} + u^2_{c,shot,dark} + u^2_{c,Rp}} \tag{13}$$

SNR and DR of the integrated voltage are functions of flicker noise coefficients, pixel area, dark current density, photocurrent density and the integration time:

$$SNR = 10\log \frac{J_{ph}^2 \cdot A}{\left[c_{dr}\left|J_{dr}\right|^{2\beta_{dr}} + c_{ph}\left|J_{ph}\right|^{2\beta_{ph}}\right] \cdot \left[1 - C - \ln(2\pi t_{int}f_{min})\right] + \dfrac{e \cdot \left(\left|J_{dr}\right| + \left|J_{ph}\right|\right)}{t_{int}} + \dfrac{2kT}{t_{int} \cdot R'_p}} \quad (14)$$

$$DR = 10\log \frac{J_{ph}^2 \cdot A}{c_{dr} \cdot \left|J_{dr}\right|^{2\beta_{dr}} \cdot \left[1 - C - \ln(2\pi t_{int}f_{min})\right] + \dfrac{e \cdot \left|J_{dr}\right|}{t_{int}} + \dfrac{2kT}{t_{int} \cdot R'_p}} \quad (15)$$

SNR AND DR OF a-Si:H PIN DIODE PIXELS OF IMAGERS

Noise components, SNR and DR of the diode pixels are presented in Fig. 2 to Fig. 7 for two TFA sensors [5, 6]: The HIRISE II (**Hi**gh **R**esolution **I**mage **Se**nsor) and the LARS II (**L**okal-**A**utoadaptiver **S**ensor) employ an a-Si:H based photodetector on top of an ASIC. One well proved option for the detector is a pin diode with 10^{-9} A/cm² dark current, a halogen bulb light referred specific current density of 67 nA/(lx cm²), a specific capacitance of 8.5 nF/cm² and a specific parallel resistance of about 10 GΩcm². The HIRISE II is designed with 10^{-6} cm² pixel size and 3.1 V maximum voltage swing at the integration capacitance of 19 fF. The equivalent LARS II data are A=15.3·10^{-6}cm², $U_{C\,max}$=3 V and C_{int}=357 fF. The calculations are based on pixels with a source follower amplification of 0.8.

The full range illumination level of the HIRISE II is a function of the integration time. The noise of the dark current depends on the integration time and the noise of the photocurrent rises with increasing illumination. At long integration time the noise voltage on the integration capacitance is dominated by flicker noise of the dark current at low illumination levels and by the shot noise of the photocurrent at high illumination levels (Fig. 2). At short integration time the noise voltage on the integration capacitance is dominated by the shot noise of the photocurrent (Fig. 3). The limitation of the SNR and the DR of HIRISE II according to Fig. 2 and Fig. 3 is shown in Fig. 4. The maximum SNR is above 50 dB and determined by the shot noise of the photocurrent. The maximum value of the pin diode DR strongly depends on the maximum illumination level.

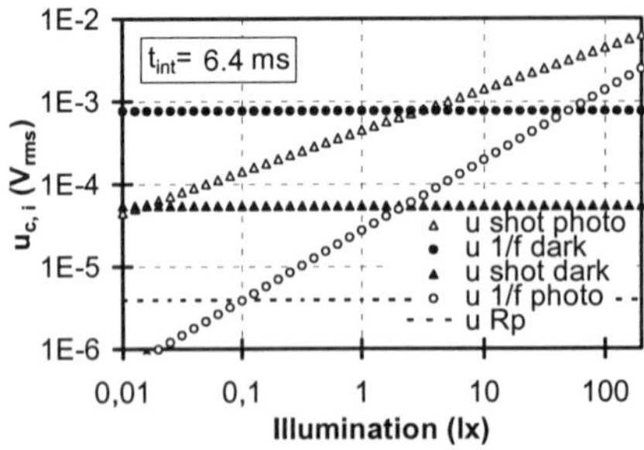

Fig. 2: Noise components of pin diode at low illumination levels (HIRISE II).

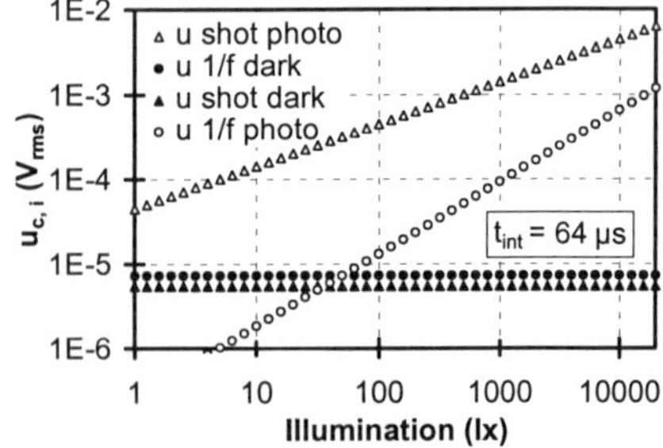

Fig. 3: Noise components of pin diode at high illumination levels (HIRISE II).

The LARS II imager chooses the integration time of each pixel according to the local illumination level (Fig. 5) to cope with an illumination range of more than 6 orders of magnitude. The LARS II covers a time stamp range (TR) from 10 µs to 10 ms and outputs a time stamp signal in addition to the analog signal of the voltage on the integration capacitance. The noise components of the integrated voltage of LARS II are shown in Fig 6. Flicker noise of the dark current dominates at low illumination levels and shot noise of the photocurrent dominates at high illumination levels. Above 100 lux the noise of the dark current decreases and the noise of the photocurrent remains about constant due to the decrease of the integration time. Fig. 7 shows the SNR and DR of the pin diode of LARS II and the corresponding TR. With up to 68 dB SNR and 130 dB DR of the diode pixels, 60 dB TR and an imager noise floor of less than 2 mV the imager output reaches a SNR and a DR of more than 120 dB each.

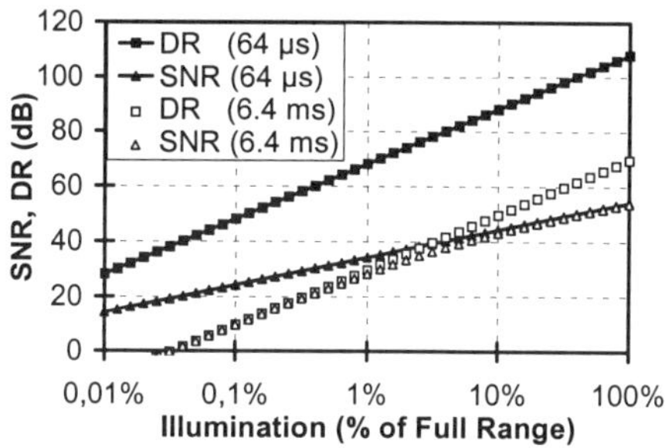

Fig. 4: SNR and DR of pin diode (HIRISE II).

Fig. 5: Integration time of LARS II.

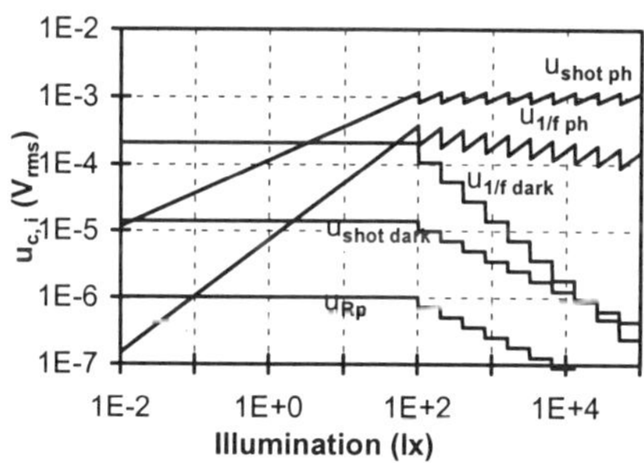

Fig. 6: Noise components of pin diode (LARS II).

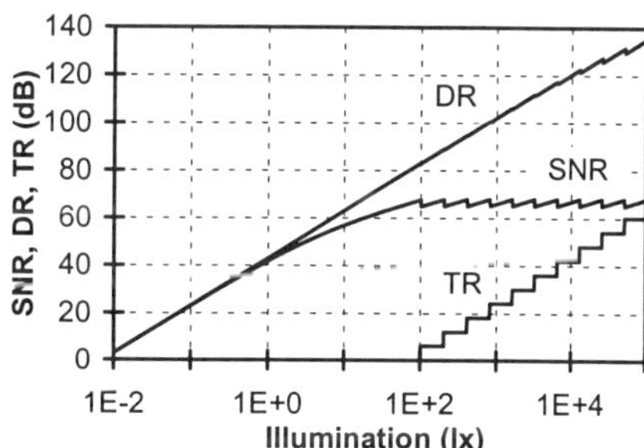

Fig. 7: SNR, DR and TR of pin diode (LARS II).

DISCUSSION

The noise of the integrated voltage due to the dark current rises with increasing integration time. According to equation (11) the shot noise component is proportional to the square root of the integration time whereas the flicker noise component is approximately proportional to the integration time. Therefore flicker noise dominates the noise of the dark current at long integration times which are required to detect very low illumination levels. However, we did not attempt to optimize the diodes with respect to flicker noise yet.

The total noise voltage on the integration capacitance due to the dark current is about 1 µV at 10 µs integration time for LARS II and below 10 µV at 64 µs integration time for HIRISE II.

This low noise level of the a-Si:H detector is superimposed by noise of the ASIC which dominates the total noise of the imagers, if the integration time is extremely short. However, the kTC reset noise values of 0.47 mV for the HIRISE and 0.11 mV for the LARS are not included in the analysis of this paper. KTC reset noise can be eliminated by correlated double sampling (CDS), if the signal voltages at the start and the end of one integration period are subtracted. Furthermore fixed pattern noise is eliminated and flicker noise of the ASIC is reduced by CDS.

The maximum SNR of LARS II and HIRISE II is limited by shot noise of the photocurrent. From equation (14) we derive equation (16) which is a good approximation for the calculated maximum SNR of LARS II and HIRISE II. However, the neglect of the shot noise reduction in this paper leads to an overestimation of noise, especially at high frequencies which correspond to short integration times.

$$SNR\big|_{\text{full range}} \approx \lim_{J_{ph} \to \infty} SNR\big|_{\text{full range}} = 10\log\frac{U_{Cmax} \cdot C_{int}}{e} \tag{16}$$

CONCLUSIONS

A detailed quantitative description of the noise current spectral density in pin diode pixels at different operating points and the superposition of noise components have been presented. The calculation of noise voltage components on the integration capacitance has been presented for imagers which are operated in the charge storage mode. This allows the characterization and optimization of noise related performance characteristics of TFA image sensors and other sensor systems which use a-Si:H pin diodes. The calculations of noise components, SNR and DR in pin diode pixels as a function of flicker noise coefficients, pixel area, dark current density, photocurrent density and integration time have been presented for the first time and applied for new TFA imagers.

ACKNOWLEDGMENTS

This work was partially funded by the DFG (Deutsche Forschungsgemeinschaft) under contract Bo 772/3.

REFERENCES

1. B. Schneider, H. Fischer, S. Benthien, H. Keller, T. Lulé, P. Rieve, M. Sommer, J. Schulte and M. Böhm, Technical Digest of International Electron Devices Meeting *1997*, 209-212.
2. F. Blecher K. Seibel and M. Böhm, Photo and Dark Current Noise in a-Si:H pin Diodes at Forward and Reverse Bias, Mat. Res. Symp. Proc. **507**, (1998).
3. F.N. Hooge, IEEE Trans. Electron Devices **41**, 1926-1935 (1994).
4. F. Blecher, K. Seibel, J. Sterzel, M. Hillebrand and M. Böhm, DFG Forschungsbericht, contract Bo 772/3-2, Universität-GH Siegen, 1998.
5. T. Lulé, H. Keller, M. Wagner, M. Böhm, C.D. Hamann and L. Humm, to be published in *Advanced Microsystems for Automotive Applications 99*, edited by D.E. Ricken and W. Gessner, (Springer Verlag, Berlin, 1999).
6. S. Benthien, M. Wagner, M. Verhoeven, M. Böhm, B. Schneider, B. van Uffel and F. Librecht, A Vertically Integrated High Resolution Active Pixel Image Sensor for Deep Submicron CMOS Processes, to be presented at 1999 IEEE Workshop on Charge-Coupled Devices and Advanced Image Sensors, Nagano, Japan, (1999).

PHOTOINDUCED SPACE CHARGE EFFECTS IN a-Si:H SOLAR CELLS

T. UNOLD, T. BINNEWIES, R. BRÜGGEMANN*, G.H. BAUER
Fachbereich Physik, Universität Oldenburg, 26111 Oldenburg, Germany
*Laboratoire de Genie Electrique, Supelec, 91192 Gif-sur-Yvette Cedex, France

ABSTRACT

We have investigated charge collection in thin amorphous silicon solar cells under light bias illumination, both experimentally and by numerical simulation. In such charge collection experiments, space charge due to trapped bias-light generated carriers leads to an enhancement of a small signal probe beam charge collection. It is found that this enhancement of the small signal charge collection is strongly dependent on the diode thickness and the defect density in the samples. In particular for thin diodes (d < 0.5 microns) the charge collection enhancement can be shown to increase with light-induced degradation of the devices. The effect of these material parameters as well as other experimental parameters, such as light bias and probe beam photon flux, will be demonstrated by means of numerical simulation.

INTRODUCTION

Light bias charge collection measurements have been performed by a number of different research groups on different device structures. Hou and Fonash showed by numerical simulation [1] and experimentally [2] that under certain conditions charge collection enhancement in reasonably thin a-Si:H diodes (d = 0.7 μm) should be observed. They named this phenomenon of charge collection enhancement „photogating", because of the analogy to gating in bipolar transistors[1]. Large photogating effects, with possible application in photodetectors, have also been observed in thick a-Si:H diodes (d = 3.5 μm) [3].

In several numerical simulations it was shown that the internal field in p-i-n diodes can become zero or even locally reversed in the presence of strongly absorbed bias illumination [1,3,4]. This lowering of the field is due to the local creation of space charge due to trapping of the photogenerated carriers. The increased collection efficiency of the small-signal „probe" illumination then arises from the additional collection of carriers which had been generated by the bias light. Photogating is easily observed in thick diodes (d > 3 microns) but, as will be shown, is much more difficult to be measured in thin diodes (d < 0.6 microns) such as those used in solar cells. Measurement of such space charge effects in thin solar cell diodes could be useful for the evaluation of defect densites and internal field distribution in these devices. In this paper we present a detailed study of the conditions that lead to photogating in thin pin devices.

EXPERIMENTAL DETAILS

The charge collection measurements were performed using a setup with a tungsten halogen lamp as a light source and a 1/8 m monochromator. The bias light from a tungsten halogen light source was directed with an optical fiber onto the sample, and the wavelength of the bias light was selected with optical bandpass filters. The probe light was chopped at 10 Hz and the photocurrent of the pin-diodes detected via a lockin-technique. In principle, there are several possibilites to illuminate the diodes in the experiment. We chose to illuminate bias light onto the p-side and the probe light onto the n-side, consistent with the experimental procedure previously carried out for

Mat. Res. Soc. Symp. Proc. Vol. 557 © 1999 Materials Research Society

thick diodes. [3] However, we would like to point out that some of the results in the literature have been obtained with a different geometry, e.g. illumination of bias and probe beam from the p-side[1].

The samples investigated in this study were several amorphous silicon pin diodes fabricated at Research Lab Siemens AG with thicknesses between 0.4 – 1.8 microns, and TCO and semitransparent Paladium as contacts. Numerical simulations of the steady-state response of the pin-diodes was carried out by numerically solving Poisson´s equation and the full set of continuity and rate equations as already described in previous publications [5].

RESULTS AND DISCUSSION

External Charge Collection Efficiency

The external charge collection efficiency (equivalent to external quantum efficiency) is defined as the ratio of charge carriers collected, normalized by the photon flux, ϕ_p incident on the sample

$$\eta_{ext} = j_{probe}/q\phi_p. \tag{1}$$

In this study we are mainly concerned with the effects of a strongly absorbed ($\alpha \cong 2 \times 10^5 \mathrm{cm}^{-1}$) light bias ($\phi_b = 10^{15}\text{-}10^{16}\mathrm{cm}^{-2}\mathrm{s}^{-1}$), on the small signal monochromatic probe beam (typically $\phi_p = 10^{13}$ $\mathrm{cm}^{-2}\mathrm{s}^{-1}$) of the quantum efficiency measurement. For such a situation we can define an apparent collection efficiency

$$\eta_{ext}^* = (j_{probe+bias} - j_{bias})/q\phi_p \tag{2}$$

where again the measured current is normalized by the small-signal probe beam light flux, ϕ_p. Because, as will be shown below, the enhancement of the charge collection under bias irradiation can be quite small for thin diodes we also introduce a so-called gating factor

$$\Gamma = \eta_{ext}^* / \eta_{ext} . \tag{3}$$

Hence, a gating factor of unity indicates no change of the charge collection, and gating factors larger than one indicate improvement of charge collection under light bias and hence, space charge effects.

Influence of Diode Thickness

The gating factor is shown for spectral quantum efficiency measurements at different bias voltages in Figures 1 and 2. It can be seen that for the 1.8 μm thick diode there is a clear deviation from unity for short circuit and small bias voltage conditions. In contrast no photogating at all is observed for the 0.5 micron thin diode shown in Figure 2. As will be shown below this is due to the low defect density and high electric field in the i-layer which does not allow for charge trapping even under strong inhomogeneously absorbed bias irradiation. The effect of the thickness on the

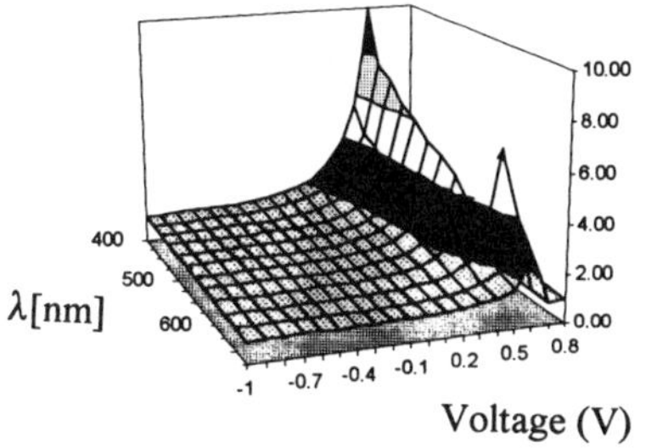

FIG 1: Gating factor Γ for n-side probe illumination and p-side bias-illumination (λ = 450 nm) as a function of probe wavelength and voltage bias, diode 4 (d = 1.8 μm).

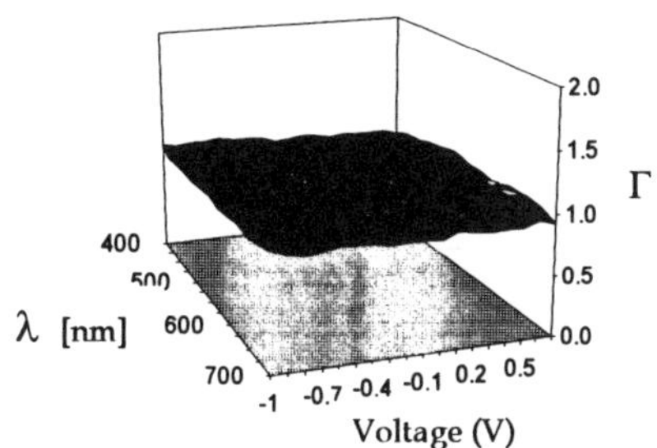

FIG 2: Gating factor Γ for n-side probe illumination and p-side bias-illumination (λ = 450 nm) as a function of probe wavelength and voltage bias, diode 5 (d = 0.52 μm).

photogating effect was investigated by numerical simulation and the results are shown in figure 3. Depicted are collection efficiencies for 3 different diode thicknesses with blue bias light incident from the p-side and red bias light incident from the n-side. The defect density was fixed at $N_d=5 \times 10^{16} cm^{-3}$ for all three calculations. It can be seen that the effect is extremely sensitive to the thickness of the diode. Whereas at d= 0.7 μm typical experimental parameters show very large values for the apparent charge collection, almost no effect at all is observed for the diode with d=0.3 μm. Note that predicted values at large reverse voltage correspond to full collection of carriers, which, however, decreases with decreasing thickness of the diode because of incomplete absorption of the probe beam. We also note that the maximum in the apparent collection efficiency occurs toward larger reverse bias for thicker diodes. This means that in order to observe the maximum in the apparent collection efficiency in diodes of different thickness, one has to make sure to measure the collection efficiency as a function of bias voltage.

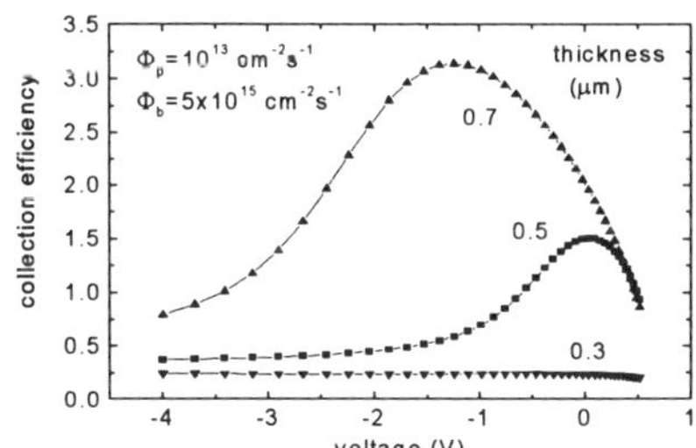

FIG 3: Simulated dependence of the photogating effect on the device thickness.

<u>**Influence of Defect Density – Degradation**</u>

We have photodegraded a 0.5μm thin sample that initially did not show any change of the collection efficiency under light bias. The result is shown in Figure 4. Shown is the photocurrent with and without blue bias illumination for 2 probe beam wavelengths, at

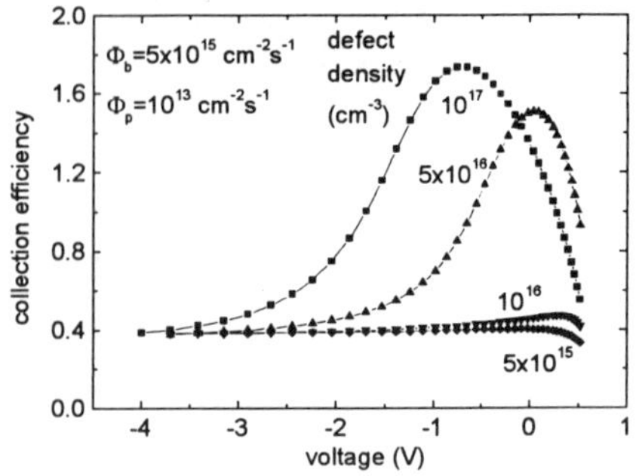

FIG 5: Simulated dependence of the photogating

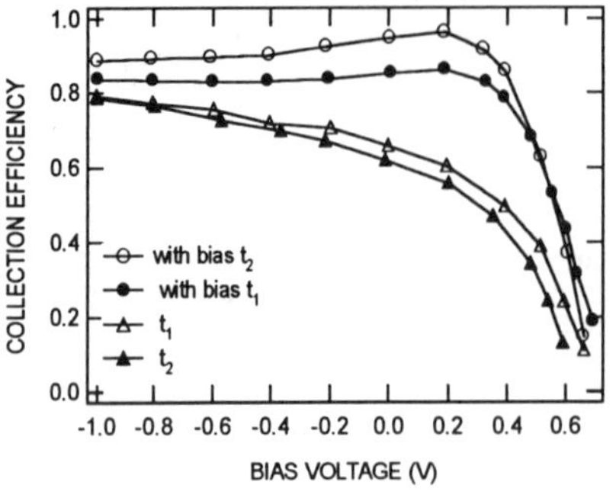

FIG 4: Photocurrent with and without light bias after two different degradation times (t_1= 7 h, t_2= 11 h) for Diode 3 (d = 0.41 μm)

effect on the defect density in a 0.5 micron thin pin solar cell.

two different degradation times. As can be seen, the „normal" charge collection decreases with increasing degradation time, whereas the charge collection under light bias increases with degradation time. This can be understood by the light-induced creation of defects in the diode. Figure 5 shows simulated collection efficiencies for 4 different assumed defect densities and it can be seen that for defect densities smaller than 10^{16}cm^{-3} (which are typical for AM1.5 degradation < 10h) a slight increase in the charge collection is expected. It also can be seen that for defect densities larger than 5×10^{16}cm^{-3} a larger effect should be observed. Indeed, we have observed such behaviour for the sample after degradation times longer than 24 hours as is shown in Figure 6. There, a pronounced increase in the charge collection even to values greater than unity can be observed.

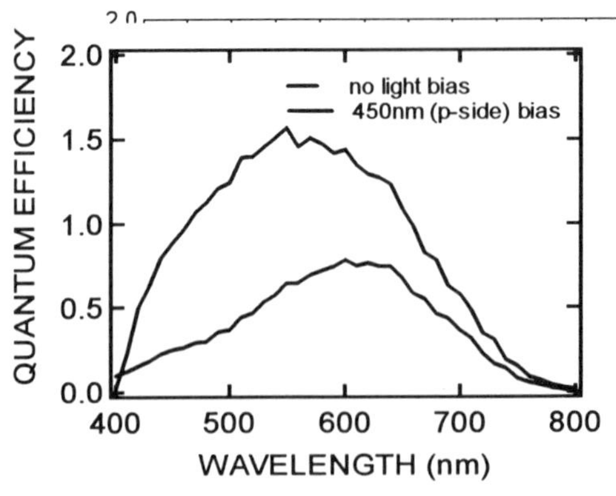

FIG 6: Collection efficiency for diode 3 after 24h of white light degradation (approx. AM1.5).

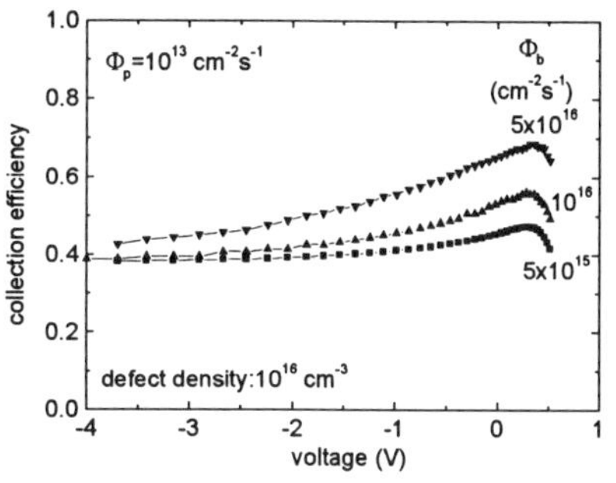
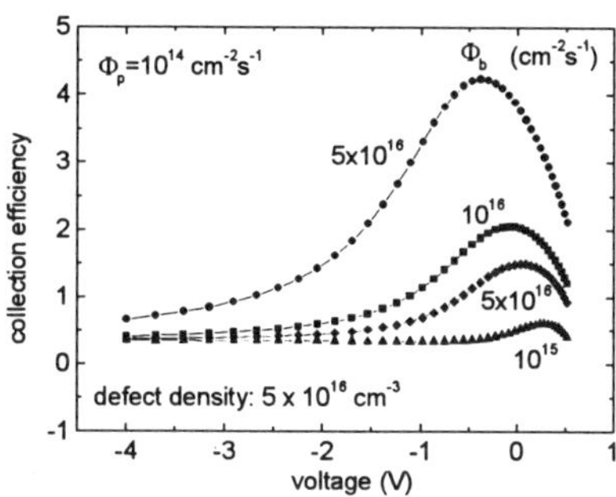

FIG.7 (a): Simulated collection efficiencies for fixed defect density and different light bias photon fluxes, d=0.5 micron. **(b):** Same as Fig.7(a) but for higher defect density of 5x10^{16}cm^{-3}.

Influence of Experimental Parameters: Bias flux and Probe Flux

So far we have demonstrated the sensitivity of the experiment to the material parameters of device thickness and defect density. However, it is also interesting to investigate the effects of experimental parameters on the experimental outcome. Experimental conditions that can be varied are bias light wavelength, bias light flux, and probe beam flux, and illumination sides. Here we focus only on the variation of ϕ_b and ϕ_p. It is reasonable to assume that increasing the magnitude of ϕ_b will increase charge trapping and thus photogating in the experiment. As shown in Figures 7(a) and 7(b), this is indeed confirmed by the numerical simulation. However, it is important to note, that for low defect densities ($N_d < 5 \times 10^{16}$cm^{-3}) the effect of increasing the bias light flux is quite small, whereas the influence is much larger for higher defect densities. At variance to some of the studies reported in the literature [1,2] we have set our probe beam intensities in the small signal regime that is, at least 1-2 orders of magnitude lower than the bias beam. For such a situation one expects linearity of the charge collection enhancement and η^*_{ext} becomes independent of the ϕ_p. That is, the improvement of the internal field by the probe beam increases linearly with ϕ_p which is exactly compensated by the normalization in equation (2). Indeed, this also has been confirmed by our numerical simulations. Deviations from linearity can be observed for higher probe beam intensites, when probe beam and bias flux become comparable.

Internal Field Distribution

To illustrate how the strongly absorbed bias light affects the internal field distribution of the device, the calculated field profiles of a 0.5µm thick device assuming several different defect densities are shown in Figure 8(a). It can be seen that for defect densities of 10^{17}cm^{-3} the internal field can be collapsed using blue bias light with $\phi_b = 5 \times 10^{15}$cm^{-2}s^{-1}. The effect of the bias light on the collected photocurrent, however, is complicated, because the trapped space charge of the bias light and therefore the field distortion, as well as the recombination of photogenerated carriers plays a role. Therefore for different defect densities and also different device thicknesses, the peak in the apparent collection efficiency occurs at different bias voltages. The actual increase in

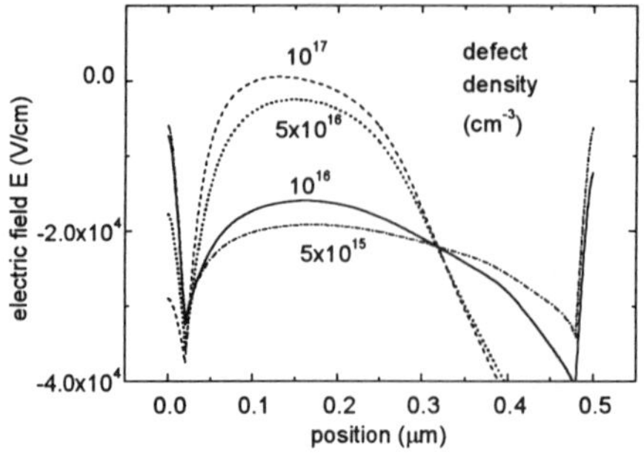

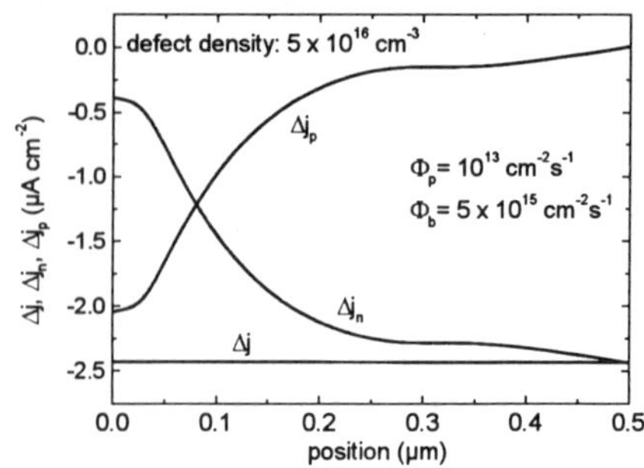

FIG. 8: (a) Simulated internal electric field for 0.5 □m thick solar cells with different defect densities at ϕ_b = 5e^{15}cm^{-2}s^{-1} (p-side on the left). (b) Change in photocurrent caused by probe-beam.

the collected photocurrent by switching the probe beam on and off, is shown in Figure 8(b). It can be seen that in the front of the device (on the left) where the absorption of the bias light occurs, the change in the current is predominantly a hole current, which means that through the probe beam excitation, holes previously trapped in this region can be collected. However, we note that the current enhancement in most of the device is dominated by electrons that have also escaped from the low field region.

CONCLUSIONS

Light-Bias induced space charge effects in thin solar cells are found to be extremely sensitive to diode thickness as well as to the defect density in the sample. Degradation on 0.4 micron thick devices confirms that the internal field can be distorted by photoinduced space-charge even in such thin devices. Simulation indicates that for very thin (d < 0.3 μm) devices with typical maximum saturated defect densities in a-Si:H (N_d =10^{17}cm^{-3}) no significant charge trapping should occur. Hence light bias enhanced quantum efficiency measurements could be an interesting diagnostic tool for solar cells.

ACKNOWLEDGEMENTS

The authors thank W. Kusian, Siemens AG, for the diodes used in this study and W.Gao for the simulation program.

REFERENCES

1. J. Y. Hou and S. J. Fonash, Applied Physics Letters **61**, p. 186 (1992).

2. S. J. Fonash, J. Hou, F. A. Rubinelli, *et al.*, Optical Engineering **33**, p. 2065 (1994).

3. J.-H. Zollondz, R. Brüggemann, W. Gao, *et al.*, in *Mat. Res. Soc. Symp. Proc*, **420**, p. 697, (1996).

4. P. Chatterjee, Journal of Applied Physics **75**, p.1093 (1994).

5. W. Gao, C. Main, S. Reynolds, *et al.*, J. of Non-Cryst. Solids **198-200**, p. 1221(1996).

AUTHOR INDEX

SUBJECT INDEX

etching, 731
ethylene, 603
exchange, 275
excimer laser
    annealing, 689
    crystallization, 177

Fermi energy dependence, 305
fill factor, 809
first-principles calculations, 275
Fourier transform infrared, 311, 317

gas
    effusion spectroscopy, 317
    jet technology, 567
gate
    dielectric, 695
    model, 695
geometric considerations, 97
germanium, 591
    carbide, 561
glass substrate, 799
glow discharge, 115, 121, 299
graded bandgap structure, 591
grain
    boundary, 171, 201
    size model, 641
growth
    epitaxial, 207, 555
    high rate, 25, 105, 513
    mechanism, 13
    rate, 127, 139, 151
    temperature, 163

helium dilution, 151
heterojunction(s), 463, 585, 863
heterostructures, 463
high rate technique, 567
hole conduction, 671
hot wire, 79, 91, 251, 299
    cell method, 531
    chemical vapor deposition, 85, 97, 519,
        525, 573
    deposition, 537
    thin film transistor, 659
hydrogen, 275, 377
    atomic, 13
    bonding, 269
    dilution, 139, 145, 239, 251, 269, 525
    diffusion, 299, 329
    exchange model, 299
    flip, 371
    in silicon, 555
    low, 97
    mobile, 383
    molecular, 287, 293, 323, 383
    molecules, 275
    plasma, 629
    properties, 317
    transport, 305

hydrogenated amorphous silicon, 13, 19, 31,
    55, 61, 105, 145, 163, 263, 269, 323, 337,
    359, 389, 409, 433, 463, 501, 543, 799
    doped, 457
    films, 121
    germanium, 189
    photodetectors, 445
    solar cell, 507, 725
hydrogenation, 635

image sensors, 827
imager, 869
imaging, 809
inductive diodes, 863
infrared, 839
    absorption, 371
    detector, 591
    spectroscopy, 269
interface(s), 457, 779
    roughness, 725
internal friction, 323, 389
interstitial, 287
intrinsic stress, 365
ionic
    bombardment, 115
    energy distribution, 115

junction, 703
    recombination, 761

lag, 827
laser
    annealing, 171
    crystallization, 623
    doping, 677
lateral grain growth, 171, 177
lattice disorder, 389
leakage current, 851
lifetime, 543
light
    bias, 875
    induced changes, 263
    scattering, 725
    soaking, 371, 389
    switching, 833
lithography, 821
localization, length, 433
long range potential fluctuations, 543
low-temperature, 665, 749, 799
    deposition, 91, 531
    fabrication, 653
    growth, 513

manufacturing, 61
medium-range order, 251
metastability, 377
metastable defects, 371